Essential Human Anatomy and Physiology

Second Edition

Barbara R. Landau
University of Washington

Essential Human Anatomy and Physiology

Second Edition

Scott, Foresman and Company

Glenview, Illinois
Dallas, Tex.   Oakland, N.J.
Palo Alto, Cal.   Tucker, Ga.   London, England

Text illustrations drawn by Cecile Duray-Bito and Leonard Morgan.

Cover photograph by Howard Sochurek.

Library of Congress Cataloging in Publication Data

Landau, Barbara Ruth, 1923–
    Essential human anatomy and physiology.

    Bibliography: p.
    Includes index.
    1. Human physiology. 2. Anatomy, Human. I. Title.
QP34.5.L36 1980   612   79–24264
ISBN 0–673–15249–9

1 2 3 4 5 6 7 8-VHJ-86 85 84 83 82 81 80 79

Preface

There is a measure of satisfaction in undertaking the preparation of the second edition of a textbook, for it suggests that the aims and objectives of the first edition have been acceptable to users and that those aims have been realized to a certain extent. More important, however, a new edition provides the opportunity for adjustments that will make the achievement of those goals more certain and for updating the contents to reflect new information and interpretations. The many changes and innovations in this edition are designed specifically to encourage and aid the student's mastery of the material.

Foremost among the innovations in this edition is a completely new illustration program. All of the figures have been redrawn and relabeled, and some new figures have been added, including light and electron micrographs. Color has been used throughout to make the illustrations more effective. The liberal use of high quality illustrations in this book should help students develop the ability to visualize structures and their relationships to one another—an ability essential to understanding anatomy.

The main objective of the second edition of *Essential Human Anatomy and Physiology* remains the same as that of the first: to provide a well-balanced and carefully integrated introduction to the anatomy and physiology of the human body, an introduction that is suitable for students of varying needs and interests. Since the essential facts of anatomy and physiology are the same regardless of field of study, this book is not tailored to meet the needs of any particular group. It should accommodate students seeking careers in health fields that require a basic understanding of the human body, as well as students who are simply interested in the subject—including those whose scientific experience and background is limited. Since some knowledge of science,

especially chemistry and/or biology, is desirable for the course, the introductory chapters provide a review of the necessary background information.

The *balance* in the text is achieved by giving anatomy and physiology equal coverage and emphasis, and by treating the several organ systems in equal depth. The *integration* of anatomy and physiology is carried out in several ways. Whether structure and function are discussed together or treated separately, the relations between them are stressed. In addition, there are frequent references to related sections of the book. The human body is a unified whole, and events or processes occurring in one part affect, and are affected by, events in other parts. It is therefore important to present each part or process as an essential element of that whole.

Whenever possible, descriptions are presented as a logically developing story, with each new piece of information fitting into its own place and adding a little more to the story, rather than being another item in a list of seemingly unrelated facts to be committed to memory. The emphasis throughout is on relationships, on similarities and differences, on causes and effects, and on hows and whys rather than merely on whats. It is hoped that the student will come away with some understanding of the human body based on facts, not just with a body of facts.

As a textbook of anatomy and physiology, this book describes normal structure and function. Because this involves a very large body of information, no attempt has been made to cover pathology as well. Abnormal and pathological conditions have generally been omitted unless they contribute to the understanding of the normal. For example, the physiological role of the hormone insulin is well illustrated by noting the effects of its absence, as in diabetes mellitus, but the point of focus is the absence of insulin, not the disease itself.

In several areas updating involves shifts in emphasis or approach that are incorporated more effectively by telling the story a bit differently. The resulting reorganizations in the text improve the sequence in developing a topic and make the relationships more apparent and understandable, as well as provide more current information. Part III, "Coordination and Control," for example, now includes hormonal as well as neural control. Introducing hormone actions early not only brings the two control systems together but also makes possible a more effective and integrated presentation of the regulation of processes that have both neural and endocrine controls. The contents of the neural chapters have been redistributed into a more functional arrangement. Thus, they deal with neurons and impulse conduction, the spinal cord and spinal reflexes, structure of the brain, sensory functions (both general and special senses), and, finally, motor functions (both somatic and autonomic controls).

At the risk of making Chapter 2 on cell function somewhat lengthy, two additional sections have been added to it: specifically, the action of enzymes and an introduction to electrical phenomena. Because enzyme action and electrical phenomena are as much a part of normal cell function as membrane transport and protein synthesis, basic cell activities cannot really be explained without involving them.

Several other sections have been reorganized, rewritten, or added. Chapter 23 on digestion has been reorganized, and Chapter 24 on metabolism of foodstuffs has been almost entirely rewritten to provide better integration of the information. Among the topics that have been added to the second edition are sections on skin, immune systems, and prostaglandins, as well as a more thorough treatment of membrane potentials. There are also numerous minor revisions in almost every chapter.

A new feature in this edition is the inclusion of a short Focus essay in most chapters. Each Focus essay highlights an interesting application (or implication) of some aspect of the material in that chapter. These essays often call attention to a clinical problem, some of which are very common and some quite unusual, but all build upon the information discussed in that chapter.

Each of the eight major parts or divisions opens with a statement that puts that part in the proper perspective. An introduction to each chapter previews that chapter and indicates how the topic discussed fits into the overall organization and operation of the body. A summary at the end of each chapter brings together the important points of the chapter. The summary is presented in narrative rather than outline form, so that relationships and explanations can be included, rather than just listing headings and names.

Features that have been retained from the first edition include study questions at the end of each chapter and a list of suggested readings for each chapter (now located at the back of the book). An improved and slightly expanded glossary, preceded by a list of prefixes, suffixes, and combining forms, aids students in the important task of vocabulary building.

A Study Guide, prepared by Jacob E. Wiebers and Wallace J. Rogers of Purdue University, provides exercises with answer keys. The Laboratory Guide, prepared by Elizabeth Arthur, contains twenty-nine class-tested experiments that teach the basics of morphology, physiology, and biochemistry. An Instructor's Manual is also available.

Preparation of this edition has been aided immeasurably by the support of colleagues and students. Their comments and suggestions, together with those of users of the first edition and reviewers of both editions, have been most helpful. Special thanks are due to the artists, Cecile Duray-Bito and Leonard Morgan, to Dr. Thomas A. Easton for the Focus essays, and to the superb book team from Scott, Foresman and Company—all competent professionals and a real pleasure to work with.

Barbara R. Landau

Contents

Part 1
Introduction and Orientation
1

1 Introduction to Anatomy and Physiology   3

An Approach to the Study of the Body   The Cell as the Basic Unit of Life
The Internal Environment—Homeostasis   The Cell—Its Structure and Function

2 Concepts Important for Understanding Cell Function   17

Chemical Organization   Translocation of Materials
Metabolism and Energy Exchange   Cell Growth and Reproduction
Electrical Phenomena

3 Organization of the Body   57

Tissues   Skin   Anatomical Orientation

Part 2
The Musculoskeletal System
85

4 Skeleton—The Framework   87

General Osteology   Bones of the Skeleton

5 Articulations   134

Fibrous Joints   Cartilaginous Joints   Synovial Joints

6 Skeletal Muscle System   148

Gross Anatomy of a Skeletal Muscle   Principles of Skeletal Muscle Action
Muscles of the Body—Their Attachments and Action

7 Muscle Contraction   192

Microscopic and Submicroscopic Anatomy of Skeletal Muscle
Excitation and Contraction of Skeletal Muscle Fiber
Contraction of Skeletal Muscles   Smooth Muscle   Cardiac Muscle

Part 3
The Nervous System
217

8 Introduction to the Nervous System —
Its Organization and Components *219*

Cells of the Nervous System   Excitation and Transmission of Impulses

9 The Spinal Cord and Reflexes *243*

Anatomy of the Spinal Cord   Nerves of the Spinal Cord   Spinal Reflexes

10 Anatomy of the Brain and Related Structures *261*

Structure of the Brain   The Cranial Nerves
Protective and Supportive Structures

11 Sensory Aspects of the Nervous System *286*

The General Senses   The Special Senses — Vision
The Special Senses — Hearing and Equilibrium   The Special Senses — Smell
The Special Senses — Taste

12 Motor Aspects of the Central Nervous System,
and Higher Brain Functions *322*

Control of Skeletal Muscle   Control of Visceral Structures — The Autonomic
System   Higher Function of the Central Nervous System   Brain Metabolism

13 The Endocrine Glands as a Control System *352*

Nature of Hormones and Hormone Action   The Pituitary Gland and
Hormone Control   Endocrine Glands and Their Secretion

Part 4
Systems of Transport
383

14 Blood *385*

The Nature of Blood   Formed Elements   Plasma   Blood Clotting
The Basis of Blood Groups   Blood in Acid-Base Balance

15 The Cardiovascular System — The Heart *409*

Introduction to the Circulation   Structure of the Heart
Excitation of the Heart   Extrinsic Innervation of the Heart
Events in the Cardiac Cycle   Cardiac Output and its Control

16 Anatomy of the Blood Vessels and Lymphatics   433

Structure of the Blood Vessels   The Pulmonary Circuit
The Systemic Circuit   The Lymphatic System

17 Circulation of the Blood and Lymph   454

Blood Flow—Hemodynamics   Arterial Pressure
The Microcirculation—The Capillary Bed   Venous Circulation
Lymphatic Circulation

18 Cardiovascular Adaptations and Adjustments   477

Circulation Through Regions with Special Needs
Cardiovascular Homeostasis—Adjustments to Some Physiological Stresses
Cardiovascular Effects of Some Abnormal Stresses

Part 5
Respiration
493

19 The Respiratory Apparatus—Its Structure and Function   495

Structure of the Respiratory Apparatus   Ventilation of the Lungs

20 Gas Exchange and Transport   515

Properties of Gases   Diffusion of Gases   Transport of Oxygen
Transport of Carbon Dioxide   Role of Respiration in Acid-Base Balance

21 The Control of Respiration   527

The Respiratory Cycle   Pulmonary Ventilation   Respiratory Problems

Part 6
Metabolic Processes and Energy
539

22 Anatomy of the Digestive System   541

General Structure of the Digestive Tract   Organs of the Digestive System
Blood Supply   Innervation of the Digestive Tract
Peritoneum

23 Digestive and Absorptive Functions
of the Gastrointestinal Tract   564

Control of the Digestive Tract
Motor and Secretory Functions of the Digestive Tract
Absorption

24 Metabolism of Foodstuffs   *589*

Some Basic Considerations   Carbohydrate Metabolism   Fat Metabolism
Protein Metabolism   Endocrine Regulation of Metabolic Processes
Some Conditions that Require Metabolic Adjustments
Metabolism of Other Substances

25 Energy Metabolism   *620*

Energy Balance   Sources of Energy   Energy Expenditure

26 Body Temperature and its Regulation   *635*

Normal Body Temperature   Heat Gain   Heat Loss
Regulation of Body Temperature

Part 7
The Urinary Systems
647

27 The Urinary System   *649*

Homeostasis and Excretion   Anatomy of the Urinary System
The Formation of Urine   Micturition

28 Regulation of the Extracellular Fluid   *666*

Reabsorption of Sodium and Potassium   Reabsorption of Water
Tonicity and Osmolarity of the Urine   Renal Regulation of Acid-Base Balance
The Regulatory Role of Renin and Angiotensin

Part 8
The Reproductive System
679

29 Reproduction and the Male Reproductive System   *681*

Origin of Gametes and the Individual   The Male Reproductive Organs
Endocrine Function of the Testes

30 The Female Reproductive System   *696*

The Female Reproductive Organs   Endocrine Function of the Ovary
The Menstrual Cycle and its Control   Pregnancy
Mammary Glands and Lactation   Control of Fertility

Further Readings   *R–1*

Some Commonly Used Metric Units and Conversions   *M–1*

Glossary   *G–1*

Index   *I–1*

Part 1

Introduction and Orientation

From TISSUES AND ORGANS—a text-atlas of scanning electron microscopy, by Richard G. Kessel and Randy H. Kardon. W. H. Freeman and Company, copyright © 1979.

Dense irregular connective tissue.

You are about to begin the study of the human body. If you are one who thinks that sciences are dull, abstract, or far removed from your everyday world, perhaps you should put those thoughts away for a while. The study of human anatomy and physiology ought to be an exciting adventure, because it is *your body* that you are studying. The structures exist, and the processes are happening in your own body.

The first chapter will introduce you to anatomy and physiology—what they are and what they include. It will also introduce you to cells and the "world" they live in. Cells are the tiny living units that are actually responsible for all the reactions and processes that occur in the body. The properties of their world, or environment, are very important to their ability to function and, indeed, to survive.

Chapter 2 will introduce you to some of the things cells do. But before you can understand what cells do and how they do it, you need to be aware of some of the chemical and physical laws that determine how cells accomplish their tasks. Accordingly, some chemical and physical concepts are described to make it easier to understand what goes on in cells and to explain their activities.

The final chapter in this section will introduce you to the way cells are organized into tissues and into the organ systems that carry out particular tasks for the whole body. You will also be introduced to the whole body, and to the relationships of its parts to one another. Finally, you will be introduced to the skin which covers the cells, tissues, organs, and parts. Learning how skin protects them from the external world and performs its other functions will provide an introduction to the way cells and tissues and organ systems work together to perform their particular tasks.

1

An Approach to the Study of the Body

The Cell as the Basic Unit of Life

The Internal Environment—Homeostasis

The Cell—Its Structure and Functions

Introduction to Anatomy and Physiology

The human body has always intrigued and challenged curious minds, and for centuries attempts have been made to learn the body's secrets. Early scholars were severely hampered both by the restraints imposed by the societies in which they lived, and by the lack of the tools and instruments of investigation. Thus progress in uncovering these secrets was slow. Concepts and ideas in human anatomy and physiology that we now take for granted were relatively late in coming, relative to the intellectual development of our civilization. Even the concept of the cell as the basic unit of living things was not firmly established until about 150 years ago, although cells had been described nearly 200 years earlier when the first microscope was developed. More recently, concerted efforts, creative investigators, and modern instruments have greatly expanded our knowledge of the human body, and it continues to expand at what seems to be an ever increasing rate. We now know a great deal of what cells are like, what they do, and how they do it, but there are still many things we do not know about them.

The cell is appropriate for introducing you to the human body. This chapter considers the role of the cell in the body, including the structure of a cell, and the environment in which it lives.

AN APPROACH TO THE STUDY OF THE BODY

Biology is the study of life and living things. Its two major categories are **zoology,** the study of animal life, and **botany,** the study of plant life. In each of these areas there are numerous subdivisions, among them such diverse fields as *taxonomy* (classification), *ecology* (relationships between organisms and their environments), and *genetics* (heredity), as well as *anatomy* (structure) and *physiology* (function). This book is about structure and function in one particular animal—the human.

Human beings are unique in some respects, but in many ways they resemble the other members of the animal kingdom. For this reason certain aspects of their anatomy and physiology can often be studied more conveniently in nonhuman species. One must use caution, however, in applying to human beings conclusions based on information obtained from other animals.

People have always been interested in the workings of their bodies. Crude drawings found on the walls of caves suggest an awareness even in prehistory of the importance of certain organs; they show, for example, that a spear was most effective in killing an animal when it pierced the heart. Several thousand years later (about 400 B.C.), **Hippocrates,** the "father of medicine," wrote extensively about the body even though he actually knew little more about the internal organs than that the heart is a muscle and that the pulsation of the blood has something to do with the beat of the heart. **Aristotle** (384–322 B.C.), the "father of biology," also wrote at great length about the body, but most of what he knew about the human body was conjecture based upon his study of other animals. Actually it had to be, since dissection of the human body was at that time (and, off and on, for centuries afterward) not considered right and proper; in fact, experimentation was regarded as degrading and unnecessary by all self-respecting philosophers. Aristotle's conclusions may strike the modern observer as an odd combination of fact and fantasy. For example, he not only considered the heart (correctly) to be the center of the system of blood vessels, but he also considered it (incorrectly) to be the seat of intelligence and the source of body heat. However, considering the lack of an experimental approach in Aristotle's day, perhaps we should be surprised at how often he was correct.

The first known human dissections were not carried out until nearly a century after Aristotle by **Erasistratus,** who made several significant observations. He described, for example, the valves in the heart but failed to realize their significance. He also found the arteries to be collapsed and nearly empty of blood after death; however, he concluded incorrectly that arteries normally contain air.

About four hundred years later, a Roman court physician named **Galen** (133–200 A.D.) made anatomical studies, mostly on apes, and performed some simple experiments. He proved that arteries contain blood. He further postulated (incorrectly) a bodily system in which digested material was absorbed from the digestive tract and carried to the liver, where it was next made into venous blood with "natural spirits" added, the blood then being distributed in the veins to all parts of the body. Among his other errors were that he believed the circulation to be an ebb and flow of the blood separately in the veins and arteries, with no connection between them, and the nerves to be hollow tubes which served to distribute the "animal spirits" manufactured in the brain to the body. The significance of Galen's views on circulation and, indeed, upon all of physiology lies in the fact that they were accepted so completely; few questions were asked and little criticism was offered for nearly fourteen centuries. Only in the sixteenth century did anatomists, including the versatile **Leonardo da Vinci,** begin to dissect the human body systematically. Working from careful dissections, **Vesalius** published in 1543 an anatomical treatise whose detailed illustrations are still classics today.

One of the first great physiological advances came in 1628 when **William Harvey** used a new experimental approach to prove that the circulatory system is a continuous circle, in which the blood is pumped from the heart to the arteries to the veins and back to the heart. His theory eliminated the dependence on Galen's ideas and provided a major breakthrough in understanding, since it treated the body as a material "machine" whose workings could be understood, not as a mysterious supernatural thing controlled by spirits and vapors.

As direct and logical and simple as Harvey's approach seems to us now, there was much contemporary criticism of his methods and reluctance to accept his conclusions. Galen's teachings had dominated scientific and medical thought for 1400 years and were not to be discarded very readily. But as Harvey's new ideas and approaches began to spread, more discoveries and applications were made and significant advances occurred at an ever increasing rate. These advances continue today.

Considering the body as the complex and wonderful machine that it is, **anatomy** is the study of how that machine is put together. It deals with the structure of the parts, ranging from the molecular components of the tiniest cells to the whole individual, and their relationship to one another and their environment. The subject matter of anatomy is sometimes divided into those structures visible to the unaided eye (*gross anatomy*) and those visible only with a microscope (*histology*). A smaller order of magnitude includes those components which can be visualized only with the electron microscope (*fine structure* or *ultrastructure*).

Physiology, on the other hand, is concerned with the mechanics of the body machine, how it works, what makes it go, and what regulates, limits, and protects the machinery. Physiology also covers a broad range, from the smallest of cellular components to the whole animal, and its many different approaches have led to the development of whole new disciplines, such as *biochemistry* and *biophysics*. In our study of anatomy and physiology we will thus be concerned with normal structure and function, including the normal adjustments to changing conditions. We will discuss *pathology*, which deals with structure and function in the abnormal conditions of disease states,

only when it affords a meaningful contrast with normality.

Questions about the normal operation of the body machinery and its adjustments can only be answered if one has a thorough grasp of the structure of the components involved. Structure and function are so closely related that it is impossible to understand one without the other, since the effectiveness with which a function can be carried out depends largely upon the structure of the parts. An organ whose function is to detect sound would be an utter failure at pumping blood. To understand the pumping of blood, one must know how the heart is constructed, and to understand the manner by which sound is heard, one must know something about the anatomy of the ear. To understand a nerve impulse requires a knowledge of the microscopic, and even submicroscopic, structure of a nerve cell. How does a cell that secretes saliva differ from a cell that conducts nerve impulses or from a cell that shortens? These cells must be different, since each performs a different role. They cannot exchange functions.

The types of questions appropriate to the study of normal structure and function are virtually unlimited. One might consider the long-lasting (chronic) adjustments in an athlete during the training season, or the immediate (acute) changes during a contest. One might properly inquire into what determines whether the heart rate should be increased, into whether an increase should be of ten or fifty beats per minute. What signals that the exercise is completed and the need for increased heart action is ended? What causes the heart to beat in the first place? Similar questions can be asked about activity at other levels, about cells and cellular components rather than intact (whole) animals and organ systems. What is the nature of the message sent by a nerve cell? What is actually sent, and what happens when the message arrives? How are things remembered and, perhaps of more immediate concern to you, the student, why are things forgotten?

The importance of questions can scarcely be overemphasized, since in any scientific endeavor it is necessary to ask the right questions before one can hope to find significant answers. This book should answer some questions and provide clues for other answers. However, it should also raise questions, many of which so far are, and may remain, unanswered. We hope that you will develop a questioning attitude, the habit of wondering how and why, and a desire to seek out better answers. And we hope that this desire will persist beyond the brief exposure provided here.

It is sometimes revealing, when undertaking a study of a new organ or system, to sit back a moment and imagine that you are a master designer or engineer charged with the task of developing an apparatus to perform a particular function. Think first what the requirements and problems are, and then see if you can devise a system to meet these requirements. For example, if the problem concerns removal of wastes, an adequate excretory system must be able to sort out and remove the unwanted wastes, and yet salvage the nonwastes. It must do this effectively and economically, without upsetting any of the other delicate balances that exist in the body. What kind of a machine could do this? (One solution is the artificial kidney, sometimes used when the body's own excretory apparatus has failed.) Now, as you study the human kidney, see how it meets these requirements. Such an approach should help in two ways. First, it should encourage you to state the problem and to see the role that organ system plays in the life of the whole organism. Secondly, it should enable you to see a physiological process as a logical solution to a particular problem and, hopefully, help you put things in the proper perspective. Some of the solutions may seem unnecessarily complex, but backup systems are needed in case of mechanical failures and adequate checks and balances must ensure control under many different conditions.

To gain a thorough understanding of a structure or mechanism, it should be investigated under varied, but controlled, conditions. For anatomists, this study usually means looking at a structure with aids to observation such as microscopes and photographic equipment. Physiologists, on the other hand, need some way to measure and record events that occur. Most of the advances in biology, particularly in recent years, have come directly on the heels of the development of instrumentation permitting more accurate observation or measurement of structures and events. Early scientists learned a great deal simply by watching and keeping an accurate record of what they saw, but the days of acquiring significant new information solely by direct observation have largely disappeared. Contemporary scientists need more detailed and quantitative information. It is not enough for them to know that a given event occurs. They want to know how this event or structure fits with what is known about other events or structures and how these relationships are affected by changing conditions. In many cases quantitative relationships can be expressed mathematically, but the precise quantitative data necessary for mathematical statements are often difficult to obtain from biological preparations; there are too many variables, and some of them have not yet even been identified, much less controlled. It has only been within recent years that sophisticated instrumentation has begun to provide biologists with the tools for obtaining the types of information needed for this approach.

THE CELL AS THE BASIC UNIT OF LIFE

It has been said that the simplest functional unit of life is the **cell.** Although recent studies in *cytology* (cell biology) have raised serious questions about its simplicity, it is safe to say that the cell *is* the basic structural and functional unit of living things. The word "cell" was first used biologically in 1665 by Robert Hooke to describe the tiny, empty compartments he saw in a thin sheet of cork. Others soon found similar compartments in many materials, but it was not until 1839 that Schleiden, a botanist, and Schwann, a zoologist, independently arrived at the **cell theory.** They really contributed no new observations, but they did recognize among the many previous observations a relationship which they expressed in the revolutionary theory that all living things are composed of tiny compartments, or cells. The resulting conclusion that microscopic examination of any living or formerly living thing should reveal the presence of cells has been universally confirmed. The cell theory is today accepted as correct.

It is now known, however, that a cell is not an empty compartment. Not long after Hooke's early description, it was discovered that cells are composed of living material, which we call **protoplasm.** This chemically complex substance is contained in all living things. The term *cell* was then broadened to include these contents as well as the enclosure, the cell membrane, or cell boundary.

The cell may also be defined as the minimal structural unit of protoplasm that can carry on all of the vital functions characteristic of living things. A definition of life itself is difficult, however, since one can easily get bogged down in problems of semantics and philosophy. It is usually sufficient to use an operational definition—to say that something is alive if it can do certain things and that it is not alive if it cannot do these things. Most of us recognize that loss of consciousness is not loss of life. We are more likely to associate loss of life with cessation of the heartbeat or of breathing, although with the advent of modern life-support systems the boundary line is not at all clear. But life is not a feature only of intact organisms; a complex living organism contains many kinds of living cells, and these cells can remain alive, even if they are removed from that organism. For example, in a technique known as tissue culture, a tiny bit of tissue is removed from an animal and the cells are separated and placed in a suitable nutrient medium. If this cell suspension is kept in a suitable environment, the individual cells will survive. We say that they are *alive* because they carry on all of their vital functions, those activities by which we define life.

Reduced to the simplest terms, the so-called vital functions may be identified by four properties:

1 metabolism
2 growth
3 irritability and adaptability
4 reproduction

Metabolism covers the processed involved in energy exchange. It is used here in its broadest sense, to encompass all of the processes associated with the use and storage of the energy extracted from ingested food. Growth and metabolism involve many of the same processes, but they are two different things. *Growth* occurs when the metabolic balance is tipped slightly in favor of building processes over breakdown processes, but growth also involves other mechanisms and controls. (The growth of a young individual to an adult is not the same as transforming a small adult to a large one by overeating!) *Irritability* denotes the ability to respond to a change in the environment, that is, to a stimulus. *Adaptation* is a long-range response to environmental change, as observed in evolutionary changes over many generations. The nature of the responses vary with the structure or cell stimulated, as well as with the nature of the stimulus. *Reproduction* is the ability to perpetuate the individual's own kind. Growth, metabolism, and irritability are concerned primarily with maintaining and protecting the individual, while adaptation and reproduction ensure the continuity of the species.

In a general way, these characteristics apply both to intact animals and to the individual cells in an animal. In fact, the simplest animals contain only one cell. Each cell, therefore, must be able to carry on independently all of the vital functions. The amoeba, for instance, appears to be little more than a shapeless bag of protoplasm, but it can, because it *is* protoplasm, carry on the vital functions (see Figure 1–1). It responds to the presence of a food parti-

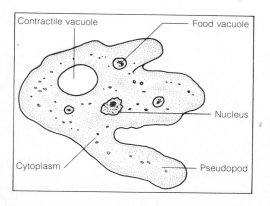

FIGURE 1–1   An amoeba, a one-celled animal.

cle by sending out footlike protoplasmic extensions (pseudopods) and literally flowing around the particle until it surrounds and engulfs the particle. The amoeba demonstrates irritability by the fact that it will move away from an "unpleasant" stimulus, such as prodding, and it will move toward a "pleasant" stimulus, such as the presence of food nearby.

In more complex and multicellular animals, the cells are neither all alike nor able to do everything for themselves, and they demonstrate different types of response. The sizes and shapes of cells vary, as in Figure 1–2, and are often characteristic, suggesting the relationship between the structure and the role of the cell. Some cells are specialized, usually in the sense of showing a well-developed or unique response to stimulation. Muscle cells have developed the ability to change shape rapidly (contraction); nerve cells are particularly adept at transmitting impulses from one end of the cell to the other (conduction); and other cells may respond to a stimulus by releasing a specific substance which they have synthesized (secretion). When cells do become specialized they often lose some other abilities; for example, muscle cells reproduce very little, if at all, and nerve cells (except in embryonic stages) cannot reproduce at all. Some less specialized cells, however, such as those in the skin, are constantly replacing themselves.

THE INTERNAL ENVIRONMENT— HOMEOSTASIS

The numerous acitivities of cells are carried out by the chemical substances of their protoplasm. These substances are dissolved or suspended in fluid—an aqueous solution—in the cell. This is true both for simple organisms like the amoeba and for multicellular organisms like human beings. In complex organisms there is fluid or water not only in the cells but also in the "spaces" between cells and in blood vessels. Water is the chief component of the body, accounting for nearly 60 per-

FIGURE 1–2   Cells of different shapes. (All cells are not drawn to the same scale.) A. A nerve cell (the long process has been cut to keep it in the figure). B. Columnar epithelial cells. C. Red blood cells and a white blood cell. D. Part of several skeletal muscle cells. E. A connective tissue cell. F. Secretory cells of the thyroid gland.

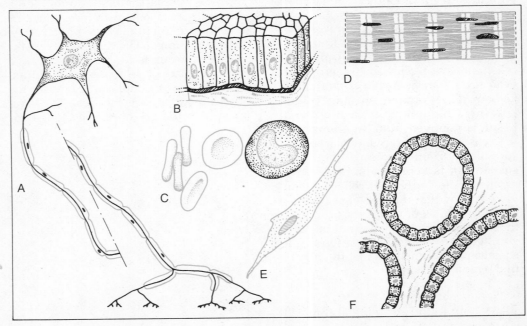

cent of the total body weight in humans. The fluid in the body is divided into several compartments by barriers such as cell membranes and the walls of blood vessels. Most of the fluid, however, is inside the cells and is called the **intracellular fluid (ICF).** Other fluid, the **extracellular fluid (ECF),** is outside cells. It includes both the fluid portion of blood, known as **plasma,** and the fluid in the "spaces" between cells, the **interstitial** or **intercellular fluid** (see Table 1–1 and Figure 1–3).

The composition of the fluid in the various compartments is quite similar in many respects, but there are some important differences. The specific concentrations of certain substances, both in extracellular and intracellular fluids are of great importance to normal cell function. The one-celled amoeba, for example, is capable of carrying on all essential cell processes, but its ability to do so depends upon the presence of all necessary elements in its immediate environment. A slight change in the composition of that environment may be crucial for its survival. It is, however, only in the simplest animals that each individual cell is bathed directly by water from the outside world. In more complex animals, nearly all of the cells are quite removed from the external environment. These cells must then either adapt to life away from water and the things contained in it or have the fluid environment brought to them. The first alternative can be eliminated, since every cell must live in a fluid environment, regardless of whether the animal lives in the sea, in a tropical forest, or in the desert. For every cell a fluid environment (the extracellular fluid) is the medium from which the cellular nutrients and respi-

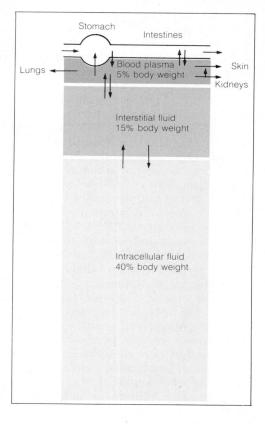

FIGURE 1–3
The body fluid compartments. The arrows indicate movement of fluid across compartment boundaries.

(Adapted and reproduced, with permission, from Gamble, *Chemical Anatomy, Physiology, and Pathology of Extracellular Fluid,* 6th ed., Harvard University Press, 1954.)

ratory gases are obtained and to which cellular wastes are discharged. If individual cells are to survive and function within an animal, therefore, an internal fluid environment must supply all of their needs.

In 1852 a French scientist, Claude Bernard, first proposed the concept of an internal environment, which he called the *milieu intérieur.* He wrote:

". . . *animals have really two environments: a milieu extérieur in which the organism is situated, and a milieu intérieur in which the tissue elements live. The living organism does not really exist in the milieu extérieur (the atmosphere if it breathes, salt or fresh water if that is its element) but in the liquid milieu intérieur formed by the circulating organic liquid which surrounds and bathes all the tissue elements; this is the lymph or plasma, the liquid part of the blood which, in the higher animals, is diffused through the tissues and forms the ensemble of the intercellular liquids and is the basis of all local nutrition and the common factor of all elementary exchanges. . . .*

"*The stability of the milieu intérieur*

TABLE 1–1 BODY FLUID COMPARTMENTS

Compartment	Percent of body weight
Intracellular fluid	**40**
Extracellular fluid	**20**
interstitial fluid	15
blood plasma	5
Total body fluid	**60**

is the primary condition for freedom and independence of existence; the mechanism which allows of this is that which insures in the milieu intérieur the maintenance of all the conditions necessary to the life of the elements."[1]

In spite of the fact that his evidence was extremely meager, Bernard's concept has successfully withstood the test of time. He did not overestimate its importance: The major task confronting the animal organism *is* that of **homeostasis.** It means a *steady state,* and is used to denote the maintenance of stability of the internal environment. The term was first used in the 1920's by Walter B. Cannon, an eminent American physiologist who further developed Bernard's ideas, and focused on what he called the "integrated cooperation" of a wide range of organs to maintain that constancy. Homeostasis involves keeping the concentration of many individual components of the internal environment within rather narrow limits. The content of nutrients, oxygen, and wastes must be kept within appropriate bounds. The volume of fluid in the internal environment, as well as its acidity, osmotic pressure, and the concentration of specific substances such as sodium and calcium, must be maintained. Many species, including humans, also keep their body temperature fairly constant. Failure to maintain homeostasis, by allowing any of these factors to fall outside of the "normal range," will interfere with the ability of the cell to carry out one or more of its activities. But when an animal is able to maintain the constancy of its own internal environment, it is free to explore external environments that are very different from its internal cellular environment.

Various organ systems have evolved in order to support homeostasis. The digestive system is a processing plant that by physical and chemical breakdown prepares ingested material for entrance into the body and removes unabsorbed material from the body. The respiratory system transfers oxygen from the air to the blood and carbon dioxide from the blood to the air. The excretory system removes wastes

from the bloodstream and carries them from the body. The circulatory system is a delivery system, operating between the other organ systems and the immediate local environment of each individual cell. There is effective exchange of materials between the blood and the interstitial fluid, so that changes in blood composition are quickly reflected in cellular activities. A case can be made for homeostatic contributions of the muscular and skeletal systems as well, although the importance of locomotion of the whole animal to the well-being of individual cells is somewhat less than direct. The activities of all these systems are coordinated by the nervous system and the secretions of the endocrine glands. Thus all the organ systems, directly or indirectly, serve the purpose of maintaining the proper fluid environment for the individual cells. Homeostasis then becomes a major objective of the organ systems and of the body as a whole. It would be good to bear this in mind when considering the intricate and seemingly devious means by which the various systems actually do this.

THE CELL—ITS STRUCTURE AND FUNCTIONS

It is the internal processes of the individual cells that you must keep in mind when considering the activities of the whole animal, since it is only as a result of these cellular functions that the activities of the whole animal can come about. We will approach this necessary knowledge by examining the workings of a "typical cell," even though one of the first things you must realize is that a "typical cell" does not exist. What is really meant by such an expression is a composite cell that possesses the general characteristics of most cells, but none of the unique properties of specialized cells.

The concept of the hypothetical "typical cell" has changed drastically in the last two or three decades, largely the result of improved methods and tools of investigation, such as the electron microscope. A modern light microscope is capable of magnifications on the order of 1000× and an effective resolution of about 200 nanome-

[1]J. F. Fulton, *Selected Readings in the History of Physiology,* 2nd ed. Charles C Thomas, Springfield, Ill. (1966), p. 326.

ters (nm),[2] but the electron microscope permits magnifications on the order of 300,000× and resolution of biological structures of about 1–1.5 nm. This permits detailed study of structures scarcely detectable with the light microscope and borders on the visualization of molecules.[3] Other advances that have been helpful to cell study include biochemical techniques for separation of various cellular components and identification of their chemical structure. With facts gleaned by using these and other methods, it has been possible to put together a remarkable picture of many of the internal processes of average and specialized cells. The picture is by no means complete yet, but more pieces are being added continually.

In its simplest form, a cell consists of a **nucleus** and **cytoplasm,** enclosed by a **cell membrane** (see Figure 1–4). From early evidence, the nucleus was known to contain a darkly staining material called *chromatin* and a small clump of material that stained differently, called the *nucleolus.* A number of structures in the cytoplasm had been described, some of which could be seen only with the aid of special techniques or stains. Details were lacking, and in most cases the function of these intracellular components remained obscure. The membrane of animal cells was known to be very thin, scarcely visible in many preparations, but it was considered to be essential to the integrity of the cell.

The current picture of the "typical cell" is much more detailed. Not only have many of the intracellular components been described structurally, but their functional roles have, to some extent, been identified.

Cell Membrane

The cell surface is covered by a thin, flexible restraining sheath or membrane, known as the **cell membrane.** It has a unique and well-defined structure, but it arises from, and is considered to be a constituent of, the protoplasm. In fact, the cell membrane is often referred to as a **plasma membrane,** which recognizes the protoplasmic origin of the membrane. Under the electron microscope the plasma membrane appears as a double-layered structure, but it is actually a single "unit" membrane. The surfaces of the membrane have a greater *electron density* than the central region, which does not show distinctly in electron micrographs. Most unit membranes are about 7.5 nm in thickness.

The function of the cell membrane is not only to outline the cell and keep its protoplasm together but also to "regulate" the flow of substances in and out of the cell. Since all exchanges between the cell and its environment occur through the membrane, the membrane must permit the passage of certain substances, prevent the passage of others, and at times actually "pump" some substances through (in one direction or the other). Membranes that perform these tasks are said to be **selectively permeable.** The selective qualities of all cell membranes are not identical but vary, depending on the specialized nature of the cell.

From evidence based on physical

FIGURE 1–4   An electron micrograph of a liver cell showing many of the intracellular structures described. 6100× magnification. A. Nucleus. B. Golgi apparatus. C. Mitochondria. D. Rough endoplasmic reticulum. E. Glycogen granules. F. Lysosome.

Robert L. Kaplan

[2]A *nanometer* (nm) is the same as a millimicrometer, and is equal to 10^{-9} meters, or 10^{-6} millimeters.

[3]A molecule is the smallest unit of an element or compound that retains the chemical properties of that substance.

and chemical properties, plus that gained by electron microscopy, the plasma membrane is generally regarded as being composed of lipid (fatlike substances) and protein. For several decades the membrane was thought to be formed of a layer of lipid two molecules thick with a thin layer of protein covering the two surfaces in a sandwichlike arrangement. The presence of tiny openings or pores was proposed to account for the passage of substances whose transfer could not otherwise be explained.

The lipid molecules were, and still are, visualized as being shaped like clothespins, oriented with the heads toward the surfaces of the membrane. The head ends, which are in contact with the fluid on either side of the membrane, are water soluble, while the tails, which lie within the membrane, are insoluble in water.

Recent evidence, however, suggests that the protein component does not really exist as a layer covering the lipid center. Rather, elongated protein molecules (integral protein) are embedded in the lipid layer with their ends projecting on either side (Figure 1–5). Other (peripheral) proteins may be attached to the ends of these molecules, usually on the inner surface of the membrane. Small carbohydrate chains may be attached to the protein or lipid, usually on the outer surface of the membrane. The embedded proteins are an integral part of the membrane and cannot be dislodged without disrupting the membrane, but the peripheral proteins can be altered or removed with relatively mild treatment. There is also evidence that the protein can actually migrate within the lipid layer. Electrically charged sites on the protein molecules can, and in fact do, react readily with chemicals in their immediate environment. Some of the sites, called **receptor sites,** react specifically with a particular substance whenever it is present. The reactions at these sites may change the shape and properties of the protein and the membrane as well. Such changes may lead to opening of temporary channels through which specific particles can penetrate the membrane. Some of the receptor sites react specifically with a particular substance whenever it comes along.

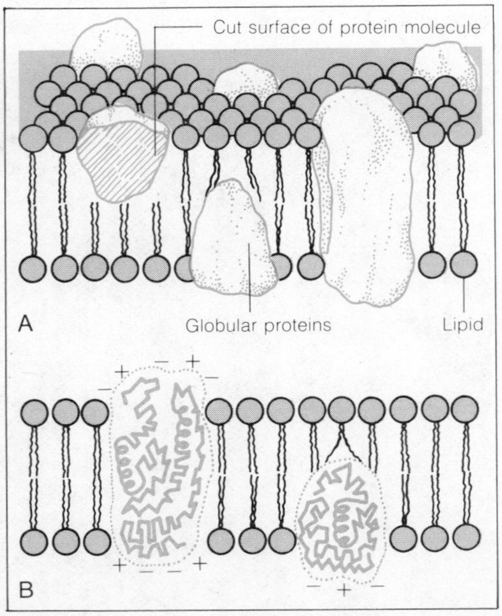

These electrochemical reactions no doubt play a major role in the functioning of the cell membrane by altering its ability to transfer (pump) substances across the membrane or to prevent or permit passage. This is a particularly important role since (1) the cell's "supplies" (or nourishment) must pass through the membrane to enter the cell, (2) its products and wastes must pass through to leave the cell, and (3) hormones or nerve impulses that might regulate the cell's activities must affect the membrane in some manner in order to influence activities inside the cell. The fact that the proteins differ from one cell to another probably explains the differing properties of various membranes.

This concept of the plasma membrane pictures a living, dynamic, ever changing fluid layer rather than an inert fence or a fixed structure. With this model it is not necessary to propose the existence of pores fixed in the membrane because the reactive potential of the protein serves the same purpose.

Cytoplasm

The **cytoplasm** is simply all the cell's protoplasm outside the nucleus. Within the semifluid substances that constitute the cytoplasm, a number of distinct structures have been identi-

FIGURE 1–5 The plasma membrane. A. The arrangement of lipid and protein molecules. B. The protein molecules are diagrammed to show the presence of electrical charges that may affect passage of substances through the membrane.

Cut surface of protein molecule

A   Globular proteins   Lipid

B

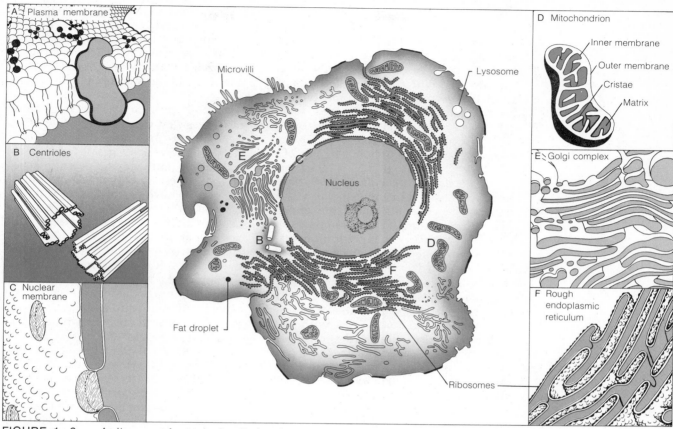

FIGURE 1–6   A diagram of a typical cell showing details of some of its components.

fied. They have been grouped either as organelles or as inclusions on the basis of their role in the cell's activity. **Organelles** are integral parts of the cell, produced by it and involved in its metabolic processes. **Inclusions** are nonliving, often temporary components that at any one moment may or may not be present; they include stored nutrients, pigments, and granules of material about to be released from the cell. See Figure 1–6 for a view of the cell and its contents.

Organelles. The mitochondria, endoplasmic reticulum, ribosomes, Golgi apparatus, lysosomes, centrosomes, and fibrils, as well as the plasma membrane, are all organelles.

Mitochondria. The **mitochondrion** is a tiny, capsule-shaped body found in all cells. Some cells contain only a few, but others, particularly those that are

metabolically quite active, may contain several thousand. Its wall resembles the cell membrane, but it is a double membrane, that is, two unit membranes. The inner one has a number of folds extending into the center of the mitochondrion forming shelflike partitions called *cristae*. The interior contains a fine granular substance known as *matrix*.

Mitochondria are important because of the types of enzymes they contain. *Enzymes* are chemical compounds that hasten or bring about particular chemical reactions (see Chapter 2). Virtually every reaction in the body requires the presence of one or more specific enzymes. The enzymes found in mitochondria are those that bring about the final steps in the metabolism of foodstuffs—the steps that account for most of the energy released in the body. The individual enzymes are believed to be arranged sequentially along the cristae of the mitochondria, and the material to be metabolized is passed along from enzyme to enzyme until it is completely metabolized. The energy released is then stored in a readily available form until needed. The mitochondria, therefore, contain the energy-releasing machinery of the cell; they are the furnaces where the fuel is actually burned. For this reason mitochondria are frequently referred to as the "powerhouses" of the cell. The metabolic reactions are discussed very briefly in Chapter 2 and more fully in Chapter 24.

Endoplasmic reticulum and ribosomes. Throughout the cytoplasm is a network of channels called the **endoplasmic reticulum.** It usually appears as groups of parallel tubes with interconnecting branches; these groups are actually flattened *vesicles* or sacs. The membranes forming the walls of the endoplasmic reticulum (ER) are identical with the cell membrane. Some of the channels are continuous with the membrane surrounding the cell nucleus, and some open through the cell membrane at the surface. Because of its numerous interconnections, the endoplasmic reticulum can provide an effective means of intracellular communication and transport.

There are two varieties of ER—rough and smooth. Most of the ER has tiny granules along its surface, giving it a beaded appearance. This is the *rough* or *granular ER*. The granules are called **ribosomes** (ribose bodies) because of their high content of ribonucleic acid (RNA). RNA contains the instructions, or code, for the synthesis of protein molecules, and ribosomes are the site of protein synthesis, which is discussed in Chapter 2. Much of the material produced by cells is protein; this includes both intracellular enzymes and many of the substances produced for "export" from the cell (secretions). Formed at the ribosomes, this protein apparently moves into the tubules of the endoplasmic reticulum and migrates in the direction of the Golgi apparatus. Some ribosomes are scattered throughout the cytoplasm, unattached to ER. However, they too are believed to be involved in synthesis, particularly of protein for use inside the cell.

The *smooth* or *agranular ER,* without ribosomes, is less abundant than rough ER. It is known to be involved in the production of some nonprotein substances that will be secreted from the cell. In the case of skeletal muscle, the smooth ER is associated with the uptake and release of calcium, which is important in initiating contraction and relaxation of the muscle cell.

Golgi apparatus. The **Golgi apparatus** was first described by the Italian histologist whose name it bears. It consists of a network of sacs or vesicles that are usually flattened and often enlarged at the edges. The vesicles commonly appear to have been stacked on one another. Associated with the flattened vesicles are a number of large rounded vesicles and many tiny ones. They are all enclosed by smooth (no ribosomes) membranes. It is likely that the protein portion of the secretions formed at the ribosomes moves along the endoplasmic reticulum toward the Golgi apparatus and eventually becomes pinched off to form tiny vesicles. These tiny packets merge with the flattened vesicles of the Golgi apparatus. Here, in other words, material to be secreted is probably "packaged for export" by the cell. It then seems to gravitate toward the edge of the Golgi vesicle and eventually breaks off into the cytoplasm as a large rounded sac or *secretion granule.* The secretion granules congregate at the free edge of the cell until the proper stimulus causes their release.

Lysosomes. The **lysosomes** are also membrane-enclosed vesicles within the cytoplasm. They contain enzymes that would be destructive to the cell if released into the cytoplasm. The exact role of these potentially dangerous organelles is not known, but they have been implicated in several processes. When a lysosome comes in contact with bacteria or other foreign material that may have gained entry into the cell, the lysosome membrane fuses with that of the bacterium. Soon the bacterium lies completely within the membrane of the lysosome, where it is exposed to the action of the enzymes in the lysosome and is soon destroyed by them. An important role of certain white blood cells is the destruction of bacteria that have entered the body. The cytoplasm of the white blood cell contains numerous lysosomes. Lysosomes also "digest" worn-out cell parts, and when a cell dies, the membranes enclosing its lysosomes probably disintegrate to release the enzymes.

Centrosome and centrioles. A **centrosome,** containing a pair of **centrioles** is usually found in an area adjacent to the nucleus. With the light microscope the centrioles appear as two dots, but in electron micrographs they are seen to be a pair of tiny cylinders lying at right angles to one another. They are composed of nine triple rods or tubules, arranged around an open center. The centrioles are most prominent at the time of cell division (mitosis), in which they play a role (see Chapter 2).

Fibrils, filaments, and tubules. The **fibrils** are cellular fibers found in several types of cells. They are formed by bundles of very slender protein threads known as **filaments.** They are probably best known in muscle cells, as *myofibrils* and *myofilaments,* where they are directly involved in the contractile process. The tiny **microtubules** found in the cytoplasm of some cells are believed to help maintain the shape of the cell. Microtubules appear in all cells during the process of cell division, and in association with centrioles, they play an important part in that process.

Inclusions. Intracellular deposits are often referred to as inclusions. Among the most frequently occurring inclusions are *stored nutrients:* carbohydrate, lipid. Carbohydrate is stored in the form of glycogen and may be present in any cell as tiny cytoplasmic granules, although the amount varies with the type of cell and with its nutritional state. Lipid too may be found at times as tiny fat droplets in the cytoplasm of certain cells, chiefly adipose cells (fat cells). These cells have a great affinity for lipid; they may acquire such an enormous lipid droplet that the nucleus and the rest of the cytoplasm are squeezed to the edge of the cell.

Secretion granules are also considered to be inclusions. They are formed in and by organelles, but at some point they separate and are no longer part of the organelle. Secretion granules are found in cells that secrete, and they are present in greatest numbers during the period between synthesis of the granules and their release from the cell.

Pigments are also found in certain cells. Some are naturally occurring and others have been taken in by the cell. Hemoglobin, the oxygen-carrying pigment of red blood cells, and melanin, the pigment of the skin, are the major naturally occurring pigments. Cells may take up colors from a wide variety of sources, however, ranging from the carotenes (yellow) of certain vegetables to foreign materials such as tattoo dye.

Nucleus

The **nucleus** plays an essential role in the life of the cell, and it is often assumed that every cell has one, but this is not true. All cells have a nucleus at one stage of development, but some types of mature cells, notably human red blood cells, have lost their nucleus. There are some cells, such as muscle cells, that have more than one nucleus.

The nucleus is essential for two vital processes of the cell, *cell division* and *protein synthesis*, both discussed in Chapter 2. Obviously, cells with no nucleus have lost the ability to reproduce themselves; also, their ability to synthesize protein is severely restricted.

The nucleus is composed of the fluid protoplasm (called *nuclear sap* or *nucleoplasm*), a sizable mass of *chromatin material,* and one or more *nucleoli.* The nucleus is enclosed by a *nuclear membrane* formed by a double layer of unit membrane with a space between the two layers. At fairly regular intervals are large windowlike openings in the nuclear membrane, and occasionally channels of endoplasmic reticulum open into the intramembranous spaces of the nuclear membrane. It should not be surprising that the nuclear membrane permits passage of larger molecules than do other membranes.

The **chromatin material** in a nondividing cell appears as a diffuse mass of darkly staining substance. The shapeless conglomeration is actually highly organized, since it consists of the **chromosomes,** which contain the heredity-determining **genes.** The chromatin (and hence the chromosomes) is composed largely of nucleoprotein and deoxyribonucleic acid (DNA).

The **nucleolus** is an area within the nucleus that is almost unique among cellular components, since it is not enclosed by a membrane. It seems to be an accumulation of nucleoprotein, but with a different nucleic acid, ribonucleic acid (RNA), which is also found in ribosomes. The nucleolus is the site of formation of RNA needed for protein synthesis.

CHAPTER 1 SUMMARY

Human anatomy and **physiology** are the studies of the structure and the function of the human body, respectively. Both have been studied since the time of the ancients, but early studies were mostly anatomical. Physiological advances have been much more recent and have coincided to a great extent with the development of measuring and recording technology.

As the structural and functional unit of living things, the cell receives much attention. A cell is said to be living when it is able to carry on the necessary energy exchanges (*metabolism*), to grow (*growth*), to respond to stimuli (*irritability*), and to adjust to long-term changes in its environment (*adaptability*). It can also reproduce its own species, although *reproduction* is greatly restricted in some kinds of cells.

Cells live in a fluid environment from which they must obtain everything they need to carry on their activities. Because cells are so dependent on their fluid environments, many properties of their environment must be carefully regulated. The maintenance of the constancy of the internal environment is known as **homeostasis.** In humans, the *interstitial fluid* is the immediate internal environment. Interstitial fluid and blood plasma make up the *extracellular fluid compartment*. The fluid inside cells is the *intracellular fluid compartment*. Fluids account for about 60 percent of the body weight.

A **"typical cell"** consists of a *nucleus* and *cytoplasm* enclosed by a cell membrane. The *membrane* determines what can and what cannot enter or leave the cell, and is said to be selectively permeable. It is composed of a double layer of lipid molecules with protein molecules distributed throughout the membrane in a particular fashion. The *cytoplasm* contains numerous specialized organelles that are responsible for many of the cell functions. Among them are *mitochondria* for energy release, *ribosomes* and *endoplasmic reticulum* for protein synthesis, the *Golgi apparatus* for secretion of substances from the cell, and *lysosomes* for the destruction of bacteria and certain other materials that may enter the cell. The *nucleus* contains the genetic material, which consists largely of the "directions" for assembling the kinds of protein molecules that are characteristic of that cell.

STUDY QUESTIONS

1 Define anatomy and physiology. How do they differ? What is their relationship?

2 What is meant by the statement that the cell is the functional unit of life?

3 How would you define life, or living? How would your definition apply in a situation involving an organ transplant (a kidney, for example)? If the organ donor is "dead," is the organ "alive"?

4 What is meant by homeostasis? Why is it important to you?

5 What are the fluid compartments? Do they have any relation to homeostasis? If so, what? How might a substance in the blood fluid find its way into a cell?

6 The requirement to maintain a certain constancy of the internal environment imposes some demands upon us, but it also gives us some advantages. What are some of the demands, and some of the advantages?

7 Of what importance is the cell membrane? What would happen to a cell if its membrane disappeared? What, if any, cell functions would be disrupted? Explain.

8 Name the important structures that make up the cell. What is the role of each? How would the life and activity of the cell be altered if any one of these structures was absent?

2

Chemical Organization

Translocation of Materials

Metabolism and Energy Exchange

Cell Growth and Reproduction

Electrical Phenomena

Concepts Important for Understanding Cell Function

In Chapter 1 cells were introduced as the smallest functional units of living things. Cells and organisms are living because they are able to carry on the metabolic activities that provide energy to sustain themselves, to grow and reproduce, and to respond to short- and long-term changes in their environment. In this chapter we will take a closer look at how cells do some of these things.

Cells are composed of chemicals, so first we will consider some basic facts of chemical organization, paying particular attention to the chemicals found in cells. The cell membrane, cytoplasm, and nucleus contain the same chemical building blocks, but they are organized differently in each structure.

Most metabolic reactions occur in the cytoplasm and generally involve substances that in some form crossed the cell membrane to get into the cell. Therefore, we will look at the properties of the cell membrane and the forces that act on chemical substances to learn how they are transported across membranes. We will learn how the cell nucleus guides such activities of the cell as the synthesis of new proteins. We will also see how the cell transmits the genetic information in its nucleus to the next generation as it reproduces to form new cells.

CHEMICAL ORGANIZATION

The cell has been described as the simplest functional unit of life, but it is not the smallest unit of structure or of function. Each of the organelles within a cell performs certain vital functions for that cell, and each organelle is a complete functional unit, even though not one of them can carry on for any length of time outside the cell. But organelles are not the smallest structural units either; they are composed of chemical compounds which in turn are made up of atoms and molecules. Each of these subcomponents plays its own role in the life of the cell and of the whole organism. Therefore, to understand how the structural components of the cell react and interact, we should know something about the nature of the most basic components of matter.

Atoms and Molecules

A **molecule** is the smallest unit of an element or compound that retains the chemical properties of that material. Thus a molecule of water, salt, or carbon dioxide is the smallest fragment distinguishable as water, salt, or carbon dioxide; anything less is no longer one of those substances. Molecules are formed by the linking together of two or more **atoms** which are the structural units of matter. For example, a molecule of water consists of two atoms of hydrogen and one atom of oxygen; table salt molecules each contain one sodium and one chlorine atom; carbon dioxide molecules simply consist of two atoms of oxygen, and one of carbon.

Substances composed of one type of atom are called **elements.** Carbon, oxygen, and iron are examples of elements. Substances that are composed of two or more different atoms are either a compound or a mixture. In a **compound** the atoms are combined in definite ratios to form molecules which are distributed evenly throughout the substance. Salt, acid, proteins, carbohydrates, and sugars are examples of compounds. In a **mixture** the atoms are not combined in the same way chemically and may be spread unevenly throughout the substance. Wood and brass are common examples of mixtures.

Even atoms, however, are composed of smaller units or particles known as electrons, protons, and neutrons (Figure 2–1). **Electrons** carry a negative charge, **protons** carry a positive charge, while **neutrons** are neutral and carry no charge. The center or *nucleus*[1] of each atom consists of a tightly bound cluster of neutrons and protons and thus has a positive charge. The amount of charge in the nucleus depends upon the number of protons it contains. The negatively charged electrons surround the positively charged nucleus as a cloud of electrical charge, although they are often said to orbit the nucleus like satellites around a planet. Because *unlike* charges attract one another and *like* charges repel one another, the orbiting negative electrons are attracted by the positive nucleus. The number of electrons surrounding the nucleus usually equals the number of protons in the nucleus, so that the atom as a whole normally carries no net charge; like the neutron, it is electrically neutral.

Although the positively charged protons would be expected to repel one another, the charged protons and uncharged neutrons are held together in the nucleus by extremely powerful forces. When these bonds are broken, as in an atomic explosion, tremendous amounts of energy are released. Ordinary chemical reactions—the ones we will consider—are concerned only with the energy of the outermost orbiting electrons and do not involve the nucleus or nuclear energy.

The number of protons, neutrons, and electrons varies from atom to atom and is one feature in which the various elements differ from one another. The **atomic number** of an element designates the number of protons (and hence the number of electrons) associated with each atom of that element. Any two different elements must have dif-

---

[1]The nucleus of an atom is not to be confused with the nucleus of a cell, although in both cases the nucleus is the center or core about which other parts are organized.

ferent atomic numbers. There are at present 106 known elements whose atomic numbers run sequentially from 1 to 106. The hydrogen atom is the smallest, with a single proton in its nucleus and one electron. Lawrencium, a very large atom, has 103 protons and 154 neutrons in the nucleus, and 103 orbiting electrons.

Atoms are so unbelievably small that thousands of them would not deflect even the most sensitive scale. They do, however, have a weight, expressed as the **atomic weight;** this number is equal to the sum of the protons and neutrons in the nucleus of the atom. It is proportional to the actual weight of the nucleus, as well as to the weight of the whole atom, since the weight of the electrons is insignificant.

Table 2–1 gives the atomic weights and other basic facts about some biologically important elements. These elements are the major components of the body and of the substances which are ingested, produced, and excreted by the body. Our attention will be directed almost exclusively to these elements and the compounds containing them.

A certain fraction of the atomic nuclei may contain a different number of neutrons than is typical of that element, and they have different atomic weights than typical atoms. This explains the decimal points for some atomic weights in Table 2–1. Atoms that have different weights are known as **isotopes.** Some isotopes are stable atoms, but many are unstable, or *radioactive,* and decay or disintegrate at a rate that is characteristic of that isotope. In so doing, these unstable isotopes emit radiation that can be detected even when present in very small amounts. Although radioactive isotopes occur naturally, they can also be produced artificially and used to "tag" chemical compounds. Since such isotopes have the same chemical properties as the other atoms of that element, they participate in chemical reactions in the same way, but because they are radioactive and can be detected readily, they can be used to study such important biological processes as the synthesis and breakdown of various substances and to establish rates and sites of transfer. Compounds containing radioactive material

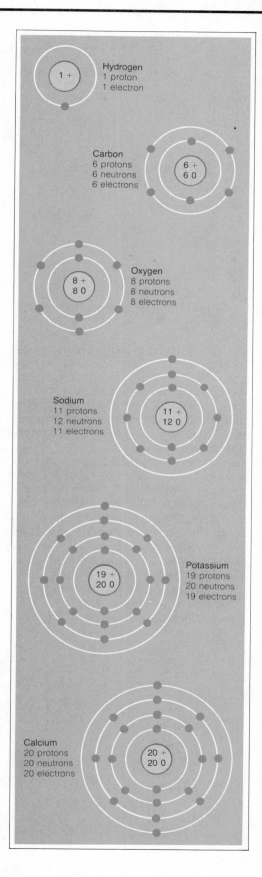

FIGURE 2–1 Some atomic structures.

TABLE 2-1   IMPORTANT ATOMS FOUND IN THE BODY

Atom	Symbol	Atomic Number	Approximate Atomic Weight	Valence*
Hydrogen	H	1	1	1
Carbon	C	6	12	4
Nitrogen	N	7	14	3 or 5
Oxygen	O	8	16	2
Phosphorus	P	15	31	3 or 5
Calcium	Ca	20	40	2
Potassium	K	19	39.1	1
Sodium	Na	11	23	1
Chlorine	Cl	17	35.5	1
Sulfur	S	16	32.1	2, 4, or 6
Magnesium	Mg	12	24.3	2
Iron	Fe	26	55.9	2 or 3

*see text below

can also be used to treat certain disease conditions, since the radiations emitted from some isotopes can damage living tissues. Diseased tissue can be destroyed without fatally damaging other parts of the body if the radioactive material is in some way selectively concentrated in that tissue.

The orbits of the electrons moving about an atom's nucleus lie in layers or *shells* at specified distances from the nucleus. Each shell has a known electron capacity; for example, the innermost shell can accommodate at most two electrons and the second shell can hold (if necessary) eight electrons. If an atom has more than two electrons, some of the orbits lie in the second shell, and if it has more than ten electrons, a third shell is needed. Electrons in the outer orbits are farther from the nucleus and are not as strongly bound by its attractive forces; they are, therefore, more likely to escape or to be removed by outside forces. In fact, these outer-orbit electrons are the ones that can bind one atom to another atom, forming a molecule.

The outer electron shell can have *at most* eight electrons (in which case it is said to be *filled*).[2] An atom whose outer electron shell is not filled may be able to gain electrons from, lose electrons to, or share electrons with other atoms (Figure 2-2). Oxygen, with six electrons in its outer shell, has room for two more electrons and can share two electrons with two hydrogen atoms to fill that shell. These two electrons are then shared by the hydrogen and the oxygen atoms to form a molecule of water, H_2O. The molecule is held together by the forces of attraction which arise from the shared electrons by the hydrogen and oxygen nuclei. When atoms share electrons the attracting forces are known as **covalent bonds.**

Ionization. The attractive forces that hold atoms together in molecules may arise not from a sharing of electrons, but rather from a giving up and a taking on of one or more electrons. For example, when sodium and chlorine are brought together, the sodium atom gives up an electron which is then captured by the chlorine atom. The chlo-

[2]Hydrogen with one electron and helium with two electrons are unique in that they have only one electron shell, which is then both the inner and outer shell. Since this first shell can have at most two electrons, their outer shell is filled when only two electrons are present.

FIGURE 2-2   Some molecular structures.

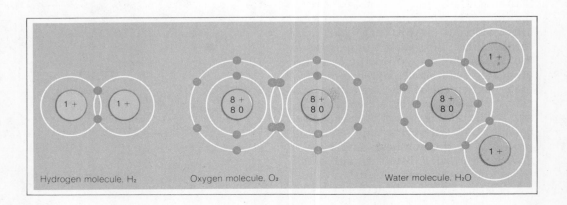

Hydrogen molecule, H_2 Oxygen molecule, O_2 Water molecule, H_2O

rine atom now has one more electron than it has protons, and it carries one negative (−) charge (see Figure 2−3). Sodium now has one less electron than it has protons, so it carries one positive (+) charge. The sodium atom and the chlorine atom are held together because opposite charges attract one another. The attractive force between these charged atoms is called an **ionic bond.**

Many atoms can gain or lose one or more electrons. Such atoms are called **ions,** and are written Na^+, Cl^-, H^+, Ca^{++}, etc. All ions are charged atoms. The process by which a molecule (especially one held together by ionic bonds) separates into ions is known as **ionization** or **dissociation.** The charge carried by a particular ion is characteristic of that ion and is indicative of the number of electrons it will donate (if it has a negative charge) or accept (if it has a positive charge). The transfer of electrons underlies chemical reactions that occur between atoms, ions, and molecules in all but atomic or nuclear reactions.

Valence. Elements differ in the number of bonds they can form as they combine with other elements. This difference is represented by the **valence,** which compares the bonding or combining capacity of an element to that of hydrogen. It often corresponds to the number of vacancies in the outer electron shell of the atom. An atom of hydrogen has a valence of one, while oxygen has a valence of two and can combine with two atoms of hydrogen. Nitrogen, which usually has a valence of three (or sometimes five), can combine with three hydrogen atoms; and carbon, with a valence of four, can combine with four hydrogen atoms:

H—   H—H H_2 hydrogen

—O—   H—O—H H_2O water

—N—   H—N—H NH_3 ammonia
  | |
                           H

          H
          |
—C—   H—C—H CH_4 methane
  | |
          H

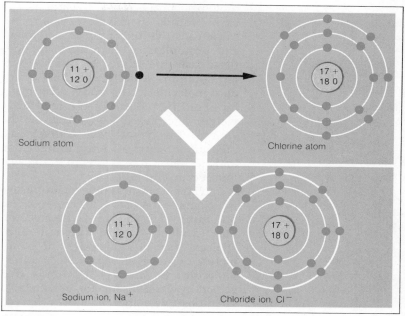

FIGURE 2−3   Ionization. An atom becomes an ion when it acquires a charge. It carries a positive (+) charge when it has lost an electron and a negative (−) charge when it has gained an electron.

Substances that ionize in solution bear a charge in proportion to their valence (Na^+, Ca^{++}). The compounds formed from these ions are determined by the combining capacity or valence of the participating ions:

$$Na^+ + Cl^- \longrightarrow NaCl$$
$$Ca^{++} + Cl^- + Cl^- \longrightarrow CaCl_2$$

Ions such as sodium, potassium, and chloride form single bonds and, like hydrogen, are said to be *monovalent.* Calcium, and magnesium can form two bonds and are said to be *divalent.*

Water and Solutions

When a compound or element dissolves homogeneously in another substance (usually a liquid), a **solution** results. The material to be dissolved is called a **solute;** the liquid into which the solute is dissolved is called the **solvent.** For example, a sucrose solution consists of sucrose and water; sucrose is the solute and water is the solvent.

All three states of matter—gas, liquid, and solid—play essential roles in biological processes. Our emphasis is upon liquids, particularly water, because cells live in a fluid environment and protoplasm is semifluid; in both these cases the fluid is an *aqueous* (water) solution. Water is the most abundant molecule in the body, accounting for about 60 percent of the total body weight and nearly 80 percent of the weight of a typical cell. Water is the medium in which all living things exist, and all chemical reactions in living systems occur in solution.

Water has a number of properties which make it suitable for this role. For one thing, it is the most effective solvent known; many different substances—gas, liquid, and solid—completely dissolve in water. For another, molecules of many compounds dissolved in water dissociate, or ionize, to yield positively and negatively charged ions. Ions respond to an electric current, a fact important to biology. If a current is passed through an ionized solution, the negatively charged ions, called **anions,** are attracted to the positive electrode (the anode) (Figure 2–4). The positively charged ions, called **cations,** migrate toward the negative electrode (the cathode). A solution of ions is therefore capable of carrying an electric current, and a compound that gives rise to ions is called an **electrolyte** (sodium chloride is an electrolyte). Substances that do not ionize in solution do not respond to an electric current and are hence called **nonelectrolytes.** Body fluids contain small but vital amounts of electrolytes.

Concentration. It is frequently necessary to know the amount of solute in a solution, that is, the **concentration** of a solution. The concentration is the amount of solute in a given amount of solvent, and it may be expressed in a number of ways. A commonly used method is **percent,** which, in a solution, is usually expressed as grams of solute in 100 milliliters of solution. Thus a 10 percent salt solution contains 10 grams (g) of salt in each 100 milliliters (ml) of solution. Most physiological fluids involve more dilute solutions, and the concentrations are often expressed as milligrams percent, or milligrams (mg)

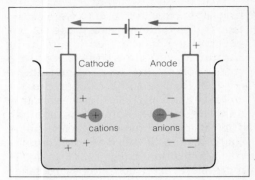

FIGURE 2–4 Ionized solutions can carry an electric current because charged particles migrate toward the electrode with the opposite charge.

of solute in 100 ml of solution. A glucose solution of 80 milligrams percent contains 80 mg (0.08 g) of glucose in each 100 ml of the glucose solution. When the solute is a gas, the concentration is often expressed as volume percent (ml per 100 ml).

Molarity. In many cases it is more meaningful to express concentration in terms of the solute's *molecular weight.* The molecular weight is the sum of the atomic weights of all the atoms in one molecule. The molecular weight of water (H_2O) is 18 (1 + 1 + 16), and the molecular weight of table salt (NaCl) is 58.5 (23 + 35.5). A **gram molecular weight** or **mole** of a substance equals numerically the molecular weight of that substance given in grams. Thus a mole of sodium chloride contains 58.5 *grams* of sodium chloride. A liter (1000 ml) of solution that contains a mole or gram molecular weight of solute is known as a **one-molar solution** (1 *M*). For reasons of ready comparison, solution concentrations are often cited in terms of a fixed volume, usually a liter. Also, because of the relatively low concentration of physiological solutions, concentration is commonly expressed in millimoles (1000 millimoles = 1 mole).[3]

Since a mole of any substance contains the same number of molecules of

[3]A variant method of expressing concentration is **molality.** A **one-molal solution** contains one mole of solute in 1000 g of solvent. At low solute concentrations, molar and molal solutions are very similar.

that substance, a liter of a molar solution of any given substance should contain the same number of molecules as a liter of a molar solution of any other substance. However, because molecular weight is proportional to the weight of the molecule in question, this means that actual weight of solute in the two solutions may be very different (Table 2–2). This is important in chemical reactions, since molecules participate in reactions only in proportion to their numbers, not to their weights or sizes.

Equivalence. Molar solutions of various substances are equal in terms of the number of molecules in solution, but if the molecules ionize, they may not necessarily be equal in terms of electrically charged particles. A one-molar solution containing calcium (Ca^{++}) yields more charges in solution than a one-molar solution containing sodium (Na^+), since calcium ions combine with twice as much chloride (or other anion) as sodium ions. For this reason, it is often preferable to speak of **equivalent** concentrations, those which are equal in terms of electrical charge. One equivalent of an ion is its molecular weight divided by its valence or ionic charge (Table 2–3). Equivalent weights are usually expressed as *equivalents per liter* or as *milliequivalents per liter.* Equivalence is not applicable to substances that do not ionize in solution.

Acids and Bases.   Substances that yield hydrogen ions upon dissociation are known as **acids.** They are called strong acids or weak acids according to the extent of their ionization. A *strong acid* dissociates to a high degree and releases many hydrogen ions into solution. A *weak acid* does not dissociate as completely and therefore yields fewer hydrogen ions. Hydrochloric acid (HCl) and sulfuric acid (H_2SO_4) are well-known strong acids, while carbonic acid (H_2CO_3) is a weak acid.

A substance that combines with hydrogen ions is a **base.** A *strong base* is highly ionized and combines readily with hydrogen, whereas a *weak base* combines less readily. The hydroxyl ion (OH^-) is a strong base and readily

TABLE 2–2   COMPOSITION OF SOME MOLAR SOLUTIONS

Substance	Molecular Weight	Gram Molecular Weight (weight of 1 mole)	Grams per Liter in 1 M Solution
NaCl (table salt)	58.5	58.5 g	58.5
Glucose (a sugar)	180	180 g	180
HCl (hydrochloric acid)	36.5	36.5 g	36.5

binds hydrogen ions. In combination with sodium it forms sodium hydroxide (NaOH). The so-called *conjugate bases* of strong acids (such as Cl^- and $SO_4^=$ of hydrochloric and sulfuric acids) are weak bases and do not readily take up hydrogen ions. The conjugate bases of weak acids (such as HCO_3^- of carbonic acid) are strong bases.

When an acid reacts with a base, water and a salt are formed. Common table salt, sodium chloride, may be formed in this manner:

$$HCl + NaOH \rightarrow NaCl + H_2O$$

It is important to realize that the term "salt" includes a whole class of compounds, of which sodium chloride is just one. *Salts* have neither acidic nor basic qualities, but, like water, are neutral.

Since biological reactions are extremely sensitive to the acidity or basicity (alkalinity) of the fluid medium, it is necessary to control the hydrogen ion concentration of the body fluids. Regulation of the acid-base balance is a major aspect of homeostasis.

Normality. It was stated above that acids can be neutralized by a base,

TABLE 2–3   SOME EQUIVALENT WEIGHTS

Ion	Gram Mol. Wt.	Valence	Equiv. Wt.
Na^+	23	1	23
Ca^{++}	40	2	20
Cl^-	35.5	1	35.5

TABLE 2-4   COMPOSITION OF SOME NORMAL SOLUTIONS

Substance	Molecular Weight	g/l in a 1 M Solution	g/l in a 1 N Solution
HCl	1 + 35.5	36.5	36.5
NaOH	23 + 16 + 1	40	40
H_2SO_4	$(1 \times 2) + 32 + (16 \times 4)$	98	49

with the formation of a salt. A liter of 1 M HCl (containing 36.5 g HCl) would exactly neutralize a liter of 1 M NaOH (containing 40 g NaOH) because they contain equal numbers of solute molecules. Both are highly ionized; and as they involve monovalent ions, they are also equivalent. A liter of 1 M H_2SO_4, however, is not equivalent to a liter of 1 M NaOH, since each molecule of sulfuric acid yields two hydrogen ions upon ionization. **Normality** is used to express the combining power of acids and bases. For HCl and NaOH, a molar solution is also a normal solution. However, a normal solution of H_2SO_4 would contain half as much sulfuric acid as a molar solution by virtue of its hydrogen content. Thus a 1 M solution of H_2SO_4 is 2 N. A **normal solution** (1 N) contains one gram of replaceable hydrogen (or equivalent base) per liter of solution, that is, one equivalent of solute per liter of solution (see Table 2–4).

pH and Acidity. Normality is determined by the amount of replaceable hydrogen (or base) present. It indicates the amount of base required to neutralize the acid (or the amount of acid required to neutralize the base). True acidity, on the other hand, is determined by the concentration of free hydrogen ions in the solution and depends upon the strength of the acid, that is, upon its degree of dissociation. All acids dissociate to a slight degree. Even water, which is composed of acidic H^+ and basic OH^-, dissociates, though only very slightly:

$$HOH \rightleftharpoons H^+ + OH^-$$

The hydrogen ion concentration (written $[H^+]$) of pure water is 1×10^{-7} grams of hydrogen per liter of water, which is 1×10^{-7} or 0.0000001 moles per liter. The hydroxyl ion concentration, $[OH^-]$, is also 1×10^{-7} moles per liter, but that is 0.0000017 grams per liter. Water is both a 0.0000001 N acid and a 0.0000001 N base, as it contains 1×10^{-7} moles of hydrogen and hydroxyl ions per liter. The addition of acid or base to water will affect the relative number of hydrogen and hydroxyl ions. If acid is added, the $[H^+]$ increases while the $[OH^-]$ decreases proportionately; if base is added the $[OH^-]$ increases while the $[H^+]$ decreases. Numbers expressing the actual concentrations are awkward to use, and it is more convenient to express hydrogen ion concentration as pH. The **pH** is the negative logarithm (to the base 10) of the hydrogen ion concentration.

If: $[H^+] = 1 \times 10^{-7}$ moles per liter
and: $pH = -log [H^+]$
then: $pH = -log 10^{-7}$
and: $pH = 7$

Table 2–5 shows the relationship of $[H^+]$ and pH to normality. You will note that as the hydrogen ion concentration increases (from 1×10^{-7} to 1×10^{-6}), the pH goes down. Thus the lower the pH, the more acid the solution. Since pH is a logarithmic scale, the difference between pH units represents a tenfold difference in hydrogen ion concentration. Thus the hydrogen ion concentration of a solution at pH 6 is ten times greater than one at pH 7; and the $[H^+]$ of a solution at pH 5

TABLE 2-5   NORMALITY AND pH

Normality of Acid	$[H^+]$	pH	$[OH^-]$	Normality of Base
1.0	1×10^{1}	0	1×10^{-14}	
0.1	1×10^{-1}	1	1×10^{-13}	
0.01	1×10^{-2}	2	1×10^{-12}	
0.001	1×10^{-3}	3	1×10^{-11}	
0.0001	1×10^{-4}	4	1×10^{-10}	
0.00001	1×10^{-5}	5	1×10^{-9}	
0.000001	1×10^{-6}	6	1×10^{-8}	
Neutrality 0.0000001	1×10^{-7}	7	1×10^{-7}	0.0000001 Neutrality
	1×10^{-8}	8	1×10^{-6}	0.000001
	1×10^{-9}	9	1×10^{-5}	0.00001
	1×10^{-10}	10	1×10^{-4}	0.0001
	1×10^{-11}	11	1×10^{-3}	0.001
	1×10^{-12}	12	1×10^{-2}	0.01
	1×10^{-13}	13	1×10^{-1}	0.1
	1×10^{-14}	14	1×10^{1}	1.0

is 100 times greater than that of one at pH 7.

Water at pH 7 is neutral, and the hydrogen and hydroxyl ion concentrations are equal. Solutions with a pH below 7 have a higher $[H^+]$ than water and are acidic; those solutions whose pH is above 7 have a lower $[H^+]$ than water and are said to be alkaline. The pH of the blood is very close to pH 7.4 and is therefore slightly alkaline, as are most body fluids. Some body fluids, however, have a different pH. Urine is usually slightly acidic and the gastric juice in the stomach is very acidic, although both fluctuate. Other secretions, such as saliva, are usually slightly more alkaline than the blood.

Organic Compounds

The structures of the body are formed for the most part of very large and very complex molecules built almost entirely of hydrogen, oxygen, and nitrogen, combined in some fashion with carbon. These *organic compounds* (that is, carbon-containing compounds) can be grouped into four principal categories: the carbohydrates, the lipids (fats), the proteins, and the nucleic acids, each with distinct properties and functions.

Carbohydrate. **Carbohydrates** are composed of carbon, hydrogen, and oxygen, the latter two elements combined usually in the ratio of 2:1, as in water (H_2O). The carbohydrates are the sugars and starches that are important energy sources for the cells of the body. The smallest are the **monosaccharides,** or **simple sugars,** which contain from three to seven carbon atoms. The most common simple sugars are the **hexoses,** with the formula of $C_6H_{12}O_6$. Because different internal arrangements of the molecule are possible, there are a number of hexoses. The three that are most often encountered in the diet are *glucose,* *fructose,* and *galactose,* with glucose being the main sugar in the body. Their formulas can be written in several forms:

glucose   fructose   galactose

glucose

Disaccharides, or **double sugars,** are formed by combining two monosaccharides, usually two hexoses:

$$C_6H_{12}O_6 + C_6H_{12}O_6 \rightarrow C_{12}H_{22}O_{11} + H_2O$$

Since there are many monosaccharides, there are also many disaccharides, but the most important biologically are *maltose, sucrose,* and *lactose:*

glucose + glucose → maltose (malt sugar)

glucose + fructose → sucrose (cane or beet sugar, table sugar)

glucose + galactose → lactose (milk sugar)

maltose

With the addition of water, the above reactions can be reversed as disaccharides are split into their component simple sugars. This process of splitting a substance by the addition of water is known as **hydrolysis:**

$$C_{12}H_{22}O_{11} + H_2O \rightarrow$$
$$C_6H_{12}O_6 + C_6H_{12}O_6$$

Polysaccharides are formed when many monosaccharides are linked together into a chain, often of considerable length. A molecule of water is removed from each molecule of monosaccharide as it is added to the chain. **Starches** are relatively long-chained polysaccharides found in plants. The particular polysaccharide produced by animal cells is called **glycogen.** It is composed of glucose molecules and is the form in which carbohydrates are stored in the body:

part of a glycogen molecule

Lipids. Lipids, like carbohydrates, are composed of carbon, hydrogen, and oxygen, but they contain much less oxygen. Lipids are relatively insoluble in water, but they dissolve readily in organic solvents such as ether and acetone.

The **neutral fats** are the most abundant lipids in the body, and they are what is usually referred to as "fat." They are also known as **triglycerides** because they are composed of glycerol and three fatty acid molecules:

glycerol fatty acids triglyceride water

Glycerol (not to be confused with glycogen) contains three carbon atoms and three hydroxyl (—OH) groups; it is with the hydroxyl groups that the fatty acids combine. A **fatty acid** is a chain of carbon atoms, ending in the acidic carboxyl group (—COOH).[4] The chains vary in length, but those with sixteen and eighteen carbons are most common in the body. Those in which all available bonds are occupied by hydrogen atoms are called *saturated fatty acids* and those in which some carbon atoms are joined by a double or triple bond are called *unsaturated fatty acids:*

saturated fatty acid

unsaturated fatty acid

Because the addition of hydrogen to an unsaturated fatty acid tends to raise its melting point, it was once customary to saturate the fatty acids in cooking fats to ensure that they would be solid rather than liquid at room temperature. Since saturated fats have been implicated in contributing to certain cardiovascular problems, manufacturers of cooking fats and oils now make much of the fact that their product is highly *un*saturated *(polyunsaturated).* Fortunately, neutral fats in the body are liquid at body temperature; if they were not, we would all be a little stiff.

Fats can be hydrolyzed to glycerol and fatty acids. Partial hydrolysis may remove only one or two of the fatty acids, leaving diglycerides or monoglycerides instead of glycerol.

Compound lipids are formed when one of the fatty acids of a triglyceride is replaced by another substance. *Phos-*

---

[4]Carboxyl groups are acidic because they yield hydrogen ions when they dissociate.

pholipids and *glycolipids,* which contain a phosphate or a carbohydrate, are among the most widely distributed lipids in the body. The lipids found in cell membranes are phospholipids, and glycolipids are found around nerve fibers. *Sterols* are lipid derivatives rather than true fats, but they share the general properties of lipids, particularly that of solubility in organic solvents but not in water. Among the sterols that have great physiological importance are *cholesterol* and its derivatives, including several hormones, some components of bile, and vitamin D. Steroid molecules are characterized by a complex ring structure rather than a long chain of carbon atoms.

Protein. **Proteins** are among the most complex molecules known. Although they can be used by cells to provide energy, they are more important in their roles as structural material and as enzymes. *Structural proteins* are part of the substance of the body, and are found both in the cell and in the intercellular material. **Enzymes** are *catalysts,* chemical compounds that cause a particular chemical reaction to "go" faster. Enzymes are of great importance in the body because virtually all biological reactions would occur too slowly to support life without their aid.

Protein molecules may be very large, for some have a molecular weight in excess of a million, and they are extremely varied. In addition to carbon, hydrogen, and oxygen, proteins also contain nitrogen and a small amount of sulfur.

Amino acids are the building blocks of protein. They are so named because each contains at least one acidic carboxyl group ($-COOH$), and at least one amino group ($-NH_2$). Proteins are formed by linking amino acids together by means of **peptide bonds.** When a carboxyl group of one amino acid reacts with the amino group of another, a compound called a *dipeptide* is formed. In the reaction below, R_1 and R_2 represent organic groups that are specific for each amino acid. They may be a single hydrogen, a carbon or carbon chain, or a simple ring structure:

More amino acids can be added to form a *polypeptide,* and adding many more makes a whole protein. Biochemists generally hold that there are twenty amino acids, although some feel that the figure is higher. The potential combination of twenty amino acid "building units" into large protein molecules provides an almost unlimited number of possibilities for different proteins with extremely varied properties. Some proteins, such as the *albumins* and *globulins* found in blood plasma and many other places are quite soluble. Relatively insoluble proteins are found in the outer layer of skin and in the hair and nails. Compound proteins are formed when proteins combine with carbohydrates *(glycoprotein),* lipids *(lipoprotein),* or any of several other substances.

In a polypeptide chain, the carboxyl group at one end of the chain and the amino group at the other end are still free to ionize and to react with other substances. Some amino acids have more than one carboxyl or amino group, and these side groups, too, can react with other substances. The protein chains may be simple or branched, and there may be cross-links between chains or parts of the same chain. Usually a long chain is folded into a coil or spiral (helix).

By difficult and time-consuming procedures, the amino acid sequence of a protein molecule can be determined. It is possible to break off the end amino acid of a chain and identify it, and to repeat the procedure for each new amino acid until all have been identified. The first protein to be described in this way (in 1955) was the hormone *insulin,* a relatively small protein. Since then the structures of many biologically important proteins, especially small ones, have been worked out.

Nucleic Acids. Nucleic acids are among the largest of molecules. In addition to carbon, hydrogen, oxygen, and nitrogen, nucleic acids also contain phosphorus and usually exist in combination with a protein as *nucleoprotein.* As the name suggests, they were first identified in the cell nucleus and, in fact, a large part of the nucleus consists of nucleoprotein. But there are important nucleoproteins in the cytoplasm as well.

There are two important nucleic acids in our cells, **ribonucleic acid (RNA)** and **deoxyribonucleic acid (DNA).** RNA is found both in the nucleus and in the cytoplasm. The chromatin material in the cell nucleus is largely DNA, which is particularly important because it is the substance of genes; it contains the genetic material that determines our heredity. Thus, the nucleic acids play vital roles in cellular activity.

The building blocks of nucleic acids are **nucleotides.** Each nucleotide consists of a phosphate group (PO_4), a sugar, and a small nitrogen-containing group called a **base.**[5] Any one of about half a dozen different bases may be found in nucleotides. The sugar is a 5-carbon sugar (a pentose), either ribose or deoxyribose. Ribose is the sugar in RNA and deoxyribose (which contains one less oxygen atom) is the sugar in DNA. Nucleic acids are formed when the phosphate group of one nucleotide binds with the sugar of the next, and so on, forming a potentially long phosphate-sugar chain. The bases extend out as side groups from the phosphate-sugar chain, and are capable of pairing with certain bases on other phosphate-sugar chains. The significance of this arrangement will become more apparent when we discuss cell division and protein synthesis later in this chapter.

Some Important Reactions

Law of Mass Action. The *law of mass action* relates the rates of reversible reactions in dilute solutions to the concentrations of the reacting substances.

[5]The base in a nucleotide is not to be confused with the base that combines with hydrogen ions.

Also, this phenomenon is behind the extreme versatility of all biochemical reactions. The rate of a reaction is proportional to the product of the concentrations of the reactants. In a *reversible* reaction in which two substances, A and B, react to form two new substances, C and D:

$$A + B \underset{\text{reaction II}}{\overset{\text{reaction I}}{\rightleftharpoons}} C + D$$

the velocity of the reaction to the right (reaction I) depends upon [A] × [B], and the velocity of the reaction to the left (reaction II) depends upon [C] × [D]. This means that reaction I cannot go to completion because reaction II begins as soon as substances C and D begin to accumulate. At some point a balance or *equilibrium* will be reached in which the *rates* of reactions I and II are equal. The two reactions will not actually have stopped, but since their rates are equal and opposite, there will be no further change in the concentrations of the individual reactants. Also, this does not necessarily mean that at equilibrium the *amounts* of A and B will be the same as the *amounts* of C and D, since they probably will not be.

An equilibrium can be upset by changing the concentration of any one of the reactants. For example, the addition of more of substance A or B will increase the rate of the reaction to the right (I), but as end-products C and D accumulate, reaction I slows and a new equilibrium is established. If the end-products C and D are removed as fast as they are formed, reaction II to the left will be negligible; the net effect then will be continued production of more C and D. This is an important relationship, because many biological processes involve a sequence of reactions, several of which are reversible:

$$A + B \rightleftharpoons C + D$$
$$C + D \rightleftharpoons E + F$$
$$E + F \rightleftharpoons G + H$$

In the sequence of reactions shown above, the end-products of the first reaction (C and D) are the initial reactants of the second. If, for some reason, the second reaction is blocked, the first reaction to the right will soon slow and stop because of the accumulation of end-products C and D. If, on the other hand, the second reaction is accelerated, the first reaction proceeds at an increased rate because of the removal of end-products C and D. In addition, the last reaction, in which E and F are initial reactants, is also accelerated.

Oxidation-Reduction Reactions. Among the most important chemical reactions in biological systems are **oxidation-reduction reactions,** since they release the energy upon which the biological engine runs. Originally, oxidations were those reactions in which a substance combined with oxygen, as, for example, when a substance is burned in air. It was later realized that this definition is too narrow. The term **oxidation** now includes all those reactions in which one of the following occurs:

(1) combination with (or addition of) oxygen
(2) removal of hydrogen
(3) loss of an electron (which also occurs in either addition of oxygen or removal of hydrogen)

The removal of electrons by any of these means implies the presence of something to accept the electrons that are removed—a source of oxygen or an acceptor of hydrogen. This process by which an atom acquires an electron is known as **reduction.** Oxidation and reduction are always coupled, in that removal of an electron from one substance (oxidation) is always accompanied by the addition of an electron to another (reduction). Oxidation-reduction reactions can therefore be thought of as *electron transfer reactions.*

Oxygen is a very good hydrogen acceptor, but it is not the only one. A substance may be oxidized by removal of hydrogen; if oxygen takes up the hydrogen, the oxygen is reduced:

$$RH_2 + \frac{1}{2}O_2 \rightarrow R + H_2O$$

where R represents the remainder of the molecule. Such oxidation-reduction reactions are commonly written in a manner that indicates the nature of the hydrogen transfer:

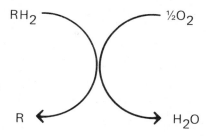

Oxidation reactions are important to biological systems because they involve the release of energy. It is by oxidation that foodstuffs are made to release the energy stored in the chemical bonds of their molecules. Many reactions in the body require energy, and this energy is obtained from energy-releasing oxidative reactions.

Although many oxidative reactions can occur without oxygen, human beings cannot live without oxygen for very long. This is because biological oxidations occur in sequence. The electrons are passed from one compound (which is oxidized) to another (which is reduced) until eventually oxygen is required as the final acceptor.

Enzyme-assisted Reactions. At body temperature the rates of chemical reactions in the body are much too slow to sustain life. They require some sort of assistance, which is provided by specific cell-produced proteins known as **enzymes.** Virtually every chemical reaction in both plants and animals is enzymatic, with a different enzyme for nearly every step of every reaction.

Enzymes, as organic catalysts, accelerate reactions without being permanently affected by those reactions. Since they are not consumed in the reaction, and do not appear among the products, theoretically they should be present and available indefinitely. In practice, however, some of the enzyme is destroyed or inactivated, so that cells must continually renew their supplies.

An enzyme acts upon a substance known as the *substrate,* and hastens a particular reaction involving that substrate. Because each enzyme is able to act only on a certain linkage or molecular configuration, they are highly spe-

FIGURE 2–5
The "lock-and-key" hypothesis of enzyme action. The enzyme fits a specific substrate and forms a complex with it. When they separate the substrate has been altered, but the enzyme remains unchanged.

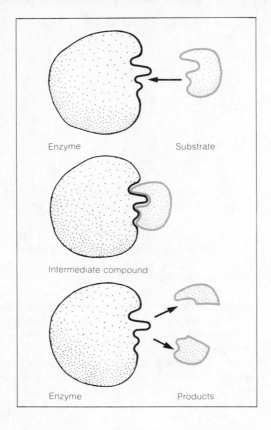

Enzyme   Substrate

Intermediate compound

Enzyme   Products

cific. The enzyme sucrase,[6] for example, catalyzes the hydrolysis of sucrose, but it cannot split maltose or lactose; those cleavages require maltase or lactase. The reason for the specificity is not completely understood, but a "lock-and-key" hypothesis provides a possible explanation (Figure 2–5). Enzymes are protein molecules whose chains are folded and coiled. The side groups projecting from it give this part of the enzymes a characteristic configuration. These groups form the reactive part of the enzyme, or "active site," which serves as the "key" that fits a "lock" of complementary shape on the substrate. When the two have combined into an enzyme-substrate complex, the key "turns" in the lock, and when the complex breaks up, parts of the lock (substrate) have been altered, but the key (enzyme) remains unchanged.

---

[6]An enzyme is named by adding the suffix *-ase* to a stem that tells its action and/or substrate. For example, succinic dehydrogenase oxidizes succinic acid by removing hydrogen. Many enzymes split their substrates by hydrolysis, and are identified only by the substrate. Sucrase, amylase, and peptidase act upon sucrose, starch, lipids, and peptides, respectively.

The effectiveness of an enzyme in increasing the rate of a reaction varies with conditions. Within rather broad limits the rate of the reaction is proportional to the amounts both of enzyme and substrate present. Extremes of pH will destroy an enzyme, but between the extremes there is an optimum pH. It is usually near the pH of the environment in which the enzyme normally functions, and hence is not the same for all enzymes, as shown in Table 2–6. Temperature also affects enzyme function since, within limits, rates of chemical reactions in general approximately double for every 10°C rise in temperature. When the temperature is below 10°C however, enzymes are virtually inactive, and when it is above about 60°C the protein is changed and the enzymes are irreversibly inactivated. In humans the optimum temperature is around 45°C, which is a few degrees above body temperature (37°C). Enzymatic reactions are slowed if their products are allowed to accumulate (Law of Mass Action), and they can be inhibited by substances that interfere with the active sites on the enzyme.

A number of enzymes require the action of another agent in order to be effective. Some enzymes are produced in an inactive form and must have their active sites exposed before they can participate in a reaction. Others require the presence of a metal ion to help bind the enzyme and substrate together during the reaction. Some enzymes require the presence of a nonprotein group, called a *coenzyme*, to serve as an intermediate carrier or transfer agent. Coenzymes combine with groups that are split off in the reaction and transfer these groups to another compound. One of these, coenzyme A, is a carrier of two-carbon (acetyl) groups, and there are several important coenzymes that are carriers of the hydrogen transferred in oxidation-reduction reactions.

If a chemical reaction requires a particular enzyme, the ability of a cell or organism to carry out the reaction depends upon its ability to produce that enzyme. What a cell can do depends upon its enzyme content, and synthesis of enzymes therefore becomes a critical element in cell function. How well a given chemical reaction "goes," however, depends upon the presence of the necessary cofactors as well as synthesis of the enzyme. A given enzymatic reaction can be accelerated by increasing the rate of enzyme activation or the availability of cofactors and coenzymes, and also it can be inhibited if any of these are reduced, or if conditions of temperature or pH are unfavorable. As we shall see, many activities of cells in the body are regulated to a great extent by controlling the effectiveness of the enzymes in this way.

TRANSLOCATION OF MATERIALS

Every activity of every cell in the body is related to and dependent upon transport of substances across membranes. All things that enter the cell—nutrients, oxygen, electrolytes—and all things that leave the cell—waste products, carbon dioxide, electrolytes, and secretions—must pass through the cell membrane. Even entry into the body entails crossing the membranes in the lung or intestine.

The movement of materials across membranes in living systems is subject to the same physical laws that control the movement of individual particles (molecules or ions) in a solution. The movement of the particles may be restricted by the presence of a barrier (a membrane) and can be drastically altered if that barrier is living tissue such as cell membrane. The physical laws and the effect of a simple barrier are purely physical factors and can be demonstrated adequately in nonliving systems. However, the cell membrane is unique to living organisms, and its presence introduces many new possibilities in the movement of particles.

Movement of Particles

The molecules of a gas are in constant random motion. The distance any one molecule travels before striking something depends upon the number of molecules per unit volume. The velocity of the molecules varies directly with the temperature. These statements apply equally to liquids and solids, but in them the extent of movement is

TABLE 2–6   OPTIMUM pH FOR SOME ENZYMES

Enzyme	Source	Optimum pH
Amylase	saliva	6.7–6.8
Pepsin	stomach	1.5–2.5
Lipase	pancreas	7.0
Trypsin	pancreas	8.0–11.0
Phosphatase	bone	8.4
Succinic dehydrogenase	mitochondria	7.4
Carbonic anhydrase	red blood cell	5.0–9.0
Adenosine triphosphatase (ATPase)	brain, muscle	7.5

greatly restricted due to the greater density and the bonds of attraction between particles. Biologists are concerned most often with either gases or solids in a liquid, and to a lesser extent with mixtures of gases, although other combinations are possible. In any case, it must be remembered that the constituent particles (molecules or ions) move constantly and randomly; only their ranges and velocities vary.

Diffusion of Gases. Samples taken from different parts of a chamber containing a mixture of gases should, when analyzed, show identical composition throughout the chamber. The random movement of gas molecules has brought about complete mixing. If it were possible (it isn't) to remove a wall separating two different gases (call them gas X and gas Y) without disturbing the contents of either side, an initial analysis of several gas samples would show gas X to be entirely on one side and gas Y to be entirely on the other. This situation would, however, exist only at the very moment of removal of the barrier. As the gas molecules move at random, some of those near the line of demarcation, or interface between the gases, quickly move through that line to the other side of the container. In time, the line of demarcation becomes progressively less distinct as more and more molecules of each gas move farther into the other side of the chamber. Successive samplings would then reveal a continuous drop in the concentrations of each gas at their original ends of the chamber. Eventually, the gases are completely mixed and evenly

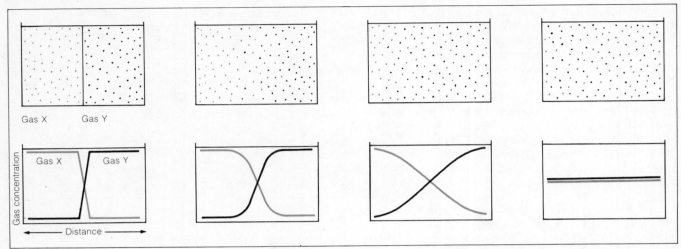

FIGURE 2-6   Diffusion of gases in a closed chamber. Upon removal of the barrier separating two gases, there is an initially steep gradient, but the random movement of the gas molecules carries more molecules of each gas to the opposite side. With continued movement the gradient diminishes until, at equilibrium, it has disappeared completely.

distributed throughout the container. This mixing process, illustrated in Figure 2-6, is called **diffusion.**

Several important points are illustrated by this example. Upon the initial removal of the barrier, a graph of the concentration of gas X (or Y) would show a sharp difference in concentration between the right and left sides of the line of demarcation, that is, a steep gradient. **Gradient** expresses the degree of difference between two sites. In this case it is the difference in concentration at two points adjacent to the line of demarcation but on opposite sides of it. The term *gradient* is applied to any situation in which there is a suitable difference, or slope, such as a difference of temperature, pressure, electrical charge, or height (as with two points on the side of a hill).

The rate of diffusion depends not only on the initial concentrations, but on the temperature, pressure, area of interface, and certain physical characteristics of the gases as well. Since these factors can be measured, it is possible to calculate the rate of diffusion. In effect, however, the rate of diffusion depends mainly upon the gradient. When diffusion occurs, the boundary between the gases becomes less distinct and the gradient less steep, until finally there is no longer any gradient for ei-

ther gas. At this point diffusion is complete.

Molecules of each gas tend to move down their own concentration gradient. Actually, of course, the molecules of each gas are moving in all directions, but more of them move *down* the gradient, that is, there is a net movement in that direction. Initially, more molecules of gas X will move from left to right than right to left simply because there are more of them on the left at the beginning. All else being equal, the steeper the gradient, the greater the **net movement,** that is, the greater the rate of diffusion. When the gradient has been eliminated, an **equilibrium** has been established. At equilibrium the molecules are still in motion, of course, but there is no net movement in any direction. At equilibrium in the above example, as many molecules of gas X are moving to the left as to the right, and as many molecules of gas Y are moving to the right as to the left. The composition of the gas in the chamber is identical (or homogeneous) throughout.

Solids, liquids, or gases may diffuse in liquids in much the same manner as do gases alone, and the rate of diffusion is determined in the same manner. A familiar example might be that of cream and sugar in a cup of coffee. The molecular movement is usually not rapid enough to suit us, however, and it is often desirable to hasten a state of equilibrium by stirring the coffee.

Movement Across the Cell Membrane

In any transport situation the rate of transfer is affected by the gradients and the membrane characteristics. When the barrier is a cell membrane, some additional factors are introduced. In many cases the barriers must be considered individually, since cell membranes do not all have the same characteristics of permeability. Transfer across an artificial membrane is visualized as passage through holes or pores which are quite abundant. Cell membranes probably do not actually contain pores as such, but the protein molecules in the membrane function much as if they contained tiny channels. Water passes readily, small molecules and ions pass with more difficulty, and many large molecules do not penetrate at all. In addition, there is an electrical charge on the cell membrane. The external surface carries a positive charge which may interfere somewhat with the passage of positively charged particles (cations), but negatively charged particles (anions) and uncharged particles are not restricted in this manner.

The processes by which materials move across cell membranes can be divided into two basic categories: **physical transport processes** and **physiological transport processes.** Physical transport processes are those that are governed entirely by physical laws and, although they are important in living organisms and cells, they can also occur in artificial systems. *Diffusion, osmosis,* and *filtration* are the physical transport processes with which we will be concerned. They are often referred to as **passive transport** because the net movement of particles (ions and molecules) is in the direction of the gradient, that is, "downhill." Such transfer requires no energy expenditure by the cell (hence, "passive"). It will occur if the membrane is permeable and if there is an appropriate gradient.

Physiological transport occurs only in living systems. The cell membrane is involved in the transport process in some way. In many cases it provides a "carrier" that helps the particles to penetrate the membrane. One type of carrier-mediated physiological transport is *facilitated diffusion,* which is a form of passive transport since no energy is required and the particles diffuse down a gradient. Most carrier-mediated physiological transport systems, however, transport particles against a concentration or electrical gradient, and require energy to do so. Such systems, known as **active transport systems,** "pump" particles across the membrane. There are also other specialized membrane processes which enable large molecules (to which cell membranes are not permeable) to enter or leave the cell.

Physical Transport Processes. *Diffusion.* The introduction of a membrane between two solutions of different composition immediately creates a barrier to the diffusion of the particles. How much it interferes depends upon the nature of the membrane and of the diffusing particles. This can be illustrated by a few simple examples.

1 A sheet of aluminum foil separating two solutions, such as water and saline (salt solution), creates a most effective barrier; any crossover by either water or saline is blocked. Days or weeks later there still will be no change (assuming no leakage or evaporation). Aluminum foil is an *impermeable membrane,* at least to water and saline.

2 If the foil separating the two solutions is removed, diffusion occurs in the manner already described for two gases, but at a somewhat slower rate.

3 A piece of wire screen between the two solutions is almost no barrier at all, and diffusion occurs almost as rapidly as if there were none. There are so many open spaces in the screen that molecules which move in that direction are more likely to pass through a hole than to strike wire and bounce back. Water molecules move into the saline and dilute it, while salt moves into the water, increasing

its salinity. Equilibrium is reached when there is a uniform concentration of salt on the two sides.

4   If some plastic beads are added to the saline in the above example, water and saline are still free to pass through the wire screen, and do, but the beads are blocked. This "membrane" is permeable to water and salt, but impermeable to plastic beads.

The permeability of a membrane is not absolute; it depends greatly on the relative sizes of the openings in the membrane and of the particles (ions or molecules) that might pass through. The size of the openings in membranes varies, but not as greatly as particle size. In biological systems the membranes are not likely to embrace the extremes of permeability shown by the foil and the screen, but they do vary in their permeability to particles found in the body fluids. They are often said to be *selectively permeable,* because they are able to "select" which particles pass and which do not.

In **simple diffusion,** the only criterion for passage mentioned so far is that of the relative size of particle and pore, but some substances may diffuse through the substance of the membrane. This route is limited to lipid-soluble substances that can dissolve in the lipid portion of the cell membrane. Oxygen and carbon dioxide are transferred in this way.

Osmosis.   The presence of nondiffusible solutes creates an interesting situation, resulting in a special type of diffusion known as **osmosis** (Figure 2–7). A cellophane membrane can be set up to separate water from a sucrose (table sugar) solution. Such a membrane is permeable to the solvent (water), but not to the solute (sucrose). Since the sucrose cannot cross, there is no diffusion of sucrose, in spite of a concentration gradient. Water molecules are free to move, however, and do, moving down the water gradient into the sucrose solution. As water dilutes the sucrose, equilibrium is approached; it cannot be achieved, because no matter how much the sucrose is diluted there is always some sugar on one side and none on the other. Osmosis is the diffusion of a solvent (usually water) through a semipermeable membrane when there is a concentration gradient of a nondiffusible solute.

Students often are troubled by the concept of osmosis, but the problems can be lessened if it is remembered that the laws are exactly the same as for other types of diffusion. A net movement of particles occurs when there is a concentration gradient, and the direction of the net movement is always down the gradient of the particles that move. In osmosis, however, only the solvent penetrates the membrane; so it shows the only net movement. In our example, one side has a "concentrated solution of water"; on the other side the water is "diluted" with solute.

As with simple diffusion, the rate of osmosis is directly related to the magnitude of the concentration gradient. As time passes, the gradient and the rate of osmosis both decline. One might expect that all of the water would enter the sucrose solution, but it does not, since new conditions are created as the volume in the sucrose compartment

FIGURE 2–7   Osmosis. The presence of a nondiffusible solute on one side of a membrane creates a concentration gradient, causing water to move into the solute. The net movement of solvent (osmosis) dilutes the solute, reduces the gradient, and increases the volume in the solute compartment.

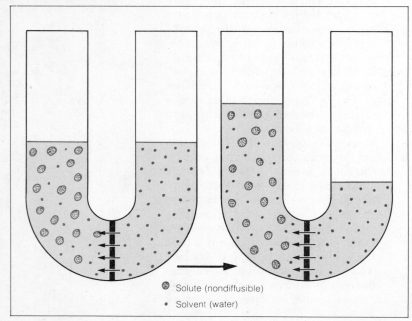

⊕ Solute (nondiffusible)

• Solvent (water)

increases. If the sucrose is in a container of fixed volume, the pressure inside rises as osmosis forces more water molecules into the limited space. This pressure is the force exerted against the membrane and other walls of the container by the molecules inside. As the pressure rises, more and more randomly moving water molecules strike the membrane and pass through back into the water where the pressure is not as great. After a time the pressure in the container is such that the number of water molecules being driven out by the pressure gradient is equal to the number being forced in by the osmotic (concentration) gradient. There is then no further net movement of water and an equilibrium is established.

For this reason, osmotic pressure is sometimes expressed in terms of the opposing pressure required to prevent osmosis. If the chamber containing sucrose has an opening in the top with a tube attached, the fluid will rise in the tube as water enters the chamber and its volume and pressure increase. The height to which the fluid rises in the tube is a measure of the osmotic pressure.

Filtration. The introduction of a mechanical force, the *fluid pressure,* as a factor in transport through a membrane leads to a consideration of **filtration** (Figure 2-8). The use of filter paper or a strainer to separate certain particles from a solution is probably familiar. The filter paper or strainer serves as a selectively permeable membrane by permitting passage of the solvent and whatever solutes can penetrate. In filtration the net movement of fluid and diffusible solutes does not depend upon a concentration or osmotic gradient, but rather upon a mechanical pressure gradient or filtration pressure which, in this case, is the weight of the solution due to the pull of gravity. In a cell the force of gravity is not the only way to cause filtration, however, since any mechanical force tends to squeeze solvent and diffusible solutes through the membrane. In fact, the resistance to osmosis that developed in the closed chamber described above is a filtration pressure.

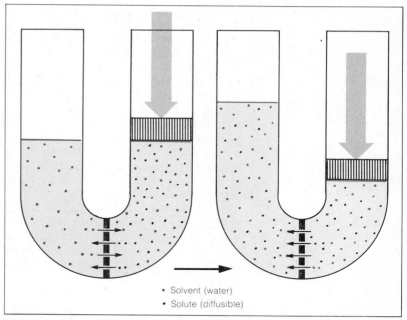

- Solvent (water)
- Solute (diffusible)

FIGURE 2-8   Filtration. The membrane is permeable to the solute, and an equilibrium exists. Pressure applied to one side forces both solute and solvent to the other side and changes the volumes in the two compartments.

Response to electrical gradients. Besides the concentration and mechanical pressure gradients that cause diffusion, osmosis, and filtration, there may also be electrical forces acting on the solute particles. When the molecules in a solution ionize, one must deal with charged particles rather than uncharged molecules. If there is an excess of cations (+) on one side of the membrane and an excess of anions (−) on the other, the resulting charge gradient is a force causing diffusion of ions to bring about an electrical equilibrium (if the membrane is permeable to the ions). The equilibrium sought is **electroneutrality,** in which there is an equal number of positive and negative ions on the two sides, and thus no charge gradient. A net positive charge on one side of the membrane would tend to drive cations to the other side and attract anions to it until the total charge is the same on both sides. When an electrical gradient causes movement in one direction, and a concentration gradient causes diffusion in the other direction, the equilibrium finally reached will be a compromise, or a balance, between the two opposing forces.

Cells in solution. In the previous chapter it was pointed out that survival of cells depends to a great extent upon the composition of the internal environment, and we have just seen some ways in which that environment can affect cells. Let us now look more closely at a representative cell, the red blood cell, in its environment, the blood plasma, and see how diffusion and osmosis affect that cell. (Filtration need not be considered here because there is no filtration pressure gradient between blood cells and plasma.)

To do this, we must know something about the permeability of the cell membrane and the gradients that exist across it. Both plasma and intracellular fluid (the cytoplasm) contain water, certain electrolytes, and certain organic substances such as glucose and protein. The membrane of red blood cells, like that of most cells, is impermeable to proteins, only slightly permeable to glucose and electrolytes, but freely permeable to water. Although there is little diffusion of solutes, you could predict that the total concentration of nondiffusible solutes is very nearly the same inside the cell as out, for if there were a gradient, water would move. Normally there is no osmotic gradient between red blood cells and plasma (or between other cells and their environments), and we say that plasma is *isotonic* to the cells. An **isotonic solution** is one that will not cause osmosis; it has the same concentration of osmotically active particles and the same osmotic pressure as the fluid in the cells.

If red blood cells are placed in pure water, there will be sizable concentration gradients, since there are solutes inside and none outside. Water will enter the cells, increasing their volume until the cell membranes can no longer withstand the pressure, and they will rupture. Cells placed in a slightly diluted plasma will swell, but probably not burst. These solutions are **hypotonic solutions.** Their osmotic concentrations are less than that in the cells, and osmosis occurs, causing cells to swell and perhaps to rupture if the gradient is great enough (see Figure 2–9).

If cells are placed in a solution whose osmotic concentration is greater than that in the cell, osmosis will occur in the opposite direction. Water will leave the cells and they will shrink, a condition known as *crenation* in red blood cells. Such a solution is a **hypertonic solution.**

Osmotic pressure is related to the total number of particles in solution, not their size or species. Most of the particles in body fluids that contribute to osmotic pressure are inorganic ions, the electrolytes obtained from dissociation of various salts. Although their concentration is low, the particles are very small and they are ionized, so the number of particles (ions) is very large. A 0.9 percent salt (saline) solution is isotonic with the body fluids of humans, and is sometimes called a *physiological saline* or *normal saline*. Red blood cells placed in such a solution neither swell nor shrink. When fluids, or medications in solution, are administered to patients intravenously (injected into a vein), an isotonic solution is used, for it does not harm the cells. (One would not inject water into a vein!) A number of standard physiological solutions can be prepared that contain not only sodium and chloride, but also other electrolytes and some of the other constituents of body fluids in their appropriate concentrations.

Physiological Transport. So far the transport processes we have discussed do not require the presence of living tissues. Physical transport processes, although of great importance to cells, can also occur in systems that do not include cells. Because they require a gradient and are limited to small particles, these processes cannot account for the transport of all things needed by cells to carry on their activities. The cell

FIGURE 2–9   Red blood cells in an isotonic solution are biconcave discs. In a hypertonic solution they shrink (crenate), and in a hypotonic solution they swell.

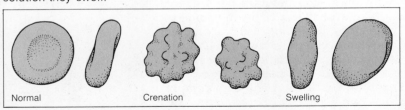

Normal   Crenation   Swelling

membranes are part of a living organism and can participate in the transfer process; they can actually help particles through.

Carrier-mediated transport. Much of the assistance is due to the presence in the membrane of a substance appropriately called a **carrier.** Solutes that are too large to diffuse through "pores," or too lipid-insoluble to pass through the membrane can combine with something on the surface of the membrane—the carrier—to form a carrier-solute combination. **Facilitated diffusion** is such a process; the carrier-solute combination diffuses across the membrane to the other side where the solute is released (Figure 2–10). Facilitated diffusion is very specific, since a given carrier can successfully combine with only one or a few closely related substances. A small amount of carrier can transport a considerable number of particles, however, because it can be "used" many times. Particles may be transferred in either direction, but movement is always down the concentration gradient. The rate at which the carrier can function is limited, however, because the carrier system can become saturated. Glucose is the most important substance transported by this mechanism.

Facilitated diffusion, because it requires the production of the carrier molecule by the cell, differs from simple diffusion and other passive processes. But facilitated diffusion is considered to be passive transport since the cell expends no energy in the transfer process; once the carrier has combined with the solute, the transfer process proceeds like simple diffusion, moving particles down the gradient.

Active transport, on the other hand, is essentially a pumping process, since particles are caused to go where they would not otherwise. Substances may be transferred that normally cannot penetrate the membrane or they may be transferred against a gradient. A gradient may be created or maintained, not reduced as in passive transport. The mechanisms by which the actions are brought about are not well understood, but they resemble facilitated diffusion in that both are be-

lieved to require specific carrier molecules located in the cell membrane. The substance to be transported is combined in some way with the carrier, transported across the membrane, and then released (see Figure 2–10). The process differs from that of facilitated diffusion in that it involves the expenditure of energy. The energy is required because the transport is against a gradient, and this is work in the physical sense. The needed energy is obtained from the metabolic processes of the cell. Therefore, active transport can be blocked by anything that interferes with cell metabolism, such as a lack of oxygen or agents that block some step in the sequence of energy-releasing reactions. There are several hormones which may alter the rate of transfer, and the cell itself can exert some control over the availability of its energy supply. In addition, even under optimum conditions the rate of active transport has an upper limit at which the carrier system becomes saturated.

FIGURE 2–10   Carrier-mediated transport. A. Facilitated diffusion. A solute in high concentration outside the cell combines with a carrier molecule in the cell membrane, diffuses down its concentration gradient, and is released to the inside of the cell. B. Active transport. A solute combines with a similar carrier and is transported across the cell membrane and into the cell against its concentration gradient. The cell must provide energy to transport a solute against a gradient.

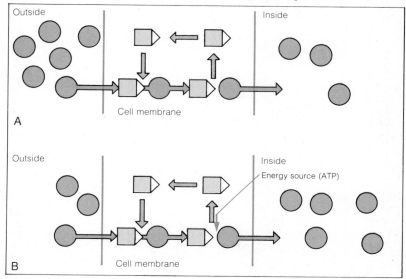

Transport of large molecules. Even active transport cannot move large molecules across cell membranes. Under certain conditions it is desirable for proteins, for example, to be taken into the cell (endocytosis) or out of it (exocytosis). This mechanism is a rather clever one, for the material gets into or out of the cell without actually passing through the cell membrane.

Endocytosis, the taking in of large molecules, involves two similar processes known as phagocytosis and pinocytosis. **Phagocytosis** ("cell eating") is the transfer of solid particles, and **pinocytosis** ("cell drinking") is the transfer of fluid droplets into the cell. These processes have long been known to occur, but it has only been relatively recently that the extent of their occurrence has become apparent. As shown in Figure 2–11, either process begins when a small portion of the surface membrane indents and invaginates around a foreign substance, forming a small pocket which closes over and becomes pinched off. This last step leaves a tiny membrane-enclosed vesicle or vacuole in the cytoplasm of the cell. The material ingested is still separated from the remainder of the cell contents by membrane.

Although pinocytosis involves the intake of fluid, it is much more likely to occur when the fluid contains certain solutes, particularly protein molecules. It has been suggested that some form of reaction between the protein solute and the cell membrane helps initiate the process. Pinocytosis is therefore viewed as a mechanism by which large molecules, such as protein, can be taken into the cell. The fate of pinocytotic vesicles in the cytoplasm is not clearly understood. They may break up into smaller vesicles whose contents eventually leak out into the cytoplasm, or perhaps they merge into the endoplasmic reticulum.

The solid particles taken into the cell by phagocytosis may include bacteria and other materials potentially dangerous to the cell. It has been suggested that when vesicles formed by phagocytosis come in contact with lysosomes in the cytoplasm their membranes fuse, permitting the enzymes of the lysosomes to reach and destroy the phagocytized material. In humans, phagocytosis is most pronounced in certain white blood cells (neutrophils) and certain connective tissue cells (macrophages). The blood phagocytes move throughout the body in the blood, but are able to leave the blood vessels at the site of an inflammation. The connective tissue phagocytes are found in large numbers in several organs such as the spleen, where they help remove worn-out red blood cells.

Exocytosis, a kind of reverse pinocytosis, is best illustrated by the process of secretion. Recall that substances to be secreted by cells are eventually packaged into secretion granules, which are tiny membrane-enclosed sacs that are pinched off from the Golgi apparatus. To be secreted, a granule moves toward the edge of the cell and its membrane fuses with that of the cell. The fused membrane soon opens at that site and the contents of the granule are spilled out into the surroundings. The secretion is now outside the cell, and the cell membrane remains intact.

METABOLISM AND ENERGY EXCHANGE

One of the fundamental properties of living matter is **metabolism.** It is concerned with the processing of foodstuffs by the cells, specifically the release and utilization of energy from

FIGURE 2–11   Phagocytosis and pinocytosis. Specific substances in the immediate environment cause the cell membrane to invaginate and enclose the substance, which soon is in a vacuole within the cell. A. In phagocytosis a particle, such as cell debris or a microorganism, is taken into the cell. B. In pinocytosis a droplet of fluid is taken into the cell. Pinocytosis is usually triggered by the presence of certain proteins in the fluid.

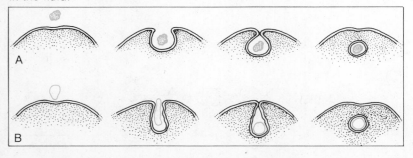

those foodstuffs. The other properties that characterize living tissues also depend upon the energy exchanges associated with metabolism. The growth of a cell or organism, its ability to respond to changes in the environment, and its reproduction all require energy. It is appropriate, therefore, to consider energy and how it is important to the living organism.

Energy and Work

We can define **energy** as the ability to do work—to produce a change or to move something. Since neither the type of work nor the nature of the change is specified, we may assume that energy can do any kind of work or produce any kind of change. There are, however, several kinds of work which are particularly important in biological processes. *Mechanical work* involves the movement of mass, *chemical work* the movement of atoms within and between molecules, and *electrical work* the movement of electrons. Each type of work is performed in response to the application of mechanical, chemical, or electrical energy, respectively. A fourth type, *thermal energy* (heat), associated with the random movement of particles, is also important in biological reactions, but the body is unable to use thermal energy for the performance of work.

These different kinds of energy may be in the form of either potential or kinetic energy. **Potential energy** is stored energy, while **kinetic energy** is the energy of motion. Kinetic energy is always associated with the movement of matter. It exists when a rock rolls down a hill, water rushes along a stream bed, blood flows in a blood vessel, or the legs of a jogger move. Potential energy, as stored energy, does not do work until it is released. It is the potential of a coiled spring or a rock at the top of a hill (both represent mechanical energy). It is the energy stored in a flashlight battery (electrical energy) or in the chemical bonds between the atoms of a molecule of glucose (chemical energy). Release of this energy means that it is no longer stored potential energy; it becomes kinetic energy, and work is done and heat is produced.

The fact that it is possible to convert energy from one form to another is basic to all energy exchanges, including those in the body.

The total amount of energy available in the universe is constant; energy can be neither created nor destroyed. When movement occurs and potential energy is converted to kinetic energy, all of the potential energy can be accounted for—no more and no less. It does not *all* appear as movement of objects and performance of work, however, since a certain portion of the energy appears as heat. Heat is produced in all types of energy conversions, and it is evident as the heat produced by a motor, by chemical reactions in test tubes, by an electric light bulb, or by an active human being. Since the heat produced cannot be converted to usable form in the body, it is energy lost from the system. If the system is to continue operating, that lost energy must be continually replaced.

Machines can obtain their energy from fuels such as gasoline and oil and from electric power, but the entire energy source for humans and other animals is the chemical energy obtained from the foods eaten. Chemical energy incorporated in the chemical bonds of the molecules of foodstuffs is released as these materials are metabolized. This energy is then stored as chemical potential energy, later to be retrieved and converted to other forms of energy. The energy released may be used to do chemical work, or it may be converted to electrical or mechanical energy to do electrical or mechanical work. However, the portion that is converted to thermal energy is lost from the body.

The amount of work performed in any energy exchange is expressed in terms of the movement produced. An example from a mechanical system will illustrate the relationship. A box placed on a shelf has acquired potential energy by virtue of its elevated position, since work had to be done *on* the box to get it there. The energy for that work came from the machine or individual that raised the box. The amount of energy required and the amount of work done depend upon the height of the shelf and the weight of the box. Work may thus be defined in terms of displacement and load:

Work = Displacement × Load
$$W = D \times L$$

The displacement in this case is the shelf height. In order to put a five-pound box on a shelf three feet high, fifteen foot-pounds of work must be done. Thirty foot-pounds of work would be done in lifting a ten-pound box to a three-foot shelf, or a five-pound box to a six-foot shelf.

If the box is very heavy, the individual or machine may not be able to lift it at all, and the displacement would be zero. By definition no work has been done, although anyone who has tried unsuccessfully to lift a piano knows that he or she "worked" hard in the attempt. A clear distinction must be made between *mechanical work,* as described by displacement and load, and what might be called *physiological work,* as described by effort. Energy is expended in the unsuccessful effort, but no mechanical work is done and all of the energy can be accounted for as heat. Heat is an ever-present and sizable component of all energy transformations. In some cases it accounts for all of the energy released. It is not surprising, therefore, that the units used to express quantities of energy are those of heat, namely calories and kilocalories (1000 cal = 1 kcal).

Energy Exchange

The energy required by an organism comes from the foodstuffs it takes in. The energy is released when the chemical bonds of the food molecules are broken. However, not all chemical reactions involve the release of energy; many of them require or consume energy. In a gross oversimplification, it could be said that energy-releasing reactions tend to be those of breakdown and *degradation,* in which relatively large and complex molecules are broken down to simpler ones with an accompanying release of energy. Reactions that require the addition of energy include those of *synthesis,* in which relatively small and simple molecules are combined to form larger and more complex ones, such as enzymes, with the incorporation of chemical energy into the bonds formed.

In biological systems the energy-releasing breakdown reactions are grouped as **catabolism;** the energy-consuming synthesizing reactions are known as **anabolism.** The energy consumed in the anabolic reactions is the energy released in the catabolic reactions. It would make a fine perpetual motion machine, except for the fact that some of the energy is lost in each exchange, primarily as heat which the cells cannot use; so new energy must be constantly supplied to the system.

Besides obtaining new energy, the organism must also deal with the problem of coupling catabolic and anabolic reactions. Energy released in the metabolism of food must be captured and stored so as to be available whenever it is needed, such as for synthesis or for other forms of work. If all of the energy in a molecule of sugar were to be liberated in one step, only a fraction of it could be used, leaving an immense quantity of energy to be dissipated as heat. If additional energy were needed later, more sugar would have to be oxidized with still more heat production. Such an inefficient operation would be rapidly self-defeating, since the energy sources would soon be depleted and animal cells could not function at the high temperatures that would result.

Adenosine Triphosphate (ATP) and High-Energy Phosphate (HEP) Bonds. There are two characteristics of the catabolic process that enable the energy released in oxidation to be coupled to energy-consuming reactions. First, a molecule of sugar (or other energy source) is oxidized in a stepwise fashion, with its energy released in a number of small "packets" rather than all at once. Secondly, cells contain compounds capable of forming a type of chemical bond that incorporates a larger quantity of energy than most. They "soak up" and hold the packets of energy released in the oxidative reactions. Later, when these bonds are broken, that energy is available for use in energy-consuming reactions.

Much of the energy transfer is associated with oxidation-reduction reactions in which hydrogen atoms or electrons are passed from one substance to another. The energy liberated in these hydrogen atom or electron transfers is taken up by various intermediaries, the most important of which is **adenosine triphosphate (ATP),** a universal *intracellular* carrier of chemical energy:

3 phosphates

adenine   ribose

The ability of ATP to take up and release energy is chiefly due to the presence of high-energy phosphate bonds. Of the three phosphate groups in ATP, two have high-energy bonds, designated by ~. These bonds contain several times as much energy as the usual phosphate linkage. Breaking one of them also yields phosphate and adenosine diphosphate (ADP), which still has one high-energy bond. The phosphate that is released in the discharge of ATP may be transferred, with its high-energy bond, to another substance in a process known as *phosphorylation.* Alternatively, the energy of the high-energy bond may be used elsewhere, leaving inorganic phosphate without an energy-rich bond.

Oxidation and Energy Release. The catabolic aspect of metabolism is usually thought of in terms of oxygen consumption and carbon dioxide production, that is, as the oxidation of foodstuffs. Oxidation involves the loss of electrons, either by the addition of oxygen or by the removal of hydrogen. Oxidation is always coupled with reduction, because oxidation of one reactant results in the reduction of another, which receives or takes up the elec-

trons. In biological oxidations, most of the energy is released in a series of oxidation-reduction reactions in which electrons from hydrogen are transferred from one intermediary compound (enzymes and coenzymes) to another and eventually to oxygen. The energy released in certain of these transfer reactions is used to form ATP from ADP and phosphate (PO_4).

Oxidation of foodstuffs in the body may be divided into an initial phase and a final phase. In the initial phase the various energy sources are metabolized individually along separate pathways. The products are then channeled into a common chemical pathway or sequence, so that the final phase, which accounts for the greatest energy release (ATP formation), treats nearly all foodstuffs in common. For now, the general, simplified scheme indicated in Figure 2−12 will be sufficient.

For carbohydrates, the initial phase involves the breakdown of sugar (mostly glucose), and is called **glycolysis.** Occurring in the cytoplasm, glycolysis involves many more steps (14 in all), than shown in the diagram, and each step is catalyzed by a different enzyme (12 in all). High-energy phosphate bonds are used in certain steps and are formed in others. Thus a small amount of energy can be obtained from the partial oxidation of glucose.

Oxidation of fats begins with their breakdown to glycerol and fatty acids. Glycerol enters the glycolytic pathway, but fatty acids must be further oxidized before they enter the common pathway. Proteins are broken into amino acids and, after removal of the amino group (NH_2), some of them also enter the glycolytic pathway. Thus, by the end of the initial phase, the carbohydrates and parts of the fat and protein molecules have entered a common metabolic pathway.

The final phase is represented by a cyclic sequence of reactions known as the **Krebs cycle** (Krebs worked it out). It is also referred to as the *citric acid cycle* (as citric acid is the first acid in the cycle), or as the *tricarboxylic acid cycle* (several of the components are acids with three carboxyl groups). The entire sequence of events is sometimes aptly referred to as a "metabolic mill," since

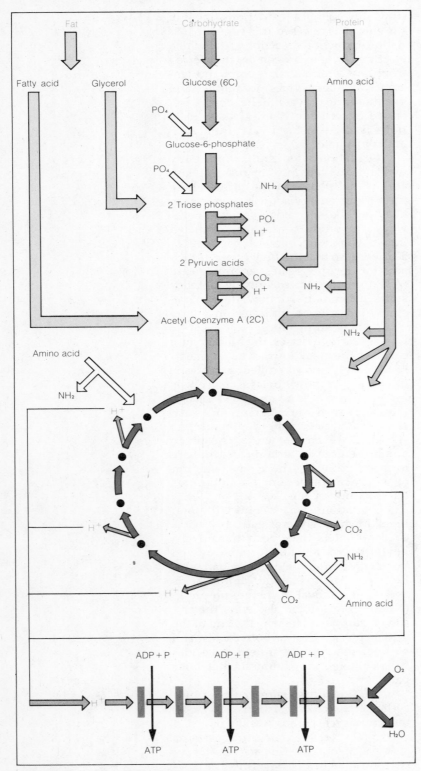

FIGURE 2-12   A summary of cellular oxidations.

substances thrown into it are "ground down" to the final end-products of carbon dioxide and water. A three-carbon molecule such as pyruvic acid first loses a molecule of carbon dioxide, and the remaining two-carbon (acetyl) group is carried into the cycle by coenzyme A. As the acetyl group passes through the cycle, two carbon dioxide molecules and several hydrogen atoms are "spun off." At the end of the cycle the pyruvic acid molecule has been completely catabolized. The carbon dioxide molecules formed diffuse out of the cell and into the bloodstream and eventually leave the body through the lungs.

As the hydrogens are released, they are promptly taken up by a coenzyme, the first of a series of hydrogen acceptors (thereby reducing the acceptor and oxidizing the donor). They are passed along from one acceptor to another (as in a bucket brigade). The last hydrogen acceptor is oxygen, and water is formed as the final end-product. Thus oxygen, which is so necessary to life, is not "used" until the last step in the metabolic sequence. If the oxygen supply is inadequate, the hydrogen cannot be removed at the end of the line; the hydrogen tends to back up until the entire process, including the Krebs cycle, is brought to a halt. Additional oxygen quickly relieves the congestion.

Three particular sites in the hydrogen transport chain involve the release of a considerable amount of energy which, in each case, is used to "recharge" a molecule of ATP. Thus for every pair of hydrogens transported along the line, three ATP molecules are formed. For every molecule of glucose that is completely oxidized to carbon dioxide and water, there is a net gain of 38 molecules of ATP, which is considerably more than is obtained from glycolysis alone.

The Krebs cycle and glycolysis, as well as the metabolism of lipids and proteins, are discussed in much greater detail in Chapter 24.

Protein Synthesis

The process of protein synthesis involves the chromatin material found in the nucleus of every cell. The chromatin material contains chromosomes in a dispersed form, and the chromosomes are made up of genes that carry the genetic information. The genetic information contains the directions for assembling a particular protein, that is, the sequence of amino acids in a given protein molecule, often an enzyme. These directions are faithfully reproduced every time a cell divides. This accuracy is critical because even a slight change in the structure of an enzyme—substituting one amino acid for another, or reversing the positions of two amino acids—might render the enzyme unable to catalyze its particular reaction. Failure of that single reaction might block a whole chain of events and impair the ability of a cell to synthesize some hormone, or absorb or metabolize some foodstuff. The deficiency might be merely an inconvenience, or it could be incompatible with life itself. In recent years a number of abnormal conditions have been traced to such errors in protein synthesis. One of these conditions is *phenylketonuria (PKU)*, a condition that may result in mental retardation, among other things. It is due to a lack of an enzyme necessary for the metabolism of the amino acid phenylalanine. Without the enzyme there is an accumulation of partially metabolized intermediates that, in high concentrations, are detrimental to brain development.

The genes contain **deoxyribonu-cleic acid (DNA),** which is an extremely large molecule consisting of a helical double chain or ladder of nucleotides (the double helix) (see Figure 2–13). Each nucleotide consists of a phosphate, a sugar (deoxyribose in DNA), and one of several nitrogen-containing bases. The sides of this twisted ladder are the alternating phosphates and sugars, and the "rungs" are the linkages between the bases of the two sides of the ladder. In DNA the bases are *adenine, guanine, cytosine,* and *thymine.* The linkages are such that adenine from one side is always paired with thymine from the other, and guanine is always paired with cytosine. The specific sequence of bases in the ladder is the code for protein synthesis.

Since the actual production of protein is carried out in the ribosomes of the cytoplasm, the DNA blueprint must be transmitted from the DNA in the nucleus to the ribosomes in the cytoplasm. This involves the other nucleic acid, **ribonucleic acid (RNA).** Hence, the first step in protein synthesis is the formation of RNA. RNA, like DNA, is composed of a chain of nucleotides, but the sugar is ribose (not deoxyribose as in DNA) and the chain is usually just a shorter single strand. The bases and base pairings are similar to DNA except that in RNA thymine is replaced by *uracil,* and therefore in RNA adenine from one side will always pair with uracil from the other side.

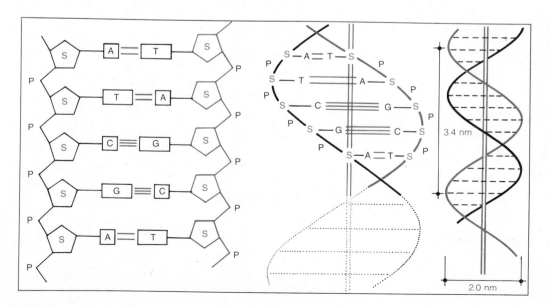

FIGURE 2–13 Schematic representation of the structure of DNA. The helical pattern is shown on the right. On the left, the sides of the chain are shown to be alternating phosphate (P) and sugar (S), and the cross-links are formed by paired bases (A, T, C, and G).

When a particular protein is to be synthesized, the appropriate part of a DNA molecule unwinds and the base-to-base cross-links are broken. The base sequence of the exposed portion of one strand of the DNA serves as a model or template for assembling an RNA molecule. Nucleotides whose bases are complementary to those of the exposed DNA pair up, and their phosphate and ribose portions link to form a strand of RNA with a specific base sequence. This process of transferring the information, or code, from DNA to RNA is known as *transcription*. The newly formed RNA chain "peels off" from the DNA template and moves through the openings of the nuclear membrane into the cytoplasm where it will become associated with ribosomes. Because it carries the genetic "message," this RNA is known as **messenger RNA (mRNA).**

As their name implies, ribosomes also contain RNA, but it is a different kind. **Ribosomal RNA (rRNA)** is produced from DNA in the cell nucleus, and the molecules are combined with protein to form RNA-protein granules. Two of these form a ribosome, which moves to the cytoplasm where it either remains free in the cytoplasm or comes to lie along the surface of the endoplasmic reticulum. Ribosomes are involved in bringing together the mRNA with the specific enzymes that cause the assembly of amino acids into a chain.

The actual synthesis of a protein molecule is thought to occur when a ribosome binds with a strand of mRNA and moves along it, "reading" the genetic code which consists of the sequence of nitrogen bases on the mRNA (as shown in Figure 2–14). Each set of three bases *(triplet)* on the mRNA represents a code-word *(codon)* that identifies a particular amino acid. If each base is thought of as a letter, then a four-letter alphabet is available from which words can be formed. It follows then that if

FIGURE 2–14   Protein synthesis. A molecule of mRNA carrying the instructions for synthesis of a particular protein moves to the cytoplasm and becomes attached to a ribosome. The ribosome "reads" the codon on the mRNA that calls for a particular amino acid. A molecule of tRNA bearing that amino acid moves into place, and its anticodon base-pairs with the codon on the mRNA. The ribosome moves to the next codon and another tRNA with its amino acid base-pairs with that codon. The two amino acids are joined by a peptide bond and the first tRNA is released. The ribosome moves along the mRNA trailing a growing chain of amino acids.

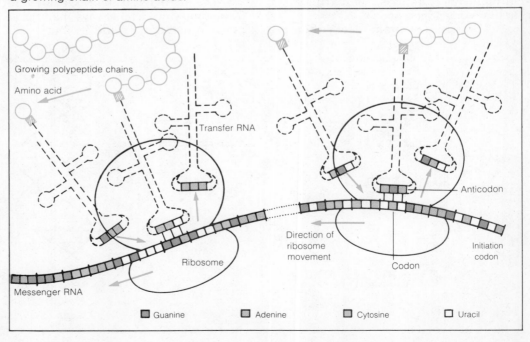

each code-word consists of three letters, then a 64-word vocabulary can be formed. Thus, each triplet or codon represents a word, and each word stands for a particular amino acid. The genetic code has been deciphered to the extent that the codon is known for the twenty amino acids commonly found in animal tissues. There are several code words for some amino acids, as well as codons that serve as "punctuation marks," by calling for the initiation and termination of an amino acid chain.

Translating the code into protein (a process known as *translation*) involves finding the appropriate amino acids and joining them with peptide bonds. The ribosome is the site of formation of peptide bonds, but delivering the amino acids requires a third kind of RNA known as **transfer RNA (tRNA)** (see Figure 2–15). Transfer RNA molecules are quite small, as RNA goes, and are found in the cytoplasm. The molecules take on a cloverleaf shape that provides for attachment of an amino acid at one end and exposes a triplet of bases on the opposite end. There is a different tRNA for each amino acid, and its triplet of bases (called an *anticodon*) aligns with a complementary triplet (codon) for that particular amino acid on mRNA.

The means by which the transfer RNA picks up its amino acid is rather complex since the amino acid molecule must first be activated; this requires an enzyme and transfer of energy from ATP. The activated amino acid is bound to the appropriate tRNA, which then moves with it to the ribosome.

When synthesis begins and the ribosome binds to the mRNA, the first mRNA codon "recognizes" the complementary anticodon of a tRNA. That molecule of tRNA, with its activated amino acid on the other end, base-pairs with the codon of the mRNA. The ribosome then moves to the next mRNA triplet and another tRNA with its activated amino acid moves into place and base-pairs with that codon. The first amino acid is attached to the second by a peptide bond and, consequently, the second tRNA has two amino acids attached to it. The first tRNA is released to "hunt for" another amino acid molecule. The ribosome moves to the next

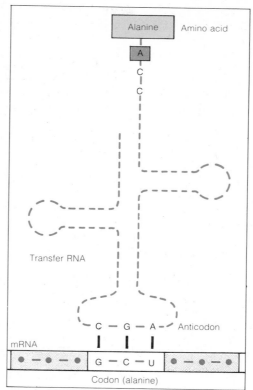

FIGURE 2–15   A molecule of transfer RNA. It carries a particular amino acid and an anticodon that base-pairs with the corresponding codon on the mRNA molecule.

triplet, and the process is repeated. As the ribosome moves along, there is a growing string of amino acids trailing from it. The chain is released as a protein molecule when the ribosome reaches the end of the mRNA. The synthesis of a molecule of protein can be accomplished in less than a minute. Since several ribosomes may be moving along a single mRNA at the same time, each mRNA can be used to form a number of protein molecules.

If the newly formed protein is to be secreted from the cell, it enters the endoplasmic reticulum upon its release from the ribosome. It then moves to the Golgi apparatus for further processing and "packaging" into secretion granules in preparation for release from the cell. Proteins formed on ribosomes that are not associated with the endoplasmic reticulum probably are intracellular proteins, such as enzymes that stay in the cytoplasm.

CELL GROWTH AND REPRODUCTION

Like whole organisms, cells grow and age, and there are controls and limits for both. For one thing, there seems to be a limit to the amount of cytoplasm which a cell nucleus can control. Because the needs of the cell must be met by exchange across its cell membrane, there is a cell size or volume beyond which the needs cannot be met through surface exchange. Irregularities and folds in the surface membrane increase the surface area and extend the limit somewhat, but not greatly.

When a cell has reached a certain size, it may stop growing but continue to carry on its other activities as it ages, or it may divide into two new daughter cells. The course taken varies with the type of cell and its location. Cell division is more likely to occur among cells that are relatively *undifferentiated* (unspecialized). In certain tissues a new generation of cells may appear every few hours or days. Cells which have become specialized divide less frequently; some cells, in fact, have lost completely the ability to divide. The latter cells seem to withstand the aging process well, however, since many of them survive and function throughout the life of the organism.

Virtually all cells that divide do so in a process known as **mitosis.** It consists of several well-recognized steps or phases, followed by an interval of varying duration before there is another mitotic cell division. For many years the nondividing phase of cell life was almost entirely overlooked, and the cell was said to be in a "resting stage" during this time. This is inaccurate, since the cell is actively carrying on all of its normal activities, as well as preparing for the next cell division. Since the nondividing cell is not in any of the phases of mitosis, it is more appropriate to refer to this period as **interphase** than as a resting stage. A cell that divides fairly often might take an hour to go through the stages of mitosis, with the interphase lasting about 20 hours. In other cells the interphase may last for weeks.

Interphase — The Growth Phase

During interphase the cell is carrying on its metabolic activities, growing, and preparing for subsequent mitosis. Each of the new cells to be formed by mitotic division will receive complete and identical sets of chromosomes. Furthermore, it is during the period of interphase that the additional set of chromosomes is produced. During this time the chromatin material of the nucleus appears to be diffuse and without definable structural features, but it is actually made up of strands too fine to be resolved by the light microscope. It is said that if the chromatin of a single nucleus were stretched out, it would be about one meter in length.

When the time comes to replicate the DNA in the chromosomes of a cell nucleus, the strands of the DNA molecule are believed to unwind and separate longitudinally. At this point the linkages between the bases are broken. Each half of the DNA strand then serves as a pattern or template against which a new complementary strand is formed. The newly formed strand is identical with the one that was broken away. Therefore, the result is two complete and identical sets of chromosomes in the cell nucleus. The DNA content of the nucleus is doubled during interphase, and when this has been completed the cell is ready to divide.

Mitosis — Cell Division

For convenience of description, mitosis is arbitrarily divided into four stages or phases; however, the process actually is continuous and there is no sharp demarcation between one phase and the next (Figure 2–16).

The first stage, **prophase,** is characterized by nuclear changes. The chromatin material appears to condense and individual strands become discernible. They become thicker and shorter, probably as the coiling of the chromatin threads becomes tighter. Meanwhile, the nucleolus and nuclear membranes disappear, and the two centrioles also separate and move toward opposite sides of the cell. Tiny microtubules (astral rays) appear to radiate from each centriole. Some of the microtubules extend from one centriole to the other, and as the centrioles pull apart they appear to "spin out" more microtubules, forming what is known as the *spindle*. The centrioles divide late in prophase.

In **metaphase,** a nucleus is not recognizable, but the spindle is stretched across the cell between the centrioles at opposite poles. The chromatin material is condensed into definite chromosomes lined up across the center of the spindle (the equatorial plate). Each chromosome consists of two paired threads joined by tiny *centromeres*. The centromeres appear to be attached to the microtubules of the spindle.

Anaphase is marked by some sort of attraction—or perhaps a contraction of the spindle microtubules—since the centromere of each chromosome appears to be pulled away from the midline toward the nearest centriole. Soon the chromosomes are separated into two identical groups, clumped at opposite sides of the cell. The centrioles are divided, and a hint of cleavage between the two chromosome clusters begins to appear in the cytoplasm.

In the final phase, **telophase,** mitosis is completed with the formation of two separate cells. The cytoplasm cleaves, the chromosomes disperse into a diffuse mass of chromatin material, and the nuclear membrane and nucleolus reappear. The result is two new cells, identical with one another and with the cell from which they arose except for being slightly smaller in size.

ELECTRICAL PHENOMENA

One of the fundamental properties of cells is their electrical activity, for it is as essential as the energy-releasing activities. Electrical activity is so characteristic of living things that it could well have been included as one of the properties of life. In fact, absence of electrical activity in the nervous system is an important criterion for determining clinical death. It is therefore appropriate to consider some of the electrical phenomena that underlie so many activities of cells.

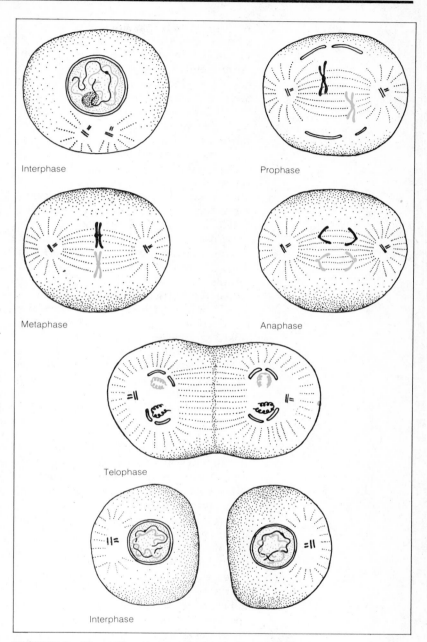

Interphase

Prophase

Metaphase

Anaphase

Telophase

Interphase

FIGURE 2–16   The stages of mitosis in a typical cell.

Electrical Concepts

When we speak of electricity we are talking about electrons and the movement of electrons, and when we speak of electrical charge we are talking about an amount or quantity of electricity. The electrical charge associated with electrons is said to be negative, and the electrical charge associated with protons in the nucleus of atoms is said to be positive. If we say a charge is negative, we mean that more electrons are present than protons, and if we say a charge is positive, we mean there are fewer electrons than protons.

It was noted earlier that when a molecule dissociates, it forms charged particles known as ions. Anions are negatively charged as they have one or more "extra" electrons, and cations are positively charged because they are short one or more electrons. When a molecule such as sodium chloride, NaCl, ionizes, it yields a cation, Na^+, and an anion, Cl^-. Because unlike charges attract one another, the Na^+ and Cl^- attract one another, but Na^+ and Na^+, or Cl^- and Cl^- do not, because like charges repel one another.

When electrical charges are separated by some sort of barrier so that positive charges are in one place and negative charges are in another, an electrical gradient is created, and there is an attractive force between the separated unlike charges. The greater the charge difference, the greater the charge gradient and the greater the force of attraction. This force is a type of potential energy since it is capable of doing work. It is referred to as the **voltage, electrical potential,** or **potential difference.** It is comparable to the pressure gradient that causes filtration or the concentration gradient that causes diffusion or osmosis.

Whether or not there is a flow of electricity (that is, an electrical current) when there is a charge gradient or potential difference depends upon the nature of the material that separates the charges. If the barrier is permeable to electrical charge, it is a good conductor; it has low electrical resistance and the current will flow readily. If the charges are separated by a poor conductor (a good insulator), there is a high electrical resistance and little or no current will flow. Such is the situation in a flashlight battery, where negative and positive poles are separated by a good insulator, and the potential energy is stored. When the two poles are connected, current will flow until the connection is broken or until the charge difference disappears and the battery is "dead." The amount of current flow is

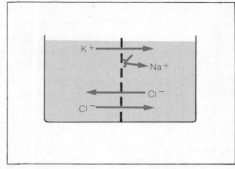

FIGURE 2–17   The presence of a nondiffusible ion (Na^+) affects the final equilibrium. As potassium ions (K^+) diffuse down their concentration gradient, an electrical gradient is developed.

therefore determined by the driving force provided by the electrical gradient and by the resistance encountered to the flow of current.[7]

Electrical activity in living cells is concerned not so much with the flow of electrons (for there are relatively few free electrons there), but rather with the movement of charged particles, the electrolytes dissolved in body fluids. The distribution and movement of these ions across cell membranes are affected just as much by the electrical forces as by the forces of concentration gradients.

As noted earlier, when two solutions are separated by a permeable membrane, diffusion will occur as solute particles move down the concentration gradient. The diffusion will continue until the gradient has disappeared. If the particles are charged and the membrane is impermeable to one class of them, however, the situation is quite different. For example, if solutions of equal concentrations of NaCl and KCl are separated by a membrane permeable to K^+ and impermeable to Na^+ (Figure 2–17), there will be no movement of the sodium because it cannot cross and no net movement of the chloride because there is no gradient, but potassium can and will diffuse.

[7]This relationship, known as **Ohm's Law,** is written as $I = E/R$, where I is the current flow, E the electrical force or potential difference, and R the electrical resistance to current flow. The current is usually measured in *amperes*, potential in *volts*, and resistance in *ohms*. In biological systems where the currents and voltages are relatively small, the first two variables are expressed more commonly as *milliamperes* and *millivolts*.

As it moves, the potassium will increase the number of positive charges on the sodium side of the membrane, thus creating an electrical gradient that opposes the chemical gradient down which it is diffusing. Potassium ions will stop diffusing when the forces exerted by the two opposite gradients are equal. This point represents an electrochemical equilibrium. The size of the electrical force, the **equilibrium potential,** can be calculated simply from a knowledge of the ion concentration at equilibrium.

Although the intracellular and interstitial fluids have almost the same total ionic concentrations, distribution of the individual ion species differs greatly. Table 2–7 shows the concentrations of some important ions in a skeletal muscle cell and the extracellular fluid around it. Although the actual figures would not be identical for all cells, they are quite similar. Note that sodium ions are about 12 times more concentrated outside than inside, while potassium ions are about 39 times more concentrated inside than out. Chloride ions are also more concentrated on the outside. Our discussion will be centered on these three ions, but it is important to remember that there are other ions, both in the intracellular and extracellular fluids. Many of these are nondiffusible organic ions inside cells, but they do not contribute significantly to the electrical activities of the cell.

The concentration differences between the inside and outside of the muscle cell cause diffusion of those particular ions, but when they diffuse electrical gradients develop across the membrane. The equilibrium potentials that result are −97 mv for potassium, +66 mv for sodium, and −90 mv for chloride. (see Figure 2–18).

Resting Potential

Equilibrium potentials are not the whole story when it comes to analyzing a cell's electrical activity. An important complication is the fact that there is a voltage difference across the membrane of all living cells, such that the inside of the cell is negative compared to the outside. For our typical inactive skeletal muscle cell the difference is about 90

TABLE 2–7   CONCENTRATIONS OF SOME IMPORTANT IONS OF EXTRACELLULAR AND INTRACELLULAR FLUIDS OF MAMMALIAN SKELETAL MUSCLE

	Extracellular (mmol/l)	Intracellular (mmol/l)	Ratio Inside/Outside	Equilibrium Potential (mv)
Cations				
Na^+	145	12	12/1	+66
K^+	4	155	1/39	−97
Others	5	—	—	—
Anions				
Cl^-	120	4	30/1	−90
HCO_3^-	27	8	3.4/1	−32
Others	7	155	—	—

mv, and we say that the cell's **resting membrane potential** (resting potential or membrane potential) is −90 mv. Resting membrane potentials are characteristic of all living cells and are present when a cell is inactive or at rest. Membranes that support such potential gradients are said to be *polarized.*

The resting membrane potential is an electrical force that affects the distribution of ions as much as does the concentration gradient. The negative charge inside the cell attracts positively charged sodium and potassium ions, but repels the negatively charged chloride ions. Thus it favors the entrance of sodium ions but impedes the outward movement of potassium ions. It also reduces the entry of chloride ions. The concentration gradients favor inward movement of Na^+ and Cl^- and an out-

FIGURE 2–18   The chemical and electrical forces across the membrane of a skeletal muscle cell affect the distribution of ions. D = diffusional forces due to a concentration gradient. They contribute to the equilibrium potential of that ion. C = charge forces due to intracellular negativity.

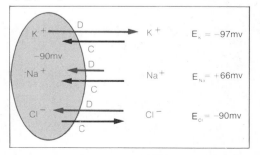

ward movement for K⁺. The magnitude of the net forces acting on each ion is determined by the resting potential of the cell and the equilibrium potential for that ion; it is the difference between them. Figure 2–19 illustrates these relationships in a muscle cell. For chloride the equilibrium potential (E_{Cl}) equals the resting potential, so that chloride is at electrochemical equilibrium; the electrical and chemical forces are balanced and chloride ions show no tendency to move either way. For potassium, with an E_K of —97 mv, the concentration gradient is great enough that at a resting potential of —90 mv there is still a slight tendency for potassium to diffuse outward. For sodium, the electrical and chemical gradients are in the same direction, both favoring entry of sodium into the cell, and the total force acting on sodium is the sum of the two. In a resting cell the chloride ion is the only one that is truly at equilibrium, while potassium has a slight net force for outward movement and sodium has a large force for inward movement.

Based on your knowledge of passive forces and the effects of gradients, you may be wondering how this condition can be maintained. Why doesn't sodium enter the cell and reduce both the Na⁺ concentration gradient and the

electrical gradient (membrane potential)? Why doesn't potassium diffuse out? And why doesn't the membrane potential "run down"? The actual values of concentrations and potentials have been measured experimentally with great accuracy and reconfirmed many times, so we know they do exist, and one must therefore find a mechanism for sustaining these gradients. There are, in fact, two factors that are responsible for maintaining the resting potential, and both have already been discussed somewhat.

The first is membrane permeability. Recall that passive processes require not only a suitable gradient but also a permeable membrane. If a particle cannot penetrate the membrane, no amount of concentration gradient will alter the situation. Permeability is not an all-or-none characteristic, however, and there are many possibilities between the extremes of unrestricted passage and total impermeability. Nor is permeability fixed, for under certain conditions the permeability of a membrane can be greatly altered.

A typical cell membrane is 50 to 100 times more permeable to potassium and chloride ions than to sodium ions. Thus, even though large forces tend to drive sodium ions into the cell, the relative difficulty of penetrating the cell membrane greatly reduces the number of sodium ions that do enter. And although the forces that tend to make potassium move out of the cell are slight, the ease with which they can cross the membrane enable some potassium ions to diffuse out. The permeability characteristics of the membrane therefore reduce the total ion movement and roughly balance the differences in chemical gradients. Because of the greater permeability to potassium and the ease with which it may enter and leave the cell, potassium is recognized as the most important ion in establishing the resting potential. This is supported by the fact that the resting potential is very near the E_K, and the membrane potential decreases if the concentration of potassium ions in the extracellular fluid is increased. These permeability differences alone would not prevent the membrane potential

FIGURE 2–19 The net force affecting ions inside and outside a skeletal muscle cell is determined by the resting potential of the cell and the equilibrium potential of each ion.

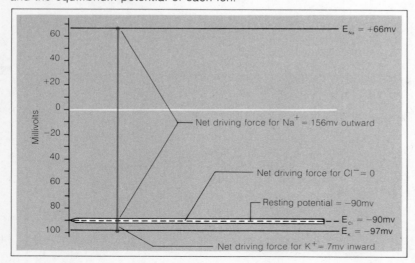

from eventually disappearing because the membrane does allow some diffusion of both sodium and potassium ions, but it certainly would prolong the process.

The second factor in maintaining the gradient, and hence the membrane potential, is an active transport system—a *sodium-potassium pump* (Figure 2–20). As an active transport process, the pump requires energy obtained from metabolic processes in the cell, and it transports or pumps ions against their gradients. Sodium ions that leak into the cell are pumped back out; and potassium ions that leak out are pumped back in. The transport of sodium was the first to be described and as a result, it often is referred to simply as the sodium pump. As long as it pumps equal numbers of cations in opposite directions, the pump does not actually create or develop an electrical gradient, but by maintaining the concentration gradient, it indirectly ensures that the membrane potential will remain.

Living cells, therefore, must continually pump sodium out and potassium in if the resting potential of the cell is to be maintained. The energy required to do this comes from ATP and accounts for a sizable portion of the energy consumed by the body at rest.

The resting potential does not have the same value in all cells, but the relationships between the various contributing factors are the same. The resting potential is a steady-state phenomenon; it remains at a constant level, but can be altered by anything that changes conditions. Changing the concentration of the extracellular fluid (perhaps increasing the K+ concentration) to alter a chemical gradient, passing a current through the membrane to alter the electrical gradient, or administering an agent that interferes with the Na-K pump all produce marked effects on the resting potential.

Action Potential

Dramatic changes occur in the resting potential of nerve and muscle cells when they are stimulated, and for this

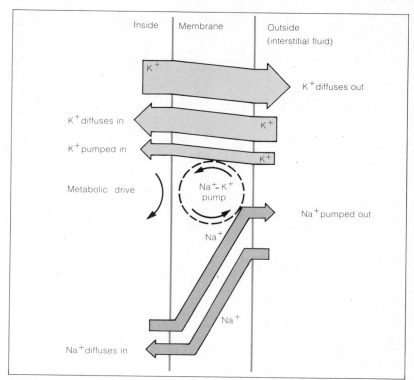

FIGURE 2–20   The resting potential is maintained by an active process that opposes passive diffusion. Net diffusion of potassium is outward and sodium inward. Sodium is actively pumped out and potassium in. The width of the arrows is proportional to the amount transported, and the slope indicates the size of the gradient.

(Adapted from John C. Eccles, *The Physiology of Nerve Cells*, Johns Hopkins University Press, Baltimore, 1957.)

reason they are called **excitable cells.** The stimulus causes extremely rapid and brief changes in the permeability of the membrane to sodium and potassium ions, which results in movement of these ions and changes in the membrane potential. Since these changes are associated with activity of the cell—contraction in the muscle cell or conduction in the nerve cell—they are known as **action potentials** (see Figure 2–21). They are of considerable importance because every muscle contraction is triggered by an action potential and a nerve impulse actually is an action potential.

A stimulus applied to the surface of an excitable cell initially causes a slight increase in permeability to sodium at that spot (see Figure 2–22). Sodium ions enter the cell, slightly reducing the resting potential; that is, they cause a

FIGURE 2–21   An action potential in a skeletal muscle cell.

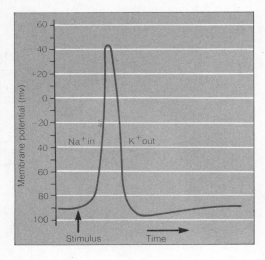

slight *depolarization*. After a weak stimulus the membrane potential returns to its resting level and nothing more happens. A stronger stimulus causes more sodium to enter and a greater depolarization. At some particular stimulus intensity, the *threshold intensity*, the membrane becomes very permeable to sodium and the resulting influx causes not only depolarization but an actual reversal of polarity. The inside of the cell becomes temporarily positive to the outside as the membrane potential approaches the equilibrium potential of sodium. The membrane quickly loses its permeability to sodium, but it now briefly becomes very permeable to potassium, and potassium ions diffuse out. Exit of the postively charged potassium ions restores the inside negativity, and the membrane potential returns to the normal resting level. In spite of the magnitude of the voltage changes, a relatively small number of ions are actually involved, and the Na-K pump will slowly remove the extra sodium ions and return the departed potassium ions.

Although each action potential is developed at one spot, it is conducted over the surface of the entire cell. If it is a nerve cell, the action potential will be conducted the length of the nerve cell and cause a response in the cell on which the nerve ends. The action potential will be discussed more fully in Chapter 9.

CHAPTER 2 SUMMARY

Chemical Organization

Matter is made up of **atoms**. Atoms consist of a nucleus that contains positively charged *protons* and uncharged *neutrons* surrounded by negatively charged *electrons*. Atoms carry no net charge because there are just enough electrons to balance the protons in the nucleus. The number of protons, neutrons, and electrons varies, and is the basis for distinguishing *elements*. Atoms can give up or gain one or more electrons, forming a charged particle or *ion*. Different atoms may link together to form **molecules**. Some molecules also can separate or dissociate into ions.

A solution is formed when a compound or element (a *solute*) dissolves in water or another fluid (a *solvent*). Ions in solution will respond to an electric current passed through the solution; the positively charged ions (*cations*) will move toward the negative electrode and the negatively charged ions (*anions*) will move toward the positive electrode. Such ions, or *electrolytes*, are present in body fluids.

A molecule that yields hydrogen ions when it dissociates in solution is an **acid**. A strong acid dissociates quite completely, releasing many hydrogen ions, and a weak acid dissociates less completely, releasing fewer hydrogen ions. A substance that will combine with hydrogen ions is a **base**. A strong base binds hydrogen ions readily. The concentration of hydrogen ions in a solution is expressed by the *pH*, which is the negative log of the hydrogen ion concentration. As the hydrogen ion concentration and the acidity of a solution increases, the pH goes down, and as the hydrogen ion concentration and acidity decrease, the pH and alkalinity go up. A pH of 7 is neutral.

The main components of cells and other biological structures are organic compounds such as carbohydrates, lipids, proteins and nucleic acids. **Carbohydrates** are sugars and starches (glycogen in animals) or *polysaccharides*, which are formed of chains of sugars. **Lipids** or fats are formed of *fatty acids* and *glycerol*. Carbohydrates and lipids

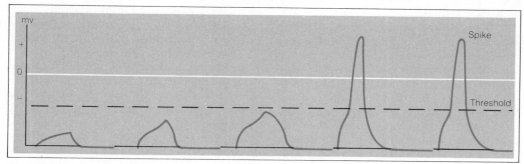

FIGURE 2-22   Response of an excitable cell to stimuli of increasing intensity. Subthreshold stimuli cause progressively greater local responses but no spikes. A threshold stimulus produces enough local depolarization to cause an action potential. A stronger stimulus produces an action potential identical to that caused by a threshold stimulus.

are important sources of energy for cellular activities. **Proteins,** whose molecules may be very large, are made up of *amino acids.* They are important primarily as building blocks for cellular and intercellular structures, but they can also be used as sources of energy. The **nucleic acids,** DNA and RNA, are the carriers of genetic information and are essential for protein synthesis.

How well a reaction goes depends upon the relative concentrations of the reactants, for when the products of a reaction are allowed to accumulate the reaction slows **(Law of Mass Action).** Many of the reactions in biological systems are **oxidation-reduction reactions** in which one substance is oxidized (by adding oxygen or removing hydrogen) and another is reduced (by adding hydrogen or removing oxygen). Oxidation and reduction reactions are always coupled. Almost all biological reactions need the assistance of a catalyst to speed the reaction. **Enzymes** are organic catalysts produced by cells, and there is a different enzyme for nearly every reaction. The ability of a cell to carry out a particular process depends upon the presence of the necessary enzymes. Since enzymes are proteins, protein (enzyme) synthesis is a very important aspect of cell function.

Translocation of Materials

Physical and physiological processes are involved in transporting solutes or water across cell membranes.

Physical transport processes are *passive* processes, dictated by gradients, and will occur whenever and wherever the necessary conditions exist. The membrane participates in physiological transport processes, so they occur only across membranes of living cells.

Physical transport processes are diffusion, filtration, and osmosis. In *simple diffusion,* a difference in solute concentration (a concentration gradient) is the driving force that brings about a net movement of solute particles, each moving down its own particular concentration gradient to an area where that kind of particle is less concentrated. This movement is possible because of the constant random motion of particles, and is independent of the movement of other particles. The final result is an equilibrium in which the concentration of all constituents is the same throughout the solvent. When there is an unequal distribution of charged particles on either side of a membrane, they will move to reduce the charge gradient and balance the positive and negative charges on the two sides; that is, to reach *electroneutrality.*

Osmosis is a type of diffusion that occurs when there is a nondiffusible solute on one side of the membrane. Inability of the solute to diffuse leaves only the solvent (usually water) as a diffusible substance. The concentration gradient created by the nondiffusible solute causes movement of solvent to the solute side of the membrane.

Filtration is caused by a mechanical

pressure which forces solvent and diffusible solutes through the membrane. The equilibrium sought is one of equal pressure, not equal concentration.

Many **physiological transport processes** involve a carrier substance located in the cell membrane. *Facilitated diffusion* is a type of carrier-mediated transport in which an otherwise non-diffusible solute combines with the carrier. This allows the solute to pass through to the other side of the membrane, where it is released. The solute always moves down its concentration gradient, as in simple diffusion. *Active transport* is a similar process, except that transport is against the concentration gradient. It is a pumping process and requires energy to operate. The energy is provided by the cell's metabolic actions.

Large molecules or particles can be taken into the cell or expelled from it by specialized processes. *Endocytosis* involves the development of an indentation in the cell membrane, which deepens and finally closes over, leaving a membrane-enclosed vacuole inside the cell. When the vacuole primarily contains fluids, the process is *pinocytosis*, and when it contains particles such as bacteria, the process is *phagocytosis*. Substances the cell has produced for export, such as secretions, are released by *exocytosis* which is essentially a reversal of endocytosis.

Metabolism and Energy Exchange

Metabolic processes in the cell are concerned with the energy the cells need to carry on their tasks. In the body, energy may be released, stored, transformed, used to perform physical work, or lost (as heat). Energy is needed to perform physical work because work involves going against a gradient or force, such as active transport across a cell membrane, or lifting a load against gravity. Energy is obtained by oxidation of the nutrients in our foods, and is released in small packets that are stored in *high energy phosphate* (HEP) bonds of *adenosine triphosphate* (ATP). The HEP bonds can be broken later, releasing small amounts of energy for use by the cell. The different foodstuffs are oxi-

dized by separate pathways to small molecules which are then metabolized along a common pathway to carbon dioxide, water, and energy. Carbohydrates are broken to sugars such as glucose for oxidation. The initial part involves a series of reactions known as *glycolysis*, which leads to pyruvic acid. From pyruvic acid a two-carbon unit is formed that combines with coenzyme A (as acetyl CoA). Fats are broken down to glycerol and fatty acids. Glycerol fits into the glycolytic pathway, and the fatty acids are broken down to two-carbon units as acetyl CoA. Amino acids also fit into the common pathways after removal of their amino (NH_2) group.

The final stage of the oxidative pathway is the *Krebs cycle*, in which the acetyl CoA and amino acid bits are oxidized to carbon dioxide and water. The hydrogen released in the course of the cycle is taken up by hydrogen carriers, then transferred from one carrier to another through a series of oxidation-reduction reactions. The hydrogen is finally taken up by oxygen, forming water. Several of these reactions yield the energy that is used to form ATP.

Protein synthesis is needed to restore structural protein of the cell and to produce the enzymes for cellular metabolism. The genetic information contained in the DNA of the nucleus includes the "directions" for assembling amino acids into a particular protein molecule.

The large DNA molecules consist of two parallel chains. *Bases* protrude from the chains and join the chains together by forming cross-links with bases on the other chain. Each base can combine with only one other base. There are four different bases arranged in a specific sequence along the chains. The genetic information is contained in the sequence of the bases. Protein synthesis begins with the formation of RNA. A RNA molecule is similar to DNA, but it is a single-chain molecule with the bases as sidechains. Strands of DNA separate, exposing the bases, which then serve as templates for assembling a molecule of RNA. The sequence of bases in the RNA formed is determined by the sequence of bases on its DNA template. The RNA moves to the cytoplasm, carrying the code for protein synthesis in its sequence of bases; it is known as *messenger RNA* (mRNA). Each set of three bases (a *triplet*) signifies a specific amino acid. The mRNA becomes attached to a ribosome, which moves along the strand of mRNA and "reads" the code of each triplet. In the cytoplasm there are small molecules of transfer RNA (tRNA), a different one for each amino acid. The transfer RNAs have a binding site for the amino acid on one end, and a triplet of bases to match a complementary triplet of the mRNA on the other end. A tRNA, with its amino acid, pairs to its particular triplet on the mRNA. The ribosome moves to the next mRNA triplet and another tRNA with its amino acid attaches to the mRNA. The two amino acids are joined and the first tRNA is released. The ribosome moves along, and yet another tRNA moves into place. A growing chain of amino acids is attached to its amino acid, and so on, until an entire protein molecule has been formed.

Cell Growth and Reproduction

Most new cells in the body are formed by a process of cell division known as **mitosis.** During the interval between cell divisions, or **interphase,** the cell conducts its normal activities, and also prepares for mitosis by synthesizing DNA to produce a new set of chromosomes. Mitosis is divided into four stages; *prophase* in which the chromatin threads of the nucleus become condensed into recognizable chromosomes; *metaphase,* in which the centrioles move to opposite sides of the cell and the chromosomes become lined up along the cell's equator; *anaphase,* in which chromosomes move away from the cell center toward the centrioles; and *telophase,* in which the cytoplasm separates and two new cells are formed.

Electrical Phenomena

Unequal concentrations of solutes on two sides of a permeable membrane lead to diffusion and an equalization of the concentrations. If the solutes are charged particles, and the membrane is permeable only to some of them, their diffusion will reduce the concentration gradient but will develop an electrical gradient. When an equilibrium is reached there will still be both a concentration and an electrical gradient. The electrical gradient present when such an electrochemical equilibrium is reached is known as the *equilibrium potential.*

Concentration gradients normally exist across cell membranes in the body. Sodium and chloride ions are more concentrated outside the cell, while potassium ions are more concentrated inside. There is also an electrical gradient, and the inside of the cell is electrically negative to the outside, which favors the entry of positive ions and the exit of negative ions. The net force acting on any of these ions is the difference between the equilibrium potential for that ion, and the electrical potential across the cell membrane. Thus, there is a large force favoring entry of sodium ions, a slight force for exit of potassium, and no force for chloride.

The membrane is relatively impermeable to sodium, but much more permeable to potassium. Some sodium does leak into the cell, however, and some potassium leaves, which tends to reduce the potential difference across the cell membrane. There is an active transport system, a *Na-K pump,* that pumps sodium ions out and potassium ions into the cell, and thus maintains the membrane in its polarized state. The electrical gradient between the inside and outside of the cell is the **resting membrane potential.**

Excitable cells, such as nerve and muscle, respond to a stimulus with changes in the permeability of the cell membrane to sodium and potassium ions. A stimulus of *threshold intensity* increases the permeability of the membrane to sodium, depolarizing it to a critical level (*threshold*). This triggers a marked increase in permeability to sodium, and enough sodium enters the cell to reduce and temporarily reverse the membrane potential. The membrane quickly becomes impermeable to sodium, but very permeable to potassium and potassium ions diffuse out of the cell. This restores both the internal negativity and the resting potential. This abrupt electrical change, known as the **action potential,** lasts only a few milliseconds, after which the membrane is ready to respond again.

STUDY QUESTIONS

1 What is the difference between an atom and a molecule? How do they combine to form chemical compounds?

2 What is an isotope? Of what value are they in health fields? Why are they dangerous?

3 How does an atom become an ion? Could a molecule become an ion?

4 How much of each ingredient would you use to make up one liter of
(a) a 5 percent solution of NaCl in water?
(b) a 1 molar solution of NaCl in water?
(c) a 0.1 normal solution of NaCl in water?
(d) a 100 mg percent solution of glucose in water?
(e) an aqueous solution containing 20 milliequivalents per liter of calcium?
(f) an aqueous solution containing 23 milliequivalents per liter of sodium?

5 Suppose you have two bottles, one of 0.1 N HCl and one of 0.1 N carbonic acid. The concentrations are the same, so why is hydrochloric acid considered to be a strong acid while carbonic acid is a weak acid? What differences would you expect in the reactions of these two solutions of equal concentration?

6 How do carbohydrate, fat, and protein differ in the way their "building blocks" are put together? What are the "building blocks" in each case, and of what are they formed?

7 When one substance is oxidized, why is another reduced?

8 What are enzymes and what do they do? How do they differ from coenzymes?

9 What is the driving force for diffusion? for osmosis? for filtration? What "moves" in each case?

10 How do active transport processes differ from passive transport processes? from facilitated diffusion?

11 How do pinocytosis and phagocytosis differ from one another, and from active transport?

12 What is the distinction between mechanical work and physiological work? Is energy expended in each case? What becomes of the energy in each case?

13 What is ATP? Where is it located? Of what value is it?

14 What is accomplished by the Krebs cycle? What are the hydrogen acceptors, and why are they needed?

15 What are cells doing during interphase?

16 What is accomplished by mitosis?

17 How do RNA and DNA differ?

18 If protein synthesis is blocked in cells, what kinds of cellular functions, if any, would be disrupted or altered? Explain.

19 An inactive skeletal muscle cell maintains a resting membrane potential. What electrical and chemical gradients are present across the cell membrane during this time? What are the mechanisms that enable the cell to maintain these gradients?

20 What is the cause of an action potential? What changes occur in the cell membrane during an action potential and the subsequent restoration of the resting potential?

3

Tissues

Skin

Anatomical Orientation

Organization of the Body

Having considered the structure of cells and some of their activities, it is time to see how they are organized into larger units that play a broader role. Cells are classified into four primary tissues, mainly on the basis of their structural characteristics. Two of these, the epithelium and connective tissue are widely distributed throughout the body, and are described in this chapter. The more specialized tissues (muscle and nerve tissues) are discussed in later chapters.

The tissues form organs (for example, the stomach), and organs make up the organ systems (the digestive system) that perform specific functions for the entire body. The skin is an excellent example of how cells and tissues are combined into an important functional unit. The description of the skin in this chapter also provides an introduction to the way in which organs and organ systems operate.

Organ systems, too, are but parts, and together they make up a unit—the body. This chapter includes an introduction to the body as a whole, its parts, and the terminology used to describe the parts and their location in the body.

TISSUES

No matter how it may seem after our discussion of cells and the processes and events associated with their activities, cells do not operate as isolated units with each performing its function independently of all other cells. They differ from one another in size, shape, and function, as well as in their relationships to other cells. Nevertheless, they are greatly dependent upon one another.

Cells are organized in various patterns and are held together (or apart) by extracellular structures. Upon examination of these patterns and structures and of the cell functions, four *primary tissues* can be recognized: epithelium, connective tissue, nerve, and muscle. In order to carry out specific functions, tissues are organized into *organs,* such as the heart, kidney, and liver. Most organs contain examples of more than one of the primary tissues, and an organ's structural and functional characteristics are determined to a large extent by the particular arrangement of its tissue elements. Those organs whose activities contribute to the same overall function or process constitute an *organ system.*

This chapter introduces some fundamental properties of epithelium and connective tissue. The discussion of muscle and nerve tissue will be deferred to Chapters 7 and 8, where they are presented as parts of the organ systems which they serve.

Epithelium

Basic Characteristics   Epithelial tissue is a covering and lining tissue. It covers the body surface (as skin) and lines the body cavities and tubular structures, including those of the digestive tract, respiratory passages, and blood vessels. Epithelium is a predominantly cellular tissue, with only enough intercellular material to hold the cells in place. In many locations it is a protective tissue, but equally important are its roles in transport and secretion. The passive exchange of materials across the walls of the tiniest blood vessels is exchange across a layer of epithelial cells. Many types of active transport, including absorption in the digestive tract or kidney, are epithelial functions, as is the glandular secretion of sweat, saliva, and hormones.

As a covering and lining tissue, epithelium always has one free surface. The other, a deep surface, is attached to a **basement lamina,**[1] the noncellular layer that anchors it to underlying connective tissue. The free surface may be modified in one of several ways. It may have tiny projections called **microvilli,** so small that they are indistinct with the light microscope. These minute structures serve to increase greatly the area of the free surface, a distinct advantage for absorptive surfaces. Other epithelia have highly developed hairlike projections called **cilia.** Cilia are *motile* (capable of motion) and can wave or beat to aid or hinder the movement of materials along the epithelial surface.

Blood vessels do not penetrate epithelial layers, which means that the needs of the epithelial cells must be met by diffusion from blood vessels in the underlying tissue. Terminal branches of nerve fibers, however, are commonly found between epithelial cells. Epithelium in general has a high regenerative capacity; this trait is particularly true of protective surfaces. Cell division occurs frequently and there is a high rate of cell turnover, a desirable property for surfaces exposed to adverse conditions.

Since epithelial cells form coverings and linings, they are almost always arranged in layers, sometimes only a single cell thick. Preservation of the functional integrity of the epithelial covering, either to protect against pressure and friction or to restrict passage of substances through it, requires that the cells be securely attached to one another.

---

[1]The terminology for this structure varies. The layer that was first described with the light microscope and known as the *basement membrane,* has been shown with the electron microscope to consist of two layers. Immediately adjacent to the epithelial cells is a sheet of extremely fine filaments in an unstructured matrix called the *basement lamina* or *basal lamina.* An outer layer consists of tiny bundles of connective tissue fibers and a condensed ground substance. Some prefer to use "basement membrane" because it includes both parts, and only the outer part is usually seen with the light microscope. Others prefer "basement" or "basal" lamina, partly because it is the most consistent feature, and partly because of the possibility of confusion with membranes of cells. Still others use the two terms synonymously.

Cell junctions. There seems to be general agreement on three different types of attachment: tight junctions, desmosomes, and gap junctions (see Figure 3–1). The firmest of these is the **tight junction,** in which parts of the plasma membranes of the two cells actually appear to be fused for a short distance. There is no intercellular space at that spot, but a barrier remains, separating the cytoplasm of the two cells. Tight junctions usually occur as continuous bands or collars around the edge of the free surface of cells, where they can block material that might otherwise diffuse between the cells.

Desmosomes are not continuous bands, but are scattered over the adjacent cell surfaces, rather like tiny spot welds. The membranes of the adjacent cells are thickened and numerous strands, or *microfilaments,* attached to the thickened regions extend out into the cytoplasm of the two cells. There is a little space between the two cell membranes, which seems to contain an as yet unidentified material.

The third type of connection is the **gap junction,** a type of connection in which the adjacent membranes are in close contact, but still there is a small amount of space between them. It contains some globular material that seems to have tiny openings or pores in it. The anatomical details of the gap junction are not entirely clear, but the pores apparently provide a more direct intercellular communication. In addition to being secure bindings, gap junctions are believed to serve as an electrical coupling between cells. They are regions of low electrical resistance, and electrical changes in one cell can be transmitted through the gap junction to the adjacent cell. Gap junctions are also found in tissues other than epithelium, most notably in smooth and cardiac muscle.

In certain types of epithelium, particularly that lining the digestive tract, there are structures which light microscopists have called "terminal bars." They appear as darkly staining areas (bars) between cells at the free surface. With the electron microscope the free border of the cells is found to have a series of several junctions, which together have been called a *junctional complex.* Such a complex consists of a

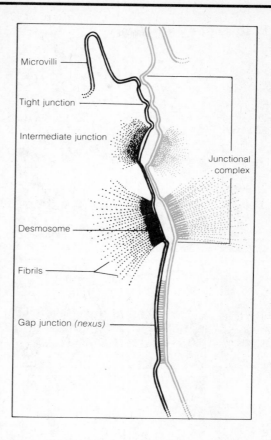

FIGURE 3–1
Diagram showing different types of cell junctions.

Microvilli

Tight junction

Intermediate junction

Junctional complex

Desmosome

Fibrils

Gap junction *(nexus)*

tight junction and a desmosome, with an intermediate type junction between them. The intermediate junction actually is intermediate both in location and structure between the tight junction and desmosome. The tight junction, and probably the intermediate junction, form a continuous band around the neck of the tall columnar cells (see below). Such a collar provides for adhesion of the cells and limits passage (leakage) of materials from the free surface through the intercellular space.

Junctions between nerve cells, or between nerve and muscle cells, are quite specialized, and are known as *synapses* and *neuromuscular junctions,* respectively. These junctions have a space between cells, but their main function is communication from one cell to another, not adhesion of one cell to the other. Synapses and neuromuscular junctions will be discussed in Chapter 9.

Classification of Epithelia As a surface tissue, epithelium is organized in sheets or layers. If the layer is one cell

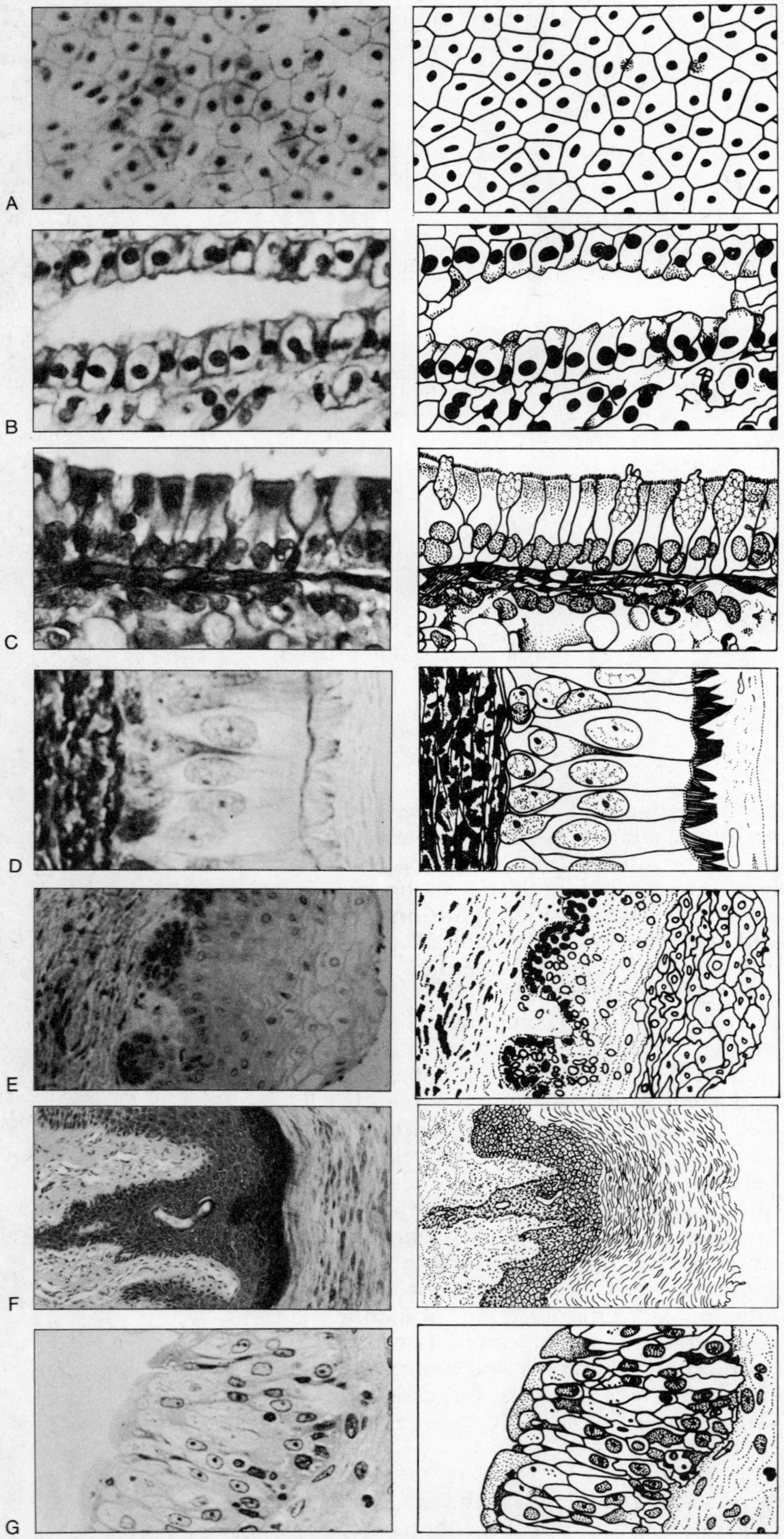

A, B, C, D, E, and G courtesy Carolina Biological Supply Company
F courtesy Manfred Kage/© Peter Arnold, Inc.

FIGURE 3–2 Photomicrographs and diagrams of several types of epithelium. A. Squamous epithelium viewed from above. B. Cuboidal epithelium in the wall of a duct. C. Simple columnar epithelium with microvilli and goblet cells. D. Pseudostratified ciliated columnar epithelium. E. Stratified squamous epithelium (nonkeratinized). F. Stratified squamous epithelium (keratinized). G. Transitional epithelium (unstretched).

thick, it is a simple epithelium; if it is two or more cells thick, it is a stratified epithelium. Simple epithelia are better suited to transport of materials, while stratified epithelia provide better protection. The shape of epithelial cells varies from very flat (squamous) to tall and slender (columnar). An overview of the types of epithelia is given in Figure 3–2.

Simple epithelia. From above, the cells of **simple squamous epithelium** appear to be rather large and roughly hexagonal. In cross section they appear flat: the cytoplasm is scarcely visible and the nucleus bulges from the surface. Simple squamous epithelium is usually found where materials are exchanged across the cell as in the smallest of blood vessels, where the blood and tissue fluid exchange occurs.

In **cuboidal epithelium** the height and width of the cells are about equal. This tissue is commonly found in the walls of ducts (tubes) and as secretory portions of glands.

Simple columnar epithelium lines much of the digestive system. Each cell is relatively tall and slender, with the nucleus located near the attached end of the cell. One-celled glands, called *goblet cells* because of their appearance, are often scattered among columnar cells. They produce a mucous secretion which, when released, protects and lubricates the epithelial surface.

Pseudostratified columnar epithelium is so named because the epithelium appears to be stratified but is not. The confusion arises from the fact that, although all cells touch the basal lamina some of them apparently do not reach the free surface and the cell nuclei therefore appear to lie in several layers. In many locations, such as the airways of the respiratory system, the cells that reach the surface are ciliated. Goblet cells, which are not ciliated, are often found among the ciliated cells.

Stratified epithelia. In **stratified squamous epithelium** only the superficial cells are squamous, while the deepest cells are small and roughly cuboidal and the middle layers are intermediate in shape. This type of epithelium is found in the epidermis of the skin. Cells on the external surface of the body (skin) are exposed to wide environmental variations. The skin has the only epithelial surface that is not kept moist. Its free surface is covered by a *cornified* or *keratinized* layer that prevents excessive fluid loss from the skin and buffers the effects of environmental fluctuations. The cornified material results from the degeneration of superficial cells. As mitosis forms new cells in the deep layers, the older cells are displaced toward the surface, further from their blood supply, until deterioration and cornification occur. Stratified squamous epithelia found in such areas as the lining of the esophagus form part of the body's moist internal surfaces.

Restricted to the lining of organs of the urinary tract, **transitional epithelium** can assume the characteristics of both simple and stratified epithelium: cells of the surface layer may extend over cells of the layer below, as in simple epithelium, or they may become balloonlike and dome-shaped. In the contracted (empty) bladder, the epithelium is many cells thick and the surface cells are domed; in the distended (full) bladder, the epithelium is stretched so as to be only a few cells thick, and the surface cells are flattened.

Glandular epithelium. Our final subgroup, **glandular epithelium,** is not truly a separate type of epithelium, but rather a modified arrangement of other forms, usually simple cuboidal or columnar. Glands are organs or cells that secrete, that is, they produce a substance to be released from the cell. There are basically two types of glands, exocrine and endocrine. *Exocrine glands* release their secretion either directly across the free surface of the cells or into special ducts for delivery to the site of action, as is the case with the salivary glands. *Endocrine glands* release their secretions *(hormones)* into the bloodstream for distribution. Under the microscope, ductless endocrine glands generally appear as clumps or cords of cells, with blood vessels nearby. Some glandular organs serve both an exocrine and an endocrine function. An example is the pancreas, in which isolated islands of endocrine tissue are scattered throughout the more abundant exocrine acini (sacs that secrete).

Secretory cells may be arranged in many different ways, and glands are classified according to their pattern of organization (Figure 3–3). Unicellular glands are individual secretory cells scattered about an epithelial surface, such as the goblet cells mentioned above. Multicellular glands may be tubular or alveolar. Tubular glands are those in which the secretory cells are situated along the walls of the tubules. In alveolar or acinar glands the secreto-

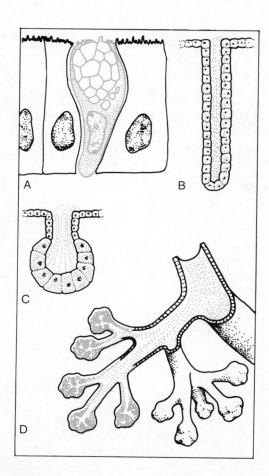

FIGURE 3–3 Types of glands. A. Unicellular (goblet cell). B. Simple tubular. C. Simple alveolar (acinar). D. Compound tubuloalveolar.

ry cells are grouped to form a rounded clump (an alveolus or acinus) at the end of the duct into which they secrete. A tubuloalveolar gland would have secretory cells along the tubular walls and also in alveoli.

Connective Tissue

Connective tissue serves as the connection between structures, between parts of structures, and between individual cells. It is the most widely distributed and probably the most varied of the primary tissues. In contrast to epithelium, the conspicuous portion of connective tissue is the *intercellular material*, consisting of fibers and a ground substance. The many types of connective tissue differ from one another chiefly in the nature and arrangement of the fibers, in the kind of ground substance, and, to a lesser extent, in the number and types of cells. The cells of connective tissue, though not abundant, are responsible in one way or another for many of the characteristics of the intercellular material, and indeed for its very existence, since they produce both the fibers and the ground substance.

Connective tissue arises from a primitive embryonic tissue known as **mesenchyme.** It has large, undifferentiated cells, *mesenchymal cells,* whose processes extend out to contact those of surrounding cells, forming a loose network. The spaces between cells are filled by an intercellular substance that is generally described as "amorphous" (without definite shape), which really means that it has no particular features by which to describe it. Such tissue is widely distributed in the embryo, and bits of the tissue, or at least the cells, remain as part of adult connective tissue. Mesenchymal cells are characterized by their ability to develop into more specialized cells, such as fibroblasts (see below), or into cartilage- or bone-forming cells (chondroblasts or osteoblasts).

Connective Tissue Fibers Three types of connective tissue fibers are recognized: collagenous, elastic, and reticular fibers. **Collagenous fibers,** composed of the protein *collagen,* are the most common of the connective tissue fibers. They are flexible, but relatively inelastic. Collagenous fibers may occur as single strands, but more often as bundles formed by a number of parallel strands. Like other proteins, collagen is altered by heat—boiling changes it to gelatin. The amount of collagen in a tissue increases with age, which is why meat from an old animal is likely to be tough. Such cuts of meat are better stewed or boiled, which converts the fibrous collagen to gelatin and makes the meat less tough.

Elastic fibers do not form bundles as does collagen, but instead branch to form a network. They are widely distributed and often interspersed with collagen fibers. Their outstanding characteristic is their elasticity. They are likely to be most abundant where this property is needed, as in the walls of large arteries. Elastic fibers are made up of a protein called *elastin* that is quite resistant to chemical action and to heat.

Reticular fibers resemble collagenous fibers in some respects. Most investigators believe that they are immature collagenous fibers. Reticular fibers are extremely fine and tend to branch to form delicate networks. They are found supporting small structures such as individual nerve and muscle cells, the smallest of blood vessels, and the terminal portions of the airways in the lungs.

Ground Substance The cells and fibers of connective tissue are embedded in a ground substance which, like connective tissue fibers, is of cellular origin. It is a complex mixture of proteins whose consistency may vary from fluid to solid, though in most locations it contains a considerable amount of water and is quite gelatinous. In fact, the tissue fluid (interstitial fluid) with which the cells carry on their exchange is intimately associated with the ground substance. The distinguishing properties of cartilage and bone are due to the deposition in the ground substance of certain inorganic salts which render it much less fluid in consistency.

Connective Tissue Cells There are few cells in connective tissues, but there are many different types of cells scattered randomly throughout the tissue, where they perform a number of important functions.

Undifferentiated cells are primitive mesenchymal cells with great potential for developing into more specialized connective tissue cells. They can be found in adult connective tissue, particularly in places where growth or regeneration and repair are likely to occur.

Fibroblasts form all three types of connective tissue fiber and the ground substance. They are the most abundant of the connective tissue cells. Fibroblasts develop from undifferentiated cells. However, they are still relatively undifferentiated and retain the potential to develop into more specialized cells, including those of bone and cartilage.

The production of collagen fibers is similar in many respects to the synthesis of other proteins. A protein (called *tropocollagen*) is synthesized on ribosomes of rough endoplasmic reticulum, then enters the ER and moves to the nearby Golgi apparatus and then it is secreted from the fibroblasts. Some tropocollagen may be released directly from the ER. Once outside the cell the tropocollagen molecules are polymerized; that is, they are joined together to form strands or microfilaments which eventually will become bundles of collagen fibers, or perhaps the more delicate reticular fibers.

Elastic fibers are formed in much the same way. Protein molecules are assembled on the ribosomes of the fibroblasts and are released from the cell. Enzymes outside the cell cause the formation of elastic strands or fibers. Fibroblasts also produce most, if not all, of the important constituents of the ground substance.

Macrophages are important because they are phagocytes. Hence they are capable of surrounding and engulfing particles in their immediate environment. Such particles, which can often be seen in the cytoplasm of the phagocyte, may consist of parts of dead cells, bacteria, or other foreign matter.

The phagocytized material is eventually destroyed by the enzymes in the phagocyte's lysosomes.

Some macrophages are fixed in a particular location, such as the liver or lung, and carry on their activities locally, while others are free or wandering cells that can migrate to a site of inflammation. Different names have been given to each variety of macrophage, but they all serve basically the same function. By helping to clean up the intercellular spaces, these scavenger cells contribute to the body defenses. The term *reticulo-endothelial system* is often used to describe these widely distributed phagocytes and their protective functions (see Chapter 14).

Fat cells, or **adipose cells,** are characterized by their ability to take up and store large quantities of fat in their cytoplasm. Once inside the cell, fat globules tend to coalesce into a large droplet that crowds the cytoplasm into a thin ring around the edge of the cell. The nucleus is forced to one side, resulting in a cell that resembles a signet ring.

Other connective tissue cells are less abundantly distributed. Among them are some large rounded cells called **mast cells,** whose cytoplasm is packed with granules. The granules contain *histamine*, which dilates small blood vessels, *heparin*, which prevents blood clotting, and probably some other substances. Under certain conditions mast cells release these substances into the interstitial fluid, causing certain allergic reactions such as hives, hay fever, and asthma, as well as a drop in blood pressure. **Blood cells,** particularly white blood cells (*leukocytes*), are found outside blood vessels in the surrounding connective tissue spaces. **Bone cells** (*osteocytes*) and **cartilage cells** (*chondrocytes*) are found in bone and cartilage, respectively.

Classification of Connective Tissue Because the components of connective tissue may be organized in many different ways, there are numerous types of connective tissue, and they have been classified in a number of ways. Table 3–1 presents one convenient and useful classification.

TABLE 3–1   CLASSIFICATION OF CONNECTIVE TISSUE

I. Embryonic connective tissue
II. Adult connective tissue
   A. Connective tissue proper
      1. Loose (areolar)
      2. Dense
         a. Irregularly arranged
         b. Regularly arranged
      3. Connective tissue with special
         properties
         a. Adipose
         b. Reticular
         c. Elastic
   B. Cartilage
      1. Hyaline
      2. Elastic
      3. Fibrous
   C. Bone
      1. Cancellous
      2. Compact
   D. Dentin
   E. Blood and lymph

Embryonic connective tissue is of interest chiefly because it is the primitive mesenchyme from which the adult types of connective tissue develop. It is diffuse and homogeneous, consisting of ground substance interspersed with primitive undifferentiated mesenchymal cells. Adult connective tissues are classified as connective tissue proper, cartilage, bone, dentin, and blood and lymph.

Connective tissue proper. When one thinks of adult connective tissue, it is of **connective tissue proper,** and particularly of **loose** or **areolar connective tissue,** often described as the "typical" connective tissue (see Figure 3–4). It contains both collagenous and elastic fibers and its cells are chiefly fibroblasts, although macrophages and other connective tissue cells are also present. The tissue is disorganized and randomly arranged, with the elements loosely packed and giving the appearance of open areas (in which the ground substance is located). Loose connective tissue can be found in most organs, extending along ducts and blood vessels, and as *septa* (partitions) between parts or segments of the organ. The tissue serves as a "general packing material," filling what might otherwise be empty spaces between or within organs. It is also found directly under the skin, where it is known as **subcutaneous tissue** or **superficial fascia.** In many areas it contains an abundance of adipose tissue.

Where the connective tissue elements are more closely packed, the tissue is called **dense connective tissue.** Obviously there is no marked separation between loose and dense, and some connective tissues could be classified in either way. In dense connective tissue the fibers, mostly collagenous, are very conspicuous and they may be *regularly* or *irregularly* arranged. In the latter case, the tissue may resemble a crowded loose connective tissue, as in the deep layer of the skin (dermis), or the fibers may lie within a single plane to form a sheet, as in the capsules that surround many organs or in the sheaths around and between muscles. Such membranes or capsules also surround bones and skeletal muscle, and may be quite tough and durable. This tissue is called **deep fascia.** It is continuous with the superficial fascia that lies between the skin and muscles or bones.

The collagen fibers of regularly arranged dense connective tissue lie parallel to one another, with fibroblasts or other cells squeezed in between the fiber bundles. This arrangement has great tensile strength and is found in tendons and ligaments where such strength is needed to attach muscles and bones.

Connective tissue with special properties is mostly loose connective tissue in which a particular type of cell or fiber predominates, imparting its properties to that connective tissue. If there are large numbers of adipose or fat cells, the tissue is **adipose tissue.** It usually develops from loose irregular connective tissue and is commonly found in subcutaneous tissue. Besides providing energy reserves, adipose tissue serves as insulation and protection.

Reticular tissue is connective tissue in which the predominant fibers are the delicate reticular fibers. It is found in many places, including lymph nodes and the spleen, where the spaces between fibers contain large numbers of *lymphocytes* (a type of white blood cell).

Elastic tissue is a form of dense connective tissue found in the walls of

FIGURE 3-4 Photomicrographs and diagrams of several types of connective tissue. A. Loose (areolar) connective tissue. B. Dense irregular connective tissue. C. Dense regular connective tissue (tendon). D. Adipose tissue. (The fat droplets have been dissolved in the preparation process.) E. Reticular tissue. F. Elastic tissue. G. Hyaline cartilage.

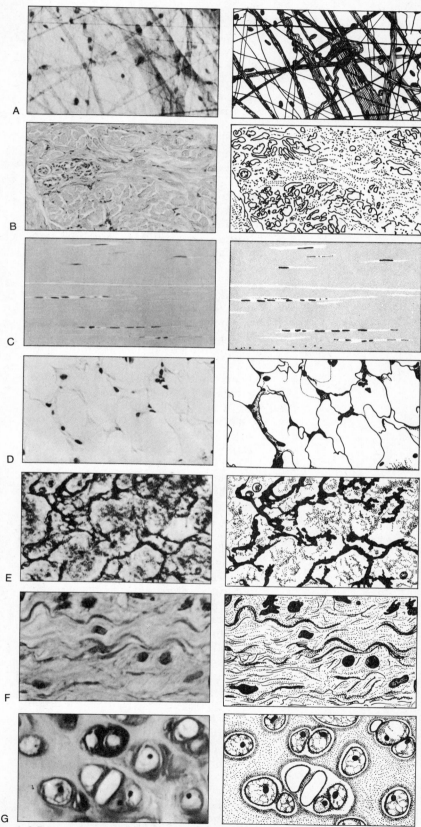

some large arteries and in some ligaments that must be able to stretch somewhat. In the large arteries (which are distended briefly with every heartbeat) there are many elastic fibers, some of which form a thick elastic membrane providing a stretchable circular layer in the wall of the artery. Parts of the heart, airways of the respiratory system, and other organs contain elastic fibers.

Cartilage. Whereas connective tissue proper is often a connective and packing material, cartilage and bone are supporting materials. The firmness necessary for the weight-bearing role is due to special properties of the intercellular substance. In the case of **cartilage tissue,** there is an abundance of fine collagen fibers, and the ground substance contains a complex of specific proteins. The proteins seem to be able to combine with the collagen in some way, to give the intercellular substance a somewhat leathery consistency—firm and solid, but not rigid.

When undifferentiated cells acquire the ability to produce the intercellular material that is characteristic of cartilage, they are called **chondroblasts** (cartilage formers). The intercellular material they produce has a homogeneous appearance that masks the fact that it also contains numerous collagen fibers. As they produce the intercellular substance, chondroblasts become separated from one another, with each cell isolated within the intercellular material in a small nest or hole called a **lacuna.** If the cell (known as a *chondrocyte* in mature cartilage) undergoes mitotic cell division, there will be two or more cells in the lacuna.

Almost all cartilage is enclosed in a sheath of dense connective tissue called **perichondrium.** The primitive cells and fibroblasts of the perichondrium retain their ability to differentiate into active

A, C, D, E, F, and G courtesy Carolina Biological Supply Company
B courtesy Manfred Kage/© Peter Arnold, Inc.

cartilage-forming cells. Cartilage is totally *avascular* (without vessels), so that the nutritive requirements of chondrocytes must be met by diffusion from blood vessels in the perichondrium through the intercellular substance. This is one reason why cartilage is always present as thin layers or plates; the diffusion distance must not be too great. Most of the bones in the body develop from cartilage, so in the embryo and in early life much of the skeleton is cartilage in various stages of replacement by bone.

Three types of cartilage are commonly recognized: hyaline, elastic, and fibrous (fibrocartilage). **Hyaline cartilage** is the most widely distributed form. It is named for its somewhat glassy translucence (hyaline means glasslike). The *articular surfaces* of bones (surfaces that move against other bones) are covered with a thin layer of hyaline cartilage that provides a virtually friction-free surface for movement. Hyaline cartilage is also found in the trachea and bronchi (the large airways to the lung), in the rib cartilages, and in the nose. **Elastic cartilage** is stiff like hyaline cartilage, but it is more flexible due to its large content of elastic fibers. The external ear contains elastic cartilage. **Fibrocartilage** contains a greater number of collagen fibers. It is found in the discs that lie between the vertebrae and in the symphysis pubis of the pelvis.

Bone. The **bone tissue** resembles cartilage in several respects, but differs primarily in that inorganic salts deposited in the intercellular material make it much more rigid than cartilage. Bone cells, called **osteocytes,** also lie in lacunae, but each cell has several long *processes* (protrusions) extending out from it. These processes extend into the intercellular material in tiny openings or canals which provide indirect contact with blood vessels. This is necessary because the more solid intercellular material of bone will not permit diffusion. The arrangement of osteocytes, intercellular material, and blood vessels is quite specific and is discussed more fully in Chapter 4.

Like cartilage, bone is surrounded by a sheath of dense fibrous connective tissue. It is identical to perichondrium, but since it surrounds bone it is called **periosteum.** The periosteum is continuous with the deep fascia that is around and between muscles.

Dentin. Often included with bone tissue, **dentin** is actually a different material. Dentin forms the bulk of substance for the teeth. It differs from bone in that dentin is harder than bone and lacks blood vessels. Also, cells are not embedded in the dentin, but rather lie along the inner surface. Only the cell processes penetrate the dentin. A harder, denser substance than dentin is the **enamel,** which is secreted onto dentin to form an especially hard covering. Enamel is composed almost entirely of calcium salts and, in fact, lacks cells entirely.

Blood and lymph. The **blood** and **lymph tissue,** whose cells are derived from connective tissue cells, differ from other types of connective tissue chiefly in the nature of the intercellular substance and in that there are no fibers (unless the blood clots). The intercellular substance of blood is a fluid known as **plasma.** In blood, the cells are of course blood cells, which are, unlike those of other tissues, completely free to travel through the blood vessels. Although some blood cells are found in many types of connective tissue, connective tissue cells are not typical of blood. Associated with blood tissue is the **marrow,** which is the blood-cell producing tissue found in the bones.

Lymph is derived from the tissue fluid (interstitial fluid). It resembles plasma in most respects, although the protein content of lymph is lower. Lymph also contains cells, but not nearly as many as blood. Organs such as the spleen, thymus gland, and lymph nodes are known as the *lymphoid organs.* Lymphoid tissue is also found in the digestive tract. Among other things, they are important in developing immunities (see Chapter 14).

SKIN

All free surfaces of the body are covered by a characteristic type of epithelium that is bound by connective tissue to the underlying structures. This is true not only for the covering on the outside of the body, but also for the linings of the major body cavities and hollow organs. The thoracic and abdominopelvic cavities are lined with membranes[2] known as *pleura* and *peritoneum.* These membranes lie on the inner surface of the chest wall and abdominal wall, and they also cover the organs in those cavities, such as lungs, stomach, and intestines. They have a smooth surface of simple squamous epithelium, with a thin layer of connective tissue underneath. Such membranes are known as **serous membranes.**

A membrane on the inner surface of a hollow organ such as the stomach is known as a **mucous membrane,** or the **mucosa.** It consists of an epithelial layer whose type varies from one organ to the next, with a connective tissue layer underneath. The connective tissue layer (*lamina propria*) also varies, for it may be quite thick, and in many locations contains a number of glands.

The exterior covering of the body is the skin, and it, too, is an epithelial layer with connective tissue underneath. But it differs from serous and mucous membranes in several ways, one being that the skin surface is not a moist surface as are the others.

The skin is an excellent example of the way in which tissues can be organized into an organ that performs certain specific functions for the entire body. In fact, the skin is considered to be an organ itself. At first glance, its most obvious function is to cover everything; to keep the "insides" in! Protection is certainly a major role, for skin protects underlying tissues and organs from various mechanical assaults, from invasions by harmful bacteria, and from the potentially damaging effect of ultraviolet light from the sun's rays. Since skin is reasonably waterproof, it greatly reduces water loss and dehydration. In addition, it plays an important role in regulating body temperature by the action of the sweat glands and the variations in the amount of blood flowing through the skin. It serves a minor ex-

cretory role, for sweat contains some waste products. The skin also is the site of vitamin D synthesis. Finally, skin is an important sense organ. There are numerous nerve endings (receptors) mostly in the deep layers of the skin that are stimulated by changes on the skin surface. These receptors send information about touch, pressure, pain, and temperature to the central nervous system. The layer of connective tissue that is directly beneath the skin (subcutaneous tissue) is an important site for the storage of fat (energy storage).

The covering of the body is made up of the skin and its specialized derivatives, namely hair, nails, and the sebaceous and sweat glands, plus the subcutaneous connective tissue that lies beneath it (see Figure 3–5). The skin itself consists of two distinct layers: **epidermis,** which is the epithelial surface layer, and **dermis** (or *corium*), a layer of dense connective tissue. The skin is attached to deeper tissues by the subcutaneous connective tissue, a loose connective tissue that is not actually part of the skin.

Epidermis

The epidermis is a layer of keratinized, stratified squamous epithelium. Its thickness varies with its location, being thickest in the palms of the hands and soles of the feet (where it may be as much as 50 cell layers thick). The keratin material is relatively impermeable to water, and it is the major reason why skin is able to limit loss of water from the surface of the body.

Epidermis grows from the inside out. Cells in the deepest layer are the only ones that regularly undergo mitosis, and they do so often enough that the entire epidermis is replaced about every 2 to 4 weeks, depending upon the location. As new cells are formed, the older ones are pushed toward the surface. As they get further from the blood vessels and closer to the everchanging conditions of the external environment, the cells undergo marked changes in appearance. On the basis of these changes, the epidermis is said to have five layers. They are best seen in the thick skin of the palm or sole because in thinner skin, some of the layers cannot

---

[2]Note the different use of the term "membrane." Our earlier use was as part of a cell, a cellular or intracellular boundary. This usage refers to a sheet of tissue, containing cells, among other things. It is more in line with the use as in basement membrane.

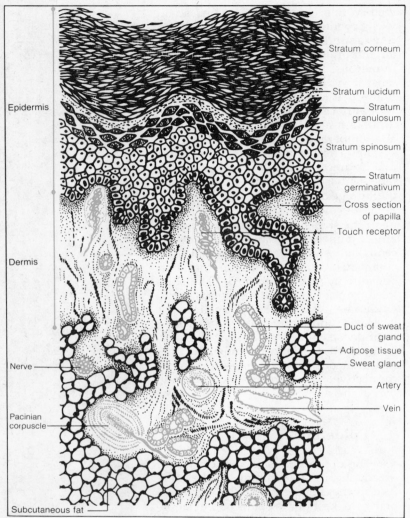

FIGURE 3–5   Diagram of a section of skin and underlying connective tissue.

be distinguished very well. As you study the layers, keep in mind that they represent different stages in the life of the epithelial cells, and that these stages occur in each cell as it moves toward the surface.

1   The **stratum germinativum** (*basal layer*) is the deepest layer. It consists of a single layer of cuboidal or columnar cells that rest upon the basement lamina. One can often find cells undergoing mitosis in this layer, for it (plus the deepest part of the next layer) is the source of replacement cells for the entire epidermis.

2   The **stratum spinosum** (*prickle cell layer*), which may be quite thick, gets its name from the fact that each cell has tiny processes protruding from the surface in all directions. The processes, which contain bundles of tiny filaments (*tonofibrils*) meet processes of adjacent cells to which they are joined by desmosomes. The bundles of filaments have been called "intercellular bridges," but there is no actual continuity between the cells. These connections increase the resistance of the skin to abrasion.

3   The **stratum granulosum** (*granular layer*) consists of a few layers of somewhat flattened cells whose cytoplasm contains characteristic granules. The granules are believed to be precursors of the keratin that covers the skin surface. Upon close examination, the cells are seen to be slightly separated from adjacent cells by a tiny space.

4   **Stratum lucidum** (*clear layer*) is easily seen in thick skin, but it is often absent from thin skin. The cells are far removed from their blood supply, having been pushed toward the surface by repeated cell divisions in the deeper layers. They are elongated, translucent, and most have no nucleus. Most of the cytoplasmic organelles have disappeared as well. The layer appears to be homogeneous, but in fact its changing cells are packed with tiny filaments, a further step in the process of keratinization.

5   The outermost layer, the **stratum corneum** (*horny layer*) consists of dead scalelike cells whose main component is the fibrous protein, *keratin*. The reactions that result in the formation of this material and the death of the cell are part of the process of keratinization, or cornification. Specialized epidermal structures, such as hairs, nails, claws, and hoofs are also formed of keratin, with their different properties due, in part, to differences in the keratin in each case. Cornified cells are continuously shed from the surface of the skin as a result of friction and abrasion. This layer of "dead skin" provides additional protection for epithelium from the external world with its great variations in temperature and moisture content, as well as pressure and friction.

In the deepest or basal layer of epithelium, (stratum germinativum) are some large rounded cells. They have long, slender armlike processes that extend up between cells of the stratum germinativum and stratum spinosum, and end by actually protruding into the epithelial cells. These large cells, called **melanocytes,** produce the dark-brown pigment **melanin,** which is a major factor in determining color of the skin.

The melanin is produced in membrane-bound vesicles in the central part of the melanocytes, and migrates out along the cell processes to enter the epithelial cells of the two deepest layers by a process much like phagocytosis. The number of melanocytes present is not determined by race or sex, but differences in skin color depend to a great extent upon the number, size, and distribution of the melanin granules in the epithelial cells. In light-skinned people the pigment granules are sparse and confined to the stratum germinativum, but in dark-skinned people there are more and larger granules, and they are not limited to the deepest layer of cells. Skin color is also influenced by *carotene,* a yellow pigment found in all skin, and by the blood in the numerous vessels in the skin. Exposure to the ultraviolet rays from the sun darkens the melanin in the melanocytes, hastens its migration to the epithelial cells, and accelerates its rate of production. Some individuals are unable to produce melanin, and these people are *albinos.* Their skin is very fair, almost pink due to the blood vessels near the surface, and they are extremely sensitive to the ultraviolet rays of the sun.

Dermis

Beneath the epidermis lies a connective tissue layer, the **dermis.** The junction with the epidermis is marked by the homogeneous basement lamina, which is typical of the epithelial layers. The cells of the stratum germinativum have tiny fringelike processes which seem to extend into the basement lamina, and fine connective tissue fibers from the dermis underneath also enter the basement membrane, thus anchoring the epidermis securely.

The surface of the junction between epidermis and dermis is irregular. If the epidermis is separated from the dermis, its undersurface clearly shows these irregularities. Often there are epidermal ridges with "valleys" in between. Bits of dermis known as *dermal papillae* protrude up into these valleys, often with several rows of papillae in a single valley. The pattern of the epidermal ridges varies in different parts of the body. In areas such as the face, the ridges are rather shallow and the valleys are wide; in other locations they are more prominent. They are best developed in the fingers and palms of the hands and soles of the feet. The pattern of these ridges, which can be detected on the external surface of the skin, is the basis of the fingerprints. Because the surface ridges in the fingertips form patterns of arches, loops, or whorls that are unique to each individual, fingerprints can be used for identification. These patterns, which first appear about the 13th week of fetal life, are probably determined by several genes. The presence of such "corrugations" on the hands and feet enables us to grip objects or surfaces more firmly, and in fact such skin is often referred to as a *friction surface.*

The dermis itself contains numerous connective tissue fibers (collagenous, elastic, and reticular), and is an excellent example of dense irregular connective tissue. The fibers in the dermal papillae are quite fine and form a network around the numerous tiny blood vessels and around the sensory nerve endings (touch receptors in particular) that are found in the papillae. In the deeper portion of the dermis the connective tissue fibers are thicker and coarser. They form a tightly interwoven network, with fibers running in various directions but generally parallel with the surface.

Subcutaneous Tissue

The **subcutaneous tissue** lies beneath the dermis. It is not a part of the skin, but it is the superficial fascia that attaches the skin (dermis) to the underlying deep fascia. Since it is beneath the dermis, it is *hypodermic;* thus a subcutaneous or hypodermic injection enters the subcutaneous tissue. Sometimes it is difficult to distinguish between dermis and subcutaneous tissue, since both are irregular connective tissue, but subcutaneous tissue is generally much looser and usually contains a considerable number of adipose cells, forming a layer of varying thickness. In regions such as the abdominal wall, *(panniculus adiposus),* the tissue may be several centimeters thick, but in other regions such as the eyelid, there is no fat. In areas such as the palm and sole, where the skin is firmly attached, the subcutaneous tissue contains tightly woven collagen fibers that are continuous with those of the dermis. Where the skin is less firmly attached, the fibers are looser and allow the skin to slide over the underlying tissues. In some animals the skin is attached very loosely and can slide over a rather wide area. The deeper layers of subcutaneous tissue are continuous with layers of deep fascia that surround muscles and with the periosteum of the bones.

Derivatives of the Skin

There are several specializations of skin. These include the nails and hair (both of which represent modifications of the cornified layer of epidermis), and the sebaceous and sweat glands.

Nails.　Nails consist of hard keratinized plates of stratum corneum and stratum lucidum. They grow out over the epithelium on the top surface of the tips of the fingers and toes. Nails are of epidermal origin and develop early in embryonic life when the epithelium invaginates and folds under to form the *nail groove* (Figure 3–6). The epithelial cells of the groove form the *matrix* from which the nail itself arises. As the *nail plate* slowly grows out from the groove, it slides over the *nail bed,* which consists chiefly of cells of the deepest layer of epidermis (stratum germinativum). Only those cells at the base of the nail produce the nail substance, however.

At the far end of the nail the epidermis of the nail bed becomes continuous with the cornified epithelium of the skin (the *hyponychium*). At the base of the nail where the epithelium folds back into the nail groove, the stratum corneum forms a fold (known as the *cuticle* or *eponychium*) over the emerging nail. The nail is semitransparent and appears to be pink in color due to the presence of many small blood vessels under the nail bed. A slight pressure applied to the free edge of the nail will cause the nail bed to blanch, or whiten, as the blood is displaced. This technique is sometimes used for a quick check of the circulation, for a delayed return of color to the nail bed after release of pressure often suggests a low blood pressure or constriction of the skin blood vessels. The significance of the curved white area at the root of the nail (the *lunula*) is not known, but it may be due in part to the thickness of the matrix, which causes the blood vessels to be deeper there, or perhaps the keratin of the nail substance is not completely matured yet.

Hair.　Most of the skin is covered by hairs, but their abundance and characteristics vary with the age, sex, and race of the individual, as well as the part of the body. Most of the hair is relatively fine, but at puberty coarser hairs appear in the axillary (armpit) and pubic areas and on the face and elsewhere in males. The function of hair is not very apparent in humans, but for many animals it provides insulation against changes in environmental temperature. In humans there are touch receptors around the base of many hairs that are stimulated when the hair is bent.

Hairs develop from *hair follicles,* and hair follicles develop from the epidermis (Figure 3–7). During fetal life epidermal buds grow downward into the dermis forming tubular epithelial extensions that become hair follicles. Some mesenchyme pushes up into the enlarged bulblike deep ends of the follicles as dermal papillae. The cells at the base of each follicle surround and fit over the papilla, forming the bulbshaped root of the hair. The epidermal cells of the root proliferate and differentiate, and become keratinized. As more cells are formed they are pushed up the follicle, and the hair—a slender strand of keratinized epidermal cells—eventually pushes its way out onto the surface of the skin. Melanocytes are found deep in the root, adjacent to the dermal papilla. Apparently here as elsewhere, they function by producing melanin and transferring it to the newly formed epithelial cells. Melanin, of course, is the major factor in determining hair color. Hairs that lack pigment are silvery white, and if some of the hairs have pigment while others do not, the hair appears to be gray.

The process of keratinization in hair is quite similar to that on the surface epithelium, except that the skin produces keratin over its entire surface, and it is continually lost from that surface. In hair, however, the process only occurs deep in the follicles, and the keratin formed is a harder material that is not so easily worn away.

The shaft of a hair is made up of three concentric layers: *medulla, cortex,* and *cuticle,* which are distinguished by the density of the keratin. Keratin is softest in the central medulla and hardest in the thin cuticle.

Hairs cut in cross section have different shapes. In general, people of the Mongol races (Asians, American Indians, Eskimos), whose hair is very straight, have round hairs. Black people, whose hair is woolly in nature, have elliptical hairs that are often flattened on one side. Individuals with wavy hair, including many Caucasians, have hairs that are oval in cross section.

Hair follicles usually lie at an angle to the skin surface. On the wide side of the angle between hair and skin surface is a bundle of smooth muscle fibers known as the **arrector pili muscle** *(piloerector).* Each tiny muscle attaches to the connective tissue sheath of the follicle and to the papillary portion of the dermis. Contraction of the muscle, controlled by the nervous system, pulls the hair into a more upright position; that is, more perpendicular to the surface. In furred animals this tends to increase the insulating effectiveness of the fur because it traps more air between the hairs. Piloerection is also a sign of anger or fear in dogs, cats, and other ani-

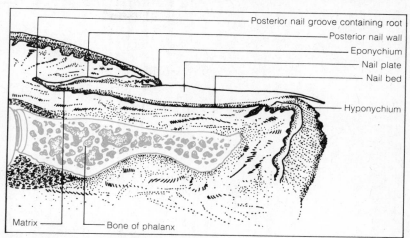

FIGURE 3-6   Longitudinal section of a fingertip showing the fingernail and related structures.

mals. The same mechanism exists in humans, but the main result is to produce "gooseflesh" or "goose bumps," the pimpling of the skin surface brought about by the movement of the hairs in response to cold or upsetting emotions.

Sebaceous Glands.   Sebaceous glands are associated with each hair follicle and are found in the angle between the piloerector muscle and the follicle shaft (Figure 3-7). They are exocrine glands, delivering their secretion through a short duct into the small space between the hair shaft and the follicle. The glands are formed of cells clustered around the ends of the branches of the ducts. As new cells are produced at the outer edges of the cluster, older cells are displaced toward the center and they acquire vacuoles containing oil droplets. Eventually their nuclei shrink, the cells die, and then rupture, spilling their contents. **Sebum,** the secretion of the gland, is composed of the oily contents of the gland cells and some of the debris of ruptured cells. It gradually moves into the duct, then into the hair follicle and toward the skin surface. Secretion of the sebaceous glands is controlled largely by the sex hormones. Disturbances in the production and flow of sebum are a major factor in the development of acne.

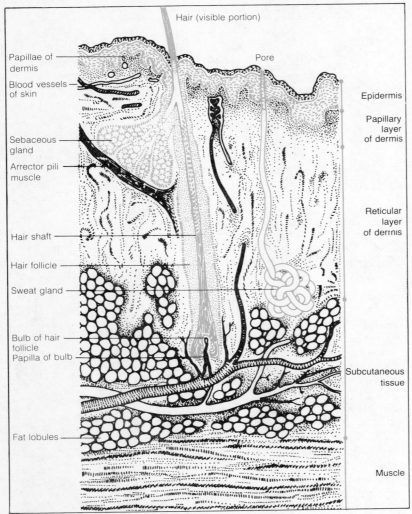

Hair (visible portion)

Papillae of dermis

Pore

Blood vessels of skin

Epidermis

Papillary layer of dermis

Sebaceous gland

Arrector pili muscle

Reticular layer of dermis

Hair shaft

Hair follicle

Sweat gland

Bulb of hair follicle

Papilla of bulb

Subcutaneous tissue

Fat lobules

Muscle

FIGURE 3–7 Diagram of a section of skin showing a hair follicle and sebaceous glands.

"Blackheads" may result if there is hardening of the material in the duct of the sebaceous gland.

Sweat Glands. Sweat glands are distributed over most of the body surface. They are especially abundant in the palms and soles and fingertips. The gland consists of a long, slender unbranched tube with the deep end coiled into a ball. This knotted end contains the secretory cells and lies in the deep part of the dermis (see Figure 3–5). From this ball of glandular cells the duct of the gland follows a wavy course through the superficial layer of dermis and penetrates the epidermis and cornified layer to reach the surface. As it passes through the epidermis, however, its cells blend with those of the surrounding epidermis and only the passageway remains identifiable.

The secretion of these sweat glands is quite watery since it contains virtually no protein or other viscous material. It contains the same electrolytes found in the plasma, particularly sodium chloride, but in lesser concentration—sweat is hypotonic to plasma. Because it also contains small amounts of some waste products, sweat glands do serve a minor excretory function. Their main role, however, is in temperature regulation. When the temperature of the environment is high, or if the body becomes heated by exercise or work, the secretion of sweat has a powerful cooling effect if it evaporates from the skin surface. Sweat glands are controlled by nerves that have connections with many centers in the brain, including those that regulate body temperature.

There are some specialized sweat glands in certain areas, notably the axillae. They are large, with their secretory portions beneath the skin in the subcutaneous connective tissue, and their ducts open into the upper portion of a hair follicle rather than on the skin surface. These sweat glands are not associated with temperature regulation and, although they do receive a nerve supply, their secretion is controlled to a certain extent by sex hormones. Their secretory activity begins at puberty. The secretion of these sweat glands is more viscous than that of other sweat glands, and contains substances which, when acted upon by ever present bacteria, acquire a characteristic odor. In our society this odor is considered highly undesirable, and the use of deodorants to destroy or mask the odor is encouraged by many businesses that manufacture and promote personal care products. But for animals, similar types of glands are a means of recognition and also serve an important function as a sexual attractant and territorial marker.

Focus

The Great Compromise: Sunburn vs. Vitamin D

The human body, lacking the typical mammalian coat of fur, is exposed directly to the benefits and dangers of sunlight. When the ultraviolet component of sunlight strikes the skin, it acts upon a steroid, 7-dehydrocholesterol, to form vitamin D_3, a substance essential to the metabolism of calcium and phosphorus. An adult human requires about 20 micrograms of vitamin D per day and can store enough in the liver for several weeks. Vitamin D_3 is converted in the kidney to 1,25-dihydroxycholecalciferol, the active form of the vitamin. This substance increases the absorption of calcium ions from the intestine, apparently by promoting the synthesis of proteins that transport or bind calcium ions in the intestinal mucosa. It also serves other functions in the regulation of calcium and phosphorus deposition in bone.

While vitamin D occurs in such foods as fish oils, humans depend on sunlight for most of their supply. A sunlight (and hence, vitamin D) deficiency results in softened, misshapen, improperly calcified bones (called *rickets* in children, *osteomalacia* in adults). Too much vitamin D can cause excessive absorption of calcium and a build-up of calcium and phosphorus in the body; bones become brittle and fragile, the skin may calcify, the appetite diminish, and the kidneys fail.

The suntan that comes with too much exposure to sunlight is part of the body's defense against sunburn and vitamin D poisoning. Ultraviolet light stimulates the melanocytes in the skin to produce the dark-colored pigment melanin, formed from the amino acid tyrosine. Melanin blocks the penetration of ultraviolet radiation into the skin, thus preventing sunburn and impeding the buildup of dangerous levels of vitamin D.

Humans vary (roughly according to the geographical origin of their ancestors) in the amounts of melanin in their skins and in the amounts they can produce at need. Blacks have a great deal of melanin; whites have relatively little, and though some of them can produce large amounts of melanin, they often use creams and lotions to prevent sunburn. Most of these anti-sunburn preparations contain substances that absorb ultraviolet light before it reaches the skin. Some lotions, usually those that are watery and transparent, do so by specifically absorbing ultraviolet light; others, more pasty and opaque, block out all light.

Blood Vessels and Nerves of the Skin. While there are no blood vessels in the epidermal layer of the skin, the dermis below is highly vascular. Blood vessels form a complex and rich network or *plexus* throughout the deep portions of the epidermis. They send branches to the sweat glands and to the base of the hair follicles. They connect to tiny plexuses in the dermal papillae that send branches to sebaceous glands and the upper portions of the hair follicles. There are also numerous blood vessels, including some larger ones, in the subcutaneous tissue. It is the control of these blood vessels that is important in regulating body temperature. If the blood vessels dilate, more warm blood is brought to the surface and the skin becomes flushed and warm to the touch. If they constrict, the skin becomes pale and cool, as the warm blood is shunted away from the surface, reducing the amount of heat lost from the body (skin) surface.

Skin also has a rich nerve supply. Some of the nerve fibers (motor fibers) go to the blood vessels to control the degree of constriction, and some cause the sweat glands to secrete or the arrector pili to contract. Many of the nerve fibers are sensory, however, for the skin is actually the largest sense organ of the body. When they are stimulated by some particular change, the ends of the sensory nerve fibers cause impulses (action potentials) to be carried along the nerve fibers to the central nervous system. There are numerous specialized endings (receptors) in the dermis — often in the dermal papillae — that are sensitive to heat, cold, or touch. Touch receptors also surround the base of hairs. Pressure-sensitive endings are found in the deeper layers of the dermis and in the subcutaneous tissue. Unspecialized bare nerve endings are the closest to the skin surface because their nerve fibers penetrate the epidermis and the naked endings lie between epithelial cells; these endings are most sensitive to painful stimuli. How the receptors provide us with the information about our environment and conditions on the body surface, and how we interpret and respond to this information are discussed in Chapters 9 and 11.

ANATOMICAL ORIENTATION

The Body as a Whole

When studying the extremely complex organism that is the human body, it is very easy to think of the structure or function of a given part as though that part were a separate and independent unit. This is not so, of course, since all parts work together in a well-coordinated fashion. Cells are grouped together to form tissues, which in turn are grouped to form organs. The organs are then organized into bodily systems. The nine major organ systems are fitted together both structurally and functionally.

The anatomical relationship of organs and organ systems is derived from a segmental pattern which appears very early in embryological development. In creatures such as the earthworm and in the human embryo, it is reasonably apparent that the body is made up of a series of similar segments. In subsequent stages of more highly developed species, however, certain segments grow and develop more extensively than others, and the animal achieves a final form in which the earlier segmentation is not very apparent. In humans, the early segmental stages can be detected in the vertebral column (backbone) and in the spinal nerves emerging from it at regular intervals. The accelerated growth of some segments often occurs in pairs, with equal and similar development on two symmetrical sides (as with the arms and legs). Organs such as those of the digestive system develop from midline structures, but they do not necessarily remain in the midline.

Terms of Direction. Well-defined terms are needed to describe the relationship of one part to another and to reduce the confusion that would otherwise result when one tries to describe the same part in two or more body positions. The head, for example, is not "on top" if the individual lies down or stands on his or her head. To avoid these difficulties, any reference to the relative positions of parts is assumed to apply to a standard position, the *anatomical position,* unless specifically stated otherwise. In this position the individual is standing erect, facing forward, and has his or her palms turned forward. Some of the terms most frequently used to locate and relate the parts, surfaces, and regions of the body are illustrated in Figure 3–8 and are defined in the following list as they apply to humans:

> *anterior*—in front of, toward the front of the body
> *posterior*—behind, toward the back of the body
> *ventral*—toward the abdominal surface
> *dorsal*—toward the back
> *medial (mesial)*—near or toward the midline
> *lateral*—to the side, away from the midline
> *superior*—above, toward the top
> *inferior*—below, toward the bottom
> *cranial (cephalad)*—toward the head
> *caudal*—away from the head, toward the tail
> *proximal*—near the point of reference
> *distal*—away from the point of reference
> *external*—outside, away from the center
> *internal*—inside, near the center
> *superficial*—on or near the surface
> *deep*—away from or below the surface

The terms *superior* and *inferior* are used particularly in reference to upper and lower parts of the body. For example, the arm is the superior extremity. The terms *proximal* and *distal* refer especially to limbs, in which the point of reference is the shoulder or hip, where the limb is attached to the body. The terms *external* and *internal* refer particularly to surfaces, as of hollow organs, body cavities, and the body wall.

These definitions, as they apply in the anatomical position, give some evidence of duplication: Anterior and ventral or posterior and dorsal can be used interchangeably. Cranial and superior or caudal and inferior have much the same connotation; either may be used in many situations. These pair-

ings break down, however, when one refers to a quadruped that does not stand erect. In this case anterior is comparable to cranial, and posterior to caudal, ventral to inferior, and dorsal to superior.

Since these terms and definitions apply to nearly all anatomical descriptions, the student should master them before proceeding.

Planes of Section. In order to visualize internal structures, it is often helpful to make use of a section, that is, a cut through the body or part in a particular plane, leaving an exposed cut surface. As shown in Figure 3–8, the three planes of section commonly used lie at right angles to one another:

> **Transverse** or **horizontal**—a plane that divides the body into superior and inferior parts.
> **Frontal** or **coronal** (corona = crown) —a vertical plane that divides the body into anterior and posterior parts.
> **Sagittal**—a vertical plane that divides the body into right and left portions; a **midsagittal** section is that sagittal plane which divides the body into right and left halves.

A *cross section* is a transverse section and a *longitudinal section* divides the part lengthwise, but these references are to a particular structure, rather than to the body as a whole. A longitudinal or cross section of a bone or a blood vessel need bear no fixed relation to the anatomical position.

Body Cavities. Most of the body's internal organs are contained in several "cavities" (Figure 3–9). The **dorsal cavity** is made up of **cranial** and **vertebral** portions which contain the brain and spinal cord, respectively. The **ventral cavity** also has two parts, the **thoracic** and **abdominopelvic cavities,** separated by a thin domelike muscle, the diaphragm. The abdominopelvic cavity is subdivided into the *abdominal* and the *pelvic* portions. Although they are really parts of a single cavity, they are often referred to as the abdominal and

FIGURE 3–8   Planes of section and terms of direction.

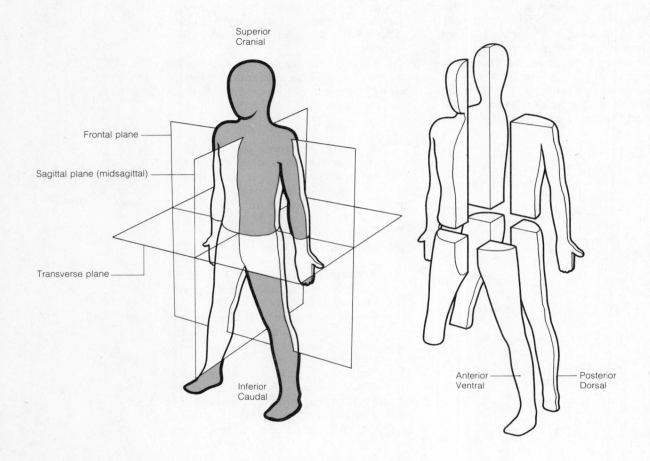

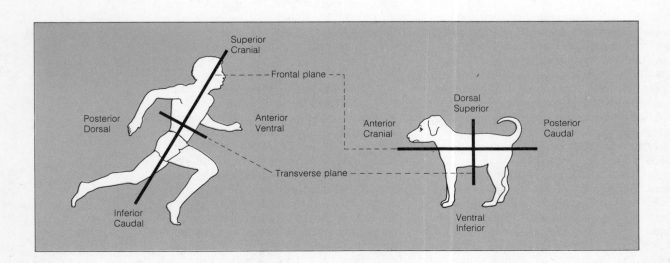

pelvic cavities. The thoracic cavity contains the heart and major blood vessels, the lungs and their airways, the esophagus, and several other structures. The abdominal cavity contains most of the organs of the digestive system (stomach, large and small intestine, pancreas, and liver) as well as the kidneys, spleen, and, of course, blood vessels and nerves. The major organs in the pelvic cavity are the urinary bladder, the rectum, and the reproductive organs. The testes of the male are outside the body cavities, and in the female of many animal species the ovary and part of the uterus are in the abdominal cavity. The organs contained in the ventral cavities are known collectively as the **viscera.**

Parts of the Body.   For descriptive purposes, the body may be divided into four parts: the head, the neck, the trunk, and the extremities. The **head** includes the face and the cranium which encloses the brain. The **neck** is the connecting link between the head and the trunk, and many important structures pass through it. The **trunk** contains the vertebral portion of the dorsal cavity and the thoracic and abdominopelvic cavities. There are four **extremities,** two superior and two inferior. The *superior extremity* consists of the pectoral girdle (shoulder girdle), arm, forearm, wrist, and hand. The *inferior extremity* includes the pelvic girdle, thigh, leg, ankle, and foot. Note that "arm" (not "upper arm") refers to the proximal portion of the upper extremity, while "leg" refers to the distal portion of the inferior extremity. Parts or regions and the scientific terms used for them are shown in Figure 3–10. You should become thoroughly familiar with these terms, since, in their adjective forms, they are widely used to locate and identify arteries, veins, and nerves, as well as to indicate general regions.

Organ Systems

The dependence of cells upon their internal environment has been pointed out in both Chapters 1 and 2. In more complex multicellular organisms the

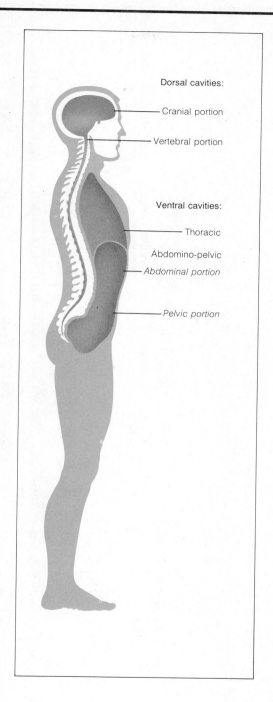

Dorsal cavities:

— Cranial portion

— Vertebral portion

Ventral cavities:

— Thoracic

Abdomino-pelvic

— *Abdominal portion*

— *Pelvic portion*

FIGURE 3–9
The body cavities as shown in a midsaggital section of the body.

immediate internal environment of most cells is far removed from the external environment, and the cells must depend upon specialized systems for the exchange between their internal and external environments. Cells and tissues are organized into organs and organ systems for this purpose. Each organ system is adapted to provide for

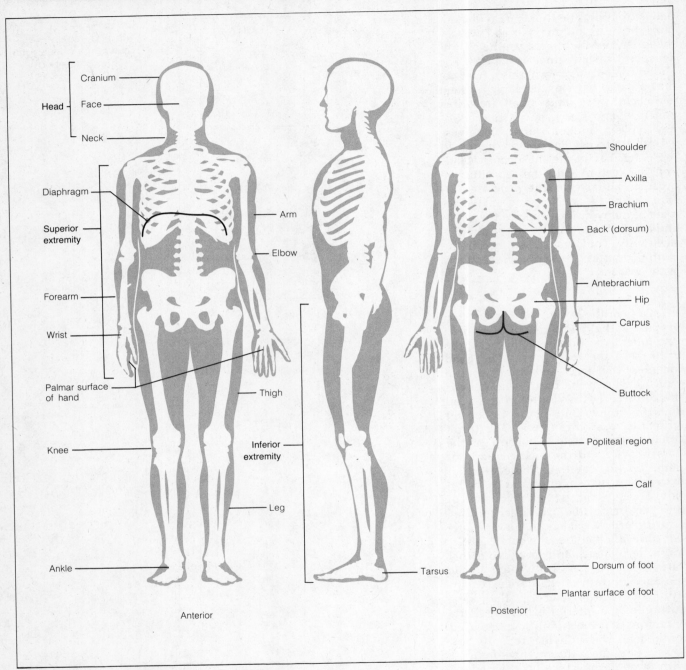

FIGURE 3-10   Body parts and regions.

a particular aspect of the internal environment of all cells (such as nutrient supply, waste removal, etc.), and to make adjustments when there are changes in the needs of groups of cells. We have already seen how the skin, which is sometimes considered to be an organ system, functions in several ways for the benefit of cells throughout the body. The major organ systems and their general functions are briefly introduced here, but they are discussed in more detail in subsequent chapters. In fact, the remainder of this book is devoted to the organ systems and how they operate to maintain the internal environment to meet the changing needs of the cells.

Skeletal System. The skeletal system consists of bones and cartilage and the joints which connect one bone to the next. The skeleton provides the framework for support of the body and its parts and provides protection for certain organs, notably the brain. The bones and joints form a system of levers which can be moved by the muscles, and this is the basis of body movement. The structure of a joint is important because it largely determines the extent and type of movement possible at that joint. In some joints the bones fit so tightly that there is no movement at all, in others there is a slight "give," and in still others there is a fairly wide range of movement. The skeleton is a strong structure, capable of supporting the weight of the body, yet permitting a considerable amount of movement of its parts.

Muscular System. The muscular system functions in such close relation to the skeletal system that the two are often considered together as the *musculoskeletal* system. The chief role of the muscles is in moving the bones of the skeleton, and they are therefore the basis of mobility for the individual. Since muscles always cross at least one joint, shortening of a muscle can produce movement at the intervening joint. However, it cannot cause the return of the part to its original position. That

requires the action of another muscle or of gravity, and we shall see that muscles are often arranged in pairs with opposing actions at a given joint. Muscles can also serve to prevent movement; while standing erect, gravity constantly pulls the body downward, and were it not for the sustained tension exerted by certain muscles, gravity would prevail.

There are several types of muscle, but only the type known as *skeletal* or *striated* muscle is associated with the movement of the skeleton and considered as part of the muscular system. The *cardiac* muscle in the heart has quite a different function and is part of the cardiovascular system. *Smooth* muscle is found in many places, usually as part of the wall of a hollow or tubular organ, such as those of the digestive tract, where its role is to modify the shape of the organ.

The roles of the skeletal and muscular systems may not seem closely related to homeostasis, but the ability to move about enables the animal to escape an inhospitable external environment or modify it. It also makes an important contribution to food-gathering activity, without which survival would be impossible.

Cardiovascular System. The cardiovascular system is a transport and delivery system. It consists of (1) a pump, the heart; (2) two sets of one-way conduits, the arteries for distribution and the veins for return; and (3) a transport medium, the blood. There is also an auxiliary return system, the *lymphatic system*. The heart, by its contractile force, generates pressure to keep the blood moving through the blood vessels. In the more peripheral parts of the vascular system, smooth muscle encircling the vessels may act as regulators to control both the pressure and flow of blood through particular vessels. Arteries and veins are connected by the very thin-walled capillaries where respiratory gases, nutrients, and wastes are exchanged between the blood and the interstitial fluid. The flow and pressure in the entire circulatory system can be altered. Because of diameter changes in many smaller vessels, flow and pressure to individual organs can also be regulated. The respiratory, digestive, excretory, and endocrine systems all depend upon the cardiovascular system for transport of the materials with which they are concerned.

Respiratory System.   The respiratory system is responsible for getting oxygen from the air into the bloodstream and carbon dioxide from the bloodstream to the air outside. It consists of the nasal cavities, larynx, trachea, and smaller airways and the air sacs within the lungs. Unlike the circulatory system, however, there is but one set of passages, and air must enter and leave the lungs by the same channels. The terminal portions of the lungs present an enormous surface area for the exchange of gases with the blood continually pumped through the lungs in the pulmonary vessels. Suitable diffusion gradients must be maintained for oxygen (which diffuses into the blood) and carbon dioxide (which diffuses into the air in the lungs). The rhythmic movement of air into and out of the lungs helps do this. The muscles of the thorax (which are part of the musculoskeletal system) change the size of the thoracic cavity to bring about the air movements that we know as breathing.

Digestive System.   The digestive system is essentially a tube running through the body from mouth to anus. Its structure is modified along the way to facilitate the specific functions of each part. Along its course various secretions are added from glands in its wall or from accessory organs nearby (such as the pancreas). The mechanical actions of the teeth and the smooth muscle in the wall of the tube physically help to break up the ingested material and mix it with the secretions. The contents of the digestive tract are then broken down physically and chemically into molecules that can be absorbed through the intestinal wall to enter the bloodstream. The task of the digestive or *gastrointestinal tract,* then, is to prepare ingested material for absorption, that is, for transport *into* the body. The fate of the absorbed substances once

they get into the bloodstream is no longer a function of the digestive system per se, but rather of all the organs that contribute to the total metabolism. One of the chief organs in this regard is the liver, which also has numerous other roles.

Excretory System.   It is the role of the excretory system to remove metabolic wastes and excesses from the bloodstream. The lungs remove much of the carbon dioxide produced by the cells, and the skin excretes small amounts of some wastes, but the bulk of the waste removal is carried out by the kidney, a remarkably ingenious apparatus. This organ receives a large blood supply and filters much of it into slender, tortuous tubules. As the filtered material passes along the tubule, fluid and nonwastes (electrolytes, glucose, amino acids, etc.) are reabsorbed from the tubule back into the bloodstream. The reabsorption of fluid greatly concentrates the waste products remaining in the tubules. There are many mechanisms to control the tubular processes in order to maintain the pH, water, and electrolyte content of the bloodstream and tissue fluids. The tubular product, urine, passes to the urinary bladder for storage until it is removed from the body.

Reproductive System.   The reproductive system consists of those organs required to produce and care for the germ cells — *spermatozoa* in the male and *ova* in the female. In the male, caring for spermatozoa involves storage, nourishment, and transport. In the female, caring for the ova involves transporting the ovum and providing a site and environment suitable for fertilization and for growth and development of the embryo. In the female, an extremely complex cycle of events involves repeatedly producing an ovum and preparing for a fertilized ovum.

Nervous System.   The activities of all of the organ systems must be coordinated and adjusted to the needs of the body at the moment. This regulation is brought about by the nervous system and the endocrine system. The nervous system consists of the *central nervous system* (the brain and spinal cord) and the *peripheral nervous system* (the nerves and nerve endings). The portions of the nervous system concerned with responses (as of skeletal muscles) make up the *motor* component by which "messages" are transmitted from some part of the central nervous system to the muscle. Information about the environment is provided by the *sensory* component. Nerve impulses are carried to some part of the central nervous system from specific nerve endings called *receptors*. Receptors are able to detect a change in their environment and are particularly sensitive to specific types of change. Receptors in the eye, for example, are stimulated by light, those in the ear by sound waves, and those in the skin by temperature, pressure, or something harmful on the skin surface. There are also receptors stimulated by stretch of a muscle or by blood pressure in an artery. The information from the various receptors is used to determine the appropriate response, and nerve impulses are then sent to those muscles that will produce the necessary response. These automatic stimulus-response mechanisms are called *reflexes,* and they account for most of our activity. The integrating, computerlike role of the central nervous system does not necessarily involve conscious thought. In fact, many reflexes can be carried out without the brain at all (the spinal cord, however, is necessary).

That part of the nervous system concerned with the control of structures other than skeletal muscle is called the *autonomic nervous system.* It regulates visceral structures through its action on cardiac muscle, smooth muscle (as in the wall of the digestive tract), and many glands. The autonomic nervous system controls structures over which we do not have voluntary control; it functions entirely as a reflex system.

Endocrine System. The endocrine glands are not connected anatomically or functionally in the same way as the elements of other organ systems. Their common bond is that they all secrete hormones that exert regulatory actions upon other structures. These actions are many and varied, affecting virtually every cell in the body. Many of the hormonal actions are concerned with regulating processes like reproduction, metabolism, and development, whereas the nervous system is often associated with responses to changes in the external environment. This is, of course, an oversimplification, since the distinctions are not clear-cut and there is much overlap.

Since its function is regulatory, the endocrine system has some things in common with the nervous system. Some hormones influence neural structures, some neural structures have secretory functions, and finally, the nervous system exerts important control over many endocrine functions.

CHAPTER 3 SUMMARY

Tissues

There are four primary tissues: epithelium, connective tissue, muscle, and nerve. The first two are described in this chapter.

Epithelium is found on surfaces. It covers the surface of the body, and lines body cavities, hollow organs, and tubular structures where it fills protective roles. It also is a secretory tissue, for glands are epithelium. Typically, epithelium occurs in sheets or layers connected to the underlying tissues by a *basement lamina.* The free surface of the cells may be specialized by the presence of *cilia* or *microvilli.* Contact between cells is maintained by modifications in the adjacent cell membranes. Among them are *tight junctions, desmosomes,* and *gap junctions,* which serve to prevent passage (leakage) of substances between cells, bind cells together, and/or allow transmission of electrical changes to adjacent cells.

Epithelia are classified by whether they are one cell thick (*simple*), or more than one cell thick (*stratified*), and as *squamous, cuboidal,* or *columnar* by the shape of the surface cells. *Pseudostratified* epithelium appears to be stratified but is not, and *transitional* epithelium may vary its thickness. *Glandular* epithelium is found in exocrine and endocrine glands. *Exocrine* glands deliver their secretions directly or through ducts. A gland may be a single cell, or arranged into tubules or cell clusters (*acini* or *alveoli*). *Endocrine* glands produce hormones and secrete them into the bloodstream for distribution.

Connective tissue contains an abundance of intercellular fibers and ground substance, in addition to its cells. The fibers are *collagen* which is flexible but inelastic and often found in bundles, *elastin* which is elastic but does not form bundles, and *reticular* which resemble very fine collagen strands. The ground substance contains proteins and certain other compounds, depending upon its location. The major connective tissue cells are *fibroblasts,* which produce both fibers and ground substance. Other cells are *macrophages* which are phagocytes, *adipose* cells which store fat, and a small number of *blood* cells.

Connective tissue is classified on the basis of its components and their arrangement. *Embryonic connective tissue* is a primitive type of tissue, whose cells are mostly undifferentiated. It is important as other types of connective tissue are derived from it. Connective tissue proper, cartilage, bone, dentin, and blood and lymph are types of adult connective tissue. *Loose connective tissue* is a typical example of connective tissue proper. Its fibers and cells are loosely arranged in a random fashion, as seen in subcutaneous tissue. When the components are closely packed, it is *dense irregular connective tissue,* as is found in the deep layers of skin and the deep fascia around and between organs. *Regularly arranged connective tissue* forms the tendons and ligaments that attach muscles and bones. Special types of connective tissue include *adipose* which has many fat cells, *elastic* with many elastic fibers, and *reticular* with numerous reticular fibers.

Other types of connective tissue have different qualities due to the special features of their intercellular material. *Cartilage* contains fine collagen fibers and special proteins in the ground substance that give it a firm consistency. The cells are trapped in tiny spaces (*lacunae*) surrounded by the intercellular material. One location of *hyaline cartilage* (the most common type) is on the ends of bones in movable joints, where its very smooth surface reduces friction during movement. Other types of cartilage are *fibrocartilage* and *elastic cartilage*, which contain large amounts of fibrous material and elastic fibers, respectively. *Bone* is hard and rigid because of minerals deposited in the intercellular material. Blood and lymph have a fluid intercellular component, which also lacks fibers.

Skin

The body surface is covered with skin, consisting of a layer of cornified, stratified squamous epithelium—the *epidermis*—over a layer of dense connective tissue—the *dermis*. Skin is attached to underlying structures by the subcutaneous tissue (*superficial fascia*).

In the **epidermis,** cells divide in the deepest layers and push older cells toward the skin surface and away from their blood supply. The displaced cells undergo progressive degenerative changes, and the layers of epidermis are described on the basis of these changes. Beginning with the deepest layer, they are: *stratum germinativum, stratum spinosum, stratum granulosum, stratum lucidum,* and *stratum corneum.* By the time cells have reached the outermost layer they have become cornified, or keratinized, and are no longer distinct nor living cells. The cornified layer is protective, but is gradually worn away (as dead skin), and is replaced by new cells from below.

Scattered large cells in the deepest epidermal layers contain the pigment *melanin.* Processes of these cells extend up between epithelial cells and distribute granules of pigment to adjacent epithelial cells, giving the skin its color.

Dermis is dense connective tissue, and contains blood vessels and nerve endings that respond to stimuli such as pain, temperature, or touch. Its junction with the epidermis forms patterns of ridges and grooves (*papillae*) which, in some areas, are well-developed and form the unique patterns that appear on the surface as fingerprints.

Nails and *hairs* are derived from epidermis. Nails consist of keratinized epithelial cells that arise from the stratum germinativum at the base of the nail. Hairs grow from hair follicles which have migrated down into the dermis. Cells at the base of the follicle divide and become keratinized as they are pushed up the follicle toward the skin surface by cells being formed below. *Sebaceous glands* located beside the hair follicle release their oily secretions into the shaft of the hair follicle. Small bundles of smooth muscle *(piloerectors)* are attached to the base of the hair follicle and to the dermis. Their contraction causes the hair to stand more erect from the skin surface. *Sweat glands* are located in the dermis of underlying connective tissue, and their ducts lead through the epidermis to the surface of the skin.

Anatomical Orientation

Relationships of body parts to one another are always described with the body in the *anatomical position*—standing erect, arms at the sides, and palms facing forward. The relationship of internal structures may be depicted in a section, or cut, through the body along one of several planes. A *transverse* or *horizontal plane* divides the body into *superior* and *inferior* (upper and lower) parts. A *frontal* or *coronal plane* divides it into *anterior* and *posterior* (front and back) parts, and a *sagittal section* divides it into *right* and *left* parts.

The main body parts are *head, neck, trunk* and the *superior* and *inferior extremities* (arms and pectoral girdle, legs and pelvic girdle). Most of the important organs are enclosed in body cavities. The *dorsal cavity* consists of a *cranial* portion that houses the brain, and a *vertebral* portion that contains the spinal cord. The *ventral cavity* includes the *thoracic cavity* and *abdominopelvic cavity*. They are separated by a muscle, the diaphragm, but the abdominal and pelvic portions of the abdominopelvic cavity are continuous with one another. The organs contained in the ventral cavities are the *viscera*.

The major organ systems of the body are:

Skeletal system—the bones, cartilage, and joints form a supporting framework for other structures, yet permit movement.

Muscular system—the muscles that attach to bones (skeletal muscle) form a system for movements of the body parts. Smooth muscle is found in viscera and blood vessels, and cardiac muscle is in the heart.

Cardiovascular system—the heart, blood vessels, and blood form a system for transporting a number of materials (such as nutrients, wastes, gases, hormones) to and from the cells.

Respiratory system—the lungs and airways provide a means for bringing oxygen to the blood and removing carbon dioxide from the blood.

Digestive system—the digestive tract and related organs prepare ingested foods for absorption into the bloodstream.

Excretory system—the kidney and related organs remove cellular wastes from the bloodstream and transport them out of the body.

Nervous system—the brain, spinal cord, and nerves form an important communications system which can monitor the external and internal environment, detect changes, and regulate accordingly the activities of many cells.

Endocrine system—a collection of glands which secrete hormones that exert important control over many cellular activities.

STUDY QUESTIONS

1 How do epithelium and connective tissue differ, in terms of their structure, components, location, and functions?

2 What is meant by an undifferentiated or primitive cell? How does it differ from a specialized cell? Would you expect to find undifferentiated cells in an adult? Why or why not?

3 How are cells, such as epithelial cells, held together?

4 What is the nature of the intercellular substance of connective tissue? How does it differ among different types of connective tissue?

5 What are glands? How do endocrine and exocrine glands differ?

6 Name the two major layers of skin. What are the structural characteristics of each?

7 What is the cornified or keratinized layer of skin? How does it arise? What function(s) does it serve?

8 Describe the glands found in the skin. What are their functions?

9 For practice, select sets of two anatomical structures (organs or parts), and use anatomical terminology to describe their relation to one another. For example, knee and ankle, lungs and heart, ear and nose, etc.

10 Name the major organ systems of the body, and the important organs of each system. What does each system contribute to the overall body economy?

Part 2
The Musculoskeletal System

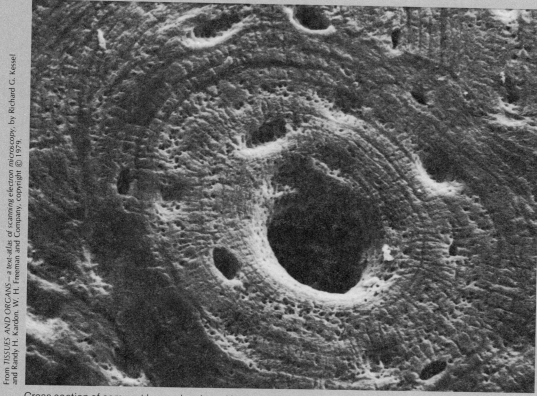

From TISSUES AND ORGANS—a text-atlas of scanning electron microscopy, by Richard G. Kessel and Randy H. Kardon. W. H. Freeman and Company, copyright © 1979.

Cross section of compact bone, showing a Haversian system (osteon).

Part I briefly introduced many types of chemicals and reactions, cells, tissues, and parts, and several kinds of cell processes. Part 2 begins a closer study of the contributions of the different parts and organ systems to the whole body.

The musculoskeletal system is an appropriate place to begin. It includes the supporting framework of the body, together with the components that permit movement and those that cause movement of the body parts. The skeleton consists of bones, cartilage, and their connections. The bones form the actual supporting system. Movement is possible because of the joints between bones, and is caused by skeletal muscles. A muscle is typically attached to the bones that form a particular joint, and when the muscle contracts it can cause movement at that joint. (Muscles can also *prevent* movement at the joints they *cross.*)

The chapters in Part 2 discuss the structure of each of these elements, including the names of the bones and their important features, the types of joints and the movements possible, the names of muscles and their attachments and actions, and finally, muscle contraction and how it brings about movements of body parts.

The anatomical portions of these chapters include a large, new vocabulary. It is recommended that you become thoroughly familiar with the lists of prefixes, suffixes, and stems at the beginning of the glossary. They will help you throughout your studies, for if you know the meaning of the word parts, it will be much easier to learn and remember the words. Familiarity with the prefixes, suffixes, and stems will clue you in on the structure or process in question.

4

General Osteology

Bones of the Skeleton

Skeleton — The Framework

The skeletal system is the first of the major organ systems to be considered in detail. Its main components are the bones, whose rigidity is essential for its support role. The joints between bones, which are also an important part of the skeleton, are described in Chapter 5.

It is easy to think of bone as being inert like a stone, but it is living tissue. It is connective tissue, and as such, it contains both cells and intercellular material. The cells produce an intercellular material that contains fibers and a ground substance or matrix in which mineral salts are deposited. The description of bone tissue, its structural organization, how it is formed and how it is maintained, make up the first part of this chapter.

The remainder of the chapter deals with the individual bones, for each has structural features that distinguish it from other bones. These include processes that serve as muscle attachments, grooves or *foramina* (holes) that provide passages for nerves or blood vessels, and surfaces and prominences that affect the movements at a joint. The size and shape of individual bones determines the way in which those bones contribute to weight-bearing or to movement, and therefore, to the total function of the musculoskeletal system.

GENERAL OSTEOLOGY

Osteology is the study of the structure, formation, and function of the bones of the body. Bones of the human skeleton differ greatly in their size, shape, and complexity, and they contribute to the functions of the skeleton in different ways. But in spite of their differences, bones have much in common. Bones are made up of living tissue, bone tissue, which is a type of connective tissue. Its cells, while unique to bone, carry on metabolic activities, respond to changes in their environment, and have homeostatic requirements similar to other cells. The intercellular material produced by its cells is what makes bone different from other kinds of connective tissue.

Bone as a Tissue

It is a characteristic of connective tissue that the intercellular material is more conspicuous than the cells, and bone is no exception. The intercellular matrix (material) of bone, however, is mineral, unlike that of most other connective tissues whose matrices are organic—the protein collagen, you may recall from Chapter 1, makes up collagenous fibers, which are the most common of connective tissue fibers. These minerals give to bone its hardness and rigidity and most of its weight—minerals account for about two-thirds of the weight of the bone. The other third—the cells, fibers, and ground substance—is largely protein and gives bone a toughness and a certain amount of flexibility. The individual contributions of these components can be demonstrated very simply. Incineration of a bone oxidizes the organic components, leaving only the inorganic material. This residue, or ash, retains the shape of the bone, but it is extremely brittle and shatters upon the slightest impact. In contrast, prolonged immersion of a bone in acid removes the mineral, leaving only organic material. The demineralized bone looks like normal bone, but it has lost its rigidity and cannot support weight. It will bend so readily that it can even be tied in a knot.

Functions of Bone

The most obvious function of bone is that it forms a framework for the body. All of the body parts are attached to and supported by the skeleton. Even most visceral structures have indirect attachments to the skeleton through other structures.

Because of the ways in which bones fit together, or *articulate*, and because the muscles attach to them, the bones of the skeleton function as a system of levers permitting movement of the body's parts. The skeleton and bones do not cause movement, however, since if the muscles did not attach to the bones as they do, there would be no movement. Articulations and muscles are discussed in Chapters 5 and 6.

The bones also serve important protective roles for certain vital but fragile soft tissues. The brain, for instance, is completely enclosed in a rigid box (the cranium), and the spinal cord is encased in a reasonably rigid tube (the vertebral column), while the heart and lungs are surrounded by the bony rib cage; though flexible, the latter does provide some degree of protection. The bones also shelter within themselves the **hemopoietic tissues,** the red bone marrow responsible for the formation of the blood cells.

Because of the deposition of inorganic salts, largely as a calcium and phosphate complex, the bones and teeth together account for about 99 percent of the calcium and 90 percent of the phosphorus in the entire body. Therefore, bone serves as a ready depot for storage or withdrawal of these minerals, since their concentration in the blood, particularly that of calcium, must be maintained within quite narrow limits. The mechanisms involved will be examined in Chapter 14.

Gross Structure of Bone

For convenience bones are classified according to their various shapes and sizes into four types: long, short, flat, and irregular bones (Figure 4–1). **Long bones** are found exclusively in the extremities. **Short bones,** such as the tarsal (ankle) and carpal (wrist) bones,

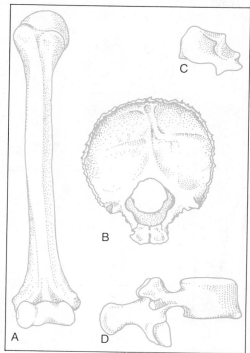

FIGURE 4–1   Types of bones. A. Long bone. B. Flat bone. C. Short bone. D. Irregular bone.

are reasonably cube-shaped. Bones of the cranium and ribs are among the **flat bones.** The **irregular bones,** such as those of the vertebral column and the pelvic and pectoral girdles and some of those of the skull, do not fit into any of the other categories.

The structure of a bone is well illustrated by a long bone such as the femur (Figure 4–2). It consists of a *shaft* and two ends or *heads* (in practice "head" is used more often for the proximal end). The shaft is the **diaphysis** and each end is an **epiphysis** (plural = epiphyses). Some part of each epiphysis has a very smooth, but rarely flat, **articular surface.** The articular surface is covered with a thin layer of hyaline cartilage that forms the surface of contact with other bones. The rest of the bone is enclosed in a sheath of connective tissue, the **periosteum,** that adheres tightly to the bone surface and is continuous with other connective tissues in the vicinity.

In longitudinal section, a long bone is seen to consist of a dense material known as **compact bone,** an apparently

fragile latticework called **cancellous** or **spongy bone,** and an open area in the center, the **marrow** or *medullary* **cavity.** The marrow cavity is lined with a thin layer of connective tissue known as *endosteum.* Compact bone is on the surface of the bone. It is quite thin at the epiphyses and over articular surfaces, but much thicker in the diaphysis. Cancellous bone in adult long bones is restricted largely to the epiphyses but extends as a thin layer between compact bone and the marrow cavity in the diaphysis. Cancellous bone is composed of small interlacing spicules or splinters of osseous material. In its interstices is found the **red bone marrow,** consisting both of primitive cells and developing blood cells in various stages of maturation and of such other cells as macrophages and fat cells. The marrow also contains some reticular fibers and, of course, a blood supply. In early life red marrow occupies all of the interior of the bone, but in the adult the cancellous bone and red marrow of the diaphysis have been replaced by fat cells to form **yellow marrow.** Under certain conditions, red marrow may reoccupy the marrow cavity and *hemopoiesis* (blood cell production) may become more extensive. Not all cancellous bone contains active hemopoietic tissue, however, since in the adult it is pretty well restricted to the sternum, vertebrae, parts of the pelvic bones, and a few others.

The structure of short and irregular bones is similar to that of the epiphyses of long bones—a thin surface layer of compact bone filled with cancellous bone without a medullary cavity. Like long bones, they are covered with periosteum, except over the articular surfaces. Flat bones consist of a sandwich formed by two plates of compact bone with spongy bone in between and periosteum covering both surfaces.

The only kind of bone not covered by periosteum is a type of short bone, called a **sesamoid bone.** They develop within the tendon of a muscle at a site where the tendon passes over and rubs on a convex surface of a bone. The best-known example is the *patella* (kneecap), but other sesamoid bones do occur, mainly in the foot and hand.

The surface markings of each bone

FIGURE 4–2
Structure of a typical long bone, as illustrated by the femur. A. Gross features of the bone. B. Diagram showing the arrangement of trabeculae in the head of the femur. C. Photomicrograph of ground bone, a cross section of compact bone showing a haversian system (osteon). D. A wedge section from the diaphysis of a long bone.

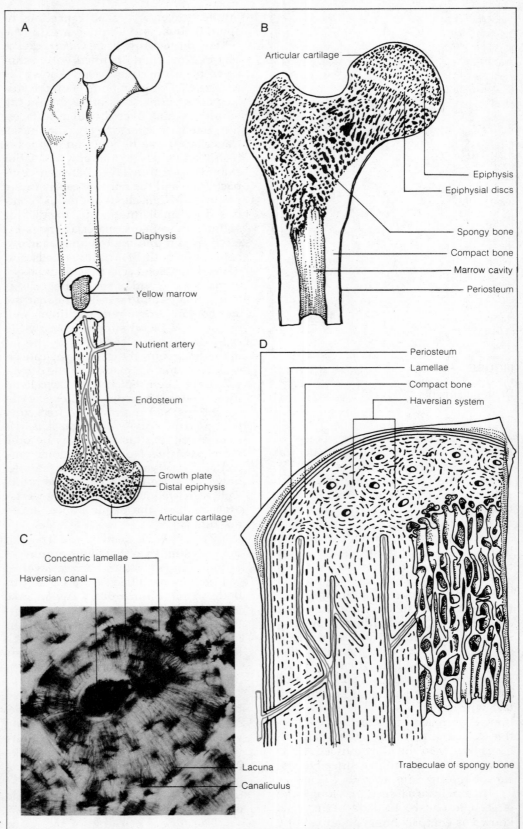

A

Diaphysis

Yellow marrow

Nutrient artery

Endosteum

Growth plate
Distal epiphysis

Articular cartilage

B

Articular cartilage

Epiphysis
Epiphysial discs

Spongy bone

Compact bone

Marrow cavity

Periosteum

D

Periosteum
Lamellae
Compact bone
Haversian system

Trabeculae of spongy bone

C

Concentric lamellae

Haversian canal

Lacuna
Canaliculus

Ray Simons/Photo Researchers

are different. They consist of various modifications—enlargements, indentations, protrusions, and holes—which each have a specific purpose. Bone is a strong material, but it is also very heavy. The problem of evolution, then, has been to produce bones that combine adequate strength with minimum weight. The two requirements are not always compatible, and in such cases there must be some compromise. In studying the individual bones, you should notice the structural adaptations that contribute to the necessary compromise. Articular surfaces are usually enlarged to provide a more secure contact. Protrusions (known as *processes*) commonly occur where ligaments or muscles attach; they can strengthen the bone at the site of application of a force or improve the leverage of a muscle. Holes may serve as passageways for nerves or blood vessels, or they may simply decrease the mass of the bone. Specific terms are used to identify these features. The student should master the names of the more common ones, which are listed here:

alveolus—small cavity, a tooth socket

canal—tubular channel through bone

condyle—rounded articular prominence

crest—prominent bony ridge

epicondyle—projection above a condyle

facet—face; particularly a smooth, flat articular surface, as on certain vertebrae

fissure—narrow, cleftlike passage

foramen—hole or perforation through a bone, the opening of of a canal

fossa—concavity, depression, or trench in a bone

fovea—small, shallow depression

head—large, rounded articular end of a bone

lamina—thin plate of bone

line, linea—line, marked by either a low ridge or a depression

meatus—canal; used only in certain cases, as external acoustic meatus

neck—narrow part of bone, between head and shaft

notch—deep indentation

process—generic term covering any prominence or protuberance

ramus—a branch

spine—sharp or pointed process

sulcus—a groove

trochanter—large blunt process, as on the femur

tubercle—small tuberosity

tuberosity—large, often rough process

Microscopic Structure of Bone

The osseous materials of compact and cancellous bone are the same. However, due to their different arrangements, their gross and microscopic appearances are somewhat different (see Figure 4–2). In cross section, compact bone is seen as tightly packed, cylindrical units known as **haversian systems** or **osteons.** Each haversian system centers around a canal or space containing blood vessels, *osteoblasts* (bone-forming cells), and undifferentiated cells. This **haversian canal** (or *central canal*) is surrounded by several rings of bone matrix called **lamellae.** Squeezed between the lamellae are tiny spaces, **lacunae,** that house the bone cells, or *osteocytes*. Tiny hairlike channels, or **canaliculi,** extend from the lacunae through adjacent lamellae to other lacunae. Slender processes of the osteocytes extend out into the canaliculi. The canaliculi provide a communication channel by which osteocytes are able to satisfy their metabolic requirements. This is important, because osteocytes have no other contact with blood vessels in the haversian canal, and nutrients and other materials do not diffuse through the intercellular matrix. The spaces between haversian systems are filled in by fragments of lamellae (interstitial lamellae), and the surface of the bone is covered by several layers of circumferential lamellae.

Although cancellous bone lacks haversian systems, its small bony spicules or *trabeculae* show a layering, with lacunae and canaliculi. Blood vessels and various bone cells occupy the interstices of cancellous bone and, where hemopoiesis is occurring, there are blood cells in all stages of development.

The blood supply of bone comes from vessels that penetrate the compact bone and then break into numerous branches. These branches provide ample blood flow to the red bone marrow in the cancellous portions of the bone and to the contents of the medullary cavity, as well as to the compact bone.

Formation of Bone— Ossification

Bone may develop in either of two environments. *Ossification* (formation of bone) within and around a fibrous membrane is **intramembranous bone formation** and its product is called **membrane bone.** Bone that develops in and replaces cartilage is called **cartilage bone,** and the production process is **endochondral bone formation.** The bone produced by the two processes is identical chemically and physically; the differences lie only in the pattern of development. Ossification begins in the second or third week of fetal life, but is not completed until about age 25.

Intramembranous Bone Formation
Membrane bone is found only in the skull, and its development occurs in and around a membrane formed when primitive mesenchymal cells differentiate into fibroblasts which form collagen fibers. The result is a loose connective tissue network or membrane. Other mesenchymal cells differentiate into **osteoblasts,** line up along the strands of the collagen network, and secrete **osteoid,** the organic component of bone. Osteoid is not bone, however, because minerals have not been deposited in it; it is not yet calcified. The osteoid develops as bars or spicules that become larger and interconnect to form a network of osteoid. Some osteoblasts become entrapped within its meshes. As more osteoid is formed, the osteoblasts come to lie in lacunae, after which they are properly called osteocytes. When environmental conditions become favorable, which is almost as soon as any osteoid has been formed, calcification begins. Inorganic minerals are deposited as submicroscopic crystals in the intercellular matrix of

the osteoid. A complex calcium salt, $Ca_{10}(PO_4)_6(OH)_2$, is the chief constituent and accounts for about 80 percent of the mineral. The remainder is mostly $CaCO_3$ with traces of other minerals.

As ossification continues, more osteoblasts line up along the spicules of bone, which then become thicker. However, ossification stops short of filling all the spaces with bone. The spaces contain blood vessels and mesenchyme, some of whose primitive cells develop into the blood-forming cells of the red bone marrow. On the surface of the bone, however, osteoblasts from the adjacent connective tissue do lay down lamellae and form a thin surface of compact bone to cover the underlying cancellous bone.

Ossification of membrane bones begins at two or three specific sites within each bone and expands from each site. These areas, called **ossification centers,** first appear during the early weeks of fetal life, but the process is not completed for some years. At birth certain areas of the skull still consist only of unossified membrane. These are the "soft spots" or **fontanels** found at the junctions of some of the bones of the cranium (see Figure 4–12). Fontanels are usually ossified by the age of two.

Endochondral bone formation
The formation of long bones and bones of the trunk begins with a cartilage model of the particular bone. As the name *endochondral* suggests, the bone actually develops within that model (see Figure 4–3).

The cartilage model. In the early embryonic stages the appearance of blood vessels in the areas of bone formation is accompanied by differentiation of the mesenchymal cells into cartilage-forming cells (*chondroblasts*), followed shortly by the appearance of a tiny hyaline cartilage miniature of the bone-to-be. This miniature grows by cell division, by *hypertrophy* (growth) of the individual cartilage cells, and by *apposition.* In apposition, new layers of cartilage are formed on the surface of the model by differentiation of primitive cells of the inner layer of the perichondrium that surrounds

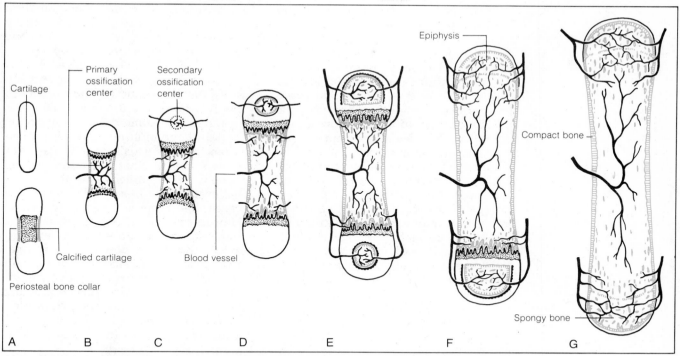

FIGURE 4–3   Stages of endochondral ossification. A. Initial cartilage model of the bone-to-be. Later, the cartilage begins to calcify in the diaphysis, and the periosteal bone collar appears around the diaphysis. B. Blood vessels and cells invade diaphysis, and the primary ossification center appears. C. Primary ossification center expands, and a secondary ossification center appears in the upper epiphyseal cartilage. D. Continued growth of bone and expansion of ossification centers. E. Secondary ossification center appears in lower epiphyseal cartilage. F. Closure of upper epiphysis. G. Closure of lower epiphysis. Growth in the length of the bone is now complete.

the cartilage. The greatest growth, however, is by cell division near the ends of the cartilage model.

As the model grows, maturing *chondrocytes* in the center undergo a marked hypertrophy, causing the intercellular material between them to become somewhat thinned out. The cells also tend to become lined up in rows parallel to the long axis of the model. Minerals then are deposited in the intercellular material, and the cartilage becomes *calcified.* Calcification triggers off a whole chain of events, since once the cartilage becomes calcified substances needed by the chondrocytes can no longer diffuse through the matrix. Unable to maintain their metabolic activities, the chondrocytes soon die. Since the integrity of the intercellular substance depends upon the chondrocytes, it too begins to disintegrate. The thin plates of calcified cartilage be-

tween the lacunae break down, leaving larger and larger holes in the center of the cartilage model.

Periosteal bone collar.   At the same time, primitive cells in the all-important perichondrium around the diaphysis of the cartilage model differentiates into osteoblasts which secrete osteoid. It soon becomes calcified and forms a thin plate of compact bone around the middle of the cartilage model. As the ring of bone grows in extent and thickness, the surrounding connective tissue is now called **periosteum.** The ring of bone is the **periosteal bone collar,** which appropriately reinforces the area where the cartilage is beginning to break down.

Primary centers of ossification.   As the bone collar forms, blood vessels from the periosteum grow through the developing bone collar into the center of the disintegrating cartilage of the model. They are accompanied by undifferentiated cells and osteoblasts; the osteoblasts begin to lay down osteoid on the remaining fragments of calcified cartilage. These cells and capillaries are the **periosteal bud** that becomes the **primary center of ossification** in the diaphysis. Spicules of bone appear and thicken, producing cancellous bone in a manner similar to that of intramembranous bone formation. The process continues as the primary center of ossification expands, and the cartilage model continues to grow, particularly at the ends. As more cartilage matures, the area of breakdown progresses toward the ends, and advancing osteoblasts increase the area occupied by the bone trabeculae or spicules.

Secondary centers of ossification. The cartilage in the epiphyses now begins to mature, initiating the sequence of events leading to its disintegration. The epiphyses are invaded by buds of osteogenic tissue — osteoblasts, undifferentiated cells, and capillaries — and bone formation proceeds as it did in the diaphysis. These are **secondary** or **epiphyseal centers of ossification.**

The centers of ossification in the diaphysis and epiphyses expand toward one another, but the epiphyseal cartilage grows rapidly enough to keep them apart for a time, and the entire bone increases in length. The cartilage becomes a thin plate, the **epiphyseal disc,** and eventually the approaching ossification overtakes the proliferating cartilage. At that point the primary and secondary ossification centers merge, and a faint **epiphyseal line** is all that remains of the epiphyseal cartilage. When the cartilage disappears there can be no further increase in the length of the bone. The time of final closure of the epiphyses varies greatly from bone to bone, but in a typical long bone it occurs somewhere around the eighteenth and twentieth year of age (see Table 4–1). The only cartilage that remains is that covering the articular surfaces of the bone, and it remains throughout life.

Remodeling of bone. The formation of bone in the cartilage model is really only half the story, however, since bone is constantly reshaped and remodeled throughout life. The process begins almost as soon as any bone is formed, since the earliest ossification occurs in the middle of the diaphysis, which is destined to become the marrow cavity. Certain large multinucleate cells called **osteoclasts** are usually found in areas of bone *resorption,* that is, in areas where bone is being eroded or removed. It is believed that these cells somehow cause the bone substance to erode. How this is brought about is not known, but osteoclasts probably secrete enzymes that break down the bone matrix. Osteoclasts cause an increase in the size of the marrow cavity by eroding the cavity's inner walls. In growing bone, osteoclasts and bone resorption follow close behind osteoblasts and bone formation, as they all spread toward the epiphyseal discs.

Osteoclasts are also found along the external surfaces of bones, where bone erosion and formation simultaneously act to reshape bone. The size of the cranial cavity, for example, is enlarged by erosion on the inside and by bone formation on the outside of the individual bones.

The haversian systems so characteristic of compact bone form as tunnels or tubes. As the calcified cartilage in the diaphysis breaks down, tubular cavities are formed in the cartilage. The invading osteoblasts line up on the surface of the cartilage fragments and secrete a layer of bone. Additional layers of bone result in a primitive haversian system, one that will undoubtedly be eroded as the enlarging marrow cavity encroaches upon it.

The outer surface of a growing bone is marked by longitudinal ridges and grooves and laced with blood vessels branching from the periosteal vessels. In growing bones, new bone is laid down along the edges of a groove until it is covered over as a tunnel enclosing a blood vessel, osteoblasts, and some undifferentiated cells. Osseous lamellae are formed within the tunnel until a complete haversian system is developed. As the bone is being thickened by the addition of new haversian systems on the surface, erosion of bone from the inside prevents the compact portion from becoming too heavy. The external surface is finally smoothed over by several circumferential lamellae parallel to the surface of the bone.

New haversian systems also are formed on the surface of adult bone. The process is quite similar to that in growing bones, except that formation of the grooves and tunnels is associated with the presence of osteoclasts. They presumably aid in hollowing out the groove by erosion of the circumferential lamellae.

Bone formation, then, does not end when ossification is completed. It is a continuing process, with new bone being laid down and old bone being resorbed throughout the life of the individual. It is a complex process requiring the coordination of a number of varied and virtually simultaneous processes—the formation of osteoid by osteoblasts, the dissolution of bone, probably by osteoclasts, and the deposition and/or withdrawal of calcium salts from the organic matrix. The rates of all of these processes must be kept within the proper limits throughout life. Most of the problems that arise are associated with the improper disposition of the calcium salts and will be discussed more fully in Chapter 13.

Regulation of Bone Metabolism

Several factors are involved in regulating the metabolism and mineral content of bone. Some of these are dietary, particularly an adequate intake of calcium and phosphorus. In addition, vitamin D is needed to ensure that the calcium ingested is absorbed into the body. Vitamin C is needed for production of the organic matrix of bone, particularly the collagen fibers, and vitamin A is involved in the proper balance of activity between osteoblasts and osteoclasts.

The *parathyroid hormone* has a number of effects on bone which together add up to resorption. It inhibits osteocytes and stimulates osteoclasts, causing both to release enzymes that erode bone; thus both the organic ma-

TABLE 4–1 AVERAGE AGE AT WHICH OSSIFICATION IS COMPLETED

Bone	Average Age at Fusion (in years)
Scapula	18–20
Clavicle	23–31 (last long bone to fuse)
Bones of arm, forearm, and hand	17–20
Os coxa	18–23
Bones of thigh, leg, and foot	18–22
Vertebrae	25
Sacrum	23–25
Sternum	23
manubrium and xiphoid	after 30, or perhaps not at all

Note: Some long bones have 2–3 ossification centers in one end. The ends of the bones that form a joint such as the knee may involve 4–5 fusions, occurring at slightly different times.

trix and the mineral content are reduced. *Calcitonin,* a hormone produced by the thyroid gland, inhibits the resorption of bone, both on the mineral and the matrix. These two hormones play a major role in regulating the level of calcium ions in the blood. This is a vital homeostatic mechanism because the concentration of calcium in the extracellular fluid is critical to many cellular activities. There are several other hormones that have effects on bone, one of which is the *growth hormone* of the pituitary gland. It stimulates the growth of cartilage in the epiphyseal discs, and therefore increases the length of long bones by delaying the final closure of the epiphyses. The actions of these hormones are discussed further in Chapter 13.

Repair of a Bone

Severe or sudden stress to a bone may cause a fracture, ranging anywhere from a tiny crack to a crush in which much of the bone is shattered. Such a fracture can be repaired, but the process is rather slow because it involves the formation of new bone (Figure 4–4).

The first effect of the injury is swelling and formation of a blood clot, both due to the tearing and destruction of blood vessels in the periosteum and to the leakage of blood from the nearby injured tissues. The interruption of their blood supply causes the death of

FIGURE 4-4
Repair of a fracture.

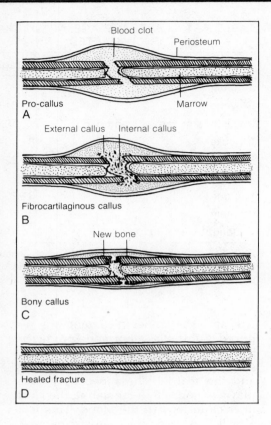

most of the osteocytes and periosteal cells in the immediate vicinity of the fracture. Within a few hours intact, undifferentiated cells from the inner layers of periosteum on either side of the injury undergo rapid cell division and differentiation to form a collar of tissue, called a *callus*, around and between the broken bone ends and through the blood clot. Some of the undifferentiated cells in the callus form fibroblasts which produce collagen, and within a week or so the callus develops into dense fibrous connective tissue. Phagocytes from the blood and locally formed macrophages help to clean up the debris of destroyed cells and matrix. Some undifferentiated cells become chondroblasts and soon the callus becomes a fibrocartilaginous structure. Osteoblasts arise from the periosteum to form bone in the outer collar portion of the callus and also from the endosteum to form bone in its central part. This is cancellous bone, formed by what is largely endochondral bone formation. The whole area is then remodeled and reshaped, with the callus being almost completely resorbed, and

eventually all or part (depending on the location) of the new cancellous bone will be converted into compact bone. Probably the most critical element in this healing process is an adequate blood supply to the area; if the supply is insufficient, repair of the bone will be greatly delayed.

BONES OF THE SKELETON

The bones of the skeleton can be divided into two groups, those of the **axial skeleton** and those of the **appendicular skeleton** (Figure 4-5). The axial skeleton is the framework of the trunk and head, and it includes the bones of the skull, vertebral column, and thorax. These bones tend to be midline structures and many are unpaired (exceptions are the ribs and some bones of the skull). The appendicular skeleton is the framework of the extremities, and it consists of the bones of the limbs and the pelvic and pectoral girdles by which they are attached to the axial skeleton. These bones are all paired.

Axial Skeleton

Skull	**29**	
cranium		8
face		14
ossicles of ear		6
hyoid		1
Vertebral Column	**26**	
cervical		7
thoracic		12
lumbar		5
sacrum (5 fused)		1
coccyx (4 fused)		1
Thorax	**25**	
sternum (3 fused)		1
ribs		24
	Total	**80**

Appendicular Skeleton

Superior Extremity	**64**	
pectoral girdle		4
arm and forearm		6
wrist and hand		54
Inferior Extremity	**62**	
pelvic girdle (3 fused)		2
thigh and leg		8
ankle and foot		52
	Total	**126**

Focus

Progress in Bone Repair Research

When a bone breaks, the normal treatment involves bringing the broken ends back into proper alignment and then immobilizing them until new bone is formed and the break is healed. The immobilization can be accomplished with splints, casts, stiff collars, straps, and wires. A break may be bridged by steel pins set into the two faces of the break. This is particularly important for the elderly, who heal so slowly that other methods are often not practical.

However, there are drawbacks to each of these methods. Casts, splints, and straps restrict mobility, and they can do so for long periods (casts can be left in place for up to six months); they therefore lead to muscle atrophy. Steel pins, although they permit the use of the bone, are not as strong as the original bone, and they are subject to rejection by the body (they may loosen as the bone around them resorbs). Researchers have therefore long sought ways to speed the healing of broken bones or to provide prompt, strong repairs. Though the clinical treatment of broken bones is still a gradual, time-consuming process, researchers are exploring new ways to speed the healing of broken bones and improve the quality of the repair. These new techniques are still in the experimental stage and are not likely to have clinical applications for quite some time, but they do offer some hope that it will someday be possible to accelerate and improve the healing of broken bone.

Bone healing has been speeded up experimentally by implanting small electrodes on either side of a break and letting a weak electrical current influence the bone cells. In some way not yet understood, this current seems to accelerate the conversion of periosteal cells to osteoblasts and the laying down of new bone. A similar result can be obtained by putting in the cast, just over the break, an apparatus that generates an oscillating electric field. Eventually, this research may lead to a device that can speed up the healing of any broken bone, perhaps even halving the time needed to regain use of the bone.

Another tack has been taken by some researchers. They have designed a glasslike material containing calcium and phosphorus, the main minerals in bone. They use this "bioglass" as a glue to fasten bone to artificial joints made of a porous ceramic, alumina. The healthy bone tissue reacts with the bioglass and grows into the pores of the ceramic, forming a bond as strong as the bone itself. The method promises a way of replacing damaged or missing bones or parts of bones. Because of the strength of the glue, it may be particularly useful for slowly healing patients such as the elderly.

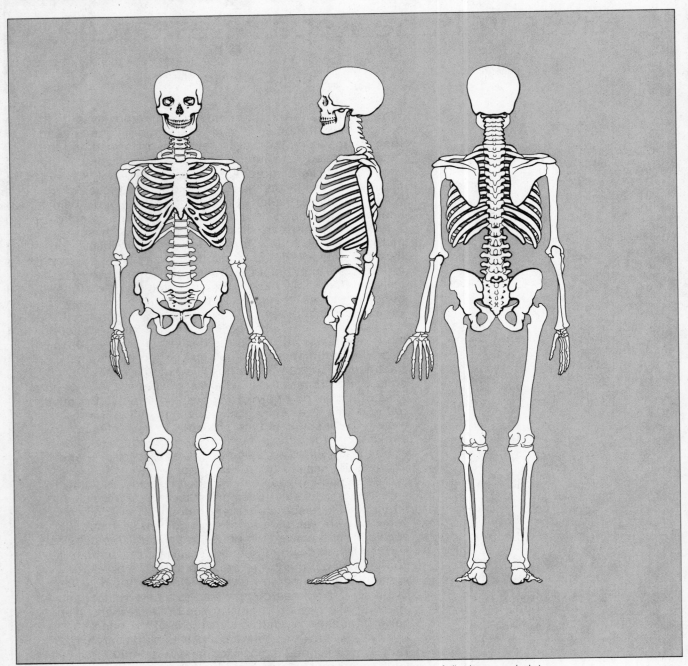

FIGURE 4–5   Anterior, lateral, and posterior views of the human skeleton.

The total of 206 bones is not absolute. The total may vary depending upon the number of sesamoid bones counted and upon how completely certain bones fuse (the coccygeal vertebrae and sternum may not fuse completely).

Axial Skeleton

The Skull The bones that make up the cranium form a large hollow vault housing the brain. They are mostly flat bones. Those that make up the face are mainly irregular bones. The division is largely one of convenience, however, since some cranial bones, such as the ethmoid, contribute very little to the wall of the cranium, and some of the bones of the face, such as the palatine, are not actually a part of the face. Also, some other bones, such as the frontal, are part of both. The ossicles of the middle ear are part of the skull, and the hyoid at the base of the tongue is considered with the skull for lack of a better category.

Bones of the Cranium
Paired: parietal
             temporal
Unpaired: frontal
             occipital
             sphenoid
             ethmoid
Bones of the Face
Paired: maxilla
             zygomatic
             nasal
             lacrimal
             palatine
             inferior nasal concha
Unpaired: vomer
             mandible
Other Bones of the Skull
Paired: ossicles of middle ear
                malleus
                incus
                stapes
Unpaired: hyoid

Any study of the skull should begin with a general consideration of the skull as a whole before looking at the individual bones, since some of the more prominent features are formed by several bones.

Anterior aspect. Among the outstanding features of the anterior view of the skull (Figure 4–6) are three large openings—the two conical **orbits,** occupied by the eyeballs and related structures, and the divided opening of the **nasal cavity**—as well as the upper and lower jaws containing the teeth. These, together with the broad expanse of the **frontal bone** (forehead) and the lateral prominences of the **zygomatic bone** (cheekbone), contribute to the general form of the face. Paired foramina (plural of *foramen*) above and below the orbit and in the **mandible** (*supraorbital, infraorbital,* and *mental foramina,* respectively) provide passage for nerves and blood vessels to superficial tissues.

Deep in the orbit are two openings, the **superior** and **inferior orbital fissures.** The superior orbital fissure and the nearby **optic canal** communicate with the cranial cavity and carry nerves and vessels to orbital structures. There is a canal at the inferior medial angle of the orbit for the *nasolacrimal duct,* which drains fluids from the eye into the nasal cavity. The **lacrimal gland** (tear gland) is located in a depression in the bone at the superior lateral angle of the orbit.

In a prepared skull, such as that shown in Figure 4–6, the opening of the nasal cavity appears large partly because the anterior portion of the nose is cartilage and is no longer present. In the midline (but perhaps off center) is the nasal *septum* (partition). The anterior cartilaginous portion is absent, but the visible portion of the bony septum is a process of the **ethmoid bone.** On each lateral wall of the nasal cavity are three curved flaps of bone, the paired **nasal conchae** (*turbinates*). The superior and middle conchae are processes of the ethmoid bone, but the **inferior nasal conchae** are separate bones. The spaces between the conchae are the **superior, middle,** and **inferior meatuses,** respectively.

Lateral aspect. Most of the bones of the skull can be identified in a lateral view (Figure 4–7). The prominent zygomatic arch is formed by processes of the zygomatic and temporal bones. Near the posterior end of the arch is the *temporomandibular joint,* where the

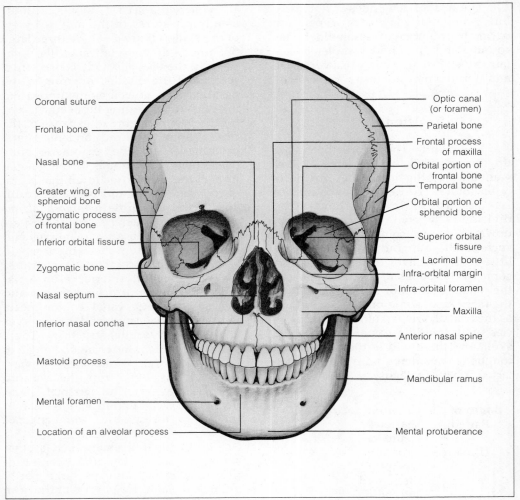

Coronal suture
Frontal bone
Nasal bone
Greater wing of sphenoid bone
Zygomatic process of frontal bone
Inferior orbital fissure
Zygomatic bone
Nasal septum
Inferior nasal concha
Mastoid process
Mental foramen
Location of an alveolar process

Optic canal (or foramen)
Parietal bone
Frontal process of maxilla
Orbital portion of frontal bone
Temporal bone
Orbital portion of sphenoid bone
Superior orbital fissure
Lacrimal bone
Infra-orbital margin
Infra-orbital foramen
Maxilla
Anterior nasal spine
Mandibular ramus
Mental protuberance

FIGURE 4–6   Anterior view of the skull.

mándible articulates with the temporal bone. It is the only movable joint in the skull. Nearby is the **external acoustic meatus,** which bores into the temporal bone, with the slender **styloid process** pointed downward just beneath it and the large, rounded **mastoid process** posterior to it.

Inferior aspect. The inferior surface of the skull (Figure 4–8) is easier to study with the mandible removed. In the midline anteriorly are the **hard palate** (formed by portions of the maxillary and palatine bones) and the **alveolar processes** of the maxillae containing the teeth. The posterior openings of the nasal cavity are visible at the posterior edge of the hard palate. The posterior

portion of the nasal septum is formed by the **vomer.** The openings to the nasal cavity are bounded laterally by the medial and lateral plates of the **pterygoid processes** of the sphenoid bone, which serve as important attachments for some of the muscles of mastication. Extending laterally from the maxillae like flying buttresses are the **zygomatic arches.**

The occipital bone occupies most of the posterior portion. Its large **foramen magnum** is flanked by the **occipital condyles** for articulation with the vertebral column. The basilar portion of the occipital bone extends anteriorly toward the sphenoid bone and meets the medial portion of the temporal bone laterally. The inferior surface of the

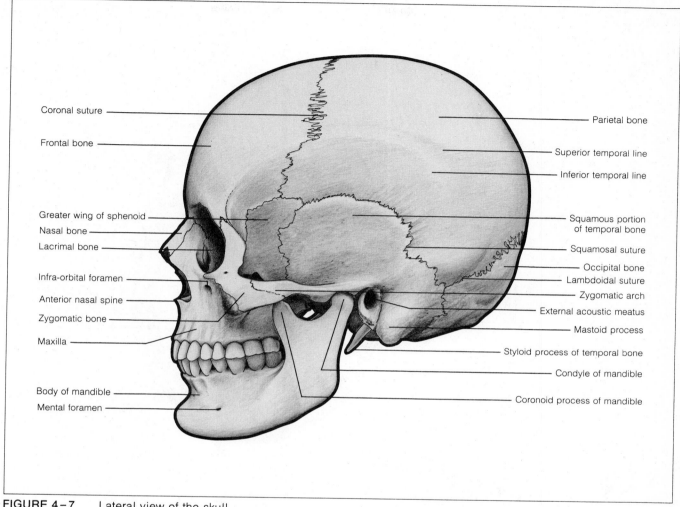

Coronal suture

Frontal bone

Greater wing of sphenoid
Nasal bone
Lacrimal bone

Infra-orbital foramen

Anterior nasal spine

Zygomatic bone

Maxilla

Body of mandible
Mental foramen

Parietal bone

Superior temporal line
Inferior temporal line

Squamous portion
of temporal bone

Squamosal suture

Occipital bone
Lambdoidal suture
Zygomatic arch
External acoustic meatus
Mastoid process
Styloid process of temporal bone
Condyle of mandible
Coronoid process of mandible

FIGURE 4–7   Lateral view of the skull.

skull is marked by several foramina that pierce or border it. The largest of these are the **jugular foramina** (for the jugular veins) and the **carotid canals** (for the internal carotid arteries). The long spindly **styloid processes** lie between these foramina and the mastoid processes.

Floor of the cranial cavity. The floor of the cranial cavity (Figure 4–9) is divided into anterior, middle, and posterior cranial fossae. The posterior fossa is the largest and is formed mainly by the occipital bone. The foramen magnum is most conspicuous and lateral to it lie the jugular foramina. The anterior boundary of the posterior fossa is the *petrous* portion of the temporal bone. (Petrous means hard or stony, and this

part of the temporal bone is very hard.)

The middle cranial fossa is bounded largely by the sphenoid and temporal bones. In the midline is the **sella turcica** (Turkish saddle), the site of the pituitary gland, and on either side of it the carotid canals enter the cranial cavity. The optic canal, carrying the optic nerve from the eye, enters the cranial cavity from under the lesser wing of the sphenoid (the anterior portion of the saddle). Several small foramina carry cranial nerves from the middle fossa to extracranial structures.

The floor of the anterior cranial fossa is the roof of the orbit. Except for a tiny midline portion that is part of the ethmoid bone, it is formed by the frontal bone.

FIGURE 4-8
Inferior view of the skull.

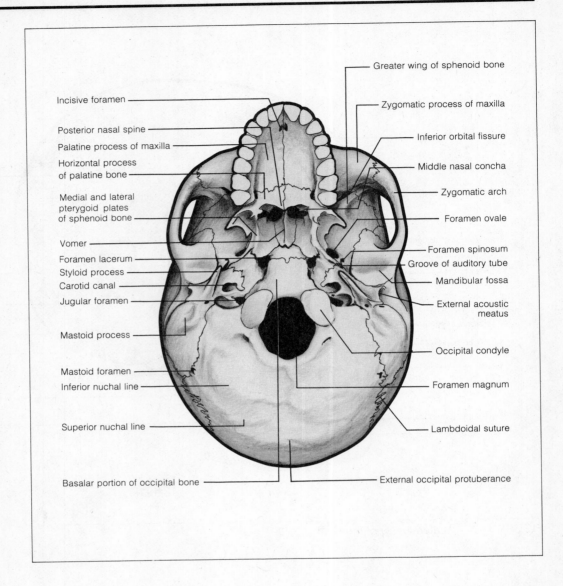

Incisive foramen

Posterior nasal spine

Palatine process of maxilla

Horizontal process of palatine bone

Medial and lateral pterygoid plates of sphenoid bone

Vomer

Foramen lacerum

Styloid process

Carotid canal

Jugular foramen

Mastoid process

Mastoid foramen

Inferior nuchal line

Superior nuchal line

Basalar portion of occipital bone

Greater wing of sphenoid bone

Zygomatic process of maxilla

Inferior orbital fissure

Middle nasal concha

Zygomatic arch

Foramen ovale

Foramen spinosum

Groove of auditory tube

Mandibular fossa

External acoustic meatus

Occipital condyle

Foramen magnum

Lambdoidal suture

External occipital protuberance

The inner surfaces of the bones of the cranial vault bear grooves or indentations marking the courses of the blood vessels. The path of a large venous sinus,[1] leading to the jugular foramen, can be traced on the occipital bone (Figures 4–9 and 4–10). On the bones of the vault, such as the parietal, the course of the *meningeal arteries* is engraved. In some areas, particularly on the floor of the anterior cranial fossa, impressions of the convolutions of the brain are plainly visible.

Sagittal section. A *midsagittal section* (Figure 4–10) shows the nasal septum formed by the perpendicular plate of the ethmoid and the vomer. It also shows the portion of the ethmoid that contributes to the cranial cavity.

A sagittal section just slightly off center (Figure 4–11) shows the lateral wall of the nasal cavity with the nasal

[1]A sinus is a hollow or space, but in practice the term is applied to several different types of spaces. The openings in the bones of the skull are sinuses—paranasal sinuses because of their proximity to the nasal cavity. Many of the veins of the brain are quite large and are also called sinuses. Venous sinuses are also found in other organs, such as the spleen. In the liver they are very small and called sinusoids. In some species (amphibians), blood returning to the heart enters a receiving chamber, the sinus venosus. Since the pacemaker of the heart is located here, the pacemaker is often called the sinus node, although the pacemaker is not itself a sinus; in many species, including humans, it is not even in the sinus venosus.

FIGURE 4-9 Floor of the cranial cavity.

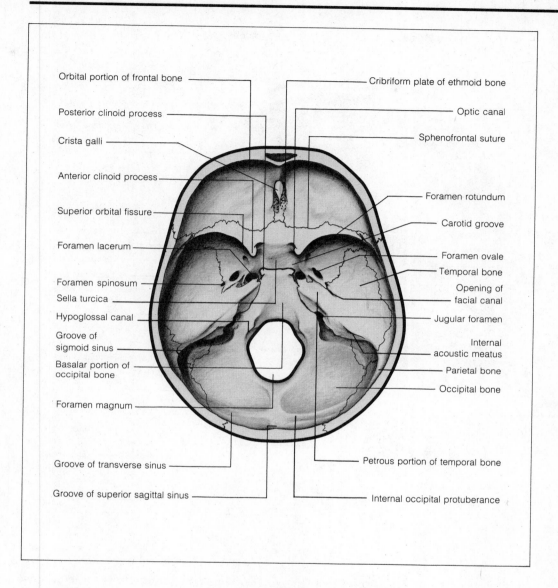

Orbital portion of frontal bone
Posterior clinoid process
Crista galli
Anterior clinoid process
Superior orbital fissure
Foramen lacerum
Foramen spinosum
Sella turcica
Hypoglossal canal
Groove of sigmoid sinus
Basalar portion of occipital bone
Foramen magnum
Groove of transverse sinus
Groove of superior sagittal sinus

Cribriform plate of ethmoid bone
Optic canal
Sphenofrontal suture
Foramen rotundum
Carotid groove
Foramen ovale
Temporal bone
Opening of facial canal
Jugular foramen
Internal acoustic meatus
Parietal bone
Occipital bone
Petrous portion of temporal bone
Internal occipital protuberance

conchae and meatuses and the openings of the paranasal sinuses. They communicate with the nasal cavity (see Chapter 19). The mastoid processes of the temporal bones have similar spaces, called air cells, but they communicate with the middle ear.

Sutures. With the exception of the mandible (jawbone), the articulations of the skull are virtually immovable joints called **sutures.** Their articular surfaces are irregular and the bones fit together rigidly and inflexibly. This is well illustrated in the cranial vault, where the lines of junction are very tortuous. The most conspicuous sutures are those that surround the parietal bones (Figure 4-7). They are:

coronal suture—between the frontal and parietal bones, in the direction of a coronal plane of section.

sagittal suture—between the parietal bones, in the line of the midsagittal section.

lambdoidal suture—between the occipital and parietal bones. It was named for its resemblance to the Greek letter *lambda*.

squamosal suture—between the parietal and temporal (squamous portion) bones.

Other sutures, especially those of the face, are less prominent. Some bones tend to fuse later in life and the

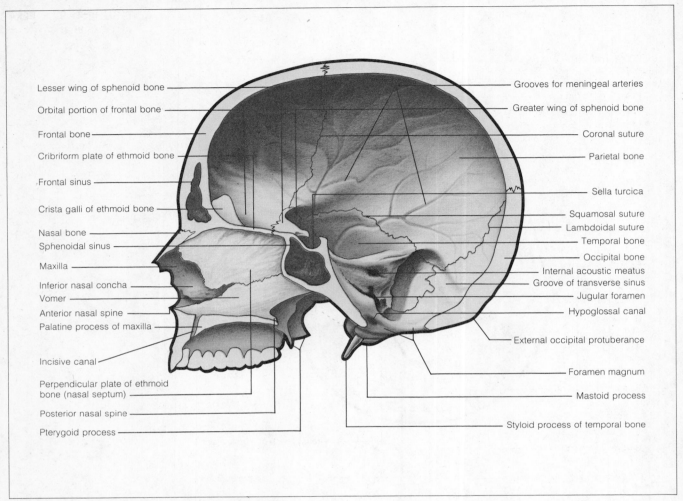

Lesser wing of sphenoid bone
Orbital portion of frontal bone
Frontal bone
Cribriform plate of ethmoid bone
Frontal sinus
Crista galli of ethmoid bone
Nasal bone
Sphenoidal sinus
Maxilla
Inferior nasal concha
Vomer
Anterior nasal spine
Palatine process of maxilla
Incisive canal
Perpendicular plate of ethmoid bone (nasal septum)
Posterior nasal spine
Pterygoid process

Grooves for meningeal arteries
Greater wing of sphenoid bone
Coronal suture
Parietal bone
Sella turcica
Squamosal suture
Lambdoidal suture
Temporal bone
Occipital bone
Internal acoustic meatus
Groove of transverse sinus
Jugular foramen
Hypoglossal canal
External occipital protuberance
Foramen magnum
Mastoid process
Styloid process of temporal bone

FIGURE 4–10   Midsagittal section of the skull.

sutures become difficult to locate. The frontal bone develops as two bones that later fuse, but occasionally they fail to do so, leaving a midline suture.

Fontanels. At the time of birth, ossification of the cranial bones is notably incomplete at the intersections of the sutures. There are six such areas, or fontanels (Figure 4–12). The **frontal fontanel,** which is the largest, lies at the junction of the sagittal and coronal sutures and is unpaired. The **occipital fontanel** lies at the junction of the lambdoidal and sagittal sutures and is also unpaired. The **mastoid fontanel** at the junction of the lambdoidal and squamosal sutures and the **sphenoidal**

fontanel at the junction of the coronal and squamosal sutures are paired. At birth the underlying structures are covered by the fibrous membrane only; hence the fontanels are also called "soft spots." The smaller, paired fontanels are closed over by ossification within a few months after birth, and the larger, midline fontanels are closed over within two years. The presence of fontanels at birth ensures that the cranium is somewhat plastic, which facilitates its passage through the birth canal. They also make possible rapid early growth of the cranium to accommodate the rapidly growing brain. Premature closure of one or more fontanels restricts growth of the cranial contents, resulting in conditions such as *microcephaly,* characterized by a small or deformed head and impaired brain development.

Bones of the cranium. The following paragraphs briefly describe the individual bones and sutures of the cranium and some of their pertinent structural features. They should be studied with frequent reference to the preceding sections and to Figures 4–6 through 4–11 in order to understand the relationships of these bones and to gain the proper perspective into their structures and interrelationships.

The **parietal bones** are approximately square and concave on the internal surface. They form a major portion of the cranial vault. The fontanels are at their corners.

Each **temporal bone** (Figure 4–13) consists of three parts: the *squamous portion,* flat and part of the lateral wall of the cranium; the dense *petrous portion,* part of the floor of the cranial cavity; and the large rounded *mastoid process.* The *external acoustic meatus,* opening on the lateral surface of the temporal bone, is a convenient reference point for locating other features. It penetrates the petrous portion as far as the *tympanic membrane* (eardrum). Anterior to the meatus is the *zygomatic process,* which forms part of the *zygomatic arch.* Near the origin of this process is the *mandibular fossa,* part of the movable articulation with the mandible (temporomandibular joint). Inferior to the external acoustic meatus is the slender *styloid process,* and posterior to the meatus is the blunt *mastoid process,* easily palpated (can be felt by the hand) behind the ear.

The superior surface of the temporal bone is marked by the ridge of the petrous portion enclosing the structures of the middle ear. The jagged medial end of the petrous portion marks the internal opening of the carotid canal, through which the carotid artery enters the cranium. Another open-

FIGURE 4–11 Lateral wall of the right nasal cavity.

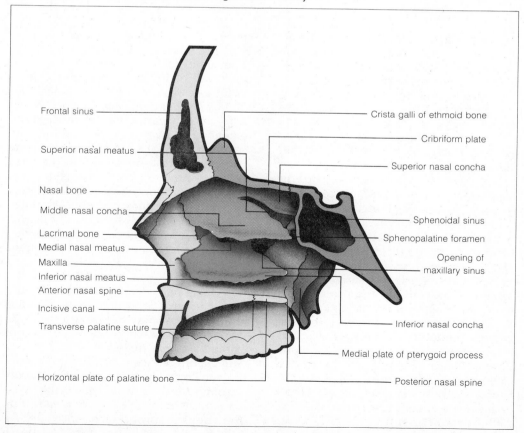

Frontal sinus

Superior nasal meatus

Nasal bone

Middle nasal concha

Lacrimal bone

Medial nasal meatus

Maxilla

Inferior nasal meatus

Anterior nasal spine

Incisive canal

Transverse palatine suture

Horizontal plate of palatine bone

Crista galli of ethmoid bone

Cribriform plate

Superior nasal concha

Sphenoidal sinus

Sphenopalatine foramen

Opening of maxillary sinus

Inferior nasal concha

Medial plate of pterygoid process

Posterior nasal spine

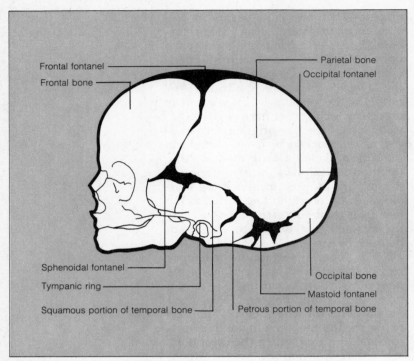

Frontal fontanel

Frontal bone

Parietal bone

Occipital fontanel

Sphenoidal fontanel

Tympanic ring

Squamous portion of temporal bone

Occipital bone

Mastoid fontanel

Petrous portion of temporal bone

FIGURE 4–12   Skull of a newborn infant, showing the fontanels.

ing in the petrous portion, the *internal acoustic meatus*, provides access for nerves and blood vessels to the parts of the ear located in the petrous portion.

The **frontal bone** is the large unpaired bone of the cranium that forms the forehead and much of the roof of the orbit. It therefore makes a major contribution to the face. The medial portion of the brow contains the rather large *frontal sinuses*, which open into the nasal cavity under the middle concha.

The **occipital bone** is a large, round, relatively heavy bone that lies at the base and back of the cranium. Its inferior part is marked by the *foramen magnum*, through which the spinal cord passes. On the external surface, lateral to the foramen magnum, are the smooth, convex *occipital condyles* by which the skull articulates with the first cervical vertebra. Slightly anterior to the condyles are canals for passage of a nerve (the hypoglossal nerve to the tongue). Internally, grooves mark the course of venous sinuses, and externally there are roughened horizontal (*nuchal*) lines representing the sites of attachment of various neck muscles. The

midline *external occipital protuberance* is sometimes quite conspicuous.

The **sphenoid bone** (Figure 4–14) at the base of the cranium is the central bone of the skull and one of the most complicated. It consists of a *body*, in which are located the *sphenoidal sinuses*, and several pairs of processes, including the *greater* and *lesser wings* and the *pterygoid processes*. When viewed by itself, the sphenoid resembles a bat with its wings spread (the greater wings) and feet dangling (pterygoid processes). The lesser wings, visible from above, separate the anterior and middle cranial fossae. As previously mentioned, the *sella turcica* on the superior surface of the body is the well-protected location of the pituitary gland. There are several openings and foramina for passage of nerves from the brain to extracranial structures, including the *superior orbital fissure* and *optic canal* medially, under the lesser wing.

The **ethmoid bone** is difficult to visualize. Although considered a bone of the cranium, the bulk of it is associated with the nasal cavity, and it is thus also a part of the face. The body of the ethmoid is roughly cubical, but it is composed of very delicate bone and riddled with air pockets, the *ethmoid sinuses*. The lateral surfaces form part of the medial wall of the orbit and give rise to two processes on each side, the *superior* and *middle nasal conchae*. The *perpendicular plate*, a part of the nasal septum, extends downward in the midline. The superior surface of the ethmoid bone contributes to the cranium in the form of the perforated *cribriform* (sievelike) *plate*. The part of the brain that overlies this surface is concerned with the sense of smell (*olfaction*). The olfactory nerves from the mucous lining of the superior portion of the nasal cavity penetrate the plate to reach the brain.

Bones of the face.   The **maxillae** articulate firmly with one another in the midline, forming what we call the "upper jaw." Each adult maxillary bone contains *alveoli* for eight teeth. Nerves and blood vessels pass through canals in the maxilla to enter each tooth through the tips of its roots. The *palatine processes* of the two maxillae form the anterior portion of the *hard palate*, which separates the oral cavity from the

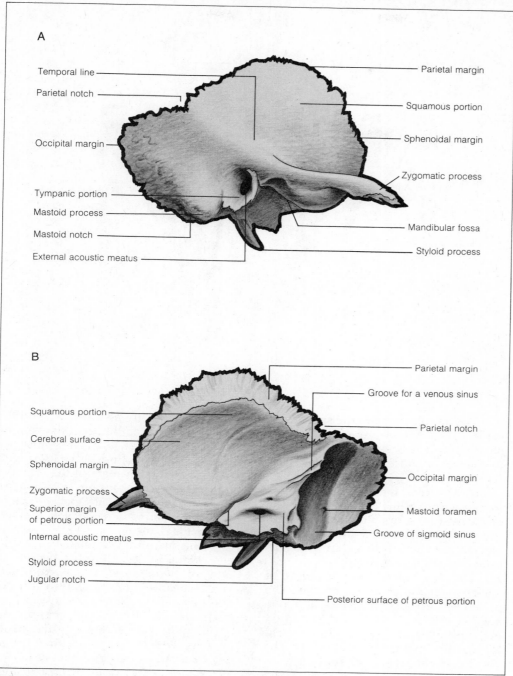

FIGURE 4-13   Right temporal bone. A. Lateral view. B. Medial view.

nasal cavity. In the *body* of each maxilla is a large space, the *maxillary sinus*, which opens into the nasal cavity (Figure 4–11). The maxillae form much of the lateral wall of the nasal cavity and the floor of the orbit.

The **zygomatic bones** (cheekbones)

articulate with the zygomatic processes of the temporal bones to form the *zygomatic arches*. Each arch protects a large muscle used in chewing (*temporalis*) which passes under it. The zygomatic bones also form much of the lateral wall of the orbit.

FIGURE 4–14
Posterior view of the
sphenoid bone.

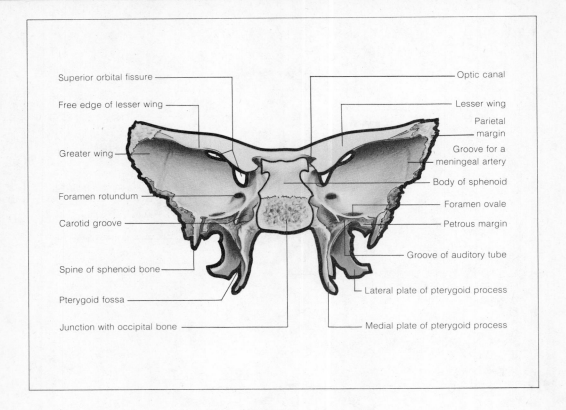

The **nasal bones** are the small elongated bones that form the bridge of the nose. They are backed and supported by the frontal processes of the maxillary bones.

The **lacrimal bones** are thin, fragile slips of bone on the medial wall of the orbit, just anterior to the ethmoid. Along their junction with the frontal process of the maxillae are the entrances to the *nasolacrimal ducts*.

The **palatine bones** each consist of a horizontal and a vertical portion. The horizontal processes articulate with one another and with the palatine portion of the maxillae, and hence form the posterior extension of the hard palate. The vertical portion extends superiorly to form part of the lateral wall of the nasal cavity and contributes a tiny portion to the orbit.

The **inferior nasal conchae,** like the other conchae, are scroll-like flaps of bone extending medially and downward from the lateral walls of the nasal cavity. The inferior conchae, however, are larger than the other two. They are separate bones and articulate chiefly with the maxillary bones.

The unpaired bones of the face are the *vomer* and *mandible*. The **vomer** is a thin flat plate of bone which joins with the perpendicular plate of the ethmoid to form the nasal septum. It rests upon the midline of the hard palate (maxillary and palatine bones) below the perpendicular plate.

The **mandible** (Figure 4–15), like the frontal bone, is formed by the fusion of two bones. The curved *body* contains *alveoli* for the sixteen teeth of the lower jaw. Extending upward on each side is the *ramus*, forming the prominent *angle of the mandible* at its junction with the body. Each ramus has two processes, a knoblike *condyloid process* for articulation with the temporal bone and a pointed *coronoid process* for the attachment of one of the muscles of mastication. On the medial surface of the ramus can be seen the *mandibular foramen,* through which blood vessels and nerves gain access to the *mandibular canal* which runs in the bone toward the midline. The vessels and nerves send *alveolar branches* to each tooth.

FIGURE 4–15
Mandible. A. Side view.
B. Posterior view.

A

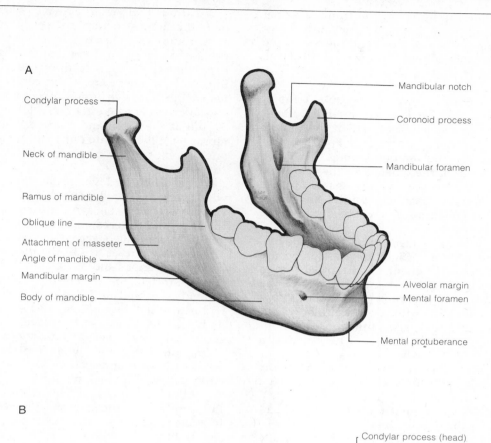

Condylar process

Neck of mandible

Ramus of mandible

Oblique line

Attachment of masseter

Angle of mandible

Mandibular margin

Body of mandible

Mandibular notch

Coronoid process

Mandibular foramen

Alveolar margin

Mental foramen

Mental protuberance

B

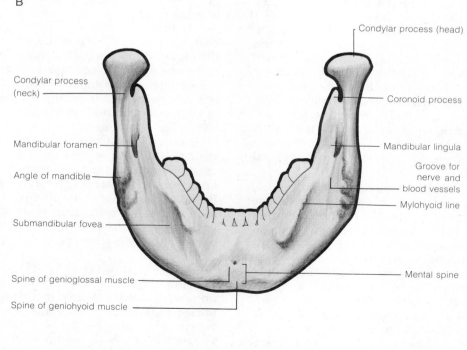

Condylar process
(neck)

Mandibular foramen

Angle of mandible

Submandibular fovea

Spine of genioglossal muscle

Spine of geniohyoid muscle

Condylar process (head)

Coronoid process

Mandibular lingula

Groove for
nerve and
blood vessels

Mylohyoid line

Mental spine

Other branches emerge from the canal through the *mental foramen* on the anterior surface of the mandible to supply the soft tissues in the region of the chin.

Ossicles of ear.   The three pairs of tiny ear ossicles—the **malleus, incus,** and **stapes**—are located in the middle ear, a small chamber in the petrous portion of the temporal bone. They are bound on the outside by the tympanic membrane and on the inside by the inner ear. Their anatomical and functional relationships will be discussed in Chapter 11.

Hyoid.   (see Figure 19–3) The **hyoid bone** is difficult to classify because it does not articulate with other bones; anatomically it is associated with the cartilaginous larynx ("voice-box"). It is considered with the skull, however, because of its close functional relationship with the tongue. The hyoid is a fairly thin horseshoe-shaped bone lying at the junction of the chin and the neck, at the top of the larynx to which it is firmly attached. The hyoid bone provides attachment for muscles, particularly those of the tongue.

Vertebral Column   Anatomically the vertebral column serves all of the functions implied in everyday references to the "backbone" and "spine." It is the support of the trunk and must bear the weight of the entire upper portion of the body—head, neck, trunk, and arms—plus whatever an individual might choose to carry. It also provides a sheltered environment for the spinal cord.

The vertebral column (Figure 4–16) is made up of a series of units, the *vertebrae*, stacked one upon the other and held in position by strong ligaments and muscles and by the interlocking nature of their articulations. This arrangement permits but slight movement between adjacent vertebrae. However, since the movements at each junction are additive, the degree of movement of the whole vertebral column is quite extensive. The interlocking structure creates regularly spaced openings, the **intervertebral foramina,** that form a protected passage for the spinal nerves leaving the cord.

Although all vertebrae have essentially the same structural pattern, there are variations in size and in some of the processes. These differences provide the basis for division of the vertebrae into groups. There are seven cervical, twelve thoracic, five lumbar, five sacral, and four coccygeal vertebrae, a total of 33, but as the sacral and coccygeal vertebrae usually fuse into one sacrum and one coccyx, there is normally a total of 26 bones.

The articulated vertebral column, when viewed laterally, exhibits several curvatures. Two, concave ventrally, are called **primary curves,** and two, convex ventrally, are called **secondary curves.** The primary curves are in the thoracic and sacral regions and are related to the fact that in fetal life the entire vertebral column is curved in this direction. After birth, when the child begins to sit up and hold its head erect, a reverse or secondary curve develops in the cervical region. Later still, when the child begins to stand and walk, a secondary curve develops in the lumbar region.

Abnormal or exaggerated curves of the spine are not uncommon. A marked thoracic curvature is called **kyphosis.** In severe form, we know it as "hunchback"; it is usually the result of some deformity or disease such as tuberculosis of the spine, affecting the body of the vertebrae. The more common and less severe forms are usually the result of muscular weakness or imbalance, or habitually poor posture. An excessive lumbar curve, **lordosis,** may result from structural defects or disease, but it is most often due to faulty posture. A certain amount of lateral curvature, **scoliosis,** is common; it is often due to such bad habits of posture as standing with the weight always on the same leg or carrying a load in the same arm.

Structure of a vertebra.   A typical vertebra (Figure 4–17) consists of a **body** with a **vertebral arch** (neural arch) affixed to its posterior surface. The superior and inferior surfaces of the weight-bearing body are covered with articular cartilage. Each articulates with the body of the adjacent vertebra through a fibrocartilage disc, called the **intervertebral disc,** interposed between

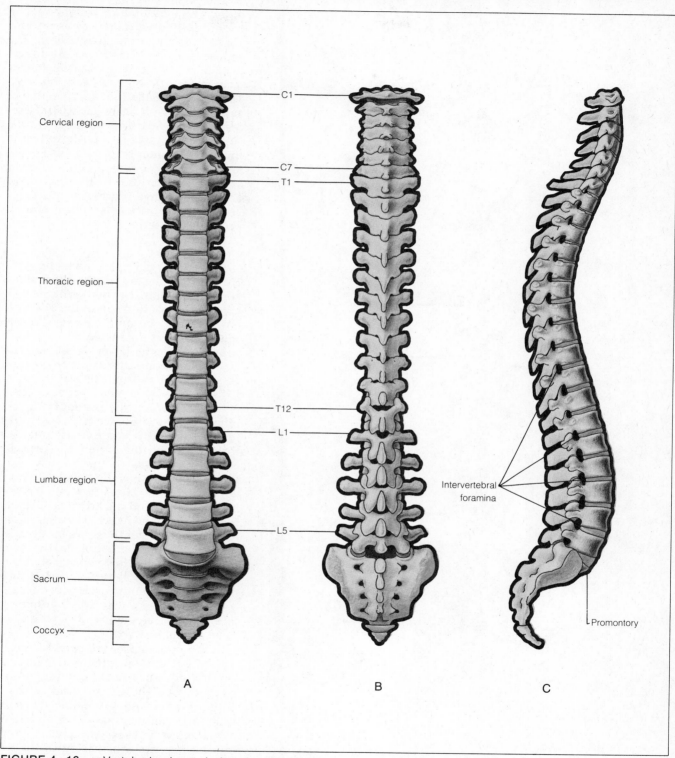

FIGURE 4–16   Vertebral column. A. Anterior view. B. Posterior view. C. Lateral view.

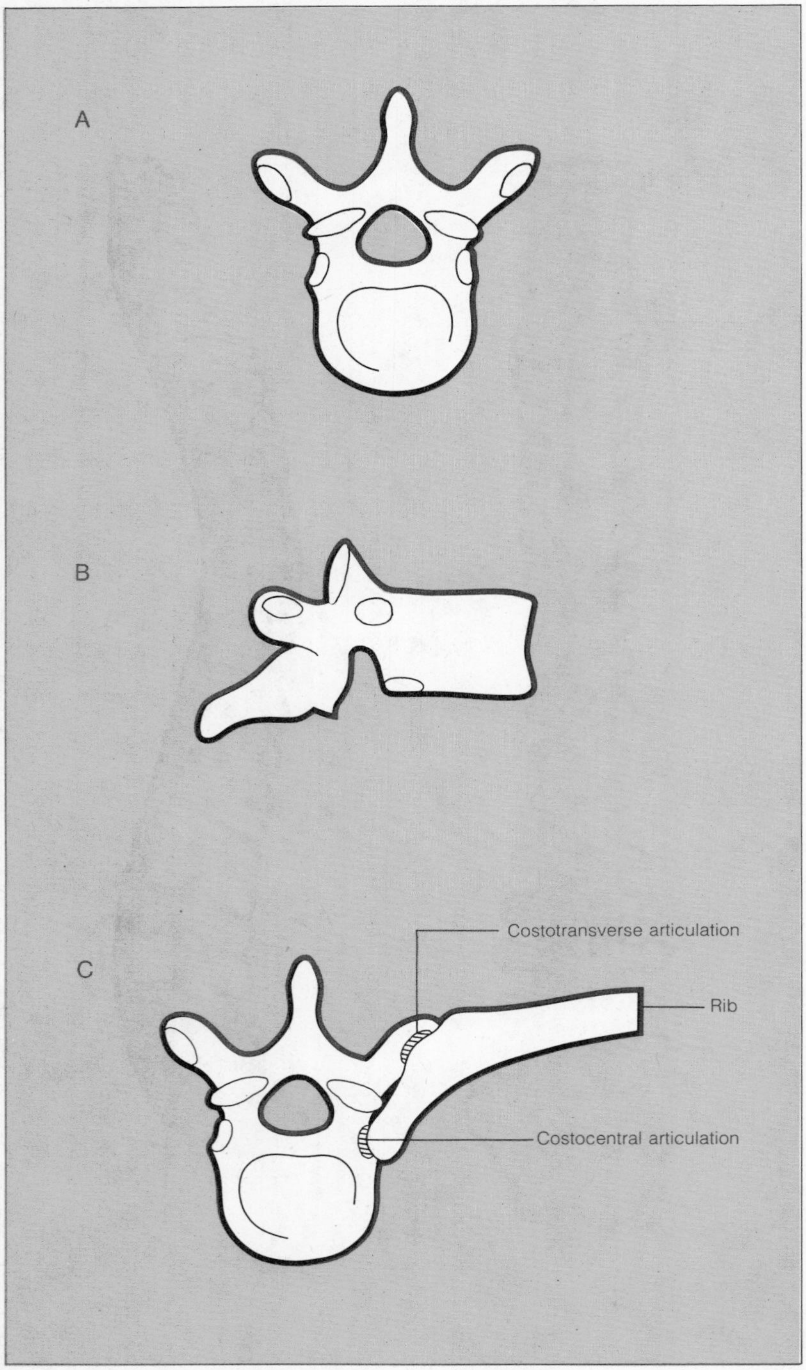

FIGURE 4-17   Structure of the vertebrae, illustrated by a thoracic vertebra. A. Superior view. B. Lateral view. C. Articulation of a rib and vertebra.

them. The vertebral arch roofs over an opening, the **vertebral foramen,** for passage of the spinal cord. From the arch several processes arise, four for articulations and three primarily for attachment of ligaments and muscles. Those for articulation are the paired **superior** and **inferior articular processes.** The articular surfaces of the superior processes face posteriorly and those of the inferior processes face anteriorly, so that when two vertebrae are properly positioned they interlock. Paired **transverse processes** extend laterally, and a single **spinous process** protrudes posteriorly from the arch.

Types of vertebrae. As one progresses from the top to the bottom of the vertebral column, the bodies of the vertebrae increase in size and the intervertebral discs become thicker. These changes reflect the increase in the load supported by the lower vertebrae. The sizes, shapes, and functions of the processes also vary in specific ways.

The typical **cervical vertebra** (Figure 4-18) is distinguished from the others by the presence of a *transverse foramen* in each transverse process, through which the vertebral artery ascends on its way to the brain. When viewed from above, the transverse processes are rather short and angular, the body is small, and the articular surfaces tend to be somewhat concave. On most cervical vertebrae, the spinous process is stubby and often split on the end. However, on the lower cervical vertebrae it is somewhat longer, particularly on C7 (the seventh cervical vertebra—see Figure 4-16), where it is very long and can be palpated readily when the head is bent forward.

The first and second cervical vertebrae merit special attention. The first cervical, the **atlas,** as its name suggests, bears the weight of the world, since it supports the skull. Anatomically, it is unique because it has no body; it is essentially a ring with two lateral masses. The superior articular processes are modified to a pair of relatively large concave articular facets which face upward and receive the condyles of the occipital bone. The second cervical ver-

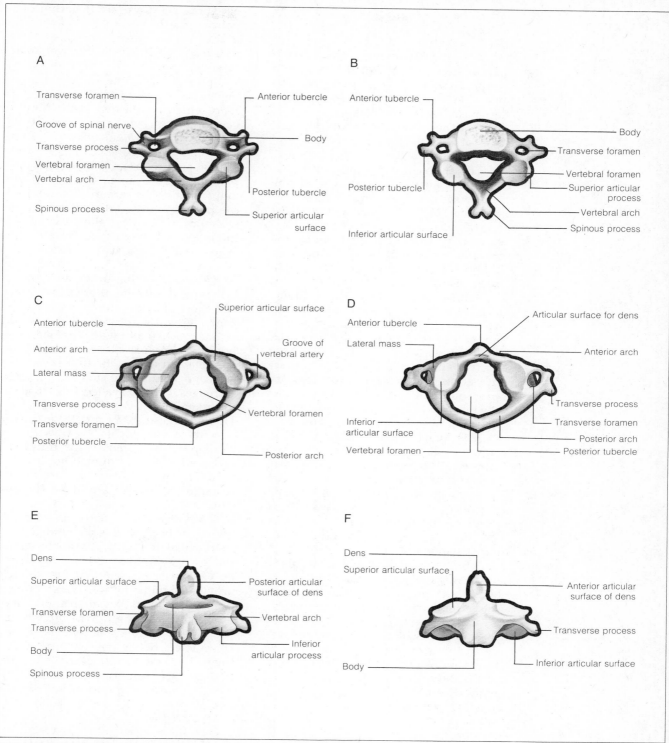

FIGURE 4–18 Cervical vertebrae. A. A middle cervical vertebra, superior view. B. A middle cervical vertebra, inferior view. C. Atlas (first cervical vertebra), superior view. D. Atlas, inferior view. E. Axis (second cervical vertebra), posterior view. F. Axis, anterior view.

tebra, the **axis,** has a prominent tooth-like process, the *dens,* protruding upward from the body. Ligaments hold the dens in place behind the anterior arch of the atlas so that it serves as a body for the atlas. This arrangement is favorable for rotation of the atlas upon the axis. Thus we shake our heads "yes" at the atlanto-occipital joint and "no" at the junction of the atlas and axis.

The **thoracic vertebrae** lack transverse foramina, but the tips of the transverse processes are usually enlarged and exhibit a small articular surface, the *rib facet.* Since the thoracic vertebrae must provide attachment for the ribs, the presence of rib facets, or *costal pits,* on the body and on the transverse processes is to be expected (Figures 4–17 and 4–19). Most thoracic vertebrae have three facets on each side, one on the transverse process and a superior and inferior pair on the body. The head of the rib actually articulates between two vertebrae and has some contact with each. The spinous processes are quite long and extend downward, but they become shorter and more horizontal in the lower thoracic vertebrae.

The body of a **lumbar vertebra** (Figure 4–20) is massive. However, its transverse processes are small and relatively unimpressive, and its spinous process is blunt and squared off. The surfaces of the articular processes tend to face medially and laterally, instead of posteriorly and anteriorly as they do in higher segments.

The five **sacral vertebrae** are fused into a single bone, the **sacrum** (Figure 4–21), but some of the features of a typical vertebra remain. There are foramina, along the lines of fusion on the anterior and posterior surfaces, corresponding to the intervertebral foramina. On the posterior surface, the *middle sacral crest* is the remnant of the spinous processes. The *sacral canal* is a reduced and flattened vertebral canal. The fused transverse processes are heavy, forming wings that present a large and sturdy surface for firm articulation with the pelvic bones (*os coxae*).

The **coccygeal vertebrae** fuse more or less completely to form the **coccyx** and are suspended from the fifth sacral vertebra. They are small and present none of the processes typical of vertebrae, with the exception of the first, which has rudimentary transverse processes.

Thorax The thorax is approximately conical and is defined by the rib cage, which is attached to the thoracic vertebrae. The rib cage (Figure 4–22) includes the *sternum,* twelve pairs of *ribs,* and the *costal cartilages.* This framework protects the thoracic organs, particularly the lungs and heart, and helps anchor the superior extremity—the pectoral girdle articulates with the sternum. In addition, the rib cage permits the movements of the thoracic cage that are necessary for breathing.

Ribs. The **ribs** articulate posteriorly with the twelve thoracic vertebrae and most of them are connected anteriorly with the sternum. They do not actually articulate with the sternum, however, because the **costal cartilages** are between them. The first seven ribs are connected to the sternum by costal cartilages. They are called **true ribs,** and the remaining five ribs are **false ribs.** Three of these (8, 9, and 10) articulate (via their costal cartilages) with the cartilage of the rib above and thus indirectly with the sternum. The last two ribs (11 and 12) have no anterior articulation at all and are called **floating ribs.**

A typical rib is a curved and twisted rod of bone. It has a rounded head that articulates with the rib facets on the bodies of two vertebrae. The rib curves posteriorly, so that the tubercle articulates with the transverse process of the higher vertebra, then angles rather sharply, slanting downward, so that the anterior end of the rib is several centimeters below its posterior attachment. Each rib is rotated so that a flat surface faces anteriorly. The floating ribs are shorter and have relatively little curvature.

Sternum. The **sternum** is a flat bone located anteriorly in the midline. It consists of three parts, the **manubrium,** the **body,** and the **xiphoid process,** which may fuse into a single bone or remain as separate bones thoughout life. The superior angles of the manubrium are marked by notches for articulation with the clavicle (collarbone) and between them is an indentation, the **jugular notch.** The costal cartilage of the first rib articulates with the manubrium, that of the second rib with the junction of the manubrium and the body, and the costal cartilages of the remaining true ribs articulate with the body of the sternum. The xiphoid process is essentially an appendage suspended from the body of the sternum. Costal cartilages do not attach to it, but it may contain a foramen.

The sternum contains functional red bone marrow in the adult and therefore is an important site of hemopoiesis. The relatively exposed position of the sternum makes it a valuable source of hemopoetic tissue, which is needed for certain diagnostic tests.

Appendicular Skeleton

The appendicular skeleton includes the framework of the appendages and their connections to the axial skeleton. For the superior extremity this connection is the *pectoral girdle,* and for the inferior extremity, the *pelvic girdle.* Structural differences between them reflect their different roles. The inferior extremity is weight-bearing; in humans it is both the only support and the major participant in locomotion. These functions require a connection that is exceedingly strong, yet flexible enough to permit free limb movement. The human superior extremity, however, need not bear weight, and range of movement need not be compromised for strength. Differences of function are also reflected in other skeletal adaptations, particularly in the specialization of the hand and foot as manipulative and supportive structures, respectively. The type of activity in which the bones must participate should be borne in mind as you study the parts of the appendicular skeleton.

Superior Extremity The superior extremity is made up of the following paired bones:

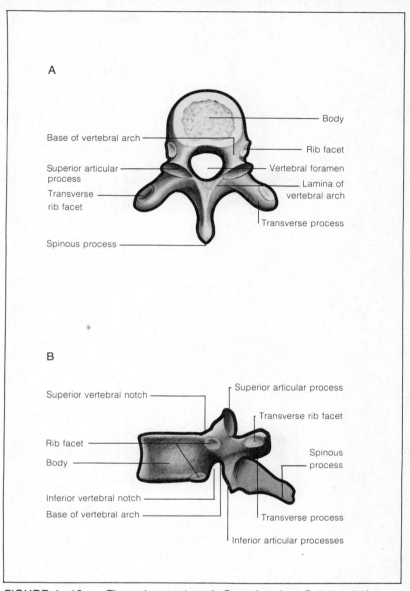

FIGURE 4–19   Thoracic vertebra. A. Superior view. B. Lateral view.

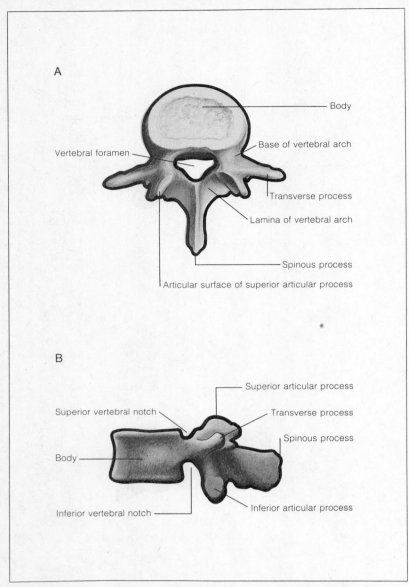

A

Body

Base of vertebral arch

Vertebral foramen

Transverse process

Lamina of vertebral arch

Spinous process

Articular surface of superior articular process

B

Superior articular process

Superior vertebral notch

Transverse process

Spinous process

Body

Inferior vertebral notch

Inferior articular process

FIGURE 4–20   Lumbar vertebra. A. Superior view. B. Lateral view.

Pectoral Girdle
   scapula
   clavicle
Arm and Forearm
   humerus
   ulna
   radius
Wrist and Hand
   carpals (8)
   metacarpals (5)
   phalanges (14)

Pectoral girdle. Composed of two scapulae and two clavicles, the pectoral girdle (Figure 4–23) seems structurally to be very insecure. It is not a complete girdle because the scapulae do not articulate posteriorly, and the anterior attachment is not very stable. The girdle articulates with the axial skeleton only at the sternum, and the articulation of clavicle and scapula is only at the tips of processes held together by ligaments. The only other bone with which the scapula articulates is the humerus, which it must support. It is kept on the posterior surface of the rib cage by the action of muscles and by the clavicle, which acts as a brace or strut.

The **scapula** (shoulder blade) (Figure 4–24) is a flat, roughly triangular bone. The base of the triangle is directed upward and is known as the *superior border.* The other edges are the *medial* (vertebral) *border* and the *lateral* (axillary) *border.* The corners are designated as the *lateral, inferior,* and *superior angles.* Its slightly concave *costal surface* fits against the rib cage. The dorsal surface is marked by a prominent ridge, the *spine of the scapula,* that extends laterally and terminates in the *acromion process,* with which the clavicle articulates. On the lateral angle, beneath the acromion, is a rounded, slightly concave articular surface, the *glenoid fossa,* for articulation with the humerus. Along the superior border just medial to the glenoid fossa, the hooked *coracoid* (beaklike) *process* projects forward. On the dorsal surface the region above the spine is the *supraspinous fossa* and that below is the *infraspinous fossa.* The *subscapular fossa* occupies the costal surface.

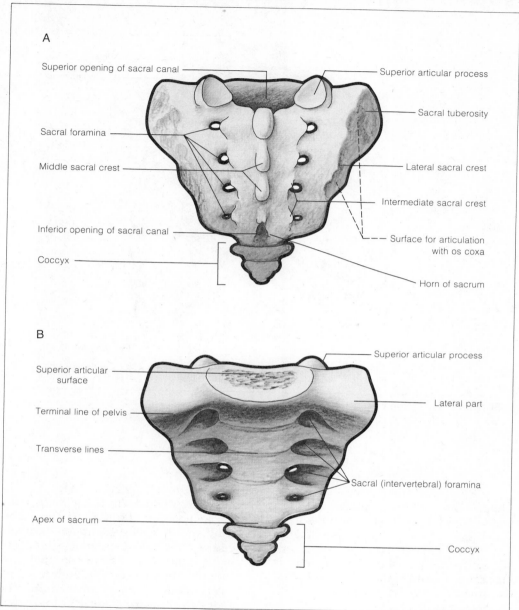

A

Superior opening of sacral canal

Superior articular process

Sacral foramina

Sacral tuberosity

Middle sacral crest

Lateral sacral crest

Intermediate sacral crest

Inferior opening of sacral canal

Surface for articulation with os coxa

Coccyx

Horn of sacrum

B

Superior articular surface

Superior articular process

Terminal line of pelvis

Lateral part

Transverse lines

Sacral (intervertebral) foramina

Apex of sacrum

Coccyx

FIGURE 4–21   Sacrum. A. Posterior view. B. Anterior view.

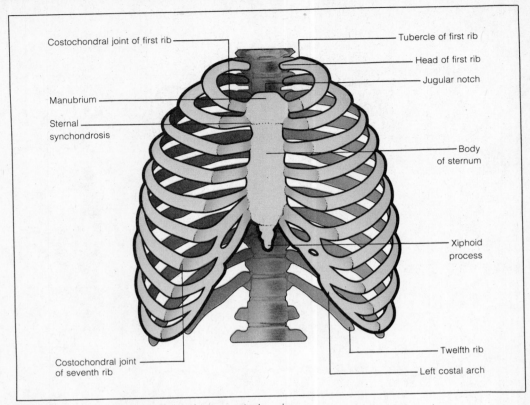

Costochondral joint of first rib

Manubrium

Sternal synchondrosis

Costochondral joint of seventh rib

Tubercle of first rib

Head of first rib

Jugular notch

Body of sternum

Xiphoid process

Twelfth rib

Left costal arch

FIGURE 4–22   Bones of the thorax, anterior view.

The **clavicle** (collarbone) is the S-shaped bone that articulates with the manubrium of the sternum medially and the acromion of the scapula laterally. The sternal end is larger and blunted, while the lateral or acromial end is flattened. The medial half of the clavicle is convex anteriorly and the lateral half is concave anteriorly.

The articulations of humerus and clavicle with the scapula permit extreme freedom of movement of the pectoral girdle and upper extremity, but they also have some disadvantages. The clavicle, being superficial and the only real support for the arm, is particularly vulnerable, since it is subjected to all forces transmitted to the pectoral girdle through the arm (as in a fall), as well as those striking it directly. As a result, the clavicle is one of the most frequently broken bones of the body. Its role as a brace to keep the scapula in place is well illustrated in those few individuals

born without clavicles. These people must rely exclusively upon their muscles to position the scapula, with the result that scapular excursions are extensive indeed. The scapulae tend either to slide downward, so that the shoulders appear very rounded, or to slide laterally and anteriorly until the acromion processes nearly touch in front.

Arm and forearm. The **humerus** (Figure 4–25), a typical long bone, is the largest bone of the upper limb. Its proximal end is dominated by the smooth, rounded *head of the humerus,* which articulates with the glenoid fossa of the scapula. The narrowed region below the head is the *surgical neck,* a frequent site of fracture. Lateral to the head is a large process, the *greater tubercle,* and slightly anterior is a smaller one, the *lesser tubercle.* Between them is the *intertubercular* (bicipital) *groove,*

FIGURE 4–23   The pectoral girdle. A. Anterior view, left side. B. Posterior view, right side. C. The pectoral girdle, superior view.

which descends along the shaft for some distance. The medial and lateral lips of this groove are important attachments for the muscles that move the humerus. Midway along the lateral surface of the shaft is a roughened area, the *deltoid tuberosity,* which is also a site of muscle attachment. The distal end of the humerus is flattened antero-posteriorly and ends in rounded condyloid articular surfaces. On either side above the condyles are the *medial* and *lateral epicondyles,* and extending up the shaft above them are the *medial* and *lateral supracondylar ridges.* On the posterior surface distally is a deep depression, the *olecranon fossa,* and on the anterior surface is the shallow *coronoid fossa.*

The **ulna** (Figure 4–26), the more medial of the bones of the forearm, tapers distally from the proximal end, marked by the conspicuous *olecranon process,* the anterior surface of which is gouged out as the *semilunar notch.* The notch has a smooth articular surface and ends in a lip known as the *coronoid process.* The proximal end of the ulna articulates with the humerus, so that its condyle fits into the semilunar notch, the tip of the olecranon process digs into the olecranon fossa, and the coronoid fossa receives the coronoid process. On the lateral surface of the ulna, just below the semilunar notch, is an articular surface, the *radial notch,* for the head of the radius. Distally, the ulna presents a smooth rounded end and a pointed *styloid process.* The ulna articulates primarily with the distal end of the radius.

In the anatomical position, the somewhat shorter **radius** (Figure 4–26) is lateral and parallel to the ulna. Its head is symmetrical, cylindrical, and concave at the end. A ligament holds the head against the radial notch of the ulna, and the concave portion articulates with the humerus. On the medial surface of the shaft a short distance below the head is the prominent *radial tuberosity.* The larger distal end of the radius has a pointed *styloid process*

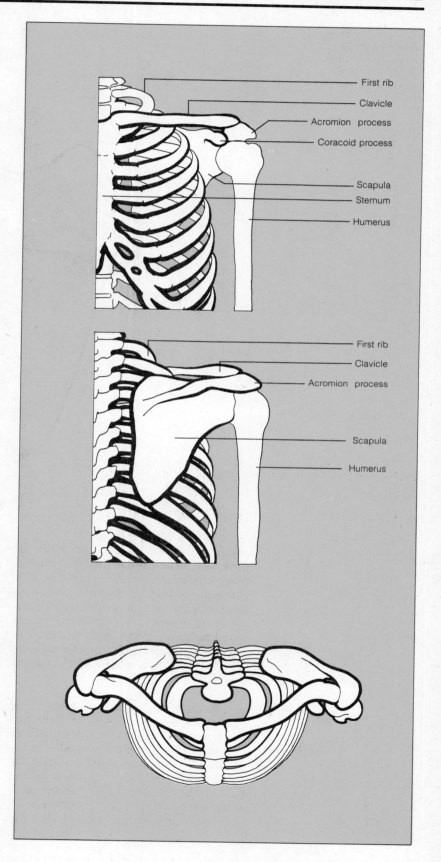

FIGURE 4–24
Right scapula.
A. Posterior view.
B. Lateral view.
C. Anterior view.

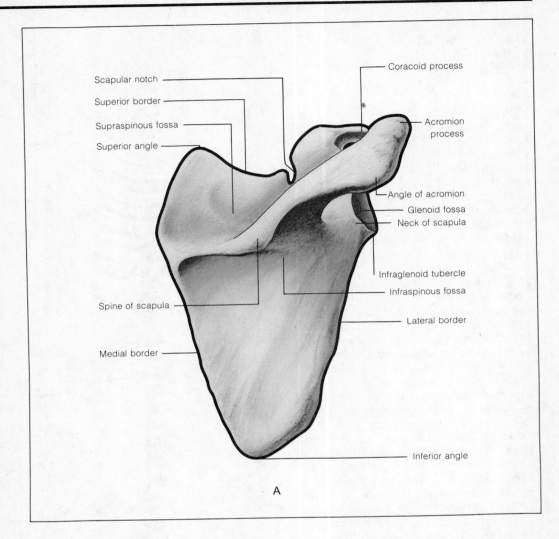

Scapular notch

Superior border

Supraspinous fossa

Superior angle

Coracoid process

Acromion process

Angle of acromion

Glenoid fossa

Neck of scapula

Infraglenoid tubercle

Infraspinous fossa

Lateral border

Spine of scapula

Medial border

Inferior angle

A

and a notch for articulation with the ulna, as well as a broad surface for articulation with the carpal bones. The ulna is the major forearm bone at the elbow joint, but the radius is the more important at the wrist. A connective tissue membrane called the *interosseous membrane* arises from ridges on the adjacent edges of the two bones and connects them securely but flexibly.

Bones of the wrist and hand. The bones of the wrist consist of eight **carpal bones,** individually identified in Figure 4–27. They are arranged roughly in two rows of four, and their articular surfaces are molded to fit those of the adjacent bones. They are bound together tightly by a network of ligaments.

The true bones of the hand are the five **metacarpal bones.** The proximal end or base of each is adapted for articulation with the carpal bones and with one another. The distal end or head of each is enlarged and rounded to articulate with the *proximal phalanx.*

The bones of the fingers are the fourteen **phalanges,** designated as the proximal, middle, and distal phalanges, except in the first digit (the thumb), which lacks a middle phalanx.

Inferior Extremity The inferior extremity is made up of the following paired bones:

Pelvic Girdle
   os coxa (innominate bone)
     ilium
     ischium
     pubis

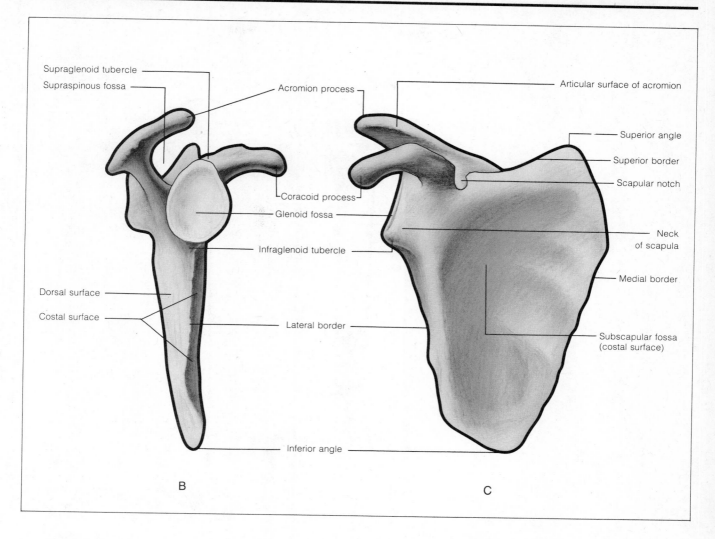

Supraglenoid tubercle
Supraspinous fossa
Acromion process
Coracoid process
Glenoid fossa
Infraglenoid tubercle
Dorsal surface
Costal surface
Lateral border
Inferior angle

Articular surface of acromion
Superior angle
Superior border
Scapular notch
Neck of scapula
Medial border
Subscapular fossa (costal surface)

B

C

Thigh and Leg
  femur
  patella
  tibia
  fibula
Foot
  tarsals (7)
  metatarsals (5)
  phalanges (14)

Pelvic girdle. The **os coxa,** or **innominate bone** (Figure 4–28), is a single bone formed by the fusion of three separate bones, the *ilium, ischium,* and *pubis,* whose names are retained as those for the parts of the os coxa. On the lateral surface is a rounded cup, the *acetabulum,* for articulation with the head of the femur. Each of the three bones contributes to it. Below the acetabulum

is the *obturator foramen,* bounded by the ischium and pubis.

The **ilium** is marked by broad wings that flare out above the acetabulum. Its superior border, known as the *iliac crest,* terminates anteriorly at the *anterior superior iliac spine.* Slightly below this is the more rounded *anterior inferior iliac spine.* Posteriorly the iliac crest terminates in a *posterior superior iliac spine,* and below that is a *posterior inferior iliac spine.*

On the posterior portion of the medial surface of the ilium is a rough surface for articulation with the sacrum. Below this region is a deep indentation (part of which is ischium), the *greater sciatic notch* (the sciatic nerve passes through it). The *arcuate line* runs from the region of the articular surface di-

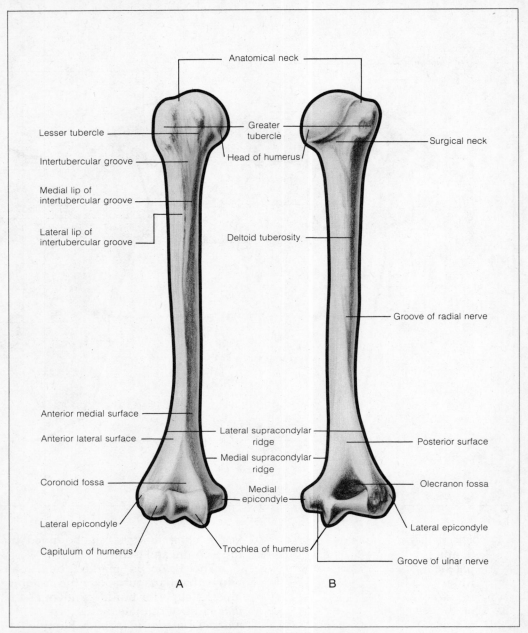

FIGURE 4-25   Right humerus. A. Anterior view. B. Posterior view.

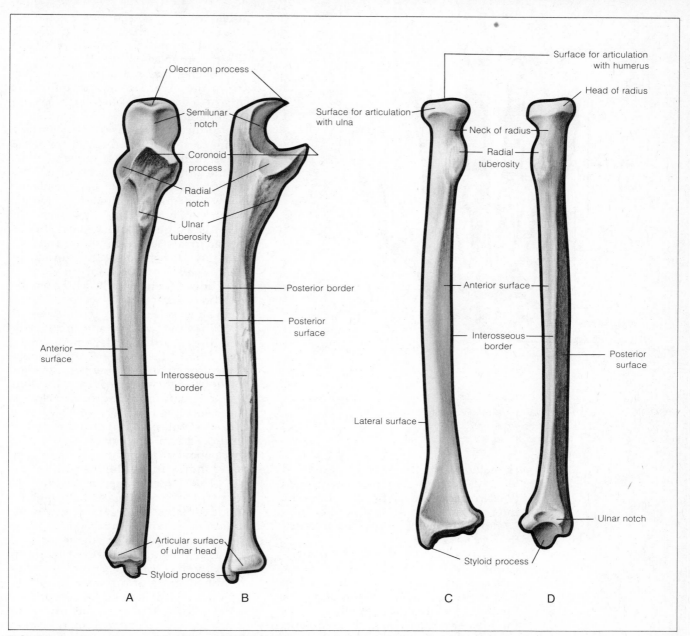

FIGURE 4–26 Bones of the forearm. A. Right ulna, anterior view. B. Right ulna, lateral view (radial surface). C. Right radius, anterior view. D. Right radius, medial view (ulnar surface).

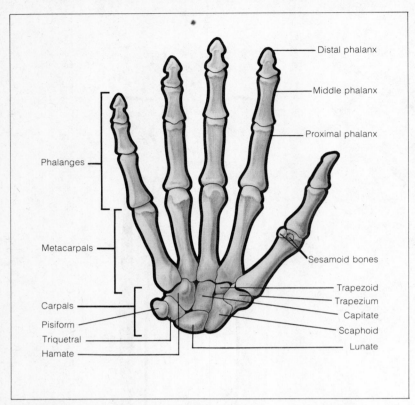

Phalanges

Metacarpals

Carpals

Pisiform
Triquetral
Hamate

Distal phalanx

Middle phalanx

Proximal phalanx

Sesamoid bones

Trapezoid
Trapezium
Capitate
Scaphoid
Lunate

FIGURE 4–27   Right hand, palmar view.

agonally to the body of the pubis. It looks as though the ilium had been twisted backward along this line. The interior (anterior) surface of the ilium is above the arcuate line and is called the *iliac fossa.*

The **ischium** makes up the posterior-inferior portion of the os coxa, where it forms part of the acetabulum and part of the boundary of the obturator foramen. On its posterior edge, below the greater sciatic notch, its landmarks are the *spine of the ischium,* and the *body* of the ischium, upon which is the large, rough *ischial tuberosity.* The ischial contribution to the inferior border of the obturator foramen is the *inferior ramus* of the ischium, which reaches out to join the inferior ramus of the pubis.

The **pubis** is the smallest of the three bones, consisting of a *body* and two *rami.* The inferior ramus is continuous with the inferior ramus of the ischium, and the superior ramus extends up toward the acetabulum. An articular surface for articulation with the other

pubic bone is present on the medial side of the body. A small *pubic tubercle* is usually found on the superior edge near the articular surface.

Several anatomical factors contribute to the great strength of the *pelvic girdle:* (1) The axial skeleton (sacrum) is an integral part of it, which makes it a true girdle in the sense that it forms a complete and closed ring; (2) the three bones of the os coxa fuse completely; (3) the articulation of sacrum and os coxa (the *sacroiliac joint*) is a broad surface with firm ligamentous attachments; (4) the anterior joint, the *symphysis pubis,* is a tightly bound joint; and (5) the femur fits securely into the acetabulum, though not too securely for movement. This construction provides a good line of transfer of the load from the vertebral column and sacrum to the thigh and leg.

When the two os coxae and sacrum are properly articulated, some interesting relationships appear. The previously mentioned arcuate line is quite conspicuous and serves as an important line of demarcation. The space below it is a region of relatively small volume enclosed within the confines of the pubis, ischium, and lower sacrum. The space above this line is much broader, being indicated by the breadth of the ilia. The larger superior portion is known as the **greater** or **false pelvis,** and it contains the organs of the lower abdominal cavity. The region below this line is the **lesser** or **true pelvis,** and the organs contained within it are those of the pelvic cavity. Because of the prominent roles of certain of these organs in the female in childbearing, it is not surprising that the structure of the female pelvis is adapted to this role. Gross comparison of the male and the female pelvises shows that the female pelvis and its openings are generally broader and shallower than the corresponding parts of the male pelvis. Specific differences are tabulated in Table 4–2 and illustrated in Figure 4–29.

Thigh and leg. The **femur** is the longest and largest bone in the body (Figure 4–30). It features a long shaft and greatly enlarged ends with complex, specialized contours. On the proximal end the rounded *head* is directed

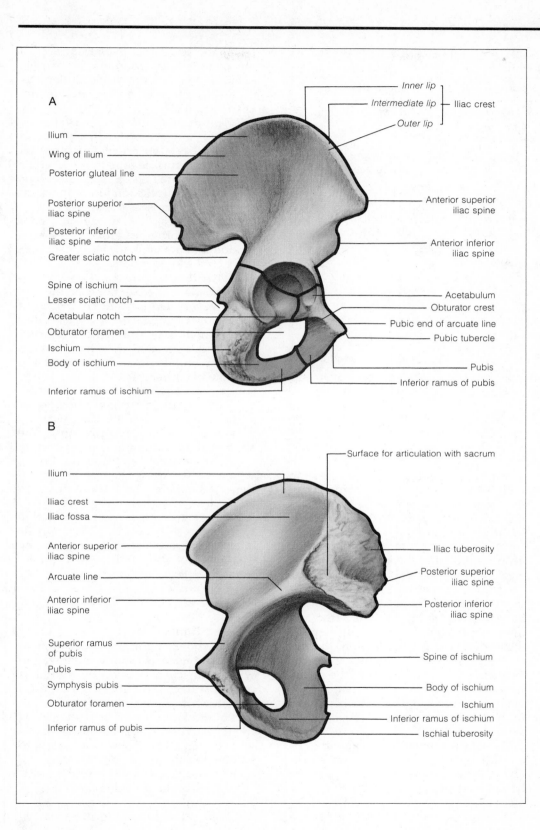

FIGURE 4–28
Right os coxa.
A. External surface.
B. Internal surface.

TABLE 4–2 SOME STRUCTURAL DIFFERENCES BETWEEN MALE AND FEMALE PELVISES

Characteristic	Male Pelvis	Female Pelvis
Bones of the pelvis	larger, heavier, more prominent marks of muscle attachments	smaller, more delicate
Sacrum	long, narrow, and curved	shorter, broader, and less curved
False (greater) pelvis	narrow	wide
True (lesser) pelvis	deep and narrow, funnel shape, less capacity	shallow and wide, more tubular shape, greater capacity
Aperture of pelvic cavity (pelvic inlet)	heart-shaped	round or oval
Greater sciatic notch	narrow	wide
Ischial spine	sharper, turned in	less sharp, not turned in
Obturator foramen	oval	triangular
Symphysis pubis	deeper, longer	shallower, shorter
Pubic angle or arch (between inferior rami of pubic bones)	narrow, pointed, acute angle	wide, rounded arch or obtuse angle
Acetabulum	faces laterally	faces more anteriorly

medially and superiorly for articulation with the acetabulum. The depression in the center of the head marks one attachment of a ligament whose other end is in the acetabulum.

Below the head of the femur is the *neck,* which joins the *shaft* at an angle. Near the junction are two large prominences which are sites of muscle attachment. The lateral *greater trochanter* can be palpated readily as it is large and quite superficial. It extends up beyond the neck, leaving a depression, the *trochanteric fossa,* on its medial side. The smaller *lesser trochanter* is on the posterior medial surface of the femur. The posterior surface of the shaft is marked by a double-lipped line, the *linea aspera.* The two lips converge from the region of the trochanters, where the lateral lip is broader and known as the *gluteal tuberosity.* Distally the lips of the linea diverge to enclose a flat triangular area known as the *popliteal surface.*

The distal end of the femur is enlarged to form the two massive *medial* and *lateral condyles,* and between them on the posterior surface is a deep *inter-condylar fossa.* The articular surfaces of the condyles are rounded for sliding over the top of the tibia. Anteriorly there is a depression for articulation with the patella.

The **patella,** or kneecap, is slightly flattened and somewhat triangular in shape. It bears an articular surface on its posterior surface for contact with the femur. The patella, as a sesamoid bone, develops within the tendon of the muscle group that crosses the knee anteriorly. It improves the leverage of these muscles and provides protection for the tendon and for the knee joint, especially when the knee is bent.

The **tibia** (Figure 4–31) is the major and more medial bone of the leg. It has a very large head, composed of the flat *medial* and *lateral condyles.* Between them are prominences and depressions associated with ligaments to the intercondylar fossa of the femur. On the lateral surface beneath the lateral condyle is a small articular facet for the head of the fibula, and on the anterior surface proximally is the *tibial tuberosity,* which may be very prominent. The shaft of the tibia is marked by a sharp edge, the *anterior crest* (the shinbone), which, as we know from painful experience, is covered only by periosteum and skin. Distally the tibia has a flattened articular surface for articulation with a tarsal bone (the *talus*). On its lateral edge is the *fibular notch* for the distal end of the fibula, and medially the *medial malleolus* extends below the articular surface.

The **fibula** (Figure 4–31) is a long slender bone with a rather angular shaft. The head is a blunt enlargement with an articular facet on the superior-medial surface. The fibula does not articulate with the femur and thus is not part of the knee joint. The distal end of the fibula, or *lateral malleolus,* is somewhat flattened, with surfaces for articulation with the tibia and the talus. An interosseous membrane connects adjacent edges of the tibia and fibula.

Foot. The seven **tarsal bones,** individually identified in Figure 4–32, are larger and more specialized than the carpal bones. Two of them merit special

A

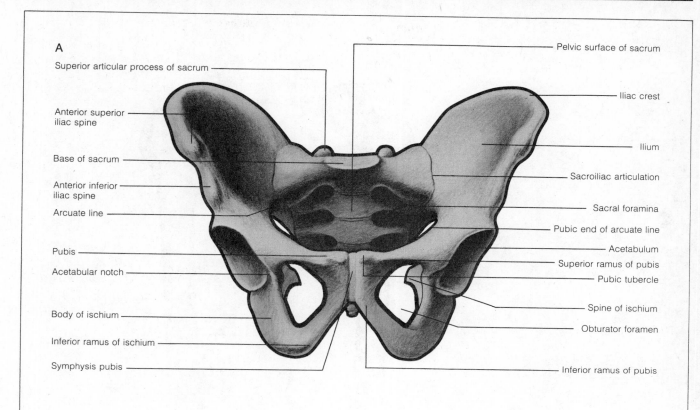

Superior articular process of sacrum

Anterior superior iliac spine

Base of sacrum

Anterior inferior iliac spine

Arcuate line

Pubis

Acetabular notch

Body of ischium

Inferior ramus of ischium

Symphysis pubis

Pelvic surface of sacrum

Iliac crest

Ilium

Sacroiliac articulation

Sacral foramina

Pubic end of arcuate line

Acetabulum

Superior ramus of pubis

Pubic tubercle

Spine of ischium

Obturator foramen

Inferior ramus of pubis

B

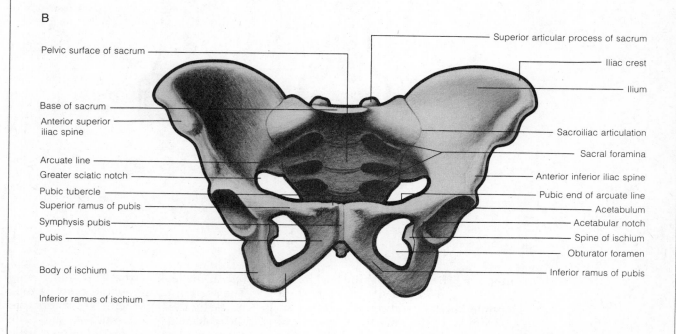

Pelvic surface of sacrum

Base of sacrum

Anterior superior iliac spine

Arcuate line

Greater sciatic notch

Pubic tubercle

Superior ramus of pubis

Symphysis pubis

Pubis

Body of ischium

Inferior ramus of ischium

Superior articular process of sacrum

Iliac crest

Ilium

Sacroiliac articulation

Sacral foramina

Anterior inferior iliac spine

Pubic end of arcuate line

Acetabulum

Acetabular notch

Spine of ischium

Obturator foramen

Inferior ramus of pubis

FIGURE 4–29   A. Male pelvis. B. Female pelvis.

FIGURE 4-30
Right femur. A. Anterior
view. B. Posterior view.

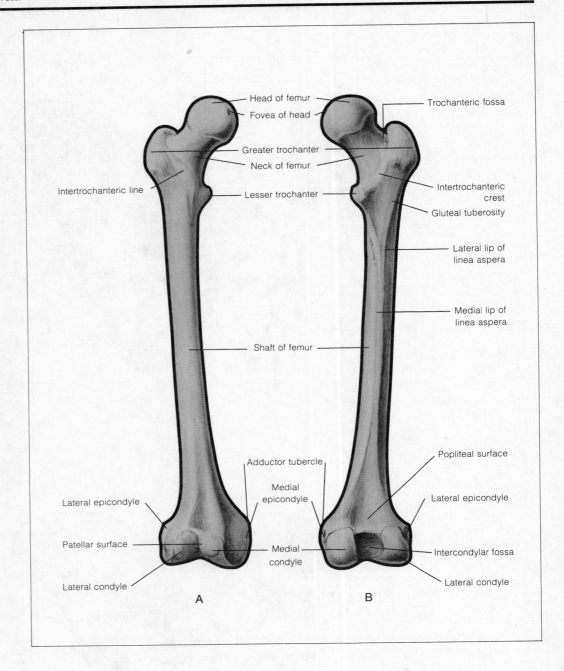

attention. The **calcaneus** (the heel bone) is the largest, and it appears as a blunted knob extending posteriorly for attachment of the muscles of the calf of the leg. The **talus** rests largely atop the calcaneus. Its superior surface is rounded for articulation with the tibia, and its sides are embraced by the medial and

lateral malleoli of the tibia and fibula. The weight of the whole body is thus borne by the talus, which transfers it to the calcaneus and other tarsals. The other tarsal bones, like the carpals, have articular surfaces on several sides for articulation with adjacent bones.

The **metatarsal** bones are similar to

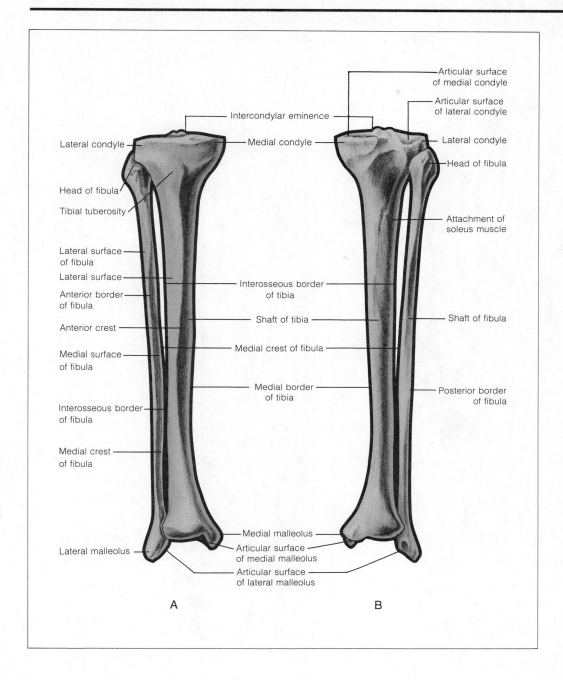

the metacarpals. They have rounded condylar surfaces distally and irregular surfaces on the proximal bases.

As in the hand, there are fourteen **phalanges,** three in each digit except the great (first) toe. The phalanges of the foot, however, are much shorter than those of the hand.

The bones of the foot are so shaped as to form several arches when properly articulated (Figure 4–33). The arches are supported and maintained by fascia, numerous reinforcing ligaments, and the muscles of the foot with their tendons. A *lateral* and a *medial longitudinal arch* are recognized, the latter being

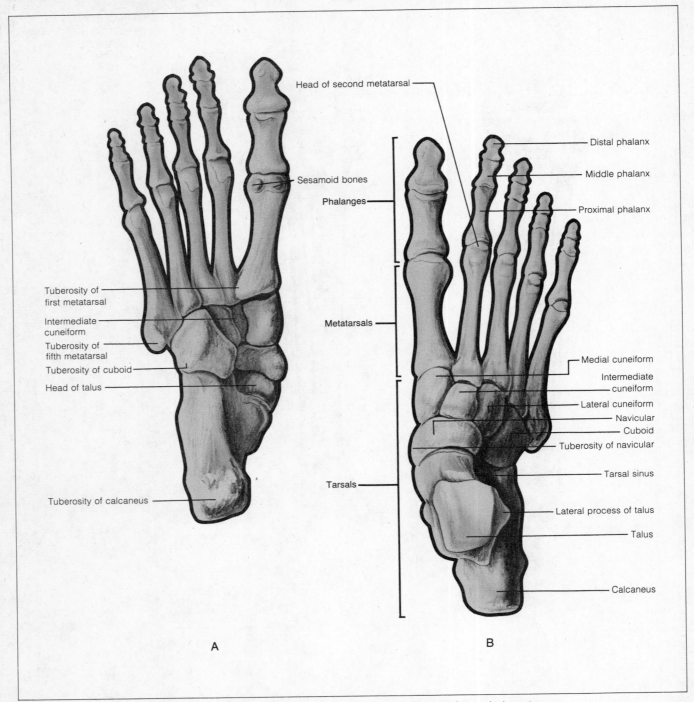

Head of second metatarsal

Sesamoid bones

Phalanges

Distal phalanx

Middle phalanx

Proximal phalanx

Tuberosity of
first metatarsal

Intermediate
cuneiform

Tuberosity of
fifth metatarsal

Tuberosity of cuboid

Head of talus

Metatarsals

Medial cuneiform

Intermediate
cuneiform

Lateral cuneiform

Navicular

Cuboid

Tuberosity of navicular

Tarsals

Tarsal sinus

Lateral process of talus

Talus

Tuberosity of calcaneus

Calcaneus

A

B

FIGURE 4–32   Bones of the right foot. A. Plantar (inferior) surface. B. Dorsal (superior) surface.

the higher arch. The medial arch is slightly asymmetrical, with the talus at the top as the keystone. Weight applied to the talus is distributed to the two pillars—the calcaneus on the one side and the tarsals and metatarsals on the other. The *transverse* or *metatarsal arch* is less obvious, but just as real. It is formed by the distal tarsal bones and the bases of the metatarsals.

Not only are the arches of the foot very strong architecturally, but they also have some of the characteristics of a spring. When a foot suddenly bears weight, the arches, especially the longitudinal arches, spread slightly. This cushions the impact and provides a certain degree of recoil (as the ligaments, tendons, and muscles are pulled taut) that aids in pushing off.

Walking is an act that requires active participation by the feet. First contact is on the heel—the posterior pillar of the longitudinal arch—and as the weight shifts over the foot, the load is transferred forward to the anterior pillar, whose contact with the underlying surface is the heads of the metatarsals. As the weight is shifted forward from heel to toe in normal walking, the greatest load is borne by the head of the first metatarsal (as shown by the effect of collapse of the transverse arch: The second and third metatarsals then bear a greater share of the burden, and a large callus may develop on the sole of the foot beneath the heads of these bones). The toes are important for balance in walking, and they aid in the push-off as weight is transferred to the other foot.

Several common foot problems are related to altered function of the arches. An individual whose longitudinal arches have collapsed (flat feet) does not have the springy cushion, and each time weight is transferred the jolt of impact is transmitted through the bones and joints of the lower extremity to the vertebrae themselves. Such a person is subject to foot and back fatigue and is not a prime candidate for any occupation that requires much walking or prolonged standing. The direction in which the toes are pointed determines the medial-to-lateral weight distribution—toeing out throws the weight medially and burdens the medial longitudinal arch. The situation is further aggravated by shoes with heavy inflexi-

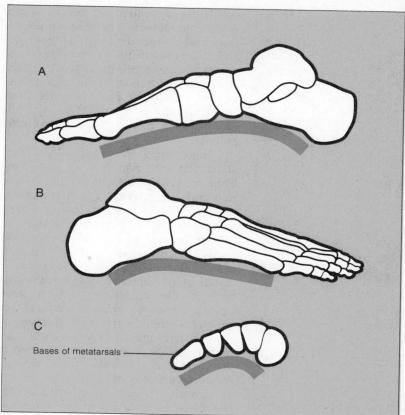

FIGURE 4–33   Arches of the foot. A. Medial longitudinal. B. Lateral longitudinal. C. Transverse (metatarsal).

ble soles, which discourage active participation of the foot muscles in walking. Wearing very high-heeled shoes also strains the arches because more of the weight is thrown onto the anterior pillar of the longitudinal arch, a situation that taxes the transverse or metatarsal arch. Walking barefoot on unyielding concrete also stresses the arches of the foot.

CHAPTER 4 SUMMARY

General Osteology

Bone is a type of connective tissue distinguished by the presence of mineral salts in the intercellular matrix. Because of their rigidity, bones are able to provide support and a framework to anchor muscles and other body parts. Bones and the joints between them constitute a system of levers which makes movements possible. Bones enclose and protect the brain, spinal cord, and structures in the thorax (lungs and heart). Bone is an important mineral storage site, for most of the calcium and phosphorus in the body is in its intercellular material. Bone is also the site of formation of the blood cells (hemopoiesis).

Bones may be classified as long, short, flat, or irregular. A typical long bone consists of a shaft (diaphysis) and two ends (epiphyses). The diaphysis is mainly compact bone enclosing a marrow cavity (yellow marrow). The epiphyses are formed of spongy or cancellous bone, which contains red bone marrow, the hemopoietic tissue. There is a thin layer of compact bone over the surface of the epiphyses, which have smooth areas, or articular surfaces, for contact with other bones. A layer of hyaline cartilage actually covers the articular surface, however, but the rest of the bone is covered with fibrous connective tissue, or periosteum.

Compact bone is organized into Haversian systems. Such a system consists of a Haversian canal that contains blood vessels and is surrounded by intercellular material arranged in concentric rings, or lamellae. Bone cells, osteocytes, lie in lacunae between the lamellae and communicate through tiny canaliculi with blood vessels in the haversian canal.

Cancellous bone lacks complete haversian systems, but spicules of bone contain osteocytes in lacunae. The spaces in cancellous bone are occupied by the red bone marrow and contain blood cells in various stages of development.

Bones of the skull are formed by the process of intramembranous bone formation, for they develop in and upon a fibrous membrane. Some primitive cells differentiate into fibroblasts, which form the collagen fibers that make up the membrane. Other primitive cells become osteoblasts and form the organic portion of bone (osteoid) on and around the membrane. Osteoblasts become trapped in the osteoid and are then osteocytes. Osteoid becomes bone when calcium and phosphorus salts are deposited in it (calcification). The process of bone formation or ossification typically begins at several sites in a bone, and spreads until the entire bone is completely ossified. Certain areas, known as fontanels, are not ossified by the time of birth. When ossification is completed, there is a thin plate of compact bone on the inner and outer surfaces, with cancellous bone in the center.

Other bones are formed within a tiny cartilage model of the bone by endochondral bone formation. As the cartilage model grows, the cartilage cells (chondrocytes) in the middle of the shaft mature and begin to undergo changes that lead to calcification of the cartilage matrix, followed by its eventual breakdown. Meanwhile, some cells of the perichondrium around the cartilage model, differentiate into osteoblasts and produce osteoid in which minerals are deposited. The result is a periosteal bone collar around the middle of the cartilage. (The perichondrium in this area is now periosteum.)

A periosteal bud, consisting of primitive cells, osteoblasts, and blood vessels invades the disintegrating cartilage from the periosteum. The osteoblasts locate along the remaining spicules of the calcified cartilage and secrete osteoid, which soon becomes calcified. This is a primary center of ossification, and immediately begins to expand into the deteriorating cartilage. The process is soon repeated for periosteal buds invade the epiphyses, forming secondary centers of ossification.

The bone continues to increase in length as long as the cartilage between the expanding centers of ossification continues to grow. Ossification finally overtakes the growing cartilage, the cartilage disappears, and ossification is complete. The bone cannot grow any longer, but it may become thicker and/or reshaped.

The process of reshaping a bone begins early and continues. The invading periosteal buds contain osteoclasts, which erode the bone that has been formed (resorption). Other osteoclasts remain on the surface of the bone. Their actions are important in forming the marrow cavity and in remodeling the bone as it grows.

Bones of the Skeleton

The skeleton is formed of two portions. The **axial skeleton** includes the bones of the skull, vertebral column, and thorax. The **appendicular skeleton** includes the superior extremity and pectoral girdle, and the inferior extremity and pelvic girdle.

The *skull* includes the bones of the cranium and the face. The cranium encloses the brain and internal structures of the ear. The bones of the face form cavities that contain the eyes (*orbits*) and the nasal and oral cavities, and sockets for the teeth. The *vertebral column,* containing the spinal cord, consists of seven cervical, twelve thoracic, five lumbar, five sacral, and four coccygeal vertebrae. The sacral vertebrae are fused into a single bone (*sacrum*), and the coccygeal vertebrae are also fused.

The appendicular skeleton consists of the bones of the arms and legs and their attachments to the axial skeleton. The *pectoral girdle* provides the attachment for the superior extremity, but the only bony connection (joint) is with the *sternum;* its other attachments are muscles and ligaments. This provides great freedom and range of movement, rather than stability. The *pelvic girdle* is much more stable, for part of it, the *sacrum,* is also part of the axial skeleton. It is designed for security and weight-bearing, but sacrifices some range of movement.

The general structural pattern of the bones of the two extremities is similar, although those of the superior extremity are smaller and lighter in weight. The hands and fingers provide for much more dexterity than the foot and toes. The pelvis of the female differs in shape from that of the male. It tends to be broader and not as deep, whereas the male pelvis is narrower and deeper.

STUDY QUESTIONS

1 Describe the structure of a typical long bone. How does the structure of a flat bone differ?

2 What is the significance of each of the following in endochondral bone formation: perichondrium (periosteum), the bone collar, calcification of cartilage, the periosteal bud, primary and secondary centers of ossification, osteoblasts, osteoclasts, closure of epiphyses?

3 How does intramembranous bone formation differ from endochondral bone formation?

4 How do osteocytes in mature compact bone receive nutrients and get rid of wastes?

5 How is a bone reshaped and re-formed as it grows? How does the marrow cavity develop? Can a bone be reshaped after growth is completed? Explain.

6 What bones form the cranium? the face?

7 What bones form the walls of the nasal cavity? the orbit?

8 Name one or more characteristics by which you could distinguish cervical, thoracic, lumbar, and sacral vertebrae. How would you recognize the atlas and axis?

9 Compare and contrast the bones that make up the superior extremity with those of the inferior extremity.

10 Compare and contrast the structure of the pectoral and pelvic girdles.

11 What are some of the anatomical differences in the structure of the male pelvis and female pelvis?

5 _____

Fibrous Joints

Cartilaginous Joints

Synovial Joints

Articulations

The most obvious and necessary function of the bony skeleton is support, but the skeleton is also essential to movement. You do not float rigidly from place to place; you walk and run and flex your waist. Every movement involves bending some part of the limbs or trunk, and every movement occurs at the junctions between the bones, the **joints** or **articulations.** Some joints are important only for strength and support, and some mainly for movement. Still other joints serve both functions.

The manner in which two bones are connected determines the type and range of movement, if any, possible at the joint between them. Joints, like bones, differ in their structure according to their particular roles. They vary in the types of connecting tissue, in the kinds of movement permitted, and in the shapes of the articular surfaces of the bones involved. Joints can be classified on the basis of any of these criteria.

The two bones involved in an articulation do not actually touch one another, since the connective tissue that holds them together also holds them apart. The type and amount of tissue between the bones determine to a great extent the strength of the joint, and how much and what kinds of movement can occur there; these criteria are also a convenient way to classify joints.

134

FIBROUS JOINTS

After stating that all joints are held together by connective tissue, which is a fibrous tissue, it may seem redundant to say that the connecting material of a particular class of joints is fibrous connective tissue. You will, however, recall that there are several varieties of connective tissue, so that a fibrous joint must be one, for example, which has no cartilage between the articulating bones. In general, fibrous joints are tight joints, since the fibrous connective tissue is strong and inelastic, permitting little or no movement. A fibrous joint may be either a *suture* or a *syndesmosis*, depending partly upon the length of the connecting fibers.

Suture

Sutures (Figure 5–1) are found only between membrane bones and are thus limited to the skull. They are characterized by interdigitations of the irregular edges of the bones which form rigid and inflexible junctions. The surfaces of the bones are covered with periosteum (also a fibrous connective tissue), and the very small space between the articulating bones is spanned by fibers of connective tissue that blend into the periosteum.

Syndesmosis

A syndesmosis (Figure 5–2) is literally "joined with fiber." Such a joint is, like a suture, held by fibrous tissue; however, the fibers are usually longer and there are more of them than in sutures. Specifically, the fibers of this joint are *ligaments*. The connection is usually firm enough that any movement is more in the nature of a "give" than of a true movement. Examples include the distal tibiofibular articulation and the coracoclavicular articulation.

CARTILAGINOUS JOINTS

Cartilaginous joints are those in which the bones are connected by cartilage. Movement is also very slight at

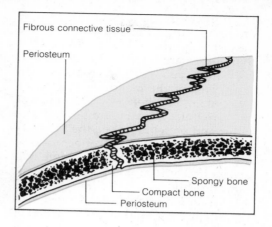

FIGURE 5–1   A suture between bones of the skull is a type of fibrous joint.

these articulations. There are two types of cartilaginous joints, *synchondroses* and *symphyses*.

Synchondrosis

Synchondroses have hyaline cartilage between the articulating bones. Many synchondroses are temporary, disappearing when ossification is completed (as in growing long bones, before the epiphyses close). The junctions between the ribs and sternum have large cartilaginous portions, and some of them are synchondroses, but many of them have a tiny joint cavity and are thus considered to be synovial joints.

Symphysis

In a symphysis the articular surfaces of the bones are covered with a thin layer of hyaline cartilage, but there is also a fibrocartilage pad between the bones. In addition to the abundant col-

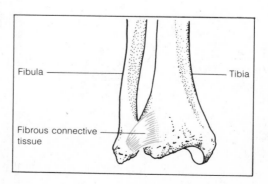

FIGURE 5–2   A syndesmosis. The distal tibio-fibular joint is also a type of fibrous joint.

lagen fibers in the fibrocartilage, there are reinforcing ligaments that ensure a very strong union between the bones. Symphyses are found between the pubic bones (pubic symphysis), between the halves of the mandible in the human infant, and between the bodies of the vertebrae (as in Figure 5–3). In the pubic and vertebral cases, the fibrocartilage serves not only as connections of the bones, but also as a cushion that gives a little under stress, thus allowing some flexibility.

The intervertebral discs must also support weight, and their structure is modified accordingly. The center of the disc contains a mass of fibrogelatinous pulp (the *nucleus pulposus*), which is a shock absorber. The outer portion is composed of fibers that are wound about the core in concentric rings (the *annulus fibrosus*). When subjected to pressure, as when you stand erect, the discs flatten and spread out a bit. Sudden severe pressure may cause the disc to rupture and the nucleus pulposus may break through the outer fibrous rings of the annulus fibrosus, cause pressure on nearby spinal nerves, and produce severe pain, or it may burst through into the body of the vertebra.

SYNOVIAL JOINTS

Synovial joints are the most complex and most widely distributed, since they are the movable joints. They are characterized by the presence of a joint cavity filled with fluid.

In a synovial joint (Figure 5–4), the smooth articular surfaces of the bones are covered with a thin layer of hyaline cartilage, the **articular cartilage.** The joint is enclosed by the **articular capsule** (joint capsule), which is continuous with the periosteum. The inner surface of the articular capsule is lined with a membrane known as the **synovial membrane.** The capsule and membrane enclose the area immediately surrounding the ends of the bones, forming the **articular** or **joint cavity,** which is filled with **synovial fluid,**[1] a clear, viscous substance secreted by the synovial membrane. The constituents of synovial fluid are similar to those of blood plasma, except for some notable differences in the protein content.

The articular cartilage serves as a shock absorber, and because of its very smooth surface there is virtually no friction with movement at the joint. The synovial fluid also provides a cushion, but it is especially important as a lubricant. Synovial fluid serves much the same function in a joint as oil between the moving parts of a machine. Both synovial fluid and oil minimize heat buildup and wear and tear by reducing friction between the moving parts. In addition, since there are no blood vessels in cartilage, synovial fluid is the source of nourishment for the cartilage cells. Changes in the synovial fluid often lead to breakdown of the articular cartilage, such as occurs in rheumatoid arthritis or in degenerative joint disease.

In addition to the above characteristics, which apply to all synovial joints, there are other, less constant properties. Nearly all synovial joints are reinforced by ligaments, either as separate bands of fibrous connective tissue or simply as thickenings of the articular capsule in appropriate places. These ligaments serve both to hold the articular surfaces in position and to limit or prevent movement in certain planes. The synovial membrane may extend into the joint cavity as **synovial folds,** increas-

FIGURE 5–3   A symphysis. The bones are united by fibrocartilage and supported by ligaments. The fibrocartilage disc in intervertebral joints contains a softer material which acts as a cushion.

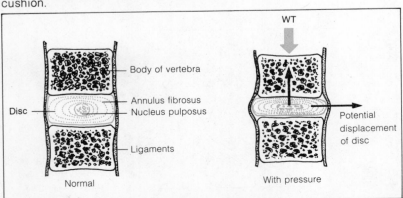

Body of vertebra

Disc

Annulus fibrosus
Nucleus pulposus

Ligaments

Normal

WT

Potential displacement of disc

With pressure

---

[1] Note the form "ova" (OVA ≡ EGG) in the term synovial. Thus it should not surprise you to learn that synovial fluid resembles raw egg white.

ing the membrane area and thus providing a greater secreting surface area. In certain joints there are additional bits of cartilage, as rings, discs, or wedges. Their presence usually improves the fit of the articular surfaces of the two bones, such as around the edge of the glenoid fossa or around the condyles of the tibia. Tendons of some muscles pass over the joint in such a way that they act as ligaments (Figure 5–4B), or the muscles themselves may contribute to keeping the articulating bones in position. Occasionally fat pads are deposited in and around the joint cavity.

Modified Synovial Cavities

Cavities lined with synovial membranes (synovial cavities) are not restricted to the enclosures around joints, however. A **bursa** (Figure 5–5A) is also a synovial cavity—it is a sac, lined with synovial membrane and filled with synovial fluid, found where some protection is needed. Bursae may be subcutaneous, lying between the skin and a bone, as over the olecranon process of the ulna; they may lie between a tendon and an underlying bone; and they may be found around some articulations that do not have a true joint cavity.

A **tendon sheath** (Figure 5–5B) resembles a tubular bursa, in which the tendon has come to be surrounded by a synovial sheath. Tendon sheaths are found particularly in "bony" regions (such as the wrist and ankle) where there is pressure on the tendon from several sides (as from a ligament on one side and a bone on the other), and in the fingers where the tendons slide great distances whenever the fingers are moved.

Analysis of a Synovial Joint

Because the synovial joints are the movable joints, they are most reasonably classified on the basis of their movements. This is reasonable because the degree and type of movement at any joint are chiefly determined by, and correspond to, the bony structure of the joint and its articular surfaces. The

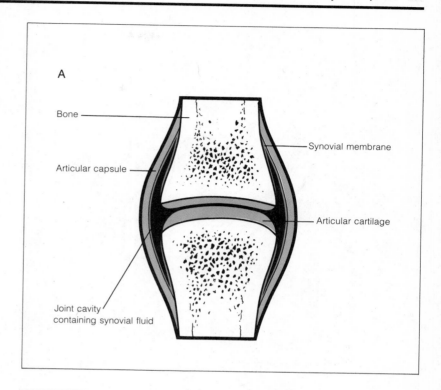

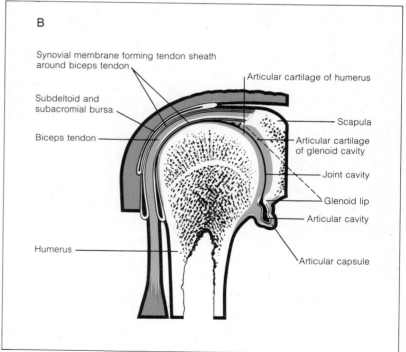

FIGURE 5–4   A. A typical synovial joint. B. The glenohumeral (shoulder) joint is a synovial joint with some special features.

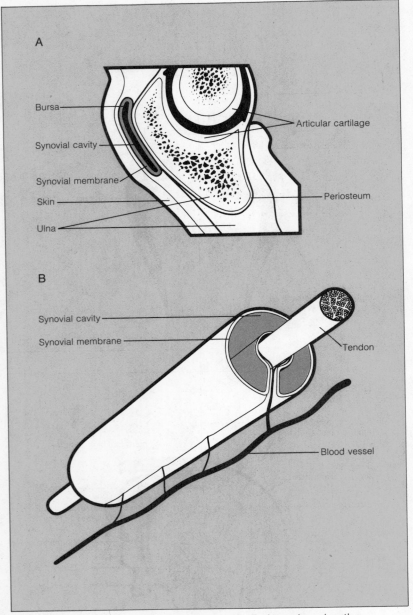

A

Bursa

Synovial cavity

Synovial membrane

Skin

Ulna

Articular cartilage

Periosteum

B

Synovial cavity

Synovial membrane

Tendon

Blood vessel

FIGURE 5–5   A. A subcutaneous bursa. B. A tendon sheath.

serves as a fulcrum about which the bone rotates (Figure 5–6). The axis for that rotation is a line passing through the fulcrum and perpendicular to the plane of the movement.[2] For the elbow joint (starting from the anatomical position), the proximal end of the ulna moves around the distal end of the humerus as the forearm swings forward. The axis of this movement is a horizontal line passing through the joint from side to side. It is a transverse axis, and its position is roughly indicated by the placement of the pin at the elbow joint of a laboratory skeleton.

A joint such as the elbow, which can have movement in only one plane and has only one axis, is called a **uniaxial joint** (Figure 5–7). Some joints, such as the metacarpophalangeal joints, permit movement in two planes and are called **biaxial joints** (Figure 5–8). The fingers may move about a transverse axis (as in making a fist) and in a frontal plane (as in fanning the fingers). The latter involves an antero-posterior (A-P) axis.

Joints that permit movement in more than two planes and have more than two axes are called **multiaxial joints** (Figure 5–9). The hip joint is an example, since it permits movement about a transverse axis, an antero-posterior axis, and a longitudinal axis which extends down the length of the femur.

Synovial Joints and Their Movements

To consider the movements at synovial joints, it is helpful to classify the joints according to their axes. A further division on the basis of the shapes of the articular surfaces is also useful, because the movements possible are largely determined by these factors. Special terms are used to describe accurately each movement that can be

movements possible are many and varied, but they are readily described and analyzed by reducing them to their simplest forms.

Virtually all movement is a form of rotation. While one end of the moving bone, usually the proximal end, remains relatively fixed at the joint, the other end (the distal end) rotates around it. The joint at the fixed end

[2]When a complex movement is reduced to its simplest components, each component occurs in a single plane. A plane is defined by three points. For a simple movement such as bending the elbow, the points are (1) the position of the joint (the fulcrum), and the position of the moving part (the hand) at the (2) beginning and (3) end of the movement. The axis is always perpendicular to the plane of the movement.

made. These movements, their names, and their relations to the joint axis are described in the following paragraphs. It is important to understand the movement at each joint before trying to learn how the muscles produce these movements. Remember that all of these movements are assumed to begin from the anatomical position, unless it is stated otherwise.

Uniaxial Joints

Hinge joints. The elbow, knee, and interphalangeal joints are typical **hinge joints.** They all have a single transverse axis, and the movements about that axis are known as flexion and extension. **Flexion** is defined as a decrease in the angle formed by the two bones involved, as a "bending" of the elbow or knee. (Note that flexion at the elbow is a forward movement, while flexion at the knee is a backward movement.) **Extension** is an increase in the angle formed by these bones, "straightening" the elbow or knee. In some joints extension may go beyond the anatomical position, so that the angle is greater than 180°. When this occurs, the position is often termed *hyperextension.*

Pivot joints. A **pivot joint** is a rather special type of joint, since the only example is the radio-ulnar joint (proximal and distal ends). In the anatomical position, the radius and ulna lie parallel and the palm of the hand is forward. To turn the palm backward, the head of the radius turns in place, while the distal end of the radius moves around to the medial side of the ulna, so that the bones become crossed (see Figure 5–10). The axis for this movement runs through the head of the radius and the distal end of the ulna. In the anatomical position, the forearm is *supinated* and the movement that brings it about is **supination.** When the bones are crossed (palm back), the forearm is *pronated,* and the action is **pronation.**[3]

[3]If the forearm is flexed, pronation and supination are indicated by palm down and palm up (supplication). These terms are also applied to the whole body; lying prone is face down, and lying supine is face up.

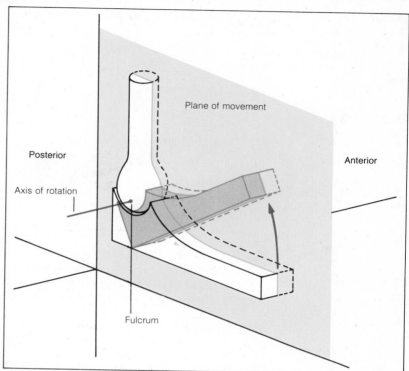

FIGURE 5–6   The relationship between the axis and plane of a movement.

The act of pronation involves movement of the distal end of the radius about a stationary ulna.

Biaxial Joints

Condyloid joints. Most biaxial joints are **condyloid joints,** with the articular surface of one bone somewhat concave and the surface of the other somewhat convex. Examples are the wrist (radiocarpal) and metacarpophalangeal joints. In addition to flexion and extension about a transverse axis, such a joint also permits movement about an antero-posterior axis. Movement away from the midline is **abduction,** and movement toward the midline is **adduction.** The midline reference is generally to the midline of the body, except in regard to the digits, where it refers to the midline of the hand or foot. Whenever it is possible to flex, extend, abduct, and adduct at a joint, it is also possible to move from one to the other without returning to the anatomical position. Such a movement is not a separate movement, but merely a com-

FIGURE 5–7
Movements of a uniaxial joint, about a transverse axis.

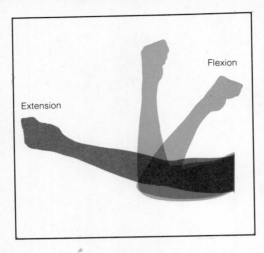

bination of others, and it is called **circumduction.** It is best illustrated by swinging the arm in a circle (although the joint involved is not a biaxial joint).

Saddle joints. The **saddle joint** is a special joint, since it is found only at the carpo-metacarpal joint of the thumb. The articular surface of each bone is concave in one direction and convex in the other (i.e., saddle-shaped). The two bones can then lock together while remaining able to slide freely. The movements are the same as for condyloid joints.

Multiaxial Joints

Ball-and-socket joints. The shape of the articular surfaces of the bones forming the **ball-and-socket joint** is aptly described by its name, as is clear from Figure 5–9C. In the human body there are only two such joints, the hip and the shoulder, but they are very important joints. In addition to the movements already described (flexion and extension about a transverse axis, abduction and adduction about an antero-posterior axis, and circumduction), there is movement about a longitudinal axis, known as medial and lateral rotation. **Medial rotation** involves rotating the limb so that the anterior surface (of femur or humerus) moves medially, and **lateral rotation** turns the anterior surface laterally, (Figure 5–9C).

Movements at the shoulder (glenohumeral joint) require some clarification because of the extreme range of movements possible. Due to the structure of the pectoral girdle, it is difficult to visualize flexion and extension as a decrease or increase in the angle formed by the participating bones. Nevertheless, flexion and extension are the terms applied to forward and backward movement of the arm. Either may be continued until the arm is raised overhead (although some rotation is also involved). There is movement of the pectoral girdle as well whenever the arm is raised above the horizontal (see below under "Special Movements"). Since it is a multiaxial joint, abduction and adduction, and medial and lateral rotation also occur at the glenohumeral joint.

FIGURE 5–8
Movements of a biaxial joint. A. Flexion and extension about a transverse axis. B. Abduction and adduction about an antero-posterior axis.

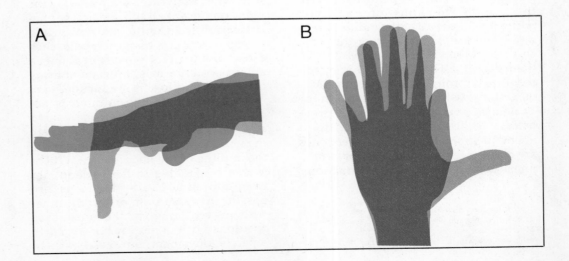

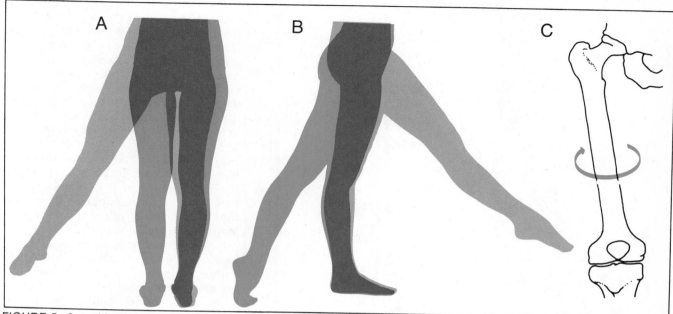

FIGURE 5–9   Movements of a multiaxial joint. A. Abduction and adduction about an antero-posterior axis. B. Flexion and extension about a transverse axis. C. About a longitudinal axis.

Gliding joints. The **gliding joints** are found between carpal bones, between tarsal bones, and in other locations where movement is limited to a slight sliding or gliding in any of several directions. The axes and specific movements are not easily defined.

Special Movements. The construction of some joints is such that the "standard" axes and movements do not fit, and special ones must be invented to describe their actions. A good example is the pectoral girdle. When one speaks of movement at the "shoulder," this is movement at the glenohumeral joint. Movements of the pectoral girdle occur chiefly at the sternoclavicular and to a lesser extent at the acromioclavicular joints. The movements are described mainly in terms of the movements of the acromion process of the scapula, and hence of the glenoid fossa as well. **Elevation** raises the acromion, as in shrugging the shoulders, and **depression** lowers it, both actions occurring about an antero-posterior axis through the sternoclavicular joint. **Protraction** moves the acromion forward, and **retraction** brings it back, as in "good" posture. This axis is roughly vertical through the sternoclavicular

joint. Protraction and retraction of the pectoral girdle causes the scapulae to move away from or toward the midline (vertebral column). For that reason they are sometimes referred to as abduction and adduction of the scapula—they are, however, also movements of the clavicle. The only exclusively scapular movement is a slight rotation in which the inferior angle moves laterally and the superior angle moves downward.

The *ankle joint* is a hinge joint (tibia-fibula with talus), but because of the way we use our foot (the anatomical position is really one of hyperextension), the usual terms of flexion and extension are not appropriate. The solution has been to use flexion for both. **Dorsiflexion** is flexion in the direction of the dorsal surface (raising the foot) and **plantar flexion** is flexion toward the plantar surface (lowering the toes; standing on tip-toe).

The movements of the foot (intertarsal and metatarsal joints) are described differently by almost every author. Since the ankle is a hinge joint, there is no abduction and adduction there. Any such movement occurs due to gliding between the bones of the foot. The result is a slight twisting such that the sole of the foot is turned medially, as **inversion,** or laterally, as **ever-**

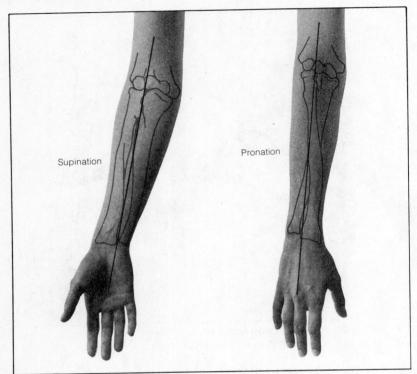

FIGURE 5-10   Supination and pronation of the forearm.

sion (Figure 5-11). The ability to swing the foot medially or laterally when seated is because a certain amount of medial and lateral rotation can occur at the knee when it is flexed.

Unique or Important Joints.   Some joints have special features of construction that enable them to serve particular functions or give them special strengths—or special weaknesses. The **temporomandibular joint,** (Figure 5-

12), the only synovial joint in the skull, is unique in that there are two joint cavities, separated by an articular disc (cartilage). Both are enclosed within a single articular capsule, and ligaments reinforce the joints laterally and medially. Their movements include depression and elevation of the mandible (opening and closing of the mouth in the manner of a hinge joint), protraction and retraction (jaw forward and backward), and a lateral rotatory motion. The latter motion is made possible by the presence of the articular discs; they have some freedom to move and thus permit the grinding type of movement necessary for mastication.

The **sacroiliac joint** is partly a synovial and partly a fibrous joint. The rather extensive articular surfaces are covered with cartilage. There is a small joint cavity, but it encloses only a part of the joint. The remainder is held firmly by interosseous ligaments, and the many additional ligaments reinforce and virtually immobilize the joint. Any movement that does occur at the sacroiliac joint is in the nature of a gliding action. The size of the articular surface, the circular arrangement of the bones of the pelvic girdle, and the ligamentous support make this a strong joint indeed. However, because of the severe stresses imposed upon it, strains of the sacroiliac joint are not uncommon.

The shape of the articular surfaces of the humerus and scapula make little contribution to the stability of the **glenohumeral joint,** even though there is a cartilage ring around the glenoid fossa to deepen the socket somewhat (see Figure 5-4b). Because of the great

FIGURE 5-11
Inversion and eversion
of the foot.

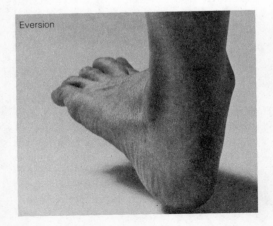

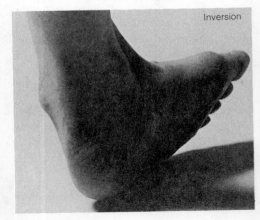

range of movement, the articular capsule must be fairly loose, but there are several ligamentous thickenings of the capsule, as well as additional ligaments. A number of muscles, particularly those that extend from the scapula to the head of the humerus, are important in maintaining the articulation, and there are several bursae around the joint. Even so, however, the head of the humerus can be displaced from its socket rather easily.

The **hip joint** (Figure 5–13) has a deep socket (acetabulum), also with a cartilage ring around the edge. The articular capsule has thickenings (ligaments), and in addition there is a ligament *(ligamentum teres)* inside the joint capsule, from the head of the femur to the fossa in the acetabulum. The joint is also held together by the cohesion resulting from the viscosity of the synovial fluid. All of this means that the hip joint is very firm, as it must be to support the body's weight.

The **knee joint** (Figure 5–14) is probably the most vulnerable and abused joint in the body. It serves as the principal weight-bearer of the body, while still permitting free movement. Although it is a hinge joint, its articular surfaces are relatively flat, unlike those of the elbow, and ligaments are almost entirely responsible for preventing movement in the wrong planes (i.e., for withstanding stresses from the side). There are two **semilunar cartilages** (*medial* and *lateral meniscus*) that lie around the edges of the medial and lateral condyles of the tibia. When the leg is flexed, a small amount of medial and lateral rotation can occur. It is possible because the menisci are able to slide a bit on the flat condyles of the tibia. These cartilages are frequently damaged or torn loose in athletic injuries and must be surgically repaired or removed. Also within the articular capsule are two **cruciate ligaments** between the tibia and femur. They are so arranged that one is taut when the leg is flexed and one when the leg is straight. The articular capsule is reinforced by very strong medial and lateral collateral ligaments on the sides, a posterior ligament, and the anterior patellar ligament. The latter is actually part of the tendon of the anterior thigh muscles and encloses the patella. There are also

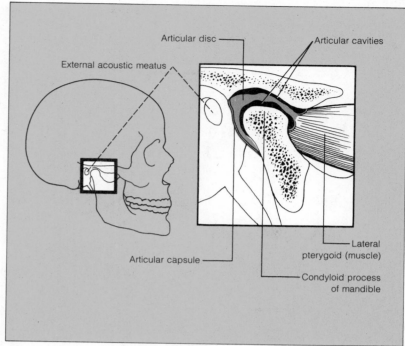

FIGURE 5–12   The temporomandibular joint.

several bursae associated with the knee joint, lying over and under the patella.

Some Joint Problems. Joints are extremely vulnerable to injury because they are capable of movement and because the bones involved are held together by relatively inelastic ligaments. The most common injuries are *dislocations* and *sprains*. A **dislocation** is the displacement of the articular surfaces of the bones, usually caused by a blow or fall or by a sudden violent contraction of the muscles crossing the joint. Dislocation is more likely to be associated with joints whose articular surfaces do not interlock in any way, and it often involves damage to the surrounding tissues or to the bones themselves. The muscles, tendons, or ligaments may be stretched or torn, or the articular cartilage displaced. Realignment of the displaced articular ends of the bones is often complicated by this damage as well as by spasm of the muscles around the joint.

Sprains are the result of twisting or wrenching a joint. The consequent strain may easily exceed the elasticity of the ligaments and tear them from their

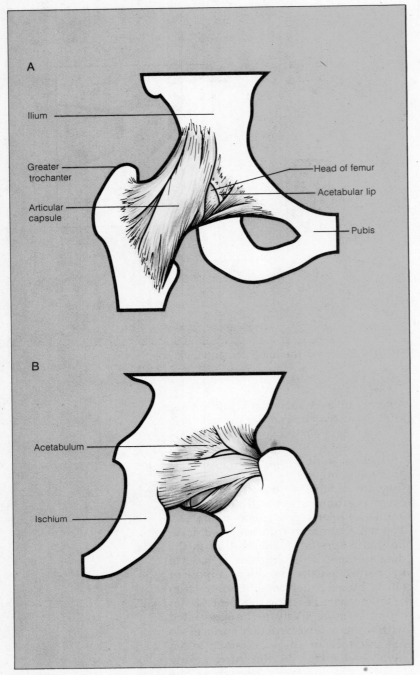

Ilium

Greater trochanter

Articular capsule

Head of femur

Acetabular lip

Pubis

Acetabulum

Ischium

FIGURE 5–13   The hip joint.

bony attachments. Like a dislocation, a sprain is painful, and there is swelling and discoloration due to the accumulation of tissue fluid and of blood from blood vessels that may have been ruptured.

Joint action is also affected by a number of diseases, most of which fit under the broad umbrella of **arthritis,** inflammation of the joint. Joint ailments may involve the bone, cartilage, or synovial membranes of the joint and result in impairment of movement, some swelling, and pain upon movement. Although a number of conditions may cause arthritis, the cause is more often not known. One form, *gout,* is due to a metabolic problem. One of the products of protein breakdown (uric acid) is formed in greater amounts than usual and, as its concentration in the blood rises, some may be deposited in the tissues. Typically, a single joint is affected, but the severe inflammation usually subsides after a few days. Because attacks often follow a period of intemperate eating or drinking, gout has long been viewed as an affliction of the wealthy.

The most common disorder, however, is not due to inflammation, but is a degenerative condition of the joint. Because weight-bearing joints, particularly the knee, hip, and spine, are subjected to stress, strain, and abuse from the time we learn to walk, they are the most likely candidates. Whenever the surface of the articular cartilage is broken, the surface becomes rough and erodes. This increases friction, and the surface erosion accelerates, so that the deterioration is progressive. Since the cartilage lacks a blood supply, regeneration is insignificant. The condition is often precipitated by injury to the joint and progresses over the years. A certain amount of degenerative joint disease is the rule in individuals of advanced age.

CHAPTER 5 SUMMARY

The structure of joints varies according to whether they are primarily to provide for strength and support, to permit movement, or a combination of all three. In all joints the bones are sep-

Focus

Artificial Knee Joints

In 1976, 30,000 human knees, mostly those of chronically disabled, elderly arthritis patients, were replaced by artificial joints in this country. These artificial knees represented more than 80 different designs, each of which relieves pain and restores function, but all of which have drawbacks. All artifical joints represent drastic, surgical solutions to problems that may one day be prevented with anti-arthritis vaccinations or cured with drugs. At present, however, artificial joints may be the only option available, even though the function they restore is limited to greater or lesser extent, and they sometimes fail. Since these artificial knee joints replace the ends of the femur and tibia and are fastened to the shafts of these leg bones by steel pins cemented into the center of the shafts, failure usually means a loosening of the bond between joint and bone. It may also mean corrosion of the metal of the joint, rejection of the joint by the patient's immune system, or, because the human knee is subject to enormous stresses, actual breakage.

A very promising artificial knee appears to be the Spherocentric knee, developed at the University of Michigan. Made of a noncorrosive alloy that does not stimulate the body's immune system, this artificial joint is so designed that it provides its wearer with virtually full freedom of movement, in all the directions of normal knee movement. Movement is limited by gradual deceleration, instead of by the abrupt impact seen in many other artificial joints. Where the two halves of the joint meet, they are covered with polyethylene, nearly as smooth and tough as the hyaline cartilage it replaces; these polyethylene bearing surfaces can be replaced fairly easily if they become worn. The joint is fastened to bone by polymethyl methacrylate cement, which provides a mechanical bond to the bone.

Comparison of a normal knee joint with the Spherocentric knee reveals that the form of the artificial joint is a near duplicate of the natural one. This mimicry means that the knee does not undergo the excessive mechanical loads of other artificial knees and is likely to last longer. It is less likely to break or loosen, and it has proved to work well in those patients on whom it has been tested. However, the Spherocentric knee is hardly the last word in aritificial joints. Work is being done with new materials, such as porous ceramics to which bone can form a living bond, with the use of electrical currents to hasten and strengthen the formation of bone-implant bonds, and with new designs. Eventually, the researchers hope to have artificial joints suitable for young, active people who put much greater stresses on their joints than do the elderly.

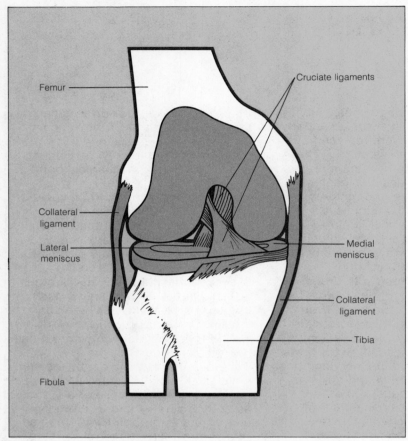

Femur

Cruciate ligaments

Collateral ligament

Lateral meniscus

Medial meniscus

Collateral ligament

Tibia

Fibula

FIGURE 5–14 The knee joint.

arated by, as well as held together by, connective tissue. Joints may be classified by the nature of this material as fibrous, cartilaginous, or synovial joints (fluid between the bones). In **fibrous joints** the bones are firmly bound by connective tissue fibers. Examples are the tightly bound *sutures* between bones of the skull, and *syndesmosis* such as the distal tibiofibular joint, where the fibers are a little longer. Sutures do not permit movement, but syndesmoses permit limited movement.

Cartilaginous joints include *synchondroses*, in which the bones are joined by hyaline cartilage, and *symphyses*, which have a pad of fibrocartilage between the bones. Joints between the pubic bones and between the vertebrae are symphyses. These are very strong joints, and although they are not

truly movable joints, they allow a certain amount of "give" or flexibility.

Synovial joints are the movable joints, although the amounts or types of movement vary greatly. The articular surfaces of the bones are covered with a thin layer of hyaline cartilage. The joint is enclosed by a fibrous *articular* or *joint capsule* lined with a *synovial membrane*. The space formed, the *articular* or *joint cavity*, is filled with *synovial fluid* secreted by the synovial membrane. The joint capsule is often reinforced by fibrous bands or *ligaments*, which limit movement.

In most movements, one end or part of a bone moves much more than the other. The bone is a lever, and rotates about a fulcrum near the fixed end of the bone. The axis of that rotation is an imaginary line through the fulcrum and perpendicular to the plane of that movement. Synovial joints may be classified as uniaxial, biaxial or multiaxial depending on whether they have movement about one, two, or more axes.

Uniaxial joints are hinge joints and pivot joints. *Hinge joints* such as the elbow or knee, have a transverse axis, and the movements are *flexion* (bending) and *extension* (straightening). *Pivot joints* are rather special joints, the radioulnar joints, in which the axis is longitudinal, extending from the head of the radius to the distal radio-ulnar junction. The movements also have special names; *supination* (palms forward) and *pronation* (palms backward).

Biaxial joints are condyloid joints and saddle joints. *Condyloid joints*, such as the wrist, are capable of flexion and extension about a transverse axis, and also movement about an antero-posterior axis. These movements are *abduction* (away from the midline) and *adduction* (toward the midline). *Saddle joints* are named for the shape of the articular surfaces. The best example is the carpometacarpal joint of the thumb. The movements are the same as other biaxial joints.

Multiaxial joints are ball-and-socket joints and gliding joints. The hip and shoulder (glenohumeral) joints are *ball-and-socket* joints. In addition to flexion and extension, and abduction and adduction, they are capable of movement about a longitudinal axis. These are *medial rotation* (anterior surface moves medially) and *lateral rotation* (anterior surface moves laterally). *Gliding joints* permit only a small amount of sliding or gliding between the bones. Specific movements and axes are not described for these joints.

The temporomandibular joint is the only synovial joint in the skull. It has two joint cavities with a small cartilage disc between them. This arrangement makes possible the lateral grinding movements of the mandible in chewing. The sacroiliac joint is part of the pelvic girdle and also the connection between the pelvic girdle and the axial skeleton. It is partly a synovial joint, and partly a fibrous joint.

The glenohumeral and hip joints, both ball-and-socket joints, illustrate the structural difference between a joint designed primarily for movement and one in which strength and stability are of prime importance. The glenohumeral joint has a very shallow socket, and the bones are held in place by ligaments and skeletal muscles. The hip joint has a deep socket, and a ligament within the socket as well as ligaments around it to provide stability and strength.

The knee joint is a hinge joint, but the articular surfaces do not interlock in any way to prevent movement in other planes. Ligaments hold the bones together and restrict the movement, and are assisted by cartilages that ring the articular surfaces of the tibia. The joint permits a wide range of movement, but also supports the weight of the body during that movement. As a result, the knee is a very vulnerable joint.

Joints are extremely vulnerable to injury because of their movement and the relative inelasticity of the ligaments. Common joint injuries are *dislocations* (displacement of articular surfaces) and *sprains* (twisting of the joint).

Joints are also susceptible to diseases, such as *arthritis* (inflammation of the joint), and to degeneration due to advancing age.

STUDY QUESTIONS

1 Describe the structure of a typical synovial joint. How is it like fibrous and cartilaginous joints? How is it different?

2 What are the structural elements of a joint that determine or limit the amount of movement possible at that joint?

3 In a synovial joint, what is the importance of the articular cartilage? of the synovial fluid?

4 How do bursae and tendon sheaths resemble synovial joints?

5 What is meant by the axis of rotation of a joint? Why is it important to understanding movement at a joint?

6 Describe the axes of rotation for movements at the hip, the elbow, and the wrist.

7 Compare and contrast the stability and range of movement of the pectoral and pelvic girdles, including the attachments of their respective extremities.

8 Since the ankle joint is a hinge joint (uniaxial), how do you explain such movements as inversion and eversion?

9 What is unique about the movements of the radio-ulnar joint?

10 What is unique about the temporomandibular joint?

11 Why is the knee joint so susceptible to injury?

6

Gross Anatomy of a Skeletal Muscle

Principles of Skeletal Muscle Action

Muscles of the Body—Their Attachments and Actions

Skeletal Muscle System

It is the skeletal muscles acting upon a system of levers formed by the bones and joints that cause movement. Muscle, which is one of the primary tissues, is characterized by the ability of its cells to change shape, that is, to contract and shorten. Skeletal muscles are attached to bones, and when the muscle contracts and shortens, the bones are caused to move. Just what that movement will be depends in part upon the movements possible at that joint, and in part upon the position of the muscle in relation to that joint. For example, if a muscle on one side of a joint causes flexion, a muscle on the opposite side of the joint would cause extension.

In this chapter we consider the anatomy of skeletal muscles, specifically their attachments and their relation to the joints they cross. This information is necessary for understanding the actions of individual muscles and their contributions to the rather complex movements of some of our daily activities.

GROSS ANATOMY OF A SKELETAL MUSCLE

The primary role of the muscles of the body is to move the parts of the skeleton, but they also make a major contribution to the shape and contour of the body. Skeletal muscle accounts for approximately 40 percent of the total body weight; it is thus the largest structural entity both by weight and by bulk. This rather formidable mass is distributed throughout the entire body as many separate and distinct organs. The individual muscles, which vary greatly in size, shape, power, and function, also differ in some of their physiological properties.

Connective Tissue Coverings

Each skeletal muscle is enclosed by a sheath of connective tissue which is part of the deep *fascia* and is continuous with the fascia of adjacent muscles, with subcutaneous connective tissue, and with the periosteum of the bone (see Figure 6–1). The fascial layer that covers a given muscle is the chiefly collagenous **epimysium.** Within the muscle, bundles of fibers are enclosed by thinner sheaths of connective tissue, the **perimysium,** and the individual muscle fibers are wrapped in very thin layers of **endomysium** that hold them together and keep them in place. The fascial sheaths carry the major nerves and blood vessels, as well as the smaller branches, to the individual muscle fibers. Each skeletal muscle cell receives a terminal nerve ending that causes it to contract. Between the muscle cells are numerous tiny blood vessels, mostly capillaries, and, of course, interstitial fluid.

Modes of Attachment

Skeletal muscles vary in their modes of attachment to bone,[1] their shapes, and the arrangements of their muscle fibers. **Fleshy attachments**

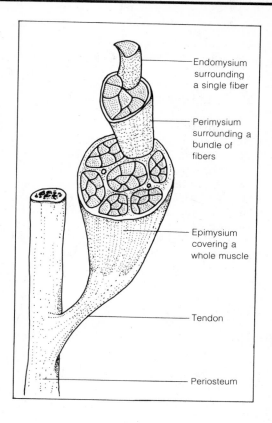

FIGURE 6–1
The connective tissue coverings of a skeletal muscle.

Endomysium surrounding a single fiber

Perimysium surrounding a bundle of fibers

Epimysium covering a whole muscle

Tendon

Periosteum

are those in which the muscle fibers arise directly from bone. They are usually fairly extensive and serve to distribute the application of force over a wide area. Fleshy attachments are generally not as strong as fibrous attachments, and they are more likely to be at the proximal end of the muscle. In **fibrous attachments** the connective tissue fibers surrounding the muscle cells converge at the end of the muscle, where some of them merge with the periosteum of the bone and others penetrate into the substance of the bone itself. Fibrous attachments, or **tendons,** are often in the form of a cord or strap, which provides a more restricted or localized connection, with all the force brought to bear at a particular point. Not only does this form a strong union, but tendinous tissue itself is very strong. It has been calculated that a tendon can support approximately five to nine tons per square inch of cross-sectional area (no human tendon is actually that large, of course). The connective tissue fibers of some tendons do not converge that much, but form a thin fibrous layer that provides a

---

[1] A few skeletal muscles do not attach to bones at both ends. Among them are some of the muscles of facial expression, which attach to skin, the extrinsic eye muscles, which move the eyeball, and the muscles of the tongue.

widespread attachment. Such a sheet-like tendon, or **aponeurosis,** is found in the lumbar region of the back.

Arrangement of Muscle Fibers

Muscle cells, which are generally known as muscle fibers, tend to lie parallel with the long axis of the muscle but there are numerous variations. As shown in Figure 6–2, some muscles are long and straplike. Some are rectangular with parallel fibers and broad attachments. In others the muscle fibers converge from a broad origin to a restricted insertion in the manner of a fan *(radiate)* In many muscles, oblique fibers run at an angle to the tendon, converging on one side *(unipennate)*, or both sides *(bipennate)* in the manner of a feather. Sometimes the fibers converge toward the end as well as the

FIGURE 6–2   Arrangements of fibers in skeletal muscles.
A. Longitudinal or parallel. B. Bipennate. C. Unipennate. D. Multipennate.
E. Radiate.

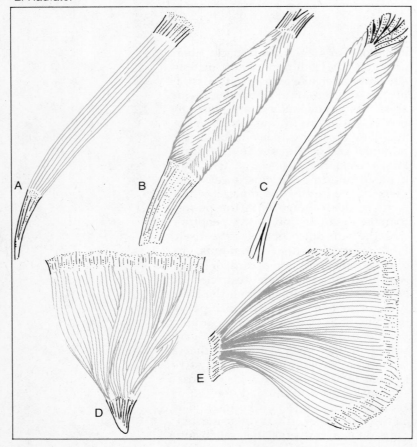

sides of the tendon *(circumpennate)*, and others have a complex interwoven arrangement of fibers *(multipennate)*. In general, pennate muscles are relatively powerful, but cannot shorten as much as the straplike muscles.

PRINCIPLES OF SKELETAL MUSCLE ACTION

Skeletal muscles may act either to produce movement of parts of the skeleton, to prevent movement, or to limit movement by steadying joints subjected to an outside force. The way in which a given skeletal muscle actually does this is largely determined by the position of the muscle and its attachments in relation to the joints involved. Its action is modified by other muscles that act upon that joint. The details of these relationships differ for each muscle, but certain general principles apply to all muscles and should be borne in mind when considering the actions of a particular muscle and its role in the production and control of a movement. Some of these considerations are discussed in the following subsections.

Functional Attachments and Joint Axes

Each muscle has at least two attachments. The proximal attachment is generally termed the **origin,** and the distal attachment the **insertion.** The origin may be quite extensive, perhaps arising in several sections or heads, or even from more than one bone. This is rarely the case with the insertion, which ordinarily has a smaller area of attachment to the bone. The origin is usually the fixed part in a movement, while the insertion is the end that is moved.[2] The origin and insertion are never on the same bone. Since movements occur at joints, a muscle must cross a joint to produce movement, and it can cause movement only at joints it crosses.

[2] These conditions may be reversed. For example, lie supine and raise your legs. The fixed ends of the muscles that produce this action are on the os coxae and lumbar vertebrae while the moving ends are on the limbs. Now, lying supine, do a sit-up. The attachments on the trunk become the moving ends of the muscles.

When a muscle contracts, it shortens or exerts tension; that is, it pulls its two ends closer together by moving (or trying to move) the insertion toward the origin. When contraction ceases, the muscle relaxes and the pull stops. A muscle can only contract and relax—it can pull and not pull, but it cannot push. While a muscle contraction may produce a movement, relaxation does not cause a return to the initial position. This requires an external force such as another muscle or gravity.

The movement that is produced by shortening a particular muscle depends upon the position of that muscle in relation to the axes of the movements at the joints it crosses (see Figure 6–3). In a uniaxial joint such as the elbow a muscle that crosses anterior to the axis

FIGURE 6–3   The movement produced at a joint depends upon the position of the muscle in relation to the axis of the movement. If the muscle is anterior to the axis, it moves the distal part forward; if it is posterior to the axis, it moves the part to the rear. Note that at the elbow an anterior muscle causes flexion, but at the knee an anterior muscle causes extension.

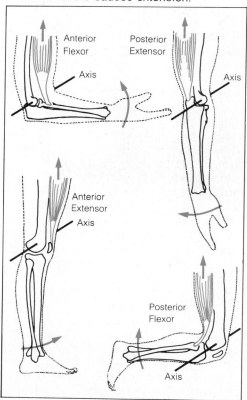

is a **flexor,** and one that crosses posterior to the axis is an **extensor.** A muscle that is medial or lateral to the axis would be expected to cause adduction or abduction, but these movements do not occur at the elbow joint; there are no muscles that cross this joint in this manner. There are ligaments that do, however, and they help to prevent such movement. The same statements apply to the knee, except that the flexors are posterior to the axis and the extensors are anterior to it.

The principle is exactly the same for multiaxial joints, but the position of the muscle must be considered in relation to each of the axes for that joint. Muscles often cross a multiaxial joint in such a way that they can exert force about each axis and produce a complex movement. Furthermore, their action may vary with the position of the part. For example, a muscle that goes from the anterior chest wall to the humerus (pectoralis major) could flex at the shoulder when the arm is extended posteriorly, but it could extend when the arm is flexed at the shoulder.

Levers and Muscle Action

The roles of bones, joints, and muscles in producing a movement are those of the parts of a lever system (see Figure 6–4). They can be illustrated by a seesaw, which rotates about its point of support, its fulcrum. A child on one end is a weight, or load, balanced by a downward force applied to the other end. The amount of force required to lift the child depends upon the distances of the child (the load) and of the force from the fulcrum. Every youngster knows that to seesaw successfully with a heavier friend, either he or she must move back or the friend must move forward. The relationship may be expressed as follows:

$$D_L \times L = D_F \times F$$

where L is the load, F the force, and D_L and D_F the distances from the fulcrum of the load and the force, respectively. When the load is very heavy, the fulcrum should be located so that the load arm, D_L, is very short and the force arm, D_F, is very long. In this way a

FIGURE 6-4　Three classes of levers, showing the mechanical relations of the load (L), force (F), and fulcrum (f), with examples from everyday life and in the body.

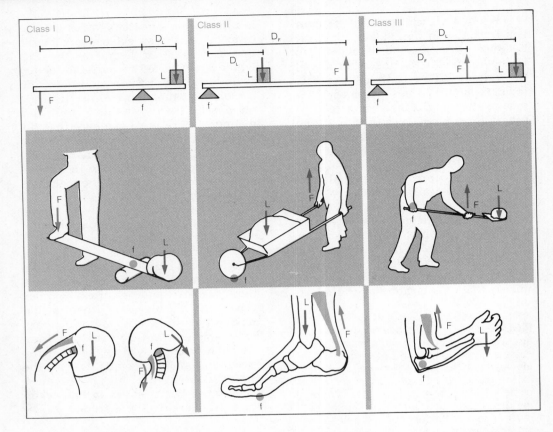

relatively weak force can lift a much heavier load (as in using a crowbar).

The fulcrum need not be central; it may be at one end with either the load or the force in between. All three types of lever occur in the body, but the most common is the one with the fulcrum at one end, the load at the other, and the force in between. The quantitative relationship between the distances from the fulcrum of the load and the force is the same in all cases. If the load arm is ten times longer than the force arm, the force of contraction would have to be ten times greater than the weight of the load in order to lift it. A muscle in such a position sacrifices power (for it must develop a force ten times the weight of the load), but it gains in range and speed of movement. While the muscle shortens one inch, the load moves ten inches and in exactly the same period of time, so the load moves ten times faster. For fine precision movements (such as writing), however, a lower ratio of part movement to muscle shortening is desirable.

In summary, a long force arm favors greater strength, decreased range of movement, and finer control of tiny movements. A short force arm permits a wide range of movement, but at the sacrifice of power, so that a larger more powerful muscle would be required to do the job than would be needed with a long force arm. One might add that certain practical considerations also apply. A long force arm in the forearm would be cumbersome and very inconvenient. The placements of the muscles about a joint and the specific locations of the attachments therefore determine not only their actions, but also their mechanical advantages in terms of strength, range, speed, and precision of movement.

Roles of Muscles

Muscles work in groups rather than individually. Several muscles cross a joint to produce a given action; when that movement is called for, all of these muscles participate in proportion to their strength and the mechanical effectiveness of their position. The muscles that produce the movement in question are called **prime movers** (or **agonists**). Muscles that produce the opposite movement are **antagonists.** Antagonists are stretched when the prime movers shorten, and they should relax when the prime movers contract. When you chin yourself, the flexors of your forearms are the prime movers and the extensors of your forearms are the antagonists. The roles of the muscles, however, can change. For example, when you do push-ups, your forearm extensors are prime movers and your flexors are antagonists.

Synergists are muscles that contribute indirectly to the movement. Their action may be either of two types, one of which is to prevent movement at an intermediate joint. When a muscle that crosses two joints contracts, it causes movement at both of the joints it crosses. Another muscle that crosses one of these joints can act synergistically as an antagonist at that joint. By contracting it can prevent movement there and ensure that all the force of the prime mover is exerted at the other joint. For example, many of the muscles involved in flexion of the fingers also cross the wrist. When you make a fist without wrist flexion, muscles that extend at the wrist but do not cross the phalangeal joints contract to immobilize the wrist and prevent the finger flexors from causing flexion of the hand. These muscles are synergists because they stabilize an intermediate joint.

A **fixator** is a special kind of synergist. Its role is to stabilize or fix one end of the muscle so that all of the force it develops is exerted at the other end. The best examples are the muscles that move the arm. Most of their origins are on the scapula, but that bone is so freely movable that their contraction would be as likely to move the scapula as the humerus. If these muscles are to be effective, the scapula must be held steady or fixed in position when they contract. The fixators are the muscles that run from the axial skeleton to the scapula and stabilize it so that when the muscles of the arm contract their insertions move instead of their origins.

Although the prime mover produces a movement and determines certain qualities of that movement, the action of antagonists is necessary for the smooth, well-coordinated movements that normally occur. Antagonists ordinarily relax when the prime mover contracts and, by so doing, decrease the resistance to the prime mover. Failure of the antagonist to relax necessitates a stronger contraction by the prime mover if the desired movement is to occur. By relaxing partially, or by terminating its relaxation, the antagonist can provide a resistance that serves to regulate and control the movement, to stop it, or to prevent an "overshoot." This type of action—a slight contraction that resists stretching of a muscle—is a very important muscle action.

Movements are often produced by gravity, however, rather than by prime movers. If you sit down slowly, for instance, the prime movers are not the flexors of the thigh and leg ; gravity is. The extensors of the thigh and leg contract and prevent sudden collapse into the chair. Certain muscle groups are acting to oppose gravity and to maintain a particular posture at all times, except when the body is completely supported as when lying down. Such muscle groups are called *antigravity muscles.* They show some functional adaptations that permit sustained activity, since gravity never relaxes. The importance of antigravity muscles therefore lies not in the movements they can produce, but in the movements they can prevent. They function primarily as antagonists.

MUSCLES OF THE BODY— THEIR ATTACHMENTS AND ACTIONS

Upon undertaking the study of skeletal muscles and their actions, you are faced with a formidable list of names of muscles. However, nearly every name tells something about the muscle and can be of some use in identifying it. It may suggest the *shape* of the muscle (trapezius, rhomboid); the *location* (tibialis anterior); the sites of its attachments (sternocleidomastoid); the

FIGURE 6–5
Anterior view of the muscles of the body.

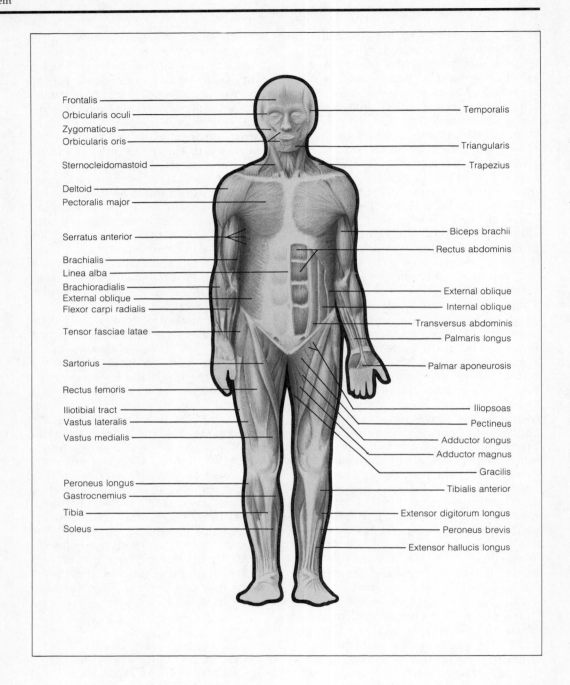

Frontalis
Orbicularis oculi
Zygomaticus
Orbicularis oris
Sternocleidomastoid
Deltoid
Pectoralis major
Serratus anterior
Brachialis
Linea alba
Brachioradialis
External oblique
Flexor carpi radialis
Tensor fasciae latae
Sartorius
Rectus femoris
Iliotibial tract
Vastus lateralis
Vastus medialis
Peroneus longus
Gastrocnemius
Tibia
Soleus

Temporalis
Triangularis
Trapezius
Biceps brachii
Rectus abdominis
External oblique
Internal oblique
Transversus abdominis
Palmaris longus
Palmar aponeurosis
Iliopsoas
Pectineus
Adductor longus
Adductor magnus
Gracilis
Tibialis anterior
Extensor digitorum longus
Peroneus brevis
Extensor hallucis longus

number of heads of origin (quadriceps femoris, biceps brachii); its *action* (flexor digitorum, adductor magnus); or some other feature (semitendinosus).

Since the action of a muscle or group of muscles depends upon the line of pull, i.e., the position of the muscle(s) in relation to the axis of movement, the specific details of muscle attachment are important only insofar as they determine this relationship.

The following discussion treats muscle attachments only in a general way. It tries to point out patterns of arrangements and actions and to explain relationships that may be important, confusing, or merely interesting. Information about specific attachments is presented in the accompanying muscle tables, which should be used for reference rather than memorized. For most muscles that have simple actions (that

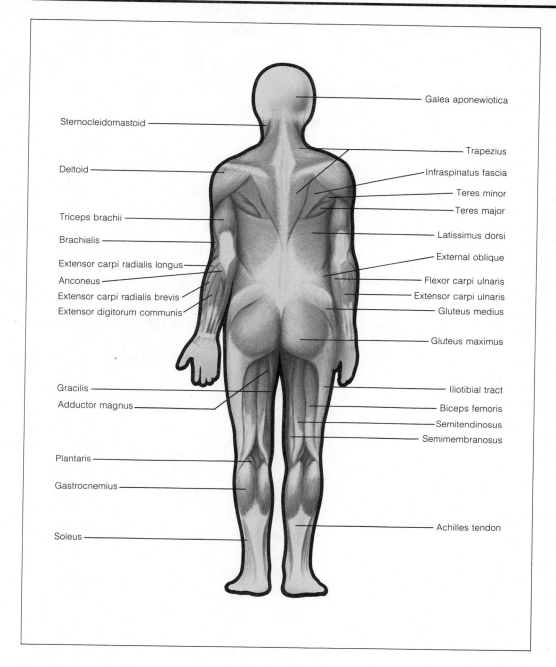

Sternocleidomastoid

Deltoid

Triceps brachii

Brachialis

Extensor carpi radialis longus

Anconeus

Extensor carpi radialis brevis

Extensor digitorum communis

Gracilis

Adductor magnus

Plantaris

Gastrocnemius

Soleus

Galea aponewiotica

Trapezius

Infraspinatus fascia

Teres minor

Teres major

Latissimus dorsi

External oblique

Flexor carpi ulnaris

Extensor carpi ulnaris

Gluteus medius

Gluteus maximus

Iliotibial tract

Biceps femoris

Semitendinosus

Semimembranosus

Achilles tendon

FIGURE 6–6
Posterior view of the muscles of the body.

is, cross one uniaxial joint), it is easy to determine the action by studying an illustration of the muscle (provided the characteristics of the bones and joints are borne well in mind). For muscles whose actions are more complex, reference to the skeleton is helpful. A piece of string or rubber tubing stretched from origin to insertion establishes the approximate line of pull and can be used to simulate the movement. When a skeleton is not available, your own body or that of a colleague can serve very well.

The locations of the major superficial muscles of the human body are shown in Figures 6–5 and 6–6. These muscles help determine the body contour, and many can be easily identified on the body surface. When studying muscles you should locate as many as possible on yourself or a colleague.

TABLE 6-1   MUSCLES OF FACIAL EXPRESSION

Muscle	Origin	Insertion	Action
Epicranius frontalis	Cranial aponeurosis	Skin, eyebrows, and root of nose	Raises eyebrows, wrinkles skin of forehead
occipitalis	Occipital bone and mastoid process	Cranial aponeurosis	Moves skin of scalp
Orbicularis oculi	Bones (frontal and maxillary) and ligaments around orbit	Fibers encircle orbit, cross eyelid, and interdigitate with one another	Closes eyelid, compresses lacrimal sac, and pulls eyebrow down
Orbicularis oris	No true origin; fibers of other oral muscles blend with it	Surrounds mouth and attaches chiefly to mucosa and skin at angles of mouth	Closes mouth, purses lips
Levator labii superioris	By three heads from zygomatic bone, infraorbital margin of maxillae, and root of nose	Skin and muscles of upper lip	Raises upper lip and wing of nose
Zygomaticus major minor	Zygomatic bone	Skin and mucosa at angle of mouth	Raises angle of mouth
Risorius	Skin and fascia over parotid gland	Skin and mucosa at angle of mouth	Draws angle of mouth laterally
Depressor labii inferioris	Base of mandible on either side of midline	Skin and mucosa of lower lip	Draws lower lip down
Depressor anguli oris	Body of mandible below incisor teeth	Skin and mucosa at angle of mouth	Draws angle of mouth down and laterally; draws up skin of chin, protrudes lip
Buccinator	Mandible and maxillae in region near molars	Skin and muscles around angle of mouth	Draws corner of mouth laterally; pulls lips against teeth and flattens cheek; used in mastication, swallowing, whistling
Platysma	Fascia of pectoralis major and deltoid	Lower border of mandible and skin around angle of mouth	Wrinkles skin of neck, draws angle of mouth downward and back, depresses mandible

Muscles of the Axial Skeleton

Muscles of the Head and Neck Some of the muscles of the head and neck belong to groups that serve very different and often specialized functions, such as those that move the tongue and the eyes.

The **muscles of facial expression** (Table 6-1 and Figure 6-7) differ from most skeletal muscles because their insertions and some of their origins are on skin and connective tissue rather than bone. As a result, their action is to move the skin and underlying tissues. The eyes and mouth are each surrounded by a circular sphincterlike muscle. The *orbicularis oculi* around the eye controls the eyelid and closes the eye, while the *orbicularis oris* encircles the mouth, a highly versatile orifice. A number of facial muscles converge into the orbicularis oris, particularly at the corners of the mouth. A cheek muscle, the *buccinator,* passes back from the angle of the mouth and is sometimes called the trumpeter's muscle because it serves to compress the cheek against the teeth; it is used more frequently, though, in manipulating food during mastication. Other muscles move the skin of the chin, the nose, and the forehead.

There is a very important group of small muscles, the *extrinsic muscles of the eye,* that move the eyes to determine the direction one is looking. They are

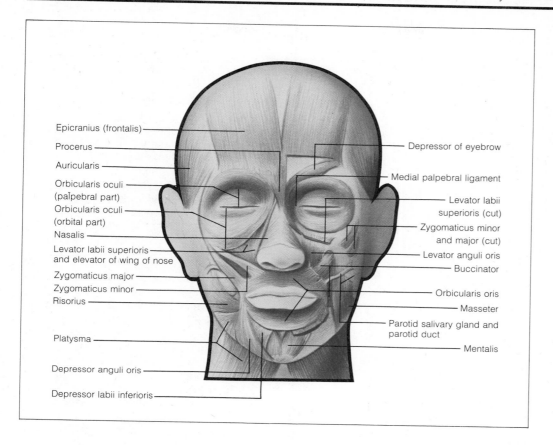

FIGURE 6-7
Muscles of facial expression.

Epicranius (frontalis)

Procerus

Auricularis

Orbicularis oculi (palpebral part)

Orbicularis oculi (orbital part)

Nasalis

Levator labii superioris and elevator of wing of nose

Zygomaticus major

Zygomaticus minor

Risorius

Platysma

Depressor anguli oris

Depressor labii inferioris

Depressor of eyebrow

Medial palpebral ligament

Levator labii superioris (cut)

Zygomaticus minor and major (cut)

Levator anguli oris

Buccinator

Orbicularis oris

Masseter

Parotid salivary gland and parotid duct

Mentalis

paired, with six muscles for each eye. Their origins are on the bony portions at the back of the orbit, and their insertions are on the eyeball. These muscles are described in Chapter 11.

The **muscles of mastication** (Table 6–2 and Figure 6–8) extend from the skull to the mandible, and their contraction closes the jaw. The largest is the *temporalis,* which arises from the temporal bone and passes under the zygomatic arch to reach the coronoid process of the mandible. Its leverage is such that it closes the jaw very forcefully. The *masseter* covers the ramus of the mandible and can be readily felt when the teeth are clenched. The two *pterygoid muscles,* medial and lateral, are located on the medial (internal) surface of the mandible, where they arise from the pterygoid plates of the sphenoid bone. They are used in side-to-side grinding movements of the jaw.

The **muscles of the tongue** include extrinsic and intrinsic muscles. The latter are entirely within the tongue and are interwoven with one another, run-

ning in all directions. They determine the shape and position of the tongue in speaking and swallowing. The extrinsic muscles of the tongue provide an anchor for it, connecting it to the hyoid bone, the mandible, and the skull (the styloid process). The *suprahyoid* and *infrahyoid muscles* (Figure 6–9), above and below the hyoid bone, elevate and depress that bone. Since the hyoid is bound to the larynx by ligaments, the larynx is also moved, as occurs in swallowing. When the hyoid bone is prevented from moving, these muscles can lower the mandible. (See Table 6–3, under "Anterior Muscles of the Neck.")

Beneath the mucous membrane lining the posterior portion of the oral cavity and pharynx (throat) is a complex interlacing of skeletal muscles (Figure 6–10) whose contraction causes specific changes in the shape of the pharynx and the openings into and out of it. They function chiefly in swallowing, which is discussed more completely in Chapter 23. The movements normally occur in a sequence that serves to direct

TABLE 6-2   MUSCLES OF MASTICATION

Muscle	Origin	Insertion	Action
Temporalis	Temporal fossa	Coronoid process and ramus of mandible	Closes jaw (elevates mandible)
Masseter	Zygomatic bone and zygomatic arch	Ramus of mandible, lateral surface, near angle of mandible	Closes jaw (elevates mandible)
Medial pterygoid	Medial surface of lateral pterygoid plate of sphenoid bone, palatine bone, and tuberosity of maxillae	Medial aspect of mandible	Elevates and protracts mandible; with lateral pterygoid, aids in grinding movements
Lateral pterygoid	Greater wing of sphenoid and lateral surface of lateral pterygoid plate	Condyle of mandible	Protrudes jaw; with medial pterygoid, aids in grinding movements

a mass of material from the mouth into the esophagus.

The **muscles that move the head** (Table 6-3) belong to two groups. Some are *superficial* and attach to non-axial structures, primarily the pectoral girdle. Others are *deep* and originate from the vertebral column, either anteriorly or posteriorly.

The superficial neck muscles (Figure 6-9) include the trapezius, sternocleidomastoid, and suprahyoid and infrahyoid muscles. The hyoid muscles acting together help flex the head and neck. The *trapezius* is a muscle of the upper extremity whose midline origin extends up as far as the occipital bone. Its insertion is on the pectoral girdle and its main action is on the scapula (Table 6-7), but when the scapula is fixed the trapezius can exert its action upon the skull and cervical vertebrae. The *sternocleidomastoid* extends from the sternum and clavicle to the mastoid process and is readily apparent on the side of the neck when the head is turned. Its actions are complicated by the fact that it crosses so many joints. Since the line of pull passes behind the atlanto-occipital joint (first vertebra and skull), the action there is extension. However, because the sternocleidomastoid passes in front of most of the cervical vertebrae, it produces flexion in the neck. Contraction of the muscle on one side will abduct the head and turn it to the opposite side.

The deep anterior muscles, for the most part, extend from the anterior surface to the first two or three vertebrae to

FIGURE 6-8   Muscles of mastication.

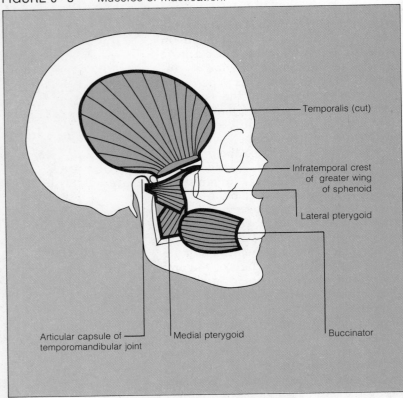

TABLE 6–3   MUSCLES THAT MOVE THE HEAD

Muscle	Origin	Insertion	Action
Lateral Muscles of the Neck			
Sternocleidomastoid	Manubrium and superior medial portion of clavicle	Mastoid process of temporal bone	Singly—abducts head and neck, rotates head to opposite side. Both—extend at atlanto-occipital joint (elevate chin), flex cervical vertebrae, or, with head fixed, elevate sternum and thus rib cage
Scalenes: anterior, medial, and posterior	Transverse processes of second to fifth cervical vertebrae	Laterally and posteriorly on first and second rib	Elevates first ribs (aids in inspiration); flexes and slightly rotates neck
Anterior Muscles of the Neck			
Suprahyoid muscles			
Stylohyoid	Styloid process, temporal bone	Hyoid bone	Pulls hyoid bone up and back
Digastric	Posterior belly—mastoid process Anterior belly—lower border of mandible	Hyoid bone, lateral tips	Elevates hyoid
Mylohyoid	Mylohyoid line of mandible (medial surface)	Hyoid bone and median raphe	Elevates hyoid bone and tongue
Geniohyoid	Inferior mental spine of mandible	Body of hyoid	Moves hyoid and tongue forward
Infrahyoid muscles			As a group, they depress larynx and hyoid, but with mandible fixed, they can flex skull (with aid of suprahyoids)
Omohyoid	Superior border of scapula	Hyoid	
Sternohyoid	Manubrium and medial end of clavicle	Hyoid	
Sternothyroid	Manubrium	Thyroid cartilage	
Thyrohyoid	Thyroid cartilage of larynx	Hyoid	
Longus		Occipital bone (capitis); bodies of upper cervical and transverse processes of fifth and sixth cervical vertebrae	Singly—abduct and rotate head and neck. Both—flex head and cervical region
capitis	Transverse processes of third to sixth cervical vertebrae		
colli	Bodies of first three cervical vertebrae		
Posterior Muscles of the Neck (and Vertebral Column)			
Trapezius—see Table 6–7			
Suboccipital Muscles			
Rectus capitis posterior major	Atlas and axis	Occipital bone	Extend and rotate head
Rectus capitis posterior minor			
Obliquus capitis superior			
Obliquus capitis inferior			

the occipital bone (anterior to the foramen magnum). They tend to flex when acting together and abduct when acting individually. The deep muscles on the posterior surface are more numerous and generally are larger and stronger. The center of gravity of the head is anterior to its line of connection to the vertebral column, so that continuous action of the posterior muscles is necessary to hold the head upright. The deepest of these muscles are analogous to those on the anterior surface; they run from the first and second vertebrae to the occipital bone. They are covered superficially by muscles that extend from the lower cervical vertebrae to the occipital bone and mastoid process. The posterior muscles are separated in the midline by a highly developed ligamentous band, known as the *ligamentum nuchae,* connecting the spinous processes and extending to the skull. It contains more elastic tissue than most ligaments and is particularly well developed in quadrupeds, such as the horse, cow, and dog, where it plays an important role in support of the head.

Muscles of the Vertebral Column

The muscles that move the vertebral column (Table 6–4 and Figure 6–11) are more numerous on the posterior surface and, as with the head, the upright posture requires that gravity be resisted. Some of the anterior muscles have both attachments on the vertebrae and some have one attachment on the ribs or pectoral girdle. The latter include some of the muscles that act upon the head. Other muscles of this group include the more laterally placed *scalene* muscles, inserting on the first two ribs antero-laterally. In the lumbar region, the *psoas major* (the portion of iliopsoas that arises from the lumbar vertebrae) flexes the lumbar spine. This muscle's major role is at the hip joint, however, and it is tabulated with those muscles

FIGURE 6–9   Side view of the superficial neck muscles.

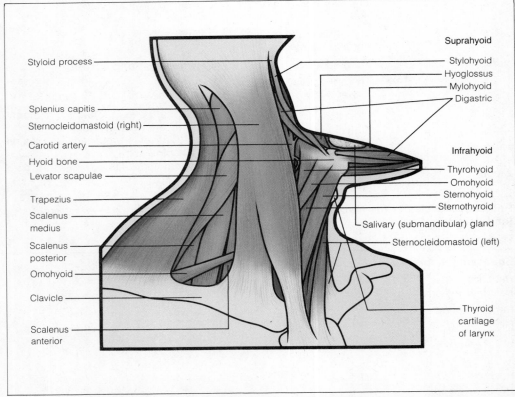

in Table 6–10. The muscles that flex the thoracic and lumbar regions of the vertebral column are muscles of the thorax and abdomen and do not attach directly to the vertebral column (see Tables 6–5 and 6–6).

The posterior vertebral muscles are divided into superficial and deep muscles, and the superficial group may be further divided into medial and lateral portions. They are all covered by the large back muscles associated with the upper extremity. The more superficial of the posterior vertebral muscles insert several segments above their origins and act over several joints, while the deepest ones insert on the segment immediately above their origins. The major muscles may have several components, as lumbar, dorsal (thoracic), cervical, and capitis (head).

The *erector spinae* is the major superficial back muscle. It has a lateral portion *(iliocostalis)* and a medial portion *(longissimus)*, both of which run superiorly and laterally, from one rib to the rib about six segments above (e.g., rib 12 to 6, rib 11 to 5, rib 10 to 4, etc.). In addition there is a more medial component *(spinalis)* that connects the spinous processes, also skipping about six vertebrae.

The deep muscles of the back are all medial, consisting of muscles that extend superiorly and medially. Some run from transverse process to spine, skipping five or six segments *(semispinalis)* or three segments *(multifidus)*. Others run between adjacent transverse processes *(rotatores)* or between adjacent spinous processes *(interspinales)*. All of these muscles extend the back and, acting unilaterally, abduct to that side; many have some action in rotation as well.

Muscles of the Thorax—Respiration

The muscles attached to the thorax (Table 6–5) include muscle groups that act upon the upper extremity and upon the vertebral column, but the chief muscles of the thorax are those of respiration. Their action is to aid in inspiration and expiration by increasing or decreasing the size (capacity) of the thoracic cavity.

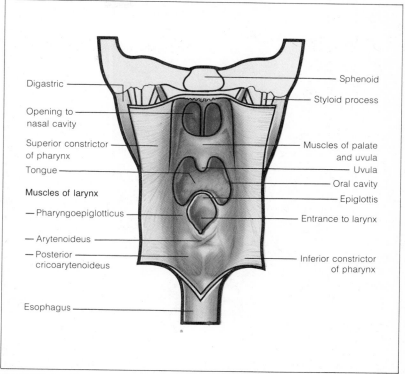

FIGURE 6–10   Muscles of the region of the pharynx and larynx, as seen from behind with the pharynx cut open.

The single most important muscle used in quiet breathing is the *diaphragm,* a dome-shaped muscle separating the thoracic cavity from the abdominal cavity (see Figure 6–23, also 16–7). It originates around the lower edge of the rib cage and inserts into the central tendon, a flat fascialike region at the top of the dome—the muscle virtually inserts into itself! When the diaphragm contracts, it pulls the top of the dome downward, thereby increasing the length of the thoracic cavity, while the abdominal muscles relax to accommodate the displacement of the abdominal organs. When the diaphragm relaxes, it returns upward to its resting position, assisted by the pressure exerted by the now contracting abdominal muscles.

The articulations of the ribs with the vertebrae permit some rotation, and since the ribs swing diagonally forward and down, a slight rotation raises the anterior ends and increases the diame-

TABLE 6–4   MUSCLES THAT MOVE THE VERTEBRAL COLUMN

Muscle	Origin	Insertion	Action
Superficial Muscles of the Back			
Splenius			Singly—abducts and rotates. Both—extend head and cervical region
capitis	Lower half of ligamentum nuchae and transverse processes of seventh cervical and first four thoracic vertebrae	Mastoid process and occipital bone	
cervicis	Spinous processes of third to sixth thoracic vertebrae	Transverse processes of first three cervical vertebrae	
Erector spinae			

iliocostalis
   lumborum
   thoracis
   cervicis
	Laterally—angle of rib and/or pelvic girdle	Angle of rib or cervical vertebrae six segments above	Extends vertebral column

longissimus
   thoracis
   cervicis
   capitis
	Medially—transverse processes of lumbar, thoracic, and cervical vertebrae	Transverse processes of vertebrae six segments above (or mastoid process for capitis)	All extend and abduct the vertebral column, but only capitis acts on the head

spinalis
   thoracis
   cervicis
	Medially—spinous processes of thoracic and lowest cervical vertebrae	Spinous processes six segments above	Extend vertebral column

Deep Muscles of the Back			

Semispinalis
   thoracis
   cervicis
   capitis
	Transverse processes of thoracic and seventh cervical vertebrae	Spinous processes six segments above, and occipital bone (capitis)	Extend and rotate vertebral column and head (capitis)

Multifidus	Pelvic girdle, lumbar vertebrae, transverse processes of thoracic and lower cervical vertebrae	Spinous processes three segments above	Extends vertebral column; each side may abduct and rotate
Rotatores	Transverse processes of all vertebrae	Spinous processes of vertebrae above	Extend and rotate vertebral column
Interspinales	Spinous processes, especially in cervical and lumbar regions	Spinous processes of vertebrae above	Extend
Intertransversarii	Transverse processes, especially cervical and lumbar regions	Transverse processes of vertebrae above	Extend and abduct

ter of the thoracic cavity. The spaces between adjacent ribs are occupied chiefly by two sets of intercostal muscles, whose fibers run at right angles to one another. The more superficial *external intercostals* are somewhat shorter when the thorax is expanded, so they are said to aid inspiration. The *internal intercostals* are somewhat shorter when the thorax is depressed, and they are said to aid in expiration, although this has been disputed (see Figure 6–12, also 19–13.) Other muscles, which are deeper or on the internal surface, also help to change the diameter of the thorax. Several muscles in the neck region (scalenes, sternocleidomastoid) can aid in inspiration by raising the first ribs. In quiet breathing, expiration occurs without assistance, due to the natural recoil of the rib cage and lungs and the pull of gravity. The importance of the neck muscles in normal quiet breathing is therefore somewhat debatable, since most people do not change the diameter of the thoracic cavity very much when breathing quietly. There is no question, however, but that all of these muscles have some role in forced respiration, as during vigorous exercise or when breathing against a resistance (sucking or blowing a wind instrument).

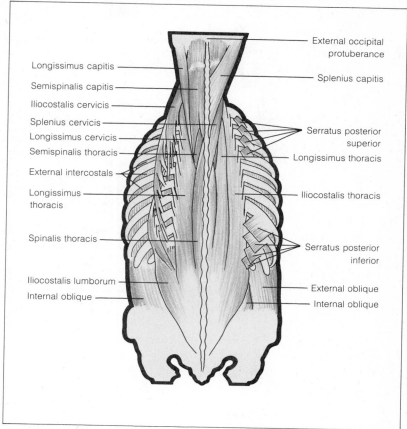

FIGURE 6–11   Superficial layers of intrinsic muscles of the back.

Muscles of the Abdominal Wall

The abdominal muscles (Table 6–6 and Figure 6–12) form the major portion of the body wall with virtually no skeletal assistance. They provide support and protection to the abdominal organs and aid the diaphragm in respiration. They flex the vertebral column and depress the ribs. Their contraction raises the intra-abdominal pressure, which is necessary for defecation, and contributes to the stability of the thorax, for straining during heavy lifting. When raising the legs from a supine position, these muscles are important fixators or stabilizers of the pelvis, from which the hip flexors originate.

Fibers of the most superficial abdominal muscle, the *external oblique,* run downward and medially as a continuation of the external intercostals. Beneath it is the *internal oblique,* whose fibers run at right angles to it, corresponding to the internal intercostals. In the lower part of the abdominal wall, however, fibers of the internal oblique tend to fan out and run more or less parallel with those of the external oblique. Beneath the obliques is the *transversus abdominus,* whose fibers run horizontally, encircling the abdominal cavity like a girdle. The fibers of these three muscles all terminate several centimeters lateral to the midline, but their fascial layers continue medially to interlace at the midline. The fourth abdominal muscle, the *rectus abdominus,* extends from the pubis to the rib cage on either side of the midline. For most of its length it is ensheathed by the fascia of the internal oblique. The peritoneum, which lines the entire abdominal cavity, lies deep to the abdominal muscles.

TABLE 6-5   MUSCLES OF THE THORAX—RESPIRATION

Muscle	Origin	Insertion	Action
Scalenes—see Table 6-3			
Sternocleidomastoid—see Table 6-3			
External intercostal	Inferior edge of rib and costal cartilage	Superior border rib below	Elevates ribs, aids inspiration
Internal intercostal	Inner surface of rib	Superior border rib below	Draws ribs together to aid in respiration
Diaphragm	Internal surface xiphoid, lower ribs and costal cartilage, and lumbar vertebrae; i.e., around the inferior border of the rib cage	Central tendon	Lowers (flattens) dome, thereby increasing size of thoracic cavity and causing inspiration
Serratus posterior inferior	Spinous processes of last two thoracic vertebrae and first three lumbar vertebrae	Inferior border of ninth to twelfth ribs	Depresses lower ribs
Serratus posterior superior	Ligamentum nuchae and spinous processes of seventh cervical to third thoracic vertebrae	Upper border of second to fifth ribs	Raises upper ribs, expands thoracic cavity

As support for the abdominal contents, the fascial layers (Figure 6-13) are almost as important as the muscles. The superficial fascia contains a layer that is prone to store fat (*panniculus adiposus*). The deeper fascial layers have complex relations with the abdominal muscles, most of which arise and terminate within them; they thus serve as fibrous origins and tendinous insertions. At their inferior edges the fascial sheaths blend to form the *inguinal ligament* (Figure 6-12), running from the anterior superior iliac spine to the pubic tubercle. The fascial sheaths of the muscles from the two sides fuse in the midline to form a tough fibrous band, the *linea alba*, into which most of the abdominal muscles are considered to insert.

Muscles of the Pelvic Floor and Perineum These muscles occupy the area of the *perineum,* or pelvic outlet, at the bottom of the true pelvis (Figure 6-14). They support the pelvic organs and provide passage for the anal canal, urethra, and, in the female, the vagina. The two major muscles originate laterally and insert into connective tissue along the midline and into the walls of the three canals. Anteriorly, the *levator ani* is the important muscle, with the canals passing through it. The *coccygeus* occupies the posterior portion of the outlet, between the coccyx and the ischia. Several smaller muscles, including those around the orifices of the canals, may also, among other things, exert some control over these openings.

Muscles of the Upper Extremity

Muscles That Act upon the Scapula The muscles of the scapula (Table 6-7 and Figure 6-15) originate from the axial skeleton, namely the ribs and vertebrae. Although contraction of any of these muscles can move the scapula and pectoral girdle, their major role is that of fixator, or stabilizer, of the scapula. Recall that the scapula does not articulate directly with the axial skeleton, but is able to move freely over the

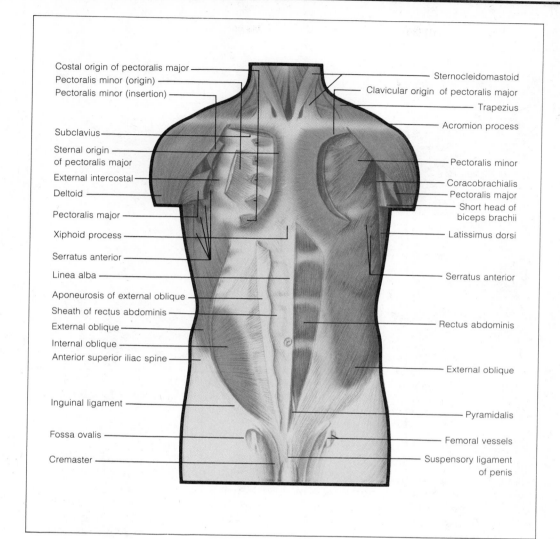

FIGURE 6–12
Muscles of the chest
and abdomen.

Costal origin of pectoralis major
Pectoralis minor (origin)
Pectoralis minor (insertion)

Subclavius
Sternal origin
of pectoralis major
External intercostal
Deltoid
Pectoralis major
Xiphoid process
Serratus anterior
Linea alba
Aponeurosis of external oblique
Sheath of rectus abdominis
External oblique
Internal oblique
Anterior superior iliac spine

Inguinal ligament
Fossa ovalis
Cremaster

Sternocleidomastoid
Clavicular origin of pectoralis major
Trapezius
Acromion process
Pectoralis minor
Coracobrachialis
Pectoralis major
Short head of
biceps brachii
Latissimus dorsi
Serratus anterior
Rectus abdominis
External oblique
Pyramidalis
Femoral vessels
Suspensory ligament
of penis

TABLE 6–6   MUSCLES OF THE ABDOMINAL WALL

Muscle	Origin	Insertion	Action
External oblique	Anterior inferior edge of last eight ribs	Linea alba and iliac crest	Depresses ribs, flexes vertebral column; compresses abdominal contents
Internal oblique	Lumbodorsal fascia, iliac crest	Inferior border last three ribs, linea alba, pubic crest	Depresses ribs, flexes vertebral column; compresses abdominal contents
Transverse abdominus	Inguinal ligament, iliac crest, lumbodorsal fascia, last five ribs	Linea alba and pubic crest	Compresses abdominal contents
Rectus abdominus	Pubic crest	Xiphoid process and costal cartilages fifth to seventh ribs	Depresses ribs, increases abdominal pressure, flexes vertebral column and/or pelvis
Quadratus lumborum (posterior wall)	Lower edge twelfth rib	Iliac crest	Flexes lumbar portion vertebral column; depresses twelfth rib

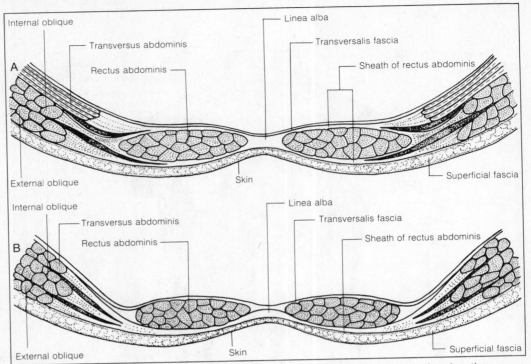

FIGURE 6–13   Schematic cross section of the anterior abdominal wall, showing the relations of the abdominal muscles and their fascial sheaths. A. At the level of the umbilicus. B. Below the level of the anterior superior iliac spine.

posterior surface of the rib cage. Since most of the muscles that move the arm (at the glenohumeral joint) originate on the scapula, this immobilizing action of the scapular muscles is very important.

The *trapezius* is the large superficial muscle on the upper part of the back. It extends from the midline of the body to the lateral portion of the pectoral girdle, where it inserts along the spine of the scapula, the acromion, and the lateral part of the clavicle. Its obvious action is retraction of the scapula. Due to its extensive origin, some fibers have a line of pull that elevates the lateral angle of the scapula (shrugging the shoulders), and some of the inferior fibers have a less effective depressing action. With the scapula fixed, contraction of the superior fibers can also cause extension of the head.

Beneath the trapezius are the *rhomboids* (major and minor), extending from the vertebrae diagonally downward to the medial border of the scapu-

la. They help hold the scapula against the rib cage and cause retraction (Figure 6–16).

The *serratus anterior*, with its serrated slips of origin from the anterolateral surface of the ribs, passes between the rib cage and the scapula to insert on the medial border of the scapula with the rhomboids. These two muscles have virtually the same line of pull, but in opposite directions. They act together as antagonists in positioning the scapula and holding it against the rib cage (Figure 6–12).

The *pectoralis minor*, a smaller muscle on the anterior surface of the thorax, inserts on the coracoid process of the scapula. It protracts the scapula and depresses the lateral angle. With these four muscles, the scapula and clavicle can be moved or fixed in nearly any position.

Muscles That Act upon the Humerus Of the many muscles that cross the glenohumeral joint, most originate on the scapula and insert on the humerus. However, two important muscles, one anterior and one posterior, bypass the scapula and extend from the axial skeleton to the humerus, thus crossing several joints (see Table 6–8). These muscles indirectly influence the pectoral girdle and its movements as well as those of the arm.

The *latissimus dorsi* is the broad flat muscle with extensive superficial origins in the lower back. Its fibers converge to pass below the inferior angle of the scapula (with a few fibers sometimes arising from it) and form the posterior border of the axilla (armpit); they then insert into the anterior surface of the humerus. The chief action of this muscle is extension of the arm, particu-

larly when the arm is flexed. The latissimus dorsi is the most powerful "pulling muscle" (used, for example, in rowing). Because it wraps around the humerus medially to reach its anterior insertion, it also rotates the arm medially (Figures 6–15 and 6–16).

The *pectoralis major*, the large chest muscle, takes its origin from the clavicle and sternum and the fascia along the inferior edge of the rib cage; it therefore covers the pectoralis minor. Its fibers pass laterally to the humerus to form the anterior border of the axilla and insert slightly lateral to the insertion of the latissimus. Its main action is adduction, to draw the arm across the chest, and it can do this from any starting point, meaning that it may either flex or extend the arm, as well as adduct.

FIGURE 6–14   Muscles of the pelvic floor (perineum), as seen from below. A. Female. B. Male.

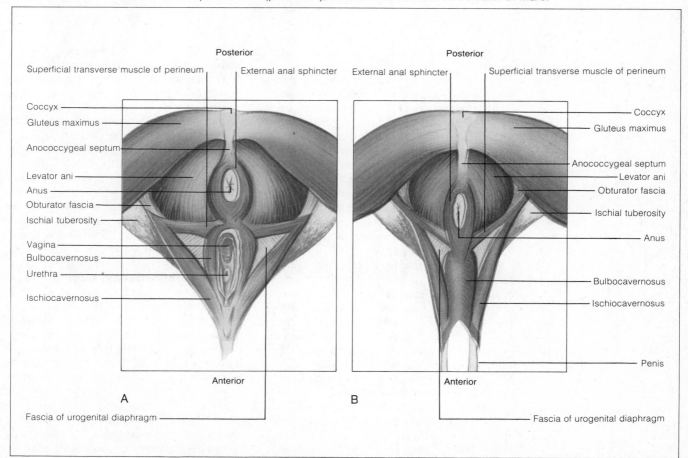

TABLE 6-7   MUSCLES THAT ACT UPON THE SCAPULA

Muscle	Origin	Insertion	Action
Trapezius	Occipital bone, midline of ligamentum nuchae, spines of thoracic and last cervical vertebrae	Lateral third of clavicle, acromion, and spinous process of scapula	Extends head; retracts scapula, elevates (upper fibers) or depresses (lower fibers) tip of shoulder; stabilizes scapula
Levator scapulae	Transverse processes of first four cervical vertebrae	Medial border of scapula near superior angle	Elevates scapula
Rhomboids major minor	Spinous processes last cervical and first four thoracic vertebrae	Medial border scapula	Elevate, retract scapulae; rotate scapulae to depress acromion
Serratus anterior	Lateral surface of first nine ribs	Medial border scapula	Protracts scapula; stabilizes and helps to hold it against rib cage
Pectoralis minor	Anterior surface second to fifth ribs	Coracoid process of scapula	Protracts scapula; depresses lateral angles and aids in inspiration

Like the latissimus, it is also a medial rotator.

The *deltoid* is the large triangular muscle that caps the shoulder (Figures 6–15 through 6–18). It originates on the clavicle and scapula along the edge of the trapezius' insertion, and inserts midway down the lateral surface of the humerus. Its obvious action is abduction. However, since some fibers pass in front and some in back of the joint, they may flex an extended arm or extend a flexed arm, as well as medially or laterally rotate (anterior and posterior fibers, respectively).

Several muscles reach from the scapula to the humerus, inserting fairly close to its head (Figures 6–17 and 6–18). These muscles, aside from the movements they cause, help hold the head of the humerus in place. Three of them—the *subscapularis*, the *supraspinatus*, and the *infraspinatus*—are named for the scapular fossae from which they arise. The subscapularis lies on the costal surface between the scapula and the ribs (actually between the scapula and the serratus anterior), and inserts on the anterior surface of the humerus.

The supraspinatus reaches over the top of the joint to the greater tubercle of the humerus, and the infraspinatus passes posterior to the joint axis, inserting just behind the supraspinatus. The supraspinatus is in a better position to initiate abduction than the deltoid, but it is a less powerful muscle. The subscapularis and infraspinatus rotate the humerus medially and laterally, respectively.

Two other scapular muscles, the *teres major* and the *teres minor*, originate along the lateral border of the scapula. The teres major is larger and longer, coming from the region of the inferior angle. Its fibers parallel those of the latissimus dorsi and insert adjacent to them on the humerus. Its actions on the arm are therefore the same as those of the latissimus. The teres minor originates adjacent to the infraspinatus and inserts on the greater tubercle of the humerus just inferior to it. The teres muscles act like a set of reins, since one inserts on the anterior surface and one on the posterior surface of the humerus to rotate it medially or laterally. (It might be interesting at this point to

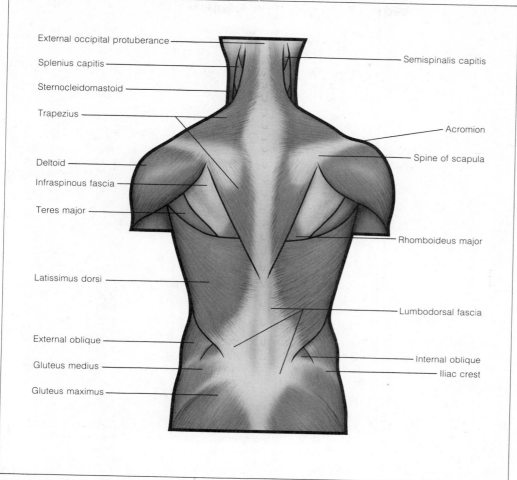

External occipital protuberance

Splenius capitis

Sternocleidomastoid

Trapezius

Deltoid

Infraspinous fascia

Teres major

Latissimus dorsi

External oblique

Gluteus medius

Gluteus maximus

Semispinalis capitis

Acromion

Spine of scapula

Rhomboideus major

Lumbodorsal fascia

Internal oblique

Iliac crest

FIGURE 6–15   Superficial muscles of the back and shoulder.

turn to Figure 19–11A and see how many of the muscles you can identify.)

Muscles That Act upon the Forearm   The muscles that move the forearm (Table 6–9 and Figures 6–17 and 6–18) produce action at the elbow and at the radio-ulnar joint (pronation and supination). They are located in both the arm and the forearm, but the muscles in the arm are more powerful.

The *triceps brachii* of the arm is alone on the posterior surface of the humerus and its insertion into the olecranon process of the ulna makes it the extensor of the elbow. Two of its three heads arise from the humerus, but the third (the long head) arises on the scapula below the glenoid fossa; its action at the shoulder, however, is minimal. The *brachialis* on the anterior surface is a major flexor of the forearm, extending from the distal half of the humerus to the ulna. The *biceps brachii* covers the brachialis and although it is the most familiar muscle of the arm, it has no attachments on the humerus. The short head of the biceps arises from the coracoid process of the scapula; the tendon of the long head runs in the intertubercular (bicipital) groove of the humerus, over the top of the joint to the scapula just above the glenoid fossa. This tendon passes through the articular capsule and contributes to keeping the

TABLE 6–8   MUSCLES THAT ACT UPON THE HUMERUS

Muscle	Origin	Insertion	Action
Axial Muscles			
Pectoralis major	Clavicle, sternum, costal cartilages of second to sixth rib, aponeurosis of external oblique	Lateral margin of intertubercular groove of humerus	Adducts and medially rotates arm; slight flexion
Latissimus dorsi	Spinous processes of lower six thoracic and the lumbar vertebrae, sacrum, and iliac crest by way of lumbodorsal fascia	Medial margin of intertubercular groove of humerus	Extends, medially rotates arm, also adducts; tends to retract and depress scapula
Scapular Muscles			
Deltoid	Lateral third of clavicle, acromion, and spinous process of scapula	Deltoid tuberosity of humerus (lateral surface, middle third of shaft)	Abducts arm; anterior fibers flex and medially rotate; posterior fibers extend and laterally rotate
Supraspinatus	Supraspinous fossa of scapula	Greater tubercle of humerus	Initiates abduction of humerus; stabilizes head of humerus
Infraspinatus	Infraspinous fossa of scapula	Greater tubercle of humerus, posterior to supraspinatus	Laterally rotates; stabilizes head of humerus
Teres minor	Lateral border of scapula	Greater tubercle of humerus, posterior to infraspinatus	Laterally rotates; stabilizes head of humerus
Teres major	Dorsal surface of inferior angle of scapula	Crest of lesser tuberosity of humerus (above latissimus dorsi)	Adducts, extends, and medially rotates (similar to latissimus dorsi)
Subscapularis	Subscapular fossa	Lesser tubercle of humerus	Medially rotates; stabilizes head of humerus
Coracobrachialis	Coracoid process of scapula	Medial surface, midshaft of humerus	Flexes and adducts humerus
Biceps brachii } Triceps brachii	These muscles exert their major action at the elbow and radio-ulnar joints and are listed in Table 6–9		

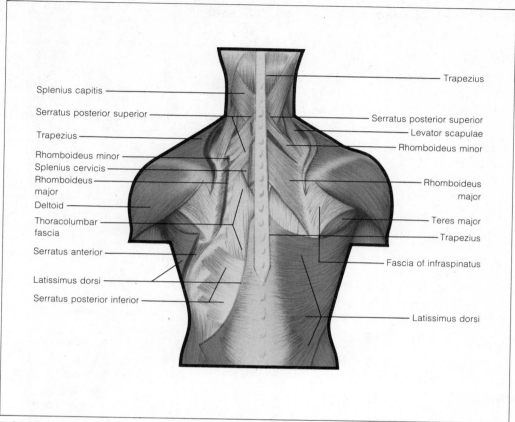

Splenius capitis

Serratus posterior superior

Trapezius

Rhomboideus minor

Splenius cervicis

Rhomboideus major

Deltoid

Thoracolumbar fascia

Serratus anterior

Latissimus dorsi

Serratus posterior inferior

Trapezius

Serratus posterior superior

Levator scapulae

Rhomboideus minor

Rhomboideus major

Teres major

Trapezius

Fascia of infraspinatus

Latissimus dorsi

FIGURE. 6–16   Deep muscles of the back and shoulder.

head of the humerus in position (see Figure 5–4B). The biceps crosses the elbow and flexes there, but its insertion on the radial tuberosity is such that it also acts upon the radio-ulnar joint. In the anatomical position (supination), the biceps is in a good position to flex the forearm. However, when the forearm is pronated, the radius, and the radial tuberosity with the biceps insertion, is rotated almost 180°. From this position the biceps is a strong supinator, but a weak flexor.

Most of the muscles in the forearm cross both elbow and wrist and act at both of these joints, as well as, possibly, at the radio-ulnar joint. These muscles may be divided into an antero-median group and a postero-lateral group. The two groups are separated medially by the superficial edge of the ulna and laterally by the *brachioradialis muscle*,

which runs along the lateral surface of the forearm from the humerus to the distal end of the radius.

The **antero-median group** (Figure 6–19) for the most part, originates from the medial epicondyle of the humerus and flexes at the elbow. Some of the muscles insert on the radius, some on the carpals or metacarpals, and some on the phalanges. In a general way, they are flexors (where flexion is possible) at all the joints crossed, including those of the wrist and fingers. The muscles in this group have been subdivided into superficial and deep muscles; for example, there are two sets of flexors to the fingers, the *flexor digitorum superficialis* and the *flexor digitorum profundus*. Those muscles that run diagonally to the radius or radial side of the wrist pronate as well as flex.

Most of the muscles of the **postero-**

TABLE 6-9   MUSCLES THAT ACT UPON THE FOREARM, HAND, AND FINGERS

Muscle	Origin	Insertion	Action
Muscles of the Arm			
Biceps brachii	Long head—superior margin glenoid fossa; short head—coracoid process	Tuberosity of radius and deep fascia of forearm	Supinates; or flexes the supinated forearm; long head stabilizes head of humerus
Triceps brachii	Long head—inferior margin glenoid fossa; short heads—lateral and posterior surfaces of shaft of humerus	Olecranon process of ulna	Extends forearm; also slightly extends and adducts humerus and stabilizes head of humerus
Brachialis	Anterior surface of distal portion of humerus	Coronoid process of ulna	Flexes forearm
Muscles of Forearm			
Brachioradialis	Supracondylar ridge of humerus	Styloid process of radius	Flexes forearm
Antero-median Group of the Forearm			
Pronator teres	Medial epicondyle of humerus and coronoid process of ulna	Midshaft of radius	Pronates and flexes forearm
Pronator quadratus	Distal end of shaft of ulna	Distal end of radius	Pronates forearm
Flexor carpi radialis	Medial epicondyle of humerus	Base of second metacarpal	Flexes and abducts hand; flexes and pronates forearm
Flexor carpi ulnaris	Medial epicondyle of humerus and olecranon process of ulna	Base of fifth metacarpal	Flexes and adducts hand; flexes and pronates forearm
Flexor digitorum superficialis	Medial epicondyle of humerus, medial ulna, and anterior border of radius	Middle phalanges of second to fifth fingers	Flexes middle phalanges; flexes forearm and hand
Flexor digitorum profundus	Anterior surface of ulna and interosseous membrane	Base of distal phalanges of second to fifth fingers	Flexes hand and distal phalanges
Flexor pollicis longus	Anterior surface of radius and medial epicondyle of humerus	Terminal phalanx of thumb	Flexes hand and phalanges of thumb
Palmaris longus	Medial epicondyle of humerus	Palmar aponeurosis	Flexes forearm and hand and tenses aponeurosis
Postero-lateral Group of the Forearm			
Supinator	Lateral epicondyle of humerus	Proximal shaft of radius	Supinates forearm
Extensor carpi radialis			Extends and abducts hand and extends forearm
longus	Lateral supracondylar ridge	Base of second metacarpal	
brevis	Lateral epicondyle of humerus	Base of third metacarpal	
Extensor carpi ulnaris	Lateral epicondyle of humerus	Base of fifth metacarpal	Extends and adducts hand
Extensor digitorum communis	Lateral epicondyle of humerus	Phalanges of second to fifth fingers	Extends fingers and hand
Extensor pollicis longus	Dorsal shaft of ulna	Distal phalanx of thumb	Extends thumb

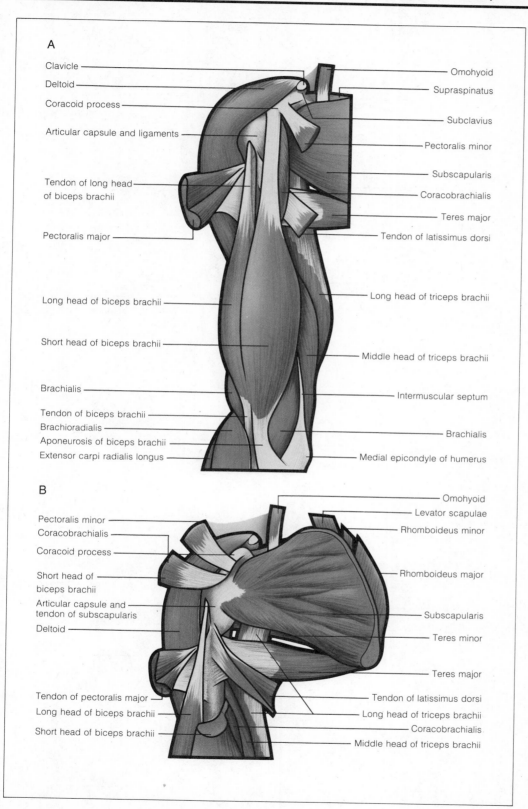

A

Clavicle

Deltoid

Coracoid process

Articular capsule and ligaments

Tendon of long head of biceps brachii

Pectoralis major

Long head of biceps brachii

Short head of biceps brachii

Brachialis

Tendon of biceps brachii
Brachioradialis
Aponeurosis of biceps brachii
Extensor carpi radialis longus

Omohyoid

Supraspinatus

Subclavius

Pectoralis minor

Subscapularis

Coracobrachialis

Teres major

Tendon of latissimus dorsi

Long head of triceps brachii

Middle head of triceps brachii

Intermuscular septum

Brachialis

Medial epicondyle of humerus

B

Pectoralis minor
Coracobrachialis
Coracoid process

Short head of biceps brachii

Articular capsule and tendon of subscapularis

Deltoid

Tendon of pectoralis major
Long head of biceps brachii
Short head of biceps brachii

Omohyoid
Levator scapulae
Rhomboideus minor

Rhomboideus major

Subscapularis

Teres minor

Teres major

Tendon of latissimus dorsi
Long head of triceps brachii
Coracobrachialis
Middle head of triceps brachii

FIGURE 6–17
A. Superficial anterior muscles of the shoulder and arm. B. Deep muscles of the shoulder and the costal surface of the scapula.

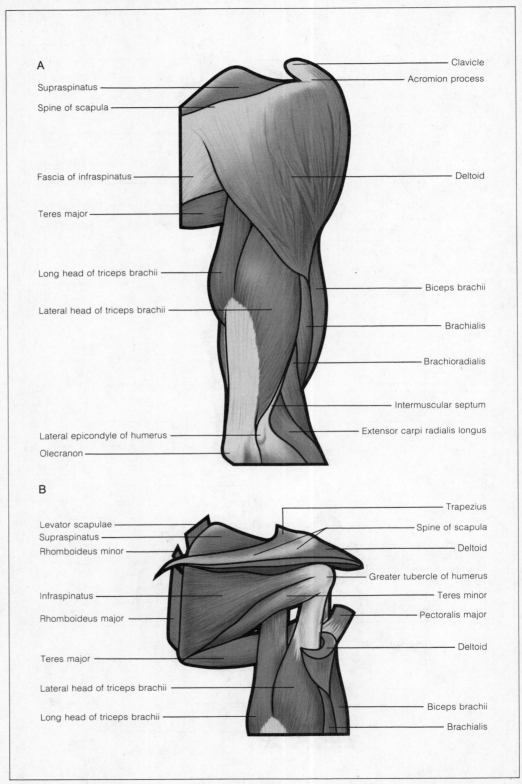

FIGURE 6–18   A. Superficial posterior and lateral muscles of the shoulder and arm.
B. Deep muscles of the shoulder and the posterior surface of the scapula.

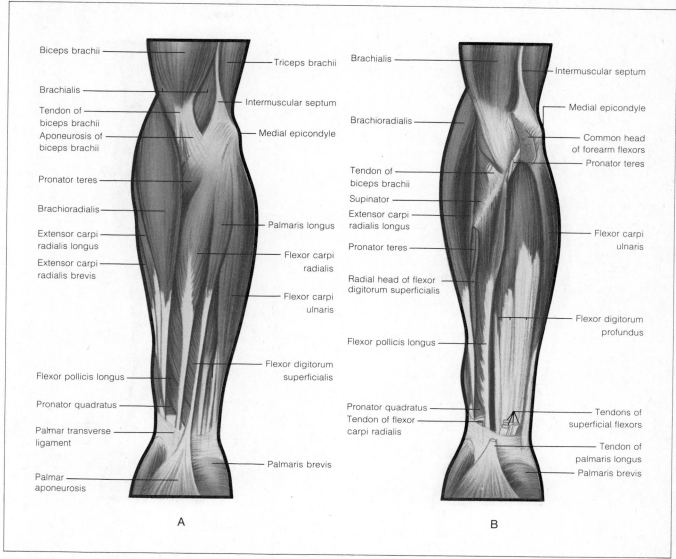

Biceps brachii

Brachialis

Tendon of
biceps brachii

Aponeurosis of
biceps brachii

Pronator teres

Brachioradialis

Extensor carpi
radialis longus

Extensor carpi
radialis brevis

Flexor pollicis longus

Pronator quadratus

Palmar transverse
ligament

Palmar
aponeurosis

Triceps brachii

Intermuscular septum

Medial epicondyle

Palmaris longus

Flexor carpi
radialis

Flexor carpi
ulnaris

Flexor digitorum
superficialis

Palmaris brevis

A

Brachialis

Brachioradialis

Tendon of
biceps brachii

Supinator

Extensor carpi
radialis longus

Pronator teres

Radial head of flexor
digitorum superficialis

Flexor pollicis longus

Pronator quadratus
Tendon of flexor
carpi radialis

Intermuscular septum

Medial epicondyle

Common head
of forearm flexors

Pronator teres

Flexor carpi
ulnaris

Flexor digitorum
profundus

Tendons of
superficial flexors

Tendon of
palmaris longus

Palmaris brevis

B

FIGURE 6–19   Anterior muscles of the forearm. A. Superficial muscles. B. Deep muscles.

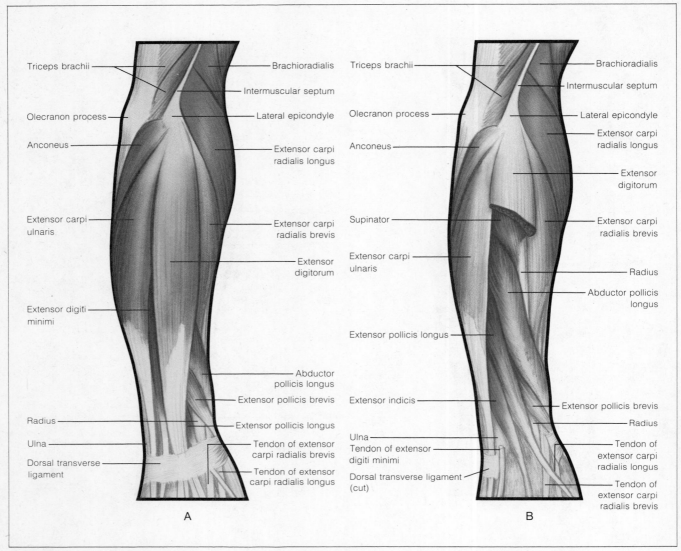

Triceps brachii
Olecranon process
Anconeus
Extensor carpi ulnaris
Extensor digiti minimi
Radius
Ulna
Dorsal transverse ligament

Brachioradialis
Intermuscular septum
Lateral epicondyle
Extensor carpi radialis longus
Extensor carpi radialis brevis
Extensor digitorum
Abductor pollicis longus
Extensor pollicis brevis
Extensor pollicis longus
Tendon of extensor carpi radialis brevis
Tendon of extensor carpi radialis longus

A

Triceps brachii
Olecranon process
Anconeus
Supinator
Extensor carpi ulnaris
Extensor pollicis longus
Extensor indicis
Ulna
Tendon of extensor digiti minimi
Dorsal transverse ligament (cut)

Brachioradialis
Intermuscular septum
Lateral epicondyle
Extensor carpi radialis longus
Extensor digitorum
Extensor carpi radialis brevis
Radius
Abductor pollicis longus
Extensor pollicis brevis
Radius
Tendon of extensor carpi radialis longus
Tendon of extensor carpi radialis brevis

B

FIGURE 6–20   Posterior and lateral muscles of the forearm. A. Superficial muscles. B. Deep muscles.

lateral group (Figure 6–20) originate from the lateral epicondyle of the humerus and are extensors at the elbow. There are not as many muscles in this group, but their pattern is similar to that of the antero-median group, since they too spread somewhat to insert on the phalanges, the metacarpals, and/or the radius. Their actions are extension at those joints at which extension occurs, and supination where they insert on the radius or radial side of the wrist.

Of special interest are four muscles, two from each group, that insert on the carpals or metacarpals (Figure 6–21). Of each pair, one inserts on the radial side and one on the ulnar (medial) side. They are the *flexor carpi radialis,* the *flexor carpi ulnaris,* the *extensor carpi radialis* (longus and brevis), and the *extensor carpi ulnaris.* Their arrangement provides four "corner muscles" whose contraction in different combinations can produce varied movements. The two antero-median muscles flex the hand at the wrist, the two postero-lateral muscles extend, the two on the ulnar side adduct, and the two on the radial side abduct. In order to make a fist, the antero-median flexors of the fingers contract, but they cross several joints and are unable to limit their action to the fingers, so other muscles must prevent flexion at the wrist. The extensor corner muscles act as synergists by stabilizing the intermediate wrist joint.

Muscles That Act upon the Hand and Fingers   The forearm muscles that cross the wrist move the whole hand, but there are also muscles that act upon parts of the hand, chiefly the fingers (see Table 6–9). Among them are those powerful forearm muscles whose long tendons insert on the phalanges. The finger flexors and extensors terminate in four tendons, one each to the second, third, fourth, and fifth digits. The thumb is moved by similar but separate muscles, the *pollicis muscles.*

The intrinsic muscles of the hand (Figure 6–22) are those with both origin and insertion in the hand. They can be divided into those of the thumb, those of the fifth finger, and the deep

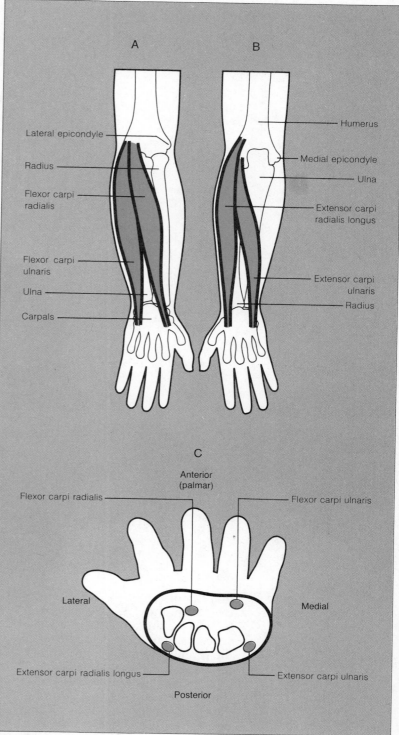

FIGURE 6–21   The "corner" muscles of the forearm. A. Postero-lateral muscles. B. Antero-median muscles. C. Cross section through the carpal bones, showing location of the tendons of the "corner" muscles.

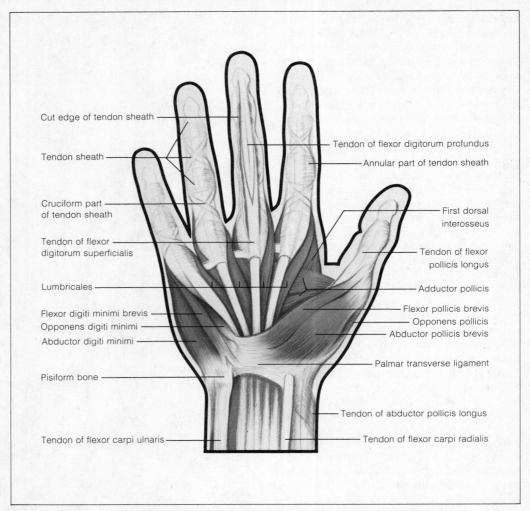

Cut edge of tendon sheath

Tendon sheath

Cruciform part of tendon sheath

Tendon of flexor digitorum superficialis

Lumbricales

Flexor digiti minimi brevis

Opponens digiti minimi

Abductor digiti minimi

Pisiform bone

Tendon of flexor carpi ulnaris

Tendon of flexor digitorum profundus

Annular part of tendon sheath

First dorsal interosseus

Tendon of flexor pollicis longus

Adductor pollicis

Flexor pollicis brevis

Opponens pollicis

Abductor pollicis brevis

Palmar transverse ligament

Tendon of abductor pollicis longus

Tendon of flexor carpi radialis

FIGURE 6–22 Intrinsic muscles of the hand. Note also the digital tendon sheaths and palmar transverse ligament. The palmar aponeurosis and palmaris longus muscles have been removed.

muscles of the hand. The thumb muscles form the *thenar eminence* at the base of the thumb. They provide for the special and independent movements of the thumb, including opposition (as in touching the thumb to the tip of the little finger). The little finger also has some special muscles, forming the *hypothenar eminence*, although they are not as well developed as those of the thumb. Between the metacarpal bones and around the metacarpophalangeal joints are several small muscles, the *interossei* and the *lumbricales*.

The varied uses of the hand require great flexibility, precise control, and a minimum of bulk; yet there must also be adequate protection for the structures in it. The intrinsic muscles of the hand are small and provide for fine and precise movements of the fingers. Gross movements are produced by the forearm muscles, represented in the hand only by their tendons. Movements of the individual joints of the fingers are possible because of the unique arrangement of the tendons of the muscles that insert on the digits. The movement of the long tendons, which slide several millimeters with flexion and extension of the fingers, is facilitated by the tendon sheaths surrounding them at the wrist and by individual digital tendon sheaths in the fingers. The tendons are held down at the wrist by the **palmar** and **dorsal transverse ligaments** and by similar ligamentous bands at the phalangeal joints.

The blood vessels and nerves in the palm of the hand are covered by a tough sheet or pad of fibrous connective tissue, the **palmar aponeurosis.** This aponeurosis, and a similar one on the sole of the foot, are much thicker than other aponeuroses, and there is a fair amount of adipose tissue embedded in it. The palmar aponeurosis is more or less continuous with the deep fascia of the hand. One of the forearm muscles (the *palmaris longus*) actually inserts into it.

Muscles of the Lower Extremity

Since movement of the pelvic girdle, unlike that of the pectoral girdle, is limited to a slight gliding and rotation at the sacroiliac joint, no special group of muscles is necessary to control or stabilize it. The other joints of the lower extremity, however, do move freely, and two-joint muscles are almost the rule. Several muscles originate on the pelvis and insert not on the femur but on the tibia and/or fibula. Others arise from the distal end of the femur and insert on the bones of the foot. It should be noted that the designation of "fixed" or "movable" attachments of muscles is often interchanged in the lower extremity. When the femur is moved as in walking, the femoral attachment moves, but when the femur remains stationary, as in bending over, it is the pelvic attachment that moves.

Muscles That Act upon the Femur
The *iliopsoas* (see Table 6–10 and Figure 6–23) is the major flexor at the hip. It is formed by two muscles, the *psoas major* and the *iliacus,* which have a common insertion on the lesser trochanter of the femur. The psoas major arises from the axial skeleton while the iliacus arises from the iliac fossa. The muscle lies beneath the femoral nerve

and vessels as they pass under the inguinal ligament. The iliopsoas is a powerful flexor at the hip, either flexing the thigh to raise the whole leg, or flexing the pelvis to bend over. Since the lesser trochanter is on the posterior medial surface of the femur, one might expect the muscle to rotate the thigh laterally as well. But because of the relation of its insertion to the longitudinal axis of the femur, there are long-standing differences of opinion about whether it causes lateral or medial rotation. In view of this, it probably is safe to assume that the iliopsoas does not have a strong action in rotation of the thigh.

The buttocks are shaped by the massive gluteal muscles (Figure 6–24). The largest and most superficial, the *gluteus maximus,* crosses the hip joint in such a way that it can exert action about each of the axes of that joint. Its major action is to extend and laterally rotate the thigh, but the upper fibers aid abduction while the lower fibers may contribute to adduction of the thigh. The gluteus maximus is a powerful extensor at the hip, but it is most effective when the hip is flexed; it is important, for example, in climbing stairs and in running, but not in walking. Beneath it are the *gluteus medius* and the *gluteus minimus,* whose attachments give them a line of pull causing abduction and medial rotation of the thigh.

The only other muscle capable of effective medial rotation, the *tensor fascia lata,* is also an abductor. This muscle is enclosed between two layers of the fascia of the thigh and actually inserts into the fascia so that contraction puts tension upon it. This fascia, called the **fascia lata** (broad fascia), invests all of the muscles of the thigh. However, it is thicker on the lateral aspect, where it extends as a tendinous band from the crest of the ilium toward the knee. This part of the fascia lata is called the **iliotibial tract,** and it is this band upon which the tensor fascia lata exerts its pull. The iliotibial tract thus functions as a long tendon.

On the posterior surface of the hip joint are six small deep muscles extending laterally from various parts of the os coxae and sacrum to the posterior surface of the femur. They are primarily

TABLE 6–10   MUSCLES OF THE HIP THAT ACT UPON THE FEMUR

Muscles	Origin	Insertion	Action
Anterior Muscles			
Iliopsoas		Lesser trochanter of femur	Flex thigh, or may flex pelvis (trunk) on femur
psoas major	Transverse processes and intervertebral cartilage of lumbar vertebrae		
iliacus	Iliac fossa and crest		
Posterior and Lateral Muscles			
Gluteus maximus	Gluteal line of ilium, sacrum, and coccyx	Gluteal tuberosity of femur and iliotibial tract	Extends and laterally rotates thigh
Gluteus medius	Posterior surface of ilium, between posterior and anterior gluteal lines	Greater trochanter of femur	Abducts thigh; anterior fibers medially rotate
Gluteus minimus	External surface ilium, between anterior and inferior gluteal lines	Greater trochanter of femur	Abducts thigh; anterior fibers medially rotate
Tensor fascia lata	Anterior portion iliac crest	Iliotibial band of fascia lata	Flexes, abducts, and medially rotates thigh
Lateral rotators			All these muscles are primarily lateral rotators of the thigh
piriformis	Anterior surface sacrum and edge of greater sciatic notch	Superior border of greater trochanter	
quadratus femoris	Ischial tuberosity	Intertrochanteric crest, femur	
obturator internus	Inner surface of membrane of obturator foramen, and bony margin	Greater trochanter	
obturator externus	External surface of membrane on obturator foramen, and bony margin	Trochanteric fossa	
inferior gemellus	Spine of ischium	Greater trochanter	
superior gemellus	Ischial tuberosity	Greater trochanter	

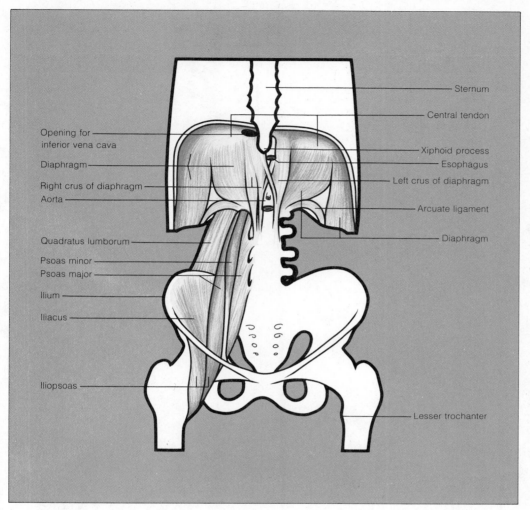

Opening for inferior vena cava

Diaphragm

Right crus of diaphragm

Aorta

Quadratus lumborum

Psoas minor

Psoas major

Ilium

Iliacus

Iliopsoas

Sternum

Central tendon

Xiphoid process

Esophagus

Left crus of diaphragm

Arcuate ligament

Diaphragm

Lesser trochanter

FIGURE 6–23   The diaphragm and iliopsoas, internal view of posterior abdominal wall.

lateral rotators of the thigh (see Figures 6–24 B,C). The large muscle mass on the medial side of the thigh is the *adductors* (Figure 6–25A). From their origin along the inferior ramus of the pubis, most of them reach to various parts of the linea aspera on the femur (see Table 6–11). Because their insertion is on the posterior surface of the femur, they are also lateral rotators.

Muscles That Act upon the Leg (Knee)   The three posterior thigh muscles known as the *hamstrings* originate mainly from the ischial tuberosity and insert into the tibia and fibula (see Table 6–11 and Figure 6–24). Except for the short head of the *biceps femoris,* they have no attachment on the femur. The

tendon of the biceps femoris passes laterally and those of the other two pass medially, forming the cords for which the group is named. These tendons can be felt on the posterior surface of the knee where they enclose a region called the **popliteal fossa.** Of the two medial muscles, one (the *semitendinosus*) has a very long tendon of insertion and the other (the *semimembranosus*) originates by a membranous band of fibrous tissue.

The anterior thigh muscles consist of one group plus one muscle, the *sartorius* (Figure 6–25B). From its origin on the anterior superior iliac spine, the sartorius crosses the thigh diagonally and wraps around behind the medial condyles of the femur and tibia to insert on the tibia anteriorly. This roundabout

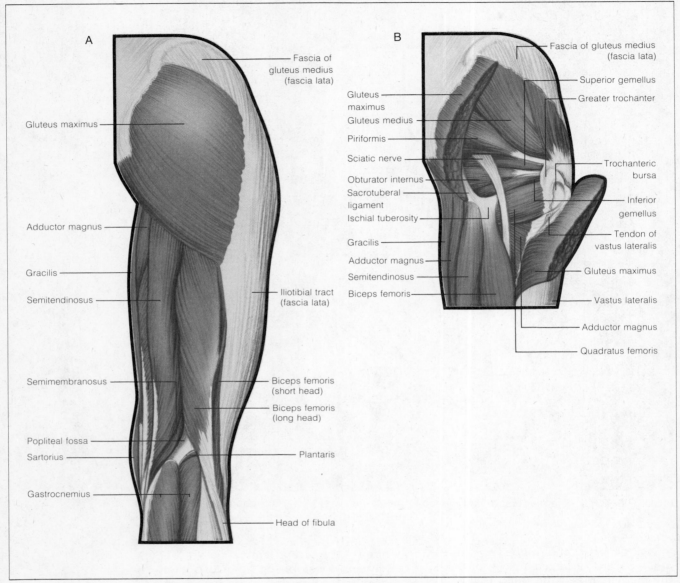

FIGURE 6–24 Posterior muscles of the hip and thigh. A. Superficial muscles. B. Intermediate layer of muscles. C. Deep muscles.

course makes it the only muscle to flex at both hip and knee; it also laterally rotates. It is known as the "tailor's muscle" because its action puts one in the cross-legged position that tailors are supposed to favor.

The *quadriceps femoris* is the prominent muscle mass on the anterior and lateral surfaces of the thigh. Three of the four heads (the *vastus muscles*) arise from the femur, occupying most of the surface of its shaft (except for the linea

aspera). The fourth head, the *rectus femoris,* arises from the os coxa and is the only one of the four to cross the hip joint. These muscles insert into the tibial tuberosity by a common tendon known as the **patellar tendon** because the patella is embedded in it.

Muscles That Act upon the Foot and Toes The posterior surface of the leg (the calf) is shaped largely by one

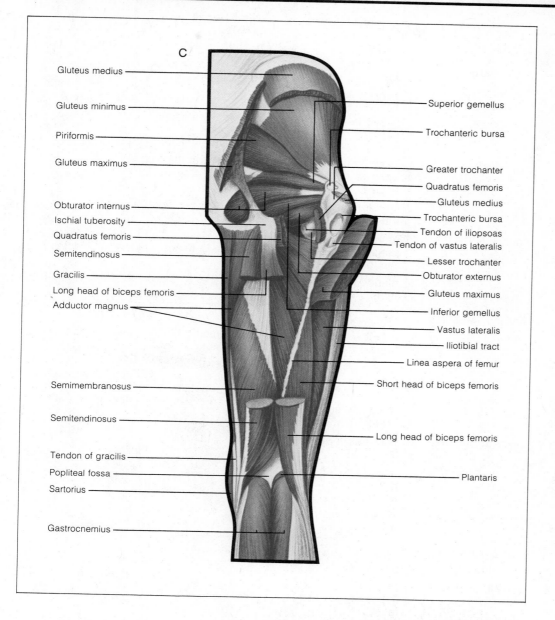

C

Gluteus medius

Gluteus minimus

Piriformis

Gluteus maximus

Obturator internus

Ischial tuberosity

Quadratus femoris

Semitendinosus

Gracilis

Long head of biceps femoris

Adductor magnus

Semimembranosus

Semitendinosus

Tendon of gracilis

Popliteal fossa

Sartorius

Gastrocnemius

Superior gemellus

Trochanteric bursa

Greater trochanter

Quadratus femoris

Gluteus medius

Trochanteric bursa

Tendon of iliopsoas

Tendon of vastus lateralis

Lesser trochanter

Obturator externus

Gluteus maximus

Inferior gemellus

Vastus lateralis

Iliotibial tract

Linea aspera of femur

Short head of biceps femoris

Long head of biceps femoris

Plantaris

pair of muscles, the *gastrocnemius* and the *soleus* (sometimes grouped as the *triceps surae*); see Table 6–12 and Figure 6–26. They have a common insertion, the **tendon of Achilles,** into the calcaneus, and they are the most powerful plantar flexors. The gastrocnemius originates by two heads from the femur, so that it also acts at the knee. The soleus arises from the tibia and fibula and so acts only over the ankle.

As in the arm, some of the muscles pass all the way to the phalanges while others terminate on the metatarsals.

The deep muscles on the posterior surface of the leg pass behind the medial malleolus to the plantar surface of the foot. They are all plantar flexors, some are weak invertors, and some are flexors of the digits. On the lateral surface of the leg (Figure 6–27) are the *peroneal muscles,* two of which pass behind the lateral malleolus to reach the plantar surface of the foot. In addition to being plantar flexors, they are evertors. The anterior leg muscles are located in the depression lateral to the sharp anterior margin of the tibia (Figure 6–28). They

TABLE 6-11   MUSCLES OF THE THIGH THAT ACT UPON THE FEMUR AND LEG

Muscle	Origin	Insertion	Action
Medial Muscles			
Adductors			Adduct and laterally rotate thigh
magnus	Ramus of ischium and pubis	Extent of linea aspera and medial epicondyle of femur	
longus	Crest of pubis near symphysis; inferior ramus of pubis	Middle third of linea aspera	
brevis	Inferior ramus of pubis	Proximal third of linea aspera	
Pectineus	Pectineal line of pubis, superior ramus of pubis	Below lesser trochanter of femur	Adducts, flexes, and laterally rotates thigh
Gracilis	Inferior ramus of pubis	Medial surface, head of tibia	Adducts thigh, flexes leg
Anterior Muscles			
Sartorius	Anterior superior iliac spine	Anterior median aspect, head of tibia	Flexes and laterally rotates thigh; flexes leg
Quadriceps femoris		Tibial tuberosity, by way of patella (ligamentum patellae)	Extends leg at knee Rectus femoris also flexes thigh at hip
rectus femoris	Anterior inferior iliac spine and superior margin of acetabulum		
vastus intermedius	Anterior surface of femur		
vastus lateralis	Greater trochanter and outer lip of linea aspera		
vastus medialis	Medial lip of linea aspera		
Posterior Muscles			
Hamstrings			Extend and adduct thigh, flex leg
biceps femoris	Ischial tuberosity and lateral lip of linea aspera	Lateral surface, head of tibia and of fibula	Biceps: weak lateral rotator
semimembranosus semitendinosus	Ischial tuberosity	Medial side of head of tibia	Semis: weak medial rotators

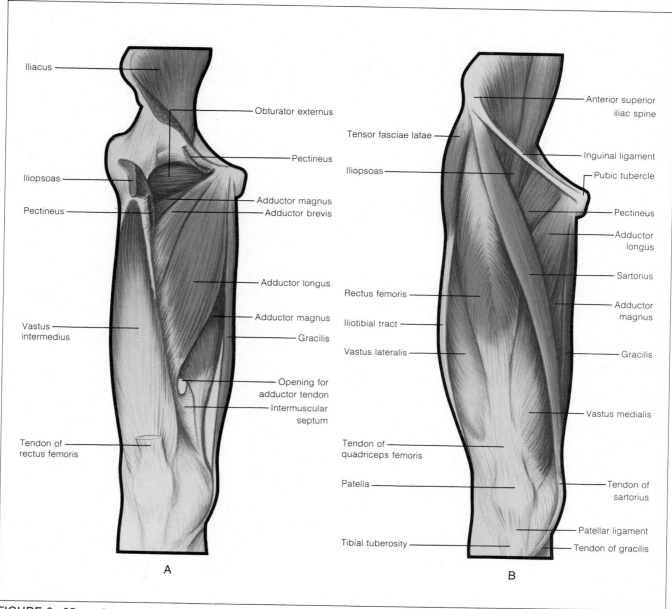

FIGURE 6–25 Anterior muscles of the thigh. A. Superficial muscles. B. Deep muscles, including the adductor muscles.

consist of the extensors of the toes and the *tibialis anterior*, which inserts on the first metatarsal. They are all dorsiflexors, and the tibialis anterior also inverts.

The general pattern of structure of the foot, while similar to that of the hand, reflects its role of support and locomotion rather than of grasping and manipulation. Like the hand, however, the dorsal surface has few muscles, and the long tendons from the leg are bound in place by **transverse ligaments**

at the ankle. The plantar surface (Figure 6–29), like the palmar, contains most of the intrinsic muscles. These are covered by a tough fibrous **plantar aponeurosis** (only part of it is shown in the figure) extending from the calcaneus to the phalanges. It is a protective cushion and contributes to the support of the longitudinal arches of the foot. The intrinsic muscles on the plantar surface include flexors to the digits, some special flexors and abductors for the first and fifth toes, an adductor for the great

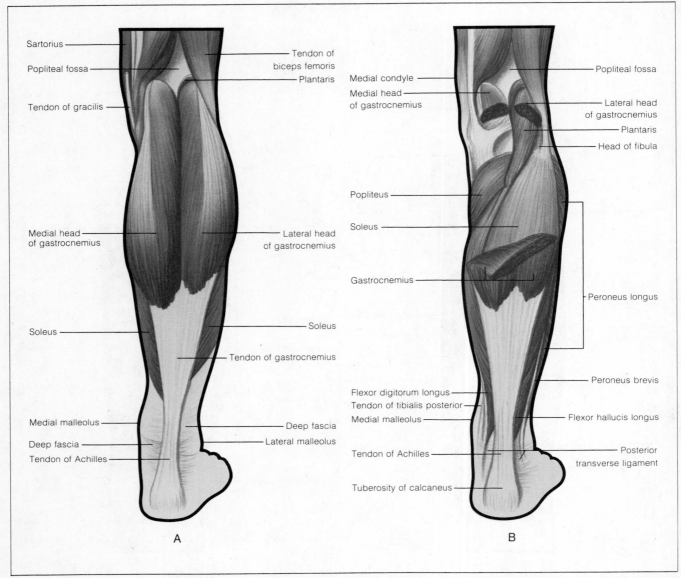

FIGURE 6-26   Posterior muscles of the leg. A. Superficial muscles. B. Intermediate layer of muscles. C. Deep muscles.

toe, and small muscles between the metatarsal bones. The muscles that move the great toe are known as *hallucis muscles.*

CHAPTER 6 SUMMARY

Skeletal muscles are enclosed by connective tissue; single cells by a fine cover of fibrous tissue, *endomysium,* bundles of cells by *perimysium,* and whole muscles by *epimysium,* which blends with the deep fascia. Muscles may be connected to bone directly (fleshy attachments) or indirectly by fibrous attachments such as a straplike *tendon* or a broader sheetlike *aponeurosis.* The proximal attachment of a muscle is the *origin.* It is usually at the fixed end, and may be quite extensive. The distal connection, or *insertion,* is usually on the end that moves. It is often by way of a rather long tendon, with a small area of attachment to the bone.

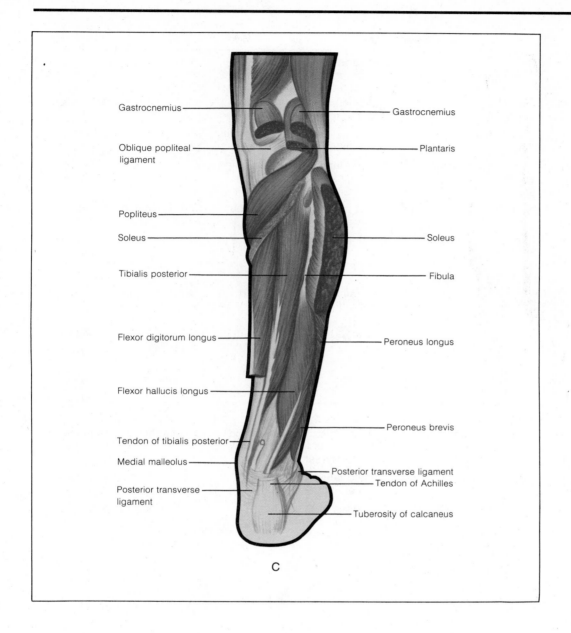

Gastrocnemius

Oblique popliteal ligament

Popliteus

Soleus

Tibialis posterior

Flexor digitorum longus

Flexor hallucis longus

Tendon of tibialis posterior

Medial malleolus

Posterior transverse ligament

Gastrocnemius

Plantaris

Soleus

Fibula

Peroneus longus

Peroneus brevis

Posterior transverse ligament

Tendon of Achilles

Tuberosity of calcaneus

C

Muscles act by exerting force on a lever (the bone) which moves about its axis. The amount of force required of the muscle and its mechanical effectiveness in producing a movement, depend upon the weight of the part to be moved (the *load*) and the distance from the axis *(fulcrum)* of both the load and the muscle insertion.

Muscles that produce a given movement, such as flexion, are *prime movers*. Muscles that oppose that movement and produce the opposite (exten-sion) are *antagonists*. *Synergists* are muscles that indirectly assist in a movement. Some muscles cross two joints, and synergists may prevent movement at one of those joints so that all of the force of the muscle contraction is brought to bear at the other joint. *Fixators* are synergists that stabilize or fix the site of origin, so that all of the force of contraction is exerted at the insertion. Fixators are particularly important for muscles that arise on the scapula and insert on the arm.

TABLE 6-12   MUSCLES THAT ACT UPON THE FOOT AND TOES

Muscle	Origin	Insertion	Action
Posterior Muscles			
Gastrocnemius	Medial condyle and lateral condyle of femur	Calcaneus, with soleus, via Achilles tendon	Flexes leg and plantar flexes foot
Soleus	Head of fibula and medial portion of proximal end of tibia	Calcaneus, with gastrocnemius, via Achilles tendon	Plantar flexes foot
Tibialis posterior	Posterior shaft of tibia and fibula, and interosseous membrane	Several tarsal bones, including calcaneus	Plantar flexes, everts foot
Flexor digitorum longus	Posterior shaft of tibia	Distal phalanx of second to fifth toes	Flexes phalanges, plantar flexes, inverts foot
Flexor hallucis longus	Shaft of fibula	Distal phalanx of great toe	Flexes great toe, plantar flexes, inverts foot
Anterior Muscles			
Tibialis anterior	Lateral condyle and upper 2/3 of shaft of tibia and interosseous membrane	First cuneiform (tarsal) and base of first metatarsal	Dorsiflexes and inverts foot
Extensor digitorum longus	Lateral condyle of tibia, upper 2/3 of anterior portion of fibula, and interosseous membrane	Phalanges of second to fifth toes	Dorsiflexes and everts foot, extends toes
Extensor hallucis longus	Anterior mid-fibula and interosseous membrane	Phalanx of great toe	Dorsiflexes foot, extends great toe
Lateral Muscles			
Peroneus longus	Head, upper shaft of fibula	Behind lateral malleolus and under sole of foot to base of first metatarsal	Plantar flexes, everts foot
Peroneus brevis	Distal shaft of fibula	Behind lateral malleolus to base of fifth metatarsal	Plantar flexes, everts foot
Peroneus tertius	Distal anterior shaft of fibula	Passes in front of lateral malleolus to dorsal surface of fifth metatarsal	Dorsiflexes and everts foot

Muscles of the Head and Neck

Muscles of facial expression insert on the skin of the face and cause movement of the skin and soft tissues of the face.

Muscles of mastication run to the mandible from the skull and cause movements of the mandible, especially in chewing (*mastication*).

The intrinsic muscles of the tongue lie within the tongue and change its shape and position as needed for speaking, chewing, etc. Extrinsic muscles of the tongue anchor it to the skull, mandible, or hyoid bone.

Among the muscles that move the head are superficial ones, including the suprahyoid and infrahyoid muscles, and muscles that extend to the skull from the pectoral girdle. The deep muscles, which are mainly posterior, include upper muscles of the vertebral column, as well as suboccipital muscles. Anterior deep muscles lie mainly on the anterior surface of the vertebral column.

Muscles of the vertebral column have origins and insertions on one of the processes of the vertebrae or the ribs. There are many such muscles, some that run between adjacent vertebrae or ribs, and others that span several vertebrae or ribs. They are all on the posterior surface.

Most of the muscles of the thorax are muscles of respiration. The majority of these muscles attach to the ribs and act to raise (expand) or lower the rib cage, thus changing its volume. The diaphragm is an exception, for it is a dome-shaped muscle that arises around the lower edge of the rib cage. When it contracts the dome flattens and the floor of the thoracic cavity is lowered, which also increases the volume of the thoracic cavity.

Muscles of the abdominal wall consist of several muscles whose fibers all run in different directions. The muscles and their interlacing fascial sheaths provide support and protection for the abdominal organs.

Muscles of the pelvic floor and perineum form a kind of sling that supports the pelvic organs and forms the floor of the pelvic cavity.

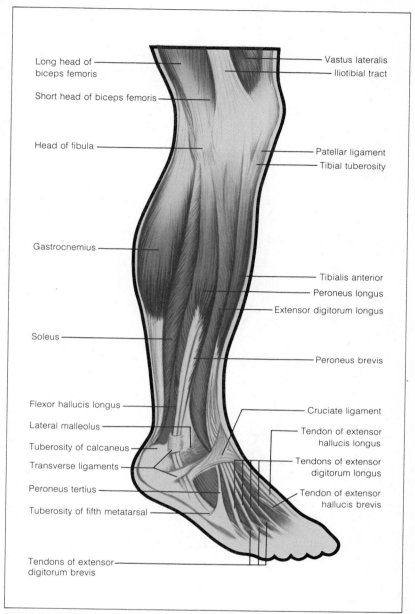

FIGURE 6–27   Lateral muscles of the leg and foot.

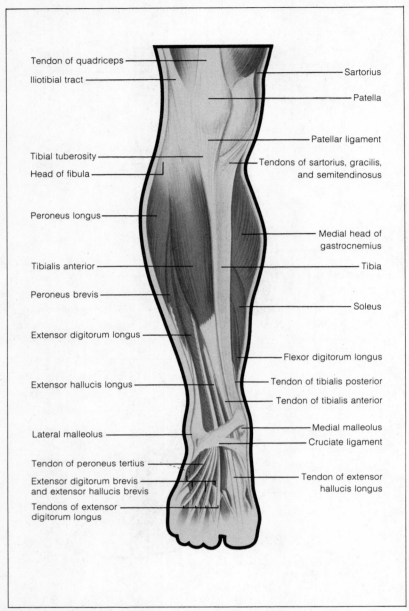

Tendon of quadriceps

Iliotibial tract

Tibial tuberosity

Head of fibula

Peroneus longus

Tibialis anterior

Peroneus brevis

Extensor digitorum longus

Extensor hallucis longus

Lateral malleolus

Tendon of peroneus tertius

Extensor digitorum brevis
and extensor hallucis brevis

Tendons of extensor
digitorum longus

Sartorius

Patella

Patellar ligament

Tendons of sartorius, gracilis,
and semitendinosus

Medial head of
gastrocnemius

Tibia

Soleus

Flexor digitorum longus

Tendon of tibialis posterior

Tendon of tibialis anterior

Medial malleolus

Cruciate ligament

Tendon of extensor
hallucis longus

FIGURE 6-28   Anterior muscles of the leg and foot.

Muscles of the Upper Extremity

Muscles that act on the scapula are those that move or stabilize the pectoral girdle. They originate on the axial skeleton — the skull, vertebrae, or rib cage — and insert on the scapula and/or clavicle. They are important fixators as well as prime movers.

Muscles that act on the humerus and move the arm arise mainly from the pectoral girdle, but two important exceptions originate on the trunk, from the anterior chest wall and the lower back.

Some of the muscles that act on the forearm arise on the scapula and humerus, and insert on the proximal ends of the radius and ulna. Those with insertions on the radius also affect pronation and supination at the radio-ulnar joint. Other muscles arise on the distal part of the humerus and insert on the forearm, carpal bones, or bones of the hand. These muscles affect movements of the hand and fingers as well as the forearm.

Muscles of the Lower Extremity

Muscles that act on the femur include those that arise on the anterior surface of the vertebral column and pelvis, the gluteal muscles of the buttocks and the hamstrings on the posterior surface, adductor muscles on the medial surface, and a small muscle in the iliotibial tract on the lateral surface.

Many of the muscles that act on the knee are in the thigh, including the hamstrings on the posterior surface and the quadriceps femoris on the anterior surface.

Muscles that act on the foot and toes are located in the leg. Those on the anterior surface dorsiflex at the ankle and extend the toes. The powerful muscles of the calf plantarflex the foot. Tendons of deeper muscles pass behind either the medial or lateral malleolus and contribute to inversion or eversion, respectively.

STUDY QUESTIONS

1 If a muscle contracts (shortens) and causes movement, how can that part be returned to its original position?

2 Of what importance is the axis of rotation of a joint in determining the action of a muscle?

3 Why are fixators so important in movements of the arm and shoulder girdle?

4 Many muscles, such as the gastrocnemius, cross two joints. If you use the gastrocnemius to stand on tip toe, how can you do so without bending your knee?

5 Suppose your sternoclavicular joint was somehow immobilized. Approximately how far could you raise your arm in abduction? Explain.

6 Why can you clench your fist tighter when the wrist is straight than when it is flexed?

7 What muscles do you use to rise from a sitting position? What muscles do you use to sit down very slowly? (Try it and see.)

8 Suppose the insertion of the biceps brachii was moved to the distal end of the radius. How would this affect the load that could be lifted? the range of movement it could produce? the speed of movement of the part? the precision of control?

9 What is unusual about the insertion of the diaphragm? What happens when the diaphragm contracts?

10 For abdominal surgery, what site(s) would you recommend for the incision, in order to cut across as few muscle fibers as possible?

11 How is it possible that the vertebral column is able to make so many different kinds of movement when virtually all of the muscles that act on the vertebral column are of its posterior surface?

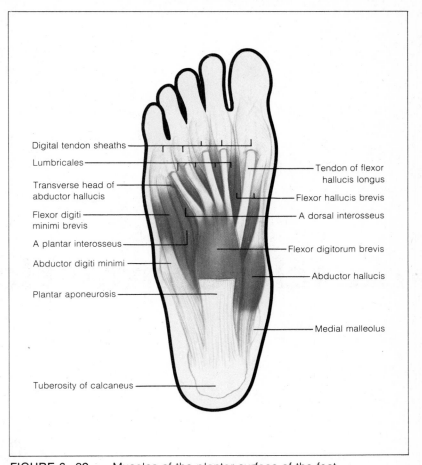

FIGURE 6–29   Muscles of the plantar surface of the foot.

7

Microscopic and Submicroscopic Anatomy of Skeletal Muscle

Excitation and Contraction of a Skeletal Muscle Fiber

Contraction of Skeletal Muscles

Smooth Muscle

Cardiac Muscle

Muscle Contraction

Muscle is distinguished from other tissues by the property of contractility, which is the ability to change shape. It involves shortening of the muscle cells (fibers). If both ends of the muscle are fixed however, the muscle cannot shorten, but it does exert considerable force at its attachments.

In this chapter we will take a careful look at the complex intracellular organization of skeletal muscle fibers, to find out how a muscle contracts, how a contraction is initiated, and what happens in the cell during contraction and relaxation.

Muscle contractions are not all alike; some are stronger, some last longer, and some do mechanical work while others do not. Because the nature of the muscle response is greatly influenced by properties of the stimulus, the effect of altering the stimulus on the contraction of muscles and the work they do is considered in this chapter.

In addition to skeletal muscle, there are smooth and cardiac muscle. They cause movements within organs of the viscera and the beat of the heart. They differ in their structure and many of their properties, but their actions, too, are based on their ability to contract.

MICROSCOPIC AND SUBMICROSCOPIC ANATOMY OF SKELETAL MUSCLE

Skeletal muscle is so named because of its attachments to the skeleton. It also is known as *striated* muscle because its fibers show alternating light and dark stripes, or as *voluntary* muscle because its contractions can be voluntarily initiated (although not all contractions are consciously willed). As a tissue, skeletal muscle is composed of large, elongated cells whose length varies between muscles, from a few millimeters to many centimeters (see Figures 7–1 and 7–2). Each cell has many nuclei located in the cytoplasm **(sarcoplasm)** directly under the cell membrane **(sarcolemma)**. The ability of the muscle fiber to contract is due to the arrangement of tiny fibers or **myofibrils** within it. They are longitudinally oriented, and so nearly fill the cell that mitochondria and other organelles and inclusions are squeezed into the spaces between them.

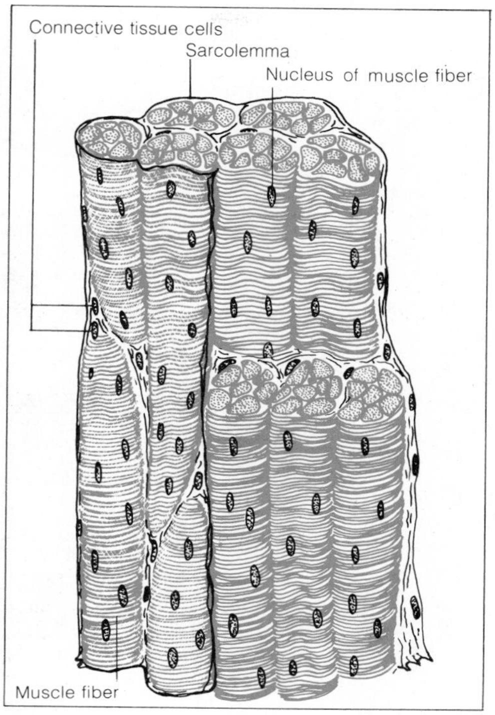

The myofibrils within each cell show a repeating pattern of light and dark areas. They are aligned in such a way that light and dark areas match up, and the striations appear to cross the whole muscle fiber. The light areas are called **I bands** and the dark areas **A bands**. There is a pale area, the **H band,** in the middle of the A band, and a thin dark line, the **Z line** in the middle of the I band.[1] The region from one Z line to the next defines a repeating unit called a **sarcomere.** It is only about 2.0–2.5 microns in length, but there may be several hundred of them in a single myofibril. Because all of the events of contraction occur in each sarcomere, contraction is usually described in terms of what happens in a single sarcomere.

Electron micrographs (Figure 7–3) reveal that the myofibrils in the muscle cell are composed of two kinds of parallel strands, or **myofilaments,** and that

FIGURE 7–1   Diagram of some skeletal muscle fibers.

the striations are due to the placement of the myofilaments in the sarcomere. The A band contains relatively thick filaments composed of the protein **myosin.** The I band contains relatively thin filaments whose main component is the protein **actin.** They are attached to the Z line at one end, and extend through the I band and into the A band to partially overlap the myosin strands. In a cross section through a myofibril, the myofilaments are seen to be packed in such a way that each thick filament is surrounded by six thin filaments, and each thin filament is surrounded by three thick ones (see Figure 7–2).

The endoplasmic reticulum of skeletal muscle cells, called **sarcoplasmic reticulum** (SR), differs from that in most cells (see Figure 7–4). It lacks ribosomes along its surfaces and forms a network of tubules around and between the myofibrils. Near the junctions of A and I bands the tubules widen out into tiny sacs *(lateral sacs)* which have a high content of calcium ions.

Numerous invaginations of the sarcolemma (cell membrane) extend into the substance of the muscle cell as

---

[1]A and I bands designate *anisotropic* and *isotropic* bands, respectively. The names describe the properties of the bands under polarized light. The H and Z bands are named for German *helle* (light) and *zwischen* (between).

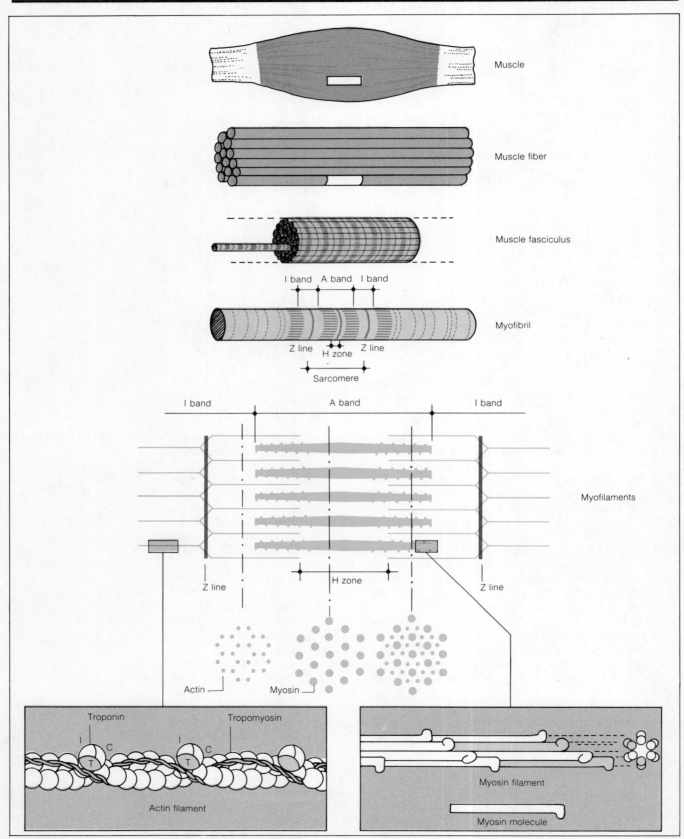

Muscle

Muscle fiber

Muscle fasciculus

I band   A band   I band

Myofibril

Z line   Z line

H zone

Sarcomere

I band   A band   I band

Myofilaments

Z line   Z line

H zone

Actin   Myosin

Troponin   Tropomyosin

Actin filament

Myosin filament

Myosin molecule

FIGURE 7-2   Structure and organization of skeletal muscle.

a second set of tiny canals known as **transverse tubules** or **T-tubes.** They wrap around the myofibrils at the level of the A and I junctions, where they lie between the lateral sacs of the sarcoplasmic reticulum. In cross sections there appears to be a *triad* of three channels, with the T-tube in the middle. These relationships were originally worked out in frog muscle which, as shown in Figure 7–4, has the triads located at the Z lines rather than the A-I junctions.

EXCITATION AND CONTRACTION OF A SKELETAL MUSCLE FIBER

When a muscle fiber contracts and shortens, the banding pattern of each sarcomere changes (Figure 7–5). The evidence suggests that the thick and thin filaments slide over one another in such a way that the region of overlap widens, the sarcomere shortens, the I band shortens, and the H band disappears. When the muscle is stretched, the filaments slide the other way, the area of overlap narrows, the I and H bands broaden, and the entire sarcomere is lengthened. The length of the thick and thin filaments, however, does not change; it is only the amount of overlap that changes. Because the amount of sliding is very slight in a single sarcomere, each one contributes very little to the total contraction, but the eventual shortening of the muscle cell is the sum of the tiny changes in the hundreds of sarcomeres in its myofibrils.

Contractile Proteins

Individual myosin molecules are shaped like small rods with rounded heads attached at an angle, rather like stubby golf clubs (see Figure 7–2). The thick filaments are bundles of myosin molecules with their heads protruding laterally from the bundle in a staggered fashion, directed toward the nearest thin filament. The central portions of the thick filaments contain only the rod portions of myosin molecules, but are thickened somewhat by some fine cross connections which presumably keep

FIGURE 7–3   Electron micrograph of a longitudinal section of skeletal muscle. 25,000× magnification. A. Z line. B. I band. C. H zone. D. A band. E. Mitochondria. F. Actin filaments. G. Myosin filaments. H. Muscle glycogen granules.

Robert L. Kaplan

the filaments in place. The myosin heads have sites that can react with actin to form the cross-bridges between actin and myosin. They also have enzyme properties that enable them to split ATP and release the energy necessary for contraction; that is, the myosin heads can act as an *ATPase.*

The thin filaments contain two proteins, tropomyosin and troponin in addition to actin. Actin filaments, which are the major component, consist of two strands of globular bits of actin twisted together like a double strand of beads. **Tropomyosin** is also arranged in double-stranded filaments which lie in the grooves between actin strands and run parallel with them. **Troponin** consists of small spheres attached to the tropomyosin at specific intervals. Troponin contains three subunits, designated T, C, and I. *Troponin* T binds the troponin to tropomyosin; *troponin C* binds calcium ions; and *troponin I* is bound to actin in such a way that the sites on actin for binding myosin are covered.

Excitation

In Chapter 2 it was pointed out that all cells maintain an unequal distribution of sodium and potassium ions

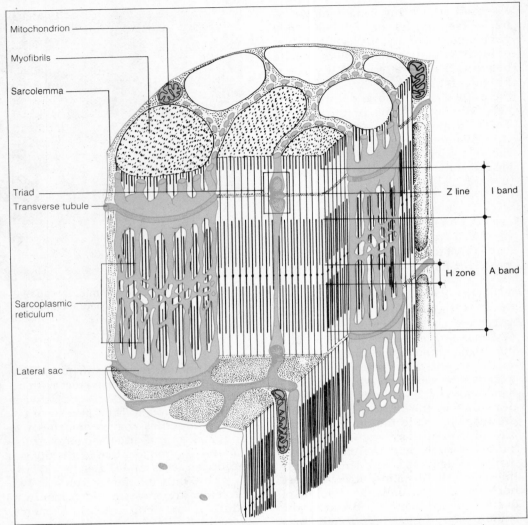

Mitochondrion

Myofibrils

Sarcolemma

Triad

Transverse tubule

Sarcoplasmic
reticulum

Lateral sac

Z line   I band

H zone   A band

FIGURE 7–4   The sarcoplasmic reticulum and transverse tubules (T-tubes) around myofibrils of an amphibian skeletal muscle cell. (In mammalian skeletal muscle the T-tubes are located at the A-I junctions.)

across their cell membranes by means of an active process, the Na-K pump. The pump creates and sustains an electrical gradient—the **resting potential.** Stimulation of a muscle cell causes changes in its resting potential. A stimulus of sufficient intensity (threshold) temporarily increases permeability of the membrane, first to sodium and then to potassium, which causes a brief depolarization of the membrane at that site—the **action potential.** The depolarization serves as a stimulus to the adjacent region of the membrane which

then depolarizes and causes depolarization of the next area, and so on. In this way an action potential is conducted over the entire surface of the muscle cell membrane; it is the event that actually initiates contraction in a muscle fiber.

Muscle cells normally develop action potentials when they are stimulated by impulses (action potentials) in nerve fibers (nerve cell processes). Each skeletal muscle cell has a nerve ending on its surface, and it can also be stimulated directly. In either case, the effect of the stimulus is depolarization of the sarcolemma (cell membrane) and development of an action potential that is conducted over the cell membrane and sets off the contraction process. If all myofibrils in a muscle cell are to contract as a unit, it is essential that the impulse be delivered throughout the muscle cell almost instantaneously. This is possible because of the system of T-tubes, which are continuations of the sarcolemma, for as the action potential is conducted over the cell surface it also travels down the T-tubes. Depolarization of the T-tubes affects the lateral sacs of the sarcoplasmic reticulum (SR) which lie on either side of it, causing them to release some of the calcium stored there. The calcium ions diffuse into the sarcoplasm in the region of thick and thin filaments and bind to troponin C. This causes some rearrangement within the thin filaments and exposes the myosin binding sites on the actin. Heads of myosin molecules quickly bind to the actin at those sites, forming cross-bridges between the actin and myosin.

As stated above, the myosin heads are capable both of binding to actin and of splitting ATP to release energy. The enzymatic reaction, however, is not very effective until cross-bridges are formed by the binding of myosin heads to the actin. But such binding cannot occur until calcium ions have exposed the binding sites. When this happens and binding occurs, the enzyme becomes more effective and ATP is split, releasing the energy used for the actual contraction—sliding of thick and thin filaments.

Contraction—Sliding Filaments

The changes that bring about sliding of the filaments are the critical events in muscle contraction, but they are not very well understood. The energy released from ATP is thought to exert a force that causes a rotation or movement of the myosin head and cross-bridge (which is bound to actin), causing a slight displacement between the thick and thin filaments. The thin filaments are moved toward the center

of the sarcomere, increasing the area of overlap of thick and thin filaments. If more ATP is present it will bind to the myosin, and the cross-bridges are quickly broken when this occurs. If calcium is still present and bound to troponin, (so that actin binding sites are available), the myosin heads will form new cross-bridges at new actin binding sites. Cross-bridges are formed and broken many times in very rapid succession, and the thin filaments are moved past the thick filaments in a rapid ratchetlike action.

Relaxation

When an action potential causes the initial release of calcium ions from the sarcoplasmic reticulum (SR), the calcium concentration in the sarcoplasm rises, and the SR starts an active transport mechanism that pumps cal-

FIGURE 7–5   The relationship of the thick and thin myofilaments varies with the contractile state of the muscle.

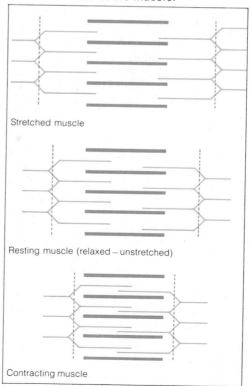

Stretched muscle

Resting muscle (relaxed—unstretched)

Contracting muscle

cium ions into the longitudinal tubes of the SR. They eventually diffuse down to the lateral sacs where they are stored until another action potential arrives. Removal of calcium from the vicinity of the thin filaments causes dissociation of calcium from the troponin C, and troponin and tropomyosin return to their resting positions. Binding sites of actin are now hidden, so cross-bridges are not formed, and the muscle relaxes. Arrival of additional action potentials (in rapid succession) can keep the calcium level in the sarcoplasm quite high, and thus maintain a prolonged contraction.

Because tropomyosin and troponin block the formation of cross-bridges, and hence prevent contraction, they have been called "relaxing proteins." More recently they have been visualized as an on-off switch that regulates contraction.

Summary

The sequence of events involved in contraction of a skeletal muscle cell might be summarized as follows:

1 An action potential on the sarcolemma travels down the T-tubes.

2 Depolarization of the T-tubes causes release of calcium ions from sarcoplasmic reticulum.

3 Calcium ions bind troponin of thin filaments, causing changes in troponin and tropomyosin that lead to exposure of the myosin binding sites on actin.

4 Myosin heads bind to actin, forming actomyosin, thus increasing the ATPase activity of the myosin. ATP is split, releasing energy for contraction.

5 Myosin heads and cross-bridges are caused to move so as to shift the thin filaments a short distance toward the center of the sarcomere.

6 The sarcomere, and hence whole muscle cell, shortens or develops tension.

7 ATP binds to the myosin molecule, which results in breaking of the cross-bridge.

8 The process is repeated as a new cross-bridge is formed by myosin binding to the "next" binding site.

9 The increased calcium concentration in the sarcoplasm activates a calcium pump that returns calcium to the SR.

10 Low sarcoplasmic calcium causes dissociation of calcium from the troponin, and the proteins of the thin filament return to their original position, thus blocking binding sites on actin. The muscle relaxes.

CONTRACTION OF SKELETAL MUSCLES

For muscle, the property of irritability, the ability to respond to a stimulus, takes the form of the ability to contract. We have just seen how a stimulus causes changes within the sarcomere that lead to shortening of the sarcomere and the whole muscle cell. Skeletal muscle contracts only when it has been adequately stimulated; it does not do so spontaneously. Our present task, then, is to determine both how a whole muscle contracts and how changes in the stimulus might modify that contraction. But first let us examine the stimulus.

Nature of the Stimulus

In the broadest possible sense, a stimulus can be defined as a change in the environment that in some way modifies the activity of the unit (cell, organ, or individual) to which it is applied. The nature of the response is determined to a certain extent by the properties of this unit (muscles contract and glands secrete), but it is modified by the qualities of the particular stimulus.

Stimuli may be mechanical, thermal, chemical, or electrical, but for study purposes, we prefer stimuli that do not damage the tissue or cause irreversible changes. Electrical stimuli are most often used, not only because they are similar to normal physiological stimuli in many ways, but also because they can be readily quantified and precisely controlled, so that stimuli with exactly the same characteristics can be applied repeatedly. Properties of the stimulus that may be varied independently are easily identified in terms of electrical stimuli.

1   The **intensity** or strength of the stimulus is the magnitude of the change. Electrically, it is measured in volts or millivolts (1 mv = 0.001 v).

2   The **duration** of the stimulus is the length of time that the stimulating current is applied. For physiological stimulation it is usually such a brief interval that it is measured in milliseconds (1 msec = 0.001 sec).

3   The **rate of change of intensity,** or the "rise time," is a measure of the length of time between the application of a stimulus and its peak intensity.

4   The **frequency** is the number of times per second a single stimulus may be repeated.

Nature of the Response

The contraction of a skeletal muscle in response to a stimulus occurs very quickly and lasts but a brief period of time. To analyze the contraction and see how it is affected by changes in the characteristics of the stimulus, it is therefore helpful to have a permanent record of the response.

Historically, and in the student laboratory, the response of a muscle to electrical stimulation has been studied with the frog gastrocnemius muscle. For a typical experiment (see Figure 7–6), a muscle is suspended from a rigid support. The Achilles tendon is attached to a response lever that moves when the muscle shortens. In most recording systems, the movement of this response lever causes corresponding deflections upon a moving strip of paper; the speed of movement of the paper can be controlled. A stimulator, with suitable controls for varying the stimulus, is connected to the muscle so that when it is turned on a current flows through the muscle and causes it to contract. Most stimulators also have connections for an event marker, a separate writing pen that marks the time and duration of the stimulus.

FIGURE 7–6   A simple setup for recording the response of an isolated muscle to electrical stimulation.

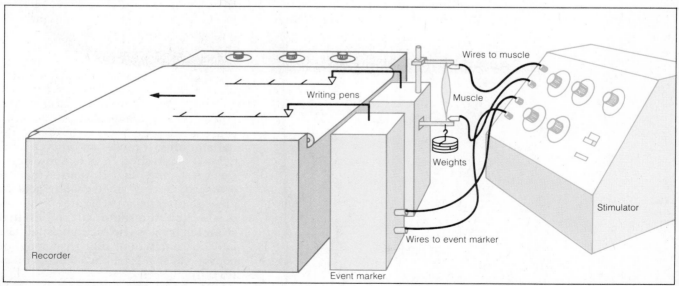

A single stimulus produces a single action potential and a single brief contraction known as a **muscle twitch.** If the recording paper is not moving, the record of the response is a vertical line, but when the paper is moving a curve is traced (Figure 7–7). This curve reveals that there is a slight delay after the stimulus (the *latent period*) before any shortening can be seen, followed by a **period of contraction** and then by a **period of relaxation.** Relaxation occupies about half the duration of the entire contraction curve. The duration of a twitch varies in mammalian muscles from about 120 msec (0.12 sec) in "slow" postural muscles to 7.5 msec (0.0075 sec) in the extrinsic muscles of the eye. In the frog muscle these times are all much longer.

Stimulation of an Isolated Skeletal Muscle

Effects of single stimuli. Skeletal muscle is an excitable tissue as described in Chapter 2. In fact, a skeletal muscle cell was the example used to illustrate the responses of excitable cells. Each muscle cell has a resting potential; its membrane is polarized and a stimulus causes depolarization, resulting in an action potential.

Recall that a stimulus increases the permeability of the membrane to sodium ions, permitting a few sodium ions to diffuse into the cell, thus reducing its internal negativity. This slight depolarization lasts only briefly, and the original membrane potential is promptly restored (see Figure 2–22). A stronger stimulus causes a greater permeability change and entry of more sodium ions, and more depolarization. If the changes are great enough, there is a very large, but extremely brief, increase in permeability to sodium (sodium activation), and the ensuing influx of sodium ions causes complete depolarization and a partial reversal of polarity. It is followed immediately by increased permeability to potassium ions; they leave the cell and restore the intracellular negativity (the membrane potential) as shown in Figure 7–8. These electrical changes, which last only a few milliseconds, produce the action potential or *spike* that is conducted over the surface of the muscle cell and down the T-tubes. It affects all the myofibrils in that cell, but it is not conducted to other cells in the muscle; each cell must be stimulated to develop its own action potential.

The experimental setup described above is designed to record the actual shortening of the muscle, its mechanical response but not its electrical response. A different setup would be needed to record action potentials. It is also important to remember that the contraction of a whole muscle (such as the gastrocnemius) is being recorded, not that of a single muscle cell. By studying contractions of a whole muscle when the stimulus is varied, we can learn a number of things about muscles and also about muscle cells.

Stimulus intensity. In order to cause contraction of a muscle, the stimulus must cause an action potential in at least some of the muscle cells. In a whole muscle, the weakest stimulus to cause a response is a **threshold** or **liminal stimulus.** It causes an action potential and contraction in only a few muscle cells, and the response that is recorded is very small (a *threshold* or *liminal contraction*) (Figure 7–9). A weaker **subthreshold** or **subliminal stimulus** causes no action potentials and no contraction, though it may, of course, cause a brief partial depolarization in some cells.

Each muscle cell has a threshold

FIGURE 7–7 Contraction curve for skeletal muscle (frog).

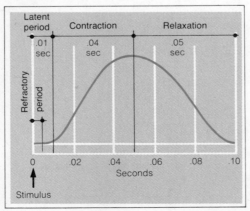

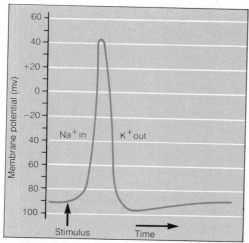

FIGURE 7–8   An action potential in skeletal muscle.

stimulus, but all fibers do not necessarily have the same threshold, even in the same muscle. Furthermore, when the stimulus is delivered this way (as an electrical current flowing through the muscle rather than from impulses in its own nerve fiber), some cells may not receive as strong a stimulus as others. Thus, what is a threshold stimulus for some fibers may be a subthreshold stimulus for others. If the intensity is increased, however, the height of the recorded contraction becomes greater as more and more muscle fibers participate. A **maximal stimulus** causes all of the fibers in the muscle to contract. The response to such a stimulus is a *maximal contraction*, and it is the greatest response that can be elicited under these conditions. Further increase in stimulus intensity (*supramaximal stimulus*) will still produce a maximal response. The actual stimulus intensities (voltages) at which threshold and maximal responses are obtained vary from one muscle to another, and in the same muscle with time, because there are other factors that also affect the irritability of the cells. These other factors may not remain constant, particularly in an isolated muscle.

Duration of stimulus. Just as a minimum intensity is required, the stimulus must also be of a certain duration in order to be effective. A weak stimulus requires a longer duration to cause contraction than a strong stimulus does. If the stimulus duration is very short, contraction will not occur even with extremely high intensity. Generally, the time required is very short, a fraction of a millisecond for some tissues.

Rate of change of intensity. The rise time of a stimulus involves a very short time interval and requires more sophisticated equipment for study than described above. Physiological stimuli usually have an adequate rate of change, but if the time is lengthened a slightly greater intensity is needed to stimulate. If the stimulus rises to peak intensity relatively slowly, the tissue may accommodate to the change and not respond at all.

Effects of Repetitive Stimuli   The response of a muscle to successive stimuli depends primarily upon the length of the interval between the stimuli (the frequency of stimulation).

Summation of subliminal stimuli. When a stimulus is just subthreshold, no response is detected. However, if several such stimuli are repeated rapid-

FIGURE 7–9   Effect of increasing stimulus intensity on response of isolated skeletal muscle.

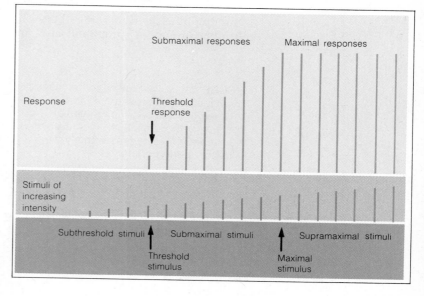

ly, a twitch may occur. Repetition of the stimulus within some brief period of time can cause an adding or **summation** of the partial depolarizations caused by each stimulus until the sum reaches threshold and produces an action potential in a few cells, resulting in a weak contraction.

Summation of twitches—tetanus. Two successive stimuli of maximal intensity cause two separate maximal contractions. However, if the interval between the two stimuli is reduced, the second contraction can be made to begin before the first is completed (Figure 7–10). The height of the second contraction is then somewhat greater than that of the first. Upon further reduction of the interval, the second response comes still earlier, and the height of its contraction is even greater—there is a **summation of twitches.** Measurement of the tension or force developed by the muscle shows that the second twitch also develops greater tension.

In order to understand how it is possible to produce a contraction that is greater than the maximal response, it is necessary to realize that the muscle contains an elastic component as well as the contractile elements (see Figure 7–11). Muscle is an elastic tissue; it resists stretch, but it can be stretched, and upon release it will return to its original length. The exact nature and location of the elastic component are not known for sure, but it undoubtedly includes the connective tissue elements that hold the muscle fibers together. The elastic component probably also in-

cludes some parts of the muscle fiber itself, perhaps Z line structures. Together these elastic components behave something like a spring. When tension is applied they stretch and then recoil.

Upon stimulation of a muscle, the contractile elements respond rapidly to the calcium released from the sarcoplasmic reticulum, but much of the shortening goes to stretch the elastic components. The elastic recoil applies tension to the muscle tendon, and there is movement—in the frog muscle experiment the writing pen records a contraction. The response of the contractile elements is, however, very rapid, and the *active state*, the actual shortening of the sarcomeres, is nearly over when the recorded contraction curve occurs.

When a second stimulus is applied, the contractile elements respond a second time before the recorded external contraction from the first stimulus is completed. The tension developed by the second active state is applied to an already-stretched elastic system; there is no "slack" for it to take up. In addition, if the interval is very short, the second stimulus may cause the release of more calcium ions before those released by the first stimulus have been returned to the sarcoplasmic reticulum. The result is a prolongation of the effect of the calcium; that is, the active state lasts a little longer and the total shortening is greater. Additional stimuli, if applied rapidly, could produce a further summation of responses. The summation is not a numerically proportional addition, however, since the contractions are not two, three, or four times that of the initial contraction.

Several series of stimuli applied at increasing frequencies illustrate the increased degree of summation (Figure 7–12). At higher frequencies the responses tend to fuse into a smooth curve that is maintained for the duration of the stimulation, as the calcium level in the sarcomere remains high enough to maintain the active state. This is **tetanus**[2], characterized by *sum-*

FIGURE 7–10   Summation of twitches. The effect of decreasing the interval between two stimuli. As the interval decreases, the summation increases until the second stimulus falls within the refractory period of the first.

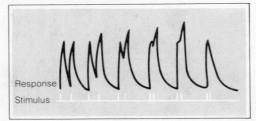

[2]This type of tetanus is a normal physiological response and should not be confused with abnormal conditions such as the tetanus of lockjaw or the tetany of low blood calcium—both of which can cause uncontrolled tetanic contractions.

mation and *fusion* of individual twitches. Tetanic contractions are important because muscle twitches do not occur physiologically. Skeletal muscle contractions are all tetanic, even the blink of an eye. The stimulus frequency required to produce tetanus varies. For the frog gastrocnemius, a frequency of 50 per second is sufficient. In humans, it ranges from about 30 per second for the "slow" soleus to about 350 per second for the extrinsic muscles of the eye. The average is about 100 stimuli per second.

It is important to realize that tetanus is a function of *frequency*, not of intensity. A maximal stimulus was used in the example so that all muscle fibers would be participating initially and the summation could not be attributed to recruiting of additional fibers. It is quite possible, and indeed very common, to have tetanic contractions involving only a few of the fibers in a muscle.

Refractory period. When the interval between two stimuli becomes short enough, the summation of responses disappears, the height of contraction declines to that of a single stimulus, and there is but a single response to the two stimuli. This happens because the second stimulus was applied so soon that the muscle had not repolarized after the action potential caused by the first stimulus. Since a membrane cannot be depolarized unless it is polarized, the second stimulus will not cause an action potential. The muscle, of course, will not respond to the second stimulus because, without an action potential, it is not actually stimulated. During this brief interval, the **refractory period,** the threshold is infinitely high, and the muscle cannot respond to a stimulus, regardless of its intensity.

Prolonged stimulation. If maximal stimuli are applied at low frequency (1–2 per second), the muscle contracts and relaxes completely after each stimulus. In the isolated muscle, deprived of a blood supply to provide nutrients and oxygen and remove wastes, fatigue begins to show after a few minutes. Relaxation takes progressively longer and is not completed by the time the

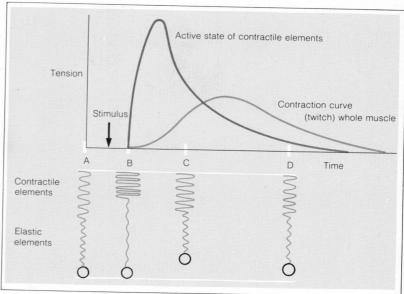

FIGURE 7–11   Role of contractile and elastic components in a skeletal muscle twitch. The curves show the time course of the tension developed by the contractile elements and by the whole muscle. The diagrams represent the status of the elastic components at the times shown.

next contraction begins, resulting in a condition often called *contracture* (Figure 7–13). It may be caused by insufficient ATP to pump calcium back into the SR. Eventually the ability to contract is also impaired and the muscle maintains a slowly declining state of contraction. This fatigued state is due

FIGURE 7–12   Effect of increasing stimulus frequency on the contraction of an isolated skeletal muscle. A., B. Individual twitches. C., D., E. Partial or incomplete tetanus. F. Complete tetanus, with summation and complete fusion of twitches.

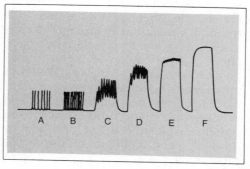

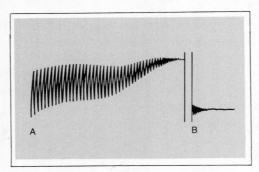

FIGURE 7-13   Fatigue in an isolated skeletal muscle, produced by repeated stimulation. A. Typical fatigue curve. B. Fatigue curve obtained in the same muscle after a 5-minute rest.

to the accumulation of metabolic waste products and the depletion of energy sources. It develops much faster if tetanic stimulation is used. A period of rest brings little recovery to the isolated muscle. However, an intact muscle with an adequate blood supply can contract for a very long time; if it does fatigue, there is a marked recovery after a brief period of rest.

Stimulation of a Motor Unit

Physiologically, a muscle fiber or a whole muscle is stimulated through its motor nerve rather than by direct application of electrodes from a stimulator. The motor nerve cell extends from the spinal cord to the muscle, where it divides into a number of terminal branches, each of which ends on a muscle fiber. The motor nerve cell (*motoneuron*) and the muscle fibers that it innervates comprise the physiological unit of muscle function, called a **motor unit.**

Stimulation of a single motor nerve fiber causes an action potential that travels over all of its branches, and all the muscle fibers of that motor unit respond. Gradations in strength of contraction of a whole muscle result from altering the number of active motor nerve fibers, and hence the number of motor units activated. Repetitive stimulation of the nerve fiber, of course, produces summation and tetanus in the muscle cells of the motor unit, just as does direct stimulation of the single muscle cell.

Types of Contraction

Although contraction is the response of muscle to a stimulus, contraction does not always involve shortening. It may involve the development of tension instead. The contractions described above, in which the muscle does shorten and moves the recording lever, are called **isotonic contractions** (isotonic = equal tension). If the load is so heavy that the muscle cannot shorten, stimulation produces an **isometric contraction** (isometric = equal length) and the development of a considerable degree of tension (isometric exercises are often favored by those who wish to develop muscular strength). Muscle contractions that result in movement of parts (especially limbs) are isotonic contractions. Isometric contractions are common among the antigravity, or postural, muscles, whose role is more to prevent movement than to produce it. Most contractions are a combination of the two, since isotonic contractions develop some tension and isometric contractions involve a slight shortening. In a third type of contraction, the muscle develops tension, but not enough to prevent stretching of the muscle, and it actually is lengthened as it contracts. An example is the action of the extensor muscles of the thigh and leg when sitting down slowly.

Contraction and Mechanical Work

A muscle that contracts isotonically and lifts a load does mechanical work. A muscle that shortens without a load, or a muscle that contracts isometrically, does not perform appreciable work. Mechanical work is defined in terms of the load and the height that load is raised (Chapter 2) and if either the load or the height is zero, there is no mechanical work.[3] As the load lifted increases, the height of contraction decreases, but the work done also increases, up to some *optimum load* beyond which the work decreases until the load is so

---

[3] Contraction involves the expenditure of energy—physiological work—whether a load is lifted or not.

Focus

Muscle Cramps

Most people are familiar with muscle cramps—spasmodic, painful, involuntary muscle contractions, according to a standard medical dictionary. They accompany and follow exercise, strike during sleep, and provide one of the more agonizing symtoms of certain disorders. Yet we understand neither the mechanisms of cramping nor the cause of the pains.

The best understood cramps are those associated with the condition known as contracture. When a patient who lacks certain glycolytic enzymes (muscle phosphorylase or phosphofructokinase) exercises vigorously, the muscles become hard, shortened, and painful. They go into a state of prolonged contraction, apparently because the sarcoplasmic reticulum fails to retrieve promptly the calcium ions it releases in response to muscle action potentials. These ions help catalyze the contractile interaction of actin and myosin in the muscle fiber; the fiber remains contracted as long as they are present. The failure to retrieve them may be due to a shortage of ATP caused by the lack of glycolytic enzymes.

The stiff-man syndrome, first described in 1956, is marked by disabling muscular rigidity and painful spasms. It usually strikes adults and is typically sporadic, progressing slowly for months and years. The muscles of the trunk grow taut and boardlike, walking becomes laborious, and painful cramps, lasting for several minutes, may be frequent. The cramps are often touched off by voluntary movement, noise, startle, and cutaneous stimuli, and they may be severe enough to break bones. Permanent joint deformities may result from this disorder.

The phenomenon behind the stiff-man syndrome is not known, although there is persistent motor unit activity and the symptoms disappear in sleep and during anesthesia and peripheral nerve block. The syndrome resembles tetanus (lockjaw) in that it may be due to a defect in spinal postsynaptic inhibition, for it is relieved by Diazepam, a drug that depresses spinal excitability and inhibits strychnine convulsions, also caused by interference with postsynaptic inhibition.

Less well understood are the so-called "ordinary" muscle cramps, which may strike almost anyone after vigorous exercise, after a sudden forceful contraction of untrained muscles, or during sleep, particularly in the legs and feet. During such cramps, a muscle or a portion of a muscle forms a hard, painful knot; though the knot can be removed by stretching the muscle, the area may remain sore and tender. The mechanism of these cramps may involve the distal portion of peripheral nerves, for though nerve block can prevent the voluntary initiation of a cramp, in some people the cramp can still be induced by electrical stimulation of the nerve distal to the block.

The reason for the pain associated with cramps is not known, but it is believed that contracting muscle releases an unidentified, slowly diffusing substance that stimulates the endings of pain nerves in muscle. It appears that this unidentified substance accumulates and exceeds a threshold level when muscle contractions are constant or when the circulation of blood is, for some reason, deficient. The soreness and tenderness that often follow a cramp may be due to actual muscle damage.

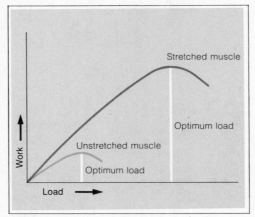

FIGURE 7–14   Effect of load on work done by stretched and unstretched muscles.

great that the muscle contracts isometrically (see Figure 7–14).

The amount of tension a muscle can develop and the amount of work it can do depend, among other things, upon the length of the muscle at the time of stimulation. Within limits, a stretched muscle can contract more forcefully and develop greater tension than an unstretched muscle. When a muscle is slightly stretched, the amount of overlap of the actin and myosin filaments permits the formation of many cross-bridges, and it is the interaction of the cross-bridges that develops the tension. If the muscle is stretched too far, the filaments overlap very little so that few cross-bridges can be formed and less tension is developed. On the other hand, if the muscle is initially short, the thin actin filaments overlap one another; upon further contraction the filaments interfere with the action of the cross-bridges. A muscle in the body, at its so-called *resting length*, is slightly stretched to very near the optimum length for development of maximum tension.

Source of Energy for Muscle Contraction

The rapid activation of the contractile mechanism that occurs when a muscle is stimulated requires the immediate release of some of the energy stored in the high-energy bonds ($\sim$P) of **adenosine-triphosphate** (ATP). Energy is released from ATP when cross-bridges are formed between actin and myosin. It is used to move the myosin heads and cause the filaments to slide; thus it provides the immediate source of energy for contraction. Energy from ATP is also needed to activate and run the calcium pump that returns calcium to the sarcoplasmic reticulum, without which cross-bridges would not break, and the muscle would not relax.

Maintained or repeated contractions require continuing energy release, but the amount of ATP present in the muscle fiber is very small and soon depleted (Figure 7–15). Another substance, **creatine phosphate** (CP), which also contains energy-rich bonds, is present in the muscle fibers in a concentration several times that of ATP. Its breakdown to creatine and phosphate provides energy that is used to recharge ADP to ATP. The creatine phosphate level diminishes during repeated contractions, but like ATP it can be reformed from creatine and phosphate if energy is available. The energy needed for this comes primarily from the breakdown of glycogen, which can be stored in skeletal muscle in considerable amounts and metabolized by glycolysis and the citric acid cycle (see Chapters 2 and 24) to fuel muscular contraction for prolonged periods of time.

Oxygen is not involved until the final step, but if there is not enough oxygen to accept all of the hydrogen ions produced in a "turn" of the citric acid cycle, that pathway will become overloaded. The pyruvic acid from partially metabolized carbohydrate, which is unable to proceed through the citric acid cycle, will be reduced to *lactic acid* instead. Should muscle activity continue, lactic acid will accumulate in the active muscles, while the energy for the continuing muscle contraction comes from the anaerobic phase of glycolysis (glucose to lactic acid). This is an advantage for the organism in that it permits continued contraction in the face of inadequate or limited oxygen. However, it also has a disadvantage, since it

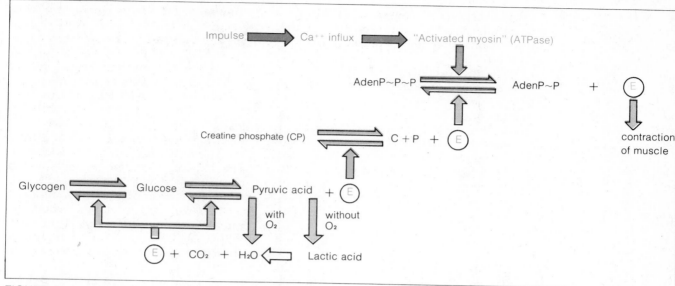

FIGURE 7–15   Release of energy for muscle contraction.

involves incomplete metabolism of the glucose — not all the available energy is released from the molecule. In order to provide a given amount of energy, many more glucose molecules must be partially metabolized than if complete oxidation through the citric acid cycle (to CO_2 and H_2O) were available.

The lactic acid formed slowly diffuses into the bloodstream and is carried to the liver. When oxygen supplies become adequate once more, about a fifth of the lactic acid is oxidized completely, and the energy released is used to reverse the reactions and form glycogen from the remaining four-fifths of the lactic acid. The metabolism of the lactic acid requires additional oxygen over and above that of current needs.

Oxygen Debt   It is possible for a skeletal muscle to contract for a time without oxygen, since although oxygen is vital to muscle activity, it is used more in recovering from, and preparing for, contraction than in releasing energy for contraction. In the absence of oxygen, contraction does not continue until all of the glycogen is reduced to lactic acid, or until all of the ~P has been released from ATP and CP. Con-

traction is limited by the ability of the tissue to tolerate the lactic acid produced. Since it is an acid, the increasing concentration of lactic acid in muscle cells tends to lower the pH of the intracellular fluid, which inhibits the action of intracellular enzymes. Individuals differ in their tolerance to lactic acid, probably because of differences in the rapidity with which lactic acid diffuses out of their muscle cells and into their circulations.

By accumulating lactic acid, you literally go into debt for oxygen, and you must repay that debt later with an increased oxygen uptake. The ability to incur an **oxygen debt** is particularly important in short bursts of strenuous exercise. For example, you might run a 100-yard dash in about ten seconds. As you are likely to hold your breath during part of the run, the entire requirement of about six liters of oxygen becomes an oxygen debt. It is repaid after the race by heavier breathing and increased oxygen consumption until all of the accumulated lactic acid is removed and the oxygen debt repaid. Some oxygen debt is incurred in light or moderate exercise as well, in the form of a lag in increasing the uptake of oxygen. The respiratory and circulatory systems take a few minutes to shift to a new level of supply. Even though the oxygen supply may soon meet the new needs, the initial deficit is not repaid until the exercise is completed. The availability of such a mechanism has great survival value for the organism. Limitation to a strictly pay-as-you-go metabolism would rule out the sudden exertion needed by a predator to capture its needed source of energy, or that needed by the prey to escape the predator. Humans too have equally urgent needs for immediate high rates of energy expenditure to meet their emergencies.

Efficiency and Heat of Skeletal Muscle Contraction Skeletal muscle is not mechanically perfect, since not all the energy released appears as useful work. Only 20–25 percent of the energy released when a muscle shortens can be accounted for as work, while the remaining 75–80 percent is dissipated as heat. The overall efficiency $\left(\dfrac{\text{work done}}{\text{energy released}}\right)$ is therefore 20–25 percent, which is comparable to that of most of our machines. When a muscle contracts isometrically, however, there is no shortening and no external work is done. All the energy appears as heat, and the efficiency is zero.

Careful analyses of the heat produced as a result of contraction show that about half the heat is produced during the contraction, and the other half is produced after the contraction. That generated during contraction is probably associated with structural changes in the muscle during shortening. The heat produced during and after relaxation is derived from the metabolic processes involved in restoring the muscle to its previous state.

Summary of Intrinsic Properties of Skeletal Muscle

In the preceding discussion, several properties of skeletal muscle have been discussed and others alluded to. It is therefore appropriate at this time to identify these properties, since together they mark the special qualities of contraction that distinguish skeletal from smooth and cardiac muscle, to be discussed below (Table 7–1).

We have seen that skeletal muscle is both extensible and elastic; that is, it can be stretched and it will return to its original length when the stretch is released. These properties arise both from the muscle tissue itself and from the elastic connective tissue within it. The tension a contracting muscle can develop varies with the extent to which it is stretched. Maximum contractile tension can be developed when the muscle is stretched about 20 percent, which is very nearly the degree of stretch applied to resting muscles in the body.

The duration of a single contraction of skeletal muscle is brief, and although it varies from muscle to muscle, it averages about a tenth of a second for the entire contraction and relaxation. To produce tetanus requires a frequency of stimulation in the range of 100/sec. The refractory period is very short, lasting only a few milliseconds, so that it is over before actual shortening begins.

Skeletal muscle is dependent upon its motor nerves, since it cannot contract automatically or spontaneously. In the intact animal, a skeletal muscle deprived of its motor nerve is paralyzed and cannot contract unless it is artificially and directly stimulated. Such a denervated muscle is not only useless, but it also soon undergoes other changes. It shrinks (*atrophies*) and eventually degenerates to be replaced by a connective tissue scar. Apparently muscle must be used if it is to maintain its integrity, since muscle unused for any reason atrophies. A temporary disuse atrophy is seen, for example, in muscles immobilized by a cast. On the other hand, excessive use has the opposite effect, enlargement of the muscle (*hypertrophy*), as is seen in well-trained individuals and body builders. Hypertrophy involves an increase in the size of each cell, but not in the number of cells. Skeletal muscle cells do not divide (it is difficult to imagine how mitosis could occur in a cell with so many nuclei).

TABLE 7-1   SOME PROPERTIES OF MUSCLE

Property	Skeletal Muscle	Smooth Muscle	Cardiac Muscle
Duration of contraction	brief, 0.1 sec (100 msec) average, (7.5–120 msec range)	long, several seconds	intermediate, about 0.3 sec
Frequency required to produce tetanus	high, about 100/sec (30–350/sec)	low, several/sec	cannot be produced
Duration of refractory period	very brief, less than 0.01 sec	brief	very long, about 0.2 sec (until beginning of relaxation)
Automaticity—rhythmicity	none—depends entirely upon nerves	single unit—more like cardiac, highly rhythmic multiunit—more like skeletal; more dependent on nerves	high—spontaneous rhythmicity
Innervation	somatic motoneurons, about 100–150 fibers per motor unit	autonomic nervous system, syncytial conduction within muscle, especially in single unit smooth muscle	autonomic nervous system, syncytial conduction with aid of specialized conducting system of heart
Mechanical properties	extensible, elastic	extensible, plastic	extensible, but with both elastic and plastic qualities

SMOOTH MUSCLE

Smooth muscle (Figure 7–16), also known as *visceral muscle* and as *involuntary muscle*, is called "smooth" because under the microscope the cells lack the striations found in skeletal and cardiac muscle. It is sometimes called "visceral" because it is found in the walls of the hollow organs of the viscera, such as the stomach, intestine, and urinary bladder. However, it also occurs in many nonvisceral structures, including the walls of most blood vessels. It is called "involuntary" because it is not under voluntary control; one cannot will the stomach wall to contract. This property is partly characteristic of the muscle cells and partly due to the fact that smooth muscle is innervated by the autonomic nervous system, an involuntary system (see Chapter 12).

Cells of smooth muscle are often arranged in sheets or layers, with the individual cells packed together like elongated sandbags. Most often the layer encircles a hollow or tubular structure, with the cells bound continuously by connective tissue to one another rather than to bone. Many organs have a second layer of smooth muscle, whose fibers run longitudinally.

Like skeletal muscle, smooth muscle is extensible, but it is characterized by plasticity rather than by elasticity. That is, it accommodates itself to a stretch without developing tension: with the cessation of the pressure or stimulus that has caused a smooth muscle to stretch, the muscle returns very slowly to its previous length: it does not "snap" back as do the more elastic muscles. A good example is the urinary bladder, which tends to "give" and expand as it is filled with liquid. There is very little increase of pressure within the bladder until it becomes quite distended.

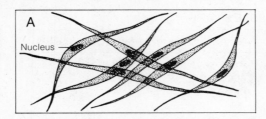

Nucleus

Manfred Kage/Peter Arnold, Inc.

Robert L. Kaplan

FIGURE 7-16   Smooth muscle.
A. A few isolated smooth muscle cells.
B. Photomicrograph of smooth muscle.
C. Electron micrograph of smooth muscle cells. 21,000× magnification.

Microscopic Anatomy of Smooth Muscle

Smooth muscle cells are very small spindle-shaped cells (0.02–0.5 mm in length), with a centrally placed nucleus. The cell membrane, as usual, is thin and virtually undetectable. The cytoplasm appears clear, but it contains mitochondria and other organelles, as well as contractile proteins. Actin and myosin are present, but there is no evidence of a systematic pattern of organization; there certainly are no striations. The sarcoplasmic reticulum is sparse, not organized, and there are no T-tubes. Presumably the intracellular mechanism of contraction is similar to that of skeletal muscle, but the absence of some sort of pattern of organization has hampered efforts to study it.

The nerve fibers to smooth muscle do not have specific endings on each muscle cell, but instead, run in the spaces between them so that each nerve fiber may establish contact with several different muscle cells as it passes. When the muscle cells are stimulated, their action potentials cause both the release of calcium ions from the sarcoplasmic reticulum and the diffusion of calcium into the cell from the interstitial fluid. The calcium initiates contraction and some sort of sliding of filaments. The mechanism of excitation, like that of contraction, is not as well-developed as in skeletal muscle, but it doesn't have to be because smooth muscle cells are so much smaller and the entire contractile process in smooth muscle occurs in slow motion. The action potential is longer, the latent period is longer, and the contraction develops more slowly and lasts longer; it is measured in seconds instead of milliseconds. Upon repeated stimulation, even at a low frequency, the individual contractions fuse into a smooth, sustained tetanic contraction in which a certain amount of force or tension is developed and maintained.

Single Unit Smooth Muscle

Functionally, there are two types of smooth muscle, which have different properties, distributions, and function.

They are single unit and multiunit smooth muscle. **Single unit** (or *visceral*) smooth muscle is organized to function as a single unit. The entire layer of cells responds together because the individual cells are connected by gap junctions. Since gap junctions are sites of low electrical resistance, any electrical change in one cell is transmitted through the gap junctions to adjacent cells. (Thus it is not so important for each cell to have direct contact with a nerve fiber.)

A particularly important property of single unit smooth muscle is that of **automaticity,** or *inherent rhythmicity,* which means that the muscle cells are able to contract without any known outside stimulation. Because of certain properties of the cell membrane, they are able to depolarize spontaneously.

The cell membrane is polarized, of course, with a resting potential which is smaller than that of skeletal muscle. Surprisingly, it is not constant—there is a fluctuation or oscillation, with rhythmically alternating phases of slight depolarization and hyperpolarization (Figure 7–17). It does not necessarily involve action potentials, but if the depolarization reaches threshold level, one or more action potentials are likely to occur. Because of the connections through the gap junctions, the oscillations (and any action potentials) are conducted to the other cells. Thus, electrical activity of all cells in that mass of muscle shows the same changes. The exact cause of this slow-wave fluctuation is not known, but presumably the action of the sodiumpotassium pump waxes and wanes.

Some smooth muscle cells behave as if the cell membrane leaks, allowing sodium ions to enter the cell, and the resting membrane potential declines (depolarizes) steadily until it reaches threshold when an action potential is discharged. After repolarization, the steady depolarization resumes and there is another action potential. Such cells fire rhythmically, and since the action potential is transmitted to other cells, they all fire together and rhythmically. The cell that generates the action potential is known as a **pacemaker**—it sets the pace for all the others. There is

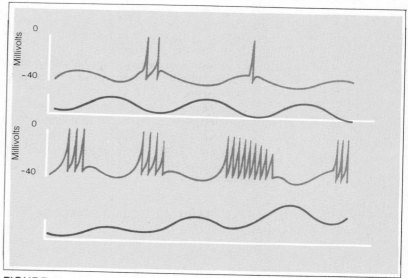

FIGURE 7–17   Electrical activity of single unit smooth muscle cells, showing spontaneous fluctuations in resting potential and action potentials (upper trace) and corresponding variations in tension of the muscle (lower trace).

nothing unique about the pacemaker cell, for all of these smooth muscle cells are capable of spontaneous depolarization. In fact, the role of pacemaker shifts randomly from one cell to another.

Single unit smooth muscle receives a nerve supply, but it also is influenced by certain chemicals and hormones, and it is particularly sensitive to stretch. Some of these factors, including stretch, tend to depolarize the membrane, increasing the likelihood that action potentials will occur during the depolarization phase of the resting potential. More action potentials mean stronger rhythmic contractions, or they may cause a continual discharge of action potentials that would result in a sustained contraction lasting many seconds. Other agents tend to hyperpolarize the membrane, decreasing the number of action potentials produced, and weakening the strength of contraction, or they may completely inhibit the production of action potentials resulting in a totally relaxed muscle. The frequency of rhythmic contractions is determined by the frequency of the slow-wave oscillations or the pattern set by the pacemaker cell, unless there is a sustained contraction. Because of its automaticity, single unit smooth muscle is not dependent upon its nerve supply, hormones, or other agents. Their effect is to increase or decrease the strength of a contraction.

Single unit smooth muscle is widely distributed, being found in organs throughout the body, and perhaps the best example is in the wall of the digestive tract.

Multiunit Smooth Muscle

Multiunit smooth muscle is more like skeletal muscle in several respects. It is organized into groups or bundles that respond individually when stimulated by their nerve, much like motor units in skeletal muscle. Multiunit smooth muscle is quite dependent upon its nerve for stimulation. It does not show automaticity, its resting membrane potential does not oscillate, and pacemaker activity is lacking. This type of smooth muscle is not stimulated by stretch, but it does respond to certain chemicals. One important location of multiunit smooth muscle is in the walls of blood vessels, where it regulates the vessel diameter.

The specific features and functions of single unit and multiunit smooth muscle will become more apparent when they are discussed as the organ systems of which they are a part. See Table 7–1 for a comparison of the properties of smooth muscle with those of skeletal and cardiac muscle.

CARDIAC MUSCLE

In some respects the properties of cardiac muscle are intermediate between those of smooth and skeletal muscle. Cardiac muscle is both involuntary and striated, but it is found only in the heart. Microscopically, the fibers appear to be long and cylindrical, with centrally placed nuclei and numerous cross-striped fibrils that give the whole fiber its striated appearance (Figure 7–18). The striations show the same banding pattern as in skeletal muscle and change similarly with contraction, indicating that the arrangements of the filaments and the mechanisms of contraction are nearly identical. The system of tubules and sarcoplasmic reticulum is present but less well developed. At intervals there are wider bands, **intercalated discs,** which appear microscopically as unstructured discs, but now are known to represent the boundaries between one cell and the next. A strand of cardiac muscle may consist of several

FIGURE 7–18   Cardiac muscle. A. Diagram of cardiac muscle cells. B. Photomicrograph of cardiac muscle.

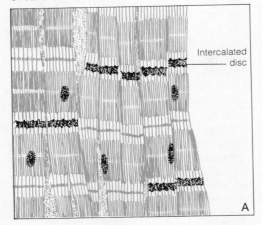
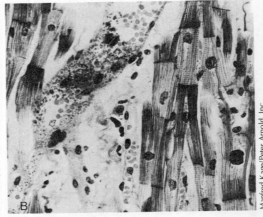

Intercalated disc

Manfred Kage/Peter Arnold, Inc.

cells in series. With the high resolution of the electron microscope, the ends of these cells can be seen to interdigitate with one another at the intercalated disc.

Cardiac muscle fibers tend to branch, so that they all seem to be connected to one another. At one time, cardiac muscle was thought to form a continuous sheet or *syncytium*, but when the intercalated disc was accurately described, it was recognized that cardiac tissue is composed of distinct cells. They are intimately connected however, for there are gap junctions between the cells at the intercalated discs. Because of this and the branching fibers, cardiac muscle behaves very much as if it were a continuous sheet. This is particularly true with regard to excitation and certain contractile properties. Hence cardiac muscle is sometimes said to be a "functional syncytium" if not an anatomical one (see Table 7–1).

Within limits, cardiac muscle is both extensible and elastic. It responds to stretch, as does skeletal muscle, by increasing the tension developed in contraction rather than by accommodating to stretch. But there is a certain element of plasticity involved as well, because between beats when it is relaxed, the heart readily distends as it fills with blood. Within a fairly wide range, the heart is able to distend to accommodate incoming blood without an increase in pressure—until it contracts.

One of the outstanding characteristics of cardiac muscle is its *automaticity*. An isolated heart—or part of a heart—can beat independently of any outside stimulus. The nerves serve mainly to control the frequency of contraction (the heart rate). As in smooth muscle, the automaticity is due to the presence of pacemaker cells that depolarize spontaneously and develop an action potential.

The role of the heart as a pump imposes additional requirements on the cardiac muscle. It must function as a unit, alternately contracting and relaxing in order to empty and fill its chambers. These requirements are met in part by the syncytial properties of the

heart muscle, and in part by the existence of a specialized conducting system within the heart. The action potentials developed by the pacemaker cells are rapidly conducted to the whole heart by this specialized conducting system, and all cardiac muscle cells contract as a unit. Because all of the cardiac muscle cells contract with each heartbeat, it is not possible to increase the strength of contraction by recruiting additional fibers as in skeletal muscle. (There are other means of doing it, however.)

The action potentials in cardiac muscle cells last longer than in either skeletal or smooth muscle cells, and so cardiac muscle has a very long refractory period. This means that it cannot respond to a second stimulus for a rather prolonged interval, and a tetanic contraction is not possible in cardiac muscle. Such a contraction, of course, would not be compatible with the pumping action required. The significance of these functional properties is discussed more fully in Chapter 15.

CHAPTER 7 SUMMARY

Anatomy of a Skeletal Muscle

Skeletal muscle fibers are large multinucleated cells marked by cross striations. The cell membrane (*sarcolemma*) encloses cytoplasm (*sarcoplasm*) that contains numerous *myofibrils*. The myofibrils consist of myofilaments, whose arrangement is responsible for the pattern of alternating light I bands and dark A bands. Thin, dark Z lines cross the I bands and define the *sarcomere*, the repeating unit of the myofibril. The A bands contain thick filaments of *myosin* and the I bands contain thin filaments composed chiefly of *actin*. Thin filaments extend part way into the A bands.

Sarcoplasmic reticulum (the ER of muscle) consists of tubules which lie along the surface of the myofibrils and form enlargements (*lateral sacs*) near the A-I junctions. Invaginations of sarcolemma extend into the muscle as transverse tubules or *T-tubes*. They wrap around the myofibrils near the A-I junctions, between lateral sacs.

Excitation and Contraction of Skeletal Muscle Fiber

A stimulus applied to a muscle cell causes an action potential which is conducted over the sarcolemma and down the T-tubes where it triggers release of calcium stored in the lateral sacs of SR. The calcium causes changes in the thin filaments that lead to binding of heads of myosin molecules to actin with the formation cross-bridges. Myosin also causes release of energy from ATP that moves the myosin heads and causes thick and thin filaments to slide over one another, increasing the amount of overlap. Rapidly repeated formation of cross-bridges, movement of myosin heads, breaking of cross-bridges and forming new ones causes each sarcomere to shorten (and hence also the myofibril and muscle fiber). The calcium is returned to the SR by active transport, the actin-myosin cross-bridges no longer are formed, and the muscle relaxes.

Contraction of Skeletal Muscles

A stimulus of sufficient intensity applied to a single muscle cell causes an action potential followed by a single contraction or muscle twitch that lasts for about 0.1 second.

A weak (*subthreshold*) stimulus to a whole muscle does not cause action potentials in any cells of that muscle, and there is no contraction. A stronger (*threshold*) stimulus causes action potentials in a few cells and a very small (*threshold*) muscle twitch. Increasing strengths of stimulus excite more muscle cells and the muscle twitches become progressively larger until all muscle fibers are contracting (*maximal contraction*).

A second stimulus applied before the twitch is completed may cause a *summation* of twitches, in which the second contraction is greater than the first (even though the first may have been a maximal contraction). This is possible because of the elastic properties of the muscle (the elastic elements are already stretched when the second contraction occurs), and a higher concentration of calcium around the myofilaments (the second stimulus causes release of calcium before that released by the first stimulus has been retrieved by the SR). Rapidly repeated stimuli cause not only a much greater contraction than a single stimulus, but also individual twitches fuse into a smooth, sustained contraction. This is a *tetanic contraction,* and is the physiological response of skeletal muscle.

If the interval between two stimuli is extremely short, the second stimulus may be applied before the sarcolemma has repolarized from the first stimulus. There is a response to such a stimulus, for it is delivered during the *refractory period* and does not cause an action potential.

Prolonged stimulation of an isolated muscle (with no blood supply) quickly causes fatigue. First the muscle fails to relax, then loses the ability to contract as well. An intact muscle fatigues very slowly, and quickly recovers when it does.

When a muscle contracts and shortens (*isotonic contraction*), it may lift a load and do mechanical work. If the load is too great and the muscle contracts but does not shorten (*isometric contraction*), it does no work but does develop tension (force). The amount of work a muscle does depends upon the weight of the load and the distance it is lifted.

Energy for muscle contraction is derived from ATP, which is replenished by energy obtained from *creatine phosphate* (CP) which, in turn, is replenished by energy released in the oxidation of glycogen stored in the muscle. Oxygen is not used until the final step, but if the oxygen supply is inadequate, the Krebs cycle is blocked, as is energy release. Some additional energy can still be obtained if lactic acid is produced, but when the lactic acid is finally oxidized, at some later time, extra oxygen above current needs must be taken in. This additional oxygen repays the *oxygen debt.* The ability to incur an oxygen debt allows the cycle to function even when the oxygen is inadequate.

Smooth Muscle

The involuntary muscle found in the walls of hollow and tubular organs and structures lacks striations and is called *smooth muscle.* The cells are very small and, although they contain actin and myosin filaments and SR, there is no detectable pattern of organization. The cells are joined by gap junctions and are usually arranged in sheets or layers. The response of smooth muscle is much slower than skeletal muscle, as a single contraction may last for several seconds.

Smooth muscle may be classified as a *single unit* or *multiunit.* Single unit muscle cells have an unstable resting membrane potential. In certain cells, called *pacemaker cells,* the resting potential fluctuates rhythmically, and whenever a depolarization reaches threshold, one or more action potentials will be generated. The electrical changes are conducted through the gap junctions to adjacent cells so that cells in that layer respond as a single unit. Stimuli such as nerve impulses or stretch of the muscle alter the strength of contraction, but do not affect its rhythm; that is determined by the rhythm of the pacemaker cells. Because of its *automaticity,* single unit smooth muscle can function after its nerves have been cut.

Multiunit smooth muscle is more like skeletal muscle in that it does not have pacemaker cells and does not demonstrate automaticity. It is more dependent upon its nerve supply, and the cells respond in small groups rather than as a single unit.

Cardiac Muscle

Cardiac muscle is found only in the heart. It is striated like skeletal muscle, which is an indication that the intracellular organization and the events of contraction are similar to those in skeletal muscle. Like smooth muscle, cardiac muscle cells are joined by gap junctions which are located in the *intercalated discs.* The muscle fibers also branch, so that electrical events are conducted throughout the heart muscle. There are pacemaker cells that initiate depolarization and a specialized conducting system to conduct impulses rapidly to all the muscle cells, so that cardiac muscle fiber contracts as a unit. Because of a very long refractory period, cardiac muscle cannot make tetanic contractions.

STUDY QUESTIONS

1 State the role of each of the following in skeletal muscle contraction: sliding filaments, T-tubes, calcium, lateral sacs of SR, ATPase, and cross-bridges.

2 Why do you suppose that tropomyosin and troponin are sometimes called relaxing proteins?

3 What changes occur in a muscle cell when the muscle contracts? When it is stretched? What causes these changes?

4 How does the response of an isolated skeletal muscle vary as the intensity of the stimulus changes?

5 If two maximal stimuli are applied to a muscle in quick succession, how is it possible that the response to the second stimulus is greater than the response to the first?

6 Could you produce a tetanic contraction in a muscle (isolated or intact) with stimuli of less than maximal intensity? Explain.

7 The refractory period in skeletal muscle is commonly demonstrated by noting the response to a quickly applied second stimulus. Would there also be a refractory period associated with a single stimulus? Explain.

8 What are the differences between isometric and isotonic contractions? With which one would you be able to do the most mechanical work? Explain.

9 If ATP supplies energy for skeletal muscle contraction, why are creatine phosphate and glycogen important for contraction?

10 What is oxygen debt? How is it incurred, and how is it repaid? How would your activities be affected if oxygen debt was impossible (strictly pay-as-you-go)?

11 What are some of the structural and functional characteristics which distinguish smooth muscle and cardiac muscle from one another and from skeletal muscle?

12 What is the significance of gap junctions in smooth and cardiac muscle? How do gap junctions relate to the functions of the muscles?

Part 3
Coordination and Control – The Nervous and Endocrine Systems

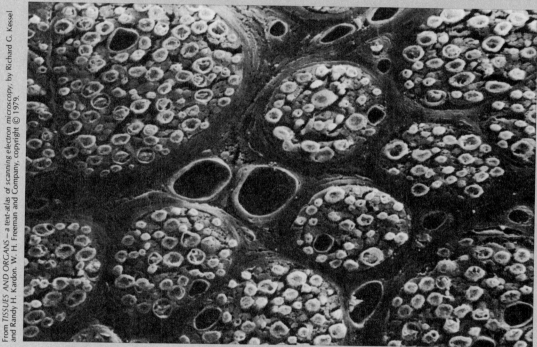

From *TISSUES AND ORGANS – a text-atlas of scanning electron microscopy*, by Richard G. Kessel and Randy H. Kardon. W. H. Freeman and Company, copyright © 1979.

Cross section of a nerve, showing bundles of nerve fibers, blood vessels, and connective tissue.

The survival of any complex organism requires coordination of the activities of the myriad cells of which it is composed. The activity of each cell is a response to a particular change in its immediate environment, and the activity of a complex organism results from the cooperative action of its many individual cells. This coordination of responses requires a communication system within the organism, to integrate the responses of the individual cells into a unified and appropriate reaction. This communication system connects cells sensitive to changes in their environment with those responsible for carrying out the organism's responses.

Information gathered by the many sensitive cells is relayed to centers where it is sorted out and evaluated to determine the best response, and directions are sent to the cells that carry out the response.

The two systems that share responsibility for communication are the nervous system and the endocrine system. In that it controls skeletal muscle activity, the nervous system senses and responds to changes that occur outside the organism. Insofar as it controls involuntary structures such as smooth muscle, the nervous system tends to regulate organ systems and help maintain the internal environment in response to signals arising within the organism. The communication medium is the vast number of nerve fibers connecting cells that detect change, and cells that respond, with the spinal cord and brain where the integration centers are located; the messages are nerve impulses, the action potentials conducted along the nerve fibers.

The endocrine system helps regulate such cellular activities as membrane transport and synthesis, particularly of proteins (enzymes) and secretory products. These actions are the bases of functions essential for maintaining the integrity of the whole organism as well as the cell—reproduction and the metabolism of energy sources and of minerals. The communication medium for the endocrine system is the blood; the messages are carried in the blood by hormones.

The distinction between the nervous and endocrine systems is not at all clear-cut. The two systems are not separate, unrelated, or antagonistic systems; rather, they work together as parallel and reinforcing systems of communication that are mutually dependent upon one another. They form a continuum of control whose actions are designed to maintain a suitable environment for the cells of the organism, (homeostasis).

The nervous system is covered in the next five chapters. The structure and function of nerve cells, the spinal cord, and the brain are discussed, as are the special senses and the regulation of responses—both of skeletal and involuntary muscle and of glands. The final chapter in Part 3 deals with the endocrine system and its regulatory role.

8

Cells of the Nervous System

Excitation and Transmission of Impulses

Introduction to the Nervous System – Its Organization and Components

The nervous system is a communications system based on the conduction of impulses that provide information and give directions. Structures called **receptors** are sensitive to certain changes in their immediate environment. Receptors in the skin, for example, generate impulses when stimulated by touch or pressure. Receptors are located on the ends of nerve fibers, which transmit the impulses to the spinal cord or brain. Other nerve fibers carry impulses from the spinal cord or brain to **effectors** such as skeletal, smooth, and cardiac muscle, and glands, which carry out the directions sent from the spinal cord or brain. Still other nerve fibers transmit information to other nerve fibers.

The operation of the nervous system involves receiving and processing information (impulses) generated by many receptors in many parts of the body, and directing the response of many effectors to bring about appropriate responses. This complex action depends upon the nerve cells. Chapter 8 presents the basic facts about nerve cells and their structure—how they conduct impulses, how they transmit information to effectors and to other nerve cells, and how receptors are stimulated. These parts are put together into a functional unit in Chapter 9.

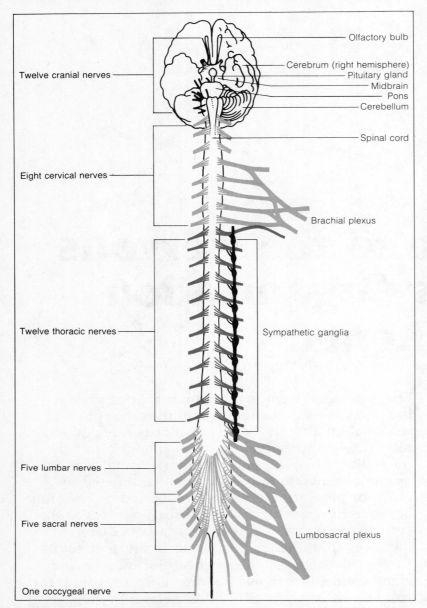

Twelve cranial nerves

Eight cervical nerves

Twelve thoracic nerves

Five lumbar nerves

Five sacral nerves

One coccygeal nerve

Olfactory bulb

Cerebrum (right hemisphere)
Pituitary gland
Midbrain
Pons
Cerebellum

Spinal cord

Brachial plexus

Sympathetic ganglia

Lumbosacral plexus

FIGURE 8–1   The central nervous system and the origin of the peripheral nervous system.

The nervous system can be and has been divided into many different fragments, each one of which is often referred to as a "nervous system." There is, however, only one nervous system, though it does have several parts.

Structurally, the nervous system may be divided into the central nervous system and the peripheral nervous system (Figure 8–1). The **central nervous system** (CNS) consists of the brain and the spinal cord. It contains the centers

where connections are made between nerve fibers bringing information from receptors and nerve fibers carrying information to effectors. The brain, which will be described in Chapter 10, is the highly specialized superior end of the spinal cord. It is capable of the sophisticated activities including consciousness, which in turn is necessary for perception and awareness, willed movements and complex movement patterns, learning, and memory. The brain exerts control over many spinal cord activities and even over certain endocrine glands.

The **peripheral nervous system** is made up of the nerves connecting the peripheral parts of the body with the central nervous system. Twelve pairs of nerves arise from the brain and are called the **cranial nerves.** Thirty-one pairs of **spinal nerves** arise from the spinal cord. There are eight cervical, twelve thoracic, five lumbar, five sacral, and one coccygeal nerve, named on the basis of the level at which they emerge from the vertebral canal.

Functionally, the nervous system may be divided into a somatic and an autonomic nervous system. The **somatic** component is what is usually meant by "the nervous system," since it innervates (supplies) the general body structures (*soma* = body). In particular, it innervates the extremities and the body wall, including the skeletal muscles and the skin. The **autonomic nervous system** supplies the involuntary effectors, such as smooth and cardiac muscle and the glands. Both the somatic and the autonomic systems are represented in the central nervous system, and both have peripheral components. They may even be represented in the same nerve.

CELLS OF THE NERVOUS SYSTEM

What is commonly known as a "nerve," such as one might find in an extremity, is made up of many nerve cell processes held together by connective tissue (Figure 8–2), in much the same way that muscle cells are held together. The outermost layer is the **epineurium,** which ensheaths the whole nerve and is continuous with the deep

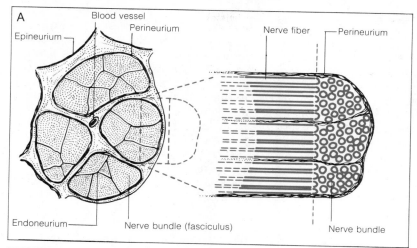
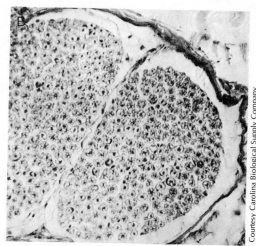

Courtesy Carolina Biological Supply Company

FIGURE 8-2   A. Cross section of a nerve trunk showing the connective tissue coverings and an enlargement of part of a single bundle of fibers. B. Photomicrograph of a cross section of a nerve trunk.

fascia. Groups of nerve fibers are surrounded by **perineurium,** and individual nerve fibers are enclosed by **endoneurium.** Small blood vessels also run in the connective tissue. As large nerve trunks approach the periphery they divide into progressively smaller branches, until finally only a single fiber remains.

At least three types of cells, other than connective tissue cells, are found in the nervous system. These are nerve cells or *neurons, Schwann cells,* and *neuroglia.* Schwann cells are found in association with peripheral nerves, and neuroglia (or glial cells) are found in the central nervous system.

Neurons

Nerve cells are among the most specialized of cells, and, as is typical among highly specialized cells they have lost the ability to reproduce. Once a nerve cell has been destroyed, it cannot be replaced, although a severed process may be regenerated. It follows that neurons normally have a long life span, since they must last throughout the lifetime of the individual.

Nerve cells are extremely irritable, and they readily respond to stimulation by initiating and transmitting one or more impulses; that is, they respond by conduction. Other kinds of cells, including muscles, can also conduct impulses, but neurons conduct impulses

more rapidly; their shape makes them particularly effective as conductors. Most neurons are elongated, and some are very long indeed. A single neuron may have one end in the spinal cord and the other in the toe, so that a single cell can be more than a meter in length.

A typical neuron consists of a **cell body** (or *soma*) and its processes which consist of a single **axon** and one or more **dendrites.** The dendrites are relatively short processes, often branching rather freely and they are specialized to receive stimuli (impulses) from other neurons. The axon is a long process that carries impulses to the site of contact with an effector or another neuron.

The cell body contains the nucleus which is fairly large and often off center and has a prominent nucleolus. The cytoplasm is marked by large easily stained granules called **Nissl bodies.** They are found throughout the cytoplasm except along the axon and in the region from which the axon arises, (the *axon hillock*) Nissl bodies are formed by layered bits of endoplasmic reticulum with an abundance of ribosomes (containing ribonucleic acid—RNA) along the surface and between the layers. Nerve cells contain more RNA than nearly any other cell in the body, and their rate of protein synthesis is correspondingly high. The Nissl bodies are the site of this protein synthesis. Mitochondria are also found both in the cell body and along the axon.

Types of Neurons Functionally, neurons are classified as motoneurons (motor neurons), sensory neurons, and interneurons (see Figure 8–3). **Motoneurons** carry impulses away from the central nervous system toward an effector. Somatic motoneurons have large cell bodies and many short dendrites. They are said to be *multipolar cells* because they have many processes. Their cell bodies are in the spinal cord or brain, and their axons extend all the way to the effector (skeletal muscle). This is not entirely true for autonomic motoneurons. The differences are described in Chapter 12. **Sensory neurons** extend from receptors in the periphery to the central nervous system. Structurally, sensory neurons are quite different from motoneurons and most other nerve cells. The cell bodies lie outside the central nervous system and have only one process; hence, sensory neurons are said to be *unipolar cells*. The single process splits into two parts, one of which goes to the central nervous system and is called the *central process*. The other extends toward the periphery and is called the *peripheral process*. The peripheral process arises from a receptor and carries impulses toward the cell body and the central process carries it on to the central nervous system. **Interneurons** resemble motoneurons in shape and form, but they are usually smaller and shorter. They lie entirely within the central nervous system where they transmit information from one neuron to another. In the central nervous system axons or central processes often divide, sending side branches (*collateral branches*) in several directions.

Nerve endings Motoneurons end at **neuroeffector junctions.** A somatic motoneuron branches many times, with each branch ending on a different skeletal muscle cell; these are **neuromuscular junctions.** Some motoneurons have endings on less than a dozen muscle cells while others have endings on (and exert control over) more than a hundred muscle cells. The motoneuron and its muscle cells is a *motor unit.* The terminal branches lose their coverings (myelin) as they near the muscle cells, and end as flattened knobs on the surface of each muscle cell. The sarcolemma is specialized in this region,[1] but the nerve ending does not actually touch it; a tiny space remains

FIGURE 8–3   Types of neurons. A. Motoneuron. B. Sensory neuron. C. Interneuron. D. Section of an axon.

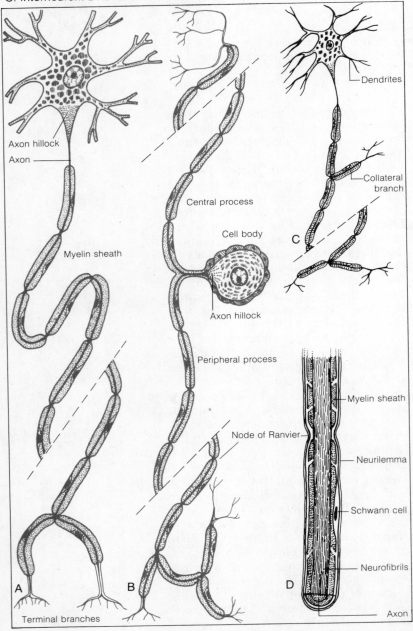

Axon hillock
Axon
Myelin sheath
Central process
Cell body
Axon hillock
Peripheral process
Dendrites
Collateral branch
Myelin sheath
Node of Ranvier
Neurilemma
Schwann cell
Neurofibrils
Axon
Terminal branches
A   B   C   D

[1]This area is commonly referred to as the **motor end plate,** but there is some confusion about just what is included. Some people use it interchangeably with *neuromuscular junction,* meaning the whole area; others consider the motor end plate to be the modified portion of the muscle cell membrane at the junction.

between them. Terminal branches of autonomic motoneurons also end on the surface of the effector cell, whether it be smooth or cardiac muscle, or a gland. As noted in the previous chapter, the nerve may not have specific branches to each smooth muscle cell, but may run between them, establishing contact with several muscle cells along the way. There is also a small space between the nerve and muscle cell in these junctions.

Interneurons and the central process of sensory neurons have similar terminals in that their axons branch profusely and end in tiny knobs, called **end-feet** or **presynaptic terminals,** on the surface of the cell body or dendrites of other nerve cells.[2] The junction between nerve cells is known as a **synapse.** One neuron may have end-feet on many different nerve cells, and many neurons may have end-feet on a single nerve cell. Again, there is a tiny gap (the *synaptic cleft*) between these endings and the next neuron. Because there is a small space between the two cells in all cases, synapses are physiological (functional) rather than anatomical (structural) connections.

---

[2]There are some exceptions to this, for in some locations there are junctions between two axons, or between two dendrites.

The peripheral endings of sensory neurons are receptors. They are part of the nerve cell and not separate organs. Their function is to detect changes in their environment and initiate impulses in the nerve fiber, thereby conveying information to the central nervous system about some aspect of their environment. Most of them are modified to be especially sensitive to particular kinds of stimuli, (Figure 8–4). For example, receptors in the eye have a low threshold for light energy, while those in the ear are most sensitive to sound waves. Of the several types of receptors in the skin, some are sensors for touch, some for temperature, and some for pain.

Receptors may be classified according to the stimulus to which they respond. **Exteroceptors** respond to stimuli arising outside the body. They include *telereceptors,* whose stimuli, such as light and sound waves, come from a distance, and *cutaneous receptors,* which monitor conditions on the body surface. **Interoceptors** are concerned with internal conditions, and they may be found in the walls of tubular and hollow structures. They may, for example, provide information about the fullness of an organ (such as the urinary bladder). However, the information from many interoceptors, such as those that

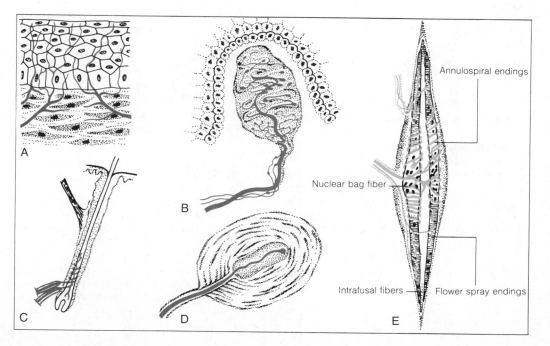

FIGURE 8–4
Examples of receptors. A. Free nerve endings in epithelium (pain). B. Meissner's corpuscles in skin (light touch). C. Touch receptors around the base of a hair. D. Pacinian corpuscle in skin and elsewhere (pressure). E. Muscle spindle in skeletal muscle (stretch).

Annulospiral endings

Nuclear bag fiber

Intrafusal fibers   Flower spray endings

monitor the pressure, oxygen content, or temperature of the blood in the vicinity of the receptor, does not reach the level of consciousness.

Proprioceptors also monitor internal conditions, but chiefly in those parts of the body under voluntary control. They are located within skeletal muscle, in tendons, and in the connective tissue surrounding joints, where they respond primarily to changes in stretch and tension. Some information from proprioceptors reaches the level of consciousness as the "kinesthetic sense" by which we are aware of the position of the body parts. Proprioceptive information that does not reach the conscious level is used to adjust the timing and intensity of muscle contractions and to ensure coordinated movements.

Receptors vary greatly in their complexity. Some, such as those associated with pain, are simply free nerve endings. Others are encapsulated structures that respond when some force stretches or presses upon them and so deforms them. The cutaneous receptors for touch and pressure are of this type, as are many interoceptors. They are sometimes called mechanoreceptors because a mechanical force, pressure (for pressoreceptors) or stretch, is the effective stimulus. Other receptors are sensitive to particular chemicals (chemoreceptors) or to temperature (thermoreceptors).

Among the more interesting and complex receptors are the **muscle spindles,** the proprioceptors in skeletal muscle (Figure 8-4E). They consist of small bundles of several striated muscle fibers called **intrafusal fibers.**[3] Most of these fibers have an enlarged area in the center with a number of nuclei and no striations. The striated ends of the fibers are capable of contraction and are supplied with their own motor nerves. (However, they are not capable of developing enough tension to contribute to muscle shortening.) The sensory nerve endings wrap around the central noncontractile part of the fiber (the *nuclear bag*) in a spiral fashion, and are called *annulospiral endings.* They are

stimulated when the nuclear bag region is stretched. Other endings, called *flower-spray endings,* are scattered along the surface of other intrafusal fibers, and they are stimulated when those fibers are stretched.

The connective tissue capsule that encloses the muscle spindle is attached to the muscle tendon and other connective tissue of the muscle, so when the whole muscle is stretched (an extensor muscle is stretched when flexion occurs), the muscle spindle is stretched and nerve endings are stimulated. When the whole muscle contracts, tension on the spindle decreases and stimulation of the receptors ceases. If, however, the intrafusal muscle fibers contract, the central portion of the fibers (where the nerve endings are) are stretched and the receptors are stimulated; they cannot distinguish between stretch caused by stretch of the whole muscle and stretch caused by contraction of their own intrafusal fibers.

Proprioceptors in the tendons (Golgi tendon organs) are also stretch receptors, and they respond whenever tension is put on the tendon. They fire (discharge impulses) when the muscle is stretched, and when the muscle contracts. The functional role of the muscle spindle and the tendon organs is further discussed in Chapter 12.

The special senses of smell, taste, vision, hearing, and equilibrium have not only unique and often intricate receptors, but also separate pathways and connections up to and including the brain. They involve truly specialized sensory mechanisms. The equilibrium receptors of the inner ear, however, are considered to be proprioceptors, probably because of their intimate relationship to muscular control.

Schwann Cells

Peripheral nerve trunks are made up of axons of motoneurons and peripheral processes of sensory neurons. But since they are indistinguishable in the nerve trunk, they are all called axons, and their cytoplasm is called **axoplasm.** The fibers vary greatly in diameter, from about 20 micrometers ($20\mu m$, 0.02 mm) down to less than $1\mu m$ (0.001 mm).

---

[3]*Intrafusal* means within the spindle-shaped capsule, as contrasted with the regular muscle fibers which are outside the spindle and thus are *extrafusal.*

The Coverings of Nerve Fibers

Beneath the fiber's connective tissue layer of endoneurium lie more specialized coverings. Most axons are covered with a chiefly lipid layer of **myelin,** which is responsible for the glistening whitish appearance of nerves in a fresh preparation. Individual fibers so enclosed are said to be *myelinated,* while those that lack this covering are said to be *unmyelinated.* The myelin sheath is interrupted at intervals by the **nodes of Ranvier** (see Figure 8–5). A thin membrane, the **sheath of Schwann** or *neurilemma,* appears to enclose the axon and its myelin. This sheath is formed by Schwann cells (or neurilemma cells), whose nuclei can often be seen along the axis cylinder; the nodes of Ranvier represent the junctions between the Schwann cells.

The electron microscope provides a clearer view of the relationships of these coverings to one another and to the nerve fiber (Figure 8–5). Schwann cells are quite large and have a great affinity for nerve fibers, which they tend to grow around and envelop. Several nerve fibers may become embedded in the cytoplasm of a single Schwann cell. Often the Schwann cell does not stop with simply enclosing the nerve fiber, but continues to grow and wrap around and around until a layer of considerable thickness is formed. This covering, made up of many layers of Schwann cell membrane, is the myelin sheath. The lipid content of myelin is due to the prominent lipid component of cell membranes. (The unmyelinated fibers are only enclosed by the Schwann cell, without the multiple wrappings.) The myelin sheath around large-diameter nerve fibers is thicker than that around small fibers, which are more likely to be unmyelinated. The myelin sheath is absent from the tiny terminal endings of nerve fibers and from the vicinity of cell bodies. There are no Schwann cells as such in the central nervous system, but certain of the neuroglial cells take over the function of Schwann cells. The myelin sheath contributes to neuron function in several ways. It is an electrical insulation that reduces the possibility of impulses in one nerve fiber affecting an adjacent fiber. Impulses travel faster in myelinated fibers (to be explained later in this chapter). Also the covering provided by the Schwann cells is necessary for any regeneration of an injured nerve.

Degeneration and Regeneration of Neurons

Although nerve cells do not undergo mitosis, they are able to recover from damage such as crushing or cutting of the nerve processes. Any damage that interrupts the continuity of a nerve fiber anywhere along its course is followed by marked changes in that neuron. There is little effect on the nerve fiber proximal to the damage, but changes known as *retrograde degeneration* occur in the cell body. The prominent Nissl granules seem to disperse or dissolve (*chromatolysis*), and the cell body swells. The magnitude of these changes is related to the proximity of the damage to the cell body. If it is too close, the cell may die, but ordinarily after some weeks the cell body regains a more normal appearance.

The portion of the neuron distal to the cut has been deprived of its necessary communication with the cell body and nucleus, and it undergoes complete disintegration (see Figure 8–6). The axoplasm deteriorates first and then the myelin, and macrophages move into the area and digest the remaining debris. All of this may take weeks to months, depending upon the extent of

FIGURE 8–5   Section of a myelinated nerve fiber showing a Schwann cell as the source of myelin.

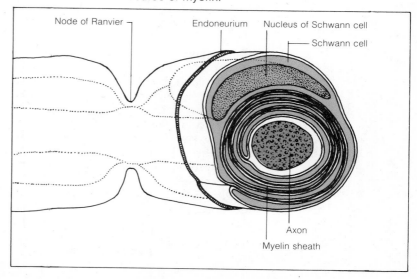

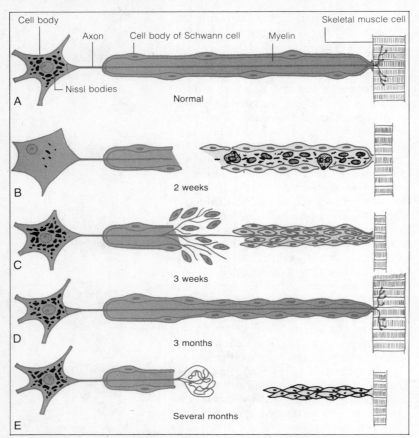

FIGURE 8–6 Degeneration and regeneration in an injured nerve fiber. A. Normal motoneuron. B. After the fiber is cut, the cell body undergoes chromatolysis, the severed portion of the fiber degenerates, and the muscle atrophies. C. Recovery. The nucleus begins to recover, the axon sprouts, and one branch finds its way through the cord of proliferating Schwann cells. D. The sprouting axon has reestablished contact with the muscle, which recovers once it can be stimulated again. E. The sprouting axon was unable to establish contact with the muscle. The sprouts grow in an unorganized manner, and the muscle remains atrophied.

the damage. In the peripheral nervous system the Schwann cells themselves survive and proliferate, forming columns of Schwann cells. Almost immediately the cut ends of the neurons in the proximal nerve stump develop tiny sprouts, some of which find their way into the channels formed by the Schwann cells, and slowly grow toward the periphery at rates of up to 1 or 2 millimeters per day. Functional connections are possible only if the Schwann cells direct the growing neurons to the right place—a sensory neuron that grows out to a skeletal muscle will not

be able to stimulate that muscle cell. The advancing nerve fibers are initially of small diameter, but they thicken as they mature and myelin develops around them once more. In the central nervous system the neuroglial cells that are responsible for myelinization do not survive, so there is no pathway to help direct the growing nerve sprouts. For this reason nerve fibers in the central nervous system do not reestablish functional connections after they have been severed.

Neuroglia

Neuroglia are found only in the central nervous system, where they account for about 90 percent of the cells and more than half of the cellular substance. But despite their abundance, their function is not well understood. They resemble connective tissue and, as the name implies, were once thought to serve a supporting function for the neural elements (neuroglia is literally "nerve glue"). But most neuroglia are not of connective tissue origin, and they do not provide the amount of support once attributed to them. Neuroglia do not conduct nerve impulses.

For the most part, the neuroglia are aptly described by their names. **Astrocytes** (see Figure 8–7) are star-shaped cells. They have many processes that often end in contact with blood vessels. They are the most numerous of the neuroglia and are believed to be involved in the transfer of substances between blood vessels and central nervous system cells. It has long been known that the mechanisms of transfer are different in the brain (the "blood-brain barrier") than in other parts of the body, and astrocytes have been implicated in controlling this transfer.

Oligodendroglia are cells with only a few processes. Some of the processes extend toward neurons, however, where they myelinate nerve fibers in the central nervous system much as Schwann cells do peripheral nerve fibers.

Ependymal cells line the tiny central canal of the spinal cord and the larger cavities, or ventricles, of the brain. In the ventricles some of the

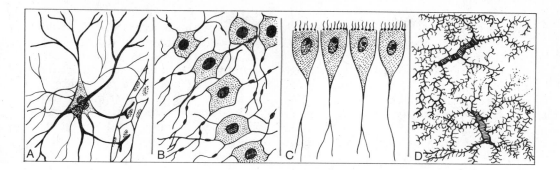

FIGURE 8–7
Types of neuroglia.
A. Astrocytes.
B. Oligodendroglia.
C. Ependymal cells.
D. Microglia.

ependymal cells are modified to produce the cerebrospinal fluid that fills these cavities and surrounds the brain and spinal cord. The formation of the cerebrospinal fluid is discussed in Chapter 10.

Microglia are the smallest of the neuroglial cells, and they are unique among them, since they develop embryologically from a connective tissue origin rather than a neural origin. They possess phagocytic properties and may become very active under certain conditions.

EXCITATION AND THE TRANSMISSION OF IMPULSES

The nervous system, in its role as a communications system, transmits information in the form of impulses or action potentials. Action potentials arise when receptors detect a particular condition in their environment. They must be transmitted to the central nervous system where the information has to be transferred to other neurons. Finally, impulses in neurons have to be translated into action potentials in effectors. (We have already seen in Chapter 7 how action potentials in a skeletal muscle lead to contractions.) Thus, we must be concerned not only with development of an action potential and its conduction along a neuron, but also with the initial stimulation or excitation of receptors and with the transmission of impulses across junctions between neurons (synapses) and between neurons and effectors (neuroeffector junctions or neuromuscular junctions when the effector is skeletal

muscle). The mechanisms by which these cells are excited is really the heart of nervous system function.

The Nerve Impulse

The neuron is the real conductor in the nervous system, for it conducts action potentials from one end of the neuron to the other. We have seen that the plasma membrane of all resting or inactive cells maintains a resting potential, which is a reflection of the electrochemical gradients that exist across that membrane. A stimulus to an excitable cell elicits changes in the cell membrane that may or may not evoke an action potential.

The nerve impulse has been defined in many ways, but one of the most concise and oldest describes it simply as a "self-propagated physicochemical disturbance."[4] In early descriptions it was compared to a spark traveling along a fuse, since it did not diminish, die out, or change perceptibly as it traveled along the nerve fiber. But, though the neuron contributes actively to conducting the disturbance, it is not consumed in the process as is a fuse. Calling it a "disturbance" recognized its transient nature, since the nerve impulse is a brief alteration of the potential difference across the nerve cell membrane. It may take many milliseconds to sweep the whole length of the fiber, but it lasts only a few milli-

---

[4]It is also, of course, the message passed along the nerve fiber communication medium. An impulse is therefore the characteristic *response* of a neuron to stimulation, although when the impulse reaches an effector or another neuron, it becomes the *stimulus* that causes a response in the effector or second neuron. An impulse may thus be both response and stimulus.

seconds at any one spot on the membrane, after which the neuron recovers swiftly and is soon able to conduct a second impulse.

The properties of nerve cells and the changes that occur in them and other excitable cells were not described much more precisely than the spark and fuse analogy until methods became available to measure and record accurately the extremely small and rapid changes that occur. The usual recording instrument now used is an *oscilloscope,* which resembles a television tube in many ways (Figure 8–8). It consists of an electron gun which shoots a beam of electrons at a fluorescent screen. The beam is focused as a single dot on the screen that can be observed and photographed. A pair of plates located on either side of the electron beam deflects it to the right or left when the potential difference between the plates is changed. The electronic circuit is usually arranged to make the beam sweep repeatedly across the face of the oscilloscope (from left to right) at a predetermined velocity. A second set of plates located above and below the beam deflects it vertically. These plates are connected to the tissue or cell under study, and the charge on them is controlled by the electrical events in that cell. Thus, the vertical deflection of the "spot" is an indication of the amplitude of the voltage change in the tissue, and the horizontal deflection or sweep provides information about the time course of the change. The inertia of this system is very slight, so that the responses are

extremely fast and very sensitive, and they provide an accurate representation of electrical changes in the cell. To produce electrical changes, the cell can be stimulated with electrical pulses delivered by an electronic stimulator. The characteristics of the stimulus (intensity, duration, frequency, etc.) can be controlled precisely and repeated exactly at will. With such instrumentation one can demonstrate some of the electrical properties of cells that were discussed in Chapters 2 and 7.

With one of a pair of electrodes placed on the external surface of the membrane of a resting nerve fiber and the other carefully inserted into the fiber, the oscilloscope will record the resting potential the instant the tip of the intracellular electrode enters the cell.[5] The inside of the axon is about 70–90 millivolts negative to the outside and holds steady. The magnitude of the voltage varies from one axon to the next, and it can also be altered by changing the external conditions.

If, instead, both electrodes are placed at two points on the external surface of the membrane, the oscilloscope will show no voltage difference between electrodes. The gradients of the resting potential lie *across* the membrane, not *along* it.

We have seen that a resting potential exists because of electrical and concentration gradients, primarily sodium and potassium, and of the relative impermeability of the membrane to those ions. Although electrical and chemical gradients are not identical to those described for muscle cells (in Chapter 2), the relationships and directions of the forces are the same. Both electrical and concentration gradients favor the entry of sodium into the cell, but for potassium the concentration gradient favors outward diffusion while the electrical gradient opposes it. The resting membrane is not very permeable to either

FIGURE 8–8   Diagram of a setup using an oscilloscope for recording electrical activity in a nerve fiber.

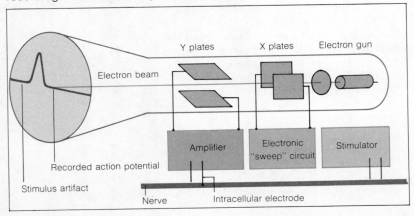

[5]Intracellular recording is done with a microelectrode made by drawing a bit of glass tubing out to a very fine tip (a few microns or less across) and filling it with a concentrated electrolyte solution, from which a wire is connected through an amplifier to an oscilloscope. Such studies were first done on the giant axon of the squid, which has a diameter of about 1 mm, many times that of any mammalian axon. More recently, using microelectrodes less than 1μ in diameter, the findings have been confirmed in mammalian axons.

one, but much less so to sodium. But since the membrane is not totally impermeable, some ions (particularly sodium) diffuse across and tend to reduce the resting membrane potential. It is maintained, however, by an active transport system, the sodium-potassium pump, which pumps sodium ions out and potassium ions in, thus preserving the concentration and electrical gradients and the membrane potential.

Action Potentials Although a nerve fiber is normally stimulated at one end by the action of another neuron or excitation of a receptor, it *can* be stimulated artificially by mechanical or electrical stimuli applied anywhere along its course. Both natural and artificial stimuli can produce an action potential or impulse. As previously described, the stimulus produces a slight localized depolarization, whose degree is related to the intensity of the stimulus. The stimulus increases the membrane permeability to sodium ions, allowing them to diffuse inward and causing the depolarization (Figure 8–9). If it is of threshold intensity, the permeability increase allows entry of enough sodium to bring about *sodium activation* (roughly a 500-fold increase in permeability to sodium) (Figure 8–10). The ensuing sodium influx results in total depolarization and partial reversal as the membrane potential approaches the sodium equilibrium

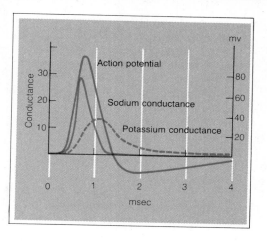

potential. At this time sodium activation ends and *potassium activation* occurs, allowing potassium ions to diffuse outward, which restores the internal negativity and the resting potential. The potassium activation develops more slowly and is longer lasting than that of sodium. These changes in permeability represent the opening and closing of specific pathways or "channels" through the membrane for sodium and potassium. The end result is the *action potential* or *spike*. It is very brief, and after 1–2 milliseconds the membrane is repolarized once more and ready for another stimulus.

After recovery from the action potential there are a few more sodium ions inside and a few more potassium ions outside the cell than before, but this is soon corrected by the sodium-potassium pump. Since the number of ions involved in one action potential is such a small fraction of the total, hundreds of action potentials could be produced without the Na-K pump before the resting potential would be seriously compromised.

If a stimulus is sufficient to trigger an action potential, a stronger stimulus does not change the potential. The size of the action potential is independent of the stimulus intensity; it is a property of the neuron, not the stimulus. In a given nerve fiber, spikes are all the same size (all else being equal), and action potentials (in muscle cells as well as neurons) are said to obey the **All-or-None Law**. In other words, there is no graded response of the action potential; if the stimulus is strong enough to elicit

FIGURE 8–9   Resting and action potential in a nerve fiber.

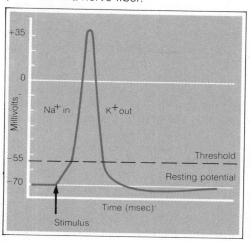

an action potential, it will be a maximum response.

We have spoken of a threshold stimulus solely in terms of its intensity, but the duration and rise time must also be sufficient (see Chapter 7 and Fig. 8–11). As in muscle, there is a relation between intensity and duration of the stimulus. A weak stimulus must be of longer duration to excite than a strong one, but there are limits at both extremes. These relationships are similar for nerve and muscle cells, except that the absolute values for both duration and intensity are much less for nerve cells.

A spike or action potential entails complete depolarization and temporary reversal followed by repolarization. The cell cannot undergo a second depolarization until it has repolarized after the first. This interval is the **refractory period.** It can be divided into an absolute refractory period and a relative refractory period, which reflect the fact that the excitability does not return instantaneously. The *absolute refractory period* extends from the beginning of the spike through the first part of repolarization. During this time the threshold is infinitely high, the excitability is zero, and the cell cannot be stimulated. The excitability returns as repolarization progresses, and during this interval the cell could be stimulated only by a stronger than normal stimulus. At the end of this, the *relative refractory period,* the membrane is repolarized and the excit-

ability and threshold are normal. The distinction between absolute and relative refractory periods is more significant in tissues that have a long refractory period which thus wears off more slowly (cardiac muscle). The duration of the refractory period determines the upper limit for the frequency of impulse conduction in a nerve fiber. If the refractory period is 1 msec, the maximal impulse frequency would be less than 1000 impulses/sec., but if the refractory period is 10 msec, the maximal possible frequency would be less than 100/sec.

Propagation of the Impulse

Conduction in a neuron. So far we have described the production of the action potential, but have said nothing about how it is conducted or propagated along the axon—other than to state that it is conducted without diminishing. The permeability changes associated with the action potential do not involve the entire neuron at once. Rather, at any instant only about a millimeter of the axon is involved. The polarity is reversed in that segment of a nerve fiber occupied by the action potential, so that it is relatively negative on the outside and positive on the inside. This creates a charge gradient with adjacent areas of the membrane and sets up local current flows (see Figure 8–12). On the outer surface of the membrane positive charges move into the active area from the membrane adjacent to it, and on the inside of the axon positive charges move from the active area to the adjacent area. Such current flows tend to depolarize that area, and soon an action potential develops there. That action potential sets up local current flows which depolarize the next segment, and so it goes, the length of the axon.

An axon is capable of conducting impulses in either direction. If it is stimulated midway along its length, the action potential developed at that site would set up local circuits of current flow that would depolarize the areas on both sides of it, and an action potential would be produced and conducted in both directions from the site of stimulation. Since a neuron is normally stimulated at one end, however, impulses are normally transmitted in only one direc-

FIGURE 8–11 Relationship between stimulus intensity and duration and the threshold of a nerve fiber.

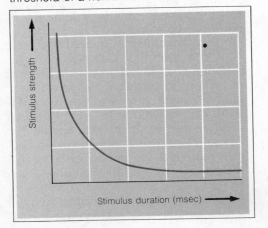

Stimulus strength

Stimulus duration (msec)

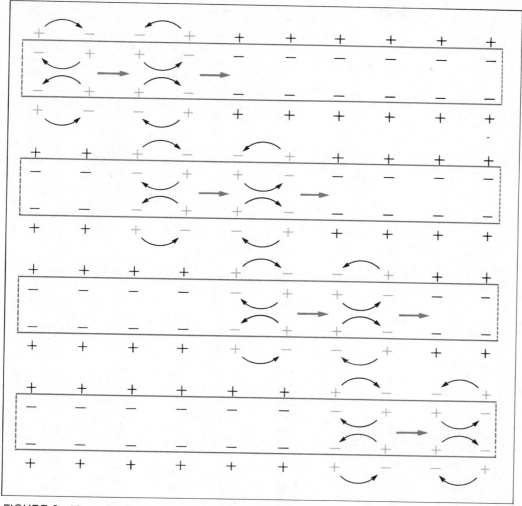

FIGURE 8–12   Impluse conduction in an unmyelinated fiber. An action potential sets up local current flows that cause depolarization of the adjacent membrane, and an action potential develops there.

tion. The local current circuits depolarize the membrane ahead of the action potential, while the area behind does not respond because it is refractory.

It is important to recognize that conduction is not a case of producing a single action potential that slides along the axon, but rather that depolarization at each site causes an action potential to develop at that site. That action potential causes depolarization of the next segment and an action potential develops in that segment, and so on. In effect, however, the segments are so short, and the events so rapid that we tend to think of the action potential as sweeping along the fiber. The rate at which an impulse travels depends upon the di-

ameter of the axon. Because of their electrical characteristics, conduction is much faster in larger neurons.

In myelinated fibers, the impulse is conducted somewhat differently. The myelin sheath is an effective insulator; it not only protects the axoplasm from the environment, but the flow of ions and electric current through it is negligible. Therefore, impulses in myelinated fibers cannot be conducted along the fiber by simple local current (and ion) flows depolarizing the adjacent area of the fiber, as described above. However, an action potential existing at a node of Ranvier, where there is a break in the myelin, sets up a potential gradient along the fiber between that node and the next one. Current and ion flows are concentrated in the region of the adjacent node, and an action potential is generated there. The action potential thus jumps from one node to the next. The myelin sheath, by enabling the impulse to skip along the fiber, reduces the magnitude of the ion shifts and greatly increases the conduction velocity. This is known as **saltatory conduction,** and it is typical of conduction in large nerve fibers. Without saltatory conduction our neurons would have to be much larger to achieve the conduction velocities of human axons. (In the squid an axon of $500\mu m$ in diameter conducts at about the same velocity as a mammalian myelinated fiber of $4\mu m$ in diameter.)

Conduction in a nerve trunk. A nerve trunk such as the sciatic or median nerve is made up of many fibers, large, small, myelinated, unmyelinated, sensory, and motor, which differ in many respects, but the basic mechanisms of resting and action potentials are the same. Many of the differences are associated with the diameter of the fiber, which in humans may be anywhere from $20\mu m$ down to less than $1\mu m$. In general, large fibers conduct more rapidly and have a lower threshold, a briefer action potential, and a shorter refractory period than small nerve fibers. Somatic motoneurons are large, but the autonomic motoneurons that supply smooth muscle are very small and unmyelinated. Most sensory neurons are quite large, but some pain fibers are very small.

The action potential recorded when a nerve trunk is stimulated is called a **compound action potential** because it is the sum of the action potentials of all fibers stimulated (Figure 8–13).[6] A weak stimulus will produce a small compound action potential because it causes action potentials in only a few fibers; a stronger stimulus will result in a larger compound action potential because more fibers are activated. This compound response is comparable in some respects to the graded response of a skeletal muscle.

When the recording electrode is placed on the nerve trunk far from the stimulus site, so that impulses must travel some distance to reach it, the contour of the recorded potential is changed. Instead of a single large spike, the deflection is made up of several component spikes representing fibers with different conduction velocities— the greater the distance of conduction the greater the separation of the peaks, just as in a race where all the contestants leave the starting line at the same time but are spread out at the finish line.

The action potential recorded at a distance from the stimulus site may show several discernible waves, indicating groups of fibers with distinct conduction velocities. These waves are designated alpha, beta, gamma, and delta waves. The fibers, however, are grouped as A, B, and C fibers, with the A group composed of the alpha (α) beta (β), gamma (γ), and delta (δ) subgroups. The A α fibers are the largest and fastest conducting; the C fibers are the smallest and slowest. The large diameter A fibers are found in muscle nerves (both motor and sensory neurons) and conduct at velocities in excess of 100 m/sec, while the small C fibers are found in autonomic nerves and some sensory nerves. C fibers are unmyelinated and conduct at velocities in the range of 1 m/sec or less. B fibers are small, myelinated autonomic fibers.

All living cells maintain a resting membrane potential, but only nerve and muscle, the *excitable tissues,* are capable of producing action potentials upon stimulation. Action potentials can be recorded from individual nerve and muscle cells but, outside of the laboratory, compound action potentials are more commonly recorded. They can give important information about the activity of the cells in question, since abnormal function is often accompanied by abnormal action potentials. Commonly recorded compound potentials include the electroencephalogram (EEG, "brain waves"), the electromyogram (EMG, from skeletal muscle), and the electrocardiogram (ECG, from the heart).

[6]To record from the many fibers of a nerve trunk, the recording electrodes are placed at two sites on the outside of the nerve trunk. Such electrodes do not record a resting potential, but can detect action potentials from many fibers.

Neuromuscular Conduction

Physiologically a skeletal muscle cell is stimulated by its motor nerve, but the impulse must first be transferred from the nerve ending to the muscle. Transmission of impulses across the neuromuscular junction is similar in some ways to stimulation and conduction in nerve fibers, since both involve depolarization of the cell membrane. The important differences are due largely to the fact that there is a small space between the nerve and the muscle.

The Neuromuscular Junction As the motoneuron reaches the periphery, it branches into many fine terminals, each ending on a different muscle fiber. The myelin is not present here, leaving only a thin layer of Schwann cell cytoplasm around the nerve fiber. The flattened nerve ending is covered only by a lidlike flap formed by the Schwann cell. The nerve terminals contain mitochondria and many tiny sacs or vesicles which contain a chemical, **acetylcholine** (ACh), that is synthesized by the neuron.

The muscle membrane portion of the junction, called the **postjunctional membrane** or **motor end plate** (Figure 8–14), is thrown into folds under the nerve endings, which greatly increases its surface area. The nerve and muscle membranes do not actually touch; there is a small gap of 30–50 nanometers (five or six times the thickness of the membrane) which must be bridged in order to excite the muscle cell.

Neuromuscular   Transmission Transmission across the neuromuscular junction probably involves the following sequence of events. The arrival of an impulse (an action potential) at the nerve ending causes a release of acetylcholine from several vesicles into the gap. The action potential apparently increases the permeability of the nerve cell membrane to calcium ions in the extracellular fluid and they enter the nerve terminals to trigger the ACh release. The ACh rapidly diffuses across the gap and combines with certain components of the postjunctional

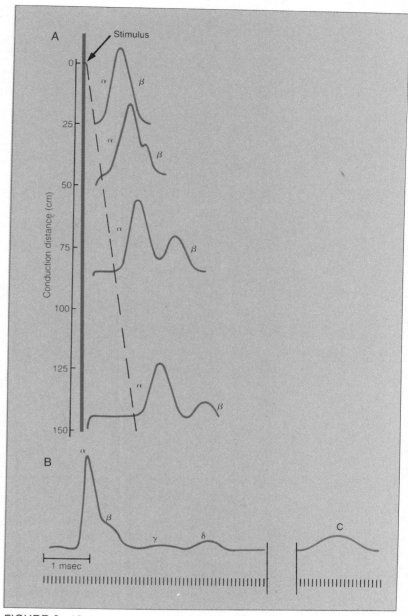

FIGURE 8–13   A. The action potential as recorded from progressively greater distances from the site of stimulation. B. The compound action potential recorded at a great distance from the stimulus site.

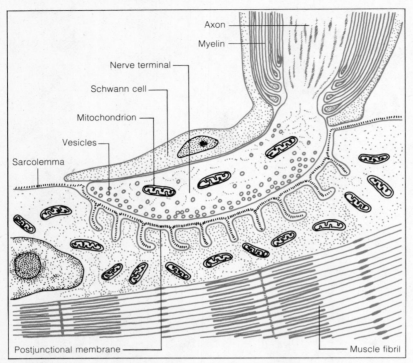

Axon
Myelin
Nerve terminal
Schwann cell
Mitochondrion
Vesicles
Sarcolemma
Postjunctional membrane
Muscle fibril

FIGURE 8–14   Diagram of a neuromuscular junction.

membrane called receptor sites.[7] The ACh-receptor combination alters the permeability of the muscle cell membrane to Na^+ and K^+, causing a slight depolarization of the end plate region, called the **end-plate potential** (EPP). In some respects, the EPP resembles the partial depolarization in a nerve fiber after a subthreshold stimulus, for it, too, is a local response that is not propagated and will disappear in a few milliseconds. The EPP is not an all-or-none response, and its magnitude varies with the amount of ACh that binds to the receptor sites. If the EPP is great enough, it will trigger an action potential that is propagated over the muscle cell membrane, setting off the events in the muscle that result in a muscle twitch. After repolarization of its sarcolemma, the muscle fiber can be stimulated once more.

---

[7]This is not to be confused with the receptors of sensory nerve endings. The term here refers to an unidentified substance that is receptive to the acetylcholine. It is probably part of a protein molecule in the muscle cell membrane with which acetylcholine forms a loose combination.

Normally, each impulse arriving at the nerve ending causes the release of sufficient acetylcholine to produce a large enough EPP to trigger one action potential in the muscle, and one contraction. Each nerve impulse triggers only one action potential (and muscle twitch), because the end-plate region contains an enzyme known as **acetylcholinesterase** (AChE), which inactivates acetylcholine. The AChE is located at sites on the postjunctional membrane very near, and perhaps between, the receptor sites for ACh, and it hydrolyzes the ACh almost as soon as the acetylcholine complex is formed. Failure to inactivate the transmitter could result in sustained depolarization of the muscle cell membrane.

Additional evidence favoring this mechanism comes from experiments with certain drugs. **Curare,** the active agent in a South American Indian arrow poison, apparently has an affinity for the same receptor sites as acetylcholine and successfully competes for them. After application of curare, nerve stimulation still causes ACh release, but the ACh-receptor complex cannot form, the EPP does not develop, and the muscle does not contract. By blocking neuromuscular transmission, curare causes paralysis of skeletal muscles, and it can be fatal if the muscles of respiration are paralyzed. With proper controls, however, the drug can be very beneficial; preparations containing curare derivatives are used clinically as muscle relaxants during surgery. There are also drugs that block or counteract the action of cholinesterase. The end result is also paralysis, but the mechanism is quite different from that of curare. Anticholinesterases permit depolarization but prevent the inactivation of ACh. Depolarization is maintained and so there can be no repolarization and subsequent stimulation.

Although nerve conduction and neuromuscular transmission have many things in common, the latter is more complex due to the introduction of a chemical transmitter. Both involve a local depolarization followed by a propagated spike or action potential. Because of the action potential, both obey the All-or-None Law, and both have refractory periods lasting until

Focus

Electroanesthesia

Most chemical anesthetics are lipids, which seem to work by dissolving in cell membranes and thus gain entry to nerve cells, where they interfere with synaptic transmission. This is true both of local anesthetics, applied by injection to peripheral nerve fibers, and of general anesthetics, usually inhaled as gases that enter the blood to reach the brain.

Unfortunately, chemical anesthetics often have assorted undesirable side-effects. They can cause nausea and headaches, and they may inhibit heart action and respiration. One intriguing alternative that researchers employed is the process of electroanesthesia.

Electroanesthesia was first conceived in 1902 in France, but because of technological limitations, nothing could be done with the idea until about 1940, when the French and Russians began to develop it. Today, it is increasingly regarded as a useful alternative to traditional anesthetics, especially in Europe. It has few side-effects beyond an occasional slight increase or decrease of heart rate, it can be turned on and off rapidly, it has a wide margin of safety, it is technically simple, and it can be used to produce both local and general anesthesia. Its main current drawbacks include some difficulties in initiating and maintaining the anesthetic effect and in achieving a suitable depth of anesthesia, although the researchers involved expect that these problems can be overcome.

The idea is not complicated. In one version of electroanesthesia, two electrodes are used, one on the forehead and one on the back of the neck. About 75 bursts of high-frequency electric current are then passed between the electrodes each second. The patient promptly enters a state indistinguishable from normal sleep. As long as the current is left on, the patient stays asleep and a surgeon can work uninterruptedly. When the current is turned off, the patient wakes up within about half a minute. Even then, the patient may remain insensitive to pain for several minutes.

Just how electroanesthesia works is not yet known. Various researchers have suggested that it activates the brain's sleep mechanisms or inhibits its waking processes. Others have suggested that as the current flows through nerve cells, it depolarizes them and holds them in a state which cannot produce nerve signals. It does work, however, and it has been used by dentists, obstetricians, and surgeons. Its simplicity has made it suitable for use on the battlefield, even as far back as World War II. The resemblance of its effect to sleep has even prompted its use as a treatment for insomnia.

repolarization has occurred. Both normally exhibit a one-to-one relation of stimulus to response (one stimulus produces one response). Because of the chemical mediator, however, the neuromuscular junction can transmit impulses in only one direction—an impulse cannot go from muscle to nerve. The increased complexity of the junction makes it more susceptible to all types of "adversity," such as fatigue, oxygen lack, and the action of many drugs.

Synaptic Transmission

The connections between neurons govern the flow of "information" through the nervous system. This channeling of information is determined by two things: the circuitry or "wiring" (that is, which neurons connect to which neurons) and the effectiveness of the connection of one neuron to the next. The anatomical circuits are often complex, but the basic patterns of neural circuits are completely dependent upon the connections between nerve cells.

The Synapse The junction between two neurons is called a **synapse.** Most synapses are in the central nervous system, although some autonomic synapses are outside of it and, because of their accessibility, have been studied in greatest detail. Synapses are structurally and functionally similar in many ways to neuromuscular junctions. In fact, the neuromuscular junction may be considered a special type of synapse.

Anatomically, the typical synapse is formed by the end of an axon and the cell body and/or dendrites of the next cell, typically a multipolar cell such as a motoneuron or interneuron. The many terminals of each incoming axon end in tiny knobs, the **end-feet** or **presynaptic terminals** (Figure 8–15). Each terminal has a number of tiny vesicles which presumably contain a chemical transmitter substance. Acetylcholine is the transmitter at some synapses, but several other chemicals are also known to be synaptic transmitters. There is a small but important space or *synaptic cleft* between the *presynaptic membrane* of the end-feet and the *postsynaptic membrane* of the stimulated cell.[8] Each postsynaptic cell receives many presynaptic terminals—one ending, or many, from any single neuron. Likewise, each incoming or presynaptic neuron has one or more presynaptic terminals on a number of postsynaptic cells. The phenomenon of one neuron (such as a motoneuron) receiving input from many incoming neurons is called **convergence;** the phenomenon of one neuron (such as a sensory neuron) branching to connect with many other neurons is known as **divergence** (Figure 8–16). If several interneurons and synapses are interposed between a sensory neuron and a motoneuron, the number of potential connections is almost infinite. It could be said that, given enough synapses, every sensory neuron is potentially connected with every motoneuron. The fact that stimulation of a sensory receptor does not actually excite every motoneuron, and hence every muscle, suggests that there must be some selectivity at the synapses to determine which motoneurons are to be excited. Each incoming impulse does not cause an impulse in its post-

FIGURE 8–15 A. A nerve cell body showing some of its numerous presynaptic endings. B. Diagram of a typical synaptic junction.

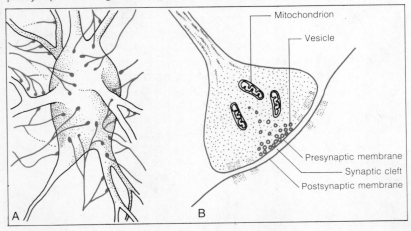

Mitochondrion

Vesicle

Presynaptic membrane

Synaptic cleft

Postsynaptic membrane

A

B

---

[8]There are exceptions to this however, for in some parts of the central nervous system, there are connections between dendrites, and between dendrites and axons. There are also gap junctions between some neurons similar to those between smooth and cardiac muscle cells. Gap junctions do not require a chemical transmitter.

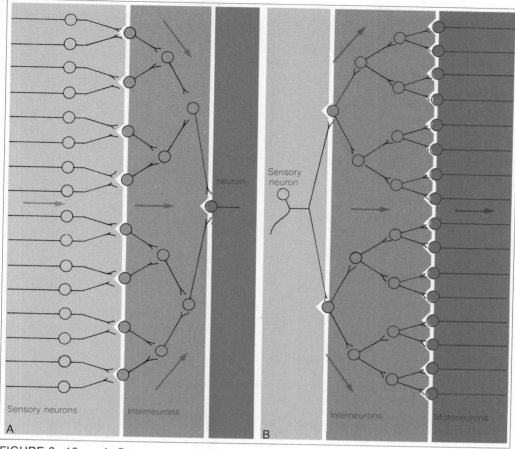

FIGURE 8–16   A. Convergence. B. Divergence.

synaptic cell because synaptic transmission is not on a strictly one-to-one basis as it is in the neuromuscular junction.

Excitation at the Synapse   The arrival of an impulse at a presynaptic terminal is believed to cause the release of a few packets of a transmitter substance. In autonomic synapses (they lie outside the CNS) the transmitter is acetylcholine. It probably is the transmitter at some synapses in the central nervous system but certainly not at all of them. As at the neuromuscular junction the transmitter diffuses across the gap and combines at certain sites on the postsynaptic membrane, where it causes a local increase in permeability of the membrane to sodium and a slight depolarization, which results in a local **excitatory postsynaptic potential** (EPSP).

Unlike the EPP, however, the EPSP, the local response produced by one impulse at the nerve terminal, is insufficient to cause an action potential in the postsynaptic membrane. Each EPSP is subthreshold, and in order to make the postsynaptic neuron "fire", an action potential summation is required. Enough depolarization can be developed by simultaneous excitation of a number of presynaptic terminals so that many sites on the postsynaptic membrane are slightly depolarized. This is called **spatial summation** because it involves summation from many sites. Summation can also be attained by a volley of impulses arriving at a single presynaptic terminal to build up the EPSP before it wears off. This is **temporal summation**—summation at one site over a period of time. Thus in order to make a neuron fire, the neuron must be excited by a certain number of impulses in a short period of time. There must be enough depolarization to reach threshold, just as in a nerve fiber or a neuromuscular junction. If threshold is not reached, the cell does not fire, but its excitability is increased until the EPSP is dissipated. The threshold is temporarily reduced and the neuron can be excited by a stimulus that ordinarily would not excite. We say that the neuron is **facilitated** during this time.

Facilitation is illustrated in Figure 8–17. Assume that each neuron requires summation of two active sites in order to fire. One impulse on neuron A would provide a threshold stimulus for neurons 1 and 2, but neurons 3 and 4 would only be facilitated; their stimulation would be subthreshold. Neurons 1 and 2 are said to be in the *discharge zone* as they discharge action potentials. Neurons 3 and 4 are said to be in the *subliminal fringe* as they remain at rest. One impulse on neuron B would fire neurons 5 and 6 and facilitate 3 and 4. If A and B were activated simultaneously, all six postsynaptic neurons would fire; the amounts of depolarization produced on neurons 3 and 4 by the two incoming neurons would sum to threshold.

Each neuron in the spinal cord probably receives a hundred or more presynaptic terminals, instead of the two shown in Figure 8–17, and some of the incoming neurons may have many more than two presynaptic endings on a single neuron. Those incoming neurons with many endings on a particular cell body have a much greater effect on the firing pattern of that cell.

Inhibition at the Synapse Not all presynaptic endings cause the development of an excitatory postsynaptic potential or contribute to facilitation; some may be inhibitory. A different chemical transmitter is released at these junctions, and when it reacts with sites on the postsynaptic membrane, it produces *hyperpolarization* and an **inhibitory postsynaptic potential** (IPSP). This transmitter causes different permeability changes, specifically, increased permeability to potassium and chloride

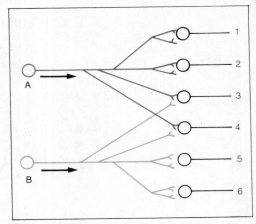

FIGURE 8–17   Facilitation. Stimulation of either neuron A or neuron B excites two neurons and facilitates the two neurons in the subliminal fringe. Stimulation of neurons A and B excites all six neurons.

ions. This allows positively charged potassium ions to cross the postsynaptic membrane and leave the cell body, causing it to become more negative inside the cell. This excitation requires depolarization, and the change in the membrane potential (hyperpolarization) is in the opposite direction. The IPSP is a local response, like the EPSP, and has a similar time course. It does not cause a different type of impulse, but simply makes depolarization of the cell more difficult. Many cells are probably being subjected to IPSPs as well as EPSPs at any given time, so that their final excitability is the algebraic sum of the inhibitory and facilitatory influences at that particular instant.

It is because of the ever-present effect of excitatory and inhibitory inputs that a neuron is less likely to fire a single impulse than a volley of impulses at a frequency dependent upon the level of excitatory and inhibitory activity. Excitatory input would increase the rate of firing, and inhibitory input would decrease it. The sources of input are many and varied, including cutaneous receptors, proprioceptors, and higher centers in the brain. The dual control—the balancing of excitation and inhibition—is characteristic of biological systems, where two opposing mechanisms counterbalance one another. You will see it in the neural control mechanisms of all organ systems.

Synaptic transmission has much in common with nerve conduction and neuromuscular transmission; it involves a chemical transmitter, one-way conduction, and depolarization of the postsynaptic membrane for stimulation. But there are two important additional considerations in synaptic transmission. One is that summation is required to reach threshold and cause the postsynaptic cell to fire. There is not a one-to-one relation between stimulus and response. As a result the response in the postsynaptic membrane does not follow closely the pattern of the input. The other major difference is the presence of inhibitory endings which tend to diminish the excitatory response. The synapse, as a more highly specialized junction, is more susceptible than the neuromuscular junction to fatigue, lack of oxygen, and the action of a number of drugs. Some synapses, especially those of the higher brain centers, are even more susceptible to such conditions. The characteristics of transmission, particularly at the synapse, should be kept in mind because many of the properties of reflex action are determined by synapses.

Excitation of Receptors

In the laboratory we can stimulate excitable tissues by applying a stimulus directly to the nerve or muscle, but physiologically there are only two ways by which they are stimulated. Most receive their stimulation from other nerves, across neuromuscular junctions or synapses. Sensory neurons, however, are excited when receptors on their peripheral ends are stimulated by some change in the immediate environment. Each type of sense organ is structurally adapted to convert the effects of a particular stimulus into nerve impulses. Each has a low threshold to one type of change, but it can also respond to other types of stimuli, although at a much higher threshold. Since the nature of the impulse or action potential is the same in all nerve fibers, the sensation (pain, touch, etc.) that is caused when a receptor is stimulated depends upon the connections that the sensory

neuron makes in the central nervous system. A given sensory neuron transmits only one type of sensory information, depending upon the type of receptors it has.

Generator Potentials Although receptors are the peripheral endings of the sensory neurons, with no gap or space between receptor and neuron, the excitation of a receptor in some ways resembles transmission across synapses and neuromuscular junctions. The process has been studied in some very large pressure receptors found in the dermis of the skin known as *Pacinian corpuscles* (see Figure 8–4). They have concentric rings of connective tissue around the nerve endings, so that they look rather like a sliced onion. Pressure applied to the surface of the receptor causes a slight depolarization of the nerve ending that resembles an EPP or EPSP; it is called a **generator potential** (see Figure 8–18). Like EPPs or EPSPs, it is not propagated, and the ion movements associated with it set up local current flows that cause an action potential to develop at the first node of Ranvier of the sensory neuron and which is then conducted along the neuron. The action potential does not "wipe out" the generator potential, however, for it persists if the stimulus continues, and as soon as the neuron repolarizes, it may be depolarized again producing a second action potential. Since a stimulus, and hence the generator potential, usually lasts much longer than the action potential, impulses in sensory neurons generally occur in bursts or volleys. The number of impulses in a burst and their frequency is related to the size of the generator potential, which, in turn, is related to the intensity of the stimulus.

Adaptation in Receptors One of the most interesting properties of receptors is **adaptation,** illustrated by the fact that while you are not normally aware of the collar against the back of your neck, you perceive it easily if you turn your head slightly. Adaptation, a decreasing response to a constant stimulus, is not due to fatigue, since the re-

FIGURE 8-18   The generator potential and the action potential in a sensory neuron.
A. A pressure receptor (pacinian corpuscle) and its sensory neuron.
B. Electrical responses to increasing increments of pressure. In a, b, and c, the stimulus causes subthreshold generator potentials. At d, the generator potential reaches threshold and causes an action potential. At e, the generator potential is strong enough to cause a series of action potentials.
C. Stimulation of the nerve ending in the receptor causes local circuit currents and development of an action potential at the first node of Ranvier.

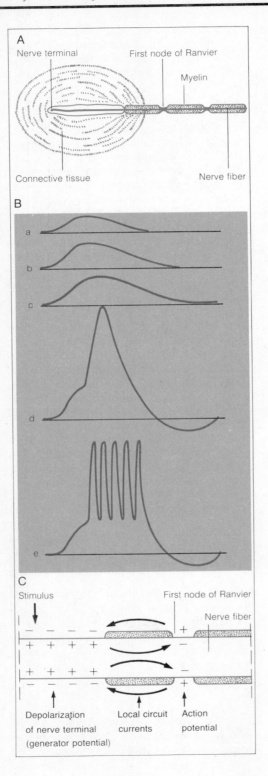

ceptor will fire readily if the stimulus is changed. Adaptation comes about because the generator potential in the receptor declines even with a steady stimulus. That decrease means a decrease in the frequency of action potentials in the sensory neuron. Adaptation by receptors is a convenient way to avoid cluttering the neural pathways and centers with unimportant information, while still allowing any changes in conditions to be immediately recognized.

Receptors differ greatly in the rapidity and degree of adaptation (Figure 8–19). Three of the receptors illustrated represent different degrees of touch, with the lightest touch adapting most rapidly. Muscle spindles adapt quite slowly, which is important because they provide information necessary for maintaining posture. Pain receptors (not shown) adapt very slowly, if at all. This may seem undesirable, but since pain receptors respond to unpleasant or harmful stimuli, their failure to adapt makes it difficult to ignore a potentially dangerous situation.

CHAPTER 8 SUMMARY

There are three kinds of cells in the nervous system; neurons, Schwann cells, and neuroglia. **Neurons** are true nerve cells and conduct impulses. *Motoneurons* arise in the central nervous system (CNS) and carry impulses to peripheral effectors. Each motoneuron consists of a cell body, a single long axon, and many short dendrites. The dendrites and cell body receive input from other cells, and the axon carries impulses to an effector. *Sensory neurons* carry impulses to the CNS from receptors. Their cell bodies, which are in a spinal ganglion, have a single process which divides into a peripheral process that carries impulses from receptors, and a central process that carries impulses into the CNS. *Interneurons* have one axon and many dendrites, and lie entirely within the CNS.

The axon of a motoneuron ends at a neuroeffector (neuromuscular) junction. Interneurons and central processes of sensory neurons terminate at synapses. The peripheral ending of a

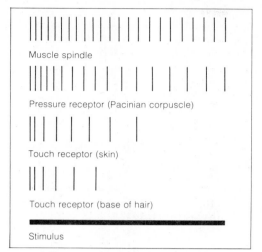

Muscle spindle

Pressure receptor (Pacinian corpuscle)

Touch receptor (skin)

Touch receptor (base of hair)

Stimulus

FIGURE 8-19   Adaptation. Typical firing patterns in sensory fibers during sustained stimulation of several types of receptors.

sensory neuron is a receptor. Receptors are sensitive to a certain stimulus, such as touch, pressure, stretch, chemical changes, or light or sound. Receptors may be classified as *exteroceptors, interoceptors,* or *proprioceptors,* depending on whether the stimulus arises outside (or on the surface) of the body, within the body, or in skeletal muscles, joints, or tendons.

Schwann cells are found only in the peripheral nervous system, where they surround nerve fibers. Their cell membranes wrap around some axons many times, forming a thick covering, the *myelin sheath*. The sheath is interrupted at the *nodes of Ranvier*, which represent the boundaries between two adjacent Schwann cells. Such nerve fibers are *myelinated*. Other nerve fibers are merely enclosed by the Schwann cell without multiple layers of wrapping and are *unmyelinated fibers*.

A nerve cell process severed from its cell body will degenerate, but it may grow back again (regenerate) if the channel formed by its Schwann cells and myelin sheath remain intact. The cut end of the nerve fiber sprouts and, if it grows into the proper channel, may eventually re-establish a functional connection.

Each neuron maintains a resting membrane potential, and a stimulus of threshold intensity causes it to depolarize, producing an *action potential* or *spike*. A fiber cannot respond to another

stimulus until it has repolarized; that is, until the refractory period is over. During the action potential there is an electrical gradient between the depolarized segment of the axon and the inactive portion adjacent to it. The gradient causes local current flows that lead to depolarization and development of an action potential in the adjacent segment, which leads to action potentials successively all along the nerve fiber. If stimulated midway, a nerve fiber can conduct an impulse in both directions but since neurons are normally stimulated at one end, they normally conduct impulses in only one direction. Impulses travel fastest in neurons that are myelinated because the local currents flow between one node of Ranvier to the next, with action potentials appearing only at the nodes. In general, large diameter neurons (which are usually more heavily myelinated) conduct impulses more rapidly than smaller unmyelinated fibers.

A *nerve trunk* contains axons of different diameters and conduction velocities. Nerve fibers may be classified in terms of diameters and conduction velocities. Large, rapidly conducting A fibers are subdivided into alpha, beta, gamma, and delta fibers. B fibers are smaller and slower, and C fibers are smallest, slowest, and unmyelinated.

Each terminal branch of a motoneuron ends on a different muscle cell where it lies close to, but not in direct contact with the sarcolemma, which is modified in the region (the motor end plate region). Each ending has many tiny vesicles that contain *acetylcholine* (ACh). An impulse in a motoneuron causes release of some ACh. It diffuses across the space and binds with proteins on the surface of the postjunctional membrane (sarcolemma) in the end plate region, which leads to a local depolarization known as the *end plate potential* (EPP). Normally the EPP is large enough to trigger an action potential in the muscle cell. An enzyme, acetylcholinesterase (ACE), present on the postjunctional membrane inactivates the ACh so rapidly that only one action potential develops. Each motoneuron impulse causes only one action potential and twitch in the muscle cell. Neuromuscular conduction is one-way only —from nerve to muscle.

The conduction between neurons occurs at a synapse, which also involves a tiny space between neuron membranes. Central processes of sensory neurons and axon terminals of interneurons synapse with cell bodies and dendrites of interneurons and motoneurons. The cell body and dendrites of a neuron (the postsynaptic membrane) may receive terminals from a great number of presynaptic nerve endings (*convergence*). A sensory neuron or interneuron may have synaptic endings on many postsynaptic cells (*divergence*).

Action potentials arriving at a presynaptic ending cause release of a chemical transmitter (which may or may not be ACh) which causes a local depolarization of part of the postsynaptic membrane called an *excitatory postsynaptic potential* (EPSP). It is similar to an EPP, but is never of threshold intensity. Excitation of the postsynaptic cell requires summation of many EPSPs, either by activation of many different presynaptic endings (*spatial summation*) or repeated stimulation from a few presynaptic endings (*temporal summation*). With so many presynaptic endings on a postsynaptic membrane, some of them are probably active at any given moment, raising the excitability of the cell, which is then said to be *facilitated*.

Some presynaptic endings release a chemical transmitter that causes a brief local hyperpolarization of the postsynaptic membrane known as an *inhibitory postsynaptic potential* (IPSP). It counteracts the effects of EPSPs, lowering the excitability of the cell and causing it to discharge impulses less frequently, or inhibiting it entirely.

Receptors on the peripheral ends of sensory neurons have a low threshold for a particular type of stimulus. Pressure applied to a pressure receptor causes a depolarization of the receptor membrane known as a *generator potential*. Its magnitude and duration depend upon the intensity and duration of the stimulus. The generator potential is not an impulse, but it does create an electrical gradient that results in local current flows with the first node of Ranvier of the sensory fiber. If the generator potential is great enough, an action potential or a series of action potentials is generated in the sensory neuron.

Many receptors may adapt to continuous stimulation and generate fewer action potentials in the sensory fiber, and then cease entirely. *Adaptation* is typical of touch receptors, but muscle spindles and pain endings show very little adaptation.

STUDY QUESTIONS

1 How do motoneurons and sensory neurons differ, in terms of their structure and the location of their cell bodies and endings?

2 What is myelin and how does it develop? How does an unmyelinated neuron differ from a myelinated neuron?

3 What happens to a neuron whose axon is severed?

4 What do neuroglia and Schwann cells have in common? How do they differ? What distinguishes them from neurons?

5 Distinguish among a stimulus, an impulse, and a response, as applied to a nerve fiber.

6 What is the resting potential, and how is it maintained?

7 What is the action potential? How does it develop, and how is it conducted along a nerve fiber?

8 What is the significance of the refractory period?

9 What is a compound action potential? How does its appearance differ if it is recorded centrally (near the central nervous system) or peripherally?

10 In what ways are the local response of an axon, the end plate potential at a neuromuscular junction, and the EPSP at a synapse alike?

11 What is the effect of an EPSP and IPSP? How are they alike, and how are they different? How are they related to facilitation?

12 What are some of the advantages and disadvantages of having a chemical transmitter for transmission at neuromuscular junctions and synapses?

9

Anatomy of the Spinal Cord

Nerves of the Spinal Cord

Spinal Reflexes

The Spinal Cord and Reflexes

Receptors generate impulses, neurons conduct them, and skeletal muscles contract, but none of them acts alone. The impulses cause effectors to respond in some way to conditions detected by receptors. Connections between the impulses coming from receptors and those going to effectors are made in the central nervous system. The nature of these connections is simpler in the spinal cord than in the brain, so we turn our attention first to those in the cord.

Anatomically, the spinal cord is highly organized, but the structural patterns established are continued throughout the spinal cord and, with modifications, on into the brain. Nerves that emerge from the spinal cord are specifically distributed throughout the body to supply all its parts. They carry sensory fibers from receptors and motor fibers to effectors.

In the second part of the chapter these parts, from receptor to effector, are put together into a unit, the reflex arc, which carries out reflex actions. The reflex arc, as the connection between stimulus and response, is the functional unit of the nervous system. Most of our actions involve reflexes of one kind or another, and here we concentrate on the most important spinal cord reflexes.

The **central nervous system** (CNS) consists of the brain and spinal cord. All nerves are connected to the central nervous system; their sensory neurons *terminate* in the central nervous system, nearly all of their motoneurons *originate* in the central nervous system (exceptions are some autonomic neurons, described in Chapter 12), and interneurons lie *entirely* within the central nervous system. The CNS is the site of the synapses and connections which are the basis of the integrating functions of the central nervous system. We begin our study of these functions in the spinal cord.

ANATOMY OF THE SPINAL CORD

General Structure

The brain and spinal cord are enclosed by connective tissue and encased by the bony skull and vertebral column. Actually a single organ rather than two separate ones, they blend into each other where the spinal cord passes through the foramen magnum into the cranial cavity. During growth the brain continues to fill the skull, but the spinal cord does not increase in length as much as the vertebral column, so that in the adult the cord ends at a level between the first and second lumbar vertebrae. Spinal nerves emerge all along the cord, but they descend within the vertebral canal to the intervertebral foramina through which they exit. Below the end of the cord, the descending trunks of these nerves are known collectively as the **cauda equina** (horse's tail) because of their appearance.

The anterior midline of the spinal cord is marked by a deep fissure, the **anterior median fissure.** A slight indentation in the posterior surface is the **posterior median sulcus.** The cord broadens in the region where the spinal nerves to the superior and inferior extremities emerge from the cord. The **cervical enlargement** (see Figure 9–1) extends from the level of C3 to that of T2, and the **lumbar enlargement** begins at about T9. Just below the lumbar enlargement the cord tapers abruptly (as

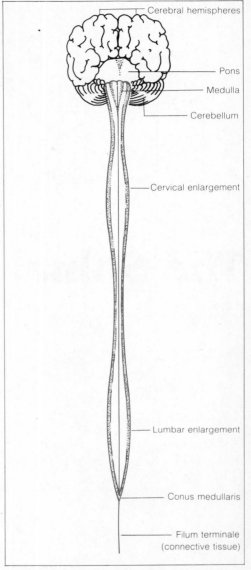

FIGURE 9–1   Anterior view of the central nervous system, showing the general shape and outline of the spinal cord and brain.

the *conus medullaris*) and ends in a slender fibrous thread, the **filum terminale,** by which it is attached to the coccyx.

Coverings of the Spinal Cord—The Spinal Meninges

The irreplaceable neural substance of the spinal cord is not only encased by the bony vertebral canal, but it is also provided with durable connective tissue coverings and a protective cushion of fluid. The three connective tissue layers are known collectively as the **meninges,** and individually, from the outside in, as the *dura mater,* the *arachnoid,* and the *pia mater* (Figure 9–2).

The **dura mater** is a tough fibrous tubular sheath which is continuous through the foramen magnum with the dura mater of the brain. Inferiorly, it extends beyond the end of the cord and encloses the cauda equina. The posterior (dorsal) and anterior (ventral) roots of the spinal nerves pass through sleeves of dura mater which extend laterally and blend into the connective tissue (fascia and epineurium) that covers the spinal nerves.

Beneath the dura mater lies the thinner and more delicate **arachnoid membrane.** It follows the course and contours of the dura, and it also enters the cranial cavity, where it is continuous with the cranial arachnoid.

The **pia mater,** the innermost layer, is quite thin and adheres closely to the surface of the cord. It follows all the fissures and grooves of the spinal cord and continues both along the nerve roots and into the cranial cavity.

Between the pia mater and the arachnoid is an appreciable space, the **subarachnoid space.** It is bridged by trabeculae formed from both pial and arachnoid tissue. These trabeculae give the arachnoid the spider-webby appearance suggested by its name. Laterally, on either side of the cord, are the **denticulate ligaments,** which extend from the pia to the dura and serve to keep the spinal cord centered in its tubular encasement. The subarachnoid space is filled with the **cerebrospinal fluid,** which provides the protective cushion. Cerebrospinal fluid around the spinal cord is continuous with that surrounding the brain.

Cross Section of the Spinal Cord

Gray and White Matter The outstanding feature of the spinal cord in cross section is the presence of a centrally placed H-shaped area, the **gray matter,** surrounded by bundles of nerve fibers ascending or descending within

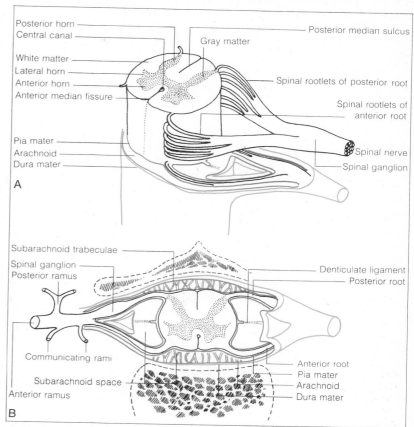

FIGURE 9–2   The spinal cord and meninges. A. Three-dimensional view with spinal roots and meninges. B. Cross section through a vertebra, showing the spinal cord and meninges in the vertebral canal.

the cord (see Figure 9–2). Because most of these fibers are myelinated, they well deserve their name of **white matter.** Gray matter contains cell bodies and interneuronal junctions (synapses), neither of which is myelinated, hence the gray appearance. The projections of the gray matter in cross section are known as the **posterior** (or dorsal) and **anterior** (or ventral) **horns,** although many anatomists prefer to call them the posterior and anterior *gray columns,* because they are actually columns extending the length of the spinal cord.

The relative size and shape of the gray and white matter in spinal cord sections vary with the cord level. The amount of white matter decreases with distance from the brain; the amount of gray matter changes less evenly. There is more gray matter in the regions of the cervical and lumbar enlargements, where neurons to and from the limbs must make their connections, and there is relatively little in the thoracic region, although a **lateral horn** is present in the thoracic and upper lumbar regions.

Anterior and Posterior Roots

Emerging from the cord, just off the tips of the posterior and anterior horns, are the **posterior** and **anterior roots,** which unite to form nerves. The anterior roots contain only axons of motoneurons, both somatic and autonomic. The neurons that supply skeletal muscles are large multipolar cells whose cell bodies and dendrites are located in the anterior horn and whose axons extend all the way to the muscle. In fact, skeletal muscle motoneurons are often called anterior (or ventral) horn cells.

The posterior roots contain only axons of sensory neurons. Their cell bodies lie outside the central nervous system in a conspicuous enlargement on each posterior root known as the **spinal ganglion** *(posterior root ganglion).* A ganglion is a collection of cell bodies outside the central nervous system, so in a sense it is a small island of gray matter. Spinal ganglia, therefore, contain the cell bodies of sensory neurons whose processes extend from the periphery to the gray matter of the posterior horn via the posterior root.

Fiber Tracts The nerve fibers in the white matter of the spinal cord are organized into bundles of fibers, or **fiber tracts.** All the fibers of each tract have a similar origin and destination (Figure 9–3). Ascending tracts serve a sensory function, as they carry impulses toward the brain. Descending tracts serve some aspect of motor function, since they carry impulses from the brain to the motoneurons. The white matter immediately surrounding the gray matter consists of short fiber tracts that provide connections between different levels of the spinal cord. Fiber tracts are usually named according to their origin and termination. Thus a *spinothalamic tract* is made up of ascending fibers running from the spinal cord to the thalamus in the brain. The *corticospinal tracts* descend from the cortex of the brain to the spinal cord. There are a great many tracts, some of which are shown in Figure 9–3.

NERVES OF THE SPINAL CORD

Nerves are formed when the anterior and posterior roots unite and leave the spinal cord and vertebral canal to become part of the peripheral nervous system, rather than the central nervous system. They are known as **spinal nerves,** but when fibers from several spinal nerves join, the nerves formed are called **peripheral nerves.**

Spinal Nerves

Spinal nerves do not have names; they are identified instead by the level at which they leave the vertebral canal. In general, spinal nerves emerge below the vertebra for which they are named. Cervical nerves, however, emerge above, which explains why there are more cervical nerves than cervical vertebrae. C8 emerges between the seventh cervical and first thoracic vertebrae.

Tiny anterior and posterior rootlets emerge from the cord in a continual line, but several combine to form the roots for each nerve (see Figure 9–2A).

FIGURE 9–3 Some important ascending and descending fiber tracts in the spinal cord. Ascending tracts are shown only on the left, and descending tracts only on the right.

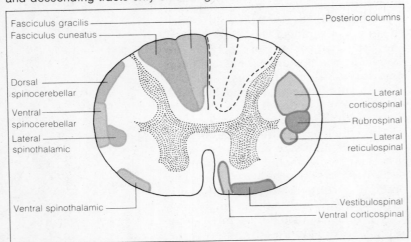

Fasciculus gracilis
Fasciculus cuneatus
Posterior columns
Dorsal spinocerebellar
Ventral spinocerebellar
Lateral spinothalamic
Ventral spinothalamic
Lateral corticospinal
Rubrospinal
Lateral reticulospinal
Vestibulospinal
Ventral corticospinal

The spinal nerve is formed by the junction of the anterior and posterior roots, which usually occurs somewhere in the intervertebral foramen or within the vertebral canal. Near the junction of the roots, there are one or two tiny branches, the *communicating rami*, for connection with the autonomic nervous system.

In spite of irregularities in the shape of the human body, the courses of the spinal nerves are remarkably similar (Figure 9–4). Each nerve very quickly divides into an anterior and posterior ramus (branch). The smaller **posterior ramus** supplies the intrinsic muscles of the back and the skin covering them. The much larger **anterior ramus** follows around the body wall toward the midline anteriorly and branches in such a way that the entire body wall is completely innervated.

The systematic pattern of distribution of the spinal nerves is most apparent in the cutaneous innervation, since the area of skin supplied by each spinal nerve forms a reasonably horizontal band, called a **dermatome** (Figure 9–5). Although the area for adjacent spinal nerves overlaps, the pattern is followed quite closely, even in the extremities. With the aid of such maps, it is possible to determine with a fair degree of accuracy which spinal nerve innervates a given cutaneous area. Such knowledge is important to neurologists when testing to determine the level and extent of damage in the spinal cord.

Innervation to skeletal muscles develops in much the same manner, but the pattern is less obvious because developing muscles often migrate from their site of embryological origin. The diaphragm, for example, is innervated by cervical nerves.

Peripheral Nerves and Plexuses

Spinal nerves maintain their identities fairly well in the thoracic region as intercostal nerves, but in the extremities they cannot be identified as such (see Figure 8–1). In these regions anterior rami of several spinal nerves unite, divide, and recombine, so that the **peripheral nerves** finally formed contain

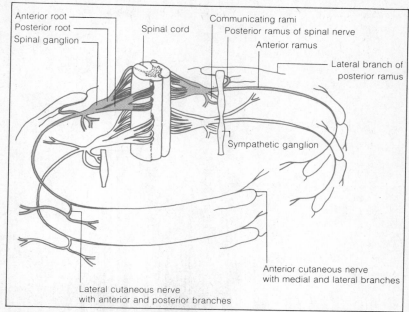

FIGURE 9–4   The distribution of a spinal nerve.

fibers from several spinal nerves. The network of nerve trunks in which this recombination occurs is called a **plexus,** and the nerves that arise from one are designated by names rather than numbers. The major plexuses are the *brachial plexus* and the *lumbosacral plexus,* whose peripheral nerves supply the superior and inferior extremities, respectively. By following the course of individual spinal nerves through a plexus and into the peripheral nerves, one can see how a segmental distribution of the spinal nerves is maintained in the periphery.

Cervical Plexus   The cervical plexus (Figure 9–6) is a small plexus formed by the upper cervical nerves. The major peripheral nerve formed by it is the **phrenic nerve,** which supplies the diaphragm. Other branches supply skin and muscles in the neck.

Brachial Plexus   The brachial plexus (Figure 9–6) is formed by spinal nerves C5–C8 and T1. Each contributes an anterior and posterior portion (of the anterior rami only). The anterior portions form a medial and a lateral cord, while the posterior portions form a pos-

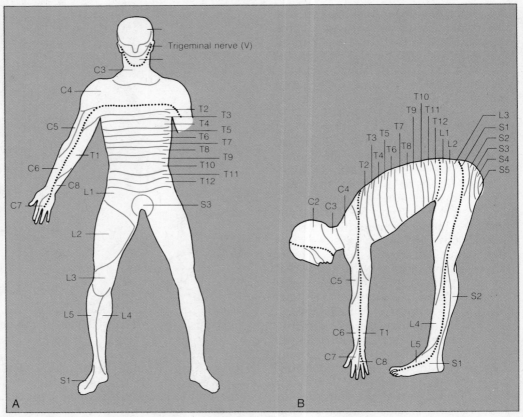

FIGURE 9-5   Cutaneous distribution of spinal nerves (dermatomes). A. Anterior view. B. Lateral view to emphasize the segmental pattern of the distribution.

terior cord. (The cords are named for their anatomical relation to the axillary artery.) The cords form four major nerves to the upper extremity and many smaller cutaneous and muscular branches.

1 The **musculocutaneous nerve** serves the muscles on the anterior surface of the arm and the skin on the lateral surface of the forearm.

2 The **median nerve** passes with the major blood vessels to the anterior surface of the forearm. It supplies the anterior flexor muscles of the forearm and the skin and muscles of the palmar surface of the hand.

3 The **ulnar nerve** passes behind the medial epicondyle of the humerus (a blow on the "crazy bone" stimulates the ulnar nerve) and supplies the skin and muscles along the ulnar (medial) surface of the forearm and hand.

4 The **radial nerve** passes posteri-

orly beneath the triceps brachii to supply the skin and muscles of the posterior surface of the arm, forearm, and hand.

Other nerves include several branches to the scapular muscles, the *axillary nerve* to the deltoid region, and several cutaneous branches to the anterior and posterior surfaces of the thorax.

Lumbosacral Plexus The lumbosacral plexus (Figure 9-7) contains two overlapping sections. The lumbar part is formed by spinal nerves L1-L4, and a small part of T12, and the sacral portion by spinal nerves L5 and S1-S3. A number of peripheral nerves arise from the plexus to supply structures in the pelvic and hip region, plus two major nerves to the inferior extremity.

1 The **femoral nerve** arises from the lumbar portion of the plexus. It passes beneath the inguinal ligament, with the femoral blood vessels, to supply the skin and muscles of the anterior surface of the thigh. A superficial branch, the **saphenous nerve,** supplies an extensive area of skin, extending all the way to the foot.

2 The **sciatic nerve,** the largest nerve in the body, arises from the sacral portion of the plexus. It passes through the sciatic notch of the pelvis and down the posterior surface of the thigh to supply the posterior thigh muscles. It ends by branching into the **common peroneal** and **tibial nerves,** which innervate most of the leg and foot.

The lumbosacral plexus also gives rise to the *obturator nerve* to the medial surface of the thigh and the adductor muscles, the *pudendal nerve* to the region of the perineum, the *gluteal nerves* to the gluteal muscles, and several cutaneous branches.

All of these nerves, except possibly their smallest terminal branches, are mixed nerves containing both sensory and motor fibers. Muscle nerves carry sensory fibers from the proprioceptors, and skin nerves carry motor innervation for sweat glands, piloerectors, and the smooth muscle of blood vessels. The cutaneous motor fibers, however, are autonomic neurons.

FIGURE 9-6 A. Major branches of the cervical and brachial plexuses. B. Major peripheral branches to the anterior surface of the arm. C. Major peripheral branches to the posterior surface of the arm.

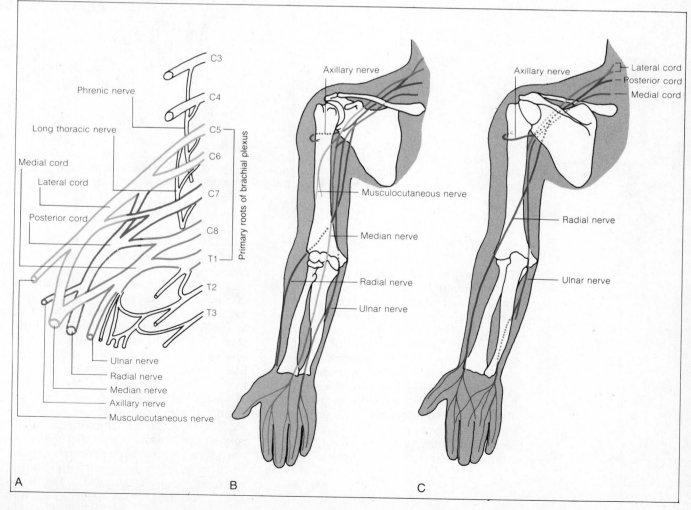

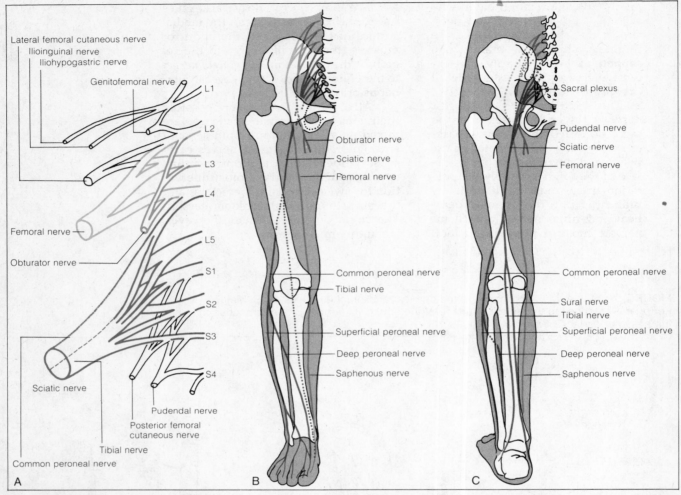

FIGURE 9–7   A. Major branches of the lumbosacral plexus. B. Major peripheral branches to the anterior surface of the leg. C. Major peripheral branches to the posterior surface of the leg.

SPINAL REFLEXES

Everyone has some concept of what is meant by reflexes and reflex action, but a good definition is difficult to achieve, and is usually replaced by descriptive adjectives. Reflexes are responses to stimuli; they are not spontaneous. They are involuntary, even though voluntary skeletal muscle is often involved—one does not "decide" to have a reflex. The response is predictable and stereotyped, and a given stimulus will always provoke the same response. There is no originality in it, although the intensity and duration may vary. And reflexes are often said to be purposeful (as, for example, when you withdraw your hand from a painful stimulus), although this is not always true, since some responses are not totally appropriate.

Reflex action is not simply a means of causing a muscle to contract, however. It is a mechanism for control and, as such, it regulates the strength, duration, and timing of the contraction of one or a group of muscles, and it very likely inhibits the contraction of other muscles at the same time. The response of a skeletal muscle (or lack of it) depends upon impulses delivered by its motoneurons. Each motoneuron receives input from many other neurons, both sensory and interneurons, and all of them communicate with the motoneurons through synapses. The synapse is so important to reflex action because transmission of impulses through synapses can be regulated to such a remarkable degree. Properties such as facilitation and the existence of inhibitory as well as excitatory connections determines excitability of the postsynaptic cells (the motoneuron) and determines the messages sent to the skeletal muscle. Since we have discussed transmission in neurons and across junctions we already have the basic elements of reflex action.

Reflexes underlie nearly everything we do. Virtually all neurally controlled responses — including most voluntary or willed movements — are reflex actions. Reflexes control not only the responses of skeletal muscles, but also those of cardiac and smooth muscle and those of glands. Some reflexes are inherent or unlearned responses, while others are conditioned or learned; some of the latter are very complex, since they can involve such intricate movement patterns as those used in signing one's name. If it were necessary to think specifically about crossing every "t" and dotting every "i," the student note-taker would be so busy writing that there would be no listening. Indeed, reflex actions free us from the need to make many individual decisions. We are "programmed" in such a way that a given input or stimulus elicits a certain response pattern, and our decision-making machinery is left for more important things than the dotting of an "i."

Upon application of a stimulus, impulses follow a particular pathway to an effector whose response is the reflex action (Figure 9–8). In order for a reflex response to occur, the entire pathway from receptor to effector must be intact and functioning. The anatomical substrate for reflex action, the **reflex arc,** must therefore contain the following components:

1 The **receptor** is the peripheral ending of the sensory neuron. It is usually a specialized ending with a low threshold for a particular type of stimulus (such as touch or light).
2 The **afferent neuron** is the sensory neuron. It provides the sensory input to the central nervous system.
3 The **center** is the site of the synapse. It is always in the central nervous system, where there may also be additional synapses and interneurons between sensory and motoneurons.
4 The **efferent neuron** is the motoneuron that carries impulses from the center in the central nervous system to the effector. For skeletal muscle it is the anterior horn cell.
5 The **effector** is the responding tissue — skeletal, smooth, and cardiac muscle or gland.

The structure and properties of the individual components of the reflex arc have already been discussed, including conduction and transmission in and between neurons, receptor responses earlier in this chapter, and skeletal muscle as a typical effector in Chapter 7.

FIGURE 9–8   Components of a spinal reflex arc.

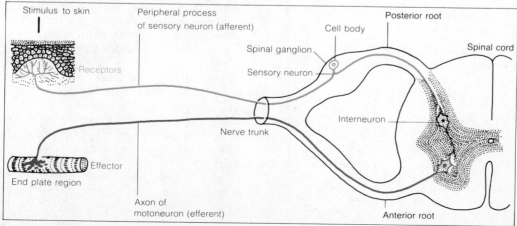

Connections of Reflex Components

When a receptor is stimulated, a volley of impulses travels along the afferent neuron at a frequency that depends upon the intensity and duration of the stimulus. A strong stimulus is likely to excite a number of receptors and thus activate more afferent neurons. Since excitation of an efferent neuron depends upon the total number of active presynaptic endings on its cell body, the strong stimulus evokes more impulses in the motoneuron. The response occurs when these impulses finally reach the effector.

Contraction of a muscle, however, is not the only effect of receptor activation. The afferent neuron gives off several branches when it enters the central nervous system (Figure 9–9). One of them enters an ascending fiber tract in the white matter that goes up toward the brain. Most of these pathways reach areas concerned with conscious perception, so that there is a sensation. (Some information from proprioceptors does not reach the level of consciousness and does not cause a sensation.) The simple reflex arc itself contains no provisions for sensation, and sensation is not a necessary part of reflex action. If you touch a hot iron two things will happen: immediate reflex withdrawal of the finger, and an acute sensation of pain. Conduction in the two pathways occurs simultaneously and the actual withdrawal has begun by the time you feel the heat—fortunately. If the spinal cord is severed in the cervical region (which disrupts all connections between the brain and the spinal cord), the simple reflex arcs remain intact and the response to a painful stimulus is prompt and undiminished, but there is no sensation. The stimulus is no longer painful because the brain is needed for sensation and awareness, and that connection has been destroyed.

Other branches of the afferent neuron terminate in the posterior horn to synapse with interneurons which may go to neurons in the anterior horn, cross over to the other side of the cord, or enter the white matter and ascend or descend to other levels of the cord where they reenter the gray matter and synapse with motoneurons on either side. These possibilities form the basis for divergence. A weak stimulus to the receptor would be likely to reach threshold only in those motoneurons with many presynaptic terminals from the sensory neuron stimulated. Other motoneurons, particularly those separated by several synapses from the afferent, might be facilitated in varying degree, or perhaps inhibited. Stronger stimuli would excite more motoneurons because of additional facilitation.

Each motoneuron receives input

FIGURE 9–9   Some of the possible connections of an afferent neuron.

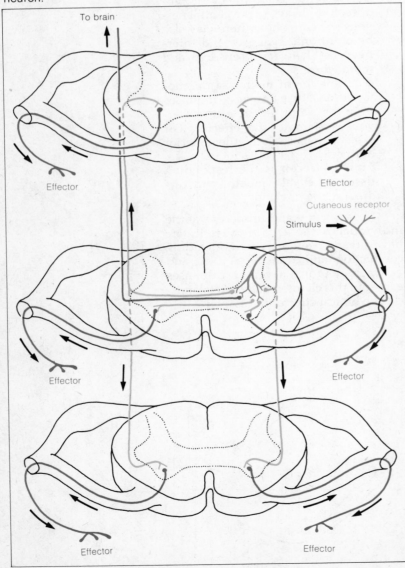

from many places (Figure 9–10); this is convergence. Those afferent neurons with many terminals on a single cell body are the ones likely to excite it with a weak stimulus. Afferents from the other side of or different levels of the spinal cord probably have less direct effect. In addition, each motoneuron is influenced by pathways descending from the brain (see Chapter 12). All of these, by contributing to facilitation or inhibition, determine the excitability of that motoneuron at that particular moment in time. And because any and all afferents that would affect a muscle must reach and stimulate the only path to that muscle, the motoneuron is sometimes called the **final common path.**

Reflex Action

The Spinal Animal   Certain reflex actions are best studied in an animal that is both unable to make voluntary responses and devoid of sensation (that is, unconscious), yet is not depressed by drugs (such as anesthetics). A convenient preparation is the spinal frog, in which the brain (only) has been destroyed. The spinal cord and spinal reflexes are intact, but the motoneurons have been deprived of any input from the brain and, of course, there is no sensation after the brain is destroyed.

For the first few minutes after destruction of the brain, the frog remains totally unresponsive, but gradually reflex responses return. This period of depression, called **spinal shock,** lasts longer in phylogenetically higher animals. Dogs and cats require hours to recover, and in humans several weeks may pass after severing the spinal cord (as when the neck is broken) before the depressed reflexes recover. This is probably because in higher animals motoneurons depend to a greater extent upon input from the brain; any disruption of the brain connections is therefore more serious and the effects last longer. While it may be difficult to be specific about the actual cause of spinal shock, it is even more difficult to explain the recovery from it. Upon recovery, however, the characteristics of reflex action can be clearly demonstrated.

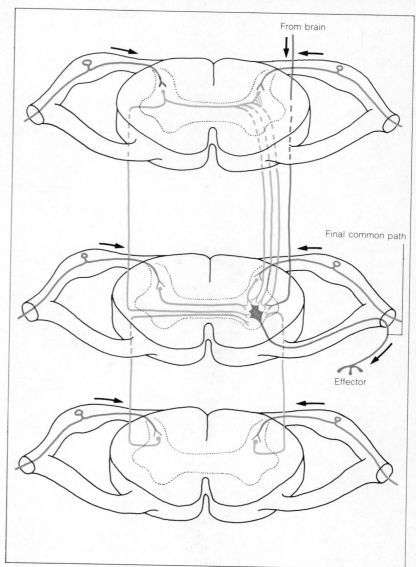

FIGURE 9–10   Some of the possible inputs to a motoneuron.

Properties of Reflex Action   Most of the properties of reflex action can be attributed to the presence of synapses in the pathway and to the arrangement of neuronal circuits in the central nervous system (Figure 9–11). The minimal anatomical requirement for reflex action involves only one synapse and two neurons, an afferent and an efferent, but each afferent has many terminal branches, and each efferent receives dozens of presynaptic endings. Thus even the minimal reflex arc has other connections. There are almost unlimit-

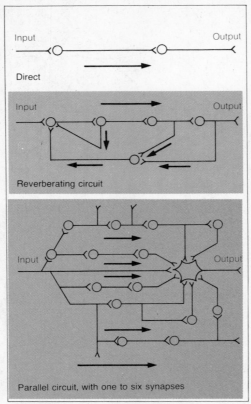

Input   Output

Direct

Input   Output

Reverberating circuit

Input   Output

Parallel circuit, with one to six synapses

FIGURE 9-11   Several types of
neuronal circuits.

ed possibilities for the afferent to influ-ence other motoneurons, interneurons, or the sensory pathways to the brain. And each motoneuron receives input from afferent neurons, interneurons, and motor pathways from the brain. These connections form the circuits necessary for convergence and diver-gence, described previously.

Because of interneurons, there may be several alternative pathways, called *parallel circuits*, by which impulses could travel from one place to another. One route may be longer than another and have more synapses, so that im-pulses taking that route would arrive at a motoneuron later than those that had followed a shorter path. Other routes may bend back upon themselves, with a branch of an interneuron serving to stimulate an earlier neuron in the path-way. An incoming signal may thus cause sustained firing of a postsynaptic neuron because of the excitatory feed-

back from such a branch. These are sometimes called *reverberating circuits* because, once started, discharge may continue until fatigue or inhibition occurs. In certain cases the collateral branches may activate inhibitory rather than excitatory neurons; the feedback is then inhibitory and firing is reduced.

The neuronal interconnections de-termine the presynaptic input to a motoneuron, and it is this input that controls its discharge of impulses. Those properties of reflexes that de-pend upon the synapses in the pathway have been discussed in the previous chapter as characteristics of synaptic transmission. Briefly, they are:

1. *One-way conduction.* Impulses pass from receptor to effector, but not the other way. Although neurons can conduct in both directions, synapses can conduct in only one direction be-cause of the release of the transmitter.[1]

2. *Summation.* Because each arriv-ing impulse is subthreshold, a certain amount of summation is required, ei-ther spatial, temporal, or both, to cause discharge in the postsynaptic neuron. Because of this, there is not a one-to-one relationship between impulses in the afferent and efferent neurons. Im-pulses in the efferent neuron tend to come in volleys which do not follow exactly the discharge pattern of the af-ferent neurons.

3. *Inhibition.* There is the possibili-ty of inhibition as well as excitation at synapses. Some neurons, by releasing a different transmitter, cause the devel-opment of IPSPs instead of EPSPs and thus inhibit the development of the excitatory state and reduce discharge of the postsynaptic neuron. Both inhibi-tory and excitatory endings on a given postsynaptic membrane may be active at the same time. The amount of inhibi-tion that is normally present in reflex action is well illustrated in the action of certain drugs. Strychnine is believed to block the action of the inhibitory trans-mitter at synapses, so there is no inhi-bition, but excitation is not affected. The result is widespread, prolonged

[1]An exception is the reciprocal junction found in several parts of the brain. In this case the two mem-branes both act as a presynaptic membrane. One is usu-ally excitatory and the other inhibitory.

Focus

Hemiplegia and Paraplegia

Human beings depend upon the integrity of their nervous systems. As long as the brain, spinal cord, and nerves are intact and functioning normally, people are able to perceive, think, talk, and move. Their limbs never seem to strive for independence.

But the story changes when the nervous system is damaged. The brain loses control over parts of the body, and local reflexes may become the only means of making a muscle contract. This is particularly true when the voluntary motor pathways are damaged.

Hemiplegia is defined as paralysis of one side of the body. It generally results from a stroke or other injury to the brain or spinal cord, especially one that interrupts the nerve fibers that carry the main voluntary commands to the muscles. These fibers cross from one side of the body to the other in the medulla (lower brain) and upper spinal cord, so that one side of the brain commands the opposite side of the body. Damage to this pathway in the brain therefore results in paralysis of the opposite side of the body. Damage in the cord may cause paralysis on the same or opposite side, depending on its location with respect to the crossing of the pathway.

Because there are other, indirect motor pathways from the cortex to the cord, the brain may still be able to evoke movement in the paralyzed portion of the body if the damage is not too extensive. At times, these movements can seem almost as competent as those of an unparalyzed person, but they do involve mainly the proximal muscles, of affected limbs; they do not involve distal muscles, such as those of the hand and fingers. The capability for these movements increases with time after the original injury and can be strengthened by training, so that hemiplegics can hope for some restoration of function.

The picture is less optimistic for paraplegics. Paraplegia is defined as paralysis of the lower half of the body. It results when the spinal cord is completely severed by such injuries as a broken back. The portion of the cord caudal to the break no longer has any connections with the brain, and it can no longer be controlled by the brain.

This lack of control is total and permanent, but the portion of the body caudal to the break in the cord remains completely nonfunctional only during the weeks or months immediately after the injury. During this period of "spinal shock," the nerve cells of the cords seem unable to respond to stimuli. The spinal reflexes cannot be elicited, even though all the paths from sensory nerve cells to interneurons to motor neurons are intact; only the paths from the brain and portions of the cord on the other side of the break have been destroyed.

In time, however, the spinal reflexes usually reappear. A paraplegic's legs may flex in response to pinches and pin pricks; they may even produce steplike movements, though without enough strength to support the body. The bladder and rectum spontaneously empty when full, since there are reflexes to produce these actions in response to distension of these organs, reflexes that are normally kept in check by the brain. Even sexual arousal and activity are possible, despite the lack of sensory feedback to the brain.

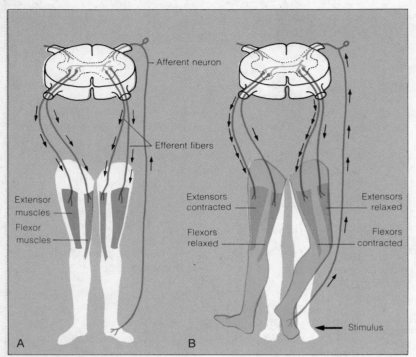

FIGURE 9–12   Reciprocal innervation. A. Normal resting state: The tonic discharge in motoneurons to flexors and extensors is balanced (indicated by the number of arrows along each fiber), and the posture is maintained. B. A painful stimulus to the left foot excites motoneurons to flexor muscles and inhibits those to extensors (antagonists) in the left leg. In the right leg extensor motoneurons are stimulated and flexor motoneurons are inhibited.

convulsive contractions of all skeletal muscles. It is also the ultimate demonstration of convergence and divergence, since every stimulus stimulates everything!

The properties of reflex action that depend upon the neuronal connections as well as the behavior of the synapses include:

1. *Irradiation.* The spreading of response that occurs with increasing stimulus intensity is known as **irradiation.** It is due to the fact that more and more facilitated neurons are recruited into the ranks of the actively firing neurons. A weak stimulus may cause reflex movement of, for instance, the foot, but with a stronger stimulus the whole leg, then both legs, and perhaps the forelimbs might become involved.

2. *After discharge.* The response may continue for some time after stimulation stops. This is due to parallel

pathways with varying numbers of synapses by which impulses may reach a motoneuron. The more synapses crossed, the more delay there is in reaching the motoneuron, and some impulses may arrive after stimulation has stopped. It is also possible that reverberating circuits may keep impulses traveling about for some time.

3. *Reciprocal innervation.* Neuronal connections in the spinal cord are such that excitation of the motoneuron to one skeletal muscle is accompanied by inhibition of the motoneuron to its antagonist. It may be recalled from Chapter 6 that one of the characteristics of skeletal muscle action is that contraction of a prime mover is accompanied by relaxation of the antagonist, thereby reducing its tone.

The effects of reciprocal innervation are not restricted to the *ipsilateral* (same) side, since there are effects on the *contralateral* (opposite) side as well. Suppose we assume an equal degree of tone in the flexors and in the extensors of both legs. Then if one foot is given a painful stimulus (Figure 9–12), that foot will be withdrawn from the stimulus by excitation and contraction of the flexors and inhibition and relaxation of the extensors of that leg. In the contralateral leg, the flexors will be inhibited and the extensors excited, producing an extension that serves to support the individual. The value of this arrangement to a biped should be clear—if you step on a tack, you will not sit on it as well!

As important as it is to skeletal muscle action, the concept of reciprocal innervation is not limited to muscle. It is intimately involved in the control of many functions, including heart rate, blood pressure, and respiration, to name but a few. It is the basis of many of the systems of checks and balances so prevalent among the mechanisms of physiological control.

4. *Tonus.* The resistance shown by a muscle to passive stretch is known as muscle tone or **tonus.** A denervated skeletal muscle is not only paralyzed, but it has no resistance to stretch and is completely *flaccid* or limp; it has no tonus. Tonus is not a property of the muscle, but it is due to the activity in the motoneurons. The continual bom-

bardment of motoneurons necessary to produce muscle tonus comes from many sources, but muscle spindles and certain areas of the brain make major contributions (see Chapter 12).

We usually think of tonus in terms of a whole muscle rather than a single fiber or motor unit. We must then think of muscle function in terms of a large population of motoneurons, or a *motoneuron pool.* At any given instant, some of the motoneurons may be subjected to considerable excitatory input; they are in the *discharge zone* and are firing off action potentials. Other motoneurons are receiving less input; they are in the *subliminal fringe* and are only facilitated to a certain extent. Any additional excitatory input will bring more of the motoneurons into the discharge zone, and any inhibitory input will decrease the number of firing neurons. Even small changes in the input to a motoneuron pool can have marked effect on the number of neurons firing and the frequency of their discharge, and will therefore affect the amount of tension developed by the whole muscle.

Types of Reflexes

A better understanding of reflex actions and their role in everyday life can be obtained by applying the principles of reflex action to some of the simple reflexes that account for a large share of spinal reflex activity.

Stretch Reflexes Stretch reflexes are also known as **myotatic, tendon,** or **proprioceptive reflexes.** They are anatomically the simplest of all, because they are the only monosynaptic reflexes in the body (see Figure 9–13A). They involve no interneurons, just an afferent and an efferent neuron. The stimulus is stretch of a muscle, which activates the muscle spindles—the proprioceptive stretch receptors—in that muscle; that same muscle is then the effector. When the patellar tendon is tapped, the quadriceps femoris is slightly but abruptly stretched. Its reflex contraction results in extension of the leg, the familiar knee jerk. Stretch reflexes are sometimes called *extensor reflexes* because they are easily demonstrated in extensor muscles. However, they can also be elicited from flexors, such as the biceps brachii.

The functional contribution of the stretch reflexes is to the maintenance of posture, and they are particularly well developed in antigravity muscles (which are mostly extensors). They help us maintain the rigidity necessary to support the weight of the body, at rest and in motion. If a supporting muscle (such as the quadriceps femoris) should yield to gravity, it would be stretched, eliciting a myotatic reflex that would increase its ability to resist gravity. Since gravity must often be resisted for prolonged periods of time, it is most helpful that muscle spindles do not adapt readily.

FIGURE 9–13   A. Stretch reflex. B. Flexion reflex, with a crossed extensor reflex.

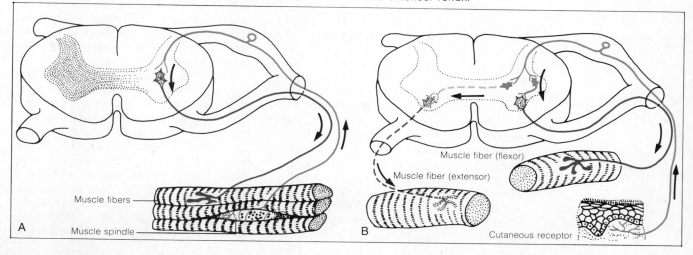

Flexion Reflexes The **flexion** or **withdrawal response,** involving one interneuron and two synapses, is at least as widespread as the stretch reflex (see Figure 9–13B). It is elicited by a noxious (unpleasant or harmful) stimulus such as a painful stimulus to a cutaneous receptor. The effectors are the flexor muscles nearby—just *which* and how many muscles depend upon the intensity and location of the stimulus. The obvious purpose of flexion reflexes is protection.

Closely associated with the flexion reflex, and actually a part of it, is the *crossed extensor reflex.* It is extension of the contralateral limb when a flexion reflex is elicited, and it is the result of reciprocal innervation.

Multineuronal, Multisynaptic Reflexes Multisynaptic pathways do not necessarily involve completely different mechanisms, since they may be based upon stretch or flexion reflexes. There are any number of intersegmental reflexes (including those that are the result of irradiation) in which, for example, strong stimulation of a hindfoot elicits a response in the foreleg as well. An interneuron's axon passes in the white matter up the cord to reenter the gray matter and synapse with appropriate motoneurons to the forelimb muscles. Many of the postural reflexes, including the righting reflexes, are intersegmental reflexes. Righting reflexes are those that enable an animal to make the proper postural adjustments to orient itself in space. There are also positioning reflexes, as when a touch on the dorsal surface of the foot of a spinal animal causes a stepping movement as if to step over an object. The grasp reflex, in which touching the palm of the hand causes the fingers to close, is conspicuous in infants.

Patterns of movement that involve a high degree of coordination of the contractions of many muscles depend upon the brain for much of their coordination, though not at the conscious level. The scratch reflex, so well developed in dogs and cats, is an example, as is the sucking reflex in young mammals. There are a number of visceral reflexes that control the secretion of some digestive juices, the emptying of the urinary bladder, and the constriction of the blood vessels. These are autonomic reflexes. In many of them we are not aware of either the stimulus or the response (emptying the bladder is an exception). These reflexes are described in Chapter 12.

Some of the most complex reflexes are the learned reflexes. Such common skills as walking, writing, and speaking, and the motor skills needed for sports—bicycling, skiing, swimming, etc.—are all primarily reflex activities. Only their initiation—the decision to make that movement—is entirely conscious and voluntary. You "think" in terms of walking across the room, not in terms of contracting one muscle to swing one foot forward, another to support your weight, another to shift your weight, etc. Reflex action frees the brain from "administrative" decisions to allow concentration on "policy" decisions.

CHAPTER 9 SUMMARY

The **spinal cord** is the part of the CNS that is enclosed within the bony vertebral canal. It is ensheathed in several layers of fibrous connective tissue, the *meninges.* The outermost layer, the *dura mater,* is the heaviest, and covers the entire spinal cord and extends a short distance along the spinal nerves. Directly under it is the *arachnoid membrane,* and beneath that the *pia mater,* which lies on the surface of the cord. The *subarachnoid space,* between the pia mater and the arachnoid membrane, contains cerebrospinal fluid. These protective coverings are continuous with similar layers that enclose the brain.

A cross section of the spinal cord shows the centrally located H-shaped *gray matter* which contains cell bodies and synapses (unmyelinated). The gray matter is surrounded by *white matter* composed of bundles of nerve fibers (many of them myelinated). The gray matter forms paired *anterior* and *posterior horns,* and in the thoracic and lumbar regions, *lateral horns* as well. *Anterior* and *posterior roots* emerge from the cord near the tips of the anterior and posterior horns. Anterior roots contain axons of motoneurons, and posterior roots contain central processes of sensory neurons. A *spinal ganglion,* containing cell bodies of sensory neurons, lies on each posterior root just before it joins the anterior root to form a spinal nerve.

Fiber tracts are bundles of nerve fibers in the white matter, and include *ascending tracts* carrying information to the brain, *descending tracts* carrying impulses from the brain to motoneurons, and bundles of fibers that connect one level of the cord with another.

Spinal nerves emerge from the cord at each vertebral level. They divide into a smaller *posterior ramus* that supplies skin and muscle of the back, and a larger *anterior ramus* that follows the body wall around to the anterior midline and sends branches to skin and muscles of the body wall. The segmental pattern of distribution of spinal nerves is clearly shown by the areas of skin innervated by each spinal nerve (dermatome).

Nerves to the extremities are distributed as **peripheral nerves** rather than as spinal nerves. Several spinal nerves come together, forming a *plexus,* in which there is an intermingling of nerve fibers. Peripheral nerves that arise from a plexus contain fibers from several spinal nerves. The *cervical plexus* sends nerves to the cervical region and gives rise to the phrenic nerve. Nerves from the *brachial plexus,* including the musculocutaneous, median, radial, and ulnar nerves supply the arm and shoulder region. The *lumbosacral plexus* gives rise to nerves that supply the leg, pelvic and hip regions, notably the femoral and sciatic nerves.

Reflexes are the basis of nervous system function. The *reflex arc* is the structural unit, consisting of a receptor, sensory or afferent neuron, a center in the CNS, an efferent or motoneuron, and an effector. The center may be a single synapse, or it may involve one or more interneurons and several synapses.

Upon entering the spinal cord, afferent neurons divide, sending a branch to the brain. A connection with the brain is necessary for sensation. Other branches of the afferent neuron synapse with many interneurons and motoneurons *(divergence).* Likewise, each motoneuron is influenced by many afferent and interneurons, including those descending from the brain *(convergence).* The motoneuron is the only connection with the effector, and is sometimes called the *final common path.*

If the spinal cord is cut, destroying the pathways to and from the brain, reflexes can still occur. That is, if all elements of the reflex arc are intact, although they may lack whatever modification the brain imposed. However, there will be no sensation.

Because of the properties of synaptic transmission, **reflex action** is characterized by one-way conduction and by summation and/or inhibition. The neuronal circuits in the CNS contribute to irradiation (greater and more widespread response with increasing strength of stimulus), after discharge (response continuing after cessation of stimulation), reciprocal innervation, and tonus. *Reciprocal innervation* is the means by which an antagonist relaxes when the prime mover contracts. The sensory neuron has connections, via inhibitory interneurons, with the antagonist's motoneurons. It also has reciprocal connections with neurons to muscles in the opposite limb. *Tonus,* the resistance of skeletal muscle to stretch, is related to the level of excitability of the neurons in a motoneuron pool. When excitability is high due to facilitation, a given stimulus will excite more motoneurons than when excitability is low.

Among the most important spinal cord reflexes are **stretch** and **flexion reflexes.** Stretch reflexes require only the afferent and efferent neurons, and a single synapse. The receptor is a proprioceptor (muscle spindle) stimulated by stretch of a muscle. The response is contraction of the stretched muscle (knee jerk). Stretch reflexes are particularly well developed in antigravity muscles and are important for postural reflexes.

Flexion or **withdrawal reflexes** involve at least one interneuron and two synapses. They are elicited by a potentially harmful stimulus to the skin, and the response is a withdrawal of the part stimulated, usually by action of flexor muscles. It is accompanied by extension of the opposite limb (crossed extensor reflex), an example of reciprocal innervation.

STUDY QUESTIONS

1 Make a diagram of a cross section of the spinal cord showing the major structural features. What cells or cell parts are found in each part?

2 What is the difference between gray and white matter?

3 What are the meninges? Describe them in relation to the spinal cord.

4 What is the significance of the fact that the spinal cord does not extend to the end of the spinal cord?

5 Based on the dermatomal distribution of spinal nerves, what spinal nerve would you expect to innervate the skin over the palm of the hand? the patella? the ischial tuberosity? the posterior surface of the calf of the leg? the acromion process?

6 What happens to spinal nerves that enter a plexus such as the brachial plexus? What happens in the individual neurons as they enter the plexuses?

7 Can a reflex response occur in the absence of sensation? If yes, by what mechanism? If no, why not?

8 What is meant by reciprocal innervation? Do you think it would be involved in performing skilled movements, such as swimming or other sports activities? Explain.

9 Describe a flexion reflex in terms of the adequate stimulus, the type and location of the receptors, and the response and its contribution to the well-being of the animal.

10 Describe a stretch (myotatic) reflex in the same terms as above.

11 It is said that most of the characteristics of reflex action are due to the neuronal circuits and the properties of synapses. How is this so?

Structure of the Brain

The Cranial Nerves

Protective and Supportive Structures

Anatomy of the Brain and Related Structures

The brain, as a continuation of the spinal cord, also contains neurons and neuroglia. The brain does the same things as the spinal cord, and more. It receives and processes information from more places, carries on more sophisticated integrations, and coordinates more diverse activities than the spinal cord. It acts on a much grander scale and, in fact, exerts important control over the spinal cord.

Predictably then, the organization of the brain is more complex than the spinal cord, but there are many similarities in pattern. This chapter discusses the structure and organization of the brain, the cranial nerves, and related structures. Little reference is made to function in this chapter, except in a very general way. That is done in Chapter 11, which deals with sensory functions, and in Chapter 12, which covers motor functions.

You will notice that many of the structures are labelled on several illustrations. This makes for heavily labelled figures, but it should help you to visualize the structures in three dimensions and to see their relations with one another. You may even want to refer back to some of the figures in Chapter 4 (bones of the skull).

Very early in the development of the human embryo a groove appears on its posterior surface. This groove soon deepens and becomes roofed over to form the **neural tube,** and the cells of its walls rapidly divide and differentiate into nerve tissue. The tube does not grow evenly, however, for the head end shows much more rapid and extensive development. It will become the **brain,** while the remainder will form the **spinal cord.** Growth within the brain region is also uneven, and soon three swellings can be recognized there; as shown in Figure 10–1, they will be- come the **forebrain, midbrain,** and **hindbrain.** The forebrain and hind- brain each soon develop two enlarge- ments of their own, giving a total of five swellings. As growth continues, the first swelling enlarges very rapidly un- til it overshadows all the other parts. It develops into the two **cerebral hemi- spheres,** which in humans are so large that the rest of the forebrain, the mid- brain, and parts of the hindbrain are nearly covered by them. One of the swellings of the hindbrain also grows very rapidly and becomes the **cerebel- lum.** The canal in the neural tube per-

FIGURE 10–1   Development of the central nervous system. A. Three stages in early embryonic development of the brain. B. The adult brain and spinal cord.

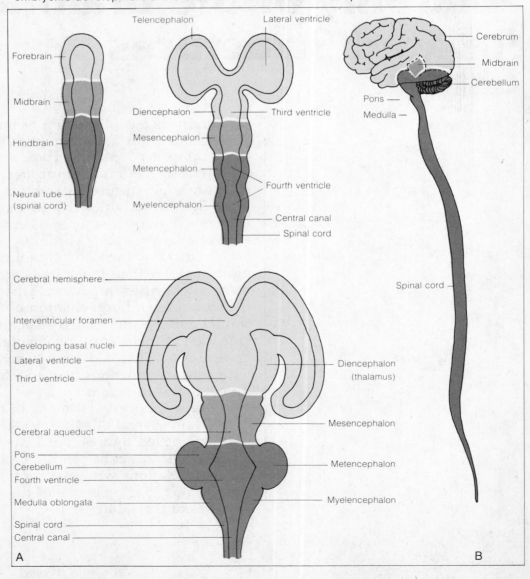

sists; in the rapidly developing parts of the brain, it enlarges to form large openings, the brain **ventricles,** and it continues through the spinal cord as the tiny **central canal.**

Although greatly modified by subsequent development, the major divisions and subdivisions remain, even in the adult brain. The major brain structures maintain a distinct relationship with the subdivision from which they developed. The subdivisions and the structures that arise from them are listed in Table 10–1, and you can see many of these relationships by comparing Table 10–1 with Figure 10–1 and then with Figures 10–2 and 10–3.

The central nervous system is a continuous structure that becomes progressively more highly developed, both anatomically and in the kinds of functions it performs, as you go from the spinal cord to the cerebral cortex. You have already studied the spinal cord, and now you will study the structure of the brain, beginning with the lowest portion, the hindbrain, and working up to the cerebral cortex.

STRUCTURE OF THE BRAIN

When the spinal cord passes through the foramen magnum in the occipital bone, it becomes brain and begins to show the structural characteristics of the medulla. The lower portion of the brain—the medulla, pons, and midbrain—is a modified continuation of the spinal cord; it is called the **brainstem.** Because of their similarities, many of the principles established for spinal cord structure and function can be applied to the brainstem with only a little adjustment.

Instead of forming a single H-shaped structure, as in the cord, the gray matter of the brainstem is broken into clumps by the white fiber tracts ascending and descending between the spinal cord and brain. The clumps of gray matter, called **nuclei,** may be defined as collections of cell bodies in the central nervous system.[1]

[1]A *nucleus* of the central nervous system is not to be confused with the nucleus of a cell. Nuclei in the CNS resemble ganglia, except that nuclei are *in* the CNS and ganglia lie *outside* the CNS.

TABLE 10–1   DERIVATION OF MAJOR STRUCTURES OF THE CENTRAL NERVOUS SYSTEM

Primary Division	Subdivision	Brain Structures	Canal
Forebrain	Telencephalon (cerebral hemispheres)	Cerebral cortex Basal ganglia Rhinencephalon	Lateral ventricles Part of third ventricle
	Diencephalon	Thalamus Hypothalamus	Most of third ventricle
Midbrain	Mesencephalon	Corpora quadrigemini Cerebral peduncles	Cerebral aqueduct
Hindbrain	Metencephalon	Cerebellum Pons	Fourth ventricle
	Myelencephalon	Medulla oblongata	
Spinal cord		Cervical, thoracic, lumbar, sacral, and coccygeal portions	Central canal

Twelve pairs of cranial nerves emerge from the brainstem at irregular intervals (see Figure 10–2). The motor fibers in those cranial nerves that supply skeletal muscle arise from nuclei that correspond to the anterior horn of the spinal cord. Motor fibers to smooth or cardiac muscle arise from separate nuclei. Sensory fibers terminate in gray matter that corresponds to the posterior horn, and they have their cell bodies in ganglia near the brainstem (the ganglia have special names, but they are equivalent to spinal ganglia).

The extensive development of the **telencephalon** (cerebral hemispheres) is especially marked in primates and humans. It is tempting to relate the greater mental capacity of these species to cerebral size, but perhaps more important is the relative surface area of the cerebrum. The entire external surface of the cerebral hemispheres is covered by a layer of gray matter, the **cerebral cortex.** Beneath the cortex is a mass of white matter, although there are some important areas of gray matter deep within the hemispheres. The extensive overgrowth of the cerebrum is well illustrated in Figure 10–3.

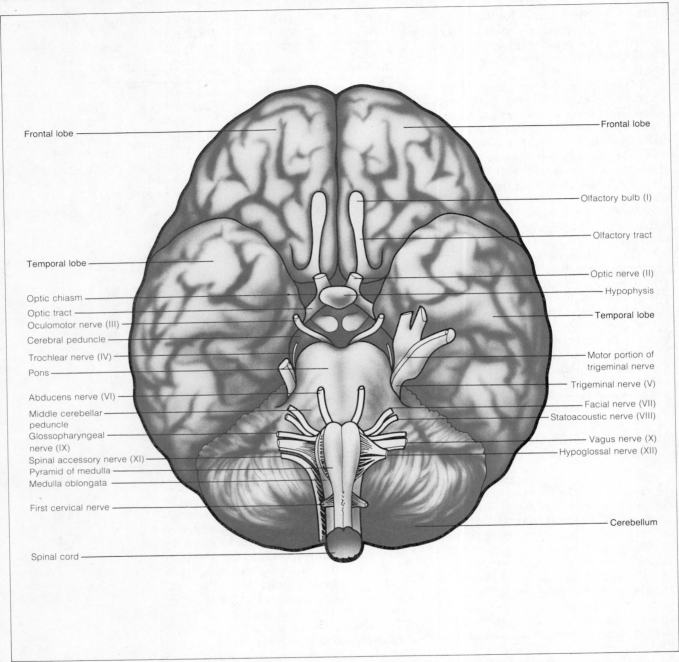

Frontal lobe

Frontal lobe

Olfactory bulb (I)

Olfactory tract

Temporal lobe

Optic nerve (II)

Optic chiasm

Hypophysis

Optic tract

Oculomotor nerve (III)

Temporal lobe

Cerebral peduncle

Trochlear nerve (IV)

Motor portion of
trigeminal nerve

Pons

Trigeminal nerve (V)

Abducens nerve (VI)

Facial nerve (VII)

Middle cerebellar
peduncle

Statoacoustic nerve (VIII)

Glossopharyngeal
nerve (IX)

Vagus nerve (X)

Hypoglossal nerve (XII)

Spinal accessory nerve (XI)

Pyramid of medulla

Medulla oblongata

First cervical nerve

Cerebellum

Spinal cord

FIGURE 10–2   Inferior surface of the brain, showing origins of the cranial nerves.

Brainstem

Medulla Oblongata   The medulla oblongata is the site of the transition from the spinal cord to the brain, and its structure reflects that fact. All of the ascending and descending fiber tracts between spinal cord and brain pass through it. On the ventral surface are the **pyramids,** (Figure 10–2) through which pass the fibers from the cortex that become the corticospinal tracts in the cord and end on or near the motoneurons. In the deeper portion of the medulla there is an extensive network of intermingled fibers and gray matter, forming what is known as the **reticular formation.** It extends throughout the entire brainstem and up into the diencephalon. There are many important synaptic connections in the medulla; among them are the so-called vital centers, which include the centers for control of heart rate, blood pressure, and respiration, as well as for salivation, coughing, vomiting, and skeletal muscle tone. It also contains nuclei for the last six cranial nerves (VII-XII), most of which have major connections with these centers.

Pons   The pons lies above the medulla and contains many of the same types of components. A large bundle of fibers extends back to the cerebellum from each side of the pons (see Figure 10–2). This band, from which the pons

FIGURE 10–3   Midsagittal section of the brain.

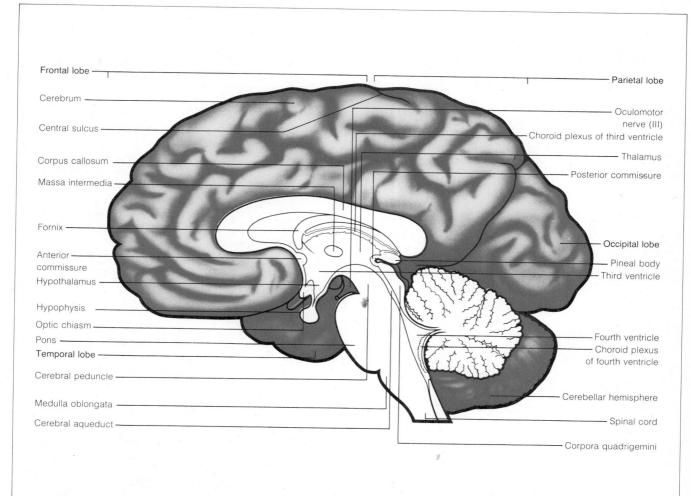

Frontal lobe

Cerebrum

Central sulcus

Corpus callosum

Massa intermedia

Fornix

Anterior commissure

Hypothalamus

Hypophysis

Optic chiasm

Pons

Temporal lobe

Cerebral peduncle

Medulla oblongata

Cerebral aqueduct

Parietal lobe

Oculomotor nerve (III)

Choroid plexus of third ventricle

Thalamus

Posterior commissure

Occipital lobe

Pineal body

Third ventricle

Fourth ventricle

Choroid plexus of fourth ventricle

Cerebellar hemisphere

Spinal cord

Corpora quadrigemini

(or "bridge") gets its name, forms a major connection of the cerebellum with the brainstem. In addition to these fibers, ascending and descending tracts pass through the pons, and it contains the central connections of cranial nerves V, VI and some of those for VII. It also contains centers for respiration and skeletal muscle tone.

Midbrain The midbrain or **mesencephalon** is a small structure containing ascending and descending pathways and the central connections for cranial nerves III and IV. The midbrain reticular formation is continuous with that of the pons and medulla. The **cerebral aqueduct** is a tiny canal that connects the fourth ventricle with the third. Posterior to the aqueduct are four rounded prominences, the **corpora quadrigemini.** The superior pair, the **superior colliculi,** are concerned with reflex responses to visual stimuli. The inferior pair are the **inferior colliculi,** and they are concerned with responses to auditory stimuli.

In a midsagittal section (Figure 10–3) the brainstem appears to end abruptly at the midbrain, but it actually divides into right and left portions, the **cerebral peduncles,** which carry fiber tracts to and from the cerebral hemispheres.

Cerebellum

The cerebellum is separated from the brainstem by the fourth ventricle, but is connected to it by three paired fiber bundles, the **superior, middle,** and **inferior cerebellar peduncles,** which connect it with the midbrain, pons, and medulla, respectively. These are shown in Figure 10–4, a view of the brainstem with the cerebrum and cerebellum removed.

The cerebellum consists of two **hemispheres** and a tiny central portion, the **vermis** (*vermis* = worm) (Figure 10–2). The surface of the cerebellum is covered by a thin mantle of gray matter, the **cerebellar cortex,** which dips into all the surface indentations, thereby greatly increasing the surface area. The white matter inside reaches to all parts of the surface. (In a midsagittal section, the cerebellar white matter is sometimes called the *arbor vitae,* because of its resemblance to the branchings of a tree.) (Figure 10–3)

The cerebellum, whose functions remain below the level of consciousness, is concerned primarily with several aspects of skeletal muscle activity, including the timing, intensity, and general coordination of the contractions of individual muscles. The spinocerebellar tracts that mediate unconscious proprioception end in the cerebellum and provide much of the information it needs to carry out those functions.

Diencephalon

This part of the forebrain is difficult to visualize, as it is surrounded by the cerebral hemispheres and only a tiny portion can be seen from the surface of the brain. Most of its components are paired, so that a midsagittal section passes through the third ventricle between them, but they can be identified in coronal sections (Figures 10–6 and 10–10).

The thalamus and hypothalamus are the major structures of the diencephalon. The **thalamus,** an egg-shaped structure, contains several nuclei and is therefore almost entirely gray matter. It is the chief sensory relay center, and all sensory fiber tracts (except the olfactory) synapse in one of the thalamic nuclei en route to the cerebral cortex. Several thalamic nuclei have cortical connections which are important in arousing the brain to the conscious state.

The **hypothalamus,** composed of many small nuclei, is located below the thalamus along the lateral walls of the third ventricle (Figure 10–5). It contains centers for control of many homeostatic processes, including body temperature, water balance and osmotic concentration of body fluids, and food and water intake. In addition, it helps control the secretions of certain endocrine glands through its action on the pituitary gland.

The **hypophysis** or **pituitary gland** is actually two different organs that are joined. The part of it known as the

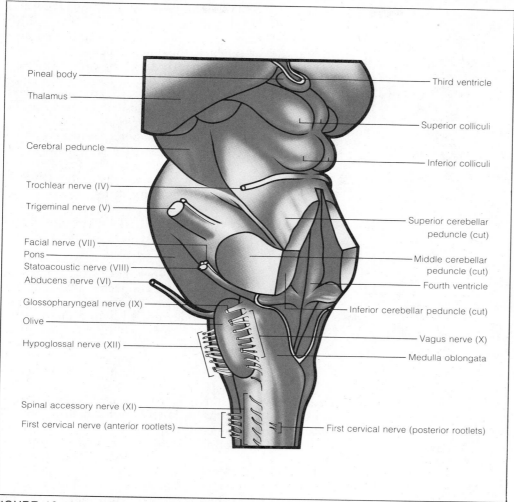

Pineal body
Thalamus
Cerebral peduncle
Trochlear nerve (IV)
Trigeminal nerve (V)
Facial nerve (VII)
Pons
Statoacoustic nerve (VIII)
Abducens nerve (VI)
Glossopharyngeal nerve (IX)
Olive
Hypoglossal nerve (XII)
Spinal accessory nerve (XI)
First cervical nerve (anterior rootlets)

Third ventricle
Superior colliculi
Inferior colliculi
Superior cerebellar peduncle (cut)
Middle cerebellar peduncle (cut)
Fourth ventricle
Inferior cerebellar peduncle (cut)
Vagus nerve (X)
Medulla oblongata
First cervical nerve (posterior rootlets)

FIGURE 10–4   Dorso-lateral view of the brainstem (cerebellum removed).

neurohypophysis, or *posterior lobe* of the pituitary, is of neural origin and is attached to the hypothalamus by a stalk, the **infundibulum.** The neurohypophysis and its stalk are thus part of the diencephalon. The **adenohypophysis,** or *anterior lobe* of the pituitary, is an endocrine gland and is not part of the nervous system. Just anterior to the pituitary stalk is an X-shaped structure, the **optic chiasm.** It is formed by fibers of the optic nerve (II) from the eye and is the site where some fibers from each eye cross to go to the other side of the brain.

The tiny **pineal body** protrudes into the posterior portion of the third ventricle between the superior colliculi (Figure 10–4). It has an endocrine func-

tion in amphibians but, although it was once thought to be the "seat of the soul," its function in humans is not understood.

The Cerebral Hemispheres

The cerebral hemispheres are the most superior portions of the brain and, in humans, the largest. They contain the cerebral cortex, many white fiber tracts, the basal ganglia, and the lateral ventricles. The gray outer surface of the hemispheres, the **cerebral cortex,** is several millimeters thick and dips down into every crevice so that the total amount of cortex is much greater than that visible on the surface (see Figure

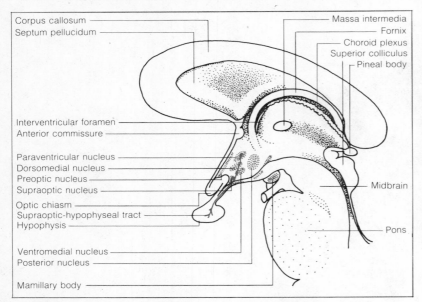

FIGURE 10-5 Midsagittal section of the region of the diencephalon, showing some hypothalamic nuclei and connections between the neurohypophysis and the hypothalamus.

10-6). White matter lies beneath the cortex. It consists of fiber tracts that connect parts of the cortex with one another and with other parts of the central nervous system. The **basal ganglia** include several large and small nuclei that surround the structures of the diencephalon, deep within the hemispheres.

Cerebral Cortex The surface of the cerebral cortex is a layer of gray matter marked by many convolutions (Figure 10-7). The ridges are called *gyri* (singular = gyrus) and the indentations are *sulci* (singular = sulcus) or *fissures.* Prominent fissures divide the hemispheres into lobes, which bear the name of the bone that covers them. The **longitudinal fissure** divides the cerebrum into right and left hemispheres. The **central sulcus** separates the *frontal* and *parietal lobes,* and the **lateral fissure** separates the *temporal lobe* from the frontal and parietal lobes. The boundaries of the *occipital lobe* are less well marked. At the bottom of the lateral fissure is an area of cortex, the **insula,** completely hidden from the surface by the adjacent cortical areas. Important

gyri are the **precentral** (frontal lobe) and **postcentral** (parietal lobe) gyri, which lie just anterior and posterior, respectively, to the central sulcus.

Several areas of the cortex are related to specific functions. Sensory pathways project to localized cortical sites, called **primary sensory receiving areas.** The postcentral gyrus contains the *somatosensory area,* and pathways arising on the body surface terminate there. The *visual area* is in the occipital lobe and the *auditory area* is in the temporal lobe. The *primary motor area* is in the precentral gyrus, and it is the origin of some important fibers descending to motoneurons.

White Fiber Tracts The parts of the cortex are connected with one another and with other parts of the brain and spinal cord by white matter consisting of bundles of fibers. The bulk of the interior of the hemispheres is composed of these bundles, which may be roughly divided into three types of tracts.

1. *Projection fibers* (Figure 10-6) project to the cortex from other parts of the central nervous system or from the cortex to these areas. They include bundles of fibers from the cord and connections with the brainstem.

2. *Commissural fibers* (Figure 10-8) connect the two hemispheres. The **corpus callosum** at the bottom of the longitudinal fissure forms the largest commissure. The others are the tiny anterior and posterior commissures (see Figure 10-3).

3. *Association fibers* (Figure 10-8) connect two cortical sites of the same hemisphere. They include connections between adjacent gyri, as well as longer tracts between the lobes.

Basal Ganglia Not all the gray matter in the cerebrum is in the cortex, however, since deep inside the hemispheres are several gray structures known collectively as the **basal ganglia** (the name is a misnomer; they are really basal *nuclei*) (Figures 10-9 and 10-10). Generally included among the basal ganglia are the *caudate nucleus* (named for its long tail which arches over the thalamus), the *globus pallidus,* the *puta-*

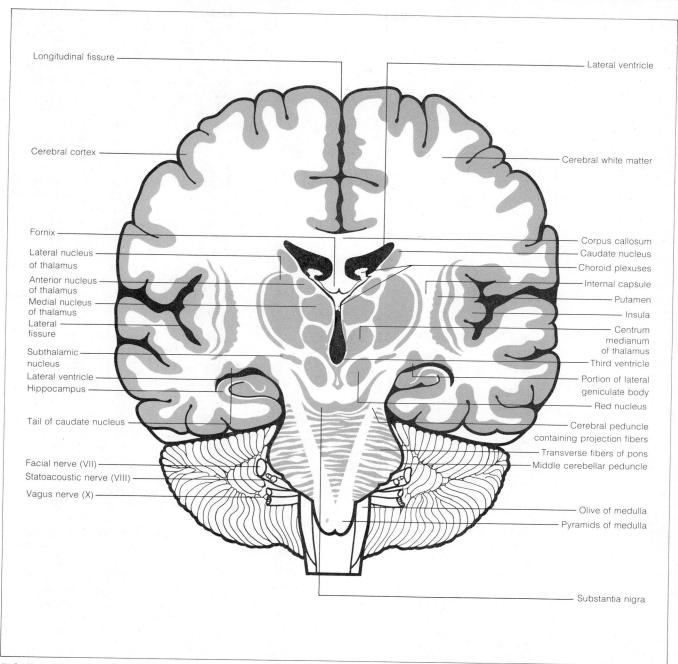

FIGURE 10-6   Coronal section of the brain at the level of the cerebral peduncles (containing projection fibers).

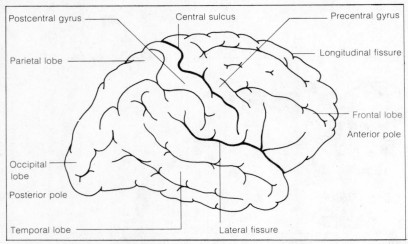

FIGURE 10–7 Major sulci and gyri of the cerebral hemispheres.

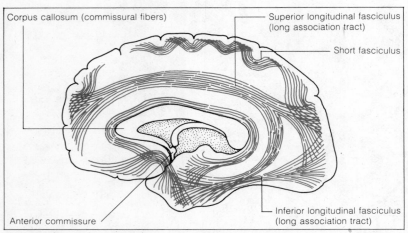

FIGURE 10–8 Major association tracts of the cerebral hemispheres.

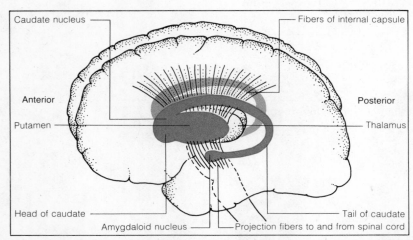

FIGURE 10–9 Three-dimensional representation of the location of the basal ganglia and their relation to projection fibers.

men, the *subthalamic nuclei,* and the *substantia nigra* (named for the fact that some of its neurons contain the pigment melanin). The caudate nucleus and putamen are closely associated structurally (they are connected anteriorly) and developmentally, and no doubt functionally. In a coronal section of the brain (Figure 10–10) these two, plus the globus pallidus and caudate, with their fiber interconnections, present a rather striped appearance, and are often collectively referred to as the *corpus striatum* (striped body). The narrow region between the putamen and globus pallidus and the thalamus, known as the **internal capsule,** contains major projection fibers to and from the cortex. It is a common site of cerebral vascular accidents (ruptured blood vessels and strokes), and because fibers serving many functions are closely packed in this region, such accidents can produce a wide variety of symptoms and problems. The **amygdaloid nucleus** lies at the end of the tail of the caudate nucleus. Functionally it is associated with the rhinencephalon.

Rhinencephalon The **rhinencephalon,** phylogenetically a very old part of the brain, contains those structures associated with the sense of smell. It includes the **olfactory bulb** and **tract** located on the inferior (orbital) surface of the frontal lobes, parts of the cerebrum, and the fiber tracts that connect them (Figure 10–11). The olfactory tract extends posteriorly from the bulb and then divides and enters the brain substance where it establishes connections with several groups of nuclei, including the amygdaloid nucleus and several thalamic nuclei. The **fornix** carries fibers from the **hippocampal gyrus** (in the temporal lobe) to the **mammillary bodies.**

The rhinencephalon is a poorly understood complex whose function is not limited to olfaction. In some species including humans, whose olfactory sense is not well developed, the olfactory bulb and tract which are associated with smell are relatively small. The other parts of the system, however, are more highly developed and make up what is called the **limbic system.** This

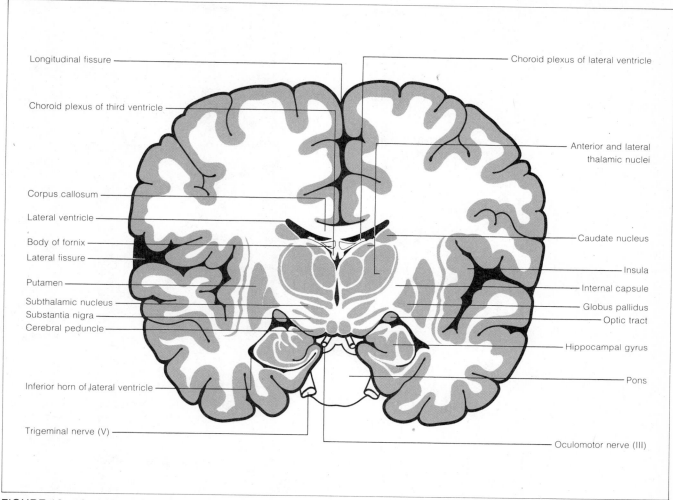

Longitudinal fissure

Choroid plexus of third ventricle

Corpus callosum

Lateral ventricle

Body of fornix

Lateral fissure

Putamen

Subthalamic nucleus

Substantia nigra

Cerebral peduncle

Inferior horn of lateral ventricle

Trigeminal nerve (V)

Choroid plexus of lateral ventricle

Anterior and lateral thalamic nuclei

Caudate nucleus

Insula

Internal capsule

Globus pallidus

Optic tract

Hippocampal gyrus

Pons

Oculomotor nerve (III)

FIGURE 10–10   Coronal section of the brain at the level of the diencephalon, showing the basal ganglia. (This section is anterior to Figure 10–6.)

system has connections with certain nuclei in the thalamus and hypothalamus and cerebral cortex. The limbic system has an important role in emotional responses, such as fear, rage, and motivation.

THE CRANIAL NERVES

Most cranial nerves are mixed nerves that contain both motor and sensory fibers, and four of them also carry autonomic efferents (Figure 10–2 and Table 10–2). Several, such as those from the eye and the ear, are wholly sensory nerves, with no efferent fibers at all. The nerves to the muscles that

move the eyes are primarily motor nerves, but they may also carry a few sensory fibers from proprioceptors in those muscles.

I. Olfactory Nerve

The first cranial nerve is a sensory nerve with very short thick fibers that arise in the specialized olfactory mucous membrane found in the superior portion of the nasal cavity. The axons pass through the perforations in the cribriform plate of the ethmoid bone to end in the olfactory bulb, an extension of the brain that lies directly above the cribriform plate.

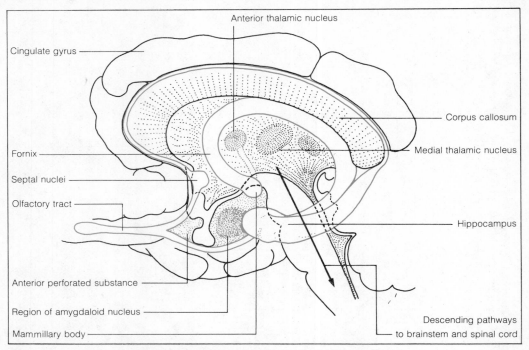

Anterior thalamic nucleus

Cingulate gyrus

Corpus callosum

Medial thalamic nucleus

Fornix

Septal nuclei

Olfactory tract

Hippocampus

Anterior perforated substance

Region of amygdaloid nucleus

Mammillary body

Descending pathways to brainstem and spinal cord

FIGURE 10–11   Structures of the rhinencephalon and limbic system.

II. Optic Nerve

The second cranial nerve is necessary for vision, since it carries impulses from the light receptors in the retina of the eye. It is not a true nerve, however, since, like the olfactory bulb, it is an outgrowth of the brain. There are also cell bodies and synapses within the retina itself. Nevertheless, the first part of the visual pathway is called the optic nerve. It emerges from the eyeball and enters the cranial cavity through the optic canal. At the optic chiasm some of the fibers cross to the other side to continue as the optic tract.

III. Oculomotor Nerve

As the name implies, this is a motor nerve to the eye. It carries motor fibers to four of the six extrinsic eye muscles which move the eyeball (superior, inferior, and medial rectus, and inferior oblique muscles). It also carries autonomic efferents to the intrinsic eye muscles which are smooth muscles that control pupillary size and adjustments for near and far vision (accomodation).

The central connections of the oculomotor nerve are in the midbrain.

IV. Trochlear Nerve

The fourth cranial nerve is a small nerve that supplies one of the extrinsic eye muscles (the superior oblique). It is the only cranial nerve that emerges on the posterior surface of the brainstem.

V. Trigeminal Nerve

One of the largest of the cranial nerves, the trigeminal nerve emerges conspicuously from the lateral portion of the pons (Figures 10–2 and 10–12). It is the major sensory nerve to the face, carrying sensory fibers from the skin of the face and the mucous membranes of the nasal and oral cavities. The large semilunar ganglion contains the cell bodies of the sensory neurons. Distal to this ganglion the nerve forms three major branches or divisions, the **ophthalmic, maxillary,** and **mandibular nerves,** which are distributed roughly to the superior, middle, and inferior portions of the face, respectively. The maxillary and mandibular branches carry the sensory fibers from the teeth of the upper and lower jaws. For dental work, branches of these nerves are frequently blocked by injection of a local anesthetic.

The trigeminal nerve also has a small but important motor component, the innervation to the muscles of mastication that control movements of the mandible.

VI. Abducens Nerve

The sixth cranial nerve emerges from the lower border of the pons, near its junction with the medulla. It carries somatic motor fibers to one of the extrinsic eye muscles (the lateral rectus). Cranial nerves III, IV, and VI on both sides must work together to produce the highly coordinated and exacting movements of the two eyes.

VII. Facial Nerve

The facial nerve is the chief motor nerve to the face. It arises from the brainstem at the junction of the pons and medulla, and innervates the somatic muscles of facial expression. It also carries autonomic efferents to two salivary glands (the submandibular and sublingual) and sensory fibers from some of the taste buds of the tongue. The nerve leaves the skull through a canal in the temporal bone and passes under the mastoid process before breaking into smaller branches on the side of the face. Damage to the facial nerve results in paralysis of the facial muscles (excluding the muscles of mastication) on that side of the face.

VIII. Statoacoustic Nerve

The eighth cranial nerve is a sensory nerve in two parts. The **cochlear portion** arises in the cochlea, which contains the organ of hearing. The **vestibular portion** arises in the vestibular apparatus, which contains the receptors for equilibrium and motion detection. Both of these receptor mechanisms are part of the inner ear located in the petrous portion of the temporal bone.

IX. Glossopharyngeal Nerve

The glossopharyngeal nerve is a mixed nerve. Its motor component contains autonomic fibers to a salivary gland (the parotid) and supplies a muscle of the pharynx. Its sensory fibers are from some of the taste buds of the tongue, the mucous membrane of the pharynx, and some visceral receptors.

X. Vagus Nerve

The vagus is also a mixed nerve, although its motor functions are more familiar because they provide part of the autonomic innervation to the thoracic and abdominal viscera (Figure 10–13). It supplies both sensory and motor innervation to the pharynx and larynx (skeletal muscles), and it carries afferents from some important visceral receptors.

FIGURE 10–12   Branches and distribution of the trigeminal nerve.

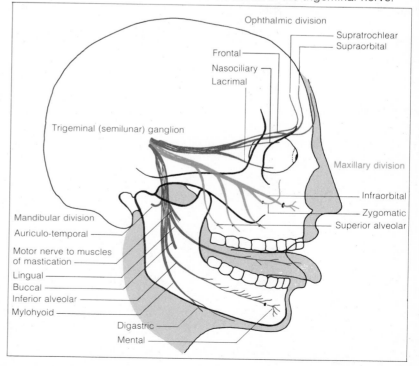

TABLE 10-2   THE CRANIAL NERVES

Nerve	Components	Central Connections	Exit from Skull	Peripheral Connections	Function
I. Olfactory	Afferent	Olfactory tract and bulb	Cribriform plate of ethmoid bone	Olfactory epithelium in upper nasal cavity	Smell
II. Optic	Afferent	Optic chiasm and tract	Optic foramen	Retina of eye	Vision
III. Oculomotor[1]	Efferent (somatic)	Midbrain	Superior orbital fissure	Extrinsic eye muscles (superior, inferior, and medial rectus, inferior oblique)	Movements of eyes
	Efferent (autonomic)	Midbrain		Intrinsic eye muscles (ciliary and pupillary muscles)	Accomodation for near and far vision, constriction of pupil
IV. Trochlear[1]	Efferent (somatic)	Midbrain	Superior orbital fissure	Extrinsic eye muscle (superior oblique)	Movement of eyes
V. Trigeminal	Afferent	Pons		Skin and mucous membrane of face, nose, and mouth regions	General sensations
Ophthalmic division			Superior orbital fissure	Face, eye, nasal cavity	
Maxillary division			Foramen rotundum	Face, oral cavity, teeth	
Mandibular division			Foramen ovale	Face, mandible, teeth	
	Efferent (somatic)	Pons		Muscles of mastication	Mastication
	Afferent	Midbrain		Proprioceptors in muscles of mastication	Muscle sense
VI. Abducens[1]	Efferent (somatic)	Pons	Superior orbital fissure	Extrinsic eye muscle (lateral rectus)	Movement of eyes
VII. Facial	Efferent (somatic)	Caudal pons	Facial canal	Muscles of facial expression	Facial expression
	Efferent (autonomic)	Medulla		Salivary glands (sublingual, and submaxillary and lacrimal glands.	Secretion of saliva and tears
	Afferent	Medulla		Taste buds, near tip of tongue	Taste

TABLE 10-2   THE CRANIAL NERVES continued

Nerve	Components	Central Connections	Exit from Skull	Peripheral Connections	Function
VIII. Statoacoustic					
Cochlear	Afferent	Medulla	Internal acoustic meatus	Receptors in spiral organ of inner ear	Hearing
Vestibular	Afferent	Medulla	Internal acoustic meatus	Receptors in semicircular canals, utricle, and saccule (vestibular apparatus in inner ear)	Equilibrium (position, linear and rotatory acceleration
IX. Glossopharyngeal	Afferent	Medulla	Jugular foramen	Taste buds in tongue	Taste
	Afferent	Medulla		Mucous membrane pharyngeal region, also carotid sinus	Sensory from visceral structures[2]
	Efferent (autonomic)	Medulla		Salivary gland (parotid)	Secretion of saliva
	Efferent (somatic)	Medulla		Muscles of pharynx	Swallowing
X. Vagus	Efferent (autonomic)	Medulla	Jugular foramen	Smooth muscle and glands of thorax and abdomen	Regulation of visceral structures (parasympathetic)
	Efferent (somatic)	Medulla		Muscles of pharynx and larynx	Swallowing, control of larynx
	Afferent	Medulla		Viscera of thorax and abdomen	Sensory from visceral structures[2]
XI. Spinal Accessory	Efferent (somatic)	Medulla	Jugular foramen	Muscles of pharynx and larynx (distributed with vagus)	Swallowing and control of larynx
	Efferent (somatic)	Upper cervical segments of spinal cord	Enters through foramen magnum leaves through jugular foramen	Sternocleidomastoid and trapezius	Movement of head and shoulder
XII. Hypoglossal[1]	Efferent (somatic)	Medulla	Hypoglossal canal	Extrinsic and intrinsic muscles of tongue	Movements of tongue

[1]Nerves III, IV, VI and XII also contain afferent fibers from proprioceptors in the extrinsic eye muscles and muscles of the tongue. They serve muscle sense.

[2]Some visceral afferents reach the level of consciousness and cause sensation, but others do not, and hence do not cause a sensation (such as afferents from receptors in blood vessels that detect blood pressure).

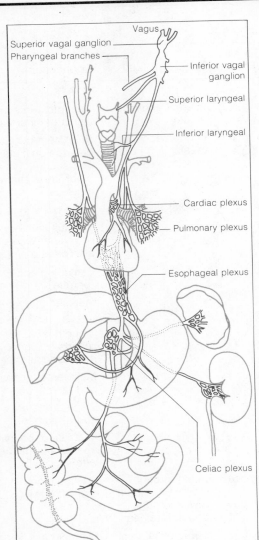

FIGURE 10–13 Distribution of the vagus nerve.

Superior vagal ganglion
Pharyngeal branches
Vagus
Inferior vagal ganglion
Superior laryngeal
Inferior laryngeal
Cardiac plexus
Pulmonary plexus
Esophageal plexus
Celiac plexus

XI. Spinal Accessory Nerve

The eleventh cranial nerve is really two nerves, one of which is a true cranial nerve and one a spinal nerve. The cranial portion arises from the medulla near the vagus nerve, and its fibers mingle with those of the vagus and are distributed with them to muscles of the pharynx and larynx. The spinal portion of the nerve arises from the cervical spinal cord, but passes up through the foramen magnum to join the cranial portion. Its fibers are motor; they innervate the trapezius and sternocleidomastoid muscles.

XII. Hypoglossal Nerve

The hypoglossal nerve is also a motor nerve. It arises from the ventral surface of the medulla and innervates the skeletal muscle of the tongue.

PROTECTIVE AND SUPPORTIVE STRUCTURES

Ventricles

As the parts of the developing brain enlarge, so does the canal in its center, and four cavities, the ventricles, appear (Figure 10–14). The first two, the **lateral ventricles,** are in the cerebral hemispheres and thus are not seen in a midsagittal section of the brain. They lie beneath the corpus callosum, each extending from the frontal lobe and around in a large "C" to terminate in the temporal lobe, though with a small spur extending posteriorly into the occipital lobe. The lateral ventricles are separated from one another by a thin partition, the **septum pellucidum.** They communicate with one another and with the third ventricle through an opening, the **interventricular foramen.** The **third ventricle** lies in the midline and is only a few millimeters wide. It is bounded laterally by the thalami and hypothalami. Posteriorly, it is channeled into the **cerebral aqueduct,** which passes through the midbrain between the corpora quadrigemini and cerebral peduncles and then opens into the fourth ventricle. The **fourth ventricle** lies between the medulla and pons on one side and the cerebellum on the other. The central canal of the spinal cord opens into it from below. There are openings in the roof of the fourth ventricle inferior to the cerebellum so that the ventricles do not form a completely closed system.

Meninges

The connective tissue coverings of the brain are the same as those of the spinal cord (see Figure 9–2) and are continuous with them. The innermost, the **pia mater,** is thin and delicate and adheres closely to the brain surface, dipping into every sulcus and fissure. The **arachnoid** is more closely associated with the dura mater than with the pia, leaving a **subarachnoid space,** between the pia mater and the arachnoid, bridged by numerous strands or trabeculae of arachnoid tissue. The cranial **dura mater,** like that around the cord, is tough and heavy, but in the skull it consists of two layers of tissue (Figure 10–15). The outer (periosteal) layer adheres to the bone and follows its contours closely. The inner (meningeal) layer also follows the shape of the skull except at the sites of certain major fissures. It dips into the longitudinal fissure and then is reflected back on itself, forming a septum between the two hemispheres known as the **falx cerebri** (*falx* = sickle). This septum is actually a double thickness of the inner layer of dura. A similar though smaller fold, the **tentorium cerebelli,** extends between the cerebellum and the cerebral hemispheres in such a way as to form a roof or tent over the cerebellum. Some of the large veins that drain the brain, the venous sinuses, are located at the edges of the dural folds. These folds serve an important protective function because they help hold the brain in place within the cranial cavity.

Cerebrospinal Fluid

The ventricles, subarachnoid space, and the central canal of the spinal cord are filled with **cerebrospinal fluid (CSF)** (Figure 10–14). Protruding into all the ventricles are tufts of capillaries,

FIGURE 10–14   The ventricles of the brain, lateral and anterior views.

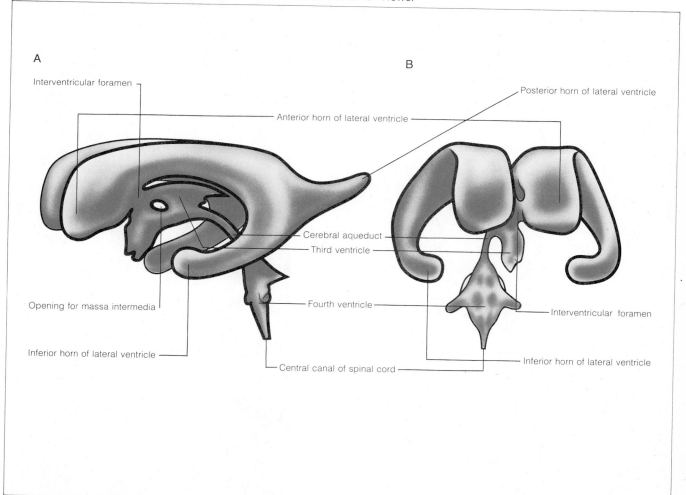

A

Interventricular foramen

Anterior horn of lateral ventricle

Posterior horn of lateral ventricle

B

Cerebral aqueduct

Third ventricle

Opening for massa intermedia

Fourth ventricle

Interventricular foramen

Inferior horn of lateral ventricle

Central canal of spinal cord

Inferior horn of lateral ventricle

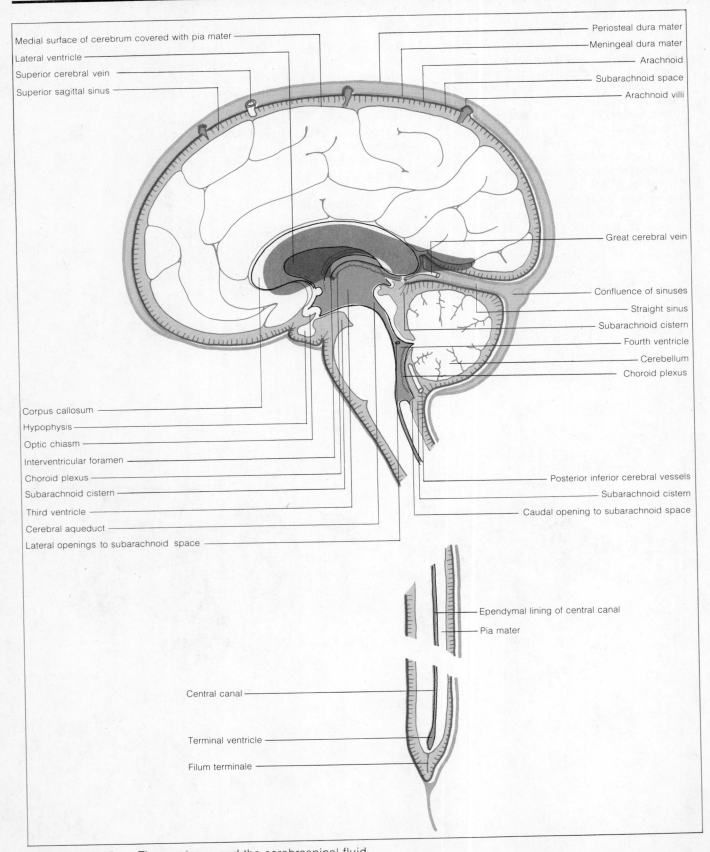

Medial surface of cerebrum covered with pia mater

Lateral ventricle

Superior cerebral vein

Superior sagittal sinus

Periosteal dura mater

Meningeal dura mater

Arachnoid

Subarachnoid space

Arachnoid villi

Great cerebral vein

Confluence of sinuses

Straight sinus

Subarachnoid cistern

Fourth ventricle

Cerebellum

Choroid plexus

Corpus callosum

Hypophysis

Optic chiasm

Interventricular foramen

Choroid plexus

Subarachnoid cistern

Third ventricle

Cerebral aqueduct

Lateral openings to subarachnoid space

Posterior inferior cerebral vessels

Subarachnoid cistern

Caudal opening to subarachnoid space

Ependymal lining of central canal

Pia mater

Central canal

Terminal ventricle

Filum terminale

FIGURE 10-15   The meninges and the cerebrospinal fluid.

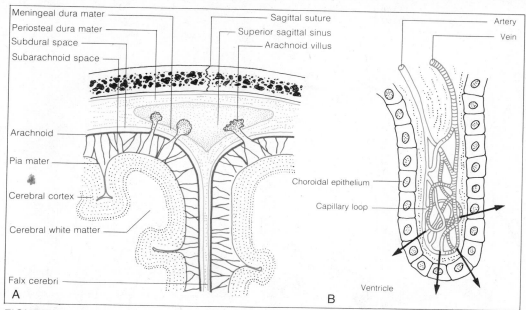

Meningeal dura mater
Periosteal dura mater
Subdural space
Subarachnoid space
Arachnoid
Pia mater
Cerebral cortex
Cerebral white matter
Falx cerebri
A

Sagittal suture
Superior sagittal sinus
Arachnoid villus

Artery
Vein
Choroidal epithelium
Capillary loop
Ventricle
B

FIGURE 10–16   A. Arachnoid villi protrude into venous sinuses for return of cerebrospinal fluid to the bloodstream. B. Cerebrospinal fluid is formed by diffusion and active transport in the choroid plexus.

covered by a layer of ependymal cells, forming the **choroid plexuses.** The cerebrospinal fluid is produced partly by diffusion and filtration, but chiefly by active transport (secretion) by the ependymal cells. Fluid is transferred from the blood in the capillaries of the choroid plexuses into the ventricular cavities. Some CSF is also formed by ependymal cells around cerebral blood vessels and by the lining of the central canal of the spinal cord. It circulates slowly from the lateral ventricles to the third and fourth ventricles, and reaches the subarachnoid spaces around the brain and spinal cord through lateral and caudal openings in the roof of the fourth ventricle. The composition of cerebrospinal fluid resembles that of the blood plasma and interstitial fluid. However, CSF is almost completely devoid of protein, and the concentration of some ions also differs slightly.

Since the cerebrospinal fluid is enclosed with the brain in a rigid container, it must be removed from the cranial cavity at about the same rate as it is formed. Tiny thin-walled projections of arachnoid tissue, called **arachnoid villi** (or *arachnoid granulations*), project into the venous sinuses of the dura (Figure 10–16). The villi are quite permeable, and the fluid readily filters

through the villi into the venous sinuses, thereby returning to the bloodstream from which it came.

The CSF is normally under only a slight pressure (10–20 mm Hg[2]), but if its movement from the ventricles into the subarachnoid space is blocked, the intracranial pressure may reach as much as 400 mm Hg or more. Chronically inadequate CSF removal may result in a condition known as *hydrocephalus,* in which fluid continues to form and the ventricles become enlarged. In children whose cranial bones have not yet firmly united, the skull may become greatly enlarged as this pressure forces the bones to spread. The brain may suffer damage from the elevated pressure, but the condition can sometimes be treated by insertion of a tube to bypass the blockage and drain the CSF into a vein outside the skull.

Tumors or damage to the brain can often be located by injecting air into the ventricles (a *ventriculogram*) or subarachnoid space (a *pneumogram*). It

[2]mm Hg = millimeters of mercury. Pressure is commonly expressed in terms of the pressure exerted by a column of mercury. In this case, the CSF exerts a pressure on the walls enclosing it equal to the pressure at the bottom of a column of mercury 15-20 mm high. This is not a very high pressure, but 400 mm Hg is three or four times the pressure in arteries (blood pressure).

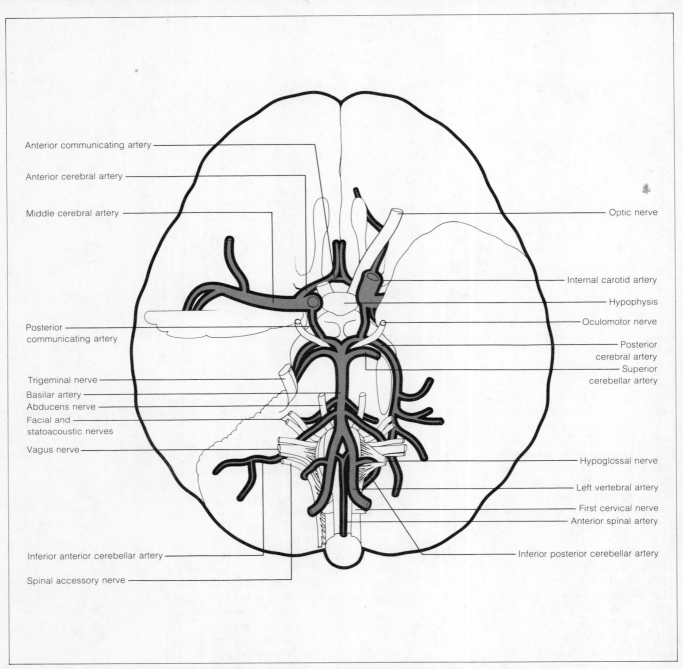

Anterior communicating artery

Anterior cerebral artery

Middle cerebral artery

Optic nerve

Internal carotid artery

Hypophysis

Oculomotor nerve

Posterior communicating artery

Posterior cerebral artery

Superior cerebellar artery

Trigeminal nerve

Basilar artery

Abducens nerve

Facial and statoacoustic nerves

Vagus nerve

Hypoglossal nerve

Left vertebral artery

First cervical nerve

Anterior spinal artery

Inferior anterior cerebellar artery

Spinal accessory nerve

Inferior posterior cerebellar artery

FIGURE 10–17 Major arteries of the base of the brain, including the circle of Willis. The left cerebellar hemisphere and the tip of the right temporal lobe have been removed.

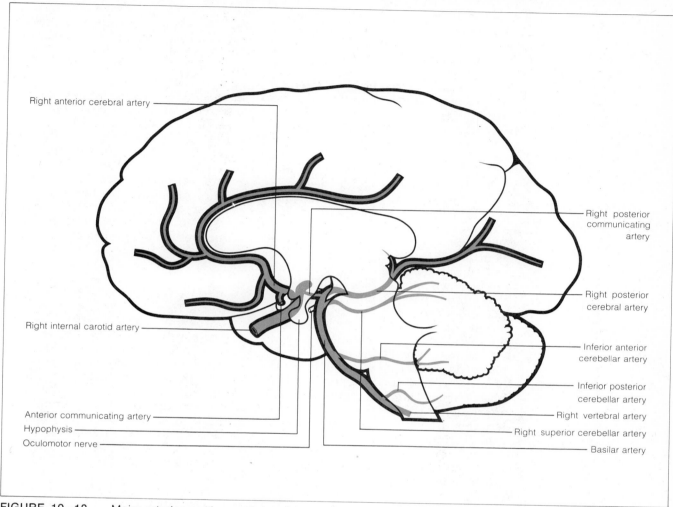

Right anterior cerebral artery

Right posterior communicating artery

Right posterior cerebral artery

Right internal carotid artery

Inferior anterior cerebellar artery

Inferior posterior cerebellar artery

Right vertebral artery

Anterior communicating artery

Right superior cerebellar artery

Hypophysis

Oculomotor nerve

Basilar artery

FIGURE 10–18   Major arteries on the medial surface of the brain.

makes the outlines of the ventricles distinguishable in an X-ray, and distortions can then be detected. Samples of CSF may be analyzed for diagnostic purposes, and sometimes anesthetics are administered via the CSF (as spinal anesthesia). For these purposes access to the CSF is usually gained by lumbar puncture, entering the spinal column below the end of the spinal cord.

The importance of cerebrospinal fluid as a protective agent can scarcely be overemphasized. Neural tissue in general, and the brain in particular, is extremely soft and fragile. An uncushioned brain would not be able to withstand blows to the skull from the outside, or the bumps and bruises caused by sudden movements of the head. The CSF lends a buoyancy that reduces the effective weight of the central nervous

system from about 1500g to 50g. It also acts as a water jacket to support and cushion the brain. The meninges and dural folds help in this regard by restricting the movement of the brain within the cranial cavity.

Blood Supply to the Brain

The blood supply to the brain is crucial, since if it is interrupted for even a few minutes, some parts of the brain may be irreparably damaged. The major arterial supply comes from two sources, the vertebral arteries and the internal carotid arteries (Figures 10–17 and 10–18). The **vertebral arteries** ascend through the transverse foramina of the cervical vertebrae and pass through the foramen magnum into the cranial cavity. Branches are given off to the spinal cord, medulla, and cerebellum before the vertebrals join to form the **basilar artery** on the ventral surface of the pons. The basilar artery sends branches to the cerebellum, internal ear, and pons before it terminates in the **posterior cerebral arteries,** which supply the inferior and posterior portions of the cerebral hemispheres.

The **internal carotid arteries** enter the cranial cavity through the carotid canal of the temporal bone. They have two major branches, the anterior and middle cerebral arteries. The **anterior cerebral arteries** lie at the bottom of the longitudinal fissure, where they follow along the surface of the corpus callosum and supply the medial portion of the cerebral hemispheres. The **middle cerebral arteries** pass laterally through the lateral fissure to supply cerebral tissue on either side of the fissure. Branches of the surface arteries penetrate the brain substance to provide a rich blood supply to deep structures in the cerebral hemispheres and brainstem. One tiny branch, the **anterior communicating artery,** joins the anterior cerebral arteries, and paired **posterior communicating arteries** connect the posterior cerebral arteries (from the basilar) with the internal carotids, forming a ring around the pituitary gland and optic chiasm known as the **circle of Willis.** Theoretically, impaired blood flow in either the basilar or one of the internal carotid arteries could be compensated by blood from the other arteries through the communicating vessels. It is no doubt effective in conditions that develop slowly, but since the communicating arteries are quite small, they would probably be inadequate in the face of a sudden *occlusion* (blockage) of one of the major arteries.

After flowing through the brain, the blood is collected in small veins that carry it to surface veins in the subarachnoid space. The latter veins eventually drain into relatively large venous sinuses located in the dura mater (Figure 10–19). Among these are the **superior sagittal sinus** at the junction of the falx cerebri and the cranial dura, and the **inferior sagittal sinus** along the deep free edge of the falx cerebri. The two are joined by the **straight sinus,** which runs along the junction of the falx cerebri and the tentorium cerebelli. **Transverse sinuses** pass laterally from the junction of the straight and superior sagittal sinuses and curve downward to leave the cranium through the jugular foramina and become the *jugular veins*. The **cavernous sinus** is a large, rather diffuse sinus located around the sella turcica. It receives blood from the deep cerebral veins and drains eventually into the transverse sinuses and jugular veins. Vascular problems in this region are not uncommon, but they are serious because of the inaccessibility of the vessels and the importance of the brain structures in their immediate vicinity.

CHAPTER 10 SUMMARY

The brain is that part of the CNS located in the skull. Three distinct segments appear very early in its development, the **forebrain, midbrain,** and **hindbrain.** The midbrain and hindbrain, as a continuation of the spinal cord, form the *brainstem*. The hindbrain consists of the *medulla oblongata, pons,* and *cerebellum.* The cerebellum is an outgrowth from the posterior surface of the pons. The midbrain includes the *cerebral peduncles* and the *corpora quadrigemini* (*superior* and *inferior colliculi*).

The brainstem provides passage for fiber tracts that connect the spinal cord and various parts of the brain, and it contains the central connections for ten of the twelve cranial nerves. It also contains centers for regulation of many processes, such as breathing and heart rate. Many of these centers are in an area called the *reticular formation* that extends throughout the length of the brainstem. At its superior end the midbrain divides into right and left parts as it connects with the forebrain, whose structures are paired.

The forebrain consists of the diencephalon and cerebral hemispheres. The *diencephalon* is located on either side of the third ventricle. Its main parts are the *thalamus* and *hypothalamus,* both of which contain many nuclei and centers. The thalamus is important in sensory pathways, and the hypothalamic centers help regulate many kinds

Focus

The Pineal Eye

The bottom-dwelling ancestor of the vertebrates—the group that includes the mammals, including humans—may have had four eyes, two on the sides of the head and two on top, to watch for predators approaching from above. The first fish apparently had three eyes, for their fossils show a hole in the top of the skull similar to the hole visible in the skulls of such reptiles as the New Zealand tuatara, which does have a third eye. In higher vertebrates, however, these extra eyes are purely vestigial. Their remnants constitute two small parts of the brain known as the pineal and parietal (or parapineal) bodies. Humans have only the pineal.

The human pineal body lies between the cerebral hemispheres and below the back of the corpus callosum. It is apparently an endocrine gland, for it secretes the hormone melatonin, but until fairly recently it was not thought to play any very great role in the body. Melatonin had no known function in humans, although in amphibians it causes contraction of pigment-containing skin cells and makes the skin turn pale. Too, the human pineal tends to calcify after puberty.

The seeming lack of a function for the human pineal has provoked a fair amount of curiosity about this tiny organ. Researchers have suggested that it is involved in controlling the pituitary gland, particularly its effects on the hormones of the adrenal cortex. There is evidence that it may help control the secretion of aldosterone, the adrenal hormone that controls salt and water excretion by the kidney. There is also evidence that it may serve to inhibit growth and delay puberty.

The evidence seems strongest for the effect on puberty. It is now believed that the pineal retains a link with its primordial function: it is no longer an eye, but it does mediate an effect of light on the body. In animals that have a breeding season, the pineal secretes hormones—probably including melatonin—that activate the gonads at the right time of year. It does so by measuring day length according to the amount of light available to stimulate it, either via the eyes or directly as light filters through the skull and into the brain. Humans do not have a breeding season, but the pineal may play a role in their sexual cycle too. It has been observed that blind girls, whose brains are less stimulated by light, enter puberty earlier than sighted girls. Children with pineal tumors, which secrete excess melatonin, do so later. Melatonin is therefore thought to inhibit the hormonal changes that initiate puberty. Just how it does so is not known.

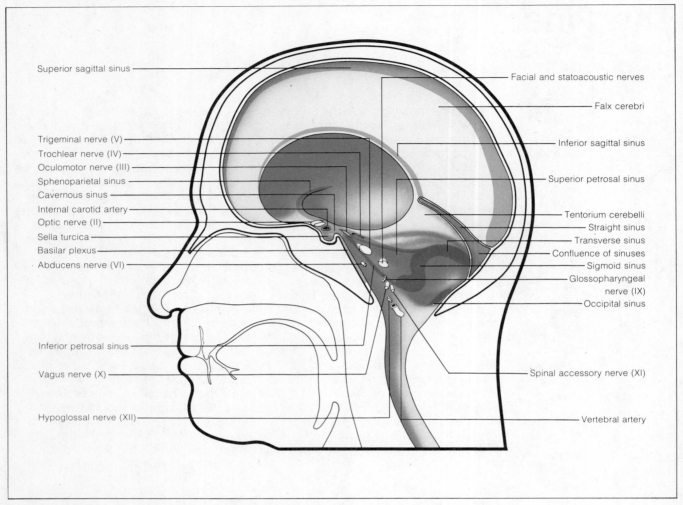

Superior sagittal sinus

Trigeminal nerve (V)
Trochlear nerve (IV)
Oculomotor nerve (III)
Sphenoparietal sinus
Cavernous sinus
Internal carotid artery
Optic nerve (II)
Sella turcica
Basilar plexus
Abducens nerve (VI)

Inferior petrosal sinus

Vagus nerve (X)

Hypoglossal nerve (XII)

Facial and statoacoustic nerves

Falx cerebri

Inferior sagittal sinus

Superior petrosal sinus

Tentorium cerebelli
Straight sinus
Transverse sinus
Confluence of sinuses
Sigmoid sinus
Glossopharyngeal nerve (IX)
Occipital sinus

Spinal accessory nerve (XI)

Vertebral artery

FIGURE 10-19   The dural folds and major venous sinuses.

of body processes. The hypothalamus is the connection for the *neurohypophysis* (the neural part of the hypophysis).

The *cerebral hemispheres* are the most highly developed part of the human brain, both anatomically and functionally. They completely cover the diencephalon and part of the brainstem. The two hemispheres are separated by the *longitudinal fissure,* and each is divided into *frontal, parietal, temporal,* and *occipital lobes,* mainly by the *central sulcus* and *lateral fissure.* On the surface of each hemisphere is the *cerebral cortex,* a layer of gray matter that is greatly expanded by numerous ridges *(gyri)* and indentations *(sulci).* Beneath the cortex are white fiber tracts that connect one gyrus or lobe with another *(association fibers),* one hemisphere with

another *(commissural fibers),* or the cortex with other parts of the brain or the spinal cord *(projection fibers).* The *basal ganglia* are centers of gray matter deep inside each hemisphere. They have connections with the cortex and other parts of the brain and spinal cord.

The visible parts of the *rhinencephalon* lie on the inferior surface of the frontal lobe, and are associated with the sense of smell. Other parts of the rhinencephalon, or *limbic system,* are associated with certain emotional responses.

Twelve pairs of cranial nerves arise from the brain. They are:

1   **Olfactory**—smell
2   **Optic**—vision
3   **Oculomotor**—movement of eyes;

adjustments for near and far vision, and size of pupil

4 Trochlear—movement of eyes

5 Trigeminal—sensory from face; motor to muscles of mastication. Three major divisions: **ophthalmic, maxillary,** and **mandibular**

6 Abducens—movement of eyes

7 Facial—motor to muscles of facial expression; secretion of saliva and tears; taste

8 Statoacoustic Coclear division—hearing and **vestibular division**—equilibrium

9 Glossopharyngeal—taste; secretion of saliva; pharynx (motor and sensory)

10 Vagus—thoracic and abdominal viscera (motor and sensory); motor to muscles of pharynx and larynx

11 Spinal Accessory—pharynx and larynx (motor, with vagus); motor to sternocleidomastoid and trapezius

12 Hypoglossal—movement of tongue

There are four openings or *ventricles* in the brain; the lateral ventricles (one in each hemisphere), the third ventricle in the diencephalon, and the fourth ventricle in the hindbrain between the cerebellum and the brainstem. It is connected to the third ventricle by the *cerebral aqueduct*.

The brain is enclosed by the same meningeal layers as the spinal cord, namely, the *dura mater, arachnoid membrane,* and the *pia mater. Cerebrospinal fluid* (CSF) is found in the subarachnoid space between the arachnoid membrane and pia mater. The CSF is produced by *choroid plexuses* in the ventricles. It is derived from blood plasma and slowly circulates through the ventricles and moves out into the subarachnoid space from the fourth ventricle. It returns to the blood from *arachnoid villi* which protrude from the subarachnoid space into the large venous sinuses of the brain. Folds of dura mater lie between the cerebral hemispheres (*falx cerebri*) and between the cerebrum and cerebellum (*tentorium cerebelli*).

The blood supply to the brain comes from the *vertebral* and *internal carotid arteries*. There are connections between these arteries (*circle of Willis*). Anterior, middle, and posterior cerebral arteries supply the hemispheres, and numerous short arteries supply the cerebellum and brainstem. Veins converge into large venous sinuses which are located along the junctions of dural folds. Among them are the *superior* and *inferior sagittal sinuses* at the top and bottom of the falx cerebri, and the *cavernous sinus* on the inferior surface of the brain. The sinuses converge to leave the skull as the *jugular vein*.

STUDY QUESTIONS

1 What are the five main divisions of the brain? How did they come about?

2 In a sense, the brainstem can be considered to be a region of transition between the spinal cord and the cerebral hemispheres. Do you agree? Explain your answer using examples of structural characteristics that support your position.

3 Name one anatomical feature by which each of the following could be identified: medulla, pons, midbrain, cerebellum, hypothalamus. Name at least one important connection of each.

4 What is included in the cerebral hemispheres? How are they connected to the brainstem?

5 In general, how do cranial nerves resemble spinal nerves, and how do they differ?

6 Which cranial nerves are entirely motor (except for proprioceptor afferents)? Which are entirely sensory? Which supply structures of the face? Which supply structures not in the head?

7 Describe the circulation of cerebrospinal fluid, from its formation to its return to the bloodstream.

8 Describe the blood (arterial) supply to the brain. Of what significance is the circle of Willis?

9 Describe the venous drainage from the brain. What is the relationship between veins of the brain and the dural folds?

11

The General Senses

The Special Senses — Vision

The Special Senses — Hearing and Equilibrium

The Special Senses — Smell

The Special Senses — Taste

Sensory Aspects of the Central Nervous System

When one speaks of sensory aspects of brain function, or of sensation, the notion of conscious awareness is implied, and both consciousness and awareness require the cerebral cortex. The central nervous system receives information from many kinds of receptors. The mechanisms for processing their varied information are similar in many ways, but there are important differences. We therefore discuss sensory function in terms of the general and special senses. The general senses include those that arise from cutaneous receptors (touch, pain), some visceral and proprioceptors. The special senses (vision, hearing) involve complex receptors and highly specialized pathways. Information from some receptors never reaches the cortex, and they do not cause a sensation.

Sensation results from impulses in nerve pathways that ascend through the spinal cord and brainstem to the cerebral cortex, but unless the cortex is sufficiently aroused to respond and process the information, there will be no sensation. The state of cortical arousal or activation depends upon additional input from other areas, including the brainstem reticular formation. Arousal enables the cortex to respond to incoming information and is necessary for what we know as the conscious state.

The sensory role of the central nervous system begins with the arrival of impulses from receptors throughout the body. Incoming information is sorted, evaluated, and integrated in the CNS. Stimulation of a receptor has several possible results: the sensory function, which involves the cerebral cortex, may register touch or pain; reflex responses, which involve only the spinal cord or medulla, may cause a somatic response like the withdrawal of a limb, or an autonomic response such as a brief increase in heart rate. Receptor stimulation may also cause movement away from a stimulus, thus involving the cortex in a decision making process.

The CNS does more than simply detect stimuli: it identifies the stimulus site (localization), recognizes the type of stimulus or modality (touch, pain, warmth), and intensity of stimulation, as well as the more complex qualities of sensation such as weight, texture and shape.

THE GENERAL SENSES

Although the general senses involve stimulation of many kinds of receptors, they can be considered together because their characteristics and pathways have much in common. References to general sensations usually brings to mind somatic sensations, which are those arising from receptors in the skin—touch, pressure, heat, cold, and pain. The sense of movement and position (conscious proprioception) depends on receptors in tendons and around joints, in addition to touch and pressure receptors. Visceral organs contain receptors that respond to stretch, giving a sensation of fullness (as in stomach and bladder), and to pain. Sensations from visceral receptors are usually not as clearly defined as those from somatic receptors.

The afferent pathways from some receptors do not reach the cerebral cortex, and they do not cause sensations. Impulses that arise from muscle spindles (unconscious proprioception) fall in this category, as well as impulses from receptors that monitor blood pressure or the oxygen content of blood.

They provide essential information for regulatory processes, and their actions are described with the processes they help to regulate. The discussion that follows describes mainly the sensations that arise from stimulation of cutaneous receptors.

Primary Sensory Pathways

Sensory pathways carry information from the receptors and end in the cerebral cortex. Peripheral endings of a given afferent neuron all have the same type of receptor. Each afferent fiber therefore carries information about only one modality (such as a pain fiber). Although the different modalities follow different pathways as they ascend through the spinal cord and brainstem, some generalizations can be made about pathways for the general senses.

1 All of them reach the cerebral cortex.
2 All ascending pathways cross to the other side of the cortex. Some cross at the level of entry into the cord, while others ascend a few segments before crossing, and still others cross in the medulla. Thus the right side of the body is represented on the left side of the cerebral cortex, and vice versa.
3 There is a certain amount of modality separation in the spinal cord, and ascending fiber tracts in the spinal cord contain "like" fibers. The posterior portion of the cord (Figure 9–3), between the posterior horns, contains two fiber tracts (*fasciculus gracilis* and *cuneatus*) which, for convenience, are often called the **posterior white columns.** Fibers in these tracts carry impulses concerned with conscious proprioception (muscle sense and position), touch and pressure, and certain perceptive functions of touch, such as localization and discrimination. The **lateral spinothalamic tracts** are associated with pain and temperature (Figure 11–1). Much attention is given to the lateral spinothalamic tracts because of the importance of

FIGURE 11–1 The sensory pathway for pain and temperature (lateral spinothalamic tract).

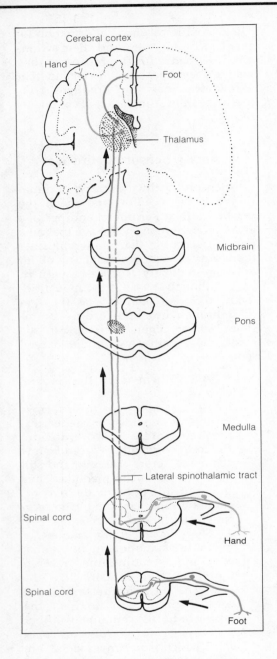

Cerebral cortex

Hand

Foot

Thalamus

Midbrain

Pons

Medulla

Lateral spinothalamic tract

Spinal cord

Hand

Spinal cord

Foot

pain in clinical situations.[1] The **anterior spinothalamic tracts** are associated with touch. Muscle sense depends upon the posterior white columns, and pain and temperature depend upon tracts in the antero-lateral portion of the cord, but touch and pressure are represented in both. The laterally placed **spino-**

[1]In cases of intractable pain, such as may occur in terminal cancer, the antero-lateral quadrant of the cord is sometimes severed to destroy ascending pain pathways.

cerebellar tracts are the pathways for unconscious proprioception and, as such, they go to the cerebellum rather than the cerebral cortex and are not involved directly in sensation, (Figure 11–2).

4 Ascending pathways involve three neurons (and two synapses) between receptor and cortex.

The direct pathways to the cortex are sometimes called the **primary sensory pathways.** The details of the location in the spinal cord, the site of crossing and synapses are quite specific and predictable for each pathway and are of diagnostic use in locating and describing injuries of the central nervous system.

The first neuron in this pathway is the familiar sensory (afferent) neuron whose cell body is in the spinal ganglion or its cranial nerve equivalent. It arises at the receptor and the central end is in the gray matter of the spinal cord or brainstem.

The cell body of the second neuron is in the cord or brainstem at the synapse with endings of the first neuron. It crosses to the other side, ascends to the thalamus, and terminates. Only a few of the thalamus's multiple nuclei are part of the relay for the primary sensory pathway. The ascending pathways for all general senses from spinal nerves synapse in one nucleus, general senses from cranial nerves in another, and the special senses have their relays in others. At the synapses, there is the possibility of connections and interactions with other neurons. This happens in the thalamus, where interneurons connect from these specific relay nuclei to some of the other (nonspecific) thalamic nuclei, some of which are part of the cortical activating system.

The third neuron in the primary sensory pathway has its cell body in the thalamus and passes to the somatosensory area of the cerebral cortex.

Somesthetic Cortex and the Primary Sensory Pathways

The sensory pathways end in one of the primary receiving areas of the cerebral cortex. The visual pathway terminates in the occipital lobe, the auditory pathway in the temporal lobe, and pathways for the general senses end in specific parts of the somatosensory area in the postcentral gyrus of the parietal lobe. The topographical representation or "mapping" of the body on the surface of the cortex is illustrated in Figure 11–3. Because it provides representation of senses from the body, this region is often called the **somesthetic cortex.** The amount of cortex devoted to a particular region of the body is related more to the number of fibers from that region than to the actual size of the region. Thus the fingertips, tongue, and lips, which have many cutaneous receptors and a high degree of sensory acuity, have a much larger cortical representation than the back or the thigh.

Electrodes placed on the exposed surface of the postcentral gyrus can record cortical action potentials as they arrive. Stimulation of a certain peripheral area evokes a spike potential in the cortical receiving area for that part. If a point in the sensory cortex of a conscious patient is stimulated,[2] the patient reports a general tingling sensation or a numbness in the appropriate peripheral region.

The primary sensory pathways to the cortex described and shown in Figures 11–1 and 11–2 provide an adequate means to recognize and interpret a stimulation. The arrival of action potentials in some part of the somesthetic cortex signals the stimulation of a certain part of the body surface. The number of action potentials and their frequency suggests the intensity of the stimulus. The modality is determined by the type of receptor and the pathways involved.[3]

We have repeatedly pointed out that the cortex is necessary for sensation, but it is difficult to say just what it is that the cortex does for sensation. If

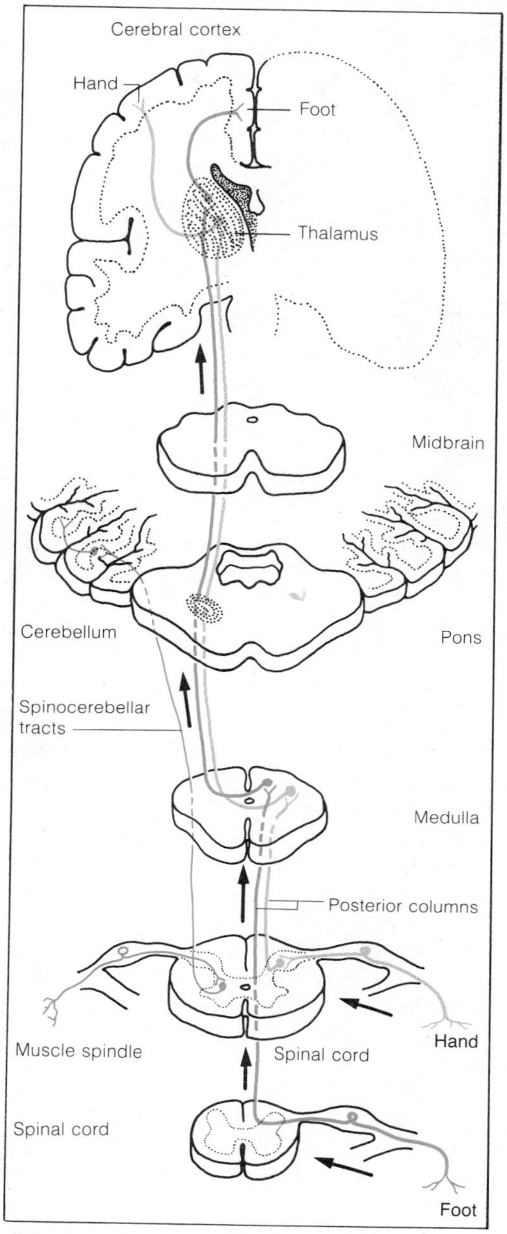

[2]Brain tissue does not contain pain or other receptors, and it is thus insensitive to painful stimuli as such. It is often desirable for a neurosurgeon to operate on a conscious patient (using local anesthesia to penetrate the skull) because the patient's responses, both verbal and nonverbal, may be very helpful to him.

[3]Since action potentials are alike, you might ask how the cortex distinguishes between action potentials from touch and pain receptors. The answer is not really known, but it must have something to do with a coding system, firing patterns, and/or interactions at synapses.

FIGURE 11–2   Pathways for touch, pressure, and conscious proprioception (posterior columns), and for unconscious proprioception (spinocerebellar tracts).

part of the somesthetic cortex is destroyed, there is no blank spot in our sensory perception of a part of the body. Unless the lesion (injury) is very extensive, the effect will be slight and the recovery good. The end result might be only an elevated threshold rather than an area of *anesthesia* (loss of sensation).

Visceral pain is vexing, as we often

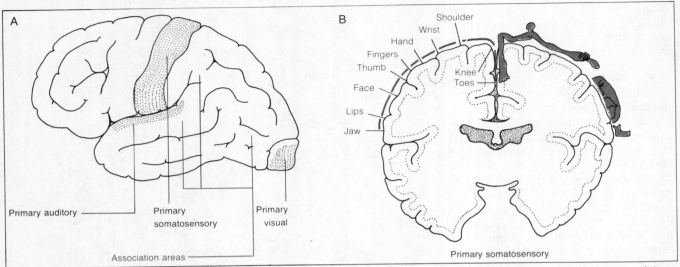

FIGURE 11–3   A. Primary sensory receiving areas and sensory association areas of the cerebral cortex. B. Representation of the body surface on the primary somatosensory area.

cannot locate the exact locus—a stomach ache rarely seems localized in the stomach. Relatively few nerve fibers carry visceral pain signals to the brain, and these signals are received in the somatosensory area of the somesthetic cortex, rather than in a special visceral area. The pain fibers tend to run with autonomic visceral fibers and enter the spinal cord at the level at which the autonomic fibers emerge. The pain may then seem to come from the region of the body wall supplied by the spinal nerve at that level. Such pain is called referred pain. Heart pain has referred and unreferred components: the referred component seems to come from the left arm and shoulder and is called angina pectoris; the unreferred component seems to come from below the breastbone.

To understand more fully the cortical contribution to sensation, we have to look beyond the somesthetic cortex and the primary sensory pathways.

Association Areas and Other Pathways

The primary receiving areas occupy only a portion of the cortical surface. Adjacent to each is a larger area known as an **association area.** The association area for the somesthetic cortex and the general senses is in the parietal lobe, just behind the postcentral gyrus. It does not receive projection fibers directly from the periphery, that is, the primary sensory pathways do not end there. It does, however, have indirect connections with the peripheral receptors, and it has important connections with other parts of the brain.

The primary receiving area is therefore not the only area receiving information from the cutaneous receptors, and the primary sensory pathways do not provide the only input to the cortex. All parts of the cortex, both somesthetic cortex and association areas, have numerous connections with other parts of the cortex—with adjacent and distant gyri and with other lobes—via association fibers. There are also very important noncortical sites that provide input to the cortex.

One of these is the thalamus. Neurons run from the thalamus to the association area in the parietal lobe, and from the association area to the thalamus. In addition, the thalamus has extremely important and very extensive connections to and from all parts of the cortex.

The other source of input is indirect, by way of the *reticular formation*, but that also involves the thalamus.

Recall that the posterior portion of the brainstem, from the medulla through the midbrain, contains an area of neurons and synapses that forms a diffuse network. It is the site of important interconnections of many kinds, and contains the so-called *vital centers* (for heart rate, blood pressure, and respiration). Our present concern, however, is the very important role the reticular formation plays in sensory function. It receives collaterals (branches) from neurons in the primary sensory pathways as they ascend through the brainstem. The reticular formation relays much of this information to the thalamus.

When it is all put together, as in Figure 11–4, we see that there is an alternate pathway to the cortex by way of the reticular formation and thalamic connections with the association areas and all the rest of the cortex. Whenever a receptor is excited, impulses travel to the cortex in both pathways. The primary pathways are the direct routes that permit recognition of such qualities as modality and localization. With few synapses, it is a simple, direct, and rapid route. The other connections and indirect pathways provide mechanisms to make certain discriminations and associations, to relate incoming information to that being received from other receptors, and to plug into some sort of "memory bank" in order to relate information to previous inputs. (The nature of the "memory bank" is highly speculative at this time.) Connections with the association area enables us to make the finer discriminations and judgments based on information from several sources.

Identification "by feel" of an object—a key, a golf ball, or a wad of cotton—placed in the hand requires consideration of the weight, shape, texture, and relative temperature of the object. These judgments utilize most of the cutaneous receptors and some proprioceptors. Identification also requires a past experience—you cannot recognize a key if you do not know what a key is. The ability to integrate in this way information received from several types of receptors and to interpret it in terms of past experiences is known as **stereognosis.** It is an important part of most of our sensory activities, and it uti-

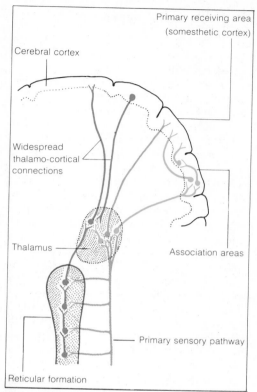

FIGURE 11–4   Schematic representation of sensory connections of the reticular formation and the thalamus and cortex.

lizes complex interconnections between many neurons and areas.

In view of all this, one might be forced to conclude that the brain is not so compartmentalized that each tiny area serves but a single role, or that one area is exclusively concerned with that role. This tends to hold true for all aspects of brain function. It is a valuable characteristic, since it means that after damage to one part of the brain, as with a stroke, other parts may be able to take over at least part of the lost function. However, it does complicate efforts to understand somatic sensory brain function. Until very recently it almost completely frustrated efforts to unravel the processes of learning, intelligence, emotion, and personality.

Although vision and hearing are special senses, many aspects of their mechanisms are similar to those of somatic senses. Visual and auditory receiving areas display a topographical representation of the areas of stimulation, and there are important association areas for integrating the data derived from their receptors.

An additional benefit of the numerous interconnections is the ability to focus attention, to screen the incoming information so as to suppress extraneous information and "tune in" on important things. Some of this is already done at the receptor, by the phenomenon of adaptation, whereby the unimportant information is simply not forwarded to the central nervous system. Recall that synaptic transmission is characterized by the possibility of inhibitory as well as excitatory input, and that summation is required to fire a postsynaptic neuron. So, at each synapse there is the possibility of facilitating some neurons and inhibiting others. The intricacies of these mechanisms are beyond the scope of our consideration here, but you can see that with selective facilitation and inhibition, the desired input or signal can be amplified, and extraneous information—the noise and static—can be suppressed and "tuned out."

The Reticular Activating System and Sensory Perception

There are times when sensory perception—and thus the electrical stimulation of the cortex by nerve impulses—is reduced or almost absent. In order for the brain to respond to incoming information, there must be a certain background of electrical activity against which arriving impulses can be measured. When this background activity is lacking, perception is impaired. It should come as no surprise that the brain is continually active. The excitability of neurons in the spinal cord is determined by the continual activity of excitatory and inhibitory endings (EPSPs and IPSPs) affecting that neuron, and one would not expect the brain to be simpler than the spinal cord. The background of electrical activity in the cortex is the basis of the alert state that is required for sensory perception.

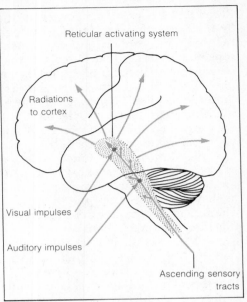

FIGURE 11–5   The reticular activating system.

Connections of the Reticular Formation   The background activity is largely due to connections from the reticular formation via the thalamus. The reticular formation receives information from and exerts important actions upon many different systems, somatic and autonomic, and motor as well as sensory. The portion of the reticular formation that we are concerned with here is a multisynaptic pathway that conducts impulses upward through the reticular formation to the thalamus. Because the elements of this system are involved in cortical arousal, it is commonly known as the **ascending reticular activating system** (*ARAS* or *RAS*) (Figure 11–5).

Although the primary sensory pathways all synapse in the thalamus, only a few of the thalamic nuclei serve as relay stations for these pathways. Other nuclei are anatomically and functionally associated with the brainstem reticular formation. Many of these nuclei have connections with the association areas, such as those of the general senses in the parietal lobe that were discussed above (see Figure 11–4). Other thalamic nuclei have diffuse connections, both to and from virtually all areas of the cortex. These nonspecific thalamic nuclei contain cell bodies

Focus

Acupuncture and Endorphins

An old principle of folk medicine, "homeopathy" holds that a disease can be treated effectively by administering something that causes the same symptoms as the disease does. Symptoms are seen as the body's response to whatever is causing the disease: reinforcing the symptoms is thought to help the body defeat the disease; alternatively, it is felt that activating the body to combat the treatment also activates it to combat the disease.

At first glance, the ancient Chinese practice of acupuncture may seem like an example of homeopathy. Needles are inserted deep into the tissues beneath the skin at specific sites to treat specific problems: headaches, muscular aches and pains, arthritis, rheumatism, and emotional disorders; they are even used as anesthetics. The needles may be left to sit, stimulating the system by their mere presence, or they may be twirled or attached to wires and electrified.

At one time, acupuncture sites were chosen according to tradition. The best site for the treatment of a particular problem was thought to have been determined long ago by trial and error, and it fitted into a more or less mystic scheme of "flows of life force." Some of these traditional sites are still used, but most are not. Regardless of the basis of site selection, acupuncture works. While its effectiveness could be demonstrated in the laboratory, its mechanism of action remained a mystery until a few years ago. The recent discovery of endorphins may provide the clue to solve the mystery.

Endorphins appear to be the body's own pain-killers. Certain brain cells release them when the body is injured, for example. These peptides inhibit the nerve cells involved in the brain's response to pain, thus dulling the sense of pain. Because these same receptor sites also combine with opiates (derivatives of opium such as morphine), researchers believe the endorphin molecule has certain similarities with the opiate molecule. They therefore call the endorphins "endogenous opiates."

The effects of endorphins, as well as opiates, can be blocked by opiate antagonists, such as naloxone, which combine with the receptor molecules and prevent them from combining with the pain-killers. The effects of acupuncture are also blocked by naloxone and by treatments that block the release of endorphins. This leads to the intriguing implication that acupuncture might work by stimulating the release of endorphins by the appropriate brain cells. Since endorphins are some 200 times as potent in stopping pain as morphine, it is hardly surprising that acupuncture enjoys the reputation it does.

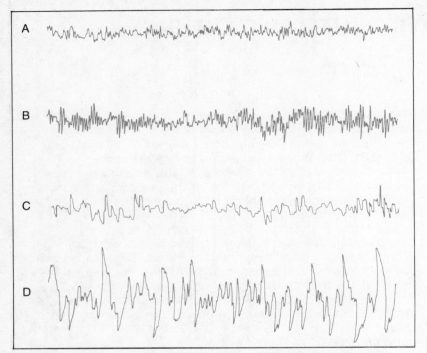

FIGURE 11-6   Samples of the electroencephalogram in various states of consciousness. A. Alert. B. Awake, relaxed, eyes closed. C. Drowsy. D. Asleep, slow waves.

of neurons that terminate in all parts of the cortex, and they contain endings of neurons that come from all parts of the cortex. The widespread thalamocortical interconnections are stimulated by input from the reticular formation. Whenever the reticular formation is stimulated, either through the branches it receives from neurons of the primary sensory pathway or from other input, there is increase activity in the neurons of the extensive thalamocortical system, and the cortex is aroused. The RAS, therefore, drives the thalamocortical connections to arouse the cortex.

Since the RAS is a multisynaptic pathway, and since synapses are vulnerable to many things, the RAS is quite susceptible to agents that affect synaptic transmission. The action of many anesthetics (including ether and barbiturates) is due at least in part to their depressing effect on conduction in the reticular formation. There are also agents that increase conduction in the reticular formation and cause varying amounts of cortical arousal. One of these is *epinephrine* (adrenalin), which

is a hormone and also a transmitter at some nerve endings.

Electroencephalogram It has been known for many years that electrodes placed on the surface of the brain, or even on the surface of the scalp, will show the presence of the continuous electrical activity of the cortex. The record of such activity, which is called an **electroencephalogram** or **EEG,** shows the electrical potential changes on the surface of the brain. It is our best indication of the level of cortical arousal or activation. The cortex, of course, is gray matter, but its surface consists primarily of processes of neurons whose cell bodies are in deeper cortical layers. They are mostly dendrites strung out along the surface and make contact with innumerable excitatory and inhibitory endings. Activity in these endings causes local hyper- and/or hypopolarizations in the dendrites. The activity recorded in an EEG is believed to be the sum of the local potential changes, rather than of individual action potentials in cortical neurons.

Although the EEG record may seem to be disorganized and without a pattern, marked differences occur in the character of the EEG with different conditions (Figure 11-6). When you are quiet and relaxed, with your eyes closed, your EEG shows a pattern of reasonably regular fluctuations or spikes. If you open your eyes and increase the amount of sensory input, or focus your attention (as by doing mental arithmetic), the pattern immediately changes to smaller but higher frequency spikes. If you fall into a deep sleep, the pattern changes to large slow waves. Large waves are believed to occur when the local potential changes are relatively in unison. When you (and your brain) are alert, the activity in the dendrites is more random, and the large spikes of the record are replaced by many small ones. The EEG thus indicates the degree of synchrony of the surface activity. The brain activity in the "sleeping" brain tends to be synchronized; that of the alert brain unsynchronized.

Abnormal electrical activity may be recorded in the EEG after brain damage or injury, or sometimes spontaneously. Epileptic seizures are accompanied by characteristic changes in the EEG, often indicated by marked synchrony of discharge. Absence of waves on the EEG (a flat line) indicates no activity of cortical cells. Repeated records showing no activity signifies "brain death," and is an important criterion of clinical death.

The dependence of the EEG and cortical arousal upon input from the RAS has been demonstrated quite conclusively by experiments in which electrodes were permanently implanted in a cat's reticular formation. At some later date, after recovery from the operation, the reticular formation could be stimulated without any need to restrain or anesthetize the animal. When the cat was asleep and had a "sleeping" EEG, stimulation of the reticular formation alerted the EEG and aroused the cat. Sensory stimulation by a sudden noise also aroused the EEG and wakened the cat. If the primary receiving areas for sound in the cortex had been destroyed previously, the cat would be deaf. But noise still aroused the cortex, the EEG, and the cat, by way of the reticular activating system. Only in the alert or aroused state is the brain able to respond effectively to a stimulus. When connections of the reticular system with the cerebral cortex are permanently interrupted, the EEG shows a sleep pattern, and the animal sleeps permanently; it never regains consciousness.

THE SPECIAL SENSES—VISION

The special senses are sight, hearing, equilibrium, smell, and taste. The receptors for these senses are located in the head and are innervated by cranial nerves. They are highly developed and some are extremely complex. The special senses each have separate neural pathways, and most have their own cortical receiving areas. By contrast, receptors for the general senses, such as touch and pain, are widely distributed throughout the body. They are relatively simple end organs whose afferent fibers are found in both spinal and cranial nerves.

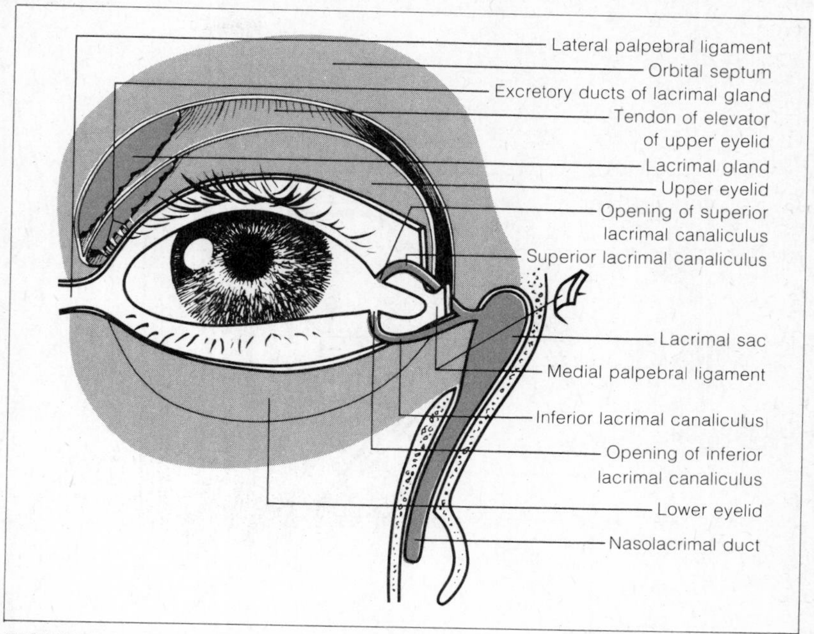
Lateral palpebral ligament
Orbital septum
Excretory ducts of lacrimal gland
Tendon of elevator of upper eyelid
Lacrimal gland
Upper eyelid
Opening of superior lacrimal canaliculus
Superior lacrimal canaliculus
Lacrimal sac
Medial palpebral ligament
Inferior lacrimal canaliculus
Opening of inferior lacrimal canaliculus
Lower eyelid
Nasolacrimal duct

FIGURE 11–7 The lacrimal apparatus.

Structure of the Eye and Accessory Organs

The eyeball, located near the opening of the orbit, contains the receptors for vision. Protected from the external world by the eyelid and the bony brow, nose, and cheekbones, its exposed surface is nevertheless easily damaged by dryness and must be constantly moistened by fluid. This is achieved by the **lacrimal** (tear) **gland,** tucked into a depression in the supero-lateral corner of the orbit (Figure 11–7). Its secretion—tears—washes diagonally across the front of the eye and drains into tiny **lacrimal canaliculi** leading to the larger **nasolacrimal duct,** which opens into the nasal cavity under the inferior nasal concha. The lacrimal gland secretes continually, but the rate may be greatly increased by certain emotions or by the presence of a foreign body or other irritant in the eye. A thin protective layer of epithelium, the **conjunctiva,** covers the anterior portion of the eyeball and lines the eyelid.

Extrinsic Muscles of the Eye Movements of the eyes are controlled by six small skeletal muscles, four of

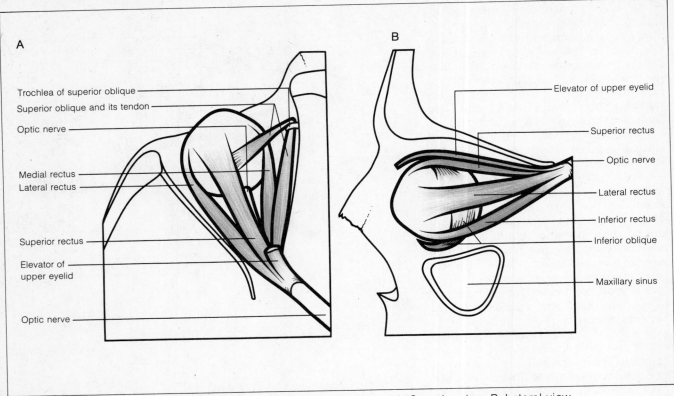

A

Trochlea of superior oblique
Superior oblique and its tendon
Optic nerve
Medial rectus
Lateral rectus
Superior rectus
Elevator of
upper eyelid
Optic nerve

B

Elevator of upper eyelid
Superior rectus
Optic nerve
Lateral rectus
Inferior rectus
Inferior oblique
Maxillary sinus

FIGURE 11–8   Extrinsic eye muscles. A. Superior view. B. Lateral view.

which arise posteriorly near the apex of the orbit (Figure 11–8). They insert on the superior, inferior, medial, and lateral surfaces of the eyeball and are named accordingly: **superior rectus, inferior rectus, medial rectus,** and **lateral rectus.** They cause the eyeball to turn up, down, medially, and laterally, respectively. The two remaining muscles are the **superior** and **inferior obliques.** The inferior oblique originates on the medial wall of the orbit and inserts on the inferior surface of the eyeball. The superior oblique originates with the rectus muscles but passes through a medial pulleylike loop, the *trochlea,* to insert on the superior surface of the orbit, (it is innervated by the trochlear nerve). Acting alone, the oblique muscles rotate the eyeball. The six muscles acting together are responsible for moving the eyes and for ensuring that the two eyes look at exactly the same object.

The eye movements are among the quickest and most precise movements the body can make, and the responsible

muscles have some interesting properties. They are the fastest-acting of all skeletal muscles; they have the shortest refractory period; and they can respond to very high frequency stimulation. It requires about 350 impulses per second to produce tetanus in these muscles as compared with about 100 per second for the average skeletal muscle. Also, each motor unit contains very few muscle fibers, and this fact contributes to their great precision.

Structure of the Eyeball   The eyeball (Figure 11–9) is a slightly flattened sphere about an inch in diameter. It has a tough, three-layered coat enclosing a semifluid gelatinous substance. The outermost layer is leathery, white, and relatively thick. It is known as the **sclera.** Anteriorly the sclera forms a transparent rounded bulge, the **cornea,** through which light passes. The middle layer, the pigmented **choroid coat,** carries the blood vessels for the eyeball

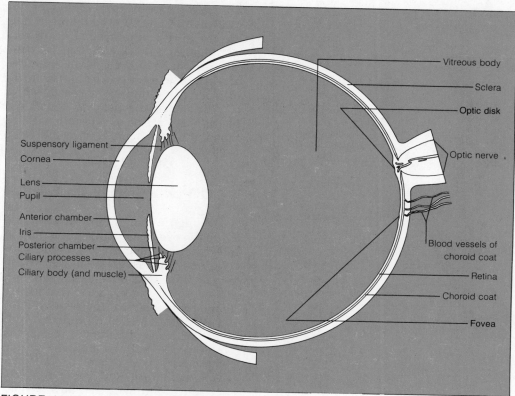

Suspensory ligament
Cornea
Lens
Pupil
Anterior chamber
Iris
Posterior chamber
Ciliary processes
Ciliary body (and muscle)

Vitreous body
Sclera
Optic disk
Optic nerve
Blood vessels of choroid coat
Retina
Choroid coat
Fovea

FIGURE 11–9   Horizontal section of the eye.

itself. Around the edge of the cornea the choroid coat forms the **ciliary body,** a thickened structure containing smooth muscle. A thin muscular diaphragm with an opening in the center is attached around the anterior margin of the ciliary body. This sheet of tissue, known as the **iris,** contains the pigment responsible for the color of the eye. The hole in the iris is the **pupil,** which permits light to enter the eye. Some of the smooth muscle fibers in the iris encircle the pupil and others radiate from it. Contraction of the radial muscle dilates the pupil and contraction of the circular muscle constricts the pupil. By their control of pupil diameter these muscles regulate the amount of light entering the eye.

Extending from the ciliary body behind the iris is the fibrous **suspensory ligament,** which holds the **lens** in position directly behind the pupil (see Figure 11–10). The lens, a transparent crystalline structure, is elastic. An increase in the tension exerted on the lens

by the suspensory ligament flattens it, while release of tension allows it to thicken once more due to its elasticity. The amount of tension on the lens is controlled by the smooth muscle in the ciliary body, the **ciliary muscle,** whose line of pull is such that contraction releases tension and allows the lens to thicken (see Figure 11–11). A thicker lens is needed for near vision, which therefore requires more effort (muscle contraction) than far vision. If the lens becomes opaque, light rays cannot penetrate it to reach the receptors. This condition is known as **cataract.**

The innermost layer of the eye, the **retina,** contains the receptors for light (photoreceptors). An extremely complex structure, the retina contains several layers of nerve cells and their processes, including two types of receptor, the **rods** for black and white vision and the **cones** for color vision. Nerve fibers from these receptors converge, after several synapses within the retina, and leave the eye slightly medial to its axis.

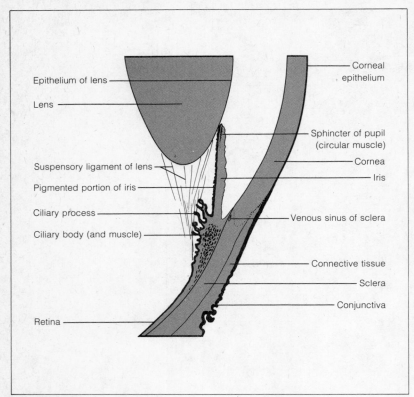

FIGURE 11–10   Detail of the anterior portion of the eye, showing the ciliary body, lens, and iris.

There are no receptors in this small rounded area which is sometimes called the **optic disc.** Because light rays falling on it have no effect, it is also known as the **blind spot** (Figure 11–12). The nerve formed by these fibers is the optic nerve (II). Blood vessels to the retina lie among the optic nerve fibers and enter the eyeball with the nerve.

In the center of the retina is a tiny depression known as the **fovea centralis.** All of the receptors in the fovea are cones, and they are very closely packed. It is therefore the most sensitive portion of the retina, and when one looks directly at an object, the light rays from that object fall upon the fovea.

The interior of the eyeball is filled with a transparent semifluid material known as the **vitreous humor.** The area between the cornea and the lens contains the **aqueous humor.** The aqueous humor is produced by cells of the ciliary body and absorbed into tiny canals near the junction of the cornea and iris.

The rate of absorption should be the same as the rate of production to hold the pressure in the eyeball quite constant and at a level sufficient to maintain the shape of the eyeball. Impaired absorption may result in a rise in the intraocular pressure *(glaucoma),* which could damage the delicate neural elements of the retina or interfere with the blood flow to them if it is not controlled.

Optical System of the Eye

Properties of Light and Lenses
Since the eye is normally stimulated by light, an understanding of its function requires some discussion of the properties of light and of lenses. Light is a wave form of energy, and can cause the generation of nerve impulses in the retina. Visible white light contains light waves of many different wavelengths. When white light is broken into its components by a prism, the different wavelengths are separated into the spectrum, forming a rainbow (Figure 11–13). The long waves of red light are at one end of the visible spectrum and the shorter violet at the other, with a continuous gradation in between.

Light rays normally travel in straight lines. However, suppose they

FIGURE 11–11   Mechanical model to illustrate the action of the ciliary muscle in adjusting the lens.

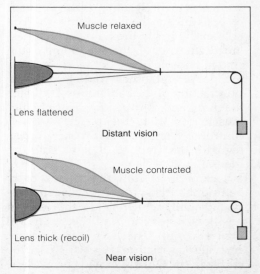

pass from a medium of one optical density (such as air) to one of a different optical density (such as glass or water). If the light rays strike the surface or interface at an angle, there is a bending or refraction of the light rays —the greater the angle, the greater the refraction. When light rays pass through a piece of glass, as in Figure 11–13, they are refracted at the near surface and again at the far surface. If the glass is a symmetrical lens the refraction is also symmetrical. When parallel light rays strike a symmetrical *convex* lens (see Figure 11–14), they are thus bent so as to converge at a single point, and a bright spot is formed on a paper or card held at the proper distance from the lens. If the card is moved closer to or farther from the lens, the spot blurs and goes out of focus. The point at which a clear focus occurs is the **focal point** (the *principal focus*) and its distance from the lens is the **focal length** of that lens. The focal length is less in a lens with a greater curvature[4] since it bends the light rays more sharply. Light rays come to focus at a greater distance when the light source is closer to the lens because the light rays are diverging as they strike the lens. A lens

[4]The focal length is therefore a measure of the strength of the lens, and is expressed in *diopters*, + for a convex lens and − for a concave one.

FIGURE 11–12   Blind spot. Close your left eye, and focus your right eye on the X. You will probably see the circle in your peripheral vision, but as you move the page away from or toward you the circle will disappear. At that distance, light rays from the circle fall upon your blind spot.

with a *concave* surface also bends light rays, but it causes them to diverge and they do not come to focus. The focal point of a concave lens can be determined by extrapolating the diverging light rays to an imaginary point. (Figure 11–14).

When the light source is an object instead of a point, light rays from each point of the object come to a focus at a particular distance from the lens, forming an image of that object (Figure 11–15). The image is inverted, with right and left reversed, and it may be reduced in size or enlarged.

Formation of an Image on the Retina   The principles of image formation described above apply to the eye as well as to separate lenses. The refractive system of the eye focuses the image on

FIGURE 11–13   White light can be broken into its component parts—the colors of the visible spectrum—by passage through a prism. As it passes through a prism, the light is bent, or *refracted*, away from its initial path.

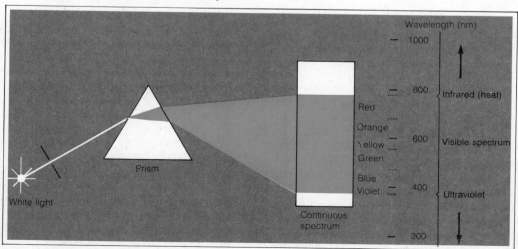

FIGURE 11–14
A. A convex lens brings parallel light rays to focus at a specific distance from the lens. B. A convex lens brings divergent light rays to focus at a greater distance from the lens. C. A stronger lens brings parallel light rays to focus nearer the lens. D. A concave lens causes the light rays to diverge. The focal point is an imaginary point on the same side of the lens as the light source.

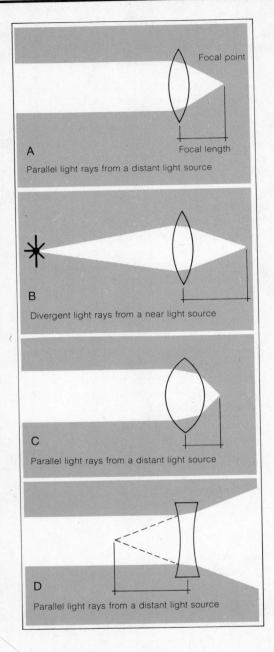

A
Parallel light rays from a distant light source

B
Divergent light rays from a near light source

C
Parallel light rays from a distant light source

D
Parallel light rays from a distant light source

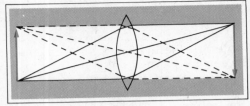

FIGURE 11–15   Formation of an image.

changing the distance between the lens and the film. In the eye, this distance is fixed, but the strength of the lens can be adjusted by the action of the ciliary muscles.

The lens is only one part of the refractive system of the eye, however, since light rays are bent whenever they pass from a substance of one optical density to one of another. There are several interfaces at which this occurs before the retina is reached: the front of the cornea (air-to-cornea), the back of the cornea (cornea-to-aqueous humor), the front of the lens (aqueous humor-to-lens), and the back of the lens (lens-to-vitreous humor). There is little difference in the optical density of the structures within the eye, so the greatest refraction occurs at the air-to-cornea interface. The role of the lens, then, is to make the final adjustment in the strength of the refractive system as needed for objects at different distances.

Lenses, even good ones, are not perfect, and light rays passing through the edges do not all come to a focus at exactly the same point as those passing through the central portion of the lens. This difficulty is minimized in the eye by the iris, which restricts the passage of light to the center of the lens. The shutter diaphragm on a camera does the same thing. The size of the opening also serves the important role of controlling the amount of light permitted to enter the eye (or camera).

Accommodation. A normal eye is one in which parallel rays come to a focus on the retina without adjustment of the lens. In practice, twenty feet is the distance from which light rays entering the pupil are considered to be parallel. Light rays entering the eye from a near light source would be di-

the retina. As with any lens, the image formed is upside down and reversed, but it is also very small. The brain nevertheless interprets it properly.

The eye resembles a camera in many respects. Both consist of a darkened chamber with a small adjustable opening through which light rays enter. Both have a complex lens system to focus the image upon the light-sensitive surface, in one the film, in the other the retina. In a camera, adjustments for distance from the object are made by

vergent, and the same degree of bending would bring them to a focus behind the retina. The image formed on the retina would be blurred unless adjustments were made.

The eye is, however, continually confronted with the need to focus on objects closer than twenty feet. The problem is solved with three simultaneous reflex adjustments which together are known as **accommodation.** The first is *contraction of the ciliary muscles* to release tension on the lens and allow it to bulge just enough to focus the image on the retina. The second is *constriction of the pupil* to eliminate the more divergent light rays that otherwise would pass through the imperfect edges of the lens. This produces a sharper image and reduces the amount of light entering the eye. The third adjustment is *convergence of the eyeballs,* so that light rays from close objects can strike the fovea of each eye (you cross your eyes to see a very near object, such as the tip of your nose). These adjustments, or their reverse, are made each time vision is shifted between far and near—from blackboard to notebook to blackboard, or between the newspaper and the television screen.

When ophthalmologists examine the refractive system of the eye, they often temporarily eliminate the first two parts of the accommodation reflex with one of the several drugs available for this purpose. The lens thickening and pupillary constriction are both brought about by smooth muscle controlled by the same part of the autonomic nervous system. Any agent that blocks the action of the transmitter released by these nerve endings will therefore prevent these responses (see Chapter 12).

Defects in the refractive system. Contraction of the ciliary muscle permits thickening of the lens for near vision, but it is the elasticity of the lens that *causes* thickening. An indication of the degree of elasticity of the lens can be obtained by determining the **near point,** the smallest distance at which an object such as a pencil point can be seen clearly. Gradual loss of lens elasticity is a normal consequence of aging. It begins at an early age and continues

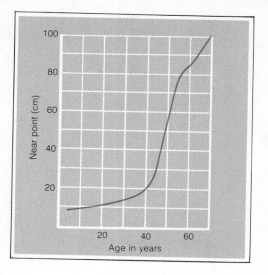

FIGURE 11–16
The near point recedes with increasing age.

throughout life but proceeds at a more rapid rate beginning at about age 40 (see Figure 11–16). When the near point has receded and accommodation diminished enough that reading becomes difficult (*presbyopia*), corrective lenses become necessary for near vision. The correction may be reading glasses (a convex lens) or bifocals (a more convex insert in regular glasses).

In the normal eye the strength of the refractive system and the distance to the retina are such that the image of a distant object comes to a focus on the retina without accommodation. But if the refractive system is too weak or the eyeball too short, the image comes to a focus behind the retina (see Figure 11–17). This condition is called **hyperopia,** or farsightedness. A hyperopic individual must accommodate to bring distant objects into focus, and he has difficulty in seeing near objects clearly because he cannot accommodate enough. The condition can be corrected by placing a convex lens before the eye, which strengthens the total refractive system.

The opposite situation is called **myopia,** or nearsightedness. In this condition the refractive system is too strong or the eyeball too long, since the image comes to a focus in front of the retina. A myopic individual can see near objects, but the lens will not flatten enough to focus distant objects on the retina. Myopia is treated with a concave

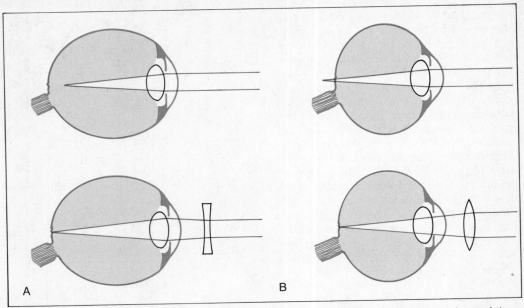

FIGURE 11-17   A. Myopia. The refractive system brings light rays to focus in front of the retina. It can be corrected by a concave lens. B. Hyperopia. The refractive system brings light rays to focus behind the retina. It can be corrected by a convex lens.

lens, which causes light rays from a distant object to diverge slightly as they enter the eye. The image then falls on the retina.

So far in our discussion of refractive errors we have assumed that elements of the refractive system are symmetrical and that light rays passing through the central portion of the lens come to a focus at a point. Unfortunately, this is not always the case, and the condition called **astigmatism** is fairly common. One might imagine the ideal symmetrical cornea as having a curvature like the surface of a basketball, with the same curvature in all planes. But if the surface of the cornea is curved like that of a football instead, the curvature is much greater along one axis than along the axis at right angles to it (Figure 11-18). Light rays in the plane of the long axis come to a focus behind those in the plane of the short axis. For an individual with an asymmetrical cornea, lines in one plane appear sharp and clear while those at right angles to it are fuzzy and poorly focused and those at intermediate angles are of intermediate sharpness (see Figure 11-19). The situation can be corrected by a lens with the same degree of astig-

matism, but at right angles to that of the cornea, so that the effects cancel one another.

Visual Acuity   The acuteness of vision, or the degree of detail that the eye can distinguish, is known as the **visual acuity.** It involves the ability to distinguish contours and edges and depends upon both the refractive system and the retina. Assuming that the image-forming apparatus is normal, the acuity is determined by what has been called the *retinal grain* (like the grain of photographic film). The grain of the retina is determined by the size of the receptors and the distance between them. The more closely they are packed the smaller the separation that can be detected between two objects. (Similarly, sensory acuity of the fingertip is greater than on the thigh because of the greater receptor density in the fingertip.) Receptors are most tightly packed in the area of the fovea centralis; therefore, this is the region of greatest visual acuity.[5]

[5]It is possible to gain some idea of the degree of acuity when it is calculated that at a distance of twenty feet an object that is half an inch high forms an image on the retina that is approximately 0.0001 inch high.

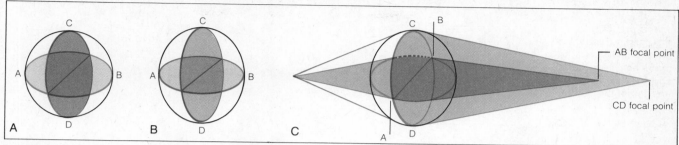

FIGURE 11–18   A. A symmetrical surface has the same curvature in planes AB and CD. B. An asymmetrical, or astigmatic, surface has a greater curvature in one plane, AB, than in the other, CD. C. In an astigmatic lens, light rays in the plane of AB come to focus nearer the lens than do light rays in the plane of CD.

Specifically, visual acuity is the minimal distance between two lines that can still be perceived as two lines. The ability to distinguish two distant points as separate is related to the angle formed by the light rays entering the eye from them. Visual acuity may be measured with the aid of eye charts whose letters are of the dimensions of this minimal angle. The letters are of different sizes for different distances, but at the prescribed distance, each letter subtends the same visual angle (Figure 11–20).

At a distance of twenty feet from the chart, a normal eye can read the lines down through the twenty-foot line, but none smaller. Such an individual is said to have 20/20 vision because at twenty feet the smallest letters that

can be read are on the twenty-foot line. This is normal vision and it requires a normal refractive system as well, since if there is defective refraction the visual acuity is reduced. If at a distance of twenty feet the smallest line that can be read is that for 100 feet, the person is said to have 20/100 vision. At twenty feet the best he or she can do is what the normal eye reads at 100 feet. Such a person has subnormal vision. A person with better than normal vision may read the fifteen-foot line and have 20/15 vision, meaning that at twenty feet he or she can read what a normal eye can read only at fifteen feet.

Physiology of Vision

Visual Fields and Binocular Vision
The *visual field* is that portion of the external world that can be seen, without moving, by one eye. Its extent is limited chiefly by the bone structure around the eye. In humans the two eyes are a few inches apart and they see very much the same view of the external world; that is, the visual fields overlap.[6] It is this overlap that makes binocular vision possible. **Binocular vision** is due to the reception of two slightly different images, one by each eye. Light rays from each part of the visual field come to a focus at a particular point on the retina, those from the object that one is looking at come to focus upon the fovea centralis, those from surrounding areas

FIGURE 11–19
Chart for testing astigmatism.

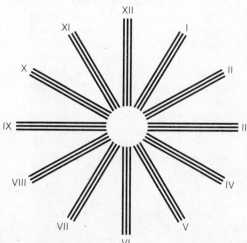

[6]Contrast this with a rabbit or some fishes in which there is little or no overlap of the visual fields of the two eyes.

FIGURE 11–20
At the prescribed distance, all letters on an eye chart subtend the same visual angle and form retinal images of the same size.

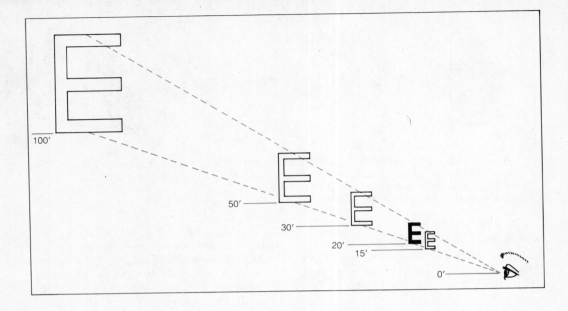

upon nearby retinal sites. Light from the temporal portion of the visual field strikes the nasal portion of the retina, while that from the nasal field strikes the temporal retina (see Figure 11–22).

Corresponding points. When one looks at the tip of a pencil, the light rays from it fall upon the fovea of each retina; yet one sees but one pencil tip. The brain is able to fuse the images from the two eyes because the neural pathways from the receptors end at the same place in the visual area of the cortex. The retinal sites stimulated are therefore said to be *corresponding points.* All sites in the binocular field of vision have corresponding points in the two retinas, and when these points are stimulated a single image is seen. When light rays are made to fall upon noncorresponding points, two images are seen. This is readily demonstrated by the simple experiment shown in Figure 11–21.

An imbalance of the extrinsic eye muscles may mean that the two eyes do not work together properly. An example is *strabismus* (cross-eyes or walleyes) in which there is a weakness of one eye muscle. The image then does not fall on corresponding points and two images are seen. This condition of double vision is known as *diplopia.* The individual learns to "ignore" the aberrant image, and if the situation is not corrected, the ignored eye may eventually become functionally blind. Early corrective surgery is very successful in restoring the muscular balance and avoiding loss of the use of one eye.

Depth perception. The fact that the two eyes do not see exactly the same visual field is the basis of meaningful depth perception. With only one eye the visual field appears to be two-dimensional; the important third dimension—depth—is missing. An individual with only one eye must judge depth or distance by learned cues, such as the fact that an object decreases in apparent size as it gets farther away. Two eyes add a much greater perception, a stereoscopic or 3-D view, because the two eyes see very slightly different views of the object. Aside from general orientation, depth perception is an important part of judging the speed of moving objects. This ability is required in many situations, as in judging the course and speed of an oncoming car in order to avoid a collision or of an approaching baseball or tennis ball in order to hit it squarely.

Visual Pathways As shown in Figure 11–22, light from objects in the left half of the visual field reaches the

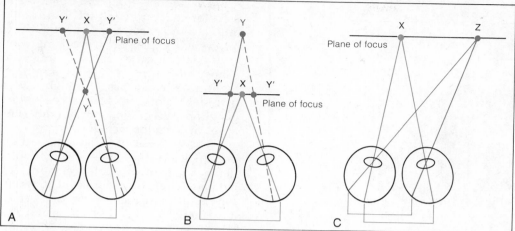

FIGURE 11–21   Corresponding points. A. If you focus on a distant point X, light rays from X will fall on corresponding points; but light rays from a near point Y fall on noncorresponding points, and you see two images of Y, one on either side of X. B. If you focus on a near point X, light rays from a farther point Y fall on noncorresponding points, and you see two images of Y, one on either side of X. C. When you focus on X, light rays from point Z, to one side of X but in the same focal plane, fall on corresponding points, and you see one image of Z.

right side of the retina, and light from objects at the top of the visual field falls on the inferior portion of the retina. Nerve fibers from all of these receptors converge to leave the eyeball as the **optic nerve.** At the **optic chiasm** there is a partial crossing, so that all fibers arising from the nasal half of each retina cross to the other side. As a result the left **optic tract** carries fibers from the left half of each retina, and this represents the right half of the visual field of both eyes. The optic tract passes to the *lateral geniculate body*, a part of the thalamus, and from there to the cerbral cortex. The **primary visual receiving area** or **visual cortex** is mainly along the medial surface of the occipital lobe. The region of central vision, around the fovea centralis, is represented more posteriorly and the peripheral parts of the retina more anteriorly. The upper half of the retina is represented on the lower portion of the cortical area and the lower half of the retina on the upper part. Each occipital lobe receives projections from one half of each retina (representing the opposite visual field). There is thus a topographical representation of the visual field in the cortex which is similar in many ways to that of the somatosensory area.

Several important reflexes depend upon the retina and optic pathway for their afferent input. One of these is the *light reflex*. When light is shined into one eye, there is a reflex constriction of both pupils. Branches from the afferent fibers in the optic tract pass to the midbrain nuclei for the autonomic efferents of the oculomotor nerve (III), and their excitation causes the pupils to constrict. Similar reflex pathways involve the extrinsic eye muscles, the pupillary constrictors, and the ciliary muscles as the components of the *accommodation reflex* previously discussed.

Other branches from the main sensory pathway go to the superior colliculi (in the midbrain) to synapse. From there some fibers lead to the nuclei of cranial nerves III, IV, and VI, which innervate the extrinsic eye muscles, and other fibers go down the spinal cord to end on anterior horn cells. These pathways through the superior colliculi are important for movements, especially of the eyes and head, but also of other parts of the body, in response to visual stimuli. Such reflexes are needed for visually following a moving object or for the hand-eye coordination involved in such activities as catching a ball or driving a car.

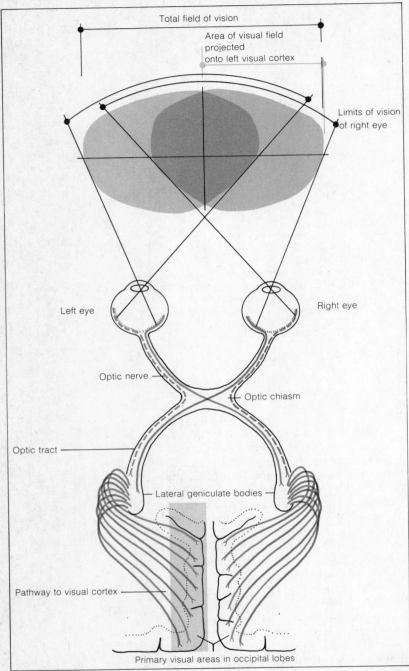

FIGURE 11-22   Visual fields and the visual pathway.

The Retina—Rods and Cones   The receptor-containing retina is a fragile multilayered sheet whose neural architecture is complex in a surprising way. The rods and cones, for one thing, are almost hidden from the light stimulus. Light rays must pass through layers of retinal blood vessels, nerve fibers, ganglion cells, bipolar cells, and synapses in order to reach the receptors.

The process of producing action potentials in the fibers of the optic nerve is more complicated than it is for cutaneous sensory neurons. The receptors, rods and cones, are separate cells, and they synapse with bipolar cells, which synapse with ganglion cells (Figure 11-23). It is the ganglion cell whose central processes are found in the optic nerve. Not only are there several cells and synapses within the retina, but also there are other cells there (horizontal and amacrine cells) that synapse with the bipolar and ganglion cells. All of these connections are sites of interaction, depolarization or hyperpolarization, that tend to facilitate or inhibit impulse production. There is a considerable degree of convergence in the pathway from the receptor to the optic nerve. Many receptors synapse with each bipolar cell. This is especially true of the rods, but even cones converge somewhat. (There are a little over a million fibers in each optic nerve, but there are well over a hundred million receptors in each retina.) The greater degree of convergence in the rod pathway provides more summation and facilitation, and may be one reason for the greater light sensitivity of the rods (as well as the poorer quality of the image formed).

Stimulation of a rod or cone with light leads to a *hyper*polarizing generator potential that does not cause an action potential. Bipolar cells develop either a hyper- or depolarizing potential, but no action potential. Ganglion cells, however, develop action potentials, but their responses vary. Some respond by discharging impulses upon stimulation while others cease firing until the light is turned off. Horizontal and amacrine cells modify the discharge of ganglion cells.

The quality of the image formed at the receptors is poor, but the processing within the retina due to inter-

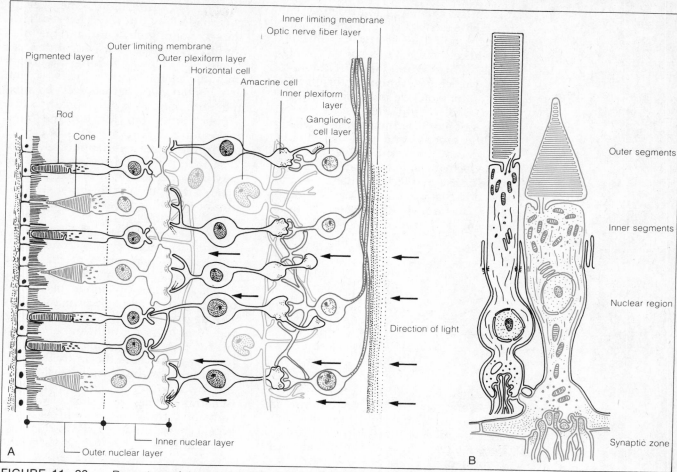

FIGURE 11-23   Receptors of the retina. A. Diagram of the retina, showing its layers and cells and some of their relationships. B. A rod (left) and a cone (right).

connections of the cells there sharpens the image that is finally transmitted in the optic nerve. The effect might be likened to the image enhancement by computers in processing photographs transmitted from space.

It has been said that the retina contains two receptor mechanisms, since although both rods and cones are receptors for light, they differ greatly in their responses to it. Cones are designed to function in daylight, while rods are designed for use at night or when the level of illumination is low. Rods have a low threshold and it takes very little light to excite them, while cones have a relatively high threshold. Vision in poorly lighted areas depends upon rods and is nearly devoid of color because the stimulus is subthreshold for cones.

The greater visual acuity in the region of the fovea centralis is due to several structural modifications. In the fovea the blood vessels and superficial layers are displaced to the side, leaving the receptors more exposed to light. The receptors themselves are packed very closely, and there is less convergence in the pathway for cones. Also, the fovea contains only cones for color vision, although both rods and cones are found in the areas immediately surrounding the fovea. The number of cones in the retina diminishes sharply with distance from the fovea.

Stimulation of rods. The mechanism for stimulating visual receptors has been described more completely for rods, but the general principles are quite similar for cones. Light produces a chemical change in the receptor, but the relationships between the chemical changes and the electrical events are not entirely clear. Rods contain a pigment known as **rhodopsin** *(visual purple)*. The configuration of the protein part of the molecule is changed when it is exposed to light, and the rhodopsin breaks down to **retinene** *(visual yellow)* and the protein, and it is bleached in the process. Retinene is very closely related chemically to vitamin A, and some of it is reduced to vitamin A and stored in that form. In the dark rhodopsin is re-formed, either directly from retinene or indirectly by way of vitamin A.

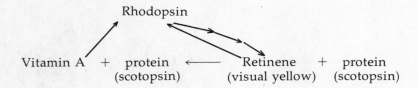

Rhodopsin

Vitamin A + protein   ←   Retinene + protein
              (scotopsin)   (visual yellow)   (scotopsin)

Dark and light adaptation. Vision is severely reduced upon entering a dark room from bright sunlight, but after a period of time objects gradually become discernible as more rhodopsin is formed. The process, known as *dark adaptation,* takes about twenty minutes. The reverse, *light adaptation,* occurs upon return from darkness into bright light. A lack of vitamin A may interfere with the prompt formation of rhodopsin and result in much less effective dark adaptation, or "night blindness." When most of the visual pigment is in the rhodopsin state, the rods are very sensitive to light, and respond to very small amounts of it. In the bleached state (retinene), the sensitivity is reduced considerably and a much higher level of illumination is required to elicit nerve impulses.

Cones and color vision. With cone vision we are able to detect more than one hundred different colors, but there is not a different type of cone for each identifiable color. There are believed to be, in fact, only three types of cones, each one containing a different visual pigment. Each pigment is affected by light of a specific wavelength (color) in much the same way that rhodopsin is

affected by white light. One type of cone has a low threshold for red light, the second for green light, and the third for blue-violet. Just as a painter can obtain any color by mixing the proper amounts of the primary hues, so can any visible color be achieved by mixing stimulation of the cones in the proper proportions. Much of the "mixing" and blending of colors is carried out in the retina. Horizontal cells synapse with cones of different colors and may be hyperpolarized by one and depolarized by the other. Equal stimulation of the three types of cones is perceived as white light.

A lack or inadequate number of any one (or more) type of cone causes problems in perception of that color. The several different types of *color blindness* involve inadequate perception of one, two, or all three primary colors. The degree of deficiency varies, ranging from minor confusion of certain shades to a total lack of color perception. In most cases the individual still sees the color, but through color mixing, by virtue of its stimulation of the remaining types of cones. The most common form of color blindness is red-green color blindness. It is due to a problem with the red receptors, resulting in confusion between certain shades of red and green, since they stimulate the same combination of cones.[7]

SPECIAL SENSES— HEARING AND EQUILIBRIUM

The ear contains two important organs. One is the mechanism for hearing, or audition, which converts sound waves into nerve impulses. The other is the mechanism for equilibrium, the vestibular apparatus, which detects position and motion of the head. Although the two receptor organs are connected anatomically, they are functionally separate. They are both located

[7]One explanation is that the red receptors respond to the same wavelength as the green receptors, making distinctions impossible. The defect is strongly sex-linked; it appears most often in males (8–9 percent of males are afflicted to some extent).

The Special Senses—Hearing and Equilibrium **309**

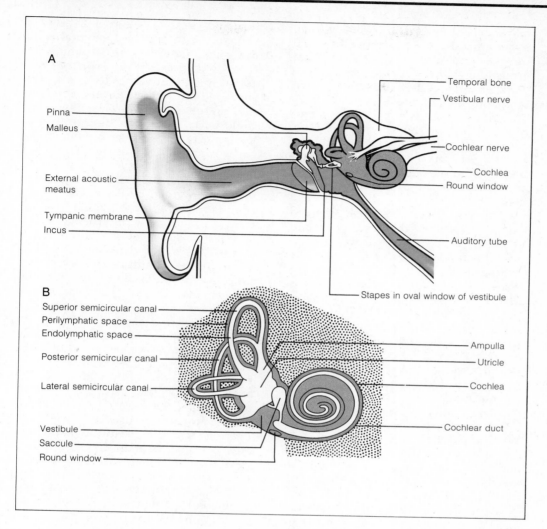

FIGURE 11-24
A. Structure of the ear.
B. Inner ear, showing the relationship between the membranous and bony labyrinths.

in the petrous portion of the temporal bone in an area no larger than a marble. However, they send their impulses to the brain over the separate cochlear and vestibular portions of the eighth cranial nerve and serve very different purposes.

Hearing

Properties of Sound The receptors for hearing are stimulated by sound waves, the alternating *compressions* and *rarefactions* of the air (or other medium) generated by a vibrating object. Two properties of sound, intensity and frequency, are of particular concern here. **Sound intensity** is related to the *amplitude* of the sound waves and is perceived as the loudness of the sound.

The **frequency** is determined by the number of waves or cycles per second and is perceived as *pitch*. High frequency sounds have a high pitch and low frequency sounds have a low pitch. The hearing apparatus must convert these changes in pressure to nerve impulses and yet retain a way of distinguishing intensity and frequency.

Structure of the Ear and the Conduction of Sound The ear is divided into three parts, the external, middle, and internal ear (Figure 11-24). The **external ear** includes the pinna, the external acoustic meatus, and the tympanic membrane. The **pinna** is what is usually meant by the word "ear." It is a flexible cartilage-containing structure located on the side of the head. In humans it is not very functional, but in many animals it can be moved to help direct sound waves into the **external acoustic meatus,** a curved passageway whose medial end is sealed off by the **tympanic membrane** (the eardrum). The latter is a slightly conical membrane separating the external and middle ears. Sound waves entering the external acoustic meatus strike the tympanic membrane and set it to vibrating at an intensity and frequency related to that of the sound.

The **middle ear** is a small air-filled chamber medial to the tympanic membrane. It contains three tiny ear ossicles, the **malleus,** the **incus,** and the **stapes** (the hammer, the anvil, and the stirrup). The handle of the malleus is attached to the center of the tympanic membrane and moves when the membrane vibrates. The head of the malleus articulates with the incus, and the stapes is fixed to a process of the incus. The footplate of the stapes fits against an opening in the bony medial wall of the middle ear. This opening, known as the **oval window,** provides communication with the internal ear.

The ossicles form a lever system that transmits sound energy across the middle ear. The area of the tympanic membrane is about 22 times that of the oval window, and it is capable of much greater excursions. The system of levers formed by the ossicles reduces the amplitude of the vibrations and concentrates the sound energy to the smaller area (that is, increases pressure). The pressure of the movement at the footplate of the stapes is about 22 times what it would be if the sound waves were applied directly to the oval window. Without the reduction in amplitude, the delicate mechanism in the inner ear would be damaged by the high-amplitude excursions of the tympanic membrane. The increase in pressure ensures that the vibrations set up at the oval window will be sufficient to reach the receptors in the fluid-filled inner ear. Two small muscles, one attached to the tip of the tympanic membrane and the other to the stapes, can contract to protect the hearing apparatus from extreme displacements caused by loud sounds or concussions.

The air in the middle ear, like any gas, responds to changes in temperature and pressure. Normally the pressure in the middle and external ear is the same. However, for example, atmospheric pressure is reduced with ascent to a higher altitude, and a pressure gradient is then developed across the tympanic membrane. Not only could such a pressure be painful and temporarily impair hearing, but it also could rupture the tympanic membrane if there were no way to equalize the pressure across the membrane. A safety valve is provided by the **auditory tube** (the *Eustachian tube*), which extends from the middle ear into the nasopharynx. It is closed most of the time, but it is briefly opened by a yawn or by swallowing. Any condition in which the mucous lining of the tube becomes swollen may prevent its opening and interfere with equalization of pressure.

The **internal ear** is located in the temporal bone medial to the middle ear. It is a maze of chambers hollowed out of the bone, and it is aptly called the **bony labyrinth.** Within it are three areas, the cochlea, the vestibule, and the semicircular canals. The snail-shaped **cochlea** houses the organ for hearing, while the **vestibule** and **semicircular canals** contain the receptors for position and movement. The vestibule is an open area; the larger end of the cochlea opens on one side and the three semicircular canals on the other.

The bony labyrinth contains the thin **membranous labyrinth,** which in general follows the contours of the bony maze. The membranous labyrinth is surrounded by one fluid, the **perilymph,** and contains another, the **endolymph.** Perilymph is related to cerebrospinal fluid, while endolymph is chemically more like intercellular fluid. The internal ear is thus fluid-filled, whereas the external and middle ear are air-filled. The bony labyrinth communicates with the middle ear through two openings: (1) the oval window, occupied by the footplate of the stapes, and (2) the round window. Both are covered by a membrane that prevents loss of perilymph into the middle ear.

The Cochlea and Hearing The spiral cochlea (Figure 11–25) winds around a bony central pillar in approximately two and a half turns. A cross

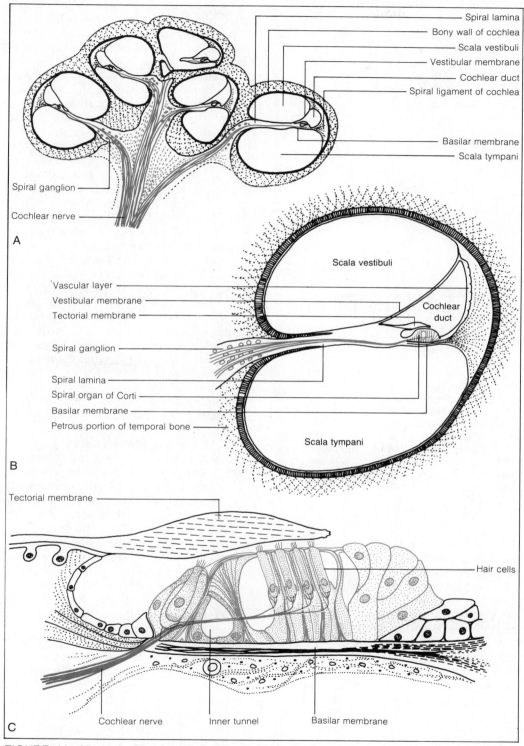

Spiral lamina
Bony wall of cochlea
Scala vestibuli
Vestibular membrane
Cochlear duct
Spiral ligament of cochlea

Basilar membrane
Scala tympani

Spiral ganglion

Cochlear nerve

A

Scala vestibuli

Vascular layer
Vestibular membrane
Tectorial membrane

Cochlear duct

Spiral ganglion

Spiral lamina
Spiral organ of Corti
Basilar membrane
Petrous portion of temporal bone

Scala tympani

B

Tectorial membrane

Hair cells

Cochlear nerve   Inner tunnel   Basilar membrane

C

FIGURE 11–25   A. Cross section of the cochlea. B. Cross section of one turn of the cochlea. C. The spiral organ.

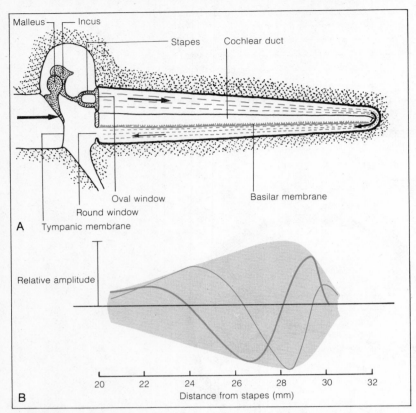

A

B

FIGURE 11–26   A. Movement of the stapes creates pressure waves in the perilymph (scala vestibuli), which causes displacement of the basilar membrane. B. A traveling pressure wave. The two lines represent a pressure wave at two moments in time. The shaded area encloses the wave at successive moments. The wave begins at the oval window and builds to a peak amplitude (pressure) at a specific distance from the oval window. The greatest displacement of the basilar membrane occurs at that distance.

section of one turn shows the cochlea to be divided into three sections. A bony shelf extends into the cochlea from the central core like the thread of a screw and partially divides it. The **basilar membrane** extends across the cochlea from the edge of the bony shelf, and a thin **vestibular membrane** divides the upper section into two parts. The two larger canals, the *scala vestibuli* and *scala tympani,* communicate with one another at the apex of the cochlea. They are filled with perilymph. The oval window connects with the upper canal (the scala vestibuli), and the round window with the lower canal (the scala tympani). Between them is the smaller *cochlear duct,* filled with endolymph. It

contains the **spiral organ** *(organ of Corti),* which is the receptor for hearing.

The spiral organ consists of a uniquely arranged mound of cells resting on the basilar membrane. Some of the cells (the *hair cells*) have tiny hairlike processes extending upward from their free surface. Nerve fibers emerge from the base of the hair cells and pass along the bony shelf to the center of the spiral. A flexible flaplike structure, the **tectorial membrane,** is suspended over the top of the hair cells. The cochlea is larger at the base than at the top, but the basilar membrane is narrower at the base and more rigid. Toward the apex of the cochlea the basilar membrane is wider, heavier, and less stiff.

Vibrations of the stapes at the oval window set up pressure waves in the perilymph of the cochlea, displacing the basilar membrane. The round window also vibrates with the pressure changes and, since the fluid is incompressible, it acts as a sort of ballast. As the basilar membrane vibrates, the hair cells are brushed against the overhanging tectorial membrane, which bends the hairs and causes the hair cells to discharge impulses along their nerve fibers.

The higher the pitch of a sound, the higher the frequency and the shorter the length of the waves produced. A pressure wave begins with movement of the oval window and reaches its peak pressure (amplitude) at some distance from it (Figure 11–26). The peak pressure of a high frequency wave is near the oval window, and for a low frequency wave it is nearer the apex of the cochlea. The greatest effect on the basilar membrane is in the region of the peak amplitude of the pressure wave. Thus sound of a given pitch causes maximum vibration of a particular segment of the basilar membrane and thereby activates specific receptors (hair cells). The afferent neurons from these receptors have their cell bodies in a ganglion in the central pillar of the cochlea (the *spiral ganglion*), and their central processes form the **cochlear nerve.**

The intensity or loudness of a sound is determined by the magnitude (amplitude) of the pressure changes produced by each wave. Loud sounds cause greater displacement of the basilar membrane and stronger stimulation of the hair cells. Thus, pitch is determined by which cells are stimulated, and loudness by the number of action potentials produced by them.

Sound can also be transmitted to the inner ear via the bones of the skull. This pathway bypasses the external and middle ear and transmits vibrations directly to the spiral organ. Bone conduction is not a very efficient method of transmitting sound, however, since much of the energy of the sound waves is lost in the transfer.

Auditory Pathway The pathway from the spiral organ to the auditory cortex is puzzling, to say the least. It has synapses where one would expect, and more as well. There are several alternate routes, involving both crossed and uncrossed pathways, and perhaps some that cross twice (see Figure 11–27). All synapses, of course, are sites of interconnections; some are concerned with connections for reflexes initiated by auditory stimulation (turning the head and eyes upon hearing a sound), and others, perhaps, are concerned with the sorting out of auditory information. But as in other sensory pathways, the final synapse is in the thalamus (the medial geniculate body) and the final neuron runs to the primary auditory receiving area, which is in the temporal lobe of the cortex along the wall of the lateral fissure. High-pitched sounds project to the anterior portion and low-pitched sounds project to the posterior part. Thus there is a topographic (tonotopic) representation of receptors on the auditory cortex, as was true for the visual and somesthetic areas.

Because of the crossed and uncrossed pathways, there is a bilateral cortical representation; that is, each side of the cortex receives input from both ears. It is also interesting to note that most of the sites of synapse in the auditory path also show a tonal mapping of the receptors. The significance of all the redundancy is unclear, but it is true that damage to the auditory path in the central nervous system is not

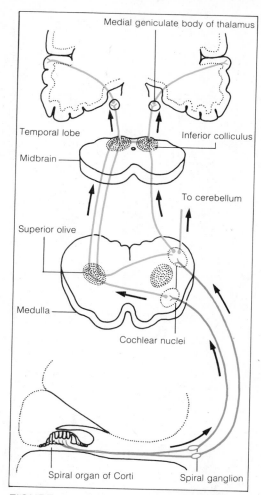

FIGURE 11–27   The auditory pathway.

likely to have great effect, since there are alternate routes available.

One function served by the bilateral representation is that of localization of sound. We apparently determine the direction of a sound by the fact that sound waves strike one ear slightly before the other, and with greater intensity. (When your eyes are closed you cannot localize very accurately sounds directed at your midsagittal plane.)

The hearing process includes much more than knowing that there is a sound of a certain pitch and intensity. We are able to identify many qualities of that sound, and the rhythm patterns of continuing sound. Making use of auditory association areas of the cortex and the multiple connections in the

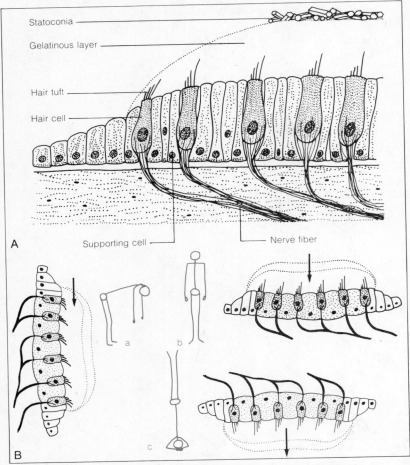

Statoconia

Gelatinous layer

Hair tuft

Hair cell

A

Supporting cell — Nerve fiber

B

a b c

FIGURE 11–28 A. Receptors in the utricle. B. Changes in the orientation of the head alter the direction of gravitational pull on the statoconia.

sensory path, we are able to recognize the sound as a friend's voice, or as a favorite melody, or perhaps, even that one note in the tune is "off-key." The mechanism by which such interpretations are made is not understood any better for hearing than it is for sight or touch.

Hearing Defects Good hearing requires that the entire mechanism be intact, and impairment results whenever there is a failure of any link in the auditory pathway. Damage to the tympanic membrane (rupture and subsequent formation of scar tissue) reduces its response to sound waves. The ossicles may fuse or the footplate of the stapes may become fixed at the oval window, and either will block transmission to the internal ear. These are essentially middle ear or conduction problems. They can be treated in several ways, including the use of hearing aids, which amplify sounds to compensate for the reduced transmission, and by surgery.

But defects of the neural apparatus, either of the receptors or of the nerve itself, do not benefit from hearing aids or improvement of conduction. There is some normal deterioration of the receptors with age, particularly at the higher frequencies. Persistent exposure to very loud sounds damages the receptors that respond to the frequencies involved. The damage may affect only a portion of the spiral organ (as in "boilermaker's disease," in which sound of a specific frequency is the offender), or it may be more widespread, caused by loud sounds in general, including highly amplified music. Such acoustic trauma is becoming an increasingly serious problem in modern society.

Equilibrium

The vestibular apparatus includes receptors in both the semicircular canals and the vestibule. The membranous labyrinth in the region of the vestibule forms two adjoining enlargements, the **utricle** and the **saccule.** The semicircular canals are continuous with and connected to the utricle, and the saccule communicates with the cochlear duct (see Figure 11–24B).

Utricle and Saccule The utricle and saccule each contain a region of hair cells from which the afferent neurons emerge (Figure 11–28). Resting on the hair cells is a gelatinous mass containing numerous small particles of calcium carbonate called **statoconia** (*otoliths*). In the upright position the receptors in the utricle are subjected to a certain stimulation due to the weight of the statoconia bearing down upon them. When the position of the head is changed, the pull of gravity on the statoconia in relation to the hair cells is altered. Because of the gelatinous sub-

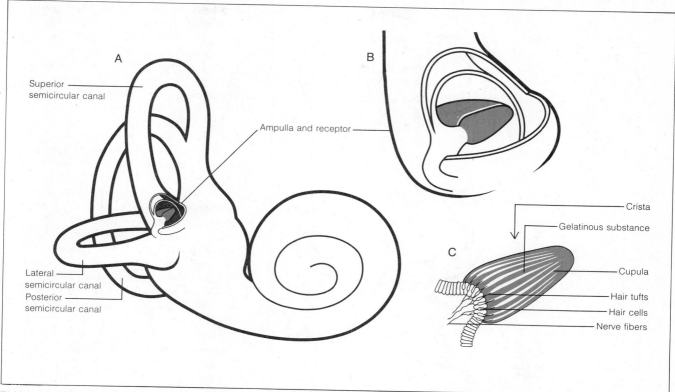

FIGURE 11-29   Receptors in the semicircular canals. A. Bony labyrinth, with a cutaway of one ampulla to show the receptor. B. Enlargement of the crista in the ampulla. C. Structure of the crista.

stance, this pull is transferred to the hair cells and they are stimulated. Each position of the head produces a different pattern of hair-cell stimulation. In this way the brain constantly receives information about the position of the head in space.

The utricle is stimulated by linear (horizontal or vertical) acceleration as well as position (Figure 11-28). At the onset of forward movement, for example, the inertia of the statoconia causes a lag, thereby bending and stimulating the hair cells. If movement continues at a constant velocity, the inertia is overcome and there is no particular sensation of movement. If the movement terminates abruptly, momentum carries the statoconia forward, stimulating once again. A slow change in velocity produces little or no sensation.

The role of the saccule is less clear, since the evidence that exists is inconclusive. Though its structure is similar to that of the utricle, its function seems to be different, at least in part. It has been suggested that it may be concerned with detection of slow vibrational stimuli.

Semicircular Canals   The semicircular canals (Figure 11-29) are concerned with angular acceleration, that is, with *rotational movement.* Each of the three endolymph-filled membranous semicircular canals has an enlargement, the **ampulla,** at one end. In each ampulla a tuft of hair cells embedded in a gelatinous mass, the **cupula,** protrudes into the endolymph. The hair cells are the receptors, and they are stimulated by movement of the endolymph. The canals are so positioned in the head that one pair is horizontal and the other two pairs are vertical, but at right angles to one another.

The semicircular canals are stimulated by rotation of the head, such as occurs when you turn a corner or swivel

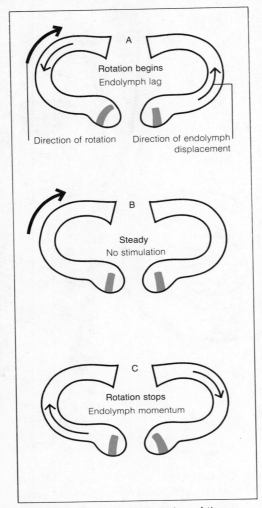

FIGURE 11–30   Stimulation of the horizontal semicircular canal. A. As rotation begins, the endolymph lags and puts pressure on one crista, causing a sensation of rotation. B. With continued rotation, the fluid catches up, and there is no sensation. C. If rotation is stopped abruptly, the fluid's momentum causes strong stimulation of the other crista, and there is a sensation of rotating in the opposite direction.

in an office chair. As indicated in Figure 11–30, the endolymph has inertia, so that the onset of rotation causes a momentary endolymph displacement. This movement of endolymph increases the pressure on the cupula and on the hair cells in one of the horizontal canals and decreases the pressure on the cupula and hair cells in the other horizontal canal. As rotation continues,

the endolymph "catches up" and the sensation of rotation decreases. Stopping, particularly if it is sudden, causes a pronounced stimulation and sensation. The head stops, but the momentum of the endolymph carries it on, producing pressure and strong stimulation of the hair cells of the horizontal canal that was unstimulated at the onset. It is this terminal or post-rotatory stimulation that causes the effects that are most noticeable to the subject. Among the effects of such stimulation are vertigo, nystagmus, and past-pointing or falling. They occur after the abrupt cessation of rotation and last for some seconds. **Vertigo** is dizziness, the sensation of spinning. **Nystagmus** consists of a rapidly repeated drift of the eyes to one side, followed by a very fast return. It represents a reflex mechanism by which the eyes can focus on a fixed object as the head rotates, with a rapid advance to fix the gaze on a new object. **Past-pointing** and **falling** are degrees of response to an error in judgment induced by semicircular canal stimulation. The error produces incorrect reflex muscular adjustments. For example, an individual who wishes to move forward after horizontal rotation may think he is angling to the left and correct for it, but since he is not going off to the left, the correction sends him to the right.

Stimulation of the vertical semicircular canals causes similar effects, but in different plane. If one puts the head on one shoulder (abducts the head 90 degrees) *during rotation*, the movement of the head simulates that produced by rotation in a sagittal plane (head over heels or somersaults). When the head is returned to the upright after stopping abruptly, the post-rotatory responses are similar to those of horizontal rotation, but in the sagittal plane. Nystagmus is up and down, and vertigo is a sensation of falling forward or back. The compensation is in the opposite direction and the subject falls backward or forward. (Whether it is backward or forward depends upon the direction of rotation and the side to which the head was tilted.) When the head is placed forward on the chest or tipped back, rotation simulates movement in a frontal plane (cartwheels). The nystagmus is rotatory about an antero-posterior axis, and vertigo is a sensation of falling to one side. The compensatory response is a fall to the opposite side.

Stimulation of the semicircular canals produces very powerful reflex responses of the postural muscles which cannot be voluntarily inhibited (they can, however, be inhibited by other reflexes). Because of this inseparable association with postural reflexes and the control of muscle tone, the vestibular apparatus is often discussed in terms of postural control. Although one is very conscious of sensations due to vestibular stimulation, the vestibular reflexes that help control skeletal muscles function below the level of awareness. This is simply one more illustration of the interdependence of the motor functions and the various sensory elements.

SPECIAL SENSES—SMELL

Olfaction, the sense of smell, is one of the least understood of the special senses. It involves perception of the presence of volatile and water-soluble substances dispersed in the air we breathe. The receptors are located in the upper reaches of the nasal cavity, extending a slight distance down along the nasal septum and the lateral wall of the nasal cavity (Figure 11–31). The receptor area consists of elongated ciliated epithelial cells liberally distributed throughout the mucous lining of that portion of the nasal cavity. The cilia are embedded in the mucus covering the epithelial surface. Nerve fibers from the receptor cells pass upward through the perforations in the cribriform plate of the ethmoid bone to the olfactory bulb. From here the olfactory tract carries fibers to the olfactory area of the cortex, located on the inferior surface of the frontal lobe, near the optic chiasm, and to connections in other parts of the brain.

Hundreds of odors can be recognized, but an adequate description of how the receptors are stimulated and the discrimination made is totally lacking. Attempts have been made to classify odors into several categories with the

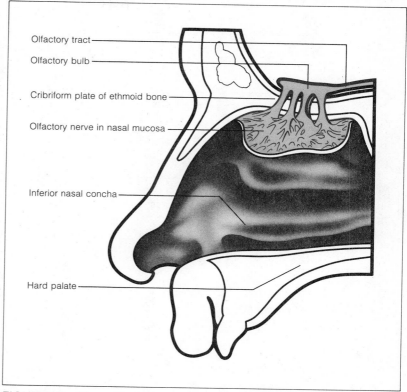

FIGURE 11–31   Sagittal section of the nasal cavity, showing the olfactory nerve, olfactory bulb, and tract.

idea that there is one type of receptor for each category. The attempt has not been very successful, partly because of the inability to classify the odors precisely. Other problems are the difficulty in accurately controlling the stimulus delivered and the rather subjective nature of the olfactory sense. What is pleasant to one person may be unpleasant to another. There seems to be general agreement, however, that there is not a special type of receptor for each odor. The several thousand odors distinguishable are the result of some sort of mixing process, perhaps as occurs in color perception.

Olfactory receptors adapt to a specific odor very quickly. However, this does not seem to impair the ability to detect a different odor. Also, it is possible to hide or mask an offensive odor quite effectively with another odor. Many of the air-fresheners on the market use a pleasant or clean odor to mask the undesirable odors.

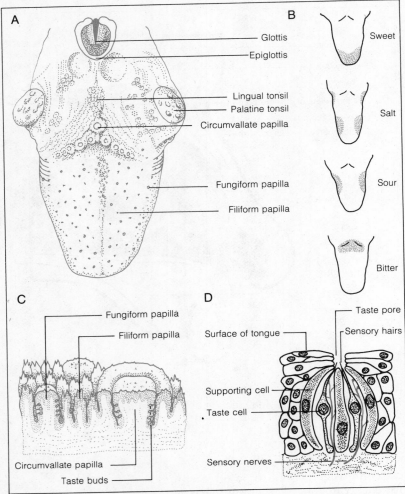

FIGURE 11-32   A. Dorsum of the tongue, showing major features.
B. Distribution of taste buds for specific tastes. C. Section of the tongue surface, showing papillae. D. A taste bud.

SPECIAL SENSES—TASTE

The sense of taste, like that of smell, is a chemical sense. The receptors are stimulated by the presence of certain chemicals in the solution bathing them. The receptors, the **taste buds,** are located chiefly on the dorsum of the tongue (see Figure 11-32), with a few on the roof of the mouth and in the pharyngeal mucosa. The dorsal surface of the tongue is covered by tiny projections, or **papillae.** The rounded *fungiform papillae* have taste buds mostly on the top, and the large *circumvallate papillae* have numerous taste buds around their sides. The latter are arranged in a "V" on the posterior dorsal surface of the tongue.

Taste buds consist of clusters of cells embedded in the epithelium. There are supporting cells and several receptor cells in each taste bud. The latter are hair cells with the tips of the hairs protruding into the porelike opening of the taste bud. Afferent nerve fibers enter the taste bud and end around the bases of the receptor cells.

Four specific tastes—salt, sweet, sour, and bitter—may be detected, and there are four types of taste receptors. Each receptor has a low threshold for one taste, but can presumably respond to each of the other basic tastes at a higher threshold. Receptors for the different tastes are distributed unequally over the tongue, since each pure taste can be detected best in certain specific regions. The many flavors that can be recognized are partly the result of blending by stimulation of various proportions of the four basic taste receptors and partly the result of olfaction. The two are sometimes confused, but the contribution of smell to normal taste becomes apparent when a head cold produces congestion of the nasal mucosa and blocks olfactory stimulation. Loss of the sense of smell leaves the impression that the sense of taste is impaired.

Nerve fibers from the taste buds are carried in three cranial nerves: (1) the facial (VII), from the anterior two thirds of the tongue; (2) the glossopharyngeal (IX), from the posterior third of the tongue; and (3) a few fibers from the pharyngeal region in the vagus (X). The pathway synapses in the medulla and in the same thalamic nucleus as the somatosensory pathways. It terminates in the inferior portion of the postcentral gyrus, which is the part of the somesthetic cortex that represents the face. Taste therefore differs from most special senses in that it does not have a single cortical area devoted exclusively to it.

CHAPTER 11 SUMMARY

The **general senses** include *touch, pressure, heat, cold,* and *pain,* plus sensations from certain proprioceptors and visceral receptors. **Special senses** include *vision, hearing* and *equilibrium, taste,* and *smell.*

Each modality has its own type of receptor. Their afferent neurons branch upon reaching the spinal cord and make synaptic connections for spinal reflexes and pathways that ascend to the brain. Unconscious proprioception is carried in the spinocerebellar tracts to the cerebellum. Since they do not reach the cerebral cortex, they do not cause sensation. Conscious proprioception and some touch are carried in the posterior columns to the medulla where they synapse and ascend to the thalamus. Pain and temperature pathways ascend in the lateral spinothalamic tracts, and other touch pathways are in the ventral spinothalamic tracts. Neurons from the thalamus terminate in the primary receiving area, the somesthetic cortex. There is a topographical representation of the body surface, with more cortical area devoted to parts that have more cutaneous receptors. This system provides for recognition of modality, stimulus intensity, and localization of the stimulus. The nearby association areas receive input from the primary receiving areas and also from the brainstem reticular formation, which provide the basis for sensory processes that make use of information from numerous receptors, discriminations, and recognition of objects (which requires memory of past experience).

The brain response to incoming sensory information requires a certain level of cortical arousal, which is dependent upon the *reticular activating system* (RAS), a system of multisynaptic pathways in the brainstem reticular formation. It receives input from collaterals from the primary sensory pathways and terminates in the other nuclei of the thalamus. Neuronal connections from the thalamus to all parts of the cortex and back again cause arousal of the cortex whenever the RAS is stimulated. Cortical arousal is indicated by changes in the electroencephalogram (EEG), and is a necessary prerequisite for brain response to sensory information arriving along the primary sensory pathways.

Highly specialized receptors in the retina of the eye respond to light rays that enter the eye. Movements of the eyes are controlled by six tiny skeletal muscles for each eye. The eyeball is formed by three layers of tissue: the tough, durable sclera; the vascular choroid; and the retina, which contains the receptors. Anteriorly the sclera is modified as the transparent cornea, through which light rays enter, and the choroid is modified as the ciliary body, to which the iris is attached. The pupil is an opening in the center of the iris whose diameter is determined by smooth muscles of the iris. Fibers of the suspensory ligament attached to the ciliary body support the lens in a position directly behind the pupil.

Light rays from a distant object (20 feet or more) are bent (refracted) as they pass through the cornea and lens, and normally come to focus at a point exactly on the retina. *Accommodation* is the adjustment necessary to focus on near objects. It involves constriction of the pupil, contraction of ciliary muscles to release tension of the lens and allow it to thicken, and convergence of the eyeballs. As the lens ages, it loses its elasticity and fails to thicken, and the individual has difficulty seeing near objects. In *hyperopia* (farsightedness) light rays come to focus behind the retina, and in *myopia* (nearsightedness) they come to focus in front of the retina. *Astigmatism* is due to an asymmetrical corneal surface, so that light rays do not all come to focus the same distance from the lens.

The *visual field* is the area seen by the eye. The visual field for the two eyes overlap, but are not identical, which enables us to have the three-dimensional vision necessary for good depth perception and judgment of speed and distance. Light rays entering the pupil from the right strike the left side of the retina. When you focus on an object, light rays from it fall on the fovea, the part of the retina that has the most acute vision. Light rays from a distant object stimulate corresponding points in the two retinas. Visual pathways from corresponding points converge in the cortex and you see a single point.

Nerve fibers from the retina leave the eye in the optic nerve. At the optic chiasm there is a partial crossing so that each optic tract carries half the fibers from each eye (those from the opposite half of the visual field). The optic tracts synapse in a part of the thalamus and continue to the primary visual cortex in the occipital lobe of the cortex, where there is a topographical representation of the visual field.

Cones, the receptors for color, are found in the fovea and the area around it, while rods for black and white vision are widely distributed but are absent from the fovea. Because rods have a lower threshold than cones, we have only black and white vision in very low light levels. Rods contain a pigment, rhodopsin, which is bleached by light and darkens in the absence of light. With low light levels most of the pigment is dark and the rods are very sensitive (dark adaptation). There are cones for blue, green, and red light, each having a different pigment.

Stimulation of rods or cones causes changes in the membrane potential of those cells and, after convergence and interaction with several other cells in the retina, action potentials are set up in the cell whose fibers are found in the optic nerve and tract. Branches from the visual fibers synapse in the superior colliculus and nuclei of cranial nerves III, IV, and VI which control eye movements and accommodation and other responses to visual stimuli.

The ear contains receptors for hearing and equilibrium. Those for hearing respond to sound waves which are pressure waves or vibrations. Loudness depends upon amplitude of the waves, and pitch depends upon their frequency or wavelength.

The external ear includes the pinna and external acoustic meatus, separated from the middle ear by the tympanic membrane. The middle ear, an air-filled chamber, contains the malleus (hammer), incus (anvil), and stapes (stirrup). The malleus is attached to the tympanic membrane, and the foot of the stapes is in contact with the inner ear. When sound waves cause vibration of the tympanic membrane, these tiny bones transmit the vibrations to the inner ear. The auditory (Eustachian) tube extends from the middle ear to the pharynx and serves to equalize the pressure across the tympanic membrane.

The inner ear is fluid-filled and contains receptors for both hearing and equilibrium. A maze of tunnels in the temporal bone, the bony labyrinth, contains a membranous labyrinth. The latter is surrounded by a fluid (perilymph) and contains a fluid (endolymph). The membranous labyrinth includes the cochlea, vestibule, and semicircular canals.

The snail-shaped cochlea is associated with hearing. The oval and round windows are openings in the bony labyrinth near the lower end of the cochlea. The oval window is occupied by the stapes, whose movements cause vibrations in the cochlear fluid. A cross-section of the cochlea shows it to be divided by the basilar membrane and a bony shelf. The spiral organ (organ of Corti) rests on the basilar membrane, with the tectorial membrane above it. Vibration of the fluid causes vibration of the basilar membrane, and hair cells of the spiral organ are brushed against the tectorial membrane, which generates impulses in afferent nerve fibers. High-pitched sounds have the greatest effect on hair cells near the oval window (base of the cochlea) and low-pitched sounds stimulate those near the apex of the cochlea.

The primary auditory pathway is unique in that it is both crossed and uncrossed in the brainstem. It terminates in the auditory cortex in the temporal lobe, where there is a cortical mapping of the basilar membrane. High-pitched sounds project to one end of the area and low-pitched sounds to the other. Each ear is represented in both sides of the cortex. The afferent neurons have branches that synapse in the inferior colliculus for reflex responses to auditory stimuli.

Receptors for equilibrium detect position, linear acceleration, and rotatory movement. The fluid-filled utricle and saccule in the vestibule contain hair cells which have otoliths in a gelatinous substance resting upon them. Change in position or acceleration alters the pull of gravity on the otoliths and stimulates hair cells. The three semicircular canals on each side lie at right angles to one another. A tuft of hair cells extends across one end of each canal. Rotation stimulates hair cells in the horizontal canals as it causes the fluid to push against the hair cells. Rotation in other planes (somersault or cartwheel movements) stimulates receptors in the other canals. Afferent neurons synapse in the vestibular nuclei of the brainstem leading to important adjustments of the postural (antigravity) muscles, and cause movements of the eyes and head.

The sense of smell is due to stimulation of chemoreceptors in the upper part of the nasal cavity. They are the endings of the olfactory nerves which enter the cranial cavity and synapse in the olfactory bulbs. The olfactory tract leads back to the olfactory cortex on the inferior surface of the frontal lobe.

Taste receptors, located in the taste buds of the tongue, consist of chemoreceptors that respond to salt, sweet, sour, and bitter stimuli. Flavors are the result of stimulation of combinations of these receptors, plus the effects of olfactory stimulation. The central pathways lead eventually to the region of the somatosensory cortex for the tongue and face.

STUDY QUESTIONS

1 Trace the general pathways of impulses from a pain receptor in a finger to their termination in the cerebral cortex. Where are the synapses in this pathway?

2 Similarly, trace the pathway of impulses from a pressure receptor in the sole of your left foot.

3 What is the role of the association areas of the cortex?

4 Why is the reticular activating system important for sensory function?

5 How is the lens of the eye attached, and what is the mechanism for adjusting its thickness?

6 What adjustments are made to accommodate for near vision?

7 What is meant by 20/20 vision? Why is it considered to be "normal" vision?

8 How is it that light rays striking corresponding points on the retinas of the two eyes are seen as a single point?

9 What kind of visual problem would you expect if the right optic nerve were cut? if the right optic tract were cut? if the optic chiasm (only) were cut? if the occipital cortex were damaged?

10 How are sound waves transmitted to the inner ear? to receptors in the inner ear?

11 What kinds of problems might you have if your auditory (Eustachian) tubes were blocked?

12 How does the inner ear distinguish pitch (frequency) and intensity of sound?

13 How does the utricle detect information on static position? on linear acceleration?

14 How do the semicircular canals respond to rotation? to the cessation of rotation? to movements in a plane other than horizontal (somersault or cartwheel movements)?

Control of Skeletal Muscle

Control of Visceral Structures—The Autonomic Nervous System

Higher Functions of the Central Nervous System

Brain Metabolism

Motor Aspects of the Central Nervous System, and Higher Brain Functions

The control of motor function involves those neural systems that regulate the activity of effectors; the somatic nervous system that controls skeletal muscle, and the autonomic nervous system that controls smooth and cardiac muscle and glands. Just as these effectors are different, so the systems that control them have different properties.

Reflex action is the basis of activity in both systems. Although we think of skeletal muscle action as voluntary, it is surprising how little of what we do is a truly voluntary action. Once the decision to act has been made, the action is carried out largely by movement patterns of which we are unaware. Autonomic function, on the other hand, has no voluntary component; it is totally reflex, and is carried out below the level of consciousness.

Somatic control of skeletal muscle involves brain mechanisms that bring about movement, regulate motoneuron excitability, and coordinate the contractions of muscles. The afferent component of autonomic reflexes is similar and often identical to that of somatic responses, but the efferent side is unique, and our attention will be directed mainly to its peripheral efferent pathways and their actions, and to a lesser extent to the effect of the central nervous system on them.

The neural control of motor systems involves primarily the control of muscles, particularly skeletal muscles. In the description of muscle contraction in Chapter 7, we saw that skeletal muscle fibers are totally dependent upon their nerve supply. Their response to stimulation is simply shortening of sarcomeres, but the contraction can be varied greatly by changing the number and frequency of impulses in the motoneurons to those muscle fibers, or the number of muscle fibers that are activated. Now let us consider how the central nervous system regulates the impulse traffic in those motoneurons, and how it correlates contractions of literally hundreds of individual muscles.

The central nervous system plays a different role for the involuntary effectors, for they are less dependent upon the nervous system. Cardiac muscle and much of smooth muscle possess automaticity, and the nervous system, therefore, serves to regulate rather than initiate their activity. Smooth muscle in particular, and many glands respond to other types of stimuli, such as hormones. Much of the activity of the involuntary effectors is geared to specific or local conditions, and the brain is not called upon to coordinate responses in the same way as for skeletal muscle.

CONTROL OF SKELETAL MUSCLE

Controlling skeletal muscle contraction involves much more than simply sending a signal through a motoneuron to the muscle cells. The motoneuron is a key element, for it is the final common path, the only physiological connection with the muscle cell, and it governs what the muscle does. Of equal importance, however, are the inputs to the motoneuron from the spinal cord and the brain, for they govern what the motoneuron does. Motoneurons do not act alone, however, and they are not stimulated individually. The inputs to them affect groups of motoneurons, and the responses (which may be contraction or inhibition) occur in whole muscles and in groups of muscles.

The input to the motoneuron from the spinal cord is the afferent limb of spinal reflexes such as the stretch and flexion reflexes described in Chapter 9. They cause such responses as a knee jerk or withdrawal of a limb. They also increase or decrease the excitability of other motoneurons, as demonstrated by reciprocal innervation.

Impulses from the brain may stimulate the motoneuron and cause a response—contraction of muscle groups and movement, or they may alter the excitability of the motoneuron. In fact, motoneurons are continually being affected by impulses (EPSPs and IPSPs) from many parts of the nervous system. We shall approach the control of skeletal muscles by considering some of the important sources of input to the motoneurons, and the qualities that each contributes to our overall performance of simple movements and complex skilled acts.

Local Segmental Control of Skeletal Muscle

The most basic control of skeletal muscle is that which occurs in the spinal cord at the level of the motoneuron. In addition to their direct control, the cord-level components are the means through which various parts of the brain exert their influence on muscle contraction.

Innervation of a Skeletal Muscle Although muscle contraction depends upon the motoneuron and all the input to it, that motoneuron is only part of the nerve supply to a skeletal muscle. Before proceeding further, it is appropriate to review the components of a muscle nerve (Figure 12–1).

1. The anterior horn cells. These are the typical motoneurons, with cell bodies in the anterior horn of the spinal cord (or corresponding nuclei in the brainstem for those in cranial nerves), and they terminate at neuromuscular junctions on extrafusal muscle fibers (those outside of muscle spindles). These are the large diameter, rapidly conducting **alpha motoneurons.**

2. Motoneurons to the intrafusal

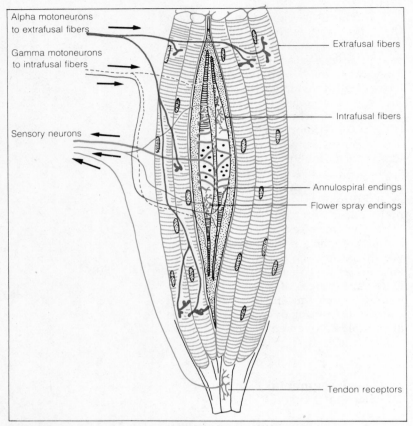

FIGURE 12-1 Innervation of a skeletal muscle.

input for unconscious proprioception. Branches of the sensory neuron form ascending pathways in the spinal cord (the spinocerebellar tracts), which end in the cerebellum instead of the cortex.

Role of the Muscle Receptors and Gamma Efferents The anatomical arrangement of the muscle receptors is such that when the whole muscle is stretched, both the tendon organ and muscle spindle receptors are stretched and discharge bursts of impulses. When the muscle actively contracts, the tendon organs are stimulated, but the tension on muscle spindles is relieved and the spindle afferents cease firing. If the fusimotor fibers (the gamma efferents) are activated, the striated ends of the intrafusal muscle fibers will contract, stretching the spindle receptors and the sensory neurons will carry impulses, just as if the whole muscle had been stretched. Since intrafusal fibers cannot develop any real contractile tension in the whole muscle, the tendon organs are not affected by their contraction (see Figure 12-2).

Impulses in the fusimotor neurons bring about an increase in the excitatory input to alpha motoneurons—a facilitatory influence. This circuit from gamma efferent and spindle afferent back to the alpha motoneuron and muscle, is sometimes called the *gamma loop* (Figure 12-3). When alpha motoneuron excitability is high due to activity in this circuit, a weak stimulus that would ordinarily be subthreshold might well elicit a muscle contraction. By their action on intrafusal fibers, the fusimotor neurons serve to control the sensitivity of the muscle spindle receptors. The input from the muscle spindles is a major determinant of the excitability of alpha motoneurons, and therefore, the gamma efferents are important in controlling both the excitability of the alpha motoneurons and the stretch reflex.

The fusimotor fibers are controlled by nerve fibers from the same areas of the brain that provide input to alpha motoneurons. In fact, most descending pathways carry impulses to both alpha and gamma motoneurons, and usually the effect is the same, that is, to stimulate both or to inhibit both.

fibers of the muscle spindles (see Chapter 8). They also have cell bodies in the anterior horn of the spinal cord (or cranial nerve equivalent in the brainstem), but they have a small diameter and slow conduction speed. They are called **fusimotor fibers** because they innervate the intrafusal fibers, or **gamma efferents** because of their small diameter.

3. Afferent fibers from stretch receptors in the muscle spindle, both from annulospiral and from flower-spray endings.

4. Afferent fibers from Golgi tendon organs. Although tendon afferents are not, strictly speaking, part of the innervation of the muscle, the response of tendon receptors helps to regulate skeletal muscle contraction, and their afferents run in muscle nerves.

The afferents from muscle spindles provide direct input to the motoneuron as the afferent limb of the stretch reflex (see Figure 9-13), but they also provide

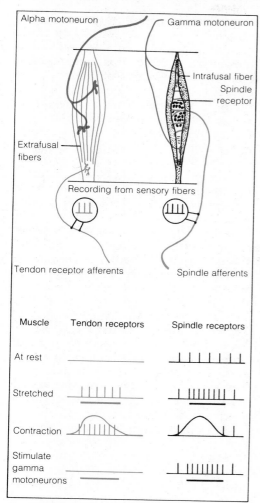

FIGURE 12-2 Responses of muscle spindles and tendon organs.

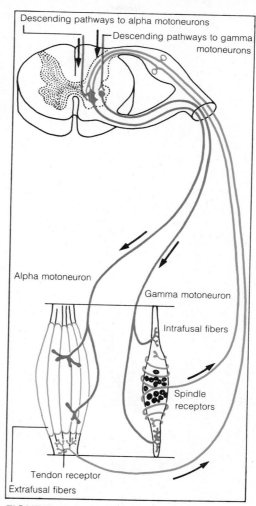

FIGURE 12-3 The gamma loop. Stimulation of gamma efferents facilitates alpha motoneurons through their effect on muscle spindle receptors.

Descending Pathways from the Brain

The brain exerts its influence on skeletal muscle contraction through a number of descending fiber tracts that end on alpha and gamma motoneurons. Some pathways are direct and some are indirect, some are excitatory and some inhibitory, and some affect mainly flexor muscles while others affect mainly extensor muscles. The input from the brain is superimposed upon the spinal mechanisms, particularly the stretch reflex and gamma loop. On the basis of structural and functional characteristics, these pathways are conveniently divided into direct and indirect pathways.

Direct Pathways The direct motor pathways (Figure 12-4) are so named because they are the only ones that project from the cerebral cortex directly to the motoneurons in the spinal cord (or brainstem for somatic pathways in cranial nerves). Most of the fibers arise in the frontal lobe of the cerebral cortex, many of them from the precentral gyrus (the primary motor cortex), and quite a few arise from the area in front of it (the premotor cortex). Nearly a third, however, arise from the parietal lobe, both the somesthetic cortex in the postcentral gyrus and association areas just behind it.

The primary motor area in the precentral gyrus contains a topographical representation of the body similar to

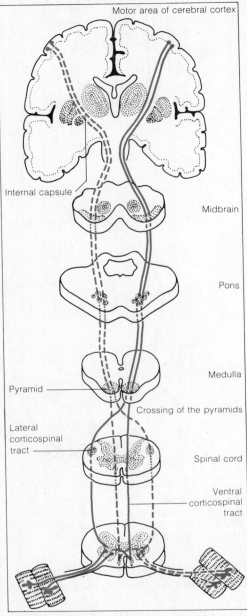

Motor area of cerebral cortex

Internal capsule

Midbrain

Pons

Pyramid

Medulla

Crossing of the pyramids

Lateral corticospinal tract

Spinal cord

Ventral corticospinal tract

FIGURE 12-4   Direct motor pathway (pyramidal tract).

that of the somatosensory area in the postcentral gyrus. Regions of the body capable of precise, complex, and varied movements (fingers, lips, tongue) have relatively large areas of motor cortex devoted to them (Figure 12–5). The brain is concerned with movements rather than regions, however, and the cortical representation in the motor area is one of muscles instead of skin surface. The premotor area and its projections are not so clearly defined, but it is believed to be associated more with muscles of the trunk and proximal parts of the limbs.

From the cortex these fibers descend through the internal capsule and brainstem to the medulla. They lie on the inferior surface of the medulla, where they form a structure known as the *pyramids* (see Figure 10–2), and for this reason the direct pathways are known as the **pyramidal tracts.** At the lower border of the medulla most of them cross over to descend in the spinal cord as the lateral *corticospinal tracts.* The remaining fibers form the *ventral corticospinal tracts* which cross at their level of termination in the spinal cord. Most of the descending fibers end on tiny interneurons, but some may end directly on motoneurons. The terms *corticospinal* and *pyramidal* are both used to identify the direct motor pathways.

Nearly all of these fibers are small, and many of them are unmyelinated. The largest are believed to arise mainly from the primary motor areas of the precentral gyrus and to be the ones that project to the motoneurons that innervate muscles of the fingers and other structures that demonstrate great dexterity. This leaves the majority of pyramidal fibers—those small neurons, many of which originate from other parts of the cortex—without a clearly defined or specific function.

Indirect Pathways   There are a number of indirect pathways from the brain to motoneurons. They are multisynaptic, multineuronal pathways (hence "indirect") with synapses in the basal ganglia, the red nucleus, the thalamus, the reticular formation, and the cerebellum. They arise from several

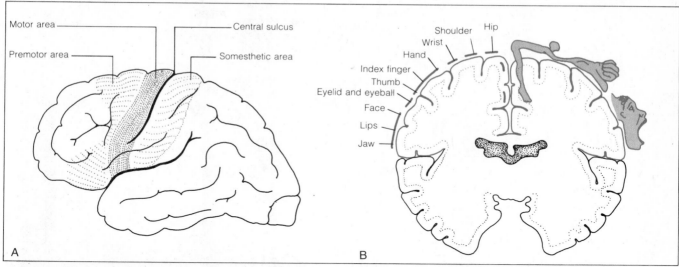

FIGURE 12-5   A. Primary motor areas of the cerebral cortex. B. Representation of various parts of the body on the primary motor area.

parts of the cortex, including the precentral gyrus, and from several subcortical areas (Figure 12-6). The indirect pathways have very little in common, except that none of them pass through the pyramids in the medulla, and for this reason they are often called the **extrapyramidal tracts.** It is not a very good term, because they are so varied, both anatomically and functionally, that lumping them together is not very meaningful.

Although several areas of the forebrain contribute to the indirect pathways, the fiber tracts in the spinal cord all arise in the brainstem. There are a number of such tracts, but three of them are of particular interest and are shown in Figure 9-3.

1. The *rubrospinal tracts* arise from the red nucleus in the midbrain. The red nucleus, named for its appearance in a fresh preparation, receives input mainly from the cerebellum and cerebral cortex. The rubrospinal fibers cross to the other side and descend to alpha and gamma motoneurons in the spinal cord. Their main effect is facilitation, especially to those that supply flexor muscles.

2. The *vestibulospinal tracts* arise from the vestibular nuclei in the brainstem and descend to alpha and gamma motoneurons in the cord on the same side. These nuclei receive input from the vestibular apparatus in the inner ear, and also from the cerebellum. The vestibulospinal tracts are facilitatory, especially to antigravity (extensor) muscles.

3. The *reticulospinal tracts* descend from the reticular formation in the brainstem. There are several tracts on each side, and they end on alpha and gamma motoneurons on the same side. The reticular formation receives input from many centers, particularly the cortex and basal ganglia, and from the ascending primary sensory pathways (the same ones that feed the reticular activating system). The reticulospinal tracts carry both excitatory and inhibitory impulses to motoneurons of antigravity muscles.

Brain Control of Skeletal Muscle

The information from the brain that is carried in the descending pathways affects all aspects of muscle activity, from the regulation of muscle tone to the initiation of voluntary movements. There are several components of these controls, however, and they involve several different brain mechanisms.

Tonus in muscle is the resistance of the muscle to passive stretch. An intact

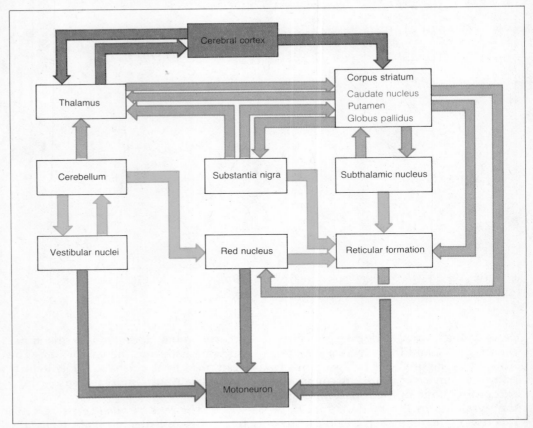

FIGURE 12–6   Schematic diagram showing some indirect (extrapyramidal) pathways and connections.

muscle resists stretch by contracting, which is the stretch reflex. Tonus, therefore, is dependent upon the nerve supply to that muscle rather than being a property of the muscle itself. The sensitivity of a muscle to stretch (how much it resists) depends upon the excitability of its motoneurons, and that is determined by impulses from the brain as well as from its muscle spindles.

The initiation of a movement may result from a decision to perform a movement requiring precision control of individual muscles, such as when you write your name. Or, it may be a less precise action, perhaps the incidental movements of swinging your arms as you walk. Movements like walking that involve many muscle groups require accurate coordination of the contraction and relaxation of each group. Walking is also an example of an activity that is a voluntary movement,

but is carried out in movement patterns without need to specifically will each portion of it.

Specific areas or centers of the brain seem to have special roles for these different aspects of movement. But as these areas also have abundant connections with one another, it is not possible to tie each functional component exclusively to a particular area of the brain for motor functions, any more than it was for sensory functions. In the sections that follow, some of the important centers for motor control are discussed and related to specific motor functions wherever possible.

Reticular Formation and Motoneuron Excitability   The brain affects the excitability of alpha and gamma motoneurons through indirect pathways from the brainstem. The critical inputs for alpha motoneuron excitability are from the muscle spindles (stretch receptors) and from the brain by way of the indirect pathways. Many of them synapse in the reticular formation and descend in the reticulospinal tracts. Stimulation of certain areas of the reticular formation may bring about a change in muscle tone, particularly of the antigravity muscles. There is a small center, an **inhibitory area,** located in the reticular portion of the medulla (Figure 12–7). Stimulation of this area decreases extensor tone and raises the threshold for stretch reflexes of the antigravity muscles. The inhibitory area receives important input from several extrapyramidal pathways, including some from the basal ganglia and cortex. Destruction of these incoming fibers effectively inactivates the reticular inhibitory area and eliminates its tonic inhibition of motoneurons.

There is also a brainstem **facilitatory area,** which is much larger, as it extends from the medulla up through the reticular formation of the pons and midbrain. It receives input from the vestibular nuclei and probably branches from the ascending primary sensory tracts. Stimulation of the facilitatory area causes a marked increase in the tone of antigravity muscles and in the sensitivity of their stretch reflexes.

Impulses from the facilitatory and inhibitory areas are carried to the anterior horn of the spinal cord in the reticulospinal tracts. Their action upon motoneuron excitability is brought about by the inhibitory and excitatory endings on alpha and gamma motoneurons. (Compare the reticular areas shown in Figure 12–8 with that of the reticular activating system shown in Figure 11–5.)

Excitability of the motoneurons is the main element in the subtle adjustments of muscle tonus and contraction that are involved in the maintenance of posture. Postural mechanisms usually involve resisting gravity, and the extensor or antigravity muscles are primarily involved. The reticulospinal tracts from the reticular centers affect primarily motoneurons to extensor muscles. Stretch reflexes are also more easily elicited from these antigravity muscles. Resisting gravity is a long-term thing, and the excitability of these motoneu-

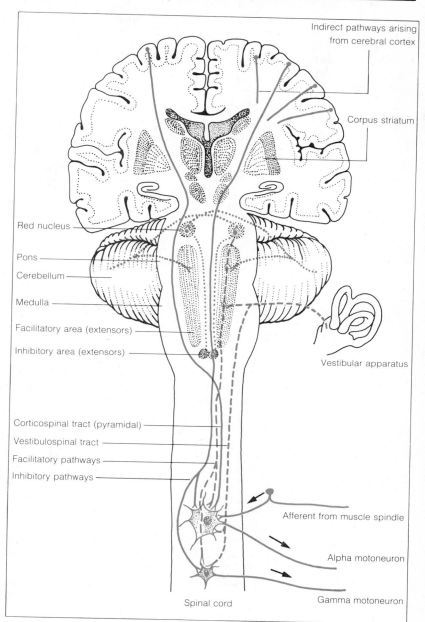

Indirect pathways arising from cerebral cortex

Corpus striatum

Red nucleus

Pons

Cerebellum

Medulla

Facilitatory area (extensors)

Inhibitory area (extensors)

Vestibular apparatus

Corticospinal tract (pyramidal)

Vestibulospinal tract

Facilitatory pathways

Inhibitory pathways

Afferent from muscle spindle

Alpha motoneuron

Gamma motoneuron

Spinal cord

FIGURE 12–7   Cerebral and brainstem pathways affecting motoneurons.

rons must be sustained for long periods of time. Impulses from the reticular formation provide an appropriate background of facilitation for those subtle adjustments.

Cerebellum and Muscle Coordination   The brain "thinks" in terms of movement rather than individual muscles, and movement of any part of the

FIGURE 12-8
Areas of the brain whose stimulation produces facilitation (blue) and inhibition (red) of stretch reflexes. These effects are mediated through the reticular formation.

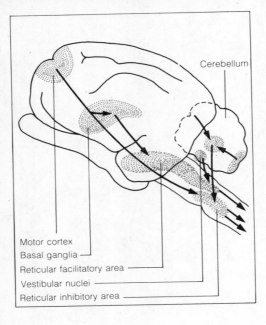

Cerebellum

Motor cortex
Basal ganglia
Reticular facilitatory area
Vestibular nuclei
Reticular inhibitory area

body involves contraction of several muscles—one or more prime movers, perhaps some synergists or fixators—and relaxation of antagonists. Because the things we do involve multiple movements and several joints, and perhaps several parts of the body, the total movement requires coordination of the actions of many muscles. This coordination is carried out primarily through the **cerebellum.**

So far as is known, the role of the cerebellum is exclusively motor. Removing it causes no sensory deficits, although it receives ample sensory input. But stimulation of the cerebellum does not cause movement either. Its role is not to initiate movement, but to smooth out the response of the muscles.

The cerebellum receives sensory input from a number of sources, including the spinocerebellar tracts from muscle spindles (unconscious proprioception) (see Figure 12-9). It also receives afferents from cutaneous receptors (as branches from the posterior white columns) and from visual and vestibular nuclei. In addition, it receives important input from the motor areas of the cerebral cortex by way of the pons. Fibers leaving the cerebellum go to nuclei in the brainstem (red nucleus, vestibular nuclei, reticular formation) as well as to the thalamus and cerebral cortex. Therefore, the cerebellum indi-

rectly affects motoneurons in the spinal cord by modifying the discharge in the descending fiber tracts to them from the brainstem centers.

The cerebellum has a thin cortex of gray matter covering its mass of white fibers, and it has several small nuclei deep inside. The internal circuitry of its fibers is complex, and just what happens between the arriving afferent and departing efferent fibers is not clear. An attractive theory of cerebellar function suggests that it operates as a "comparator." It receives information from sources such as the motor areas of the cortex that indicate what the muscles *should* be doing, and it compares this with information from proprioceptors that tells what the muscles *are* doing. It notes any discrepancy and directs the cortex and brainstem centers to make necessary corrections in the discharge to the motoneurons. This is a continuous process for as long as movement is going on, and it ensures that muscle contractions are of the proper strength and that muscles begin and end contraction (or relaxation) at exactly the right time. The constant monitoring and adjusting brings about the harmony and smoothness that lends beauty as well as effectiveness to the movements of the body.

Normal function can often be illustrated by looking at some of the problems that arise when function is not normal. It is interesting to note that the effects of cerebellar lesions are apparent only when the muscles are being used, not when the individual is relaxed and at rest. Among the problems encountered are: errors in range of movement, undershoot and overshoot when reaching for an object; breaking a movement into its individual parts and performing them sequentially; inability to perform rapid alternating movements such as pronation and supination; tremors that occur only when movements are made or attempted; and incoordination in movement of a part, as to rate, range, force, and direction of movement. All of these represent some aspect of the breakdown in the coordination process.

Basal Ganglia and Patterns of Movement The basal ganglia are large centers in the cerebral hemispheres.

The corpus striatum, consisting of the caudate nucleus, putamen, and globus pallidus, is the largest ganglion. Others are the substantia nigra and several subthalamic nuclei (see Figure 10–10). They have numerous connections with one another and almost all other brain structures that have anything to do with motor function (see Figure 12–6). In spite of this, or perhaps because of it, the role of the basal ganglia is not well understood. One investigator has aptly called it an "enigma." Neither stimulation nor removal of the basal ganglia alone produces marked effects. Part of the problem is that they do not have a direct projection to motoneurons, but instead affect other centers in the brain. Recent evidence suggests that much of their output leads indirectly back to the motor area of the cortex rather than to the brainstem centers of extrapyramidal pathways. The basal ganglia also receive fibers from the cortex, and in fact there is a multisynaptic loop or circuit from motor cortex to several basal ganglia, to the thalamus, and back to the cortex (see Figure 12–6). This suggests that their action is on the pathways arising from the motor areas of the cortex and, therefore, hints at possible involvement in voluntary activity.

In species whose motor cortex is poorly developed, the basal ganglia are the highest motor centers, and they control what might be called voluntary movement. In humans they are probably involved in the initiation of some movements, as they are in birds, but to a lesser extent. It has been suggested that, in humans, the basal ganglia are concerned with the "planning and programming of movement" in some way. They are known to be involved in the generation of mannerisms and incidental movements, such as facial expressions and swinging the arms while walking. The basal ganglia seem to stabilize voluntary movements in some way and to reduce oscillations and after-discharge. They also have some facilitatory effect on muscle tone and the stretch reflex.

Much of the information about the role of the basal ganglia in humans has been obtained from patients with disorders of these structures. One of the best known conditions is *Parkinson's disease*, which is characterized by a

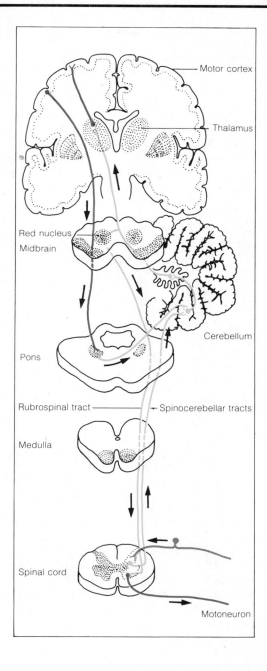

FIGURE 12–9
Some of the major cerebellar connections.

Motor cortex

Thalamus

Red nucleus
Midbrain

Pons

Cerebellum

Rubrospinal tract

Spinocerebellar tracts

Medulla

Spinal cord

Motoneuron

general stiffness and rigidity of the muscles. Incidental movements (arm-swinging) are absent and all movements are slow, and there is often a tremor at rest that disappears when movement is undertaken.[1] Other disor-

---

[1]The tremor is apparently caused by a lack of a specific neurotransmitter substance (dopamine) in certain basal ganglia. In patients with Parkinson's disease the concentration of this substance is much lower than normal, and many of them respond to administration of a precursor *(l-dopa)*.

ders of the basal ganglia have different symptoms, often accompanied by one of several kinds of uncontrolled movements of the extremities and/or head.

Mechanisms for Voluntary Movement Traditionally, the pyramidal tracts have been associated with the initiation of voluntary or willed movements, particularly those involving the fine, discrete movements of the more distal parts. For example, pyramidal tracts are more likely to be concerned with movements of the fingers than of the whole arm.

The largest neurons in the corticospinal tracts arise in the motor area in the precentral gyrus and project directly to the alpha and gamma motoneurons, primarily those to muscles in the distal part of the extremities. The effects are excitatory, and stimulation of the motor cortex is likely to produce movements of those parts. Recall that the largest areas of the motor cortex are for parts of the body that are capable of fine skilled movements such as the fingers.

There must be other pathways for voluntary movement, however, because such movement is still possible after the pyramidal tract has been cut. In experiments on monkeys, the pyramidal tract was cut on one side. After an initial period of paralysis, the affected arm could be moved, although there was a general reluctance to use it. The monkey used the "good" arm to reach for food and to manipulate it, but if that arm was restrained the monkey used the affected arm. The most obvious permanent effects of such a procedure were the awkward and clumsy movements of the fingers—lack of the precision control. It is not certain what pathways are used when the direct pyramidal path is gone, but it is known that if the rubrospinal tract is also cut, the distal muscles of the limb are almost completely paralyzed.

The implication of the rubrospinal tracts in certain aspects of voluntary movement is somewhat surprising because they do not arise from the cerebral cortex. They come from the red nucleus in the midbrain, which has connections with the cerebellum and basal ganglia, and the rubrospinal tracts also receive some branches from the descending corticospinal tracts. The corticospinal (pyramidal) and rubrospinal (extrapyramidal) tracts do have several things in common, however. Both facilitate alpha and gamma motoneurons, especially those to flexor muscles of distal parts. They also inhibit motoneurons to extensor muscles, but this is not limited to distal muscles of extremities. Flexor muscles are probably involved more often in movements and manipulations of the fingers, while extensor muscles are primarily antigravity muscles involved in the maintenance of muscle tonus and posture. Inhibition of extensor muscles accompanying stimulation of flexors seems to be a means to override postural reflexes in order to carry out a specific movement. This possibility contradicts the long-held view that the corticospinal pathways are only excitatory. It probably represents a role for the majority of pyramidal fibers which are smaller and arise from other parts of the cortex.

It is interesting to note that the pyramidal pathways are most highly developed in humans and primates. They are much less important in dogs and cats, and absent in birds, yet those species are all capable of skilled movements, though not of the kind that are characteristic of human and primate hands and fingers. These nonpyramidal movements could be considered parts of movement patterns, and so are probably controlled at least in part by the basal ganglia. The extent of the involvement of basal ganglia in voluntary movement in humans is not known.

Sensory Input for Postural Mechanisms The concept of posture conjures up the caricature that we associate with "good posture," but any body position, whatever it may be, is also a posture. Any position, except one of complete support as in repose, involves a sustained resistance to gravity. This resistance is maintained reflexively through the tonic activity of the antigravity muscles in response to stimulation of the stretch reflexes and impulses in the indirect motor pathways from the brainstem. We have discussed several areas of the brain that play important roles in determining the passage of impulses in these descending motor pathways. Information from all available sources contributes to the discharge in these pathways, including input from many types of receptors. Although we may not think of touch receptors as having much to do with postural mechanisms, all primary sensory pathways give off collateral branches into the reticular formation, and they all have synaptic connections with other pathways in the thalamus. Information from these receptors provides a constant updating of information about the orientation of the body in relation to its surroundings, and in this way modifies the discharge in the descending pathways to motoneurons. Some of these sources of information are familiar receptors.

1. In addition to being part of the basic stretch reflex and gamma loop, *stretch receptors* in the muscle spindles provide information through the spino-cerebellar tracts (unconscious proprioception) and cerebellum that modifies the discharge in some of the motor pathways.

2. Signals from such *cutaneous receptors* as touch and pressure receptors contribute to postural adjustments in many ways. Those in the sole of the foot, for example, provide information about the center of gravity and the distribution of weight over the foot. When descending a stairway, you feel more secure if you touch the handrail. You need not grip it for support, but just touching it gives you additional cues about your position in relation to your surroundings.

3. The contribution of *visual cues* can be demonstrated readily by closing your eyes while standing on one foot. In doing so, you remove those visual cues about your relation to your surroundings. We seldom realize how important the visual information is, for example, in adjusting the contraction of leg muscles to support the body weight at a particular instant, until that information is lacking or in error. It is quite a "jolt" to suddenly step off a curb that you did not see.

4. The *vestibular apparatus* in the inner ear provides essential information about the position of the head in space. In addition to effects on motoneuron excitability by way of the vestibular nuclei and vestibulospinal tract,

stimulation of these receptors may set off a complex chain of postural adjustments known as the *righting reflexes*. These are well illustrated by an experiment that nearly every youngster has tried—dropping a cat to see it if always lands on its feet. The animal first turns its head, then its forequarters, and finally its hindquarters. The vestibular receptors, perhaps aided by vision, detect the inverted position of the head and provoke an immediate rotation of the head to the proper position in relation to gravity (the ground). Twisting the head excites receptors in the joints of the neck, which in turn cause the contractions of the limb and trunk muscles that rotate the body into its proper relation to the head. The animal is then ready to land on its feet. Reflex postural adjustments and righting reflexes are not, however, limited to cats.

Levels of Motor Integration

Perhaps the easiest way to gain an understanding of the roles of the various pathways and centers that participate in the control of motor activity would be to see what each portion adds to the total picture. This can be done experimentally by studying preparations in which certain parts of the nervous system have been stimulated or removed, or clinically by observing the effects of injury or disease in human patients.

To begin, we can assume that a normal intact animal or individual can move voluntarily and in a coordinated manner, and has "normal" reflexes and muscle tone. Thus, stimulation of the skin promptly elicits a withdrawal response, a tap to a tendon produces a reflex contraction of the muscle stretched, and the muscles all demonstrate some degree of tone.

The Motor Nerve If the motor nerve is damaged, the axons of all neurons to the muscle are destroyed. Since that muscle receives no impulses, it has no tone and no contraction, either voluntary or reflex. The muscle is limp, a condition known as *flaccid paralysis*, and after a time the muscle atrophies and degenerates (unless there is some regeneration of axons). *Poliomyelitis* is a disease in which anterior horn cells are destroyed, leaving the muscle fibers permanently paralyzed.

The Spinal Cord   Transection of the spinal cord produces a spinal animal (Chapter 9). Immediately after the cut there is a period of time, a period of *spinal shock*, during which all responses are suppressed. In humans, spinal shock lasts several weeks. After it wears off the individual still has no sensations or voluntary movements in those parts below the cut, since their connections with the brain have been severed. Sensations and movements of parts above the cut, and brain function in general, are unimpaired, however. Spinal reflex arcs below the level of transection also remain intact, and withdrawal and stretch reflexes can be elicited.

The spinal reflexes in a spinal animal are not, however, quite "normal." The flexion reflexes return first and are often exaggerated; responses of extensor muscles (stretch reflexes) appear later. A spinal human may develop what is sometimes known as a "mass reflex," in which stimulation of the skin (perhaps bumping one leg) may set off a prolonged response, primarily withdrawal of the limb stimulated and flexion-extension responses of the other limb as well; it may also include autonomic responses (defecation, urination, skin flushing, or sweating) in more extreme cases. After many months hyperactive extensor reflexes and sustained stretch reflexes develop in addition to flexion responses.

Transection of the cord interrupts pathways from the brain, including those from the reticular formation (Figure 12–7). The interruption of these pathways undoubtedly deprives many neurons of a considerable amount of subthreshold depolarization (facilitation) and thereby lowers the excitability of the spinal neurons. The return of excitability is not well understood. One suggested mechanism is a *denervation hypersensitivity*, since it is known that denervated cells (nerve or muscle) become hypersensitive to the transmitter that normally activates them (although this would not explain the extremely

rapid recovery from spinal shock of animals such as the frog).

The Brainstem   Lesions in the brainstem can produce a great variety of effects. A ruptured blood vessel might destroy selected pathways and alter the reflex response or tone of certain muscle groups. The result might be only altered tone, or it might well include the development of one of several types of *tremor*. Since brainstem lesions are more likely to damage one of the extrapyramidal tracts than the pyramidal, the effects on tone and reflex responses are more pronounced than on voluntary movements, which may be totally unaffected.

Severing the brainstem through the midbrain between the superior and inferior colliculi leaves most of the brainstem connected to the spinal cord. The cerebrum, including the diencephalon, is removed and the animal is said to be *decerebrate*. Many of the connections from the cerebellum to the brainstem are also interrupted. Those cranial nerves that emerge below the transection are intact, but those from above are nonfunctional.

The reflex picture in a decerebrate animal is quite different from that in a spinal animal. Stretch reflexes and extensor tone are greatly exaggerated. The limbs, in fact, are extended so rigidly that the decerebrate animal can be balanced upon them—the condition known as *decerebrate rigidity*. Flexion reflexes are often difficult to elicit.

The differences between spinal and decerebrate preparations are due in large part to differences in the remaining connections of the facilitatory and inhibitory areas of the reticular formation. In the spinal animal, reticular connections with the brain are intact, but those with the spinal cord have been destroyed. In the decerebrate animal, the reticular areas still have their connections with the spinal cord, but they have been deprived of some of their input from the brain. After decerebration, the facilitatory area retains its important input from the vestibular apparatus and collaterals from the primary sensory pathways ascending through the brainstem, so that it continues to function reasonably well. The smaller inhibitory area, on the other hand, no longer receives inputs from the cortex, cerebellum, or basal ganglia, so that its function is severely impaired. The decerebrate animal thus has an imbalance; there is too much extensor facilitation and not enough inhibition. In a normal animal the input to the inhibitory area from other parts of the brain enables it to counterbalance the facilitatory output, resulting in what we call "normal" muscle tone and reflex movements.

The Cerebral Cortex The effect of lesions in the motor cortex is difficult to demonstrate experimentally. In general, lesions limited to the areas from which the pyramidal tracts arise result in difficulty in initiating voluntary movements, clumsiness, and weakness, particularly in performing skilled movements. Extensive cortical damage causes greater impairment, especially in the patterning of movement. All primates, but notably young ones, recover to a remarkable degree from the deficits caused by such lesions.

The types of lesion produced experimentally are rarely seen in patients, in whom damage is usually the result of accident or *stroke* (ruptured brain blood vessel). Either may produce damage directly and/or indirectly by pressure caused by localized swelling (there may be at least partial recovery when the swelling subsides and pressure is relieved). Vascular accidents are particularly common in the region of the internal capsule, where there are many tightly packed fiber tracts, both ascending and descending. Such lesions may cause widely varied effects, depending upon which particular pathways are involved.

Clinically, lesions in the motor pathways are often described as due to **upper motoneuron** or **lower motoneuron** damage. The latter, of course, refers to the alpha motoneuron, the anterior horn cell. The upper motoneuron designation came into use before the motor pathways were understood. It refers to lesions in the descending motor tracts in the central system and therefore includes both extrapyramidal and pyramidal pathways. The effects of upper motoneuron

lesions vary, but generally involve muscle tone, reflex responses, and tremors. Upper motoneuron lesions are classically considered to cause spastic paralysis characterized by resistance to stretch or passive movement. Most other signs, including exaggeration of tendon reflexes, are related to damage to extrapyramidal pathways.

CONTROL OF VISCERAL STRUCTURES—THE AUTONOMIC NERVOUS SYSTEM

The **autonomic nervous system** (ANS) consists of those portions of the nervous system that innervate the involuntary effectors—cardiac muscle, smooth muscle, and glands. The effectors are located in organs that are part of many different organ systems, and serve varied functions. The skeletal muscles innervated by somatic neurons, on the other hand, function to move or maintain a position of the body and its parts, and thus have a somewhat common purpose. These muscles also have very sophisticated control systems to coordinate their activities, to make them responsive to conditions in the external environment based on information from many receptors, and to allow for voluntary initiation of contractions. The control of autonomic effectors does not have to be coordinated to quite the same degree, for their actions are determined primarily by systems that regulate the organ system in which they are located. They respond, for example, to factors that regulate the cardiovascular system rather than contraction of smooth muscle in general.

Since it controls involuntary effectors, the ANS does not have a mechanism for voluntary action. It is totally a reflex system, and most of it is carried out below the level of consciousness.[2]

---

[2]You may become aware of your heartbeat when it speeds up, but you do not directly decide to reduce it, or to reduce your blood pressure. You can voluntarily initiate some activities and avoid others so that you will not get angry or lose your temper, both of which would raise your heart rate and blood pressure. These are indirect behavioral effects, however. Some individuals have been "trained" through conditioned reflexes and biofeedback techniques to exert some voluntary control over some autonomic functions of the digestive and cardiovascular systems.

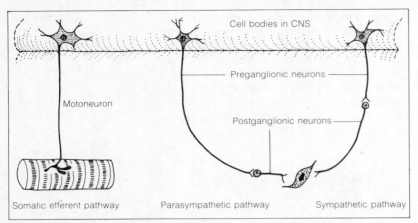

FIGURE 12–10   Comparison of somatic, parasympathetic, and sympathetic efferent pathways.

characteristic of cardiac and visceral smooth muscle, and some of the effectors exhibit a marked tone which, unlike that of skeletal muscle, is independent of the innervation. The action of autonomic nerves, therefore, is to alter ongoing activity—to change the degree of tone (as in constriction of the pupil of the eye) or to change the rate of rhythmicity (as with the heart rate). This is in contrast to somatic neurons, which tend to initiate and produce an on-off action, whose maintenance is dependent upon continued nerve activity.

The elements of autonomic reflexes are the same as those for somatic reflexes, however, and the characteristics of somatic reflexes discussed in Chapter 9 apply equally well to them.

The CNS exerts its influence on autonomic function at the reflex centers in the spinal cord and brainstem. Because these centers are associated more with a particular function (heart rate), the overall control is not as well defined as for somatic functions. The input that determines the discharge of impulses from these centers depends to a great extent upon afferents from receptors throughout the body. The receptors may provide information for specific centers (about pressure in an artery or distension in the stomach), or they may be the same receptors that provide information for somatic reflexes. In fact, single sensory neurons may synapse with both somatic and autonomic efferents.

Anatomically it is only the autonomic efferent pathway that differs from that of the somatic. Because special characteristics of the peripheral autonomic efferent pathways account for some of the properties of autonomic function, much of our discussion centers on these pathways.

Some generalizations about autonomic activity can be made at the outset, based upon our knowledge of the properties of the neurons and effectors. Autonomic responses tend to be relatively slow, in terms of both latency and duration of response. Automaticity is

Structure of the Autonomic Nervous System

The *autonomic reflex arc* has the same basic components as a somatic reflex arc. The receptors and afferent path are similar and may even be identical. Exclusive autonomic afferents may differ in the nature and location of their endings, and their fibers may travel in an autonomic nerve. However, their cell bodies lie in the spinal ganglia and they enter the central nervous system through the posterior roots (or their counterparts for cranial nerves). Autonomic reflex centers in the central nervous system contain synapses at which typical synaptic phenomena occur. Aside from the effectors, which have been discussed, most of the unique characteristics of the autonomic nervous system are related to the efferent pathway.

Each autonomic efferent pathway (with one exception) has two neurons between the central nervous system and the effector. The first neuron, called the **preganglionic neuron,** arises in the central nervous system and terminates in a ganglion where it synapses with the cell body of the **postganglionic neuron.** This neuron lies entirely outside the central nervous system (see Figure 12–10).

The autonomic nervous system is composed of two divisions anatomically and physiologically. Each division innervates most (but not all) of the viscera. Many organs are supplied by both divisions; where this occurs, the two divisions usually (but not always) have opposing actions on that organ (for example, one increases and one decreases the heart rate). The two divisions are known as the **craniosacral,** or **parasympathetic, division** and the **thoracolumbar,** or **sympathetic, division** (see Figure 12–11). The anatomical names represent the sites of origin from the central nervous system.

Parasympathetic Division The central connections of the parasympathetic, or craniosacral, division are with the brain and the sacral spinal cord. Only four of the twelve cranial nerves contain autonomic fibers as part of their outflow.

The **oculomotor nerve** (III) supplies the eye, specifically the ciliary muscles (which adjust the lens for near vision) and the pupillary constrictors (which reduce the amount of light entering the eye).

The **facial nerve** (VII) and the **glossopharyngeal nerve** (IX) supply the salivary glands and the mucous membranes of the head (in the nasal and oral cavities).

The **vagus nerve** (X), the "wanderer," supplies the thoracic and most of the abdominal viscera, including the heart, bronchioles, esophagus, stomach, small intestine, and the first part of the large intestine.

Branches of sacral nerves 2, 3, and 4 unite to form the **pelvic nerve** to pelvic structures, including the lower part of the digestive tract, the urinary bladder, and the genitalia.

Parasympathetic preganglionic neurons are relatively long. Their cell bodies are in nuclei in the brainstem or in the lateral portion of the gray matter of the sacral cord. The preganglionic fibers end in **terminal ganglia** located very close to their destination and, in some cases, actually within the walls of the organ. The postganglionic neurons are very short. Their cell bodies are in the terminal ganglia or walls of the organ (they are sometimes called *ganglion cells*). Their peripheral endings are on nearby muscle or secretory cells. All parasympathetic pathways have a synapse, but none is known to have more than one.

Sympathetic Division The thoracic and upper lumbar segments of the spinal cord give rise to sympathetic fibers. (Notice that several regions of the cord do not contribute to the autonomic nervous system.) The preganglionic neurons have cell bodies in the gray matter of the lateral horn, which is quite prominent in this part of the cord. The postganglionic neurons are relatively longer than those of the parasympathetic; their cell bodies lie in one of the ganglia of the sympathetic chain or in one of the outlying or collateral ganglia. The **sympathetic chain** is a paired string of ganglia lying on either side of the vertebral column and extending from the cervical to the sacral region. There is a ganglion by nearly every spinal nerve, which is connected to it by **communicating rami.**[3] Many sympathetic preganglionic neurons terminate in the chain and synapse with a postganglionic neuron, while others pass through the chain without synapsing and proceed to one of the collateral ganglia to synapse. The **collateral ganglia** are found near certain arteries.

In the cervical region the sympathetic chain has only three ganglia. The uppermost is the *superior cervical ganglion,* which is quite large and is located behind the ramus of the mandible. Fibers that synapse there arise in the thoracic region and ascend in the chain. The postganglionic fibers go to structures in the head. Among fibers that synapse in the other cervical ganglia are those to the heart and other thoracic structures.

A cross section of the spinal cord in the region of sympathetic outflow shows the relationships of the components (Figure 12–12). The preganglionic neuron arises from the lateral horn gray matter, leaves the spinal cord through the anterior root, and passes by way of a communicating ramus to the nearby sympathetic chain. Beyond the sympathetic ganglion several courses

---

[3]Sometimes the communicating rami are identified as gray rami and white rami. This distinction is based upon the fact that the preganglionic neurons are myelinated and form the white ramus, while postganglionic fibers are unmyelinated and form the gray ramus. The distinction, however, is largely academic, since it is not possible to tell them apart with the naked eye.

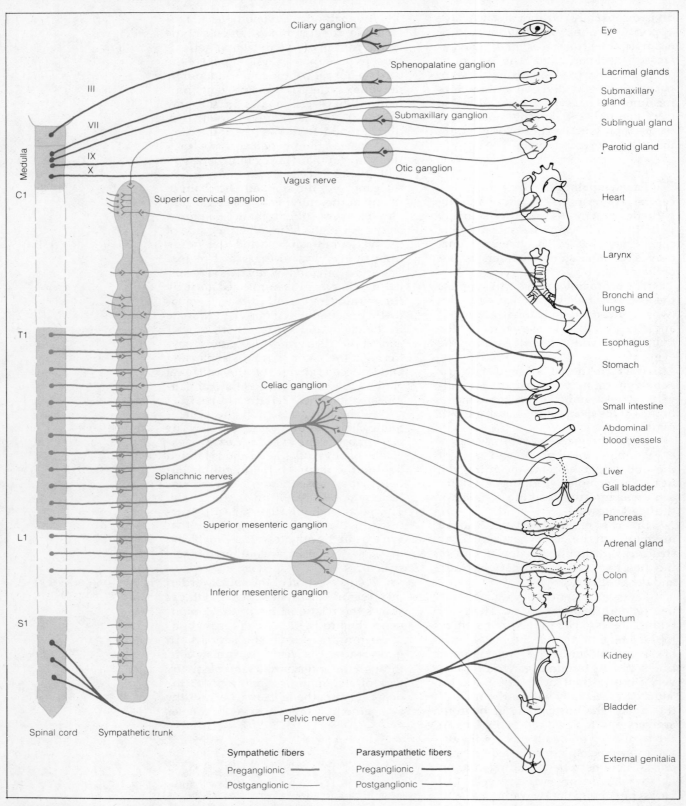

FIGURE 12-11   Distribution of autonomic efferent fibers.

are possible. The preganglionic neuron may:

1. Synapse in the chain. It may ascend or descend a few segments in the chain before synapsing, whereupon the postganglionic neuron returns to the spinal nerve via another communicating ramus and is distributed with that nerve. These fibers go to structures in the skin (blood vessels, sweat glands, and the erector muscles of the hairs).

2. Synapse in the chain, usually after ascending several segments. Many of the postganglionic fibers in this pathway form separate autonomic nerves that go directly to their destination. Others form plexuses along the arteries as they proceed to their effectors. This group includes the sympathetic innervation of structures in the head and thorax.

3. Pass through the chain without synapsing. The preganglionic fibers that emerge from the chain come together to form the **splanchnic nerves,** which go to one of several **collateral ganglia** to synapse. They are *celiac, superior mesenteric,* and *inferior mesenteric ganglia,* and are located at the origin of the artery with the same name. The postganglionic neurons form plexuses along the arteries as they proceed to their destinations. Parasympathetic fibers may intermingle in these plexuses, but there is no functional (synaptic) connection.

Sympathetic pathways also have only one synapse (with one exception, that to the adrenal medulla), and none is known to have more than one. All sympathetic pathways go through the chain, whether or not they synapse there (see Figure 12–11).

Function of the Autonomic Nervous System

The activities regulated by the autonomic nervous system are not the same as those governed by the somatic nervous system. The somatic innervation regulates the animal in relation to the external environment, while the autonomic nervous system is associated with functions related to regulation of its internal environment. In Chapter 1 it was pointed out that the

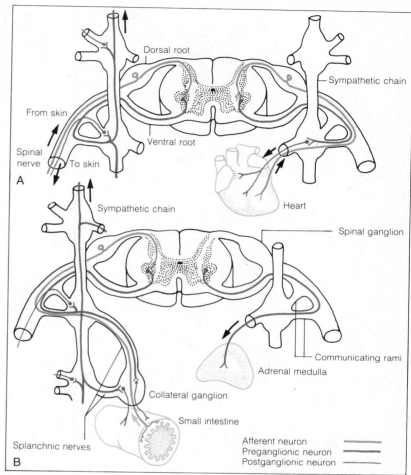

FIGURE 12–12 Pathways of sympathetic efferent fibers: A. To structures in the skin and in the thorax (heart). B. To structures in the abdominopelvic cavity and to the adrenal medulla.

individual cell, whether a one-celled organism or a cell of the quadriceps femoris, has certain needs which must be met by the environment in which it lives, and that the needs of all types of cells have much in common. The one-celled animal meets its needs by exchange across the cell membrane with its watery environment; the muscle cell meets its needs by exchange across the cell membrane with the interstitial fluid. Interstitial fluid in turn is serviced by exchange with the blood across the capillary membrane. The substances that the blood must deliver or take away are prepared and added, or removed and disposed of, by the organ systems. The autonomic nervous system, by exerting control over the organ systems, plays an important role in

TABLE 12-1   ACTIONS OF THE AUTONOMIC NERVOUS SYSTEM

Effector	Sympathetic (Adrenergic)	Parasympathetic (Cholinergic)
Cardiac muscle and pacemaker	Stimulate, accelerate	Inhibit, decelerate
Smooth muscle		
Skin (piloerectors)*	Contract	—
Digestive system (except sphincters)	Inhibit	Increase tone and motility
Urinary bladder	Inhibit	Contract (empties)
Bronchioles	Dilate	Constrict
Pupil	Dilate	Constrict
Blood vessels		
*Skin**	Vasoconstrict	—
Coronary	Vasodilate	Vasoconstrict
Skeletal muscles	Vasodilate (a few vasoconstrict and some are cholinergic).	—
Viscera	Vasoconstrict	Vasodilate (some, e.g., genitalia)
Glands		
Digestive	Secrete (generally low enzyme content)	Secrete (generally high enzyme content)
Salivary		
Pancreas		
Stomach		
Intestine		
Liver	Increase blood glucose	—
Adrenal medulla	Secrete	—
Sweat glands*	Secrete (cholinergic)	—
Lacrimal glands	—	Secrete

*Note that skin structures have no parasympathetic innervation.

maintaining the constancy of the internal environment; it is deeply involved in *homeostasis*. (This does not, however, mean that the autonomic nervous system is the only regulatory factor, since, as we shall see, endocrine controls are also of great importance.) This type of role illustrates the earlier statement that the autonomic nervous system is a *regulatory* or *integrative* system rather than an initiating one.

The regulatory role is carried out by different and often opposing actions of the two divisions, as can be seen from the actions of each division on specific organs shown in Table 12-1. As previously indicated, most structures are innervated by both divisions, except for those in the skin. And where there is a dual innervation, the actions

of the two divisions are opposing, except in some digestive glands. Even there, however, the actions are not identical, since the nature of the secretion produced differs in each case. It should also be noted that smooth muscle does not all respond in the same manner, not even all vascular smooth muscle.

After examining the kinds of actions produced, you might assume that the sympathetic division always stimulates and accelerates, and the parasympathetic division inhibits and slows activity. This common misconception should be avoided, because the facts simply do not support it, as will become evident upon further study of Table 12-1 and the following paragraphs.

Role of the Sympathetic Division

That the sympathetic nervous system is not essential to life is demonstrated by removing it surgically. Cats *sympathectomized* by severing the connections of the sympathetic chain manage very well in the controlled environment of the laboratory. They can carry on vital processes, run about, bear young, and lactate. Their problems seem to be related to stressful conditions, such as the presence of a strange dog or a room that is cold. They cannot erect the hair on their backs to register anger and fear or to improve insulation, and they seek out the radiator for warmth more consistently than do normal cats.

Examination of the list of actions of the sympathetic nervous system in Table 12–1 indicates that it produces responses that would be helpful in a stressful or emergency situation. The heart rate is increased. The bronchioles of the lungs are dilated, which facilitates the movement of air in and out of the lungs. The digestive system and urinary bladder are inhibited (in a time of acute emergency these activities can be postponed). Additional energy is made available by the release of glucose from the liver into the bloodstream. The blood vessels of the skin and viscera constrict, thus reducing the delivery of blood to inactive tissue, while blood vessels to the active skeletal muscle and heart dilate. Because its actions are associated with the mobilization and expenditure of energy necessary to meet an emergency, the phrase "fight or flight" has been applied as an overall description of what sympathetic nervous system activity prepares the organism to do.

The sympathetic division exerts the major control over blood vessels. The vessels in the skin and viscera, which make up the bulk of the neurally controlled vessels, lack a parasympathetic innervation. The tone of these vessels, and thus their resistance to blood flow, depends heavily upon the sympathetic nervous system, particularly since they contain multiunit smooth muscle which has little automaticity. Blood vessels to skeletal muscle make up a large portion of the vascular tree, but they receive few sympathetic and no parasympathetic fibers.

The sympathetic nervous system is capable of discharging as a unit, which would be desirable in an emergency situation. This is brought about through central connections in the brain and spinal cord, and it is reinforced by the secretion of the **adrenal medulla.** This is an endocrine gland which is stimulated by sympathetic fibers to secrete the hormone **epinephrine** (or *adrenaline*).[4] Epinephrine released by the gland enters the bloodstream and is circulated throughout the body. Its actions mimic those of the sympathetic nervous system in almost every organ. In addition, epinephrine causes dilatation of the blood vessels in skeletal muscle. Thus epinephrine secretion is a part of the sympathetic discharge, and it tends to reinforce the action of the nerves. This interrelationship is recognized in references to the **sympathoadrenal system.** Drugs and chemicals with sympathetic or epinephrine-like actions are known as *sympathomimetic* agents.

In view of the importance of the sympathetic nervous system in maintaining resistance in blood vessels, and hence in maintaining blood pressure, there was a time when it was quite popular to treat certain types of hypertension (high blood pressure) by surgically performing a partial *sympathectomy* (usually in one limb), to lower vascular resistance and therefore blood pressure. The measure is effective for a time, but the denervated vascular smooth muscle develops a heightened sensitivity to circulating epinephrine. It soon regains its tone and the hypertension returns.

Role of the Parasympathetic Division

The parasympathetic nervous system is not organized for a mass discharge, either centrally or by circulating chemicals. A quick look at its actions will make it apparent that this would be highly undesirable. The parasympathetic system, instead, is organized for

---

[4]In the United States, this hormone is called *epinephrine,* but in Britain and Canada it is *adrenaline.* In the U.S. "Adrenalin" is the brand name for a particular preparation of epinephrine. Both names recognize the gland as one located on or above the kidney.

more localized actions, relatively distinct from those of other parts of the division. The parasympathetic system is probably not essential to life, although this has not been strictly verified, because it cannot be surgically disrupted or severed as surely as the sympathetic. Most of the smooth muscle that is excited by the parasympathetic nervous system is of the visceral (single unit) type, which has a high degree of inherent rhythmicity and is less dependent upon neural control.

Central Connections Specific control mechanisms for various facets of autonomic control will be considered in more detail with the particular organ systems involved. However, there are some features, particularly the applications of some of the general properties of the reflexes, that are common throughout the autonomic nervous system. Because it is a reflex system, the reflex center determines the activity in the efferent pathway. Most of the autonomic reflex centers are located in the brainstem, many of them in the reticular formation. There are centers for the control of a specific process, and they are often coupled (as with the sympathetic center that increases heart rate and the parasympathetic one that decreases it). Neurons from these centers either descend as interneurons in the spinal cord and synapse with a preganglionic neuron in the lateral horn (sympathetic) or connect with neurons of the appropriate cranial nerve (parasympathetic).

There is reciprocal innervation between the paired sympathetic and parasympathetic centers, just as there is between the motoneurons to corresponding flexor and extensor muscles. An increase in heart rate is therefore due to facilitation of the sympathetic and inhibition of the parasympathetic fibers. There is often a certain degree of tonic discharge from one of the centers, even though the effector may have an automaticity of its own. This means that the tonically active center is the dominant one and is more important in the control of that particular activity. The parasympathetic center is tonically active and more important in the regulation of heart rate, while the sympathetic center is almost exclusively responsible for control of blood vessels.

Autonomic centers receive input from many sources, including other parts of the brain and receptors located in the walls of organs such as the urinary bladder and arteries.

Peripheral Connections The dual innervation in the autonomic nervous system introduces a new complication in transmission of impulses to an effector. It means that a single muscle or gland cell may receive innervation from two neurons with antagonistic effects. If impulses in one neuron excite the cell, impulses in the other must inhibit it, and this requires a different type of neuroeffector transmission.

In a somatic reflex, inhibition occurs centrally; the motoneuron is prevented from discharging impulses and there is no contraction. Central inhibition also occurs in the autonomic nervous system, where inhibition at a center results in fewer impulses in certain efferent neurons. But in the autonomic nervous system, there is also **peripheral inhibition,** a direct inhibition of the effector organ. *Stimulation* of the parasympathetic nerve to the heart (increasing the number of impulses in the vagus nerve) causes the heart to *slow*. This is in marked contrast to impulses in the sympathetic nerves to the heart, which increase its rate of contraction. The two kinds of neurons exert different actions on the cardiac effector cells. Since nerve impulses are all alike, that difference must be at the neuroeffector junction.

The mechanism of peripheral inhibition was described in 1920 by a German investigator, Otto Loewi, with some simple but significant experiments (for which he received the Nobel prize). He *perfused* (poured or pumped through) a frog heart with a physiological solution and then pumped the same fluid through a second heart (Figure 12–13). Stimulation of the vagus nerve to the first heart slowed and stopped the heart as expected, but after a short time the second heart also stopped. The only explanation was that the first heart produced some inhibitory substance which then entered the perfusing fluid to stop the second. Subsequent investigations showed the substance to be acetylcholine released at the vagal post-ganglionic endings; acetylcholine will inhibit cardiac muscle when applied directly. This discovery was important for two reasons: It was the first proof of peripheral inhibition, and it was also the first proof of a chemical mediator anywhere. (Identification of acetylcholine at skeletal neuromuscular junctions and of chemical transmitters at synapses was still years away.)

Other investigators studied transmission in the sympathetic pathway. They perfused the superior cervical ganglion, which contains the synapses between pre- and postganglionic neurons; they found that stimulation of the preganglionic fibers yielded acetylcholine, indicating that it was the transmitter at these synapses. Other studies in which chemicals or drugs were used as blocking agents identified acetylcholine as the transmitter at parasympathetic ganglia as well. Thus acetylcholine is secreted by the terminals of somatic motoneurons, sympathetic and parasympathetic preganglionic neurons, and parasympathetic postganglionic neurons. For this reason they are all termed **cholinergic fibers** (see Figure 12–14).

The only endings left, the sympathetic postganglionics, must be different, since if they released acetylcholine they would have the same effect as the parasympathetic endings. They were found to release a transmitter substance very much like epinephrine (or adrenaline), and these neurons are known as **adrenergic fibers.**[5] It is now known that there actually are two substances, epinephrine and norepinephrine, which are chemically very similar (see Chapter 13); they differ chiefly quantitatively in their effectiveness at certain sites. Epinephrine has a greater effect on heart rate, while norepinephrine has a greater effect on blood vessels. Although both are secreted by sympathetic postganglionic endings and by the adrenal gland, the proportions of each in the two secretions differ. In humans, sympathetic endings release mostly norepinephrine; the adrenal medulla secretes

---

[5]Exceptions are sympathetic fibers supplying skeletal muscle blood vessels and sweat glands in the skin; they are cholinergic.

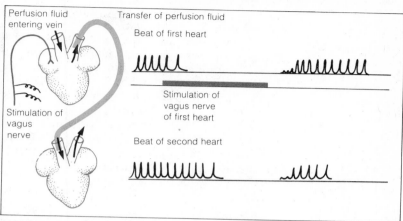

FIGURE 12–13   Demonstration of the existence of a chemical mediator. Stimulation of the vagus nerve slows the first heart and, later, the second heart. The effect could only be due to an inhibitory substance released by the first heart into the perfusing fluid reaching the second heart.

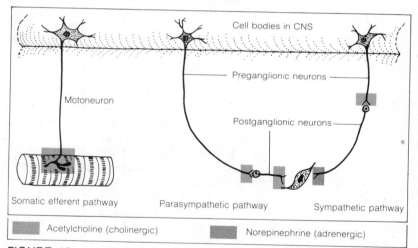

FIGURE 12–14   Chemical mediators at peripheral endings of somatic and autonomic neurons.

mostly epinephrine. The close relationship between the adrenal medulla and the sympathetic nervous system and the similarity of the effects of their secretions become readily understandable in view of the transmitter.

We have seen that all effectors—skeletal, smooth, and cardiac muscle and glands—are activated by the release of a chemical transmitter, and that the response of skeletal muscle can be altered by blocking the action of the transmitter or by interfering with its inactivation. The same things can be done in autonomic ganglia and in neuroeffector junctions. Curare is not very effective in the autonomic nervous system, but there is a group of drugs that effectively blocks transmission at sympathetic postganglionic junctions, another group that blocks parasympathetic postganglionic junctions, and a third group that blocks at autonomic ganglia. In addition, there are drugs that mimic the effects of the transmitters at the various junctions. The selective blocking is possible because the receptors that bind acetylcholine are not identical in skeletal muscle and autonomic ganglia or effectors.

Many of these drugs have important clinical uses. Those with parasympathetic action (parasympathomimetic) can be used when the digestive tract or urinary bladder lack tone, or if the heart suddenly begins to beat too fast. Drugs that block parasympathetic action at neuroeffector junctions, such as atropine and related compounds, are widely used before surgery, for example, to reduce the secretion of saliva and mucus. These secretions could be dangerous to the anesthetized patient, whose cough reflex (among others) is suppressed—the secretions could get into the lungs. Another use is during eye examinations, to block constriction of the pupil and accommodation. It is interesting to note that certain mushrooms are poisonous because of their parasympathomimetic action, and that one group of insecticides (the organophosphates) is a potent anticholinesterase; both can be counteracted by atropine.

There are innumerable sympathomimetic drugs, including epinephrine, and they, too, have many clinical uses. One use is to relieve the constriction of the airways that occurs during an asthmatic attack. Many sympathetic blocking agents have also been used, with varying success, to treat hypertension and other circulatory problems.

These are only a few of the many uses of drugs to modify transmission at peripheral junctions. Some of these agents also affect transmission at junctions in the central nervous system. Their actions are not as well understood, but many are widely used—and misused. They modify sensory and motor functions as well as behavior and perception, and range all the way from aspirin, tranquilizers, and pep pills to the hard drugs.

Hypothalamus and Visceral Control The hypothalamus is a relatively small area located in the walls of the third ventricle (see Figure 10–5). It consists of many nuclei which have connections with virtually every part of the central nervous system. Although it is small, the hypothalamus is involved in many different activities; more functions have been attributed to it than to any other structure of comparable size. Its actions can be said to fall into three general areas, although there is a great deal of overlap.

1. The hypothalamus is an important link in the pathways involved in emotional behavior. Destruction of certain portions of the hypothalamus, for example, produces an animal which, upon harmless stimulation, manifests all the signs of rage (growling, scratching, piloerection, and other signs of discharge of the sympathetic nervous system). Many behavioral responses to this and other emotional stimuli involve autonomic functions. The hypothalamus has important connections with the limbic system, a part of the brain that is particularly involved in emotions. Some of these aspects are discussed in the next section of this chapter.

2. The hypothalamus controls the pituitary gland, both the anterior and posterior portions, and therefore exerts a major control over most of the endocrine system. Much of this, too, is associated with autonomic functions. These relationships are discussed in Chapter 13.

3. The hypothalamus is the chief integrative center for the autonomic nervous system. It was once thought to be specifically an autonomic control center with sympathetic and parasympathetic sections. This was partly because stimulation in various parts of the hypothalamus could produce some autonomic responses. But rather than being organized around the autonomic nervous system, hypothalamic regulation is organized around certain processes or functions, such as the regula-

Focus

Are the Two Hemispheres Identical?

The two halves of the human brain, the cerebral hemispheres, appear at first glance to be identical masses of convoluted tissue joined by the corpus callosum and the commissures. The two hemispheres do, in fact, differ in structure and function: The patterns of convolution and of cellular organization differ, as do certain specialized functions served by each half of the brain. The two hemispheres work in the same way: Each one receives sensory data from, and sends motor commands to, structures on the opposite side of the brain. Yet people show distinct preferences for using one side of the body (generally the right) for any task that requires intricacy and coordination.

Other functional differences between the two hemispheres, such as those that underlie speech, are better understood than those of hand preference. In most people the left hemisphere contains two areas that are larger than their counterparts in the right hemisphere. These are Broca's area, in the frontal lobe just anterior to the motor area for the face and organs of speech, and Wernicke's area, in the auditory association area of the temporal lobe. They are connected by a bundle of fibers.

Broca's area is involved in the production of speech, in the appropriate and coordinated use of the speech mechanism. A lesion in this area causes difficulty in speaking and putting together a grammatical sentence; speech is slow and inarticulate, but surprisingly, there may be no difficulty in singing. An individual with lesions in Wernicke's area is able to speak with ease, but the words and sentences may not make much sense.

In other words, damage to Broca's area affects the act of speaking but has less effect on comprehension, while an injury to Wernicke's area interferes less with speaking than with the meaning of the words. Wernicke's area is also important in interpreting auditory and visual language stimuli and passing that information along to Broca's area, which then directs vocalization of the response.

The right hemisphere seems to be more important for the perception and production of music, for visual patterns and spatial relationships. This shows up when a person whose corpus callosum has been severed (a "split-brain" patient) is asked to identify an object presented to only one half of his or her visual field, and hence to only one hemisphere. When the left hemisphere sees the object the patient can name it; when the right sees it the patient can only match it with a similar object. The right hemisphere (left hand) is also better able to assemble blocks or copy a simple drawing (even in right-handed patients).

The right hemisphere also seems to be more involved in emotional responses. A person who suffers damage to the right hemisphere shows the same sense of loss and depression that one associates with any serious injury, but a person who suffers such damage to the right hemisphere may not even seem to care and fails to recognize emotions in others.

The differences between the two hemispheres are not normally apparent, as long as the corpus callosum and commissures are intact and the hemispheres can and do communicate with one another.

tion of body temperature. There are centers in the hypothalamus, for example, that control heat loss and heat gain (or production), which, through their balanced actions, maintain body temperature. Much of this control is exerted through participation of the autonomic nervous system, such as sweating or constriction and dilation of blood vessels. Note that temperature control is not exclusively autononomic, however, for shivering (heat production) is a skeletal muscle activity. Therefore, in spite of its importance to autonomic function, the hypothalamus also exerts some somatic control.

Other important hypothalamic centers control the volume of the body fluids, thirst, food intake, and certain aspects of sexual behavior.

Hypothalamic regulation is mediated through fibers that lead to centers in the brainstem, such as those in the medulla that control blood pressure (constriction of blood vessels) or heart rate. Fiber tracts from these centers end on sympathetic and parasympathetic preganglionic neurons in the spinal cord and brainstem. Hypothalamic control is also exerted through its regulation of the pituitary gland and the hormones it produces.

The hypothalamus relies on a wide variety of afferent inputs. Some come from various visceral and somatic receptors. There are specialized receptors within the hypothalamus that detect certain things about the blood: its temperature (for temperature regulation), its osmotic pressure (for fluid balance), and glucose utilization by the receptor cells (for control of food intake). It is also influenced by the concentration in the blood of the hormones whose secretion it controls. In addition, and perhaps more importantly than we may realize, much of the afferent input for regulation of visceral function comes from pathways mediating emotional responses.

Many of the autonomic, somatic, and endocrine mechanisms under hypothalamic control are those concerned with maintaining the integrity and stability of various aspects of the internal environment. For this reason the hypothalamus is closely identified with homeostasis, perhaps more than any other single structure or organ.

HIGHER FUNCTIONS OF THE CENTRAL NERVOUS SYSTEM

The central nervous system activities covered by the designation of "higher functions" are neither simple nor simply described. They include the neural mechanisms that enable an animal to evaluate environmental signals and to develop complex, goal-oriented behavioral responses. They involve far more than the perception of sensory stimulation or the performance of skilled motor activity, since implicit in evaluating signals, developing responses, and reaching a goal (even such a basic one as finding and eating food) are the commonly recognized higher functions of learning, remembering, planning, problem solving, and decision-making. And these activities are colored by the emotions, which also come under the umbrella of higher functions. The ability to carry out all these tasks requires the existence of a state that we call "consciousness." Consciousness is closely associated with the diurnal sleep-wakefulness pattern, but it is not identical with it.

In general, the higher functions are more developed in primates, particularly in humans. They are extremely difficult to study, however, partly because of their complexity and the high degree of interaction between them and between their underlying neural structures. Furthermore, it is in these areas that the differences between humans and other animals appear the greatest. Certain types of controlled experiments that could provide needed data are morally or ethically impossible to do on humans, yet it is difficult to design animal experiments in which behavioral responses unquestionably reveal the state of some rather abstract parameter. Fortunately, there are now some tools and techniques that can provide objective information upon which to base conclusions.

Sleep

Considering how behavioral and electroencephalographic patterns are related to consciousness and arousal inevitably leads to questions about sleep. According to the EEG descriptions, sleep might seem to be unconsciousness, but the two are not identical. An animal can be aroused readily from sleep, but not at all from a coma.

If you are average, you spend about 30 percent of your life sleeping. Sleep is a puzzling phenomenon, and a great deal of research has been done on it, but it is still difficult even to give an adequate definition. We know neither exactly what sleep does for the individual nor exactly what the lack of it does. A person deprived of sleep suffers impaired efficiency, but this could be due to lack of physical rest.

When you become drowsy and then fall asleep, your EEG pattern shifts to one of larger, slower waves which, in deep sleep, may become quite prominent, suggesting a synchronized discharge pattern. This is sometimes called **slow-wave** or **synchronized sleep.**

At certain times during sleep the EEG shows signs of cortical activation or desynchronization of the EEG. There is an overall reduction of muscle tone at this time, although there occasionally are muscle twitchings. The most characteristic feature is bursts of rapid movements of the eyes. Such sleep is known as **desynchronized** or **rapid eye movement (REM)** *(paradoxical)* **sleep,** in contrast to slow-wave sleep, which is sometimes called **nonrapid eye movement (NREM) sleep.** These bouts last for some minutes and recur several times during the night. Apparently it is during the periods of REM sleep that dreaming occurs, although individuals rarely remember their dreams unless awakened at this time. Individuals awakened each time they show REM sleep, however, tend to become anxious and irritable and, once allowed to sleep without interruption, show an increased amount of REM sleep for a time. It has been said that the amount of REM sleep, and the accompanying dreaming, is more important to the well-being of the individual than the total amount of sleep, but in fact both kinds are necessary.

The actual cause of sleep is not known. Certainly the reduction of sensory input and the reduced cortical stimulation from the reticular activating system contribute, but they cannot be the whole story. Nearly everyone has gone to sleep under the most unfavorable conditions, with a great deal of sensory input (noise, light, movement, etc.), and has as well, perhaps, experienced a frustrating failure to sleep under optimal environmental conditions. It is known that the neural genesis of the slow-wave sleep and REM sleep are not the same. A synchronizing mechanism associated with slow-wave sleep has been identified in the lower brainstem, while periods of desynchronized or REM sleep begin with bursts of activity in nuclei in the pontine (i.e., of the pons) portion of the reticular formation.

Emotions

The emotions, like most of the activities that come under the heading of the higher functions, cannot be precisely localized; no specific brain structures can be identified as necessary for them. This may be due to the fact that of their two principal aspects—the internal awareness of the "feeling" and the external manifestation of that feeling—only the latter can be easily studied. Particularly with animals, experimental efforts must deal largely with the external indicators of emotions, in terms of behavior in particular situations. Stimulation or destruction of any of several areas of the brain can alter emotional behaviors, but there are two areas or systems that seem to be particularly involved. One is the hypothalamus, a few of whose nuclei seem to be concerned largely with emotions. The other is the chiefly cerebral part of the rhinencephalon known as the **limbic system** (Figure 10–11). Limbic structures are quite well developed in species such as humans, in whom the sense of smell is not of great importance. They include medially placed cortical and subcortical structures and the fiber tracts connecting them with one another and with the hypothalamus.

Stimulation or ablation (removal) of these structures can produce or eliminate a great variety of emotion-associated behavior patterns, such as rage or docility, fear or lack of fear, hypersexuality, hunger, and satiety. The hypothalamus seems to be a central component through which the effects of the limbic system are funneled. It is probably significant that the hypothalamus strongly influences autonomic activity and the secretions of a number of hormones, since these are important parts of the behavioral manifestations of emotion.

Learning and Memory

Learning, intelligence, and memory are even less well localized than the emotions. Learning represents a change in behavior as a result of previous experience, and it thus involves memory of that previous experience. Intelligence lies essentially in the ease with which the learning occurs, but we usually also include the abstract qualities of reasoning, planning, and problem solving in our concept of intelligence. Learning is generally considered to be a cortical activity, although some of the simplest types of learning can eventually be acquired by a decorticate animal. Neither the process of learning nor the retention of a learned behavior seems to be associated with any one part of the cerebrum. Based on ablation experiments, memory seems related more to the amount of cortex retained than to the presence of specific areas.

Intelligence was once thought to be a function of the prefrontal cortex. This idea was probably due to the fact that this area of the brain is most highly developed in those species considered to be most intelligent. Experimental evidence and clinical experience do not bear this out, however. Removal of the frontal lobe (*frontal lobectomy*) or severing certain of its connections (*frontal lobotomy*) does not produce a moron or an idiot. Rather, it usually produces personality changes often marked by lack of concern for others or for the consequences of an act and by an inability to plan and organize.

One is thus left with the frustrating realization that the most complex higher functions of the brain are likely to be whole-brain activities. Not only do the myriad interconnections make it difficult to determine the contribution of any one part, but the often remarkable recovery from even massive lesion-induced deficits indicates that more than one location can carry out a particular operation and that each location is involved in many functions. Such overlap and duplication of function allow many interpretations of the experimental data, and it will no doubt be many years before the discrepancies are resolved and a single theory settled upon.

BRAIN METABOLISM

The central nervous system is never truly at rest, although the level of activity may fluctuate widely. Indeed, the absence of electrical activity in the brain (as indicated by the EEG) is a more reliable criterion of death than is cessation of heartbeat or breathing.

As one ascends through the central nervous system, the metabolic rate in general increases; cells of the cerebral cortex have a greater requirement for oxygen and nutrients and a greater sensitivity to certain drugs and anesthetics than cells of the medulla. When the brain is deprived of oxygen (either by insufficient oxygen in the blood or by inadequate blood flow), the first areas to be affected are those necessary for consciousness, perception, voluntary movement, and the higher functions. It may seem to be a poor arrangement to lose consciousness first, but it is important to note that the vital centers controlling heart rate, blood pressure, and breathing are in the medulla and are not threatened immediately. In this way, it is possible to anesthetize an individual without inactivating these essential mechanisms. Prolonged deprivation (for just a few minutes) of oxygen or nutrients can result in irreversible damage from which the cortical cells may never recover, although the heartbeat and breathing continue, and the individual is said to be alive.

The high metabolic rate of brain cells emphasizes the great importance of maintaining the cerebral blood flow. But there are two other properties of brain tissue that make the blood flow even more critical:

1. Not only does the brain need more oxygen than many tissues, but it needs it continually. Unlike muscle, brain cells cannot go into oxygen debt. Skeletal muscles can contract for a time in the absence of oxygen and repay the debt later with increased oxygen con-

sumption, but brain tissue cannot do this. Without adequate oxygen, it promptly ceases to function.

2. Brain tissue obtains its energy almost exclusively from the oxidation of glucose. But unlike muscle, neural tissue stores very little glycogen for conversion into glucose. Glucose, like oxygen, must be supplied continually by the blood. Failure to do so results in impaired function of the neurons involved, and a low level of glucose in the blood can cause loss of consciousness, though it is relieved quickly by glucose administration. Some of the metabolic mechanisms responsible for maintaining a suitable level of glucose in the blood are discussed in Chapter 24.

CHAPTER 12 SUMMARY

Skeletal muscles are controlled by the somatic nervous system. A skeletal muscle nerve contains *alpha motoneurons* to extrafusal fibers, *fusimotor neurons (gamma efferents)* to intrafusal fibers, afferent fibers from the muscle spindle receptors, and probably afferents from Golgi tendon organs. Afferents from muscle spindles synapse with alpha motoneurons for stretch reflexes, and send branches to the cerebellum (unconscious proprioception). Stimulation of gamma efferents causes contraction of intrafusal fibers, which increases the response of muscle spindle receptors and raises the excitability of alpha motoneurons. This is the *gamma loop,* important in the control of alpha motoneuron excitability.

Nerve fibers in most pathways descending from centers in the brain synapse with both alpha and gamma motoneurons. The corticospinal or *pyramidal tracts* form a direct pathway from the motor cortex to motoneurons in the spinal cord. Indirect pathways (*extrapyramidal tracts*) include the rubrospinal tract from the red nucleus mainly to motoneurons of flexor muscles; the vestibulospinal tract from vestibular nuclei to motoneurons of antigravity (extensor) muscles; and the reticulospinal tracts from the brain stem reticular formation mainly to motoneurons of antigravity muscles. The brain controls the various aspects of skeletal muscle activity by its influence on these descending pathways.

The reticular formation is of major importance in determining the excitability of motoneurons to the postural (antigravity) muscles. The formation contains a small *inhibitory area,* stimulation of which generally reduces excitability of these motoneurons and inhibits the stretch reflex. The inhibitory area receives input from higher motor centers and is ineffective if these connections are cut. A larger *facilitatory area* increases the excitability of the motoneurons and stretch reflex; it is less dependent upon input from higher centers.

The role of the cerebellum is the coordination of muscle contractions. It receives information from muscle spindles and the cerebral cortex, and it sends information to nuclei of the descending fiber tracts. It probably is a center that compares what *is* with what *should be* in muscle contraction, and brings about adjustments to correct any errors.

The basal ganglia are believed to be involved in movement patterns and incidental movements. They receive information from all parts of the cortex and other areas as well, and send information back to the cortex and extrapyramidal centers.

The larger neurons in the corticospinal tracts are responsible for the finely controlled voluntary movements of distal parts (fingers). Rubrospinal tracts may contribute to control of movements of more proximal muscles and inhibit antigravity muscles.

Postural reflexes are based on stimulation of motoneurons of antigravity muscles based on input from muscle spindles (and the gamma loop), from cutaneous touch and pressure receptors, from visual pathways, and from the vestibular receptors, which are involved in the righting reflexes.

Smooth and cardiac muscle and glands are controlled by the **autonomic nervous system** (ANS). As it has no voluntary component, the ANS operates on strictly a reflex basis. The afferent side of ANS reflexes is similar to that of somatic reflexes.

The ANS efferent pathway has two divisions: the *parasympathetic division* which arises from the brainstem and sacral spinal cord, and the *sympathetic division* which arises from the thoracic and upper lumbar portions of the spinal cord. In each division the peripheral pathway consists of a preganglionic and a postganglionic neuron connected by a synapse located in a ganglion outside the CNS. Many effectors receive innervation from both divisions, and their actions are often opposite—one division stimulates the effector and the other inhibits it. Opposing actions are possible because of different chemical transmitters at the postganglionic endings. The parasympathetic postganglionic transmitter is *acetylcholine,* and these fibers are said to be **cholinergic.** The sympathetic transmitter is mainly *norepinephrine,* and sympathetic postganglionic fibers are said to be **adrenergic.** Norepinephrine is very similar to epinephrine, a hormone secreted by the adrenal medulla. The actions of the sympathetic nervous system are virtually identical with those of epinephrine, and in fact the adrenal medulla is stimulated by the sympathetic nerves and is an integral part of sympathetic action.

The sympathetic nervous system may be thought of as a mobilizing system whose actions aid in meeting emergencies; it increases heart rate and blood pressure, raises blood glucose, and frequently discharges as a unit. The parasympathetic division is not organized to function as a unit, and its actions are more restorative in nature. It is not correct to think of the sympathetic division as excitatory and the parasympathetic as inhibitory, for parasympathetic fibers increase activity of the digestive tract and cause contraction of the urinary bladder.

Sleep is closely related to cortical arousal. During sleep the EEG typically shows the large, slow waves of an unaroused brain and reticular formation. This is interrupted periodically by intervals in which the EEG is that of cortical arousal and there are rapid movements of the eyes. This is *rapid eye movement* (REM) sleep, in contrast to regular sleep, which is *nonrapid eye movement* (NREM) or slow-wave sleep. A certain amount of REM sleep seems to be necessary, perhaps because dreaming seems to occur then.

Emotions are difficult to study experimentally, but they seem to involve pathways from the *limbic system* (rhinencephalon), and certain nuclei in the hypothalamus. Learning, memory, and intelligence are not localized in a particular part of the brain, although they are mostly cortical functions.

The brain, especially the cortex, has a high rate of metabolism and a high oxygen requirement. The brain needs a continual supply of oxygen because it cannot incur an oxygen debt. It depends almost exclusively upon glucose as a source of energy and needs a continuous supply of it since it stores very little glycogen.

STUDY QUESTIONS

1   What would be the effect on motor function if the anterior roots were cut? the posterior roots?

2   What is the role of the gamma efferents and the gamma loop in control of skeletal muscle?

3   What are the pyramidal and extrapyramidal tracts? How do they differ anatomically and functionally?

4   What aspects of motor function do you associate with the brainstem reticular formation? the cerebellum? the vestibular apparatus and nuclei? the basal ganglia? the motor cortex?

5   Why is an animal likely to have increased muscle tone and exaggerated stretch reflexes when the brainstem is sectioned at the level of the midbrain?

6   Compare and contrast the autonomic and somatic nervous systems. What kinds of effectors are innervated by the autonomic nervous system? What kinds are not?

7   What effectors are innervated by only one division of the autonomic nervous system? Do you see any important implications in this?

8   Make a diagram of a cross section of the spinal cord and draw the pathway of sympathetic fibers that would supply the heart, the small intestine, and a blood vessel in the skin.

Could you draw the parasympathetic pathway to the heart or intestine on this same cross section? Explain.

9 To inhibit skeletal muscle contraction, the number of impulses in the motor nerve is reduced, but in cardiac muscle and intestinal smooth muscle inhibition can be brought about by stimulating certain neurons. How can

this be? How does inhibition come about in smooth muscle of blood vessels?

10 The sympathetic division is often associated with meeting emergencies, and thus is often viewed as an excitatory system that increases or stimulates most processes. Comment on the validity of this view.

13

Nature of Hormones and Hormone Action

The Pituitary Gland and Hormone Control

Endocrine Glands and Their Secretions

The Endocrine Glands as a Control System

Preceding chapters have been devoted to the role of the nervous system in the control of body functions. As important as it is, and as widespread as its actions are, the nervous system does not, however, control everything. There is another control system, the endocrine system, whose actions are also vital to our well-being and, indeed, to our very existence.

Most of the body's systems consist of several anatomically connected organs and structures organized to perform important and specific functions for the entire body. The organs of the endocrine system seem to be relatively independent of one another, anatomically and functionally. Though they are not an organ system in the usual sense, the endocrine glands do share a common mode of action: instead of conducting impulses, they secrete hormones which affect many cell functions that are not under neural control. Together the endocrine glands constitute a control system that parallels, but does not duplicate, the nervous system.

In this chapter we discuss how hormones act and examine the characteristics common to all endocrine controls. The individual glands and their hormones are discussed in a rather general way, with the specific details added later in the sections covering the processes that each hormone helps to regulate.

TABLE 13–1 THE ENDOCRINE GLANDS

Pituitary gland (hypophysis)
anterior lobe (adenohypophysis)
posterior lobe (neurohypophysis)
intermediate lobe

Adrenal glands
adrenal medulla
adrenal cortex

Thyroid gland

Parathyroid gland

Pancreatic islands (islets of Langerhans)

Gonads
testis
ovary

Placenta

Digestive tract (stomach, small intestine)

Kidney

Pineal gland (?)

The endocrine system includes the glands listed in Table 13–1. All **endocrine glands** act by secreting their hormones into the bloodstream for distribution throughout the entire body. In contrast, **exocrine glands** such as the salivary glands or sweat glands release their secretions directly (or via a duct) onto the surface of the body (or one of its inward extensions). An endocrine gland is therefore considered to be a specific organ whose bloodborne secretion produces a specific action on an organ (the target organ) at some distance from the gland itself.[1]

NATURE OF HORMONES AND HORMONE ACTION

Hormones affect all types of cells, not just the usual effectors. They do so chiefly by altering a basic cell activity in some way, and this affects the rate of a particular process that may be an essential part of the body economy. The results from hormonal actions are many

and varied, but they can be fitted into four broad categories:

1. Hormones help regulate the energy supply, including processes by which the energy is obtained from foodstuffs (or stored reserves). Thus they help control the digestive system, and they are of major importance in regulating metabolism and energy release from foodstuffs after they have been digested and have entered the bloodstream.

2. Hormones help control certain properties of the extracellular fluid (the internal environment). Specifically, they help control the metabolism of water and electrolytes; that is, they determine the fate of much of the sodium, potassium, calcium, phosphorus, and water in the body.

3. Hormones help us cope with adverse conditions or stresses, including such adversities as cold, heat, dehydration, trauma, blood loss, and emotional stress, in terms of withstanding the stress, fighting it, or escaping from it.

4. Hormones help regulate such basic aspects of life as growth, development, and reproduction.

Formation and Transport of Hormones

The chemical nature of many hormones is known, including their molecular structure. Many of those identified are protein derivatives, either *amines*, *peptides*, or *polypeptides*. Amines are small molecules, essentially an amino acid with the carboxyl (acid) group removed and perhaps a small side chain added. Peptides consist of a few amino acids, and polypeptides are longer chains of amino acids. They have many of the properties of proteins, in terms of the kinds of reactions they can enter into, but, of course, they are much smaller molecules.

Protein-related hormones are produced in the manner described in Chapter 2 for general protein synthesis. The short amino acid chains are assembled on ribosomes of rough endoplasmic reticulum and are transported in the ER to the Golgi apparatus where they are packaged into secretion gran-

---

[1]By this definition, such important chemical regulators as acetylcholine or carbon dioxide are excluded, since acetylcholine is not carried in the blood to a distant site and carbon dioxide is not secreted by a specific organ. They remain chemical regulators, but they are not hormones.

ules to be held until the proper stimulus causes their release from the cell.

The other major group of hormones are *steroids*. These hormones contain the basic steroid nucleus (a specific molecular configuration), so they are all quite similar, but they differ in slight but important specific groups attached at certain sites (see Figure 13–7). One important property of steroids is that they are lipid derivatives so they are fat soluble and thus can diffuse through cell membranes. Steroid hormones are synthesized from cholesterol (a widely distributed steroid) partly in mitochondria and partly in smooth ER. Important steroid hormones are those of the adrenal cortex, ovary, testis, and some of those produced by the placenta.

Since hormones reach their destination by traveling in the bloodstream, much attention has been given to the concentration of hormones in the blood. Many of the advances in endocrinology have come on the heels of better ways to measure the hormone levels in the blood. This is often a difficult problem for, although hormones are potent substances, they are present in very small amounts.[2]

Even knowing the exact concentration of a hormone in the blood does not tell us very much, because hormones do not exert their action in the blood; the blood is merely a transport system for most of them. Hormones in the blood are on their way either *to* some place or *from* some place. The amount of hormone in the blood at any given time depends upon the rate at which it is entering the blood (rate of secretion)

and the rate at which it is leaving the blood, either by excretion or inactivation. Many hormones are inactivated in the liver or elsewhere, and many are excreted rather rapidly in the urine. In fact, the half-life (the length of time for one half of a given dose to disappear from the blood) varies from a few minutes to a few hours, and averages only 10–30 minutes.

The fact that hormones are removed so rapidly suggests that in order to maintain a blood level, production must be continual. This is the case, for they are produced continually, but not necessarily at a constant rate. Stimulation of an endocrine gland increases the ongoing rate of production, and inhibition reduces it. The need for continual production is all the more important when one realizes that, with one or two exceptions, we do not store hormones to any great extent. Except for the thyroid hormones, only a one- or two-day supply is on hand.

Most hormones are bound to proteins as they are transported in the blood. It is a reversible condition, and an equilibrium is maintained between the free hormone and the bound hormone:

Free hormone + plasma protein
$\rightleftarrows$ Protein-bound hormone
Action

Some hormones are nearly 95 percent bound while others are more nearly half and half. Only the free portion can act, or be acted upon. When some of the free hormone is inactivated or leaves the bloodstream, more is released from the protein and the free/bound equilibrium is maintained. Such binding serves a purpose, for bound hormones are not available to diffuse out of the blood vessel or to be excreted in the urine.

Mechanisms of Hormone Action

Depending upon how one chooses to classify or count, there may be 40 to 50 different hormones with many different actions, but surprisingly, of those whose mechanism of action is known, all seem to involve only two or three different mechanisms.

---

[2]Some, such as amines (epinephrine), can be measured directly by chemical analysis. Some are detected by *bioassay*, a method used to measure a hormone when its chemical nature is not known. A bioassay consists of comparing a biological effect (such as the increase in the weight of an organ) of a sample containing an unknown amount of a hormone sample with the effect of a known amount of that hormone. A newer, extremely sensitive method is the *radioimmunoassay*, which makes use of the fact that proteins are bound firmly by specific antibodies. A sample containing an unknown amount of a hormone would interfere with the binding of a known amount of a pure sample. By determining the extent of the interference, the amount of hormone in the unknown sample can be determined. To do so, the pure sample carries a radioactive label, which can be measured very accurately. This method is applicable to protein and to peptide hormones.

As far as is known at present, all hormone actions begin with the combining or binding of the hormone with a specific receptor site of the target cell, in a manner somewhat similar to the binding of transmitter substances at neuromuscular junctions and synapses. Therefore, a hormone can only act on a cell that has receptor sites specifically for that hormone. This is the way in which the hormone "recognizes" its target cell (or vice versa), and it is the mechanism that limits the action of a hormone to certain target cells. The receptors have been described as either fixed receptors or mobile receptors.

Fixed Receptors For hormones that bind to **fixed receptors,** the receptors are on the outer surface of the cell membrane. The hormone binds there and never enters the cell. The hormone-receptor combination serves to activate an enzyme that is also in the cell membrane, but presumably on its inner surface (Figure 13–1A). This enzyme, **adenyl cyclase,** acts upon adenosine triphosphate (ATP) in the cell cytoplasm, converting it to a cyclic form, **cyclic AMP (cAMP):**

cAMP——Cyclic Adenosine 3' 5' Monophosphate

Cyclic AMP is inactivated rather quickly, but before that happens, cAMP causes an action which, in most if not all cases, is the activation of a **protein kinase.** Protein kinase describes a class of enzymes that activates another enzyme by transfering phosphate (from ATP) to it. The action that occurs when a protein kinase is activated depends upon the action of the enzyme it activates. Therefore, the effect of a hormone (which is specific) depends upon the action of a particular protein kinase (which is also specific). The link between them is the adenyl cyclase and cyclic AMP, and they are the same in many hormone actions.

Because cAMP is inactivated rather quickly, more hormone is needed and

FIGURE 13–1 Mechanisms of hormone action. A. Fixed receptor model. B. Mobile receptor model.

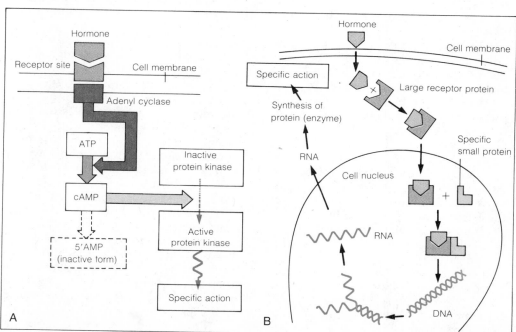

more cAMP must be produced to continue the action (in much the same way that additional packets of acetylcholine must be continually released if a muscle contraction is to be maintained). Epinephrine is one of about a dozen hormones known to act by way of cAMP. In fact, studies on how epinephrine causes glucose to be released from the liver led to the discovery of the role of adenyl cyclase and cAMP in the first place. Cyclic AMP is sometimes called the "second messenger," the hormone being the "first messenger."

Mobile Receptors **Mobile receptors** are involved in the action of steroid hormones. Recall that as lipid substances, steroids can pass through the cell membrane. The receptor site for steroid hormones is in the cytoplasm of the cell, and they diffuse into the cell and bind to the site (Figure 13–1B). The hormone-receptor combination then moves into the nucleus of the cell where it combines with a smaller protein which enables it to have an action on the DNA of the nucleus. The DNA of a particular gene unwinds leading to the synthesis of mRNA that carries the code for a particular protein. The mRNA moves to the cytoplasm, and that protein is synthesized at the ribosomes. The new protein is generally an enzyme to catalyze a particular reaction.

In either case, the varied effects of a hormone, whether brought about by fixed or mobile receptors, result in the presence of more active enzymes, thus accelerating (or blocking) a particular cellular reaction, which we describe as the action of that hormone.

Other Possible Mechanisms of Action A third mode of action involves an alteration in the properties of the cell membrane, so as to increase the permeability of the membrane to a specific substance. One of the best-known examples is the hormone insulin, which increases the rate of entry of glucose into the cells. Some hormones that act in this way do so by way of cAMP, but others do not and their mechanism of action is not understood.

A final possible mechanism of action has been postulated. It is not understood yet, but it could explain the fact that some hormones alter the effectiveness of another hormone. It is suggested that the action may come about by an effect on the receptor sites. If hormone A destroys receptors for hormone B, or occupies them leaving no place for hormone B to bind, the effectiveness of hormone B will be greatly reduced. Some hormones have what is called a *permissive action.* They do not bring about a particular effect by themselves, but their presence is necessary for another hormone to be fully effective. One possible explanation (there are others) might be that the permissive action of hormone A could have a favorable effect on the binding sites for hormone B.

Prostaglandins and Hormone Actions Prostaglandins are among the most confusing substances we will encounter in the body! The confusion begins with their name. Prostaglandins were first discovered in semen, the secretion of the male reproductive tract, and were presumably produced by the prostate gland, hence the name. It turned out that the seminal vesicles are the chief source of prostaglandins in semen, but the name has stuck. Prostaglandins are not hormones, but they are often considered with them (partly for lack of a better category), and their actions are related to hormone action in several ways.

Prostaglandins are synthesized by most, if not all, cells of the body. There are more than a dozen different prostaglandins, differing only slightly in their molecular configuration (and in some actions). They are all derivatives of a 20-carbon fatty acid and are therefore lipid-soluble and able to diffuse through cell membranes. Prostaglandins are extremely potent substances, but are present in very small amounts, more so in some tissues or at certain times. They are present in the blood, probably entering by diffusion rather than actually being secreted into it, but they are rapidly inactivated, especially in the lungs, liver, and kidney.

Prostaglandins are almost as

widely distributed as ATP, but their actions are so varied that it is almost impossible to find a common thread among them. Since they are produced by virtually all cells, and since their actions seem to be local (that is, within "diffusing distance" rather than requiring a transport medium), it is assumed that they exert their effects locally—in the cell or its immediate vicinity. They have actions on some part of nearly every organ system, but their actions on the reproductive, cardiovascular, and digestive systems seem to have received the most attention. One of their actions, for example, is to inhibit smooth muscle in blood vessels, which dilates the vessels and reduces the blood pressure. But they also cause contraction of other smooth muscle, such as in the digestive tract and uterus. These and other actions have prompted investigations into their possible value for a number of applications ranging from treatment of high blood pressure to inducing labor.

When all the effects of the prostaglandins are tabulated, it is noted that most if not all actions are those in which cAMP is involved. Prostaglandins increase the content of cAMP in most cases, but there are a few tissues in which they decrease the amount of cAMP. By altering the formation of cAMP in cells the prostaglandins can affect the response of those cells to a hormone whose action involves cAMP. Prostaglandins may turn out to be local modulators of cAMP-induced reactions. It remains to be seen what controls the production of the prostaglandins.

Control of Hormone Secretion

The rate of secretion of an endocrine gland is regulated by neural and/or chemical means. A number of endocrine glands are innervated by autonomic fibers, chiefly parasympathetic, but for the most part these nerves exert only secondary control. Cutting the nerve does not have much effect on the secretion of the gland, and stimulation may only raise the level of secretion slightly.

One endocrine gland, however,

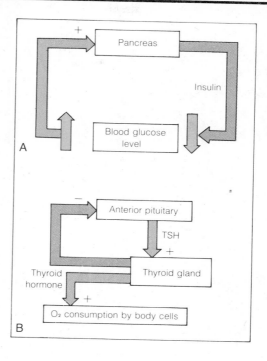

FIGURE 13-2
Negative feedback control of hormone secretion. A. Direct action, as in control of insulin secretion by the blood glucose level. B. Indirect action, as in control of thyroid hormone secretion by its effect upon secretion of thyroid stimulating hormone (TSH) by the anterior pituitary.

receives an important nerve supply. That is the adrenal medulla, which receives sympathetic innervation by a preganglionic neuron, and secretion by the adrenal medulla is part of the response of the sympathetic nervous system. (Recall that epinephrine, the hormone produced by the adrenal medulla, resembles norepinephrine, the sympathetic postganglionic transmitter, both chemically and in the action produced.) In addition, the release of hormones from the posterior lobe of the pituitary gland is directly controlled by neurons from the brain in a unique manner, to be discussed below.

The major control of endocrine glands is chemical in nature. It is exerted by what is known as a **negative feedback mechanism.** The simplest form is a direct feedback such as in the control of the secretion of insulin. The overall effect of insulin is to lower the blood glucose level. Elevation of the blood glucose concentration causes increased production of insulin, which causes glucose to leave the blood and enter cells. As the level of glucose in the blood falls, there is less stimulation of the pancreas, insulin secretion declines, and the blood glucose level begins to rise once more (Figure 13-2A). When

tion is a rather complex subject that requires further consideration for two reasons. First, the anterior pituitary gland produces hormones that stimulate secretion by other endocrine glands, and it is regulated by the secretions of those glands (by negative feedback). Second, the pituitary is an important link between the nervous system and the endocrine system.

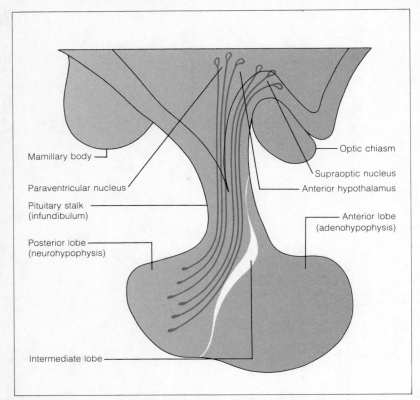

Mamillary body

Paraventricular nucleus

Pituitary stalk (infundibulum)

Posterior lobe (neurohypophysis)

Intermediate lobe

Optic chiasm

Supraoptic nucleus

Anterior hypothalamus

Anterior lobe (adenohypophysis)

FIGURE 13-3   The pituitary gland and its innervation.

THE PITUITARY GLAND AND HORMONE CONTROL

The pituitary gland occupies a unique position among endocrine glands because of its role in the control of secretion by other endocrine glands. It also has a unique association with the nervous system—part of it (the neurohypophysis) is actually part of the diencephalon. The anterior portion (adenohypophysis) also has close ties with the brain. These connections regulate secretion by the pituitary, and therefore also the secretion by those glands controlled by the anterior pituitary. Because the pituitary gland has a special role in the overall regulation of hormone secretion, it is considered in some detail at this time.

Anatomy of the Pituitary Gland

The pituitary gland **(hypophysis),** safely hidden in the sella turcica of the sphenoid bone (see Figure 10–3), is three essentially separate glands, the anterior, posterior, and intermediate lobes (Figure 13–3). The **posterior lobe,** or **neurohypophysis,** is an outgrowth of the brain. It is connected directly to the brain by the **pituitary stalk** *(infundibulum),* and a tiny extension of the third ventricle of the brain can often be traced into the stalk and into the neurohypophysis itself. The posterior lobe contains relatively few cells and does not look much like glandular tissue. The **anterior lobe,** or **adenohypophysis,** develops embryologically as an isolated outgrowth of the primitive gut and migrates toward the neurohypophysis; it has a glandular appearance, with cords of several types of cells inter-

blood glucose is made to fluctuate by other factors (such as eating or exercise), the secretion of insulin is adjusted to reduce the magnitude of the fluctuations of glucose concentration. Negative feedback mechanisms are designed to maintain the status quo, to prevent change or, if change occurs, to bring things back to "normal." Thus, negative feedback is a homeostatic mechanism.

Several endocrine glands are controlled by a more indirect version of the same mechanism, as illustrated by the thyroid gland (Figure 13–2B). Thyroid secretion is stimulated by a hormone from the anterior pituitary known appropriately as the **thyroid-stimulating hormone (TSH).** A high level of thyroid hormone in the blood inhibits the pituitary secretion of TSH, which results in less stimulus for secretion of the thyroid hormone. Soon there is less of the latter to inhibit the production of TSH, and as TSH rises, the thyroid gland secretes more of its hormone.

The implication of the pituitary gland in the control of endocrine secre-

spersed with networks of vascular sinusoids. The **intermediate lobe** arises where the anterior and posterior portions of the developing pituitary come into contact with one another.

The neurohypophysis receives a prominent nerve supply in the several short fiber tracts that enter it by way of the pituitary stalk. These bundles arise in specific nuclei of the hypothalamus where the neurons have their cell bodies, and end near capillaries in the posterior lobe. Despite many persistent attempts to demonstrate secretory nerves, the adenohypophysis has no known innervation other than a few autonomic fibers to the blood vessels.

The posterior lobe receives its blood supply from tiny branches of the internal carotid artery (Figure 13–4). Arteries for the anterior lobe first form a network of capillary loops in the inferior portion of the hypothalamus. These capillaries then converge to form the several vessels of the **hypophyseal portal system** that descend along the stalk and drain into the sinusoidal capillaries in the adenohypophysis. Blood reaching the anterior lobe has therefore already passed through one capillary bed in the nearby hypothalamus.

The hormones secreted by the pituitary gland and their main actions are shown in Table 13–2. Over the years, other "factors" or "principles" have been attributed to the anterior lobe, but these effects are now generally believed to belong to the known pituitary hormones. Of those produced by the anterior pituitary, only the *growth hormone* and *prolactin* (in humans) do not have another endocrine gland as their target organ. FSH and LH, the gonadotropins, affect the gonads (ovary and testis), but they do much more than stimulate hormone secretion by the gonads. Prolactin has sometimes been considered to be a gonadotropin because in female rodents (especially the much-studied laboratory rat) it affects the gonad (luteotropic action), but this action is lacking in humans.

Control of Pituitary Secretion

Because the adenohypophysis controls the activity of several other endocrine glands, it has often been called the

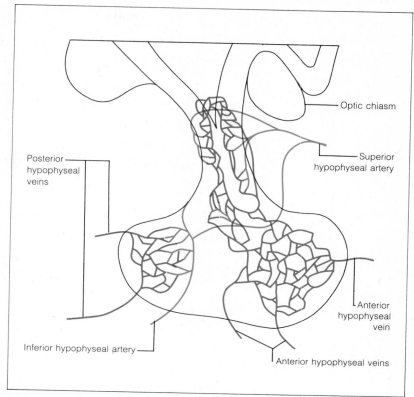

FIGURE 13–4   The blood supply of the pituitary.

"master" gland. It is not, however, really in decisive control of the "lesser" glands; it relays messages, and it is itself controlled by the hypothalamus, which secretes minute amounts of specific substances called **releasing hormones** into the blood of the hypophyseal portal system. Upon reaching the anterior pituitary, each releasing hormone causes secretion of one of the anterior lobe hormones (Figure 13–5).

Releasing hormones that have been identified include a *growth hormone-releasing hormone (GRH)*, a *prolactin-releasing hormone (PRF)*, a *thyrotropin-releasing hormone (TRF)*, a *corticotropin-releasing hormone (CRF)*, a *follicle-stimulating hormone-releasing hormone (FRH)*, and a *luteinizing hormone-releasing hormone (LRH)*, although there is some question as to whether FRH and LRH are actually two different substances. There is also a hormone that inhibits the release of growth hormone (*GIH* or *somatostatin*) and prolactin (*PIH*). There may also be a releasing hormone for melanocyte-stimulating hormone (MSH) produced

TABLE 13–2   THE PITUITARY HORMONES AND THEIR ACTIONS

Hormone	Other Names	Major Action
Adenohypophysis		
Growth hormone	GH; somatotropin, somatotropic hormone	stimulates growth of body
Thyroid-stimulating hormone	TSH; thyrotropin	stimulates thyroid growth and secretion
Adrenocorticotropic hormone	ACTH; corticotropin	stimulates growth and secretion by adrenal cortex
Follicle-stimulating hormone	FSH; gonadotropin	stimulates growth of ovarian follicle in female and spermatogenesis in male
Luteinizing hormone	LH; interstitial cell-stimulating hormone, ICSH; gonadotropin	stimulates ovulation, formation of corpus luteum, and hormone secretion in female; stimulates secretion by interstitial cells in male
Prolactin	lactogenic hormone, luteotropic hormone, LTH	stimulates secretion of milk; maintains corpus luteum in female rodents
Neurohypophysis		
Antidiuretic hormone	ADH; vasopressin	promotes water reabsorption from collecting tubule of kidney
Oxytocin	—	stimulates milk ejection, and contraction of pregnant uterus
Intermediate lobe		
Melanocyte-stimulating hormone	MSH; melanotropin, intermedin	expands melanophores (changes skin color); no known function in humans

by the intermediate lobe. One wonders if there is any significance to the fact that inhibitory hormones exist for the hormones whose target organ is not another endocrine gland.

A rather different arrangement exists for the neurohypophysis. The neurons in the fiber tracts that run from the hypothalamus to the posterior lobe release substances from their terminals upon stimulation, but the nerve endings are near capillaries in the posterior lobe. These capillaries are not part of the hypophyseal portal system, and the substances entering them are carried throughout the body. The hormones of the posterior pituitary are actually produced by neurons whose cell bodies are in the hypothalamus. Since so-called posterior lobe hormones are released from it, but not produced by it, one may rightfully question whether the posterior lobe is, in fact, an endocrine gland.

The arrangement by which the hypothalamus exerts control over the pituitary gland is not as unique as it might seem. We have seen that all nerve fibers produce their actions by releasing a chemical, which we call a transmitter, from their terminals when they are stimulated. The hypothalamic neurons do the same. The difference is that they release their "transmitter substances" near capillaries instead of postjunctional membranes. In the posterior lobe the substances are released near and enter capillaries of the general circulation of distribution throughout the body; the substances are called *hormones*. Other hypothalamic fibers release their "transmitter" near capillaries of the hypophyseal portal system to be carried to the anterior and intermediate lobes where they diffuse into the interstitial fluid and act on the pituitary cells; these substances are called *releasing hormones*. Each kind of releasing hormone "recognizes" the cell that produces "its" hormone because the membrane of that cell has receptor binding sites for that releasing hormone. The releasing hormone then binds to it and triggers the sequence of events for which that cell is programmed. The hormone produced as a result is released from the pituitary cell, enters the bloodstream, and is carried to the target organs.

Hypothalamic control of pituitary function is the means by which the nervous system exerts control over many hormonal systems. The nervous system has a much more highly developed sensing mechanism, and the hypothalamus, in a sense, gives the endocrine system a "digest" of the sensory information obtained by the nervous system. This information ensures that endocrine responses are appropriate to internal and external conditions. The hypothalamic connections explain the effects of stresses (emotional and otherwise) upon a number of hormonally controlled processes.

Neural Versus Hormonal Control

Most of the activities of the body are in response to a stimulus, with controls built into the system to ensure its effectiveness. A control system for stimulus-response behavior requires the following elements:

1 A mechanism to produce a response (the efferent and effector).
2 A mechanism to detect or monitor conditions (receptor and afferent).
3 A mechanism to bring the two together in order to assess and evaluate incoming information and to determine the appropriate response (a center).

We have two systems by which responses can be directed, the nervous system and the endocrine system. The nervous system sends messages from one place to another in the form of action potentials which travel along nerve fibers that extend from here to there. The action is exerted by the release of small packets of chemicals (transmitters) at the neuroeffector junction. The endocrine system sends messages in the form of chemicals (hormones) released into the bloodstream to be car-

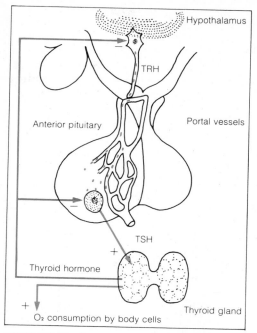

FIGURE 13–5   Control of hormone secretion by hypothalamic releasing hormones.

ried from here to there. They exert their action at some distant target organ.

The nervous system has all the required elements of a control system. It has a well-developed efferent system. It can get messages to all sorts of effectors (excitable cells) to produce a response that may be brief and instant or continuous, precise and localized, or generalized and widespread. The nervous system has a well-developed array of monitors and sensors, which includes receptors on the body surface that provide information about the external environment, and receptors in skeletal muscle and viscera that tell about conditions inside the body. The nervous system has marvelous centers to integrate all this information and to send out directions. The centers are organized in a hierarchy of command with the minimal elements for a simple response (a stretch reflex) on the spinal cord level. Centers in the brainstem exert control over the cord centers, and centers above that (cerebellum, basal ganglia) exert a higher level of control, while areas of the cerebral cortex exert control over all of those below.

The endocrine system, on the other

hand, has a good efferent component, but lacks a true integrating center of its own. It also does not have a very good mechanism for sensing and monitoring the environment, either external or internal. There is no means for providing information about what is happening "out there," except for the negative feedback mechanism, which is limited in scope.

Further contrasts are apparent if we consider characteristics of the responses produced by the two control systems. We tend to think of neural responses as rapid and quick, as in a knee jerk, while the endocrine response is slower and with a longer latency and more prolonged action. Perhaps it would be better to make comparisons in a different way, so as to show the two systems less as opposite or antagonistic systems. This may be generalizing too much, but it might contribute some perspective to your overall view.

Consider the somatic nervous system. It involves responses to changes in the external environment. The stimuli come from outside the body, applied to the body surface or to visual or auditory receptors that detect changes at a distance. The response is of the "body." Skeletal muscle, as the effector, brings about movement of parts, primarily the limbs, resulting in movement of the body. The response is quick (measured in milliseconds to seconds), usually brief and precise, and well localized.

Consider the autonomic nervous system. It involves primarily responses to changes in the internal environment (pressure of the blood, distension of the stomach or urinary bladder). The responses are carried out by smooth muscle, cardiac muscle, and glands of certain organ systems and bring about adjustments that tend to support and maintain homeostasis. The responses are of longer latency, slower and more prolonged, being measured in seconds and minutes.

Consider the endocrine system. Here we see a system that responds to metabolic needs, involving stimuli such as changing levels of substances in the blood (glucose, calcium, etc.). The

responses are carried out at the cellular level and involve metabolic functions, such as increased synthesis of a particular enzyme, changes in the permeability of a cell membrane to a substance (glucose), or activation of some enzyme. These responses are generally of still longer latency and are slower, more prolonged actions. They are measured for the most part in minutes and hours.

The nervous system and endocrine system should not be considered as two opposing systems, but rather as elements of a continuous spectrum of control with considerable overlap between the various parts. The autonomic nervous system seems to be a link between neural and hormonal controls through its association with the hypothalamus, and with the adrenal medulla and its secretion of epinephrine which has both neural and endocrine functions.

There are certain reflexes, known as *neuroendocrine reflexes,* in which the receptor and afferent are neural and the efferent limb is hormonal. Such mechanisms are involved in control of one of the hormones released from the posterior pituitary (see below). And finally, the nervous system, which has the better sensory system and better integrative centers, directs and coordinates much of the endocrine function through the action of hypothalamic releasing hormones[3] on the anterior pituitary and its endocrine target organs.

ENDOCRINE GLANDS AND THEIR SECRETIONS

The remainder of this chapter is devoted to the individual endocrine glands and the hormones they produce. The discussion of the action of these secretions and their significance, however, will be more in the nature of an introduction than a complete coverage. Hormones affect the actions of specific cells that carry out specific processes of

---

[3]Evidence is accumulating that the releasing hormones may have other actions. Some have been found in other parts of the nervous system and may be synaptic transmitters, and at least one (GIH), which has been found in other organs, has some effect on the digestive system.

the individual organ systems, and you cannot fully understand the hormone actions until you have some understanding of the organ systems. The actions of most of the hormones identified here are discussed more fully in the chapters that follow, as part of the regulation of a particular organ system.

The Pituitary Gland

The anatomy and control of the pituitary gland have been described in earlier sections, and some of its hormones have been mentioned. At this time we need only consider those hormones and their actions.

Adenohypophysis The action of most of the hormones of the adenohypophysis is very simple—they stimulate secretion by their target organs. Some have other actions as well, and these will be discussed with their target organs.

The target organ for the **growth hormone,** however, is not another endocrine gland. As indicated by its other name, **somatotropin,** it stimulates many parts of the body, although its major effects are on the metabolic reactions of skeletal muscle and on developing bone. In the latter case, it increases the rate of cell division among cartilage cells in the growing bones, which increases cartilage growth in the epiphyses of long bones and delays the completion of ossification. By postponing the final closure of the epiphyses the bone is able to grow for a longer period of time and attain a greater length. Besides helping cartilage formation in bone, growth hormone also has a positive effect on synthesis of the organic component of bone, which is related to its effect on protein metabolism.

In all tissues, but especially in skeletal muscle, growth hormone has several actions that favor protein anabolism, that is, the use of amino acids to form protein rather than for energy or formation of glucose or fat. This leads to an increase in the amount of structural protein and to a general increase in muscle mass. Such use of amino acids is aided by other actions of the hormone,

because growth hormone also causes mobilization of fat from storage, releasing more fatty acids into the circulation, and it raises the blood glucose level by causing the release of glucose from the liver. The way in which these metabolic actions fit into the total body economy will be made more apparent in Chapter 24.

Although the pituitary gland is not essential for life, an animal deprived of it (by *hypophysectomy*) has serious problems, due largely to absence of the adenohypophysis. Target organs that depend upon the drive of the pituitary tropic hormones tend to atrophy, and there is a reduction both of their hormone secretion and of their other functions. Most of the symptoms of hypophysectomy (removal of the gland) can be traced to the resulting *hypofunction* of the various target organs. Since several of the hormones of the pituitary and its target organs have important effects on metabolism, the hypophysectomized animal is susceptible to many things because it cannot properly adjust its metabolic processes. Similarly, *hyperfunction* of the pituitary shows up largely as increased activity of the target organs. Pituitary malfunction usually involves all the pituitary hormones, but selective malfunction is most likely to involve growth hormone.

Hyposecretion of growth hormone in a young child results in an individual of small stature, the **pituitary dwarf** or midget. Such people are well proportioned, but they are small. They are usually of normal mentality, but hyposecretion by other pituitary target organs, particularly the thyroid and adrenal cortex, may cause other problems. Also, the pituitary dwarf often fails to attain sexual maturity or fertility because of insufficient gonadotropin to cause development of the gonads.

Hypersecretion of growth hormone during the growth years, before ossification is completed, produces a condition known as **gigantism,** in which the individual is tall but has the musculature appropriate to the large frame. If the secretion of growth hormone becomes excessive only after the epiphyses have closed, there is no increase in the length of the long bones; instead there is a thickening of certain bones.

The lower jaw grows and the brows become prominent, giving a characteristic and striking facial appearance. The hands and feet become enlarged and there are changes within the bone tissue. This condition is known as **acromegaly.**

Thyroid-stimulating hormone (TSH) and adrenocorticotropic hormone (ACTH) cause secretion of the thyroid and certain adrenal cortical hormones, respectively. The effects of malfunction are those of hypo- or hypersecretion of the target organ and are discussed with those organs. Likewise, gonadotropin and prolactin are discussed with the reproductive systems whose function they help control.

Neurohypophysis   Two hormones known to be released from the neurohypophysis are the **antidiuretic hormone (ADH)** and **oxytocin.** They have almost identical chemical structures, each one being composed of nine amino acids, seven of which are the same. They are released from the neurohypophysis by stimulation of the nerve fibers from the hypothalamus rather than by releasing factors. As stated above, hormones released from the neurohypophysis are synthesized in the nuclei of the hypothalamus (ADH in the supraoptic nucleus and oxytocin in the paraventricular nucleus), from which the nerve fibers arise (see Figure 13–3). The hormones migrate down the nerve fibers as secretory granules and accumulate at the nerve endings near blood vessels in the neurohypophysis.

Antidiuretic hormone is released when receptors in the hypothalamus (*osmoreceptors*) detect an increase in the osmotic pressure of the plasma of the blood flowing through them, as might occur when you are dehydrated. A decrease in the plasma osmotic pressure inhibits the release of ADH. The hormone acts almost exclusively on cells of certain ducts of the urine-producing apparatus of the kidney (collecting ducts). The fluid in these ducts is destined to be excreted from the body as urine. By themselves, they are not permeable to water, and nearly all of the fluid passing through the ducts is excreted. ADH increases the amount of

cAMP in the cells; this increases their permeability to water. Therefore, in the presence of ADH, much of the water in the ducts is transferred through these cells and is returned to the blood (reabsorption). ADH thus reduces the volume of urine excreted but makes it more concentrated. A lack of ADH causes *diabetes insipidus,* a disease characterized by excretion of excessive amounts of very dilute urine (see Chapter 28).

In large amounts (*pharmacological* rather than *physiological* doses), ADH causes constriction of blood vessels which raises blood pressure. This effect is not very prominent in humans, but in some species it is important, and it is the basis for the other name for ADH, *vasopressin.*

Oxytocin has two actions, both upon reproductive structures: the mammary glands and the uterus. (It has no known function in the male, although it is secreted.) Once the mammary glands have been primed by the actions of several other hormones, which cause development of a duct system and secretory apparatus within the gland, oxytocin causes ejection of the milk by a neuroendocrine reflex. The suckling infant stimulates touch receptors in the nipple whose afferent fibers have connections with the hypothalamus and lead to oxytocin release. After a short latency, milk is ejected from the alveoli in the mammary gland.

The second action of oxytocin is upon the smooth muscle of the uterus. During pregnancy the uterine musculature becomes progressively more sensitive to oxytocin until, in late pregnancy, strong uterine contractions can be elicited by very small amounts of oxytocin. It is doubtful that oxytocin plays an important part in the induction of labor, but it is released during the course of labor. The mechanism seems to be another neuroendocrine reflex, this one apparently initiated by stimulation of receptors in the lower part of the uterus (the cervix) as the fetus passes along the birth canal. Pituitary extracts containing oxytocin are sometimes administered to induce labor, but this must be done with great care, since extremely powerful uterine contractions may result.

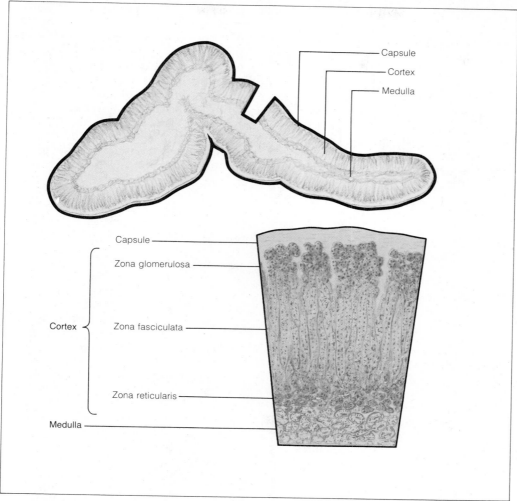

FIGURE 13-6   Structure of the adrenal gland.

Intermediate Lobe The skin of amphibians, fish, and reptiles contains melanocytes whose melanin pigment granules can move about within the cytoplasm. When the melanin granules are congregated around the nucleus they are less visible and the skin appears light, but when they are dispersed throughout the cytoplasm the skin looks dark. In these species, the hormone of the intermediate lobe, **melanocyte-stimulating hormone (MSH),** causes dispersion of the pigment granules and darkens the skin. The mechanism is probably a neuroendocrine reflex. Receptors in the retina of the eye have connections with the hypothalamus, which may either cause or inhibit the release of MSH. There is an MSH-RFH as well as nerve fibers to the intermediate lobe. This response does not occur in mammals, since the melanocyte granules cannot move. MSH therefore has no known effect in mammals, although there is some evidence of its involvement in certain conditions of hyperpigmentation in humans.

Adrenal Glands

Because of their location atop the kidneys, the adrenal glands are also known as the **suprarenal glands.** Like the pituitary, each adrenal gland consists of separate entities, an outer **adrenal cortex** and an inner **adrenal medulla** (see Figure 13-6). The two are very different glands in terms of their embryological origins, their controlling mechanisms, the hormones they produce, and the processes these hormones affect.

Anatomy of Adrenal Glands   The adrenal gland receives a greater blood flow per gram of tissue than any other organ in the body. The adrenal arteries enter the gland through the capsule that encloses it. Some of them break into capillaries in the cortex, while others pass directly to the medulla, but all vessels eventually drain through the medulla and empty into the adrenal vein, which thus drains both cortex and medulla.

The adrenal cortex develops near the kidney and remains in close association with it. Its characteristic three layers include the **zona glomerulosa,** whose cells are arranged in globular clumps (glomeruli), a thicker middle **zona fasciculata,** whose cells are arranged in long parallel cords separated by sinusoids (small venous sinuses), and the innermost and relatively narrow **zona reticularis.** The latter zone lies adjacent to the adrenal medulla, but there is no sharp line of demarcation between them, even though they are different organs. The adrenal cortex does not receive a nerve supply other than to its blood vessels.

The smaller adrenal medulla develops from neural tissue, but early in fetal life it migrates laterally until it comes to be embedded in cortical tissue. It is composed of irregularly interconnecting cords of cells interspersed with capillaries and venules. It has been likened to a sympathetic ganglion whose neurons have lost their axons, since it is innervated by sympathetic preganglionic neurons, and its secretion is very similar to the transmitter released by sympathetic postganglionic fibers.

Adrenal Medulla   The anatomical tie between the adrenal medulla and the sympathetic nervous system is further reflected in their functions. Medullary secretion is governed by sympathetic neurons and is an important part

of sympathetic discharge. The adrenal medulla, like the rest of the sympathetic nervous system, is not essential to life, but it is important to the ability of the organism to meet emergencies.

The human adrenal medulla secretes a mixture of about 80 percent epinephrine and 20 percent norepinephrine. The latter is also the chief component of the transmitter substance released by sympathetic postganglionic endings. Epinephrine and norepinephrine are chemically the same, except that epinephrine has an extra CH_3 (methyl) group.[4] They belong to a class of compounds known as the **catecholamines,** and are frequently referred to by this name because they are often found together, and because many analytical tests cannot distinguish between them. For the most part, the actions of epinephrine and norepinephrine are similar to one another and to those of the sympathetic nervous system. The differences between them are in part quantitative and in part due to the fact that epinephrine, as a true hormone, is widely distributed throughout the body, while the action of norepinephrine is more restricted to the immediate vicinity of the adrenergic (sympathetic postganglionic) nerve endings. Once released, both epinephrine and norepinephrine have a very short half-life.

Norepinephrine   Epinephrine

[4]They are synthesized from the amino acid phenylalanine, to which the two hydroxyl (OH⁻) groups are added, forming *dihydroxyphenylalanine (DOPA).* Removal of the carboxyl group (COOH⁻) leaves *dopamine,* and addition of another OH⁻ gives *norepinephrine,* while addition of a methyl group yields *epinephrine.*

Dopamine is probably a transmitter in its own right. It is present in high concentrations in certain parts of the brain. Dopamine is the substance that is in low concentration in the basal ganglia of patients with Parkinson's disease. DOPA is the substance that relieves the symptoms in some of these patients.

Both epinephrine and norepinephrine increase the heart rate, but epinephrine has a greater effect. Epinephrine also causes the cardiac muscle to shorten faster and contract more forcefully. It therefore can cause a marked increase in the amount of blood pumped by the heart (the cardiac output).

Norepinephrine and epinephrine are both effective constrictors of the blood vessels of the skin and viscera, which receive important sympathetic innervation. This action increases the resistance to the flow of blood and raises the blood pressure. Epinephrine, however, is a dilator of skeletal muscle blood vessels (where the role of the sympathetic innervation is still unsettled), and because so much of the body's mass is muscle, the overall effect of epinephrine may be a fall in total resistance and a consequent slight fall in pressure. These effects on blood pressure are also influenced by the response of the heart. The cardiovascular effects of the catecholamines are also discussed in Chapters 15, 17, and 18.

Epinephrine has greater effects on metabolic processes than norepinephrine, such as raising the blood glucose level by causing its release from the liver. However, both facilitate the use of glucose by skeletal muscle and other tissues, raise the metabolic rate and heat production, and stimulate the central nervous system.

Epinephrine and norepinephrine, like other hormones, must first bind with a receptor on the target cell. Observations that certain drugs would block some of the actions of the catecholamines, but not others, led to the discovery that there are two kinds of receptors to which they may bind. They have been designated *alpha* and *beta* *receptors.* In general, binding with alpha receptors leads to constriction of blood vessels (and elevation of blood pressure), while binding to beta receptors leads to increased rate and force of the heartbeat as well as metabolic and other effects. Beta receptors activate adenyl cyclase resulting in the formation of cAMP. Drugs that block actions mediated by the alpha receptors have been helpful in treating certain types of hypertension (high blood pressure).

There are no particular problems due to hyposecretion of the adrenal medulla, and only rarely may a tumor cause hypersecretion of epinephrine. The symptoms of this condition can readily be predicted.

Adrenal Cortex   The adrenal cortex is essential to life, and its removal is fatal within about a week unless certain hormones are administered. This fact has been known for a long time, but it was not until shortly after World War II that the substances produced by this gland were isolated, identified, and made available for use.

Approximately forty compounds have been isolated from the adrenal cortex. About a fourth of them are biologically active, but in humans only three or four are actually secreted by the gland in significant amounts. The other compounds are all chemically related and many are intermediates in the synthesis or degradation of the secreted hormones.

All of these substances belong to a class of compounds known as the **steroids.** The steroids, which include male and female sex hormones, bile acids, vitamin D, and cholesterol, are a special group of lipids characterized by the presence of the "steroid nucleus," a four-ring framework (Figure 13–7). For convenience, each carbon is numbered, and individual steroids are described by the sidechains attached to certain carbon atoms.

The adrenal cortical steroids are synthesized from cholesterol and differ only slightly in their molecular configurations. They are divided into the following three groups principally on the basis of their major actions:

1   The **mineralocorticoids** primarily affect electrolyte metabolism.
2   The **glucocorticoids,** named for their effect on glucose metabolism, also importantly affect protein and fat metabolism.
3   The **androgens** have male sex hormone activity.

In general, glucocorticoid activity has been associated with the presence of oxygen (as =O or —OH) at C11 (the eleventh carbon atom), and mineralo-

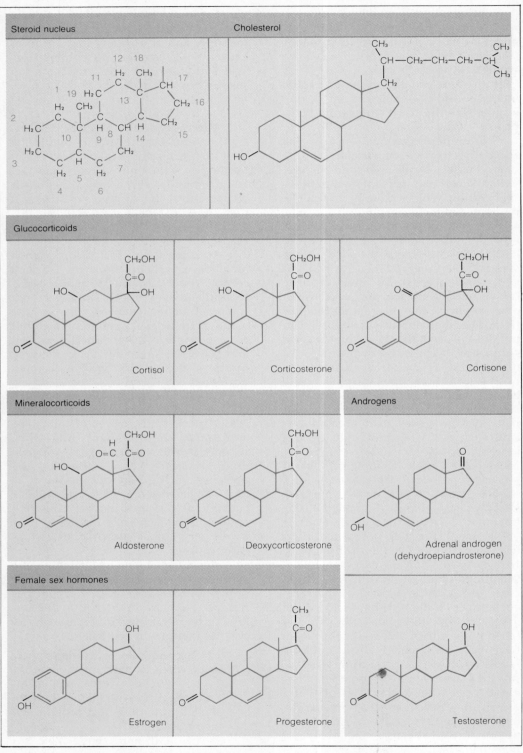

FIGURE 13-7   The steroid nucleus and some biologically important steroids.

corticoid activity with steroids lacking oxygen at that site. Mineralocorticoids and glucocorticoids have 21 carbons, but steroids with androgenic activity tend to have only 19 carbons. These "rules" are not, however, hard and fast, since aldosterone, the most potent mineralocorticoid, has an —OH at C11. Aldosterone also has some glucocorticoid activity as well, and this overlapping of actions is actually typical of most of the adrenal cortical steroids; their actions are not one *or* the other, but varying degrees of both. Certain chemical configurations enhance or impair particular actions, and knowledge of these structure-activity relationships has been applied by organic chemists in the search for synthetic compounds with the desired actions, but without undesirable side-effects.

As steroids, the adrenal cortical hormones act in the nucleus of the target cell. They increase formation of the RNA needed for synthesis of the enzymes that carry out the specific actions attributed to the steroid homones.

Mineralocorticoids. The two important mineralocorticoids are **aldosterone,** which is effective in extremely small amounts and is the most important physiologically, and **deoxycorticosterone,** which is much less potent. Deoxycorticosterone is not produced by the adrenal cortex in significant amounts, but because it can be synthesized easily and inexpensively, it is often used clinically (as deoxycorticosterone acetate, DOCA, DCA) when mineralocorticoid treatment is called for.

Aldosterone is secreted only by the zona glomerulosa, and it is produced continuously, for unlike the rest of the adrenal cortex, it does not depend upon the pituitary command (ACTH) for secretion to occur. ACTH is not totally ineffective, however, since it can cause an increase in the rate of aldosterone secretion. In the absence of ACTH, the aldosterone secretion in response to stressful stimuli is greatly reduced, but the zona glomerulosa does not atrophy.

The rate of aldosterone secretion is influenced by many conditions, including blood volume, plasma electrolyte concentration, and certain types of stress, but its major control is apparently the renin-angiotensin mechanism described in Chapter 28. The angiotensin II produced in response to renin release is an important stimulus to aldosterone secretion. All the necessary stimuli for this mechanism are not completely understood, but renin release is known to follow both reduction in blood flow to the kidney and reduction in sodium transport in the kidney. Low sodium and increased potassium in the plasma are also known to increase aldosterone production, but whether directly or through the renin-angiotensin mechanism is unclear. At any rate, aldosterone secretion is increased when blood volume and pressure are down or when the electrolyte content is disturbed. The resulting secretion of aldosterone then brings about adjustments that help relieve these situations.

The mineralocorticoids, in general, and aldosterone in particular, reduce the amount of sodium lost in the urine. Sodium is actively transported from the urine across the kidney cells and is returned to the bloodstream. The reabsorption of sodium secondarily increases reabsorption of chloride and, by diffusion and osmosis, water. Aldosterone helps retain sodium by reducing its content in sweat, saliva, and gastric juice. Mineralocorticoids also increase the loss of potassium in the urine by reducing its reabsorption.

In the absence of mineralocorticoids, sodium reabsorption is reduced. Although the excretion of sodium causes some increase in water excretion, the sodium loss is greater, and the extracellular fluids become slightly hypotonic. Water then tends to enter the cells and the extracellular fluid volume (including plasma) is further reduced. This causes circulatory problems, beginning with a fall in blood pressure that reduces the flow of blood to all organs, including the kidney. The ability of the kidney to excrete wastes is impaired and waste concentration in the blood rises. Continued lack of mineralocorticoids eventually leads to circulatory collapse and death. The adrenal cortex is essential to life, chiefly because of the mineralocorticoids.

Glucocorticoids. The glucocorticoids are concerned with metabolism of protein, fat, and carbohydrate. Metabolic effects are associated with steroids that have an oxygen at C11, and the most potent have oxygen at C17 as well. The most important glucocorticoids produced by the human adrenal cortex are **cortisol** and **corticosterone** (see Figure 13–7), with cortisol being produced in much greater amounts. **Cortisone** is not actually secreted by the gland, but it is biologically active and commercial preparations are widely used for conditions that respond to glucocorticoid therapy.

The actions of glucocorticoids on the three major foodstuffs are so interrelated that it is difficult to determine which are primary actions and which are secondary. The glucocorticoids tend to elevate the blood glucose level by increasing *protein catabolism,* which raises the amount of amino acids available for conversion to glucose. They also mobilize fat in adipose tissues, thus increasing the free fatty acids in the blood, and they interfere with the utilization of glucose in the tissues, especially in skeletal muscle. An important, though poorly understood, action of the glucocorticoids is their "permissive action." There are a number of metabolic reactions which the glucocorticoids do not cause, but which will not occur in their absence.

The glucocorticoids have several specific actions that do not appear to be direct results of their metabolic effects. One of these is to maintain *vascular reactivity,* since without glucocorticoids the blood vessels are unable to respond to circulating epinephrine or to norepinephrine liberated by sympathetic nerve fibers. This inability undoubtedly contributes to the circulatory difficulties encountered in adrenal insufficiency.

Glucocorticoids (primarily cortisone) are often administered in massive (pharmacological) doses because at very high levels they inhibit the *inflammatory reaction.* The inflammatory reaction is the normal response of cells to tissue injury (see Chapter 14), and typically includes three phases:

1 local swelling, redness, and pain due to dilation of blood vessels and movement of fluid from blood vessels into the tissue spaces of the injured area;

2 invasion of the area by macrophages from the blood (some white blood cells are phagocytes) and from surrounding tissues;
3 increased synthesis of connective tissue, which walls off the injured area and is the first step in repair of the injured tissue.

Glucocorticoids exert an inhibitory action on all three phases of the inflammatory reaction. They are believed to inhibit the local release of some of the substances that cause dilation of the blood vessels, thereby reducing the accumulation of fluid in the affected area. Part of the normal inflammatory reaction is due to the action of hydrolytic enzymes (similar to enzymes in the digestive tract) released from the lysosomes of injured cells, including overworked phagocytes. Glucocorticoids are believed to stabilize the membranes of the lysosomes and thus prevent the release of these enzymes and the breakdown they cause. Fibroblasts produce the collagen fibers and ground substance that surround and wall off the inflamed area and begin the repair process. Glucocorticoids inhibit the fibroblasts, which slows the process of healing and tissue repair.

Because they suppress the inflammatory reaction, the glucocorticoids are effective in alleviating the symptoms of the so-called *collagen diseases,* such as rheumatoid arthritis. But they also interfere with wound healing in these patients (as after surgery). Furthermore, the absence of the inflammatory response may mask the presence of dangerous infection and allow it to spread unchecked.

The glucocorticoids also have beneficial effects on certain allergic reactions because of some of their anti-inflammatory effects.

The glucocorticoids are believed to be secreted chiefly by the middle layer of the adrenal cortex, the zona fasciculata. Secretion is the result of stimulation by ACTH from the adenohypophysis, and this portion of the adrenal cortex is quite dependent upon ACTH. Without it the cells atrophy and fail to secrete adequate amounts of hormone. Circulating glucocorticoids inhibit the release of ACTH, probably by direct action upon the pituitary gland and upon hypothalamic production of the cortico-tropin-releasing hormone (CRH).

ACTH secretion is increased by a number of stimuli commonly considered stressful. They probably increase the secretion of CRH. Such diverse factors as trauma, infection, hemorrhage, exposure to cold, heavy muscular exercise, and such psychological traumas as fear, pain, and noise all quickly increase ACTH production, often enough to elicit a maximum secretion of glucocorticoids. The ability to respond to stress situations with an increase in glucocorticoids seems to be essential for the "nonspecific systemic reactions" that enable the body to survive long exposure to such stresses. Many of these stresses also cause a response of the sympathetic nervous system, and it is probable that part of the role of the glucocorticoids is related to this. Glucocorticoids are needed for the catecholamines to be fully effective, since they maintain the vascular reactivity that is so much a part of the catecholamine effects, and they enhance the fatty acid mobilizing action. Since sympathectomized animals can tolerate stress, the glucocorticoids must also have other important roles that are not yet understood.

Androgens. The adrenal cortex also produces steroids with sex hormone activity. The actions of these C19 steroids are predominantly those of male sex hormones (androgens), but there is probably some secretion of female sex hormones as well. Both are secreted by males and females under the stimulus of ACTH rather than of gonadotropins. In normal amounts these androgens have no known physiological function, but symptoms may occur when they are produced in abnormally high amounts. Androgens cause development of male characteristics, and when adrenal androgen secretion is excessive in young males, it may bring about an early puberty. The effects are less apparent in adult males. If adrenal androgen secretion is elevated in the female, the effect is masculinization, commonly including a receding hairline and facial hair. The actions of the sex hormones are further discussed in Chapters 29 and 30.

Abnormal secretion of adrenal cortical hormones. Removal of the adrenal glands (*adrenalectomy*) eliminates the source of both medullary and cortical hormones and, without replacement therapy, results in death. In humans adrenal cortical hypofunction is known as **Addison's disease.** The most conspicuous effects are the electrolyte disturbances due to lack of aldosterone, leading to reduced blood pressure and the related cardiovascular problems. Other symptoms include muscular weakness, low blood sugar (glucose), gastrointestinal disturbances, including loss of appetite and weight loss, and a progressive spotty pigmentation of the skin. These patients can be maintained with mineralocorticoids (DOCA), but they tolerate stress very poorly unless they also receive glucocorticoids.

Hypersecretion of cortical hormones may result from a tumor of the adrenal cortical tissue or from a pituitary malfunction. **Cushing's disease** is an example of the effects of excessive production of ACTH. The disease is characterized by obesity of the face, neck, and trunk, but *not* of the extremities. The skin is usually thin and the muscles weak, due to protein catabolism, and wounds heal poorly. There is high blood sugar, a certain amount of salt retention, frequently hypertension, and sometimes *hirsutism* (excessive hairiness). Many of these symptoms also occur as side effects of treatment with large doses of glucocorticoids (cortisone) or ACTH, but disappear once the treatment is discontinued.

Thyroid Gland

Anatomy of the Thyroid Gland
The thyroid gland is an H-shaped organ located in the neck; it sits astride the trachea at its junction with the larynx (Figure 13–8). Two lobes, one on either side of the trachea, are joined by a narrow band of tissue, the *isthmus.* The thyroid has a blood supply second only to that of the adrenal glands, but its nerve supply is meager, mostly to the blood vessels.

The microscopic structure of the thyroid gland is unique. It consists of numerous closely-packed spheres, or

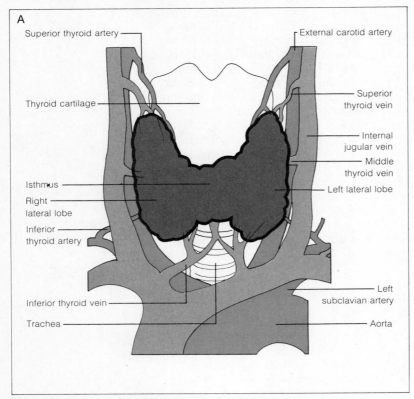

A
- Superior thyroid artery
- External carotid artery
- Thyroid cartilage
- Superior thyroid vein
- Internal jugular vein
- Middle thyroid vein
- Isthmus
- Right lateral lobe
- Left lateral lobe
- Inferior thyroid artery
- Inferior thyroid vein
- Left subclavian artery
- Trachea
- Aorta

B

Courtesy Carolina Biological Supply Company

FIGURE 13–8   A. Gross anatomy of the thyroid gland.
B. Photomicrograph of the thyroid gland.

follicles, formed of a single layer of cuboidal epithelial cells. These are the major secretory cells of the gland. The follicles contain a homogeneous protein substance called **colloid.** Interspersed between the follicles are small clusters of rather clear cells, often called **C-cells,** which also have a secretory function. Also between the follicles are numerous blood and lymph vessels, as well as the connective tissue necessary to support the gland.

Thyroid Hormones   The function of the thyroid follicle is to produce, store, and release **thyroxin and tri-iodo-thyronine,** the major thyroid hormones. The cells that make up the follicle are responsible for the production of the thyroid hormones, but the actual synthesis occurs in the colloid of the follicle. The follicle cells are the source of the necessary enzymes, and they also produce the protein found in the col-

loid, which is known as **thyroglobulin.** The thyroid gland is unique in that the hormone produced in the follicles is stored there until it is needed, when it is transported across the follicle cells and released into the circulation.

To make the hormone, however, the cells require iodine, which occurs in the blood as *iodide* (I^-). Cells obtain it by actively pumping it from the blood into their cytoplasm, from which it diffuses into the colloid and is oxidized to "active iodine." This trapping process is so effective that nearly half the iodine in the body is in this small gland.[5]

The basis for the hormone molecules is the amino acid *tyrosine* (Figure 13–9). It is bound to the thyroglobulin

[5]The ability to trap iodine is the basis for one method of treating certain hyperthyroid conditions. Radioactive iodine (^{131}I) is administered in doses so small that only in the thyroid gland, which traps it so efficiently, is the concentration of radioactive iodine great enough to destroy any tissue. This permits tumors to be destroyed without surgery.

in the colloid in such a way that the ring portion of its molecule protrudes. Iodine becomes attached to the tyrosine ring at one or two sites to form mono-iodotyrosine (MIT) or di-iodotyrosine (DIT). These molecules are condensed to form tri-iodothyronine (T_3) or tetra-iodothyronine (T_4). The latter, T_4, is the chief hormone of the thyroid gland, thyroxin. These iodinated compounds are stored within the follicle, bound to the thyroglobulin (Figure 13–10).

Release of the hormone into the bloodstream involves transporting it across the follicle cell. The follicle cell takes a "bite" of the edge of the adjacent colloid substance, ingesting thyroglobulin in a process much like phagocytosis. The thyroglobulin, with iodinated compounds still attached, now lies in a vesicle in the thyroid cell. The membrane of the vesicle merges with that of a lysosome, whose enzymes break off the iodine compounds. T_3 and T_4 are released into the bloodstream as thyroid hormones, but the iodine is removed from MIT and DIT and all of these components are recycled back into the colloid.

The thyroid gland also produces another hormone, **calcitonin.** It is not produced by follicle cells, but rather by the C-cells which lie between follicles. Calcitonin acts to lower the calcium level in the blood. Its role is discussed with the parathyroid glands whose secretions are most important in calcium metabolism (below).

Action of thyroid hormones. The specific effects of thyroxin and T3 on several organ systems can usually be traced to their more general effects, an increase in the oxygen consumption and metabolic rate of nearly all cells. The major action appears to be exerted on the mitochondria, since the cells of hyperthyroid animals have a larger number of mitochondria. It has been suggested that thyroxin stimulates the mechanism for forming mitochondria. The mitochondria contain the enzymes for breaking down fatty acids, for the citric acid cycle and the hydrogen transport system, and it is here that the energy released by oxidation is used to form ATP. It has been suggested—and dis-

puted—that thyroid hormones cause more energy to be released by oxidation, but that less of it is used to form ATP and more is released as heat. Thyroid hormones do seem to increase RNA synthesis and hence protein formation, but this does not completely explain their effect.

Regulation of thyroid secretion. Thyroid secretion—the trapping of iodine and hormone synthesis and release—is increased by the thyroid-stimulating hormone (*thyrotropin,* TSH) from the adenohypophysis. Thyrotropin is in turn controlled by the thyrotropin-releasing hormone (TRH) from the hypothalamus, and TSH and TRH are both inhibited by circulating thyroxin and T_3. Other controlling factors are those that influence the release of TRH. Continued cold and various neural stimuli, including stress of various types, increase it, while thyroxin, T_3, and a warm environment inhibit it (the effects of temperature are beneficial, in view of the effect of thyroxin on heat production). In addition, several chemical agents can reduce thyroxin secretion by blocking synthesis at one stage or another.

FIGURE 13–9   Compounds important in thyroid hormone synthesis.

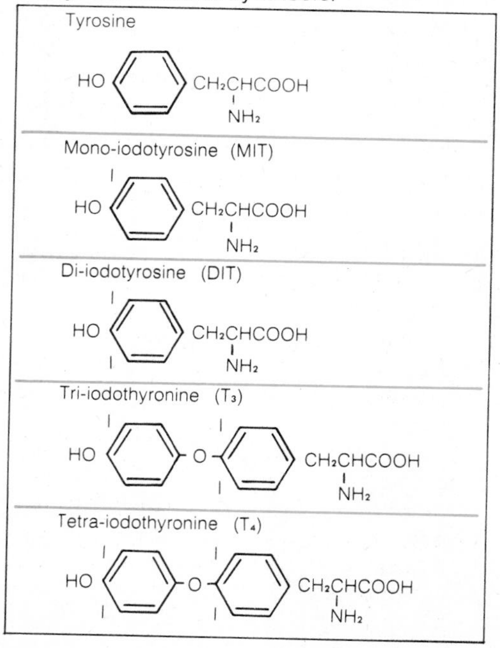

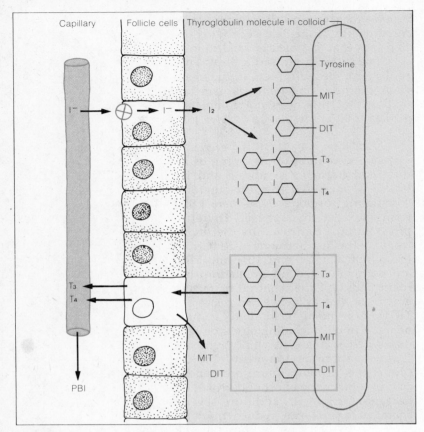

FIGURE 13-10   Synthesis of thyroid hormones.

In addition to the above symptoms, the hypothyroid child suffers from arrested growth and development and may be mentally retarded. Such a condition is known as **cretinism,** and the patient is a *cretin.* Both myxedema and cretinism can be remedied by the administration of thyroid hormone. Since both T_3 and thyroxin are relatively small molecules, and can be absorbed from the digestive tract, they can be given orally.

Hyperthyroid persons have increased basal metabolic rates. They use more oxygen and produce more heat and eat more food, while nevertheless losing weight. They are jumpy, irritable, and cannot tolerate heat. Their skin is warm due to dilation of cutaneous blood vessels, and the rate and pumping action of the heart are increased. In severe cases (*thyrotoxicosis*), many of the important symptoms are those of an overloaded cardiovascular system.

Hyperthyroid conditions may be due to a tumor of the thyroid gland (*toxic goiter*), in which case treatment may be surgical removal, destruction of part of the gland by radioactive iodine, or the use of antithyroid drugs to block hormone synthesis. Hyperthyroid symptoms may also be due to excessive stimulation of the gland (*exophthalmic goiter*). It is due to production of an abnormal protein that causes hypertrophy and increased thyroid secretion, but it is not inhibited by the negative feedback of T_3 and T_4. This substance also causes the accumulation of fluid in the soft tissues behind the eyes, which, in turn, causes the eyeballs to protrude (*exophthalmos*). The eyelids are opened widely, and the individual appears to be staring.

Goiter is the term referring to an enlarged thyroid gland. It may reflect either a hypo- or a hyperthyroid condition. *Simple goiter* is often due to an iodine deficiency that reduces hormone output. TSH production is then not inhibited, and the gland responds to the continual stimulation by hypertrophy as it "tries harder." The growth of the gland, by increasing the iodine-trapping tissue, may enable it to produce enough hormone so that the only sign is the increased size of the gland—and it may become very large. If it cannot produce enough hormone, symptoms

It should be noted that TSH release alone may not result in an increased hormone secretion. The gland must also have sufficient iodine. If it does not, the stimulated gland will appear histologically active (that is, the cells will hypertrophy), but no hormone will be forthcoming.

Abnormal secretion of thyroid hormones. The symptoms of thyroid hypofunction are almost entirely the direct result of the reduced metabolic rate. Afflicted individuals have a diminished rate of oxygen consumption and the body temperature, blood pressure, and heart rate may be reduced. They are mentally and physically sluggish and do not tolerate cold very well. Their hair tends to be coarse, dry, and rather sparse, and their faces appear bloated and puffy due to an accumulation of protein and fluid in the subcutaneous tissues. This last symptom provides the name of the condition in the adult: **myxedema.**

Focus

Addison's Disease

The adrenal cortex produces a number of hormones essential to the normal functioning of the body. When something goes wrong with this organ, the effects can therefore be serious and far-reaching. Hypersecretion of the adrenal cortical hormones, caused by an adrenal tumor or excessive pituitary secretion of ACTH, causes the disorder known as Cushing's disease. Hyposecretion, caused by a lack of ACTH, atrophy of the adrenal cortex, or damage to the adrenal cortex, causes Addison's disease, which was first described over one hundred years ago, before any of the hormones were known.

Addison's disease is marked by two kinds of symptoms because both mineralocorticoids (including aldosterone) and glucocorticoids (including cortisol) are in short supply. The lack of mineralocorticoids increases the excretion of sodium and water; as a result, blood volume and blood pressure fall. The lack of glucocorticoids leads to depletion of glycogen stores—they are not adequately replenished because synthesis of glucose from amino acids is impaired. This causes a tendency to hypoglycemia, an increased sensitivity to insulin, and a lack of physical energy. Without cortisol's negative feedback action, there is increased release of melanocyte stimulating hormone (MSH), as well as ACTH. The MSH causes an increase in skin pigmentation, especially in portions of the body exposed to light, pressure, or friction. The lack of glucocorticoids, which are also involved in the body's response to injury, infection, and other forms of stress, decreases the body's ability to adapt and defend itself.

A person with a severe case of Addison's disease may be listless and apathetic, have a small heart with a weak beat, and have very low blood pressure. The metabolic rate and body temperature may be low, and the body is slow to recover from damage and overexertion. Fortunately such cases are not common. More often the symptoms are less severe, owing to a relative lack rather than near or total absence of adrenal cortical hormones. Most individuals can manage quite well on replacement therapy consisting of drugs with mineralocorticoid and with glucocorticoid actions. Synthetic drugs are usually used rather than the natural hormones because they are less expensive. Salt intake usually has to be increased as well, in order to replace the sodium lost in urination. Since these hormone actions are necessary for a person's well-being, it is essential to continue the medication for life; and in times of stress, the dose, particularly of glucocorticoids, must be increased.

of hypothyroidism appear. The disease was formerly quite common in inland areas where the iodine content of the soil (and food) is low, but availability of iodized salt has largely eliminated this problem. Hyperthyroid conditions also produce goiter, since the gland is enlarged due to a tumor or excessive stimulation.

Parathyroid Glands

For many years it was believed that the thyroid gland was essential to life, since when it was removed, the patient died. However, it is now known that death resulted from the inadvertent removal of the parathyroids during the *thyroidectomy*. Two or three of these tiny glands are embedded in the dorsal surface of each lobe of the thyroid gland. They produce **parathormone,** which is necessary to maintain the calcium level of the blood. Thus, the parathyroid gland is essential to life.

Calcium Metabolism The body contains about 1200 g of calcium Ca^{++}, more than any other cation, and about 99 percent of it is in the bones and teeth. The remaining 1 percent is in the extracellular fluid and soft tissues, but only a small part of that, less than 1 g, is in the blood. The calcium in the blood, about 10 mg per 100 ml of blood, exists as a roughly fifty-fifty equilibrium between diffusible calcium and calcium bound to a plasma protein, with the diffusible calcium being the active portion.

The level of calcium in the blood must be maintained within a very narrow range (10 ± 1 mg percent), since calcium is essential for normal contraction of cardiac and skeletal muscle, for normal nerve function, and for blood coagulation. Too much or too little calcium causes serious difficulties. Too high a level of calcium in the blood depresses muscle excitability and causes cardiac irregularities. Too little raises the excitability of both nerve and muscle and results in muscle twitches, which may develop into convulsions and tetany (tonic spasms), particularly of the muscles of the limbs and larynx. If low enough, the end result will be

asphyxiation from laryngeal spasm.

The 99 percent of the body's calcium that is in the hard tissues serves as a storage depot. Some of it is readily exchangeable, and bone is continually resorbed and rebuilt as its calcium repeatedly departs and returns. Calcium enters the body (and the blood) from the digestive tract and leaves by excretion from the kidney. While the level of calcium in the blood is the resultant of all processes by which it enters and leaves the blood, it is determined primarily by the shift of calcium to and from bone. Because the bone content is so great, a slight change in bone calcium can make an extremely great change in the blood calcium level.

Parathormone The natural equilibrium between the reactions shown in Figure 13–11 is such that without parathormone the serum calcium levels off at about 7 mg percent (instead of 10 mg percent), which is low enough to produce signs of tetany. The effect of parathormone is to cause more calcium to be absorbed from the digestive tract (and hence less to be lost in the feces) and less to be excreted in the urine. More calcium is removed from the urine being formed in the kidney, and is returned to the bloodstream (a process known as *reabsorption*). Its major action, however, is to raise blood calcium by removing calcium from the bone, perhaps by stimulating the osteoclasts, the cells important in bone resorption. One could therefore summarize the action of parathormone on calcium as increased *absorption* from the gut, increased *reabsorption* in the kidney, and increased *resorption* from bone.

Parathormone also affects phosphorus, whose metabolism is closely associated with that of calcium. The mineral in bone is a complex salt composed mostly of calcium and phosphate. When calcium is withdrawn from bone, the level of phosphate in the blood is therefore also raised. The fact that the calcium and phosphate levels usually vary in opposite directions is explained by noting that the solubility of calcium phosphate depends upon the product of calcium and phosphate ion concentrations in the blood. When this

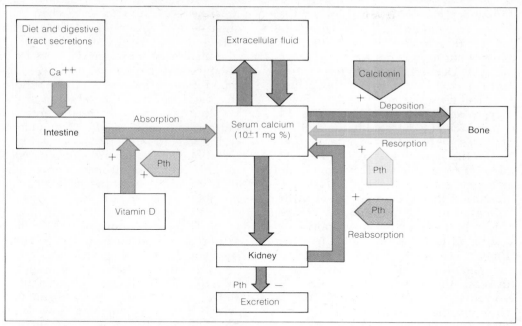

FIGURE 13–11   Hormonal regulation of serum calcium levels. (Pth = parathormone)

product increases, less calcium phosphate can remain in solution and it precipitates out; when it falls, the solubility is greater and more calcium phosphate can remain in solution in the blood. Parathormone keeps the product low, since it reduces the level of phosphate in the blood by increasing the renal excretion of phosphate, which is necessary if more calcium is to be withdrawn from the bone.

There is no evidence of pituitary or hypothalamic involvement in the control of parathormone secretion. The only determining factor appears to be the level of free calcium in the blood; a fall elicits an increase in parathormone secretion, followed by a rise in blood calcium, which in turn depresses parathormone secretion.

Hypo- or hyperfunction of the parathyroid gland is rare. Hypofunction may occur as a result of thyroid surgery and is indicated by signs of tetany, which can often be controlled by administration of calcium and vitamin D (see below). Hyperfunction may eventually cause bone diseases due to excessive removal of calcium from the bone, and calcium-containing kidney stones are common due to the high level of calcium in the blood.

Calcitonin   In the early 1960s, a disorder marked by excess calcium in the blood (*hypercalcemia*) was traced to a hormone other than parathormone. This new hormone was called **calcitonin** until it was found to be produced by the thyroid gland and renamed thyrocalcitonin. Since then, cells that produce it have been found in the parathyroid and thymus glands as well, so it is known as calcitonin once more. Calcitonin is secreted when the level of calcium in the blood is elevated, especially in young individuals. Its action is primarily to prevent the resorption of bone and therefore to lower the blood calcium level by counteracting one of the actions of parathormone.

Vitamin D   A discussion of calcium and its metabolism is hardly complete without mentioning the role of vitamin D. For many years this vitamin has been highly regarded as a preventive against *rickets*, a disease in which the bones do not calcify properly. The beneficial effect of vitamin D is an increased deposition of mineral in the bones, thanks in part to improved absorption of calcium from the intestine. Without vitamin D, so little calcium and phosphate are absorbed that there is not enough available for normal calcification of the bones. When calcium absorption is improved with vitamin D, calcification becomes normal once more.

The term *vitamin D* covers several closely related steroid compounds. Vitamin D can be ingested in the diet, but it can also be formed from a steroid present in the skin by the action of ultraviolet light from the sun. The compound is converted to its active form by actions first in the liver and then in the kidney. Since the compound is produced in the body, and transported by the blood to a distant site, it is sometimes considered to be a hormone.

Vitamin D's action seems to be to increase the synthesis of protein involved in the active transport of calcium in the intestine, and probably other cells as well.

Pancreas

The pancreas lies on the posterior wall of the abdominal cavity inferior to the stomach (see Figure 22–9). It is both an exocrine and an endocrine gland. Its exocrine function, the secretion of important digestive enzymes, is discussed in Chapter 23, and its anatomy in Chapter 22. The endocrine function is performed by small isolated clumps of tissue, the **pancreatic islands** or **islets of Langerhans,** scattered throughout the substance of the gland. The cells of these islands are arranged in irregular cords with numerous interspersed capillaries, and they consist of two main types of cells: The alpha cells are believed to secrete **glucagon,** while the more numerous beta cells produce **insulin.** A few delta cells are present, and they produce *somatostatin* (another name for the growth hormone-releasing hormone, GH-RH, of the hypothalamus).

Insulin   Insulin is associated primarily with carbohydrate metabolism and blood glucose regulation, although it has important effects on fat and protein metabolism. A number of other hormones contribute to regulation of the same processes, and their actions

must be taken into account when considering the actions of insulin. These actions are discussed in Chapter 24.

Insulin is *hypoglycemic*. That is, it enhances those reactions that lower the blood glucose level and inhibits those that raise it. This is largely due to the fact that insulin increases the entry of glucose into cells. Insulin stands alone in this regard, since most of the other metabolic hormones are *hyperglycemic*. Because it enables glucose to enter cells more readily, insulin facilitates all metabolic reactions of glucose, including glycogen formation and use of glucose. These reactions are particularly important in skeletal muscle, where much of the glucose is used.

Insulin has effects on fat metabolism which are almost as important as those upon glucose metabolism, and which indirectly tend to lower the blood glucose. It increases the formation of fat and its storage in adipose tissue and, by increasing glucose entry into cells (including adipose cells), more glucose is available for conversion to and storage as fat.

Insulin also increases the transport of amino acids into cells, with a consequent increase in protein synthesis and reduced protein catabolism. The actions of insulin favor removal of glucose, amino acids, and fatty acids from the blood; they are anabolic, for they lead to synthesis of glycogen, fatty acids and fats, and protein.

A lack of insulin causes a disease known as *diabetes mellitus*, in which there are many metabolic disturbances. The most characteristic is the accumulation of glucose in the bloodstream (*hyperglycemia*) because it cannot enter the cells readily. The hyperglycemia leads to the excretion of glucose in the urine. The role of insulin and the other "metabolic" hormones are discussed in more detail in Chapter 24.

Glucagon   Glucagon is a polypeptide produced by the alpha cells of the pancreatic islands. Glucagon's action is exerted mainly on the liver, because it is rapidly inactivated there and reaches the general circulation in reduced amounts. The polypeptide's actions favor energy release; it increases the breakdown of fats, which raises the supply of fatty acids that can be used as a source of energy. But more importantly, glucagon is the most potent agent known for increasing the breakdown of glycogen in the liver, which it does by activating an enzyme necessary for the reaction. The glycogen breakdown yields more glucose which is released into the circulation. It is interesting to note that epinephrine causes a similar breakdown of glycogen in the liver as well as in skeletal muscle, and that epinephrine also causes the release of fatty acids from fat stores.

Secretion of Pancreatic Hormones

The secretion of insulin is controlled primarily by the glucose content of the blood to the pancreas, since an elevation of the blood glucose level acts directly and promptly to increase the production and release of insulin. When blood glucose falls, insulin secretion also falls. The pituitary has no known effect, but epinephrine, glucagon (the other pancreatic hormone), and secretin and pancreozymin (hormones that increase the exocrine secretion of the pancreas) all increase insulin secretion to some extent. Several drugs have similar effects, and in those cases of diabetes in which the pancreatic islands are functional but simply do not produce enough insulin, these drugs can be used to drive the insulin-producing beta cells. This is an advantage, as these agents can be taken orally; they are the so-called "oral insulins." Unfortunately, they are ineffective in any patient whose pancreas cannot secrete insulin.

The secretion of glucagon also depends largely upon the blood glucose level, except that glucagon secretion is increased by hypoglycemia and diminished by hyperglycemia. Stimulation of sympathetic nerves to the pancreas increases glucagon release, and so do several "stressful" stimuli.

Gonads

The hormones produced by the gonads are the so-called sex hormones. They have important roles in reproduc-

tion and in the development and maintenance of the sex characteristics. These hormones and the pituitary gonadotropins that control them are discussed in Chapters 29 and 30 on reproduction. The hormones produced by the placenta during pregnancy will also be discussed in that section.

The Pineal and Other Endocrine Glands

The pineal has been called a mere *vestigial* remnant, a controller of aldosterone secretion, an antagonist of ACTH, and an inhibitor of puberty and of growth, but to date its function in mammals remains unknown. The primary reason for considering it an endocrine gland is precisely that it has no known function. It has therefore been assumed that it must produce an unidentified hormone of unknown action. There is some evidence that in rats it serves as some sort of connection between the retina of the eye and various endocrine glands, thus relating the diurnal rhythm of some glands to the photoperiod (periods of light and darkness). There are nerve connections with the retina, and the indication is that light causes a hormonelike substance to be released by the pineal gland. It should be added, however, that there is also evidence which contradicts this.

Several other hormones, such as those produced by the organs of the digestive tract (gastrin, secretin, pancreozymin-cholecystokinin) and by the kidney (erythropoietin and renin) are discussed in later chapters.

CHAPTER 13 SUMMARY

The endocrine glands secrete into the bloodstream specific substances (hormones) which have a specific effect on some distant organ (target organ).

Hormones secreted by the endocrine glands, their major controls and actions, are listed in the table below. Some of these hormones are not discussed in this chapter, but are covered in later chapters and are included here for completeness.

TABLE 13–3   THE ENDOCRINE GLANDS AND THEIR SECRETIONS

Endocrine Glands and Hormones	Main Stimulus for Secretion	Major Actions
Pituitary gland (hypophysis)		
Anterior lobe (adenohypophysis)		
growth hormone	growth hormone-releasing hormone (GRH); inhibited by inhibitory hormone (GIH)	stimulates body growth, protein anabolism, and other metabolic effects; stimulates growth of epiphyseal cartilage of long bones.
thyroid-stimulating hormone (TSH)	thyrotropin-releasing hormone (TRH)	stimulates growth and secretion by thyroid gland
adrenocorticotropic hormone (ACTH)	corticotropin-releasing hormone (CRH)	stimulates growth and secretion by adrenal cortex, especially secretion of glucocorticoids
follicle-stimulating hormone (FSH)	follicle-stimulating hormone-releasing hormone (FRH)	stimulates growth and development of ovarian follicle in female; stimulates spermatogenesis in male
luteinizing hormone (LH) or interstitial cell-stimulating hormone (ICSH)	luteinizing hormone-releasing hormone (LRH)	stimulates ovulation, formation of corpus luteum and secretion of estrogen and progesterone in female; stimulates secretion of testosterone by interstitial cells in male
prolactin	prolactin-releasing hormone (PRH); inhibited by prolactin inhibitory hormone (PIF)	stimulates secretion by mammary glands
Posterior lobe (neurohypophysis)		
antidiuretic hormone (ADH)	increased osmotic pressure of plasma via hypothalamic osmoreceptors	decreases volume of urine excreted, increases volume of water reabsorbed in kidney
oxytocin	neuroendocrine reflexes initiated by stimulation of receptors in uterus and mammary glands	causes contraction of smooth muscle of uterus and ejection of milk from mammary glands
Adrenal medulla		
epinephrine and norepinephrine	sympathetic nervous system	same actions as sympathetic nervous system (sympathomimetic); metabolic actions, including increased blood level of glucose and release of fatty acids into bloodstream
Adrenal cortex		
glucocorticoids (cortisol)	ACTH	metabolic actions, including protein catabolism, increased levels of glucose and fatty acids in blood; aids in withstanding stress; suppresses inflammatory reaction
mineralocorticoids (aldosterone)	renin-angiotensin mechanism, also ACTH	decreases sodium excretion by kidney, increases retention of sodium; increases excretion of potassium, decreases retention
androgens	ACTH	no significant effects at normal levels
Thyroid gland		
thyroxin and tri-iodothyronine (T_3)	TSH	increases oxygen consumption and metabolic rate of all cells
calcitonin	increased calcium in blood	lowers calcium in blood, causes calcium deposition in bone
Parathyroid gland		
parathormone	decreased calcium in blood	raises calcium in blood, causes increased absorption from digestive tract, withdrawal from bone, and decreases excretion; increases excretion of phosphate

Endocrine Glands and Hormones	Main Stimulus for Secretion	Major Actions
Pancreatic Islands		
insulin	increased blood glucose level, and other metabolic stimuli	decreases blood glucose levels, increases entry of glucose into cells; metabolic actions include storage of fat and synthesis of protein
glucagon	decreased blood glucose level and other metabolic stimuli	raises blood glucose level, breakdown of glycogen; causes release of fatty acids into bloodstream
Gonads		
Testis		
testosterone	ICSH	stimulates growth and development of male sex organs and sex characteristics; aids spermatogenesis
Ovary		
estrogen	LH	stimulates growth and development of female sex organs and sex characteristics
progesterone	LH	stimulates further development and differentiation of sex organs
relaxin	—	loosens (relaxes) pubic symphysis and aids in dilation of cervix of uterus at end of pregnancy
Placenta		
human chorionic gonadotropin	—	maintains corpus luteum and stimulates it to secrete estrogen and progesterone in early pregnancy
estrogen progesterone	—	stimulates continued growth and development of sex organs, especially endometrium, during pregnancy
Digestive tract		
Stomach		
gastrin	vagus nerve, and presence of certain substances in stomach	causes secretion of hydrochloric acid and digestive enzyme (pepsin) by lining of stomach
Small intestine (duodenum)		
cholecystokinin-pancreozymin (CCK)	protein digestion products and fat in duodenum	causes secretion of digestive enzymes by exocrine pancreas; causes contraction of gall bladder
secretin	acid (low pH) in duodenum	causes secretion of high-bicarbonate fluid by exocrine pancreas
Kidney		
erythropoietin	hypoxia (low oxygen) in blood to kidney	increases rate of production of red blood cells by bone marrow
renin	decreased blood pressure and blood flow to kidney; decreased sodium transport in kidney	activates angiotensin, which causes constriction of blood vessels and increases blood pressure; increases secretion of aldosterone

Hormones help control processes associated with regulation of metabolism and energy supply, regulation of the extracellular fluid, responses to stress, and with growth, development, and reproduction.

Most hormones are derived from amino acids, as amines or peptides, or are *steroids*. Most are rapidly inactivated and/or excreted and must, therefore, be produced continually.

Many hormones affect target cells without entering them. The hormone "recognizes" and binds a specific *fixed receptor* on the surface of the target cell membrane. This activates adenyl cyclase, an enzyme in the membrane which causes formation of cAMP from

ATP in the cell. The cAMP then activates a specific enzyme in the cell to bring about the hormone's action. Steroid hormones diffuse into the cell to find the receptor *(mobile receptor)*. The receptor-hormone complex moves to the cell nucleus and causes formation of the RNA needed to direct synthesis of a particular protein—the enzyme needed to catalyze a certain reaction. Some hormones act by altering the permeability of the cell membrane to a specific substance.

Some endocrine glands are controlled by the nervous system, but most are controlled by a negative feedback mechanism in which an increase in blood concentration of the hormone inhibits its production by the endocrine gland. The anterior lobe of the pituitary gland *(adenohypophysis)* is important in regulating secretion of many endocrine glands because it produces several hormones whose target organ is another endocrine gland. The adenohypophysis also has direct connections with the hypothalamus by way of the hypophyseal portal system. The blood supply to the anterior lobe passes through a capillary bed in the hypothalamus before reaching the adenohypophysis. Neurons that arise in various hypothalamic nuclei release chemicals (neurotransmitters) near these capillaries, and the neurosecretions are carried to the adenohypophysis where they stimulate secretion of the pituitary hormones. There is a specific neurosecretion, or releasing hormone, for each hormone of the adenohypophysis, plus inhibitory hormones for two of them. The negative feedback control of the target organs of the pituitary hormones is exerted both at the pituitary gland and at the hypothalamus.

The posterior lobe of the pituitary *(neurohypophysis)* receives nerve fibers that arise in hypothalamic nuclei. These fibers release their secretions—ADH and oxytocin, which are synthesized in the hypothalamus—into the bloodstream in the neurohypophysis and become the hormones of the posterior lobe; the neurohypophysis itself does not produce the hormones.

STUDY QUESTIONS

1 In what ways are hormonal and neural actions alike? In what ways are they different?

2 How do hormones exert their actions on cells? Explain the fixed receptor and mobile receptor models for hormone action.

3 What is the hypophyseal portal system? What is its significance?

4 What is meant by negative feedback control?

5 How does the production and release of thyroid hormone differ from that of other hormones?

6 What is the general effect of thyroid hormones? Based on this, what symptoms would you expect to find in a hyperthyroid condition?

7 Compare and contrast the actions of epinephrine and norepinephrine. How do these compare with the actions of the sympathetic nervous system?

8 Do the metabolic actions of growth hormone and glucocorticoids augment or interfere with one another? Explain.

9 What processes determine the calcium level in the blood? How does parathormone affect these processes?

10 What effect does insulin have upon blood glucose level? What processes does it alter to bring about the effect?

11 What effect does insulin have upon fat metabolism?

12 Absence of which hormones is fatal? What controls the secretion of these hormones?

Part 4

Systems of Transport

Disclike red blood cells and rounded white blood cells in an arteriole.

The dependence of cells upon the composition of their immediate environment has been mentioned repeatedly. The maintenance of that local environment (homeostasis) depends upon the transport systems that deliver necessary materials to cells, remove substances from them, and transport components that are needed for conditions favorable for normal cell function.

The transport systems are the cardiovascular and lymphatic systems. The cardiovascular system circulates blood to all cells and to those organs that supply the things to be delivered to the cells, or accepts those things removed from cells. Thus the cardiovascular system is also an integral part of the digestive, respiratory, and excretory systems.

Blood is the transport medium, for materials transported are either suspended or dissolved in its fluid, the blood plasma. The cardiovascular system is a mechanism for distributing the blood. It consists of a pump (the heart), with vessels (arteries) branching from it into progressively smaller and more numerous vessels and finally, to the tiny thin-walled capillaries, where the exchange between blood and interstitial fluid actually occurs. The capillaries are so numerous that every cell is within diffusing distance of one. From capillaries the vessels converge into fewer but larger vessels (veins), which return the blood to the heart. The lymphatic system is an auxiliary system, for it returns fluid (interstitial fluid) and the substances in it to the blood.

The chapters in Part IV deal with the components of the transport systems, beginning with the blood as the medium of transport. Other chapters discuss the heart and its pumping action, the blood vessels and their distribution and control, and the lymphatic system. Special attention is given to mechanisms for controlling the action of the heart and blood vessels in order to adjust the volume of blood delivered to particular groups of cells when their needs change, and to maintain their blood supply when other conditions change.

14

The Nature of Blood

Formed Elements

Plasma

Blood Clotting

The Basis of Blood Groups

Blood in Acid—Base Balance

Blood

Blood is a unique type of connective tissue, for it is a fluid tissue. Great numbers of cells are suspended in the fluid (plasma), and a variety of solutes are dissolved in it. Most of the substances carried by the blood are going somewhere or coming from somewhere, and their importance lies in what happened before they entered the bloodstream or what happens when they leave the blood.

Although many of the blood's constituents act outside the circulation, blood is vital for transport because (except for nerve impulses) it is the only means the body has for distributing anything for some distance within the body. Blood provides the contact between the internal environment of the cells and the external world. It carries nutrients, wastes, electrolytes, proteins, water, hormones, and numerous other substances. It also distributes body heat and elements of the body's defense mechanisms. In addition, blood contains special components to transport gases such as oxygen, to deal with the invasion of harmful bacteria or other "foreign" material, to prevent loss of blood if a vessel is broken, as well as to maintain the pH, osmotic pressure, and other elements of homeostasis.

This chapter describes these roles of the blood and some of the mechanisms by which they are carried out.

THE NATURE OF BLOOD

The blood carries oxygen from the lungs to the tissues and carbon dioxide from the tissues to the lungs. It brings absorbed material from the intestine to the liver and other organs for further processing and storage, and thence to the cells for use. It carries cellular wastes to the kidney for excretion. It carries hormones from the endocrine glands to their sites of action throughout the body. It plays an important part in the regulation of the temperature of the body and the amount, distribution, and pH of the body fluids. Because both its fluid and certain of its cells have properties that enable the body to combat bacteria, viruses, and other potentially dangerous agents which may invade the body, it serves to protect the body.

Blood is therefore the keystone of homeostasis. It is the medium through which the internal environment is maintained, since it is only through the blood that most of the organ systems contact the cellular environment and the interstitial fluid and cells communicate with the external environment. It is the body's transport system, and it is thus an essential link in the functional chain for most of the body's systems.

Blood Volume

With such a varied and vital task to perform, the blood must have a sizable volume to "service" all the cells properly. The amount of blood circulating through the vessels varies between individuals, and within an individual from time to time. However, it rarely falls below a certain volume; otherwise the flow to the tissues would be endangered. The total blood volume is usually 5–6 liters in males and 4–5 liters in females, of which at least a tenth, or about a pint (the volume of a blood donation), can readily be spared. In either case, the amount of blood averages out to about 7.7 percent of the total body weight.

Although total blood volume may vary, there are numerous mechanisms that help prevent extreme deviations in either direction. These mechanisms include those concerned with the regulation of water balance and with the concentration of the individual constituents of the blood. Loss of water from the blood concentrates all of the substances it carries, while increased water content of the blood dilutes them.

Composition of the Blood

Blood is a type of connective tissue. It has cells and intercellular material, but no fibers unless it clots. In appearance, freshly drawn blood is a red, sticky, rather viscous fluid that clots readily. The cells can be separated from the fluid portion by centrifuging the blood in a special test tube, after which the cells are packed at the bottom of the tube, leaving the fluid or **plasm** at the top. The cells are mostly **red blood cells** (RBCs'), or **erythrocytes** (erythro = red), although upon careful examination a thin whitish layer may be seen resting on top of the red column. This is the "buffy coat," formed chiefly of **white blood cells** (WBCs), or **leukocytes** (leuko = white). The red cell volume percentage is known as the **hematocrit,**[1] and it is about 45 percent of the blood volume (Figure 14–1). It is usually slightly higher in males (47 percent) than in females (42 percent) and varies with a number of conditions. The remaining fluid portion of the blood, about 55 percent, is plasma.

FORMED ELEMENTS

The major formed (nonfluid) elements of the blood are the red and white blood cells, with the red cells making up by far the greatest portion. The term "formed elements" is used because it also includes tiny fragments, known as **platelets,** which are not actually whole cells.

---

[1]*Anemia,* for example, is a condition in which red blood cell formation is impaired and their numbers may be reduced. *Polycythemia,* on the other hand, is the result of increased red blood cell formation. An increased hematocrit also occurs in dehydration, where there may be a normal number of cells but a decrease in the plasma volume.

Red Blood Cells

Erythrocytes are small cells, without nuclei, whose chief role is the transport of the respiratory gases, particularly oxygen. They are shaped like biconcave discs, which increases their surface area. This permits them to take in a considerable amount of fluid without undue strain on the cell membrane. This unusual shape is an important adaptation because the fluid environment for erythrocytes is less constant in its composition than the fluid environment for many other cells. The shape is also adapted to the constant tumbling to which the cells are subjected, since it allows them to twist and fold as they squeeze through the smallest blood vessels.

Red blood cells, like other cells, are enclosed by a typical unit membrane. The membrane is selectively permeable and readily permits the passage of water and of some small ions. Plasma normally has an ionic concentration of 0.9 percent, which is *isotonic* with the intracellular fluid of cells (see Chapter 2). If blood cells are placed in a fluid whose ionic concentration is greater—a *hypertonic* solution—the osmotic gradient drives fluid from them and they shrink about their cellular frameworks, a phenomenon known as **crenation** (see Figure 2–8).

When red cells are placed into a *hypotonic* solution, the osmotic gradient causes fluid to enter the cells and they swell to resemble biconvex discs. Both crenation and swelling due to changes of the tonicity of the surrounding fluid are reversible within limits, but if the solution is very hypotonic, the gradient is so great that enough fluid may enter to rupture the cell membrane. This is **hemolysis;** it is not reversible. When the cell membrane bursts, the cellular contents are spilled out into the plasma, and the cell ceases to function as a cell. Hemolysis can also be produced by any agent that destroys the integrity of the cell membrane; fat solvents such as ether and chloroform are examples.

The number of erythrocytes in the blood is extremely high. Each cubic *millimeter* (mm³) contains about 5,000,000 (5 × 10⁶) red blood cells. Males have a

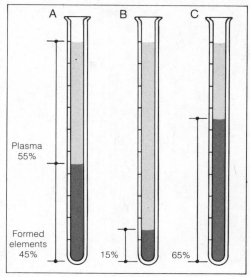

FIGURE 14–1   Hematocrit. A. Normal blood. B. Anemia. C. Polycythemia.

slightly higher red cell count, 5.4×10^6 cells/mm³, and females have about 4.8×10^6 cells/mm³.

Hemoglobin   Hemoglobin is the most important component of the red blood cell, both quantitatively and functionally. The amount of hemoglobin present in the blood is about 15 g of hemoglobin in every 100 ml of blood. In males it is about 16 g (14–18) and in females 14 g (12–16).

Hemoglobin has a great affinity for oxygen and forms a loose, reversible association with it. When hemoglobin is combined with oxygen, it is said to be **oxygenated** and is called **oxyhemoglobin (HbO₂).** When it is not carrying oxygen, it is called **deoxyhemoglobin (Hb)** or reduced hemoglobin, a misleading term because the uptake and release of oxygen by hemoglobin is not an oxidation-reduction reaction in the chemical sense.

Virtually all the hemoglobin in blood which is en route to the tissues is oxygenated, and the high content of oxygenated hemoglobin in *arterial blood* is responsible for its bright red color. About a quarter of the hemoglobin in blood returning from the tissues is in reduced form; the remaining three quarters is still oxygenated. Blood with

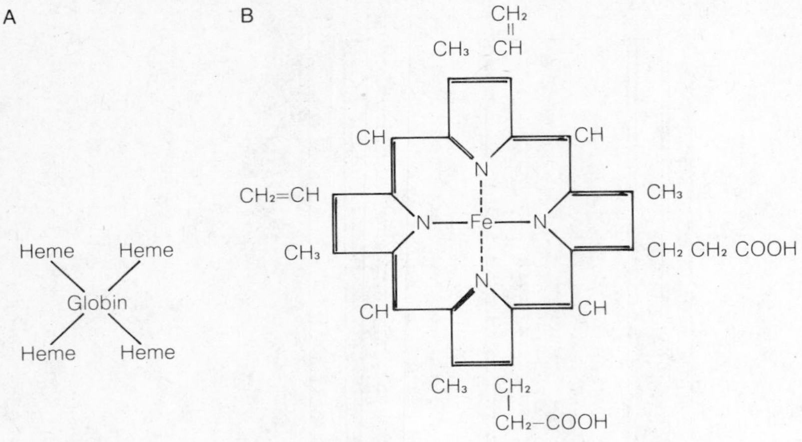

A

B

CH₃, CH₂=CH, CH₃, CH₃, CH₂CH₂COOH, etc. structure with Fe center and four N atoms (heme structure diagram)

Heme   Heme
Globin
Heme   Heme

FIGURE 14-2   A. Structure of hemoglobin. B. Structure of heme.

a significant content of reduced hemoglobin is darker in color, as is *venous blood*.

Hemoglobin contains a red pigment, **heme,** which contains iron (Figure 14-2). The iron (Fe) is a very important constituent of the heme and hemoglobin, since without it the hemoglobin is not complete and cannot carry oxygen. The iron-containing heme is bound to a protein, a **globin,** to form the hemoglobin molecule. The globin consists of several polypeptide chains which have specific amino acid sequences. Sometimes there are genetically determined variations in the sequence, resulting in different hemoglobins.

Formation of Red Blood Cells

Because they have no nucleus, erythrocytes do not form new cells by mitosis as many types of cells do. They do not reproduce themselves, but instead arise from the *hemopoietic* (blood forming) *tissue* located in the red bone marrow. In the fetus and newborn all of the bone marrow is red, and the potential for blood cell formation is very great. In the adult, however, much of it has been replaced with yellow marrow and red cell production is restricted to such bones as the sternum, vertebrae, ribs, bones of the skull, and possibly the heads of some long bones. Under special conditions yellow marrow may revert to active hemopoietic tissue.

Together with many of the white blood cells, erythrocytes arise from a common primitive cell, called a **stem cell,** although the red and white cells develop along different lines. The pathway for erythrocytes involves a number of steps. In successive mitotic cell divisions the large nucleated stem cell is replaced by smaller cells. Their cytoplasm takes on the pinkish appearance of red blood cells as they synthesize increasing amounts of hemoglobin. At this stage the cells look like rather large nucleated erythrocytes and are called **normoblasts.** Their nuclei are then *extruded;* with them goes the capability of further mitotic division and protein (hemoglobin) synthesis, but some RNA remains in the cytoplasm for a time. Circulating blood normally contains a few of these cells called *reticulolytes,* but their concentration may rise when red cell formation is accelerated. The process of forming red blood cells is known as **erythropoiesis.**

Fate of Red Blood Cells   The life of an erythrocyte is a difficult one. It is almost constantly on the move, sometimes traveling at speeds of many millimeters per second and at other times slowly traversing vessels so narrow that the cell must bend in order to pass through. Without a nucleus, the ability to synthesize protein for self-renewal is severely limited, and after a time the erythrocyte simply wears out. The average life span of a red blood cell is about 120 days (4 months). At the end of this time, the cell may simply disintegrate and the fragments be picked up by the cells of the *reticuloendothelial system.* This system consists of phagocytic cells (macrophages) that are found in a number of places throughout the body, notably in the spleen and liver.

Disintegration of the erythrocyte releases the hemoglobin, which is then broken down into components that are salvaged for reuse whenever possible (see Figure 14-3). Globin, the protein portion of hemoglobin, is reduced to amino acids and used by the tissues for protein synthesis. The iron from the heme may be returned to the bone marrow for reincorporation into red cells or it may be stored, chiefly in the liver, for future use. The remainder of the heme

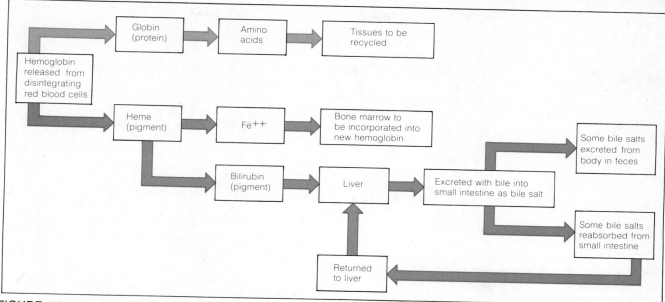

FIGURE 14-3   Catabolism of hemoglobin.

is converted to a substance known as *bilirubin*, which is excreted by the liver in the bile. The fate of the bile, which enters the small intestine, is discussed in Chapter 23.

Control of Erythropoiesis The rate of red blood cell formation normally equals the rate of destruction, but there are conditions in which this is not so. For instance, any reduction in the oxygen delivered to the tissues will stimulate erythropoiesis, perhaps as much as sixfold. The oxygen lack, or *hypoxia,* may be due to a decrease in the number of red blood cells, the blood flow, or the amount of oxygen in the air. It can be caused by such varied conditions as hemorrhage (loss of blood), hemolysis, anemia, reduced pumping by the heart, or high altitude. There is less stimulation of erythropoiesis when the red blood cell count is high.

A reduced oxygen delivery to the tissues triggers the release into the bloodstream of a hormonelike substance, **erythropoietin.** The kidney is generally believed to be the source of erythropoietin, but recent evidence suggests that the kidney produces a substance that acts upon a protein in the plasma to form erythropoietin. In

any case, renal hypoxia leads to the formation of erythropoietin which is transported to the bone marrow where it stimulates stem cells to differentiate into red blood cells. A small amount of erythropoietin is normally present in circulating blood, and there is a direct relationship between its level and the degree of tissue hypoxia.

People who live at high altitudes, where the air contains fewer molecules of oxygen per liter than at sea level, generally have an elevated red cell count. Indians in the Andes Mountains of South America, who may live their entire lives above 15,000 feet, may have red cell counts of 7-8 million cells/mm^3. A sea-level resident will also have a higher count after two or three weeks at a high elevation.

Red Cell Abnormalities The most common problem associated with red cells is that of **anemia.** Literally, anemia means lack of blood. However, the basic problem is a lack of hemoglobin in the blood, which may be due to an inadequate number of red blood cells, to a lack of hemoglobin in each cell, or to both. Anemias due to insufficient hemoglobin in each cell include the nutritional anemias, particularly iron-deficiency anemias. Since iron is an essential part of the hemoglobin molecule, its availability is essential for hemoglobin production. Iron-deficiency anemias are fairly common among women whose dietary intake of iron is not enough to make up for the iron lost in the menstrual flow. Other causes may be related to problems with the complex mechanism for absorption and utilization of iron (see Chapter 24).

Anemias due to lack of red blood cells may be the result of hemorrhage, or loss of blood. Following such a loss, the blood volume is quickly restored by compensatory mechanisms that bring fluid into the blood vessels from the tissues. However, this fluid contains no cells and the hematocrit and red cell count fall. If the loss is not prolonged or repeated, the red cell count can be reinstated by flushing cells from reservoirs (such as the spleen) and by increased erythrocyte production. Repeated or chronic hemorrhage can produce severe anemia, even if there is an increase in erythropoiesis.

A small amount of *vitamin B₁₂* is needed to produce normal red blood cells. It is widely distributed in the diet, but it cannot be absorbed from the digestive tract in the absence of an unidentified substance produced by the mucous lining of the stomach. This substance known as the **intrinsic factor,** binds the vitamin B_{12} in a way that facilitates its absorption and then releases the vitamin to enter the circulation. When vitamin B_{12} reaches the red bone marrow, it causes the proper development and maturation of the red blood cells. A lack of the intrinsic factor, and the resulting lack of vitamin B_{12}, causes a disease known as *pernicious anemia,* which is characterized by faulty red cell formation. The erythrocyte count may be extremely low (1–2 million cells/mm³), and the cells may have abnormal sizes and shapes. The disease can be treated by injections of vitamin B_{12}, thus bypassing the absorption problem.

Hemolytic anemias are those in which the rate of red cell destruction is increased. *Sickle cell anemia* is one example. It is a hereditary condition found almost exclusively among black people. It is due to a slightly different amino acid sequence in one of the polypeptide chains of the hemoglobin molecule. It changes some of the properties of the hemoglobin so that the red cells take on an unusual crescent shape (they "sickle") when oxygen is low, which renders them more susceptible to destruction.

The drugs administered for treatment of many conditions have side-effects upon red cells. They may inhibit erythropoiesis due to action on the bone marrow, or they may have a hemolytic effect, either of which can be very serious.

Polycythemia is a condition in which the red cell production is accelerated, producing a very high red cell count and hematocrit. Although the oxygen-carrying capacity is greater, the viscosity of the blood is increased. This causes problems of clogging and blockage of small blood vessels and makes it more difficult for the heart to keep the blood circulating.

White Blood Cells

White blood cells (WBCs) or leukocytes are not really white. Rather, they are colorless unless stained. In a sense they are not really blood cells either because they must leave the circulatory system to perform most of their functions. They are found both in the bloodstream and in great numbers outside the circulatory system in the tissues, especially in loose connective tissue and in many organs. Usually there are about 5000–9000 white blood cells per cubic millimeter of blood (5–9 × 10^3/mm³). There are, therefore, more than 500 red blood cells for every white blood cell. The leukocyte count is closely regulated but is highly variable, for it may increase severalfold with an infection and may reach extremely high levels in some disease conditions.

Unlike erythrocytes, leukocytes are nucleated and have the typical organelles with which to carry on cell functions such as protein synthesis. A suitably stained preparation examined under the microscope shows them as flat circular cells, most of which are somewhat larger than red cells. This may be misleading because in the body they are neither flat nor round. Leukocytes are amoeboid cells, which means that they possess varying degrees of motility and change their shape constantly. Like amoebas, they can squeeze through spaces too small for a spherical cell—this is how they move from the bloodstream into the tissues.

Leukocytes have been divided into two groups, **granulocytes** and **agranulocytes,** depending upon whether or not there are prominent granules in their cytoplasm. Figure 14–4 shows the different types of white blood cells, and Table 14–1 summarizes their characteristics, along with those of the other formed elements.

Granulocytes Granulocytes are formed in the red bone marrow, along with erythrocytes. They develop from the same stem cell but along a different pathway. A potential granulocyte passes through many stages as it develops along one of three pathways leading to its release into the circulation. There are three types of granular leukocytes named for the staining characteristics of their cytoplasmic granules.

Neutrophils are the most common of the leukocytes, accounting for 65–75 percent of all white blood cells. Their granules, which are lysosomes, are numerous, small, and usually stain a neutral purple. Their nuclei are so irregular in shape, especially in older cells, that a cell may appear to have several nuclei. However, upon careful examination, a tiny thread can be seen connecting the lobes. For this reason they are often called *polymorphonuclear leukocytes* (or simply "polys").

Two outstanding characteristics of neutrophils are their prominent amoeboid tendency and their phagocytic ability. They act as scavengers, engulfing bacteria and other foreign material, as well as cellular debris from the tissue spaces.

Eosinophils normally make up only about 2–4 percent of the white blood cell population. Their distinguishing feature is their granules which fill the cytoplasm and stain red with the dye eosin. The granules are larger but less numerous than those of neutrophils, and they are also lysosomes. Eosinophils are also capable of amoeboid movement and they are known to increase in number in certain

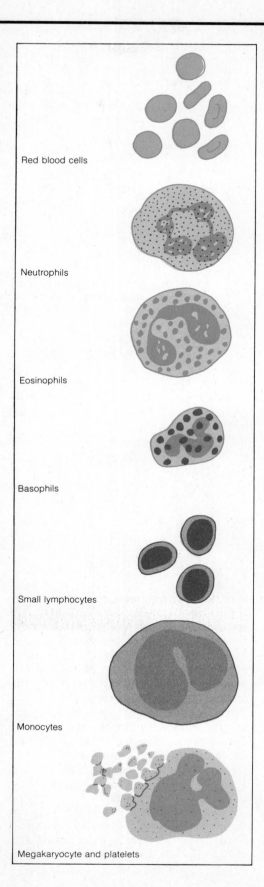

FIGURE 14–4
Formed elements of the blood.

Red blood cells

Neutrophils

Eosinophils

Basophils

Small lymphocytes

Monocytes

Megakaryocyte and platelets

TABLE 14–1   THE FORMED ELEMENTS OF THE BLOOD

Characteristics	Red Blood Cells	Neutrophils	Eosinophils	Basophils	Lymphocytes	Monocytes	Platelets
Number	$4.5–5 \times 10^6$ per mm^3	65–75% of all white cells (total white cells $= 5–9 \times 10^3$ per mm^3)	2–4% of all white cells	1% of all white cells	20–25% of all white cells	2–6% of all white cells	$2.5–3 \times 10^5$ per mm^3
Average size (diameter, in μm)	7–8	10–12	10–14	9–12	6–12	12–15	2–3
Appearance of nucleus	none	dark, up to 5 lobes	pale, 2–3 lobes	pale, 2–3 lobes	very dark, generally rounded	light, indented	none
Appearance of cytoplasm	red	clear, many tiny purple granules	pale, moderate number, red granules	pale, a few coarse dark blue granules	blue, agranular, scant	blue-gray, agranular	very dark
Source	bone marrow	bone marrow	bone marrow	bone marrow	bone marrow	bone marrow	bone marrow (megakaryocyte)
Motility	none	marked	moderate	marked	marked	marked	none
Phagocytic ability	none	marked	some	negligible	negligible	marked	none
General function	O_2 and CO_2 transport	phagocytosis	??? allergy?	???	cell-mediated and humoral immunity	phagocytosis	blood clotting
Miscellaneous properties	shape of biconcave disc				found in connective tissue in great numbers	become tissue macrophages	cell fragment

allergic diseases and to decrease in conditions of stress.

Basophils are the rarest of blood cells, accounting for less than 1 percent of all leukocytes. Their granules stain dark blue with a basic stain and are quite large but rather sparse. Basophils are motile, but not very phagocytic. Their function is not known, although their granules contain *heparin*, an agent that prevents clotting of blood (that is, an anticoagulant), and *histamine*, a powerful dilator of blood vessels.

Agranulocytes   Agranulocytes are so named because they do not ordinarily have granules in their cytoplasm, although with special stains a few tiny granules can often be seen. **Lymphocytes** and **monocytes** are the agranular leukocytes. They are also formed in the bone marrow, but after release into the circulation, they tend to leave it rather quickly. Some lymphocytes go to organs that have been called *lymphoid organs* (thymus gland, spleen, lymph nodes), as well as to other organs, and some may re-enter the circulatory system. Monocytes enter the connective tissue and become tissue macrophages.

Lymphocytes make up 20–25 percent of all leukocytes. Most lymphocytes are small, about the size of a red blood cell, but some are nearly twice as large. They have a fairly round, dark-staining nucleus. The small lymphocytes have a narrow ring of pale blue cytoplasm, while large ones have a greater amount. Lymphocytes are very active as far as motility goes, but they are not phagocytic. Lymphocytes are important in specific immune systems (see below).

Monocytes account for only 2–6 percent of the total leukocyte population, but they are the largest of the blood cells. They look something like oversized lymphocytes and are sometimes difficult to distinguish from them. Usually the monocyte nucleus is pale (when stained) and indented, and it is surrounded by a large amount of less distinctly colored cytoplasm than that of lymphocytes. Monocytes are motile and markedly phagocytic, especially when they leave the bloodstream and enter the tissues.

Leukocytes and Body Defenses

Leukocytes function in the body's defense against foreign materials. There are two aspects to this role; one is a nonspecific defense against bacteria and other microorganisms that enter the body. It is carried out largely by the phagocytic white blood cells (neutrophils and the other granular leukocytes, and monocytes). The other defense involves specific immune systems which are mediated through lymphocytes. They protect against certain microorganisms and are also involved in responses to (and rejection of) "foreign" substances or tissues, as in allergic reactions and rejection of tissue transplants.

Nonspecific defenses. There are many barriers to the entry of microorganisms into the body. The skin protects the external surface and the mucous membranes protect the inner surfaces. Both the secretions that keep mucous membranes moist and the environment in some of the organs (such as the acidity in the stomach) are inhospitable to invaders. But in spite of this, microorganisms do occasionally invade the tissues. Their presence sets off the **inflammatory response,** which is a general tissue response to many different kinds of assault. It is triggered by substances released from the invading

bacteria and from the cells in the invaded area. Both histamine from the tissue cells (and mast cells) and some of the other substances released dilate blood vessels and increase their permeability. This increases the flow of blood through the area and also increases the amount of plasma (including even some protein) that leaks into the tissues. These effects account for the local redness, heat, and swelling, and the pain (caused by stretching of the tissues) that occurs. Other substances are released into the bloodstream and react with proteins in the plasma to form chemicals that attract phagocytes to the area. The movement of the phagocytes is called *chemotaxis.* Neutrophils are the first to respond, usually arriving in significant numbers within an hour or so; they are therefore the first line of defense. They stick to the wall of the inflamed blood vessel and, although normally they are not very active in the circulation, they now become very active. By their amoeboid action they are able to leave the circulation and enter the tissue where they begin to engulf bacteria. Inside the neutrophils, the vacuoles containing bacteria merge with the granules (lysosomes) and the bacteria are then destroyed. The neutrophils do not return to the bloodstream, but remain in the tissue for the remaining days of their short life span. Some bacteria produce *toxins* (poisons) that cause the lysosomal membranes to rupture and release their destructive enzymes within the neutrophil, resulting in its death.

Monocytes respond in much the same manner as neutrophils, but they are much slower. They will ingest about anything around—bacteria and other microorganisms, dead cells and debris, even dirt and dust particles. They are believed to be of greater importance in long-term infections, since they are slower to arrive, and since neutrophils die off rather quickly. Monocytes normally leave the circulation within about 24 hours, whether there is an infection or not. They enter the tissues and become the macrophages of the connective tissue and part of the reticuloendothelial system, the clean-up system of the tissues. Many macrophages leave the connective tissues and become lodged in various organs, such as the lungs, liver, lymph nodes, and many others, where they carry out the same function.

Eosinophils respond more slowly and more selectively. They are likely to be present in relatively large numbers in allergic reactions and inflammatory responses that involve *antigen-antibody* reactions (see below). The role of basophils is unknown, but one suggestion has been that by releasing the histamine from their granules, they may contribute to the development of the inflammatory reaction.

The number of leukocytes in the blood seems to be controlled rather carefully, but the factors involved are not well understood. Neutrophils have been studied most, and it is known that their production is increased if their number in the blood is reduced. Production is greatly increased in the presence of infection, probably due to the effect of some of the substances released into the circulation by bacteria and damaged tissues as part of the inflammatory reaction. Neutrophils remain in the circulation only a few hours, then leave by "oozing" through capillary walls to the tissues. They can also penetrate mucous membranes, into the digestive tract and lungs, for example, where they die. (They are digested in and/or excreted from the digestive tract; in the lungs, they take care of the debris from dead cells.)

Specific immune systems. The immune systems are those by which the cells of the body are able to recognize those materials which are foreign to it, and to inactivate or otherwise remove any threat they might present. These systems provide a means by which the body can recognize and distinguish "self" from "nonself."

Proteins are extremely varied, and they differ between organisms, species, and individuals of the same species. When a protein from another source is introduced into the bloodstream, the native tissues recognize it as a foreign protein and react against it. The foreign substance acts as an **antigen** because it stimulates the tissues to initiate an immune response against it.

There are two types of immune responses. One is a system of **humoral immunity** in which certain lymphocytes, called **B lymphocytes,** or **B cells,** are responsible for the production of **antibodies** that are released into the circulation. These are specific proteins produced in response to a particular antigen; they combat the structure or microbe of which the antigen is a part. The other is a **cell-mediated system,** in which a lymphocyte known as a **T lymphocyte,** or **T cell,** may become "sensitized" to a specific antigen. B and T cells cannot be distinguished under an ordinary microscope, but with the electron microscope both are seen as small spheres. B cells have many tiny finger-like processes, while the T cells have very few. Both types of lymphocyte have receptor sites on their surface membrane for binding a particular antigen.

When lymphocytes that are destined to become T cells are released from the bone marrow where they are formed, they migrate to the **thymus gland** (hence the name T cell). Here they proliferate and in some way mature into cells that are able to carry out cell-mediated immunity. Upon leaving the thymus, they move to lymphoid organs and lymphoid tissue scattered throughout the body, but they can also return to the circulation. The thymus gland is believed to exert some control over them through hormones that it produces.

Other lymphocytes, the B cells, develop the ability to produce and release antibodies into the circulatory system. It is not known where they go to acquire this capacity in humans, but in chickens it occurs in a lymphoid organ known as the *bursa of Fabricius* (hence the name B cell). It is assumed that there is an organ that performs this function in humans, but until it is identified, the lymphocytes are still called B cells. They also move to lymphoid tissue where they take up permanent residence; they do not generally return to the circulation.

B cells are responsible for antibodies that confer particular immunities against most bacterial infections. T cells are major causes of specific immunities against fungi, viruses, and some types of bacteria. They also mediate destruction of malignant (cancerous) cells and the rejection of tissue transplants. It is possible for an individual to have a deficiency of one or the other type of lymphocyte. B cell-deficient individuals cannot produce antibodies and so are very sensitive to most any bacterial infection, but they are fairly resistant to nonbacterial invasions, and they can reject foreign tissues normally. T cell-deficient individuals have a normal resistance to most bacteria, but they are prone to cancer and they tolerate grafts and transplants of foreign tissue.

Humoral immunity: When an antigen reaches an organ where B cells are located, the antigen binds to B cells that have appropriate receptor sites on their surface. The antigen stimulates proliferation of those B cells to form a cluster or *clone* of such cells. Some of these cells will differentiate into *plasma cells*, which do the actual antibody producing. Other cells of the clone become "memory" cells instead of antibody producers. They are ready to respond promptly if that antigen returns in the future: the memory cells will divide and form new cells, leading to new plasma cells and antibody production, and to new memory cells.

A different antigen would bind different B cells, and stimulate proliferation and differentiation of a different clone, resulting in formation of another antibody. Each group of B cells apparently can produce only one kind of antibody, which means that there must be many thousands of B cell types to respond to all the possible antigens. This is is apparently the case. It takes a few days for B cells to be stimulated and an antibody to be produced the first time, but because of the memory cells a second exposure to an antigen provokes an almost immediate production of antibody.

The immunity conferred when there is a direct contact with a microorganism (that is, an antigen) that causes antibody production is known as **active immunity.** It is the sort of immunity one acquires by having a disease. Active immunity can also be acquired by exposing the individual to a small amount of weakened antigen, enough to stimulate antibody production but not enough to cause full-blown illness

and all the symptoms of the disease. This is the effect of most vaccinations and immunizations. Some antigens, however, do not leave any memory cells, and the response to each exposure is like the first. A **passive immunity** can be conferred, however, by injecting antibodies produced by someone else, that is, literally injecting the immunity, rather than inducing the body to produce its own immunity (antibodies). This provides immediate protection, but it lasts only as long as the injected antibodies remain intact in the circulatory system, usually only a few weeks.

Antibodies provide protection because of events triggered when they combine specifically with the antigen that stimulated their production. The antibodies, which are part of the **gamma globulin** fraction of the plasma proteins, are also called **immunoglobulins (Ig).** A segment of the antibody molecule is able to bind a system of enzymes found in the plasma (known as *complement*), which initiates the reactions by which invading microorganisms are destroyed. The antibody, therefore, does not actually do the destroying, it recognizes a microbe as "foreign" and then binds and activates complement, which does the job. Various members of the complement family are responsible for different aspects of the process. Some cause dilation of blood vessels and increase the permeability of capillaries, some cause release of histamine (which reinforces the above effects), some attract phagocytes (chemotaxis), and some aid the entry of phagocytes into the tissues and stimulate their phagocytic activity. In other words, complement enhances and amplifies the inflammatory reaction. In addition, some elements of the complement system act directly to kill the invading microorganisms.

Cell-mediated immunity: A T cell, when exposed to a particular antigen, recognizes and binds the antigen to its surface. In the process, the T cell somehow becomes "sensitized" to that antigen. In so doing, it acquires the ability to release a variety of chemicals whenever it binds that antigen. As with B cells, only a few of the T cells become sensitized by a given antigen, and some of them become "memory" cells that will hasten the response to a second exposure to that antigen.

When sensitized T cells bind antigen, they release their chemicals, and these chemicals stimulate all the actions that enhance the inflammatory response and some that directly kill the microorganisms or foreign cells. In addition, the chemicals released by T cells attract phagocytes, especially monocytes, keep them in the area, and stimulate phagocytosis in the monocytes—now macrophages. They become very active and vigorously attack almost anything in the way.

Thus we see many similarities between the action of B cells and T cells in spite of their differences. The B cells (or plasma cells) remain wherever they take up residence and send their antibodies into the circulation and to the site of infection. The antibodies bind antigen and complement, causing components of complement to amplify the inflammatory response and destroy the invader. Sensitized T cells bind antigen and then release chemicals that attack the microbes or foreign tissue. T cells return to the circulation and move to the site of infection. From the action of these agents on the inflammatory response, we see that the nonspecific response mediated by phagocytes is also an important part of the response of the specific immune systems mediated by the lymphoctyes.

The presence of symptoms of the inflammatory reaction signals to us that some sort of "foreign" matter is present, and that our defenses are responding to it. In the previous chapter it was mentioned that the glucocorticoids (cortisol) suppress the inflammatory reaction by reducing the breakdown of lysosomal membranes and the subsequent release of their destructive enzymes. Glucocorticoids also inhibit the release of histamine, which normally increases the response. While it is often highly desirable to avoid the discomfort of the symptoms of inflammation, it may be a mixed blessing to do so, because the presence of an infection may go undetected if there are no symptoms.

Sometimes we are, or become, very sensitive to a particular antigen, or antigens, and the response to them may actually involve damage to the tissues. This is a case of *hypersensitivity* or *allergy*, which may involve antigens that act through either a humoral or a cell-mediated system. The response is often associated with excessive histamine release and may be local or systemic (widespread); it may even be life-threatening, as is the case for those who are hypersensitive to the sting of bees and similar insects.

When an organ transplant is made, the recipient's cells will recognize the donated organ as foreign and eventually will react against it. Transplants stand the best chance of success when the donor is a near relative (ideally an identical twin), because the cells would be less foreign. Even so, the recipient must be treated to suppress the cell-mediated immune systems in order to minimize the chances of rejection of the organ. Of course, extreme care must be taken to avoid, and protect against, any infection by an organism that would normally be controlled by a cell-mediated immune system.

White Cell Abnormalities The leukocyte count fluctuates much more widely and readily than the red cell count. **Leukopenia** is a reduction in the number of circulating leukocytes, usually as a result of damage to the leukocyte-forming tissue in the bone marrow. The stem cell is extremely sensitive to ionizing radiation (such as nuclear radiation or X rays). Bone marrow is also affected by various poisons, and some drugs used clinically may have toxic side-effects on the marrow. Leukopenia creates a serious situation because it leaves the individual without adequate protection against microorganisms that require response of phagocytes or immune systems.

A mild to moderate increase in circulating leukocytes is known as **leukocytosis.** It is generally the result of an acute infection, but it may also follow a physiological stress such as strenuous physical exercise. The increase is chiefly in granular leukocytes and the white count may reach 50,000 cells/mm.[3]

Focus

Gnotobiosis and Bubble Babies

It has proven useful at times to raise laboratory animals in isolation from the bacteria, viruses, and allergens to which all living things are normally exposed. The animals are born by Caesarean section in a sterile operating room, immediately washed, and promptly transferred into a closed chamber where they can be given sterile foods and sterile air and handled with sterile instruments. Where normal animals acquire populations of millions of bacteria on their skins and in their guts soon after birth, these animals acquire none. They are germ-free, and as such they enjoy peculiar problems. They have no intestinal bacteria to provide vitamin K or other nutrients. Their stools, which in humans contain large amounts of dead bacteria, are small. Their immune systems, which in other animals are constantly stimulated by foreign cells, lie dormant.

These animals are known as gnotobiotic or germ-free animals. Among other things, they have made it possible to study the body's responses to single infections and organ transplants without the confusion introduced by extraneous immune system activity. By using them, we have acquired the technology necessary for handling one rare human ailment, the lack of a functioning immune system.

Very occasionally, a baby is born who lacks the B cells that produce antibodies and the T cells that attack foreign cells. He or she has no defenses against infection and is likely to become sick and die upon the first exposure to the world and its bacteria. If the child is to survive, he or she must be kept in a germ-free environment. Thanks to the work on gnotobiotic animals, we have learned how to build and maintain such environments.

A few years ago, a boy was born without an immune system. To protect his and his family's privacy, he is known only by the pseudonym "David." As soon as his problem was diagnosed, he was put in a sterile room and kept there, where he could not be attacked by microorganisms. As a baby, he could be cuddled only through heavy rubber gloves of the sort used for working in a microbiologist's "glove box," a chamber used for keeping dangerous materials out of contact with an experimenter. He could be talked to only through loudspeakers; he therefore had to be taught everything from how to speak to his elementary school lessons by remote control. He lived in a plastic bubble, with no real, intimate contact with the people around him.

Despite years of effort, the researchers have not yet found a way to give David an immune system. They have, however, been able to give him a portable bubble. NASA scientists devised a variation on the spacesuit that allows David to walk around and still stay germ-free. But until they find an immunological solution, David will never be able to smell a flower, go swimming, or pet a puppy. His plastic bubble, even the portable one, will never let him touch the world.

Mononucleosis is the presence of an abnormally large number of monocytes.

Leukemia, literally "white blood," is a form of cancer in which the rate of leukocyte production is greatly accelerated and the white count may reach several hundred thousand per cubic millimeter. It may be a myeloid or lymphocytic leukemia, depending upon whether there is an excess of granular or agranular leukocytes. Excessive numbers of leukocytes may result in such problems as: (1) anemia, because the red cells are literally crowded out; (2) heart strain, because the blood becomes more viscous; (3) actual clogging of some blood vessels with the cellular mass; and (4) severe overburdening of the organs that normally remove worn-out blood cells. In addition, the leukocytes themselves are likely to be abnormal and therefore to fail to carry out their tasks (such as phagocytosis) in spite of their great numbers.

In certain conditions or diseases a particular type of leukocyte is increased or decreased more than others. The reasons for these deviations are not always understood, but recognition of the condition is often of diagnostic value. Relative numbers can be determined by doing a *differential white count,* in which 100 white cells are classified and the results are compared with the normal distribution of white cells.

Platelets

Platelets are tiny disclike bodies with a diameter that is usually slightly less than half that of a red blood cell. They stain very darkly and show few clear structural features. They tend to stick together and in blood slides are usually found in groups. Platelets are not cells and have no nuclei. They are fragments of the cytoplasm of a type of large cell of the bone marrow (a megakaryocyte). There are about 250,000–300,000 platelets per cubic millimeter of blood, which is 30–50 times the number of leukocytes. On account of the important function that platelets serve in initiating the formation of a blood clot, they are sometimes called *thrombocytes* (thrombo = clot).

TABLE 14–2   COMPOSITION OF PLASMA

Constituent	Percent of Plasma
Water	90–92%
Protein	6–8
Inorganic ions	<1 (0.9)
Organic substances	<1 (0.5–0.9)

PLASMA

The fluid portion of the blood, in which the formed elements are suspended, is the blood plasma. As shown in Table 14–2, it is mostly water, but it also contains proteins and other organic and inorganic constituents. The latter two are present in small but very important amounts.

Plasma Protein

The plasma proteins are in solution in the plasma, as distinct from the structural proteins in cells or parts of intercellular fibers. The plasma proteins, unlike many constituents of the blood, are important *in* the blood, and if they should leave it to enter the tissue, serious problems would arise.

Most of the protein in the plasma belongs to one of the categories listed in Table 14–3. *Albumin* is the most abundant, accounting for slightly more than half the total protein content. The higher concentration and smaller molecular size of albumin mean that there are many more particles of albumin in the plasma than of any of the other proteins. Albumin is therefore the major protein contributor of the osmotic pressure of the plasma.

TABLE 14–3   PLASMA PROTEINS

Protein	Plasma Concentration	Molecular Weight
Albumin	4.5%	70,000
Globulin	2.5%	150,000
Fibrinogen	0.25%	400,000

Protein in the plasma is important because it is a nondiffusible solute. Since there is very little protein in the interstitial fluid, its presence in the plasma is responsible for the osmotic gradient across the capillary wall. If plasma protein content falls, the osmotic pressure of the plasma falls and there is a danger of fluid leaking from the blood vessels into the tissues. Albumin molecules are just barely too large to pass through the walls of capillaries. However, with only a slight change in the permeability of the vessel walls, albumin can, and does, leak out.

Plasma protein both increases the viscosity of the blood and contributes to the regulation of acid—base balance, that is, to maintaining the pH of the blood. In severe malnutrition or starvation, plasma protein may be used as a source of energy. However, it is not an efficient source and it is used only when other sources have been depleted.

Some plasma proteins have additional specific functions. *Fibrinogen* can be converted into the insoluble protein clump, the blood clot. The *globulins*, especially those known as the gamma globulins, are important in immunities and antigen-antibody reactions.

Inorganic Ions

The inorganic ions of the plasma are the same as those found in the extracellular fluid. The total concentrations are the same, and the concentrations of most of the individual ions are also similar. These ions are the electrolytes, chiefly sodium, potassium, calcium, magnesium, chloride, phosphate, and bicarbonate. Although the total concentration is relatively low (0.9%), the small size of the ions means a great number of osmotically active particles in the plasma. Since the electrolyte concentration in interstitial fluid does not differ notably from that of the plasma, and the walls of the smallest blood vessels permit passage of electrolytes with relative ease, the only osmotic gradient between blood and interstitial fluid is due largely to the proteins.

Changes in plasma electrolyte concentrations are followed by similar changes in interstitial fluid, which in turn affect the cells, and vice versa. Shifts in potassium affect the membrane potential of cells, shifts in calcium affect the excitability of cells, and shifts in sodium affect the movement of water in and out of blood vessels. Some of these specific actions are considered elsewhere.

Organic Substances

There are several nonprotein organic substances present in the plasma, though in relatively small amounts. The least abundant—although still extremely important—are the *hormones, vitamins,* and certain *enzymes.* The most abundant is the *blood sugar,* which is almost exclusively glucose, a major source of energy for cellular metabolism. Elaborate homeostatic mechanisms keep the blood glucose level within a fairly narrow range, even while fasting or after the ingestion of a large meal.

Other organic substances in the plasma include those grouped as *nonprotein nitrogen.* Nitrogen is an essential and constant component of protein, and virtually all nitrogen taken into the body is associated with protein in some way. Nonprotein nitrogen is the nitrogen associated with substances that will become, or have been, protein, including amino acids and products formed by the metabolism of proteins. The latter, found in the plasma because they are being transported to the kidney for excretion, include urea, uric acid, creatine, and creatinine.

Plasma also has a significant content of *lipid substances.* They may be fats on their way to the tissues for storage or to a cell for utilization, or they may be specific lipid-containing substances—phospholipids or cholesterol. The cholesterol may be ingested, but it is also produced in the body. Its role in the body economy is not completely understood, although a number of functions are known. It is perhaps best known for its tendency to be deposited in the walls of blood vessels of certain individuals, contributing to a condition known as *atherosclerosis.*

BLOOD CLOTTING

The blood is kept within a closed system (except for exchanges through the walls of the smallest vessels), and throughout much of this system the blood is under fairly high pressure. To ensure that the blood stays in the system, a means is needed to stop leaks and to minimize blood loss if a vessel is broken. Stoppage of the flow from blood vessels, or *hemostasis* (*not* homeostasis), is brought about chiefly by the formation of a blood clot, a process known as *blood coagulation*. It involves the formation of an insoluble mass that stops the flow of blood as it leaves the blood vessel.

Fresh blood drawn into a test tube will clot in a few minutes into a red, semisolid, gelatinous mass. The clot is formed by a meshwork of strands of a protein called *fibrin*. As the clot forms, blood cells are trapped within the meshes, and the clot appears to be red. If freshly drawn blood is whipped, however, the fibrin strands will form on the beater without trapping red cells; the clot then appears as white or ivory in color. The remaining fluid, which contains the cells and other constituents of plasma, will not clot.

A clot that is allowed to stand undisturbed tends to retract after a time, to shrink and withdraw from the edges of the container, squeezing out a clear straw-colored fluid known as **serum** (Figure 14–5). The distinction between serum and plasma is that plasma is the fluid portion of *unclotted* blood. The difference is the absence from serum of those constituents which contribute to clotting.

Hemostasis

The first thing that happens when a blood vessel is damaged is a marked constriction of the wall of the damaged vessel. It is due to a spasm of the muscle in the vessel wall, to a reflex constriction, and to the effects of substances that are released from the injured tissue. The constriction reduces the size of the opening, and may be enough to completely occlude or block even a fairly large artery, at least for a time.

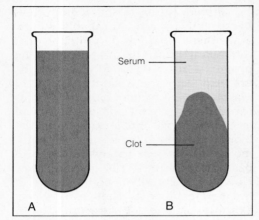

FIGURE 14–5   Retraction of a blood clot. A. Fresh clot. B. After retraction.

Damage to the blood vessel interrupts the smooth epithelial surface of the vessel and is likely to expose the connective tissue (especially collagen) that lies under the epithelium. Both a rough surface and collagen attract platelets, which adhere to the vessel wall and to one another. The platelets, or thrombocytes release substances that make them very sticky, and more platelets become attached. They also release substances (including epinephrine) that increase and prolong the vasoconstriction. The result is the formation in a rather short time of a clump of platelets firmly stuck to the vessel. This **platelet plug** blocks the opening in the vessel and reduces the loss of blood until a clot can be formed.

Formation of the Clot

The essential reaction in blood clotting is the conversion of the soluble plasma protein fibrinogen to the insoluble protein fibrin. However, the process involves many steps and is extremely complex. A dozen or more substances, called factors, participate. Most of them have several names and so, to reduce the confusion, they have been arbitrarily designated by Roman numerals. To ensure that clotting always occurs at the right time and never occurs at the wrong time, there are many checks and balances. All of the clotting elements are present in the circulating plasma, but in inactive form; all that is needed is a trigger to set it off. Obviously the trigger is not present in the plasma or clotting would occur continually.

The basic events in clotting, however, are quite straightforward:

1 Prothrombin, a protein in the plasma, is converted to thrombin (an enzyme).
2 Thrombin catalyzes the conversion of fibrinogen to fibrin.

In both reactions an inactive substance present in the blood is activated, and the factor activated in the first step is what causes the second to occur. The real complexities of the process are those that lead to the first reaction, the activation of prothrombin.

There are two mechanisms or pathways by which this can be done, an *intrinsic* or intravascular pathway, in which the trigger comes from within the blood vessel, and an *extrinsic* or extravascular pathway, in which it comes from the damaged tissue (see Figure 14–6).

The intrinsic mechanism is initiated when blood comes in contact with the damaged vessel wall, particularly the collagen. It seems to activate plasma factor XII, which activates factor XI, which activates factor IX, which (in the presence of platelets and factor VIII) activates factor X, and factor X (in the presence of platelets and factor V) activates prothrombin. Finally, prothrombin is converted to the active agent thrombin. The sequence is often described as a cascade, with each factor activating the next in line.

Not only are platelets and factors V and VIII necessary for certain steps, but calcium ions must also be present at many of the steps. In fact, calcium is officially factor IV.

The extrinsic pathway is not as long or complex. The mechanism is initiated when a substance, called **tissue thromboplastin,** is released from injured cells of the blood vessel or tissues. Tissue thromboplastin activates factor VII, which activates factor X, which then activates the prothrombin.

The thrombin formed when prothrombin is activated catalyzes the final and critical reaction in the clotting

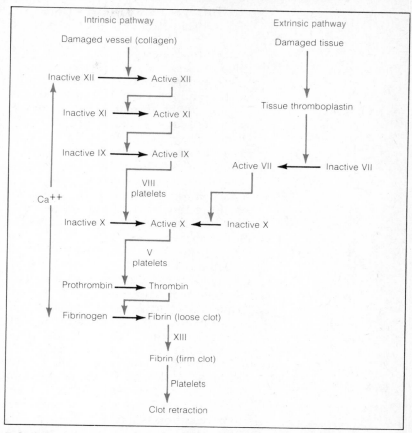

FIGURE 14–6   Summary of the blood clotting mechanism.

process. It causes the conversion of the soluble plasma protein, fibrinogen, to the insoluble protein, fibrin. As the conversion develops, the fibrin forms a network of threads and strands that entraps blood cells. The clot is rather loose at first, but it is soon stabilized by the action of factor XIII which causes the formation of cross-linkages between the fibrin strands.

Within a few minutes after its formation, the clot begins to shrink. Platelets trapped in the clot are responsible for this process, which is known as *clot retraction.* The platelets, which contain a contractile protein and ATP for energy, attach themselves to the fibrin strands. The action of their contractile proteins draws the surfaces of the wound together and reduces the size of the opening. It also squeezes out any fluid caught in the clot. This fluid, of course, is serum, since fibrinogen and many other clotting elements have been removed (see Figure 14–5).

Intravascular Clotting and Anticlotting Mechanisms

A roughness in the lining of a blood vessel may cause some platelets to stick there, where they tend to disintegrate and release substances that contribute to the formation of thrombin. The consequent small clot, called a **thrombus,** impedes the blood flow and may become large enough to occlude or block the vessel completely. A thrombus in a blood vessel of the heart (a coronary thrombosis) may result in the death of the cardiac muscle fibers that are no longer receiving an adequate blood supply.

A thrombus that breaks loose is called an **embolus.** It is swept along in the blood vessel until it finally lodges in a vessel that is too small for it to pass. This may occur in the tissues, but it is more likely in the lungs (a pulmonary embolism), where it may cause serious damage to the delicate lung tissue.

Clotting at the wrong time can be as undesirable as a failure to clot, and there are several anticlotting mechanisms to prevent it. They not only help to prevent formation of unwanted clots in vessels, but also remove any unwanted clots that may form, as well as any clots that have served their hemostatic function.

The normal mechanism for clot disintegration is a slowly developing process that is initiated by the same trigger as clotting. Vessel damage acts on factor XII to start the cascade of events that lead to clot formation. That same factor XII also initiates a series of steps that lead to activation of an enzyme in the plasma (plasmin) which then dissolves the clot by breaking down the fibrin. Needless to say, this reaction is quite slow, so that the clot has served its purpose by the time it disintegrates.

There are several agents that tend to prevent or inhibit intravascular clotting. One of these anticoagulants is *heparin,* found in the cytoplasmic granules of certain connective tissue cells (the mast cells) and basophils. Heparin prevents activation of factor IX, and may be administered clinically to prevent formation of a thrombus or embolus. Another anticoagulant used clinically is *dicumarol,* first discovered in spoiled sweet clover, in a search undertaken because Wisconsin dairy cattle were suffering from a severe hemorrhagic disease.

Failure of the Clotting Mechanism

If any of the essential elements of the clotting mechanism are absent, a clot will not form. There are therefore several reasons for clotting failures other than the presence of anticoagulants. Some of them are listed below.

1. *Calcium lack.* This is not physiologically important because a calcium level low enough to interfere with coagulation would cause other serious problems first. *Oxalate* and *citrate* are anions with a great affinity for calcium, and if added to blood they will prevent coagulation by making calcium ions unavailable. Clotting could occur, however, if excess calcium were added later. The addition of citrate to blood samples is a common method of preventing clotting in the test tube or during storage.

2. *Lack of vitamin K.* Prothrombin and several other clotting factors are produced in the liver and require vitamin K for their synthesis. Vitamin K is a fat-soluble vitamin, and its absorption from the intestine depends upon the presence of bile from the liver. If an obstruction blocks bile from reaching the intestine, vitamin K is not absorbed and clotting problems develop because of a lack of the vitamin. (The anticoagulant dicumarol inhibits the action of vitamin K.)

3. *Hemophilia.* The so-called bleeder's disease is a hereditary condition affecting males almost entirely. There are several forms, all of which are due to a lack of one of the factors involved in the release or activation of thromboplastin from the platelets. The danger often lies not so much in bleeding from a wound (for there may be tissue thromboplastin), but in the prolonged oozing of blood from ruptured tiny blood vessels within the tissues. A simple bruise may result in a slow loss of blood in the tissues for hours, producing a large painful lump. In addition, bleeding in and around joints may interfere with movement.

THE BASIS OF BLOOD GROUPS

Early efforts to transfuse blood from one individual to another were sometimes successful. However, more often than not they met with disastrous failure because the transfused cells would clump together, or *agglutinate*, and eventually hemolyze (rupture). This undesirable reaction has been found to be due to an antigen-antibody reaction.

Blood groups are based upon two particular types of antigen and two types of antibody, which may or may not be present in the blood. Since the antigen-antibody response that occurs in this case is a clumping or agglutination of the transfused red blood cells, the antigens are called **agglutinogens** and the antibodies are called **agglutinins.**

The antigens or agglutinogens are in the membranes of the red blood cells, and they are designated as A or B. A person may have either, both, or neither, and his blood type is said to be A, B, AB, or O, accordingly. Whichever agglutinogen is lacking is a foreign protein to that person, and his serum contains antibodies against it. These are naturally occurring antibodies and do not depend upon previous contact with the agglutinogen. The antibodies, or agglutinins, are designated as α (anti-A) and β (anti-B), and a person may have either, both, or neither of these, depending upon which antigens are on the cells. A summary of blood groups is given in Table 14-4.

Blood type can be determined by a simple test. A drop of the blood to be typed is added to serum containing α (anti-A) agglutinin and to serum containing β (anti-B) agglutinin. Any clumping of the added cells can be readily detected without the aid of a microscope or other instrument (Figure 14-7 and Table 14-5).

When a transfusion is to be given, one is concerned only about the effect of the recipient's plasma (containing agglutinin) upon the donor's red cells (agglutinogen), since agglutinins in the recipient's plasma would cause clumping of the transfused donor cells. The plasma of the transfused donor blood

TABLE 14-4 BLOOD GROUPS

Blood Type	Occurrence (percent)	Agglutinogen (on cells)	Agglutinin (in serum)
O	45	none	α(anti-A) and β(anti-B)
A	41	A	β(anti-B)
B	10	B	α(anti-A)
AB	4	A and B	none

would not affect the recipient's cells because it would be greatly diluted by the recipient's plasma. Blood type O, with no agglutinogens on the cells, could thus be mixed with blood of any type without causing agglutination. A person with type O blood is therefore known as a **universal donor.** Blood type AB, on the other hand, with no agglutinins in the serum, could receive blood of any type without clumping, and a person with type AB blood is known as a **universal recipient.**

The A and B antigens are not the only antigenic proteins in the blood which must be reckoned with. Another is the **Rh factor,** named for the rhesus monkey in which it was first identified. There are several Rh subgroups, but the presence of any of them—as in about 85 percent of the white population—marks the individual as Rh-positive. Those individuals who have no Rh factor at all are Rh-negative. Rh-negative blood does not contain antibodies for the Rh factor, but they are produced upon exposure to it, as by transfusion of Rh-positive blood. Routine blood typing includes determination of the Rh type as well as of the A-B type.

A problem may arise if an Rh-negative mother should bear an Rh-positive child. Blood of the mother and fetus do not normally mix, but some Rh-positive red blood cells from the fetus may diffuse into the maternal circulation and cause the production of antibodies against them. Some of these antibodies will diffuse back into the bloodstream of the fetus and cause damage and destruction of the fetal blood cells, which can be fatal if enough antibodies enter. The antibody concentration of the mother's blood is likely to increase with exposure to Rh-positive blood, thus increasing the danger with

TABLE 14–5   TRANSFUSION REACTIONS

Donor's cells	Recipient's serum			
	anti-A&B (type O)	anti-B (type A)	anti-A (type B)	None (type AB)
O	−	−	−	−
A	+	−	+	−
B	+	+	−	−
AB	+	+	+	−

+ = agglutination; − = no reaction.

each successive pregnancy. The condition (erythroblastosis fetalis) has been treated by transfusion of the newborn, or even by total replacement of its blood to rid it of the antibodies from the mother's blood (which will not be produced by the Rh-positive infant).

There are a number of other antigens in the blood, but they do not ordinarily cause transfusion problems. Because their presence, or absence, is inherited, determination of some of these is sometimes used as an aid in cases of disputed parenthood.

BLOOD IN ACID–BASE BALANCE

One of the most critical factors in maintaining the internal environment is the acidity of the body fluids. The acidity of the cells, the interstitial fluid, and the plasma are all interdependent, although most of our regulatory mechanisms are directed at the plasma. Acidity is determined by the concentration of the hydrogen ions that are formed when acids ionize and is expressed as the **pH.** Hydrogen ion concentration, [H⁺], must be maintained within rather narrow limits because all the reactions of the body occur in solution, and they are adversely affected when it deviates from the optimum concentration. Proteins are particularly sensitive to pH changes and, as enzymes, they are involved in virtually every reaction. Amino acids (and hence proteins) can ionize as either an acid (release H⁺) or as a base (accept H⁺), depending upon the pH. Plasma proteins bind many substances (including hormones) as they are transported in the blood, and

the extent of the binding varies with the pH. The effectiveness of substances such as hormones, ions, and drugs is influenced by acidity. Changes in pH cause some ions to shift between the intracellular and extracellular fluid compartments, affecting vital (electrical) gradients.

Recall that hydrogen ions are released when an acid dissociates in solution, but that all acids do not ionize to the same extent. Those which are highly ionized and release many hydrogen ions into the solution are "strong" acids; those which are weakly ionized and release relatively few hydrogen ions are "weak" acids.

Aqueous solutions contain both hydrogen and hydroxyl ions, derived from the ionization of any acid or base in the solution and from the ionization of water (it is very slight, but it does occur). The hydrogen and hydroxyl ions will combine to form water until an equilibrium is reached:

$$[H^+] + [OH^-] \rightleftarrows HOH$$

At equilibrium the product of hydrogen and hydroxyl ions is a constant:

$$[H^+] \times [OH^-] = 1 \times 10^{-14} \text{ moles/liter}$$

For water, which forms equal numbers of hydrogen and hydroxyl ions upon ionization,

$$[H^+] = 1 \times 10^{-7}, \text{ and } [OH^-] = 1 \times 10^{-7}.$$

In acid solutions the [H⁺] increases, while the [OH⁻] goes down. In alkaline solutions the [H⁺] falls and [OH⁻] rises (see Table 2–5). The acidity of solutions is expressed in terms of the concentration of hydrogen ions, specifically the negative log of the [H⁺], or pH. For water, the pH is 7, and it decreases when something is added that causes the [H⁺] to rise. A change from pH 7 to pH 6 means a tenfold increase in [H⁺], and a change from pH 7 to pH 5 means a hundredfold increase.

pH of the Blood

The pH of the blood is very close to 7.4, with normal variation between 7.35 and 7.45; the range compatible with life is 7.0–7.8. The blood pH is therefore slightly basic (alkaline), and this is also generally true of the fluid of the internal environment. To keep the pH of the internal environment within the proper limits, the pH of the blood must be carefully controlled. The mechanisms for this are complex, involve many different systems, and are in continual operation.

Continuous regulation is needed because the bloodstream is almost constantly subjected to an "acid load"; that is, most of the substances which enter it as a result of cellular metabolism are acidic. Carbon dioxide in solution forms carbonic acid (H_2CO_3), which, although it is a weak acid, is present in large amounts. Also transported in the blood are the acidic intermediate products of carbohydrate and fat metabolism and several acidic endproducts of protein metabolism. The rates at which these substances enter the bloodstream vary with metabolic activity, and without regulatory mechanisms the pH of the blood would fluctuate beyond the tolerable range.

There are three major means of coping with an acid (or alkaline) load: the lungs, the kidneys, and the blood. The first two are concerned with excreting the acid substances from the body, and the third with handling them safely until that time. The lungs remove carbon dioxide from the blood. The kidneys both remove other metabolic wastes and regulate the loss of water and certain inorganic ions, which in turn helps maintain the pH of the blood. The blood must carry the acidic substances from the tissues where they are formed to the lungs or kidneys for excretion, but must do so without change in the blood pH. The roles of the lungs and kidneys in acid-base balance are discussed more fully in Chapters 20 and 28, respectively.

Buffers

In a sense, the blood is the first line of defense against pH changes in the internal environment, since it serves as a buffer between the acidity of the metabolites and the cells. It does so by means of substances that we call *buffers*, which are found both in plasma and in

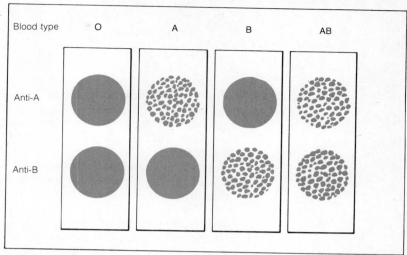

FIGURE 14–7 Determination of blood groups by agglutination with sera of a known type.

red blood cells. Buffers are present in all cells and in the interstitial fluid, as well as the blood, for pH changes must also be minimized there. Those buffers in the blood cells and plasma, however, are particularly important in acid-base regulation throughout the body.

Buffers work in pairs, usually a weak acid and a salt of that acid with a strong base, or a weak base and a highly ionized salt of that base. The important buffers are the bicarbonate, protein, and phosphate buffers. In the extracellular fluid (interstitial fluid and plasma), the bicarbonate buffer, for example, consists of the weak acid, carbonic acid (H_2CO_3), and a salt with a strong base, sodium bicarbonate ($NaHCO_3$); the buffer pair is conventionally written as

$$\frac{NaHCO_3}{H_2CO_3}.$$

The protein and phosphate buffers in the extracellular fluid are:

$$\frac{Na\ proteinate}{H\ proteinate} \quad and \quad \frac{Na_2HPO_4}{NaH_2PO_4}.$$

In extracellular fluid the proteins are largely albumin and globulin.

The same buffers are present in cells, except that the cation is K^+ instead of Na^+, and their relative concentrations may differ. These buffer pairs are:

$$\frac{KHCO_3}{H_2CO_3}, \frac{K\ proteinate}{H\ proteinate}, \quad and \quad \frac{K_2HPO_4}{KH_2PO_4}.$$

In red blood cells the protein is hemoglobin (KHb/HHb).

The action of buffers can be illustrated by the bicarbonate buffer system, which is quantitatively the most important system. The protein and phosphate buffers operate in much the same manner.

If a strong acid, such as hydrochloric acid (HCl), is added to a solution containing no buffer, the [H^+] rises markedly and the pH drops. But if HCl is added to a solution containing the bicarbonate buffer, the strong, highly ionized acid reacts with the salt of the buffer pair (sodium bicarbonate), to produce sodium chloride and carbonic acid. Bicarbonate (HCO_3^-) is a stronger base than Cl^-, and binds H^+ more firmly, so hydrogen displaces sodium. The carbonic acid formed is a weak acid that does not dissociate readily and therefore provides few hydrogen ions. It lowers the pH much less than the same amount of HCl. Thus a strong acid is converted to a weak acid, and the effect on pH is minimal. Under certain conditions the carbonic acid then releases CO_2 to be excreted from the lungs, leaving only water:

$$HCl + NaHCO_3 \rightarrow NaCl + H_2CO_3$$
$$\downarrow$$
$$H_2O + CO_2$$

There are times when there is an alkaline load instead of an acid load, but the buffer system can cope with this type of load also. The highly ionized sodium hydroxide (NaOH) reacts with the carbonic acid of the buffer pair to produce the bicarbonate. In this case OH^- is a stronger base than bicarbonate, and hydrogen displaces sodium. Again, a highly ionized substance is converted to a weakly ionized substance:

$$NaOH + H_2CO_3 \rightarrow NaHCO_3 + H_2O$$

In a buffering action the concentration of one of the buffers is reduced and that of the other is increased. With an acid load, bicarbonate falls and carbonic acid increases. With an alkaline load, bicarbonate increases and carbonic acid decreases.

Normally the plasma ratio of $NaHCO_3:H_2CO_3$ is 20:1. With a heavy acid load, the amount of $NaHCO_3$ in the blood decreases and the amount of H_2CO_3 increases, reducing the ratio and lowering the pH slightly. An alkaline load brings a shift in the opposite direction, increasing the ratio and raising the pH slightly. The magnitude of the ratio gives an indication of the saturation of the buffering mechanism and of whether there is adequate bicarbonate to buffer acids. When the buffered substances, acids or bases, have been removed from the blood by the lungs or kidneys, the buffer systems are ready to be used once more.

CHAPTER 14 SUMMARY

The blood vessels normally contain about 5 liters of blood consisting of *formed elements* and *plasma*. The formed elements, which make up about 45 percent of the blood volume *(hematocrit)*, include erythrocytes, leukocytes, and platelets.

Formed Elements

Erythrocytes (red blood cells, RBCs) are tiny biconcave discs without a nucleus. In a hypotonic solution osmosis occurs and the cells swell and perhaps rupture *(hemolysis)*; in a hypertonic solution fluid leaves the cells and they shrink. The main component of RBCs is *hemoglobin*, which consists of a protein, globin, and an iron-containing pigment, *heme*. It can form a loose combination with oxygen (oxyhemoglobin, HbO_2) and it can release the oxygen (deoxyhemoglobin).

RBCs are formed in red bone marrow from primitive stem cells in a many-stepped process. After a life span of about 120 days they are broken down and phagocytized by macrophages. The hemoglobin is largely excreted in bile by the liver, the iron is recycled, and globin is metabolized. RBC formation *(erythropoiesis)* is stimulated by low oxygen *(hypoxia)* of blood to the kidney, resulting in formation of *erythopoietin,* a hormone that stimulates the bone marrow.

Leukocytes (white blood cells, WBCs) are also formed in the red bone marrow, but they eventually leave the circulation to perform their functions. The granulocytes are *neutrophils, eosinophils,* and *basophils,* and they can be distinguished by the staining properties of their cytoplasmic granules. Agranulocytes, which lack prominent granules, are the *lymphocytes* and *monocytes.*

Leukocytes are important for body defenses. The neutrophils and monocytes are involved in nonspecific defenses. Invading microorganisms release substances that initiate the inflammatory response in which blood vessels dilate and fluid moves from blood vessels into the area. Neutrophils, attracted to the site by chemotaxis, leave the blood vessel and promptly attack the invaders. Neutrophils are phagocytic, and their granules contain powerful enzymes to destroy the microorganisms. Monocytes respond in a similar manner, but more slowly and are longer lasting.

Specific immune systems depend upon lymphocytes. They are also formed in the bone marrow, but come to rest at least temporarily in lymph nodes and other lymphoid organs. Foreign substances *(antigens)* elicit an immune response from tissues. **Humoral immunity** involves *B lymphocytes,* and **cell-mediated immunity** involves *T lymphocytes.* Certain B lymphocytes respond to a given antigen by binding it, and then proliferating. Some of the new cells become plasma cells and produce antibodies against the antigen and release them into the circulation. Other cells become memory cells which will hasten the response to a second exposure to that antigen. *Antibodies* are proteins that are specific for a particular antigen. They bind to certain proteins in the plasma *(complement)* which enhances all phases of the inflammatory response and thus increases destruction of the harmful foreign substance.

Cell-mediated immunity depends upon T lymphocytes, which acquire their specific properties in the thymus gland before they take residence in one of several lymphoid tissues. Upon exposure to an antigen a few T cells recognize and bind the antigen and become sensitized to it. They proliferate and form some memory cells and some cells that produce and release chemicals that both enhance the inflammatory response and directly attack the offending foreign substance.

Plasma

Substances transported in the blood are dissolved or suspended in the plasma. These include the plasma proteins (chiefly *albumins, globulins,* and *fibrinogen*) which are responsible for the colloid osmotic pressure of the plasma, inorganic ions (electrolytes), and organic substances such as glucose and other nutrients, metabolic products, hormones, and many other substances.

Blood Clotting

When a blood vessel breaks, hemostasis is established by several events, chiefly clotting of the blood. Initially the injured vessel constricts. Contact with damaged tissues and/or roughened surfaces causes platelets to aggregate and form a temporary plug at the injured site. Clot formation involves converting fibrinogen to fibrin threads which form a tight network, the clot. The formation of fibrin is catalyzed by thrombin, which is derived from prothrombin in the blood. Thrombin may be formed by either of two mechanisms. An *intrinsic (intravascular) pathway* occurs entirely within the blood vessel and is triggered by exposure of blood to the damaged vessel wall (probably collagen under the endothelium). This sets off a cascade of events involving a number of "factors" and reactions, several of which require calcium, and eventually leads to thrombin formation. An *extrinsic (extravascular) pathway* is initiated by release of thromboplastin from damaged tissues. It also triggers a sequence of reactions that leads to thrombin formation. The clot formed soon shrinks (*clot retraction*), squeezing out *serum* (plasma minus clotting elements) from the clot.

Basis of Blood Groups

Blood groups are based on the presence (or absence) of certain antigens (A and B) associated with red blood cells, and antibodies (anti-A and anti-B) in the plasma. Each person's plasma naturally contains antibodies (*agglutinins*) for the antigen (*agglutinogen*) that their own RBCs do *not* carry. If a transfusion of blood containing antigen (A) is given to a person whose blood contains the antibody for it (anti-A), the added (donor) cells will clump or agglutinate.

The *Rh factor* is an antigen that induces antibody formation in the plasma of a person lacking this factor (Rh-negative) whenever that person is exposed to it. A problem may develop if an Rh-negative mother has an Rh-positive child. If there is any mixing of fetal and maternal blood, the mother will produce antibodies against the Rh antigen, and if the antibodies get into fetal blood, there will be destruction of fetal erythrocytes.

Blood in Acid-Base Balance

Buffers in the blood allow the blood to transport acidic (or alkaline) substances without a marked fall (or elevation) of the blood pH. Buffers are substances that combine with highly ionized (strong) acids and form weakly ionized acids. They reduce the increase in hydrogen ion concentration that would otherwise occur and hence reduce the fall in pH. Alkaline substances are buffered in a similar manner, preventing a rise in pH. Buffers function in pairs, most commonly a weak acid and its salt with a strong base. The most abundant buffer system is carbonic acid and sodium bicarbonate (in the extracellular fluid) or potassium bicarbonate (in cells). Other important buffers are protein buffers found in the extracellular fluid and in cells (in RBCs the protein is hemoglobin), and the phosphate buffers.

STUDY QUESTIONS

1 What are some of the functions performed by blood?

2 What kinds of problems would you be likely to encounter if your red blood cell count were cut in half? If your leukocyte count were cut in half, would the effect be the same if some types of leukocytes were reduced more than others? Explain.

3 What is involved in the body's nonspecific defense systems? What is the role of neutrophils and monocytes in these mechanisms?

4 What is the difference between B and T lymphocytes? Explain the role of each in humoral and cell-mediated immune systems.

5 Name four important constituents of plasma, and indicate what might happen if the concentration of each of them were elevated or reduced.

6 Distinguish between serum and plasma; homeostasis and hemostasis; vitamin K and vitamin B_{12} (as related to blood).

7 Since most of the substances involved in forming a blood clot are normally present in the blood, one would think blood should clot spontaneously. Why doesn't it?

8 Platelets seem to be involved in several aspects of the overall process of hemostasis and blood clotting. Identify and describe these actions.

9 If you know your blood type, to which blood types could you safely donate blood? From which types could you safely receive blood?

10 How do buffers act, and of what importance are they?

15 _____

Introduction to the Circulation

Structure of the Heart

Excitation of the Heart

Extrinsic Innervation of the Heart

Events in the Cardiac Cycle

Cardiac Output and Its Control

The Cardiovascular System — The Heart

The heart supplies the driving force that causes blood to move through the blood vessels. Without the heart's action, the blood and all the substances it carries would not reach the tissues. The heart is a remarkable organ in many respects, not the least of which is the fact that it must work constantly. The only rest it gets is that fraction of a second between one beat and the next.

When we think of the resting level of cardiac function, we are thinking about the heart when the body is at rest. Although the heart's work load is then at a minimum, the load is still significant and, of course, vital to our very existence. When the body is not at rest and the cells need more of the substances carried in the blood, the heart is able to meet those increased needs. It can step up its action to increase by severalfold the amount of blood it sends out to the tissues each minute.

In this chapter we consider what makes the heart beat, and what actually happens in the course of a single beat. The other important aspect considered is the means by which the heart is controlled, so that its pumping action is adjusted to meet the needs of the tissues, whatever their level of activity.

409

INTRODUCTION TO THE CIRCULATION

As the keystone of homeostasis, blood is vital to the survival of all the cells of the body. To serve this role, it is transported to and from the cells by the cardiovascular system, consisting of a pump (the heart), large and small tubes (the blood vessels), and assorted valves, shunts, and other controls. The flow of blood through the blood vessels and, indeed, the behavior of the entire cardiovascular system are determined by the same physical laws that regulate any system of fluids flowing through tubes.

The cardiovascular system is also called the circulatory system, since to all intents and purposes it is a closed "circular" system. Every drop of blood continually retraces a circular path from the heart into the arteries, through the capillaries, and thence through the veins back to the heart. Any fluid that is lost from the capillaries is returned to the bloodstream by the lymphatic system, and so, unless there is damage to the vessels, it is essentially a closed system. Blood flow in this circle is one-way flow: There are valves in the heart and in the veins to ensure that it does not flow in the wrong direction. All arteries carry blood *away* from the heart, and all veins carry blood *toward* the heart.

The simplified schema of Figure 15–1 illustrates some of the basic facts about the circulatory system.

1.   The heart is actually two pumps, though they are physically joined. Each controls the flow of blood into one of the two loops of the blood's circuit. The right side of the heart sends blood through the **pulmonary circuit** (the lungs) and receives it from the **systemic circuit** (the rest of the body). The left side of the heart receives blood from the pulmonary circuit and sends it out through the systemic circuit.

2.   Each pump consists of a receiving chamber, or **atrium,** and an ejection chamber, the **ventricle.** The ventricles must develop enough pressure to drive the blood through the entire circuit and back to the proper receiving chamber. As might be expected, the ejection chambers and arteries have strong thick walls, and the receiving chambers and veins have thin walls. Because there is more resistance in the longer systemic circuit to move blood through it, a greater force is required. Therefore, the left ventricle and the largest systemic arteries have thicker walls than the right ventricle and pulmonary artery. The right and left sides of the heart must, however, pump equal amounts of blood (over a period of a minute or so), or else sizable quantities of blood will be shifted from one circuit to the other. In addition, each ventricle must pump all of the blood it receives—inability to do so is a sign of a failing heart.

FIGURE 15–1   Schematic diagram of the circulation.

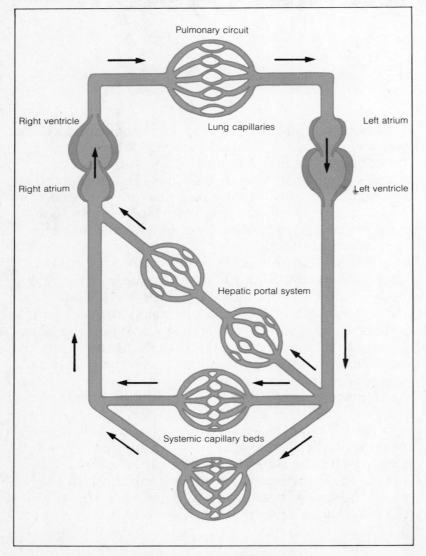

The ultimate role of the circulatory system is to deliver to, or collect from, the cellular environment whatever is necessary to maintain the constancy of that environment, that is, to maintain homeostasis. What exactly *is* necessary to accomplish this, however, varies with the activity of the tissues. Active cells require a greater blood flow than resting cells, and the cells of certain organs, such as the brain, cannot tolerate a reduction in their blood supply. It is essential, therefore, that the blood flow be adjustable, both in the total circulation and in the circulation to specific organs. The heart, by increasing its pumping action, can greatly increase the total blood flow, and local flow can be changed by altering the resistance in blood vessels supplying the individual organs. Enlargement of a vessel *(vasodilatation)* reduces the resistance to flow and increases the volume of blood traveling through that vessel. Narrowing the vessel *(vasoconstriction)* diminishes the blood flow.

In traversing the systemic circuit, a blood cell progresses from larger to smaller vessels (from arteries to arterioles) and into the capillaries, where the vital exchange of substances occurs. There is a capillary bed in every route through the circuit, and in some special regions a blood cell may have to traverse two capillary beds. The blood is then gathered into larger and larger vessels (venules and veins) for return to the heart. One might generalize and say that the large arteries and veins are merely conduits for quick transfer of blood from one site to another. However, the large arteries must also absorb the intermittent spurt of blood ejected by the pulsating pump. Because their diameter is readily variable, the smallest arterial vessels (arterioles) serve as valves to regulate the flow of blood into the capillaries "downstream." The capillaries are the smallest vessels in diameter ($8-10$ μm), length ($1-2$ μm), and in wall thickness (a single squamous cell), but it is in the capillaries that the cardiovascular system carries out its function—the exchanges that maintain the internal environment. Due to the specific roles of the different components of the cardiovascular system, and the varying conditions under which

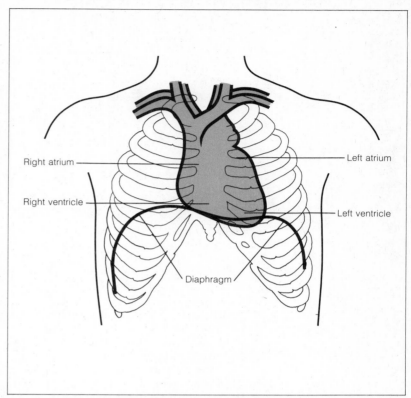

FIGURE 15–2   Position of the heart in relation to the ribs, sternum, and diaphragm.

they must fulfill their assignments, the structure, mechanisms, modes of actions, and means of control also vary. We are concerned in Part 4 with identifying and understanding some of the problems that are overcome in fulfilling these roles, and in this chapter we are concerned specifically about the role of the heart in cardiovascular function.

STRUCTURE OF THE HEART

The heart is a hollow muscular organ about the size of a clenched fist. It is situated between the lungs in the part of the thoracic cavity known as the **mediastinum.** Viewed anteriorly, it is somewhat triangular in shape, with the tip or **apex** extending to the left and resting on the diaphragm (see Figure 15–2). The **base,** which is located behind the sternum, is where the great vessels enter or leave the heart. The

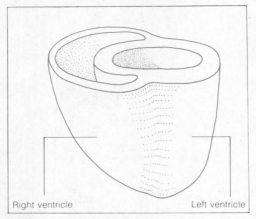

Right ventricle   Left ventricle

FIGURE 15-3   Relative sizes and shapes of the ventricles.

heart usually extends between the levels of the second and fifth interspaces (the intercostal spaces below the second and fifth ribs).

The mammalian heart consists of four chambers, of which the two atria are receiving chambers and the two ventricles are ejection chambers. Each atrium has a small earlike appendage, or *auricle*. The exit from each atrium and ventricle is guarded by a one-way valve. The heart is lined with a single layer of squamous cells, the **endocardium,** which is continuous with the endothelium that lines the blood vessels. It is enclosed in a double-walled sac, the **pericardium,** whose inner layer, the **epicardium** or *visceral pericardium,* is also very thin and adheres closely to the external surface of the heart. The outer layer, the *parietal pericardium,* is thicker and contains fibrous tissue which makes it quite tough and inelastic. The two layers of pericardium are continuous with one another around the bases of the great vessels, forming a closed sac or *pericardial cavity.* There is no real cavity, however, since the two pericardial layers are in contact with one another. A small amount of fluid lubricates the two surfaces and permits them to slide upon one another with very little friction.

The heart itself, the **myocardium,** is composed of cardiac muscle, whose structural and functional characteristics

were discussed in Chapter 7. A fibrous structure known as the "skeleton of the heart" is located approximately at the junction of the atria and ventricles (see Figure 15-6). It is semirigid and consists of four connected rings surrounding the openings into and out of the ventricles. It serves as a site of attachment for the muscle bundles of the atria and ventricles and for the departing arteries. Atrial walls are relatively thin and ineffective as pumps. Ventricular walls are much thicker as they are formed by several layers of spiraling bundles of fibers which wrap around the central cavity (the *lumen*) in a complex pattern. The left ventricle is roughly cylindrical in shape, with a cone-shaped apex (Figure 15-3). The right ventricle is a flattened sac applied to one side of the left ventricle. The *interventricular septum* is anatomically a part of the left ventricle. Ventricular contraction results in a squeezing or wringing of the left ventricle and a flattening of the right ventricle.

The right atrium receives blood from the systemic circuit. Blood returning from the head, upper extremities and much of the trunk enters it from above as the large and thin-walled **superior vena cava.** The **inferior vena cava,** carrying blood from the inferior extremities and abdominal viscera, enters it posteriorly (Figures 15-4 and 15-5). The **coronary sinus,** which returns blood from the myocardium, drains directly into the right atrium. The left atrium receives blood from the pulmonary circuit, usually through two pulmonary veins from each lung. There are no true valves at the entrances of any of these atrial vessels, although the inferior vena cava and coronary sinus sometimes have a flap of atrial lining tissue near the orifice, which *may* deflect or direct the flow of blood somewhat. The absence of functional valves indicates that there is nothing to guarantee one-way flow at this point or to regulate the inflow.

The other atrial orifices are the **atrioventricular valves** (Figure 15-6A), through which blood enters the ventricles from the atria. These valves are flaps of fibrous tissue covered with endocardium and attached around the

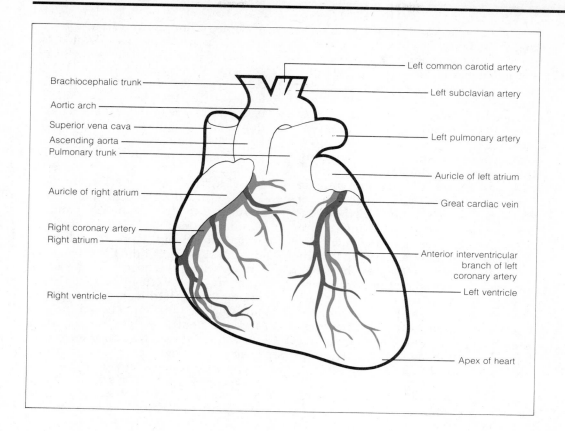

FIGURE 15–4
Anterior view of the
heart.

Left common carotid artery

Brachiocephalic trunk

Left subclavian artery

Aortic arch

Superior vena cava

Ascending aorta

Left pulmonary artery

Pulmonary trunk

Auricle of right atrium

Auricle of left atrium

Great cardiac vein

Right coronary artery

Right atrium

Anterior interventricular
branch of left
coronary artery

Left ventricle

Right ventricle

Apex of heart

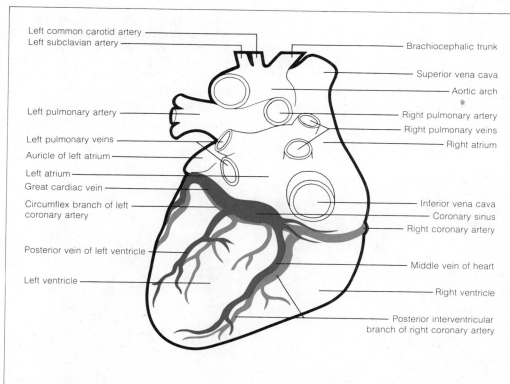

FIGURE 15–5
Posterior view of the
heart.

Left common carotid artery

Left subclavian artery

Brachiocephalic trunk

Superior vena cava

Aortic arch

Left pulmonary artery

Right pulmonary artery

Right pulmonary veins

Left pulmonary veins

Right atrium

Auricle of left atrium

Left atrium

Great cardiac vein

Circumflex branch of left
coronary artery

Inferior vena cava

Coronary sinus

Right coronary artery

Posterior vein of left ventricle

Left ventricle

Middle vein of heart

Right ventricle

Posterior interventricular
branch of right coronary artery

rim of the orifice. It is important to note that the valve leaflets, or *cusps,* contain no muscle of any kind and are therefore incapable of any contraction. The right atrioventricular valve has three cusps and is known as the **tricuspid valve.** The left valve has but two cusps and is called the **bicuspid** or **mitral valve** (the latter because of its resemblance to a bishop's miter). In the open position, the cusps extend down into the ventricles (see Figure 15–6B) and provide little interference with the flow of blood, but as the ventricles fill they float up and effectively block the return route. When the ventricles contract, considerable pressure is applied to the valves from below. However, the valves are held in position by fine, strong tendinous strands called the **chordae tendineae.** Several such cords run from the free edge of each cusp to the cone-shaped **papillary muscles** found along the inner wall of the ventricle. The papillary muscles are part of the myocardium and exert tension on the valve flaps, thus preventing them from reversing when the ventricles contract.

The only other openings of the ventricles are those of the arteries, by which blood leaves the heart, the *pulmonary artery* from the right ventricle and the *aorta* from the left ventricle. These vessels, unlike the entering veins, have thick walls and a valve at their entry. The valves are often called **semilunar valves** due to their shape, or **pulmonary** and **aortic valves** after the arteries in which they are located (see Figure 15–7). Each consists of three small pockets attached around the inside wall of the artery. As blood leaves the ventricle, the pockets are flattened back against the artery wall. When ejection is completed and blood tries to re-enter the ventricle, it catches in the pockets, causing them to fill and abut against one another so as to close the vessel tightly.

The blood supply to the myocardium is provided by the right and left **coronary arteries.** They arise from the aorta just beyond the semilunar valves (Figure 15–6A), behind the valve cusps. They encircle the heart, passing to the right and to the left in an indentation, known as the **atrioventricular groove,** marking the junction between the atria and the ventricles. On the posterior surface of the heart the two arteries *anastomose* (join together). Along the way, branches are given off to descend toward the apex, particularly in the region of the interventricular septum. The peripheral branches of these arteries form an abundant network of vessels throughout the entire myocardium, and there are numerous anastomoses between them. Blood can thus be supplied to a region in case of blockage of the vessel normally supplying that region. Unfortunately, these anastomoses may be too small to suffice if the interruption is sudden.

FIGURE 15–6   A. Valves of the heart viewed from above, after removal of the atria. B. Tricuspid valve viewed from the side.

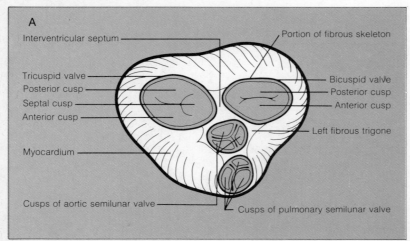

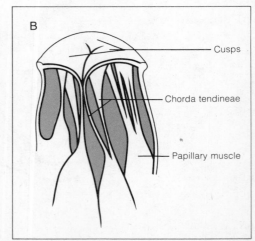

From the capillaries in the myocardium, the blood enters the **coronary veins,** whose course parallels that of the coronary arteries. They converge at the **coronary sinus** (see Figure 15–5), a large venous channel located on the posterior surface of the heart and opening directly into the right atrium. The vicinity of the coronary vessels is a prime location for the deposition of adipose tissue, and it is not uncommon for the vessels to be almost completely obscured by it.

EXCITATION OF THE HEART

Functional Characteristics of Cardiac Muscle

Some of the properties of cardiac muscle discussed in Chapter 7 should be reviewed at this time because of their bearing upon the manner in which the heart functions as a pump:

1. Cardiac muscle possesses *automaticity;* a heart removed from the body can continue to beat for some time if the proper environmental conditions are maintained. (These conditions are much easier to achieve for hearts of frogs and turtles, which do not maintain a constant body temperature, than for mammals.)

2. Cardiac muscle, and indeed, the whole heart follows the *all-or-none law,* which means that any contraction is maximal for the conditions at that time. In skeletal muscle, a weak stimulus reaches threshold for only a few muscle fibers, and increasing its intensity recruits more fibers and produces a stronger contraction. In cardiac muscle, stimuli of graded intensity cannot produce a graded response. If a stimulus is at or above threshold, all cardiac muscle fibers respond; there are none left to recruit. Though it is well known that the force of contraction of the heart does vary greatly, this need not be a violation of the all-or-none law, since the conditions may change from one beat to the next.

3. Cardiac muscle has a very long *refractory period,* lasting throughout contraction and into the early relaxation

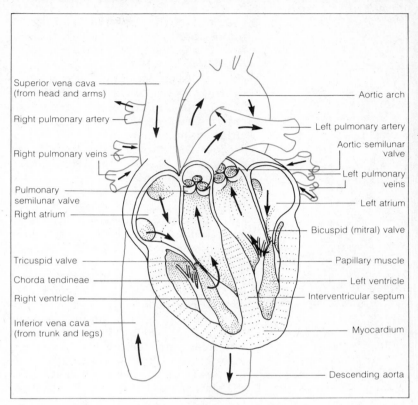

FIGURE 15–7   Interior view of the heart, showing the path of blood flow through it.

phase. Because of this, cardiac muscle neither exhibits tone, as does skeletal muscle, nor can it be thrown into tetanic contraction.

The Conducting System of the Heart

The heart contains some tissue that is specialized to initiate impulses, deliver them rapidly to the entire heart, and ensure a coordinated contraction of all its fibers. The membranes of cardiac muscle cells, like those of all cells, are polarized; they show an unequal distribution of sodium and potassium ions on either side of them, and excitation involves conduction of ionic and electrical changes along the cell membranes. The branching ("syncytial") nature of the myocardium permits impulses to be conducted to all the muscle fibers. However, muscle cell membranes conduct slowly and, due to the complex geometry of the ventricle, it would be impossible to get an effective contraction if this were the only means of excitation. The solution lies in the specialized tissue that is known as the **conducting system of the heart** (Figure 15–8). The conducting system consists of modified cardiac muscle tissue. It has striations and intercalated discs, but no T-tubes, and lots of sarcoplasm. It can conduct impulses more rapidly than regular cardiac muscle.

Directly under the endocardium in the wall of the right atrium, near the entry of the superior vena cava, is an area containing such tissue, called the **sinoatrial node** (*SA node*). It possesses the highest degree of rhythmicity and is the **pacemaker** of the heart. ("Sino-" comes from the fact that in some species the pacemaker is located in a structure known as the sinus venosus, which is formed by the confluence of the great veins before they enter the right atrium.) In the *interatrial septum*, just above the atrioventricular junction, is a similar mass of tissue, the **atrioven-**

tricular node (*AV node*). A bundle of this specialized tissue extends from the AV node as the **atrioventricular bundle** (*AV bundle, bundle of His*). In the ventricle the AV bundle splits into right and left branches, descending on the right and left sides of the interventricular septum beneath the endocardium. Each bundle branch gives off terminal branches to the surrounding myocardium as it descends toward the apex of the heart and then passes up the sides of the ventricles. The terminal branches (*Purkinje fibers*) ramify throughout the myocardium and reach every part of it.

Intracellular recording of the electrical activity of the pacemaker cells in the SA node reveals similarities with that of nerve fibers and other excitable tissues (Figure 15–9). There is a resting membrane potential which is interrupted by an action potential spike each time an impulse is generated. The nature of the trigger for the spike potential is not clear at this time, but it is related to the fact that pacemaker cells do not maintain a steady resting membrane potential. Perhaps the membrane "leaks" ions, since at some point the membrane potential has been reduced (depolarized) enough to reach threshold, and the spike is triggered. Heart rate can thus be altered by changing the rate at which pacemaker cells depolarize.

There is not a well-defined bundle of conducting tissue from the SA to the AV node. There is some evidence for three different pathways between them, but it has been questioned, and a functional role is disputed. Impulses from the SA node probably are conducted along the cell membranes from one atrial muscle cell to the next. A wave of depolarization thus passes out from the SA node over the right and left atria, rather like a ripple in a pond when a pebble is dropped in. This mode of conduction is relatively slow, but the distance is not great and it is adequate for the thin-walled atria. The wave of depolarization reaches the AV node, whose spontaneous rhythm is much slower, and after a slight delay the AV node is excited and the impulse is then conducted very rapidly through the AV bundles and terminal branches to reach all of the ventricular muscle.

FIGURE 15–8   Conducting system of the heart.

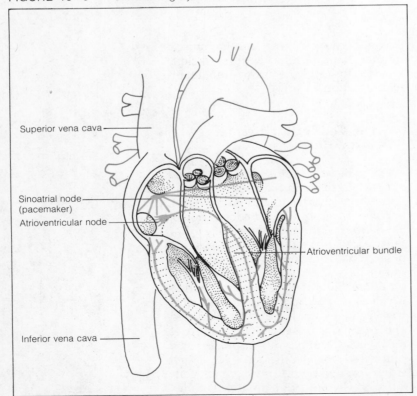

Superior vena cava

Sinoatrial node (pacemaker)

Atrioventricular node

Atrioventricular bundle

Inferior vena cava

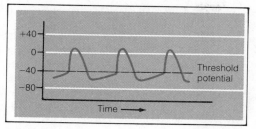

FIGURE 15–9   Membrane potential in cardiac pacemaker cells, showing the spontaneous depolarization that leads to an action potential.

The wave of depolarization is then followed by repolarization, after which the myocardium is ready for the next impulse. Since excitation must precede the actual muscle contraction, it is apparent that a disturbance in transmission of the wave of depolarization could profoundly alter the heart's contraction, and hence its pumping action.

Conducting Abnormalities

Pacemaker cells are not restricted to the SA node, however, since if the node is removed the heart will usually continue to beat. (This is easier to demonstrate in a frog or turtle where the SA node is not in the atrium.) The new rate is slower, since the AV node has now taken over the role of pacemaker, and the atria and ventricles contract almost simultaneously. If both nodes are removed, as in an isolated strip of cardiac muscle, the cells may continue to contract spontaneously, thus demonstrating a true automaticity. Mammalian cardiac muscle is not so likely to demonstrate its automaticity, although under certain conditions it may. Such contractions are not particularly regular, however, and the rate is much slower even than the AV rhythm.

The rhythm of a normal healthy heart is not always perfectly regular (see Figure 15–10). The heart rate may vary with breathing movements, or there may even be occasional premature beats. Abnormal rhythms may be due to irregular impulse generation by the SA node, or by pacemaker activity at some other location, or to disturbances of the conducting system. Pres-

sure or other damage to the conducting system may delay conduction at the AV node, resulting in failure of some or all of the impulses to be transmitted to the ventricles. The condition, known as **partial heart block,** occurs when the atria beat normally, but only every second (or third or fourth, etc.) atrial beat is followed by a ventricular contraction, producing a 2:1 (or 3:1, or 4:1, etc.) block. If no impulses get through, the ventricle may stop completely (*ventricular arrest*) for a time, or continue beating at its own slow rate. Such a ventricular beat is completely independent of the atrial beat, which is still following the SA nodal rhythm. This is **complete heart block.** Conduction problems, blocks, or very slow sinus rhythms can often be corrected by an artificial pacemaker. This is an electronic device implanted under the skin, with wires to the area of the pacemaker cells. It generates impulses that stimulate the heart to beat at a normal rhythm.

Under certain conditions an abnormal site, or *ectopic focus,* may become excitable and initiate a beat or beats on its own. If the site is in the ventricle, the ventricle responds with a **premature beat,** but it will probably then be refractory when the next impulse arrives from the SA node. It will miss that beat, resulting in a prolonged interval, or *compensatory pause,* before returning to the normal rhythm of the SA node. The atrial rhythm remains undisturbed, since the impulse of the ectopic beat does not travel backward from the AV node. If the excitable site is in the atrium, the premature beat and compensatory pause appear in both the atria and the ventricles. If the ectopic focus fires repetitively, it may usurp the role of the SA node as pacemaker.

Conduction of the impulse over the atria may be abnormal, causing the atrial contraction to be uncoordinated, and neither atrium can contract as a unit. This condition, known as **atrial fibrillation,** is very distressing, but it is not fatal as long as the ventricle contracts as a unit, even though irregularly. **Ventricular fibrillation** is another story, since asynchronous ventricular contractions neither develop any pressure nor pump any blood. This is incompatible with life!

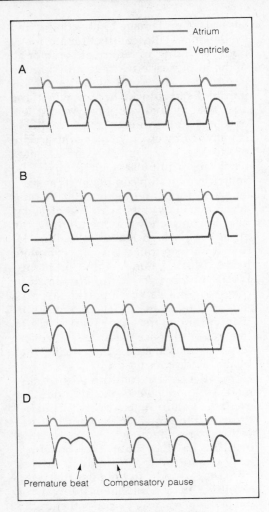

FIGURE 15–10   The relationship
between atrial and ventricular contractions.
A. Normal beat. The atrium contracts
shortly before the ventricle. B. A 2:1 partial
heart block. The ventricle misses every
other atrial contraction. C. Complete heart
block. The ventricle contracts at its own
rate, independent of atrial contractions.
D. Ectopic focus. An impulse arises at an
abnormal site in the ventricle and
generates a premature beat. This leaves
the ventricle refractory to the impulse from
the pacemaker, resulting in a long
compensatory pause until the next impulse
arrives.

The Electrocardiogram

The action potential of a cardiac
muscle cell, like that of a skeletal muscle
cell, consists of depolarization and repolarization; but unlike skeletal muscle,
there is a delay of many milliseconds
before cardiac muscle repolarizes. Because the heart contracts as a unit, we
are concerned with the action potential
of the whole heart rather than that of a
single fiber. The heart is a large and
complex organ, and the wave of depolarization associated with its excitation
sweeps over it in a specific sequence.
It should, therefore, not be surprising that the action potential of the
heart, recorded as the **electrocardiogram
(ECG),** is not a simple single spike such
as seen in nerve or skeletal muscle fibers. The differences are dictated by
certain features of the anatomy of the
heart and its conducting system.

1   The specialized conducting system distributes the impulse rapidly,
but not instantaneously.
2   The impulse in the ventricle
passes down the interventricular
septum to the apex and back up the
sides of the ventricles.
3   The heart is placed neither centrally nor symmetrically in the thoracic cavity.
4   There is a larger mass of cardiac muscle to be excited on the left
side of the heart than on the right.

An electrode placed directly on the
surface of the heart should detect only
the electrical events in its immediate
vicinity; it would not detect all of the
heart's electrical behavior. But because
the body fluids contain electrolytes,
and hence can conduct electrical signals, electrodes can be placed on the
surface of the body where they can
monitor the course of the entire depolarization of the heart. This fact is a key
to the electrocardiogram, for it is recorded from electrodes placed on the skin at
sites far removed from the heart, such
as the wrists and ankles. When they are
connected through a suitable amplifier
and recorder, one can record the passage of the wave of depolarization over
the heart. Because of the factors men-

tioned above, the potential changes picked up by electrodes at different locations are neither simultaneous nor identical; each electrode has a slightly different "perspective" as the wave of depolarization and repolarization sweeps over the heart. When connected so that the "view" seen by one electrode is compared with that seen by another, the record obtained shows the potential differences between those two electrodes. The ECG is a record of the changes in the potential difference between specific electrodes as the impulse passes over the heart. It should not be surprising that the deflections recorded have a complex, though predictable, contour, as shown in Figure 15–11.

The individual waves or peaks of the cardiac action potential are designated by the letters P, Q, R, S, and T, each of which is associated with a specific electrical event (the letters themselves are not significant). The P-wave is associated with depolarization of the atria, and the QRS complex is caused by depolarization of the ventricle. The T-wave, which appears a short time later, represents the repolarization of the ventricles. There is said to be an atrial T-wave as well, but it is small and is masked by the large QRS deflection. The duration of the PR interval (from the beginning of the P-wave to the beginning of the QRS) represents the time required for the impulse to pass from the pacemaker through the AV node and bundle to the ventricular muscle and is referred to as the conducting time. When this interval is prolonged, conduction is delayed and heart blocks may occur.

It is customary to record the ECG from several locations, or *leads*, to get several different "views." The three so-called standard limb leads (Figure 15–12) are designated Lead I (right arm-left arm), Lead II (right arm-left leg), and Lead III (left arm-left leg). The deflections are similar in the three leads, but the amplitudes of the component waves vary slightly from one to the other. Clinical electrocardiography makes use of a series of additional leads, the chest leads, which provide still other "views."

Many types of cardiac malfunction can be detected by the electrocardiogram, including *arrhythmias* and conduction defects such as heart block. If a portion of the cardiac muscle has been damaged by interruption of its blood supply (as by rupture or occlusion of a coronary blood vessel), it will not be normally polarized, and the course of depolarization will be abnormal. Such defects in the pattern of depolarization and repolarization are very apparent in the ECG, and an experienced analyst can assess the severity of the damage and its location with remarkable precision. It is important to remember that the ECG is a record of the action potential, the excitation of cardiac muscle, and that it must occur before each contraction. But the ECG itself is not a contraction; it is possible (though highly unlikely) to have an action potential without a contraction.

FIGURE 15–11   A. Normal electrocardiogram of a single beat. B. Electrocardiograms in a 2:1 partial and a complete heart block.

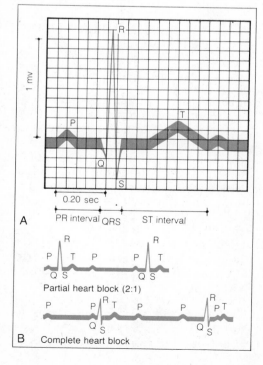

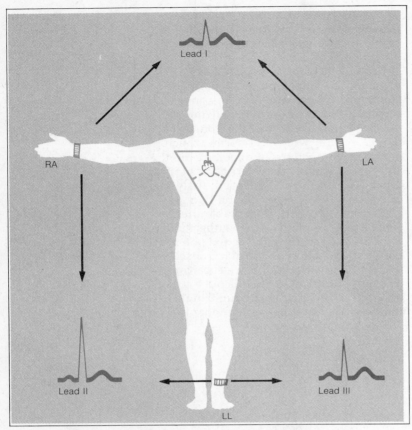

Lead I

RA

LA

Lead II

Lead III

LL

FIGURE 15-12   Standard limb leads for the electrocardiogram.

aptic) connections. The vagal fibers synapse within the heart and terminate in the region of the SA node (especially those from the right vagus) and also in the cardiac musculature (see Figure 12–11).

The sympathetic nerves are adrenergic and excitatory. They are often called the *accelerator nerves* because of their action on the pacemaker. The vagal fibers are cholinergic and inhibitory. Vagal effects are dominant due to the tonic discharge of these fibers, by which they exert a constant braking action on the pacemaker. Acetylcholine released by the vagal endings tends to retard the process of spontaneous depolarization of pacemaker cells so that it takes longer for them to reach threshold and fire. Cardiac arrest can be produced by either vagal stimulation or application of acetylcholine to the pacemaker. Acetylcholine also slows AV conduction, which may result in some form of heart block or arrhythmia. Norepinephrine from the sympathetic endings hastens the slow depolarization of pacemaker cells, so that the threshold for firing is reached more rapidly. Epinephrine in the circulating blood increases both the heart rate and the strength of contraction.

EXTRINSIC INNERVATION OF THE HEART

Although the pacemaker is capable of initiating beats independently of outside neural influence, there is an important nerve supply to the heart. It comes from both divisions of the autonomic nervous system, and its role is to control heart rate and to help regulate the strength of contraction. The sympathetic innervation arises from the upper thoracic portion of the spinal cord, synapses in the sympathetic chain, and continues as the cardiac nerves, ending chiefly in the vicinity of the pacemaker, though some fibers supply the ventricular muscle. The parasympathetic innervation is carried by the vagus nerve. There is a rich nerve plexus around the esophagus and adjacent blood vessels involving both sympathetic and vagal fibers, but there are no functional (syn-

EVENTS IN THE CARDIAC CYCLE

The heart receives blood at·low pressure and sends it out at high pressure. Contraction of a ventricle raises the pressure of the incompressible fluid in it until a valve opens and the fluid (blood) is ejected with sufficient impetus to travel through the vessels and back to the heart. An understanding of the action of the heart in developing this driving force can best be obtained by careful analysis of the pressure changes within the various chambers during the course of a single heartbeat.

Obtaining precise information about intracardiac pressures in a beating heart is a major undertaking. Early approximations were made by inserting a needle connected to a pressure-sensitive device directly into the heart of an anesthetized animal. This information is now obtained in humans by means of *cardiac catheterization,* in which a very narrow tube is inserted into an arm vein and threaded back into the heart. On one end of the catheter is a pressure-sensitive tip, and on the other a suitable recording device. With the subject before a fluoroscope, the catheter tip can be guided into the right atrium, right ventricle, or even into the pulmonary artery. More recently, techniques have been developed to catheterize the less accessible chambers of the left side of the heart. Catheterization is a most useful, and virtually essential, tool for diagnosing congenital defects in the heart.

From information provided by this method, accurate diagrams have been constructed to define clearly the events and relationships during a single beat. It may seem improper to relate venous return in the right atrium to contraction in the left ventricle as in Figure 15–13, and, of course, it is not strictly accurate. The action of the right atrioventricular valve depends upon the pressure in the right atrium and right ventricle. From a practical point of view, however, it can be defended on the basis of the fact that our chief concern is with the systemic circuit. The important homeostatic mechanisms of the cardiovascular system are those that maintain the pressure and flow in the systemic circuit. A quick glance at Figure 15–14 confirms that the pressures in the two atria are very nearly the same, and that the timing of events on the two sides are also very nearly identical. The only real differences are the pressures in the aorta and pulmonary artery, and the pressures developed by the two ventricles.

Between beats the heart is relaxed, a phase known as **diastole.** Contraction is called **systole.** During diastole the semilunar valves are closed, but the atrioventricular valves are open. The pressure of the venous blood returning to the atria is only a few millimeters of mercury (mm Hg), but it is even lower in the relaxed ventricles. Blood therefore flows continuously into the atria and on through the open AV valves into the ventricles. The ventricles are almost filled before atrial contraction,

which develops only a few millimeters of additional pressure. It forces some blood into the ventricles, but some blood also moves back into the veins since there is no valve to prevent it. When a person is at rest the rather ineffective atrial contraction does not contribute much to the pumping action of the heart.

When the ventricle begins to contract, the intraventricular pressure rises and the blood seeks to escape by the only available route—back into the atrium. The result is immediate closure of the AV valves, shutting off that route. Pressure then builds up in the ventricles, but there is no change in ventricular volume until the intraventricular pressure exceeds that in the aorta (or pulmonary artery). When it does, the semilunar valves are suddenly forced open and blood is rapidly ejected into the artery, whose pressure rises accordingly. The arterial (aortic) pressure at the peak of the pressure curve is

FIGURE 15–13 Events of the cardiac cycle.

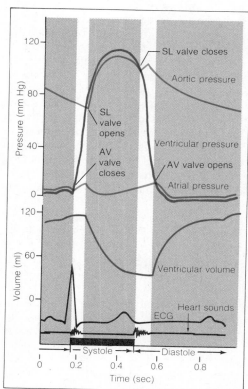

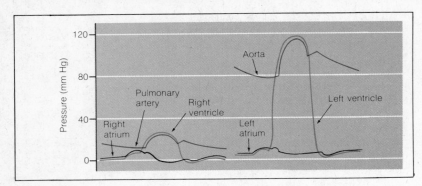

FIGURE 15–14   Comparison of the pressure changes in the right and left sides of the heart.

the **systolic pressure** and is the maximum pressure attained in the artery during the cardiac cycle. The rate of ventricular outflow tapers off as the contraction reaches its peak and less blood remains in the ventricles. With relaxation, the intraventricular pressure falls precipitously. When it drops below that in the artery the pressure gradient is reversed, but as arterial blood begins to return to the ventricle it quickly fills the pockets of the semilunar valves and they snap shut. Closure of the semilunar valves produces a brief secondary rise in arterial pressure (the *dicrotic wave*), followed by a steady decline in pressure interrupted only by the next ventricular systole. Pressure in the artery is lowest just prior to the opening of the semilunar valves, and this is the **diastolic pressure.** (Systolic and diastolic pressure can be measured in any artery, but unless otherwise stated, it is assumed to refer to the pressure generated by the left ventricle as it is customarily measured in the brachial artery.)

Blood continues to enter the atria throughout the ventricular contraction, but only when pressure in the relaxing ventricle falls and the AV valves open does the accumulated blood enter the ventricles.

The arterial pressure determines the ventricular pressure required to eject blood. If the arterial (and hence aortic) pressure (Figure 15–13) were elevated to a diastolic pressure of 150 mm Hg, no blood would be ejected at all unless the force of ventricular con-

traction were also raised. The ventricle is quite capable of the required increase in force of contraction, and it adjusts to marked changes in arterial pressure within a few beats. Prolonged elevation of arterial pressure increases the work load of the heart and could cause an overload or strain. A healthy heart, however, responds like any muscle to a chronic increase in work load—it increases in size and work capacity.

The ventricle can eject blood only when the semilunar valves are open, and it can receive blood only when the AV valves are open, as they are throughout most of the cardiac cycle. The ventricles fill most rapidly, however, immediately after the AV valves open. Blood accumulates in the atria throughout systole, when the AV valves are closed, but as soon as they open, that blood flows into the ventricles. Ventricular filling is nearly complete by mid-diastole and continues more slowly as blood enters from the veins. Contraction of the atria adds very little. It is fortunate that most of the filling occurs early in diastole, because when the heart rate increases the period of diastole is shortened. Diastole can therefore be shortened very considerably before it actually encroaches upon filling of the ventricle. In the average adult human, only heart rates in excess of 160–180 beats per minute begin to interfere with adequate filling of the ventricles.

The right and left sides of the heart contract in unison and the events of the cardiac cycle occur more or less simultaneously on the two sides (Figure 15–14). The chief difference lies in the pressures in the pulmonary artery and the aorta, the latter being about six times higher, necessitating greater ventricular pressure on the left side to eject blood. This accounts in large part for the difference in the thicknesses of the walls of the left and right ventricles. Atrial pressures are about the same on both sides and are never very high.

Heart Sounds

The contraction of the ventricles is an abrupt and powerful event, and the consequent sudden closing of the valves causes distinct vibrations of the heart, arteries, and blood. These vibrations are transmitted through the chest wall and can be detected as sounds with the aid of a stethoscope (*auscultation*). The two sounds commonly heard are often described as "lub-dupp." The first heart sound is primarily due to the closure of the AV valves, although ventricular systole and the opening of the semilunar valves probably also contribute to it. It is therefore heard at the onset of ventricular contraction. The second heart sound is due largely to the closure of the semilunar valves and the ensuing vibrations in the heart and chest wall, and it is heard as the ventricle is relaxing.

Heart Murmurs and Valvular Defects

Heart murmurs are a type of heart sound heard when sounds are not expected, and they are often the result of valvular defects. **Incompetence** or **insufficiency** of a valve is a condition in which the cusps do not provide a secure seal when closed. Insufficiency may be caused by scarring of a valve, resulting in regurgitation of blood back into the chamber it just left, which produces a sound when that valve should be closed. Any valve may be affected, but the condition is especially critical at the aortic valve because of the high pressure gradient across it. Aortic insufficiency can severely impair the effectiveness of the heart, since much of the ejected blood then returns and must be pumped a second time. This places an extra load upon the left ventricle and reduces the amount of blood sent to the tissues.

A **stenosis** is the result of valvular damage that narrows the opening and interferes with the flow of blood. A higher pressure is then required to force an adequate volume of blood through the valve. A stenosis produces a sound or murmur when the valve is supposed to be open. Not uncommonly, a valve may exhibit both stenosis and insufficiency, which compounds all the problems. Not all murmurs are abnormal, however, since turbulence in the flow of blood also causes sounds. The occurrence of turbulence is related chiefly to the velocity of the flow and the radius of the vessel. Such sounds are not uncommon, particularly in children after vigorous exercise.

CARDIAC OUTPUT AND ITS CONTROL

The effectiveness of the heart must be gauged by the volume it pumps in relation to the body's needs. For some pumps, adjustment is a simple matter of turning the pump on and off as needed, but the heart must operate continually. It must pump all of the blood that comes to it, adjusting both its rate and stroke to the returning blood, whose volume is determined by noncardiac factors. The volume of blood pumped by the heart, the **cardiac output,** is the amount of blood ejected per minute, or the **minute volume of the heart.** It is expressed in terms of the amount of blood pumped into the systemic *or* pulmonary circuit (they should be equal), and it thus refers to the output of a single ventricle. The cardiac output is the product of the amount pumped with each beat, the **stroke volume,** and the number of beats in a minute, the **heart rate.** Using conveniently typical resting values for each:

Heart rate × stroke volume = cardiac output
70 beats/min × 70 ml/beat = 4900 ml/h
= 4.9 L/m

Thus the resting cardiac output is in the range of 4.5–5 liters every minute of every day. You will recall that the total blood volume is about 4.5–5 liters, so that even at rest just about all the blood is pumped through the heart each minute. An average individual is capable of increasing his or her cardiac output two or three times, to 10–15 liters/min, and a trained athlete may achieve a fivefold increase.

The problem is to determine what sets the level of cardiac output in a given situation. How is it controlled? Heart rate and stroke volume are independently variable and are controlled by different mechanisms to a large extent. Some of the contributing factors are discussed in the following paragraphs, and their relationships are summarized schematically in Figure 15–15.

Control of Heart Rate

Although 70 beats/min may be considered as a typical "normal" resting heart rate, this is one characteristic that is widely variable, with resting heart rates between 40 and 100 beats/min considered normal. It is usual-ly higher than this in infants, and the extremely low rates are usually found only in trained athletes, particularly those who specialize in endurance sports (distance runners). In the average healthy adult, cardiac output begins to decline at heart rates above 160–180 beats/min. Approximately a 1.5–4-fold increase in heart rate is thus possible, depending upon the initial rate. Maximum rates well in excess of 200 beats/min have been reported, however, usually in young people.

The heart rate is largely under reflex control by the autonomic nervous system (Figure 15–16). The chief center for cardiac control is in the reticular portion of the medulla. Stimulation in

FIGURE 15–15   Factors contributing to the control of cardiac output.

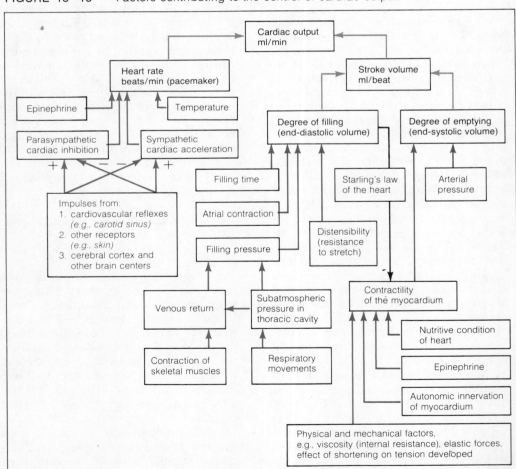

Focus

Cardiopulmonary Resuscitation (CPR)

Over the years, first aid techniques have become more elaborate and more useful as more has been learned about the human body and what it needs (and can be given) in emergencies. In the process, first aid has come to mean more than bandaids, splints, warm blankets, and artificial respiration. Modern first aid actually includes the restarting of hearts stopped by heart attack, electric shock, drug overdose, or suffocation. The process is known as cardiopulmonary resuscitation, or CPR. Unlike bandaids, it requires enough skill that one who has not been trained in CPR should not attempt it. Fortunately, free courses in CPR are offered frequently throughout the country, through the American National Red Cross and the American Heart Association.

CPR is appropriate whenever heart action has stopped or is greatly reduced, and the nature of the technique is specified by that of the problem. The heart is a pump. When is contracts, it decreases its volume and squeezes blood into the arteries that lead to the lungs and the rest of the body, including the brain. When it fails to contract, the blood does not flow and oxygen is neither absorbed in the lungs nor delivered to the body. The entire body suffers, but the brain suffers most of all. It can tolerate only a very brief oxygen deprivation before its cells begin to die. It is therefore crucial that a stopped heart be restarted as soon as possible and that the flow of blood and oxygen to the brain be maintained in the meantime. This is the function of CPR.

CPR can be described simply as rhythmic compression of the heart combined with artificial respiration. The giver of CPR positions the victim flat on his or her back and, using both hands, presses downward on the lower half of the sternum, at a rate of about 80 times per minute. The pressure used is such as to depress the sternum 4 to 5 centimeters and compress the heart. The cardiac compression squeezes the blood in the heart into the arteries and establishes a flow of blood through the lungs, back to the heart, and to the brain, keeping the nervous system alive. At the same time, artificial respiration is performed by another person, if possible, or a single operator must alternate cardiac and respiratory activity. Artificial respiration ensures that the blood can absorb oxygen as it flows through the lungs.

The effect of CPR is twofold. By supplying an external motive force for the blood flow, it gives the heart a little time to recover from the immediate effects of whatever stopped it. In addition, because the heart is an excitable tissue, it provides a mechanical stimulus that can spur the heart back into action. Done properly, it can be amazingly effective, keeping the victim alive until medical professionals can arrive on the scene.

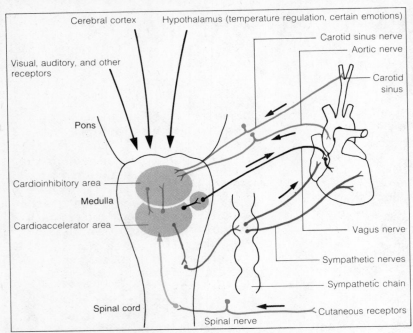

FIGURE 15-16   Neural control of heart rate.

one part of this cardiac center, designated the **cardioaccelerator center (CA),** causes an increase in the heart rate. It is the sympathetic center, since its effects are mediated through fibers which synapse with the sympathetic accelerator fibers arising from the thoracic region of the spinal cord. In other parts of the cardiac center, stimulation causes a decrease in heart rate; this region is called the **cardioinhibitor center (CI).** Its effects are mediated through connections with the vagus nerve, so it is a parasympathetic center. These centers are not discrete and readily identifiable entities. They are quite diffuse, and the cell bodies tend to be somewhat intermingled with one another and with those of other "centers" in the vicinity.

The activity of the CA and CI centers is controlled, in part, by numerous interneuronal connections between them providing a reciprocal action. Any input which excites CA neurons will inhibit CI neurons, and vice versa. The inhibitory neurons are tonically active under resting conditions and send a continual stream of impulses along the

vagus nerves to the heart. The bulk of the neural regulation of the cardiac pacemaker is mediated through alteration in the rate of discharge of the vagal fibers. Cutting these fibers results in a chronic elevation of heart rate. The CA (sympathetic) center, on the other hand, is not tonically active, and cutting the accelerator nerves does not change the resting heart rate.

Afferent fibers from many areas impinge upon the medullary cardiac centers—each afferent augments the activity of one center and inhibits the activity of the other. Among the important inputs to the cardiac centers are those from receptors in various parts of the cardiovascular system. There are, for example, receptors (located in the **carotid sinus** and in the **aortic sinus**) which monitor blood pressure and cause the heart rate to adjust accordingly; a fall in arterial pressure results in an increase in heart rate. As these reflexes are of prime importance in control of blood pressure, they are discussed more fully in Chapter 17. Other cardiovascular reflexes may arise from receptors in the heart itself, in the great veins, in the pulmonary vessels, and at other sites. In addition, stimulation of almost any cutaneous nerve will elicit a change in the heart rate. Pain is particularly effective, but heat, cold, and touch stimuli may also elevate the heart rate for a short time; severe pain will lower it. Other sensory inputs, such as sight or sound, may do the same, especially if the element of surprise is involved. Such stimuli are closely related to input from higher centers, including emotional inputs (joy, exhilaration, fear, and anxiety).

The medullary cardiac centers do not provide exclusive control of cardiac activity, since these centers are influenced by other areas of the brain. The arrangement may be likened to that for the control of skeletal muscle tone. It is well known, for example, that stimulation of certain areas of the brain (such as the hypothalamus) can reproduce the pattern of cardiac response associated with physiological activities such as exercise. It is probable that the cells in the cardiac centers are normally subjected to a background of activity from other parts of the brainstem, which determines the degree of response (excitability) of those cells to sensory input.

Heart rate may also be altered by nonneural factors, such as temperature, drugs, and chemicals. If the temperature of the pacemaker is increased, the heart rate is also increased, as would be expected. This effect is not often of physiological significance in humans, since we have a reasonably effective mechanism for maintaining a constant temperature. It does, however, account for the elevation in heart rate during fever and for the reduction during *hypothermia* (which is used for some cardiac surgery). Of the more important pharmacological factors, epinephrine is undoubtedly of the greatest physiological significance. It is released into the bloodstream by the adrenal medulla upon sympathetic stimulation, and it is therefore often present in the blood perfusing the pacemaker area. Both epinephrine and norepinephrine tend to increase the rate of pacemaker discharge, but epinephrine is several times more potent in this regard. It is not within the province of this text to consider in detail the actions of pharmacological agents. However, you should be aware of the fact that the pacemaker is susceptible to stimulation or depression by a variety of agents, often as an unwanted side-effect of other actions.

Control of Stroke Volume

The physiological contribution of stroke volume to the regulation of cardiac output is more difficult to assess than that of heart rate. There is no unanimity about either the changes in stroke volume that occur under varying conditions or the relative importance of possible regulating mechanisms. Some of the problems which have created this dilemma will come to light in the discussion that follows.

The stroke volume, the volume of blood ejected in a single beat, is the difference between the volume of the ventricle when the semilunar valves close at the end of contraction (the **end-sys-**

tolic volume or **ESV**) and that of the ventricle as the AV valves close at the onset of contraction (the **end-diastolic volume** or **EDV**). The end-diastolic volume is an indication of the degree of ventricular filling during diastole; the greater the filling, the greater the EDV. The end-systolic volume, on the other hand, is a measure of the degree of emptying during systole; the greater the ESV, the *less* blood was ejected. Filling and emptying are not governed by the same factors, and they may therefore be varied independently. Both are greatly affected by cardiac factors, but filling is also dependent upon conditions on the receiving (venous) side of the circulation, and emptying is related more to conditions on the ejection (arterial) side.

Filling — End-Diastolic Volume

Filling of the ventricle, as measured by end-diastolic volume, is not controlled directly by neural or hormonal influences, but rather by several mechanical factors:

1. *Filling time.* As the heart rate increases, the duration of the filling period becomes shorter. At high heart rates the ventricle may contract before it is adequately filled, thus reducing stroke volume. Filling is greatest in early diastole, and it is not greatly increased by an unusually slow heart rate alone.

2. *Atrial contraction.* Normally, ventricular filling is virtually completed before atrial systole begins; the weak atrial contraction adds little. If, however, inflow to the ventricle is limited by reduced filling time or resistance at the AV valve, the atrial contraction may add measurably to the ventricular volume. It is particularly helpful at high heart rates during exercise.

3. *Distensibility.* As blood enters the ventricle, the musculature "gives," or distends, so that very little pressure is required to cause an appreciable increase in ventricular volume. Since the pressure of blood returning to the heart is normally quite low, the resistance of the ventricular muscle to stretch is a critical factor in filling. A relatively empty ventricle fills rapidly because of its distensibility. However, as it be-

comes filled, and as the muscle fibers become stretched to the limit of their elasticity, further distension requires a greater pressure increment.

4. *Filling pressure.* The pressure difference between the inside and outside of the ventricle, known as the **transmural pressure,** is the effective filling pressure. The pressure inside the ventricle is directly related to the amount of blood returning to it, the **venous return.** The venous return is quite dependent upon the contraction of skeletal muscles in the limbs, as well as upon the smooth muscle in the blood vessels. Loss of blood, or pooling of blood in the lower extremities, interferes with the venous return and results in reduced filling pressure. Some of the situations in which this occurs are discussed in Chapter 18.

The other half of transmural pressure is the pressure outside the ventricle, in the thoracic cavity. It is an effective aid to filling the ventricle because it is several millimeters of mercury below atmospheric pressure. This reduced pressure facilitates distension of the heart and the return of blood into the thorax by virtually sucking it in. Pressure in the thorax is subject to fluctuations associated with the respiratory movements, and it may have profound effects on the blood returning to the thoracic cavity and heart.

Emptying — End-Systolic Volume
The end-systolic volume indicates the effectiveness of the ventricle in pumping out the blood that has entered it. It is obvious that the ventricle cannot pump more blood than it contains, but this should not be construed to mean that in any one beat it ejects *all* that it contains. The degree of emptying in any given beat is determined by the arterial pressure that the ventricle must overcome to open the semilunar valve, and by the pressure developed by ventricular contraction.

1. *Arterial pressure.* The pressure in the aorta or pulmonary artery just prior to ventricular contraction is the pressure which must be exceeded before the semilunar valve will open. It already has been pointed out that if arterial pressure is markedly elevated without other adjustment, the duration of ejection, and hence the stroke volume, will be reduced. The reverse will hold if the arterial pressure is lowered.

2. *Contractility of the myocardium.* All kinds of factors have been included under the heading of contractility, but it is still a convenient term to reflect the multitude of factors that contribute to the force of ventricular contraction. The subtle changes of these factors help keep the stroke volume relatively constant during changes in arterial pressure and diastolic volume. Everyone knows that the heart beats "harder" during exercise and excitement, in spite of the fact that it is supposed to obey the all-or-none law. The apparent contradiction is due to changes in the conditions which alter one or more of the factors associated with myocardial contractility. Some of the more pertinent of these are mentioned below.

Metabolic conditions of cardiac cells. The coronary blood flow is a critical element in meeting the needs of the cardiac muscle cells for oxygen, nutrients, and waste removal. Normally the blood flow is adequate to provide the metabolic needs of contracting cells, but sudden or chronic circulatory problems which limit the effectiveness of the coronary blood flow will quickly impair the ability of the ventricular muscle to develop pressure. Conditions arising elsewhere, as in the lungs, may also interfere with one or more of the metabolic requirements of the heart and may similarly hinder pumping action.

Neural effects. The majority of nerve fibers to the heart end in the vicinity of the conducting system (SA and AV nodes). Some nerve fibers, however, do end in the muscle. They are primarily sympathetic, and their excitation strengthens the force of contraction and increases the rate of shortening and the power developed. The increased force of contraction is an important element in all cardiovascular reflexes.

Epinephrine. Even though vagal influences dominate in the control of heart rate, sympathetic transmitters play an important role in determining contractility. Epinephrine from the adrenal medulla is of particular importance because it is carried in the bloodstream and, as a hormone, reaches more muscle fibers than the sympathetic nerve endings. (Recall that epinephrine has a greater effect on cardiac muscle than norepinephrine has.) In physiological concentrations, epinephrine increases the force of contraction and the rate of both contraction and relaxation. Many other chemical agents, therapeutic and otherwise, also affect the force of contraction.

Physical and mechanical factors. Some of the numerous structural features that modify the effectiveness of myocardial contraction include, briefly:

(a) As the ventricle contracts, the cellular components must change their physical relationship to one another. There is an internal resistance to this rearrangement which must be overcome, however, and in so doing some energy is dissipated that otherwise would be used for shortening. More energy is consumed in this way when the rate and amount of shortening are great.[1]

(b) Contraction in the well-filled heart is aided by the forces of elastic recoil, which are greater with increased distension. The distended heart incurs a mechanical disadvantage, however, in that at a greater volume (or radius) more tension must be developed by the muscle in order to produce a given pressure inside the chamber (Law of Laplace). Thus an overly distended ventricle contracts less effectively.

Starling's Law and Stroke Volume For many years it has been taught that stroke volume is determined largely, if not exclusively, by what is known as *Starling's Law of the Heart.* In 1914,

Starling stated that "The law of the heart is therefore the same as that of skeletal muscle, namely that the mechanical energy set free on the passage from the resting to the contracted state depends . . . on the length of the muscle fibers.[2] In other words, the strength of contraction depends upon the degree of diastolic filling. Such a mechanism would provide an effective means of autoregulation of stroke volume without need for extrinsic neural or hormonal control, and there is much evidence to support the concept. In skeletal muscle we have seen that greater tension is developed in an isometric contraction as the initial length of the muscle is increased. In cardiac muscle too, greater tension is developed in an isovolumetric contraction (one with no ejection of blood and hence no shortening of the muscle fibers) when the initial volume is greater — within limits.

Starling worked with what is known as a heart-lung preparation, in which the heart and lungs of an anesthetized animal are virtually isolated by cutting the nerves and directing the blood through an artificial systemic circuit in which the volume and pressure of the "venous return" and the "arterial" pressure can be measured and controlled. From data obtained in this manner, a curve, sometimes referred to as a Starling curve, can be plotted to explain the changes in stroke volume with varying conditions in the artificial systemic circuit (Figure 15–17). Decreased venous return results in reduced filling, less stretch of the ventricle, a weaker contraction, and a smaller stroke volume. Elevation of the arterial pressure reduces the volume ejected, but leaves a larger residual volume (a larger ESV) which leads to a raised end-diastolic volume producing greater stretch and greater stroke volume. In this way, after a few beats for adjustment, the force of contraction adapts to new conditions. These relationships have been extrapolated to the hearts of intact humans and other animals, in conditions of varying cardiac output, notably exercise. The dilation of a fail-

---

[1] In order to eject a given volume of blood (say, 100 ml), a heart containing 200 ml would decrease its volume by a smaller fraction than a heart containing only 110 ml. The fibers of the latter would have to undergo greater shortening and thus would lose more energy in overcoming viscous resistance.

[2] S. W. Patterson, H. Piper, and E. H. Starling, "The regulation of the heart beat," *Journal of Physiology (London),* **48** (1914):465–513.

FIGURE 15-17
A. Curve relating stroke volume to diastolic filling. B. Family of ventricular function curves relating stroke work (stroke volume × pressure gradient) to diastolic pressure under different conditions.

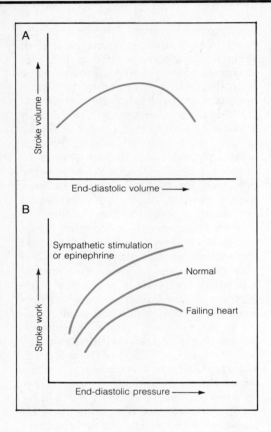

A

Stroke volume

End-diastolic volume ⟶

B

Stroke work

Sympathetic stimulation or epinephrine

Normal

Failing heart

End-diastolic pressure ⟶

ing heart has been viewed as a situation in which the heart is "over the hump" of the Starling curve.

In recent years, with the development of more sophisticated data-gathering methods, experimental results have been reported which are not totally consistent with Starling's Law:

1. If other conditions are altered slightly, for example, by stimulation of the sympathetic nerves, the diastolic volume—stroke volume relationship remains, but the position of the curve varies with each new condition. Such ventricular function curves indicate that diastolic filling (atrial pressure) is not the only factor controlling myocardial contractility.

2. Studies on the length-tension relationships in cardiac muscle have shown that greater stretch or filling does not increase the force of contraction nearly as much if the muscle is allowed to shorten (to pump blood). This is due chiefly to the viscous resistance encountered during shortening.

3. Other studies have shown that in an intact animal the end-diastolic

volume is actually reduced during exercise, rather than enlarged as Starling's Law would predict for a high cardiac output. Furthermore, a maximal end-diastolic volume seems to occur at rest in a reclining position.

The re-evaluation of concepts that followed the publication of these data was based on the realization that the cardiac response in an intact unanesthetized animal is quite different from that of a heart which:

(1) does not pump blood (that is, an isometric or isovolumetric contraction);

(2) is deprived of a nerve supply (and is thus unresponsive to reflex directions);

(3) is subjected to an anesthetic (a depressant);

(4) is exposed to atmospheric pressure (loss of the negative intrathoracic pressure).

Therefore, the relationship described by Starling's Law is an important part of stroke volume control in intact animals, including humans, but superimposed upon it are the neural, hormonal, and mechanical factors described above. Because of this mechanism, the ventricle is very sensitive to changes in the venous return and can adjust its stroke volume accordingly. Thus an increase in venous return to the right side of the heart from the systemic circuit distends the right ventricle, which then contracts with enough force to expel a somewhat greater stroke volume. It takes only a few seconds for the effect of the greater stroke volume to pass through the pulmonary circuit and increase the venous return to the left side of the heart and cause a corresponding increase in the output of the left ventricle. Starling's Law is therefore a very precise mechanism for adjusting the stroke volumes of the right and left ventricles so that over a period of time their outputs are equal and the proper distribution of blood between the pulmonary and systemic circuits is maintained. An unequal output of as little as 1 ml/beat for even a half hour would shift at least 1.8 liters of blood (1 ml × 60 beats × 30 min). This is roughly the amount of blood in the pulmonary circuit and half that in the systemic circuit. Such a shift cannot be tolerated.

The response of the heart—its rate, force, and output—as part of the overall cardiovascular response to some common physiological situations is discussed in Chapter 18.

CHAPTER 15 SUMMARY

Structure Of The Heart

The heart is a four-chambered muscular organ lined with *endocardium,* covered by *epicardium,* and enclosed by the *pericardium.* The thin-walled right and left *atria* receive blood from the venae cavae and systemic circuit, and from the pulmonary veins, respectively. The thick-walled right and left *ventricles* receive blood from the right and left atria and pump it into the pulmonary artery and the aorta, respectively. The *AV valves* between atria and ventricles contain three cusps on the right side (tricuspid) and two on the left (bicuspid, mitral). The cusps are attached by chordae tendineae to papillary muscles of the ventricle. The AV valves are open except during ventricular contraction. The *semilunar* (pulmonary and aortic) *valves* at the outflow from the ventricles resemble tiny pockets on the inner wall of the arteries, which fill and close if blood begins to flow back into the ventricles.

The *coronary arteries* which supply the myocardium arise from the aorta just above the aortic valve, and circle the heart in the AV groove, giving off branches to the myocardium. *Coronary veins* empty into the coronary sinus which drains into the right atrium.

Excitation of the Heart

Cardiac muscle demonstrates automaticity, follows the all-or-none law, and has a very long refractory period, and hence does not develop tetanic contractions. It contracts as a single unit because of communication between cells and the presence of a specialized conducting system.

Pacemaker cells in the *sinoatrial (SA)* node in the right atrium depolar-ize rhythmically. The depolarization spreads over the atrium which then contracts. After reaching the AV node the impulse is conducted rapidly through the AV bundles whose terminal branches reach all parts of the myocardium. Blockage or interruption of the impulse passage may lead to heart block. In partial heart block a ventricular beat follows only every second or third (or more) atrial beat. In complete heart block the atria and ventricles beat independently of one another and at different rates. Cells other than the pacemakers may depolarize and initiate a premature beat. *Fibrillation* occurs when the impulse conduction is abnormal and the muscle does not contract as a unit. Ventricular fibrillation is fatal because blood is not being pumped.

The action potential of the whole heart can be recorded from various sites on the skin surface. Such a record, an *electrocardiogram (ECG)* shows several deflections: a *P wave* denoting atrial depolarization, a *QRS wave* for ventricular depolarization, and a *T wave* for ventricular repolarization. Irregularities in these waves or in the time intervals between them, indicates changes in the course of depolarization of the heart which may affect the heart's pumping action.

Extrinsic Innervation

The heart is innervated by parasympathetic fibers carried in the vagus nerve and sympathetic fibers carried in cardiac nerves. The vagus nerve (and acetylcholine) slow the heart by inhibiting the spontaneous depolarization of the pacemaker cells and it also slows conduction. Sympathetic fibers (norepinephrine and also epinephrine) accelerate the heart by causing faster depolarization of the pacemaker cells. They also increase conduction and cause faster and more forceful contraction by the myocardium.

Events in the Cardiac Cycle

Contraction of the heart is *systole,* and relaxation is *diastole.* During diastole the semilunar valves are closed, but the AV valves are open and the ventricles are filling. Atrial contraction adds little more blood. Ventricular contraction quickly raises pressure in the ventricles, closes the AV valves, and soon forces the semilunar valves open, allowing blood to flow into the artery (pulmonary artery or aorta). As the ventricles begin to relax, and ventricular pressure begins to fall, blood in the arteries begins to return to the ventricles, but catches in the pockets of the semilunar valves and closes them. Ventricular pressure falls rapidly, and when it is less than that in the atria, the AV valves open allowing blood in them to enter the ventricles. Pressure in the arteries falls slowly until the next ventricular contraction. The highest pressure in the aorta during systole is the systolic pressure, and the lowest pressure, at the end of diastole, is diastolic pressure.

A stethoscope placed on the chest wall over the valves allows one to hear the heart sounds, which are caused primarily by closing, first of the AV valves and then the semilunar valves. Failure of the valves to close properly (*insufficiency*) or to open properly (*stenosis*) leads to sounds (*murmurs*) other than normal heart sounds.

Cardiac Output and Its Control

The *cardiac output* is the volume of blood pumped by the heart each minute. It is determined by the heart rate and stroke volume, and is about 5 liters a minute in a resting individual, but can be increased greatly by activity. The *heart rate* is controlled chiefly by reflexes involving sympathetic and vagal fibers. Cardioinhibitor and cardioaccelerator centers in the medulla receive afferent input from many receptors, and cause excitation or inhibition of parasympathetic and sympathetic fibers. There is reciprocal innervation between the two centers, but the parasympathetic is the dominant influence, as the vagus nerves are tonically active.

Stroke volume depends upon factors that affect both the filling and the emptying of the ventricle. Filling, as measured by the end-diastolic volume, is determined by the filling time (heart rate), the atrial contraction, distensibility of the ventricle, and the venous return. Emptying, as measured by the end-systolic volume, depends upon the pressure in the arteries and the contractility of the myocardium. Epinephrine increases the force of contraction, as does greater filling of the ventricle (*Starling's Law*).

STUDY QUESTIONS

1   Trace the course of an erythrocyte through the heart from vena cava to aorta, naming all the chambers and openings through which it passes.

2   What are some of the special properties of cardiac muscle that are particularly important in cardiac function?

3   Trace a cardiac impulse from its development in the pacemaker, through the conduction system and to its final disappearance. How is this related to the electrocardiogram?

4   Consider the aorta, atria, and ventricles during the cardiac cycle. Which of these structures always has a low pressure? Which always has a high pressure? Which has a pressure that varies between high and low? Into which can blood enter throughout the cardiac cycle? Into which can blood enter during most of the cycle? From which can blood leave throughout the cycle?

5   If the heart has automaticity, what is the role of the cardiac nerves? What, specifically, do they innervate?

6   What is the chief mechanism for controlling heart rate?

7   What effect does epinephrine (the hormone) have upon the heart? Is it the same as the action of the sympathetic nervous system?

8   What is meant by the contractility of the myocardium? What might alter this contractility, and how might this affect cardiac output?

9   Suppose, for one reason or another, that the right ventricle pumped out more blood each beat than did the left ventricle. What would be the consequence after a couple of hours?

16

Structure of the Blood Vessels

The Pulmonary Circuit

The Systemic Circuit

The Lymphatic System

Anatomy of the Blood Vessels and Lymphatics

In some respects the blood vessels are like a system of roads and highways. If you wish to travel to a particular place you must travel along certain roads. And for blood to reach a given organ it must flow through specific blood vessels. In a few cases there are alternate routes, but not often. The vascular highway system is entirely a one-way system, although the "going" and "coming" vessels often lie side-by-side and follow similar branching patterns.

Just as you may consult a map to decide how to get to your destination, you may use the following pages to determine how a blood cell gets to an organ. By the time you have completed your study of the blood vessels, you should be able to trace the course of a blood cell to any part of the body and back again.

From your map you can learn something about the kinds of roads you will encounter. Freeways are constructed differently from country lanes because each are used differently. Blood vessels also differ in their structure in accordance with their specific roles in the fluid transport system.

433

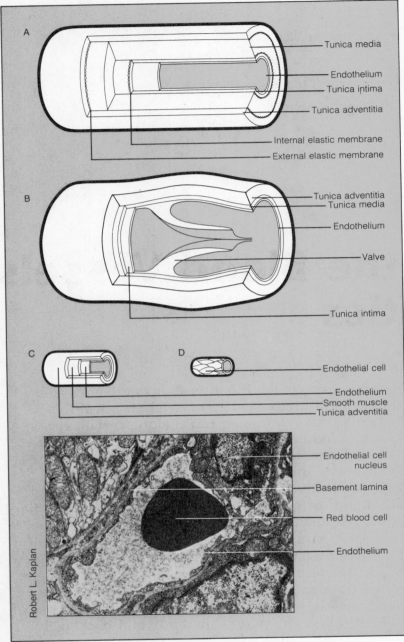

Robert L. Kaplan

| Tunica media |
| Endothelium |
| Tunica intima |
| Tunica adventitia |
| Internal elastic membrane |
| External elastic membrane |
| Tunica adventitia |
| Tunica media |
| Endothelium |
| Valve |
| Tunica intima |
| Endothelial cell |
| Endothelium |
| Smooth muscle |
| Tunica adventitia |
| Endothelial cell nucleus |
| Basement lamina |
| Red blood cell |
| Endothelium |

FIGURE 16–1   Structure of the blood vessels. A. Artery. B. Vein. C. Arteriole. D. Capillary. E. Electron micrograph of a cross section of a capillary. 13,000× magnification.

STRUCTURE OF THE BLOOD VESSELS

Blood vessels, the conduits that carry blood from the heart to the tissues and back to the heart again, vary in size and wall thickness according to their function and position in the vascular circuit. Arteries transport blood from the ventricles under relatively high pressure and have thick walls to withstand that pressure. They soon divide into progressively smaller arteries, into arterioles, and finally into the capillaries that are so abundantly distributed throughout the tissues. As the diameter of the vessels decreases, so does the thickness of their walls, down to the capillaries, whose walls are only one cell thick. The capillaries then converge to venules, then to veins and larger veins, and the largest veins return the blood to the heart. Although the walls of the veins increase in thickness as the vessels increase in size, they never become as thick as the walls of the corresponding arteries. Venous blood is under lower pressure than that in arteries, and the walls of veins do not need to withstand the same strain.

The structure of vessel walls, as well as their thickness, varies with the type of vessel (see Figure 16–1). There are three layers, but only the larger vessels contain them all. The innermost, the **tunica intima,** consists of a layer of simple squamous epithelium, the *endothelium,* together with its basement lamina. Endothelium lines all blood vessels and is continuous with the endocardium in the heart. The middle layer, the **tunica media,** often quite thick, is made up of circularly arranged smooth muscle fibers and/or connective tissue fibers. The outermost layer, the **tunica adventitia,** is chiefly connective tissue and is continuous with the surrounding connective tissue or fascia. Large blood vessels have their own blood supply, tiny vessels known as *vasa vasorum,* which are found in the tunica adventitia. In the larger vessels the boundaries between layers are often marked by sheets of elastic tissue, the **internal elastic membrane** between the tunica intima and the tunica media, and the **external elastic membrane** between the tunica media and the tunica adventitia.

Arteries

The largest arteries are called **elastic arteries** because the thick tunica media contains a large portion of elastic connective tissue, and the elastic membranes are also well developed. The elastic arteries, the aorta and its major divisions and the pulmonary artery, are able to distend to accommodate the volume of blood suddenly ejected at each heartbeat. Their elastic recoil helps maintain pressure in the artery between heartbeats.

Elastic arteries gradually give way to **muscular arteries,** in which the largely elastic tunica media is replaced by one which is predominantly smooth muscle. Large muscular arteries may contain many layers of circular smooth muscle fibers.

Arterioles, with a diameter of 0.5 mm or less, are the smallest of the arterial vessels. Their most prominent feature is a muscular tunica media, which may be several cells thick, although in the smallest arterioles it consists of only a few muscle cells spiraled around a relatively small central cavity or lumen.

The smooth muscle in small arteries and arterioles permits selective narrowing *(vasoconstriction)* or enlarging *(vasodilatation)* of these blood vessels, permitting the blood to be directed toward one organ and away from another. Some arterioles have rings of smooth muscle at the entrance to capillaries, called **precapillary sphincters,** which effectively regulate the flow of blood into those particular capillaries.

Capillaries

Capillaries are the tiniest of vessels; their diameter is scarcely greater than that of a red blood cell. They also have the thinnest walls, since the only significant layer is the endothelium. The junction between endothelial cells is such that the capillary walls are easily penetrated by nearly all constituents of the plasma. Most of these substances

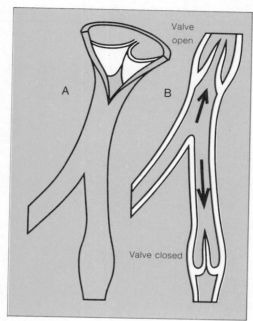

FIGURE 16–2   Valves in veins. A. The vein has been cut open to show the appearance of the valves. B. Longitudinal section through several valves to show how they permit flow in one direction only.

pass between cells, but some are probably transported by pinocytosis. We have already noted that white blood cells can squeeze through the openings because of their amoeboid properties.

Veins

From the capillaries the blood passes into venules and thence into progressively larger veins. **Venules** are a little larger than capillaries and have a layer of connective tissue and maybe a few muscle cells around the endothelium. Veins usually have a thin but recognizable tunica media containing some smooth muscle cells and elastic fibers. The thickest layer, however, is the tunica adventitia, about half the thickness of the entire wall. A characteristic of many larger veins is the presence of valves at intervals to ensure the one-way flow of blood (Figure 16–2). Valves are found most frequently in those veins that must transport blood against gravity; they are thus more

TABLE 16-1   STRUCTURAL CHARACTERISTICS OF BLOOD VESSELS

Type of vessel	General features	Tunica intima	Tunica media	Tunica adventitia
Elastic arteries	Large, at least 1 cm in diameter, with thick walls; carry blood at high pressure and high velocity	Endothelium with underlying connective tissue; prominent internal elastic membrane	Very thick; conspicuous elastic fibers with smooth muscle fibers; external elastic membrane present, but indistinguishable from tunica media	Relatively thin; mostly connective tissue with some smooth muscle fibers; contains vaso vasorum
Muscular arteries	Diameter less than 1 cm; relatively high pressure and velocity of blood	Endothelium with underlying connective tissue; internal elastic membrane very conspicuous	Thick, mostly smooth muscle; external elastic membrane conspicuous	Thinner than tunica media; contains vaso vasorum and smooth muscle fibers
Arterioles	Diameter less than 0.5 mm	Endothelium	Muscular, thickest layer	Thin, connective tissue
Capillaries	Smallest vessels, 8–10 μm; thinnest walls; intermediate pressure (30 + mm Hg), and slowest velocity (less than 5 mm/sec)	Endothelium; entire wall only one cell thick	None	None
Venules	Diameter up to 1 mm; very thin wall	Endothelium	Present in larger venules	Relatively thick, mostly collagenous fibers
Veins	Larger than corresponding arteries, but with thinner walls; valves in many veins; pressure low and velocity intermediate	Endothelium	Relatively thin, some smooth muscle and elastic fibers	Thickest layer; mostly collagenous, with some smooth muscle (longitudinal); contains vaso vasorum

numerous in the veins of the extremities, but absent from those of the brain and certain viscera.

Table 16–1 provides a summary of the blood vessels' structural characteristics.

THE PULMONARY CIRCUIT

The vessels that transport blood from the right ventricle to the lungs and back to the left atrium make up the **pulmonary circuit** (Figure 16–3). The pressure in the pulmonary circuit is much lower than that in the systemic circuit, and the walls of the pulmonary arteries are thinner than those of systemic arteries.

The **pulmonary trunk** arches over the top of the heart and divides into right and left **pulmonary arteries** which pass directly to the right and left lungs. As they enter the lungs, the right pulmonary artery divides into three lobar branches and the left into two lobar branches, one lobar artery to each lobe of the lungs. These branches then divide further, in accordance with the subdivisions of the lungs. The capillary beds form a network of vessels in the walls of the terminal air sacs, or *alveoli*, in the lungs. (The details of these relationships are discussed in Chapter 19.) This is where the blood picks up oxygen and loses carbon dioxide. The oxygen-rich blood then flows into progressively larger veins and returns to the left atrium via the **pulmonary veins.** Typically the left atrium receives two pulmonary veins from each lung.

THE SYSTEMIC CIRCUIT

The vessels that carry blood from the left ventricle to the tissues of the body and back to the right atrium make up the **systemic circuit.** It is longer and has a more complex distribution system than the pulmonary circuit.

Systemic Arteries

The arteries of the systemic circuit, shown in an overall view in Figure 16–4, all branch from the **aorta.** From the left ventricle the aorta ascends a few centimeters before it arches over the pulmonary arteries to descend along the posterior body wall through the thoracic and abdominal cavities. It is usually considered in segments for convenience in identifying the branches arising from its various parts.

The **ascending aorta** is a short segment beginning just beyond the aortic semilunar valves. Its only branches are the coronary arteries, which arise almost behind the cusps of the aortic semilunar valve and supply the myocardium (see Figure 15–6A).

The **arch of the aorta** is aptly named, since it is the portion which curves over the pulmonary arteries. It has three branches, the larger *brachiocephalic trunk* and the smaller *left common carotid* and *left subclavian arteries.* The brachiocephalic divides into the *right common carotid* and *right subclavian arteries,* which follow courses similar to those of their counterparts on the left. The common carotids are the major arteries to the head and neck, while the subclavian arteries supply the upper extremity.

Arteries of the Head and Neck
The common carotid arteries ascend in the neck on either side of the trachea (Figure 16–5). Slightly below the angle of the mandible, they *bifurcate* (divide or branch) into the *internal* and *external carotid arteries.* In the internal carotid artery, near the bifurcation, is an area containing specialized receptor cells. The part of this region called the **carotid sinus** contains pressoreceptors that monitor the blood pressure in the artery. Another part, known as the **carotid body,** contains chemoreceptors that respond to chemical changes in the blood (chiefly a low oxygen tension). Similar regions found in the arch of the

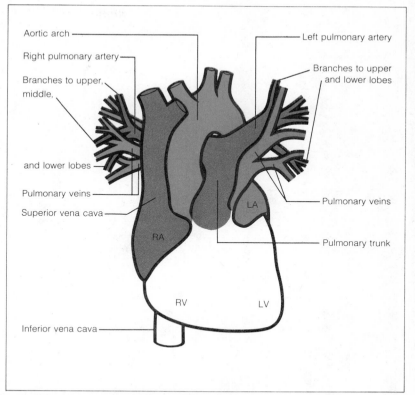

FIGURE 16–3   The pulmonary circuit.

aorta are known as the **aortic sinus** and **aortic body,** respectively.

The **internal carotid** enters the cranial cavity through the carotid canal in the temporal bone. The distribution of its major branches, the *anterior* and *middle cerebral arteries,* and their contribution to the *circle of Willis* have been discussed in Chapter 10 (see Figures 10–17 and 10–18).

The **external carotid artery** supplies those structures of the head and neck that lie outside the cranial cavity. Its branches include the *superior thyroid artery* to the thyroid gland and larynx, the *lingual artery* to the tongue, the *facial artery* to the skin and muscles of the face, and the *occipital* and *posterior auricular arteries* to the posterior portion of the scalp. The terminal branches of the external carotid artery are the *maxillary artery* and the *superficial temporal artery.* The former supplies the deep structures of the face, the muscles of mastication, the mucous membranes of the nasal cavities, the pharynx, the re-

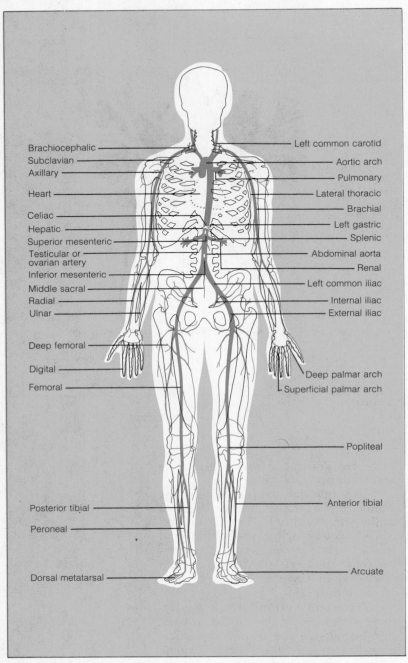

Brachiocephalic
Subclavian
Axillary

Heart

Celiac
Hepatic
Superior mesenteric
Testicular or ovarian artery
Inferior mesenteric
Middle sacral
Radial
Ulnar

Deep femoral

Digital

Femoral

Posterior tibial

Peroneal

Dorsal metatarsal

Left common carotid
Aortic arch
Pulmonary
Lateral thoracic
Brachial
Left gastric
Splenic
Abdominal aorta
Renal
Left common iliac
Internal iliac
External iliac

Deep palmar arch
Superficial palmar arch

Popliteal

Anterior tibial

Arcuate

FIGURE 16-4   Major arteries of the body.

gion of the palate, and, via its *superior* and *inferior alveolar* branches, the individual teeth of the upper and lower jaws. Branches of the superficial temporal artery supply the frontal, parietal, and temporal regions of the scalp.

Arteries of the Upper Extremity   The first branches of the **subclavian artery** are not directly associated with the upper extremity. The **vertebral artery** arises just beyond the origin of the subclavian and ascends through foramina in the transverse processes of the cervical vertebrae and the foramen magnum to the cranium. The two vertebral arteries join to form the **basilar artery** on the ventral surface of the brain (see Figure 10–17). The basilar artery gives off branches to the brainstem and cerebellum before it terminates in the *posterior cerebral arteries*. It also contributes to the circle of Willis. A second branch of the subclavian artery, the *thyrocervical trunk,* sends one branch to the thyroid gland and several to the cervical and scapular regions. The *costocrevical trunk* sends branches to the upper intercostal muscles and to muscles in the back of the neck. The *internal thoracic artery (internal mammary)* descends on the inside of the anterior body wall on either side of the sternum. Its branches include the *anterior intercostal arteries* which run in the intercostal spaces and terminate in branches to the diaphragm and to the anterior thoracic and abdominal walls.

The subclavian artery passes laterally beneath the clavicle. At the level of the first rib it becomes the *axillary artery* as it traverses the axillary region (the armpit) (see Figure 16–6). Branches of the axillary artery, including the thoracoacromial, long thoracic, and subscapular, supply the lateral thoracic wall and muscles around the scapula and head of the humerus. It continues into the arm as the *brachial artery,* which gives off numerous branches to the muscles of the arm. Proximally, it lies medial to the humerus, but it comes to occupy the anterior surface as it nears the elbow.

Just below the elbow the brachial artery divides into the *radial* and *ulnar arteries*. The radial artery continues

along the anterior surface of the forearm on the radial side. At the wrist it
lies next to the radius and is often used
for taking the pulse. The ulnar artery
approaches the hand along the ulnar
side of the forearm. Both arteries terminate in anastomosing branches to form
two arches, the *superficial* and *deep palmar arches,* from which arise tiny *digital
arteries* running down each side of the
fingers.

Arteries of the Thorax The **descending thoracic aorta** passes behind
the heart and along the anterior surface
of the thoracic vertebrae, slightly to the
left of the midline. Its branches are all
small and include the *posterior intercostal arteries,* which pass laterally in the
intercostal spaces to connect with the
anterior intercostal arteries from the internal thoracic artery, the *bronchial arteries* to the bronchi and lungs, and
esophageal branches to the esophagus.

Arteries of the Abdominal Viscera
The aorta passes through the diaphragm and continues as the **descending abdominal aorta** to about the level
of the fourth lumbar vertebra, where it
divides into the **common iliac arteries**
(Figure 16-7). The branches of this section of the aorta include a few small
vessels to nonvisceral structures, namely several pairs of *lumbar arteries* and
the terminal *middle sacral artery.*

The visceral branches of the abdominal aorta are much larger and have
profuse plexuses formed about them by
the autonomic nerves. These major arteries, in order from above downward,
are the *celiac, superior mesenteric, renal,
gonadal (testicular* or *ovarian),* and *inferi-
or mesenteric.* Only the gonadal and
renal arteries are paired.

The *celiac artery,* arising just below
the diaphragm, is very short and it divides immediately into three important
arteries, the *splenic* to the spleen, the
left gastric to the stomach, and the *hepatic* to the liver. The left gastric artery
courses along the upper edge of the
stomach (the lesser curvature) and joins
with the *right gastric,* a branch of the
hepatic artery. A branch from the
splenic artery, the *left gastroepiploic,*

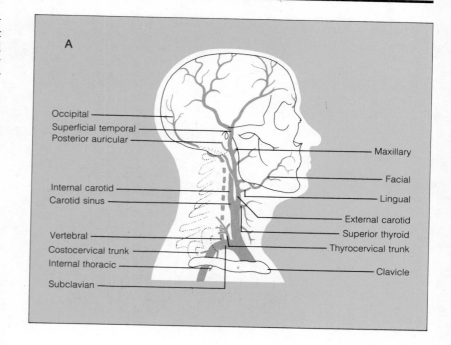

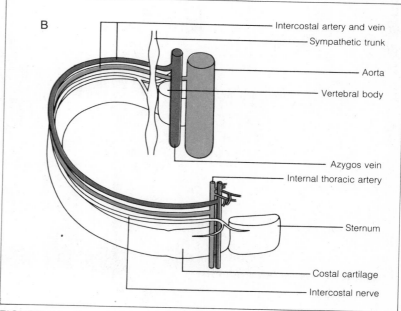

FIGURE 16-5   A. Arteries of the head and neck—the carotid
arteries. B. Arteries of the ribs and thoracic wall—the subclavian
artery.

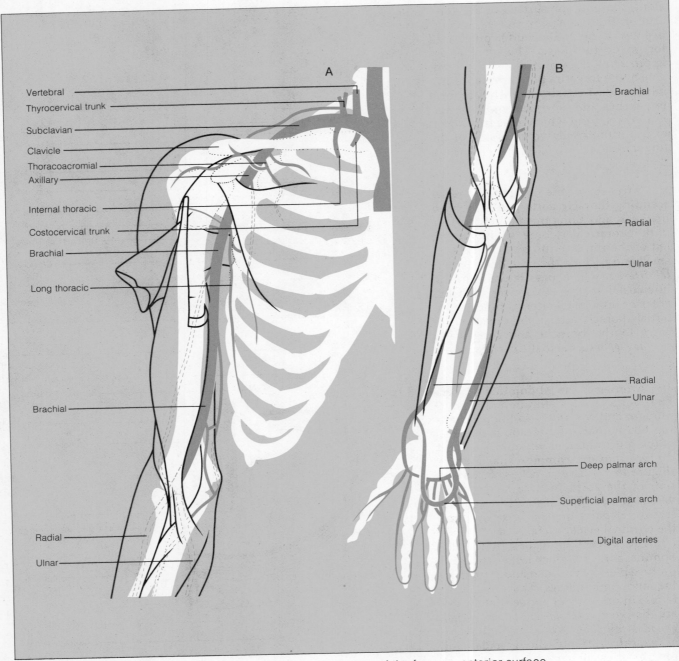

FIGURE 16-6   A. Arteries of the shoulder and arm. B. Arteries of the forearm, anterior surface.

runs along the inferior edge of the stomach (the greater curvature), and joins with the *right gastroepiploic* from the hepatic artery. Other branches from the hepatic artery supply parts of the duodenum and pancreas and the gallbladder.

The *superior mesenteric artery* arises a short distance below the celiac artery (see Figures 16–7 and 16–8). Its branches, entirely enclosed in the mesentery, terminate in numerous anastomosing loops—actually several tiers of loops are formed before the vessels fi-

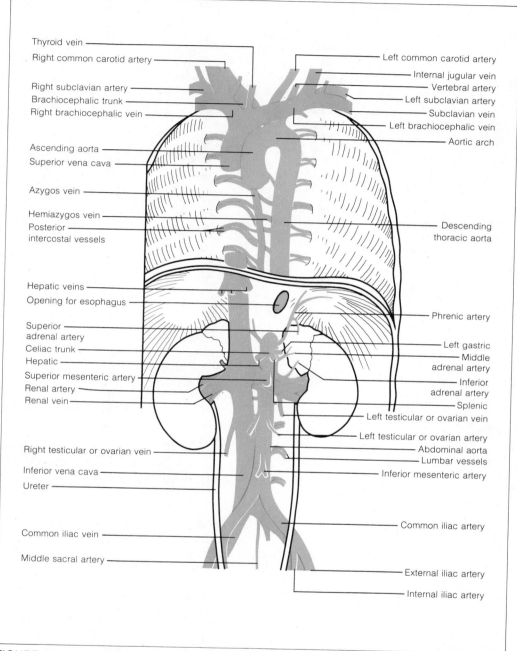

Thyroid vein
Right common carotid artery
Right subclavian artery
Brachiocephalic trunk
Right brachiocephalic vein
Ascending aorta
Superior vena cava
Azygos vein
Hemiazygos vein
Posterior intercostal vessels
Hepatic veins
Opening for esophagus
Superior adrenal artery
Celiac trunk
Hepatic
Superior mesenteric artery
Renal artery
Renal vein
Right testicular or ovarian vein
Inferior vena cava
Ureter
Common iliac vein
Middle sacral artery

Left common carotid artery
Internal jugular vein
Vertebral artery
Left subclavian artery
Subclavian vein
Left brachiocephalic vein
Aortic arch
Descending thoracic aorta
Phrenic artery
Left gastric
Middle adrenal artery
Inferior adrenal artery
Splenic
Left testicular or ovarian vein
Left testicular or ovarian artery
Abdominal aorta
Lumbar vessels
Inferior mesenteric artery
Common iliac artery
External iliac artery
Internal iliac artery

FIGURE 16-7   Great vessels of the trunk.

nally terminate in the wall of the intestine. The superior mesenteric artery supplies all of the small intestine (except the first part of the duodenum), plus the cecum, ascending colon, and transverse colon.

The *renal arteries* are short, large-diameter vessels leading directly to the kidneys, which lie on either side of the aorta.

The tiny *testicular (spermatic)* or *ovarian arteries* pass inferiorly and laterally. The ovarian arteries terminate in the ovaries, and the testicular arteries

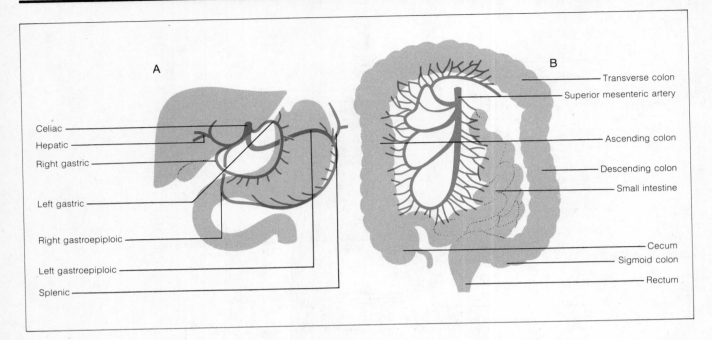

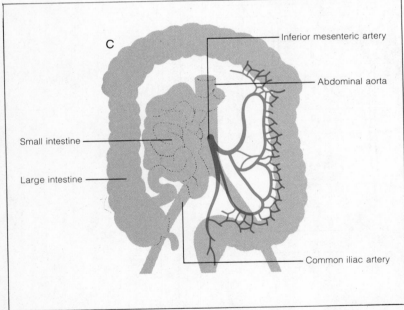

FIGURE 16–8   A. Celiac artery. B. Superior mesenteric artery. C. Inferior mesenteric artery.

pass through the inguinal canal in the inguinal ligament to enter the scrotum and supply the testes.

The *inferior mesenteric artery* is smaller than the superior mesenteric artery. It supplies the descending and sigmoid (S-shaped) portions of the colon and most of the rectum. It anastomoses with branches of both the superior mesenteric artery and the internal iliac artery (see below).

Arteries of the Pelvic Region   The common iliac artery divides into the internal and external iliacs (Figure 16–9). The **internal iliac** (*hypogastric*) **artery** forms an anterior and a posterior division to supply the organs of the pelvic cavity and the structures around the pelvic girdle. The posterior portion sends branches to the muscles of the iliac fossa, the lumbar region, and the gluteal muscles. Branches of the anterior division go to the pelvic viscera, including the uterus, vagina, bladder, and rectum; the *rectal arteries* anastomose with branches of the inferior mesenteric artery. Other branches are the *obturator artery*, which passes through the obturator foramen to supply the adductor muscles on the medial surface of the thigh, the *gluteal branch* to the gluteal muscles, and the terminal *inferior pudendal artery,* which supplies much of the perineal region and external genitalia.

Arteries of the Lower Extremity The **external iliac artery,** a continuation of the common iliac, crosses the iliac fossa, passes beneath the inguinal ligament, and leaves the pelvic cavity, after which it becomes the **femoral artery** (Figure 16–10). Numerous branches of the femoral artery supply the skin and muscles of the thigh and hip region; one large branch is the *deep femoral (profunda) artery,* whose branches supply the muscles on the posterior surface of the thigh. Distally, the femoral artery passes through a canal between the extensor and adductor muscles of the thigh and emerges on the posterior surface of the knee as the *popliteal artery.*

Branches of the popliteal artery supply the muscles and skin around the knee joint. Just below the knee the popliteal artery divides into the anterior and posterior tibial arteries. The *anterior tibial artery* penetrates the interosseous membrane between the tibia and fibula and descends along the anterior surface of that membrane, giving off branches to anterior muscles of the leg. When it reaches the ankle and foot it becomes the superficial *dorsalis pedis* and continues on to the great toe. Branches to the other toes arise from a lateral branch, the *arcuate artery.*

The *posterior tibial artery* descends on the posterior surface of the tibia beneath most of the calf muscles and supplies branches to those muscles. It passes to the medial side of the calcaneus. On the sole of the foot it forms two *plantar arteries* which join to form the *plantar arch,* from which *digital arteries* arise. A large proximal branch, the *peroneal artery,* descends parallel with the fibula and sends branches to the lateral surface of the leg and calcaneus.

Systemic Veins

Blood distributed to the tissues by the arteries is collected from capillaries and returned to the heart by the veins. The pattern of veins converging as they

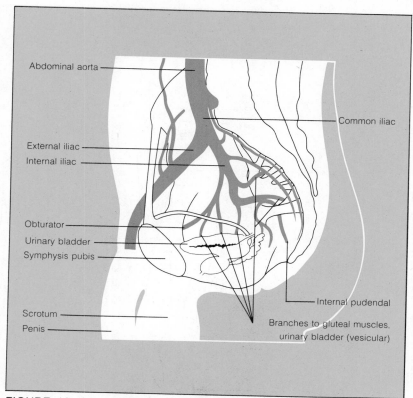

FIGURE 16–9   Internal iliac artery.

approach the heart is similar to the pattern made by the arteries as they leave the heart (Figure 16–11). In the veins, however, blood flows from smaller vessels into larger ones, so that a vein receives tributaries where an artery gives off branches. In general, arteries and veins travel side by side and bear the same names, so it is necessary to consider in detail only those of the veins that differ from the arteries.

Veins are classified as **deep veins** or **superficial veins.** The deep veins parallel the arteries and are often bound to them by connective tissue. Where veins closely accompany an artery, especially in the extremities, they are often paired as **venae comitantes.** Superficial veins lie just beneath the skin and drain the skin and subcutaneous structures. They do not have arterial counterparts, since arteries are not usually superficial vessels. Veins in general, and superficial veins in particular, are extremely variable in their course and pattern (this is readily demonstrated by

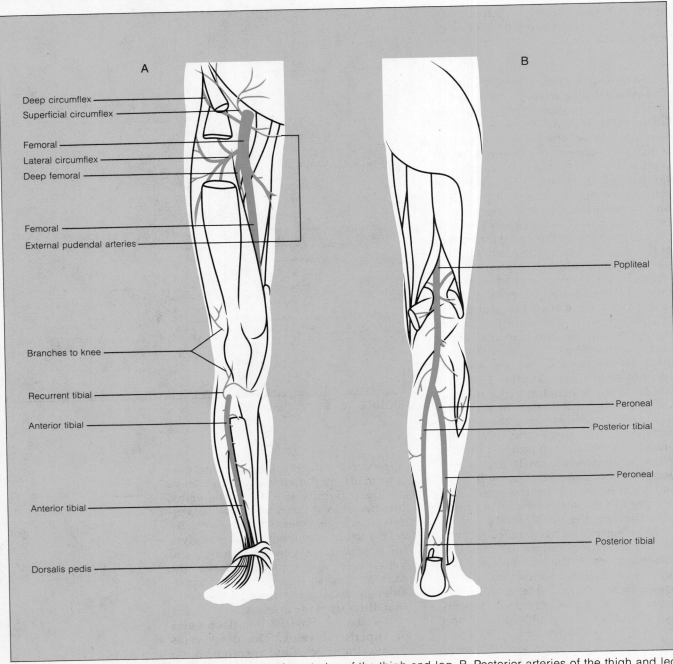

FIGURE 16–10   A. Anterior arteries of the thigh and leg. B. Posterior arteries of the thigh and leg.

comparing the superficial veins on the backs of your two hands). Names are given only to the major, more constant superficial veins.

A third type of vein is the **venous sinus,** a vessel of relatively large diameter found only in certain locations. The most notable examples are the coronary sinus of the heart and the venous sinuses in the dura mater around the brain. Venous sinuses are also found within certain organs, including the liver, spleen, and adrenal glands, but these are much smaller and are microscopic structures (*sinusoids*).

Veins of the Head and Neck

Most of the venous blood from the head finds its way into the jugular veins (Figure 16–12). Blood from the face and scalp (those structures outside the cranial cavity) empties into veins which drain into the superficial *external jugular vein*. Most of the blood from inside the cranial cavity enters the venous sinuses in the dura mater, which in turn converge and leave the skull through the jugular foramen as the large *internal jugular vein* (see Figure 10–19). The latter is enclosed with the common carotid artery and the vagus nerve in the **carotid sheath.** The external jugular vein empties into the **subclavian vein** just before the subclavian and internal jugular veins join to form the **brachiocephalic vein.** The two brachiocephalic veins join above the arch of the aorta to form the **superior vena cava,** which then empties into the right atrium.

Veins of the Upper Extremity

The arm has an extensive superficial venous drainage in addition to those deep veins that accompany the arteries (Figure 16–13). A rich venous network covers the palmar and posterior surfaces of the hand. From the lateral side of the posterior network, the *cephalic vein* emerges and courses upward along the lateral (radial) surface of the forearm and arm. It passes in the groove between the pectoralis major and deltoid muscles and then goes deep to empty into the *axillary vein* (sometimes the subclavian vein). The *basilic vein* arises

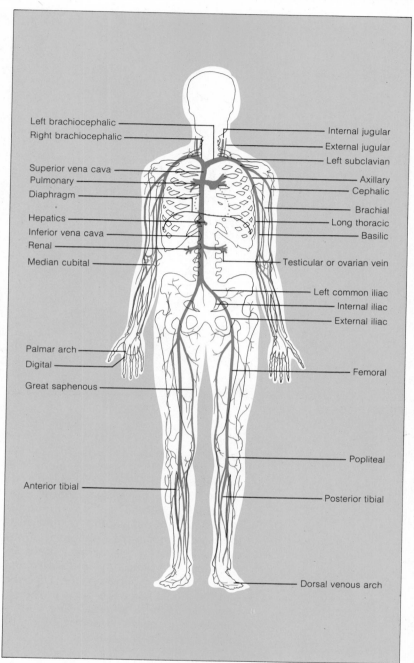

FIGURE 16–11   Major veins of the body.

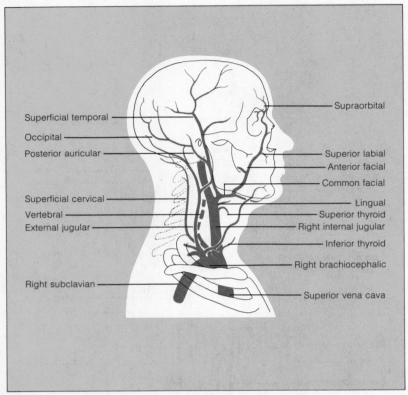

FIGURE 16–12   Veins of the head and neck.

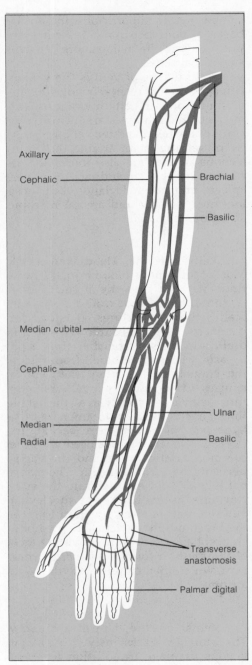

FIGURE 16–13   Veins of the arm and forearm, anterior surface.

from the medial side of the dorsal venous network and ascends along the medial aspect of the forearm. Above the elbow it goes deep and empties into the *brachial vein.* There are numerous branches and anastomoses between these veins, forming a network of superficial veins in the forearm. Very often there is a vein, the *median cubital vein,* that runs between the cephalic and basilic veins at the bend of the elbow (the antecubital fossa). This vein is commonly used for drawing blood or for intravenous injection.

Veins of the Thoracic Cavity   The thoracic veins drain the wall of the rib cage. They do not drain into the venae cavae as one might expect; the presence of the heart and the entry of the venae cavae into it from above and below interfere somewhat. Instead, there are special veins, the **azygos system,** to receive the thoracic drainage (Figure 16–14). The *azygos vein* arises in the upper right lumbar region, penetrates the diaphragm, ascends along the pos-

terior thoracic wall, and empties into the superior vena cava. On the left side, the *hemiazygos vein* follows a similar course, but crosses over to empty into the azygos vein in the midthoracic region. An *accessory hemiazygos vein* drains the upper left side of the thoracic wall and also empties into the azygos vein. The azygos veins receive the *inter-*

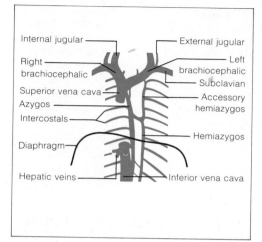

FIGURE 16-14   The azygos veins.

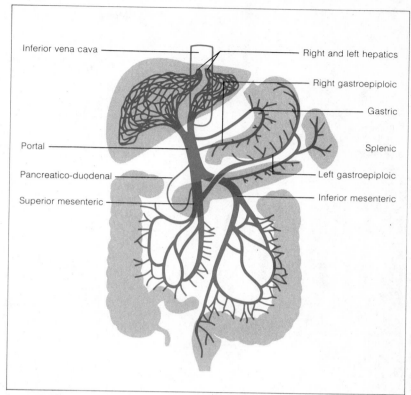

FIGURE 16-15   The portal circulation.

costal veins, which may also empty into the internal thoracic vein (a tributary of the subclavian vein) on the anterior thoracic wall. Other tributaries of the azygos system are the bronchial, esophageal, and pericardial veins.

Veins of the Abdominal and Pelvic Regions   Veins from the abdominal and pelvic organs and the lower extremity empty into the **inferior vena cava,** which is formed by the confluence of the common iliac veins. For the most part, veins from organs in the abdominal and pelvic cavities are the same as the arteries to those structures. The inferior vena cava receives blood from several *lumbar veins,* the *renal veins,* and the *testicular* or *ovarian veins* (these empty into the renal vein on the left side, however). Just before passing through the diaphragm, the inferior vena cava receives the *hepatic vein.* Veins from the pelvic region correspond to the pelvic arteries and converge to form the *internal iliac veins,* which are joined by the *external iliac veins* from the lower extremity to form the *common iliac veins.*

The Portal Circulation   Veins from structures concerned with digestive functions do not drain directly into the inferior vena cava. Instead, the *inferior mesenteric vein* drains into the *splenic vein,* which is then joined by the *superior mesenteric vein* to form the **por-**

tal vein or *hepatic portal,* (Figure 16-15). The veins from the stomach (the *gastric* and *gastroepiploic*) also drain into the portal vein. The portal vein then enters the liver and breaks up into a second capillary bed. These capillaries then converge to form the **hepatic vein,** which returns all of the blood from the digestive tract to the inferior vena cava.

This circulatory modification is functionally significant, since the venous blood in the portal vein contains nutrients and other substances absorbed from the digestive tract. The portal system thus ensures that these substances pass through the liver before being distributed throughout the body. The liver, acting as a clearinghouse, may remove these substances from the blood and store them, "preprocess" them for release to the general circulation as needed, or detoxify those which are foreign to the body.

Veins of the Lower Extremity
The venous drainage of the lower extremity, like that of the upper extremi-

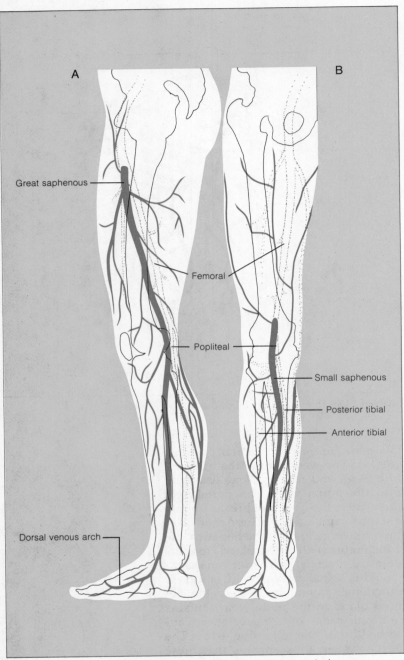

FIGURE 16–16   Veins of the thigh and leg. A. Lateral view.
B. Posterior view.

ty, follows the arterial distribution and is augmented by important superficial veins (Figure 16–16). The major superficial veins of the leg emerge from a *dorsal venous arch* in the foot. The *great saphenous vein* arises from the medial side of the arch and ascends along the medial surface of the leg and thigh to enter the femoral vein a short distance below the inguinal ligament. The *small saphenous vein* arises from the lateral side of the arch, courses up the posterior surface of the leg, and plunges deep to enter the popliteal vein behind the knee. These superficial vessels have numerous tributaries and connections with deep veins. Both superficial and deep veins of the lower extremity contain valves.

THE LYMPHATIC SYSTEM

Although the lymphatic system is not, strictly speaking, a part of the circulatory system, it is necessary for normal function of the cardiovascular system over any length of time. The lymphatic system is an auxiliary venous system, since its role is to return excess fluid from the tissue spaces to the bloodstream. It has no arterial counterpart, since the tissue fluid, which becomes **lymph** when it enters the porous lymphatic vessels, is formed when more fluid leaves the capillaries by filtration than enters by osmosis. Normally there is a slight excess of filtration, and tissue fluid, containing a small amount of protein, tends to accumulate in the tissue spaces. The lymphatic system returns this extra fluid and the protein to the bloodstream. The physiological role of the lymphatic system and the mechanism by which it operates are discussed in Chapter 17.

Lymph Vessels

Peripherally, the lymphatic system arises in a network of vessels resembling blood capillaries, equally thin but slightly larger and much more permeable, since protein molecules are able to enter. These lymphatic capillaries form rich plexuses in the connective tissue within and around the organs and in the subcutaneous tissue. Larger lymphatics, or collecting vessels, conduct the lymph to and through lymph nodes and thence to still larger lymph channels (see Figure 16–17).

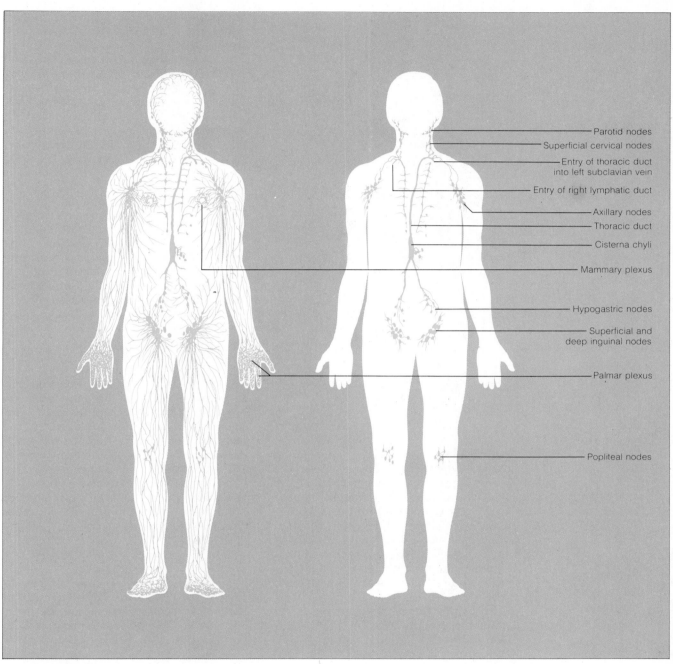

Parotid nodes
Superficial cervical nodes
Entry of thoracic duct into left subclavian vein
Entry of right lymphatic duct
Axillary nodes
Thoracic duct
Cisterna chyli
Mammary plexus
Hypogastric nodes
Superficial and deep inguinal nodes
Palmar plexus
Popliteal nodes

FIGURE 16–17   The lymphatic system.

Lymph Nodes Lymph nodes are tiny organs of lymphoid tissue, usually only a few millimeters in diameter (Figure 16–18). Each is enclosed by a fibrous capsule from which connective tissue trabeculae divide the node into compartments. A fine reticular network further subdivides the compartments and supports their contents. The outer portions of the node (the cortex) contain **germinal centers,** which are concentrations of lymphocytes. Around these aggregates of lymphoid tissue are fluid-filled sinuses that contain numerous macrophages. Although the lymph node contains numerous lymphocytes,

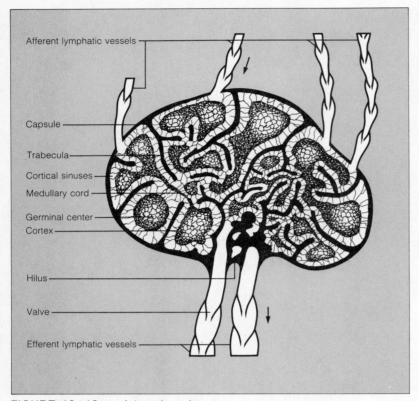

Afferent lymphatic vessels

Capsule

Trabecula

Cortical sinuses

Medullary cord

Germinal center

Cortex

Hilus

Valve

Efferent lymphatic vessels

FIGURE 16–18   A lymph node.

both T and B cells, efforts to determine exactly where they locate within the lymph node have not been successful. The lymphatic vessels (*afferent lymphatics*) enter a lymph node at the peripheral edge. The lymph then filters through the lymph node and leaves by way of small vessels (*efferent lymphatics*) that emerge from the hilus on the other side of the node and lead to larger lymphatic vessels.

Lymph nodes usually occur in groups in certain areas, as in the axillary region, the inguinal region (groin), and along the great vessels in the neck and abdominal and pelvic cavities. A few lymph nodes are also located near the antecubital fossa (elbow) and the popliteal fossa (knee).

Lymphatic Drainage   Lymph collected by the peripheral lymph capillaries may filter through several lymph nodes as it moves centrally. Lymphatic vessels from the lower extremities and

from the abdominal and pelvic viscera converge to a saclike structure, the **cisterna chyli,** which lies just anterior to the vertebral column at about the level of the celiac artery. The **thoracic duct,** the largest lymphatic channel, arises from the cisterna chyli and ascends through the thoracic cavity, receiving tributaries from the left side of the thorax along the way. In the neck it receives the lymphatic drainage from the left arm and left side of the head. The thoracic duct empties into the left subclavian vein near its junction with the jugular vein, thus returning lymph from three-fourths of the body to the bloodstream. Lymphatic vessels from the right side of the thorax and head and from the right arm converge to the smaller **right lymphatic duct,** which enters the right subclavian vein.

Lymphoid Organs

Lymph nodes are important lymphoid organs, but there are other important lymphoid organs that are neither in the direct pathway of lymphatic drainage nor involved in the transport of lymph. They are, however, associated primarily with functions of the lymphocytes. The spleen, thymus, and tonsils are the specific organs; in addition, there are large aggregates or nodules of lymphoid tissue in other organs, particularly in the walls of the intestine.

Spleen   The **spleen,** shown in Figure 16–19, is the largest lymphoid organ. In color it resembles the liver, but the spleen is much smaller and is on the other (left) side of the body. It lies under the diaphragm and posterior and lateral to the stomach. Its internal structure is much like that of a lymph node, although the spleen is much larger. It has a well-developed fibrous capsule, and prominent trabeculae penetrate the substance of the organ to divide it into compartments or *lobules*. The lobules are composed largely of **red pulp,** a tissue which resembles lymphoid tissue because of the presence of reticular cells, lymphocytes, macrophages, and other typically lymphoid cells, all supported by a network of reticular fibers. It differs, however, in that it contains numerous red blood cells as well. Whereas lymph nodes contain lymph-filled sinuses, splenic red pulp contains blood-filled sinuses. It is the presence of blood for which the red pulp is named.

Masses of densely packed lymphoid tissue known as **white pulp** are scattered throughout the red pulp. Some of them contain germinal centers with many lymphocytes. The lack of red blood cells and sinusoids in white pulp accounts for the whitish appearance.

Once lymphocytes have matured and differentiated into committed T or B cells, many of them come to rest at least temporarily in the spleen. Although the lymphocytes arise from cells in the bone marrow, there will be proliferation (cloning) in a particular group of T or B lymphocytes when those cells bind their specific antigen. The T cells will produce sensitized T cells, and B cells will give rise to plasma cells that produce the antibodies; both can produce memory cells. The spleen is, therefore, important in both cell-mediated and humoral immunities. It also contains great numbers of macrophages, so it is important in the nonspecific responses as well.

The spleen serves as a filter for blood in much the same way that lymph nodes serve as a filter for lymph. This function depends heavily on the phagocytic action of the macrophages. A related and important function of the macrophages in the spleen is the destruction and removal of worn-out red blood cells. The spleen has been said to be a reservoir, especially for red blood cells (RBCs) which tend to get trapped in the sinusoids as the blood filters through the red pulp. Theoretically, they can get flushed out into the circulation, given the proper stimulus, but this function is quantitatively unimportant in humans. In some animals smooth muscle fibers in the capsule of the spleen contract to force stored blood out of the spleen, but in humans the splenic capsule has no smooth muscle.

In fetal life the spleen serves as a site of production of blood cells—

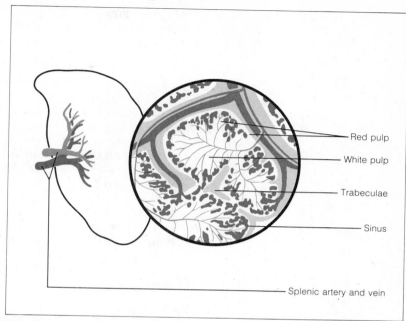

FIGURE 16-19   The spleen.

erythrocytes and granular and agranular leukocytes. This ability is lost before birth, but under certain abnormal circumstances the spleen may become hemopoietic once more. In spite of its many important roles, the spleen can be removed without the individual suffering serious effects because many of its functions can be taken over by other organs. Diseases of the spleen, however, may affect other organs.

Thymus   The **thymus gland** is located in the anterior thorax, between the great vessels and the sternum. It is relatively large during fetal life and early childhood, but after puberty there is a progressive atrophy. The thymus is composed of masses or lobules of lymphoid tissue separated by connective tissue trabeculae. Each lobule has an outer cortical region, which contains densely packed lymphocytes with little supporting connective tissue, and an inner medulla, which contains lymphocytes in looser aggregates and with a heavier connective tissue framework. The thymus is where lymphocytes go to become T cells, and where they acquire the capacity to respond to certain antigens by binding them and releasing various chemicals that facilitate their destruction (see also Chapter 14).

Tonsils Tonsils are found in several sites. The **palatine tonsils** are rather large masses of lymphoid tissue located on either side of the throat and are the structures commonly referred to as "the" tonsils. The **pharyngeal tonsils** are located on the posterior wall of the nasopharynx, between the nasal cavity and the throat. When they become enlarged in children, they are known as *adenoids*. **Lingual tonsils** are tiny clumps of lymphoid tissue on the dorsum of the tongue near its base. T and B lymphocytes and macrophages are found in tonsils.

CHAPTER 16 SUMMARY

Arteries carry blood away from the heart, and veins carry blood toward the heart. Walls of blood vessels are described as having three layers: tunica intima, tunica media, and tunica adventitia. The innermost, *tunica intima*, consists of endothelium, which is a single layer of squamous epithelial cells. The *tunica media* is smooth muscle and connective tissue, and the *tunica adventitia* is connective tissue and blends with the surrounding fascia. An internal and external elastic membrane is found between the layers in larger vessels. Larger vessels also have tiny blood vessels (*vaso vasorum*) in the tunica adventitia as the blood supply to the vessel walls. In small vessels the layers are thinner and some of them are absent.

The largest **arteries** such as the aorta are known as *elastic arteries* because their thick tunica media consists mainly of elastic tissue which provides distensibility and recoil. Smaller arteries are *muscular arteries* as their tunica media contains smooth muscle. *Arterioles* are much smaller vessels, but still have smooth muscle in their walls, with which their diameter can be varied. *Capillaries* are the smallest blood vessels, whose wall consists only of the single layer of squamous epithelial cells. Exchanges between interstitial fluid and plasma occur across capillary walls.

Veins, in general, have a larger diameter and thinner walls than arteries. *Venules,* the smallest venous vessels, are mostly connective tissue, but with some smooth muscle. *Larger veins* have a recognizable tunica media with some smooth muscle cells. In all veins the tunica adventitia is the thickest and most prominent layer. Most veins have one-way valves at frequent intervals.

The major arteries, and some of their more important branches are listed below.

MAJOR ARTERIES

Ascending Aorta
Right and Left Coronary arteries
Arch of the Aorta
Brachiocephalic Artery
　Right Common Carotid Artery
　　Internal Carotid Artery
　　　Cerebral branches
　　External Carotid Artery
　　　Numerous branches to neck and face
　Right Subclavian Artery
　　Vertebral—Basilar
　　　Branches to brainstem, cerebellum, and cerebrum
　　Thyrocervical Trunk
　　Costocervical Trunk
　　Internal Thoracic Artery
　Axillary Artery (continuation of subclavian)
　　Branches to thoracic wall, shoulder, and axillary region
　Brachial Artery (continuation of axillary)
　　Radial Artery
　　Ulnar Artery
Left Common Carotid Artery
　(Branches same as right common carotid)
Left Subclavian Artery
　(Branches same as right subclavian)
Descending Thoracic Aorta
Several intercostal, bronchial, and esophageal branches
Descending Abdominal Aorta
Celiac Artery
　Splenic Artery
　Left Gastric Artery
　Hepatic Artery

Superior Mesenteric Artery
Renal Artery
Testicular or Ovarian Artery
Inferior Mesenteric Artery
Common Iliac Artery
Internal Iliac Artery (hypogastric)
Posterior muscular branches to lumbar, gluteal, and iliac fossa muscles
Anterior visceral branches to pelvic viscera
Obturator, Gluteal, and Pudendal branches
External Iliac Artery
Femoral Artery
Deep Femoral Artery
Popliteal (continuation of femoral)
Anterior Tibial Artery
Posterior Tibial Artery
Peroneal Artery

The deep veins come together in a pattern very similar to that of the distribution of the arteries, and most of the names are the same. Deep veins often run in pairs alongside the arteries as *venae comitantes.* The venous drainage of most of the abdominal cavity (digestive tract) converges to the *portal vein* and goes through a second capillary bed in the liver before emptying into the vena cava. The veins of the thoracic wall drain into the *azygos* system which empties into the vena cava. Superficial veins are the external jugular in the head, the cephalic and basilic in the arm, and the great and small saphenous in the leg.

The **lymphatic system** drains the interstitial fluid. Lymphatic vessels arise from rich networks of lymphatic capillaries in and around organs throughout the body. Their walls are extremely thin and extremely permeable. When interstitial fluid enters a lymphatic channel it becomes *lymph.* Small lymph vessels empty into lymph nodes which are clustered in and around most organs and in the neck, axillary, and inguinal regions. They contain macrophages and lymphocytes. Lymph is usually filtered through several lymph nodes. The lymph vessels converge into larger vessels with numerous valves, and those from the lower half of the body empty into the cisterna chyli from which the thoracic duct arises. The *thoracic duct* ascends along the posterior body wall and receives lymphatic vessels from the left side of the head and thorax, and left arm before it empties into the left subclavian vein near its junction with the internal jugular vein. Lymphatic vessels from the right side of the head and thorax, and right arm form the smaller right lymphatic duct which empties into the right subclavian vein.

STUDY QUESTIONS

1 State a structural characteristic of arteries, arterioles, capillaries, and veins that is important to the particular function of that type of vessel.

2 In general, how does the structure of the vessels of the pulmonary circuit differ from that of the systemic circuit?

3 What are anastomoses? Under what conditions might they be of particular importance?

4 A tagged erythrocyte enters the aorta. Name in order all vessels it would pass through in going to: (a) the right thumb and then to the coronary sinus; (b) the small intestine and then to the cerebral cortex; (c) the azygos vein and then to the cephalic vein.

5 How does the venous drainage of the digestive system differ from that of most other viscera? What is the significance of this difference?

6 Why do you think a separate set of veins is needed to collect venous blood from the thorax?

7 What is the relationship between the lymphatic system and the rest of the cardiovascular system? What is the functional significance of the lymphatic system?

8 Compare and contrast the movement of lymph in lymphatic vessels with the movement of blood in veins.

17

Blood Flow—Hemodynamics

Arterial Pressure

The Microcirculation—The Capillary Bed

Venous Circulation

Lymphatic Circulation

Circulation of the Blood and Lymph

So far we have studied properties of the blood and of the pump that sends it through the vascular channels. In Chapter 15 it is pointed out that the force generated by contraction of the ventricle has to be sufficient to drive the blood through the entire circuit and back to the heart, and that blood leaves the left ventricle at a high pressure and returns to the right atrium at very low pressure. In this chapter we turn our attention to what happens between the time blood leaves the left ventricle and returns to the right atrium. We ask what forces act upon the blood to cause it to flow, how that flow is regulated, and how blood is distributed among organs whose needs may be very different.

To answer these questions, one needs some understanding of the physical forces that determine flow in any fluid system, and a knowledge of the effects of specific properties of this particular fluid system. Among factors that determine the flow and distribution of the blood are the arterial blood pressure and the means by which it is maintained, the venous pressure and its effect upon return of blood to the heart, and conditions in the capillaries that govern the exchange of materials between the blood and the interstitial fluid.

BLOOD FLOW— HEMODYNAMICS

The blood maintains a suitable environment for the individual cells of the body by transporting the necessary substances to and from that environment. The mechanisms that ensure an adequate flow of blood to the tissues are therefore vital parts of the homeostatic process. Since the need of any cell for blood-borne substances varies with its metabolic activity—an active muscle uses more oxygen than a resting muscle—there must be a way to increase its blood supply when necessary. The amount of blood available, however, is limited, and any increase in the blood flow to one organ must be accompanied by a decrease elsewhere. Redirecting the flow and distribution of blood requires changes in those factors that are responsible for moving the blood. The flow of blood and the factors that control it constitute what we know as **hemodynamics.**

There are three basic elements in the movement of blood: *flow, pressure,* and *resistance.* They are so closely interrelated that one cannot consider one without the others. Flow of blood is caused by pressure generated by the ventricle and is directly proportional to the pressure. Flow is impeded by any resistance along the way, and is inversely proportional to that resistance. We will first consider these basic elements, and then see how they affect— and are affected by—such factors as the velocity of the blood and the elasticity of the walls of the vessels.

The blood flow is the volume of blood passing a point in a given period of time, usually a minute, and is expressed in milliliters or liters per minute. It may be the blood flow to a particular organ, or the blood flow in the entire circulatory system (see Figure 17–1). The total blood flow is the volume of blood leaving the left ventricle each minute, and is the same as the *cardiac output.* About five liters per minute, this flow is distributed among the organs of the body according to their needs. The organs in the abdominal cavity receive nearly half the total cardiac output (see Table 17–1), while

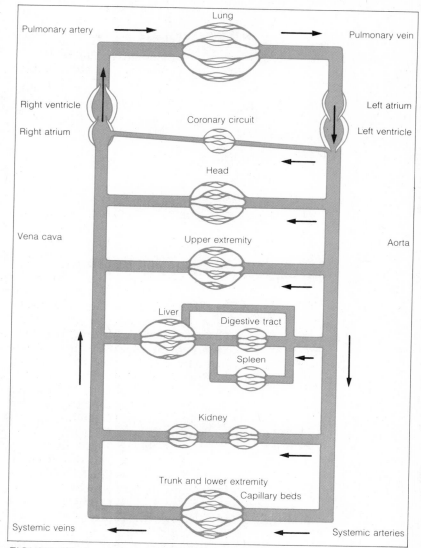

FIGURE 17–1 Distribution of the blood to the body parts through parallel circuits.

skeletal muscle, which accounts for nearly half the body weight, receives only about a fifth of the cardiac output at rest. However, during vigorous physical activity, it may receive as much as 75 percent of the entire output of the heart. Blood flow to the skin may also be increased greatly as part of the temperature-regulating mechanism. Such large increases are, however, accompanied by reductions in flow to visceral structures and/or increases in the cardiac output.

TABLE 17-1   ESTIMATED DISTRIBUTION OF CARDIAC OUTPUT IN A RESTING SUBJECT

Circulation	Blood flow (ml/min)	Percent of cardiac output
Splanchnic	1350	27
Renal	1100	22
Cerebral	700	14
Skin	300	6
Skeletal muscle	750	15
Coronary	250	5
Other organs	550	11
Total	5000	100

Pressure and Resistance

With each beat the left ventricle ejects the blood with enough force to drive it through the systemic circuit and back to the right atrium.[1] The force imparted to the blood by the heart is opposed by various types of resistance to its flow, and if the pressure developed by the ventricle is not enough to overcome the resistance, there is little or no blood flow and the circulation virtually ceases.

It is not the pressure itself that causes blood to flow, but the pressure gradient. This can be illustrated with a mechanical system in which a reservoir provides the head of pressure for a fluid-filled system (Figure 17-2). When the outlet is closed, the pressure is the same throughout, as indicated by the identical fluid heights in the vertical tubes; there is pressure, but no pressure gradient and no flow. If the outlet is opened, the outlet pressure will drop, creating a pressure gradient along the tube, and there will be flow in proportion to the size of that gradient.

A similar situation exists in the circulatory system. Blood enters the aorta at high pressure, but by the time it reaches the other end of the circuit (the

---

[1]Our discussion is directed to the systemic circuit, for conditions there determine the blood flow to the organs. Although the total blood flow through the pulmonary circuit is the same as that through the systemic circuit, pressures and resistances are much less. In general, the pulmonary circuit is not affected by some factors that are important for the systemic circuit, and is less responsive to some of the control mechanisms. Blood flow in the pulmonary circuit is discussed in Chapter 18.

right atrium) the pressure is very low. Since the right atrial pressure is nearly zero, the pressure gradient or driving force that causes blood to flow, is the pressure in the aorta.

Not all of the energy expended by the heart to develop the pressure is used to cause blood to flow. Because the blood in arteries is under pressure, it pushes outward against the artery wall. The artery resists this force by "pushing in" to balance the outward force of the blood (Figure 17-3). However, blood is incompressible, so when the ventricle contracts and abruptly forces more blood into an artery (aorta) that is already full and under pressure, the pressure rises in the artery and it be-

FIGURE 17-2   Interaction between pressure and resistance in a fluid system. A. The outlet is closed and the pressure is the same throughout the horizontal tube (as shown by the height of the fluid in the vertical tubes). B. The outlet is open. There is no resistance at the outlet, but a pressure gradient exists, and the fluid flows. C. A constriction in the horizontal tube increases resistance at that site. Upstream pressure is raised and downstream pressure is reduced.

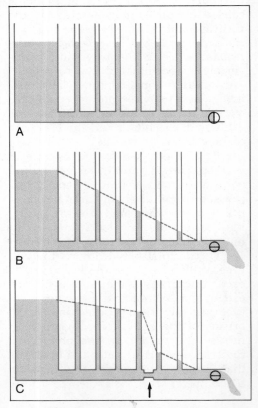

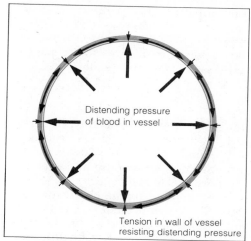

FIGURE 17–3   Relationship between the distending force of blood in a vessel and the resisting tension in the vessel wall.

comes distended. The connective tissue elements in its wall permit a certain amount of stretch and the artery is distended until its resistance balances the pressure inside. If the artery could not "push back" with equal force the blood would break through its wall. Occasionally there is a weak spot in an artery wall, and the pressure inside causes the wall to bulge outward, a potentially dangerous condition known as an *aneurysm.*

Several types of resistance to flow are found in any fluid system, including the circulatory system. These include viscosity, which is an internal friction within the fluid itself, and a frictional resistance, which is a fixed resistance that cannot be regulated. In the circulatory system there is also a variable resistance that can be regulated. It is known as the **peripheral resistance,** and is associated with changes in the diameter of the smaller vessels.

The **viscosity** is the internal resistance of the fluid to flow. The greater the viscosity the stronger are the forces that bind the particles of that fluid and resist their displacement. Water, for example, has a relatively low viscosity, whereas molasses has a relatively high viscosity. Water will flow through a tube with little pressure behind it, but molasses requires so much pressure that rapid flow may not be possible. Blood has a higher viscosity than water, though not nearly as high as molasses.

The **frictional drag** is developed by contact of the moving fluid with the stationary walls of the vessels through which it flows. Drag is proportional to the length of the vessel, and it is increased if the inner surface of the vessel is rough. Frictional resistance is greater in vessels of small diameter because of the relatively large area of surface contact (See Figure 17–4). The outer layer of fluid brushes against the stationary vessel wall which impedes its movement. The next layer of fluid comes in contact with the slowly moving outer layer and is subjected to less drag. In larger vessels the flow is considerably faster in the center than along the edges, producing what is known as *laminar flow.* It is not apparent in small vessels in which all of the fluid layers are quite close to the walls.

The **variable resistance** in the circulatory system is caused by the peripheral resistance of the arterioles and small muscular arteries. Because these vessels contribute so much to the total resistance, they are known as *resistance vessels.* Their small diameter has the same effect on pressure and flow in blood vessels as a clamp on the tube in the mechanical model. If the clamp is completely closed, the situation in the tube will resemble that in Figure 17–2A, in which there is no flow and the pressure above the clamp (upstream) is the same as in the reservoir, and below the clamp (downstream) it is the same as at the outlet. If the clamp is wide open, the conditions are as in Figure 17–2B, for the clamp has no effect. But when the clamp is partially closed (or partially open) the conditions in the tube lie somewhere between— just where depends upon how much the clamp is opened (Figure 17–2C). The pressure above the clamp is elevated, and downstream it is reduced. The

FIGURE 17–4   Frictional drag and laminar flow in large and small vessels.

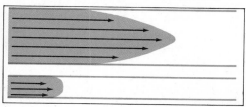

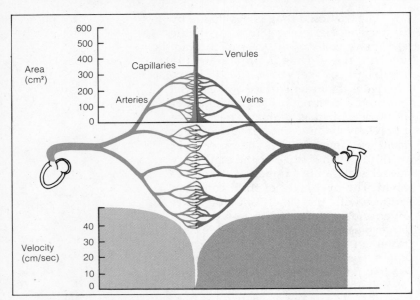

FIGURE 17–5   Relationship between velocity of flow and cross-sectional area of segments of the systemic circulation in a dog. In humans the areas are different, but the relationships are the same.

(Adapted and reproduced, with permission, from R. F. Rushmer, *Cardiovascular Dynamics*, 3rd ed., W. B. Saunders Company, Philadelphia, 1970.)

clamp is an increased resistance that dams the fluid back behind it and alters the pressure on either side of it. The total pressure drop along the tube is the same, but a large portion of it occurs over the region of the increased resistance. The flow is reduced because the pressure gradient along the tube below the clamp is small.

The importance of the peripheral resistance lies in the fact that it is so highly variable, and is therefore a major factor in regulating the blood flow and pressure. The resistance vessels—the small muscular arteries, but especially the arterioles—determine the flow of blood from arteries to veins. Constriction (vasoconstriction) raises the resistance, reduces the flow of blood into the capillaries, and raises the pressure upstream in the arteries. Decreased constriction (vasodilitation) reduces the resistance, allows more blood to leave the arteries, and lowers the arterial pressure. Smooth muscle in the walls of these vessels is responsive to stimulation by the sympathetic nervous system, epinephrine, and a variety of local conditions.

Summary. We have seen that a fluid flows when there is a force or pressure driving it and that the flow is impeded when there is a resistance in its path. This interdependence between flow, pressure, and resistance may be expressed by a simple equation:

$$F = \frac{P}{R} \text{ or } P = FR$$

where F is the flow, P is the pressure or driving force, and R is the resistance. The equation states that flow is directly proportional to pressure and inversely proportional to resistance.

As applied to the circulatory system (the systemic circuit), F is the total blood flow or cardiac output; P is the pressure drop from the beginning of the aorta (semilunar valves) to the right atrium, which for all practical purposes is the arterial pressure,[2] since atrial pressure is so low. R is the total resistance to flow along the entire circuit, including the viscosity of the blood, the frictional resistance, and the variable peripheral resistance due to the action of smooth muscle in the muscular arterioles. According to the equation, the pressure in the arteries is the product of the volume of blood flow into the arteries, and the resistance to its flow out of the arteries. Arterial pressure can be varied by changing either the cardiac output or the resistance, and if pressure is to be maintained, they both must be regulated.

Velocity of Flow

One of the factors that affects the distribution and flow of blood is the velocity or speed with which the blood moves through the vessels. Like water in a stream that rushes at high speed through a narrow chasm, but passes slowly through a placid pool, the velocity or speed of a fluid moving in response to a pressure gradient depends partly upon the size of the channel through which it flows. Stated more precisely, if the volume of flow is constant, the velocity is inversely proportional to the cross-sectional area of the channel (see Figure 17–5). The cross-

[2] In these discussions, arterial pressure refers to the average pressure in the artery, the mean arterial pressure, not systolic or diastolic pressures.

sectional area considered is that of the entire system at one level, however, not the area of the individual channel, a distinction that is important in the circulatory system. Velocity is greatest in the aorta, but least in the capillaries, whose total cross-sectional area is roughly 600–800 times greater than the 3 cm² of the aorta. Because the total flow must be the same in both, it means that the velocity in the capillaries is 600–800 times less than the 300 mm/sec in the aorta. Capillaries are generally only a millimeter or two in length, however, so it takes a blood cell only two or three seconds to pass the length of a capillary.

As the blood moves into the progressively larger veins (of progressively smaller total area), the velocity increases once more. It never reaches the speed in the aorta, however, because the venae cavae are larger than the aorta. It is important to remember that pressure falls all along the way from the aorta to the venae cavae, but that velocity increases again beyond the capillaries.

These flow characteristics enhance the effectiveness of the cardiovascular system. Blood travels faster in large vessels, where the main function is transport, but it travels very slowly in the capillaries. In addition, capillaries are small, their walls are extremely thin, and they have a large total surface area. It is in the capillaries that the all-important exchanges occur between the blood and the extravascular fluid.

Elasticity of the Vessels

A rigid (glass) tube has no "give," and if the pressure in it exceeds a certain level, the tube will burst. The walls of blood vessels, however, are not rigid. They are elastic, and they are distended as the pressure within them increases. The ability to *distend*, or to be stretched, is a particular advantage when the pressure of the fluid in the vessels is not constant.

The output of the heart, like that of most pumps, is intermittent, so that the flow in the arteries is *pulsatile* (has pulsations). When blood is ejected into the aorta (the most elastic of arteries) the vessel "gives." This accommodates part

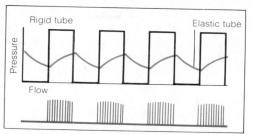

FIGURE 17–6 Pressure developed in rigid and elastic tubes by pulsatile flow.

of the blood ejected and acts as a damper to minimize the pressure rise in the artery. At the end of ejection the elastic recoil of the stretched artery squeezes down on the blood in it and helps sustain a pressure within the artery until the next ejection. If the aorta were rigid, ejection would develop an immediate and extremely high pressure, since there could be no distension, and hence no increase in capacity of the artery. However, the pressure would fall drastically almost as soon as the ejection was completed because there would be no recoil. The effects of a pulsatile flow on the pressure in a rigid and an elastic tube are compared in Figure 17–6. Fortunately, the aorta is quite elastic, and its pressure curve resembles that of the elastic tube rather than the rigid tube (see also Figure 15–13).

Arterial pressure falls slowly but continuously from the end of one ejection to the beginning of the next due to the fact that blood is continually running out the other end of the arterial system into the capillaries aided by the pressure from the recoiling arteries. This volume of blood is known as the **peripheral run-off.** If the heart did not beat again, the run-off would continue and the pressure would continue to fall toward zero.

The Arterial Pulse

The distension caused by the ejection of blood spreads over the arteries like a wave, beginning at the root of the aorta and sweeping out to the tiny peripheral arteries. The waves of distension, or **pulse waves,** are routinely counted as an indication of the heart rate. The pulse rate is commonly taken from the radial artery at the wrist, but it can be counted from any artery that is close enough to the body surface for its pulsations to be detected. Similar pressure changes or pulsations occur throughout the arterial system with each beat, although the changes vary in different parts of the system. The pulsations become smaller in the terminal arteries and are rarely seen in capillaries.

The pulse wave extends over the "arterial tree" at a speed of 5–7 m/sec. It travels more slowly in elastic vessels, but if they lose their elasticity and become more rigid, the pulse wave is transmitted more rapidly. It should be noted that the speed of the pulse wave is *not* the same as the velocity of the blood in the arteries. The pulse wave travels at least 15 times faster than the blood (at 300 mm/sec maximum). It is analogous to the waves in a stream, which are quite distinct from the current.

When blood vessels lose their elasticity, they become more like rigid tubes. They neither distend nor recoil as readily. Ejection of blood therefore produces an unusually great rise in systolic pressure. Since there is no recoil to maintain pressure between beats, the diastolic pressure falls markedly. A slight loss of elasticity and the accompanying rise of systolic pressure typically occur as one grows older. Excessive rigidity is known as *arteriosclerosis.* Frequently, however, the loss of elasticity may be due to, or aggravated by, the deposition of material in the walls of the arteries *(atherosclerosis),* which not only reduces the distensibility and recoil, but narrows the lumen and thus tends to increase the frictional resistance as well. In such a condition, diastolic pressure is also raised, while systolic pressure is raised even more, causing a potentially dangerous condition of generalized elevated blood pressure. The disease of high blood pressure *(hypertension),* may be caused by conditions other than arteriosclerosis. Whatever causes hypertension increases the danger of circulatory problems (such as *thrombosis, occlusion,* and *rupture)* in severely affected arteries (see Chapter 18).

ARTERIAL PRESSURE

It is the pressure developed by the ventricle that keeps the blood moving and ensures an adequate filtration pressure for exchange in the capillaries. If blood pressure is insufficient, flow will not be maintained. If it is too high, excessive loss of fluid from the capillaries is possible. The level of arterial pressure must be carefully regulated.

Measurement of Arterial Pressure

When we refer to "blood pressure," we refer to mean arterial pressure in the brachial artery, unless otherwise specified. Because the pressure in the arteries fluctuates during the course of each cardiac cycle, the pressure that would result if these fluctuations were smoothed out, the **mean arterial pressure,** is what we are concerned with, since this is the pressure that keeps the blood moving. The brachial artery is the usual site of measurement largely because of its convenience.

Arterial pressure may be measured directly by inserting a needle or narrow-tipped tube *(cannula)* into the artery and connecting it to some sort of pressure-detecting device. Although such methods have been extremely useful in experimental situations, they are not practical for routine determination of blood pressure, and indirect methods are preferred. These methods involve measuring the amount of externally applied pressure required to cause the artery to collapse.

Arterial pressure in humans is usually measured with a **sphygmomanometer,** which consists of an inflatable cuff connected to a bulb, with valves for controlling inflation of the cuff, and to a gauge for registering the cuff pressure. The cuff is wrapped around the arm and inflated. Whenever the pressure in the cuff exceeds that in the brachial artery, the artery is occluded and the pulse in the radial artery disappears. This is the *palpatory method* of determining systolic pressure.

A more common method is the *auscultatory method,* by which the blood flow distal to the cuff is detected with a

stethoscope placed over the brachial artery near the elbow (Figure 17–7). When the pressure in the cuff is greater than in the artery, the vessel is occluded and there is no blood flow. When the pressure in the cuff is reduced to less than in the artery, the artery is open and blood flow is uninterrupted. But when the pressure in the cuff is between the systolic and diastolic pressures, flow is intermittent; blood flows through the artery only during that part of the cardiac cycle in which the arterial pressure is greater than the cuff pressure. When arterial pressure falls below cuff pressure, the artery collapses; and each time blood spurts through the artery under the cuff, a sound can be heard with the stethoscope (the sound can be heard only when the flow is intermittent). The highest pressure (in millimeters of mercury) at which blood passes under the cuff is the **systolic pressure** (SP), and the pressure at which blood begins to flow continuously is the **diastolic pressure** (DP). The information is customarily written as SP/DP. Typical textbook values for normal resting arterial pressure are likely to be stated as 120/80, although "normal" covers a rather wide range (Figure 17–8). Among college students readings of 90/60 are not unusual (the differences may reflect their youth, as well as the fact that determinations of blood pressure by a classmate and by a physician do not have the same emotional effects on the blood pressure). Arterial pressure increases slightly and slowly with age, related, at least in part, to the loss of elasticity in the blood vessels with aging.

The difference between systolic and diastolic pressure is the **pulse pressure.** It represents the pressure change in the artery during the cardiac cycle, and is a rough approximation of the stroke volume. It also varies with the elasticity of the vessels.

Maintenance of Arterial Pressure

The arterial pressure is the driving force that keeps the blood moving through the blood vessels. It is the

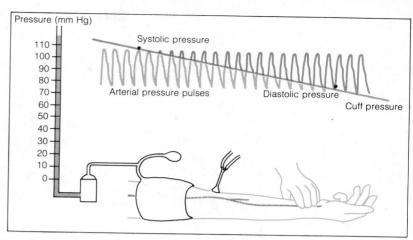

FIGURE 17–7   Measurement of arterial pressure with a sphygmomanometer.

source of the filtration pressure for the exchange of materials in the capillaries and the removal of wastes and regulation of plasma composition in the kidney, and it ensures that blood flow to a given tissue can be increased when its need increases. It must provide a sufficient driving force to distribute blood to all tissues, but a pressure any higher than that would impose a load on the heart and vessels. To achieve these goals, the arterial pressure must be carefully regulated; it must be kept within certain limits, yet there must also be the means to change it when necessary. But arterial pressure is dependent upon the blood flow and the resistance to flow, both of which change with exercise and metabolic activity. If a particular level of blood pressure is to be maintained, change in either resistance or flow in any part of the circuit must be counterbalanced by adjustments in the other in some part of the circuit (Figure 17–9).

When considering the cardiovascular system as a whole, the relationships between flow, pressure, and resistance previously discussed are very much in evidence. The blood flow is the output of the left ventricle, that is, the cardiac output. The pressure is the mean arterial pressure, and the resistance includes all those factors that impede the flow of blood, especially the peripheral resistance. Blood flow is controlled by the smooth muscle in the arterioles, which governs the diameter

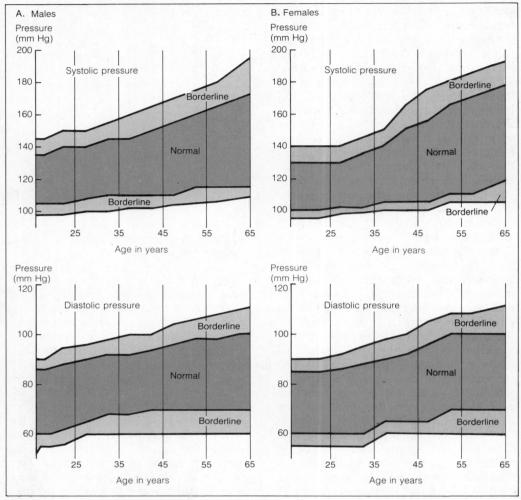

FIGURE 17–8 Range of normal blood pressures in males and females of different ages. A. Males. B. Females.

of these vessels and therefore controls the peripheral runoff and flow into capillaries. Peripheral resistance is of particular importance because, unlike other types of resistance, it can be readily changed and is thus a means of continually regulating the resistance.

The relationship between pressure, flow, and resistance assumes a constant volume of blood in the circulatory system. To have a pressure in the arteries, they must be slightly overfilled, but if the blood volume falls (as through injury), the blood pressure declines, and so does the flow.

The major factors that contribute to the mean arterial pressure are therefore the blood volume, the cardiac output, and the resistance. Of these factors,

cardiac output and peripheral resistance merit the greatest attention because they can be changed quickly and within a fairly wide range. The minute-to-minute control of blood pressure is almost exclusively due to adjustments of cardiac output and peripheral resistance. Other components of resistance to flow—vessel elasticity and blood viscosity—have important effects upon blood pressure, but neither they nor blood volume changes are means by which arterial pressure is regulated. The interrelationships of these factors are summarized in Figure 17–9.

Total Blood Volume We do not adjust the blood volume in order to

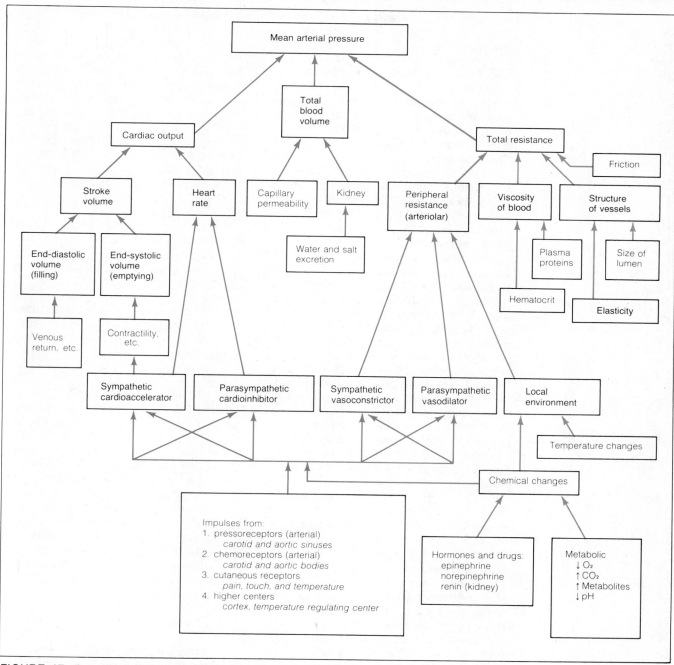

FIGURE 17-9   Factors that contribute to maintenance of mean arterial pressure.

maintain arterial pressure. Our homeostatic mechanisms are designed to maintain the blood volume and restore it whenever it deviates in either direction. If something causes the blood volume to change however, it will have an effect on arterial pressure, since the pressure depends, in part, upon there being enough blood in the system to put the vessels under some tension. Compensatory adjustments are quickly invoked to maintain arterial pressure whenever the blood volume is reduced. Of course, getting the blood volume back to normal helps to maintain the pressure, but it is not part of the minute-to-minute regulation of blood pressure. The total blood volume might be reduced by hemorrhage, by a loss of fluid from the surface of the body (as might occur with severe burns), or by a loss of fluid from the vascular system into the tissues, a condition known as **edema.** Since only about 20 percent of the blood in the systemic circuit is in the arteries, a shift or redistribution of blood from the arterial to the venous side of the circuit would have the same effect upon arterial pressure as a reduction in the total blood volume. Increases in blood volume are less likely to occur, but could result from transfusion or intravenous administration of fluid.

Most compensatory mechanisms for the restoration of blood volume are triggered by the resulting changes in arterial pressure, rather than by the actual volume change itself. The pressure changes affect blood volume because they alter the gradients that determine movement of fluid into and out of the capillaries. For instance, because the kidney's normal role of regulating the loss of water and certain electrolytes from the plasma depends upon the arterial pressure, any change in this pressure may result in the excretion of more or less water and a consequent decrease or increase in the blood volume. The role of the kidney in this matter is discussed in Chapters 27 and 28.

Cardiac Output The flow of blood into the arterial system from the left ventricle depends upon two continuously and independently variable components—the stroke volume and the heart rate. The latter is regulated largely by neural means, while stroke volume depends more upon factors related to filling of the ventricles and the contractile force of the cardiac musculature. As the input side of the circuit, cardiac output is as important to arterial pressure as is resistance on the outflow end of the arteries. The control of cardiac output was discussed in Chapter 15, and it would be appropriate to review these mechanisms at this time.

Elasticity of the Blood Vessels The contribution of the elasticity of the blood vessels to arterial pressure has been discussed above. It is a factor that affects the arterial pressure, but one cannot specifically alter vessel elasticity; it is a property that changes slowly over a long period of time, and its effect upon arterial pressure is a long-range effect.

Viscosity of the Blood The viscosity of the blood is due largely to the presence of the formed elements and the plasma proteins. Viscosity therefore varies with the hematocrit, which reflects not only the number of cells, but also the relative fluid content of the blood. Viscosity may be quite high in such conditions as polycythemia or leukemia, forcing the ventricle to develop more pressure in order to drive the blood through the vascular system. The viscosity is less when the hematocrit is reduced, as in anemia or hemodilution, or when the plasma protein content is low.

Peripheral Resistance The smooth muscle concerned with peripheral resistance is found in the small muscular arteries and arterioles, where its contraction increases the peripheral resistance by damming blood back on the arterial side, thereby raising the arterial pressure. When smooth muscle relaxes, the peripheral resistance decreases and lowers the arterial pressure by allowing more blood to leave the arteries. This is multi-unit smooth muscle and it does not have the high degree of rhythmicity that is characteristic of cardiac muscle or visceral smooth muscle. It is innervated almost exclusively by the sympathetic division of the autonomic nervous system, and the degree of vascular tone is largely dependent upon these nerves. In this respect, vascular smooth muscle is similar to skeletal muscle.

Capillary walls have no muscle cells and therefore are not capable of active constriction. There is smooth muscle in the walls of the venules and veins. Though it does vary in tone, its contraction cannot affect arterial pressure directly, since the venous smooth muscle is downstream from the arteriolar resistance. Alterations in venous tone serve more to regulate the storage capacity of the veins.

Neural control of peripheral resistance. Control of the peripheral resistance involves reflex mechanisms similar in many respects to those involved in the control of heart rate (Figure 17–10). The components of the reflex arcs include the usual effector, efferent path, centers, afferent paths, and receptors. The first of these, the effector, is, of course, the smooth muscle in the walls of the arterioles and small arteries, particularly those in the skin and viscera.

The efferent path consists almost exclusively of sympathetic vasoconstrictor fibers from the spinal cord. Those fibers to vessels in the skin synapse in the sympathetic chain and are then distributed with the spinal nerves, while those to visceral vessels pass through the chain and synapse in a collateral ganglion before being distributed to the vessels. Vessels of the genitalia and some glands of the digestive system receive a parasympathetic innervation (vasodilator), but these fibers are concerned with regulation of the blood flow to those structures rather than with the maintenance of arterial pressure.

Neurons of the **vasomotor center** in the reticular portion of the medulla oblongata are intermingled with those of the cardiac centers. Two parts of the vasomotor center, a sympathetic **vasoconstrictor center** and a parasympathetic **vasodilator center,** have been described, but since most of the vascular

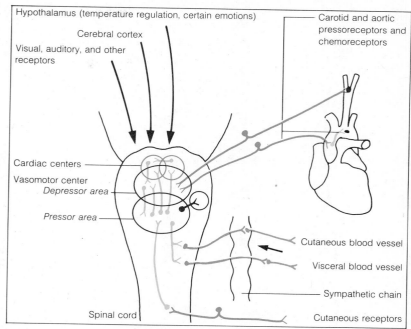

FIGURE 17–10   Neural control of blood vessels, showing some of the important inputs to the vasomotor center.

smooth muscle is innervated only by sympathetic fibers, the vasodilator center is of minimal importance. Effects mediated by the medullary centers are the result of an increase or decrease in the tonic discharge of the vasoconstrictor center. It is therefore probably more accurate to speak of a single vasomotor center which contains *pressor* and *depressor areas.* Stimulation of pressor areas causes contraction of the innervated vascular smooth muscle and raises arterial pressure (hence the term "pressor area"). Stimulation of depressor areas inhibits contraction, resulting in vasodilatation and a reduced arterial pressure (hence "depressor area"). Fibers from the vasomotor center descend in the spinal cord to synapse with cell bodies of the sympathetic preganglionic neurons.

Afferent fibers from many receptors influence the vasomotor center, and the interconnections between the vasomotor and cardiac centers are so rich that any stimulus that affects one affects the other. It should therefore be no surprise to find that the receptors and afferent paths that affect the vasomotor center are the same as those that regulate heart rate.

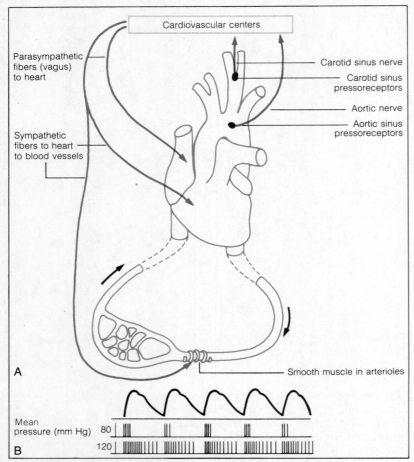

FIGURE 17–11   A. Carotid and aortic sinus mechanisms.
B. Responses in the carotid sinus nerve during the cardiac cycle at
mean arterial pressures of 80 and 120 mm Hg.

The most important receptors are probably the **pressoreceptors** located in the walls of certain large arteries (Figure 17–11). The *carotid sinus*, at the bifurcation of the common carotid into the internal and external carotid arteries, sends afferent fibers to medullary centers in a tiny branch of the glossopharyngeal nerve (IX). The *aortic sinus* is located in the arch of the aorta, and its afferents are part of the vagus nerve (X). These receptors are sensitive to the pressure of the blood within the artery, and they are tonically active; that is, pressure in the artery is normally high enough to cause them to fire. The frequency of discharge varies throughout the cardiac cycle as the pressure changes in the artery (see Figure 17–11B). A rise in mean pressure in the

artery therefore causes an increased frequency of firing in the afferent nerve and an inhibition of the vasoconstrictor center, which reduces the peripheral resistance and the mean arterial pressure (a depressor response). A fall in mean arterial pressure and the ensuing decreased rate of discharge by the pressoreceptors elicit vasoconstrictor activity from the center, causing an increased peripheral resistance and a rise in mean arterial pressure (a pressor response). You will recall that similar changes occur in heart rate, so that when pressure in the artery rises, the cardiac output and peripheral resistance are both reduced, and mean arterial pressure is rapidly returned to "normal." This is why the afferent fibers for these reflexes are often called the "buffer nerves."

The arterial pressoreceptors are another example of a negative feedback system. The pressoreceptors detect a deviation in arterial pressure and cause the cardiac and vasomotor nerves to change heart rate and peripheral resistance to bring the arterial pressure back to normal. This is a typical homeostatic mechanism because it tends to prevent or minimize changes from a desired level or "set point."

Located very near the carotid and aortic sinuses are the *carotid* and *aortic bodies*, which contain **chemoreceptors.** They are stimulated by changes in the chemical composition of the blood flowing through them, primarily by reduced oxygen levels, and to a lesser extent by elevated carbon dioxide and hydrogen ion concentrations (reduced pH) of the arterial blood. Although the major effect of stimulation of these chemoreceptors is on respiration, they also cause an increase in arterial pressure. Under normal conditions the composition of the arterial blood does not change very much (it has just come through the lungs), and the chemoreceptors are not activated. In abnormal or stressful conditions, however, under certain conditions in which the cardiovascular and respiratory systems are not meeting the needs of the tissues, these receptors can be very important.

Cutaneous receptors can affect arterial pressure by eliciting vasomotor responses. The flushing effect of warmth and the paling effect of cold are familiar examples. Touch and pain stimuli are also effective, but whether they cause vasoconstriction or vasodilation varies with the stimulus intensity. The stimulation of visceral receptors or of those for the special senses, particularly those for sight and sound, may also cause cardiovascular responses. The effects of the latter may last only a few seconds, but they may involve great increases in both heart rate and blood pressure.

In addition, the medullary vasomotor centers receive important inputs from higher centers. The hypothalamus is necessarily the source of many of those inputs, since it contains centers for the regulation of body temperature. Much of this regulation is brought about by action on the blood vessels, particularly in the skin. But areas of the cerebral cortex, including areas associated with the limbic system (rhinencephalon), which also affect arterial pressure. The limbic system is involved in emotional responses, and it has numerous anatomical connections with the hypothalamus. In fact, the hypothalamus seems to serve as a funnel through which input from cerebral structures is channeled to the cardiovascular centers. Emotional effects on vascular tone are well known. Many are familiar with blushing and some turn pale with fear or pain.

Local control of peripheral resistance. For the most part, reflex control of blood vessels is associated with the sympathetic nervous system, and the response is quite generalized; it tends to cause constriction primarily in those vessels that receive a sympathetic innervation, specifically those of skin and viscera. One large and important vascular bed, however, that of the *skeletal muscles,* receives few vasomotor fibers. These vessels do not participate in the generalized vasoconstriction associated with sympathetic discharge. During vigorous exercise, when active skeletal muscles require an increase in blood flow, vasoconstriction of the muscular vessels would be highly undesirable. The meager sympathetic innervation of skeletal muscles is mostly *cholinergic* and *vasodilator* in action.

Increased metabolic activity in tissues such as skeletal muscle brings about some striking changes in their local environment. The higher rate of oxygen consumption means that the cells remove more oxygen from their environment and add more carbon dioxide and other metabolites, such as lactic acid, to it. Since most metabolic end-products are acidic, they lower the pH slightly, and activity of skeletal muscles produces heat and tends to raise the temperature locally. As a result, the immediate environment of the vessels in the active tissue soon has lower oxygen tension and pH, a higher content of carbon dioxide and metabolites, and an elevated temperature. Each of these has its own independent vasodilator effect on vascular smooth muscle. Collectively, they bring about a marked decrease in resistance in these vessels and an increase in blood flow through that particular region. This is a local effect, which occurs only in the area where these changes are significant. There can thus be a vasodilatation and an increased blood flow through the active muscles while there is a generalized vasoconstriction elsewhere.

Chemical agents other than those associated with metabolic activity also may exert effects directly on vascular smooth muscle. In view of the action of the sympathetic nervous system on this kind of muscle, it is to be expected that norepinephrine, the sympathetic postganglionic neurotransmitter, is a vasoconstrictor (except in coronary vessels). Epinephrine, whose secretion occurs upon sympathetic stimulation of the adrenal medulla, has similar effects, with the important exception that it dilates vessels in skeletal muscle as well as in the heart. Epinephrine increases resistance in general (in the skin and visceral vessels) and reduces it in skeletal muscle vessels. This is a direct action of the drug on the muscle cells and does not involve any nerve fibers.

Among vasodilator substances are acetylcholine, the parasympathetic neurotransmitter, and histamine, which is released from a number of tissues as part of the inflammatory response and many allergic reactions. In addition, there is a group of substances known collectively as *kinins*, which are formed

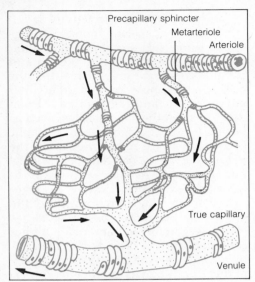

Precapillary sphincter
Metarteriole
Arteriole
True capillary
Venule

FIGURE 17–12   Portion of a capillary bed.

THE MICROCIRCULATION— THE CAPILLARY BED

Between the arterioles and the venules lie innumerable tiny capillaries in the form of a network of vessels which lie both in series and in parallel with one another. The specific details of the arrangement vary slightly from one organ to another. However, in the typical pattern, shown in Figure 17–12, several vessels branch from a terminal arteriole. Some have a few smooth muscle fibers in their walls (metarterioles), while others are true capillaries with only endothelial cells in their walls.

At the beginning of each true capillary is a tiny ring of smooth muscle, the precapillary sphincter, that controls the flow of blood into that capillary. As a result, some capillaries may be closed while the flow in others may be intermittent or even momentarily reversed. The capillaries branch and anastomose freely, and eventually they converge into a venule. In skeletal muscle some of the channels are more direct (thoroughfare or preferential channels) and probably function as shunts from arteriole to venule when the muscle is at rest and inactive.

The exchange of materials between blood and tissue fluid occurs through the walls of the true capillaries as they make their way between the tissue cells. Capillary walls are very permeable; though the endothelial cells are joined together by an intercellular "cement," there appear to be numerous openings or gaps between the cells. This permits passage between the endothelial cells of substances that ordinarily would not pass through their membranes. The size, frequency, and nature of the intercellular openings vary from one organ to another, with more and larger gaps in organs (such as the liver) where exchange is particularly important and abundant.

Capillary Exchange

A number of factors, both physical and physiological, contribute to the effectiveness of the capillary as the site of exchange between the blood and the extracellular fluid. Capillaries are of

in certain glands during active secretion. They cause dilatation of the vessels in those glands and hence increase blood flow through the gland when it is active.

Of the several vasoconstrictor agents, one is the potent angiotensin II, whose appearance in the blood may be triggered by reduced blood flow to the kidney (see Chapter 28). In addition, many agents taken into the body as medication (or otherwise) have distinct effects upon blood vessels and blood pressure—alcohol is a vasodilator and nicotine is a vasoconstrictor.

These chemical agents, whether produced in the body or taken into it, exert their effects at several different sites to alter peripheral resistance. They may act directly, only upon the smooth muscle of the blood vessels in their immediate vicinity, or they may, like some drugs and hormones, reach the brain through the circulation and act upon the vasomotor centers. The latter effects are more widespread, though they are limited to the vessels controlled by these centers. A third site of action is on specific receptors, notably the chemoreceptors of the carotid and aortic bodies, which then activate the medullary centers. Certain agents, notably the respiratory gases, may act at all three sites.

very small diameter, only a millimeter or two in length and they contain no more than 5 percent (200–250 ml) of the circulating blood at one time. Furthermore, there are so many capillaries that the total area of the capillary walls in the body (6300 m²) is many times greater than the surface area of the body (1.5–2.0 m²). Because the blood travels slowly in these vessels, requiring 2–3 seconds to traverse the capillary bed, the passive exchange processes of diffusion, osmosis, and filtration (discussed in Chapter 2) have ample time to act over short distances.

Most small particles penetrate capillary walls with relative ease. Among these substances are the respiratory gases, the electrolytes, nutrients (particularly glucose), and the metabolic wastes (such as urea and lactic acid). In fact, virtually all constituents of the plasma except plasma proteins and those substances bound to them *can* diffuse across the capillary wall. These substances *do* diffuse whenever there is a suitable gradient in the form of a concentration gradient. Oxygen and nutrients, for example, are normally in greater concentration in the plasma than in the extracellular fluid, and they diffuse out of the capillary. Carbon dioxide and waste products produced by the cells are in higher concentration in the extracellular fluid; therefore, they diffuse into the capillary to be carried away. Electrolytes diffuse, but their concentration in the plasma is essentially the same as that in the tissue fluid, so their movement is purely random diffusion with no net movement in either direction. There is very little protein in the extracellular fluid. Despite the favorable gradient for outward diffusion of protein, the capillary wall normally blocks almost all protein transport. If capillary permeability is increased, as occurs in such conditions as oxygen lack and shock, a detectable amount of protein may enter the extracellular fluid.

The fact that diffusion of protein is severely restricted is important, since protein is responsible for the existence of one vital concentration gradient. As a nondiffusible solute, existing in unequal concentrations across a barrier, plasma protein creates an osmotic gra-

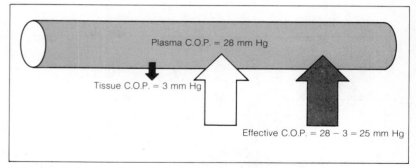

Plasma C.O.P. = 28 mm Hg

Tissue C.O.P. = 3 mm Hg

Effective C.O.P. = 28 − 3 = 25 mm Hg

FIGURE 17–13   Effective colloid osmotic pressure.

dient or osmotic pressure by which water is drawn into the capillaries. This pressure is referred to as the **colloid osmotic pressure** (COP), or **oncotic pressure,** to distinguish it from the total osmotic pressure or that osmotic pressure due to the electrolytes (*crystalloid osmotic pressure*). The colloid osmotic pressure is a very small part of the total osmotic pressure (approximately 28 mm Hg of nearly 6000 mm Hg), but its importance is all out of proportion to its magnitude. Since protein is the only significant nondiffusible solute in the plasma, it therefore contributes the only *effective* osmotic gradient.

The concentration of protein in the tissue fluid, normally quite low, contributes a tissue colloid osmotic pressure of up to 5 mm Hg. The colloid osmotic pressure of the plasma tends to draw water into the plasma, while the colloid osmotic pressure in the tissues tends to draw water into the tissues. The difference between these opposing forces, which is about 25 mm Hg, is the **effective colloid osmotic pressure** that tends to draw fluid into the capillaries from the tissue spaces (Figure 17–13).

In addition to the osmotic gradient, there is a filtration pressure, or **hydrostatic pressure,** within the capillaries caused by the blood pressure. The hydrostatic pressure tends to drive from the vessel everything that can pass through the capillary wall, fluid as well as solutes. The hydrostatic pressure is not constant throughout the length of the capillary, however, since just as blood pressure declines from artery to vein, so it declines from the arterial end of the capillary to the venous end. The magnitude of the pressure at the arteri-

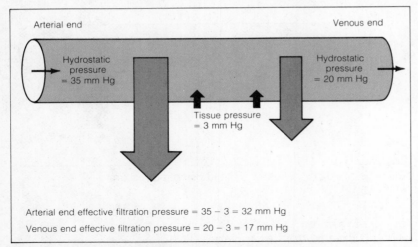

Arterial end effective filtration pressure = 35 − 3 = 32 mm Hg

Venous end effective filtration pressure = 20 − 3 = 17 mm Hg

FIGURE 17−14   Effective filtration pressure.

al end of the capillary depends upon (1) the arterial pressure and (2) the degree of constriction of the smooth muscle of the precapillary sphincter. Hydrostatic pressure in the capillary is opposed by a slight pressure in the tissues, but this amounts to only a few millimeters of mercury. The difference between the plasma and tissue hydrostatic pressures is the **effective filtration pressure** that tends to force fluid with its solutes from the capillary to the tissue spaces (Figure 17−14).

The net movement of fluid into or out of a capillary is determined by the effective filtration pressure and the effective colloid osmotic pressure (Figure 17−15). At the arterial end, the effective filtration pressure is normally greater than the effective colloid osmotic pressure, and there is a net movement of fluid *out* of the capillary into the extracellular spaces. Filtration exceeds osmosis. At the venous end, the hydrostatic pressure has fallen, so that the effective filtration pressure is less than the effective colloid osmotic pressure, and there is a net movement of fluid *into* the capillary. Osmosis exceeds filtration. Diffusible solutes will follow the water back into the capillary.

The Capillary Equilibrium

Theoretically, the influx at the venous end of the capillary should be equal in the long run to the outward movement at the arterial end. In practice, however, it often is not, perhaps due partly to a small loss of protein from the plasma. The slight excess of extracellular fluid is normally removed by the lymphatic system, so that the extracellular fluid volume does not increase.

The arteriolar smooth muscle and precapillary sphincters, by their control of blood flow and hydrostatic pressure in the capillaries, determine the balance between the filtration and osmotic gradients. A change in any one of the pressure components will alter the gradients and disturb the equilibrium, bringing about a net movement of fluid. When arterial pressure is elevated, the capillary hydrostatic pressure is raised and the equilibrium is shifted in favor of a net movement of fluid into the tissues. The result is an increase in extracellular fluid volume. The same thing would happen if the venous pressure were raised, since that would also raise the capillary hydrostatic pressure. A reduction of capillary pressure would bring about a net uptake of fluid by the capillaries.

A reduced oncotic pressure (decreased plasma protein) would produce

FIGURE 17−15   Exchange in a typical capillary. At the arterial end there is net movement of fluid out of the capillary, and at the venous end there is net movement of fluid into the capillary.

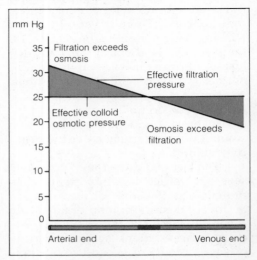

edema, while an elevation would draw fluid into the blood. The oncotic pressure may be changed by direct alteration of the amount of plasma protein. However, such a change is more likely to be the result of changes in the fluid content of the blood. Dehydration, for example, raises the relative concentration of plasma protein. The effect is movement of fluid into the capillary, which alleviates the problem. An increase in capillary permeability has the same effect as a decrease in colloid osmotic pressure, because more protein is allowed to leak into the extracellular spaces, thus reducing the osmotic gradient.

VENOUS CIRCULATION

Blood reaching the systemic veins has passed through the arteries, arterioles, and capillaries. The hydrostatic pressure has been reduced to 15–20 mm Hg by the time the blood enters the veins, and it continues to decline to about 5 mm Hg or less at the right atrium. The veins are therefore a "low pressure" system, whereas the arteries are a "high pressure" system. At any given instant, the veins contain about 75 percent of the blood in the systemic circuit, arteries about 20 percent, and capillaries the remaining 5 percent. The vessels on the venous side of the systemic circuit are appropriately referred to as *capacitance vessels*, in contrast to those on the arterial side, which are *resistance vessels*.

The veins, with their thinner walls, are more easily distended by pressure in them, and they tend to collapse somewhat when the pressure falls. Collapse of a vein does not mean occlusion, however, since blood can flow along the "edges" of the partially collapsed vessel. Larger veins normally are partially collapsed, but their capacity can be markedly increased with only a slight rise in pressure (Figure 17–16). A pressure increase of 1 mm Hg may increase the capacity of a vein threefold as the vein becomes more circular. A 10 mm Hg increase might cause a sixfold increase in capacity, but beyond that there is increasing resistance to distension. Because the capacity of the veins

is so large and so variable, the venous system is an effective blood reservoir.

The low pressure in veins is quite adequate to return blood to the heart when one is recumbent, but when one is standing erect gravity is a force that must be reckoned with. The hydrostatic pressure due to the weight of a column of blood extending from the foot up to the level of the heart is actually closer to 80 mm Hg than to 15 mm Hg. That this does not create a problem for blood flow is due to the fact that the weight of the column of blood in the arteries affects the pressure on the arterial side in the same way. To illustrate, in a U-shaped tube filled with fluid (or mercury) the fluid level in the two arms will be the same, and the hydrostatic pressures will be equal (Figure 17–17A). The higher the columns, the greater the pressure at the bottom. Fluid may be added to one arm and the level will rise to the same height in the other. Although the pressure is high at the bottom, no extra force is needed on the inflow side to lift fluid on the other side against gravity. That force is provided by the downward push, the weight of the column on the inflow side. Consequently, fluid under a certain head of pressure will come out of a tube at the same flow rate regardless of whether the tube is straight or has a loop in it (Figure 17–17B).

FIGURE 17–16   Effect of venous pressure on the cross-sectional area and capacity of veins.

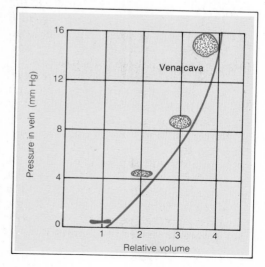

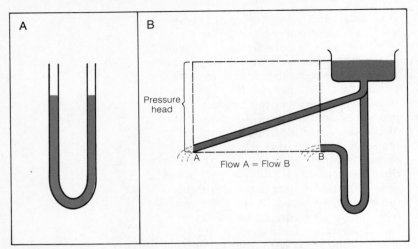

FIGURE 17–17   A. Fluid in a U-tube rises to equal heights on the two sides. B. Flow through a tube is dependent upon the pressure head, which is the same in a straight tube and a tube with a loop. Thus, the erect posture does not place an additional burden on the heart.

In the upright position the arteries and veins of the systemic circuit operate very much like the two arms of a U-tube. Because of this, we do not have to worry about providing enough pressure to force blood up to the heart against gravity, but there are two things we do have to take into account. One is the fact that the pressure at the bottom of the loop (in the capillaries of the feet) is high, which alters the gradient across the capillary walls and may lead to edema. The other is that, because the veins have such a great capacity, they may simply hold some of the blood rather than return it all to the heart. This creates a problem because an inadequate venous return leads to a reduced cardiac output. Humans, and those animals that assume an upright posture, have highly developed compensatory mechanisms to help prevent excessive pooling of the blood in the venous reservoirs of the inferior extremities and abdominal viscera. Among them are increased breathing movements, and rhythmic contractions of skeletal muscles in the legs, both of which aid the return of venous blood to the heart. Their contributions to the compensatory adjustments for gravity effects are discussed in Chapter 18.

LYMPHATIC CIRCULATION

Lymph resembles both tissue fluid and plasma in composition. Since blood capillaries normally resist the passage of protein, the tissue fluid is relatively protein free. It is, however, important to realize that a small amount of protein does leak out of the capillaries, more in some tissues than others, and that a small increase in capillary permeability may result in a significant increase in protein in the tissue fluid. Because lymphatic capillaries permit the passage of all constituents of the tissue fluid, including the protein, lymph , is actually only tissue fluid that has found its way into the lymphatic vessels. The lymph capillaries arise as blind-end vessels, networks, and plexuses in the interstitial spaces among the blood capillaries. They lead through lymph nodes to larger and larger lymphatic vessels, which eventually drain into the bloodstream.

Lymph Flow

The volume of lymph returned to the circulation is about 2–4 liters every 24 hours, certainly much lower than the blood flow in that time, but approximately equal to the total volume of plasma. If it were allowed to remain in the extracellular spaces without being returned to the circulation, problems of edema formation and reduced blood volume would rapidly develop.[3] Although only a small amount of protein leaks into the tissue spaces from the capillaries, its effect on colloid osmotic pressure would be marked if it were allowed to remain in the interstitial fluid. The role of the lymphatics in returning protein to the circulation is therefore almost as important as its role in returning fluid.

[3]There is a tropical parasite that invades and blocks lymphatic channels. The result is a sometimes extreme swelling or edema of the part from which lymph flow is obstructed, usually a leg. The appearance of the affected limb is the source of the appropriate common name for this condition — elephantiasis. Radical mastectomy, a surgical treatment for cancer of the breast, involves removal of lymph nodes in the axillary region and may result in a tendency to edema of that arm due to impaired lymph drainage.

Focus

Cancer and the Lymphatic System

The human body is sometimes its own worst enemy. Even as it provides a system of trenches and booby traps to defend against infection, it offers homegrown traitors an easy path to its vitals.

The traitors are cancer cells. The defenses are the lymphatic system, that network of channels that permeates the body and drains excess fluid from all its tissues. Lymphatic capillaries pass near almost every cell in the body. They flow into each other, the many streams of lymph merging until they are dumped into the veins near the heart. Along the way, the lymph flows through the lymph nodes, filters which trap debris and bacteria for removal by specialized cells.

Unfortunately, a characteristic of cancers is that at some point in their history, their cells become mobile and leave the mass of the parent tumor. They enter the surrounding tissues, and because lymphatic capillaries are always nearby and are very porous, they join the flow of lymph. They are then carried to the nearest lymph node, where they may be trapped. Subsidiary cancers may then grow in the lymph node. If cancer cells make it past the node, they may reach the general circulation. They may then be carried to any point in the body.

Because when cancers spread, or metastasize, their first stop is generally a lymph node, and because lymph nodes *are* effective filters, cancer surgery often involves removing a large part of the lymphatic system in the neighborhood of the cancer. In this way, the surgeon hopes to catch all of the cancerous cells trapped in nearby lymph nodes; his or her chances of success are greatest when the cancer is operated on early, since the longer the cancer cells have to move through the lymphatic system, the better their chances of reaching the general circulation. If the surgeon misses one or two cells, they can then grow into a whole new tumor. For this reason, surgery is usually followed by treatment with radiation and drugs aimed at killing any remaining cancer cells. This treatment is so successful today that breast cancer, for instance, can be cured by removing the breast, with its primary tumor, and then using radio- and chemotherapy. It used to be thought necessary to remove the breast together with the neighboring lymphatic system and large portions of the chest muscles, but many surgeons now believe that the simpler treatment is just as effective.

The means of producing a flow of lymph are not altogether clear. Since the lymphatic system is a one-way system without an arterial counterpart, there is no driving force to guarantee movement of fluid. Lymph vessels, even large ones, have extremely thin walls, and pressure within them is very low so that they are easily collapsed by external pressure. Rhythmic contractions of skeletal muscle and smooth muscle of the abdominal viscera are effective aids to lymph flow, as the vessel is closed by each contraction but is allowed to open and refill with each relaxation. The action of this "muscle pump" is enhanced by frequent valves in the lymphatic vessels which prevent backflow. The subatmospheric pressure in the thoracic cavity also helps draw lymph into the thoracic region.

Other Functions of the Lymphatic System

The lymph nodes serve as filters by trapping foreign particles in the lymph, as when wounds or infections bring bacteria and other foreign matter into the extracellular fluid. This foreign material does not normally enter the blood capillaries (unless there has been vascular damage), but if it is in the interstitial space it can enter the lymphatic capillaries. The particles are then carried to lymph nodes where the phagocytes (macrophages) destroy or inactivate the undesirable material. This is why there is often swelling in the region of lymph nodes proximal to an area of infection. The role of lymph nodes in specific immune systems has been discussed. Antigens brought to lymph nodes can initiate responses by groups of T and B lymphocytes.

A final contribution of the lymphatic system is associated with the absorption of fat from the small intestine. Much of the lipid material is absorbed into tiny lymphatic vessels, called **lacteals,** in the lining of the intestine rather than into the bloodstream. This lipid gets into the general circulation without going through the liver, whereas nutrients absorbed into the

bloodstream do go through the liver. This aspect of lymphatic circulation will be considered further in Chapter 23.

CHAPTER 17 SUMMARY

Blood Flow—Hemodynamics

Blood flow is the volume of blood flowing past a point in a given amount of time; in the systemic circuit it is the *cardiac output,* which is distributed among the tissues of the body. Flow through a vessel is directly proportional to the pressure drop from one end to the other; in the systemic circuit it can be considered to be the aortic pressure, since the right atrial pressure is near zero. Flow is inversely proportional to the resistance. In the circulatory system resistance is caused by the viscosity of the blood, the fixed frictional drag against the vessel walls, and a variable resistance, which is the peripheral resistance due to smooth muscle mostly in the arterioles. An increase in resistance along the way raises pressure above the resistance and lowers the pressure below it.

If blood flow is constant, the velocity of the blood is inversely proportional to the cross-sectional area of the vascular bed at that level. It is, therefore, fastest in the aorta and slowest in the capillaries, as the total cross-sectional area of the capillary bed is much greater than that of the single aorta. In general, velocity in veins is faster than in capillaries, but less than in arteries.

The large elastic arteries are distended by the pressure of the blood forced into them with each beat, and their elastic recoil maintains a pressure in them between beats. The pressure wave and the distension it causes pass rapidly over the arterial system as the *pulse wave.* Pulse waves are routinely counted as a measure of heart rate.

Arterial Pressure

Arterial blood pressure is determined by measuring the amount of externally applied pressure needed to collapse the artery (by convention, the brachial artery). *Systolic pressure* is the highest pressure in the artery during the cardiac cycle (peak of ejection), and *diastolic pressure* is the lowest pressure (the end of diastole). *Pulse pressure* is difference between them.

Mean arterial pressure (MAP) is the driving force that causes blood to flow, and it provides the filtration pressure for capillary exchange. When the blood volume is constant, MAP depends upon flow and resistance, and is controlled through cardiac output and peripheral resistance (the variable resistance). If blood volume is reduced, arterial pressure will fall until either the blood volume is restored or compensatory mechanisms restore the blood pressure. MAP is raised by either increased cardiac output or peripheral resistance. Cardiac output (described in Chapter 15) depends upon stroke volume and heart rate. Stroke volume is associated largely with venous return and the force of ventricular contraction, and heart rate is controlled primarily by neural mechanisms. The peripheral resistance is also largely under neural control, mainly by reflexes mediated through the sympathetic nervous system and, in fact, is initiated by the same stimuli that elicit cardiac reflexes. The most important input is from pressoreceptors in the carotid and aortic sinuses. They are tonically active and thus respond to both an increase and a decrease in arterial pressure. A pressure increase causes a decrease in both heart rate and peripheral resistance (vasodilatation). A fall in arterial pressure brings about an increase in heart rate and more constriction of the vessels with sympathetic innervation (skin and viscera). Other input comes from chemoreceptors, cutaneous receptors, as well as input from the brain, including hypothalamus and cerebral cortex.

Blood vessels are also subject to local control arising from changes in their immediate environment. This is especially important in skeletal muscle whose blood vessels receive little innervation. Changes likely to occur during exercise (including reduced oxygen and increased carbon dioxide) tend to dilate the vessels. The hormone epinephrine also dilates skeletal muscle vessels, though it constricts other blood vessels.

The Microcirculation

The exchange of materials between blood and interstitial fluid occurs in capillaries whose walls are permeable to all constituents of plasma except proteins. The exchange is passive and depends upon filtration and osmotic gradients. There is a filtration gradient because pressure in capillaries (blood pressure) is greater than in the interstitial fluid. There is an osmotic (*colloid osmotic pressure,* or *COP*) gradient because there is more protein in the plasma than in the interstitial fluid. At the arterial end of the capillary the filtration pressure is greater than the COP and fluid leaves the vessel, but at the venous end of the capillary the filtration pressure has fallen, so the COP is greater and fluid returns to the capillary. If capillary fluid loss is greater than the return, fluid accumulates in the tissues as edema fluid.

Venous Circulation

Pressure in veins is lower than in arteries, but veins have a tremendous capacity—normally they contain about half the blood in the systemic circuit. Any increase in the amount of blood in veins can seriously reduce venous return. The venous return is greatly enhanced by contraction of skeletal muscles in the lower extremities (the "muscle pump"). Blood is squeezed out of a section of a vein with each muscle contraction, and with each relaxation it refills from the capillaries. Valves in the veins prevent backflow. Respiratory movements also help return venous blood to the thoracic cavity.

Lymphatic Circulation

The lymphatic system drains fluid from the interstitial spaces and returns it to the circulation. Since lymphatic vessels are permeable to protein, any protein lost from capillaries is picked up and returned to the circulation. Lymphatics therefore help to maintain the COP gradient. Flow of lymph is very slow and is aided by contraction of smooth and skeletal muscles in the vicinity of the lymphatic vessels.

STUDY QUESTIONS

1 Define: blood flow, peripheral resistance, and pressure, as applied to the circulatory system. How are they related? What happens to one if one of the others is increased?

2 What effect does elasticity of blood vessels have upon blood pressure?

3 What is the arterial pulse, and what causes it? How is the velocity of the pulse wave related to the velocity of the blood flow in the arteries? Would you expect to find a pulse in veins? Why or why not?

4 When you measure blood pressure, what are you actually measuring? What is the significance of these numbers?

5 Distinguish between local and central control of peripheral resistance. What are the mechanisms and which vessels are affected in each case?

6 What is the function of the carotid pressoreceptors? How do they work?

7 How does venous return affect mean arterial pressure?

8 Consider arteries, capillaries, and veins. List them in order in terms of: (a) pressure in the vessels; (b) velocity of flow in the vessels; and (c) volume of blood contained at a given moment.

9 Fluid exchange between a capillary and the surrounding interstitial fluid is determined largely by the relationship between hydrostatic and osmotic pressure in the capillary and interstitial fluid. Explain these relationships. What would happen if the pressure of the interstitial fluid was higher than that in the capillary? Explain? What is meant by *effective* filtration pressure, or *effective* osmotic pressure?

10 What are some of the differences between flow in the lymphatic system and in the cardiovascular system? How do they compare as to volume and velocity of flow? How does lymph get into the lymph channels? What makes lymph move, and what happens if you block its flow?

18

Circulation Through Regions with Special Needs

Cardiovascular Homeostasis—Adjustments to Some Physiological Stresses

Cardiovascular Effects of Some Abnormal Stresses

Cardiovascular Adaptations and Adjustments

The preceding three chapters have presented the mechanisms for the delivery of blood to the tissues. Most of the discussions have been centered around maintaining circulation in the systemic circuit in a normal individual under resting conditions. It is now time to look at the delivery of blood under some rather different conditions—conditions which require some modifications in the operation of the cardiovascular system.

Several organs or regions have certain anatomical characteristics, or present special problems that necessitate adjustments to assure them an adequate blood supply.

Although we describe the circulatory system in the resting state, we are not at rest during most of our waking hours, and we do things that impose some load or stress on the cardiovascular system. The system must be able to adjust to provide ample blood flow during all levels of physical activity, or while merely quietly standing erect. In addition to physiological stresses, the system must also be able to cope with abnormal—pathological or traumatic—stresses. This chapter indicates circulatory adaptations to some special situations and the adjustments to a few of the more common physiological and pathological stresses.

CIRCULATION THROUGH REGIONS WITH SPECIAL NEEDS

Fetal Circulation and Changes at Birth

For most of one's life the circulation follows the course described at the beginning of Chapter 15 (and shown in Figure 18–2). Blood returning from the tissues of the body enters the right side of the heart and is pumped through the pulmonary circuit and lungs, where it takes up oxygen. The oxygenated blood then enters the left heart, which pumps it through the systemic circuit, where it gives up oxygen, and returns to the right heart. There is no mixing of the oxygen-rich arterial blood with the less well-oxygenated venous blood.

Before birth the circulatory system cannot operate in this fashion because the lungs contain no air. The blood obtains its oxygen instead in the **placenta,** the organ through which the developing fetus is connected with the mother. The placenta is formed in part from maternal tissue and in part from fetal tissue. In it, maternal blood, carrying oxygen and nutrients needed by the fetus, flows through large pool-like sinuses, while the fetal blood flows through the tiny **chorionic villi** which project into the sinuses (Figure 18–1). The fetal blood is separated from the maternal blood by a membrane (the *chorion*), and all exchanges occur across this membrane. The materials exchanged are: oxygen *to* the fetal blood, and carbon dioxide *from* it; all nutrients that in later life are absorbed from the digestive tract; and the fetal waste products. Hormones, vitamins, and many drugs (medications and otherwise) as well as certain immunoglobulins, can also normally cross the placental barrier. The placenta thus serves as more than lungs for the fetus; it is the lifeline, a vital link in the life-support system.

Fetal Circulation Fetal blood travels to and from the placenta in the **umbilical cord,** which contains two **umbilical arteries** and one **umbilical vein.** The umbilical arteries are branches of the internal iliac arteries. They meet the umbilical vein at the **umbilicus** (the navel) and spiral around it en route to the placenta. Blood in these arteries is "mixed" blood, not highly oxygenated as one would expect in a systemic artery. In the placenta the fetal blood circulates through the chorionic villi, where it is exposed (through the membrane) to maternal blood and the vital exchanges of gases, nutrients, and wastes take place. The newly oxygenated blood returns to the fetus in the umbilical vein. Part of the blood enters the liver and passes through its capillary bed before emptying into the inferior vena cava. The remainder is joined by the portal vein from the digestive tract and passes directly to the inferior vena cava in a vessel known as the **ductus venosus.** The blood is now mixed blood, since that in the portal vein and inferior vena cava is venous blood from capillary beds in the tissues.

This mixed blood then enters the right atrium, but very little of it follows the usual path through the right heart to the lungs. Since the lungs are collapsed and have a fairly high resistance, much of the blood follows one of two alternative paths. One is a shunt, the **ductus arteriosus,** between the pulmonary artery and the aorta. Most of the blood ejected from the right ventricle into the pulmonary artery takes this route to the aorta because it offers less resistance. The other alternative, a shunt from the right atrium to the left atrium, allows the right ventricle and pulmonary circuit to be completely bypassed. In the interatrial septum is an opening, the **foramen ovale,** covered on the left by a small flap of tissue that acts as a one-way valve. Some of the blood entering the right atrium is thus deflected straight across to the left atrium and pumped into the aorta. The high resistance in the lungs means that pressure in the pulmonary arteries and right ventricle is also relatively high, which makes these alternate routes possible. The aorta therefore carries mixed blood to the tissues, and it is a portion of this mixed blood that finds its way to the placenta in the umbilical arteries.

The umbilical vein is the only fetal vessel that carries highly oxygenated blood. All other vessels contain varying proportions of oxygenated and "venous" blood from the tissues. The flow

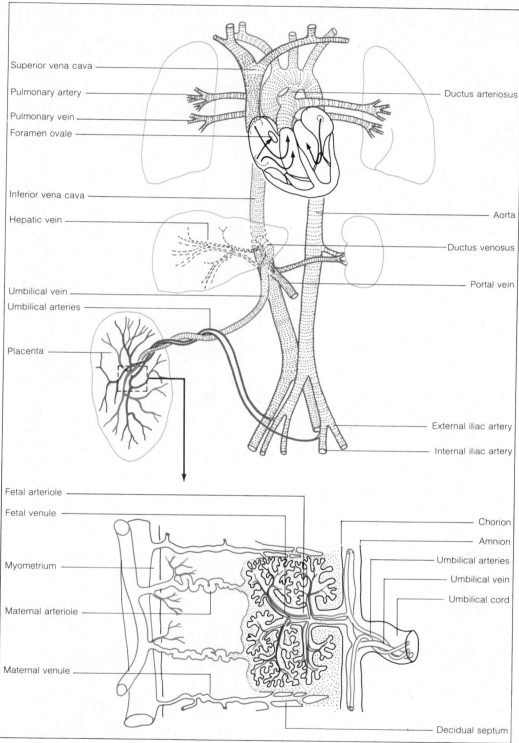

Superior vena cava

Pulmonary artery

Pulmonary vein

Foramen ovale

Inferior vena cava

Hepatic vein

Umbilical vein

Umbilical arteries

Placenta

Ductus arteriosus

Aorta

Ductus venosus

Portal vein

External iliac artery

Internal iliac artery

Fetal arteriole

Fetal venule

Myometrium

Maternal arteriole

Maternal venule

Chorion

Amnion

Umbilical arteries

Umbilical vein

Umbilical cord

Decidual septum

FIGURE 18-1   The fetal (prenatal) circulation.

patterns are such, however, that the more oxygenated blood in the inferior vena cava is likely to go through the foramen ovale and into the aorta, whereas blood in the superior vena cava is more likely to go into the right ventricle and out the pulmonary artery. This means that the arteries to the brain, which come off the arch of the aorta, are likely to get more highly oxygenated blood than those to the trunk and inferior extremities. It may seem to be a very inefficient system, with "arterial" (oxygenated) blood and "venous" (less oxygenated) blood being mixed throughout most of the circulation. It suffices, however, because the fetal oxygen requirement is not as high as the infant's or adult's and because fetal hemoglobin is able to take up and release oxygen at lower oxygen tensions than postnatal hemoglobin; it can therefore function adequately at the lower oxygen tensions that prevail in fetal blood.

Circulatory Changes at Birth At birth the first gasp expands the lungs and fills them with air. This immediately lowers the resistance in the lungs and permits an increased amount of blood to pass into and through the pulmonary circuit. Oxygenation must now occur in the lungs, since the connection with the placenta and the source of oxygen are interrupted when the umbilical cord is tied off. There is no longer any blood flow in the umbilical vessels, and after a time they become obliterated. The umbilical arteries remain as the **umbilical ligaments,** which can be seen on the inner surface of the anterior abdominal wall. The umbilical vein becomes the **ligamentum teres** (*round ligament*) of the liver, and the ductus venosus becomes the **ligamentum venosum** (Figure 18–2).

The reduction in resistance in the lung lowers the pressure in the right atrium and ventricle and in the pulmonary artery. Blood will no longer shunt to the aorta. It might even go the other way, but this is apparently prevented by constriction of the ductus arteriosus, which soon fuses shut and becomes the **ligamentum arteriosum.**

The fall in pressure on the right side of the heart and the increase on the left side also remove the pressure gradient which sent blood through the foramen ovale to the left atrium. Any reverse flow (back into the right atrium) is blocked by the valvelike flap over the foramen. In due time it grows shut and an indentation, the **fossa ovalis,** is all that remains. A few days after birth the normal adult pattern of circulation, in which blood *to* the tissues (arterial) is completely separated from blood *from* the tissues (venous), is established.

Developmental defects are not uncommon. The ductus arteriosus may fail to close, in which case blood from the aorta (where the pressure is now greater) enters the pulmonary artery to go through the lung again. The left ventricle then must pump more blood to get adequate flow to the systemic circuit, but even so, tissues supplied by the systemic circuit may not get enough blood. The septum between the right and left atria may form incorrectly, and the foramen ovale may be too large or may fail to close. There may also be a defect in the interventricular septum, leaving an opening between the ventricles. Any of these defects may prevent adequate amounts of oxygenated blood from reaching the tissues. Fortunately, many of the problems arising from faulty development of the heart can now be corrected surgically.

Pulmonary Circulation

The pulmonary circuit differs from the systemic circuit in many respects, both anatomically and physiologically, since it does not serve the metabolic needs of lung tissue (the bronchial vessels do that). Its function is rather to bring blood into contact (across a membrane) with the air in the lung so that an exchange of gases may occur. Since the output of the right ventricle is the same as the output of the left ventricle, the flow in the pulmonary circuit must equal that in the systemic circuit. The resistance in the pulmonary circuit is very low, and much less pressure is required to drive blood through it. Where arterial pressure in the systemic circuit may be 120/80, that in the pulmonary artery is more like 22/9.

Pulmonary vessels branch rapidly to form a capillary network. The vessels

Focus

Blue Babies

The mammalian (and human) heart is a relatively recent evolutionary development. It is so designed that the flow of blood through the lungs is completely separate from the flow of blood through the rest of the body. This maximizes the amount of oxygen in the blood and ensures that the mammalian body gets enough oxygen to support its intense metabolism. The hearts of reptiles, amphibians, and fishes do not separate the pulmonary and systemic flows. In these animals, the oxygen-depleted venous and the oxygen-rich arterial blood mix in the heart. The blood that leaves the heart does not carry as much oxygen as in mammals, but this amount of oxygen is enough for the slower metabolisms of these animals.

Like other organs, the mammalian heart develops in the embryo. Well before birth, the ventricles are completely walled off from each other. The atria are almost completely walled off; at birth, the only connection remaining between the atria is the small foramen ovale. Together with the ductus arteriosus, this opening shunts most blood into the systemic circuit. At birth, the shunts close off, forcing blood into the lungs to obtain the oxygen the body can no longer get from the placenta.

However, the perinatal circulatory changes do not always happen correctly. The ductus arteriosus does not always seal off. The flap over the foramen ovale does not always fall precisely into place. In addition, the wall between the ventricles may not form properly before birth, and there may be other problems with heart structure. The net effect of all these problems is that the heart comes to resemble that of a lower animal. Oxygenated and unoxygenated blood mix in the heart. Since oxygenated blood is bright red and unoxygenated blood is bluish, an infant with any heart defect that produces such a mixing has bluish blood and its skin shows a bluish tinge. It is appropriately called a "blue baby."

The effect of blue blood is not just cosmetic. Because it carries less oxygen than normal, the body's cells receive less oxygen than they need. The load on the heart is increased, both because the body tries to compensate for the lack of oxygen by increasing the flow of blood and because the physical defects reduce the cardiac output. As a result, blue babies do not long survive without surgical help. Fortunately, the surgical help is available. An open ductus arteriosus can be tied off, and most openings in the atrial and ventricular walls can be patched. A blue baby usually can be made completely normal, although there is the risk always associated with major surgery.

FIGURE 18-2   The postnatal circulation.

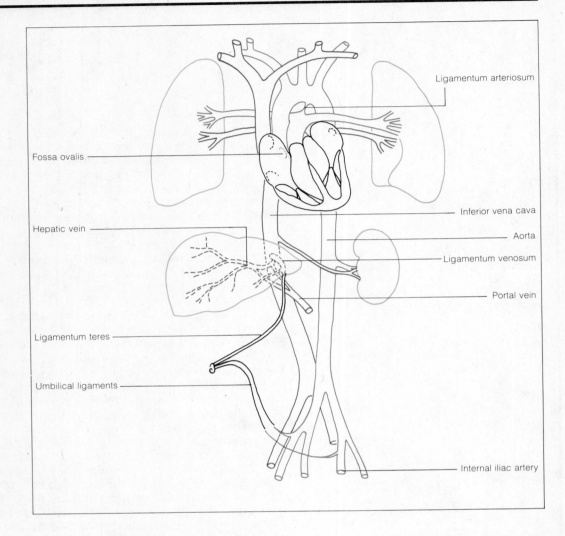

Ligamentum arteriosum

Fossa ovalis

Inferior vena cava

Aorta

Hepatic vein

Ligamentum venosum

Portal vein

Ligamentum teres

Umbilical ligaments

Internal iliac artery

have thinner walls, are larger in diameter, and contain less elastic and muscle tissue than comparable systemic vessels. Functionally, they are rather like veins since they are readily expanded by a slight increase in pressure. The pulmonary circuit typically contains about 15–20 percent of the blood volume (that is, up to about a liter of blood), but this capacity is so variable that it increases by about 400 ml when you lie down. (This is one reason why some patients with respiratory or cardiac problems have difficulty breathing when they lie down—more blood in the lungs leaves less room for air!)

There are no elaborate mechanisms to regulate the blood flow or its distribution within the lungs. Vessels in the pulmonary circuit are not affected by most of the stimuli that cause marked responses in the systemic vessels, and

they are not very sensitive to most pressor chemicals. Because of the distensibility of the vessels and the lack of reactivity, pulmonary blood pressure does not, within a wide range of blood flow, change very much. An interesting property is that the smooth muscle of the pulmonary vessels constricts in some situations that cause dilatation in systemic vessels. This is useful when portions of the lung are not well ventilated, since in these areas the oxygen content is rapidly depleted, carbon dioxide rises and pH falls, and gas exchange ceases. This elicits a constriction of the vessels in that region, which restricts blood flow there and forces blood to go to well-ventilated parts of the lung where the exchange of gases *can* be carried out.

Capillary exchange in the lungs is chiefly the diffusion of oxygen into the blood and of carbon dioxide into the lungs. The concentration gradients for these gases are normally such that the exchange is rapid. (For further discussion of the exchange of gases in the lung, see Chapter 20.) There is very little fluid exchange in the lung. The plasma colloid osmotic pressure is about the same as in the systemic circuit (25–30 mm Hg), but the filtration pressure is so much lower (probably 6–7 mm Hg in the capillaries) that any tissue fluid present would probably move *into* the capillaries. Since there are also numerous lymphatic vessels in the lung that remove any fluid forming in the tissues, there is normally little extracellular fluid in the lungs.

Coronary Circulation

Cardiac muscle, like skeletal muscle, is at times called upon to increase its work load greatly. However, unlike skeletal muscle, the heart is rarely at rest for more than a second at a time. At all times it must receive a blood supply sufficient for a working muscle.

Blood flow in the coronary arteries is complicated by the fact that when the left ventricle contracts it develops a pressure greater than that in the coronary arteries. As a result the terminal branches of these arteries, which penetrate the ventricular muscle, are occluded (blocked) during much of systole. Coronary blood flow is therefore intermittent, although it may reach very high flow rates during diastole (Figure 18–3). A rapid heart rate may significantly reduce coronary blood flow because of the shortening of diastole.

Unlike skeletal muscle, cardiac muscle cannot incur a large oxygen debt. Most of its energy is derived from oxidation of lipids (fatty acids), and the remainder comes from glucose and from lactic acid. This ability to use lactic acid for energy is both somewhat unusual and very helpful. Lactic acid is a metabolite that is formed in skeletal muscles when they contract anaerobically, so it is likely to be present in the blood in greater concentration during strenuous physical exercise when skeletal muscles incur an oxygen debt.

The heart also extracts a relatively

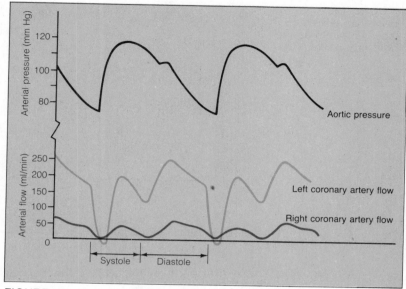

FIGURE 18–3   Blood flow in the coronary arteries.

high proportion of oxygen from the blood under resting conditions—about 75 percent, as compared with 25 percent for other tissues. When cardiac work and oxygen requirements are increased, the heart can therefore meet these needs only if the coronary blood flow is increased correspondingly. Probably the greatest factor in increasing coronary blood flow is the amount of oxygen present locally. *Myocardial hypoxia* (low oxygen in cardiac muscle) has a vasodilator action on coronary vessels and can quickly increase coronary blood flow several-fold. Epinephrine also increases the blood flow, as does sympathetic stimulation. Vagal (parasympathetic) stimulation reduces it. The actions are the reverse of those on most other vessels.

Occlusion or blockage of a coronary artery deprives the muscle tissue nourished by that vessel of its nutrition and oxygen and that tissue then soon ceases effective contraction. Although there are anastomoses between the various coronary arteries, they are quite small for the most part and probably would not be adequate to deal with sudden occlusion. A gradual blockage might, however, cause the anastomoses to enlarge and take over adequately. Coronary occlusion is quite common, and it is one of the most frequent causes of death. It often occurs as a result of deposition of material (lipid and/or

FIGURE 18-4
Effect of changes in mean arterial pressure on cerebral blood flow.

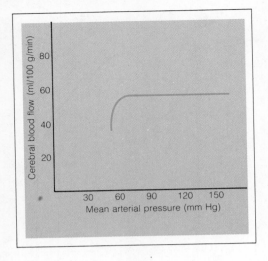

cholesterol) in the wall of the coronary arteries. The material progressively thickens the wall of the vessels, reduces the size of the lumen, and curtails blood flow through it.

Cerebral Circulation

The brain is another organ that is extremely dependent upon a continual blood flow. It was pointed out in Chapter 12 that the brain has a high oxygen requirement due to its high metabolic rate. It oxidizes glucose almost exclusively for energy. However, it cannot store either glucose of glycogen, and it cannot go into oxygen debt. The rate of oxygen consumption in the brain does not fluctuate widely—mental activity does not cost much in terms of energy expenditure. Accordingly, the mechanisms for regulating cerebral circulation are not directed as much toward increasing the blood flow as toward preventing a decrease.

Cerebral blood flow remains quite constant, in spite of changes in the mean arterial pressure, as long as the arterial pressure is above about 70 mm Hg, but it falls precipitously when the pressure is reduced (Figure 18-4). The cerebral vessels do not participate much in cardiovascular reflexes, but they do respond effectively to changes in the plasma content of carbon dioxide and oxygen (Figure 18-5). The cerebral blood flow increases more when the level of carbon dioxide increases than when that of oxygen decreases. Fortu-

nately, since the rate of brain metabolism does not vary greatly, local carbon dioxide production does not vary much either, which should help keep the cerebral blood flow steady.

The larger cerebral veins are unique, since they are the dural sinuses. They can therefore neither distend nor collapse with pressure changes as other veins do. Since the head is usually above the heart, the cerebral venous pressure is quite low. Puncture of a venous sinus (as might happen with a skull fracture) could result in air being sucked into the circulatory system—a highly undesirable situation.

The size of the cranium determines the volume of material that can be enclosed within it. This space is occupied by the brain, the blood, and the cerebrospinal fluid. A marked change in the volume of any of the three will affect the other two. For instance, if intracranial pressure is raised because of increased cerebrospinal fluid or a growth (tumor) of the brain, the cerebral blood flow will be reduced. Gravitational forces, such as acceleration upward (positive g), temporarily lower intracranial pressure and cerebral blood pressure as all fluids tend to shift toward the feet. The reduced intracranial pressure reduces resistance and helps maintain the cerebral blood flow in spite of the reduced cerebral blood pressure. Similarly, acceleration downward (negative g) has the opposite effect, and the increased intracranial pressure helps prevent the marked increase in cerebral flow which would otherwise occur.

CARDIOVASCULAR HOMEOSTASIS— ADJUSTMENTS TO SOME PHYSIOLOGICAL STRESSES

The two most common stresses imposed upon the cardiovascular system, *gravity* and *exercise,* are encountered so often, and frequently in such mild degree, that it may seem unrealistic to speak of them as stresses. But in a broad sense these are conditions to which the cardiovascular system must respond in order to maintain homeostasis. Gravity affects the hydrostatic pressure of blood with every change of posture, and the effects of exercise vary with the severity of the activity. Both are involved in nearly everything we do.

Effects of Gravity

In the reclining position, the effect of gravity upon the fluid in the vascular system is minimal, as all of the vessels are very near heart level. Arterial pressure is about the same in the head as in the feet (Figure 18–6), and stroke volume is relatively high. When the erect posture is assumed, however, gravity tends to pull the blood toward the feet. Pressures in the lower extremity, both arterial and venous, become elevated due to the weight of the column of blood from the feet to the heart (Figure 18–7). In vessels above the heart, pressures are correspondingly lower. In the foot, this might add an extra 80–85 mm Hg; in the head, it might subtract 25–30 mm Hg. Therefore, if mean arterial pressure were 90 mm Hg, the hydrostatic pressure while standing would be 170–175 mm Hg in the foot, but only 65–75 mm Hg in the head. The effect on venous pressure would be similar. If venous pressure at the right atrium were 5 mm Hg, that in the foot would be 80–85 mm Hg. In the head it would be negative, and the veins in the head (except the venous sinuses in the dura) would collapse. Blood flow through the brain would not necessarily be impaired, however. The pressure drop from artery to vein should provide sufficient driving force, and blood can flow through collapsed veins.

In the capillaries of the lower extremities, the capillary filtration pressure exceeds the colloid osmotic pressure. Therefore, there is movement of fluid from the capillaries to the tissues (your feet swell after a day of standing). Increased pressure in the veins causes them to distend, increasing their capacity, and blood pools there. This is most apparent in superficial veins, but pooling also occurs in deep veins and in veins in the abdominal cavity. That portion of the blood that stays in the veins instead of returning to the heart reduces venous return and so affects stroke volume. If cardiac output falls, so does arterial pressure. A fall in arterial pressure may mean reduced

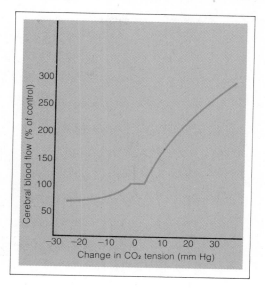

cerebral blood flow, and fainting may occur. Fainting is an effective if inconvenient solution, since it brings about the horizontal position and removes the hydrostatic pressure problem; it thus restores cerebral blood flow.

That you do not faint each time you assume the erect position shows that there are compensatory mechanisms to prevent pooling of blood in the venous reservoirs and maintain venous return, cardiac output, and arterial pressure, and hence to ensure adequate blood flow to the brain and other organs. One of these mechanisms is that a slight initial drop in arterial pressure activates the carotid and aortic pressoreceptors. The cardiac accelerators then help maintain cardiac output in the face of

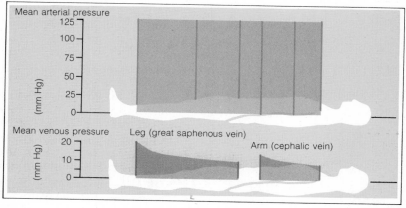

FIGURE 18–6   Mean arterial (red) and venous (blue) pressures in the reclining position.

FIGURE 18–7
Mean arterial (red) and venous (blue) pressures in the standing position.

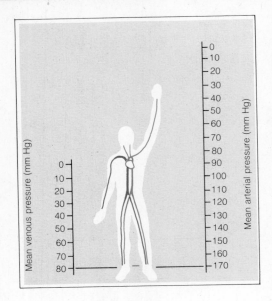

the reduced stroke volume. The vasomotor centers bring about vasoconstriction to increase peripheral resistance, particularly in the capacious vascular beds of the viscera. Constriction on the venous side helps limit the pooling of blood in the veins.

One of the most effective aids to the return of blood from the extremities—so effective, in fact, that venous pressure in the foot falls significantly after just a single step—is the so-called **muscle pump.** Leg muscle contractions apply pressure to the deep veins of the leg and squeeze the blood from these parts of the vessels. Contraction of the leg muscles applies pressure to the deep veins of the leg and squeezes blood from them. The blood is forced to move upward because the one-way valves in the veins prevent it from flowing back toward the capillaries. When the muscles relax, the veins are able to fill again and blood moves in from the capillaries, and again the valves prevent backflow in the veins. With another contraction the veins are emptied again and more blood moves upward.

Rhythmically repeated muscle contractions, as in walking, markedly lower the venous pressure in the foot for the duration of the activity. This is because the valves in the veins break up the column of blood between the heart and capillaries of the foot. When a valve is closed it supports the column of blood immediately above it, and the

hydrostatic pressure is therefore based only on the height of the column of blood up to the next valve. With each contraction the blood is moved a little higher, beyond another valve, in what has been called a "milking action."

Standing casually relaxed and occasionally shifting the weight from one foot to the other is tolerated much better than a rigid, motionless posture. The latter may result in fainting, even in a physically fit individual—as sometimes occurs at a formal military review. Sustained (isometric) contractions do not aid venous return, but rather block blood flow for the duration of the contraction.

If an individual is placed on a tilt table, so that the transition from reclining to erect posture is achieved with no muscular activity, the blood simply pours to the feet. Cardiovascular reflex mechanisms alone may not be able to compensate for it. Some individuals tend to show a drop in arterial pressure, dizziness, blurred vision, and even fainting after rising quickly. This is particularly noticeable in patients who have been confined to bed for a period of time. It also happens occasionally in perfectly healthy individuals who get out of bed or stand up from a squat too abruptly. In spite of this, however, humans are much better equipped to cope with gravity than most quadrupeds. Many species are unable to prevent the pooling of venous blood in their hindlimbs and viscera when held in an upright position.

The cardiovascular effects of gravity are greatly exaggerated by acceleration, including that which commonly develops in rotary motions, as in aircraft during sharp turns. The compensatory mechanisms require assistance (pressure suits, reclining position) to cope with such severe gravity stresses. At the other extreme is the weightlessness encountered by space explorers, since it eliminates the need for the mechanisms that combat gravity. After a prolonged space venture, the sensitivity of these cardiovascular reflexes seems to be temporarily reduced. Astronauts returning from a prolonged sojourn in space find that it takes a while before their cardiovascular compensatory mechanisms readjust to gravity.

Effects of Exercise

Marked cardiovascular adjustments are necessary during physical exertion in order to meet the tremendous increase in metabolism of the active skeletal muscles. With vigorous exercise their oxygen consumption may increase nearly a hundredfold. For this to be possible, muscle blood flow must be raised to deliver a correspondingly greater amount of oxygen. Exercising muscles remove nearly all of the oxygen from the blood flowing through them (in contrast to resting muscles which remove only about a fourth of the oxygen). Even so, at least a 25-fold increase in muscle blood flow is still necessary. The circulating blood volume increases only slightly, due to the flushing out of blood from reservoirs such as the spleen. Most of the increase in blood flow is brought about by moving the blood more rapidly and by redistributing it so that a greater share of the cardiac output goes to the active muscles and less to other tissues.

The adjustments of the cardiovascular system to physical exercise show up as changes in familiar parameters. There is a heart rate increase, a slight increase in stroke volume, and an increase in mean arterial pressure. Systolic pressure goes up roughly in proportion to the increase in cardiac work, while diastolic pressure may either rise or fall. Diastolic pressure reflects the rate of peripheral run-off and is likely to go up slightly in mild exercise where reflex vasoconstriction predominates. With more vigorous exercise the vasodilatation and the fall in peripheral resistance in active muscles are usually greater; the diastolic pressure falls, sometimes to very low levels.

The increase in blood flow through active muscles depends greatly upon the local factors—a decrease in oxygen, an increase in carbon dioxide and other metabolites, and a fall in pH—all of which reduce peripheral resistance and increase blood flow through the muscle vessels. In resting muscle many capillaries are closed, but in active muscles the dilation of arterioles and precapillary sphincters produces a 10–100-fold increase in the number of open capillaries. The resulting increase in cross-sectional area of the vascular bed reduces the velocity of the blood flow (or perhaps just prevents it from rising), which facilitates the transcapillary exchange. It also, however, decreases peripheral resistance and hence arterial pressure; the latter must be maintained if flow is not to cease completely.

The mechanisms for maintaining pressure and increasing total flow are largely systemic reactions that increase both cardiac output and peripheral resistance (exclusive of skeletal muscles). The mechanisms for regulating cardiac output are summarized in Figure 15–15 and those for controlling arterial pressure and peripheral resistance in Figure 17–9. Cardiac output is raised by increases in either heart rate or stroke volume. Stroke volume depends upon the venous return and contractility, while the heart rate is controlled through the cardiac centers in the medulla. There is also an important element of stimulation from higher centers in the brain, since heart rate may actually increase before exercise begins. Stimulation of certain sites in the diencephalon can produce the changes in cardiac and vasomotor activity that are characteristic of exercise.

Stimulation of the vasomotor centers in the medulla accompanies activity in the cardiac centers. The resulting vasoconstriction of the vessels that receive sympathetic innervation serves two important purposes: (1) It restricts flow to inactive areas, particularly to the abdominal viscera and the skin, thus reducing the size of the vascular bed; and (2) by increasing peripheral resistance in these areas, it decreases the overall peripheral run-off, which helps raise mean arterial pressure. The capacitance vessels (veins) are influenced more by vasoconstrictor fibers than by the presence of vasodilator substances in the muscles, and their reservoir functions are thus reduced by sympathetic activity. Otherwise the blood passing through the muscles might just pool in the veins and venous return would decline.

Skeletal muscle resistance vessels (arteries and capillaries) do receive some sympathetic innervation, and some of these fibers are vasodilators. It has been suggested that these fibers may hasten the redistribution of blood by causing dilation and reduced resistance before the metabolic dilators become effective.

There is usually a release of epinephrine from the adrenal gland during, or even before (in anticipation of), exercise. It reinforces all the above responses, since epinephrine dilates muscle vessels while causing constriction elsewhere. It is also a stimulus to both cardiac muscle and its pacemaker.

The systemic and local vascular responses tend to bring about a rather effective redistribution of blood. By reducing peripheral resistance in the active muscles, blood flow there is increased; by increasing peripheral resistance in the viscera and skin, the flow to inactive organs is reduced; by venoconstriction, venous pooling is prevented, and venous return is maintained. The size of the total vascular bed therefore does not increase very much and arterial pressure is maintained or raised. There are some complications, however. For example, vigorous exercise in a hot environment poses the problem of how to get rid of the extra heat. One way is to dilate skin vessels to bring the warm blood to the surface for cooling. Unfortunately, this enlarges the vascular bed and makes it difficult to maintain arterial pressure. There is no completely satisfactory solution to this problem, and we find that our ability to perform vigorous exercise in a hot environment is limited.

Reflexes of the carotid and aortic pressoreceptors serve primarily as checks to prevent excessive increases in arterial pressure. They undoubtedly contribute to returning blood pressure and heart rate to control levels after the end of the exercise. Blood pressure falls very rapidly, reaching pre-exercise levels in a few minutes. Heart rate falls more slowly, and after strenuous exercise it may require many minutes to reach the control level.

Individuals differ in their response to exercise, depending in part upon their physical endowment and in part upon its effectiveness. Chronic exercise seems to improve the overall efficiency and effectiveness of the entire system. Trained athletes are not only able to perform better (that is, to run farther and faster) than untrained people or nonathletes, but their physiological responses to exercise are different.

Trained athletes generally have slower heart rates and greater stroke volumes at rest, which enables them to achieve higher cardiac outputs with given rate increases. They can also increase their cardiac outputs to higher levels than the untrained. Therefore, they can deliver more oxygen to their active muscles each minute. Distribution of blood to muscles is further improved because of the greater vascularization of muscles that are chronically exercised.

Any number of tests have been designed to measure what is called "physical fitness." Many of these tests are for cardiovascular fitness and are based upon the response of the cardiovascular system to a specific "bout" of exercise. A "good" test score is usually based upon one or more of the following: slow heart rate and low arterial pressure pre-exercise, minimal changes due to exercise, and a quick return to control levels post-exercise. Performances on such tests are adversely affected by physical factors such as fatigue or a recent cigarette and by emotional factors such as anxiety or tension.

CARDIOVASCULAR EFFECTS OF SOME ABNORMAL STRESSES

Circulatory Shock

Shock is a term that has been applied to a number of unrelated events ranging from electrical shock to insulin shock to spinal shock, and more. One of the more common uses is in *cardiovascular shock*, a circulatory collapse. The cascade of events that occurs in cardiovascular shock, and the compensatory adjustments employed to combat them, involve the entire cardiovascular system and eventually other systems as well. It should be profitable to follow through the sequence of events because it is an excellent example of the interdependence of the many mechanisms, and it provides a good review of the cardiovascular system.

Circulatory shock has been variously defined as a bout of acute hypotension (low blood pressure), as a state of inadequate tissue perfusion, and as a condition of inadequate cardiac output.

It is all of these, since one leads to the others. The symptoms are all so closely related that it is often not possible to determine which is cause and which is effect. There is a decrease in blood volume which may be real, due to actual fluid loss, or apparent, due to increased capacity of the vascular bed (vasodilatation). In either case, it leads to poor venous return followed by a decrease in cardiac output. The reduced cardiac output contributes to the fall in arterial pressure, which means a low capillary pressure and reduced flow through the capillaries. These conditions all contribute to an imbalance between the capacity of the vascular system and the volume of blood in it.

A convenient classification of circulatory shock is on the basis of the origin of the difficulty. It may be a low volume shock, due to loss of whole blood, to a loss of fluid (as from burns, excessive sweating, or diarrhea), or to a transfer of fluid to the tissues (as in edema). It may be a low resistance shock, due primarily to vasodilatation brought about either reflexively or by vasodilator substances released into the circulation. A third cause of circulatory shock is cardiac failure (see below).

The sequence of events is probably best described in **hemorrhagic shock,** a rather common type, caused by the loss of whole blood from the body. The reduced blood volume and venous return bring about an immediate fall in cardiac output and arterial pressure. As arterial pressure falls, the carotid and aortic pressoreceptors evoke increases in heart rate and peripheral resistance. Arteriolar constriction occurs primarily in arterioles of skin and abdominal viscera, including the kidney. Venoconstriction minimizes the pooling of blood in the veins.

If these measures *are* adequate, and if the fluid loss has not been too great, venous return is sufficient, and the arterial pressure levels off and, after a time, rises once more. The arteriolar constriction, while raising pressure in the arteries, lowers it downstream in the capillaries. The reduced filtration pressure in the capillaries favors movement of fluid into the capillaries from the extracellular spaces. Over a period of hours or days the blood volume is restored (although it takes longer for the lost blood cells to be replaced).

If these measures are *not* adequate, and the period of hypotension is prolonged, a progressive deterioration sets in that sooner or later develops into irreversible shock. The arteriolar constriction, which some investigators view as a spasm of the precapillary sphincters, is so effective that capillary flow is reduced to the point of being inadequate to supply the tissues. Oxygen lack (hypoxia) and accumulation of carbon dioxide and other metabolites interfere with cellular function, particularly in the kidney and liver. The blood vessels themselves are eventually affected, and their vasoconstriction gives way, causing arterial pressure to fall even further. The capillary walls become increasingly permeable and leak even protein. At this stage, efforts to restore blood volume by transfusion are not very effective, because the fluid leaks out of the vascular system. The specific cause of these changes is unclear, but it is surely related to hypoxic changes associated with the reduced capillary flow. In addition, vasodilator substances and *proteolytic* (protein-splitting) enzymes are released from damaged cells and add to the collapse (this is particularly true of shock due to antigen reactions — *anaphylactic shock* — or infections, in which vasodilatation is a primary factor). The integrity of the cardiac muscle cells is compromised by severe reduction of the coronary blood flow, and the heart's action is weakened. As the condition progresses, each new development hastens the next step in a vicious downward cycle. Other forms of circulatory shock, if severe enough and long enough, lead along the same unrelenting course. The chief differences lie only in the part of the circuit that suffers the initial breakdown.

Hypertension

Hypertension is a condition of sustained elevated arterial pressure. Persistent pressures above 140–170/90–100 (depending on age and sex) are commonly considered to be hypertensive. Since vessels of the pulmonary circuit do not respond to the same types of stimuli or controls as the systemic vessels, hypertension does not ordinarily involve these vessels (although pulmonary hypertension is not unknown).

In hypertension, the arterial pressure is elevated by an increase in peripheral resistance rather than by an increase in cardiac output. It may result from one of several malfunctions, including ones in the vascular, neural, endocrine, and renal (kidney) systems.

Renal hypertension is perhaps the most studied of the known forms of hypertension. A permanent hypertension, which lends itself to experimental study, develops when the blood flow to the kidney is restricted *(renal ischemia)*. The ischemic kidney releases a substance, **renin,** into the circulation. Renin is an enzyme that acts upon a plasma protein fraction *(angiotensinogen)* to form **angiotensin I,** which is converted to **angiotensin II** by an enzyme in the blood. Angiotensin II acts directly upon vascular smooth muscle to raise peripheral resistance throughout the body. It is an extremely powerful vasoconstrictor, although it is short-acting because it is destroyed by enzymes in many tissues. Other more physiological roles of the renin-angiotension mechanism are discussed in Chapter 28.

The most common form of hypertension, accounting for about 90 percent of all cases, is called **essential hypertension.** Its causes are largely unknown, but innumerable suggestions have been made. The true cause probably arises from multiple factors, and may involve the carotid pressoreceptors. It has been suggested that perhaps the receptors have been "reset" to a different (higher) pressure, in a manner similar to resetting a thermostat, or that the walls of the arteries in which they lie may have become *sclerotic* (hardened—more rigid and less distensible), so that a greater pressure is required to activate the pressoreceptors.

Whatever the causes of the hypertension, the effects vary only in the many complications and secondary effects. With elevated arterial pressure the left ventricle must develop greater pressure in order to eject blood. Like any muscle subjected to an increased work load, it hypertrophies (the cells become larger) and thus needs more oxygen. The cardiac muscle thus becomes more vulnerable to any reduc-

tion in coronary blood flow. In addition, there is always the danger that the myocardium may not be able to meet the demands upon it and that the heart may fail. The high blood pressure also increases the likelihood of a ruptured vessel. In addition, resistance vessels in hypertensive patients hypertrophy, which makes them more rigid and narrows the lumen. *Atherosclerosis,* a type of **arteriosclerosis** in which calcium and lipid materials are deposited in the walls of arteries, commonly accompanies hypertension. The roughened inner surface and slowed flow in these areas raise the likelihood of thrombosis (blood clotting) in the affected vessels. Any of these complications in the coronary vessels drastically reduces the ability of the heart to pump blood and may lead to a heart attack. In a cerebral vessel, a clot or rupture may temporarily or permanently damage certain centers or pathways (cerebrovascular accident, *stroke*). If arteries within the kidney become sclerotic, kidney function is impaired and there may be kidney failure.

Cardiac Failure

Cardiac failure is said to occur whenever the pumping action of the heart becomes inadequate and cardiac output fails to meet the needs of the tissues. Of the many causes, probably the most frequent is damage to the myocardium resulting from impaired coronary blood flow. Failure may also be brought on by an increased cardiac work load due to increased resistance (hypertension) or defective heart valves.

Heart failure, or **cardiac decompensation,** usually develops first on one side of the heart. If the left ventricle fails, its output (stroke volume) declines, its end-systolic volume rises, the left atrium becomes enlarged, and congestion backs up into the lungs. The pulmonary congestion is more severe when the right ventricle continues to function normally, but its output will eventually be reduced when the venous return to the right atrium declines. Congestion in the lung may lead to pulmonary edema, which impairs respiratory function and causes difficulty in breathing.

If the initial failure is in the right ventricle, on the other hand, the congestion is in the systemic veins and the edema is in the peripheral tissues. Left ventricular output is soon reduced because of decreased inflow to the left atrium from the lungs.

As cardiac output falls, blood flow to the kidney is reduced and urine formation ceases. The consequent retention of both fluid and electrolytes (particularly of Na^+ and Cl^-) contributes to the edema. The mechanism is complicated, but it may result in a considerable increase in extracellular fluid, which is apparent as visible swelling and an increase in body weight. Because this imposes its own burdens on the heart, the objective of treatment must be both to improve the pumping action of the heart and to remove the excess fluid from the tissues. The first of these goals can be accomplished with drugs such as *digitalis*, the second by limiting the dietary intake of salt and by the use of agents *(diuretics)* that increase the excretion of fluid and electrolytes by the kidney.

CHAPTER 18 SUMMARY

Circulation Through Regions with Special Needs

During fetal life modifications of the circulation are necessary because all the exchanges with the external environment are carried out with the mother's blood in the placenta. Blood from the fetus reaches the placenta in the umbilical arteries and returns in umbilical vein. Most of the blood in the umbilical vein flows into the ductus venosus which empties into the vena cava, bypassing the liver. Upon entering the right atrium, blood may be shunted directly to the left atrium through the foramen ovale, or pumped out the pulmonary artery, with most of it shunted to the aorta through the ductus arteriosus. Very little blood goes to the lungs, for they are collapsed and resistance there is quite high. At birth the lungs expand, reducing the resistance in the pulmonary circuit, thus increasing blood flow to the lungs; the umbilical cord is tied off, stopping flow in the umbilical vessels, which raises the resistance in the systemic circuit. A valvelike flap over the foramen ovale, which prevents backflow from left to right atrium, eventually grows shut leaving only a fossa ovalis. Constriction (spasm) of the wall of the ductus arteriosus closes the shunt between pulmonary artery and aorta, and forces all the blood through the lungs. All of the ducts soon become obliterated: the ductus arteriosus becomes the ligamentum arteriosum, the ductus venosus becomes the ligamentum venosum, the umbilical vein becomes the ligamentum teres, and the umbilical arteries become the umbilical ligaments.

The pulmonary circuit exists for exchange of respiratory gases rather than as blood supply for tissues. Oxygen is taken up and carbon dioxide removed in pulmonary capillaries, which is the reverse of that in other capillaries. Flow in the pulmonary circuit is the same as in the systemic circuit, but the pressure and resistance are much less. Most of the mechanisms for maintaining arterial pressure do not affect the pulmonary circuit, and pressure and resistance do not vary much there.

The coronary circulation is intermittent because contraction of cardiac muscle occludes the coronary arteries during systole, but flow is very high during diastole. Blood flow to the heart must be increased with increased cardiac work. Coronary vessels are caused to dilate by sympathetic stimulation, epinephrine, and hypoxia, which increase coronary blood flow, even while other vessels are constricting.

Because the metabolic rate of the brain tissue does not change much, it is not necessary to be able to greatly increase blood flow to the brain. But since the brain is very dependent upon a continual oxygen and energy supply, it is important that its blood flow not be decreased. The cranium has a fixed volume and factors such as gravity, that tend to lower blood flow to the brain, are likely to reduce the pressure of the cerebrospinal fluid as well, which lowers the intracranial pressure and facilitates the blood flow to the brain. The large veins in the skull are sinuses in the dural folds, and cannot readily distend or collapse.

Adjustments to Some Physiological Stresses

The cardiovascular system responds to normal and abnormal stresses by the same mechanisms that control the normal functions of the system.

Gravity decreases venous return, as blood pools in venous reservoirs of the abdominal cavity and lower extremities. Reduced venous return decreases cardiac output and arterial pressure, which stimulates arterial pressoreceptors and leads to reflex increases in heart rate and peripheral resistance (increased diastolic pressure). The contraction of skeletal muscles (the *muscle pump*) and increased breathing movements aid venous return. If these mechanisms fail to maintain cardiac output, arterial pressure, and blood flow to the brain, cardiac output falls in spite of an increased heart rate, the pulse pressure narrows, and fainting is likely to occur.

Exercise elevates venous return, resulting in increased cardiac output, mean arterial, diastolic, and systolic pressure. Resistance increases in the vessels with sympathetic innervation (skin and viscera); decreased oxygen and increased carbon dioxide in the active muscles cause dilation of the muscle vessels, which reduces the resistance and increases the flow. Increased resistance in the skin and viscera and decreased resistance in the muscles redistribute the blood to the muscles, where it is needed.

Effects of Some Abnormal Stresses

Circulatory shock is a condition of circulatory collapse. It is a vicious cycle of events that can be triggered by any of a number of factors. There is a fall in blood pressure, in venous return, and in cardiac output, and an increase in capillary permeability that leads to edema and decreased blood volume; all contribute to a reduction in the force that causes the blood to flow.

Hypertension is due to an increased resistance requiring a greater pressure to maintain flow. It imposes a burden on the heart to develop enough pressure to overcome the resistance, and a burden on the vessels to withstand the increased pressure. Cardiac failure occurs when the heart fails to pump all the blood that comes to it: cardiac output is less than venous return, causing venous congestion and reduced pressure in the arteries. It may occur in the right or left ventricle.

STUDY QUESTIONS

1 What environmental conditions necessitate modifications in the fetal circulatory system?

2 What changes occur at or shortly after birth to adjust to conditions encountered in the world?

3 In what ways does the pulmonary circuit differ from the systemic circuit? How can the pulmonary blood flow be maintained when the pressure is so low?

4 What problems are encountered in maintaining a continuous blood flow to cardiac muscle? Why is it so important that adequate circulation be maintained?

5 What are the particular problems in maintaining cerebral blood flow? How are they met?

6 What would be the effect of each of the following? Answer with *increase, decrease,* or *no change.* (a) hemorrhage on arterial pressure; (b) lumbar sympathectomy on arterial pressure; (c) increased venous return on cardiac output; (d) ventricular fibrillation on cardiac output; (e) moderate exercise on skeletal muscle blood flow; (f) moderate exercise on blood vessels to abdominal viscera; (g) standing quietly for several minutes on pulse pressure; (h) getting out of bed too quickly on cardiac output; (i) straining (as in defecation or lifting) on venous return; (j) a very high heart rate (200 beats/min) on stroke volume; (k) a defective (leaky) aortic semilunar valve on cardiac output and arterial pressure; (l) submerging up to the neck in a hot bath on arterial pressure.

7 As a final review, see if you can explain the compensatory mechanisms with which one counteracts or accommodates the changes described above.

Part 5

Respiration

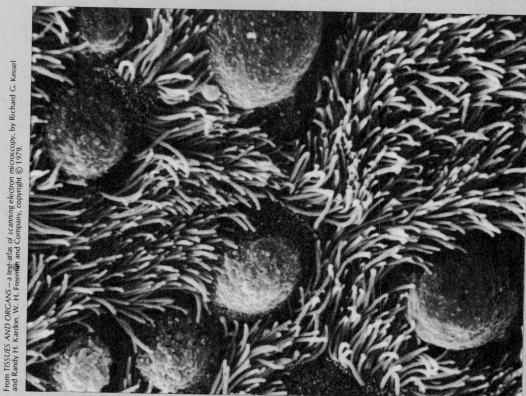

Lining of an airway in the lung; a bronchiole with ciliated cells and nonciliated secretory cells.

Now that you have studied the cardiovascular system as a transport system, it is appropriate to consider in more detail the organ system that regulates some of the substances carried by it. Part 5 deals with the system that transports oxygen and carbon dioxide between the external environment and the cells. A large portion of Part 6 deals with processes by which nutrients reach the cells from the digestive tract and from storage sites within the body. Much of Part 7 deals with removal of cellular wastes from the body and removal (or retention) of many other substances carried in the blood.

Part 5, respiration, is about oxygen and carbon dioxide in the body. Respiration is the exchange of oxygen and carbon dioxide between the cells and the external environment. The cells need oxygen to carry on their metabolic processes, and it must be delivered to the immediate environment of the cell. Carbon dioxide is produced by the cells in the course of their metabolic activities, and it must be removed from the cellular environment and then from the body.

To a biochemist or cell physiologist, however, respiration refers to the processes that occur within the cell. It includes the chemical reactions in which oxygen is used and carbon dioxide is produced as the cells break down carbohydrates and other substances to release energy and produce ATP.

Our concern in Part 5 is with those parts of respiration involved in getting oxygen and carbon dioxide to and from the cells; with the transport of these gases between cells and the external world. Some of the intracellular aspects of respiration are discussed in Part 6. There are several major elements in the respiration process to be considered in the next three chapters. They are:

1. the structure of the respiratory apparatus;
2. ventilation—the movement of air in and out of the lungs;
3. the actual exchange of gases between the lungs and blood, and between blood and the cells;
4. the transport of gases from the lungs to the tissues and from the tissues to the lungs;
5. the control of breathing.

19

Structure of the Respiratory Apparatus

Ventilation of the Lungs

The Respiratory Apparatus – Its Structure and Function

Chapter 19 examines two aspects of respiratory function, the anatomy of the respiratory system and the process of moving air into and out of the lungs. The upper part of the respiratory tract serves some important nonrespiratory functions, but the lower portion is exclusively respiratory, consisting of the channels that carry air to the tiny alveoli where gas exchange occurs.

As you study these branching airways, recall the vessels of the circulatory system. The respiratory and and circulatory ducts resemble one another in that some only provide passage, some regulate flow, while others provide for exchange. They differ in that the respiratory system moves a gas (air) and the circulatory system moves a fluid (blood). As you might expect, the mechanisms for ventilating the lungs—moving air in and out—are quite different from those for moving blood. The respiratory system has no pump to keep air moving and, since there is but one set of airways, air must move in and out through the same passages. In addition, there must be a means to prevent collapse of these passages.

This chapter describes the structure of the airways, the lungs, and the thoracic cavity, as well as certain conditions in them. With this knowledge you can understand how the lungs are ventilated.

STRUCTURE OF THE RESPIRATORY APPARATUS

The structures we usually think of as part of the respiratory system are those involved in ventilation and the exchange of gases between the lungs and the blood. Their role is to get the gases to and from the blood. Transporting those gases between the lungs and tissues is the responsibility of the cardiovascular system. The circulatory system is therefore as important to respiration as the respiratory apparatus itself. In fact, the chief difficulty in circulatory failure is due to the interruption of its respiratory function, oxygen delivery.

For convenience of presentation, the respiratory apparatus can be divided into three basic sections: (1) the upper respiratory tract, consisting of structures from the nasal cavity through the larynx; (2) the lower respiratory tract, consisting of the trachea through the lung; and (3) the thoracic cavity.

The Upper Respiratory Tract

The structures of the upper part of the respiratory tract serve as channels for moving air to and from the system of distributing airways, but they are not directly involved in the exchange of gases. These structures do serve some important nonrespiratory functions, however. The nasal cavity contains the olfactory receptors and helps protect the fragile terminal airways by warming, humidifying, and filtering the incoming air. The pharynx, shared with the digestive system, serves as passageway for food as well as air. The larynx contains the "voice box," the apparatus for producing sounds. It controls the opening into the lower respiratory tract and provides fine adjustment for the passage of air as needed for speech and other forms of vocalization. It also closes the passage for such activities as straining.

Nasal Cavity Air entering the respiratory tract first enters the nasal cavity shown in Figure 19–1, through the **external nares** (nostrils) when the mouth is closed. The nasal cavity is a narrow paired passageway leading to the pharyngeal region. Its medial wall (Figures 19–2 and 4–10) is formed by the nasal septum, made up of the vomer and the perpendicular plate of the ethmoid bone and an anterior cartilage extension. The lateral walls are made irregular by the **superior** and **middle nasal conchae,** which are processes of the ethmoid bone, and the **inferior nasal concha,** a separate bone that articulates with the maxilla. Under each concha is a recess, the **superior, middle,** or **inferior meatus,** respectively. The roof of the nasal cavity is the perforated cribriform plate of the ethmoid bone, through which the fibers of the olfactory nerve (cranial nerve I) reach the cranial cavity. The floor of the nasal cavity is formed in part by the bony hard palate and in part by the soft palate. Posteriorly, it opens into the nasopharynx through the **choanae** *(internal nares).*

The **paranasal sinuses** are openings in the ethmoid, sphenoid, frontal, and maxillary bones. Those in the ethmoid bone consist of several large air cells. The frontal sinuses are located near the midline above the orbit and open into the middle meatus. The large maxillary sinuses occupy much of the body of the maxillary bone. All the sinuses, as well as the *nasolacrimal duct,* open into the nasal cavity.

The entire nasal cavity, including the sinuses and conchae, is lined or covered by mucous membrane, which has a rich blood supply. The epithelium is *pseudostratified ciliated columnar* (characteristic of the respiratory system), containing numerous goblet cells which secrete a mucous substance that helps protect the membrane and keep it moist. The cilia of the epithelium and the hairs at the entrance to the nose serve protective functions. The cilia "beat" toward the pharynx (throat) and tend to move dust particles or other foreign material toward the oral cavity for easy removal. The hairs filter out larger particles and impede their entry into the nasal cavity.

The nasal cavity is not very large, but the scroll-like conchae extending into it increase the surface area so that

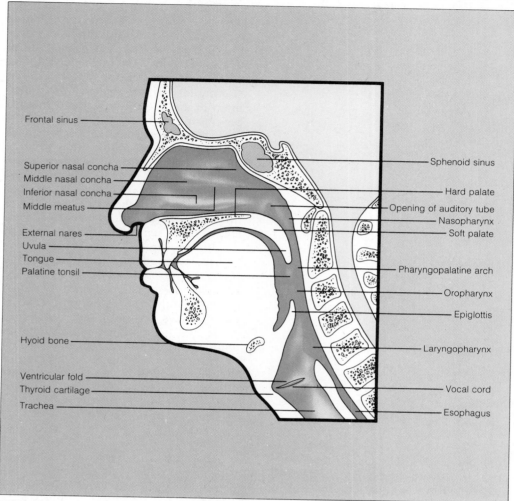

FIGURE 19–1 Midsagittal section of the head, showing the nasal cavity and related structures.

air entering the respiratory passages comes into close contact with the mucous membrane lining the nasal cavity. Because the inspired air may show extreme variations in temperature and moisture content, it must be brought to body temperature and saturated with water before it reaches the terminal portions of the lung, whose delicate membranes could quickly be dried out or frozen. The large area of moist mucous membrane and the ample supply of warm blood therefore serve to protect the lung tissues. The abundant blood supply also ensures that the mucous membranes of the nasal cavity will not be cooled too much by cold air or dried too much by dry air. When you

breathe through your mouth, the nasal passages are bypassed and the burden of warming and moisturing the incoming air falls upon the mucous membranes of the mouth and pharynx. They are not equipped for the job and quickly become dry.

Under certain conditions (irritations or allergies), the mucous membranes may become inflamed—the vessels dilate, the membranes swell, and the mucous secretion increases. Since the airways are quite narrow at best, this congestion may interfere with or completely block the movement of air through the nasal passages. Many of the familiar symptoms of the common cold can be traced directly to the mu-

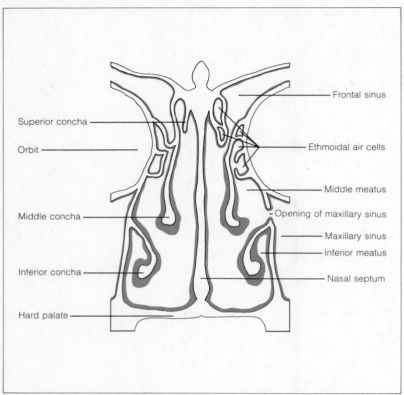

Superior concha
Orbit
Middle concha
Inferior concha
Hard palate

Frontal sinus
Ethmoidal air cells
Middle meatus
Opening of maxillary sinus
Maxillary sinus
Inferior meatus
Nasal septum

FIGURE 19–2 Frontal section through the nasal cavity, showing the conchae and meatuses.

The auditory tubes from the middle ear open into the nasopharynx on its lateral wall. A small mass of lymphoid tissue, the pharyngeal tonsil, is found on the posterior wall. Abnormal enlargement of the pharyngeal tonsil in children ("adenoids") interferes with the movement of air and often leads to mouth-breathing.

The **oropharynx** is continuous with the nasopharynx above, the laryngopharynx below, and the oral cavity anteriorly. It is the area roughly described as the throat. In accord with its close association with the oral cavity, it is lined with *stratified squamous epithelium*, which is typical of the upper part of the digestive tract.

The **laryngopharynx** is between the pharynx and oral cavity above, and the larynx and esophagus below. The oropharynx and laryngopharynx serve as passageways for both the respiratory gases and the solid and liquid material entering the digestive system.

Beneath the mucous membrane in the walls of all three parts of the pharynx are a number of skeletal muscles used in such activities as swallowing and vocalization.

cosal congestion. The loss of resonance of the voice and the feeling of pressure are due to the accumulation of mucus in the paranasal sinuses whose openings are blocked.

Pharynx The pharynx is part of both the respiratory system and the digestive system, and it will be discussed again in Chapter 22. Although it is usually considered an organ, it is easier to think of it as a space or area. It may be divided into three parts, the nasopharynx, the oropharynx, and the laryngopharynx, associated primarily with the nasal cavity, the oral cavity, and the larynx, respectively (Figure 19–1).

The **nasopharynx** is posterior to the nasal cavity and opens into it through the choanae (internal nares). It is continuous below with the oropharynx and oral cavity. Its antero-inferior boundary is the posterior part of the soft palate and its midline appendage, the **uvula.**

Larynx The larynx is the beginning of the exclusively respiratory portion of the airway. It opens into the laryngopharynx above and connects with the trachea below. The larynx contains several large cartilages held together by ligaments and fibrous membranes, with the hyoid bone bound to its upper edge (Figure 19–3). Skeletal muscles move some of the cartilages individually and the entire larynx as a unit. The larynx is located on the anterior surface of the neck. The **laryngeal prominence,** more commonly known as the Adam's apple, is quite conspicuous in the adult male.

There are three unpaired and three paired cartilages in the larynx. The largest is the **thyroid cartilage,** whose anterior portion is formed by two broad plates, or laminae, which meet at an angle to form the Adam's apple. Below it is the **cricoid cartilage,** which somewhat resembles a class ring, with the face directed posteriorly. The cricoid cartilage is connected to the trachea

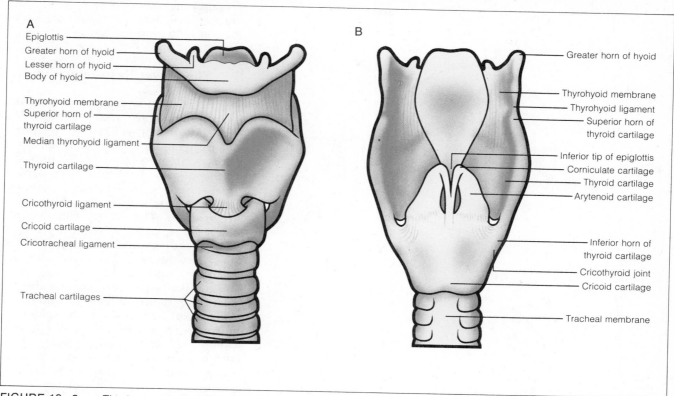

A

Epiglottis
Greater horn of hyoid
Lesser horn of hyoid
Body of hyoid

Thyrohyoid membrane
Superior horn of
thyroid cartilage
Median thyrohyoid ligament

Thyroid cartilage

Cricothyroid ligament

Cricoid cartilage

Cricotracheal ligament

Tracheal cartilages

B

Greater horn of hyoid

Thyrohyoid membrane
Thyrohyoid ligament
Superior horn of
thyroid cartilage

Inferior tip of epiglottis
Corniculate cartilage
Thyroid cartilage
Arytenoid cartilage

Inferior horn of
thyroid cartilage
Cricothyroid joint
Cricoid cartilage

Tracheal membrane

FIGURE 19–3   The larynx. A. Anterior view. B. Posterior view.

below. The **epiglottis** is a small leaflike flap of cartilage extending upward posterior to the base of the tongue. It is attached to the inside of the thyroid cartilage below, but its upper portion is free. When the larynx moves upward in swallowing, the epiglottis is tipped so as to deflect food and fluid away from the larynx and toward the esophagus. The **arytenoid cartilage** is paired. Each is a small triangular plate on the posterior wall of the larynx, above the cricoid cartilage. They are attached to the vocal cords and because they are slightly movable they are involved in the production of sound (see below). The **corniculate** and **cuneiform cartilage** are tiny bits of cartilage found at the top and in front of the arytenoid cartilage (see also Figure 19–4).

Viewed from the laryngeal cavity above it, the **laryngeal aperture** can be seen between the epiglottis and the arytenoid cartilages, with two pairs of horizontal tissue folds extending across the laryngeal cavity from the arytenoid cartilages to the inner surface of the

thyroid cartilage (Figures 19–4 and 6–10). The upper pair is the **ventricular folds** (*false vocal cords*), and the lower pair is the **true vocal cords.** The space between them on each side is the *ventricle.* The **glottis** is the opening between the two true vocal cords. It is through this opening that air must pass as it moves in and out of the lungs.

A mucous membrane lines the laryngeal cavity and covers the vocal cords. The vocal cords themselves are ligamentous bands of elastic connective tissue, and they are stretched across the opening under a certain amount of tension. The size of the glottis and the degree of tension on the cords can be adjusted by contraction of skeletal muscles to change, for example, the position of the arytenoid cartilages. As air passes through the glottis, it causes the true vocal cords to vibrate at a frequency depending upon their tension in much the same way that sound can be produced by flicking a stretched rubber band. The amplitude of the vibration is related to the force of the air movement.

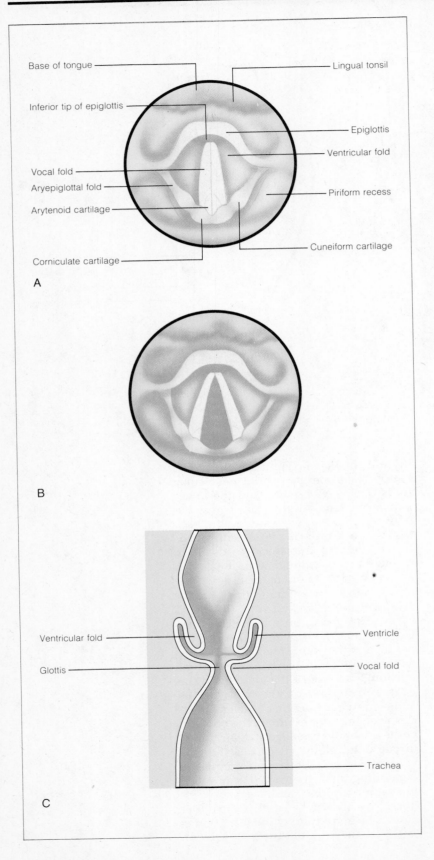

A

- Base of tongue
- Inferior tip of epiglottis
- Vocal fold
- Aryepiglottal fold
- Arytenoid cartilage
- Corniculate cartilage
- Lingual tonsil
- Epiglottis
- Ventricular fold
- Piriform recess
- Cuneiform cartilage

B

C

- Ventricular fold
- Glottis
- Ventricle
- Vocal fold
- Trachea

The skeletal muscles of the pharynx and oral cavity (cheeks, lip, tongue) shape the sound to produce speech.

The Lower Respiratory Tract

Below the larynx the respiratory system branches into smaller and smaller units, rather like the branchings of a tree (Figure 19–5). As in the circulatory system, a large trunk, here the trachea, divides into progressively smaller passageways until the sites of functional exchange, here the alveoli, are reached. The largest channels are solely for air transport and the presence of cartilage in their walls ensures that these airways do not collapse when air is drawn into the lungs. Other passages, like arterioles, are capable of changing their diameters, and have smooth muscle in their walls for this purpose. The smallest channels, like the capillaries, are very small, thin-walled, and numerous; they provide a large surface area for the rapid exchange of gases.

Unlike the circulatory system, the respiratory apparatus functions as a two-way system—air moves in and out through a single set of passages. The respiratory system is a low pressure system: the pressure of the air in it never fluctuates by more than a few millimeters of mercury during normal quiet breathing.

Trachea and Bronchi   The **trachea** is a tube approximately 2.5 cm in diameter and 10–12 cm long, lying just anterior to the esophagus. It is characterized by the presence of C-shaped rings of cartilage. The rings are joined by dense fibrous connective tissue, and there is some smooth muscle between the cartilage rings and, more prominently, across the opening of the "C." The trachea is lined with respiratory (pseudostratified ciliated columnar) epithelium containing numerous goblet cells. Other mucous-secreting glands are

FIGURE 19–4   The laryngeal cavity. A. Glottis closed. B. Glottis open. C. Diagram of a frontal section through the region of the glottis.

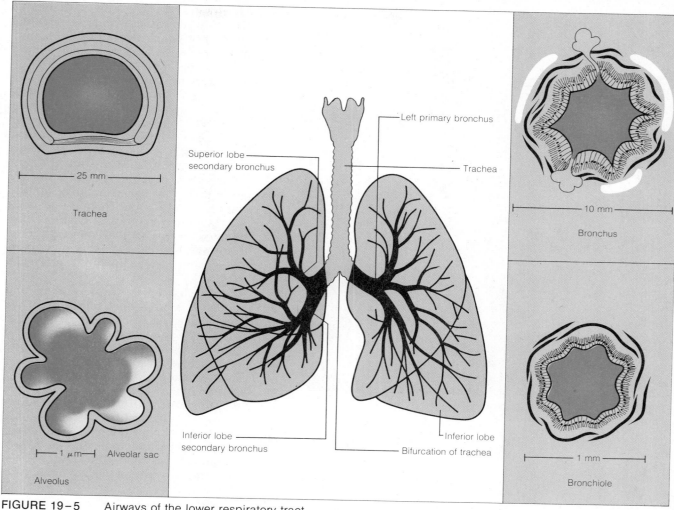

Superior lobe
secondary bronchus

Left primary bronchus

Trachea

25 mm

Trachea

10 mm

Bronchus

Inferior lobe
secondary bronchus

Inferior lobe

Bifurcation of trachea

1 μm Alveolar sac

Alveolus

1 mm

Bronchiole

FIGURE 19-5   Airways of the lower respiratory tract.

found in the deeper layers of the tracheal wall. In the lower respiratory tract the cilia beat upward toward the pharynx, which helps prevent foreign particles from entering the airways while clearing mucous secretions from those airways. The thyroid gland lies at the junction of the larynx and trachea (Figures 19–6 and 13–8).

Behind the arch of the aorta, the trachea bifurcates to form the two **primary bronchi** (see Figures 19–5 and 19–6). The left bronchus tends to go off at an angle, so that foreign objects in the trachea are more likely to lodge in the straighter right bronchus. The primary bronchi divide into **secondary bronchi,** which in turn divide into **tertiary bronchi.** There are two secondary bronchi

on the left and they are distributed to the two lobes of the left lung. There are three secondary bronchi on the right, and they go to the three lobes of the right lung. The bronchi continue to divide through more than a dozen "generations" of ever smaller airways.

The larger bronchi resemble the trachea in structure, although they are smaller. The rings of cartilage give way to cartilage plates, which become smaller and less frequent in the smaller bronchi. The epithelium also becomes thinner and the number of goblet cells decreases.

The Lung—Internal Structure   The smallest bronchi branch into **bron-**

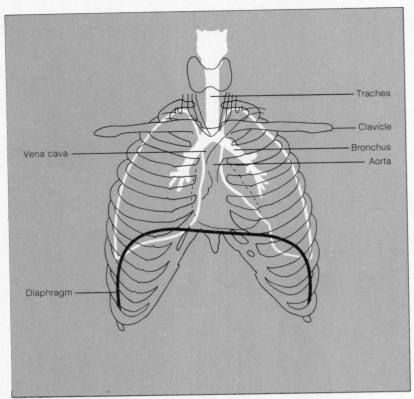

FIGURE 19-6 Relationship of the respiratory system to other thoracic structures.

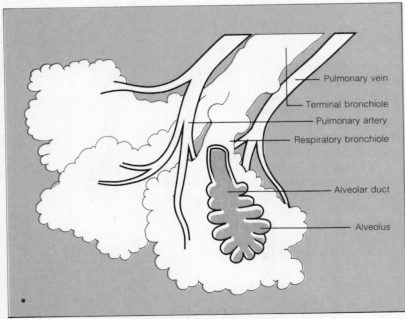

FIGURE 19-7 Terminal structures of the lungs.

chioles less than 1 mm in diameter. They have smooth muscle in their walls, but no cartilage. The epithelium is thinner still, lacks cilia, and contains no secretory cells.

From the terminal bronchioles arise **respiratory bronchioles,** which have air sacs, known as **alveoli,** scattered at intervals along their walls (Figure 19-7). The respiratory bronchioles end in **alveolar ducts,** whose walls are sometimes difficult to find because of the great number of alveoli opening from them. The alveoli are tiny rounded sacs clustered about the alveolar ducts like a bunch of grapes. Each alveolus is encased in a capillary net formed by terminal branches of the pulmonary artery and located in the interalveolar walls.

The barrier between capillary and alveolus is important, since it is the membrane across which oxygen and carbon dioxide must diffuse rapidly. It consists of the alveolar epithelium and its basement lamina, perhaps some interstitial fluid with a few scattered elastic fibers, and finally the capillary endothelium with its basement lamina (Figure 19-8). The two walls are fused in places, so that the entire membrane is very thin and delicate—the whole distance between alveolus and capillary is normally less than half a micron.

The Lung—External Structure
The lungs are somewhat conical in shape, with their apices (plural of *apex*) directed toward the top of the thoracic cavity. The bases of the lungs are concave and rest upon the diaphragm (see Figure 19-6). The lungs are bounded on three sides by the thoracic wall, which defines their **costal surface.** The medial surface, known as the **mediastinal surface,** bears the **hilus,** or root, of the lung through which the bronchi and the pulmonary and bronchial vessels enter and leave the lung.

Each of the two lobes of the left lung and the three of the right are ventilated by a secondary bronchus, and each secondary bronchus divides into two to five tertiary bronchi (a total of ten in each lung). Each tertiary bronchus supplies a region of lung tissue known as a **bronchopulmonary seg-**

ment. The specific relationships in the formation of the lung segments are of particular importance to the thoracic surgeon (Figure 19–9).

The structures of the lung—the bronchi, bronchioles, and larger blood vessels—do not receive their blood supply from the pulmonary circulation, but from tiny bronchial arteries that branch from the descending aorta and enter the lung at the hilus. The lung tissue nerve supply consists essentially of afferent fibers from stretch receptors in the lung and autonomic fibers innervating the smooth muscle of the bronchioles and blood vessels.

The Thoracic Cavity

The thoracic cavity is an airtight enclosure whose boundaries are formed by the vertebral column, rib cage, and diaphragm. The bony structure of the thorax and the muscles attached to it are discussed in Chapters 4 and 6, respectively. It is important to note here that these muscles are skeletal muscles and are therefore innervated by the somatic and not by the autonomic nervous system.

The Pleura The thoracic cavity is lined, and the lungs are covered, by a thin serous membrane known as **pleura.** The layer that lines the thoracic cavity is the **parietal pleura,** and the one that covers the lungs is the **visceral pleura.** The two are continuous with one another at the hilus of the lung and with the pericardium around the heart.

These membranes divide the thoracic cavity into several compartments. The visceral pleura encloses the two areas that contain the lungs (the **pulmonary** or **intrapulmonary spaces**). The area between the lungs is known as the **mediastinum.** Surrounding the lungs, between the parietal and visceral layers of pleura, and occupied only by a thin film of lubricating fluid, is the **pleural** or **intrapleural space.** It is only a potential *space* in that the two pleural surfaces contact one another with no gas or air between them, but it is an important space from a functional standpoint.

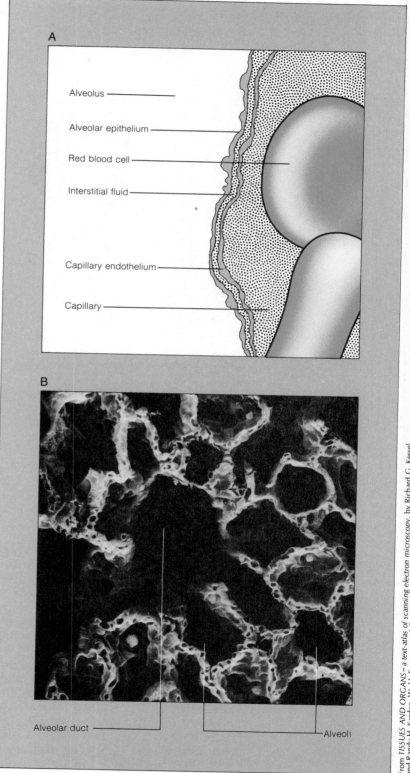

From *TISSUES AND ORGANS—a text-atlas of scanning electron microscopy*, by Richard G. Kessel and Randy H. Kardon, W. H. Freeman and Company, copyright © 1979

FIGURE 19–8   A. The alveolar membrane. B. Scanning electron micrograph of lung tissue, showing a cross section of an alveolar duct with several alveolar sacs opening into it. 615× magnification.

FIGURE 19–9
Bronchopulmonary
segments.

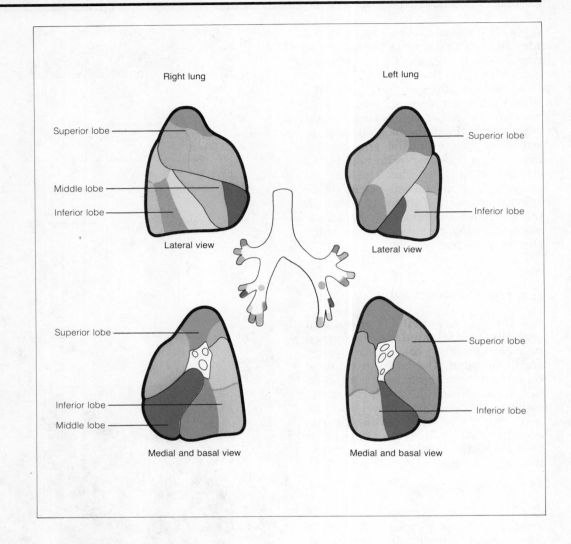

Mediastinum All the other structures of the thoracic cavity are in the mediastinum between the two lungs (Figures 19–10 and 19–11). The mediastinum is not a *structure* covered by pleura, but a *region* with pleura on either side. One of the major organs within it, the heart, is enclosed by the pericardium, which is continuous with the pleural membranes. The other structures in or passing through the mediastinum include the trachea and primary bronchi, the esophagus, the great vessels (aorta, vena cavae, and pulmonary vessels), azygos veins, several important nerves (vagus, phrenic, and sympathetics), the thoracic duct and other lymphatic vessels, and the thymus gland. These structures, with connective tissues that support them,

form an effective septum separating the right and left pleural cavities (see Figure 19–11). It is therefore possible to collapse one lung without greatly disturbing structures on the other side, although there is then some displacement of the soft tissues.

VENTILATION OF THE LUNGS

The respiratory apparatus is designed to bring fresh air into and remove "stale" air from the lungs. Only by maintaining or restoring the composition of the air in the alveoli can the respiratory system effect the necessary gas exchanges.

Mechanics of Breathing

Air must be taken into the alveoli and then forced to leave by the same way it entered. A pump, such as that employed by the circulatory system, would not be particularly effective here, where inflow and outflow are through the same passageways.

The Thoracic Cavity at Rest To understand the movement of air in and out of the lungs, certain anatomical relationships must be borne in mind. The thoracic cavity is airtight, with its flexible side walls and curved, flexible floor surrounding the lungs, which are open to the atmosphere (Figure 19–12). The lung surface and the thoracic wall are held in contact by the strong surface tension between the visceral and parietal pleura. This surface tension is increased by the thin film of lymphlike fluid that covers the pleural surfaces. Although these surfaces can slide against one another with little friction, they cannot be separated easily.

The lung itself is passive. The only smooth muscle is in the blood vessels and bronchioles, where it regulates their diameters. The lung is not capable of contracting or expanding by itself, but it does contain a significant amount of elastic tissue which tends to reduce the size of the lung by its recoil. If the lungs were removed from the body they would promptly collapse. They do not collapse in the body because the surface tension holding them against the thoracic wall is greater than the elastic forces within the lung.

Because the lung is constantly pulling away from the thoracic wall, the pressure in the intrapleural space is less than atmospheric pressure. It is a *suction*, or *subatmospheric pressure* (sometimes called a negative pressure). Measurements made during the interval between breaths have shown it to be about −4 mm Hg (or 756 mm Hg, if atmospheric pressure is 760 mm Hg). The pressure within the lungs in the intrapulmonary space is zero (or 760 mm Hg). That space is open to the atmosphere; if it were not at atmospheric pressure there would be a pressure gradient and air would move in or out of the lungs to establish equilibrium.

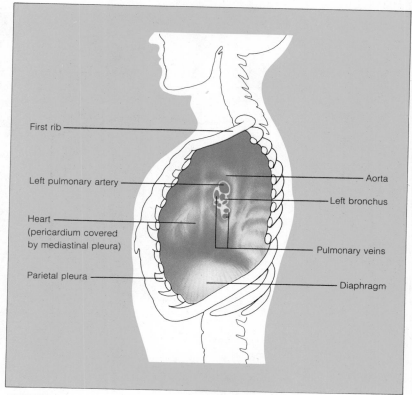

FIGURE 19–10 Structures of the mediastinum, as seen from the left (left lung removed).

The subatmospheric intrapleural pressure develops shortly after birth. During fetal life the lungs are collapsed and contain no air, and the rib cage is also partially collapsed. At birth the thoracic wall expands and the lungs are literally pulled out with it into the semistretched position which they will maintain throughout life. The subatmospheric pressure that is then created in the intrapleural space is the same pressure mentioned previously as an important aid in returning blood to the thoracic cavity and hence to the heart.

Inspiration The movement of air into the lungs is known as **inspiration.** It is accomplished by increasing the size of the thoracic cavity, which further expands the lungs and lowers slightly the pressure of the air in them to produce a pressure gradient between the atmosphere and the lungs. Air then moves into the lungs until there is equilibrium.

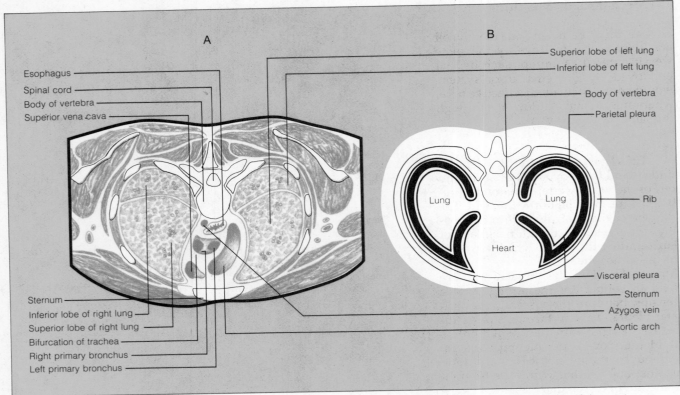

FIGURE 19-11   A. Cross section of the thorax at the level of the arch of the aorta.
B. Diagram of a cross section of the thorax, showing the pleura.

There are two ways to enlarge the thoracic cavity. The arched floor may be lowered and flattened to increase the vertical dimension, and the walls may be expanded to increase the antero-posterior dimension. The first of these is accomplished by contracting the diaphragm, the dome-shaped skeletal muscle which separates the abdominal and thoracic cavities (see Chapter 6). Contraction of the diaphragm pulls its central portion down and flattens the dome. The resulting displacement of the abdominal contents is accommodated by reflex relaxation of the abdominal muscles and bulging of the anterior abdominal wall.

In the resting position, the ribs extend diagonally down and forward from their dual articulation with the vertebral column. When they move there is a slight rotation at these junctions such that the anterior (sternal) ends of the ribs are brought forward and elevated. The volume of the thoracic cavity is then increased because the antero-posterior dimension is larger. This is accomplished by the action of skeletal muscles (Chapter 6). The external intercostals extending diagonally down and forward between each pair of ribs act to pull each rib up toward the one above it (Figure 19-13). Several other muscles, such as the scalene muscles, the sternocleidomastoid, and the pectoralis minor, also can help raise the rib cage.

Normal quiet inspiration is produced chiefly by contraction of the diaphragm. This is **diaphragmatic breathing.** Elevation of the rib cage—**costal breathing**—is added when ventilation is increased, as in exercise. Less effort is required for diaphragmatic breathing, and singers prefer it because a finer control of breathing can be achieved.

Expiration The movement of air from the lungs is **expiration,** and in quiet breathing it is a passive process. When the muscles of inspiration relax, the rib cage returns to its resting position, aided by the weight of the rib cage itself. When the diaphragm relaxes, it rises to its former position, aided by the subatmospheric intrapleural pressure pulling it upward and by the pressure of the abdominal contents below.

The decrease in the size of the thoracic cavity permits the elastic fibers of the lung to recoil accordingly, raising the pressure in the lungs (the intrapulmonary pressure) and pushing the air out of the lungs until equilibrium with the atmospheric pressure is reestablished. In quiet breathing expiration is followed by a brief pause before the next inspiration.

During heavy breathing, as in exercise, or forced expiration, as in playing a wind instrument, expiration may be aided by skeletal muscle contractions. Some of the muscles that contribute to expiration are the internal intercostals, which extend diagonally up and forward, and the anterior abdominal muscles, which help to return the diaphragm to its elevated resting position. Other muscles are also involved; in vigorous activity most of the trunk muscles participate in respiratory efforts.

Work of Breathing To bring air into the lungs, the body must increase the size of the thoracic cavity and expand the lungs. The expansion, brought about by skeletal muscle contraction, requires the expenditure of energy. The amount of energy used depends upon the resistance of the lungs and thorax to the movement. We noted earlier how the distensibility of the cardiac ventricle affects its filling; distensibility has a similar effect on inflation of the lungs. **Compliance** is the term used to describe the ease with which the lungs can be inflated, the "adjustability" of the lungs and thorax. It is defined in terms of the change in lung volume produced by a given amount of pressure. Compliance is high when a small increment of pres-

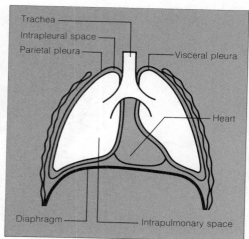

FIGURE 19–12   A "simplified" thoracic cavity, showing the spaces within it.

sure causes a large increase in lung volume; the lungs and thorax distend easily and comply readily to the pressure increase. Compliance is low when they resist stretch, and more pressure is required to inflate the lung.

Several factors contribute to lung compliance. One of these is the elasticity of the lung tissue itself. If the elasticity is lost, the lung does not stretch easily, and either more pressure is required to move air into them or the volume of air moved is greatly reduced. Another factor is the resistance to air flow, chiefly in the bronchi and bronchioles. Any obstruction of an airway or constriction of bronchioles would increase the pressure required to force air through. It would also increase the time required to move air in and out of the alveoli. Accumulation of fluid in the lungs (in the alveoli or the interstitial spaces) also interferes with inflation.

One of the most important elements of lung compliance is related to surface tension within the alveoli. Recall that surface tension is due to forces of attraction between the molecules on a fluid surface. It tends to pull the surface molecules into the smallest possible space; it is what makes a drop of water assume a spherical shape. The strength of these forces of attraction is inversely proportional to the radius of the sphere, and thus surface tension is

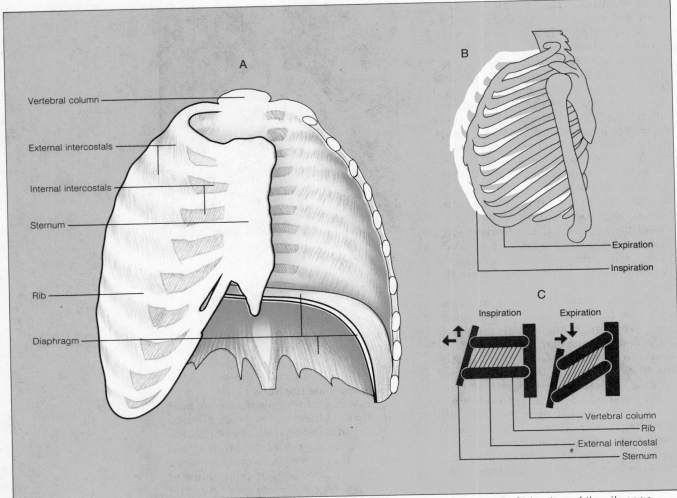

FIGURE 19-13   A. Muscles of respiration, anterior view. B. Side view of the rib cage, showing the action of the intercostal muscles. C. A mechanical model to show movements of the rib cage.

greater in small droplets. A fluid film coats the inner surface of the alveoli. This fluid exerts a surface tension which tends to collapse alveoli, particularly impairing the expansion of the smaller alveoli. If a terminal airway opens into two alveoli, one large and one small, the surface tension is much greater in the small one (Figure 19-14). Air forced into this passageway tends to enter the large alveolus, where the surface tension is less and the sac distends more readily. Such unequal expansion of alveoli in the lung is not optimal for the exchange of oxygen and carbon dioxide, because a few very large alveoli do not provide nearly as much surface area for exchange as do

many smaller ones. An even greater danger, however, is that high surface tension in small alveoli might be enough to collapse them and force all of the air out. Fortunately, there are cells in the alveolar epithelium that secrete a substance which reduces surface tension in alveoli. It is a surface-active material known as *surfactant*. It has detergent properties which decrease the collapsing tendency of the smaller alveoli, and which enable them to inflate under much less pressure. Pulmonary surfactant increases compliance and reduces the work of breathing.

Newborn infants are occasionally afflicted with a condition known as *respiratory distress syndrome*, or *hyaline membrane disease.* These infants usually have many collapsed alveoli, and the condition is believed to be due to some condition or stress that destroys the surfactant or interferes with its production.

Summary of Pressure and Volume Changes During a Breath In normal quiet breathing, the pressure in the lungs equals the atmospheric pressure during the interval between breaths (end of expiration) and momentarily at the end of inspiration, when there is no air movement in the respiratory passages. This can be seen in Figure 19-15, where the volume of air moving into and out of the lungs is shown by the tidal volume curve. The intrapleural pressure also fluctuates a few millimeters, but in quiet breathing it is always subatmospheric. It falls when the thoracic cavity is expanded, and rises toward the atmospheric pressure during expiration. In a forced expiration with the glottis closed (as in straining) or against a resistance (as in playing a wind instrument), the pressure in the thorax may become positive; it may go as high as 100 mm Hg. When pressure in the intrapleural space is greater than that in the veins, blood flow into the thoracic cavity is impaired and venous return may be severely reduced. It shows in the veins of the face, which become greatly distended. When the expiratory effort ceases, and all of the dammed blood is allowed to rush into the thoracic cavity and to the heart, the ventricle may be overloaded. Straining, as in lifting or defecation, can be dangerous to a person with a weakened heart.

Lung Volumes

Once the lungs have expanded and filled with air in the first breath, they are never again completely empty—you cannot exhale *all* of the air in your lungs. Of the air that can be moved in and out, only a small portion is actually exchanged during normal quiet breathing.

As indicated in Figure 19-16, the **total lung capacity** is the maximum

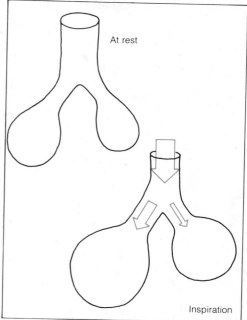

At rest

Inspiration

FIGURE 19-14 Surface tension is less in large alveoli than in small ones, and large alveoli are more easily distended, causing unequal ventilation.

amount of air that the lungs can hold. It includes the **residual volume** that cannot be exhaled, as well as all the air that can be inhaled. At the end of a maximal inspiration, the **vital capacity** of about 4800 ml has been added to the residual volume of 1200 ml, for a total lung capacity of about 6000 ml. The vital capacity is the maximum amount of air that can be moved in and out of the lungs. During normal quiet breathing, about 500 ml of **tidal air** is moved in and out with each breath. A maximal inspiration adds the **inspiratory reserve,** while a maximal expiration removes the **expiratory reserve.** Vital capacity is therefore the sum of inspiratory reserve, tidal air, and expiratory reserve. The functional residual capacity (residual volume plus expiratory reserve) is the volume of air in the lungs at the end of a normal expiration. It is the amount of air to which the tidal volume is added, and it represents the resting position of the lungs and thorax.

The actual values of these lung components vary greatly with the size and general body build and sex of the individual (Table 19-1). The tidal volume is only about 10-15 percent of the

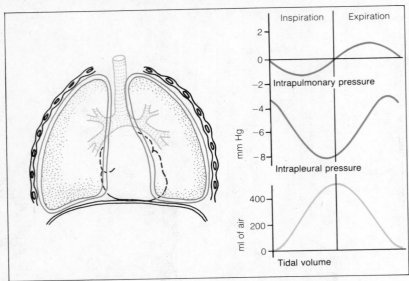

FIGURE 19–15   Pressure changes during quiet breathing.

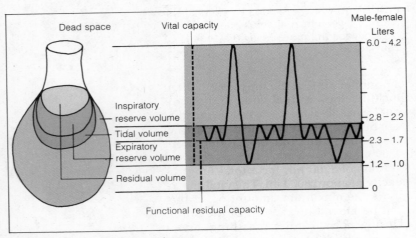

FIGURE 19–16   Lung volumes.

(Modified and reproduced, with permission, from Comroe, J. H., Jr., et al., *The Lung: Clinical Physiology and Pulmonary Function Tests,* 2nd ed. Copyright © 1962 by Year Book Medical Publishers, Inc., Chicago.)

vital capacity and lies just about in the middle of the lung capacity, leaving a great reserve for increasing the amplitude of either inspiration or expiration.

The exchange of gases occurs only in the alveoli. As inspiration begins, the first air to enter the alveoli is not "fresh" air. It is air that was left in the airways at the end of the last expiration. The last portion of an inspiration does not reach the alveoli, but remains in the airways and is the first to be exhaled. The air that does not participate in the gaseous exchange occupies about 150 ml of *dead space.* Since the tidal volume is about 500 ml, of which 150 ml is dead space, only about 350 ml of air from the outside reaches the alveoli with each breath. If the tidal air is only about 350 ml, the alveolar exchange involves only 200 ml. From this it can be seen that very shallow breathing may not provide adequate alveolar ventilation. A panting dog moves dead space air back and forth, and must frequently stop panting to take a few deep breaths to ventilate the alveoli. The 350 ml or so of "fresh" tidal air enters alveoli that already contain about 2500 ml of air—the residual volume plus the expiratory reserve plus dead space air—so that it is greatly diluted in the alveoli. A tidal volume of 500 ml and a typical breathing rate of 10–16 breaths each minute would provide a **pulmonary ventilation,** or **respiratory minute volume,** of between 5 and 8 liters of air per minute (the alveolar ventilation would be 1.5–2.4 liters per minute less). Although respiratory function depends upon the air that gets to the alveoli, pulmonary ventilation is measured more often than alveolar ventilation. Pulmonary ventilation refers to the amount of air you breathe each minute in order to meet your requirements for oxygen uptake and carbon dioxide removal.

Ventilatory Problems

Because the lungs are kept inflated by the surface tension existing between the pleural membranes, any interference with that action will end the inflation. For example, if an injury damages the thoracic wall so that air *can* enter the thoracic cavity, air *will* enter until the intrapleural pressure is the same as atmospheric pressure (the presence of air in the intrapleural space is known as *pneumothorax*). When that happens, the lung on the damaged side will be collapsed. Fortunately, the mediastinal structures form a barrier so that the other lung need not collapse. Because it is sometimes desirable to rest one lung by inactivity, a physician may inject air into one intrapleural space, thus allowing that lung to collapse.

When thoracic surgery is performed and the thorax is opened, the lungs collapse. In spite of the activity of the muscles of inspiration, the lungs cannot inflate; ventilation must then be carried out by other means.

Artificial Respiration When for any reason respiration ceases, ventilation of the lungs can be achieved artificially. The *respirator* is one effective method. With this machine, air (or oxygen) is intermittently forced into the respiratory system (through a face mask) until the lungs are inflated. The pressure is then released to allow the lungs to collapse partially (expiration). Of the many methods that can be applied when a machine is not available, *mouth-to-mouth resuscitation* is the most effective. The operator rhythmically forces air from his or her lungs into the lungs of the victim to inflate them, allowing time for expiration in between. Other, less satisfactory methods involve intermittent pressure applied to the victim's back or sternum to force air out, alternated with raising the arms to enlarge the thorax and draw air in.

In cases of paralysis of the muscles of inspiration, the *iron lung* has been used. It consists of a large airtight chamber into which the patient's body (below the shoulders) is placed. As pressure in the chamber is lowered, the thoracic cage is enlarged and air enters the lungs. As pressure in the chamber is raised, the thoracic wall lowers and air leaves the lungs.

Nonrespiratory Air Movements

There are several causes of movement of air through the respiratory passages that are not related to alveolar ventilation. Coughing and sneezing, for example, are protective reflexes which tend to dislodge or remove substances from the airways. *Sneezing* is usually the result of chemical or mechanical irritation in the upper respiratory tract; it consists of an inspiration followed by a sudden explosive expiration. *Coughing* is elicited by stimuli arising from the region of the epiglottis, larynx, trachea, or bronchial passages. It consists of an inspiration followed by a build-up of pressure within the thoracic cavity with the glottis closed. The glottis is then opened and air is expelled with great force to dislodge the offending material. Coughing is controlled and coordinated through a cough center located in the medulla oblongata. The *hiccup* is a short abrupt inspiration caused by a brief contraction of the diaphragm. It too is a reflex, usually elicited by stimulation of nerve endings in the digestive tract or abdominal cavity.

Yawns and *sighs* are both deep inspirations followed by expiration. Yawning is usually more prolonged and accompanied by a widely opened mouth. The cause is poorly understood, but it is thought to be related to reduced oxygen tension in the blood. The deep inspiration should improve ventilation of alveoli that had been poorly *aerated* in quiet breathing.

Laughing and *crying* are relatively similar, since both consist of an inspiration followed by a series of short expirations. Each is accompanied by appropriate sounds and facial expressions, although sometimes it is difficult

TABLE 19-1   TYPICAL LUNG VOLUMES IN HEALTHY YOUNG ADULTS

Volume	Male (in ml)	Female (in ml)
Total Lung Capacity	6000	4200
Vital capacity	4800	3200
Inspiratory reserve	3200	2000
Tidal volume	500	500
Expiratory reserve	1100	700
Residual volume	1200	1000
Functional residual capacity	2300	1700

to distinguish one from the other.

Talking and singing are both produced during expiration. Although sound can be produced during inspiration, there is little control over that sound or its quality. Both speech and singing require an extremely fine control over the expiratory process.

CHAPTER 19 SUMMARY

Structure of the Respiratory Apparatus

The respiratory tract carries air to and from the lungs. The upper respiratory tract includes the nasal cavity, pharynx, and larynx. The lower portion includes the trachea and airways that lead to the alveoli.

The *nasal cavity* is a small space with a large surface area. It warms and moistens air, its ciliated epithelium helps remove foreign particles, and it contains the receptors for the sense of smell. The paranasal sinuses (frontal, maxillary, ethmoidal, and sphenoidal) and the nasolacrimal ducts open into it.

The *pharynx* is made up of the *nasopharynx*, which receives the auditory tubes from the middle ear, the *oropharynx*, and the *laryngopharynx*, both of which are shared with the digestive system. The laryngopharynx opens into the esophagus and larynx.

The *larynx* contains the apparatus for producing sound. It is attached to the hyoid bone and is formed by four major cartilages (the unpaired *epiglottis, thyroid,* and *cricoid,* and the paired *arytenoids*), plus ligaments, membranes, and skeletal muscles. The *vocal folds* (vocal cords) are fibrous bands of tissue that extend across the airway at the level of the thyroid cartilage. The space between them, through which air passes, is the *glottis.* Air moving past the vocal folds causes them to vibrate and produce sounds. Contraction of laryngeal muscles can vary the size of the glottis and the tension on the vocal cords, which modifies the sounds and helps make them recognizable.

The wall of the trachea contains C-shaped cartilage rings which prevent its collapse. It branches into right and left *primary bronchi*, which branch into *secondary bronchi*, then *tertiary bronchi*, and more. Further branching leads to *bronchioles, respiratory bronchioles, alveolar ducts*, and finally *alveoli*. The amount of cartilage in the walls diminishes in the smaller bronchi, and is absent from bronchioles which contain relatively more smooth muscle. Alveoli are the tiny rounded sacs where gas exchange occurs. Some of them open into respiratory bronchioles, but most of them open into alveolar ducts. Their walls are exceedingly thin and contain numerous capillaries of the pulmonary circulation.

The *thoracic cavity* is enclosed by the rib cage and diaphragm. It is lined with *parietal pleura*, which is reflected back over the lungs as *visceral pleura.* The *intrapleural space* is an almost hypothetical space between the two layers of pleura, and the *intrapulmonary space* is within the lungs. The *mediastinum* is the region between the lungs, and it contains the heart, large arteries and veins, trachea and primary bronchi, esophagus, and nerves and lymphatics. It forms an effective septum between the right and left sides of the thoracic cavity.

Ventilation

The thoracic cavity is airtight, except for the tracheal opening. During the interval between breaths (end-expiration) the lungs are open to the outside and the pressure in them is equal to that of the atmosphere. Surface tension keeps the visceral and parietal layers of pleura in contact with one another, which means the lungs are somewhat stretched. Their natural elasticity causes them to pull away as if to collapse. The result is that intrapleural pressure is normally a few millimeters below atmospheric pressure. If air is allowed to enter the intrapleural space *(pneumothorax)* through injection or accident, intrapleural pressure will no longer be subatmospheric and the lung(s) will collapse.

Inspiration is brought about by contraction of the diaphragm which lowers the floor of the thoracic cavity,

Focus

The Heimlich Maneuver

Until the middle of 1974, it was not unheard of that a restaurant patron might get a fishbone or other food fragment caught in his or her throat and choke to death. The only help available might be no more than a companion who could pound the victim's back, and that was not very effective. Physicians, with their tools for reaching into the throat and extracting the obstacle or for performing an emergency tracheotomy, were not generally on hand.

Since that date, fatal choking cases have become much less common. The reason is that in June 1974 Dr. Henry J. Heimlich introduced a first aid technique for choking that is so effective, simple, and easy to use that since then many states have passed laws insisting that at least one member of every restaurant's staff know how to use it. This technique is the "Heimlich maneuver."

The Heimlich maneuver can be used on sitting, standing, or lying choking victims. With a standing victim, the rescuer stands behind the victim, puts his or her arms around the victim's waist, and grasps one fist with the other hand. The rescuer then places the thumb side of the fist just above the victim's navel and presses the fist into the abdomen with a quick, upward thrust, repeating as necessary. The technique is similar with a sitting victim.

With a supine victim, the rescuer kneels astride the victim's thighs, places both hands, one atop the other, on the victim's abdomen just above the navel, and presses into the abdomen, again with a quick, upward thrust. Again, this action is repeated as necessary.

The Heimlich maneuver works as well as it does because even if a victim has exhaled all the air he or she possibly can in the course of choking, there is some air left in the lungs. The quick, upward thrust into the abdomen pushes the diaphragm upward, which compresses the lungs. This forces the air in the lungs out the bronchi, trachea, and mouth, often in a quick enough surge to expel the material obstructing breathing. The maneuver does not always work, of course, since the material can be too tightly wedged in the airway and since people do choke for other reasons than wrong-way swallows. It also does not work if the thrust into the abdomen is not abrupt enough; too gentle a thrust will not develop enough air pressure to loosen and expel the blockage in the throat.

Because the Heimlich maneuver is so simple, death by choking is now often preventable. Everyone who cares about their friends and family should be familiar with the technique and ready to use it, for choking is all too common an event.

and contraction of the external intercostal and other muscles which raise and expand the rib cage. These actions increase the size of the thoracic cavity and intrapleural pressure falls as the lungs are further expanded. As the volume of the lungs is increased, the intrapulmonary pressure falls, creating a pressure gradient that causes air to enter the lungs.

Expiration is brought about by relaxation of the muscles of inspiration; the diaphragm rises and the rib cage returns to resting position. This reduces the size of the thoracic cavity and raises pressure in the lungs above that in the atmosphere, and air moves out of the lungs.

Lung Volumes

The air in the lungs can be partitioned into several volumes. The *total lung capacity* of about six liters consists of the residual volume and the vital capacity. The *residual volume* remains in the lungs at the end of a maximum expiration; it cannot be expelled. The *vital capacity* is the maximum volume of air that can be moved, from a maximum inspiration to a maximum expiration. It includes the tidal volume, inspiratory and expiratory reserves. The *tidal volume* is the air moved in each normal breath and is about ten percent of the total lung capacity. *Inspiratory reserve* is the volume that can be inspired beyond a normal inspiration, and *expiratory reserve* is the volume that can be exhaled beyond a normal expiration. The *functional residual capacity* is the air in the lungs at the end of a normal expiration, and includes the expiratory reserve and residual volume. The *dead space* is air in the airways (trachea, bronchi, bronchioles, etc.) at the end of an inspiration or expiration; it does not participate in gas exchange because it never reaches the alveoli.

STUDY QUESTIONS

1 What are some of the functions of the nasal cavity, aside from that of being a passageway? How is its structure adapted to perform these functions?

2 How does the structure of the larynx enable it to produce sounds?

3 What are some of the similarities and differences between the walls of the air passages and of the blood vessels of similar sizes?

4 What is the mediastinum? What is found in it? What are its relationships with the pleura?

5 Under normal conditions: (a) why don't the lungs collapse? (b) in quiet breathing, how do you get air to enter the lungs? (c) how do you get air to leave the lungs?

6 What changes would you expect in the intrapulmonary and intrapleural spaces when you try to blow up a balloon (forced expiration)? When you try to draw a thick mix up through a straw?

7 How is it that contraction of the diaphragm increases the capacity of the thoracic cavity? What is the role of the abdominal muscles in breathing?

8 In mystery and spy adventure stories, the hero sometimes escapes by completely submerging and breathing through a reed to the surface. Can you think of any practical limitations to such a tactic?

9 What differences (if any) would you expect if the lung volumes were measured in a reclining subject instead of a standing subject? Explain.

10 What is the significance of the functional residual capacity? of the residual volume? What would be the effect of an increase in residual volume?

20

Properties of Gases

Diffusion of Gases

Transport of Oxygen

Transport of Carbon Dioxide

Role of Respiration in Acid—Base Balance

Gas Exchange and Transport

Only a small fraction of the air we breathe is oxygen, yet our cells can use only oxygen and we must somehow extract that oxygen from the air and transport it to the cells. Oxygen is transported in the blood, of course, but as a gas; it cannot simply be carried as tiny bubbles in the blood, because gases occupy different volumes when the pressure changes. This also applies to carbon dioxide produced in the tissues.

This chapter deals with the means by which these gases get into and out of the blood, and how they are transported in it. Some of the properties of gases are discussed, particularly those properties of oxygen and carbon dioxide that affect their behavior in the body. One of the most important of these relates to the behavior of a gas in a liquid. The blood plasma by itself cannot carry enough oxygen or carbon dioxide to meet our metabolic requirements, and special mechanisms are necessary to increase the carrying capacity of the blood. Red blood cells play an essential role in providing the means to transport adequate amounts of oxygen and carbon dioxide.

Chapter 20, in dealing with the transport and delivery of sufficient amounts of oxygen and carbon dioxide, is really about those aspects of respiration that are carried out by the cardiovascular system.

PROPERTIES OF GASES

Molecules of gas are in constant random motion, colliding with one another and anything else in their paths. Enclosed in a chamber they exert a pressure or tension against its walls. The amount of pressure, measured in millimeters of mercury, is related to the number of molecules of gas per unit volume and to the temperature of the gas. If the number of molecules remains constant, pressure is greater when the gas is heated or the volume of the chamber is reduced. Likewise, the pressure is less when the gas is cooled or the chamber is enlarged. If the chamber contains a mixture of gases, the pressure exerted by each gas is related to the number of molecules of that gas alone and is independent of any other gases that may be present.[1] This pressure is the **tension** or **partial pressure** of that gas, and is written as P_{O_2} (for oxygen) or P_{CO_2} (for carbon dioxide). The total pressure of a mixture of gases is the sum of the partial pressures of all the gases in the mixture.

Air is a mixture of gases, chiefly nitrogen and oxygen. Since the volume occupied by a sample of air depends upon its pressure and temperature, gas volumes are measured under well-defined standardized conditions—standard temperature, standard pressure, and dry—STPD. Thus at sea level, the pressure of the atmosphere is 760 mm Hg when the temperature is 0°C (273°K) and the air contains no water vapor; these are the standard conditions. The partial pressures of the individual gases in the air add up to the pressure of the atmosphere.

Knowing the partial pressure of each gas in a mixture, it is possible to determine its percent concentration.

TABLE 20-1   COMPOSITION OF DRY AIR

Gas	Partial pressure in mm Hg	Volumes percent
Nitrogen	600.6	79.03
Oxygen	159.1	20.93
Carbon dioxide	0.3	0.04
Total	760.0	100.00

Thus air is mostly nitrogen[2] (600/760 or 79 percent) and oxygen (159/760 or nearly 21 percent) with very little carbon dioxide (less than 1 mm Hg or 0.04 percent).

The percentages given in Table 20–1 are volumes percent. For example, 100 liters of air would contain 79 + liters of nitrogen, nearly 21 liters of oxygen, and about 40 ml of carbon dioxide.

Changes in the atmospheric pressure bring corresponding changes in the partial pressure of each gas. For example, at 18,000 feet the pressure of the atmosphere is reduced by exactly half (Mount McKinley in Alaska is a little more than 20,000 feet high). The partial pressure of each gas is also reduced by half, but the percent concentration is not changed. The partial pressure of oxygen (P_{O_2}) would be only 80 mm Hg, but it would still account for nearly 21 percent of the air (80/380). Dry air is roughly one-fifth oxygen and four-fifths nitrogen at any atmospheric pressure.

DIFFUSION OF GASES

The uptake of oxygen and the release of carbon dioxide in the lung occurs by diffusion, as do the release of oxygen and the uptake of carbon dioxide in the tissues. Diffusion, however, is a passive process which occurs only

---

[1]The laws that define the relationships between pressure, temperature, and volume of a gas are expressed in several statements known as the *gas laws*:
1. Equal volumes of different gases at the same temperature and pressure contain the same number of molecules (*Avogadro's Law*).
2. At constant temperature, the volume of a gas varies inversely with the pressure (*Boyle's Law*).
3. At constant pressure the volume of a gas varies directly with the absolute temperature (*Charles' Law*).

[2]Included with the nitrogen are certain rare and inert gases which are present in very minute amounts. In these days of concern about air pollution, it might be added that some of these inert gases may not be so inert. Their concentrations are still relatively low, being counted in some cases in parts per million.

when suitable conditions exist. These conditions are related to the permeability of the membranes which *permit* diffusion and to the gradients which *cause* diffusion. The membranes involved are basically very permeable to the respiratory gases, but several factors can decrease gas exchange across them. In the previous chapter we saw that the alveolar walls are very thin, with pulmonary capillaries sandwiched into the delicate interalveolar septa. Both the alveolar lining and the capillary walls consist of thin squamous cells with little else between them, so that the diffusion distance from alveolus to capillary is very short. The rate of diffusion is reduced by anything that increases the diffusion distance, such as a thickening of the wall of the presence of fluid in the alveoli (which also decreases the amount of air they can hold). Alveoli are very small, but they have an extensive total surface area—about 70 m², nearly 40 times the total area of the body surface. If the interalveolar membranes break down, forming larger but fewer alveoli (as in emphysema), the gaseous exchange is impaired because of the reduced area of the diffusion surface.

The driving force for diffusion is a concentration gradient across the membrane. All else being equal, the greater this gradient the faster the diffusion. In the lung, the diffusion gradient is the difference in the partial pressures of the gases in the alveoli and in the blood. The former depend upon and are maintained by the ventilation of the alveoli. If ventilation is uneven, alveoli in certain segments of the lung may be poorly aerated. Gas tensions in these alveoli will become more like those of the venous blood, and as the gradients decrease, so will the rate of gas diffusion. But even ample ventilation can be wasted if the alveoli do not receive an adequate supply of blood for the oxygen to diffuse into and carbon dioxide out of. Only where both ventilation and perfusion are adequate do the necessary gradients exist for the exchange of gases (see Figure 20–1).

Under normal conditions, the overall gradients are such as to cause diffusion of oxygen from the atmosphere to

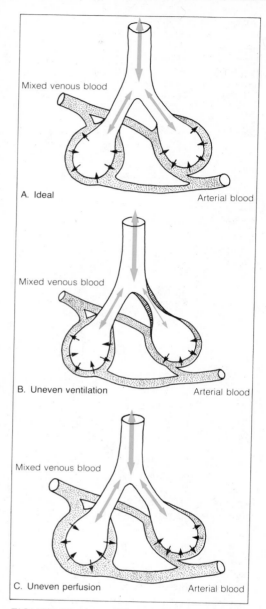

FIGURE 20–1   Effect of nonuniform ventilation or perfusion on gas exchange.

(Modified and reproduced, with permission, from Comroe, J. H., Jr., et al., *The Lung: Clinical Physiology and Pulmonary Function Tests*, 2nd ed. Copyright © 1962 by Year Book Medical Publishers, Inc., Chicago.)

the tissues and of carbon dioxide from the tissues to the atmosphere. The pressures shown in Figure 20–2 and Table 20–2 are typical normal values, although the tissue pressures vary widely. During vigorous exercise the tissue P_{O_2} is lower than that shown in the figure (it may approach zero in

TABLE 20-2   PARTIAL PRESSURE OF GASES IN THE AIR
AND BODY AT SEA LEVEL* IN MM HG

Gas	Dry air	Moist air (trachea)	Alveolar air	Arterial blood	Venous blood
Oxygen	159.1	149.2	104	100	40
Carbon dioxide	0.3	0.3	40	40	46
Water vapor	0.0	47.0	47	47	47
Nitrogen	600.6	563.5	569	573	573
Total pressure	760.0	760.0	760	760	706

*Blood values are mean values in healthy young men.

Source: J. H. Comroe, Jr., *Physiology of Respiration*, 2nd ed. p. 9. Copyright © 1974 Year Book Medical Publishers, Inc. Used by permission.

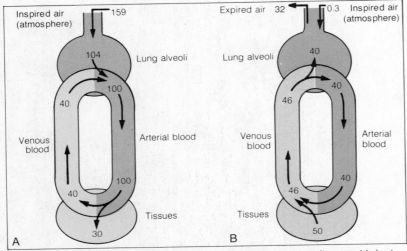

FIGURE 20-2   Partial pressures of respiratory gases (in mm Hg) at several sites between the atmosphere and the tissues. A. Oxygen. B. Carbon dioxide.

active muscles) and the P_{CO_2} is higher. These variations increase the gradients for both gases and enhance diffusion of both.

A drop of blood traverses the pulmonary capillaries in less than one second, but the diffusion is rapid. About 250 ml of oxygen and 200 ml of carbon dioxide are exchanged each minute. Enough oxygen is taken up to bring the venous oxygen tension to within a few millimeters of that in the alveoli, and enough carbon dioxide is released to bring the P_{CO_2} down to that in the alveoli. With this exchange, the "venous" blood becomes "arterial" blood.

When two gases (for example, 100 percent Gas A and 100 percent Gas B)

are allowed to come to equilibrium across a membrane, the final partial pressures are the same on both sides, as are the percentage concentrations (see Figure 20–3). The situation is different when a gas (oxygen in air) comes into equilibrium with a liquid (oxygen dissolved in water). At equilibrium, the partial pressures of the gas in the air and in the liquid are the same, but the gas is still 100 percent gas, while the liquid is obviously not 100 percent gas. The *amount* of gas (in volumes) that goes into solution in the liquid varies with the partial pressure and is greater when the partial pressure is high. Of equal importance, however, is the *solubility* of that gas in the liquid, for it determines the volume of gas that dissolves in the liquid. Carbon dioxide is about 20 times more soluble in water than is oxygen, so at equal partial pressures 20 ml of carbon dioxide will dissolve in water for every 1 ml of oxygen that enters solution.

The ability of blood to transport oxygen and carbon dioxide depends upon the ability of oxygen to diffuse into the blood in the lungs and of carbon dioxide to diffuse into it in the tissues. The conditions that exist in the body are such, however, that the solubilities and gradients are not great enough to transport the necessary amounts of either oxygen or carbon dioxide. For example, 100 ml of water can dissolve nearly 5 ml of oxygen at STPD, but at body temperature only half as much oxygen is dissolved. Since air is only 20 percent oxygen, the capacity is reduced to 0.5 ml of oxygen for every 100 ml, or 5.0 ml per liter, of water or plasma. At a cardiac output of 5 liters of blood per minute, less than 15 ml of oxygen could be delivered to the tissues each minute. At rest the body consumes oxygen at about 250 ml per minute; in exercise the body may consume up to 4 *liters* per minute. The volume of plasma that would be necessary to provide for these demands is completely out of the question. Yet these demands are met: oxygen must be transported by some means other than physical solution in plasma. Table 20–3 shows the actual content of oxygen and carbon dioxide in the blood and in the air.

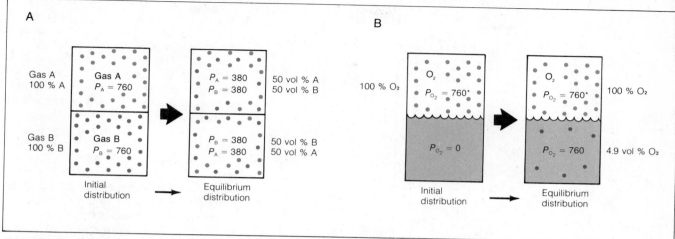

FIGURE 20–3 A. Diffusion of a gas in a gas. B. Diffusion of a gas in a liquid.
*The final P_{O_2} actually would be slightly less than 760 mm Hg because some oxygen molecules leave the gaseous phase as they dissolve in the liquid.

TRANSPORT OF OXYGEN

Transport of a sufficient amount of oxygen in the blood is made possible by the presence of **hemoglobin** in the red blood cells. Hemoglobin consists of an iron-containing pigment, **heme,** which is bound to a protein, **globin.** It has a great affinity for oxygen, with which it forms a loose reversible combination known as **oxyhemoglobin** (see Chapter 14):

$$Hb + O_2 \rightleftharpoons HbO_2$$

Each gram of hemoglobin can combine with 1.34 ml of oxygen. If 100 ml of blood contains 15 g of hemoglobin, the oxygen-carrying capacity of that blood is 20 ml of oxygen per 100 ml of blood. This is much more than the 0.5 ml of oxygen per 100 ml that can be carried dissolved in the plasma.

Before it can combine with hemoglobin, however, the oxygen must first diffuse into the plasma and then into the red blood cell where the hemoglobin is located. The greater the alveolar oxygen tension, the more oxygen can diffuse into the plasma and then into the red blood cell. As oxygen diffuses into the blood cell from the plasma, more can enter the plasma from the alveoli. Equally important, however, is the fact that the ability of hemoglobin to combine with oxygen depends upon plasma oxygen tension. When the oxygen tension is low, hemoglobin can take up very little oxygen, but when the P_{O_2} is raised, the hemoglobin approaches its oxygen-carrying capacity. Therefore, although only a small portion of the oxygen is actually carried in the plasma, it is the partial pressure of that oxygen which determines how much oxygen will be combined with hemoglobin at any given moment. The oxygen content of the hemoglobin—its degree of oxygenation—is expressed as the **percent saturation,** the percent of its oxygen-carrying capacity that is used. This expression recognizes that the ac-

TABLE 20-3   OXYGEN AND CARBON DIOXIDE CONTENT (VOLUMES PERCENT) OF AIR AND BLOOD

Medium	Oxygen	Carbon dioxide
Inspired air	20.93	0.04
Expired air	16.3	4.5
Alveolar air	14.0	5.6
Arterial blood	20.3	50
Venous blood	15.5	54

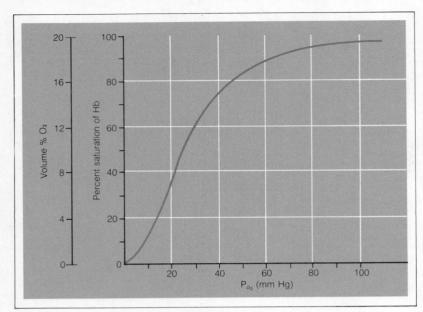

FIGURE 20-4 The oxygen-hemoglobin dissociation curve.

tual volume of oxygen being carried also depends upon the amount of hemoglobin in the blood.

The relationship between the hemoglobin saturation and the oxygen tension is not linear. Several important facts about oxygen transport are illustrated by the oxygen-hemoglobin dissociation curve, shown in Figure 20-4:

1. At a P_{O_2} of 100 mm Hg, the P_{O_2} of arterial blood, the hemoglobin is 97 percent saturated. Hemoglobin in arterial blood therefore normally carries oxygen at 97 percent of its capacity to do so. A further increase in P_{O_2}, as by breathing 100 percent oxygen, brings about virtually no increase in oxygen transport by the hemoglobin (unless the saturation has been reduced below normal for some other reason).

2. The dissociation curve flattens out at higher oxygen tensions. Because of this, arterial P_{O_2} can be reduced considerably before hemoglobin saturation is seriously affected. The hemoglobin is still about 80 percent saturated with P_{O_2} cut nearly in half.[3]

[3]This is equivalent to an altitude of about 15,000 feet. At 10,000 feet the saturation would be about 87 percent, at 7500 feet about 90 percent, and at 5000 feet about 95 percent. Functional impairment is appreciable when saturation falls below 90 percent.

3. At low oxygen tension a slight change in P_{O_2} causes a great change in saturation. This is desirable, since it favors the release of oxygen from the hemoglobin in the tissues. It is as important to the well-being of the organism that hemoglobin readily "unload" oxygen in the tissues as it is that hemoglobin readily take it up in the lungs.

The P_{O_2} of blood in the venae cavae is about 40 mm Hg, which means that the hemoglobin is still about 75 percent saturated with oxygen. Each 100 ml of blood, therefore, gives up only about a fourth of its load of oxygen—about 5 ml of oxygen—as it passes through the tissues. A cardiac output of 5 liters per minute therefore provides the 200-250 ml of oxygen needed each minute. The oxygen tension is lower in active tissues and more oxygen is given up by hemoglobin passing through capillaries there.

When arterial blood flows through a region of low tissue oxygen tension, oxygen dissolved in the plasma diffuses out to the tissues. As the plasma P_{O_2} falls, the hemoglobin gives up its oxygen, which diffuses into the plasma and thence to the tissues. The oxygen saturation of hemoglobin does not depend exclusively upon the oxygen tension, however. The position of the whole dissociation curve is shifted to the right by an increase in temperature or P_{CO_2} or a fall in the pH of the blood (Figure 20-5). A shift to the right, as occurs normally in the tissues, means that the hemoglobin releases oxygen at a higher P_{O_2}, or that it releases more oxygen at a constant P_{O_2}, than it otherwise would, a condition which favors "unloading" of oxygen. The curve is shifted to the left when temperature or P_{CO_2} is down or pH raised, as occurs in the lungs. This improves the uptake of oxygen by hemoglobin, but it impairs its release.

TRANSPORT OF CARBON DIOXIDE

Carbon dioxide is produced in, and is in greatest concentration in, the tissues, particularly in metabolically active tissues. Diffusing down its gradient takes it first to the blood and then

Focus

Fetal Hemoglobin

At birth, each human being moves from a warm, moist cradle with built-in food and oxygen supplies and waste removal to a cold, uncaring arena of constant conflict, from dependence to independence, from womb to world. In the process, humans effortlessly adapt to the change in environment. Their circulatory systems adjust to new needs, their lungs expand, and even the nature of the oxygen-carrying component of their blood changes.

This component, of course, is hemoglobin, the iron-containing protein found in red blood cells. Its function is to combine with oxygen, in the lungs after birth, in the placenta before birth. In the fetus, its chemistry is such that it combines with oxygen readily at the oxygen tension (P_{O_2}) characteristic of the maternal blood in the placenta; it releases the oxygen it carries when it is exposed to the lower oxygen tension of the fetal tissues.

Normal adult hemoglobin is a complex of two alpha and two beta protein strands, the blobins. Fetal hemoglobin is a complex of two alpha and two gamma globins. The gamma protein differs from the beta in its amino acid sequence, though both contain 146 amino acids. This difference in structure affects the ease with which the hemoglobin's hemes can combine with oxygen. Fetal hemoglobin becomes almost completely saturated with oxygen when exposed to the relatively low oxygen tensions found in the placenta; it releases the oxygen only at the low oxygen tensions found in the fetal tissues. Adult hemoglobin, on the other hand, saturates with oxygen only at higher oxygen tensions, such as those in the lungs; it also releases oxygen at higher oxygen tensions, such as are found in adult tissues. This difference in response can be seen by comparing the oxygen dissociation curves of fetal and adult hemoglobin, in which the percent saturation of the hemoglobin is plotted against the partial pressure of oxygen (P_{O_2}, or oxygen tension) in the fluid surrounding the hemoglobin molecules. The fetal curve lies to the left of the adult curve, indicating that at a given partial pressure of oxygen the hemoglobin is bound to more oxygen.

Shortly after birth, the body stops synthesizing gamma globins and starts synthesizing beta globins. Fetal hemoglobin is replaced by adult hemoglobin. The body thus adjusts to the change to a richer source of oxygen. The adjustment is progressive, becoming complete by about the age of six months.

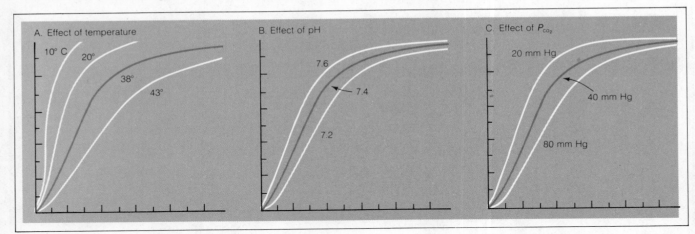

FIGURE 20–5   Factors affecting the oxygen-hemoglobin dissociation curve.
A. Temperature.   B. pH.   C. P_{CO_2}.

to the alveoli and the atmosphere. Since carbon dioxide is much more soluble than oxygen, more can be carried in solution. Even so, not enough carbon dioxide can be carried in this way, since at body temperature and at the P_{CO_2} of venous blood, each 100 ml of plasma can take up only 3 ml of carbon dioxide.

Most of the carbon dioxide is not carried in physical solution—that is, dissolved in the plasma—but the P_{CO_2} of that dissolved carbon dioxide determines the amount that is carried by two other means, both of which involve the red blood cell.

All carbon dioxide diffusing into the blood first goes into solution. Some of it—about 8 percent of the total CO_2 transported—remains in the plasma, (and contributes to the P_{CO_2}) but the rest moves on into the red blood cell, where much of it forms a loose reversible combination with hemoglobin (Figure 20–6). This combination is, however, very different from that of oxygen and hemoglobin, since it involves a reaction of the carbon dioxide with the amino groups on the globin portion of the hemoglobin molecule to form a compound known as **carbaminohemoglobin** (written as $HbNH_2$ to emphasize the amino group):

$$HbNH_2 + CO_2 \underset{\text{in lungs}}{\overset{\text{in tissues}}{\rightleftharpoons}} HbNHCOOH$$

The amount of carbon dioxide carried in this manner is related to the carbon dioxide tension, but it is not a linear relationship. In the lungs, where the P_{CO_2} is lower, the carbon dioxide is released to diffuse into the plasma and to the alveoli. About a fourth of the carbon dioxide is carried in the red blood cells as carbaminohemoglobin.

The rest of the carbon dioxide, about two-thirds, is transported as bicarbonate, mainly in the plasma. The bicarbonate-forming reaction is a very slow one in plasma, and little is formed there. But as carbon dioxide diffuses into the red blood cell, the reaction goes very rapidly due to the presence of an enzyme, **carbonic anhydrase** (c-a), which catalyzes the formation of carbonic acid, some of which dissociates to form hydrogen and bicarbonate ions:

$$CO_2 + H_2O \overset{\text{c-a}}{\rightleftharpoons} H_2CO_3 \rightleftharpoons H^+ + HCO_3^-$$

The enzyme is capable of catalyzing the reaction in either direction. It is typical of reversible reactions, in that the direction and rate of the reaction are related to the concentrations of the compounds involved. When the relative concentration of CO_2 is increased, the reaction goes to the right. When the relative concentrations of H^+ and HCO_3^- are greater, the reaction goes to the left. Thus when carbon dioxide is elevated, as it is in the tissues, more hydrogen and bicarbonate ions are formed. In the lung the reaction goes to the left, since the equilibrium is shifted by the decrease in carbon dioxide concentration as it leaves the bloodstream.

The hydrogen ions formed in car-

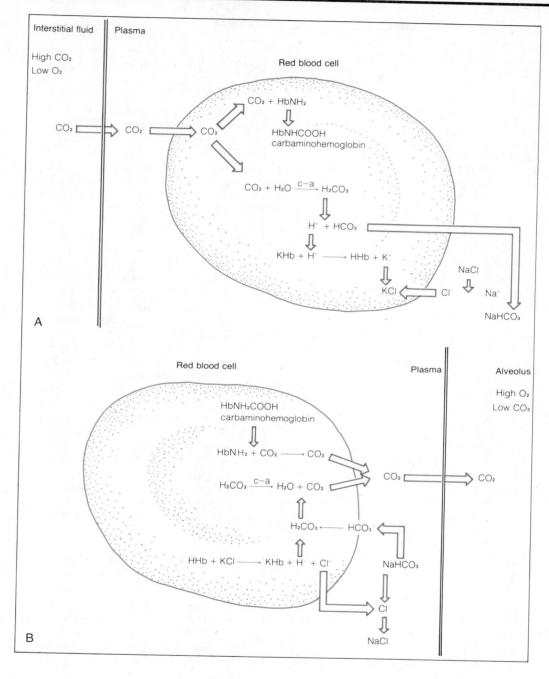

Interstitial fluid | Plasma

High CO₂
Low O₂

Red blood cell

CO₂ → CO₂ → CO₂

CO₂ + HbNH₂

HbNHCOOH
carbaminohemoglobin

CO₂ + H₂O $\xrightarrow{c-a}$ H₂CO₃

H⁺ + HCO₃⁻

KHb + H⁺ ⟶ HHb + K⁺

NaCl

KCl ← Cl⁻   Na⁺

NaHCO₃

A

Red blood cell

Plasma | Alveolus

High O₂
Low CO₂

HbNH₂COOH
carbaminohemoglobin

HbNH₂ + CO₂ ⟶ CO₂ → CO₂ → CO₂ → CO₂

H₂CO₃ $\xrightarrow{c-a}$ H₂O + CO₂

H₂CO₃ ← HCO₃⁻

NaHCO₃

HHb + KCl ⟶ KHb + H⁺ + Cl⁻

Cl⁻

NaCl

B

FIGURE 20–6
A. Carbon dioxide exchange in the tissues.
B. Carbon dioxide exchange in the lungs.

bonic acid dissociation are buffered by hemoglobin, the protein buffer in red blood cells, and the bicarbonate ions formed diffuse into the plasma where they are transported as sodium bicarbonate ($NaHCO_3$). (The red blood cell and plasma buffering mechanisms are discussed further below.) The removal of the hydrogen and bicarbonate ions as quickly as they are formed allows more carbonic acid to dissociate, which means that more carbon dioxide can enter the red blood cells. More carbon dioxide can therefore be taken up by the blood in the few seconds it takes to traverse the tissue capillaries.

The reactions described above occur in the tissues, where carbon dioxide tension is high, favoring the diffusion of carbon dioxide into the blood. In the lung, however, the carbon dioxide gradients are reversed and carbon dioxide diffuses into the alveoli, reducing the plasma P_{CO_2} and triggering a reversal of the reactions just described. In the red blood cell, carbon dioxide separates from the carbaminohemoglobin combination and diffuses outward. The bicarbonate in the plasma re-enters the cell (and chloride leaves) and carbon dioxide is released to diffuse into the plasma and to the alveoli (Figure 20–6).

ROLE OF RESPIRATION IN ACID–BASE BALANCE

The respiratory system is as important for acid–base regulation as it is for carbon dioxide excretion, and the two are inseparably bound. Carbon dioxide is quantitatively the most important acid to be dealt with, because it is produced in such large amounts. It becomes acid upon entering the red blood cell, where it is hydrated to carbonic acid (with the aid of carbonic anhydrase). Although carbonic acid is a weak acid and dissociates very little, it is formed in enough quantity that the pH would be seriously reduced if it were not buffered in the blood.

Buffer solutions minimize changes in hydrogen ion concentration by combining with the dissociated hydrogen ions to form less ionized compounds (Chapter 14). One of the most important of these is the bicarbonate buffer system ($\dfrac{NaHCO_3}{H_2CO_3}$). When hydrogen ions are added to a solution containing these buffers, hydrogen ions displace the sodium, increasing the carbonic acid content and reducing the amount of bicarbonate salt:

$$H^+Cl^- + NaHCO_3 \rightarrow H_2CO_3 + NaCl$$

The carbonic acid formed from carbon dioxide is both a substance to be buffered and a part of a major buffering system. When the carbon dioxide enters the red blood cell and carbonic acid is formed, a portion of the acid ionizes, releasing some hydrogen ions within the cell (Figure 20–6). Intracellular pH does not fall appreciably however, because these ions are buffered inside the cells by the hemoglobin buffer system, $\dfrac{KHb}{HHb}$. Recall from Chapter 14 that a buffer pair consists of a weak acid and a salt of that acid. In blood cells the protein is hemoglobin and its potassium salt (potassium is the chief intracellular cation). In the buffering reaction the hydrogen ions from carbonic acid displace the potassium of the hemoglobin salt, yielding the weak hemoglobin acid, HHb.

$$KHb + H_2CO_3 \rightarrow HHb + KHCO_3$$

Some of the potassium bicarbonate ($KHCO_3$) formed remains in the cell, but most of it ionizes, and the bicarbonate ions (HCO_3^-) diffuse out into the plasma where they combine with sodium ions and are carried as sodium bicarbonate ($NaHCO_3$). The departure of the negatively charged bicarbonate ions disturbs the electrical equilibrium, creating an electrical gradient. The cell membrane is not very permeable to cations, but Cl^-, the abundant extracellular anion, can penetrate, and it enters the cell to reestablish the electrical equilibrium. This exchange is sometimes known as the *chloride shift*. Thus, the important buffering actions occur in the red blood cells, but most of the carbon dioxide is transported in the plasma as bicarbonate.

Both oxyhemoglobin (HbO_2) and deoxyhemoglobin (Hb) are buffers, but deoxyhemoglobin is a weaker acid than oxyhemoglobin, and hence is a better buffer. The effectiveness of the hemoglobin buffers is improved by the fact that deoxyhemoglobin is formed when oxygen leaves the red cells in the tissues, which is when carbon dioxide enters the red blood cells and increases the need for buffering. Buffering is not perfect, however, since the H^+ concentration is slightly higher and the pH is slightly lower in venous blood than in arterial blood (7.37 vs. 7.40).

Carbon dioxide affects the pH in several ways, since the amount present as carbonic acid affects both the hydrogen ion and bicarbonate buffer concentrations. The P_{CO_2} of the blood also influences the diffusion of carbon dioxide and therefore the operation of the intracellular buffers. When a fall in the blood pH is due to an increase in its carbon dioxide content, the homeostatic response would be an increase in the removal of carbon dioxide, and there is a mechanism to do this.

The carbon dioxide tension of the blood of certain chemosensitive areas in the brainstem exerts a powerful, though indirect, effect on respiration (see Chapter 21). If the arterial P_{CO_2} goes up (suggesting that not enough carbon dioxide is being removed in passage through the lungs), respiration increases and more carbon dioxide is exhaled. When arterial P_{CO_2} is reduced, ventilation decreases and the carbon dioxide level is restored. It is possible, by **hyperventilating** (over-breathing), to exhale carbon dioxide faster than it is being produced, and to reduce the H_2CO_3 level and raise the pH, a condition known as **respiratory alkalosis.** Symptoms of light-headedness and dizziness accompany the upward shift of pH. If arterial P_{CO_2} is increased, the result is a **respiratory acidosis,** whose symptoms (short of coma) are not so easily recognized.

The lung therefore helps regulate pH by controlling carbon dioxide removal. This affects primarily the carbonic acid member of the bicarbonate-carbonic acid buffers. Regulation of the pH by controlling the excretion of HCO_3^- is also important, but that is largely a function of the kidney.

CHAPTER 20 SUMMARY

Since the pressure, volume, and temperature of gases are interdependent, it is customary to refer to gases under standard conditions: pressure at sea level, temperature of 0°C, and with no moisture. Under these conditions the pressure of the atmosphere is 760 mm Hg. Each gas in a mixture of gases exerts a pressure in proportion to its concentration in that mixture, which is the *partial pressure* of that gas. In air,

nitrogen accounts for about 600 mm Hg and oxygen for about 159 mm Hg. Carbon dioxide and other gases account for less than one mm Hg. Air is therefore about 79 percent nitrogen and nearly 21 percent oxygen. Changes in atmospheric pressure will change the total and partial pressures, but will not change the percent concentration of the gases.

The exchange of oxygen and carbon dioxide in the body occurs by diffusion. Each gas diffuses in response to its partial pressure gradient. The partial pressure of oxygen (P_{O_2}) is higher in air and alveoli than in blood entering the lung, and oxygen diffuses into the blood. The P_{O_2} is higher in arterial blood than in the tissues, and oxygen diffuses into the tissues. Carbon dioxide tension is highest in the tissues, and it diffuses into the blood and then into the alveoli, from which it is exhaled.

Diffusion of a gas into a fluid such as plasma depends upon the solubility of the gas in that liquid as well as on the partial pressure gradient. When a gas dissolves in fluid and reaches equilibrium, the actual volume of gas dissolved in fluid is much less than it would be in a mixture of gases at the same pressure. Not enough oxygen or carbon dioxide dissolves in plasma to meet the body's needs, and there are additional transport mechanisms for these gases.

Transport of Oxygen

Most of the oxygen is transported in red blood cells in loose combination with hemoglobin (*HbO₂*). The volume of oxygen bound by hemoglobin depends upon the P_{O_2} of the plasma. At the P_{O_2} of arterial blood hemoglobin is carrying about 97 percent of its oxygen capacity; it is 97 percent *saturated.* The actual volume of oxygen depends upon the amount of hemoglobin present. If hemoglobin is about 15g per 100 ml blood, and it is 97 percent saturated, then each 100 ml of blood contains approximately 20 ml of oxygen, an amount normally sufficient to meet the body's needs.

When oxygen diffuses to the tissues, where the P_{O_2} is low, the plasma

P_{O_2} drops, and causes the hemoglobin to give up some of its oxygen. Hemoglobin in venous blood is typically about 75 percent saturated; it has given up about a fourth of its oxygen to the tissues. The relationship between plasma P_{O_2} and hemoglobin saturation is not linear. When P_{O_2} is high as in arteries, slight changes in P_{O_2} cause little change in saturation, but when P_{O_2} is low a slight change in P_{O_2} causes a great change in the hemoglobin saturation and the amount of oxygen released. In active tissues, where the P_{O_2} may be very low, hemoglobin may give up nearly all of its oxygen.

Transport of Carbon Dioxide

Like oxygen, dissolved carbon dioxide accounts for but a fraction of the carbon dioxide transported, but it determines the amount of carbon dioxide that diffuses into red blood cells. Most of carbon dioxide does diffuse into the red blood cells and some of it combines with hemoglobin and is transported as *carbaminohemoglobin*. Most of the carbon dioxide in red blood cells reacts with water to form carbonic acid (with the aid of carbonic anhydrase) which ionizes somewhat. The resulting hydrogen ions are buffered in the red blood cell by hemoglobin (KHb/HHb), and the bicarbonate ions diffuse into the plasma to be transported. They are replaced in the red blood cells by chloride ions.

In the lungs the gradients are reversed. Carbon dioxide diffuses from the plasma and from the red blood cells, and all the other reactions are reversed.

The transport of carbon dioxide in the blood is a major factor in acid-base regulation. Because of its acid forming capacity, it is important to control carbon dioxide excretion and prevent its accumulation. This can be done to a great extent by controlling respiration.

STUDY QUESTIONS

1 If air (20 percent oxygen) comes to equilibrium with a fluid, the P_{O_2} will be the same in the air and the fluid. Will the concentration (volumes percent) of oxygen be the same in both? Explain.

2 By what process does oxygen move from lungs to blood to tissue fluid to cells? By what process does carbon dioxide move from cells to lungs? Why do they "move" in opposite directions?

3 If you hold your breath as long as you can, what changes will occur in the P_{O_2} and P_{CO_2} in the alveoli? in the arteries? in the veins?

4 Suppose you are breathing a gas that contains 10 percent oxygen, but is at a pressure of 2 atmospheres (1520 mm Hg). How will this affect the P_{O_2} in the alveoli and arterial blood, and oxygen transport in general? Would it have any effect on P_{CO_2}?

5 If oxygen is carried mainly in combination with hemoglobin, why do we pay so much attention to the partial pressure of the oxygen in the plasma?

6 What is the significance of the oxygen-hemoglobin dissociation curve?

7 Why is it more accurate to refer to the transport of oxygen by hemoglobin as percent saturation instead of simply as volumes percent of oxygen?

8 Although most of the carbon dioxide is transported in the plasma, most of it has to enter the red blood cells, both upon entering and upon leaving the blood. Why?

9 Why is the transport of carbon dioxide in the blood so closely associated with acid–base balance?

10 What is meant by respiratory acidosis and alkalosis? How might they be produced or combatted?

21

The Respiratory Cycle

Pulmonary Ventilation

Respiratory Problems

The Control of Respiration

As cells increase their metabolic activity, they increase their needs for blood-borne substances, including oxygen. Cardiac output can be increased to deliver greater amounts of these substances to the tissues, but if the blood is to carry more oxygen to the tissues, pulmonary ventilation must increase to bring more oxygen to the lungs.

It is to be expected that there are mechanisms to regulate pulmonary ventilation, but there must also be a mechanism to cause breathing. In the cardiovascular system the heart, which makes the blood move, has an autonomic innervation to control its rate. The heart also has automaticity and will beat without outside stimulation. This is not true for respiration, for breathing depends upon skeletal muscles. There is no automaticity, since skeletal muscles are totally dependent upon their nerve supply.

Respiration is therefore controlled by two different regulatory systems. One, the mechanism responsible for breathing, arises from centers in the brainstem which generate intermittent bursts of impulses in motoneurons to respiratory muscles and cause rhythmically alternating inspirations and expirations. The other system controls pulmonary ventilation in response to neural and chemical stimuli that reflexively and/or directly modify the basic breathing cycle.

THE RESPIRATORY CYCLE

The alternating inspiration and expiration of breathing resembles the alternating systole and diastole of the cardiac pump in that both must continue unceasingly throughout the life of the individual and both are rhythmic rather than sustained. There are some significant differences between them, however. Cardiac muscle is characterized by its automaticity; the beat of the heart is initiated by the activity of its pacemaker under the influence of autonomic nerves, but even without them cardiac muscle can contract independently and rhythmically. Breathing, on the other hand, is brought about by skeletal muscles with neither automaticity nor a pacemaker. They are completely dependent upon their somatic nerves. They contract when a volley of impulses arrives on these nerves, and they relax when the volley ceases. The origin of breathing movements therefore lies within the central nervous system, not in the muscles of respiration.

In quiet breathing, inspiration is the dominant phase. It is actively brought about by the contraction of the muscles of inspiration, which enlarges the thoracic cavity and draws air into the lungs. Expiration is usually passive and occurs when inspiration ceases and the muscles of inspiration relax; the expiratory muscles do not participate in quiet breathing.

The Respiratory Centers

The control of the basic respiratory cycle is largely the control of the initiation and termination of inspiration. In the intact animal the mechanism is reflexive. However, since the central nervous system is the source of the basic rhythmicity, the efferent limb is more important; the afferent limb is more important for adjustments and need not be considered until later. Some of the anatomical components in the basic respiratory cycle are already familiar:

1. The *effectors* are the muscles of inspiration, primarily the diaphragm, the external intercostals, and certain other thoracic muscles.

2. The *efferent paths* (Figure 21–1) are somatic motor fibers. The diaphragm is innervated by the phrenic nerves, which arise from the cervical cord, and the intercostal muscles are innervated by the thoracic spinal nerves. The motoneurons are excited by interneurons that descend in the spinal cord from the respiratory centers.

3. *Respiratory centers* (Figure 21–2) are located in the brainstem. By analogy to the cardiovascular system, one might predict that there would be two reciprocally related centers or areas, one for inspiration and one for expiration, and that they would probably lie in the medulla. Such a prediction is reasonably correct within limits. Although we do speak of an **inspiratory center** and an **expiratory center** located in the medulla, they are not separate and distinct entities. Within a single area there are cells which, upon stimulation, cause inspiration, and other cells which, when stimulated, bring about expiration. The cells seem to be intermingled with one another and, perhaps, with cells of the cardiovascular centers. Probably the best view is to recognize that *anatomically* there is a single respi-

FIGURE 21–1   Efferent pathways for respiration. Pathways for inspiration are shown on the right and those for expiration on the left.

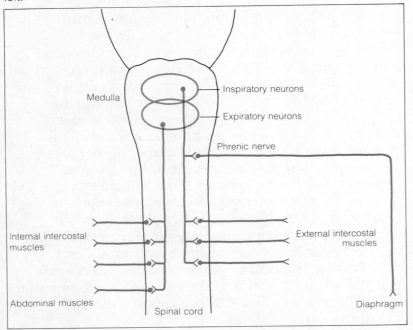

ratory center in the reticular formation of the medulla, but that *physiologically* one may think of an inspiratory and an expiratory center, because within that anatomical region there are neurons with separate functions although they are interconnected and have reciprocal relations. They are not, however, sympathetic and parasympathetic centers, as are the cardiac and vasomotor centers.

This medullary region receives input from various receptors and, in the manner of a reflex center, sends impulses along efferent pathways to the respiratory muscles. Some input comes on afferent fibers from the periphery, but a major source is from other parts of the brainstem, including at least two other areas which have specific respiratory roles and which may be considered as respiratory centers.

An **apneustic center** is located in the reticular formation of the lower pons. It is tonically active and prolongs inspiration by facilitating medullary inspiratory neurons. It receives input from other parts of the brainstem as well as from certain peripheral sources.

A **pneumotaxic center,** located in the upper pons, also exerts its influence through the medullary centers. It tends to facilitate expiration, not by causing active expiration, but by inhibiting inspiratory neurons and thus interrupting the inspiratory act.

It is generally agreed that the medullary respiratory center is capable of producing intermittent discharge in inspiratory neurons, since cyclic respiration still occurs after experimental interruption of all but the medullary centers and their connections to the inspiratory muscles. The breathing may be somewhat irregular, but inspiration and expiration still follow one another. The apneustic center exaggerates inspiration, and when it is also present the inspirations are prolonged and the infrequent expirations are only brief interruptions of inspiration.

The pneumotaxic center periodically interrupts inspiration. It does not have a rhythmicity of its own but seems to be activated by a feedback from the discharge of inspiratory neurons. Stimulation of the pneumotaxic center causes inhibition of inspiratory neu-

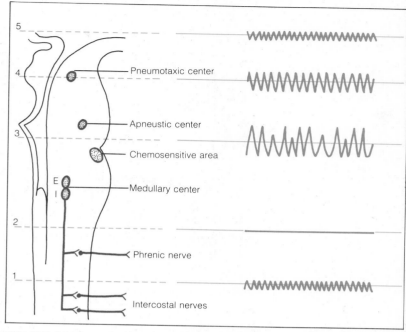

FIGURE 21-2 The respiratory centers. The relative contribution of each is shown by contrasting the different breathing patterns that result from the interruption of their motoneuron connections by transection of the brainstem at various levels.

rons. It probably acts by inhibiting the apneustic center (thus reducing inspiratory facilitation) and/or by direct inhibition of the inspiratory neurons. Inhibition of the inspiratory neurons deprives the pneumotaxic center of its source of excitation, and it soon ceases firing. Without the inhibitory influence of the pneumotaxic center, both the apneustic and inspiratory centers resume activity and another inspiration begins. The system of centers in the brainstem is thus capable of continuous oscillating action.

The relative contributions of each center to the production of normal rhythmic respiration are shown in Figure 21-2, in which the respiratory pattern is shown following transection of the brainstem at each of several levels:

Transection just below the medulla separates the respiratory muscles from the respiratory centers. There is then no breathing, a condition known as **apnea.**

If the cut is below the level of origin of the phrenic nerve, the diaphragm retains its connections with the medul-

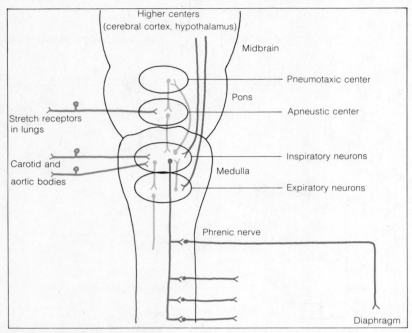

FIGURE 21–3   Some important neural connections of the respiratory centers.

lary centers and an adequate, reasonably normal respiration occurs. Survival of a person after injury to the cervical region of the spinal cord may well depend upon whether the damage is above or below the phrenic outflow. Such breathing is, of course, wholly diaphragmatic, since the intercostal muscles are no longer connected with the respiratory centers.

Transection of the brainstem between the medulla and the pons leaves intact the connections between the medullary centers and their effectors, and respiratory cycles occur, although in a somewhat gasping manner. This demonstrates that the basic oscillating pattern exists at the medullary level. However, without the modulating influences from higher levels it is not the smooth and even cycle typical of normal breathing.

After transection at the level of the upper pons, the apneustic center and its connections are retained. Breathing is regular but deep, and there is no pause at the end of each expiration. This is what would be expected with facilitation of inspiration.

When the transection is above the pons, the basic respiratory cycle is essentially normal. The pneumotaxic center is now part of the system and the basic mechanism is complete. This does not, however, mean that the cerebrum or other higher centers have no effect on respiration—it means only that while these centers may modify the basic cycle, they are not necessary for its operation.

The Inflation Reflex

One input from the periphery probably plays some role in the basic rhythmic respiratory cycle. It is the afferent limb of a reflex known as the **inflation reflex,** the **Hering-Breuer reflex,** or simply the **vagal mechanism** (Figure 21–3).

Within the airways of the lungs are stretch receptors whose afferents are carried in the vagus nerves. The fibers terminate near the apneustic center and apparently have connections with the medullary centers as well. The impulses inhibit the apneustic center and thus inhibit inspiration. As the lung inflates during inspiration, these stretch receptors are stimulated. The deeper the inspiration, the more rapidly they fire and the greater the inhibition, until inspiration stops and expiration occurs.

The contribution of the inflation reflex in humans is difficult to assess. There are marked species differences in its effectiveness, and it seems to be more active in anesthetized animals. In humans, it is apparently not of major importance in halting inspiration during normal quiet breathing. But when the tidal volume is approximately doubled (as it might be in exercise), inhibition of inspiratory neurons does occur, and overinflation is avoided. At normal tidal volumes the receptors are stimulated, and impulses are conducted along the vagus nerve; the respiratory centers seem to have a high threshold, however, or perhaps other influences override the vagal effects. In quiet breathing it may be that the inflation reflex contributes a generally inhibitory influence on the brainstem centers. Another possibility is that it helps determine the optimum tidal volume that will provide the necessary alveolar ventilation with the least effort. (Very deep inspirations require much more effort, while very shallow inspirations provide less alveolar ventilation due to the increasing effect of the dead space.)

The array of respiratory centers and vagal afferents—two of which contribute to inspiration (medullary and apneustic) and two of which contribute to expiration (pneumotaxic and vagal)—provides another example of the checks and balances on the mechanisms which control vital functions.

PULMONARY VENTILATION

Pulmonary ventilation refers to the amount of air delivered to the lung each minute, and it is determined by the breathing rate and the tidal volume. The rate and amplitude of breathing may be increased so greatly that pulmonary ventilation may exceed 100 liters per minute. At a typical resting breathing rate of 10–16 breaths per minute, and a tidal volume of 500 ml per breath, the normal pulmonary ventilation is only 5–8 liters per minute. The **alveolar ventilation** (minus the dead space) is then about 3.5 to 5.6 liters per minute. The air to the alveoli is of prime importance; due to the greater ease and accuracy of determining pulmonary ventilation, however, the latter is the volume used in most considerations of ventilation.

Chemical Regulation of Ventilation

Like the cardiovascular system, the respiratory system is charged with the delivery of certain chemicals. It should not be surprising that it too is controlled by changes in the levels of these chemicals, namely the oxygen and carbon dioxide tensions, and in the related hydrogen ion concentration. There are receptors that detect changes in one or more of these chemicals and then relay the appropriate information to the medullary respiratory center. Some of the receptors are located peripherally, while others are located within the central nervous system. Afferent fibers

from the latter are less well described than those from the peripheral receptors, but there must be connections between these areas and the medullary centers. Each receptor is stimulated by a particular aspect of its immediate chemical environment.

Central Receptors It has been known for some time that very slight changes in the arterial P_{CO_2} can markedly alter respiration. An increase of as little as 2 mm Hg has been shown to be enough to double pulmonary ventilation, but it is not due to a direct effect of the carbon dioxide tension upon the medullary center. More recently areas have been found on the antero-lateral surfaces of the medulla that are particularly responsive to chemical changes (see Figure 21–2). The areas, designated **chemosensitive areas (CSA)**, are just that. They are not true centers, but simply collections of receptor cells in the central nervous system. (We have seen other receptors in the brain, such as the osmoreceptors and temperature receptors in the hypothalamus.)

The response of the CSA to reduced arterial oxygen tension is not significant, and the increase of arterial hydrogen ion concentration required to elicit a response is much greater than would occur in normally buffered blood. The response to changes in arterial carbon dioxide tension is marked, however, although the change in the arterial P_{CO_2} itself is probably not the direct stimulus.

The cells of the CSA are particularly sensitive to changes in hydrogen ion concentration, but charged particles such as H^+ do not cross the blood-brain barrier very well. The receptor cells of the CSA, which lie close to the brain surface, are thus influenced more by changes in the cerebrospinal fluid than by changes in the arterial blood. Carbon dioxide, as an uncharged particle, diffuses readily into the cerebrospinal fluid, where it combines with water to form carbonic acid. The acid dissociates to yield the hydrogen ions that stimulate the CSA. This is a rather strange sequence of events because the cerebrospinal fluid contains virtually no carbonic anhydrase, and the formation of carbonic acid is relatively slow. Cerebrospinal fluid, unlike blood, does not circulate rapidly, which allows time for the reaction. In addition, cerebrospinal fluid contains little protein, so it has very poor buffering capacity. Therefore, any hydrogen ions released are likely to remain as unbuffered H^+ which lowers the pH of the cerebrospinal fluid. Thus it is the hydrogen ion concentration of the cerebrospinal fluid, not changes in the PH or P_{CO_2} in the blood, that stimulates the cells of the CSA. It is therefore not incorrect to continue to view carbon dioxide as the most potent stimulus to respiration, even though hydrogen ions are the true stimulus to the receptors.

Peripheral Receptors The important peripheral chemoreceptors are those of the **carotid body** and **aortic body,** small clumps of tissue located at the bifurcation of the carotid artery and along the arch of the aorta very near the carotid and aortic sinuses (which contain the pressoreceptors) (Figure 21–3). Although the bodies and sinuses are near one another and send afferents with the same nerves (glossopharyngeal and vagus, respectively), they are separate and distinct receptors.

These chemoreceptors respond to changes in P_{O_2}, P_{CO_2}, and pH of arterial blood, but the threshold for the latter two are so much higher that, for all practical purposes, the chemoreceptors may be regarded as low-oxygen detectors. They increase ventilation when the P_{O_2} is low, but they do not cause apnea when it is raised. Since the stimulus is the partial pressure, they do not respond to reduced oxygen caused either by low hemoglobin or by poor saturation. They do respond, however, to a reduced atmospheric pressure even if the oxygen concentration is normal. The threshold is relatively high, and there is little response until the arterial P_{O_2} is reduced by about half. Further reduction may increase pulmonary ventilation three- or four-fold.

These carotid and aortic chemoreceptors do not participate in the control of normal quiet respiration, but they are an effective means of increasing ventilation when oxygen supply is reduced independently of the carbon dioxide level. They are particularly important at high elevations where atmospheric pressure is low, and this is probably their major function. These receptors can also evoke a cardiovascular response, although it is less marked. Low oxygen causes vasoconstriction and an increase in heart rate.

Nonchemical Regulation of Ventilation

Reflexes A number of receptors whose stimulation causes reflex changes in respiratory activity have been identified (See Figure 21–4). The most important are probably those participating in the inflation reflex and the chemical reflexes discussed above. Most of the others have been identified under quite specific conditions and do not participate in regulation of normal respiration. These receptors are located in various parts of the thorax—the lungs, the heart, and the great vessels—and most of their afferent fibers are believed to run in the vagus nerve. The pressoreceptors of the carotid and aortic sinus, although primarily concerned with cardiovascular control, do have a respiratory effect. Increased arterial pressure causes apnea or reduced ventilation as well as decreased heart rate and blood pressure. There is also some evidence that ventilation can be increased by reflexes arising from joint proprioceptors, since passive movement of a limb, which involves minimal expenditure of energy by the individual, increases respiration significantly.

Higher Centers It is well known that parts of the brain above the respiratory centers influence ventilation. One such part is the hypothalamus which, in some species, produces the panting associated with heat dissipation. Another may be the limbic system, since pain and the emotions also modify respiration. The cerebral cortex, of course, has a regulatory effect, since we have some voluntary control over respiration. Voluntary breath-holding, however, gives way when the arterial P_{CO_2} builds up sufficiently to produce a stronger respiratory drive than any

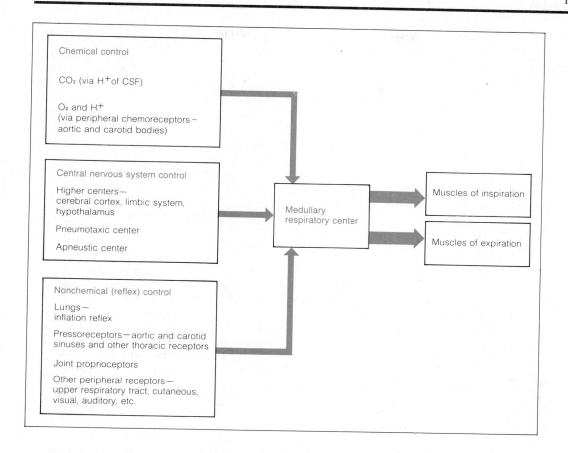

FIGURE 21–4
A few of the chemical
and neural stimuli that
affect the medullary
respiratory neurons.

cortically induced inhibition. Many other acts, including vocalization, are mainly reflex modifications of respiration, even though they are initiated from the cortex or other higher centers.

Hyperpnea of Exercise

Since exercising muscles use more oxygen and produce more carbon dioxide than do resting muscles one might expect changes in oxygen, carbon dioxide, and pH in directions that would increase ventilation. But ventilation increases immediately and well before such changes could occur. Furthermore, if one measures these parameters in the *arterial* blood, the change is noticeable only in severe exercise. The striking increase in pulmonary ventilation *(hyperpnea)* that accompanies exercise is ordinarily enough to prevent any changes in the composition of the arterial blood.

As discussed in Chapters 17 and 18, the cardiovascular system greatly increases the flow of blood to exercising muscles. This allows for the delivery of the large amounts of oxygen needed and removal of the equally great amounts of carbon dioxide produced. Also, the very low oxygen tension in the capillaries of active muscles causes the hemoglobin to give up more of its oxygen. However, because the ventilation is increased, the returning venous blood is able to take up enough oxygen in the lungs and lose enough carbon dioxide that arterial P_{O_2} and P_{CO_2} do not change. There are therefore no known chemical signals for the arterial chemoreceptors to detect. Various investigators have suggested a role for (1) reflexes from moving limbs, (2) increase in temperature, (3) the action of epinephrine, and (4) cortical (psychic) influences. All of these undoubtedly are involved in some way. There are probably other factors, such as changes in the sensitivity of the centers themselves, that contribute to this common everyday occurrence. However, the total increase in ventilation that occurs in exercise is much greater than the sum of the individual ventilatory responses produced by each known respiratory stimulus acting alone.

RESPIRATORY PROBLEMS

The function of the respiratory system, with the aid of the cardiovascular system, is to deliver oxygen to and remove carbon dioxide from the tissues. Difficulties can arise from malfunctions anywhere along the line—the composition and pressure of the air breathed, the effectiveness of the ventilatory mechanism, the gas exchange processes, or the transport by the cardiovascular system. A few of the problems and some of the solutions are mentioned in the following subsections.

Oxygen-Related Problems

Hypoxia The most common of oxygen-related problems is a lack of oxygen delivered to the tissues. This condition is known as **hypoxia** and it may be brought about by any of several defects. It was formerly referred to as anoxia, but since the problem is one of *insufficient* oxygen rather than one of the *absence* of oxygen, hypoxia is a more appropriate term. The hypoxias are classified according to the cause of the oxygen lack.

Hypoxic hypoxia refers to tissue oxygen lack due to low arterial P_{O_2} and the low hemoglobin saturation which accompanies it. The cause may be a reduced oxygen content of the air breathed or a low barometric pressure. It may be inadequate ventilation because of resistance or obstruction of the airways, paralysis of the respiratory muscles, or even pneumothorax. Exchange difficulties due to reduced alveolar surface area, fluid in the lungs, or improper balance between ventilation and perfusion of the alveoli could also cause hypoxic hypoxia. All of these conditions diminish the amount of oxygen reaching the arterial blood.

Anemic hypoxia is due to impaired ability to transport oxygen. The arterial P_{O_2} is normal, but the oxygen content is low because of a reduced amount of available hemoglobin. There may be an insufficient amount of hemoglobin (or red blood cells) as in anemia, or the hemoglobin may be unable to combine with the oxygen. Some of the hemoglobin-related problems were discussed in Chapter 14.

Stagnant hypoxia is a cardiovascular problem. The arterial blood carries enough oxygen, but it is not delivered to the tissues that need it. There may be general circulatory collapse (shock), or impaired blood flow to a specific organ or limb.

Histotoxic hypoxia occurs when the tissues have been poisoned and their metabolic activities impaired. Cyanide, for example, inhibits an intracellular enzyme system, so that the cells are unable to take up and use the oxygen, although an adequate supply is brought to the tissues by the blood.

Cyanosis When arterial blood has not taken on enough oxygen and a significant amount of reduced hemoglobin (deoxyhemoglobin) is present in arterial blood, a condition of **cyanosis** exists (this is not to be confused with cyanide poisoning). Reduced hemoglobin has a darker color than oxyhemoglobin, and its presence gives to the tissues the rather bluish tinge for which the condition is named. Cyanosis is due to the presence of reduced hemoglobin rather than to a lack of oxyhemoglobin and is therefore rarely seen in anemia.

Carbon Monoxide Carbon monoxide (CO) is a deadly poison which causes a type of anemic hypoxia and, because it is colorless and odorless, its presence is not readily detected. It is a well-known component of automobile exhaust and is present in minute amounts in cigarette smoke. It combines with hemoglobin, binding at the same sites as oxygen, so that the two compete for binding sites. Unfortunately, carbon monoxide has an affinity for hemoglobin that is 210 times that of oxygen. Thus if the carbon monoxide concentration of inspired gas is 1/210 that of oxygen, the hemoglobin binds equal amounts of oxygen and carbon monoxide. The carbon monoxide bond is a tenacious one, and very little of the bound gas is removed in the next passage of the blood through the lungs, while half of the hemoglobin that had given up its oxygen in the tissues subsequently takes on carbon monoxide. Fortunately, carbon monoxide is rarely present in such a high concentration; if it were, a few minutes would suffice for all of the hemoglobin to be occupied by carbon monoxide. Since the P_{O_2} is unchanged, the chemoreceptors do not respond and ventilation is not increased. The symptoms of carbon monoxide poisoning (headache, dizziness) appear when about 20 percent of the hemoglobin is saturated with carbon monoxide. Exposure to 0.04 percent carbon monoxide would not be safe after about an hour, and 0.5 percent would be dangerous in about 15 minutes.

Hyperoxia Breathing pure (100 percent) oxygen does not significantly increase the hemoglobin saturation but it does increase the amount of oxygen dissolved in the plasma. The intake of pure oxygen is known as **hyperoxia** and, surprisingly, can be detrimental if continued for several hours. At atmospheric pressure, pure oxygen is irritating, but at higher pressure the symptoms are more severe and convulsions and coma may result. Premature infants exposed to 100 percent oxygen in an incubator for a long period of time may suffer visual damage and blindness. The mechanism of these effects is not well understood.

Carbon Dioxide-Related Problems

The effects of increased or decreased carbon dioxide (*hypercapnia* or *hypocapnia*) are not as apparent as those of hypoxia. Hypercapnia often accompanies hypoxia, and its effects are often masked by the more dramatic hypoxic symptoms.

Two effects of increased carbon dioxide tension have already been described—the stimulus to respiration and an increase in hydrogen ion concentration leading to *acidosis* (excessive amounts of acid). If the carbon dioxide content gets too high, as from breathing air containing 10 percent carbon dioxide, it has a depressant effect and may cause a loss of consciousness. Carbon dioxide therefore can be, and has been, used as an anesthetic, but this is risky because of the danger of too much respiratory depression.

Hypocapnia is usually the result of hyperventilation in which carbon dioxide is "blown off" and the alveolar carbon dioxide tension is reduced, thus producing a condition of *alkalosis* (excessive amounts of alkali). Several minutes of voluntary hyperventilation may be followed by a period of apnea, with a gradual return to normal respiration as carbon dioxide returns to normal levels.

Changes in Barometric Pressure

When the barometric pressure is reduced, P_{O_2} and hemoglobin saturation are also reduced, less oxygen reaches the tissues, and hypoxia may result. The first signs are those of cerebral hypoxia. Sudden exposure to low barometric pressure (as might occur upon decompression of an airplane cabin at high altitude) can result in unconsciousness within a few seconds. Unacclimatized persons usually get into difficulty at altitudes much above 20,000 feet without extra oxygen. Acclimatized persons can tolerate low atmospheric pressure somewhat better: in the Andes Mountains people live permanently and do heavy work at about 18,000 feet. At very high altitudes, above 40,000 feet, the oxygen tension is so low that even breathing 100 percent oxygen is insufficient, and oxygen must be supplied under pressure. At 63,000 feet the total atmospheric pressure is, about 47 mm Hg, which is the vapor pressure of water at body temperature. At this pressure body fluids boil. Survival at such altitudes (and in space) obviously requires pressurization of the individual or his environment, as well as of his oxygen supply.

Increased barometric pressure also has important effects, not all of which are respiratory. Such conditions are encountered mainly under water, where pressure increases markedly below the surface. The maximum inspiratory effort can counteract a pressure of about 100 mm Hg, and the pressure is increased by that much at a depth of about 4 feet. Deeper than that, it would be impossible to expand the thoracic cavity to get air into the lungs. In order to ventilate the lungs at depths greater than this, the air breathed must also be under pressure, in proportion to the depth. Breathing air at high pressure cannot significantly increase the transport of oxygen by hemoglobin, but there is a marked increase in the amount of oxygen dissolved in the plasma.

At pressures of several atmospheres, a considerable amount of nitrogen dissolves in the pulmonary capillary blood and diffuses into the tissues. Large amounts of nitrogen in the tissues cause central nervous system symptoms (nitrogen narcosis) similar to hypoxia. The real problem develops when, for example, a diver returns to the surface, since then the dissolved nitrogen comes out of solution. If ascent is rapid, the nitrogen is unable to diffuse slowly into the blood and lungs, but comes out of solution rapidly and forms gas bubbles in the tissues, especially around muscles and joints. The symptoms of decompression sickness, or the bends, depends upon the severity and location of the bubbles. The treatment and prevention is slow decompression, over a number of hours, determined by the depth and duration of the dive.

Asphyxia

Suffocation or **asphyxia** is a condition of hypoxia and hypercapnia which results from inability to ventilate the lungs. It may be due to confinement in a very limited amount of air (as in cave-ins or mine disasters), to an obstruction in the airway, or, as sometimes hap-

pens, to a foreign object setting off a spasm of the laryngeal muscles—the choking that follows prevents any effective movement of air.

Some Common Respiratory Diseases

Some of the respiratory problems discussed above are related to changes in oxygen, carbon dioxide, or atmospheric pressure, and thus are caused by unfavorable environmental conditions. However, there are also numerous disease conditions which directly affect the respiratory apparatus by interfering with ventilation or gas exchange.

Asthma is due to hypersensitivity (allergy) to certain substances. The response is a spasm of the bronchiolar smooth muscle, increased mucous secretion, and, over a period of time, a thickening of the epithelium and muscle layers. The result is a narrowing of the airways and difficulty in moving air (dyspnea). Epinephrine will bring relief since it dilates the bronchioles by inhibiting their smooth muscle.

Bronchitis is an inflammation of one or more bronchi. The inflammation involves excessive production of sputum and often leads to attacks of coughing and chest pains.

Tuberculosis is an infection of the lungs caused by the tuberculosis bacillus. In fighting the bacillus, the body produces tubercles of fibrous tissue to block out the infection sites. However, this fibrosis of the lung decreases lung tissue elasticity and compliance, and thereby reduces vital capacity and interferes with diffusion.

Pneumonia is an inflammation of the lungs, often, but not always, caused by invasion of bacteria. It is characterized by the presence of fluid in the alveoli and airways.

Emphysema is a slowly developing condition in which the alveoli become enlarged. The interalveolar septa break down, reducing the surface area for gas exchange and the number of pulmonary capillaries. There is frequently a reduction in the amount of elastic tissue in the lung, and difficulty in expiration of air is typical. Emphysema is often secondary to some other respiratory problem, such as *asthma, tuberculosis,* or *chronic bronchitis,* and it is not uncommon in long time heavy smokers.

We have seen that the lung will collapse when air enters the intrapleural space *(pneumothorax),* raising the intrapleural pressure on that side. Blockage of a bronchus or bronchiole may cause collapse **(atelectasis)** of anywhere from a few alveoli to a whole lung. The lung tissue beyond the obstruction is not aerated, and the gas in these alveoli is gradually absorbed and they collapse. An increase in the surface tension within the alveoli (lack of surfactant) will also cause them to collapse.

Pulmonary congestion, leading to **pulmonary edema,** may result when there is an increase in the pressure of the pulmonary vessels. It is often due to failure of the left ventricle to pump all of the blood that is returned to it. It may also occur in those who ascend quickly to high altitude and engage in heavy physical work too soon. The edema fluid, which accumulates in the interstitial spaces and alveoli, interferes with gas exchange by increasing the diffusion distance and reducing the amount of space available to air.

The inhalation of air containing certain types of dust or chemical particles affects the alveolar-capillary membrane in any of several ways that interfere with the diffusion of gases. Several industrial diseases, including *asbestosis, silicosis,* and *black lung,* are examples, as are the effects of the ubiquitous smog and air pollution and even, apparently, the propellants used in aerosol sprays.

CHAPTER 21 SUMMARY

Respiration is controlled by two mechanisms, one for bringing about the normal breathing cycle, and the other to determine the level of pulmonary ventilation.

Respiratory Cycle

Breathing is the result of contraction of skeletal muscles, and is therefore under neural control. In normal quiet breathing inspiration involves contraction of inspiratory muscles (diaphragm and external intercostals), and expiration occurs when they relax. Expiration is passive; expiratory muscles are not active in quiet breathing.

A center in the medulla of the brain contains inspiratory and expiratory neurons whose stimulation causes contraction of inspiratory or expiratory muscles. The center operates as an oscillating system: the inspiratory neurons fire for a period of time, thus causing inspiration. This is followed by inhibition of those neurons, leading to expiration. An *apneustic center* in the lower pons is tonically active and drives the inspiratory neurons. A *pneumotaxic center* in the upper pons periodically inhibits inspiration by inhibiting the apneustic center and/or the inspiratory neurons. The pneumotaxic center is probably activated by feedback from discharge of the inspiratory neurons. Receptors in the lungs are stimulated when the lungs are well inflated and their impulses, carried in the vagus nerves, inhibit the apneustic center and inspiratory neurons. This is the *inflation reflex,* and it interrupts inspiration to bring about expiration.

Pulmonary Ventilation

Pulmonary ventilation is the amount of air breathed each minute. It is determined by the breathing rate and tidal volume. During quiet breathing it amounts to about 5–8 liters per minute, but it must be increased whenever the body's needs for gas exchange increase. Pulmonary ventilation is controlled largely by chemical stimuli, and a major element in its control is carbon dioxide in the plasma. Carbon dioxide is the most powerful physiological stimulus to respiration, although carbon dioxide itself is not the actual stimulus.

The so-called *central receptors* are chemoreceptors located on the ventral surface of the medulla, in contact with the cerebrospinal fluid (CSF). The cells of this *chemosensitive area (CSA)* are stimulated by hydrogen ions. Carbon dioxide readily diffuses into the CSF from the blood where it reacts to form carbonic acid, which then yields hydrogen ions upon dissociation. These hydrogen ions in the CSF stimulate the cells of the CSA which then stimulate the inspiratory neurons in the medulla, causing a marked increase in pulmonary ventilation.

Important *peripheral receptors* are chemoreceptors located in the *carotid* and *aortic bodies*. They respond chiefly to lowered P_{O_2} of the arterial blood and are particularly important when the atmospheric pressure (and hence P_{O_2}) is reduced, as at high altitude.

Many other receptors have connections through which they may affect pulmonary ventilation, including pressoreceptors in the carotid and aortic sinuses and cutaneous receptors. A number of centers in the brain affect the respiratory centers, for control of breathing is part of many activities such as vocalization, emotional responses (laugh, cry), and there is also voluntary control.

Respiratory Problems

Hypoxia, a condition of insufficient oxygen to the tissues, may result from reduced P_{O_2} in the blood (or air), insufficient hemoglobin to carry it, failure of the cardiovascular system to deliver it, or inability of cells to use it. Reduced P_{O_2}, as occurs at altitude, causes an increase in ventilation by stimulating the arterial phemoreceptors. At very high altitudes it may be necessary to breathe 100 percent oxygen, perhaps at an increased pressure. Reduced oxygen carrying capacity may be caused by carbon monoxide, which interferes with oxygen transport because it binds with hemoglobin more avidly and tenaciously than oxygen. Increased baro-

metric pressure raises the partial pressure of all gases in the air breathed, including nitrogen which is quite soluble in tissue fluids. Return to normal atmospheric pressure must be slow to prevent formation of nitrogen bubbles in the tissues, which will occur if nitrogen comes out of solution too rapidly, a condition known as the bends.

STUDY QUESTIONS

1 The muscles of respiration are skeletal muscles rather than smooth muscles. What does this tell you about the act of breathing?

2 What is the role of the vagus nerve in quiet breathing? If both vagi were cut, what would be the effect on quiet breathing?

3 What is the normal stimulus to the chemosensitive areas (CSA) of the medulla, and how does it come about? What is the effect of their activation?

4 How does the role of the CSA differ from that of chemoreceptors in the carotid and aortic bodies?

5 Do you think you could hold your breath long enough to cause hypoxic damage to the brain? Explain.

6 Carbon monoxide can cause severe hypoxia. Why is there no stimulation to respiration?

7 Why do changes in the barometric pressure (either a decrease as at high altitude, or an increase as in diving) pose problems if the air breathed still contains the same concentration of oxygen (20 percent)?

Part 6
Metabolic Processes and Energy

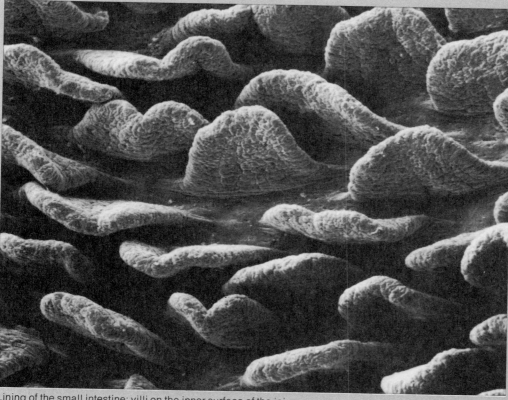

Lining of the small intestine; villi on the inner surface of the jejunum.

Energy is required to conduct all bodily activities, everything from running upstairs to producing a secretion. This energy is obtained from the foods we eat. Most of the food eaten is not in suitable form for entry into the body, however, and it must be broken down and transported to a site where it can be absorbed. It is the role of the digestive system to perform these functions and to get the energy-containing nutrients into the bloodstream, which then carries them to the cells.

Upon reaching the cells, the energy may be used to fuel the cell's activities, converted to other substances, and/or stored for future use. The regulation of these processes, and the "decision" of how to use the energy, is largely under hormonal control. It requires a careful balancing of many reactions so that we use our energy wisely.

Part 6, then, is about the energy needs of the body. It examines how we get the energy into the body (the digestive system); what we do with it after it is in the body, and how we get the energy from the nutrients (metabolism); how we use the energy released; and finally, how it all is regulated to balance our energy intake and expenditure.

22

General Structure of the Digestive Tract

Organs of the Digestive System

Blood Supply

Innervation of the Digestive Tract

Peritoneum

Anatomy of the Digestive System

Before the nutrients we ingest can be absorbed and used by the cells, they must be released from the foods that contain them. This is accomplished largely by the mechanical and chemical breakdown of food in the digestive tract, by the actions of smooth muscle and by enzymes in the secretions (digestive juices) of the glands of the digestive system.

Not only must the body provide for the breakdown (digestion) of nutrients, but it must transport them to appropriate sites for absorption into the bloodstream. The blood supply to the digestive tract is therefore important for its digestive functions, as well as for meeting the needs of the cells.

The organs of the digestive system occupy most of the abdominal cavity, and many are held in place only rather loosely. The relationship of the organs and their attachments, and the peritoneum that lines the abdominal cavity and covers all the organs is an important part of understanding the digestive tract.

Chapter 22 describes the structures of the digestive system that enables it to carry out these functions.

The functions of the digestive system are to prepare material for absorption and to excrete what is not absorbed. Its structure is admirably suited to this role. The digestive system consists primarily of a tube running through the body from mouth to anus. Its wall contains muscle, glands, and absorptive surfaces, but its precise structure varies to reflect the varying functions of its several parts. Located outside but near the tube are important glands, whose secretions are carried into it by ducts.

GENERAL STRUCTURE OF THE DIGESTIVE TRACT

The tubular digestive or **gastrointestinal tract** has well-developed layers of smooth muscle in its walls and valves of skeletal muscle at its entry and exit. The churning and mixing produced by the action of the smooth muscle aid in the mechanical breakdown of the food, ensure contact of the contents with both the digestive juices and the absorptive surfaces, and propel material through the tract.

Several types of glands provide the digestive tract with the digestive enzymes and with lubricating fluids. Some are one-celled glands in the lining of the tract, while others are more complex and located in the wall or outside the gastrointestinal tract entirely. Some of the glands are endocrine glands, which secrete hormones (into the bloodstream instead of into the gastrointestinal tract) that help control the tract's motor and secretory activity.

The gastrointestinal tract is a single continuous tubular structure, but each of its several portions is modified according to the functions performed by that portion (see Figure 22–1). The stomach, for example, is a saclike enlargement which holds ingested material and releases it gradually to the small intestine beyond. The small intestine is relatively long and has well-developed muscle layers and a number of glands, since this is where the action of digestive enzymes is greatest and nutrients are absorbed. The large intestine is less active and has less muscle and fewer glands.

The gastrointestinal tract shows a consistent structural pattern along almost its entire length (Figure 22–2). There are four layers throughout, but the details and degree of development vary in the different parts. From the *lumen* (the space inside a tubular organ) outward, these layers are as follows.

1. The **mucosa.** The mucosa or mucous membrane lining the digestive tract is the innermost layer and consists itself of three layers: the **epithelium,** the **lamina propria,** and the **muscularis mucosa.** The epithelium is on the luminal surface. The lamina propria is a relatively loose connective tissue containing blood vessels, some lymphoid tissue, and usually glands. The muscularis mucosa is a thin layer of smooth muscle that separates the mucosa from the deeper layers.

2. The **submucosa.** Beneath the mucosa is a layer of loose connective tissue, the submucosa. It contains blood vessels, lymphoid tissue, autonomic neurons and plexuses, and sometimes glands.

3. The **muscularis.** The muscle layers are well developed along most of the digestive tract. They are smooth muscle and are usually arranged in two layers, an inner circular one and an outer longitudinal one. Between them are nerve fibers and scattered ganglion cells, the cell bodies of the parasympathetic postganglionic neurons that supply the smooth muscle, and secretory cells.

4. The **serosa** (or **adventitia**). The gastrointestinal tract is surrounded by a thin connective tissue layer. In the abdominal cavity the connective tissue is covered with epithelial cells to form the **serosa.** The serous membrane that lines the entire abdominopelvic cavity is known as the **peritoneum.** It covers certain organs that adhere to the body wall, but it completely surrounds and suspends those organs that are not so firmly attached, including many of the digestive organs. Other organs (such as the esophagus and rectum) are not adjacent to a body cavity for most of their length; the connective tissue surrounding them is continuous with the deep fascia of the area and is known as **adventitia.**

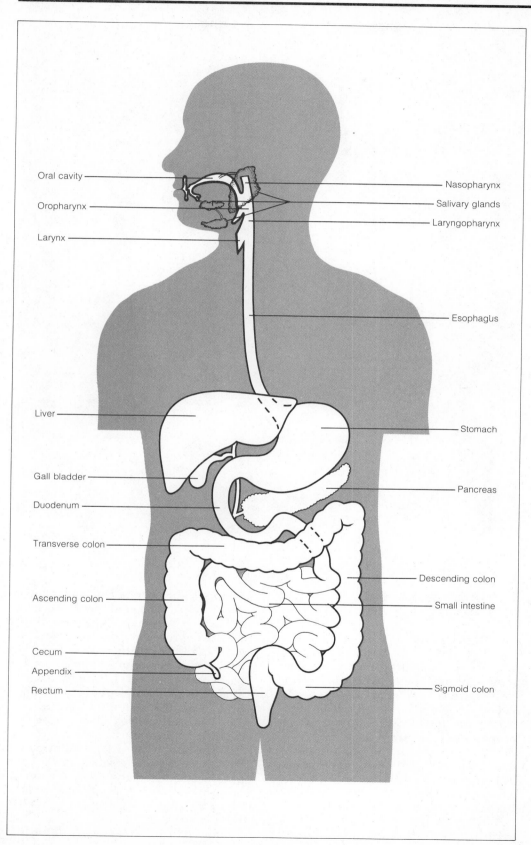

FIGURE 22–1
Organs of the digestive
system.

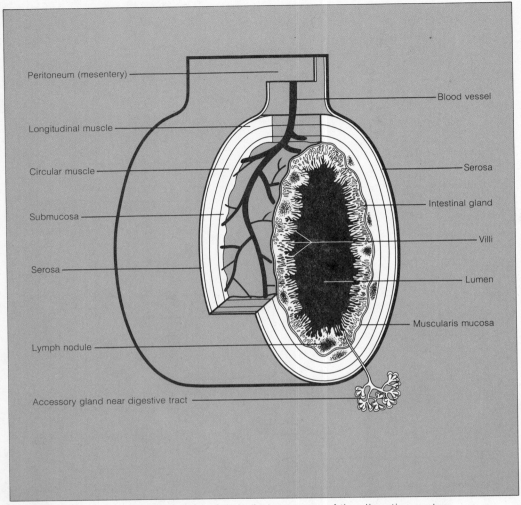

Peritoneum (mesentery)

Longitudinal muscle

Circular muscle

Submucosa

Serosa

Lymph nodule

Accessory gland near digestive tract

Blood vessel

Serosa

Intestinal gland

Villi

Lumen

Muscularis mucosa

FIGURE 22–2   Structural pattern of the tubular organs of the digestive system.

ORGANS OF THE DIGESTIVE SYSTEM

Oral Cavity

The digestive tract begins with the mouth, or **oral cavity,** which is bounded by the teeth and gums anteriorly and laterally, the hard and soft palate superiorly, the tongue inferiorly, and the oropharynx posteriorly (Figure 22–3). The space between the teeth and the lips and cheeks is known as the **vestibule;** both the vestibule and the oral cavity are lined with mucous membrane covered with a *stratified squamous epithelium* surface. Three pairs of salivary glands are located near the oral

cavity. Their secretions are carried to the oral cavity through special ducts.

The **hard palate** is formed by the palatine processes of the maxillary bones and the palatine bones. The **soft palate** is a posterior extension of the hard palate, but it is chiefly skeletal muscle and has the **uvula** suspended from it in the midline. Together the hard and soft palates separate the oral cavity from the nasal cavity.

The hard palate is formed by bones that grow from the sides toward the midline. Occasionally it does not develop properly, and one side may fail to reach the midline, leaving an opening between the nasal and oral cavities (*cleft palate*), or the two sides of the lips may not fuse together, leaving a split (*harelip*).

Teeth The teeth develop in the sockets or **alveolar processes** of the maxillae and mandible. The twenty **temporary** or **deciduous** teeth appear between the ages of 6 months and two-and-a-half years. They begin to be lost at about 6 years and are replaced over the next few years by the **permanent teeth,** of which there are 32 (Figure 22–4). Beginning at the midline, each quadrant (that is, quarter) of the jaw contains the teeth indicated in Table 22–1.

The teeth differ in shape in accord with their differing functions. The *incisors* are chisel shaped and adapted for cutting, while the *canine* teeth are pointed for tearing. The *bicuspids* (premolars) and *molars* have broad flat surfaces suitable for crushing and grinding. The *third molars,* the "wisdom teeth," do not erupt until much later, between the ages of 17 and 25. They may become impacted or otherwise fail to erupt completely, or they may never appear.

A tooth consists of a **crown** and one or more **roots,** joined by a relatively narrow neck (Figure 22–5). The crown is the visible exposed portion, and the root is that part within the alveolar process of the bone. Incisors have a single root, but molars have four.

The major portion of the tooth is made up of **dentin.** The dentin of the crown is completely covered by **enamel,** an extremely hard white material. In the center of the tooth is the **pulp cavity,** which contains blood vessels, nerves, and connective tissue. The pulp extends down through each root as the **root canal** and provides passage for the vessels and nerves to the tooth. The tooth is held firmly in its socket or alveolar process by a substance known as **cementum,** which fixes it to the connective tissue lining the socket (the *periodontal membrane*). Covering the bone around the teeth is the gum, or **gingiva,** a dense fibrous connective tissue covered with epithelium. Normally it is quite firm and hard and fits snugly around the neck of the tooth.

Teeth, like bones, have a protein

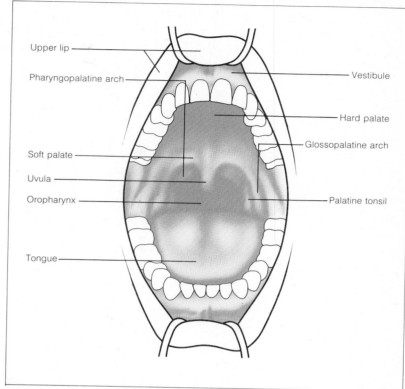

FIGURE 22–3   The oral cavity.

matrix within which a large amount of calcium and phosphorus salts are deposited. These elements are important ingredients of the diet, especially during the time that the teeth are forming. There is turnover of the mineral in dentin and cementum throughout life, but apparently very little turnover in enamel once it has been formed. Some of the bacteria that get into the mouth in the food are capable of producing or releasing protein-splitting enzymes and acids. They can destroy the protein matrix of the enamel, leading to breaks in and erosion of the enamel, forming **dental caries** (cavities). A very small amount of fluoride in the drinking water seems to promote the formation of enamel which is more resistant to the action of caries-producing agents.

Tongue The tongue is composed of skeletal muscle tissue, arranged both as *intrinsic* and as *extrinsic* muscles (see Figure 22–8). The former lie entirely within the tongue, while the latter have

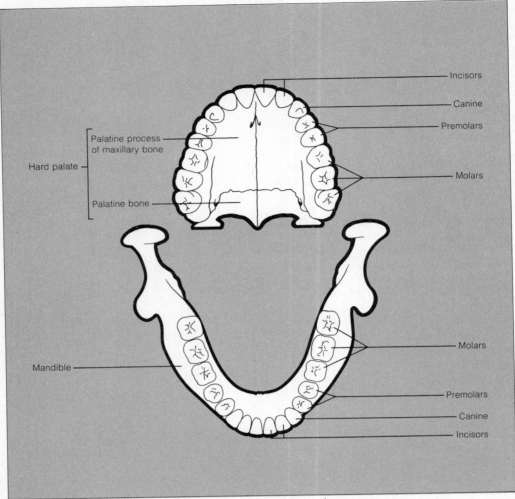

FIGURE 22-4   The permanent teeth, upper and lower jaws.

their origins on other structures, such as the hyoid bone and mandible, and are concerned with movements of the whole tongue (protraction, retraction, etc.). The fibers of the intrinsic muscles are interwoven and run in many differ-

ent directions within the tongue. They are, therefore, able to change the shape and position of the tongue in a great variety of ways. Such versatility is important because of the role of the tongue in speaking and in manipulating food, both of which require precision movements. The tongue is attached by its extrinsic muscles to the mandible, the hyoid bone, and the styloid process of the temporal bone, and by folds of mucous membrane to the floor of the mouth and to the epiglottis.

The mucous membrane covering the tongue is continuous with that lining the oral cavity and pharynx, but over the dorsal surface it is uniquely specialized, differing even from that on the ventral surface. The presence of numerous papillae gives the dorsum a

TABLE 22-1   TEETH OF EACH QUADRANT OF THE JAW

Temporary		Permanent
2	Incisors	2
1	Canines	1
—	Bicuspids (premolars)	2
2	Molars	3
20	Total (= × 4)	32

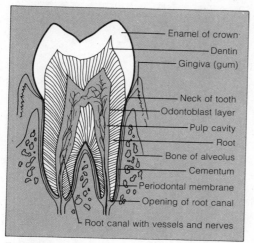

FIGURE 22-5   Longitudinal section of a tooth in its alveolus.

rough texture (see Figures 22-6 and 11-32). Slender thread-like **filiform papillae** and larger, more rounded **fungiform papillae** are distributed over the anterior two-thirds of the dorsum of the tongue, which is separated from the posterior third by a V-shaped line of about a dozen large **circumvallate papillae**. The **taste buds**, the chemoreceptors for the sense of taste, are to be found on the sides of the fungiform and circumvallate papillae. The surface of the posterior third of the dorsum has a coarser surface, due to numerous aggregates of lymphoid tissue known as *lingual tonsils*.

Salivary Glands   Three pairs of salivary glands (Figure 22-7A) lie near the oral cavity and secrete fluids which both lubricate and contribute to the digestive process. The largest is the **parotid gland**, found anterior to and slightly below the ear. Its duct, the *parotid duct*, extends anteriorly across the masseter muscle and goes deep to empty into the vestibule lateral to the second upper molar. The **submandibular gland** (sometimes called the submaxillary gland) lies medial to the angle of the mandible. Its duct penetrates the floor of the mouth to open into the oral cavity under the tongue. The **sublingual gland** lies in a fold of mucous membrane under the tongue and empties its secretion into the oral cavity through a number of small ducts.

The salivary glands are made up of spherical acini formed of single layers of pyramidal cells whose apices point toward the center of the acinus (Figure 22-7B). Branches of the salivary duct arise from the center of each acinus. The salivary glands have a rich blood supply and are innervated by both divisions of the autonomic nervous system.

Tubular Organs

Pharynx   Posteriorly, the oral cavity opens into the pharynx through an opening known as the *isthmus of the fauces*, bounded above by the **uvula** and on each side by two folds of mucous membrane, the **glossopalatine arch** and the **pharyngopalatine arch** (Figure 22-3). The arches are formed by skeletal muscles of the same names. Between the arches on each side is a fossa containing a mass of lymphoid tissue, the **palatine tonsil** (see Figure 22-8).

The pharynx serves as a passageway for both the respiratory system and the digestive system. Its three portions,

FIGURE 22-6   The dorsum of the tongue.

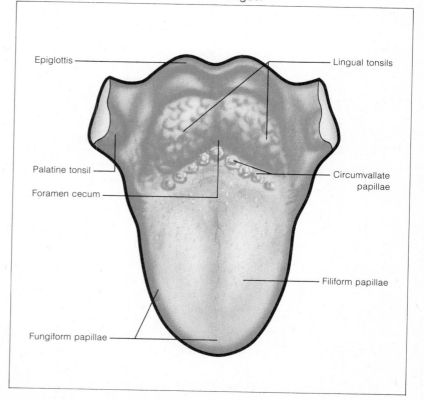

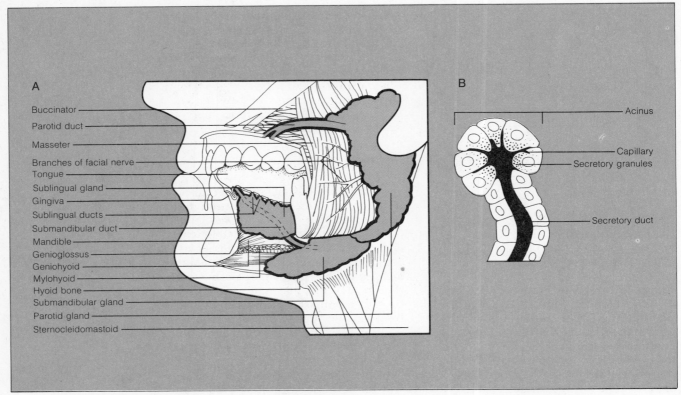

A
- Buccinator
- Parotid duct
- Masseter
- Branches of facial nerve
- Tongue
- Sublingual gland
- Gingiva
- Sublingual ducts
- Submandibular duct
- Mandible
- Genioglossus
- Geniohyoid
- Mylohyoid
- Hyoid bone
- Submandibular gland
- Parotid gland
- Sternocleidomastoid

B
- Acinus
- Capillary
- Secretory granules
- Secretory duct

FIGURE 22–7 A. Dissection of the face, showing the salivary glands and their ducts. Part of the mandible has been removed. B. Structure of a salivary gland.

the **nasopharynx,** the **oropharynx,** and the **laryngopharynx** are continuous with one another. The nasopharynx is above and behind the soft palate and uvula, and the nasal cavity opens into it. The oropharynx is directly behind the tongue and oral cavity and is continuous with the oral cavity as well as with the other two portions of the pharynx. The laryngopharynx opens inferiorly into the larynx and esophagus.

The mucous membrane of the pharynx, like that of the oral cavity, is lined with stratified squamous epithelium (except for the nasopharynx which has pseudostratified columnar epithelium). Beneath the mucous membrane are a number of muscles, mostly skeletal, whose contractions change the size and shape of the pharynx and are particularly important in swallowing.

Esophagus The esophagus is the tube that carries material from the pharynx, through the diaphragm into the abdominal cavity, and into the pouch-like stomach. It lies posterior to the trachea but, unlike the trachea, it has no cartilage rings to keep it open. It is therefore collapsed except when food is passing through. Its wall contains skeletal muscle at the upper end, which gradually gives way to smooth muscle nearer the stomach. It is lined with stratified squamous epithelium like the oral cavity and pharynx, and it contains only a few glands. The esophagus passes an opening in the diaphragm and enters the abdominal cavity. Its lowest portion is therefore covered by serosa (peritoneum) rather than adventitia.

Stomach The stomach (Figure 22–9) typically lies slightly to the left of the midline, under the diaphragm, although its position, like its shape, may vary. It resembles a swollen tube that has been bent to one side; the two

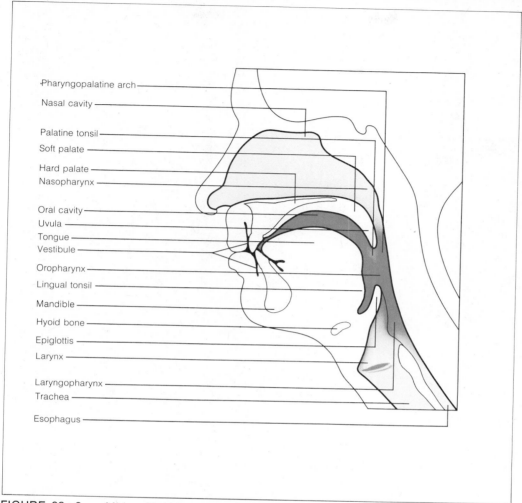

Pharyngopalatine arch

Nasal cavity

Palatine tonsil

Soft palate

Hard palate

Nasopharynx

Oral cavity

Uvula

Tongue

Vestibule

Oropharynx

Lingual tonsil

Mandible

Hyoid bone

Epiglottis

Larynx

Laryngopharynx

Trachea

Esophagus

FIGURE 22–8   Midsagittal section of the head, showing the oral cavity and pharynx.

sides of the bend are known as the **lesser curvature** and the **greater curvature.** The stomach extends above the entrance of the esophagus into a rounded portion known as the **fundus.** The main portion of the stomach is the **body,** which tapers to the **pyloric antrum,** leading to the **pyloric canal** and **sphincter.**

The stomach is a distensible sac, but when it is empty it is quite flat and longitudinal ridges, or *rugae,* on the inner surface become prominent (Figure 22–10). The mucosa lining the stomach is covered with simple columnar epithelium which dips down into frequent openings known as **gastric pits.** From the bottom of each pit long tubular glands extend down to the muscularis mucosa. These glands, so numerous that they fill the entire lamina propria, are the **gastric glands,** and they produce the gastric juice. The most common type of cell in the glands is the **chief cell,** which secretes a digestive enzyme. Large, plump triangular cells known as **parietal cells** produce the hydrochloric acid found in gastric juice. There are also mucus-secreting cells in the necks of the gastric glands.

The stomach has three muscular layers: the usual circular and longitudinal layers and an internal oblique layer. They are better developed toward the lower end of the stomach, and become quite thick as the antrum blends into the pyloric canal. The pyloric sphincter is anatomically impressive, although it

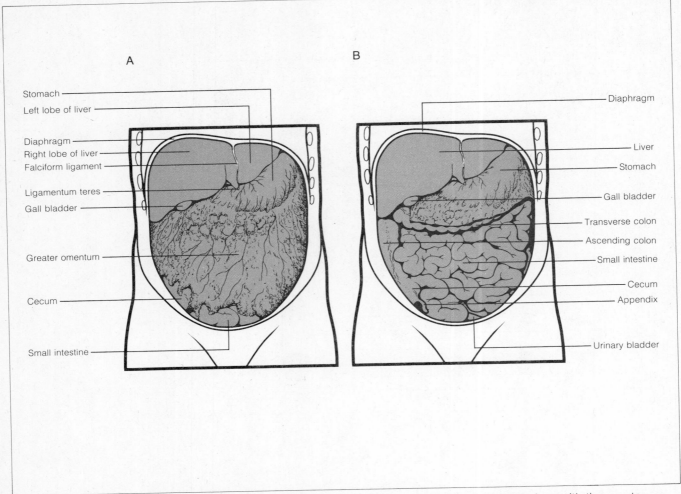

FIGURE 22–9 Organs of the abdominal cavity. A. Superficial view, with the greater omentum in place. B. With the greater omentum removed. C. With the stomach and liver raised and most of the small intestine removed.

is partially open most of the time. The stomach is covered and supported by peritoneum.

Small Intestine Most of the breakdown of foodstuffs and absorption of nutrients occurs in the small intestine, a fairly narrow tube about 10 feet long during life.[1] Structural and functional differences along its length divide it into three portions, the duodenum, the jejunum, and the ileum. They add many digestive juices to the intestinal contents, churn the resulting mixture (*chyme*), facilitate mechanical and chemical breakdown, and absorb nutrients, salts, and fluid. Although the structure of the small intestine changes along the way as the functions change, the change is gradual and there are no sharp boundaries between the three portions.

Duodenum. The first 10–12 inches of small intestine is the **duodenum.** From the pylorus it curves around in a "C" and becomes attached to the posterior body wall. The head of the pancreas is located in the curve of the duodenum (Figure 22–9).

A few centimeters beyond the pyloric sphincter, two ducts empty into the duodenum. The **common bile duct**

[1]With the loss of muscle tone that accompanies death, the small intestine relaxes to the length of 20–25 feet, which is more often cited.

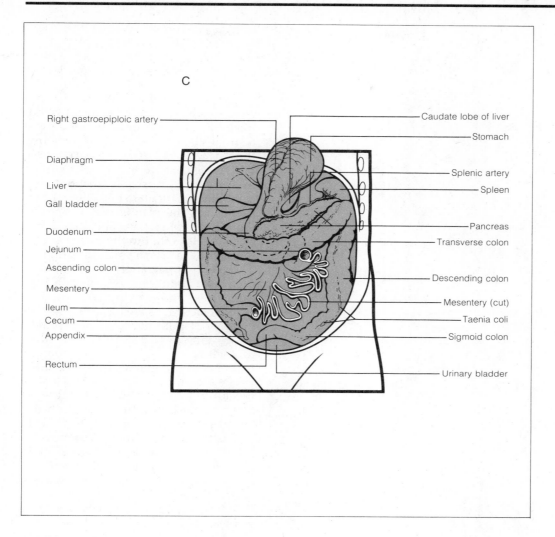

C

Right gastroepiploic artery —

Diaphragm —

Liver —

Gall bladder —

Duodenum —

Jejunum —

Ascending colon —

Mesentery —

Ileum —

Cecum —

Appendix —

Rectum —

— Caudate lobe of liver

— Stomach

— Splenic artery

— Spleen

— Pancreas

— Transverse colon

— Descending colon

— Mesentery (cut)

— Taenia coli

— Sigmoid colon

— Urinary bladder

delivers bile from the liver, and the **pancreatic duct** brings pancreatic juice. They usually join and enter the duodenum together, although occasionally they have separate openings. The opening, which is guarded by a small sphincter, is marked by a slight prominence in the intestinal lining known as the **duodenal papilla.**

Jejunum and ileum. Roughly two-fifths of the small intestine (4 feet) is **jejunum,** while the remaining three-fifths (6 feet) is **ileum.** The jejunum and ileum form a continuous, greatly convoluted tube, with no valves, sphincters, or ducts opening into it. They receive no secretions from other organs, but there are numerous glands within their walls. At the entry of the ileum into the large intestine (at the cecum) is the **ileocecal valve.** It is a somewhat

modified sphincter, as there are two flaplike folds of tissue which form a one-way valve permitting ready passage from the ileum to the cecum, but no return. Circular muscle fibers, however, by contracting or relaxing, provide a means of opening and closing the valve.

The ileum and jejunum, unlike most of the duodenum, are not bound closely to the body wall. They are rather suspended in the **mesentery,** a fold of peritoneum extending from the posterior abdominal wall. The arrangement allows the small intestine a considerable range of movement or displacement.

Wall and lining of the small intestine. The inner surface of the small intestine (Figure 22–11) is thrown into numerous shelf-like folds, the **plicae**

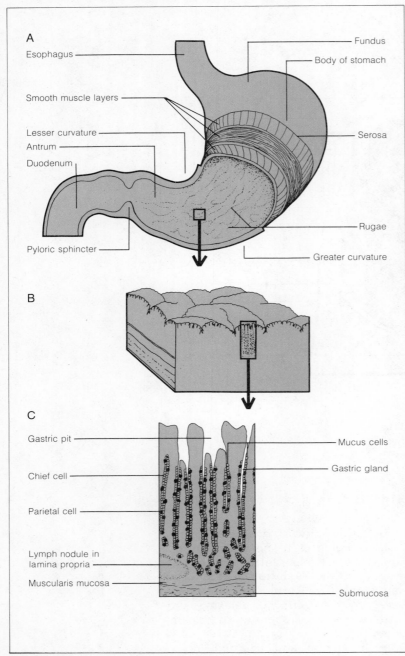

A

Esophagus

Fundus

Body of stomach

Smooth muscle layers

Serosa

Lesser curvature

Antrum

Duodenum

Pyloric sphincter

Rugae

Greater curvature

B

C

Gastric pit

Mucus cells

Chief cell

Gastric gland

Parietal cell

Lymph nodule in
lamina propria

Muscularis mucosa

Submucosa

FIGURE 22–10   A. The stomach, with cutaway to show the internal
surface and pyloric sphincter. B. The surface of the gastric mucosa.
C. A cross section through the wall of the stomach.

circulares, which make it impossible for material to pass through the intestine without frequent contact with its surface. These folds are particularly large and numerous in the jejunum.

The wall of the small intestine has all the typical layers of the digestive tract. The mucosa contains not only intestinal glands which dip down from the surface, but also tiny projections, or **villi,** which extend up from the surface. In fact, in microscopic sections it is sometimes difficult to tell where the villi end and the glands begin. The villi are slender processes, and they are so close together that to the naked eye the inner surface of the small intestine has a velvety appearance. The center of each villus contains an artery, a vein, capillaries, and a lymphatic capillary called a **lacteal,** as well as a few smooth muscle cells. Simple columnar epithelium covers the villi and lines the intestinal glands. Scattered among the epithelial cells are mucus-secreting goblet cells which increase in numbers toward the lower end of the small intestine. Most of these cells are in the mucosal layers *(lamina propria)*, but the duodenum also has some in the submucosa.

The surface area of the lumen of the human small intestine has been estimated as about 4500 square meters, which is nearly the area of a football field. This enormous surface area is achieved by several levels of surface convolutions. The lining itself has the folds of the plicae circulares, whose surfaces are covered with villi, and, in addition, the free edge of each epithelial cell has *microvilli*. All these irregularities facilitate mixing of the contents with the digestive juices and provide the large surface area needed for rapid and effective absorption.

The *lamina propria* of the small intestine contains a diffuse distribution of lymphocytes, as well as numerous aggregates of lymphoid tissue called **lymph nodules.** The size and frequency of the nodules increase with the distance from the stomach, and they are particularly conspicuous in the ileum, where they may spill over into the submucosal layer. These large lymphoid aggregates in the ileum (called *Peyer's patches*) are visible to the naked eye. They somewhat resemble tonsil tissue, which is also lymphoid tissue.

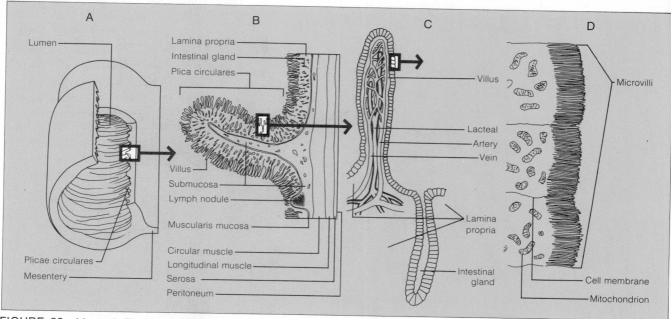

FIGURE 22-11   A. The small intestine, cut open to show the internal surface. B. A section of the intestinal wall, containing a plica circularis. C. A villus. D. Epithelial cells of the mucosal lining.

Large Intestine The large intestine, or **colon,** which is shorter in length but greater in diameter than the small intestine, is arranged rather like a border around the abdominal cavity (see Figures 22–9 and 22–12). The ileum opens into the side of a blind-ended pouch, the **cecum,** from which the **appendix** is suspended. From there, the large intestine continues upward as the **ascending colon,** turns just below the liver (the *hepatic flexure*), and crosses the abdominal cavity as the **transverse colon.** It then turns downward near the spleen (the *splenic flexure*) as the **descending colon** and curves posteriorly along the floor of the pelvic cavity as the **sigmoid colon,** which leads to the rectum.

The colon is bound more closely to the body wall than is the small intestine, with only the transverse colon having a mesentery (or *mesocolon*). Its wall includes the same layers as the rest of the digestive tract, but with some modifications. The mucosa lacks villi and plicae circulares, but it has intestinal glands and numerous lymphoid nodules. The epithelium is simple columnar with very many goblet cells.

The longitudinal layer of smooth muscle is reduced to three straplike bands, the **taenia coli.** Between the bands the intestinal wall tends to balloon out and form *sacculations* (little sacs) known as **haustra.** There are often fat-filled *tabs* or folds of peritoneum, called **epiploic appendages,** suspended from the colon. The large intestine shows much less muscular and secretory activity than the small intestine. By the time the intestinal contents have reached the colon the nutrients and most of the fluid and electrolytes have already been absorbed. The material that reaches the colon remains there longer than in other parts, and during this time some additional absorption converts it to a semisolid fecal material.

Rectum and Anal Canal The sigmoid colon swings toward the midline and just anterior to the sacrum it opens into the **rectum,** which then descends along the sacrum to the **anal canal.** The rectum's muscle layers are well developed and include a complete longitudinal layer. A sharp bend at its distal end marks the beginning of the anal canal, which leads to the external anal opening **(anus).** The anal canal is quite short and its walls contain two anal sphincters, an internal one of smooth muscle

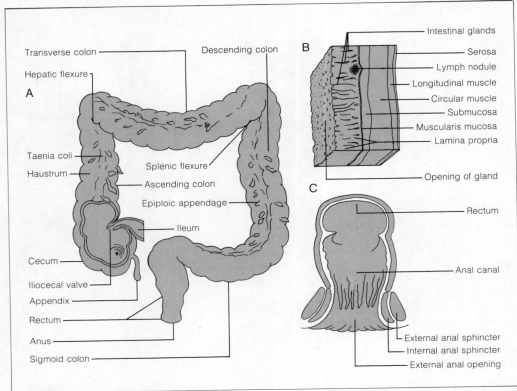

FIGURE 22–12 A. The large intestine, cut open to show the internal surface. B. A section of the wall of the large intestine. C. Longitudinal section through the anal canal.

and an external one of skeletal muscle. The external one is formed by the muscles that form the floor of the pelvic cavity; it is not as specific around the anal opening, but it provides voluntary control over the final exit from the digestive tract. Near the anal opening the epithelium changes from the simple columnar of the digestive tract to the stratified squamous typical of external surfaces (skin).

Accessory Organs of the Digestive Tract

Several important organs of the digestive system are located outside the tubular tract itself. These are the salivary glands (discussed earlier), the pancreas, and the liver. Each produces a secretion important to the digestive process and reaches the digestive tract by a duct. The gall bladder is not a secretory organ, but rather a storage site for bile produced by the liver.

Pancreas The pancreas (Figure 22–13) is a rather diffuse gland located below the stomach and behind the peritoneum. It has a **head,** located in the curve of the duodenum, a **neck,** a **body,** and a **tail** that extends almost to the spleen. It consists of closely packed acini and resembles the salivary glands under the microscope. It produces the **pancreatic juice,** containing several important digestive enzymes as well as electrolytes. Its ducts arise from the acini and empty their secretion into the **pancreatic duct,** which enters the duodenum in the company of the bile duct from the liver. Sometimes an accessory pancreatic duct enters the duodenum separately.

A unique feature of the pancreas is the presence of **pancreatic islands** *(islets of Langerhans),* isolated clumps of tissue scattered throughout the organ. The islands have a rich blood supply, but no ducts, and contain several types of cells. Two of these are known as *alpha* and *beta* cells, with the latter being much more numerous. These islet cells are functionally unrelated to the rest of the pancreas, since they constitute an important endocrine gland. They secrete several hormones into the bloodstream (see Chapters 13 and 24).

Liver The liver, with a weight of about 1500 grams, is the largest gland in the body (Figure 22–14). It is located directly under the right side of the diaphragm. It is divided into two main lobes, the smaller of which is the **left lobe.** The larger is the **right lobe,** and it has two small subdivisions visible on the inferior surface, the **quadrate lobe** anteriorly and the **caudate lobe** posteriorly.

The liver (except for a small "bare area" posteriorly) is covered by a layer of peritoneum reflected back from the inferior surface of the diaphragm. The right and left lobes are separated in front by a fold of peritoneum, the **falciform ligament,** which encloses the **ligamentum teres** (the *round ligament of the liver*) in its free (inferior) border. The round ligament (once the umbilical vein) extends from the umbilicus to a junction with the ligamentum venosum (formerly the ductus venosus) in the liver. On the inferior surface of the liver, between the caudate and quadrate lobes, is a fissure (the *porta hepatis*) through which the hepatic artery, the portal vein, the hepatic (bile) duct, lymphatics, and nerves enter the liver. The hepatic vein from the liver exits posteriorly and is very short because the inferior vena cava is adjacent to the liver. Between the quadrate lobe and the main portion of the right lobe on the inferior surface is a depression into which the gall bladder fits.

In view of the fact that the liver performs so many varied functions, one might expect to find many different types of cells in it. So far as is known, however, there is only one type of liver cell other than the blood cells and connective tissue cells, particularly macrophages, that are found both in the liver and elsewhere. This one kind of liver cell is so arranged, together with the blood vessels, as to give the liver a distinctive internal structure.

The functional units of the liver are

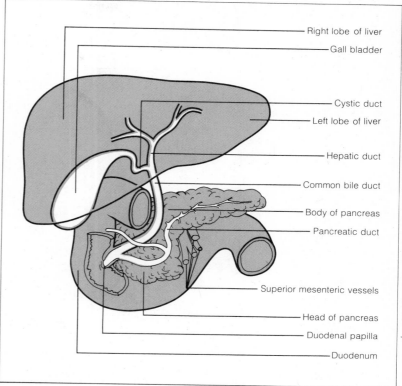

Right lobe of liver
Gall bladder
Cystic duct
Left lobe of liver
Hepatic duct
Common bile duct
Body of pancreas
Pancreatic duct
Superior mesenteric vessels
Head of pancreas
Duodenal papilla
Duodenum

FIGURE 22–13 The liver, pancreas, and the relations of their ducts. The stomach and most of the small intestine have been removed.

the **liver lobules,** which have been described as "polygonal prisms" of tissue (Figure 22–15). In cross section they appear to be roughly hexagonal in shape and separated from one another by a small amount of connective tissue. This connective tissue is prominent, however, only at the corners of each lobule. Each corner marks the intersection of several lobules. In this location there are almost invariably found three types of tube, forming the *portal triad*— at least one branch of the hepatic artery, a branch of the portal vein, and a bile duct. A central vein passes through the center of each lobule and drains into the hepatic vein.

The liver cells of each lobule are arranged in what seem to be cords of cells radiating from the central vein. In three dimensions, however, the cords become irregular plates or sheets of cells. The spaces between the plates are blood-filled sinusoids lined with squamous epithelial cells and holding a number of macrophages *(Kupffer cells)*

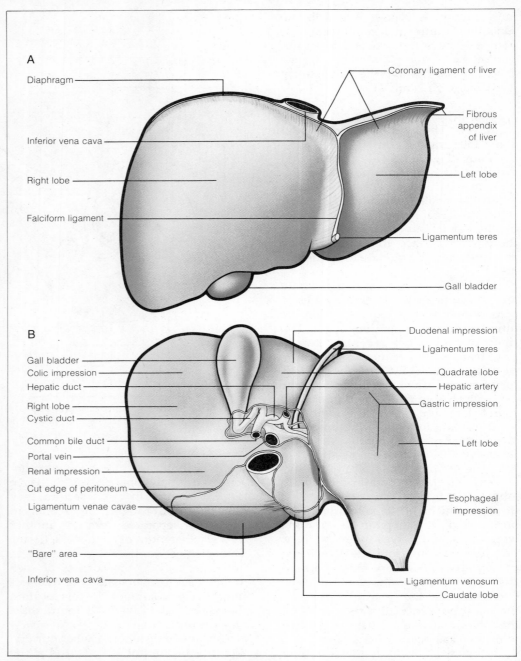

A

Diaphragm

Inferior vena cava

Right lobe

Falciform ligament

Coronary ligament of liver

Fibrous appendix of liver

Left lobe

Ligamentum teres

Gall bladder

B

Gall bladder
Colic impression
Hepatic duct
Right lobe
Cystic duct
Common bile duct
Portal vein
Renal impression
Cut edge of peritoneum
Ligamentum venae cavae

"Bare" area

Inferior vena cava

Duodenal impression
Ligamentum teres
Quadrate lobe
Hepatic artery
Gastric impression

Left lobe

Esophageal impression

Ligamentum venosum
Caudate lobe

FIGURE 22–14   The liver. A. Anterior surface. B. Posterior surface.

which phagocytize foreign material and cellular debris (such as fragments of disintegrating blood cells).

Blood from both the hepatic artery and the portal vein enters the sinusoids and moves slowly toward the central vein. As it does so, nutrients from the portal vein and oxygen from the hepatic artery are removed by the liver cells.

Some of the nutrients may be stored for a while before being released into the blood once more. In addition, the liver cells produce many substances (such as plasma proteins) which are added to the blood as it passes through the sinusoids.

Between adjacent liver cells are very tiny channels or crevices, the **bile**

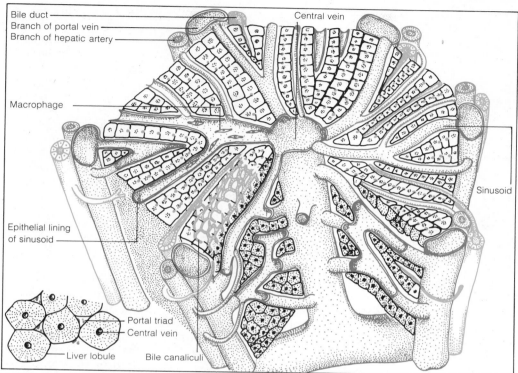

Bile duct
Branch of portal vein
Branch of hepatic artery
Central vein
Macrophage
Sinusoid
Epithelial lining
of sinusoid
Portal triad
Central vein
Liver lobule
Bile canaliculi

FIGURE 22-15   Structure of a liver lobule.

canaliculi. The bile is produced by the liver cells and secreted into the canaliculi, through which it moves toward the bile ducts at the periphery of each lobule. The bile ducts converge into larger ducts and eventually empty the bile into the **hepatic duct.**

Gall Bladder and Bile Ducts   The gall bladder is a small pear-shaped sac on the inferior surface of the liver. As a storage site for the bile, it is connected to the bile duct system. The **cystic duct** from the gall bladder joins the hepatic duct from the liver to form the **common bile duct,** which enters the duodenum with the pancreatic duct. The sphincter at the entrance to the duodenum is usually closed. Since the liver produces bile continuously, the bile backs up in the common bile duct and into the cystic duct and gall bladder, where it is held and concentrated. The gall bladder adds nothing to the bile, but does change its composition by absorbing a considerable amount of water from it. Appropriate stimuli relax the

duodenal sphincter and cause contraction of the smooth muscle in the wall of the gall bladder, expelling bile from the gall bladder into the duodenum.

BLOOD SUPPLY

As was discussed more fully in Chapter 16, the digestive tract receives its blood supply from several sources. The oral cavity and pharynx, including the deep structures of the face, the teeth of the upper and lower jaws, and the tongue are supplied by branches of the *external carotid artery,* while branches of the *thoracic aorta* supply the esophagus. The bulk of the digestive tract lies in the abdominal cavity and receives its blood supply from three abdominal arteries, the *celiac,* the *superior mesenteric,* and the *inferior mesenteric arteries.* The three branches of the celiac artery supply the digestive organs in the upper abdominal cavity — the stomach, liver, gall bladder, lower esophagus, duodenum, and pancreas — as well as the spleen. Anastomoses between these branches provide an alternative source of supply, particularly for the stomach (see Figures 22–9C and 16–8A). The intestine, from the duodenum through the transverse colon, receives blood from the superior mesenteric artery (Figure 16–8B) which anastomoses with branches of the celiac and inferior mesenteric arteries. The inferior mesenteric artery supplies the descending and sigmoid colon and part of the rectum. The rectum and anal canal also receive blood from the rectal branches of the *internal iliac* (hypogastric) *artery*.

Anastomoses between the terminal branches of the major arteries may be of great importance as alternate sources of blood in case the flow in any of them is restricted. Such anastomoses are found between branches of the esophageal and the gastric, between the gastric (gastroepiploic) and the superior mesenteric, between the superior and the inferior mesenteric, and the inferior mesenteric and the rectal branches of the internal iliac artery.

Most of the veins draining the digestive tract empty into the *portal vein*, which carries the blood to the liver before it enters the general circulation (Figure 16–15). The portal vein is formed by the confluence of the inferior mesenteric and splenic veins with the superior mesenteric vein, and it is joined by branches from the stomach. The portal vein blood passes through the liver lobules and then is collected in the hepatic vein, which empties into the inferior vena cava.

INNERVATION OF THE DIGESTIVE TRACT

The digestive tract, except for the voluntary structures at the two ends, is innervated by the autonomic nervous system. The exceptions include most of the muscles of the oral cavity, tongue, pharynx, and upper part of the esophagus, and the external anal sphincter. They are supplied mainly by somatic neurons in cranial nerves VII, IX, X, and XII. These nerves also carry sensory fibers. Most of the digestive structures innervated by the autonomic nervous system (except the blood vessels) receive fibers from both divisions.

Most of the *sympathetic* fibers to the abdominal and pelvic portions of the digestive tract pass through the sympathetic chain without synapsing and emerge as the *splanchnic nerves,* which go to one of several outlying collateral ganglia. There they synapse with the cell bodies of the postganglionic neurons, which then proceed to the effectors. If the sympathetic chain or splanchnic nerves are severed, the postganglionic neurons are deprived of their innervation and cease to function effectively. (The denervated smooth muscle, however, in time develops a hypersensitivity to epinephrine, the sympathetic mediator.)

The *parasympathetic* innervation to the salivary glands and mucous membranes of the oral cavity and pharynx is carried by cranial nerves VII and IX. The vagus nerve (X) supplies the digestive tract from the lower part of the esophagus down through the transverse colon. The remainder is innervated by sacral autonomic fibers. The long parasympathetic preganglionic neurons reach to the organ, while the short postganglionic neurons lie entirely within the organ.

Branches of these neurons synapse with one another to form networks or plexuses of interconnecting postganglionic fibers (Figure 22–16). In a histological section of the digestive tract, the cell bodies of these neurons, known as *ganglion cells,* can be seen, both in the *myenteric plexus* between the circular and longitudinal muscle layers and in the *submucous plexus* inside the circular layer. This type of arrangement is different from any neuronal connections encountered so far. Because of the plexus arrangements, ganglion cells act on other ganglion cells as well as on smooth muscle or secretory cells, and they are affected by one another as well as by preganglionic neurons. In addition, dendritic processes of ganglion cells may end in the mucosa, where they function as chemoreceptors that respond to the composition of the contents of the digestive tract, or as stretch receptors that respond to the tension in its wall. We have, therefore, all the necessary elements for a reflex-type action

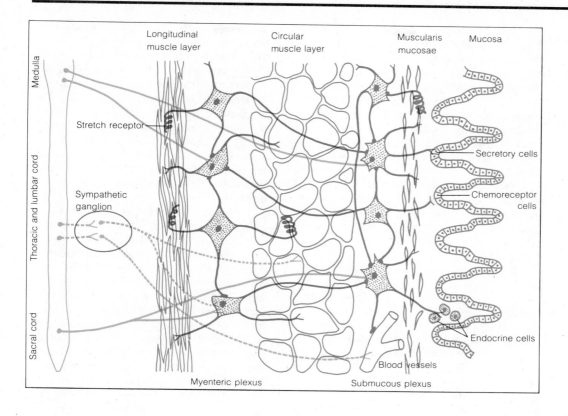

Longitudinal muscle layer   Circular muscle layer   Muscularis mucosae   Mucosa

Medulla

Thoracic and lumbar cord

Stretch receptor

Sympathetic ganglion

Sacral cord

Myenteric plexus   Submucous plexus

Secretory cells

Chemoreceptor cells

Endocrine cells

Blood vessels

completely within the wall of the digestive tract and independent of the central nervous system; it is known as the *intrinsic innervation* of the gut. Cutting the vagus nerves severs connections between the intrinsic plexuses with the central nervous system: that is, it removes their extrinsic nerve supply, but since ganglion cells receive input from one another and the local environment, the network may still function independently.

In spite of the possibility of regulation in the absence of central nervous system connections, the more conventional reflex pathways are normally present and very functional. There are numerous sensory neurons that travel to the central nervous system (in the vagus, pelvic, and sympathetic nerves) to provide the afferent limbs of typical autonomic reflexes and the input for sensations arising from the gut.

The intrinsic plexus exists throughout the digestive tract (except where there is skeletal muscle), but it becomes increasingly important in the lower portions of the tract. The smooth muscle of the gut is of the visceral type,

and as such it possesses a rather high degree of automaticity. The role of the nerve supply, therefore, is to alter this ongoing activity. In general, smooth muscle of the digestive tract is excited by parasympathetic fibers and inhibited by sympathetic fibers.

The effect of autonomic stimulation of the digestive glands varies, since in several instances secretion is increased by stimulation of either division. However, the nature of the secretion (that is, its enzyme content) in each case is different. Secretion is also facilitated indirectly by vasodilatation of the blood vessels, a parasympathetic effect (brought about by reciprocal inhibition of sympathetic vasoconstrictor fibers).

PERITONEUM

The peritoneum of the abdominopelvic cavity resembles the pleura of the thoracic cavity in several respects. **Parietal peritoneum** lines the wall of the cavity and **visceral peritoneum** covers the organs. The parietal and visceral layers of peritoneum are continuous with one another, forming the **peritoneal sac** (sometimes called the **greater sac**). It encloses an area known as the **peritoneal cavity,** which is very nearly filled by the organs within it. The epithelial cells on the surface of the peritoneum secrete a small amount of fluid that serves as a lubricant and permits the surfaces to slide freely over one another.

Some organs adhere to the body wall or floor of the pelvic cavity and are covered by peritoneum, but not surrounded by it. These **retroperitoneal** organs include part of the duodenum, the pancreas, and the rectum, as well as the kidneys, the urinary bladder, and the reproductive organs. Some organs are suspended in the abdominal cavity in the free edge of a fold of peritoneum. The support for the small intestine is called the **mesentery,** while parts of the large intestine are suspended in **mesocolon** (and the appendix has a **mesoappendix**). Arteries, veins, nerves, and lymphatics to the intestine are found between the two layers of peritoneum which form these attachments.

The arrangement of the peritoneum and its folds can be described by tracing the course of the peritoneum around the abdominopelvic cavity (Figure 22–17). The parietal peritoneum adheres to the fascia of the innermost muscle of the abdominal wall, the transversus abdominis. It continues along the inferior surface of the diaphragm and is reflected back over the top of the liver as visceral peritoneum. A fold of peritoneum, the falciform ligament, emerges anteriorly between the right and left lobes of the liver and extends to the anterior abdominal wall (Figure 22–14A). It encloses the ligamentum teres (the obliterated umbilical vein) in its inferior border. A ligament-like continuation of peritoneum extends between the inferior surface of the liver and the lesser curvature of the stomach. It forms part of the **lesser omentum,** through which pass the hepatic artery, portal vein, bile ducts, and lymphatics. Peritoneum covers the anterior surface of the stomach, and from the greater curvature it hangs down in a great fold reaching almost to the symphysis pubis before turning back upon itself to reach

the transverse colon. If the anterior abdominal wall is laid back to expose the abdominal contents, most of the organs remain covered by this long fold of peritoneum (see Figure 22–9A). It is known as the **greater omentum,** but because of its location and appearance it is sometimes called the "lace apron." Its peritoneal surfaces adhere to one another but are incomplete, since there are often holes in the greater omentum, and bits of fat are deposited within or upon it.

The peritoneum covers the inferior surface of the transverse colon and continues to the posterior body wall to form part of the **transverse mesocolon.** Along the posterior body wall the peritoneum covers the aorta, the inferior vena cava, and the kidneys. It then turns away from the body wall as the fold of mesentery that encloses and suspends the small intestine. At no place is the mesentery more than a few inches in length, and the whole length of small intestine is attached to it with many twists and turns, rather like ruffles. The ascending and descending colon lie against the body wall and do not have a true mesocolon, but the sigmoid does have a short one. Finally, the peritoneum encloses the rectum and passes up and over the uterus (in the female) and over the urinary bladder, then up the anterior abdominal wall.

Behind the stomach is a second, smaller peritoneal sac, known as the **lesser peritoneal sac,** or the **omental bursa.** It covers part of the posterior inferior surface of the liver and joins with the peritoneum from the anterior surface to form the lesser omentum. The lining of the lesser sac passes around the posterior surface of the stomach and meets that of the greater sac. It then extends down into the greater omentum and returns over the transverse colon and back to the posterior body wall as part of the transverse mesocolon. The lesser omentum and the transverse mesocolon are, therefore, formed by one layer of peritoneum from the greater sac and one layer from the lesser sac. The greater omentum is formed by four layers of peritoneum, a fold (two layers) from the lesser sac enclosed in a fold (two layers) from the greater sac. They are all fused together, however, so that there is normally no real extension of the open space of the lesser sac into the greater omentum (although such an extension is shown in Figure 22–17).

The lesser sac is not completely isolated, though, since there is a small opening from it into the greater peritoneal cavity. It is called the **epiploic foramen** and is located at the right edge of the lesser omentum, between the liver and the stomach.

CHAPTER 22 SUMMARY

General Structure

The major portion of the digestive system is a tube running through the body. It is modified in different areas to reflect differing functions, but a general pattern applies throughout. Four layers are described in the wall, beginning on the inner (lumenal) surface as *mucosa, submucosa, muscularis,* and *serosa* or *adventitia*. The mucosa has three layers, the innermost *epithelium*, a connective tissue *lamina propria* that usually contains glands, and a layer of smooth muscle, the *muscularis mucosa*. The submucosa is also connective tissue and may contain glands, blood vessels, and nerves. The muscularis layer typically contains an inner circular and an outer longitudinal layer of smooth muscle. The outermost layer of organs that are adjacent to a body cavity (such as small intestines) is known as serosa. It is part of the peritoneum, and its free surface is a layer of epithelium. Adventitia covers organ surfaces that are continuous with deep fascia.

Organs of the Digestive System

The structural pattern differs at the upper end of the digestive tract. The *oral cavity* is separated from the nasal cavity by the hard and soft palate, and bounded laterally by the teeth. There are 32 *permanent teeth,* two incisors, one canine, two bicuspids, and three molars on each side of each jaw. A tooth is mainly dentin which, like bone, con-

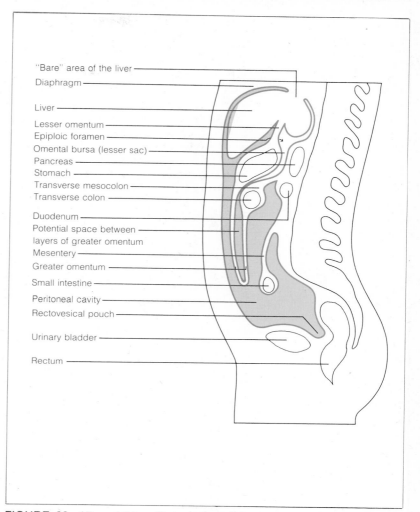

Labels (top to bottom): "Bare" area of the liver, Diaphragm, Liver, Lesser omentum, Epiploic foramen, Omental bursa (lesser sac), Pancreas, Stomach, Transverse mesocolon, Transverse colon, Duodenum, Potential space between layers of greater omentum, Mesentery, Greater omentum, Small intestine, Peritoneal cavity, Rectovesical pouch, Urinary bladder, Rectum

FIGURE 22–17   Midsagittal section of the abdominopelvic cavity to show the peritoneum.

tains a large amount of calcium. It is covered by enamel on its exposed parts. The exposed portion is the crown, and the roots extend down into the alveolar process. The pulp cavity is inside and contains blood vessels and nerves which emerge through the root canal. The teeth are held tightly in the alveolar processes of the maxillae and mandible by cementum. The gingiva, or gum, is also tightly bound to the bone.

The *tongue* occupies the floor of the oral cavity. It consists of skeletal muscle whose fibers run in all directions. Chemoreceptors for taste are located in taste buds found on many of the papillae on the dorsal surface of the tongue.

Three pairs of *salivary glands* (parotid, submandibular, and sublingual) deliver their secretion (saliva) to the oral cavity by special ducts.

The *pharynx*, like a funnel, channels material into the *esophagus*, with which it is continuous. Both are lined with stratified squamous epithelium, and there is skeletal muscle in the pharynx and upper esophagus, but it gives way to smooth muscle toward the lower end of the esophagus.

The *stomach* is an expansion of the digestive tube and lies directly under the diaphragm. Its parts are the *fundus, body, antrum,* and *pyloric canal,* ending at the *pyloric sphincter.* The mucosal surface has ridges or rugae, and gastric pits. Gastric glands, which contain *chief cells, parietal cells* and *mucus cells,* open into the gastric pits.

The *small intestine* includes the *duodenum, jejunum,* and *ileum.* The duodenum is quite short and receives secretions from the pancreas and liver. It is continuous with the jejunum which becomes the ileum which opens into the cecum. The small intestine is suspended from the posterior body wall by mesentery, which is part of the peritoneum.

The inner surface of the small intestine is thrown into transverse folds, the *plicae circulares.* The surface is covered with tiny fingerlike *villi,* and the epithelial cells contain numerous *microvilli,* all of which greatly increase the surface area of the small intestine. Each villus contains a tiny artery, vein and lymphatic capillary *(lacteal).* Numerous glands produce intestinal juice.

The large intestine consists of the *cecum* and the *ascending, transverse, descending,* and *sigmoid colon.* The cecum, the blind-ended pouch to which the appendix is attached, receives the contents of the small intestine through the *ileocecal valve.* Its mucosa lacks villi, and its glands secrete mostly mucus. The longitudinal muscle layer exists mainly as three muscular bands *(taenia coli).*

The colon empties into the *rectum,* which leads to the *anal canal* and *anus.* Glands are lacking, but the muscle layers are well-developed. Smooth muscle forms an *internal anal sphincter,* and skeletal muscles of the floor of the pelvic cavity function as an *external anal sphincter* providing for voluntary control.

The salivary glands, pancreas, and liver, are accessory digestive organs. The *pancreas* is located below the stomach in the curve of the duodenum. Its exocrine secretion is produced in the acini and reaches the duodenum in the *pancreatic duct.* Small clusters of cells, the pancreatic islands, are endocrine glands and their secretion enters the bloodstream.

The *liver,* the largest gland in the body, lies in the upper right portion of the abdominal cavity. The portal vein and hepatic artery enter, and the hepatic duct leaves from the inferior surface. The hepatic vein leaves from the posterior surface. The liver is organized into roughly polygonal *lobules* of liver cells arranged in irregular sheets or plates. A *central vein* penetrates the center of each lobule and the blood vessels and bile ducts are along the edges. The hepatic artery and portal vein empty into *sinusoids* that occupy the spaces between the sheets of liver cells. As blood moves through the sinusoids toward the central vein, the liver cells remove some substances from the blood and add others to it. There are macrophages in the sinusoids. Bile produced in the liver cells drains into tiny *bile canaliculi* located between cells. They empty into bile ducts at the edges of the lobules, which converge to form the *hepatic duct* leading from the liver.

The *gall bladder* is located on the inferior surface of the liver. Its one duct, the *cystic duct,* joins the hepatic duct to form the *common bile duct,* which empties into the duodenum with the pancreatic duct. Since the opening is guarded by a sphincter that is usually closed, the bile from the liver backs up into the gall bladder where it stays until its release into the duodenum.

Blood Supply

The blood supply for the parts of the digestive tract in the abdominal cavity comes from the *celiac axis* (hepatic, splenic, and left gastric arteries), and the *superior* and *inferior mesenteric arteries.* The esophagus is supplied by branches of the thoracic aorta and the rectum is supplied by branches of the internal iliac artery. Most of the veins converge to form the *portal vein* which enters the liver.

Innervation

The digestive tract is innervated by the autonomic nervous system, except for skeletal muscles in the upper portion and external anal sphincter. Sympathetic fibers reach abdominal organs by way of splanchnic nerves and collateral ganglia. The parasympathetic innervation is mainly by way of the vagus and pelvic nerves. The postganglionic neurons are the *ganglion cells* in the *myenteric* and *submucosal plexuses* in the walls of the hollow organs. The ganglion cells synapse with other ganglion cells, and their processes arise from stretch and chemoreceptors in the mucosa, and end on secretory cells or smooth muscle cells. They form an *intrinsic innervation* that can still function if connections with the central nervous system are interrupted. The ability to function independently is particularly important in the lower part of the digestive tract.

Peritoneum

Free surfaces in the abdominal cavity are covered by a serous membrane, the *peritoneum,* and the organs are supported by peritoneal folds such as *mesentery* and *mesocolon*. The *greater omentum* is a fold of peritoneum suspended between the greater curvature of the stomach and the transverse colon, while the *lesser omentum* lies between the liver and the lesser curvature of the stomach. Parietal and visceral peritoneum are continuous with one another and enclose the *peritoneal* or *greater sac*.

A *lesser sac,* or *omental bursa,* lies behind the liver and stomach and communicates with the greater sac through the *epiploic foramen* at the edge of the lesser omentum. Organs that lie outside of the peritoneal sac, such as the kidney and urinary bladder, are said to be *retroperitoneal*.

STUDY QUESTIONS

1 What anatomical features are common to all of the tubular structures of the digestive tract?

2 In what ways do the structures of the esophagus, stomach, small intestine, and colon differ from this common pattern?

3 Name in order the structures through which food passes as it moves through the gastrointestinal tract. Make a statement about each structure—something of its anatomical or functional significance.

4 What important glands of the digestive system are not part of the "tube"? What is their role? What about the gall bladder?

5 What vessels carry blood to the liver? Trace a drop of blood through the liver to the inferior vena cava.

6 Trace the probable course of a drop of bile from its site of production to its entry into the small intestine.

7 Distinguish between: the greater and lesser omentum; the peritoneal cavity and the lesser sac (omental bursa); mesentery and mesocolon.

8 What is unique about the innervation of the digestive tract? What is the relationship between the intrinsic and extrinsic innervation?

23

Control of the Digestive Tract

Motor and Secretory Functions of the Digestive Tract

Absorption

Digestive and Absorptive Functions of the Gastrointestinal Tract

Chapter 23 examines the processes through which energy-containing foodstuffs are broken down, mechanically and chemically, into molecules that can be transported across the intestinal wall into the bloodstream.

Materials which enter the digestive tract are moved along by the action of the muscles in its walls. As the contents progress through the tract they are exposed to the action of the digestive juices that are secreted into it along the way, and at the same time they are mixed and churned by muscle actions. Nutrients obtained from the foodstuffs are absorbed, but some things we eat are not digested very well and are excreted from the body.

What causes the muscular activity to increase after food has been eaten? What determines which of several secretions is to be produced at a particular time? How do these secretions actually break down ingested foods? How does absorption occur? What happens to the secretions, and to the undigested material? You should find some answers to these questions as we consider how the motor and secretory activities of the digestive tract are controlled and coordinated to prepare nutrients for absorption.

In the previous chapter the digestive tract was described as a tube extending through the body from mouth to anus. It is chiefly a muscular tube, but its walls also contain glands whose secretions are poured into its lumen. In addition, the lumen receives the secretions of accessory organs of the digestive system. Foodstuffs put into the upper end of the digestive tract are broken down to tiny bits and mixed with the digestive juices as they are moved along. Nearly all of the ingested material is digested, and the end products are absorbed through the intestine.

The food that enters the gastrointestinal tract is made up of six major components: carbohydrates, fats, proteins, minerals, vitamins, and water. Most of our attention will be directed to the first three because they furnish the body's energy and provide the material used for growth and repair. Vitamins and minerals are neither energy sources nor building blocks, but they play important roles in many metabolic reactions concerned with energy exchange and growth and repair. Water, of course, is involved in all reactions, since it is the major constituent of the body's cells and of many foodstuffs, and all reactions in the body take place in an aqueous solution.

The process of digestion involves the chemical and physical breakdown of ingested foodstuffs, to molecules small enough to be absorbed from the digestive tract. The chemical breakdown is brought about by action of the digestive enzymes contained in the secretions. There are a number of such enzymes, each of which splits a specific linkage in a carbohydrate, fat, or protein molecule. Several different enzymes are often required to complete the digestion of a particular foodstuff. Important enzyme actions are listed in Table 23 – 1.

The physical breakdown involves not only propelling the contents along the tube, but also chewing in the mouth and churning and mixing in the stomach and intestines. These actions mechanically break down the food, which increases the effectiveness of the digestive enzymes. They also bring the digested materials into contact with the intestinal wall, which facilitates absorption.

In addition to the production and release of the digestive secretions and their enzymes, the gastrointestinal tract also secretes several hormones that help regulate its activities.

CONTROL OF THE DIGESTIVE TRACT

Since the breakdown of ingested materials is the result of activity of muscle in the wall of the tract and of glands in the walls and in the nearby accessory organs, the story of control of the digestive tract is the story of the control of muscle and gland. Both are subject to neural and hormonal regulation. Both may also respond directly to changes in their immediate environment. Before proceeding, however, it might be well to review some of the properties of digestive tract muscle, both skeletal and smooth (see Chapter 7).

Properties of Muscle of the Digestive Tract

Skeletal muscle is found in the walls of the oral cavity, pharynx, and upper esophagus, as well as the anal region. The highly organized internal arrangement of the myofilaments of its large multinucleated cells is responsible for the striated appearance. A complex system of intracellular tubules ensures nearly simultaneous contraction of entire cells, as described in Chapter 7. Skeletal muscle cells are totally dependent upon action potentials in their motoneurons, for they possess no automaticity or tonus on their own.

Smooth muscle, on the other hand, has very different properties. The tiny cells lack the elegant internal organization of skeletal muscle cells, although they do contain myofilaments and some sarcoplasmic reticulum. Smooth muscle in the gastrointestinal tract is of the single unit type, which means that the cells are arranged in sheets or layers and large bundles of cells act together

TABLE 23–1   THE PRINCIPAL DIGESTIVE ENZYMES

Enzyme	Source	Substrate	End-product	Comment
Salivary amylase	salivary glands	starch	polysaccharides, disaccharides	inactivated by gastric acidity
Pepsin (*Pepsinogen*)	chief cells of the stomach	proteins polypeptides	polypeptides	pepsinogen activated by HCl
Pancreatic amylase	pancreas	starch	disaccharides	—
Pancreatic lipase	pancreas	fats, triglycerides	glycerides, fatty acids, glycerol	absorbable end-products
Trypsin (*Trypsinogen*)	pancreas	proteins, polypeptides	smaller polypeptides	trypsinogen activated by enterokinase
Chymotrypsin (*Chymotrypsinogen*)	pancreas	proteins, polypeptides	smaller polypeptides	chymotrypsinogen activated by trypsin
Carboxypeptidase	pancreas	small polypeptides	shorter peptide chains and amino acids	—
Ribonuclease	pancreas	RNA	nucleotides	—
Deoxyribonuclease	pancreas	DNA	nucleotides	—
Aminopeptidase	intestinal mucosa	small polypeptides	shorter peptide chains and amino acids	—
Dipeptidase	intestinal mucosa	dipeptides	amino acids	absorbable end-products
Disaccharidases *Maltase* *Sucrase* *Lactase*	intestinal mucosa	disaccharides *maltose* *sucrose* *lactose*	monosaccharides *glucose* *glucose, fructose* *glucose, galactose*	absorbable end-products
Enterokinase	intestinal mucosa	trypsinogen	trypsin	activates trypsin, no digestion per se
Nucleotideases	intestinal mucosa	nucleotides	nucleosides and phosphates	—

as single units. Unitary action is possible largely because of the low electrical resistance between cells due to the presence of gap junctions. Each cell is polarized, but its resting potential is unsteady, exhibiting varying degrees of hypo- and hyperpolarization and, therefore, fluctuations in its excitability. In certain "pacemaker" cells, the resting potential oscillates rhythmically; because of the communication between cells, the "pacemakers" drive the other cells in a bundle, whose resting potentials then oscillate in synchrony. Whenever the resting potential reaches threshold during the hypopolarization phase of the oscillation, there will be one or more action potentials followed by contractions in all of the cells affected.

The whole contraction process in smooth muscle cells is very slow when compared to that of skeletal muscle. The action potentials are prolonged (50 msec, as compared to 1–2 msec for skeletal muscle), the contraction itself (latency, shortening, and relaxation) lasts several seconds, and single contractions fuse into tetanic contractions at very low rates of stimulation. Unlike skeletal muscle, smooth muscle maintains a certain degree of tension at all times, even in the absence of action potentials and contractions. The amount of tension fluctuates with the changes in resting potential.

In the digestive tract, pacemakers are located in the longitudinal muscle layer. In the stomach they are in the fundic region along the greater curvature. Pacemaker cells generate slow waves of hypopolarization which pass over the stomach toward the pylorus. These rhythmic waves of partial depolarization constitute what is known as the **basic electrical rhythm, BER.** Whether or not there is any contraction of the muscle during the hypopolarization portion of the slow waves depends upon the excitability of the muscle cells (Figure 23–1). If it has been raised—perhaps by nerve stimulation—action potentials followed by contraction are likely to occur with each slow wave.

There are pacemakers at various sites in the intestine as well, and they are responsible for the development of a BER in these regions. The frequency of the waves is not the same in all parts (they are more frequent in the intestine for example), but the direction is always caudally, that is, toward the anus. The speed at which the wave progresses is determined by the rate of conduction of the slow wave through the cells and gap junctions. It is relatively slow—about a centimeter per second.

Smooth muscle in the digestive system, therefore, is capable of contracting without a known outside stimulus; it possesses *automaticity* or *inherent rhythmicity* because of the pacemakers and the communication between cells. A nerve supply is not essential for contraction of gastrointestinal smooth muscle; the nerve supply helps set the level of excitability of those cells and thus influences the occurrence of action potentials in the smooth muscle cells during each slow wave depolarization.

Neural Control

The motoneurons to smooth muscle do not end at a motor end plate on each cell. Rather, they wander between the cells. At intervals along the axon are areas which contain vesicles, presumably of transmitter substance. An action potential in the neuron causes transmitter to be released into the extracellular fluid where it affects the muscle cell membranes in the immediate vicinity. Acetylcholine (from parasympathetic endings) tends to depolarize the muscle cell membranes, increasing the likelihood of reaching threshold during the BER hypopolarization. Norepinephrine

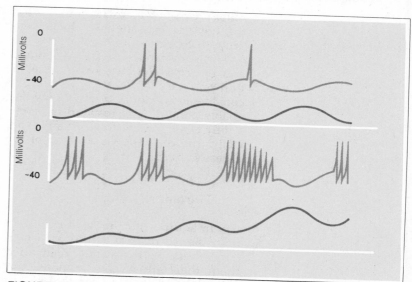

FIGURE 23–1   Electrical activity of intestinal smooth muscle. Upper trace: Electrical activity of a single cell recorded with an intracellular electrode. Lower trace: Tension developed by the muscle.

and epinephrine (from sympathetic endings and in the circulation) hyperpolarize the muscle cell membranes and reduce the likelihood of action potentials and contractions.

The motoneurons are the efferent limbs of typical autonomic reflex arcs, whose reflex centers are in the central nervous system and which respond to a wide variety of stimuli from various parts of the digestive tract as well as from other areas. It is important to remember that the digestive system has an additional neural network, the *intrinsic innervation,* formed chiefly by the interconnections of the parasympathetic postganglionic neurons within the wall of the gut. These neurons are known as ganglion cells, and their bodies lie between and under the smooth muscle layers in the wall of the gut (see Figure 22–16). Ganglion cells not only receive input from parasympathetic preganglionic neurons (vagus), but they also synapse with one another. In addition, they have receptor endings in the epithelium and in the muscle that function as chemoreceptors and mechanoreceptors (stretch). They may also have motor processes that end on smooth muscle and secretory cells (both endocrine and exocrine). Thus, there is the functional equivalent of a reflex arc completely within the wall of the gut (an exception to the general "rules" of reflex action).

The properties of smooth muscle and the characteristics of its innervation provide the digestive tract with a considerable amount of autonomy and enable parts of it to respond independently to local conditions. This is especially true of the lower portion of the digestive tract.

Hormonal Control

In addition to the neural controls, both the extrinsic and intrinsic innervation and all the stimuli that affect them, there are several hormones that are produced by the digestive system and exert their action upon it. Endocrine cells can be found in the epithelium of the mucosal glands of the stomach and intestine. Their hormones enter the bloodstream and are carried in the portal vein to the liver before entering the general circulation, but eventually they return to influence the motor and/or secretory activity of some organ of the digestive system. Release of these hormones is triggered by stimuli associated with eating, such as the presence of a partially digested foodstuff in some part of the digestive tract, and their action contributes to the digestion and absorption of that foodstuff.

The hormonelike substances are all peptides containing 11 to 44 amino acids. Some of them show similarities in structure and action and may activate the same or similar receptors on target cells. So far, however, only three have been accepted as bona fide hormones. The others are still "candidates," because it has not yet been shown that they are secreted in sufficient amounts under normal conditions to play a role in the digestive process. Some of them will probably be shown eventually to make a physiological contribution.

Gastrin is produced in the stomach, and it acts on the stomach. Its chief action is to stimulate the production of hydrochloric acid by the parietal cells of the gastric mucosa, although it also increases other secretory and motor functions of the stomach.

Many years ago it was discovered that the presence of fat in the duodenum resulted in the secretion of a substance by mucosal cells of the small intestine that caused contraction of the gall bladder and relaxation of the sphincter at the duodenal papilla. Because this caused bile to be emptied into the small intestine, the hormone was appropriately named **cholecystokinin (CCK).** Some twenty years later it was discovered that partially digested protein in the duodenum resulted in secretion of a substance that caused the pancreas to secrete a pancreatic juice with a high content of digestive enzymes. This substance was named *pancreozymin*. And twenty years after that it was discovered that these were two actions of the same substance. It was then called CCK-PZ (cholecystokinin-pancreozymin), or just CCK in recognition of its initial role. Part of the amino acid sequences of gastrin and CCK are identical, and they have similar actions. In fact, the list of the effects that can be produced by the administration of the two are virtually identical. The differences lie in those actions that are *physiological*—those that can be produced by hormone concentrations that normally occur in the body. These actions are essentially those mentioned above—acid secretion by gastrin, and contraction of the gall bladder, relaxation of the sphincter, and secretion of pancreatic enzymes by CCK.

The third gastrointestinal hormone, **secretin,** is produced by the duodenal mucosa when there is acid in the duodenum. Its main action is to elicit the secretion of pancreatic juice that has a high bicarbonate content. Secretin is structurally similar to glucagon (a hormone produced by the endocrine pancreas), whose main actions are metabolic; predictably, therefore, secretin has some rather weak metabolic actions, and glucagon is a weak stimulant to pancreatic secretion of bicarbonate.

One of the candidate hormones, known as **gastric inhibitory peptide,** or **GIP** is produced in the small intestine in the presence of fat and glucose. As the name implies, its major action is to inhibit gastric secretion. Other potential hormones include such peptides as *vaso-active intestinal peptide (VIP), motil-*

in, chymodenin, and several others.

The discovery and naming of secretin has an interesting history, because the original proof that such things as hormones existed and could cause glandular secretion turned out to be surprisingly important. In 1902 two English physiologists, Sir William Bayliss and Ernest Starling, showed that a completely denervated pancreas would secrete large quantities of pancreatic juice in response to acid placed in a denervated segment of the jejunum. They postulated that the acid provoked the release of a substance from the intestinal mucosa which traveled in the bloodstream to the pancreas and caused pancreatic secretion. They named this remarkable substance secretin. Such a mechanism does not seem very unusual now, but at the time it was greeted with much skepticism. This was the first evidence that a "chemical messenger" or hormone (in fact, Bayliss and Starling coined the word "hormone") could play a role which, until then, had been assigned only to nervous impulses. The study of hormones (*endocrinology*) has grown from its shaky introduction until now it rivals study of the nervous system in importance for understanding the control of bodily processes.

Types of Motility

Given the fundamental properties of the smooth muscle of the digestive tract, and the presence of the basic electrical rhythm (BER), plus the intrinsic nerve plexuses and perhaps gastrointestinal hormones, several different types of muscular activity or *motility* occur in the digestive tract. Visceral (single unit) smooth muscle normally exhibits a certain amount of tonus. The level of tension varies and it normally fluctuates rhythmically, in step with the slow wave oscillations of the membrane potential. Propulsive and mixing movements are superimposed upon this background or baseline of tonic activity. Two general types of contraction occur throughout much of the digestive tract, and one of them, peristalsis, also occurs in the skeletal muscle of the pharynx and esophagus.

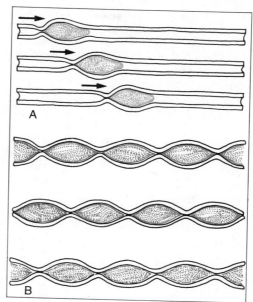

FIGURE 23–2   A. Peristalsis.
B. Segmentation.

Peristalsis is a ring of constriction produced by contraction of a section of circular smooth muscle. Each constriction moves caudally as a *peristaltic wave* and tends to push the contents of the tract ahead of it (Figure 23–2A). A given peristaltic wave, however, normally progresses only a few centimeters before it dies out. When something causes distension in one part of the intestine, the response will be contraction above and relaxation below that site, which tends to move the material downward. (This is known as the Law of the Intestine.) The frequency of such waves and the speed with which they progress is determined by the BER. Peristalsis can occur after the extrinsic nerves have been cut, but it is interrupted if the connections of the intrinsic plexus are blocked. It is most characteristic of the esophagus and stomach, and to a lesser extent of the intestine.

Segmentation, on the other hand, is not a propulsive movement (Figure 23–2B). It occurs primarily in the intestine and consists of rings of constriction that appear at intervals several centimeters apart along a portion of the intestine (thus breaking the intestinal lumen into segments). As these contractions wane other contractions develop at intermediate sites, producing alternating contractions and relaxations which serve to break up and mix the intestinal contents. Like peristalsis, the frequency of segmenting contractions is also determined by the BER, with each contraction occurring during a hypopolarization, though not necessarily with each slow wave. Segmentation also requires an intact intrinsic plexus. Whether peristalsis or segmentation occurs in a given situation probably depends to a certain degree upon local conditions, such as amount of distension, and fluidity of the organ's contents. Both peristalsis and segmentation may occur in the same section of gut, however.

MOTOR AND SECRETORY FUNCTIONS OF THE DIGESTIVE TRACT

The Oral Cavity and Esophagus

The oral cavity is the entrance to the digestive tract. Much of the material put into it is not in a form suitable for enzyme action or, for that matter, for passage through the digestive tract. The process of preparing food for the digestive actions that lead to absorption begins in the oral cavity with the powerful action of jaws and teeth.

Chewing Chewing, or mastication, is normally initiated voluntarily, but it is carried out by reflex action. It involves not only rhythmic opening and closing of the jaw, but constant manipulation of the food by the tongue and cheek muscles to improve the efficiency of the crushing and grinding movements of the jaw. Its function is to reduce ingested material to small pieces, mix it with the salivary secretions, and form it into a soft manageable mass of material called a *bolus*—a necessary preliminary to swallowing.

Salivary Secretion The secretion present in the oral cavity is that of the salivary glands. It contains water, electrolytes (notably potassium), mucin (a glycoprotein), and the enzyme **salivary amylase** *(ptyalin)*. Its composition may vary with the stimulus, from a thin, watery secretion to one that is thick and viscous, and the enzyme content differs widely among individuals. Saliva serves several important functions, most of which are related to its fluid content. Its lubricating action keeps the mucous membranes of the oral cavity moist and provides the fluid needed to moisten food and mold it into a bolus for swallowing. In addition, saliva is needed for taste, since taste buds can only detect chemicals in solution. About 1500 ml of saliva are secreted each day, a surprisingly large volume when one considers the relatively small size of the salivary glands.

Salivary amylase splits polysaccharides (starches) to disaccharides (such as maltose) if given enough time. Since it does not function in an acid medium, its action is stopped when the bolus is mixed with the highly acid gastric secretion. Food stored in the fundus of the stomach usually does not get mixed for some time, however, so salivary amylase may be able to break down much of the starch in food.

Salivary secretion is under neural control, and the salivary glands are innervated by both divisions of the autonomic nervous system. Parasympathetic stimulation produces an abundant secretion of watery saliva containing little organic material; it also causes dilatation of blood vessels to the salivary glands. The resulting increase in blood flow is due in large part to the fact that parasympathetic stimulation leads to the local release of a powerful vasodilator substance *(bradykinin)* into the interstitial fluid of the gland. Sympathetic stimulation results both in a small volume of viscous secretion with a high organic content and in vasoconstriction of the blood vessels.

Salivary secretion can be initiated reflexively by several types of stimuli. Mechanical stimuli (the presence of something in the mouth, such as food, chewing gum, or dental instruments) elicit a copious flow of saliva, as do certain chemicals (vinegar, lemon juice) and dryness of the oral mucosa. The sight, sound, smell, or even the thought of food may also evoke salivation ("make the mouth water").

Swallowing Once the ingested material has been sufficiently broken up and moistened with saliva, it is formed into a bolus and moved to the posterior portion of the tongue to be swallowed, as shown in Figure 23–3.

Swallowing, or *deglutition*, may be divided into three stages, related to the passage of the bolus through the (1) mouth, (2) pharynx, and (3) esophagus. The first stage is voluntary and involves maneuvering the bolus to the back of the tongue. Its presence there stimulates receptors which initiate the second stage, followed by the third, both of which are strictly reflex. Once the second (pharyngeal) stage of swallowing has begun, the act is no longer voluntary. You cannot stop in the middle of a swallow!

The second stage, passage through the pharynx, is a complex act since when the bolus reaches the back of the tongue and begins to descend through the pharynx it must be directed to the esophagus and prevented from going elsewhere. To do this, the tongue elevates to press against the hard palate, displacing the bolus posteriorly and preventing it from moving forward and out of the mouth. Muscles of the soft palate and pharynx move the soft palate up and back, blocking the opening to the nasopharynx and the nasal cavity. The larynx moves forward and upward and the glottis closes, while at the same time the epiglottis tips back and deflects food from the laryngeal opening. (The epiglottis does not act as a lid, however, because food does not necessarily enter the larynx if the epiglottis is absent.) As the material passes through the pharynx, it spills over and around the epiglottis into the esophagus, the only passageway still available.

The third, or esophageal, stage is a continuation of the second. Constriction of the pharyngeal muscles during the second stage reaches the upper end of the esophagus and then continues down it as a wave of peristalsis. The esophagus is normally collapsed and its upper end is kept shut by the tonic

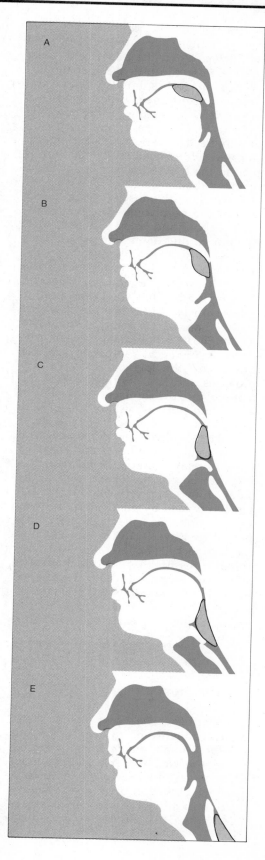

FIGURE 23–3 The sequence of events in swallowing. A. The bolus is moved to the back of the tongue. B. The soft palate and uvula cover the opening to the nasal cavity. C. The larynx is raised and the epiglottis is tipped. D. The bolus enters the esophagus as the upper esophageal sphincter relaxes E. The bolus is transported down the esophagus.

contraction of the muscles that form the pharyngeal or upper esophageal sphincter. When the pharyngeal muscles begin to contract, those of the upper esophageal sphincter relax, and this, along with the elevation of the larynx, briefly opens the sphincter allowing the bolus to pass before it closes once more. The peristaltic wave emerges as a wave of constriction that forms behind the bolus and moves down the length of the esophagus.

The smooth muscle of the lowest part of the esophagus—the few centimeters above and below where it passes through the diaphragm—is tonically active. Pressure in this part of the esophagus is higher than the pressure above it, or in the stomach below it. This region is called the lower esophageal sphincter, but it is a physiological rather than an anatomical sphincter. Although it is normally closed, the lower esophageal sphincter relaxes briefly as the peristaltic wave approaches and it allows passage of the material into the stomach.

When fluid is swallowed, gravity carries the liquid quickly to the lower sphincter, which remains closed until the first part of the peristaltic wave arrives to relax it. Rather solid materials may not be carried all the way to the stomach by a single wave, but the distension caused by a bolus left midway in the esophagus will trigger additional peristaltic waves until it is swept into the stomach. The slightly subatmospheric pressure that exists in the thorax affects the esophagus, but the presence of a sphincter at either end prevents its inflation with air from above, or the regurgitation of gastric contents from below.

The Stomach

Motility The empty stomach possesses enough tone to keep it flat and relatively small, but its smooth muscle is quite relaxed and does not generally show much motor activity. With each swallow, however, there is a reflex reduction in tone of the gastric musculature known as *receptive relaxation.* In this way the stomach enlarges to accommodate its increasing volume with virtually no rise in intragastric pressure, within rather wide limits.

The first material to enter the empty stomach lines its wall, and later additions line that, so that incoming food tends to become layered in the stomach with little mixing. As the stomach fills, however, it begins to show muscular contractions in response to the ever-present slow waves of the BER which occur about every 15–20 seconds. The fundus remains relatively inactive. Peristaltic contractions become apparent in the body of the stomach and move toward the pylorus. As each wave passes it causes the pyloric sphincter to contract briefly, but the wave does not continue into the small intestine. The peristaltic waves progress slowly, becoming stronger as they advance, accelerating as they reach the antrum. As time passes, the contractions become more intense, particularly in the antral region. Each wave tends to push the gastric contents toward the antrum. The pyloric sphincter is open most of the time, so it does not block passage of the gastric contents, but it does create some resistance. Each wave pushes the contents ahead, and a small amount squirts into the duodenum. When the contraction accelerates over the antral region, it overruns the contents of the antrum, and much of the material is pushed back toward the body of the stomach. The next wave pushes it forward again and forces a little more into the duodenum. The resulting forward and backward movements of the gastric contents are responsible for almost all of the mixing of the ingested material with the gastric secretion. This is important, since there is virtually no mixing in the fundus and very little in the body. The muscular activity continues until all of this semifluid mixture, which is called chyme, has been transferred to the duodenum.

The rate at which the stomach is emptied depends upon the pressure gradient between the stomach and the duodenum. The pressure in the stomach is related to the consistency and volume of its contents, and the vigor of the contractions of its smooth muscle.

The rate of gastric emptying is regulated by altering the vigor of the gastric smooth muscle contractions. Distension of the stomach, for example, increases the force of the peristaltic contractions, raises intragastric pressure, and hastens emptying.

The most effective stimuli for controlling gastric motility arise from the duodenum and are inhibitory, thereby prolonging emptying (Figure 23–4). Most of the stimuli act by both neural and hormonal mechanisms. Distension and the presence of acid in the duodenum inhibit gastric motility, chiefly by a reflex known as the *enterogastric reflex*. The receptors are in the duodenal mucosa, and both afferent and efferent fibers are carried in the vagus nerve. The reflex is also mediated through the connections of the intrinsic plexus, for if the vagus nerves are cut the response is weakened but not abolished.

There are receptors in the duodenal mucosa that respond to an increase in osmotic pressure in the lumen of the duodenum. They too inhibit gastric motility, but the actual mechanism is not well understood.

Fat in the duodenum apparently reduces gastric motility, chiefly by hormonal means. Cholecystokinin (CCK) and gastric inhibitory peptide (GIP) are the principle agents for this action. All three of the major gastrointestinal hormones and several of the "candidate" peptides can inhibit gastric motility and slow gastric emptying, but only CCK and GIP are produced in sufficient amounts to do so under physiological conditions.

It might be noted that stimuli such as fat, acid, increased osmotic pressure, and distension are normally present in the duodenum only when the stomach has passed chyme into it. These inhibitory mechanisms serve the purpose of preventing the stomach from delivering more chyme for processing before that already in the duodenum has been taken care of. They adjust the stomach's rate of emptying to the duodenum's ability to deal with it. It usually requires two to six hours for the stomach to empty (an average of about four hours), but sometimes material may remain in the stomach for more than

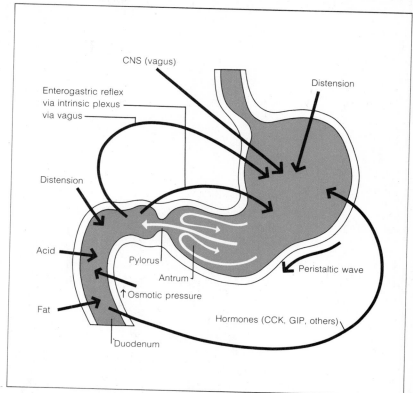

FIGURE 23–4 Factors affecting gastric emptying.

eight hours. The stomach is thus emptied just about in time for the next meal.

Gastric Secretion Every day the cells of the gastric glands produce two to three liters of a fluid known as gastric juice, which contains digestive enzymes, hydrochloric acid, and mucus, as well as salts and water. **Pepsin** is the only important enzyme in the gastric secretion. It is produced by the chief cells in an inactive form, **pepsinogen,** which is converted to the active form in the presence of acid or some already active pepsin. It functions best in an acid medium, about pH 2.0, such as is found in the stomach. Pepsin starts the digestion of proteins, splitting whole proteins to polypeptides by attacking certain amino acid linkages. Because there is very little mixing of the gastric contents until they reach the antrum, pepsin normally accounts for only a small portion of total protein digestion; one can digest proteins in its absence.

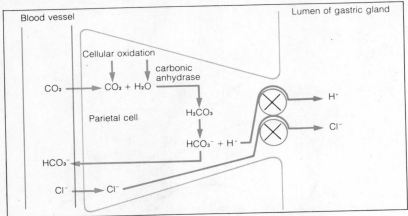

FIGURE 23–5 Secretion of HCl by parietal cells.

Gastric juice is also reported to contain *rennin*, which acts upon the protein in milk (casein). Its presence in human gastric secretion is questionable, however, and it is of little importance in humans.

The most distinctive feature of gastric juice is the presence of hydrochloric acid (HCl) in rather high concentration. The acid is produced by the parietal cells, located mainly in the upper part of the stomach—the fundus and upper part of the body. It is put out at a pH of about 1, but when it is mixed with the other constituents of gastric juice (which are formed throughout the stomach), its acidity approaches pH 2 or 3. The hydrochloric acid in gastric juice has several functions: As a strong acid it may denature proteins and break some chemical bonds, it helps activate pepsinogen, and it is an effective disinfectant, since many types of bacteria cannot tolerate high acidity. It also inactivates salivary amylase.

Secretion of acid. The ability of parietal cells to produce such an acidic substance is interesting from several aspects. First, it requires considerable metabolic work by the cells, because both the H^+ and Cl^- are secreted against sizable gradients, both electrical and chemical. The hydrogen ions apparently come from the ionization of water in the cells (Figure 23–5). They are actively pumped into the canaliculi of the cells, and then into the lumen of the gland, which would leave hydroxyl ions (OH^-) behind in the cells. There is carbon dioxide in the cells, obtained from metabolic processes in the parietal cells and by diffusion from the blood. A high concentration of the enzyme carbonic anhydrase catalyzes the formation of carbonic acid (H_2CO_3). The carbonic acid dissociates to a certain extent, releasing HCO_3^-, which diffuses back into the blood plasma, and H^+, which combines with the OH^- from the ionization of water, thus replacing the hydrogen that was secreted. Often, as the potentially acid CO_2 is removed from the plasma (P_{CO_2} becomes lower in gastric veins than in gastric arteries) and the basic HCO_3^- is added to the plasma, there is a slight rise in the pH of the blood, known as the "alkaline tide." It can be detected as an increase in the pH of the urine during gastric secretion.

As bicarbonate ions leave the cell, chloride ions diffuse into the cell from the plasma. They pass across the cell and are pumped into the lumen of the gastric gland by a process that is somehow coupled with the pumping of hydrogen ions. Oxidation of any of several intracellular energy sources provides the energy (ATP) to run the secretory process.

The most important reason why the acidity and enzymes of the gastric juice do not harm the stomach is probably the mucus secreted by the neck cells of the gastric glands. It is slightly alkaline, and its consistency is such that it normally adheres to the stomach wall to form a protective coating against mechanical abrasion, erosion by acid, and digestion by proteolytic protein-splitting enzymes. If, however, for some reason it is not secreted, or is ineffective, and the gastric secretion comes into direct contact with the mucosa, an *ulcer* is likely to develop. In the stomach, the area near the pylorus is most often affected. More ulcers actually occur in the first portion of the duodenum, above the site of entry of the alkaline pancreatic juice and bile.

There are several reasons why the protective mucous barrier may break down. Gastric irritants, such as aspirin, may reduce its effectiveness, or acid secretion may be increased by such factors as emotional stress, which increases vagal tone, or increased secretion of the hormone gastrin. The treatment for an ulcer is usually aimed at reducing the acid production.

Control of gastric secretion. The secretion of gastric juice is under both neural and chemical control (Figure 23–6). Stimulation of the vagus nerve elicits a marked increase in pepsinogen and acid production (as well as in gastric motility). It is for this reason that vagal fibers to the stomach are sometimes sectioned in cases of stomach ulcer. Gastric secretion is associated with the intake of food and occurs at three different times, known as the cephalic, gastric, and intestinal phases of secretion.

The **cephalic phase** of gastric secretion occurs before food enters the stomach. It is the result of a reflex whose stimuli and receptors are similar to those which evoke salivary secretion and whose efferent fibers are carried by the vagus nerve. The presence of food therefore elicits gastric secretion as well as salivary secretion. Psychological factors are involved here as well, of course, because unappetizing foods do not cause this type of gastric secretion.

The **gastric phase** of secretion occurs while food is in the stomach. The main stimuli are distension as the stomach fills and the presence in it of partially digested proteins, which act reflexively and through the hormone gastrin. Distension causes secretion both of pepsinogen and hydrochloric acid. It stimulates receptors in the gastric mucosa, initiating reflexes whose afferent and efferent fibers run in the vagus nerve. The receptors also have local connections within the intrinsic plexus. Acetylcholine (from vagal endings) is the most potent stimulus for pepsinogen production. Acetylcholine also causes secretion of gastrin, which stimulates gastric secretion, especially of acid. Gastrin is an important stimulus because its production is also increased by distension in the antral region of the stomach and by the presence of peptides and amino acids. Peptides and amino acids stimulate chemoreceptors in the mucosa which

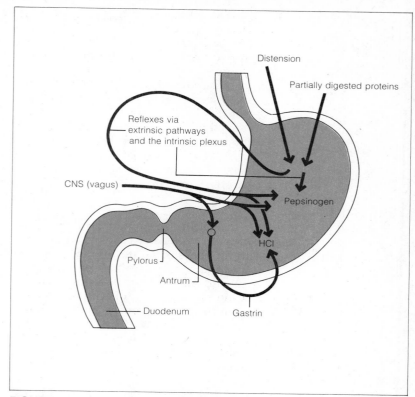

FIGURE 23–6   Control of gastric secretion.

cause local reflexes through the intrinsic plexus.

The **intestinal phase** covers the secretion of gastric juice which continues for several hours after chyme has left the stomach. It is probably at least partly due to the release of gastrin by gastrin-secreting cells in the intestinal mucosa. There are also some intestinal factors that inhibit gastric secretion, for fats, carbohydrates and acid in the duodenum cause the production of gastric inhibitory peptide (GIP) which inhibits gastric secretion as well as motility.

The Small Intestine

When chyme enters the small intestine from the stomach, the digestion of protein and starch have only begun, and that of fat has not started at all. Most of the digestion of the acidic, semifluid mixture that we call chyme therefore occurs in the small intestine and, in fact, is completed there. Digestion is carried out not only by enzymes in the intestinal secretion but also by those of the pancreas, aided by bile from the liver (which contains no digestive enzymes).

The motility of the small intestine is an essential ally of chemical digestion. Not only must the chyme be moved through the tract, but it must be thoroughly mixed with the secretions in order to come in contact with the surface-acting enzymes and with the absorptive surfaces.

Motility Segmentation is the most characteristic type of motor activity of the small intestine. Single contractions involve several centimeters of gut and last for several seconds; as they wane similar rings of constriction appear in the previously inactive intervening segments. The frequency of these contractions is determined by the slow waves of the intestinal BER which travel along the longitudinal muscle layer.

Pacemaker cells in the small intestine set a higher frequency than those in the stomach—11–12 per minute in the duodenum and 8–9 per minute in the ileum (as compared to 3 per minute in the stomach). The presence of chyme, by causing a localized distension of the small intestine, increases the likelihood of that section contracting when a slow wave passes. Segmentation has a very effective mixing action, as the chyme is moved back and forth by the alternating contractions and relaxations in a section of gut. But due largely to the higher frequency of contractions in the upper portion of the small intestine, the chyme tends to move forward more than back, and segmentation therefore contributes somewhat to progression of the chyme.

Peristaltic contractions in the small intestine are much less prevalent and of lesser intensity than in the stomach. They usually consist of only two or three waves that progress for several centimeters and then die out.

Pancreatic Secretion The secretion of the pancreas serves two important roles: Its several digestive enzymes carry out digestive functions, and its alkalinity helps neutralize the acidity of the chyme. Either of these functions can be emphasized by varying the composition of the pancreatic secretion. It may be a thin, copious, aqueous solution with a high content of electrolytes, particularly bicarbonates, or one of scant volume and high enzyme content.

The cells of the pancreatic acini produce the digestive enzymes and store them as secretion granules until an appropriate stimulus causes their release into the pancreatic duct. Cells of the tiny ducts from the acini are believed to be the source of the electrolyte secretion. How the cells can secrete so much bicarbonate is somewhat puzzling, but the mechanism may be similar to that for the secretion of H^+ by the gastric mucosa, except in the opposite direction. The pH of the pancreatic secretion is on the alkaline side (ranging from pH 7 to 8.2) and, along with the bile, is sufficient to make the intestinal contents nearly neutral. This stops the action of pepsin, but it provides a more suitable environment for the pancreatic and intestinal enzymes.

The pancreatic secretion contains enzymes to act upon all three major foodstuffs, including several proteolytic enzymes.

1. **Pancreatic amylase,** like salivary amylase, splits carbohydrates to disaccharides (chiefly maltose, sucrose, and lactose). It continues the job begun by salivary amylase, but it does not complete the digestion of carbohydrates since disaccharides cannot be absorbed.

2. **Pancreatic lipase** is the only important fat-splitting enzyme. It hydrolyzes fat to monoglycerides or fatty acids and glycerol, all of which can be absorbed from the intestine.

3. **Trypsinogen** is a protein-splitting enzyme which, like gastric pepsinogen, must be converted to an active form, in this case, **trypsin.** The conversion is brought about by **enterokinase,** an enzyme secreted by the intestinal mucosa. (Enterokinase is therefore present when trypsinogen reaches the intestine.) It is also activated by previously activated trypsin. Trypsin acts on proteins, breaking them down to smaller polypeptides.

4. **Chymotrypsinogen** must be activated to **chymotrypsin** before it can be effective. The activation occurs in the intestine in the presence of trypsin. Chymotrypsin is also a protein-splitting enzyme whose end products are polypeptides.

5. **Carboxypeptidase** acts upon partially digested proteins and polypeptides, breaking off the end amino acid (from the acid or carboxyl end) of certain polypeptide chains.

6. **Ribonuclease** and **deoxyribonuclease** break the nucleic acids RNA and DNA to their respective nucleotides.

We have seen that the enzymes that act on whole proteins are secreted in an inactive form and are not activated until they have reached the site where the appropriate protein substrate is to be found. This is a protective measure, since proteolytic enzymes would be likely to attack the secretory cells which produced them if they were formed in a ready-to-act state.

Trypsin, chymotrypsin, and gastric pepsin are all enzymes that act upon whole proteins and break them into smaller protein fragments. Their actions are not identical or overlapping, however, since each attacks the linkage adjacent to different amino acids. After a protein has been exposed to the action of these enzymes, the end-products are partially digested proteins — polypeptide chains of varying lengths.

Control of pancreatic secretion. Secretion of pancreatic juice increases in response to vagal stimulation. It can also be increased by the same stimuli that cause salivary secretion and the cephalic phase of gastric secretion: that is, by the presence of food in the mouth and probably by the sight or smell of food. These stimuli excite reflex pathways in the vagus nerve.

The major factors in the control of pancreatic secretion, however, are hormonal; they involve secretin and CCK. Secretin, produced by the mucosal cells of the duodenum, is released when there is acid in the intestine (see Figure 23 – 7). It elicits a copious flow of pancreatic juice with a low enzyme content but much bicarbonate. Such a secretion is not very effective in digestion, but it can neutralize the acid

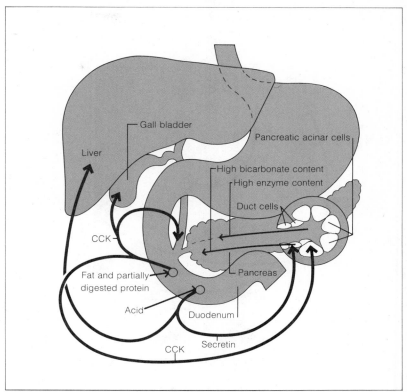

FIGURE 23 – 7   Hormonal control of the exocrine secretion of the pancreas.

which initiated its secretion in the first place. CCK is also secreted by the intestinal mucosa, but in response to the presence in the intestine of certain foodstuffs, such as fats and whole and partially digested proteins. It elicits a pancreatic juice of a somewhat lower volume and electrolyte content, but of a high enzyme concentration.

The Liver, Gall Bladder, and Bile

The liver contributes to the digestive process by the secretion each day of 500 – 1000 milliliters of bile, a viscous, yellowish fluid that is slightly alkaline and has a very bitter taste. As formed in the liver, bile is mostly water (97 percent), with mineral salts, bile salts, bile pigments, and cholesterol as important constituents. Bile salts are steroids derived from the metabolism of cholesterol; the bile pigments (chiefly bilirubin), which result from the breakdown of hemoglobin released when red blood cells disintegrate are excretory products to be removed from the body. There are no digestive enzymes in bile; its diges-

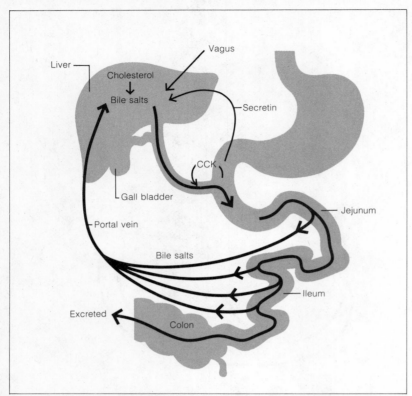

FIGURE 23–8   Enterohepatic circulation and the secretion of bile.

tive function is due to the fact that bile salts are emulsifying agents. Because of their surface tension, fat particles have a tendency to coalesce into larger and larger globules, which reduces the relative amount of surface exposed to the action of lipase. The bile salts reduce the surface tension of the fat globules, resulting in more and smaller globules and, therefore, in a greater surface area and increased effectiveness of the surface-acting enzymes. If bile is prevented from reaching the intestine, little fat digestion occurs and the ingested fats are excreted in the feces.

Control of bile secretion.   The delivery of bile encompasses two separate processes, the *formation* of bile by the liver and the *release* of bile into the intestine. The sphincter at the entry of the bile duct into the duodenum is usually closed, and bile is diverted to the gall bladder for storage until there is a stimulus for its release. The factors that increase the rate of bile formation are not the same as those that cause its release into the intestine, and they do not necessarily occur at the same time.

The liver produces bile continuously, but a potent stimulus to additional bile formation is the presence of bile salts in the blood to the liver. When bile is released into the small intestine, it is presumably destined for removal from the body. Almost all of the bile salts, however, are absorbed from the lower ileum into the portal blood and returned to the liver, where they increase the rate of bile formation and are excreted in the bile once more (Figure 23–8). This circuit is known as the **enterohepatic circulation** of bile salts; it helps regulate bile formation.

Secretin is also an effective stimulus to bile formation. Those things that cause the production of secretin, namely food and acid in the duodenum, therefore increase the rate of bile formation as well. The liver also responds to vagal stimulation by the increased production of bile, but this control is less important than the chemical controls.

When the sphincter at the end of the bile duct is closed, bile does not enter the intestine. Instead it backs up into the gall bladder where it is stored and concentrated. The capacity of the gall bladder is too small to accept all the bile produced, but the absorption of water and inorganic salts keeps the volume down to a manageable amount. Under certain conditions there may be a precipitation with formation of crystals, mainly cholesterol, in the gall bladder. Such crystals, or *gallstones,* may be very painful when the gall bladder contracts, especially if one becomes lodged in the cystic or bile duct.

The smooth muscle in the wall of the gall bladder will contract upon vagal stimulation, but denervation has little effect upon normal gall bladder function, suggesting a major nonneural control. Contraction of the gall bladder and the appearance of bile in the small intestine follow the arrival of chyme in the small intestine. Fats are particularly effective, as are acids and the products of protein digestion, for they cause the intestinal mucosa to release cholecystokinin, CCK. It brings about contraction of the smooth muscle of the gall bladder and relaxation of the sphincter, thus permitting bile to enter the duodenum. Thus the presence in the small intestine

Focus

Gall Stones and Cholesterol

Many chemistry departments have a jar of curious, brownish lumps with granular surfaces, ranging in size from sand grains to lemons. Cut open, the lumps reveal a fine-grained crystalline structure. Held in the hand, they have a greasy feel.

The lumps are gall stones obtained from human gall bladders after surgery or upon autopsy. Their principal component is cholesterol, and they were once the primary source of cholesterol for laboratory work. Pure cholesterol stones are rare, for they are produced only by persons whose cholesterol metabolism is upset. Most stones are a mixture of cholesterol, blood pigments, and calcium. Occasionally, stones may be composed of pure bilirubin, bilirubin and calcium, or calcium carbonate. The pure bilirubin stones seem to be formed when a person suffers excessive breakdown of red blood cells.

Gall stones are found in seven percent of the human population. They are more common in women than in men, and then increase in frequency in people past the age of thirty. They may be formed after injury to the gall bladder wall, disturbances of cholesterol metabolism, changes in bile pH, or a blockage of the bile duct. In each case, they are a product of the normal functions of the gall bladder. This small organ receives bile secreted in the liver and stores it until CCK, secreted by the duodenal mucosa in response to the presence of fat, stimulates it to contract and release bile to the intestine, where it serves as an emulsifier. While bile is in the gall bladder, the gall bladder wall removes water and some solutes, concentrating the bile salts and cholesterol in the bile up to

tenfold. The longer the bile resides in the gall bladder, the more concentrated it becomes (up to a point) and the more likely it is that its contents will crystallize to form stones. The residence time is increased by duct blockage and by impairments of the bladder's ability to contract; the likelihood of crystallization is increased by changes in bile pH and in the initial concentration of cholesterol.

Gall stones do not always cause symptoms, but often enough they are responsible for abdominal pain, nausea, and vomiting, especially after a fatty meal. The symptoms are due to blockage by small stones of the cystic or common bile ducts. Treatment may require a low-fat, low-seasoning diet, which minimizes the secretion of bile and the contraction of the gall bladder around the stones. It may also require surgical removal of the gall bladder.

Cholesterol was first known from gall stones because of their relative accessibility. Not long after, it was observed in atherosclerotic plaques (hardened arteries). Not surprisingly, people then thought of cholesterol as a chemical villain. Cholesterol is, however, a necessary precursor of bile pigments and steroid hormones. It enters the body in the diet, but it is also synthesized by the body's cells. It is essential to the normal function of the immune system, particularly the body's response to such invaders as cancer cells.

For all the health problems associated with cholesterol, this substance is not something we should try to eliminate from our bodies. It does far more good than harm, and the harm is generally symptomatic of other problems.

of the elements of a normal meal — protein, partially digested protein, and fat — causes the secretion of a hormone that brings about the release of the bile needed to emulsify the fat and to facilitate its digestion and of a pancreatic juice that contains the enzymes needed to digest the foodstuffs.

Intestinal Secretion As chyme passes along the first part of the small intestine it is mixed and churned, exposed to bile, and acted upon by pancreatic enzymes. The end-products of carbohydrate and protein digestion are small molecules of disaccharides, small polypeptides, and some amino acids. Only the latter can be absorbed, the others must be further broken down by enzymes provided by the intestine. However, while the volume of fluid secreted by the small intestine is large, its enzyme content is not great. The necessary enzymes are not actually secreted into the intestinal lumen, but remain attached to the surface membranes of the cells. (Enterokinase is an exception: it is believed to be secreted.) Epithelial cells of the mucosa are formed by mitosis in the intestinal glands. Subsequent cell divisions displace the cells toward the surface, and eventually they are pushed off the tips of the villi. As these extruded cells disintegrate, their enzymes are spilled into the intestinal lumen. There is a complete turnover of the intestinal epithelium about every four to five days, so the magnitude of desquamation is very great indeed. The following enzymes are among those that are produced in the intestinal mucosa and act upon the contents of the small intestine.

1. **Aminopeptidase** acts upon partially digested proteins and polypeptides to form short-chain polypeptides. It splits off amino acids from the amino end of the amino acid chain.

2. **Dipeptidases** split dipeptides (two amino acids linked) into amino acids that can be absorbed.

3. **Disaccharidases** *(maltase, sucrase, lactase)* act upon disaccharides (maltose, sucrose, and lactose, respectively) to split them into the absorbable monosaccharides.

4. **Enterokinase** activates pancreatic trypsinogen to trypsin.

5. Several **nucleotidases** hydrolyze the various nucleotides from RNA and DNA.

Most of the intestinal enzymes are produced in the upper part of the small intestine, so that chemical breakdown is virtually complete before the intestinal contents reach the ileum. The volume of secretion of the small intestine is considerable, two or three liters per day, making it quantitatively the largest volume of secretion added to the digestive tract.

The Colon and Rectum

Material enters the colon whenever a peristaltic wave passes over the ileocecal valve and pushes chyme through it into the cecum. It receives around 500 to 1000 ml of fluid each day. The nutrients and most of the fluid and electrolytes have been absorbed from the small intestine, but the colon absorbs 80–90 percent of the remaining fluid, enough to reduce the contents from a semifluid to a semisolid mass.

The secretions of the large intestine are scant and have no digestive functions. They contain no digestive enzymes and no unique substances such as hydrochloric acid, nor are any important regulatory hormones produced by the large intestine. The major constituent of the colonic secretion is mucus (recall the great abundance of goblet cells in the epithelium).

The large intestine contains a significant and flourishing population of bacteria which came into the digestive tract with the food and survived the action of acid and enzymes along the way. The bacteria make several minor contributions, including some breakdown of indigestible cellulose, some fermentation of sugars (with the formation of gas), and putrefaction of certain protein residues. The end products, especially of putrefaction, are potentially dangerous, but they are poorly absorbed from the colon. Any that are absorbed are detoxified by the liver, so they present no real threat.

It usually takes more than 24 hours

for material to pass through the large intestine. It is inactive for long periods of time, and the activity which does occur is slow and sluggish. Weak constrictions and longitudinal shortenings of the haustra that resemble segmentation have a kneading and churning action on the contents. Their frequency and intensity increase when material enters the colon. There are also some weak peristaltic contractions, and two or three times a day peristaltic waves sweep over large portions of the colon. These long slow-moving mass contractions move the colonic contents toward the rectum.

The rectum is usually empty, but when a wave of mass contraction moves material into it, the resulting distension stimulates receptors in the wall of the rectum and gives rise to a sensation of fullness and the urge to defecate. It also stimulates the center for the defecation reflex, which is located in the sacral portion of the spinal cord. The efferent fibers, which include parasympathetic fibers to the distal colon and rectum, and somatic fibers to several skeletal muscles, increase the motility of the wall of the colon and relax the anal sphincters. Contraction of such voluntary muscles as the diaphragm and abdominal muscles raises the intra-abdominal pressure and adds force to the act of defecation.

Higher centers in the brain may inhibit the defecation reflex and keep the external anal sphincter (skeletal muscle) contracted, thereby postponing evacuation until a more suitable time. Voluntary control of defecation requires intact connections between the brain and the reflex center in the cord. If these connections are severed, defecation can still occur as a completely spinal reflex but, of course, there is no voluntary control.

The material excreted from the digestive tract has taken from two to five days to travel from one end of the tract to the other: two to six hours in the stomach, six to eight hours in the small intestine, and the rest of the time in the large intestine. As the feces are largely of nondietary origin, their composition is not affected very much by changes in the diet. They are about 75 percent wa-

ter. Nearly a third of the solid portion consists of bacteria from the digestive tract. The remainder contains cellular debris from the continual replacement of the epithelial lining, residues from digestive secretions (both organic and inorganic material, including mucus, some components of bile, and some fat and fat derivatives). There is also a variable amount of indigestible fibers and cellulose (berry skins, grain hulls, etc.), but normally virtually no unabsorbed foodstuffs or nutrients are excreted.

Alterations in Motility

A typical meal is likely to require from two to five days to pass completely through the gastrointestinal tract. The speed of passage depends upon the motility of the gut. When it is altered, the length of time that chyme is subjected to the processes in a particular segment of the gut is correspondingly altered. The results are probably familiar to most of you.

Diarrhea Diarrhea is due to increased motility of the small intestine. This may hasten the passage of chyme, particularly through the lower small intestine, to the extent that there is not enough time for absorption, particularly of water and electrolytes. The volume of chyme that is delivered to the large intestine then surpasses the capacity of the colon to absorb water and electrolytes, resulting in frequent and sometimes explosive passage of watery stools. A prolonged bout of diarrhea can pose problems of dehydration and electrolyte imbalance because of the loss of water, sodium, and potassium, a situation that becomes serious more quickly in infants, where there is less reserve.

Constipation When the intestinal motility is reduced and material remains in the lower colon and rectum for long periods of time, more water is absorbed from it, increasing the difficulty of eliminating the fecal material. The symptoms, feelings of discomfort and general malaise, disappear promptly upon relieving the distension in the rectum.

Ours has been called a "bowel-conscious society," in which the amount of misinformation and apprehension about constipation is probably as great as that about any other health topic. It is often taken as self-evident that defecation once each day is essential and that failure to maintain such "regularity" endangers health and must be dealt with immediately with a particular laxative. Some individuals, however, normally defecate twice or three times a day while others do so only every two or three days. There are individuals, admittedly atypical but not exactly unhealthy, who empty the rectum much less often.

The symptoms of constipation are due to distension of the rectum, not to absorption of "toxic substances" from the rectum. The typical treatment for suspected constipation—a laxative, or *cathartic*, that irritates the lining of the intestine—increases its motility, often to the extent of cleaning out the entire tract. The result is that several days may be required to accumulate enough bulk for normal defecation, but by that time the patient is convinced that he or she is constipated once more and repeats the treatment.

Vomiting Vomiting is the rapid emptying of the stomach by passing its contents in the wrong direction. It is not, however, simply a siege of reverse peristalsis in the stomach and esophagus, since vomiting is a complex reflex involving the coordinated response of many muscles. The upper end of the stomach, the lower esophagus, and the sphincter lose tone and become flaccid, while the lower end of the stomach, the pyloric sphincter, and the duodenum demonstrate increased tone. A sudden and powerful contraction of the diaphragm and abdominal muscles raises the intra-abdominal and intragastric pressure, and, due to the altered tone, there is only one way for the gastric contents to go.

The required coordinated muscle action is controlled through a vomiting

TABLE 23–2   TYPICAL VOLUMES OF GASTROINTESTINAL SECRETIONS

Secretion	Volume (liters per day)
Salivary juice	1.0 – 1.5
Gastric juice	2.0 – 3.0
Pancreatic juice	0.2 – 0.8
Bile (liver)	0.5 – 1.0
Intestinal juice	2.0 – 3.0
Total secretion	5.7 – 9.3

center located in the medulla. Afferents converge at the center from many areas. The sensation of nausea which usually precedes the act is often triggered by something that is disgusting or repulsive to the individual; it may be the sight of blood, an unpleasant odor, or something irritating in the stomach. Stimulation of the semicircular canals (motion sickness) is a common cause of nausea. (Some of the pills used to combat motion sickness act by depressing the vomiting center, thereby rendering it insensitive to these stimuli.)

ABSORPTION

Absorption is usually understood to be the removal from the digestive tract of the 1.5 – 2.5 liters of fluids and nutrients ingested each day, and their transfer to the bloodstream and lymphatics. However, almost as important, is the necessary removal of the gastrointestinal secretions, the digestive juices. They amount to between five and ten liters every day (see Table 23–2). When one recalls that the circulating blood volume is only about five liters, the ease with which water and electrolytes could be lost and the need for retrieving them become quite obvious.

Although nutrients are not absorbed from the stomach, there is some transfer across the gastric mucosa. Water can move in either direction, but there is usually no *net* movement of water. Two substances that are absorbed from the stomach are alcohol and aspirin. They thus gain entrance to the circulation and exert their effects much more rapidly than if they had to reach the intestine for absorption.

About 90 percent of the foodstuffs entering the small intestine are absorbed in the duodenum and jejunum, and virtually none reaches the large intestine. Most of the fluid and electrolyte absorption also occurs in the small intestine. Although the large intestine absorbs water and some electrolytes, it accounts for a fairly small share of the total volume absorbed, less than a liter. The colonic absorption does, however, determine the final consistency of the fecal mass and the amount of fluid and electrolytes lost from the body by this means.

Absorption of Carbohydrates

Carbohydrates are absorbed mostly as 5- or 6-carbon monosaccharides, i.e., *pentoses* or *hexoses*. But because these sugars are absorbed at different rates, it is clear that they are not all transferred by the same mechanism. A few sugars are absorbed by simple diffusion and some, including *fructose*, are apparently transported by carrier-mediated facilitated diffusion. *Glucose*, the most abundant sugar in the body, and *galactose* are absorbed by an active process that requires the expenditure of energy by the mucosal cells and is therefore independent of any gradient across the cell membranes. Some details of the mechanism for glucose absorption are still unclear, since the transport of glucose in the gut, unlike its transfer in most other sites, does not involve phosphorylation of the glucose as it enters the cell and dephosphorylation inside the cell. Glucose transport in the gut is somehow associated or coupled with sodium transport, since increased absorption of sodium in the gut facilitates the absorption of glucose, and vice versa. The active transport seems to be a mechanism to get the sugar *into* the epithelial cells of the mucosa. Glucose apparently moves out the other side of the cell to the blood vessel passively, by simple diffusion.

Occasionally an individual will lack one of the intestinal disaccharidases, leaving that person unable to hydrolyze and absorb a particular sugar. The most common of these genetic deficiencies is a lack or absence of lactase, which causes an inability to use lactose and an intolerance of milk and milk products. The lactose in the intestine causes various symptoms, including diarrhea, due to its osmotic effect. It may cause malnutrition when it occurs in an infant. Most often the enzyme is present in infants but disappears by about age six. The condition is fairly widespread, particularly among black people; about 70 percent of adult American blacks are said to be lactase deficient.

Absorption of Lipids

Most of the lipid material entering the small intestine is in the form of triglycerides—glycerol combined with three molecules of fatty acid. Lipase splits the fatty acids from the glycerol, but chemically it is difficult for these enzymes to remove the middle fatty acid. The result is that two fatty acids are split off and absorbed, but most of the glycerol is absorbed as monoglyceride. Once inside the cell there are two possibilities, both of which are shown in Figure 23–9. Fatty acids with short carbon chains (less than 10–12 carbons) pass through the cell and enter the portal vein to be carried to the liver. Within the cell the longer-chain fatty acids are recombined with the absorbed monoglyceride to form triglycerides once more. They then enter the rough endoplasmic reticulum where they receive a coating that stays with them as they leave the mucosal cell. These tiny protein-coated lipid droplets, known as **chylomicrons,** are extruded from the cell and find their way into a lacteal, the tiny lymphatic in each villus. After a high fat meal, the fat content of the lacteals and other lymphatics of the mesentery is so great that they can be seen with the naked eye and have a milky appearance (hence the name "lacteal"). By entering the lymphatic system instead of the portal vein, the chylomicrons bypass the liver and enter the

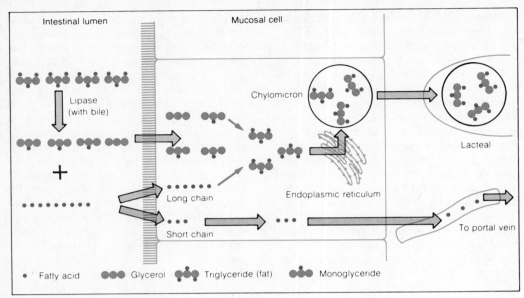

FIGURE 23-9   Absorption of fat.

general circulation to be distributed throughout the body. Since most of the fatty acids in the diet are larger than 10–12 carbons, most of the ingested fat follows this course.

Absorption of Proteins

Protein is absorbed mainly as amino acids from the small intestine, particularly the duodenum and jejunum. The amino acids are rapidly transferred into the mucosal cell by an active process involving one of probably three transport systems. Amino acids using the same transport system compete for that carrier and thus can block one another, but there is no competition between amino acids that use different carriers. Once in the cell, however, they cross the cell and leave it to reach the bloodstream by passive diffusion. A few di- and tripeptides are actively transported into the epithelial cells, most of which are hydrolyzed within these cells and released into the portal vein as amino acids.

Occasionally, whole proteins may be absorbed from the intestine. When they enter the bloodstream, they are treated as foreign proteins and antibodies are formed. Upon subsequent exposure to that protein there is an antigen-antibody reaction to the protein which may produce discomforting or even dangerous symptoms of allergy. The body is therefore protected by the fact that ingested protein is normally digested to amino acids before it enters the bloodstream. This does mean, however, that medications containing proteins will not be effective when taken orally. The gastrointestinal tract must be bypassed, usually by injection, to get the agent into the bloodstream and the tissues intact. A common example is insulin for the control of diabetes.

Absorption of Water and Electrolytes

Of the 6–12 liters of secretions and ingested fluids that enter the small intestine each day, only about 500–1000 ml reach the colon, where all but about 100–150 ml are absorbed. The actual volume of water moved across the intestinal epithelium is much greater, however, because the epithelium is very permeable to water and large volumes can move rapidly in either direction whenever there is an appropriate gradient. Gastric chyme may be hypertonic; the digestive action of pancreatic enzymes raises the tonicity further by increasing the number of osmotically active particles. The material in the intestine, however, soon becomes isotonic with the body fluids because if it is not isotonic, water will quickly move in or out until it is.

The driving force for the absorption of the water is generated by the active transport of other substances. The digestive juices contain large numbers of electrolyte ions that are actively transported across the intestinal cells to the blood vessels. The active uptake of other substances, such as glucose, amino acids, and other nutrients, also contributes to the osmotic gradients, but they are less important because there are fewer of these particles than there are of inorganic ions. The permeability of the mucosa to the water ensures that it will respond very quickly to any gradients created by the transport of these substances. The average rate of absorption of water amounts to about 200–400 ml/hr. However, the maximal rates are much higher, since if one drinks a liter of water, it will be absorbed within an hour.

The electrolytes to be absorbed are largely constituents of the digestive juices secreted into the lumen of the intestine. They are essentially the same as those that are important in the blood plasma and other body fluids. Some electrolytes, of course, are among the minerals in the ingested foods, since that is their ultimate source. Sodium is able to diffuse in and out passively, but it is also transported out of the intestine actively. Apparently, its transport has some effect upon the absorption of some other substances. Chloride absorption is probably passive, resulting from the electrical gradients established by the active absorption of sodium.

Monovalent ions such as potassium and bicarbonate are absorbed more readily than divalent ions such as calcium or magnesium. Calcium and iron have specific mechanisms for absorption, so that the amount of either absorbed is related more to the body's needs than to availability. Calcium is actively absorbed, but its absorption is impaired in the absence of vitamin D. Iron absorption is also an active process, but it is complicated by the formation in the mucosal cells of **ferritin,** an iron-protein complex. An equilibrium is maintained between ferritin in the mucosa and circulating iron. When the latter decreases, it is replaced by iron from the ferritin. More iron is then absorbed to replenish the ferritin (see Chapter 24).

Absorption of Vitamins

Vitamins, essential for the proper use of many nutrients within the body, are ingested as part of the dietary intake and must be absorbed with the nutrients, although small amounts of some vitamins may be produced by the intestinal bacteria. The water-soluble vitamins present no problems, as they are rapidly transferred with the water and nutrients in the upper part of the small intestine. However, the fat-soluble vitamins, A, D, E, and K, may not be adequately absorbed if there is any problem with fat digestion, due to lack of either bile or pancreatic lipase. Vitamin B_{12}, needed for development of adequate numbers of normal red blood cells (see Chapter 14), requires special treatment for absorption. It becomes bound to the *intrinsic factor* produced by the parietal cells in the gastric mucosa, which somehow enables it to be absorbed in the ileum, perhaps by pinocytosis.

CHAPTER 23 SUMMARY

Control of the Digestive Tract

Smooth muscle of the digestive tract is of the single unit type. In the longitudinal muscle layer of the stomach and small intestine are pacemaker cells, whose resting potential fluctuates rhythmically. These potential changes are conducted to adjacent cells through gap junctions as waves of hypo- and hyperpolarization. They pass caudally over the muscle and constitute the *basic electrical rhythm (BER)*. If the excitability of the muscle cells is raised, one or more action potentials may occur during the hypopolarization phase and cause contractions that occur with the same frequency as the BER. Excitability might be raised by the action of parasympathetic nerves (including the intrinsic innervation) and by local stimuli such as stretch. Sympathetic stimulation and epinephrine lower the excitability and diminish the response. Neither nerves nor local stimuli affect the BER.

Smooth muscle of the gut generally exhibits a certain amount of tone, which fluctuates with the BER. Peristalsis and segmentation are superimposed on the underlying tonus. *Peristalsis* is a ring of constriction that progresses caudally along the gut for a short distance, and *segmentation* is rhythmic contraction and relaxation of alternate sections, especially of the small intestine.

At least three hormones secreted by the digestive tract affect the digestive organs. *Gastrin* is produced by the stomach and stimulates the stomach, *cholecystokinin (CCK)* is produced by the duodenum and stimulates the pancreas and gall bladder, and *secretin* is produced by the duodenum and also stimulates the pancreas.

Motor and Secretory Functions

In the oral cavity chewing or mastication begins the breakdown of ingested material and mixes it with salivary secretion, which is necessary for taste and for forming a bolus for swallowing. Salivary juice contains an *amylase* that begins starch digestion. Its secretion is under reflex control and may be initiated by a number of stimuli, especially those arising in the oral cavity.

Swallowing is initiated by moving the bolus to the posterior portion of the tongue. This triggers the reflex portion; contraction of pharyngeal muscles blocks off the nasal cavity, and movement of the larynx directs the bolus into the esophagus. A wave of constriction begins with the pharyngeal muscles and continues the length of the esophagus. It is associated with a brief relaxation first of the upper and then the lower esophageal sphincters.

When food enters an empty stomach the tone diminishes (receptive relaxation), allowing the stomach to increase in size without an increase in pressure in it. With food in the stomach, peristaltic waves begin to occur with each hypopolarization of the BER (about 3 per minute). They appear in the body and become progressively stronger as they move toward the pylorus. These waves mix the contents, or *chyme*, with gastric secretions primarily in the antrum, and also move chyme into the duodenum. Stimuli arising in the duodenum tend to slow the emptying of the stomach. Among these stimuli are bulk and acid acting via the enterogastric reflex, fat and protein by hormonal action (mostly CCK), and increased osmotic pressure of the cyhme.

Gastric juice contains *pepsin*, a proteolytic enzyme secreted in an inactive form (pepsinogen) which is activated by HC1. The *hydrochloric acid* is produced by parietal cells in high concentration (low pH). The secretion of mucus cells helps protect the mucosa.

Gastric juice is secreted before, during and after food is in the stomach. The first, the *cephalic phase*, is a reflex response initiated by many of the same stimuli that elicit salivary secretion. The second, or *gastric phase*, is caused by gastrin, and the third is an *intestinal phase*.

Motility of the small intestine is characterized chiefly by segmentation and occasionally short peristaltic waves. The frequency, which is set by pacemaker cells in the small intestine, is faster than in the stomach. These movements mix the chyme with digestive secretions and facilitate both digestion and absorption.

Pancreatic juice and bile from the liver enter the duodenum a short distance below the pyloric sphincter. Pancreatic juice contains an *amylase* which acts on starches, and *lipase* which digests fats. It also contains *trypsinogen* which is activated to trypsin by *enterokinase* (an enzyme produced by the intestinal mucosa), *chymotrypsinogen* which is activated to chymotrypsin by trypsin, and *carboxypeptidase*. These act on proteins and polypeptides to produce smaller polypeptides and a few amino acids. *Ribo-* and *deoxyribonucleases* act on RNA and DNA.

Pancreatic secretion is mostly under hormonal control. Fat and partially digested proteins in the duodenum cause production of CCK, which elicits a secretion with a high enzyme content. It also causes contraction of the gall bladder and relaxation of the sphincter that allows both pancreatic juice and bile to enter the duodenum. Acid in the duodenum stimulates the production of secretin, which elicits a secretion high in bicarbonate.

Bile from the liver contains no digestive enzymes but its emulsifying action is important for fat digestion. Most of the secreted bile salts are absorbed from the ileum and returned to the liver where they are a powerful stimulant to further bile secretion. Bile produced when the sphincter is closed is stored and concentrated in the gall bladder until CCK causes its release.

Enzymes produced by the small intestine include *disaccharidases, peptidases,* and *nucleotidases,* that yield absorbable products. Most of the enzymes are not actually secreted, but enter the intestinal lumen when the surface epithelial cells are eroded. The intestinal mucosa is completely replaced every few days.

The colon and rectum are relatively inactive. Peristaltic waves move material toward the rectum several times a day. The secretion consists mostly of mucus. The colon normally contains a large population of bacteria. Distension of the rectum due to entry of material stimulates receptors which cause the urge to defecate.

Absorption

Most of the nutrients are absorbed from the small intestine, primarily the duodenum and jejunum. Carbohydrates are absorbed chiefly as monosaccharides, most by simple or facilitated diffusion, but some, including glucose, are absorbed by active transport. Proteins are absorbed as amino acids by active transport. Monosaccharides and amino acids enter the bloodstream and are carried by the portal vein to the liver. Fats are absorbed mostly as fatty acids and glycerol. Inside the mucosal cells most are recombined into triglyc-

erides and coated with protein to form *chylomicrons* which enter the lacteals and are carried by the lymphatic system to the general circulation, thus bypassing the liver.

The fluids and electrolytes absorbed include not only those ingested, but those secreted by organs of the digestive system as well. Water is transported passively in response to gradients created by active transport (absorption) of electrolytes and other substances. Much of this occurs in the lower jejunum and ileum. Some of the fluid reaches the colon and most of that is also absorbed.

Water-soluble vitamins are readily absorbed with fluid and nutrients, but absorption of fat-soluble vitamins (A, D, E, and K) depends upon adequate fat digestion, and may be reduced if that is impaired (by lack of bile or lipase).

STUDY QUESTIONS

1 Compare and contrast the physiological properties of skeletal muscle and digestive tract smooth muscle.

2 What is the basic electrical rhythm (BER)? How does it affect smooth muscle of the gastrointestinal tract? How is it related to the action of motor nerves?

3 What are the important gastrointestinal hormones? What are their physiological actions?

4 Distinguish between peristalsis and segmentation. Could they both occur at the same time in a given section of intestine?

5 What are the elements of the act of swallowing that ensure that the bolus of food is directed into the esophagus rather than elsewhere?

6 What are the major mechanisms that govern the rate of emptying of the stomach?

7 What is the significance of hydrochloric acid in the stomach, and of bicarbonate ions in the pancreatic secretion?

8   In general, how does the control of secretion in the upper part of the digestive tract (such as salivary glands) differ from that in the lower part (small intestine or pancreas)? What controls the production of each of these secretions?

9   What is bile? What is its role in digestion? What is the significance of the enterohepatic circulation?

10   What substances are actually absorbed from the digestive tract? What is unique about the absorption of lipid materials?

11   To review the digestive functions, follow a piece of food—a cube of sugar, a pat of butter, or a piece of meat—through the digestive tract, and describe the effect of each secretion on it. Indicate the enzymes involved, and the end-products formed, as well as the sites of absorption.

24

Some Basic Considerations

Carbohydrate Metabolism

Fat Metabolism

Protein Metabolism

Endocrine Regulation of Metabolic Processes

Some Conditions that Require Metabolic Adjustments

Metabolism of Other Substances

Metabolism of Foodstuffs

The action of the digestive system in breaking down the foods we eat is not the final step in the overall process of providing energy to carry out our activities. Digestion is but a preparatory process that leads to absorption. The real tasks of managing our energy are carried out in the cells.

Each cell must obtain the nutrients it needs from the blood and either oxidize them to release their energy or store them for future use. What the cell does is determined to a great extent by each cell's needs at that time, but there must be some way to coordinate the metabolic activities of cells in different parts of the body. Some organs, such as the liver and adipose tissue, help control the availability of nutrients in the bloodstream, and that process, too, must be regulated.

The metabolic processes of the cells are regulated and coordinated largely by a group of hormones which affect the blood level of nutrients and determine whether these energy sources are stored, oxidized, or converted to another substance. The hormones coordinate cellular metabolic activities to bring about efficient management of our energy resources. Most of Chapter 24 is devoted to the cellular processes of metabolism that are concerned with energy.

Metabolism deals with energy. All cells require energy to carry on their activities and, indeed, to survive. The energy comes from the foods eaten, and it is the responsibility of the digestive system to prepare these foods for absorption into the bloodstream and for delivery to the cells. Metabolic processes are cellular activities and must be preceded by digestion and absorption.

SOME BASIC CONSIDERATIONS

To study metabolism is to study the transformations of energy and matter that occur in the cells of the body. Metabolism was discussed only briefly in Chapter 2, and some of the general principles should be reviewed and perhaps expanded. Energy is stored in the chemical bonds of molecules; it can be released when those molecules are broken down and their chemical bonds severed. The energy released in such reactions must be captured quickly and used or stored, or it will be lost.

Adenosine triphosphate

adenine   ribose   phosphates

adenosine

adenosine monophosphate (AMP)

adenosine diphosphate (ADP)

adenosine triphosphate (ATP)

The breakdown of large molecules to smaller ones is known as **catabolism.** Small molecules can be assembled into larger ones by forming new chemical bonds, but energy must be provided to form the new bonds. The synthesis of larger molecules is called **anabolism.** Catabolic reactions yield or release energy, while anabolic processes require or use energy.

Certain molecules are capable of forming "high-energy" chemical bonds that incorporate greater amounts of energy than most. A common example is the **high-energy phosphate bond** ($\sim$P) which links phosphate (PO_4) to the rest of a molecule. There are several substances which can contain energy-rich bonds, but the most widely distributed is **adenosine triphosphate (ATP).** It is formed of a nucleotide, plus two additional phosphate groups that are attached by high-energy bonds. Removal of one phosphate yields **adenosine diphosphate (ADP),** and removal of the other leaves **adenosine monophosphate (AMP).** Recall that AMP in its cyclic form (cAMP) is an intracellular mediator of many hormone actions.

Every cell contains some ATP and ADP. The ability of ATP to release energy when a high-energy phosphate bond is broken makes it an ideal source of the energy required for anabolic reactions and for muscle contraction and active transport.

$$ATP \rightarrow ADP + PO_4 + Energy$$

The ability of ADP to take up energy and form high-energy phosphate bonds makes it an excellent way to store the energy released in catabolic reactions.

$$ADP + PO_4 + Energy \rightarrow ATP$$

ATP–ADP serves as the "energy carrier" needed for coupling catabolic and anabolic reactions.

In the body the breakdown or catabolic reactions are, for the most part, oxidation reactions. A substance is *oxidized* when it combines with oxygen or when it gives up hydrogen or electrons (see Chapter 2). A substance that gives up oxygen or takes up hydrogen or electrons is *reduced.* Oxidation and reduction are always coupled, because whenever one substance is oxidized another is reduced. Most biological oxidations involve removal of hydrogen or electrons and require compounds to accept the hydrogen or electrons. Such substances are often called carriers, because they accept hydrogen or electrons from one compound (and are reduced), then donate them to another compound (and are oxidized). A number of hydrogen carriers are involved in energy transfers in the body.

Metabolic processes often consist of a series of reactions, each dependent upon and controlled by a specific enzyme. Many of the enzymes that catalyze oxidation-reduction reactions require a coenzyme. A **coenzyme** is an organic compound produced by cells that must combine in some way with an enzyme in order for it to be effective. A number of coenzymes are carriers of hydrogen or electrons; two that are particularly important are **nicotinamide adenine dinucleotide (NAD)** and **flavin adenine dinucleotide (FAD)**. Both of these, as their names state, are formed of two nucleotides, and both contain one of the B vitamins. NAD contains niacin, and FAD contains vitamin B_2, riboflavin. Vitamins are constituents of several other coenzymes as well.

Not all important coenzymes are hydrogen carriers, however. **Coenzyme A,** as a carrier of 2-carbon (acetyl) groups, provides a vital link in the oxidative process. It, too, is vitamin related, for it contains the B vitamin pantothenic acid. ATP, which has already been mentioned as an energy carrier, is also a coenzyme.

In our study of metabolism we will be concerned with the way in which energy is: (1) released from foodstuffs; (2) transformed for storage as energy-containing compounds, such as fat, or in the form of high-energy bonds; and (3) eventually used by cells to carry out their activities. These three aspects of metabolism cannot be isolated and treated separately, for they are closely interrelated. The fate of one foodstuff may depend upon what is happening to another. When the carbohydrate supply is abundant, for example, fat is stored, but when carbohydrate is in short supply, fat stores are reduced and fat is used for energy.

Carbohydrates, fats, and proteins are broken down in the digestive system to relatively small and simple molecules for absorption from it and eventual entry into the bloodstream. Once inside the various cells they are broken down to even smaller molecules, following metabolic pathways that are different for each foodstuff. The resulting two- and three-carbon fragments are chemically quite similar, and their oxidation is completed by a common pathway, the **Krebs** or **citric acid cycle.** Hydrogen ions are released in several reactions of the cycle and are transported through a series of carriers in oxidation-reduction reactions. Some of these reactions yield energy, much of which is captured in the high-energy phosphate bonds of ATP. In the following sections the major metabolic pathways are discussed in greater detail.

CARBOHYDRATE METABOLISM

Carbohydrates are absorbed from the digestive tract as monosaccharides. Many of them are absorbed by active transport systems which require energy but do not depend upon a concentration gradient. The concentration of sugars in the portal vein depends upon the rate of sugar absorption from the intestine and varies greatly. In the general circulation, the sugar level varies much less, for there are homeostatic mechanisms operating to keep it relatively constant. Upon reaching the liver, most monosaccharides are converted to glucose, so that glucose is about the only sugar in the general circulation. For all practical purposes, then, the study of carbohydrate metabolism is the study of the fate of glucose in the body.

Pathways in the Metabolism of Glucose

Since glucose is used by all cells, and preferred by most, many of the oxidative reactions occur in all cells, though some conversion reactions are more or less restricted to certain organs.

The first step in glucose metabolism is the entry of a glucose molecule into the cell. Unlike absorption from the intestine, glucose uptake by the cells is brought about by facilitated dif-

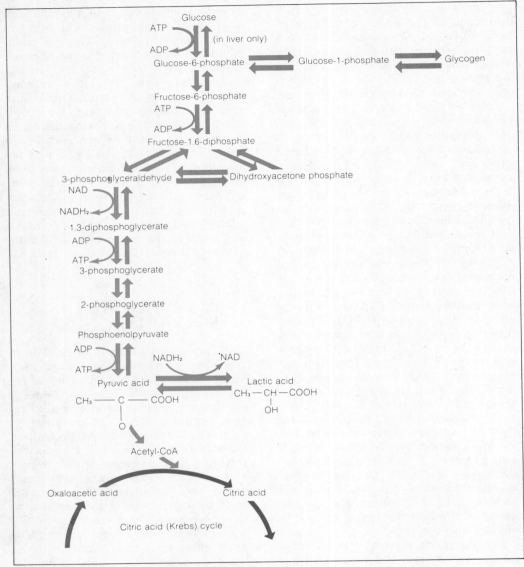

FIGURE 24-1   The sequence of reactions in glycolysis.

fusion, which is a passive process that requires no energy but does depend upon a concentration gradient.

As soon as glucose enters a cell, a phosphate group (obtained from ATP) is attached to carbon number 6 of the glucose molecule, forming **glucose—6—phosphate (Gl—6—P).** The enzymes that catalyze the reaction are present in the cytoplasm of all cells. This step, known as **phosphorylation,** is important because it is a prerequisite for any further reactions of the glucose. Prompt phosphorylation of the entering glucose keeps the concentration of

"free" glucose in the cell low, so that glucose can continue to diffuse into the cell. Gl-6-P is a key compound because it is at the intersection of several metabolic pathways. It is not an end point, however, since glucose is not stored or held as Gl-6-P. The several possible metabolic pathways for the glucose are shown in Figure 24-1 and described below.

The phosphate could be removed and the glucose returned to the blood-stream. This can occur only in the liver, because other cells lack the enzyme necessary to remove the phosphate.

This is important because it means that the liver is the only organ that can release glucose into the bloodstream. This is the main reason why the liver plays the dominant role in regulating the blood glucose level.

Glycogenesis Glucose may be stored in the cell as glycogen, a polysaccharide made up of long chains of glucose molecules linked together. The process of forming glycogen is known as **glycogenesis.** Almost all cells contain the enzymes necessary for glycogenesis, but liver and skeletal muscle store the greatest amounts.

Glycogenolysis The reactions of glycogenesis are all reversible, so glycogen may be broken down once more to Gl–6–P in a process called **glycogenolysis.** Glycogenolysis occurs in all cells that form glycogen, but the glucose may be released into the circulation only by the liver. In other cells the Gl–6–P formed must be used within the cell.

Glycolysis The oxidation of glucose in the cell is called **glycolysis.** It consists of several reactions that lead into the Krebs cycle.[1] Glycolysis begins with the addition of another phosphate (from ATP), after which the 6–carbon (hexose) molecule is split into two 3–carbon (triose) molecules. In two subsequent reactions energy is released and ATP is formed (2 ATPs for each triose, hence 4 ATPs). In addition, hydrogen is split off, which is taken up by the hydrogen carrier, NAD, forming $NADH_2$. The end result is two 3-carbon molecules of pyruvic acid.

All these reactions are reversible, so this pathway could be used to synthesize glucose (Gl–6–P) from pyruvic acid. Reversal would require the availability of energy (from ATP) and a source of hydrogen ($NADH_2$). The direction of reversible reactions is often

[1]You need not be concerned with the details of the individual reactions in glycolysis, nor the specific compounds, but you should be aware of what happens to the glucose molecule in the process.

determined by the relative amounts of such compounds, that is, by whether there is more ADP or ATP, or more NAD or $NADH_2$, as well as the relative concentrations of the reactants (such as Gl–6–P or pyruvic acid).

Note that in the glycolysis of one mole of glucose, 2 moles of ATP are "used," and 4 are formed, so there is a net gain of 2 moles of ATP for each mole of glucose oxidized. Note also that no oxygen has been consumed in this process.

Krebs Cycle and Hydrogen Transport

All the above reactions are carried out by enzymes in the cell cytoplasm. The pyruvic acid produced in glycolysis diffuses into the mitochondria where the enzymes for the Krebs cycle are located (Figure 24–2). Here carbon dioxide and a pair of hydrogens are split off, and the remaining 2-carbon acetyl group combines with Coenzyme A, forming **acetylcoenzyme A (acetyl–CoA).** This reaction differs from the glycolytic reactions because it is not reversible, which means that one cannot form pyruvic acid or synthesize glucose from acetyl-CoA.

Coenzyme A serves to carry acetyl groups into the Krebs cycle and transfer them to the 4–carbon oxaloacetic acid, forming the 6–carbon citric acid (hence the name "citric acid cycle"). Fatty acids from lipids and several amino acids can also be broken down to acetyl groups and combine with CoA. Acetyl-CoA is therefore a common ground, as it is where the metabolic pathways for the three major nutrients converge. The Krebs cycle reactions are as much a part of lipid and protein metabolism as of carbohydrate metabolism.

When the acetyl group enters the citric acid cycle, the coenzyme A is freed to pick up another acetyl group. The citric acid molecule formed when the acetyl group combines with oxaloacetic acid is then processed through the series of reactions that make up the cycle. Two more molecules of carbon dioxide and four pairs of hydrogen are split off and several molecules of water

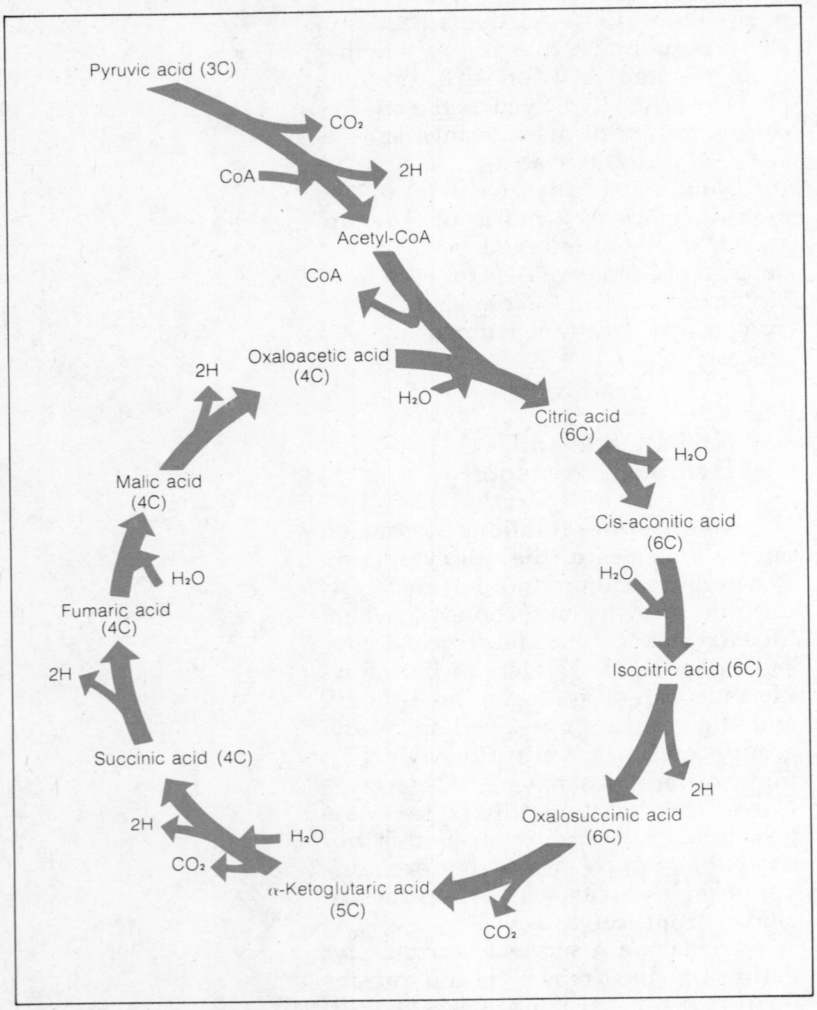

FIGURE 24–2   The Krebs (citric acid) cycle.

Energy Release   At three sites along the respiratory chain the hydrogen transfer results in the release of energy with which to form energy-rich bonds of ATP (if ADP and inorganic phosphate are available). An ATP molecule is formed at each of these sites with every pair of hydrogen ions carried through the chain. The process by which the oxidation of transport enzymes is coupled with the formation of ATP is **oxidative phosphorylation.**

In the complete oxidation of one mole of glucose to carbon dioxide and water, a total of 38 moles of ATP is formed. Of these, eight are formed in the degradation of glucose to pyruvate (glycolysis). Two of these represent the net gain of ATP previously mentioned. The other six result from oxidation of the $NADH_2$ and transport of the hydrogen ions through the hydrogen transport chain. The remaining thirty moles of ATP are formed in the passage of two moles of pyruvic acid through the citric acid cycle and in the transport of the resulting hydrogen ions through the chain of respiratory enzymes.

Oxidation of a mole of glucose yields approximately 686,000 calories of energy. If each energy-rich bond accounts for about 7600 calories, roughly 288,000 of the total calories released are captured in ATP—about 42 percent of the total yield. The remainder appears as heat. The efficiency of this system compares favorably with that of machines, in which 20–25 percent efficiency is typical.

Anaerobic Glycolysis and Oxygen Debt   A shortage of oxygen blocks the hydrogen transport system, since without oxygen the hydrogen ions cannot be removed from the final carrier in the chain. The consequent congestion eventually backs up and halts the entire Krebs cycle, preventing any more pyruvic acid from entering it. The lack of hydrogen acceptors (NAD) would also block formation of pyruvic acid, and hence also prevent the gain of energy from glycolysis. Pyruvic acid can, however, be reduced to lactic acid. This reaction requires hydrogen ions, which can be provided by $NADH_2$, thereby freeing a hydrogen acceptor and per-

are added, so that after one cycle the pyruvic acid molecule is completely "gone." The carbon dioxide diffuses from the mitochondria into the cytoplasm, then into the blood, and is carried to the lungs and excreted in the expired air. The hydrogen ions are immediately taken up by a hydrogen carrier, the coenzyme NAD, which transfers them to another carrier, FAD, which transfers them to another in a chain of linked carriers known as the **respiratory enzymes** or **hydrogen transport system** (see Figure 24–3).

Note that oxygen is the final hydrogen acceptor in the chain, resulting in the formation of a molecule of water, but note also that each of the transfer enzymes in the chain is repeatedly oxidized and reduced.

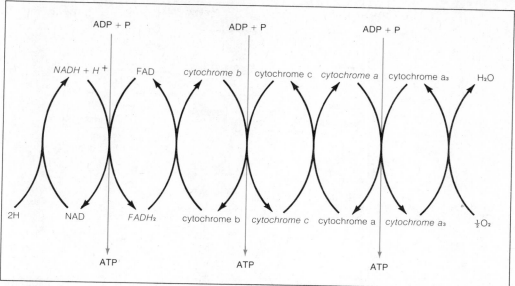

FIGURE 24–3   The hydrogen transport system, showing most of the steps and enzymes in the chain and the probable reactions that yield energy for the formation of ATP. NAD = nicotinamide adenine dinucleotide; FAD = flavin adenine dinucleotide. The cytochromes are hydrogen transport enzymes. The italicized compounds are in the reduced state.

mitting the breakdown of more glucose to pyruvic acid.

Lactic acid formation is a very important alternative in skeletal muscle, since, as was shown in Chapter 7, muscle contraction depends upon the release of energy from ATP and creatine phosphate (CP). The supply of these sources of energy-rich bonds is limited and is rapidly depleted in vigorous exercise. The ATP and CP can be "recharged" by the oxidation of glucose which, of course, requires oxygen. In vigorous exercise the supply of oxygen often cannot keep up with the need for oxygen, and the energy-rich bonds are broken faster than they can be restored.

Skeletal muscle is able to contract in spite of an oxygen shortage because pyruvic acid can be reduced to lactic acid, thus allowing more glucose to be metabolized, at least to lactic acid. Because the catabolism of glucose to lactic acid does not require oxygen, it is sometimes known as **anaerobic glycolysis.** During anaerobic glycolysis, lactic acid accumulates in the active muscles and considerably less energy is obtained from each mole of glucose (2 moles of ATP per mole of glucose). For a given energy expenditure, much more glucose must be degraded when lactic acid is formed than is necessary when oxygen is available.

Even anaerobic glycolysis has an aerobic aspect however, since eventually the lactic acid must be removed—it must be oxidized. The oxygen used for this constitutes the **oxygen debt** that is incurred during vigorous exercise. The lactic acid formed in the contracting muscle cells slowly diffuses out into the circulation and is returned to the liver. There it is oxidized to pyruvic acid, and roughly a fifth of it is then carried through the citric acid cycle. The energy released forms enough ATP to convert the remaining four-fifths of the lactic acid back to glucose or glycogen. The oxygen required for this, the oxygen debt, is additional oxygen which must be taken in, over and above that required for current needs. It is provided by the hyperventilation that occurs for a time immediately after exercise.

Fate of Glucose in the Body

Glucose is the first choice energy source for many cells, but the cells of the central nervous system depend upon it almost exclusively. They store very little glycogen, so it is important

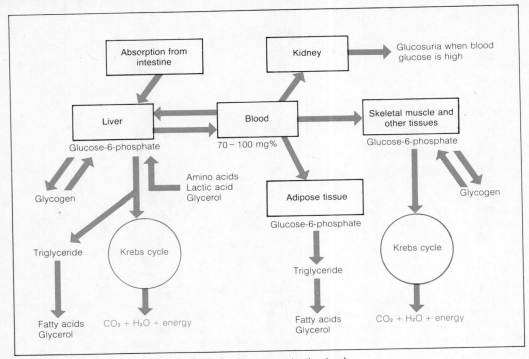

FIGURE 24-4   The source and fate of glucose in the body.

that blood always contains sufficient glucose to meet their needs. The maintenance of the blood glucose level within a fairly narrow range is an important aspect of homeostasis. There are a number of processes which contribute to it, and they are regulated by the same factors that regulate metabolism in general. It is not necessary to control the blood levels of amino acids and fatty acids as closely as glucose but, as we shall see, their levels are influenced by the same factors that control blood glucose.

The glucose concentration in the general circulation is normally maintained at about 70–100 mg percent (mg per 100 ml blood). At any given time it is a balance between the rate at which glucose is entering the bloodstream and the rate at which it is leaving (see Figure 24–4). Physiologically, glucose can enter the circulation only from the digestive tract or from the liver. All glucose and other monosaccharides absorbed from the intestine are carried in the portal vein to the liver, which releases glucose into general circulation (via the hepatic vein). All glucose in the general circulation, therefore, has

passed through the liver. Since it is the only organ that can return glucose to the circulation, all cells are totally dependent upon the liver for their glucose.

Upon leaving the bloodstream glucose enters cells of all tissues, but particularly those of skeletal muscle. They account for about half of the total cellular mass of the body, and at times their energy requirements can be extremely high. Glucose may also enter adipose cells where it is converted to fat and stored. Normally glucose is not excreted, but if the blood level is high, the kidney may not be able to retain all of it, and some glucose will be lost in the urine (*glucosuria*).

In skeletal muscle and most other tissues, glucose is stored as glycogen (glycogenesis) or oxidized for energy (glycolysis). Stored glycogen may be broken down to Gl–6–P (glycogenolysis) and oxidized, but not returned to the bloodstream. Glycolysis may continue through the Krebs cycle, or the pyruvic acid may be reduced to lactic acid. Because there is so much skeletal muscle in the body, it normally stores more glycogen than any other tissue.

In adipose tissue, the glucose (Gl–6–P) is converted to fat (triglycerides) and stored as such (*lipogenesis*). When necessary, fat can be broken down to fatty acids and glycerol (*lipolysis*) and released into the circulation, from which it can be picked up and used by other cells.

The liver is capable of carrying on all these processes, plus a few more. It stores glucose as glycogen (glycogenesis), whose concentration in the liver is ordinarily higher than in any other tissues, though the total amount is less than in skeletal muscle. It can also convert glucose to triglyceride for storage. The liver, of course, can carry on glycogenolysis and glycolysis, and it can release glucose into the circulation. Glucose can also be formed from certain noncarbohydrate sources in the liver, a process called **gluconeogenesis.** The main sources are certain amino acids and lactate, with glycerol a poor third. The liver is the only organ that is capable of gluconeogenesis.

Of the metabolic pathways available for glucose, the concentration of glucose in the blood would be lowered by glycogenesis in liver, muscle, and other tissues, and by lipogenesis in adipose tissue and liver, as these processes take up glucose. It is also lowered by glycolysis, since more glucose then tends to enter cells to replace that which has been oxidized. The blood glucose level is raised by glycogenolysis and gluconeogenesis in the liver, for they make more glucose available. The blood glucose level at any particular time depends upon which of these processes are dominant. Blood glucose is regulated, therefore, by the factors that determine which pathways are followed. These factors are largely hormonal, and they influence the pathways for lipid and protein metabolism as well.

FAT METABOLISM

You will recall that most of the fatty acids and glycerol absorbed from the digestive tract are recombined in the mucosal cells of the intestine and coated with a lipoprotein material to form droplets known as *chylomicrons*. These are extruded from the cells and absorbed into the lacteals and lymphatics. They therefore enter the general circulation without going through the liver. Short chain fatty acids and glycerol do enter the portal vein and go to the liver, but they are a small fraction of the total lipid. Following absorption of fats there is an immediate and marked elevation of the lipid content of the blood which may last for several hours. It is due chiefly to chylomicrons, although other lipid substances, particularly phospholipids and steroids such as cholesterol, are also present. Our concern, however, is with the major lipid component, neutral fats or triglycerides, and their constituents, fatty acids and glycerol.

Fate of Fats in the Body

Virtually all of the fat absorbed is stored at least temporarily. Most of it eventually reaches the tissues and is oxidized. A small amount of lipid, however, appears in secretions such as milk or sebum (from sebaceous glands in the skin), and some of it becomes part of the lipid structure of the cells and is not available for oxidation. This is largely phospholipid, an important component of cell and organelle membranes; its amount is relatively constant and does not vary with the size of energy reserves.

The site of fat storage is the fat cells in the adipose tissue distributed throughout the body. These connective tissue cells have such a great affinity for lipid that the nucleus and cytoplasm are often almost completely obscured by the immense fat globules they contain. About half the stored, or depot, fat is found in the subcutaneous tissue where it also serves as insulation. The remainder is distributed largely in the connective tissue around organs such as the kidney and heart and in the mesenteries. Body fat is liquid at body temperature.

To enter a fat cell, the triglyceride in the chylomicron must be hydrolyzed to fatty acid and glycerol, which requires an enzyme. Such an enzyme, a

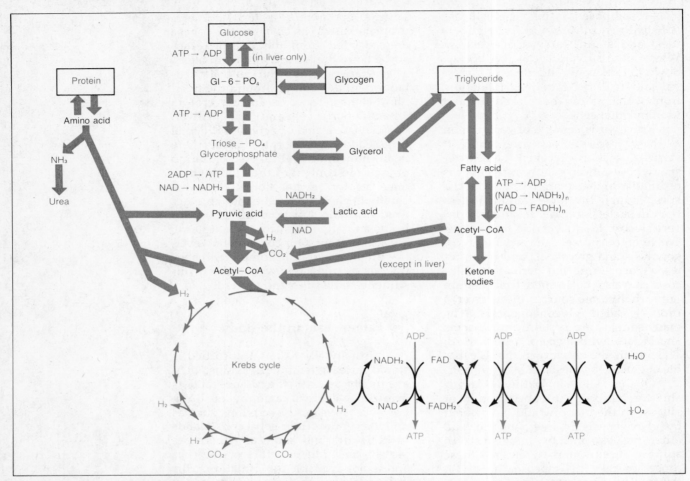

FIGURE 24-5 Summary of metabolic pathways, showing interconversions and common pathways.

lipase known as **lipoprotein lipase,** is found in the endothelium of capillaries. It frees the fatty acid and glycerol, which then diffuse into the adipose cell. Once inside the cell they are combined again into triglyceride for storage. Triglyceride can also be formed from glucose in fat cells.

The first step in the use of neutral fat is its hydrolysis to fatty acids and glycerol again, which also requires a lipase. This one is known as **hormone-sensitive lipase,** and it is found within the adipose cells. The fatty acid and glycerol are released into the circulation to be picked up by other tissues for oxidation. The energy requirement of the adipose cells is not high, and they themselves do not use much of the energy they store. Although fat may be stored for long periods of time, true

depot fat is readily available for use, and some of it is usually mobilized within a few hours of storage. There is thus a constant turnover of depot fat, and the amount in storage may vary widely. It represents the intake of energy (in calories) in excess of that expended by the individual.

Most of the glycerol formed when stored fat is hydrolyzed reaches the liver and enters the glycolytic pathway. It may be carried on through the citric acid cycle or used to resynthesize glucose. The fatty acids follow a different pathway.

Oxidation of Fatty Acids

Fatty acid oxidation is carried out by enzymes located in the mitochondria of nearly all cells, by a process known as *beta-oxidation* (because the action occurs at the second, or beta, carbon) or *alternate oxidation*. The energy from one ATP is needed to initiate the series of steps beginning when a molecule of coenzyme A is attached to one end of the fatty acid molecule, and ending when a two-carbon unit is split off the end as acetyl CoA (Figure 24–5). The steps are repeated to form more acetyl–CoAs. Since most naturally-occurring fatty acids contain 16 or 18 carbon atoms, each molecule will yield 8 or 9 acetyl–CoA molecules. The oxidation of the acetyl groups is completed through the Krebs cycle and the respiratory enzymes. The whole process has a high energy yield, for not only is energy released (and ATP formed) by oxidation of the acetyl groups (12 molecules of ATP for each acetyl group), but the cleavage of each acetyl group from the fatty acid yields 2 pairs of hydrogen ions, one of which combines with NAD and the other with FAD. When it is all totalled up, a 16–carbon fatty acid accounts for formation of 130 ATP molecules.[2]

In many tissues, particularly when the rate of fatty acid oxidation is high, the 2-carbon acetyl groups may condense to form *acetoacetic acid*, some of which is converted to *beta-hydroxybutyric acid* and/or *acetone*. These substances are known as **ketone bodies**.[3] They are normal products of fatty acid oxidation, but as most tissues are able to transfer CoA to them for entry into the Krebs cycle, their concentration in the blood normally remains quite low. The liver, however, lacks the enzyme needed to do this. When it oxidizes fatty acids, the ketone bodies diffuse into the bloodstream and are carried to other tissues for oxidation. The liver is the most important, and probably only, source of ketone bodies in the blood.

Ketone Bodies

$$CH_3-\overset{\overset{O}{\|}}{C}-CH_2-\overset{\overset{O}{\|}}{C}-OH$$

Acetoacetic Acid

$$CH_3-\overset{\overset{OH}{|}}{CH}-CH_2-\overset{\overset{O}{\|}}{C}-OH$$

β-hydroxybutyric acid

$$CH_3-\overset{\overset{O}{\|}}{C}-CH_3$$

Acetone

Fatty Acid Synthesis

Although the foods we eat are an important source of fatty acids, most of them can be synthesized in the body. Most fatty acid synthesis occurs in the soluble portion of the cytoplasm by a process quite different from that of beta-oxidation. The result is a saturated fatty acid that is nearly always 16 carbons in length. There are other pathways for synthesis, but they seem to involve chiefly lengthening of short-chain fatty acids and are probably of less significance.

The fatty acids synthesized in the body are all saturated, but there is a need for a certain amount of unsaturated fatty acids. The prostaglandins, for example, seem to be synthesized from unsaturated fatty acids. Since these fatty acids cannot be produced in the body, they must be provided in the diet; they are thus known as *essential fatty acids*. An inadequate dietary intake of the essential fatty acids produces a number of symptoms related primarily to growth, skin conditions, and kidney function.

[2]The ATP formed when a 16–carbon fatty acid is oxidized can be accounted for as follows: 8 acetyl groups × 4 $NADH_2$ per acetyl group × 3 ATP per $NADH_2$ = 96 ATP. There are 7 cleavages of the fatty acid × 5 ATP per each = 35 ATP. (Each $NADH_2$ results in formation of 3 ATP and each FAD results in formation of 2 ATP). The total is thus 96 ATP + 35 ATP − 1 ATP (used in the initial step), or 130 ATP.

[3]These substances are ketones, and except for acetone, they are acids. Since many other substances are also keto acids, these particular compounds are identified as "bodies." The term should not suggest that they are large molecules or have any special form.

PROTEIN METABOLISM

Protein, as the major organic constituent of protoplasm, is not used primarily as a source of energy, but as structural material. During growth the total amount of protoplasm, and hence of protein, in the body is increasing. However, even as new protein is being formed, existing protein—both in cell structure and in enzymes—is constantly being torn down and replaced. The amino acids released in the protein breakdown are used in the synthesis of new protein.

Amino acids from dietary protein are absorbed from the small intestine by one of several active transport systems and carried to the liver. They may remain there or be sent to one of the other tissues where, along with amino acids released in protein breakdown, they are used for synthesis of new protein. In the liver and certain other organs amino acids are used for synthesis of numerous specific substances, such as the purine and pyrimidine bases of nucleotides, histamine, melanin, and others. Amino acids also can be oxidized for energy in the liver.

Neither protein nor amino acids are actually "stored" in the same sense as glycogen or fat. While it is true that structural proteins can be catabolized for energy and thus constitute something of a reserve, they are not a true storage depot. Proteins are synthesized for a specific purpose, and that purpose is compromised when the proteins are broken down for energy. Since there is a continuous turnover of protein and amino acids, there is what has been called an *amino acid pool*, consisting of amino acids en route. This is not a true storage depot either, but it serves as a temporary "holding tank" for amino acids.

Protein Synthesis

Proteins are synthesized by linking large numbers of amino acids by peptide bonds. The nature of a particular protein is determined by the specific sequence of its amino acids, and that sequence is dictated by the genetic code carried by the DNA of the cell nucleus and expressed through the RNA, as discussed in Chapter 2. Protein is synthesized in all cells to some degree; it is severely limited in red blood cells, which have no nucleus, and some cells form only enough protein to maintain their own structure. A great deal of protein is synthesized by cells that are metabolically active, for they produce the enzymes to catalyze the reactions, or they produce secretions (hormones or enzymes) for export from the cell.

Protein synthesis imposes very specific amino acid requirements on the cells, which they meet largely by converting one amino acid to another. An important element in this conversion is *transamination*, the process by which an amino group (NH_2) from one amino acid is transferred to a keto acid, forming a new amino acid and a new keto acid. Amino acids can also be synthesized from glucose and lipids, through interconversions with common metabolic pathways (see below, and Figure 24–5).

Transamination

Amino acid #1   Keto acid #2

Keto acid #1   Amino acid #2

Of the 22 amino acids used in protein synthesis, 8 of them cannot be formed in the body. They are known as *essential amino acids* and must be provided in the diet. The ability to synthesize the proteins requiring these amino acids depends upon their availability. Protein of animal origin is more likely to contain all the essential amino

Focus

The Body's Response to Starvation

The body of a normal, sedentary, 70 kg man contains enough available metabolic fuel to live for about three months without food. He has approximately 141,000 kcal of adipose tissue (15 kg), 24,000 kcal of proteins, mainly muscle (6 kg), 900 kcal of glycogen, and 100 kcal of circulating glucose, fatty acids, and the like. An obese individual may be able to survive for more than a year without food; he or she may have 750,000 kcal of adipose tissue (80 kg), while other components remain about the same.

When the food supply stops, the body first draws upon liver glycogen to maintain the blood glucose level. The muscles meet their needs by drawing upon their own glycogen stores. This phase of the response to food deprivation lasts only a day or two, however. As soon as the easily metabolized glycogen is gone, the body increases its use of the fat reserves in the abdomen and under the skin. This increases the amount of ketone bodies in the blood and gives the starving person his or her characteristic bad breath. At the same time, the body begins to use its protein, even though its fat supply is far from exhausted. It does so because it cannot turn fatty acids into the glucose which the brain needs in order to remain functional; it can turn some amino acids into glucose, and in general 100 g of protein will yield about 57 g of glucose. The body must therefore use a large amount of protein every day to provide the 140 g of glucose the brain needs in that time.

The first proteins to be used are the stored digestive enzymes, which are no longer needed. The next to go are the nutrient-processing enzymes of the liver, followed by the contractile proteins and glycolytic enzymes of muscle. When the body starts to consume its muscle, the starving person becomes inactive, which may be regarded as an adaptation to the body's energy crisis. However, protein is used rapidly only during the first week or two of starvation. After four to six weeks, the brain becomes able to use ketone bodies, in particular the one known as D-β-hydroxybutyrate, and the body can shift back to an emphasis on using fat reserves. Protein is then conserved until the fat is gone. Then, since the protein is all that is left, the protein is quickly used up. Death soon follows.

The sequence of changes in the body through the course of starvation seems designed to do two things. The body draws first on those fuels it can most easily spare and use with the least consequence to its ability to survive if and when food becomes available again; only at the very end does the body use essential components. In addition, the sequence of changes seems designed to keep the central nervous system functional as long as possible. This highlights the crucial role played by the central nervous system in maintaining and coordinating all aspects of the body's activities.

acids than is vegetable protein. Deficiencies of the essential amino acids are commonly associated with a diet that is generally deficient both in protein and calories. Such a diet is said to be the single most serious nutritional problem in the world among young children.

Oxidation of Amino Acids

While protein synthesis and the conversion of one amino acid to another occurs in many tissues, the oxidation of amino acids is essentially a liver function. It begins with the oxidative removal of the amino group, a process known as **deamination,** which results in the formation of ammonia (NH_3),[4] which is a toxic substance that requires further treatment. Most of the ammonia is accounted for by the formation of **urea,** which is formed from two molecules of ammonia and one of carbon dioxide. Urea is normally present in the blood and is harmless in the amounts normally present. Each day the kidney excretes 25–30 grams of urea in the urine as a waste product. Its concentration in the blood may be lowered in certain types of liver malfunction marked by reduced deamination, and it may be elevated by inadequate excretion in kidney failure (uremia). Small amounts of other nitrogen-containing compounds are formed in the breakdown of related compounds and are excreted in the urine. Among them are uric acid from metabolism of purines (such as adenine) and creatinine from phosphocreatine in skeletal muscle.

The nonnitrogen portion of the amino acid remaining after deamination is a keto acid, like that shown in the transamination reaction above. Although the specific details for each amino acid are different, the keto acids produced from them can be processed for entry into the common metabolic pathway at one of four different sites. All of these keto acids can be oxidized through the Krebs cycle, and some can

be used in fatty acid synthesis, but only those amino acids whose keto acids can be converted to pyruvic acid can be used for synthesis of glucose (gluconeogenesis). Since these reactions are all reversible, many amino acids can be synthesized from components of the glycolytic and Krebs pathways.

Nitrogen Balance

Since amino acids are not actually stored, one might assume that amino acids taken into the body are excreted if they are not incorporated into new protein or related compounds. This is a correct assumption, and it underlies the concept of nitrogen balance. Nearly all the nitrogen in the body is associated with protein and its derivatives, and the nitrogen content of proteins is remarkably constant. The amount of nitrogen excreted in the urine is a good indicator of the breakdown of protein and amino acids in the body.

When the amount of nitrogen excreted is equal to the amount of nitrogen in the protein ingested, the individual is said to be in **nitrogen balance.** The turnover of protein in the body is such that the total amount of protein in the body is not changing and the amount of protoplasm is constant. This is the case in a normal healthy adult.

If, however, the amount of protoplasm in the body is increasing, more nitrogen is retained in the form of new protein, and less nitrogen is excreted than is ingested. Such an individual is said to be in *positive* nitrogen balance. It is likely to occur during growth, pregnancy, and recuperation from certain diseases, all of which are characterized by an increase in the total amount of protein in the body.

When nitrogen excretion is greater than nitrogen intake, there is a *negative* nitrogen balance. Protein is being broken down faster than it is being formed, and the amount of protoplasm is decreasing. This occurs during starvation and fever, after extensive burns, and for a brief period after surgery. These are conditions in which protein synthesis is impaired and/or its breakdown is excessive.

[4]Ammonia is also produced in the kidney and excreted as part of the renal regulation of acid-base balance. See Chapter 28.

ENDOCRINE REGULATION OF METABOLIC PROCESSES

We have seen that absorbed food-stuffs have several alternatives in terms of the metabolic pathway to be followed; generally, these are storage, conversion, or oxidation. The "choice" is not a random thing, but is influenced by the "needs" of the body. When glucose is abundant, for example, it is more likely to be stored, and when glucose is in short supply it is more likely to be withdrawn from the stores. This involves not only glucose metabolism in many different organs, but fat and protein metabolism as well.

Although glucose is the primary energy source for most cells, most of them can easily shift to fatty acids when glucose is not available. Nerve tissue, however, can oxidize fatty acids only with difficulty, and the entire nervous system is almost totally dependent upon glucose. This is particularly true of the higher centers in the brain, which also have higher rates of metabolism. They cannot function much longer without glucose than without oxygen. Maintenance of the glucose level is of importance primarily for brain function.

The processes which regulate the blood glucose level are the same ones that determine whether glucose is stored, converted, or oxidized. And because of interconversions, the metabolism of lipid and protein also influences the blood glucose level. It is appropriate, and almost essential, to consider regulation of the three together because they are controlled by the same hormones.

Careful regulation of metabolic processes and the blood glucose level is necessary for two reasons. First, we need it because our energy intake is intermittent. We are geared to replenish our energy supply three times a day (presumably!) but we can cope with fasts of more than just a few hours without ill effect. If necessary, we can survive several weeks of severely reduced energy intake (though not necessarily without ill effect). Second, we need regulation because our energy expenditure is intermittent. Periods of relative inactivity are interspersed with periods of activity, and our greatest energy outputs are usually confined to short bursts. The metabolic processes must thus adjust to cope with changes in both supply and demand.

One would not expect something as important as the management of our body's energy resources to be left to a single mechansim, and it is not. There are at least five hormones that play important roles and several others with less direct actions. Most of the hormones exert their actions by affecting the enzyme that catalyzes one of the metabolic reactions. It may be activation or inactivation of the enzyme, an increase or decrease in synthesis of the enzyme, or an increase or decrease in availability of a cofactor. Some hormones influence permeability of the cell membrane to a nutrient and thus affect the amount of substrate available. These hormones and their actions are discussed briefly in Chapter 13, and only those actions that are concerned with metabolism will be presented here. Figure 24–6 shows the metabolic actions of the hormones discussed below.

Insulin

As an exocrine gland, the pancreas is a major source of digestive enzymes, but it also has an endocrine component, the pancreatic islands or Islets of Langerhans, which are dispersed throughout the gland. Certain cells in these islands, the beta cells, produce insulin, which is probably the most important of the metabolic hormones. Insulin is secreted directly into the vessels of the hepatic portal system and carried directly to the liver where it has important metabolic actions. It is partially inactivated by an enzyme in the liver, but enough escapes into the general circulation to have widespread effects throughout the body.

Insulin is a polypeptide consisting of two peptide chains joined by cross linkages. It was the first protein whose molecular structure was completely worked out (in 1955). Now we know the

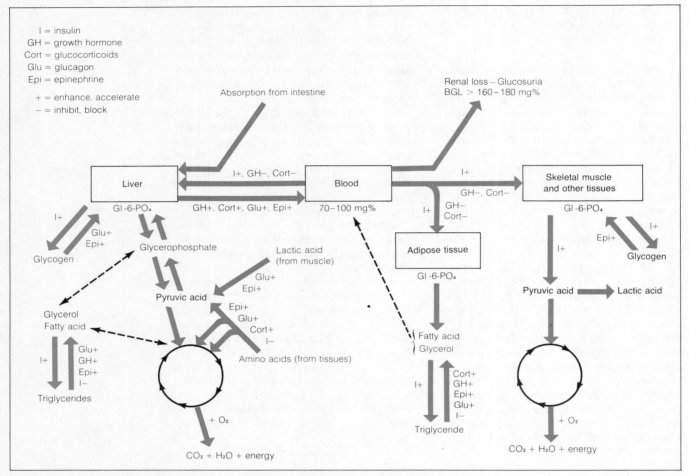

I = insulin
GH = growth hormone
Cort = glucocorticoids
Glu = glucagon
Epi = epinephrine

+ = enhance, accelerate
− = inhibit, block

FIGURE 24–6   Hormonal control of metabolic pathways and blood glucose level.

molecular makeup of many proteins, including almost all of the protein and peptide hormones.

Insulin has important actions on carbohydrate metabolism, but it affects fat and protein metabolism as well. Its actions on carbohydrates can be summarized quite well by saying that it lowers the level of glucose in the blood. The low blood glucose level, or hypoglycemia, is brought about by several different actions, but the major one (which was once thought to be the only one) is an increased permeability of the cell membrane to glucose. By this means insulin increases the entry of glucose into almost all cells, particularly those in which glucose enters by facilitated diffusion, most notably into all types of muscle and adipose tissue. (In-

sulin does not, however, influence glucose uptake in intestinal mucosa or kidney where transport is active, and it has little effect on glucose uptake in the brain or red blood cells.) An increased amount of glucose in the cells facilitates all reactions that begin with glucose, such as glycogenesis and glycolysis.

Most other actions of insulin on carbohydrate metabolism are on specific enzymes, often by inhibiting the formation of cAMP. Insulin stimulates the action of enzymes involved in glycogenesis and glycolysis and it inhibits several involved in gluconeogenesis. In general when one reaction is accelerated, its opposite is inhibited, so increased glycogenesis is accompanied by decreased glycogenolysis. Directly or indirectly, then, insulin favors all pathways that contribute to lowering the glucose in the blood, and it inhibits all those that tend to raise it.

Insulin inhibits the breakdown of fat, both in adipose tissue and liver. Recall that an enzyme, hormone-sensitive lipase, is required to hydrolyze stored triglyceride to fatty acids and glycerol. Insulin inhibits this lipase, thus preventing the formation of fatty acids and glycerol and their subsequent release into the circulation. Also, because insulin increases entry of glucose into cells, more glucose is available for synthesis of triglycerides. Insulin therefore increases the formation of triglycerides and inhibits their breakdown.

Amino acids are similarly affected, for insulin enhances their entry into cells as well, thereby reducing the amino acid concentration in the blood. It also increases protein synthesis because it promotes the incorporation of those amino acids into protein. Because of this, insulin has some important growth-promoting actions.

Taken all together, the actions of insulin lead to removal of all three foodstuffs from the bloodstream and increase the storage and/or synthesis of each. Its actions are anabolic. Without insulin, little glucose gets into the cells and the blood glucose rises, leading to *hyperglycemia*. The extracellular fluid compartment has an oversupply, but the intracellular compartment lacks glucose. The high level of glucose in the blood leads to the excretion of glucose in the urine (glucosuria). Because glucose has a powerful osmotic effect, its presence in the urine causes an excessive loss of water as well.

An insufficient supply of insulin leads to the condition known as **diabetes mellitus.** The very name of the disease emphasizes the large volume of urine and its sugar content (*diabetes* = pass through; *mellitus* = honey). Removal of the pancreas (pancreatectomy) causes diabetes in addition to problems due to the absence of the digestive enzymes. A pancreatectomized animal cannot use glucose very well in muscle, liver, or elsewhere, because so little gets into the cells. Accordingly, the animal does not readily store it as glycogen or convert it to fat either.

Without its anabolic actions to an-

tagonize the catabolic processes, glycogen is broken down to glucose, triglycerides are broken down to fatty acids and glycerol, and proteins are broken down to amino acids, all at a faster rate than if insulin were present. This is not all bad, however, because with the difficulty in oxidizing glucose for energy, other energy sources are needed. The catabolic reactions allow fatty acids and, to a lesser degree, amino acids to become the major source of energy.

Glucagon

A second pancreatic hormone is **glucagon,** produced by the alpha cells of the pancreatic islets. Like insulin, it goes first to the liver where an enzyme inactivates much of it. Glucagon is also inactivated in other tissues and has a half-life of only a few minutes. As a result, its main actions are exerted in the liver.

The actions of glucagon antagonize those of insulin, for glucagon acts to bring about a condition of hyperglycemia. It does so by increasing glycogenolysis in the liver (but not in muscle), brought about by a cAMP-mediated increase in the activity of one of the enzymes (a phosphorylase) necessary to convert glycogen back to $Gl-6-P$. This action raises the blood glucose level, since the liver can convert the $Gl-6-P$ to glucose and return it to the circulation. Glucagon also raises the glucose supply by stimulating gluconeogenesis, both from amino acids in the liver and from lactic acid produced in skeletal muscle. Glucagon causes the breakdown of triglycerides, which leads to the mobilization of fatty acids. It does so by activating the hormone-sensitive lipase needed for lipolysis.

The three main actions of glucagon—glycogenolysis, gluconeogenesis, and lipolysis—are all catabolic. They increase the supply of readily available energy, and all oppose the action of insulin. A pancreatectomized animal lacks both insulin and glucagon, but it has a less severe diabetes than one lacking only insulin. It is postulated that the severity of the diabetes in human patients may also be aggravated by the unopposed glucagon.

Growth Hormone

Unlike most other hormones of the anterior pituitary, **growth hormone** acts directly on cells in the body rather than on another endocrine gland. As its name suggests, growth hormone has widespread anabolic effects which are associated with growth of the individual, as discussed in Chapter 13.

Its actions on skeletal muscle are due largely to its stimulation of protein synthesis. This effect is not limited to skeletal muscle, though the effects are most conspicuous there. Growth hormone increases both the entry of amino acids into cells and synthesis of protein from those amino acids. The hormone probably acts on the ribosomes for the latter effect. These actions are shared with insulin, but its other actions oppose those of insulin. Growth hormone impedes the entry of glucose into cells and increases the release of glucose from liver cells, an effect probably brought about by inhibiting the phosphorylation of glucose as it enters cells. Growth hormone also causes a breakdown of triglycerides in adipose tissue, with the release of fatty acids into the circulation and increased oxidation of fatty acids. As with glucagon, the action is on the hormone-sensitive lipase, but in this case it is apparently an increase in production of the enzyme.

Growth hormone produces a hyperglycemic effect, and because of the increased use of fatty acids, it may cause ketosis; that is, it may be *ketogenic*. Because it causes symptoms like those of diabetes mellitus, it is therefore said to be *diabetogenic*. A pancreatectomized animal has diabetes because of its insulin lack, but if it is then hypophysectomized (the pituitary removed), its diabetes is much less severe. This is because an important means of *raising* glucose level has been removed as well as an important means of *lowering* it. In fact, such an animal can manage reasonably well—if it is not faced with conditions that require much regulation. A period of fasting or high energy output would cause a severe drop in blood glucose, and a large meal would be followed by a great increase. This illustrates the importance of the regulatory processes in maintaining the blood glucose level. It also shows that growth hormone has a continuing role long after growth has been completed.

Epinephrine

The actions of **epinephrine** and other catecholamines already have been discussed several times, primarily for their involvement in functions of the sympathetic nervous system and control of the cardiovascular system. Although the secretion of the adrenal medulla is mostly epinephrine, and the sympathetic transmitter is chiefly norepinephrine, both the gland and the nerve endings do produce a little of both. Epinephrine, as the catecholamine in highest concentration in the blood, has more effect than the others on organs that do not have a sympathetic innervation. Its secretion, of course, depends upon sympathetic stimulation of the adrenal medulla. Recall that one effect of sympathetic discharge is the release of glucose from the liver into the blood. Our concern at present is not so much with those responses that are associated with a full-blown fight-or-flight response, but rather with the metabolic effects, many of which may be initiated independently.

Epinephrine causes glycogenolysis in liver and skeletal muscle by increasing the amount of cAMP, which activates an enzyme causing formation of $Gl-6-P$. It was the study of this action of epinephrine that led to the original description of the role of cAMP in mediating many hormone actions. The breakdown of glycogen to $Gl-6-P$ in the liver leads to release of glucose into the circulation, but in skeletal muscle it increases glycolysis which, for some reason, results in a marked increase in lactic acid formation. Epinephrine stimulates the conversion of that lactic acid to glucose or glycogen when it reaches the liver (gluconeogenesis).

Epinephrine stimulates the hormone-sensitive lipase to cause hydrolysis of triglycerides and mobilization of fatty acids from adipose tissue. A comparison of the actions of epinephrine and glucagon shows many similarities. Both act via cAMP and both can elevate the blood glucose within a few minutes. Glucagon is about 30 times as potent for liver glycogenolysis, but it does not affect muscle glycogenolysis. Both act on adipose tissue, however.

Glucocorticoids — Cortisol

Certain hormones produced by the adrenal cortex can have profound effects on these same metabolic processes; these are the **glucocorticoids,** typified by *cortisol.* In spite of their name, they affect the metabolism of fat and protein as much as of carbohydrate. They are produced mainly in the zona fasciculata, ·the middle layer of the adrenal cortex. This portion of the gland is completely dependent upon ACTH (adrenocorticotropic hormone) from the anterior pituitary, and it atrophies after hypophysectomy.

Several of the most important actions of cortisol can be attributed to its action on proteins. Cortisol has a catabolic effect and causes the degradation of proteins into amino acids, which then enter the bloodstream. The abundance of amino acids leads to increased gluconeogenesis in the liver. The newly formed glucose may be stored in the liver as glycogen or released into the blood (cortisol increases activity of an enzyme for this). Cortisol indirectly interferes with uptake and utilization of glucose by inhibiting its phosphorylation. It also causes the breakdown of triglycerides by increasing synthesis of the hormone-sensitive lipase. Cortisol is thus both hyperglycemic and ketogenic, and therefore also diabetogenic. Its main effect, however, is probably the increased gluconeogenesis due to accelerated protein catabolism.

Glucocorticoids exert an elusive "permissive" action because their presence is necessary for the normal occurrence of certain metabolic reactions, including some actions of glucagon and epinephrine. Glucocorticoids resemble growth hormone in their effects on glucose and fatty acid metabolism, but they are opposites in their action on protein and amino acids. Aside from their permissive action, however, the glucocorticoids are not believed to contribute much to the control of metabolic processes in normal day-to-day regulation. Their secretion is associated primarily with the so-called "stressful" conditions that cause the secretion of ACTH.

Other Hormones

If one were to mention all the hormones that affect metabolic processes, nearly every hormone would have to be named. Among these are the *thyroid hormones* which increase the overall metabolic rate of nearly all cells, as shown by an increase in oxygen consumption. The thyroid hormones specifically increase the absorption of glucose from the intestine, and this contributes to an increased blood glucose level, at least temporarily. They have other, less understood actions which seem to reinforce, or potentiate, the actions of epinephrine. *ACTH* and *TSH* control the release of cortisol and thyroid hormones and therefore indirectly affect metabolic processes. *Androgens* (male sex hormones) have protein anabolic effects and cause an increase in the total muscle mass and general muscle strength. *Prolactin* has some actions that are very similar to those of growth hormone.

Summary of Hormone Regulation

Having considered the hormones that control metabolism and their actions, you are confronted with a confusing welter of similar actions that sometimes seem to be contradictory and are very difficult to sort out. When they are all brought together, as in Table 24–1, some patterns of action begin to appear. The table presents an oversimplification, because often several individual actions have been lumped together as a single effect, but by doing this it is easier to see the patterns.

Note that insulin is involved, either directly or indirectly, in all the processes. With the exception of growth hormone in the case of protein metabolism, one or more of the other hormones oppose or antagonize each effect of insulin. If you look at the processes that insulin increases or stimulates, you will see that they are all anabolic; they favor storage and synthesis. The other

TABLE 24-1   SUMMARY OF THE MAJOR ACTIONS OF HORMONES
ON METABOLISM

Process	Increased By	Decreased By
1. Blood glucose level	epinephrine glucagon growth hormone cortisol	insulin
2. Glucose uptake by cells	insulin	growth hormone cortisol
3. Glycogenesis (storage)	insulin	epinephrine glucagon
4. Glycogenolysis (mobilization)	epinephrine glucagon	insulin
5. Gluconeogenesis	epinephrine glucagon cortisol	insulin
6. Triglyceride formation	insulin	epinephrine glucagon growth hormone cortisol
7. Mobilization of fatty acid (breakdown of triglycerides)	epinephrine glucagon growth hormone cortisol	insulin
8. Amino acid uptake and protein synthesis	insulin growth hormone	cortisol
9. Protein breakdown and amino acid release	cortisol	insulin growth hormone

hormones, in one way or another, favor catabolic processes. Their actions are more in the nature of mobilizing energy resources—glycogenolysis provides glucose, breakdown of triglycerides provides fatty acids, and the breakdown of protein provides amino acids for gluconeogenesis. Because insulin is really the only anabolic hormone, a deficiency of insulin would leave the catabolic hormones unopposed, thus seriously disturbing the whole metabolic process. A lack of any one of the catabolic hormones is likely to have less serious metabolic effects since almost all the insulin-opposing actions involve more than one hormone.

The metabolic hormones act in concert to adjust the metabolic processes. Their secretion is coordinated because their controls have much in common. Insulin is controlled chiefly by the blood glucose level, with more insulin released when the glucose level is high and less when it is low. Its se-

cretion is also increased by certain amino acids (whose storage insulin facilitates) and by the intestinal hormones, especially CCK. Vagal stimulation and acetylcholine cause insulin release, but sympathetic stimulation and epinephrine inhibit it. Many of the stimuli which cause insulin secretion are related to food intake and the presence of absorbed nutrients.

Glucagon release is also stimulated by certain amino acids in the blood, but in most other respects its control is the reverse of that of insulin. A rise in blood glucose inhibits glucagon release, as does parasympathetic stimulation and acetylcholine, while sympathetic stimulation and epinephrine increase its release.

Epinephrine is controlled directly by the sympathetic nervous system, which has connections with the hypothalamus. Growth hormone, the glucocorticoids, and the thyroid hormones are controlled somewhat indirectly by the nervous system, for they all depend upon pituitary hormones which are controlled by hypothalamic releasing hormones. The hypothalamus receives almost unlimited input from the nervous system about all manner of stimuli to, and conditions in, various parts of the body. Included are such things as exercise, fasting, and low blood glucose, all of which are associated with a decrease or lack of immediate energy sources. Secretion of these hormones at such times enables us to mobilize stored energy when energy sources are not being supplied from the digestive tract.

Other stimuli which cause the release of the catabolic hormones include such diverse conditions as cold, infection, fever, and physical and emotional trauma. These factors have one thing in common—they are all stressful conditions. Low energy supplies and low blood glucose also can be stressful conditions. If all these conditions elicit the secretion of the mobilizing or catabolic hormones, we can see their role as contributing to coping with all types of stressful conditions. The special role of the glucocorticoids in this regard was pointed out in Chapter 13 and, indeed, it seems that ACTH and the glucocorticoids are much more important in coping with stresses than in adjusting to normal physiological conditions such as occur during and after each meal.

SOME CONDITIONS THAT REQUIRE METABOLIC ADJUSTMENTS

How the interplay of the metabolic hormones affects our daily lives may become more apparent if we examine some of the situations that require adjustments of the metabolic pathways— adjustments that can be produced by changes in the secretion of the various metabolic hormones. The situation that occurs most often involves the adjustments for processing nutrients just absorbed after a meal, followed by the adjustments to provide energy after absorption has been completed. Another condition is a pathological one, in which the normal hormone balance is upset by insulin lack.

Absorptive and Postabsorptive States of Metabolism

For several hours after a meal there is a period of absorption, during which the blood is enriched with nutrients (Figure 24–7A). The triglycerides enter the general circulation as chylomicrons by way of the lymphatic system. They are stored as triglycerides, briefly in adipose tissue. Amino acids pass through the liver and are distributed to all tissues for use in meeting their normal requirements for protein synthesis. Some amino acids are deaminated in the liver, and the nonnitrogen keto acid portion may be converted to fatty acid, oxidized for energy, or used in synthesis of various substances in the liver.

Much of the glucose also passes through the liver and enters cells throughout the body. In skeletal muscle it is stored as glycogen, and in adipose tissue it becomes fat. Glucose in the liver is stored as glycogen or converted to fatty acid or glycerophosphate needed for production of triglyceride. The triglycerides formed in the liver often become part of complex molecules such as lipoproteins that are eventually secreted from the liver.

During the absorptive phase, glucose is being used as the source of energy, and glycogen reserves are being restored primarily in liver and skeletal muscle, but also in other tissues. Any nutrients in excess of requirements are stored as fat in adipose tissue. These actions are all enhanced by insulin, which is the dominant hormone during the absorptive state.

The postabsorptive state gradually develops as absorption is completed and the absorbed nutrients have been taken up by cells (Figure 24–7B). The cells continue to require energy, which now must come from the newly replenished energy stores. In adipose tissue, the triglycerides are broken down. The glycerol released goes to the liver, but the fatty acids are distributed to the tissues which begin to use them as their energy source instead of glucose. Any ketone bodies that are produced are metabolized in those tissues. Skeletal muscle uses some fatty acids, but it has large glycogen stores to rely on when contraction is called for. The formation

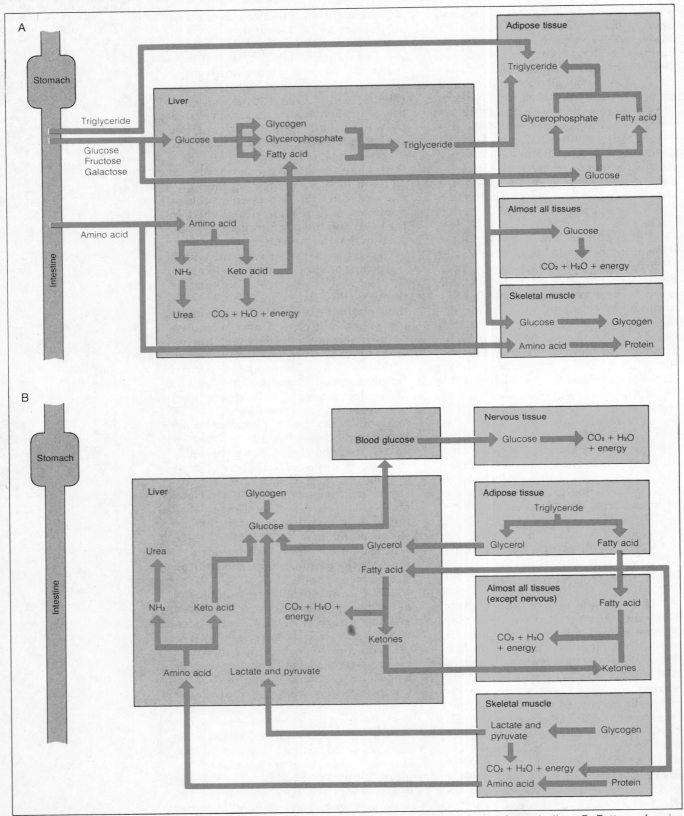

FIGURE 24–7   A. Pattern of major metabolic pathways during the absorptive phase of metabolism. B. Pattern of major metabolic pathways during the postabsorptive phase of metabolism.

of lactic acid is increased, however, and it is carried to the liver. Fatty acids are oxidized in the liver, but here the ketone bodies cannot be oxidized, so they diffuse into the circulation and reach other tissues to be metabolized. The nervous system is the exception. The central nervous system does not shift to fatty acid, and it continues to rely on glucose as its energy source.

Amino acid uptake and protein synthesis are reduced, and the amino acids in the blood plus those from normal catabolism find their way to the liver.

In the liver, everything is converted to glucose through glycogenolysis and gluconeogenesis (using glycerol from fats, lactate and pyruvate from skeletal muscle, and the keto acid portion of amino acids). The amount of glucose formed from glycerol is relatively small, but that from amino acids is potentially a significant amount. The glucose formed is released into the bloodstream to maintain the blood glucose level and provide glucose for the brain.

Glucose stores are not really very great, since the liver does not contain a large amount of glycogen. Although the concentration of glycogen in the liver is quite high, it amounts to no more than about 85 g, which will not last very long, and glucose from the large quantities of glycogen in skeletal muscle cannot be returned to the blood. After a time, therefore, the blood glucose level is maintained entirely by gluconeogenesis. In prolonged fasting (starvation), amino acids obtained by protein breakdown are the main source of glucose. It should be noted, however, that after a number of days of fasting, nervous tissue becomes able to oxidize ketones, which eases the glucose requirement. An individual can survive for some time using fat for energy and protein for maintaining the glucose in the blood, but at a lower blood glucose level than "normal."

The ability to shift our metabolic processes from a "glucose economy" to a "fatty acid economy" has obvious advantages for survival. It provides a means to carry on when the interval between meals (and energy replacement) may be very long. Normally, of course, these intervals are not so long, and the metabolic adjustments are not so extreme, but we do shift to fatty acid metabolism quite regularly. These shifts are brought about by the catabolic hormones, and very little insulin is produced during the postabsorptive state.

Diabetes Mellitus

Diabetes mellitus is a fairly common disorder, in which the basic problem is a lack of insulin, either absolute or relative. There may be no production, reduced production, or perhaps adequate production but insensitivity to it. It is known, for example, that there are individuals whose cell membranes do not have enough receptor sites to which the insulin can bind. Such persons have an elevated insulin level in the blood, but they do not have the benefits of its action.

The disease is usually thought to be a problem with carbohydrate metabolism, which it is, but since insulin normally affects fat and protein metabolism, they too are altered, especially fat metabolism. The situation is an exaggeration or caricature of the postabsorptive state. The actions caused by insulin are not happening, and those caused by the other hormones are unchecked.

In terms of glucose, there is hyperglycemia, with an excess of glucose in the extracellular fluid but a lack in the cells. As a result the cells have insufficient glucose to oxidize for energy, or to store as glycogen or triglyceride. Some of the symptoms are predictable, such as hyperglycemia, glucosuria, increased urine volume (polyuria), and great thirst (polydipsia). The energy lost from the body in the excreted glucose contributes to increased hunger, and the fluid loss may cause dehydration and the subsequent cardiovascular problems associated with a low blood volume.

Without insulin's action on fat storage, triglycerides are broken down by the unopposed lipolytic action of the other hormones. The increased fatty acid supply provides energy for the glucose-starved cells, as a shift to the postabsorptive pattern develops. The greater use of fatty acids means an increased production of ketones, and before long the ketone concentration in the blood rises as the other tissues approach their maximum rate of oxidizing ketone bodies. This situation is responsible for some additional symptoms. With increasing level of ketones (*ketosis*), some of them are excreted in the urine (*ketonuria*). One of the ketone bodies, acetone, is a volatile substance and can sometimes be detected by its characteristic odor, both in the urine and on the breath of the individual. The ketone bodies are acidic and contribute to *acidosis*, which may lead to depressed function of nerve cells. In severe cases there are also problems of dehydration and loss of electrolytes because of the large urine volume. Together, unless treated, these problems may lead to coma (*diabetic coma*) and possible death. The symptoms, and the metabolic derangements that caused them, can quickly be alleviated by administration of insulin. This causes a brief shift to absorptive state metabolism, in which glucose enters the cells and is oxidized and/or stored, and fatty acids are stored as triglycerides once more.

If too much insulin is given, however, the metabolic shifts may be too extreme. So much glucose may enter the cells that the blood glucose falls to levels that are too low to meet the needs of the brain cells. The individual becomes hyperirritable, and there may be convulsions followed by depressed function and eventually coma. This is *insulin shock*. It can be rapidly relieved by glucose, either injected or eaten (candy or fruit, for example, provide available glucose).

Insulin must be administered by injection into the bloodstream, for otherwise it would be destroyed by digestive enzymes. The individual whose pancreas produces virtually no insulin must pay close attention, both to the amount of food he or she eats and to the amount of exercise performed, in order to avoid the great swings of blood glucose and all of its consequences. Exercise, for example, increases the entry of glucose into cells and temporarily lowers the insulin requirement. Eating too much food at one sitting raises the blood glucose and increases the insulin requirement. Those individuals with a mild case of diabetes may be able to manage without insulin by careful planning and management of diet and activity. Others, whose pancreas is still capable of producing insulin, may be able to use an "oral insulin," which is not really insulin, but rather an agent that stimulates the beta cells of the pancreas to produce more insulin.

METABOLISM OF OTHER SUBSTANCES

Minerals

Inorganic constituents make up but a tiny fraction of the total body composition, but they are vital to the normal operation of many physiological processes. They are part of the medium needed for protoplasmic activity, both within the cell and in its immediate environment. They contribute to osmotic equilibrium, to acid—base balance, and to enzyme functions. The bulk of the body's inorganic matter consists of seven elements: calcium, phosphorus, sodium, potassium, chloride, magnesium, and sulfur. Other mineral elements, including iron, copper, iodine, manganese, zinc, cobalt, and fluorine, are present in smaller or trace amounts. Most of the minerals are widely distributed in the components of a normally balanced diet. Therefore, special efforts to ensure an adequate intake are not usually necessary.

Calcium   About 99 percent of the calcium present in the body is in the bones and teeth. The small fraction present in the blood plasma and other body fluids is critical in maintaining the normal irritability of the tissues. It is involved in the release of transmitters at synapses and neuromuscular junctions, and it seems to be a modulator of actions brought about by cAMP. It also is essential for coagulation of blood.

The absorption of calcium from the intestine, its deposition in (or resorption from) bone, and its excretion by the kidneys are regulated in such a way that the serum calcium level is maintained at about 10 mg percent. The absorption from the intestine is facilitated by vitamin D. The amount of calcium that is in the blood, or in bone, or is excreted is controlled by parathormone and calcitonin, as discussed in Chapter 13.

Phosphorus   Phosphorus is part of the calcium–phosphorus complex that makes up the inorganic material of bone. Phosphorus metabolism is closely associated with that of calcium, since the ratio of calcium to phosphorus in the blood is kept fairly constant by appropriate changes in the excretion of either calcium or phosphorus.

Although bones and teeth account for about 80 percent of the phosphorus in the body, it is present in all cells, often in combination with protein, lipid, carbohydrate, or other compounds. Its role in reactions involving energy transfer has been mentioned repeatedly, but although of vital importance, this does not account for a large quantity of phosphorus.

Sodium   Some sodium is found within cells, but most of it is in the extracellular fluid, where it is the chief cation. Its abundance there makes it particularly important in maintaining osmotic equilibrium and in preserving the membrane potentials which are so important for nerve and muscle irritability.

Sodium comes mostly from the salt used to season food. The amount of this intake varies greatly from one individual to another. Its concentration in the body fluids is regulated largely through the control of sodium excretion. Some is excreted in sweat, but most is lost from the kidney under the control of *aldosterone* from the adrenal cortex (see Chapters 13 and 28). Increased intake of sodium may be necessary following excessive loss by sweating, or in conditions in which aldosterone secretion is inadequate and too much sodium is excreted. Because of its osmotic effect, excessive sodium loss is accompanied by fluid loss; if prolonged, this can result in dehydration, weakness, and vascular collapse. Retention of sodium is accompanied by retention of fluid, leading to edema.

Potassium   The concentration of this, the chief intracellular cation, is much higher than in the extracellular fluid. The low extracellular potassium level is maintained by the sodium–potassium pump which pumps K^+ into the cell as it pumps Na^+ out. A rise in the concentration of potassium in the extracellular fluid reduces the resting membrane potential of the cells and affects the excitability of nerve and muscle. Elevation of serum potassium causes skeletal muscle weakness and depresses activity of the heart. Because of its abundance inside cells, potassium is an important part of the intracellular osmotic pressure.

Potassium is widely distributed in foods, especially seafood, and it is readily absorbed from the intestine. It is lost from the body, largely through the kidney, which is able to excrete it at a high rate. Its excretion is controlled in part by aldosterone, but since potassium has a role in the renal regulation of acid-base balance, its excretion is also influenced by those needs (see Chapter 28).

Chlorine   Chlorine, present mostly as chloride ions, is particularly important in the extracellular fluid. It is closely associated with and shares many functions with sodium, namely participation in osmotic pressure regulation and acid–base balance.

Chloride is normally ingested with sodium as table salt, and they are excreted together. Chloride ions generally can penetrate membranes with relative ease, and their transfer is by passive diffusion along the gradient created by the transfer of sodium or other ions (which are often transported actively). Thus, although chloride excretion is not controlled directly, it is influenced by many of the same factors that determine sodium loss.

Magnesium   Most of the magnesium in the body is bound with calcium and phosphorus in the bones. The remainder is distributed throughout the body fluids and soft tissues, particularly in muscle and nerve tissue.

Because magnesium is found in all green plants, a deficiency is difficult to produce, even experimentally. However, when one does occur, the symptoms include increased irritability of nerve cells. A high concentration of magnesium, on the other hand, has a depressant action on the nervous system. The metabolic role of magnesium seems to be as an activator of enzymes, particularly some of those involved in glycolytic reactions. Magnesium is excreted mostly from the intestine rather than by the kidneys.

Sulfur   As a component of several amino acids, sulfur is found in proteins and protein derivatives containing those amino acids. Although present in quantitatively small amounts, part of its importance is due to the chemical bonds that can be formed by sulfhydryl (–SH) groups. In addition, high-energy sulfate bonds (similar to high-energy phosphate bonds) play a vital role in tissue respiration (as in acetyl CoA-SH).

Iron   Two-thirds of the small but important amount of iron in the body is in the hemoglobin of the red blood cells; the rest appears as an essential part of several enzyme systems, including the cytochromes of the hydrogen transport system (Figure 24–3). The metabolism of iron differs from that of most other minerals in that there is no mechanism for excreting it; the amount present in the body is regulated by controlling the absorption of iron from the digestive tract.

The absorption of iron is an active process, since as it is taken up by the mucosal cells of the intestine, it is combined with a protein *(apoferritin)* to form an iron-containing protein complex known as **ferritin.** This complex is maintained in equilibrium with a second iron–protein complex in the blood. As the level of the latter falls, iron is released and diffuses into the blood, replenishing the iron protein complex there. The freed apoferritin can then combine with more iron from the intestine, but the ability of the mucosal cells and the body to absorb iron is limited by the availability of apoferritin. When it is all combined with iron, there is none left to combine with

more iron, and no iron can be absorbed.

Some iron is lost from the body as the epithelial cells of the intestinal mucosa wear away and release their contents into the digestive tract, but more is lost by bleeding. This may be serious in certain diseases but it is mostly commonly associated with menstrual blood, which is the cause of the higher iron requirements of females during the reproductive years. Iron deficiency is usually brought on by excessive loss of iron (as in blood loss) and causes the iron deficiency anemias in which the red blood cells lack hemoglobin. Inadequate intake or absorption problems have the same effect. Because of the mechanism for controlling its absorption, an excessive amount of iron in the body is not likely.

Other Minerals   Another mineral essential to the body is **iodine** because it is needed for the synthesis of thyroxin, the chief hormone of the thyroid gland. However, it is needed only in very small amounts, as this is its only known function. The thyroid gland is very efficient in removing iodine from the blood, but if the iodine supply is inadequate, formation of thyroxin is correspondingly reduced. The thyroid gland may attempt to compensate for the shortage by enlarging in order to capture more iodine and produce more thyroxin—a condition known as *goiter* —but the attempt is futile. Iodine metabolism is closely bound to thyroid function, discussed in Chapter 13.

Copper is a trace element needed for the formation of hemoglobin and the function of certain enzymes. The dietary requirement is very small, and it is so widely distributed in foods that a deficiency is unlikely.

Fluorine is found in many tissues, but particularly in bones and teeth. Small amounts, in any form, seem to reduce the formation of dental caries in children. In somewhat higher concentrations it tends to cause mottling of the tooth enamel. The mechanisms of these effects are not clearly understood. Drinking water is an important source of fluorine, but since it is often lacking, many communities add it to their water supply. It has been added to some toothpastes, and can also be applied to the teeth by the dentist.

Zinc, manganese and **cobalt** participate in several enzyme reactions in one way or another. Zinc is found in carbonic anhydrase and carboxypeptidase, and a zinc deficiency impairs the reactions in which these enzymes participate. Manganese activates cholinesterase and several phosphatases, but its exact role is not known. Cobalt is a constituent of vitamin B_{12} and is therefore needed for red blood cell formation.

Vitamins

An animal or human maintained on a diet containing the proper amounts of purified carbohydrate, fat, protein, and minerals will not prosper. In fact, such a diet is not enough to sustain life. Certain substances found in natural foods, the **vitamins,** must also be present in small amounts for life, health, and normal growth.

Vitamins are not used as a source of energy or as building blocks for protoplasm, but they do contribute to metabolic activity, each in a specific manner. As we have seen, several vitamins act as cofactors, often by serving as a carrier.

The first two vitamins to be described were the fat-soluble vitamin A and the water-soluble vitamin B. Further study eventually showed that there were a number of different substances concealed within each of these designations and another letter (or a number) was added as each new vitamin was recognized.

Fat-Soluble Vitamins   The fat-soluble vitamins are vitamins A, D, E, and K. Because of their solubility properties, they are poorly absorbed without pancreatic lipase, bile, and a certain amount of dietary fat.

One of the best-known roles of **vitamin A** is concerned with vision. *Rhodopsin*, or visual purple, the pigment contained in the rods of the eye, is split by light into a protein and *retinene* (visual yellow), which is derived from vitamin A. In the dark, rhodopsin is formed once more if there is an adequate supply of vitamin A. With a lack of vitamin A, rhodopsin formation is reduced and vision in dim light is impaired. Night-blindness is one of the first symptoms of vitamin A deficiency.

Vitamin A has other roles as well, and one is concerned with the integrity of the epithelium. Without the vitamin the skin becomes dry, and the normal function of secretory epithelium is impaired. It is also involved in carbohydrate metabolism and other metabolic functions, but the mechanisms of these actions are not well understood.

Vitamin A is widely distributed in foods, being found in dairy products such as milk, eggs, cheese, and butter, and in liver. Its precursors are found in carotene, the yellow pigment found in such foods as carrots, sweet potatoes, and apricots, and in leafy green vegetables.

The **vitamin D** group contains several related compounds formed when certain steroid compounds, such as the vitamin D provitamin found in the skin, are irradiated with ultraviolet rays. Vitamin D therefore tends to be a somewhat "seasonal" vitamin. Food sources of vitamin D are largely of animal origin—egg yolk, butter, and liver. Many foods, particularly milk, butter, and margarine, are irradiated or "fortified" to increase their vitamin D content.

The chief role of this vitamin is to permit or facilitate the absorption of calcium from the intestine. It therefore tends to raise the calcium level in the blood and promotes the deposition of calcium in bone. Lack of vitamin D eventually leads to problems of inadequate calcification of bones and teeth. The deficiency disease is *rickets*, which is characterized by a soft and often deformed bone structure.

The importance of **vitamin E** in humans is difficult to assess, even though it is known to retard the oxidation of certain compounds, notably fats. In laboratory animals it is necessary for normal reproduction and muscle and peripheral vascular function. A deficiency results in muscular dystrophy and other structural changes in muscles, as well as sterility. If a pregnancy does occur in a deficient animal, the embryo dies and is resorbed. The effects are less clear-cut in humans, however, since a lack of vitamin E is not known to cause sterility, and even large doses do not remedy sterility-related problems. Nor is there any effect of vitamin E on human muscular dystrophy. Massive doses have been recommended for certain cardiovascular conditions, but this treatment is highly controversial. Vitamin E is found in many foods, including meat and dairy products, leafy vegetables, and several vegetable oils (including wheat germ oil).

Vitamin K is necessary for the formation of prothrombin in the liver. In its absence the prothrombin content of the blood declines and blood clotting is delayed, sometimes enough to cause severe bleeding from minor injuries. Since vitamin K is widely distributed in foods a deficiency is not likely under usual circumstances.

Clotting problems may develop in the newborn, since the fetal supply is by diffusion in the placenta. This source is removed at birth, and hemorrhagic tendencies in the newborn are not unusual. Administration of large doses of vitamin K to the newborn infant, or to the mother shortly before delivery, reduces the danger until other sources become sufficient.

Water-Soluble Vitamins The water-soluble vitamins are vitamins B and C. The single vitamin C is ascorbic acid, but the B vitamins include thiamin (B_1), riboflavin (B_2), niacin, pyridoxine (B_6), pantothenic acid, and several others. These vitamins are chemically quite distinct from one another, but, largely because of their water solubility, deficiencies of them are likely to occur as multiple deficiencies. For the same reason, the water-soluble vitamins are likely to be lost from foods during cooking.

Thiamin (B_1) serves primarily as a coenzyme (thiamin pyrophosphate) in certain reactions leading to carbon dioxide formation (oxidative decarboxylation), as in the citric acid cycle. It must be supplied daily because it is not stored in the body. When it is lacking there is difficulty in oxidizing pyruvic acid, and therefore problems in carbohydrate metabolism arise. A prolonged lack causes the deficiency disease known as *beriberi*, which is character-

ized by complex symptoms, primarily of the peripheral nervous system, and disorders of the cardiovascular system and gastrointestinal tract. Beriberi and milder signs of thiamin deficiency are common in populations existing on a high carbohydrate diet (rice) with a low fat and protein intake. Thiamin is found in small amounts in many foods, but it is more abundant in unrefined cereal, liver, kidney, and heart.

Riboflavin (B_2) also plays an important role as a component of FAD, a coenzyme involved in metabolic oxidations and hydrogen transport. Lack of riboflavin produces cutaneous lesions of the lips, angles of the mouth, tongue, and eye. It is widely distributed in foods of both plant and animal origin, but it is particularly high in milk, liver, kidney, heart, and some vegetables.

Niacin (*nicotinic acid* or *nicotinamide*) forms part of NAD, an important hydrogen acceptor that participates in a number of reactions in cellular respiration. It is present in all plant and animal cells, but it is in higher concentration in heart muscle and red blood cells. It can be synthesized from the amino acid tryptophan, but not in sufficient amounts in humans. Lack of niacin results in a disease known as *pellagra* one of the most common and serious deficiency diseases in the United States, although the symptoms, which include inflammation of the skin and tongue, and diarrhea, are similar to those of some other vitamin deficiencies.

Pyridoxine (B_6) is involved in the metabolism of unsaturated fatty acids and of several amino acids and in the formation of niacin from tryptophan. It apparently also plays a role in the metabolism of the central nervous system, since in children the deficiency is often accompanied by seizures. Some of the other symptoms are associated with the skin.

Pantothenic acid, as a constituent of coenzyme A, participates in many important metabolic functions. Probably the best known is the "activation" of acetate, that is, the formation of acetyl–CoA prior to the entry of the 2–carbon acetate into the citric acid cycle. Pantothenic acid is thus involved in the metabolism of lipid and amino acids as well as of carbohydrate. It also contributes to the formation of acetylcholine and cholesterol, and hence of the steroid hormones. A lack of pantothenic acid produces symptoms in many systems, including the gastrointestinal tract, the skin, and the adrenal gland.

Vitamin B$_{12}$ (*cobalamin*) is the antianemic factor for normal red blood cell formation. The molecule is structurally unique in that it contains a molecule of cobalt in the center. Its absorption from the intestine depends upon the presence of the "intrinsic factor" from the gastric mucosa. Vitamin B$_{12}$ is a coenzyme which participates in a number of different reactions, particularly in the synthesis of certain amino acids.

Ascorbic acid, or **vitamin C,** somewhat resembles a monosaccharide in its chemical structure. It is the least stable of the vitamins, being easily destroyed by heat. Its specific biochemical role is not known, although it helps maintain the intercellular substance of connective tissue, including bone and cartilage. A lack of ascorbic acid results in a disease known as *scurvy*, characterized by painful joints and bleeding from the joints, mucous membranes of the mouth, and gastrointestinal tract. Teeth loosen, bones fracture easily, and wound healing is greatly slowed. All of these are related to poor formation and maintenance of the organic intercellular material. There is evidence that ascorbic acid is involved in the synthesis of the amino acid hydroxyproline, an important constituent of collagen.

Because vitamin C is found in fresh fruits and vegetables, particularly in citrus fruits, berries, melons, tomatoes, and cabbage, it should not surprise you to learn that scurvy was probably the first disease to be associated with a dietary deficiency. It was a serious problem to early sailors and explorers, but it disappeared as soon as they began to carry a supply of these foods with them.

CHAPTER 24 SUMMARY

Some Basic Considerations

Metabolism includes all of the chemical reactions in the body. *Anabolic* reactions involve synthesis of larger molecules and require energy, and *catabolic* or breakdown reactions release energy. The energy may be used to form chemical bonds that incorporate greater-than-average amounts of energy, such as the *high energy phosphate (HEP)* bonds. A high energy phosphate bond added to *adenosine diphosphate (ADP)* forms *adenosine triphosphate (ATP),* an almost universal energy carrier. The reaction is reversible and the energy can be released for reactions that need it.

Catabolism involves oxidation of foodstuffs, usually by the loss of hydrogen. There are several important hydrogen carriers, such as the coenzyme *nicotinamide adenine dinucleotide (NAD),* which accept the hydrogen released in one reaction and give it up in another reaction. *Coenzyme A (CoA)* is an important carrier of 2-carbon (acetyl) groups.

Carbohydrate Metabolism

Carbohydrates are absorbed as monosaccharides, and are carried to the liver where most are converted to glucose. Whenever glucose enters a cell it is *phosphorylated* to glucose-6-phosphate (Gl-6-P), then stored or oxidized. It is stored as glycogen (*glycogenesis),* and it also can be returned to Gl-6-P (*glycogenolysis).* It is oxidized in a series of reactions (*glycolysis)* in which pyruvic acid is formed and a small amount of energy released (ATP formed). The reactions are all reversible, and glucose can be synthesized from pyruvic acid. The liver can remove phosphate from Gl-6-P and release glucose into the circulation.

Pyruvic acid is broken to an acetyl group that combines with CoA (*acetyl CoA)* and enters the *Krebs cycle.* Fats and proteins also yield acetyl groups that combine with CoA to enter the Krebs cycle. The Krebs cycle is therefore the common pathway for final metabolism of all three foodstuffs. It consists of a cyclic sequence of reactions in which carbon dioxide and hydrogen are split off. The carbon dioxide diffuses from the cell and the hydrogens are taken up in a series of oxidation-reduction reactions in which the final hydrogen acceptor is oxygen, and water is formed. Three of these reactions produce energy that is used to form ATP *(oxidative phosphorylation)*. Considerably more ATP is formed by oxidation through the Krebs cycle and hydrogen transport system than by glycolysis.

If there is not enough oxygen to accept all the hydrogen released, there will be a backup in the hydrogen transport system and Krebs cycle. Glycolysis and energy release can still occur without oxygen, however, by reducing pyruvic acid to lactic acid (anaerobic glycolysis). In vigorous exercise, for example, lactic acid accumulates in exercising muscles. It is oxidized in the liver at a later time, and the extra oxygen required to do this is the *oxygen debt*.

Glucose is the energy source preferred by most cells, but those of the nervous system are most dependent upon it. The level of glucose in the blood is regulated to ensure a continual supply of glucose for them. The blood glucose level is raised by glucose entering the bloodstream from the digestive tract and the liver, and is lowered by glucose entering cells or, when it is high, glucose may be excreted in the urine. In skeletal muscle large quantities of glucose can be stored as glycogen and/or oxidized. In adipose tissue glucose is converted to fat (triglyceride) for storage. The liver can store glucose as glycogen or convert it to fat, and it is the only organ that can release glucose into the bloodstream, and the only one that can convert other substances (especially amino acids) to glucose *(gluconeogenesis)*.

Fat Metabolism

Lipids absorbed from the digestive tract enter the circulation as chylomicrons by way of the lymphatics. An enzyme in capillary endothelium (lipoprotein lipase) breaks them down to fatty acid and glycerol for entry into adipose cells where they are recombined into triglycerides. Fats must be hydrolyzed again to fatty acids and glycerol (by a hormone-sensitive lipase in fat cells) for release into the circulation and distribution to other cells for oxidation. The glycerol is oxidized through the glycolytic pathway, and the fatty acids are oxidized by *beta-oxidation*, in which acetyl–CoA is formed for entry into the Krebs cycle. Both beta-

oxidation and the oxidation of the acetyl groups yield large amounts of energy (ATP). Some acetyl groups condense to form *ketone bodies* (acetoacetic acid). The liver is unable to oxidize ketone bodies, but they can be oxidized by other cells to a certain extent. If the rate of ketone body production is high, their concentration in the blood may rise *(ketosis)*. Most fatty acids can be synthesized from acetyl–CoA.

Protein Metabolism

Proteins are absorbed as amino acids, which are then used by all cells for synthesis of enzymes and structural proteins and, if necessary, as a source of energy. Protein synthesis requires specific amino acids, and many of them can be synthesized by converting one amino acid to another (transamination).

Amino acid oxidation occurs in the liver, beginning with the removal of the amino (NH_2) group *(deamination)*. This yields ammonia (NH_3) a toxic substance that is removed by the formation of *urea* and excreted in the urine. The remaining keto-acid portion is reconverted to intermediate compounds that fit into the glycolytic pathway or the Krebs cycle.

Endocrine Regulation of Metabolic Processes

Whether nutrients are stored, oxidized, or converted to something else is determined largely by the action of several metabolic hormones. *Insulin,* produced by beta cells of the pancreas, is hypoglycemic and generally anabolic in its action. One of its actions is to increase the entry of glucose and amino acids into cells. This leads to increased storage of glucose as glycogen or as fat, and increased protein synthesis. Insulin inhibits the breakdown of glycogen, triglycerides, and protein. Lack of insulin leads to hyperglycemia and increased catabolic reactions.

Glucagon, produced by alpha cells of the pancreas, is hyperglycemic. It increases glycogenolysis in the liver, breakdown of fat stores, and gluconeogenesis.

Growth hormone, produced by the anterior pituitary is anabolic for protein (increasing amino acid uptake and protein synthesis), but catabolic for glucose and fats. It causes glycogenolysis and release of glucose by the liver and the breakdown of fats.

Epinephrine, produced by the adrenal medulla, has effects similar to those of glucagon. It is hyperglycemic, causing glycogenolysis and release of glucose by the liver, glycogenolysis in skeletal muscle, and breakdown of fat in adipose tissue.

Glucorticoids from the adrenal cortex are also hyperglycemic. Their most notable metabolic action is catabolism of protein, which yields amino acids that are used by the liver to increase gluconeogenesis.

Conditions that Require Metabolic Adjustments

During the absorptive state after a meal, glucose is used for energy by most cells, and the remainder is stored as glycogen or converted to fat. Absorbed fats are stored, and amino acids used for protein synthesis. These reactions are increased by insulin, and it is the dominant hormone during this time.

After absorption, during the post-absorptive state, energy to carry on cellular activities comes from stored glycogen and fat. Most cells shift to fatty acids as the energy source (obtained from adipose tissue). The liver obtains glucose by glycogenolysis and gluconeogenesis, and releases it into the bloodstream. This maintains the blood glucose level for cells of the nervous system which cannot readily shift to fatty oxidation. These reactions are brought about by the catabolic hormones, which are secreted in greater amounts at this time, while less insulin is produced.

Diabetes mellitus, which is the result of insufficient insulin, represents an exaggerated postabsorptive stage. The effects of the catabolic hormones are not adequately opposed. Since glucose cannot enter the cells for use or storage, the blood glucose level rises and glucose is excreted in the urine. Fatty acids are used for energy and the formation of ketone bodies may be excessive, causing ketosis and acidosis (which may lead to diabetic coma and

death). Injection of insulin quickly alleviates the symptoms by causing a temporary shift to the absorptive state of metabolism. Too much insulin may cause hypoglycemia (insulin shock).

STUDY QUESTIONS

1 What is the first thing that happens to glucose when it enters a cell?

2 Distinguish between digestion and metabolism. Where does each occur?

3 What is the role of coenzymes in the metabolic process?

4 Where does the energy come from that is released in glycolysis, and what happens to it?

5 What is the effect of a lack of oxygen on the citric acid cycle, and on the release of energy?

6 By what processes is glucose added to the blood, and by what processes is it removed from the blood? What cells can add glucose and what cells can remove it?

7 Why is it so important to maintain the blood glucose level? By what means is this done when glucose supply is low?

8 Where and in what form are fatty acids stored? How and when are they released from storage?

9 What does urea have to do with protein metabolism? Where is it formed?

10 What is the role of insulin in metabolism? Why is a lack of insulin more serious than the lack of any one of the other metabolic hormones?

11 How do the metabolic pathways in the post-absorptive state differ from those in the absorptive state? Of what significance is the ability to make this shift in metabolic pathways?

12 In general, what kinds of stimuli control the release of metabolic hormones? How are these stimuli related to the overall body needs?

13 In general, what role do vitamins play in the metabolic processes?

25

Energy Balance

Sources of Energy

Energy Expenditure

Energy Metabolism

In previous chapters we have discussed the intake of food, which is our source of energy, and we have discussed processes by which cells release and use that energy. The next step is to consider the overall use of energy by the entire body.

Energy intake, in the form of food, must be adequate to cover the energy expended in our various activities. If it is insufficient our energy reserves diminish and we lose weight, and if it is excessive our energy stores increase and we gain weight. How much energy is required to maintain a balance? What controls the energy (food) intake? For what purposes do we use our energy, and how much is required to meet these needs? How much does it actually "cost" in terms of energy to "run our machine?" We all need answers to these questions to properly balance our personal energy equation.

A common definition of metabolism states that it is the sum of all the body's transformations of energy and matter, while the metabolic rate is the rate at which these transformations occur. Metabolism encompasses both catabolic and anabolic reactions, the use, conversion, and storage of all the foodstuffs discussed in the previous chapter. It is the ability to capture and store the energy in food, and it is the ability to make that energy available when needed. Indeed, it is this ability to handle energy that enables living things to carry on all the activities we associate with life.

ENERGY BALANCE

The basic **law of conservation of energy** states that energy can be neither created nor destroyed. For this reason, to say that energy is "produced" means that it is *released* by being *converted* to a useful form of energy, and to say that energy is "used" means that it does work and/or is converted to heat. Implicit in this law is the fact that the energy already and always exists, so that after any series of reactions or energy exchanges, all the energy is still present in some form.

This brings us to the concept of *energy balance*. When a given amount of energy has been "put into" the components of a reaction, the same amount of energy should be "gotten out" of that system after the reaction, although in a different form. The energy input equals the energy output:

$$E_{in} = E_{out}$$

These statements apply equally to metabolic reactions within the body and nonmetabolic reactions outside the body. The energy input for the latter may be electrical, light, mechanical, or chemical energy, but animals can only use the chemical energy (E_{chem}) stored in the chemical bonds of the foods they eat. This energy is released when the foods are metabolized by the cells. It is released slowly, bit by bit, and it is held temporarily as ATP.

Energy made available from stores or from ATP may be used for *mechanical work,* as in muscle contraction, for *electrical work,* as in the maintenance of electrical potentials in the nervous system, or for *chemical work,* as when new compounds are synthesized. Much of the energy released, however, appears directly as *heat* and even that energy used to perform useful work is degraded to heat eventually. Heat is therefore a by-product of all energy conversions. Because the body cannot harness heat energy to do work, most of it is wasted and must be dissipated. Of the energy released in metabolic reactions, about 20 percent can be accounted for as useful work and 80 percent as heat. This is sadly comparable to the efficiency of an internal combustion engine.

The equation above may now be rewritten to consider the nature of the energy intake and output:

$$E_{in} = E_{out}$$

$$E_{chem} \text{ (food)} = E_{heat} + E_{work}$$

This equation describes a system in equilibrium, one whose energy content is constant. In the body, however, the energy content may vary over a period of time. If the energy intake (food) is greater than the output, there will be a gain in the body's energy content—that is, a storage of energy. If energy intake is less than output over a period of time, there will be a loss of energy from the body's stores. In other words, there is another factor, energy storage (E_{stor}), that must also be considered to make the equation balance under all conditions:

$$E_{chem} \text{ (food)} = E_{heat} + E_{work} + E_{stor}$$

Whenever E_{stor} is positive for a period of time, there will be a weight gain, and when E_{stor} is negative for a period of time, there will be a weight loss. Balancing this equation is one of the most serious and widespread problems in the world today. In most of the world and parts of the United States, the problem is to ensure that energy intake is sufficient to cover the output and avoid negative storage. In much of this country, the problem is excessive intake coupled with reduced output, resulting in too much storage, which may

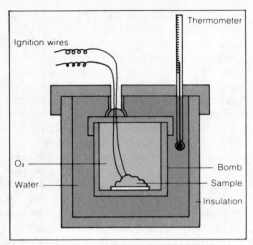

FIGURE 25–1   The bomb calorimeter, used for determining the caloric value of foodstuffs.

lead to *obesity*. For many of us there will be (or is) a persistent problem of weight control, a continual struggle to maintain an energy balance.

SOURCES OF ENERGY

The ultimate source of energy for living things is radiant energy (E_{rad}) from the sun. No animal can convert this energy into a useful form, but plants, in a process known as **photosynthesis,** use the energy of the sun to form a hexose sugar from carbon dioxide and water:

$$6CO_2 + 6H_2O + E_{rad} \rightarrow C_6H_{12}O_6$$

To obtain their necessary energy, some animals eat the plants, some animals eat other animals, and some animals, including humans, eat both plants and animals. In any case, once the energy has been incorporated into the chemical bonds of the sugar molecules, it is in a form that can be used by animals, that is, chemical energy. Animals can then transform it again into mechanical, electrical, or chemical energy in order to perform useful work.

Energy Content of Foods

The unit used to express quantities of energy is the unit of thermal energy, the calorie. A **calorie (cal)** is the amount of heat required to raise the temperature of one gram of water 1°C. It is an extremely small quantity of energy, however, and a more convenient unit is the **kilocalorie (kcal)** or **large calorie (Cal),** which is the amount of heat required to raise the temperature of one kilogram of water 1°C. A kilocalorie is thus equal to 1000 calories (small calories), and it is the unit used in expressing the caloric content of foods and the energy output of the individual.

The amount of energy that can be obtained from a food can be measured quite accurately experimentally with a *bomb calorimeter* (Figure 25–1) which consists of a tiny sealed chamber into which a weighed sample of the food in question is placed with oxygen. The chamber is surrounded by a water jacket of known volume and temperature, and the whole is thoroughly insulated. The food inside can be ignited electrically, and as it burns (is oxidized) the heat produced can be determined by noting how much the water temperature is raised. Table 25–1 shows the values obtained when various foodstuffs are oxidized by this method. Since carbohydrate and fat are completely oxidized in the body, the same amount of heat is obtained whether they are oxidized in the body or in the calorimeter. Protein is not oxidized completely in the body, and less heat is obtained when it is burned in the body. The difference is due to the energy contained in nitrogenous compounds, such as urea, which are excreted.

The amount of energy contained in the foods eaten varies tremendously. It depends upon the relative proportions of fat, carbohydrate, and protein present, as well as upon the water content and the amount of indigestible material

TABLE 25–1   AVERAGE CALORIC VALUE OF FOODSTUFFS*

Food	Bomb Calorimeter	Oxidation in the Body
Carbohydrate	4.1	4.1
Protein	5.4	4.2
Lipid	9.3	9.3

*Given in kilocalories per gram of foodstuff oxidized.

TABLE 25–2   THE CALORIC VALUE OF SOME COMMON FOODS

Food	Approximate Amount	Kilocalories	Food	Approximate Amount	Kilocalories
Breads and cereals			**Vegetables**		
bread, white	1 slice	63	asparagus, canned	6 stalks	20
bread, wheat	1 slice	60	beans, green, cooked	1 cup	25
cake, plain, without icing	1 piece (3 × 2 × 1½″)	180	cabbage, raw, shredded	1 cup	25
			carrot, raw	1 carrot	20
cookies, assorted	1 cookie (3″ diam.)	110	celery, raw	1 stalk	5
			corn, sweet	1 medium ear	65
cornflakes	1 oz	110	lettuce	2 large leaves	5
crackers, soda	2 crackers	45	onion, raw	1 medium onion	50
doughnut, cake	1 doughnut	135	peas, green, cooked	1 cup	110
pancake, wheat	1 pancake (4″ diam.)	60	potatoes, white, boiled	1 potato	105
			potatoes, sweet	1 potato	155
pie, apple	1 sector	330	tomato, fresh	1 medium tomato	30
pizza, with cheese (14″ diam.)	1 sector (⅛ of pizza)	180	tomato juice, canned	1 cup	50
spaghetti with tomato sauce and cheese	1 cup	210	**Fruits**		
			apple, raw, medium	1 apple	70
wheat flakes	1 oz	100	apple juice	1 cup	125
Dairy products			banana	1 banana	85
butter	1 pat	50	cantaloupe, medium	½ melon	40
cheese, cheddar	1 oz	105	grapefruit, fresh	½ grapefruit	50
cheese, cottage, creamed	1 oz	30	grapefruit juice, canned, unsweetened	1 cup	100
cheese, cream	1 oz	105			
cream, coffee	1 tbsp	35	orange	1 orange	60
ice cream (brick)	1 slice	145	orange juice	1 cup	110
milk, skim	1 cup	90	peach, fresh	1 peach	35
milk, whole	1 cup	165	pear, fresh	1 pear	100
1 egg, whole	1 egg	80	pineapple, canned, sliced, with syrup	1 slice with juice	95
Meat, poultry, and seafood			plum, fresh	1 plum	30
beef, hamburger, broiled	3 oz	245	prune, dried	4 prunes	70
beef liver	2 oz	120	raisins, dried	1 cup	460
beef roast, lean	3 oz	220	strawberries	1 cup	55
beef steak, lean, broiled	3 oz	220	watermelon	1 wedge (4″ × 8″)	120
pork, bacon	2 slices	95	**Miscellaneous**		
pork, ham	3 oz	290	candy, chocolate	1 oz	145
pork chop	1 chop	260	jellies, assorted	1 tbsp	50
bologna	1 slice	86	peanut butter	1 tbsp	90
frankfurter	1 frankfurter	155	sugar, white granulated	1 tbsp	50
chicken, broiled (without bones)	3 oz	185	walnuts, English	½ cup	325
			mayonnaise	1 tbsp	110
perch, ocean	3 oz	195	French dressing	1 tbsp	60
salmon, pink, canned	3 oz	120	beer (4% alcohol, avg.)	1 cup	114
shrimp, canned	3 oz	110	cola beverages	1 cup	105
tuna, canned (drained)	3 oz	170	coffee	1 cup	0
			tea	1 cup	0

Source: Home and Garden Bulletin No. 72, U.S. Department of Agriculture, 1960.

(such as cellulose) present. Table 25–2 shows the caloric content of average servings of some common foods. Much more extensive and complete tables, listing the composition of individual nutrients, minerals, and vitamins for almost any food imaginable, are available. From such tables the total intake of energy (in kcal) and of other dietary constituents can be determined quite accurately.

The amount of energy an individual obtains from the foods he or she eats may be somewhat less than the caloric content of the food. It would be reduced, for example, if absorption from the intestine were impaired and some nutrients never entered the

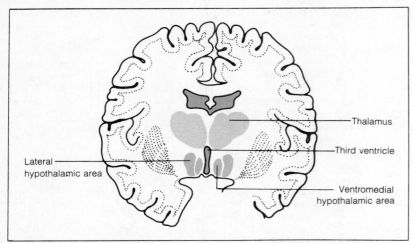

Lateral hypothalamic area

Thalamus

Third ventricle

Ventromedial hypothalamic area

FIGURE 25-2 Location of hypothalamic areas which control food intake.

bloodstream (this occurs most often in elderly people), or if biochemical irregularities, such as the lack of a particular enzyme, were to prevent digestion or metabolism of a certain food.

Control of Food Intake

Assuming that an adequate supply of food is available, what causes an animal (including a human) to eat or not to eat? On first thought it would seem that the animal eats when it gets hungry, but it is not that simple. Even such a basic act as eating is the combined result of several factors and of a complex neural control system.

A distinction should be made between appetite and hunger. **Hunger** may be considered to be a physiological need for food, that is, a need for additional energy. **Appetite,** on the other hand, represents more of a psychological need for food—a desire for food. The hungry individual is not particularly selective about what he or she eats, whereas one with an appetite may turn down food that is unappealing. The intake of food relieves hunger, but it may only stimulate appetite.

Factors involved in the regulation of food intake seem to be mediated through centers in the hypothalamus (Figure 25-2) which receive information from monitors located throughout the body. These centers were initially identified in rats by destroying or stim-

ulating relatively restricted areas of the hypothalamus with small, permanently implanted electrodes.

These techniques showed that a **feeding center,** located in the lateral portion of the hypothalamus, causes the animal to eat. When the feeding center is stimulated, the animal eats voraciously, but if it is destroyed on both sides of the brain, the animal has no urge to eat and may literally starve to death beside a full food dish.

A **satiety center** which inhibits eating is located in the medial portion of the hypothalamus, close to the walls of the third ventricle. When it is stimulated the animal is satisfied, has no urge to eat, and ignores food placed in front of it, but when the area is destroyed the animal eats continually. Rats with hypothalamic lesions that cause overeating *(hyperphagia)* may become extremely obese. A normal rat may weigh 250-300 grams, but rats with hypothalamic hyperphagia may reach weights of 1000 grams. They do not gain weight indefinitely, however; they tend to reach a plateau and to maintain that new high weight.

The bulk of evidence supports the idea that the feeding center is dominant. It is believed to be chronically active, except when periodically inhibited by the satiety center, as normally happens after eating. The fact that rats in which the satiety center has been destroyed tend to stabilize their weights at a new elevated level suggests that there may be something like a thermostat that is set at a certain level or "set point," and the regulatory mechanisms operate to maintain that level. In the case of the hyperphagic rats, the set point for weight has been raised, and the animals must then eat excessively until they reach that level. This is only one of a number of factors that serve to complicate our understanding of the way in which we control our food intake. As with other neural systems, input to the hypothalamic centers comes from many sources, not all of which are clearly understood.

There are two aspects to the control of food intake, the short-term, meal-to-meal control and the long-term regulation of the energy stores in the body, or weight control. For short-term regulation, the rate of glucose utilization is

Focus
Weight Wars

Obesity is basically a very simple problem, and fighting it is likewise simple—though by no means always easy. It is the result of taking in more food calories than the body needs for activity and growth, of the body's conversion of excess food into a form it can easily store—fat. The fat can be useful as an insulation, especially in cold climates, and as a hedge against future food shortages, but in modern America fat is usually superfluous.

In rare instances, obesity *may* result from a physical disorder, a flaw in the control of hormone secretion or in the brain's appetite controls. Most often it is due to a lack of self-control. An exercise of self-control to minimize the difference between intake and outgo, however, may not always be sufficient to eliminate long-term obesity. Enlarged fat cells increase the body's fat-making and -storing capacity; increased amounts of fat in the body increase the production of insulin, which in turn enhances the appetite because the increased insulin levels reduce the blood sugar level. At the same time, increased insulin levels induce the formation of more insulin receptors in the body's cell membranes, which further increases the body's response to insulin and the synthesis of fat. The resulting feedback cycle is particularly vicious for children who may become so adapted to their obesity that they can never escape it. Most people, however, can escape the problem of excess weight more easily.

The solution to the problem of obesity is inherent in its own basic, simple nature: intake exceeds outgo. Increasing the outgo by exercise can help, but that can be a painful and impractical option. Decreasing the intake can be done fairly painlessly, without relying on any of the many fad diets. If you have a weight problem, start by examining your present diet. Identify the high-calorie items than can be replaced with low-calorie equivalents: french fries can be switched for salad, banana cream pie for jello, whole milk for skim milk or water, cookies for carrots. Make up your own diet, and put it into practice.

The advantage of such a diet is that it changes your personal eating pattern as little as possible. You continue to eat when you're used to eating. You continue to be as full as ever. But every 600 kilocalories you remove from your daily diet in this way is a pound of fat per week you do not gain. If your weight was stable before the diet, the 600 kilocalories less per day is a pound of fat lost per week. Your ultimate weight loss depends only on your patience and persistence. If you can keep it up long enough, you can even develop new and healthier food habits.

You should, however, be careful not to chop your daily diet by too many calories. Severe dieting should be done only under a physician's supervision. If you choose to do it yourself, be prepared to take your time and be sure to consume a balanced diet with all the minimum daily requirements for a person of your height and age. Cut out only the excess.

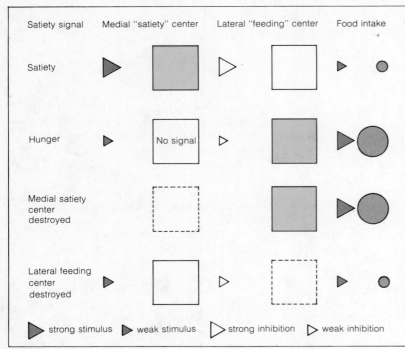

FIGURE 25-3   Role of the hypothalamus in control of food intake. The feeding center is active unless inhibited by the satiety center.

(Adapted from J. Tepperman, *Metabolic and Endocrine Physiology*, 3rd ed. Copyright © 1974 by Year Book Medical Publishers, Inc., Chicago. Used by permission.)

believed to be an important factor. Certain cells in the hypothalamus can detect the rate of removal of glucose from the blood passing through the area. When glucose is being utilized at a high rate, the satiety center is active and the feeding center is inhibited (Figure 25-3). When glucose utilization is low, however, the satiety center is inactive, the feeding center is unchecked, and the animal is hungry. This has been called the *glucostatic theory* of intake control, and the detector of glucose utilization is sometimes called a *glucostat*.

There is evidence that increases in free fatty acid levels in the general circulation lead to increased food intake, though it is not known just how this is brought about. During the postabsorptive stage of metabolism fatty acids are being mobilized from fat stores for use as an energy source, and their blood levels are elevated (compared with the absorptive state).

Reactions involved in the metabolism of foodstuffs produce some extra heat (the specific dynamic action, see

below) that raises the temperature of the blood. The increase in temperature is detected by other cells in the hypothalamus and has been suggested as an inhibitor of food intake. Stimuli arising from the gastrointestinal tract—the presence (or absence) of food in the mouth, pharynx, or stomach—affects the hypothalamic centers. There are also inputs from other parts of the central nervous system. The limbic system, for example, is involved in some way, since lesions in certain parts of it (such as the amygdaloid nucleus) can produce some hyperphagia. Psychological factors that affect how we feel about food are important elements in the control of intake, but they are difficult to assess. Among them are cultural factors, food habits and eating patterns, emotional states, and use of food as a reward (or when we feel sorry for ourselves).

The long-term regulation of intake involves integration and analysis by the central nervous system of information from all sources concerning our energy stores and reserves, as against our energy needs and expenditures. It has been suggested that information about energy stores might be based on integration of many of the short-term inputs, particularly the mobilization of fatty acids. The energy expenditure levels would have to be determined not only by physical activity and exercise, but also by other factors such as growth. Needless to say, this process is poorly understood, and the widespread problems with long-term regulation are good evidence of it.

ENERGY EXPENDITURE

We stated above that some of the energy obtained from food is used to perform useful work, but that most of it is dissipated as heat. It is now appropriate to look more closely at the functions for which this energy is released. These activities account both for all the energy expended for useful work and for the heat by-product. In fact, heat energy is an eventual end result of all work.

The energy released by the body is used for the following purposes:

(1) To maintain the body functions—to "keep the motor running." It is the energy cost of living, of staying alive, and is known as the **basal metabolism.** It is the baseline energy requirement upon which energy needs for other activities are superimposed.
(2) To do muscular work.
(3) To metabolize food.
(4) To maintain the body temperature.
(5) To cope with certain special conditions which may increase energy expenditure in any of several ways.

Control of Energy Expenditure

Contraction of skeletal muscle markedly increases the rate of energy expenditure. Because of the size of the muscle mass—about half the total body mass—and the extent of the increase possible—roughly twentyfold—muscle contraction is the most important physiological determinant of the level of energy expenditure. The energy expenditure associated with skeletal muscle activity, of course, depends upon voluntary control of the skeletal muscles and related reflex activities. There are, however, factors which influence the neural control of muscle contraction and thereby exert some control over energy expenditure.

General motor activity is known to vary rhythmically in response to several types of cycles. The patterns of activity have been studied more extensively in animals than in human beings, and although the results cannot be transferred totally, there are some implications for humans. Motor activity can be measured quite easily in rats by noting the revolutions of an exercise wheel placed in their cage. Rats show much greater activity at night (they are nocturnal animals), in a cool environment, during starvation, and just before feeding time, and female rats show greatly increased activity during estrus (when they are in heat).

There is also evidence of specific neural control over activity, since animals become very inactive and lethargic after lesions in certain areas of the hypothalamus (not, however, the same areas that control food intake). This may be associated with the role of the hypothalamus in determining sleeping and waking states. Destruction of certain parts of the frontal lobes of the cerebrum in monkeys produces hyperactive animals. They are not necessarily hyperexcitable, but they do show incessant movement.

Precisely how these modifiers of motor activity might function as means of adjusting energy expenditure is difficult to assess, however, because most of them seem to be determined by factors other than the need to maintain energy balance.

Cost of Muscular Work

Virtually anything one does involves an increase in the rate of metabolism, and one's routine activities in the course of the day add up to a considerable expenditure of energy. Merely sitting instead of lying supine, for instance, increases the metabolic rate by about 10 percent. More obvious forms of exercise, such as sports, may cause very high rates of metabolism. However, the average individual does not perform so vigorously for more than a small portion of the day.

Table 25–3 shows the energy cost of a wide range of activities. By comparing the energy cost of activity with the caloric value of foods, one is forced to the disturbing conclusion that to "burn off" the 453 kilocalories contained in three ounces of chocolate requires nearly an hour and a half of walking (at 3.5 mph) or half an hour of running (7 mph). All of these figures are for an individual of average size, which is arbitrarily taken to be a 70-kg man (154 pounds). Unfortunately, the great burst of physical activity indulged in as a weight-reducing measure may backfire, since after half an hour the runner just may eat several more pieces of chocolate, putting his energy balance in a worse state than it was before!

A 300-pound man would use about twice as much energy to walk or run at the same speed simply because he is moving a load that is twice as great. It has been suggested, however, that some obese people tend to be consider-

TABLE 25-3   ENERGY COST OF SELECTED ACTIVITIES*

Activity	Energy Expenditure (in Kilocalories per Hour)	Activity	Energy Expenditure (in Kilocalories per Hour)
Sleeping	70	Mountain climbing	600
Lying quietly	80	Snowshoeing (trail snowshoes) at 2.5 mph	620
Sitting	100	Fencing	630
Mental work, seated	105	Skating at 11 mph	640
Standing	110	Rowing at 3.5 mph	660
Singing	120	Walking up a 36% grade at 1 mph carrying a 43-lb load	680
Driving a car	140		
Office work	145	Swimming crawl stroke at 1.6 mph	700
Housekeeping	150	Horizontal running at 5.7 mph	720
Walking at 2 mph	170	Walking up a 14.4% grade at 3.5 mph	740
Riding a bicycle at 5.5 mph	190	Walking in 12–18 inches of snow	760
Walking *down* stairs at 2 mph	200	Skating at 13 mph	780
Bricklaying	205	Wrestling	790
House painting	210	Swimming backstroke at 1.6 mph	800
Carpenter work	230	Horizontal running at 5 mph carrying a 43-lb load	820
Billiards	235		
Baseball (except pitcher)	280	Horizontal running at 7 mph	870
Horizontal walking at 3.5 mph	290	Walking up a 36% grade at 1.5 mph carrying a 43-lb load	890
Rowing for pleasure	300		
Table tennis	345	Running up an 8.6% grade at 7 mph	950
Horizontal walking at 3 mph carrying a 43-lb load	350	Rowing at 11 mph	970
		Football	1000
Walking up a 3% grade at 3.5 mph	370	Rowing at 11.3 mph	1130
Baseball pitcher	390	Swimming sidestroke at 1.6 mph	1200
Pick and shovel work	400	Horizontal running at 11.4 mph	1300
Shoveling sand	405	Rowing at 12 mph	1500
Swimming breaststroke at 1 mph	410	Swimming crawl stroke at 2.2 mph	1600
Bicycle riding, rapid	415	Swimming breaststroke at 2.2 mph	1850
Swimming crawl stroke at 1 mph	420	Swimming backstroke at 2.2 mph	2000
Walking up an 8.6% grade at 2.4 mph	430	Horizontal running at 13.2 mph	2330
Chopping wood	450	Swimming breaststroke at 2.4 mph	2530
Skating at 9 mph	470	Horizontal running at 14.8 mph	2880
Sawing wood	480	Swimming sidestroke at 2.2 mph	3000
Swimming breaststroke at 1.6 mph	490	Swimming breaststroke at 2.7 mph	3690
Swimming backstroke at 1 mph	500	Horizontal running at 15.8 mph	3910
Skiing at 3 mph	540	Horizontal running at 17.2 mph	4740
Swimming sidestroke at 1 mph	550	Horizontal running at 18.6 mph	7790
Walking up an 8.6% grade at 3.5 mph	560	Horizontal running at 18.9 mph	9480
Walking up a 10% grade at 3.5 mph	580		
Walking *up* stairs at 2 mph	590		

*Calculations based on 70-kg (154-lb) man.
Source: Morehouse, Laurence E., and Miller, Augustus T., Jr., *Physiology of Exercise*, 6th ed., The C. V. Mosby Co., St. Louis, 1971; compiled from data chiefly obtained at the Harvard Fatigue Laboratory.

ably less active than the average individual. For these people the energy imbalance may be due to reduced expenditure as much as (or instead of) to excessive intake.

Cost of Metabolizing Foodstuffs

The metabolic rate is elevated for several hours after food has been eaten, with the extent of the elevation depending on the food. One hundred kilocalories of protein produces an extra 30 kcal of heat, while similar amounts of carbohydrate and fat raise the metabolic rate by 6 and 4 kcal, respectively. A mixed diet causes an average increase in the basal metabolic rate of approximately 10 percent. This elevation is known as the **specific dynamic action** (SDA) of the food.

That the SDA is not produced by the processes of digestion or absorption is shown by the fact that injection of amino acids directly into the bloodstream causes an increase in the metabolic rate even though the digestive tract has been bypassed. The SDA is instead believed to be due to the cost of the reactions involved in the metabolism of the foodstuffs. As a result most of the heat associated with the SDA is produced in the liver. The cost of "processing" is higher for proteins probably because they require more processing, including *deamination* of amino acids and the formation of urea.

Cost of Maintaining Body Temperature

Humans, as warm-blooded animals, maintain the temperature of their bodies within rather narrow limits. Except in unusual circumstances, this temperature is higher than that of the environment and there is a continuous loss of heat to the environment due to the temperature gradient. This loss represents an energy expenditure. When environmental temperature is very low, the rate of heat loss may be enough to necessitate increased heat production to maintain the body temperature. The increase is actually brought about by skeletal muscle contraction, in the form of increased muscle tone or shivering. When the environmental temperature approaches or is greater than body temperature, the problem is to prevent overheating, and additional means are employed to facilitate heat loss. Measures for adjusting and regulating heat loss and production in terms of temperature control are discussed in Chapter 26.

Other Forms of Energy Expenditure

A number of other physiological, emotional, chemical, and pathological factors may also increase heat production. Two hormones, thyroxin and epinephrine, increase the metabolic rate by their action on energy-releasing mechanisms. Pregnancy increases metabolic rate, due at least in part to hormone action. Anxiety and fear cause both the release of epinephrine and increased muscle tone, both of which are associated with an elevation of metabolic rate. The caffeine in coffee and tea is a *stimulant* which raises the metabolic rate, as are the amphetamines in some "diet pills," while other agents may be *depressants*. And many diseases are associated with an increase or decrease of heat production. Because a rise in temperature increases the rate of any chemical reaction, including those in the body, basal heat production is increased 7 percent for every 1°F rise in body temperature (or 13 percent for every 1°C rise). Lowering the body temperature has the opposite effect, and this is the rationale behind cooling the body (*hypothermia*) for certain surgical procedures. The lowered metabolic rate reduces the needs of the tissues for oxygen and nutrients and minimizes the danger of temporary interruption of the circulation.

Basal Metabolic Rate

The **basal metabolic rate** (BMR) is the minimum rate of energy expenditure (heat production) when one is awake. Basal metabolism includes the energy needed by the heart to pump blood, by the respiratory muscles to move air in and out of the lungs, by the kidneys to separate wastes from the blood, and by secretory cells to synthesize and release secretions. It also includes the energy needed to maintain central nervous system activity and muscle tone and to replace heat lost from the body surface.

The basal state exists in a person in whom all the other energy-consuming functions are absent or minimal. This means that he or she must be in the following state:

(1) At rest and relaxed—no physical activity for at least the preceding half hour.

(2) Post-absorptive—12–14 hours after the last meal, so that all food has received its initial preprocessing and the body is calling upon stored energy to function.

(3) Subjected to a comfortable environmental temperature, with no need for increased muscle tone or shivering and with a normal body temperature (i.e., no fever).

(4) Mentally at ease—not anxious or fearful.

(5) Not under the influence of stimulants or depressants.

It should be noted that the basal state is defined as the *minimum metabolic rate while awake*. It is *not* the absolute minimum rate of metabolism. In the sleeping individual there is a slowing or depression of all the metabolic processes (heart action, respiration, muscle tone, etc.), with the result that the metabolic rate during sleep is about 10 percent below that of the basal state.

Factors That Affect the Basal Metabolic Rate Basal energy expenditure varies with the size, age, and sex of an individual

Body size. Since energy release and heat production are carried on by living cells, it follows that a large number of cells must produce more heat than a small number. The total energy exchange is therefore much greater in an elephant than in a mouse. For the same reason a large person produces more heat than a small person.

There is, however, no simple linear relationship between body weight and heat production. If total heat production over a period of time is divided by the weight of the animal and expressed as kcal/kg/hr, it turns out that small animals and birds have much higher metabolic rates than large ones. The heat produced by each kilogram of tissue is greater in small animals, so apparently body weight per se is not the ideal unit of metabolic size.

Since an animal's body temperature remains fairly constant, one can assume that the body loses heat at the same rate that it produces heat. And since most of the heat is lost from the surface of the body, there should be a relationship between the area of the body surface and the amount of heat-producing tissue. Small animals (mice) have more body surface in proportion to their size (weight) than large animals (elephants). If the metabolic rates of animals are expressed in terms of body surface, the rates of species of very different sizes are moderately comparable. For humans, the body surface area in square meters is generally accepted as the best unit of metabolic size. The metabolic rates are expressed as kcal per square meter of body surface per hour.

Age. The basal metabolic rate is not constant throughout the life of an individual, since it is highest in the early years, declines rapidly through the growth years, and decreases steadily thereafter, though more slowly (Figure 25–4). The growing individual or animal has a higher energy requirement, not only because he or she is likely to be more active, but also because there is a greater rate of tissue turnover associated with increasing the amount of protoplasm.

Sex. The basal metabolic rates for men and women of the same age differ. All else being equal, that of men is about 10 percent higher than that of women. The reason for this difference is unknown. However, one factor is probably the sex hormones, since they affect several metabolic reactions. Dur-

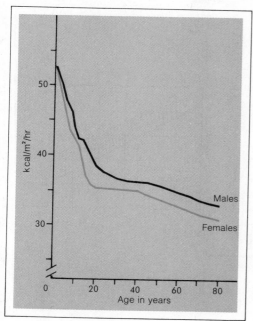

FIGURE 25-4   Normal basal metabolic rates in males and females of different ages.

ing pregnancy and lactation, however, a woman's basal metabolic rate is the higher.

Measurement of Metabolic Rate

The most obvious method of measuring heat production would be to enclose the individual in a chamber and measure it in a manner similar to that used for determining the energy content of foods in a bomb calorimeter. (It is not necessary to ignite the individual, however, because the aim is to determine the rate of heat production, not the total energy content.) This method is known as **direct calorimetry** because heat production is measured directly. In theory it is simple, but in practice it is very complicated, since the chamber needed to accommodate a human being (see Figure 25-5) and the technical staff necessary to operate it are so large and expensive that few laboratories use this method. Direct calorimetry is, however, commonly used for measuring heat production in small animals such as rats.

In humans it is more convenient to measure something else which is related to heat production and convert that to kilocalories. This method is known

as **indirect calorimetry.** Metabolic reactions in the body depend ultimately upon energy obtained from the oxidation of foods. Oxygen is required for this and carbon dioxide is produced, both in amounts proportional to the level of metabolic activity. Either oxygen consumption or carbon dioxide production can be measured, but in practice it is more convenient to measure oxygen consumption. All that is needed is a conversion factor to relate oxygen consumption to calories of heat produced.

The factor by which a volume of oxygen consumed is converted to heat produced is known as the **caloric value of oxygen**—the number of kilocalories released when a liter of oxygen is consumed in the oxidation of foods. The relation between oxygen and calories is not a simple or direct one, however, since, as shown in Table 25-1, the amount of heat produced per gram of foodstuff depends upon what is oxidized. Furthermore, the amount of oxygen required to burn a gram of foodstuff also depends upon what is being burned. A liter of oxygen liberates more heat when used to burn carbohydrates (5.05 kcal) than when used to oxidize fat (4.7 kcal) or protein (4.5 kcal). Therefore, it would seem that in order to de-

FIGURE 25-5   Apparatus for determining heat production by direct calorimetry.

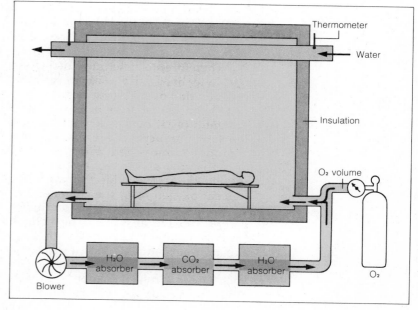

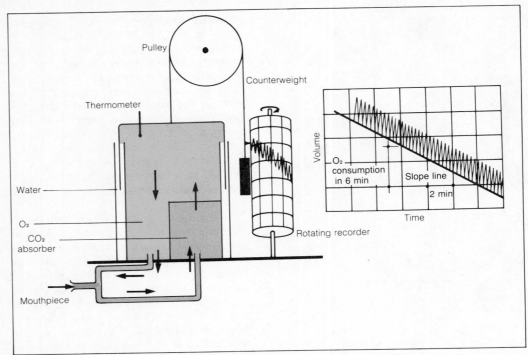

FIGURE 25-6   Apparatus for determining heat production by measuring oxygen consumption (indirect calorimetry).

termine accurately the caloric value of the oxygen used, the composition of the so-called metabolic mixture must be known. In the basal state, however, the body is "living off itself," and people are oxidizing essentially the same mixture of foodstuffs; there is actually very little difference between individuals in the caloric value of oxygen, and one liter of oxygen used to oxidize foodstuffs yields 4.83 kcal of heat in the basal state.

Therefore, to determine metabolic rate, one need only measure oxygen consumption over a period of time (corrected for temperature and pressure) and multiply it by 4.83 kcal/liter to get the heat production. The oxygen consumption is measured with the aid of an apparatus similar to a spirometer (Figure 25-6). The subject breathes 100 percent oxygen from the chamber and the expired air returns to it through a carbon dioxide absorber. The volume in the chamber fluctuates with each breath, but gradually declines as some oxygen is retained by the subject with each breath. From the slope of the decline, the amount of oxygen consumed

in the test period is determined. From the amount consumed, the heat production in kilocalories per hour is calculated, and that is the metabolic rate.

Metabolic rate can be measured during all sorts of activities. A portable oxygen supply and oxygen-metering device can be fixed to a backpack to be carried by a runner, skier, mountain climber, etc., in order to determine his or her energy expenditure. Similar methods were used to obtain much of the data in Table 25-3. (The rates given in Table 25-3 are metabolic rates, not *basal* metabolic rates however.)

Tables have been prepared (Table 25-4) which predict the heat production (in kcal/m²/hr) for an individual of a given age and sex. These tables are based on the results of thousands of measurements of the BMR in "normal" people. It is customary to compare the actual measured heat production with the predicted normal rate, and then to express the BMR as the percent deviation from that standard. For example, a BMR of +10 means that the individual produces 10 percent more heat each hour than is predicted for a person of that age and sex. A deviation of ±10 percent is usually considered to be within the normal range. Considering all the factors that can affect metabolic rate and its measurement, it is remarkable that this degree of accuracy can be achieved. A meal, especially a high-protein meal a few hours before a test, may cause an otherwise normal BMR to come out at +30 percent due to the specific dynamic action of the protein. Exercise may increase it by several hundred percent (see Table 25-3).

TABLE 25-4   NORMAL STANDARDS FOR BASAL METABOLISM*

Age	Male	Female	Age	Male	Female
1	53.0	53.0	17	40.8	36.3
2	52.4	52.4	18	40.0	35.9
3	51.3	51.2	19	39.2	35.5
4	50.3	49.8	20	38.6	35.3
5	49.3	48.4	25	37.5	35.2
6	48.3	47.0	30	36.8	35.1
7	47.3	45.4	35	36.5	35.0
8	46.3	43.8	40	36.3	34.9
9	45.2	42.8	45	36.2	34.5
10	44.0	42.5	50	35.8	33.9
11	43.0	42.0	55	35.4	33.3
12	42.5	41.3	60	34.9	32.7
13	42.3	40.3	65	34.4	32.2
14	42.1	39.2	70	33.8	31.7
15	41.8	37.9	75	33.2	31.3
16	41.4	36.9	80	33.0	30.9

*Heat production is given in kilocalories per square meter per hour.
Source: A. Fleisch, "Le metabolisme basal standard et sa determination au moyen du 'metabulicalculator,' " *Helvetica medica acta*, **18** (1951):23–44, published by Schwabe & Co., Verlag, Basel, Switzerland. Reprinted by permission.

CHAPTER 25 SUMMARY

The energy balance in the body depends upon the relationship between energy intake and energy expenditure. Energy taken into the body is in the form of chemical energy incorporated in the chemical bonds of food molecules. Upon release the energy is stored, at least temporarily, as ATP. Energy expenditure involves either using the energy from ATP to perform mechanical, electrical, or chemical work, or losing the energy as heat which the body cannot use to perform useful work. Only about 20 percent of the energy expenditure is used to do work, the remainder appears as heat. If energy intake is greater than expenditure, the energy reserves of the body will increase and there will be a gain in weight. If intake is less than expenditure, the deficit is made up from energy stores, and there will be a loss in weight.

Sources of Energy

The energy content of the foods is measured (in kilocalories, kcal), as the amount of heat produced by combustion of the food under controlled conditions. Oxidized in the body, a gram of fat yields about twice as much heat (kcal) as a gram of protein or carbohydrate.

The amount of energy taken in is controlled through a pair of centers in the hypothalamus, a *feeding center* and a *satiety center*. The feeding center causes eating, and is believed to be tonically active. When it is stimulated eating is excessive, but if it is destroyed there is no drive to eat. The satiety center reduces eating by inhibiting the feeding center. If the satiety center is stimulated food intake is reduced, but if it is destroyed the feeding center is not inhibited and eating is unchecked. An increased rate of glucose uptake by certain hypothalamic cells signals satiety and the feeding center is inhibited (*glucostat*). Other factors that may affect the centers include the level of free fatty acids in the bloodstream, heat produced in association with metabolism of foods (SDA), and impulses from various parts of the digestive tract. Long term control of food intake may involve signals that in some way indicate the size of the energy reserves.

Energy Expenditure

Energy released by cells is used to do muscular work, metabolize food-stuffs and maintain body temperature. These expenditures are in addition to the energy used to maintain the body's vital functions, the basal metabolism.

Skeletal muscle can consume great quantities of energy, for it can increase its metabolic rate many fold. All of our activities have a cost in terms of the energy expenditure (kcal) required to perform them.

Metabolism of foodstuffs (postabsorptive) also has a cost known as the *specific dynamic action (SDA).* It adds an average of about 10 percent to the basal rate of metabolism; protein adds about 30 percent. The SDA is caused by metabolic processes, primarily those that occur in the liver, such as deamination.

Whenever temperature of the environment is lower than that of the body, heat is lost from the body to the environment, and in order to maintain body temperature heat must be produced to replace it. Since the body produces ample heat, this is not a problem unless heat loss is excessive. Among other factors that raise metabolic rate are an elevated body temperature (fever), anxiety, fear, and thyroid hormone. Metabolic rate is lowered by reduced body temperature, sleep, and several disease conditions.

Basal metabolic rate (BMR) is the minimal energy expenditure in an awake individual. It therefore can be accurately measured only when the other forms of energy expenditure are absent. Basal energy expenditure varies with the size, age, and sex of the individual. The BMR is usually expressed in terms of the body surface area, as kcal/m²/hr. This is valid because most of the heat is lost from the skin and, if one assumes heat production and loss to be equal, the amount lost is proportional to the area of skin surface.

Heat production is hard to measure in humans, but since the heat is produced by oxidation which eventually requires oxygen, metabolism can be determined from the rate of oxygen consumption. The amount of heat produced when a liter of oxygen is used *(calorific value of oxygen)* depends upon the content of the food oxidized. Fat yields more calories per gram than carbohydrate or protein, but more oxygen is required to oxidize a gram of fat than of carbohydrate or protein. In practice, the metabolic mixture oxidized does not vary much and the calorific value of the oxygen is fairly constant. The measured BMR or heat production (kcal/m²/hr) is compared to normal standards for individuals of the same age and sex, and is finally expressed as a percent deviation from the predicted standard.

STUDY QUESTIONS

1 How does the law of conservation of energy apply to overall energy metabolism in our bodies?

2 What mechanisms and factors control our intake of food? How is this control exerted?

3 What conditions must be met if a determination of metabolic rate is to be a determination of *basal* metabolic rate? Why is each of these conditions a requirement?

4 What would you have to do in the way of exercise or reduced food intake (see Tables 25–2 and 25–3) to lose a pound of fat? Assume: 1 pound = 453 grams, and 1 g of fat yields 9.3 kcal.

26

Normal Body Temperature

Heat Gain

Heat Loss

Regulation of Body Temperature

Body Temperature and Its Regulation

Heat is a by-product of all chemical reactions and body activities, and we have seen that most of our energy expenditure can eventually be accounted for as heat. The management of heat produced in the body is therefore an important aspect of homeostasis, and humans have elaborate control systems which enable them to maintain their body temperature within relatively narrow limits.

Body temperature is a balance between heat produced in the body and heat lost from it. The temperature can be altered by changing either the rate of heat production or the rate of heat loss, and we must be able to adjust either of them to maintain a reasonably constant body temperature. For example, when a cold environment causes us to lose heat, we need to produce more and lose less heat. In this chapter we consider the ways in which heat production and heat loss can be altered, and how these mechanisms are regulated and balanced to maintain the constant body temperature.

Most of the energy released by metabolism appears as heat, and when metabolic rate is measured, it is assumed that the rate of heat loss from the body is the same as the rate of heat production. When this is the case, the heat content of the body and its temperature remain constant. When heat is produced faster than it is dissipated, the body temperature rises; and when heat is lost faster than it is produced, body temperature falls.

If an organism is to prevent changes in its body temperature, it must be able to change the rates of heat production and heat loss to maintain the balance between them. Only birds and mammals, including humans, are able to do this, and they are called **homeotherms.** They are therefore the only animals whose metabolic rates are not governed by the environmental temperature and who can function quite well throughout a wide range of external temperatures. Such freedom and independence, however, have one major disadvantage. Because the environmental temperature is almost always less than that of the body, the animal is continually losing heat to the external environment. It is expensive in terms of energy expenditure to maintain a gradient of any kind, and the greater the gradient the greater the energy cost.

Animals that cannot control their body temperatures are called **poikilotherms.** Their body temperatures and metabolic rates vary with the temperature of the world around them. In cold weather they cannot move rapidly even when their lives are in danger. Such animals can tolerate neither a very high nor a very low body temperature, and their sole means of temperature regulation is necessarily behavioral. On a cool morning a snake may sun itself on a warm rock, but in the heat of the day it must seek the shade under that rock. In hot weather a frog does not wander far from its pond. When it gets cold the frog buries itself below the frost line to await more favorable days. Poikilotherms require much less energy because they do not have to maintain a significant temperature gradient and therefore need to produce less heat, but they are at the mercy of their environ-

ment. Their activities, indeed their survival, depend upon the environmental temperature.

Some animals represent a unique adaptation or compromise—they hibernate. Most of the time they are homeothermic, but in winter they seek out an underground burrow and become nearly poikilothermic for long periods of time. During hibernation the body temperature and metabolic rate drop dramatically, and hence the energy requirement is greatly reduced at a time when food is difficult to find.

No animal can tolerate a very high body temperature—43°C (110°F) is about the maximum, since higher temperatures result in irreversible changes in cellular enzymes. Nonhibernating homeotherms do not usually tolerate body temperatures as low as the hibernators. In humans, consciousness is usually lost by about 32°C (90°F) and cardiac irregularities such as fibrillation may occur. At these temperatures the individual needs externally applied heat to rewarm. Several cases have been reported, however, of individuals who have survived extremely low temperatures resulting from accidental exposure.

NORMAL BODY TEMPERATURE

It is often stated that the body temperature is 37°C (98.6°F), but this is incorrect for two reasons. First, there is no *single* body temperature—various parts of the body may differ by several degrees at any particular time. Second, body temperature is regulated within a narrow *range*, not at a rigidly fixed value.

Deep body or **core temperature** is usually higher than surface temperature (Figure 26–1). The regulatory mechanisms are such that core temperature is kept quite constant, very close to 37°C, while the surface temperature is allowed to fluctuate, and may be considerably lower than the core temperature. The internal organs are important heat sources, but they are separated from the external environment by an insulating shell of skin and subcuta-

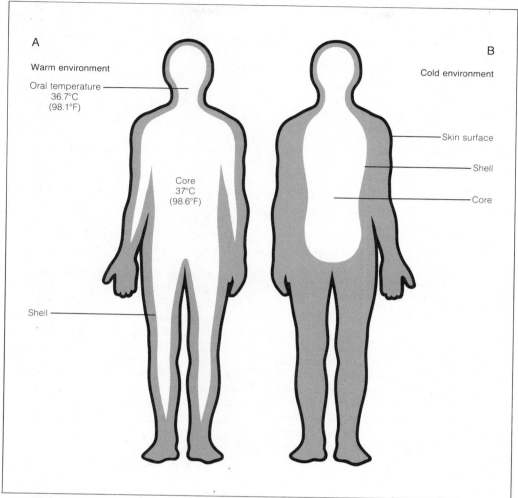

A
Warm environment

Oral temperature
36.7°C
(98.1°F)

Core
37°C
(98.6°F)

Shell

B
Cold environment

Skin surface

Shell

Core

FIGURE 26–1   Diagram to show the body as a warm core surrounded by an insulating core. A. In a warm environment. B. In a cool environment.

neous tissue. The heat they produce must therefore be carried to the surface by the blood, which is warmed as it flows through the core. When cutaneous blood flow is high, skin temperature approaches core temperature and when cutaneous blood flow is low, the skin temperature falls toward that of the environment.

Body temperature measurements are made to determine the core temperature. A thermometer bulb placed under the tongue, where there are large superficial veins, is likely to give a reasonable indication of the core temperature, although it is usually slightly lower than the rectal temperature, which is a true core temperature.

In spite of homeostatic mechanisms, there are several physiological circumstances that may cause the temperature to vary. In the first place, everyone does not have exactly the same body temperature, although most individuals fall within about 1°C of one another (Figure 26–2). The regulatory mechanisms are not as finely adjusted in young children and in young animals; among them, body temperature fluctuations are common and often of greater magnitude than in adult humans or animals. Like many other metabolic activities, body temperature shows a *diurnal* variation. The lowest temperature occurs in the early morning hours and the high adds 1–1.5°C

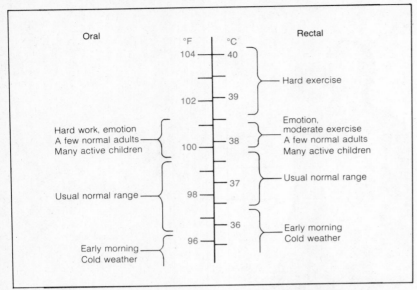

FIGURE 26-2   Body temperature ranges in normal persons.

late in the day. Activity, of course, can increase body temperature because it involves a great increase in the rate of heat production. Women show a slight rise in body temperature during the latter half of the menstrual cycle, which is probably related to the presence of progesterone, a hormone secreted at that time.

HEAT GAIN

The heat produced by the body is the result of metabolic reactions, and so the control of heat production is largely a matter of regulating metabolic activity. For this reason the regulation of heat production is often known as **chemical thermoregulation.** The amount of heat produced in carrying on basal metabolism and in metabolizing foodstuffs (specific dynamic action), however, is altered very little as a regulatory measure; it cannot be reduced below a certain minimum.

Metabolic heat production can be increased dramatically by skeletal muscle contraction. During voluntary physical exercise the metabolic rate may be increased more than fifteen to twenty times for short periods, or sustained for

longer periods at rates of up to ten times the basal rate. Involuntary skeletal muscle contraction, in the form of muscle tension or outright shivering, is also capable of increasing heat production up to five- or sixfold.

Epinephrine and thyroid hormone increase heat production because of their effects on metabolic rate. Epinephrine and norepinephrine are short acting, while thyroxin causes a slow, more prolonged increase. In many animals a cold environment causes a marked stimulation of the thyroid gland, but this effect is not prominent in humans. Skeletal muscle contraction, either voluntary or involuntary, is the main means of increasing heat production in humans.

Whenever the environment is warmer than the body, heat is transferred to the body. This may be in the form of radiant heat from the sun, from a radiator, or from an electric blanket. Heat from the external environment provides a relatively small portion of the total body heat supply, however, and only under certain conditions does it become very important.

HEAT LOSS

Processes by Which Heat Is Lost

The actual heat exchange between the body and its environment is the result of passive processes that depend upon the existence of suitable gradients. The regulation of heat loss involves mechanisms that alter the physical conditions (e.g., the gradients) and is thus known as **physical thermoregulation.** Most of these processes also operate to gain heat when the environment is warmer than the body.

The amount of heat lost by physical means depends upon the steepness of the gradient—more heat is lost when the environment is cool than when it is warm. It also depends upon the area exposed to that gradient. In cold weather dogs and cats curl up in a ball to nap, which reduces the area for heat loss, but when it is warm they sprawl out and increase their exposed surface. Humans reduce heat loss by covering more of the body surface with insulating clothing when the weather gets cold.

Radiation, Conduction, and Convection Radiation involves the exchange of heat between objects in space. All objects emit radiant energy in the same way a light bulb emits light, and this radiant energy produces heat when it strikes an object. The rise in temperature which then occurs is related to the temperature of the object and to whether it absorbs or reflects heat. Radiant energy from the sun, our greatest source, will heat a sidewalk to a temperature well above that of the air because the sidewalk absorbs heat while the air does not.

The human body is also a very good radiator, and it is constantly radiating heat in all directions. The quantity of heat radiated depends upon the temperature of the body and upon its exposed surface area. Objects also radiate heat in proportion to their temperatures and surface areas, but the net transfer of heat between the body and objects in the vicinity will be from the warmer to the cooler body or object. For humans, conditions are usually quite favorable for the loss of heat by radiation. This loss typically accounts for about 60 percent of the total heat loss from the body.

Conduction is the passage of heat directly from a warm object to a cooler one. It depends not only upon the temperature gradient, but also upon the conductivity of the cooler object, that is, upon the rate at which heat is conducted away from the warm surface. A cool metal object "feels" colder than a ball of wool of the same temperature because metal is a good conductor and draws more heat from the skin than does wool.

Convection refers to the transfer of heat from a warm object to cooler air (rather than a cooler object). Air itself is a *poor conductor* of heat, but when heated it rises and allows cooler air to replace it (that is, *convection* occurs).

Convection and its cooling effect are both greater when air movement is increased, as by the wind or a fan. On the other hand, if the air is prevented from moving, it provides an excellent insulation which reduces heat loss. Many types of commercial insulation function by trapping air. The same is true of warm clothing, which by holding air that has been warmed creates a comfortable micro-environment. Wool is warmer than cotton because it traps air better.

Conduction and convection together usually account for a relatively small portion of the total heat loss. Radiation, conduction, and convection all depend upon a temperature gradient. They become less effective as the environmental temperature rises. In fact, when it becomes greater than body temperature these processes contribute to heat *gain* rather than heat *loss*.

Evaporation Evaporation is the process of converting a liquid to a gas or vapor, and it depends upon a gradient of humidity rather than one of temperature. It is the only effective means of losing heat when the environmental temperature is greater than the body temperature.

The amount of water vapor the air can hold varies with the temperature of the air. At a high temperature a given volume of air can hold much more water (as vapor) than when its temperature is low. When the temperature is reduced, moisture condenses, perhaps as "steam" on a window or as "sweat" on the outside of a glass of cold water. The moisture content of the air is expressed as the percent of its capacity at that temperature, or the relative humidity. When it is low, the air can take up much more fluid, and evaporation occurs rapidly, but when the air holds much, little fluid can be evaporated.

To raise the temperature of a kilogram of water one degree centigrade requires one kilocalorie of thermal energy. To convert the kilogram to water vapor with no temperature change requires 580 kilocalories. Evaporation therefore requires a lot of heat, and it is a very effective method of cooling—if

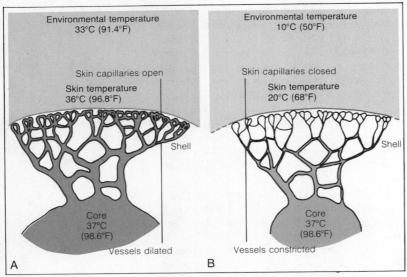

FIGURE 26-3 Effect of blood flow on heat loss from the skin. A. In a warm environment with blood vessels dilated. B. In a cool environment with blood vessels constricted.

there is an adequate humidity gradient. Perspiration is effective only if the sweat is evaporated; it has no cooling value when the sweat simply drips off.

Avenues of Heat Loss

Skin The greatest loss of heat, whether by radiation, conduction, convection, or evaporation, is from the body surface. The heat is brought to the skin from the core of the body by the blood. The gradient between skin and air can be regulated quite well by adjusting the cutaneous blood flow (Figure 26-3). If the cutaneous vessels are dilated, more warm blood reaches the surface, skin temperature approaches that of the core, and more heat is lost from the surface. Constriction of the cutaneous vessels reduces skin blood flow, less heat is then brought to the surface, and the skin temperature falls and approaches that of the air. This reduces the gradient between skin and air and the rate of heat loss falls.

The temperature gradient and rate of heat loss may also be reduced, independently of blood flow, by insulation. In winter, animals that live in cold climates develop heavy fur coats and/or thick layers of subcutaneous fat, both of which serve as insulation. Many animals increase the effectiveness of their insulation by ruffling the fur (piloerection), thereby trapping more air and reducing heat loss. Humans also try to reduce heat loss by piloerection, but it is inefficient, producing only "gooseflesh." We change our insulation by adding or removing layers of clothing. Unlike other animals, we can also regulate the surface-to-air temperature gradient by heating or cooling our environment.

Much heat is lost by evaporation of moisture from the skin surface. Some of this fluid penetrates the skin by simple diffusion from the underlying tissues, a phenomenon known as **insensible perspiration.** This transfer accounts for the loss of about 700 ml of water per day, but it does not change much to alter the rate of heat loss. We are normally unaware of insensible perspiration, but become acutely aware of it when we wear a garment such as a plastic raincoat. Because of its water-tight surface, we soon feel cold and clammy since surface moisture cannot be evaporated.

Active secretion by the sweat glands is the chief means by which humans lose excess heat. Known as **sensible perspiration,** its rate can be regulated across a wide range as part of the temperature-regulating process. Active sweating can also appear as the result of *sympathetic stimulation,* which is "cold sweat," unrelated to the need for evaporative cooling.

When the environmental temperature is higher than body temperature, heat cannot be lost by radiation, convection, or conduction, and the only means left is evaporation. Humans can perform heavy physical labor when the environmental temperature is well above body temperature—but only if the humidity is low enough to evaporate the large amount of sweat produced can the body temperature be maintained. If fluids are replaced, sweat can be produced at rates of up to 1.5 liters per hour, or 10-11 liters per day (more than double the total plasma volume). Since sweat is a saline solution, (though it is hypotonic at 0.4 percent) if water is replaced and salt is not, electrolyte problems soon develop when

TABLE 26–1	HEAT LOSS FROM THE BODY

Avenue of Loss	Percentage of Heat Lost at 21°C (70°F)
Radiation, conduction and convection	68
Evaporation	27
Respiration	3
Excretion	2

sweating is excessive. If the humidity is not low, the sweat produced cannot be evaporated, and its cooling value is lost. We cannot maintain our body temperature under those conditions, and our ability to perform physical work is greatly reduced when it is hot and humid.

Lungs Expired air has been warmed to body temperature and saturated with water vapor. The amount of heat needed to warm the inspired air is usually slight, but in very cold weather it may account for an appreciable heat loss. In a dry environment, saturating the inspired air may require a significant amount of fluid. The necessary heat and water are obtained from the mucous membranes lining the nasal cavity and respiratory passages. Many furred animals use this as a regulatory mechanism to increase heat loss. By panting, they move a large volume of air rapidly back and forth over moist surfaces, especially the tongue, and greatly increase their evaporative cooling.

Excreta The amount of heat lost through excretions of the urinary and digestive tracts is small and virtually insignificant. See Table 26–1 for a summary of the relative contributions of the different avenues of heat loss.

REGULATION OF BODY TEMPERATURE

The control of body temperature involves all available mechanisms for increasing or decreasing heat loss and

heat production. These include autonomic, somatic, endocrine, and behavioral adjustments, but none is effective beyond a certain point. Some heat is always produced as long as the organism is carrying on metabolism; the amount cannot be reduced below a certain minimum, the basal metabolic rate. Some heat is lost from the body as long as there is a surface temperature or humidity gradient. It is the role of the regulating mechanisms to balance heat loss and production so that the temperature of the body changes very little (see Figure 26–4).

The chief control of body temperature is mediated through two centers in the hypothalamus. In the anterior portion is an area primarily concerned with heat loss. Stimulation of this region activates such heat loss mechanisms as vasodilatation and sweating or panting. After its destruction the animal cannot increase heat loss to prevent a rise in body temperature. It may be able to keep body temperature within tolerable limits in a cool environment, but not when exposed to a high temperature.

In the posterior portion of the hy-

FIGURE 26–4 Summary of temperature regulating mechanisms.

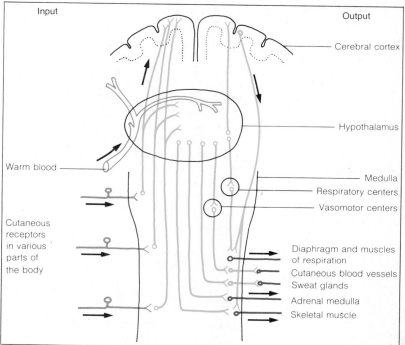

pothalamus is an area that controls heat production. Stimulation of this area results in vasoconstriction and shivering, while its destruction renders the animal unable to reduce heat loss or increase heat production as protection against a cold environment.

It is, however, not strictly accurate to say that there is a discrete heat loss (or anti-rise) center in the anterior hypothalamus, and a well-localized heat production (or anti-drop) center in the posterior hypothalamus. As is frequently the case, the centers are not distinct and isolated areas. They are instead somewhat intermixed and have numerous interconnections with other areas. Stimulation of certain spots may elicit responses typical of both centers.

Afferent Input

To allow them to order the appropriate adjustments, the hypothalamic centers have two main inputs—one which provides information about the body surface, and one which provides information about the core temperature. Surface temperature, largely an indication of heat loss, is detected by two types of cutaneous temperature receptors which are widely distributed over the body surface. These receptors have been described as heat and cold receptors, but in reality they are heat receptors that respond to different degrees of heat. Both are particularly sensitive to changes in temperature. Their afferent pathways lead to the cerebral cortex where the stimulus is recognized, perceived, and localized. Important branches of these pathways reach the temperature regulating centers and permit an assessment of total body surface temperature.

Information about the temperature or the heat production of the core is not received from nerve fibers, but directly from the blood flowing through the hypothalamic temperature centers. This arterial blood comes from the internal core regions of the body where most of the heat production occurs. When the blood temperature is raised, mechanisms that increase heat loss are activated; when blood temperature is reduced, heat conservation and production measures are activated.

Efferent Output

From the hypothalamus, efferent fibers descend through the brainstem to synapse with brainstem centers for the individual autonomic, as well as somatic, effector mechanisms (such as vasomotor and respiratory centers). From these centers fibers descend through the spinal cord to reach motoneurons controlling panting, vascular tone, shivering, and sweating. The cerebral cortex receives temperature information from the hypothalamic centers, as well as directly from the cutaneous receptors, which may lead to the initiation or restriction of voluntary activity (see Figure 26-4). In cold weather, for example, arm-swinging and foot-stamping are well-known methods of increasing skeletal muscle activity and heat production.

Some of the increased heat production in response to cold involves endocrine glands. The sympathetic nervous system, for example, controls the vascular responses and stimulates the adrenal medulla. The epinephrine released favors increased heat production by its action on glycogenolysis and skeletal muscle glycolysis. The hypothalamus exerts important control over the anterior pituitary gland, which in turn controls the secretions of other endocrine glands. Through this connection, a cold environment leads to increased production of the thyroid hormone, increasing metabolic rate and heat production in cells of all tissues. This is a long-range effect, however, and is not as important in humans as it is in other animals.

The hypothalamic temperature centers function rather like a thermostat in some respects (Figure 26-5). When the temperature falls below some predetermined level, known as the **set point,** mechanisms for heat production and conservation are activated, and the temperature is soon raised to the desired level. If the temperature rises above that of the set point, the mecha-

Focus

Keeping Cool

Many animals have special problems of temperature regulation. Animals such as gazelles, antelopes, and sheep live in hot climates where their body temperatures may exceed 46°C (comparable to a fever of 114°C in a human), while their brains must be kept much cooler. Birds that spend much of their time swimming in cold water face a large potential heat loss through their uninsulated legs and feet; whales have a similar problem with their flippers and flukes. A special arrangement of blood vessels, known as a "countercurrent exchange system," enables the gazelles, antelopes, and sheep to remain in the hot sun without succumbing to heat shock, and protects the birds and whales from excessive heat loss through their exposed extremities.

The tolerance of the hot-climate animals for hot, dry conditions is aided by their ability to tolerate high body temperatures while cooling only their brains. This cooling of the brain is due to the arrangement of the blood vessels that enter and leave the head. Many of the veins that drain blood from the head merge to form the cavernous sinus, which is drained in turn by the internal jugular veins. Among these veins are those from the interior of the nose; the blood they bring to the sinus has been cooled by evaporation of water from the nasal lining.

The blood in the sinus is thus cooler than the blood in the rest of the body and offers a way to keep the brain temperature down. In antelopes, sheep, and other animals whose main temperature control mechanism is panting, branches of the carotid arteries enter the sinus and break up into an anastomosing network, or "carotid rete." The function of the rete is to maximize the surface area for exchange of heat between the hot arterial blood from the body and the cooler sinus blood; the cooled arterial blood then enters the brain while the warmed venous blood flows toward the heart, and the delicate brain is kept from overheating. Once the arteries of the rete have traversed the sinus, they come together again to form the arteries that join the circle of Willis as the internal carotids.

Humans and other animals that have such alternate temperature control mechanisms as sweating do not have a carotid rete. In them, the internal carotid passes through the cavernous sinus with a minimum of heat loss and delivers blood at body temperature to the circle of Willis and the brain.

Sea birds and whales have similar blood vessel arrangements in their limbs, where they ensure that only relatively cool blood reaches the limbs. Networks of arteries lie next to veins, and as the two flows of blood pass each other, the warmth of the arterial blood is given up to the cool venous blood (hence the term "countercurrent exchange") and returned to the body where it can be conserved. When combined with such other temperature control mechanisms as the constriction of superficial blood vessels, this arrangement serves very well to minimize heat loss from the body and the metabolic burden of a cold environment.

In humans, too, limb arteries and veins lie next to each other. Venous blood serves to cool arterial blood and reduce the amount of heat delivered to the extremities. The cooling is not terribly efficient, but it does make a difference in cold weather.

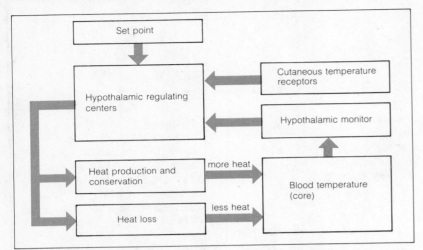

FIGURE 26–5   Schema to show temperature regulation as a control system. Peripheral and central sensors monitor temperatures in those sites and relay the information to the hypothalamic centers. The centers compare the information with the desired temperature (set point) and cause either more production and less loss, or the opposite, to bring measured temperature to set point temperature.

nisms for heat loss are called into play and heat production is minimized until the temperature falls.

Fever and Hypothermia

Fever is the most common example of malfunction of the temperature-regulating mechanism. It is not necessarily due to a breakdown of the mechanism, but is rather a case of "resetting the thermostat," often by the action of fever-inducing agents (called *pyrogens*), such as bacterial toxins, which reach the hypothalamus through the blood and act directly upon the thermoregulatory centers. When the set point is raised to some new level, perhaps 39°C, the regulating center compares the existing body temperature with the new set point and immediately determines that it is below that temperature. The temperature-raising mechanisms are then activated—you shiver and your skin is cold and pale because of vasoconstriction—the production of heat is increased, and heat loss is reduced until the body temperature rises to the new level. When the hypothalamic thermostat is set back to the "normal" temperature, the fever "breaks" and heat loss mechanisms are now

needed. You stop shivering, your blood vessels dilate, your skin is flushed and warm, and you sweat. When the temperature has been reduced, these measures are discontinued and body temperature is once again held at the proper level.

Until very recently the main interest in a reduced body temperature, or **hypothermia,** was in relation to its use in certain types of surgery. Carefully controlled reduction in body temperature lowers the metabolic rate and oxygen consumption of the tissues so that interruption of the circulation can be tolerated for short periods of time.

With the current interest in "getting back to nature,"—camping, backpacking, wilderness treks and the like—large numbers of people are putting themselves in a position to experience the effects of hypothermia, and it is becoming increasingly apparent that this phenomenon is much more common than we would wish. The environmental temperature does not have to fall very low in order to produce conditions that overload the body's heat production and conservation mechanisms, especially if one is wet and tired and has no dry clothing or shelter.

Shivering increases heat production, but it requires a great deal of energy and cannot be maintained for long periods. Vasoconstriction helps maintain the core temperature by keeping the body heat inside, but it may sacrifice the extremities. Reduced blood flow to the fingers may allow their temperature to fall to dangerously low levels, or at least to interfere with the dexterity needed to build a fire or shelter. If body temperature falls there is central nervous system depression, which may impair judgment and the ability to make the right decisions, or even to recognize that there may be a problem.

CHAPTER 26 SUMMARY

Species that maintain a fairly constant body temperature are known as *homeotherms. Poikilotherms* do not regulate their temperature. Cells of homeotherms, including humans, do not tolerate variation in body temperature very far from the "normal" range. The temperature of the body is usually higher in the *core,* where most of the heat is produced (about 37°C), and cooler on the surface where most of the heat is lost. The skin and subcutaneous tissues form a *shell* that insulates the core structures from the usually cooler environment.

Heat Gain

If the body temperature is to be held constant, heat gain must equal heat loss. Heat gain *(chemical thermoregulation)* is almost entirely due to heat produced in cellular metabolism. It can be increased by muscular activity (including shivering), SDA and by thyroid hormones. Although less important quantitatively, heat can be gained from the environment if it is hotter than the body (such as the sun or a hot bath).

Heat Loss

Heat loss *(physical thermoregulation)* occurs by physical processes which depend upon gradients. *Radiation, conduction* and *convection* involve transfer of heat from a warmer to a cooler object by radiation through the air to another object, by direct contact, or by warming the adjacent air, respectively. These processes all depend upon the temperature gradient between the body and the object or air, and upon the area of body surface involved. Converting water to water vapor *(evaporation)* requires considerable heat, which cools the evaporative surface. Evaporation depends upon the relative humidity (a humidity gradient) rather than a temperature gradient, therefore it can still be effective when environmental temperature is higher than body temperature.

Most of the heat is lost through the skin and is controlled, within broad limits, by changing conditions so as to alter the gradients. Dilation of blood vessels brings more warm blood to the surface, skin temperature rises, and more heat is transferred to the environment. Vasoconstriction reduces blood flow and keeps warm blood in the core.

Skin temperature falls and less heat is lost.

Cutaneous heat loss can also be reduced by insulation. Furred animals can make their fur stand out (piloerection) thereby trapping more insulating air; some species increase insulation by adding layers of subcutaneous fat. Humans add insulation by putting on more clothing.

Evaporative heat loss from the skin includes *insensible* and *sensible perspiration.* Insensible perspiration is simply diffusion of water through the skin and occurs at a fairly constant rate. Sensible perspiration is due to the secretion of the sweat glands and can reach very high rates of fluid secretion. Perspiration does not cool, however, unless the fluid is evaporated.

Air exhaled from the lungs is saturated with moisture and is at body temperature. In humans it accounts for a significant heat loss only if the environmental temperature is very low. For many animals, however, evaporative cooling from the tongue (panting) is a major means of losing heat.

Regulation of Body Temperature

Body temperature is controlled through a pair of centers in the hypothalamus. Stimulation of the *heat loss center* activates heat loss mechanisms such as vasodilation, perspiration, or panting, while stimulation of the *heat gain* (or conservation) center leads to vasoconstriction and shivering. The output of these centers is based on input both from the shell and the core of the body. Impulses from cutaneous temperature receptors provide information about the temperature and heat loss on the body surface. The temperature of the blood flowing through the hypothalamus monitors temperature and heat production in the core. The centers integrate this information and their output affects both smooth and skeletal muscle, and exocrine and endocrine glands to produce more or less vasoconstriction, shivering, or secretion of sweat, epinephrine, or thyroid hormone.

The centers regulate temperature to

a particular *set point* and initiate heat production and conservation measures when body temperature is found to be below the set point, much like a thermostat for the furnace in a home. When body temperature is found to be above the set point, heat loss mechanisms are activated. If the set point is raised (as in fever), the centers will regulate to the new level and maintain an elevated body temperature until the set point returns to normal once more. A reduced body temperature (hypothermia) may occur if the regulating mechanism is impaired, or if one is forced to cope with an excessive or prolonged temperature stress and is unable to keep heat production in balance.

STUDY QUESTIONS

1 What are the advantages and disadvantages to an animal of being homeothermic or poikilothermic?

2 What is "normal" body temperature? What value is considered to be a normal body temperature in degrees centigrade (Celsius)? In degrees Fahrenheit?

3 How much variation would you expect to find if you measured body temperature at several sites? Explain. How much normal variation might occur with various activities?

4 By what means can a homeotherm (such as a human being) normally lose heat? What conditions are necessary for each form of heat loss to occur (i.e., what gradients are involved)? By what means can a homeotherm gain heat?

5 What are some of the adjustments such an animal can make to increase or decrease the rate of heat loss or production?

6 Suppose the environmental temperature is about 38°C (about 100°F), and the humidity is about 80 percent. By what means would you maintain your body temperature? Explain.

7 What is meant by the "set point"? What would happen if it were raised? Explain.

Part 7
The Urinary System

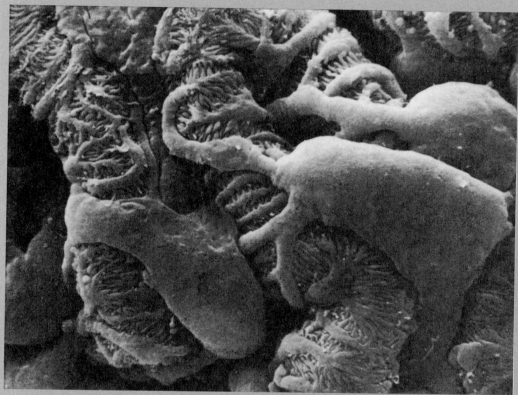

From *TISSUES AND ORGANS—a text-atlas of scanning electron microscopy,* by Richard G. Kessel and Randy H. Kardon. W. H. Freeman and Company, copyright © 1979.

Capillaries in the glomerular capsule of the kidney; fingerlike processes of large cells (podocytes) wrap around the capillaries.

The urinary or excretory system is the last of the major organ systems whose primary function is homeostasis. Other systems have been concerned with one particular aspect of the internal environment: the respiratory system with oxygen and carbon dioxide; the digestive system with nutrients and energy; and the cardiovascular system with transport of these substances to and from the internal environment. The urinary system rids the body of waste products. The kidney is the major excretory organ, and it does, indeed, remove and excrete waste products, but it does much more.

As blood flows through the kidney, the plasma is "processed" in many ways. The kidney excretes more of a substance when it is present in excess, and retains it when it is lacking. By removing—or not removing—certain normal plasma constituents the kidney regulates the composition of the plasma. Recall that the blood plasma carries on exchange with the internal environment through the relatively permeable capillary walls. Therefore, by excreting more or less of certain plasma constituents (including fluid), the urinary system regulates not only the composition of the plasma, but of the entire extracellular fluid compartment.

The first chapter in Part 7 deals with the urinary system, its structure and how it selectively removes certain substances from the blood plasma. Chapter 28 examines how the kidney is able to adjust the removal or retention of fluid, electrolytes, and other solutes in order to regulate the extracellular fluid.

27

Homeostasis and Excretion

Anatomy of the Urinary System

The Formation of Urine

Composition of Urine

Micturition

The Urinary System

Although Chapter 27 is about the urinary system, it focuses on the kidney because it is the organ that carries out most of the functions. To begin, you might refer back to the section on the approach to the study of the body in Chapter 1. The kidney was chosen as an example of an organ whose structure is very well adapted to its function. You were asked to consider the design of, and the requirements for, an excretory organ. Such an organ must be able to economically remove wastes and salvage nonwastes without upsetting any of the delicate balances in the body.

By now you have a much better idea of the significance of those requirements: that the removal is from the blood plasma; that the processes available to do it are active transport and diffusion, osmosis, and filtration; what economical processes are, in terms of energy; and an awareness of some of the delicate balances, as well as the consequences of upsetting them.

Before you proceed, give some further thought to this design problem. Consider the requirements of the system and their implications. As you study the kidney, see how its design satisfies the requirements.

HOMEOSTASIS AND EXCRETION

Throughout this book, the organ systems have been presented in terms of homeostasis, of fulfilling the needs of the individual cells. So far, we have considered those systems that provide specifically for the supply and delivery of respiratory gases and nutrients. However, an important need that remains to be considered is the removal of the wastes of cellular metabolism, that is, the excretory mechanisms.

Of the systems involved in the excretory process, we have already discussed several in relation to their other roles:

1. *Digestive tract.* Much of the material excreted from the digestive tract never gains entry into the body. It is not absorbed, and although it is discarded, it is not an excretory product produced by cellular metabolism. Of the few truly excretory (cellular) products in the feces, the most notable are several constituents of the bile.

2. *Skin.* Sweat secreted onto the skin surface contains mostly water and electrolytes, but small amounts of urea, ammonia, and other waste products are also present.

3. *Lungs.* We do not usually think of the lungs as excretory organs, but they play a major role in the removal of carbon dioxide, the most abundant single waste product of the cells. By its stimulatory effect on respiration, carbon dioxide exerts an important control over its own excretion.

The **kidney** is probably the most important excretory organ, although you could survive a little longer without renal excretion than without respiratory excretion. However, the kidney is responsible not only for getting rid of more kinds of wastes than is the lung, but also for regulating several aspects of your extracellular fluid. The kidney separates certain substances from the blood plasma and excretes them in the urine. We think of urine formation as the chief function of the kidney, but in reality the function of the kidney is to maintain the composition of the extracellular fluid. It does so by excreting (or retaining) certain substances. The formation of urine is simply the means by which this is carried out.

In performing its excretory role the kidney must be selective. The kidney must perform its excretory role selectively: while some substances need to be excreted, others must be kept in the body. The excretory function therefore encompasses a great deal more than merely getting rid of wastes. What the kidney excretes or retains determines in large part the composition of the plasma, and the properties of the plasma determine the properties of the interstitial fluid, the internal environment of the cells. The kidney helps maintain the concentration of a number of specific substances, excreting very little when the plasma concentration of a substance is low and excreting more when it is high; it regulates water loss, and hence water balance; and it varies the pH of the urine, and hence stabilizes the pH of the body fluids. In addition, it has some endocrine functions and a role in the regulation of arterial pressure.

In the absence of properly functioning renal tissue, excesses and deficiencies of various substances develop in the plasma, changing the composition of the plasma and interstitial fluid. After a few days the changes of the internal environment are great enough to be incompatible with normal cell function. The alternatives to death are then a kidney transplant or periodic cleansing of the blood with an artificial kidney machine.

ANATOMY OF THE URINARY SYSTEM

Gross Anatomy

The principal structures of the urinary system are the kidneys, ureters, urinary bladder, and urethra (Figure 27–1). The **kidneys** are paired reddish-brown organs lying against the posterior wall of the abdominal cavity, between the level of the twelfth thoracic and the third lumbar vertebrae, with an **adrenal gland** sitting on top of each one like a small cap. The kidneys are *retroperitoneal*, which means they are covered, but not surrounded, by peritoneum. They are, however, enclosed by a connective tissue capsule, and often almost entirely hidden behind a mass of adipose tissue, the **perirenal fat.** Each kidney is a bean-shaped organ four or five inches long with an indentation at the *hilus* from which the nerves, the large renal blood vessels, and the ureters emerge.

The urine formed in the kidney first empties into a saclike funnel within the kidney known as the **renal pelvis.** At the hilus, this structure narrows down into a slender muscular tube, the **ureter.** The ureters descend on either side of the aorta and vena cava, behind the peritoneum, and enter the urinary bladder on either side of its posterior inferior surface. Because their walls contain smooth muscle, the ureters are capable of peristaltic contractions (anywhere from less than one to about five per minute) which help tiny droplets of urine move toward the bladder.

The **urinary bladder** (Figure 27–2) is a hollow muscular organ resting on the floor of the pelvic cavity and, like other urinary structures, outside the peritoneal cavity. It is a relatively small organ when empty, but in keeping with its storage function, it is capable of a great increase in capacity. The bladder is lined with *transitional epithelium* and the smooth muscle of its wall is arranged in three layers oriented in different directions. Very little internal pressure is required to distend the bladder, since both the muscle layers and the lining readily and progressively decrease in thickness as it fills.

The three openings into the urinary bladder, two for the ureters and one at the base of the bladder for the urethra, enclose a triangular area known as the **trigone.** The smooth muscle of the bladder receives parasympathetic innervation and contracts reflexively, causing an increase in pressure within the bladder, forcing the urine through a sphincter (which relaxes) and out the urethra. The urine does not back up into the ureters partly because there is a valvelike flap of mucous membrane at the opening, but largely because the ureters pass through the bladder wall at an angle, so that when the muscle contracts it acts like a sphincter, closing off the ureters temporarily.

The **urethra** is a muscular tube leading from the bladder. In the female

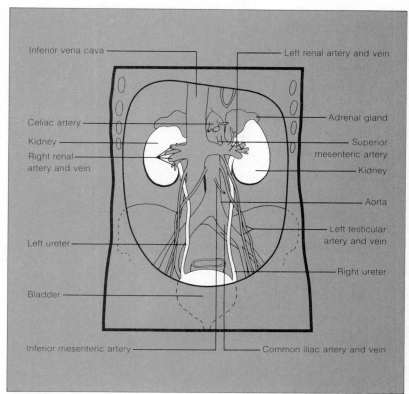

FIGURE 27–1   Organs of the urinary system. The peritoneum has been removed.

it is a short direct channel, with its own opening to the outside. In the male (see also Chapter 29), it is much longer and more complicated, and it serves both the urinary and the reproductive systems. The flow of urine through the urethra is controlled in part by a smooth muscle sphincter near its exit from the bladder. There is also a voluntary skeletal muscle sphincter where it passes from the pelvic cavity through what is known as the **urogenital diaphragm.**

Internal Structure of the Kidney

The internal structure of the kidney can be readily seen in a frontal section (Figure 27–3). The outer or surface portion of the kidney is the **cortex,** and beneath it lies the **medulla.** The **renal sinus,** in the region of the hilus, is a fairly large space occupied by the renal vessels and the renal pelvis. Any remaining space is usually filled with adipose tissue.

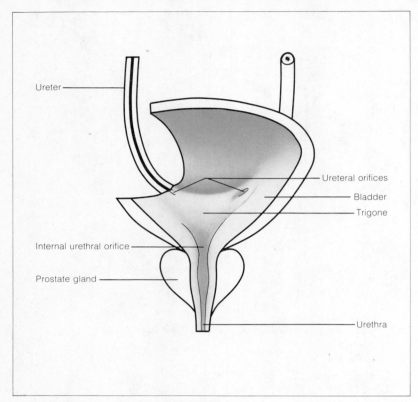

Ureter

Ureteral orifices

Bladder

Trigone

Internal urethral orifice

Prostate gland

Urethra

FIGURE 27–2   The urinary bladder.

The medulla of the kidney is somewhat paler in color than the cortex and is made up of several **pyramids.** The base of each pyramid abuts against the cortex, and the apex protrudes into the renal sinus. The pyramids have a striated appearance due to numerous tiny tubules that merge into a few channels which open at the tip or **papilla** of each pyramid.

The renal pelvis branches into several **major calyces,** which then branch into several smaller **minor calyces.** A calyx fits like a cup over the papilla of each pyramid. Urine formed in the kidney drains through the papilla to the renal pelvis and ureter for transport to the bladder.

The Nephron

The **nephron** is the functional unit of the kidney, since it performs every aspect of urine formation, including the regulation of the urine's acidity, volume, and tonicity. One need only multiply the activities of a single nephron by the million or so nephrons in each kidney to have a total picture of renal function.

The nephron's task is formidable, since it must remove unwanted substances from the bloodstream while at the same time retaining others. The nephron must be able to change the balance between excretion and retention of a single component promptly to meet current needs. And it must be able to do all this without imposing gradients its cells cannot withstand or energy demands they cannot meet.

Structure of the Nephron The nephron is basically a tube with an enlarged blind-end sac on one end, the other end leading eventually into a channel that empties into the renal pelvis and ureter (Figure 27–4). The enlarged end of the nephron, known as the **glomerular capsule** *(Bowman's capsule),* surrounds a tuft of coiled capillaries called the **glomerulus.** Together the capsule and the glomerulus make up the **renal corpuscle,** a structure similar to the one you would make by pushing your fist into a half-inflated balloon. Your fist is the glomerulus, surrounded by the two continuous walls of the capsule (balloon), except for a tiny opening through which the blood vessels (your wrist) enter and leave. The capsular space between the two walls is continuous with the lumen of the tubule.

The wall of the capsule is made up of *simple squamous epithelium,* whose cells are very thin and flat. The inner, or visceral, layer of the capsule is applied directly to the surface of the glomerular capillaries. Unlike typical squamous epithelium, or that of the outer layer the cells of the visceral layer, known as **podocytes,** have large branching and rebranching processes which interdigitate with those of other cells as they surround the capillary. Such an arrangement leaves numerous large openings in the inner capsule wall. The endothelium of the glomerular capillaries also has more and larger openings than does endothelium elsewhere. With so many openings in both walls, there are numerous places where the basement lamina is all that separates the inside of the capillary from the capsular space. This is all highly favorable to rapid exchange between glomerulus and capsule.

The tubule of a nephron is a tiny channel, about two inches long, whose wall is composed of a single layer of mostly *cuboidal cells* (Figure 27–5). It emerges from the renal corpuscle in tortuous coils, drops in a long hairpin loop toward the medulla and back. It becomes highly coiled again before it empties into the **collecting tubule** which extends to the papilla and empties into the renal pelvis. The parts of the tubule, in order from the capsule, are called the **proximal convoluted tubule,** the **loop of Henle,** which consists of a **descending limb,** a **thin segment,** and an **ascending limb,** and finally the **distal convoluted tubule.** The thin segment near the bottom of the loop consists of simple squamous epithelium. In nephrons whose glomerular capsule lies close to the medulla, the loops are long and dip almost to the papilla, but cortical nephrons have relatively short loops.

Under the microscope the cells of the proximal and distal portions of the tubule look different, suggesting that they serve different functions. Proximal cells (of the proximal convoluted tubule and descending limb) are thick and have many microvilli on the inner luminal surface, which greatly increases their free surface area. Distal cells (of the ascending limb and distal convoluted tubule) are flatter, and their internal (luminal) surfaces have few microvilli, though they are irregular.

As the ascending tubule returns to the cortex, it touches the blood vessel which enters the glomerulus (the afferent arteriole). The cells of both the arteriole and the distal tubule are modified at this site. Those of the tubule form the *macula densa,* and those in the arteriole are called juxtaglomerular cells. Together they form what is known as the **juxtaglomerular complex** or **apparatus.**

Blood Supply of the Nephron

The pattern of distribution of blood through the kidney and to the nephron is an essential feature of nephron func-

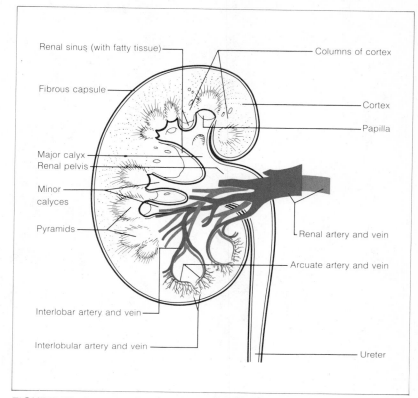

FIGURE 27–3   Frontal section of the human kidney.

Labels: Renal sinus (with fatty tissue); Columns of cortex; Fibrous capsule; Cortex; Papilla; Major calyx; Renal pelvis; Minor calyces; Pyramids; Renal artery and vein; Arcuate artery and vein; Interlobar artery and vein; Interlobular artery and vein; Ureter

tion, as is suggested by the intimate relation of the glomerulus and glomerular capsule. The short renal arteries arise directly from the aorta and divide at the hilus into several large **interlobar arteries** which pass between the medullary pyramids to reach the cortex. Near the junction between the medulla and the cortex these arteries branch and arch along the line of junction as the **arcuate arteries.** From them numerous smaller **interlobular arteries** extend upward through the cortical substance toward the surface. At intervals tiny branches, the **afferent arterioles,** each enter a nearby glomerular capsule and break into several parallel **glomerular capillaries,** instead of the usual freely anastomosing capillary network. The glomerular capillaries then converge to a single vessel, the **efferent arteriole,** which emerges from the renal corpuscle and shortly breaks into the **peritubular capillaries** around and between the tubules. Some of the peritubular capillaries near the medulla form long loops, the **vasa rectae,** which follow the course

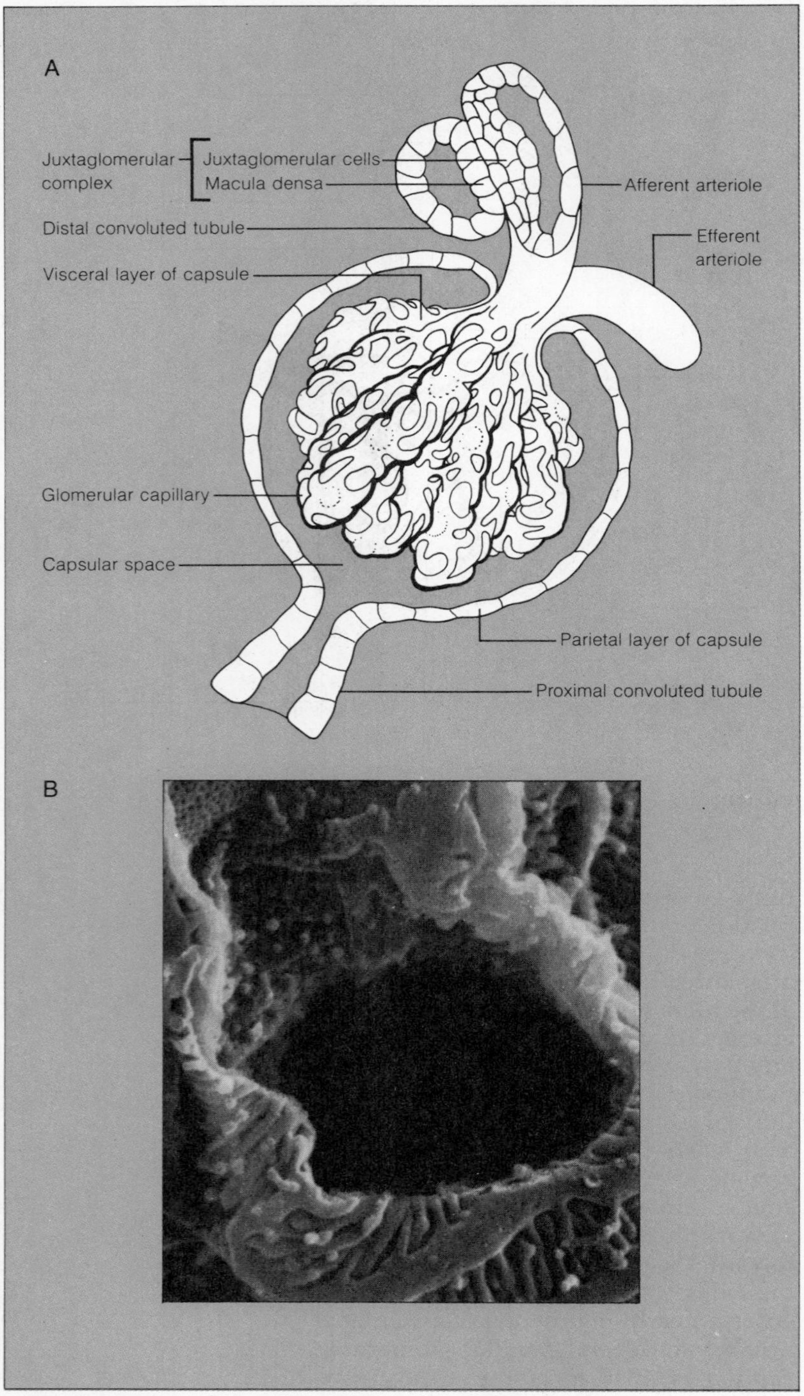

A

Juxtaglomerular complex
Juxtaglomerular cells
Macula densa
Distal convoluted tubule
Visceral layer of capsule
Afferent arteriole
Efferent arteriole
Glomerular capillary
Capsular space
Parietal layer of capsule
Proximal convoluted tubule

B

FIGURE 27–4   A. A renal corpuscle.   B. Scanning electron micrograph showing podocytes and a glomerular capillary whose wall has been partially removed to show openings in the endothelium. 15,020 x magnification.

of the loops of Henle deep into the medulla and back. The peritubular capillaries eventually converge into veins which flow into *interlobular veins, arcuate veins, interlobar veins,* and finally the **renal vein.**

Several features of the renal circulation are significant from a functional standpoint.

1. The blood flows through *two capillary beds,* that of the glomerulus and that of the peritubular capillaries.

2. The vessels leading to and from the first capillary bed (the glomerulus) are both arterioles, which implies that the blood leaving the glomerulus is still "arterial" blood. Very little oxygen is removed from the blood as it passes through the glomerulus, which tells us that there is little metabolic activity (such as active transport) carried on in the glomerulus.

3. In other parts of the circulatory system there is a considerable drop in blood pressure across a capillary bed due to the peripheral resistance. With two capillary beds, one must conclude that either the blood pressure is high or the resistance is low. Both are true in the kidney. Due to the short renal arteries and their abrupt branching, capillary pressure in the glomerulus is higher than that in capillary beds elsewhere. The peripheral resistance in the kidney is relatively low, but it is variable. The afferent and efferent arterioles on either side of the glomerulus have in their walls smooth muscle whose tone can be varied when systemic pressure fluctuates to maintain a fairly constant pressure in the glomerulus. These vessels and their sympathetic (vasoconstrictor) innervation apparently represent the only site of nervous control of the kidney.

THE FORMATION OF URINE

The formation of urine by the nephron involves three basic processes: filtration at the glomerulus and reabsorption and secretion along the tubule. **Filtration** is a passive process in which a portion of the plasma flowing through

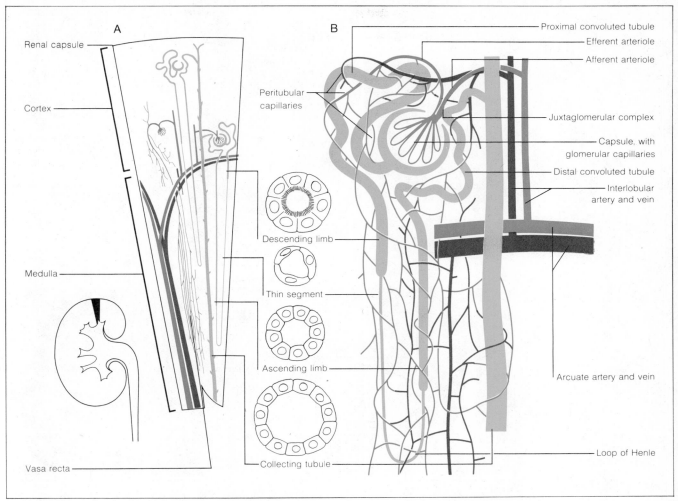

FIGURE 27-5 A nephron. A. A wedge of renal tissue, showing the orientation of a nephron. B. Basic structure of a nephron and its blood supply.

the glomerulus is transferred from the blood to the capsule. **Reabsorption** is the return of some of the filtered material (originally absorbed from the digestive tract) to the blood. Some reabsorption is passive, but the return of many substances depends upon active processes. In the main, substances that are reabsorbed are those the body must retain—nutrients, electrolytes, and fluids—while most of the waste products are reabsorbed poorly or not at all. **Secretion** is similar to reabsorption, except that materials are transferred in the opposite direction, that is, from the peritubular capillaries into the tubule. It is largely an active process, and thus can cause a substance to be excreted more rapidly than by filtration alone.

The mechanisms for active transport in either reabsorption or secretion are many and varied, and they differ for nearly every substance so handled. For this reason it is possible to alter the transport of a particular substance independently of that of other substances, while some of the passive processes may be controlled indirectly by changes in the active processes (active transport of electrolytes creates gradients that cause water transport). It is these adjustments in filtration, reabsorption, and secretion that make it possible to change the composition, volume, tonicity, and acidity of the urine.

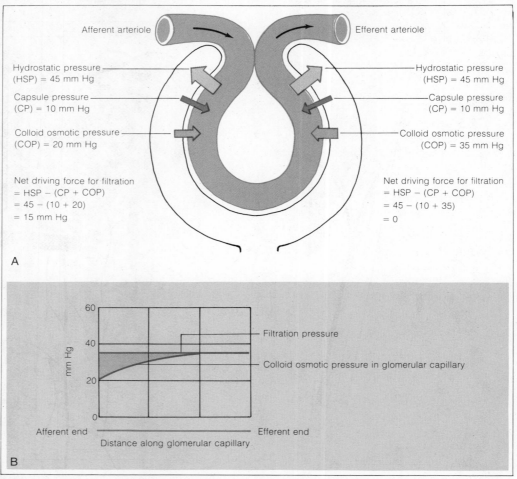

FIGURE 27–6   A. The pressure gradients between the glomerular capsule and capillaries (as measured in the rat).   B. Profile of the net driving force for filtration (shaded area) along an idealized glomerular capillary.

Glomerular Filtration

The renal corpuscle is admirably designed as a filter: (1) Its membrane is extremely porous due to the numerous openings; (2) the glomerular capillaries present a surface area of about 1.5 square meters (roughly the same as the area of the body surface); and (3) the existing pressure gradients are favorable for filtration. The pressure of the blood entering the glomerular capillaries is now believed to be about 45–50 mm Hg, which is lower than it was formerly thought to be (Figure 27–6A). It is opposed by the pressure of fluid in the capsule, about 10 mm Hg, and by the plasma colloid osmotic pressure of about 20–25 mm Hg. (The col-

loid osmotic pressure of the fluid in the capsule is considered to be 0, since protein filtration is negligible.) The net filtration pressure, or driving force, is thus about 15 mm Hg, which is not much higher than in capillaries elsewhere.

A high rate of glomerular filtration is achieved without a large filtration pressure for at least two reasons. First is the great permeability or porosity of the capillary walls. It is over 100 times greater than that of skeletal muscle capillaries, so that filtration occurs very rapidly. In a typical capillary, filtration occurs primarily at the arterial end, while at the venous end net movement is back into the capillary (see Figure 17–15). In the glomerular capillaries there is only filtration, as the return occurs in the peritubular capillaries. As blood flows through the glomerular capillary loops, rapid filtration reduces the volume of fluid left in the capillary.

The filtration pressure remains fairly constant in spite of the loss of fluid from the capillary because of a mechanism known as *autoregulation* (see below), in which the resistance of the afferent and efferent arterioles is adjusted so as to maintain the filtration pressure. The colloid osmotic pressure increases as the blood flows along the glomerular capillary because the plasma that filters into the capsule leaves its protein in the capillary. Part of the way along the glomerular capillary the colloid osmotic pressure comes to equal the filtration pressure, and there is no net gradient, and hence no more filtration in the remainder of the capillary (Figure 27–6B). Thus, part of the glomerular capillary does not ordinarily participate in filtration, but if the renal blood flow were to increase, filtration would occur along a greater length of the capillary.

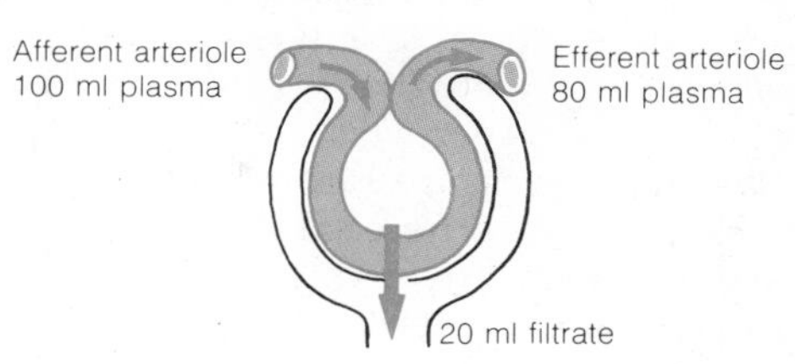

Glomerular Filtration Rate The renal blood flow is quite high, as nearly 25 percent of the total cardiac output flows through the kidneys. This is typically in the range of 1100–1200 ml of blood, or 600–650 ml of plasma, per minute. It is the latter that is of concern, since it is the plasma that is filtered; the blood cells are not involved. Nearly a fifth (15–20 percent) of the plasma that enters the glomeruli from the afferent arterioles passes into the capsule and becomes *glomerular filtrate* (See Figure 27–7). The remaining 80–85 percent stays in the blood vessels and leaves the glomeruli through the efferent arterioles to pass on to the second capillary bed. The volume of the filtrate formed is thus enormous. It amounts to about 125 ml of filtrate formed each minute, or 7.5 liters (nearly two gallons) each hour. If one assumes the total plasma volume to be a little less than three liters (55 percent of five liters of blood), then all the plasma in the body is filtered in less than half an hour, and in twenty-four hours 180 liters of filtrate are formed. From figures such as

FIGURE 27–7
Distribution of 100 ml of plasma containing 10 mg of solute as it passes through the glomerulus. The solute is completely filtered so that its concentration in the filtrate equals that in the plasma.

	Afferent arteriole	Efferent arteriole	Filtrate (tubule)
Plasma flow ml/min	100	80	20
Concentration mg solute/ml	10	10	10
Amount mg solute/min	1000	800	200

these, it becomes apparent that reabsorption must be a very important part of the whole process of urine formation. Since only 1.2 liters of urine are excreted each day, over 99 percent of the filtrate must be reabsorbed.

Composition of the Glomerular Filtrate The composition of the fluid filtered depends upon the composition of the plasma and the properties of the membrane. In general, the filtrate is a protein-free plasma; that is, all the constituents of the plasma except the protein pass through the membrane into the capsule in proportion to their concentrations in the plasma. The composition of the filtrate is the same as that of the plasma except for protein. The plasma in the *efferent* arteriole is identical to normal unfiltered plasma, except in that it contains *relatively* more protein. The composition of the filtrate can be determined experimentally by collecting and analyzing samples of glomerular filtrate. Though such experiments are simple in theory, they are exceedingly difficult technically because of the small size of the nephrons and the tiny volumes of filtrate that can be obtained for analysis.

Control of Glomerular Filtration As a passive process, glomerular filtration depends upon pressure gradients. Since arterial pressure is probably the most variable of the pressures involved, one might expect glomerular filtration rate to fluctuate with it. Glomerular filtration, however, is determined by *glomerular* pressure rather than *arterial* pressure, and there are mechanisms that maintain a relatively constant pressure of blood entering the glomerulus, as long as the mean arterial pressure is between 80 and 180 mm Hg. Glomerular filtration rate, therefore, does not change very much over the usual range of arterial pressures. If the arterial pressure falls to dangerously low levels (below 80 mm Hg), the glomerular filtration pressure is reduced and urine formation may stop completely. When arterial pressure rises above about 180 mm Hg, the limit of the protective mechanisms is exceeded, and filtration pressure and filtration rate increase.

The flow of blood *into* the glomerulus is regulated by the afferent arteriole, and the flow *out of* it by the efferent arteriole. Constriction of the afferent arteriole can reduce glomerular pressure (or prevent a rise if arterial pressure rises). Contraction of the efferent arteriole can raise the glomerular pressure (or prevent a fall if arterial pressure falls). The smooth muscle in these vessels receives sympathetic innervation, and the afferent arteriole is particularly responsive to such stimulation and to epinephrine. But a denervated and perfused kidney is still able to regulate and maintain its blood flow, so the mechanism of control must lie within the kidney itself. The process, known as **autoregulation,** is not well understood, but several suggestions have been offered to explain it. Two of them are that the arteriolar smooth muscle contracts in response to stretch caused by increased blood pressure in the arterioles, or that the juxtaglomerular apparatus (see Figure 27–7) contains some sort of pressure-sensing device that triggers constriction.

Glomerular filtration is also affected by the plasma colloid osmotic pressure, a force that tends to reduce filtration. A decrease in the plasma protein concentration lowers the colloid osmotic pressure and will increase the filtration rate, even though the glomerular pressure remains unchanged.

Tubular Reabsorption

As the fluid filtered from the plasma passes along the tubule of the nephron, it is drastically altered, largely by reabsorption of many of its constituents. In fact, the fluid known as glomerular filtrate actually exists only at the very beginning of the tubule, since reabsorption begins as soon as it begins to move along. The reabsorbed substances are actively or passively transferred from the tubule across the tubular cells to the interstitial fluid, from which they enter the peritubular capillaries. The result of reabsorption is that while certain substances are removed from the tubule, those not removed are greatly concentrated by the reabsorption of water.

Passive Transport Water is the major constituent transferred by passive reabsorption. The transfer is brought about by the gradients created by the active reabsorption of other solutes, such as sodium and glucose, as well as by the osmotic gradients due to the increased protein content of the peritubular capillaries. The conditions are such that about 80 percent of the water has been removed by the end of the proximal portion of the tubule, while another 19.5 percent is recovered in the distal portion of the tubule. Chloride ions are passively reabsorbed, due to the electrical gradient created by active sodium reabsorption. Urea, formed in the liver upon deamination of amino acids, is also passively reabsorbed. This would not seem to be very efficient, since urea is a waste product and presumably should be removed from the body. However, urea is a fairly small molecule, and about half of it is carried back to the blood with the reabsorbed water.

Active Transport More reabsorbed substances are transported actively, which involves metabolic work and increased oxygen consumption as the tubular cells transfer or pump these substances against their gradients. There is, however, a maximum rate at

Focus

The Artificial Kidney

Dialysis is a method of removing particular substances from a solution in the laboratory; it provides the basis of the artificial kidney. A solution containing material to be removed is passed through tubing made from a selectively permeable cellophanelike membrane, or "dialysis tubing." The tubing is surrounded by a fluid containing all the diffusible solutes—electrolytes (salts) and other small molecules—that are in the solution being dialyzed, except the solutes one wishes to remove from the solution in the tubing. The solutes present in both fluids are present in the same concentrations in the two fluids. There are therefore no concentration gradients and no net movements of solutes. For the solutes that are *not* present in both fluids, however, there *are* concentration gradients and net movements. Since these solutes are present in the fluid in the tubing, but not in the outside fluid, they move out of the fluid being dialyzed. Because the fluid being dialyzed is constantly moving and being exposed to fresh outside fluid, the dialysis never reaches equilibrium, and these solutes can be virtually completely removed.

This scheme can be varied. The bathing fluid does not have to contain an equal concentration of diffusible solute or none at all. It can contain an unequal concentration, in which case the solute crosses the membrane until the concentrations are equal. The concentrations of the solutes in the solution being dialyzed can thus be raised or lowered by addition or subtraction. If the concentrations of nondiffusible solutes, such as proteins, are manipulated, the movement of water can also be controlled.

In the laboratory, dialysis is used to purify protein solutions by removing electrolytes. In the body, it approximates what the kidney does in producing urine (though the kidney also uses active transport). It can therefore be used as the basis of an artificial kidney machine. Usually, cannulas are chronically implanted or placed in an artery and a vein, so they can be readily and repeatedly connected to the artificial kidney machine. The patient's blood is then routed through a length of dialysis tubing as a fluid to be dialyzed. The tubing is immersed in a fluid that is carefully designed to contain all the normal constituents of blood—the proteins, nutrients, and other substances—except the waste products normally removed by the kidneys. As the patient's blood flows through the tubing, waste materials are therefore given up to the bathing fluid. The cleansed blood is then returned to the circulation via the cannula inserted in the artery. The procedure normally takes several hours and must be performed several times a week.

Until fairly recently, artificial kidneys have been bulky, nonportable machines. Patients have had to visit hospitals periodically to have their blood cleansed. Now, however, there are experimental artificial kidneys the size of attache cases. They are portable, and if enough of them can be made to go around, kidney patients will find life much more bearable.

which the cells can function, a limit on how much they can pump in a minute. This is known as the **tubular maximum** (T_m). Just as a fruit inspector can remove all the defective fruit passing on a conveyor belt only if most of the fruit is not damaged, so the tubular cells can totally remove a substance from the tubule only if there is not too much of it. If the plasma concentration (and hence the filtrate concentration) of that substance is high enough to cause more of the substance to be filtered each minute than can be transferred, some will appear in the urine. The plasma level at which a substance begins to appear in the urine is called its **threshold.** Since nearly every substance that is actively reabsorbed involves a different transport mechanism, the thresholds and tubular maxima differ widely. For some substances the thresholds are so low that they almost always appear in the urine. For other substances, the thresholds are high enough that they are not reached under physiological conditions.

Glucose is actively reabsorbed in the proximal tubule. It has a relatively high threshold, and at normal plasma concentrations (70–100 mg percent), all the glucose is reabsorbed and none is excreted. When the plasma glucose concentration surpasses 160–180 mg percent, however, the tubular maximum for glucose (T_{m_g}) is exceeded and glucose begins to appear in the urine. Further increase in plasma glucose results in greater glucose filtration and excretion, but not in greater glucose reabsorption, since that already has reached its maximum.

Sodium is actively reabsorbed throughout much of the tubule, but the rate of transfer in distal parts of the tubule is regulated so as to preserve acid–base balance and electrolyte content. Other *inorganic* ions actively reabsorbed include potassium, calcium, phosphate, and sulfate. Other *organic* substances actively reabsorbed are amino acids (which are not all reabsorbed to the same extent), a number of organic anions, and ascorbic acid. Some of the waste products of protein metabolism are also reabsorbed to a limited extent.

Tubular Secretion

Secretion is similar to reabsorption, but it transfers substances *into* the tubules rather than *out* of them, and it is almost entirely active. It removes from the blood certain substances that are not filtered in the glomerulus, but are instead passed along the efferent arteriole and into the peritubular capillaries and interstitial fluid. Because secretion can be carried out against a gradient, it can remove all of a substance from the blood passing through the kidney (like reabsorption, however, it is limited by the tubular maximum for that substance).

Some important substances, including hydrogen ions, ammonia, potassium, and metabolic end products such as creatinine and hippuric acid, are secreted as part of the normal regulatory processes (Figure 27–8). However, many substances that can be secreted are not normally present in the plasma. They are instead foreign (nonphysiological) compounds. Although secretion is an effective means of removing them, it can be inconvenient if the substance was given as a medication, since it necessitates more frequent administration.

Other secreted substances can be used to evaluate kidney function, since if the plasma concentration of a substance and the amount excreted in the urine over a period of time are known, the volume of plasma needed to bring the substance to the kidney can be calculated. If the plasma concentration does not exceed the threshold and tubular maximum for that substance, then this volume of plasma is the total renal plasma flow, which in turn determines the total renal blood flow.

COMPOSITION OF URINE

Although the concentrated fluid that emerges from the collecting tubules was derived from blood plasma, its final composition differs very markedly from that of plasma. Virtually all constituents of urine were removed from the plasma by glomerular filtra-

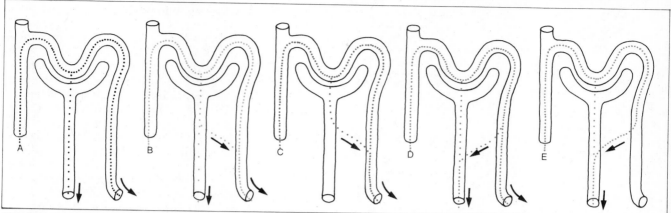

FIGURE 27–8   Tubular function.   A. Substance A is filtered only.   B. Substance B is filtered and partially reabsorbed.   C. Substance C is filtered and totally reabsorbed. None is excreted.   D. Substance D is filtered and partially secreted.   E. Substance E is filtered and totally secreted. None remains in the venous blood from that nephron.

tion, since few physiologically produced substances are secreted. Many of the filtered substances do not appear in the urine at all, however, since they are totally reabsorbed in passage through the tubule.

Normal Constituents

The volume of urine produced in a day is usually about 1200 ml, but it may be as little as 600 ml or as much as 2500 ml. A volume of 600 ml is near the minimum needed for removing a day's waste products. The urine volume may be considerably in excess of 2500 ml when fluid intake is high or when water reabsorption is impaired for one reason or another (see Chapter 28); in either case the urine is likely to be very dilute, as indicated by a *specific gravity* of about 1.003 (that of water is 1.000). When the urine volume is scant, the solute concentration is much greater and the specific gravity is more likely to be near 1.030. Dilute urine has a very pale color while concentrated urine is darker and more amber in color.

The amount of organic waste produced for excretion is related to the total metabolic activity, not to urine volume. The amounts of inorganic salts excreted are likely to vary, since they are determined by other needs. Dietary intake affects excretion of several constituents. Most of the organic constitu-

ents are the result of protein metabolism, and most contain nitrogen. The most abundant is *urea*, and even though roughly half of the urea is reabsorbed, 25–30 grams are excreted each day. Among other nitrogenous compounds in the urine are *uric acid*, which is closely related to urea, and a small amount of *creatinine*, which is associated with muscle metabolism. Small amounts of *amino acids* may also be found in the urine, although they are almost totally reabsorbed. Other organic constituents include certain vitamins, hormones, and enzymes (or their metabolic end-products), which are normally present in extremely small concentrations (see Table 27–1).

Most of the anions and cations found in the plasma and interstitial fluid are also found in the urine, but not in the same concentrations. These include Na^+, K^+, Ca^{2+}, Mg^{2+}, Cl^-, PO_4^{3+}, SO_4^{+2}, and HCO_3^-. Sodium and chloride are the major ions excreted, although they are also reabsorbed in the greatest amounts. Their excretion, like that of other constituents of urine, is affected by dietary intake.

Abnormal Constituents

The significance of urinary constituents, particularly of protein and glucose, lies as much in their absence as in their presence. The presence in the

TABLE 27–1   MAJOR CONSTITUENTS OF NORMAL DAILY URINE*

Constituent	Amount
Volume	1200 (600–2500) ml
Specific gravity	1.003–1.030
pH	6.0 (4.7–8.0)
Total solids excreted	30–70 grams/liter of urine
Inorganic constituents	
chloride	10 g
sodium	4 g**
phosphorus	2.2 g
potassium	2 g**
sulfur	2 g
calcium	0.2 g
magnesium	0.15 g
Nitrogenous organic constituents	
urea	25–30 g**
creatinine	1.4 g
ammonia	0.7 g
uric acid	0.7 g
Other organic constituents	
ascorbic acid	15–50 mg
sugar	2–3 mg/100 ml***
hippuric acid	0.1–1.0 g

*Amounts excreted are for a normal 24-hour period.
**Amount excreted varies with diet.
***This amount of sugar appears in the urine of 50% of people after a heavy meal.
Source: Modified and reproduced, with permission, from Krupp, M. A., and others, *Physician's Handbook,* 19th ed. Copyright © 1979 by Lange Medical Publications, Los Altos, California.

urine of a substance (such as glucose) that is normally absent may be due to impaired renal function (such as reduced reabsorption) or to an excessive load created by impaired function of some other organ system (such as the pancreas). Protein in the urine (*proteinuria*) occurs if there is an alteration in the glomerular membrane that permits slightly larger molecules to pass. This protein is usually *albumin* since it is the smallest protein (mol. wt. = 70,000). When excessive *hemolysis* results in large amounts of hemoglobin (mol. wt. = 68,000) in the plasma, hemoglobin can be found in the urine as well. Proteinuria is not, however, al-

ways pathological, since it may occur temporarily after a bout of vigorous exercise or a period of prolonged standing.

Similarly, the presence of glucose in the urine (*glucosuria*) when the plasma glucose level is high may be entirely normal after ingestion of a large amount of carbohydrate. *Ketones* may be present in the urine when increased metabolism of fatty acids results in excessive formation of ketone bodies. And many other substances may also be present in readily detectable amounts when, for some reason, they are produced in excess or not properly metabolized. Among the latter are products excreted in high concentrations due to inborn errors of metabolism, usually the absence of a particular enzyme. A well-known example is *phenylketonuria,* a condition in which the absence of an enzyme needed to metabolize the amino acid phenylalanine forces an alternate metabolic path that leads to large amounts of a characteristic ketone in the urine.

MICTURITION

Transport of the urine from the renal pelvis to the urinary bladder is carried out not by gravity, but by peristaltic contractions of the smooth muscle in the ureter walls. The urinary bladder itself is relatively small when it is empty, but like the stomach, it can adjust its size to its contents with very little increase in pressure. Its wall contains smooth muscle fibers which are arranged roughly in three layers, with fibers of the middle layer running approximately at right angles to the others. The muscle is stimulated by parasympathetic fibers of the pelvic nerve. Sympathetic fibers (hypogastric nerve) supply the blood vessels, internal sphincter, and some of the musculature, but their role in bladder function is not very clear. Somatic fibers from the pudendal nerve innervate the skeletal muscle that forms the external sphincter. All three nerves, but especially the pelvic nerve, carry afferent fibers.

When enough urine has collected

in the bladder to distend it sufficiently, the stretch receptors in its wall are stimulated. Their impulses are carried to the sacral portion of the spinal cord where they make synaptic connections with the parasympathetic and somatic efferents (Figure 27–9). At the same time, branches of the afferent fibers carry impulses to the brain and produce the sensation of fullness of the bladder and the urge to void. The result is contraction of the bladder wall and relaxation of both the internal sphincter and the external skeletal muscle sphincter. Contraction of the bladder muscle closes off the entry of the ureters, thereby preventing backflow of urine.

The ascending afferent fibers also make connections with several centers in the brain that exert control over the excitability of the motoneurons, and hence the threshold of the micturition reflex. In the pons and posterior hypothalamus are facilitatory centers, and in the midbrain an inhibitory area. The cerebral cortex can exert both inhibitory and facilitatory influences, but it is primarily inhibitory, as it can override facilitatory input to a certain extent.

Micturition (urination) is normally a voluntary act initiated by cortical activity. It is usually the result of impulses arising from the full bladder, but it can be initiated voluntarily before the bladder is filled. It can also be inhibited for a time after the bladder is filled. After transection of the spinal cord the modifying influences from the brain are interrupted, and urination becomes strictly a spinal cord reflex, dependent upon stimulation of stretch receptors in the bladder wall or upon other sensory input at the spinal cord level. An individual whose spinal cord has been severed can sometimes gain a measure of control by making use of some of these reflexes. The pathways mediating brain control are not developed at birth, and the infant bladder operates as a spinal reflex.

CHAPTER 27 SUMMARY

The kidney is the major excretory organ, but the digestive system, skin, and lungs also contribute to excretory

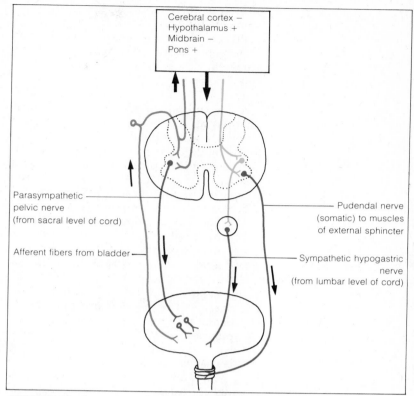

FIGURE 27–9   Neural control of micturition.

functions. The kidney removes wastes from the plasma and excretes them in the urine. It also maintains the plasma concentration of many specific substances by altering the amounts excreted.

Anatomy of the Urinary System

The urinary system consists of the **kidneys, ureters, urinary bladder,** and **urethra,** all of which are retroperitoneal. Urine formed in the kidney drains into the funnel-like renal pelvis then into the ureters to be carried to the urinary bladder. The ureters enter the bladder posteriorly and the urethra leaves from its inferior surface. The urinary bladder is a muscular organ, with the muscle arranged in three layers that run in different directions. Contraction of the bladder tends to block the ureteral openings preventing backflow of urine, the muscle around the entrance to the urethra functions as an internal sphincter, and skeletal muscles of the floor of the pelvis (through which the urethra passes) serve as a voluntary external sphincter.

Each kidney consists of an outer *cortex* and an inner *medulla*. The renal sinus, at the hilus, contains arteries, veins, and the renal pelvis and calyces. The medulla is arranged into pyramids, whose apices protrude into the renal sinus as papillae. A cuplike calyx fits over the papilla of each pyramid and receives the urine formed in the kidney.

The cortex and medulla contain the **nephrons** which produce the urine. A nephron consists of a *renal corpuscle* and a *tubule*. Renal corpuscles lie in the cortex and consist of a *glomerular capsule* enclosing a tuft of capillaries, the *glomerulus*. The inner portion of the capsular wall that covers the glomerular capillaries and the capillaries themselves have numerous openings, so that this is a very porous membrane.

The tubule begins as tortuously coiled *proximal convoluted tubules*, followed by a hairpin loop, the *loop of Henle* which dips into the medulla. The loop is made up of a straight *descending limb*, a *thin segment*, and a straight *ascending limb*. It continues as the *distal convoluted tubule* and empties into a *collecting tubule* which opens into the calyx at the papilla of the pyramid.

The renal arteries divide into interlobar, and then arcuate arteries, from which interlobar arteries penetrate the cortex. They give off *afferent arterioles* that enter the glomerular capsules to become glomerular capillaries, which emerge as *efferent arterioles*. Efferent arterioles break into the *peritubular capillary* network among the convoluted tubules, and eventually drain into the renal vein. Some peritubular capillaries form loops that dip down into the medulla as the *vasa rectae*.

Just before the afferent arteriole enters the capsule, its wall contains the specialized juxtaglomerular cells. The ascending limb of the tubule makes contact with this area, and its wall, too, is specialized at that site (macula densa). Together they form the *juxtaglomerular apparatus* (JGA).

Formation of Urine

Urine is formed by **filtration** in the glomerular capsule, and **reabsorption** and **secretion** in the tubule. The net filtration pressure is slightly above that in other capillaries. The permeability of the capillary wall is so great that fluid is rapidly filtered into the capsule and into the tubule. Nearly 20 percent of the plasma entering the glomerulus is filtered, and the remainder of the plasma (and all the blood cells) continues on in the efferent arteriole to the peritubular capillaries. The composition of the glomerular filtrate is the same as that of the plasma, except for proteins and substances bound to them. Thus, there is normally no colloid osmotic pressure in the filtrate, but that in the efferent arterioles is slightly elevated. The concentrations of other constituents are virtually equal in the afferent and efferent arterioles, and in the glomerular filtrate.

Of the 180+ liters of filtrate formed each day, only 1–1½ liters are excreted as urine; all the rest is reabsorbed from the tubules and returned to the peritubular capillaries. Much of the reabsorption is passive, notably that of water, but also urea and some ions. Active transport accounts for reabsorption of glucose, amino acids, most electrolytes, and other substances. Active reabsorption is limited by the *tubular maximum* (T_m). Glucose is normally totally reabsorbed in the proximal convoluted tubule, but when the blood glucose level is raised to a certain level (threshold), the T_m is exceeded and glucose begins to appear in the urine. Most of the reabsorption, both passive and active, occurs in the proximal convoluted tubule, including all of the glucose, and about 80 percent of the water and electrolytes. Reabsorption in the distal tubules accounts for smaller amounts, but it is here that most of the adjustment and regulation of reabsorption occurs.

Tubular secretion is the transfer of substances from the peritubular capillaries into the tubule. It includes both passive and active transport, but is quantitatively less significant since few normal constituents of plasma are processed in this way (with the exception of certain substances associated with acid–base regulation). Many nonphysiological substances (including some medications) are rapidly secreted.

Composition of Urine

In spite of reabsorption, urine normally contains small amounts of the anions and cations found in plasma, but no protein or glucose. With the exception of urea, metabolic wastes are poorly reabsorbed and, since most of the water is reabsorbed, they are highly concentrated in the urine. Urine is usually hypotonic and slightly acidic, although there is wide variation in volume, tonicity and pH, depending upon other regulatory processes.

Micturition

Urine is carried to the bladder by peristaltic contractions of the smooth muscle in the wall of the ureters. The bladder muscle readily distends to accommodate its increasing volume, until it reaches a volume sufficient to stimulate stretch receptors in its wall. Afferent fibers lead to a reflex center in the sacral portion of the spinal cord. Parasympathetic fibers from the center (via pelvic nerves) cause contraction and emptying of the bladder. Branches of the afferent neurons ascend to the brain, where they cause the sensation of a full bladder, and also establish connections with facilitatory and inhibitory centers in the brain. The cortex exerts voluntary control, which is mainly inhibitory, but it can initiate micturition by removing inhibition of facilitatory mechanisms. If the spinal cord is cut, interrupting pathways from the brain, bladder control is strictly a stretch reflex on the cord level.

STUDY QUESTIONS

1 What systems are involved in removing wastes from the body?

2 What are some of the characteristics of the renal circulation that are of particular importance in filtration?

3 What are some of the characteristics of the renal corpuscle that are of particular significance in filtration? What forces (gradients) cause filtration?

4 What do you know about the composition, osmotic pressure, etc., of the glomerular filtrate as compared with the fluid in the afferent arteriole? in the efferent arteriole? About what proportion of the plasma in the afferent arteriole leaves the glomerulus via the efferent arteriole? via the tubule?

5 How can so much water be reabsorbed when all fluid transport is passive?

6 What is tubular maximum (T_m)? What tubular processes have a T_m? What is the significance of T_m?

7 What are some of the important similarities and differences between tubular reabsorption and tubular secretion?

8 Glucose and protein are normally absent from the urine, but for entirely different reasons. Explain.

Reabsorption of Sodium and Potassium

Reabsorption of Water

Tonicity and Osmolarity of the Urine

Renal Regulation of Acid — Base Balance

The Regulatory Role of Renin and Angiotensin

Regulation of the Extracellular Fluid

Like Chapter 27, Chapter 28 examines the function of the kidney, but it is concerned not with the removal of waste products. Rather, it looks at the excretion of substances that are important constituents of the plasma and extracellular fluid. These are substances whose presence in the plasma is essential, and they must not be completely removed by the kidney. The fluid (water) content and the concentration of electrolytes and hydrogen ions in the plasma are important in themselves, but they also determine the pH and tonicity of the plasma. Regulation of the volume and concentration of the plasma, of course, is also regulation of the volume and concentration of the extracellular fluid. The mechanisms by which the kidney processes these substances and controls their excretion in order to maintain the volume and composition of the plasma and extracellular fluid are described in this chapter.

The kidneys are essential for homeostasis not only because they remove the wastes that would otherwise poison the cells, but also because they control the amounts of water and electrolytes excreted or retained in the body. In so doing they play an important role in regulating the body's water, electrolyte, and acid–base balances.

Although the kidney carries out its actions on the blood plasma, its homeostatic effects extend to fluids throughout the intracellular and extracellular compartments. The extracellular fluid compartment consists mainly of the blood plasma and the interstitial fluid, separated from one another by capillary walls that permit relatively free passage of most plasma constituents (except protein).

The extracellular fluid is of vital importance because it is the intermediary between the cells and the external environment. It is through this compartment that all the cells are supplied with nutrients and oxygen and through which all their wastes are removed. And it is the kidney that maintains the integrity of the extracellular compartment by regulating the volume and composition of the blood plasma.

The sizes of the individual fluid compartments are at best difficult to measure. It is usually done by injecting a known amount of some colored or radioactive indicator substance that is quickly and evenly distributed throughout the compartment in question. After the indicator has had time to be dispersed but not excreted, its concentration is determined. This tells how much it has been diluted, which in turn indicates the volume of fluid in which it is distributed. Different indicators, however, give different results because they do not become or remain equally distributed. In fact, some extracellular fluids, such as cerebrospinal fluid, the digestive juices, and the fluid in cartilage and bone, are virtually excluded from such measurements because they do not participate readily in intercompartmental exchanges. The volume of the intracellular compartment cannot be measured directly at all, but it can be estimated by subtracting the extracellular volume from the total body water. The composition of the plasma and interstitial fluid can be determined

TABLE 28–1   ION DISTRIBUTION IN THE FLUID COMPARTMENTS OF THE BODY*

| Ion | Extracellular Fluid | | Intracellular Fluid |
	Plasma	Interstitial	(Skeletal Muscle)
Sodium, Na^+	142	144	±10
Potassium, K^+	4	4	160
Calcium, Ca^{++}	5	2.5	—
Magnesium, Mg^{++}	3	1.5	35
Chloride, Cl^-	103	114	±2
Bicarbonate, HCO_3^-	27	30	±8
Phosphate, $PO_4^=$	2	2.0	140
Sulfate, $SO_4^=$	1	1.0	—
Protein	16	0.0**	55

*Data are given in milliequivalents per liter of fluid.
**Less than 1.0 percent in most sites, but may be higher in some regions, such as the liver.
Source: R. F. Pitts, *The Physiological Basis of Diuretic Therapy.* Copyright © 1959. Courtesy of Charles C Thomas, Publishers, Springfield, Illinois.

with reasonable accuracy, but that of the intracellular fluid can only be approximated. This is partly because the composition of the intracellular fluid is not the same in all cells.

Table 28–1 reminds us once more that sodium is the chief extracellular cation, while potassium is the chief intracellular cation. Chloride is the dominant extracellular anion, and phosphate and protein are the dominant anions within the cell. The chief difference between the plasma and the interstitial fluid is the amount of protein in the plasma.

The volume and composition of all fluid compartments ultimately depend upon the volume and composition of the plasma. The character of the plasma is, in turn, determined to a great extent by the activities of the kidney. The renal regulation of the reabsorption or excretion of water, sodium, and other electrolytes is particularly important in maintaining the plasma. The control of these processes is the subject of this chapter.

REABSORPTION OF SODIUM AND POTASSIUM

As the most abundant cation in the extracellular fluid, sodium accounts for most of the osmotic activity in both plasma and interstitial fluid. Control of its reabsorption is most important partly because of its abundance, but also partly because the return of sodium creates the electrical gradient that causes reabsorption of chloride. Sodium and chloride reabsorption together create the osmotic gradient needed for passive reabsorption of water.

The human kidney filters daily about 600 gm of sodium, of which 96 – 99+ percent is reabsorbed. The regulatory mechanisms can adjust the sodium excretion over a wide range (from less than 1 milliequivalent[1] to 400 meq per day) depending upon the dietary salt intake. Sodium is actively reabsorbed from the proximal and distal convoluted tubules and the collecting ducts, most of it from the proximal tubules. More sodium is reabsorbed from the proximal tubules when the glomerular filtration rate (and amount of sodium filtered) is high, and less is reabsorbed when the filtration rate is down, but otherwise the proximal tubules are relatively insensitive to factors that regulate sodium reabsorption. Sodium diffuses into the tubular cells and is actively pumped out the other side into the interstitial fluid and diffuses into the capillaries.

Sodium reabsorption in the distal and collecting tubules is variable and is subject to the action of the mineralocorticoids produced by the adrenal cortex, primarily **aldosterone.** Hormones control only a small portion of the total sodium reabsorption; with aldosterone present, sodium reabsorption might be 99.5 percent, and without it reabsorption might be 98 percent, but this small difference amounts to a severalfold increase in sodium loss. Such a loss is intolerable, and for this reason the mineralocorticoids are essential to life for humans.

As a steroid, aldosterone acts by the mobile receptor mechanism. It diffuses into the tubular cells to bind its receptor, resulting in the formation of RNA and protein synthesis. The nature of the protein produced is unknown, but it may act either to facilitate the entry of sodium into the cell or to increase the amount of ATP available to fuel the

sodium pump. Nevertheless, there is no doubt that aldosterone is a potent stimulant to sodium reabsorption.

An important factor in the control of aldosterone secretion is the *renin-angiotensin* mechanism (discussed below). Cells of the *macula densa* of the distal tubule (where it touches the afferent arteriole) are believed to be receptors that in some way monitor the amount of sodium transported (reabsorbed) in the distal tubules. When sodium transport is low, this information is relayed to the adjacent *juxtaglomerular cells* of the afferent arteriole, and they secrete the substance known as *renin* into the circulation. Renin initiates a series of reactions that lead to the production of aldosterone by the adrenal cortex. Thus we see that sodium plays an important role in regulating its own excretion. Aldosterone secretion is also increased by ACTH from the anterior pituitary gland. Recall that the mineralocorticoids are produced by the zona glomerulosa, the outermost layer of the adrenal cortex. It responds to ACTH stimulation but, unlike the other layers, is not dependent upon it and does not atrophy after hypophysectomy.

Although potassium content of the extracellular fluid is much less than that of sodium, one cannot really discuss sodium and aldosterone without touching on potassium. After being filtered at the glomerulus, most if not all of the potassium is actively reabsorbed in the proximal tubule. In the distal tubule, however, it is secreted into the tubular lumen in varying amounts and then excreted. All else being equal, the amount secreted is approximately equal to the amount ingested, thus maintaining a balance. The regulation of potassium excretion is brought about by changes in its secretion by the distal tubules, and there are several factors that affect it. Potassium secretion is associated with the hydrogen ion secretion that is part of the acid–base regulation (see below), since relatively more potassium ions are excreted with an alkaline load. More potassium is secreted when the potassium concentration of the tubular cells is high (which is more likely when there is an alkaline load). At least some of the potassium secretion is coupled to sodium reab-

---

[1] A milliequivalent of an ion is its molecular weight in milligrams divided by its valence or ionic charge.

sorption, for increased reabsorption of sodium is accompanied by an increased secretion of potassium (that is, decreased sodium excretion is accompanied by increased potassium excretion). The effect is due in part to aldosterone, which increases potassium secretion. Thus, aldosterone causes sodium retention and potassium loss. Prolonged treatment with mineralocorticoids can deplete potassium stores. Elevated plasma potassium levels, acting through the renin-angiotensin system, increase the production of aldosterone.

REABSORPTION OF WATER

Of the roughly 180 liters of fluid filtered from the plasma each day, more than 99 percent is reabsorbed, leaving only about 1.2 liters to be excreted. Moving so much water against a gradient by active transport would impose a tremendous work load on the tubular cells, particularly since water can penetrate most membranes with ease. Fortunately, all water reabsorption is passive in response to osmotic gradients.

Eighty percent or more of the water is reabsorbed in the proximal portion of the tubule. The main cause of this transfer is the osmotic gradient created by reabsorption from the filtrate of sodium and chloride ions and, to a lesser extent, of glucose and amino acids.

The remaining 20 percent of the tubular fluid (which still amounts to 30–35 liters per day) is regulated by several factors that modify the osmotic gradients and the permeability of the tubule membranes to water. The ascending limb of the loop of Henle is not permeable to water, so none is reabsorbed from this portion, even though there is a favorable gradient. The permeability of the walls of the distal and collecting tubules is also limited, but it can be varied by the action of a hormone.

This hormone is the **antidiuretic hormone (ADH),** released into the general circulation from the posterior pituitary gland. It acts specifically upon the cells of the distal and collecting tubules to increase their permeability to water. In the absence of ADH, reabsorption of water in this part of the nephron is

virtually nonexistent, and a large volume of fluid passes into the urine from the kidney; that is, a *diuresis* occurs. In the presence of ADH, the walls of the distal and collecting tubules become permeable enough that water can respond to the osmotic gradients that exist (see below), and the volume of urine is greatly reduced.

Since ADH conserves body fluid by increasing the reabsorption of water, it should not be surprising to find that the release of ADH is related to fluid needs. When the body is dehydrated, the solutes in the blood plasma are more concentrated due to the reduced amount of water. The greater concentration of solutes, most of which are osmotically active inorganic ions (Na^+, Cl^-, etc.), raises the osmotic pressure of the plasma and stimulates cells in a small area in the hypothalamus. When the blood is hypertonic, these **osmoreceptors** cause the posterior pituitary gland to release ADH into the circulation, resulting in increased reabsorption of water and a return of the osmotic pressure of the blood to normal. When the blood is hypotonic, ADH is not released and more water is excreted.

ADH is rapidly inactivated, so when production is inhibited, it quickly disappears from the blood. It is released continuously except when inhibited, and without it urine formation may be as much as 25 liters per day. Such a condition is known as **diabetes insipidus.** It is characterized by a large volume of insipid or dilute watery urine, as contrasted with *diabetes mellitus,* in which there is a large volume of sweet or sugary urine. As might be expected, both are accompanied by great thirst.

For a hormonal mechanism, the response of ADH is quite rapid. If a person drinks a liter of water, the urine volume soon increases due to the inhibition of further ADH release and the normally rapid inactivation of that already present. The diuresis reaches a peak in about an hour or so, and after two or three hours the excess fluid has been eliminated (Figure 28–1). Drinking a large volume of isotonic saline solution, however, does not have the same effect, since the osmotic pressure does not change and the osmoreceptors

FIGURE 28-1 The effect of drinking a large volume of water or isotonic saline on urine formation.

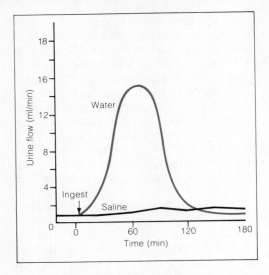

cannot detect it. It takes nearly a day to get rid of the extra isotonic fluid.

ADH is also increased as a result of a reduced blood volume without increased osmotic pressure (as, for example, after hemorrhage). There are stretch receptors in certain low pressure parts of the vascular system—the atria and pulmonary vessels—which are sensitive to changes in blood pressure in those regions. These pressures, which are always quite low, are influenced to a great extent by the fullness of the vascular system. ADH and water retention would help to alleviate a reduction in blood volume. An increased blood volume tends to inhibit ADH release and is probably the mechanism by which an isotonic fluid load is finally excreted.

TONICITY AND OSMOLARITY OF THE URINE

Another important aspect of the relation between the reabsorption of electrolytes and of water is that the metabolic wastes present in the tubules must be excreted in solution, even during severe dehydration. The urine excreted during dehydration is therefore not only of small volume, but also very concentrated. It may have a concentration of osmotically active particles several times that of the plasma, which means that the urine may be quite hypertonic. When there is excess water to be removed, the urine volume is high but very dilute, since approximately the same amount of solute is suspended in the larger volume, and the urine is hypotonic. Whenever the urine excreted is *not* isotonic with the plasma, an osmotic gradient exists across the tubular cells, even though the transport of water is always a passive process.

The kidney gets around this problem of controlling the *tonicity* of the urine by means of something called a **countercurrent mechanism.** It is based upon the effects that parallel channels have upon one another when their contents flow in opposite directions (*counter* to one another). This situation exists in the kidney where the descending limb of the loop of Henle dips into the medulla and the ascending limb returns very close to it (see Figure 28-2). The entire medullary pyramid is a closely packed mass of tubules, all reasonably parallel to one another. In addition, the capillaries in this region (the vasa recta) also form loops that dip deeply into the medulla and return.

Between the loops of tubules and capillaries is the interstitial fluid. It is the link between them. The osmotic and concentration gradients between the tubule and the interstitial fluid and between the interstitial fluid and the capillary govern passive transfer from one to the other. In the kidney cortex there is no osmotic gradient between either the capillary or the tubule and the interstitial fluid. But the osmotic pressure of the interstitial fluid increases progressively in the medulla, until near the papilla it is several times that in the cortex. As the loops of Henle and the vasa recta descend into the medulla, their immediate environment therefore becomes increasingly hypertonic.

To determine how this mechanism operates to control the tonicity of urine excreted, it is helpful to follow along the tubule and examine the changes that occur. The proximal tubule lies in the cortical region in an *isotonic* environment. The many substances that are reabsorbed in this region, including sodium and water, are quickly taken up by the peritubular capillaries. No osmotic gradients are developed and the fluid in the tubule remains isotonic with both the interstitial fluid and the plasma. The nephrons that lie close to

Focus

Diuretics

The kidneys normally maintain the body fluids at constant tonicity. When the blood and interstitial fluid become more dilute, as happens after a drink of water, the pituitary posterior lobe releases less antidiuretic hormone (ADH), and the kidneys reabsorb less water from the tubules. More water is allowed to leave the kidney in a more voluminous urine that contains relatively few solutes. The body fluids are restored to their normal tonicity.

When the body fluids become concentrated, as after a salty meal, hypothalamic osmoreceptors cause more ADH to be secreted, and the kidneys increase their reabsorption of water. The urine is less voluminous and contains more solutes. Again, the body fluids are restored to normal.

ADH is only one of the mechanisms that control kidney reabsorption of water, but it works primarily when body fluid tonicity is disturbed (otherwise, the osmoreceptors cannot "know" they need to change their secretion of ADH). It does not work when, for instance, fluid is accumulating in one part of the body, as it does with the edema caused by such factors as hypertension, inflammation, and some infections. It is in such cases, as well as in other failures of kidney control, that drugs—diuretics—must be used to rid the body of excess water.

Diuretics are substances that in one way or another cause the kidneys to decrease their reabsorption of water from the tubules. Some of them may act by interfering with the release or effect of ADH. Others are substances such as mannitol, sorbitol, or urea that increase the osmotic activity of the tubular fluid, since they are only partially, if at all, removed by the tubule cells. They decrease the amount of water that is reabsorbed, and hence the amount of water lost to the urine. The diuretic substances caffeine and theophylline, found in coffee and tea, respectively, appear to depress the reabsorption of water in the distal convoluted tubule. They inhibit an enzyme that is crucial to the action of cyclic AMP on cellular functions. Since ADH exerts its effect via cyclic AMP, caffeine and theophylline may work by stopping the action of ADH.

Most people run into diuretics both as medical treatments and in their daily lives. The caffeine in a cup or two of coffee is enough to have a diuretic effect, and caffeine and theophylline are members of a class of compounds, the xanthines, that supply many medical diuretics.

One might expect alcohol to be a diuretic, since after a beer or two urine volume is typically high, but the diuresis in this case is due more to the volume of liquid in the beer. More concentrated alcohol does not have the same effect. The increase in urine formation represents the body's attempt to keep the tonicity of the body fluids constant. However, smokers do not seem obliged to empty their bladders so often when drinking. The nicotine in tobacco may stimulate reabsorption in the kidney tubules.

FIGURE 28–2
Diagram summarizing active and passive transport of water and ions in the operation of the countercurrent system. Numbers indicate the concentration of the fluid at that site (in milliosmals/liter). Boxed numbers are the approximate percentage of the glomerular filtrate in the tubule at that level.

(Adapted from R. F. Pitts, *Physiology of the Kidney and Body Fluids,* 3rd ed. Copyright © 1974 by Year Book Medical Publishers, Inc., Chicago. Used by permission.)

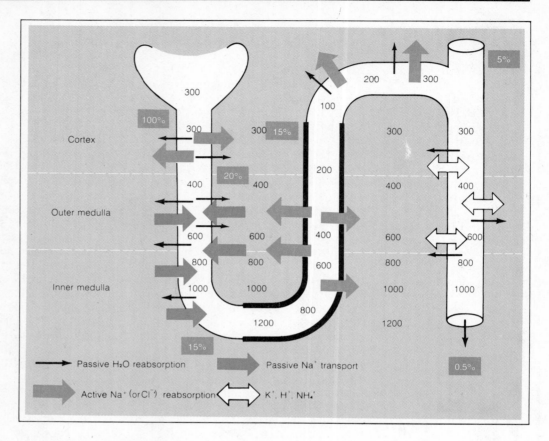

the medulla all have a long loop of Henle, and most of the descending limb is a thin segment. This is very favorable for passive transport. As the descending limb enters the increasingly hypertonic environment of the medulla, sodium and chloride ions actually diffuse into the tubules, and some water leaves them. As a result, the fluid in the tubule at the bottom of the loop is virtually in equilibrium with its immediate environment, though it is considerably hypertonic to the plasma in the cortex and the general circulation.

The wall of the ascending limb of the loop is impermeable to water, so no water can be reabsorbed from this segment. There is active transport along the ascending limb, but apparently it is chloride that is actively pumped out of the lumen and it is sodium that follows. Some of the ions that are removed from the ascending limb diffuse into the nearby descending limb, only to be reabsorbed once more as they travel up the ascending limb. Some sodium and chloride ions cycle many times between the ascending and descending limbs. This is the major factor in developing

and maintaining the hypertonicity deep in the medulla. Some ions do diffuse into capillaries of the vasa recta to be carried back to the general circulation, but if all of them were carried away, there would be no osmotic gradient in the medulla. Reabsorption of sodium and chloride from the ascending limb without reabsorption of water dilutes the fluid in the tubule. By the time this fluid has reached the distal convoluted tubule, it has actually become *hypotonic* to the cortical interstitial fluid and plasma.

In the convoluted portion of the distal tubule, more sodium is actively reabsorbed, and usually some water is reabsorbed passively, so that the tubular fluid becomes isotonic with its environment and the plasma once more. It has, however, lost much of its sodium and some other solutes, along with a corresponding amount of water; its volume is much reduced and its composition is very different from that of the glomerular filtrate that entered the proximal tubule.

The contents of the collecting tubule may become progressively more concentrated as it descends through the hypertonic medulla once more on its way to the papilla, since the permeability of its walls and those of the distal convoluted tubule depend upon the presence of antidiuretic hormone. When ADH is present, the walls of the collecting tubule may permit enough water to be reabsorbed so that the tubular fluid stays in osmotic equilibrium with the nearby interstitial fluid as it descends through the medulla. The urine released into the renal pelvis is then considerably hypertonic to the plasma in the general circulation. In the absence of ADH, the distal tubule and collecting tubules permit very little reabsorption of water, and the urine excreted may be almost as hypotonic as the fluid at the beginning of the distal convoluted tubule.

The water, sodium, and chloride removed from the distal and collecting tubules diffuse into the adjacent capillary loops, the vasa recta, in which a countercurrent exchanger also operates. As the capillary descends into the medulla, electrolytes diffuse in and water out, so that the plasma at the bottom of the capillary loop is hypertonic to the plasma at the surface. The exchange is reversed as the capillary returns to the surface. But since the blood flow is faster than the exchange across the capillary's wall there is a slight lag, and total equilibrium is never quite reached; the blood leaves the medulla with a little more sodium and chloride than it had originally. The sodium and chloride are returned to the general circulation along with the materials reabsorbed from the proximal tubules.

With this mechanism the kidney is able to vary the tonicity of the urine without requiring a large osmotic gradient across any part of the tubule and with a minimal amount of active transport. Additional transport in the distal and collecting tubules is associated with acid—base balance.

RENAL REGULATION OF ACID—BASE BALANCE

One of the most important aspects of the cellular environment is the pH of the extracellular fluid. To maintain it, the pH of the plasma must be kept close to 7.4; any deviation from this figure would soon have serious effects upon chemical reactions in the cells. This aspect of homeostasis is a continuous necessity, since metabolic wastes, including carbon dioxide, are largely acid, and many foods leave an acid or alkaline residue. It requires mechanisms both for transporting the mostly-acid waste products and for excreting just the right amounts of hydrogen ion. The first mechanism is the buffer system of the blood, which enables acid or alkaline substances to be transported without greatly changing the blood pH. Buffers are capable of combining reversibly with highly ionized substances (which would change the hydrogen ion concentration too much) to form weakly ionized compounds (which change it little). Upon reaching an excretory organ, the buffer reactions are essentially reversed to release the acid or alkaline substances for removal.

The two major excretory organs are the lungs, which excrete most of the carbon dioxide, and the kidneys, which excrete most of the other wastes. Although some waste products are excreted by the skin (in sweat) and the digestive tract, the amounts are relatively small and do not make an important regulatory contribution. The action of the blood buffers has been discussed in Chapter 14, and the role of the lungs and carbon dioxide in maintaining blood pH has been discussed in Chapter 20.

The mechanisms by which the kidney maintains acid—base balance involve many of the same mechanisms with which it regulates fluid and electrolyte balance. Acid—base regulation involves the processing of tubular fluid, plus secretion by tubular cells. Acid (or alkaline) substances that are to be excreted are brought to the kidney bound to buffers in plasma and are filtered into the tubules still buffer-bound. Excreting an acid load involves excreting H^+, and removal of an alkaline load requires the excretion of fewer H^+. The buffers that transport these substances must also be retrieved. Particularly important are the bicarbonate and sodium ions of the carbonic acid—bicarbonate buffers in the plasma, since they are present in such large quantities. Managing the buffers becomes almost

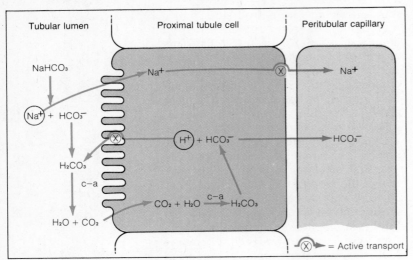

FIGURE 28–3   Secretion of hydrogen ions and retention of bicarbonate ions in the proximal tubule.

as important as excreting the acid (or alkaline) loads. The lungs play an important role in this for, by excreting more or less carbon dioxide, they control the carbonic acid member of the buffer pair. Recall that P_{CO_2} is an important regulator of pulmonary ventilation, and hence of carbon dioxide excretion. The kidney maintains the bicarbonate member of the pair by reabsorption both of bicarbonate and sodium ions. It also, of course, secretes and excretes hydrogen ions.

The basic mechanism involved is an ion exchange in which hydrogen ions are secreted into the tubular lumen in exchange for sodium ions which are reabsorbed. The hydrogen ions are derived from carbonic acid, which is formed from carbon dioxide and water in the tubular cells. (They contain carbonic anhydrase to catalyze the reaction.) Secretion of H^+ from H_2CO_3 leaves the bicarbonate ion, HCO_3^-, to be returned to the circulation. This basic mechanism is modified in different parts of the tubule to meet different requirements or loads.

In the proximal tubules large amounts of bicarbonate ions are present in the tubular fluid, and bicarbonate reabsorption is the major activity (Figure 28–3). Sodium diffuses into the tubular cell and is actively pumped out to the interstitial fluid. Hydrogen ions secreted into the tubular lumen displace the sodium, forming carbonic

acid. The cells of the proximal tubule have numerous microvilli, and carbonic anhydrase is found on the luminal surface of the microvilli membranes. It splits the carbonic acid, yielding carbon dioxide and water in the tubular lumen. The carbon dioxide diffuses back into the tubular cell and participates in the formation of more carbonic acid there. The water may do the same or be excreted in the urine.

As is true of other proximal tubule functions, this is a bulk transfer process. About 90 percent of the bicarbonate buffer is reabsorbed, and much hydrogen is secreted. The urine remains isotonic, and there is little change in pH. Other parts of the tubule remove most of the remaining bicarbonate, chiefly by the processes described below. These "fine adjustments" are carried out primarily by cells of the distal and collecting tubules.

Small amounts of strong acids are produced by metabolism of proteins, phospholipids, and other compounds. These acids are buffered and are carried in the plasma as sodium salts. The sodium–hydrogen ion exchange still operates, and a more acid salt is excreted in the urine (Figure 28–4). Sodium and bicarbonate are retained, although some sodium is lost. This process involves an increase in the hydrogen ion concentration of the urine, and thus a fall in pH. It is the excretion of these strong acids that reduces the pH of the urine. In fact, this mechanism is sometimes referred to as the "acidification of the urine."

Hydrogen ions cannot be excreted as free ions; there must be something to bind them. If the tubular fluid contains little buffer, secreted H^+ may combine with Cl^-, and since HCl ionizes so completely, the pH of the urine would fall drastically. This cannot happen, however, because the tubular cells are unable to secrete H^+ against a steep gradient (as would occur with a high H^+ concentration). This is actually what sets the lower limit of the pH of the urine. If this were to occur, many of the hydrogen ions simply could not be secreted.

There is a mechanism to cope with situations in which there is not enough buffer in the tubular fluid (Figure 28–5). Cells of the distal and collecting tubules are able to synthesize ammonia (NH_3), chiefly from glutamine (derived from glutamic acid, an amino acid). Ammonia is not a very desirable substance, but when secreted into the tubular lumen it combines with hydrogen ions there, forming the ammonium ion (NH_4^+). This effectively binds hydrogen ions, and enables the tubular cells to secrete more hydrogen ions without lowering the pH of the urine. The ammonium ion combines with an anion in the tubule, such as chloride, and is excreted. The amount of ammonia produced varies with the acid load and is likely to be greatest when there is need for excretion of increased amounts of hydrogen.

On occasion, the need is for excretion of an alkaline load. In this case retention of the bicarbonate and sodium are still important, but secretion of hydrogen ions is not. In fact, to prevent an increase in pH, hydrogen ions should be retained. In this situation, cells of the distal and collecting tubules secrete potassium ions (instead of hydrogen ions) in exchange for sodium. This reduces the hydrogen ion concentration in the urine and raises the pH.

THE REGULATORY ROLE OF RENIN AND ANGIOTENSIN

We have seen how the kidney helps to regulate important aspects of the extracellular fluid, and in discussing these roles we have touched upon some of the ways in which the kidney itself is regulated. This involves regulation of the glomerular filtration rate and of tubular function. An important element in glomerular function is the arterial pressure because of its effect on the glomerular filtration pressure. There are different regulatory mechanisms for almost every one of the tubular processes. One mechanism, however—the *renin-angiotensin system*—touches upon control of both glomerular and tubular functions. It responds to decreases in arterial pressure and in sodium reabsorption.

Renin itself does not affect either process. Renin is an enzyme produced by the juxtaglomerular cells in and around the wall of the afferent arteriole.

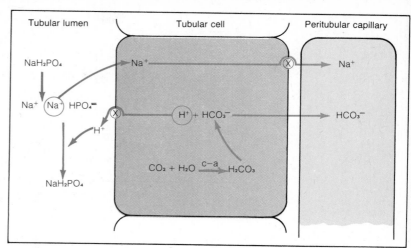

FIGURE 28-4 Secretion of hydrogen ions as acidic salts, which lowers the pH of the urine.

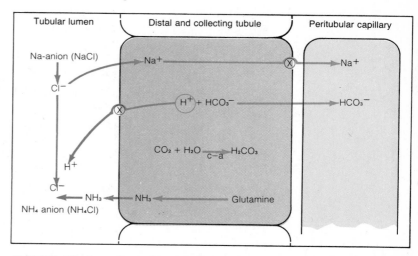

FIGURE 28-5 Secretion of ammonia, which makes possible the secretion of more hydrogen ions.

It is released into the circulation, where it splits a protein (a globulin produced by the liver known as *angiotensinogen*) to **angiotensin I** (Figure 28-6). An enzyme in the plasma converts angiotensin I to **angiotensin II,** which has two important functions. It is a very potent and generalized vasoconstrictor, but it is effective for only one or two minutes before it is inactivated (by an enzyme in the blood). Angiotensin II also stimulates the adrenal cortex to secrete aldosterone.

There are a number of stimuli that will cause the release of renin and the

FIGURE 28–6
The renin-angiotensin
system.

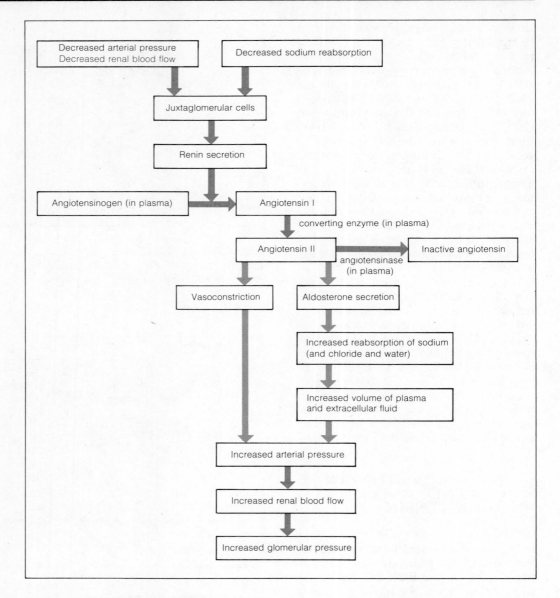

formation of angiotensin. One of these is a decrease in arterial blood pressure. A general fall in arterial pressure is also a fall in pressure in the afferent arterioles. The juxtaglomerular cells of the arteriole function as stretch receptors and detect the fall in pressure, and renin is produced. The role of reduced reabsorption of sodium across the macula densa of the distal tubule has already been mentioned as a mechanism to trigger renin release and aldosterone secretion. Thus, we have a system by which a fall in arterial pressure (which reduces filtration pressure) initiates a sequence of events leading to vasoconstriction and a general rise in arterial pressure which prevents any fall in glomerular filtration pressure and ensures continued formation of urine by the kidney. Low sodium in the plasma, or a reduction in the amount of sodium delivered to the tubule, reduces the amount of sodium being reabsorbed, and this, too, triggers the release of renin. The resulting secretion of aldosterone, by increasing sodium reabsorption, increases water reabsorption (due to the osmotic effect of the sodium). Together they help raise the arterial pressure and increase the extracellular fluid volume. The renin-angiotensin II system thus contributes to maintaining arterial pressure, to maintaining an adequate filtration pressure, and to overall fluid and electrolyte balance.

CHAPTER 28 SUMMARY

The extracellular fluid (ECF) consists of plasma and interstitial fluid, separated by the relatively permeable capillary walls. The kidney, by regulating the excretion of individual constituents of plasma, controls the volume and composition of the plasma, and hence also of the ECF.

Reabsorption of Sodium and Potassium

A large quantity of sodium is filtered each day, and nearly all of it is reabsorbed, mostly by active transport from the proximal convoluted tubules. The smaller amount reabsorbed from the distal convoluted and collecting tubules is controlled by *aldosterone* secreted by the adrenal cortex. It increases sodium reabsorption in the distal but not the proximal tubules. Although aldosterone controls only a fraction of sodium reabsorption, we cannot tolerate the extra sodium loss if it is absent. Aldosterone secretion is increased by the *renin-angiotensin* mechanism (activated by low sodium transport at the macula densa and reduced renal arterial blood flow and pressure), and by ACTH from the anterior pituitary.

Small quantities of potassium are filtered since its concentration in the ECF is normally low. Potassium is filtered at the glomerulus, reabsorbed in the proximal convoluted tubule, and may be secreted in the distal tubule. Aldosterone increases potassium secretion and excretion.

Reabsorption of Water

More than 99 percent of the filtered water is reabsorbed, most of it from the proximal convoluted tubules. It is transported passively, due to gradients created by active reabsorption of solutes. Cells of the distal and collecting tubules are relatively impermeable to water, except in the presence of *antidiuretic hormone (ADH)* from the posterior pituitary. ADH increases tubule permeability and permits water to respond to osmotic gradients. Osmoreceptors in the hypothalamus respond to increased plasma osmotic pressure with the release of ADH, which increases the absorption of water. Decreasing the plasma osmotic pressure by drinking a large volume of water inhibits ADH production and leads to an increased output of dilute urine (diuresis). Drinking isotonic saline stimulates volume receptors in the central veins, which are low-pressure receptors. When central venous pressure rises, they inhibit ADH secretion, and when the pressure falls, ADH secretion is increased.

Tonicity and Osmolarity

To excrete waste products and still retain water, the kidney must excrete a concentrated (hypertonic) urine. It does so by a *countercurrent mechanism*, in which materials can be transferred from one limb of the loop of Henle to the other, because of the hairpin shape of the loop.

Interstitial fluid in the cortex is isotonic with the plasma, but it becomes more hypertonic deeper in the medulla. Tubular fluid in the descending limb of the loop becomes progressively hypertonic in the medulla; at the bottom of the loop its tonicity is several times that in the cortex. The wall of the ascending limb is not permeable to water, but chloride is pumped out of the tubule and sodium follows, so that fluid reaching the distal convoluted tubule in the cortex is quite hypotonic. The ions removed from the ascending limb diffuse into the descending limb, and are largely responsible for the hypertonicity of the medulla.

If ADH is present, water is reabsorbed from the distal portions of the tubule. As the fluid descends through the increasing hypertonicity of the medulla, more water is reabsorbed and a smaller volume of hypertonic urine is excreted. Without ADH, the distal and collecting tubules are not permeable to water, and a large volume of dilute urine is excreted.

Blood flow through the vasa rectae also becomes hypertonic as it descends through the medulla, and blood leaving the vasa rectae is still somewhat hypertonic to the plasma of the general circulation. Changes in the blood flow through the vasa rectae alters the removal of ions from the interstitial fluid in the medulla, and thus controls its tonicity. If it were not hypertonic, there would be no gradient for reabsorption of water from the collecting tubule, and no way to excrete a hypertonic urine; if there were no active reabsorption of ions without water (in the ascending limb) there would be no way to excrete excess fluid in a hypotonic urine.

Renal Regulation of Acid–Base Balance

Maintenance of the plasma (and ECF) pH requires the removal of hydrogen ions, while the blood's hydrogen buffer, $H_2CO_3/NaHCO_3$ is retained. The lung regulates the excretion of carbon dioxide and thus controls the carbonic acid. The kidney retrieves the bicarbonate and secretes hydrogen ions. The basic process is an ion exchange mechanism: Hydrogen ions are secreted into the tubule in exchange for sodium ions, which are reabsorbed from the tubule. Sodium bicarbonate is filtered at the glomerulus and the sodium is actively reabsorbed, mostly in the proximal convoluted tubule. Carbonic acid is formed in the tubular cells and it dissociates to provide the hydrogen ions that are secreted into the tubule. The remaining bicarbonate ions enter the peritubular capillaries.

Stronger acids transported, for example, by phosphate buffers, use a similar ion exchange, but the acid salt formed yields some hydrogen ions and the pH of the urine falls. As it falls, the $[H^+]$ rises, which limits hydrogen ion secretion, because tubular cells are unable to secrete against a high concentration gradient. In this case, tubular cells are able to secrete *ammonia* (NH_3) into the tubules. By combining with the hydrogen ions to form ammonium ions (NH_4^+), it permits secretion of more hydrogen ions. When needs are for excretion of an alkaline load, potassium ions are secreted by the tubular cells rather than hydrogen ions.

Renin-Angiotensin

Renin is secreted by the juxtaglomerular cells of the afferent arterioles when they detect a decrease in the blood pressure and flow in the arteri-ole, or when the macula densa detects a reduced sodium transport in the distal tubule. Renin is released into the bloodstream and activates *angiotensinogen* (present in the plasma), to angiotensin I, which is converted to *angiotensin II*. It is a powerful vasoconstrictor, and helps to raise the arterial pressure and renal blood flow (and filtration pressure). It also causes secretion of aldosterone which increases sodium reabsorption.

STUDY QUESTIONS

1 What fluid is included in the intracellular fluid? the extracellular fluid? the interstitial fluid? the plasma? What is the relative volume of each of these, and how do they differ from one another in composition?

2 Contrast the mechanism for tubular reabsorption of water with that for reabsorption of sodium.

3 What is a countercurrent mechanism? How does it apply to the nephron? What is the advantage of such a mechanism?

4 Why is it important that the kidney be able to increase or decrease the tonicity of the urine excreted?

5 What is the mechanism by which the kidney responds to a decrease in the osmotic pressure (dilution) of the blood? To an increased blood volume?

6 The secretion of hydrogen ions by the tubular cells is an important aspect of renal regulation of acid–base balance. Where do these hydrogen ions come from?

7 How is the secretion of hydrogen ions related to sodium and bicarbonate reabsorption? Why is it important to reabsorb the sodium and bicarbonate?

8 What is the significance of the secretion of ammonia by the tubular cells?

9 Under what conditions is renin released? By what mechanism do they affect blood pressure and aldosterone secretion?

Part 8

The Reproductive Systems

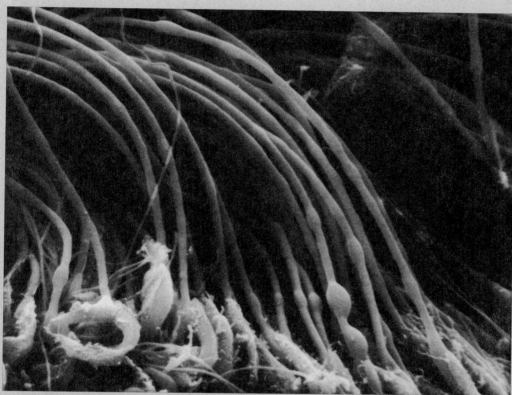

From *TISSUES AND ORGANS—a text-atlas of scanning electron microscopy*, by Richard G. Kessel and Randy H. Kardon. W. H. Freeman and Company, copyright © 1979.

Developing spermatozoa in the seminiferous tubule.

The role of the reproductive system is unique among the organ systems. Whereas the other systems sustain the individual, the reproductive system continues the species. In all but the most primitive species, this requires the participation of two individuals, a male and a female. The major differences between the two sexes lie in the organs of their respective reproductive systems and in the hormones they produce. It is appropriate, therefore, that the two chapters in Part 8 should be devoted to the reproductive systems in the male (Chapter 29) and in the female (Chapter 30)—their structure, function, and hormonal relations.

Reproductive function is concerned with the conception and development of a new individual, one who is unique in his or her genetic makeup and its manifestations. Conception occurs upon the union of a sperm from the male and an ovum from the female to form a single cell that contains chromosomes from each parent. A brief discussion of how male and female germ cells are formed, and how the new individual receives the proper number of chromosomes, half from each parent, is included in Chapter 29.

The single cell must undergo many cell divisions before it develops organ systems that can sustain it independently. That period of growth and development occurs within the mother's uterus. During that time the mother's circulatory system meets all the needs of the developing organism. Chapter 30 therefore includes some description of the mechanisms that contribute to support of the developing new individual.

29

Origin of Gametes and the Individual

The Male Reproductive Organs

Endocrine Function of the Testes

Reproduction and the Male Reproductive System

The structures and activities of the male reproductive system are geared to the production and care of spermatozoa, the male germ cells. These organs store, nourish, and transport the tiny cells, which are produced in prodigious numbers. The final role of the male reproductive system is the deposition of sperm in the female reproductive tract.

The reproductive system is under hormonal control. Hormones of the anterior pituitary regulate the production of sperm and the secretion of male sex hormones. The male sex hormones assist with sperm production, but mainly they control development and maintenance of the ducts and glands of the reproductive system, as well as sex-related characteristics.

This chapter is mainly a description of the organs of the male reproductive system, how they produce and care for sperm, and the regulation of these processes. There is also a brief description of the cell division involved in formation of germ cells. These cells are different from other cells of the body, and they are formed by a different type of cell division.

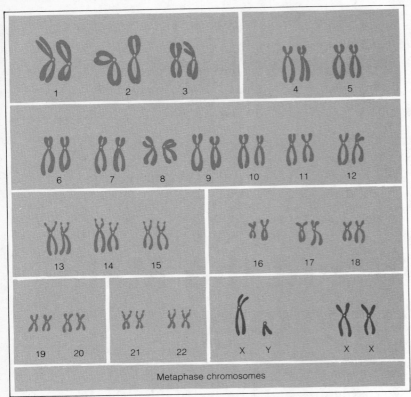

Metaphase chromosomes

FIGURE 29–1   Array of chromosomes grouped according to their sizes and shapes.

The major differences between male and female lie in the organs of their reproductive systems and in the hormones they produce. These organs, though not at all alike in the two sexes, serve comparable roles; it is convenient to divide them into three categories on the basis of their respective roles in the reproductive process:

1 The **primary** or **essential sex organs** are the gonads, the **testes** in the male and the **ovaries** in the female. The gonads serve two vital functions. They produce the germ cells or **gametes—spermatozoa** in the male and **ova** in the female. As endocrine glands, the gonads produce the sex hormones—*androgens* in the male and *estrogens* in the female. The ovary also produces at least one other important hormone, *progesterone*.

2 The **secondary** or **accessory sex organs** are directly involved in the care of the gametes while they are in the body. They protect and nourish these delicate cells and transport them from the body.

3 The **secondary sex characteristics,** those features that mark one sex from the other, play no direct role in the reproductive process or in the care of the germ cell. They include such characteristics as general body physique and muscular development, the distribution of body hair, and the pitch of the voice. Their development and maintenance depend upon the presence of the sex hormones.

ORIGIN OF GAMETES AND THE INDIVIDUAL

Meiosis and Development of the Gametes

Recall that the genetic information for all cells is contained in the pattern of bases of the DNA that makes up the chromosomes in the nucleus of every cell. Most human cells contain a total of 46 chromosomes, including two sex chromosomes. Cells of females have two of the rather large X chromosomes, while those of the male have one X and one of the smaller Y chromosomes, in addition to the other 44 chromosomes. Each gamete, however, contains only 23 chromosomes, including one sex chromosome. All ova contain an X chromosome, while half the sperm contain an X and half contain a Y.

The sex chromosomes are not the only ones that are distinctive. The various chromosomes are studied in cells growing in tissue culture. A mitotic cell division can be stopped in metaphase when the chromosomes have separated and, by applying appropriate techniques, the individual chromosomes can be identified and classified. There are twenty-two pairs, plus X and Y. (see Figure 29–1). The pairs differ slightly in size, shape, and in the banding patterns that appear following use of certain stains. Study of such preparations has made it possible to relate certain genetic problems to abnormalities of a particular chromosome.

The means by which we are able to produce cells that have only half the usual number of chromosomes needs to be considered. In Chapter 2 it was stated that virtually all cells divide by mitosis, a process which begins with the formation of an additional set of chromosomes. The two sets of chromosomes align themselves in such a way that when the cell divides they are equally distributed between the two daughter cells formed. In this way, cells of the new generation are genetically identical to each other and to the previous generation. In humans, the only exceptions to this type of cell division occur during the development of sperm and ova. They are formed by **meiosis,** a type of cell division that produces cells with half the usual number of chromosomes. It involves two cell divisions, but only one duplication of chromosomes.

Meiosis is carried out in the testes and ovaries during the processes of spermatogenesis and oogenesis (Figure 29–2). **Spermatogenesis,** or sperm formation, begins with a cell known as a **spermatogonium** (44 + XY). It undergoes a typical mitotic cell division forming two **primary spermatocytes** (44 + XY). The primary spermatocytes begin the meiotic division by first duplicating their chromosomes, forming 46 sets of doubled chromosomes, much as in mitosis. But when the cell actually divides, the original and duplicate member of each chromosome stay together, and half of the chromosomes (originals and duplicates) go on to one new cell, and the other half goes to the other cell. These daughter cells are **secondary spermatocytes,** and they contain a double set of half the chromosomes; thus half of them contain an X chromosome and half contain a Y chromosome. Secondary spermatocytes quickly divide, without additional chromosome formation. The doubled chromosomes separate in this division, forming **spermatids,** each of which contains a single set of 22 + X or 22 + Y chromosomes. The spermatids undergo maturation into spermatozoa without further cell divisions. Thus, meiosis in the male results in the production of four spermatozoa from each primary spermatocyte, half of them containing an X and half a Y chromosome.

Development of an ovum, **oo-**

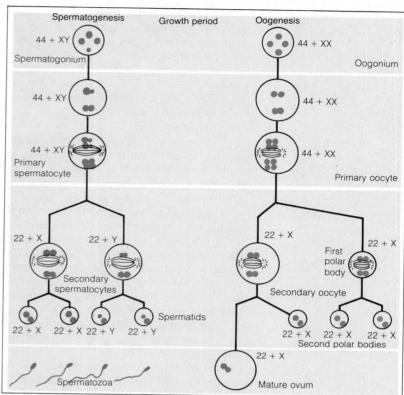

FIGURE 29–2 Meiosis and maturation of gametes.
A. Spermatogenesis. B. Oogenesis.

genesis, is similar in many respects. **Oogonia** in the ovary give rise to **primary oocytes** by mitotic division. Like primary spermatocytes, the primary oocytes duplicate their chromosomes and then undergo the first meiotic division, forming two new cells, each having a double set of half the chromosomes (22 + X). There is one important difference, however, in that the cytoplasm is not distributed equally when the cell actually divides. One cell gets most of it, and becomes a **secondary oocyte.** The other daughter cell gets its set of chromosomes, but very little cytoplasm, and is known as the **first polar body.** The second meiotic division involves distributing the duplicated chromosomes equally, and the cytoplasm unequally. The result is a single large cell, the **ootid,** and a small **second polar body,** both with 22 + X chromosomes. A mature ovum (22 + X) develops without further cell division from the ootid. In oogenesis just one ovum results from meiotic division of a primary oocyte, and the time it takes is much longer than in spermatogenesis (see Chapter 30). The polar bodies are nonfunctional and soon disappear.

Sex Determination and Early Differentiation

The conception of a new individual occurs at fertilization, when a single male sperm penetrates a single female ovum. The gender of the individual is determined at that moment. When a sperm enters an ovum the result is a cell with a full chromosome complement, 22 + X from the ovum and 22 + X or Y from the sperm. If the ovum was fertilized by a sperm containing an X chromosome, the offspring will be female (XX); if it was fertilized by a Y-containing sperm, the offspring will be male (XY).

Most cells of individuals who have two X chromosomes (genetic females) show an extra bit of chromatin material *(sex chromatin)* in their cell nuclei. This is apparently caused by the presence of the second X chromosome, which tends to become condensed against the nuclear membrane. This material can be made visible rather easily in many epithelial cells such as those lining the cheek, where it is called a Barr body. In neutrophils it appears sometimes as an appendage to the nucleus, called a drumstick. By this means it is possible to determine "chromosomal sex," and it is also useful in studying certain chromosomal abnormalities.

Although sex is determined at the time of fertilization, male and female are at first distinguishable only by the presence or absence of sex chromatin. This early embryonic stage is the **indifferent stage,** and it lasts six to seven weeks in humans. During this time a primitive or **indifferent gonad** develops near the *mesonephros,* a part of the embryonic urinary system (Figure 29–3). Connected to it are two pairs of ducts, the *mesonephric ducts* and the *Müllerian ducts.* If the indifferent gonad is to become a testis, the inner portion (medulla) then begins to develop and the outer portion (cortex) regresses. If it is to become an ovary, the cortex develops and the medulla regresses. In the male, the duct system of the mesonephros remains and develops into the ducts for transport of the sperm (epididymis, ductus deferens). In the female, the Müllerian ducts develop to become the organs and tubes that transport and care for the ovum (uterine tubes, uterus). In both cases, the undeveloped set of embryonic ducts degenerates. Thus the indifferent embryo has all the components needed to develop either male or female genitalia.

The development of gonads into testes sets the stage for the differentiation of the other male organs. The embryonic testis produces two substances, a regression factor and the male hormone testosterone. The regression factor and testosterone act together to cause development of the male duct system (the internal genitalia). Testosterone acting alone causes the development of male external genitalia. If the fetal testes fails to produce these substances at this time, female organs will develop.

The embryonic ovary of a genetic female does not produce hormones or other substances at this time. Female organs will develop unless the substances produced by the testes are present at the right time and in the right amounts.

The events described so far are complex and delicate, and they are very critical. Even a small failure can cause serious, perhaps lethal, abnormalities. Difficulties may arise from genetic (chromosomal) or hormonal abnormalities, or from the effects of other chemicals if they are present at critical times.

Improper separation and distribution of sex chromosomes during meiosis may lead to sperm or ova that have no sex chromosomes, or two (or even more if the problem occurs in both meiotic divisions). If such gametes are involved in fertilization, the cells of the individual conceived might have such chromosome patterns as XXY, XXX, or XO. Most such combinations produce abnormal development of certain reproductive organs, and in some combinations, possibly other symptoms as well.

Faulty distribution of chromosomes is not limited to the sex chromosomes; it may happen with any chromosome: in recent years a number of conditions have been traced to such an "error." Down's syndrome (mongol-

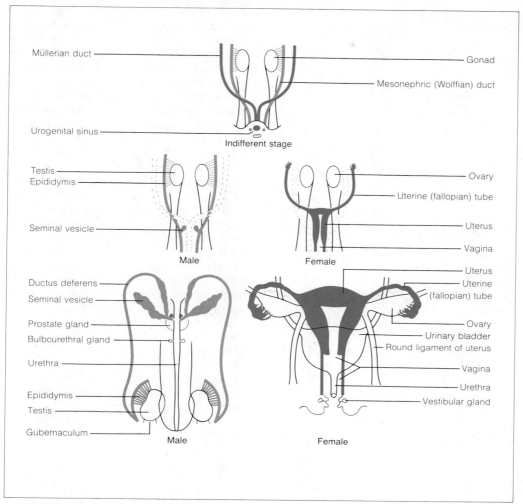

FIGURE 29–3   Differentiation of the internal genitalia.

ism), for example, has been attributed to an extra chromosome 21. There are many other examples, but fortunately most of them are extremely rare.

Since the development of the internal and external genitalia (the accessory sex organs) depends upon the presence or absence of certain hormones and "factors," any deviation from the normal pattern will have marked effects. Males whose testes fail to secrete testosterone will develop female external organs. Females who are exposed to androgens at a very early stage might show some development of male organs. (The adrenal cortical secretion of androgens may be excessive, or the mother may be receiving treatment by substances with androgenic side effects.)

Throughout development, but especially in the very early stages, the fetus is susceptible to many kinds of chemicals, including medications given to the mother, pollutants in the air, additives in the mother's food, or the effects of some diseases (measles). These factors affect not only the reproductive system, but the rapidly developing fetal organ systems, in general.

THE MALE REPRODUCTIVE ORGANS

Testes and Spermatogenesis

The primary male sex organs, the **testes,** are suspended outside the body in a sac, the **scrotum** (see Figure 29–4).

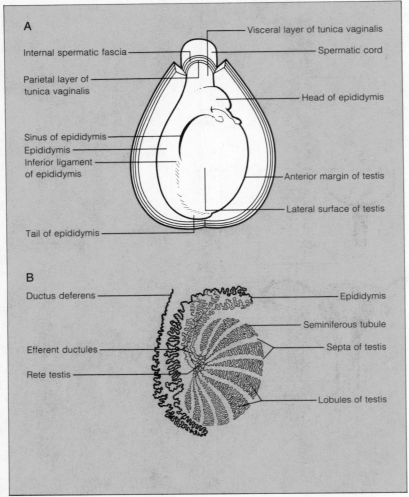

A

Internal spermatic fascia
Parietal layer of tunica vaginalis
Sinus of epididymis
Epididymis
Inferior ligament of epididymis
Tail of epididymis

Visceral layer of tunica vaginalis
Spermatic cord
Head of epididymis
Anterior margin of testis
Lateral surface of testis

B

Ductus deferens
Efferent ductules
Rete testis

Epididymis
Seminiferous tubule
Septa of testis
Lobules of testis

FIGURE 29–4   The testis. A. Lateral view, showing the fascial sheaths. B. Lateral view without the coverings and with a section removed to show the internal structure.

Each testis is a small, egg-shaped organ enclosed in a sturdy fibrous capsule. A longitudinal section shows it to be divided by connective tissue partitions into several segments, tightly packed with the coiled **seminiferous tubules,** in which the vast numbers of sperm are produced. In the tiny spaces between the seminiferous tubules are scattered clusters of cells, appropriately named the **interstitial cells** *(Leydig cells)*. These are the endocrine cells of the testes, and they produce the male sex hormones.

The seminiferous tubules themselves are filled with cells (see Figure 29–5). An outer layer of moderate-sized cells, **spermatogonia,** is the source

of the germ cells (sperm). When they undergo mitotic cell division, the daughter cells may either become new spermatogonia or move toward the lumen of the tubule and become **primary spermatocytes.** Primary spermatocytes may grow quite large before they divide to form **secondary spermatocytes** in a *meiotic* or maturation division. The second meiotic division occurs rather quickly, and the **spermatids** formed are small cells.

Sandwiched between the developing spermatocytes are large, rather ill-defined cells called **supporting cells** (or *Sertoli cells*). They extend from the basement lamina of the seminiferous tubule to the lumen. They do not give rise to sperm, but they do play an important (though not very well understood) role in spermatogenesis. The tiny newly formed spermatids tend to migrate to these cells and establish some sort of close contact with them, as they mature into spermatozoa. The supporting cells are believed to provide support and nourishment during this time.

There is considerable internal change in a spermatid as it develops into a spermatozoon (see Figure 29–6). The nucleus is flattened and condensed into a compact mass of chromatin that becomes the main component of the head of the sperm. The Golgi apparatus is reorganized and comes to fit like a helmet over the top of the nucleus. It is known as the *acrosome*, and it contains several enzymes which are believed to be important in fertilization, when a sperm penetrates an ovum. Bits of cytoplasm seem to be pinched off from the cell until virtually all of it has been lost. The discarded fragments are apparently removed by phagocytic action of the supporting (Sertoli) cells. During this time a tail (flagellum) develops from one of the centrioles. It grows to considerable length and provides for motility of the sperm. The mitochondria are organized in a spiral fashion about the proximal end of the tail.

Though motile sperm have a very high energy requirement, they unfortunately have no place to store much energy—they have been described as little more than elaborately gift-wrapped packages of DNA. They cannot survive on their own for any length of time, and the fluid in which they are suspended, the **semen,** must meet their needs for energy. This fluid, produced by several of the secondary sex organs, also maintains a favorable pH and ensures that sperm become highly active only when they are about to be ejected from the body.

Scrotum and the Descent of the Testes

As long as the testis is an indifferent gonad, it remains in the abdominal cavity. About the time it begins to differentiate, a swelling appears in the wall of the abdominal cavity. This swelling is destined to become the scrotum, and as it becomes larger, its walls come to include the muscle and fascial layers of the abdominal wall and the peritoneum. The testis, which is retroperitoneal, is connected to the inside of the scrotal swelling by a fibromuscular band, the **gubernaculum.** As the embryo grows and the scrotum enlarges, the distal end of the gubernaculum is moved downward. The gubernaculum thickens, and it has been said to shorten and pull the testis into the scrotum. The shortening, however, is probably more relative than absolute because of the great increase in size of the fetus. The gubernaculum more likely serves to guide the testis into the scrotum rather than to pull it there. The testes pass through the inguinal canal in the inguinal ligament with its appendages, the testicular (or spermatic) arteries, veins, and nerves, its lymphatics and the ductus (vas) deferens, still attached (see Figure 29–7). They are bound together by their connective tissue sheaths as the **spermatic cord.**

The testis and spermatic cord are still retroperitoneal as they enter the scrotum. The *evagination* (an outpouching) of the peritoneum and abdominal fascia that extends into the scrotum (the *vaginal process*) lies anterior to the testis. Its opening into the abdominal cavity normally becomes sealed off to leave within the scrotum an isolated sac of peritoneum and fascia known as the **tunica vaginalis.** If the vaginal process does not close off completely, some part of the abdominal viscera (usually a bit

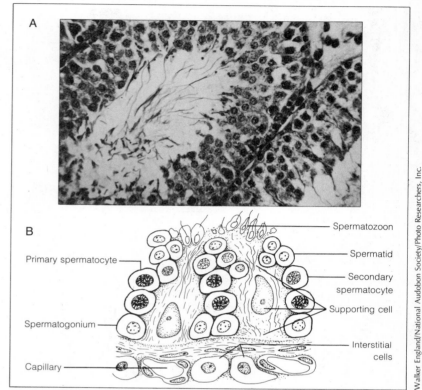

FIGURE 29–5   A. Photomicrograph of a cross section of the seminiferous tubules. B. Diagram of a cross section of a seminiferous tubule.

Walker England/National Audobon Society/Photo Researchers, Inc.

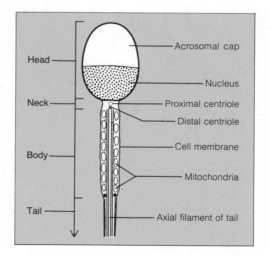

FIGURE 29–6
A spermatozoon.

of small intestine) may later protrude into it as a **hernia.** Similar accidents may occur wherever there is a real or potential weak spot in the abdominal wall, such as the inguinal canal, under the inguinal ligament (with the femoral vessels and nerve), and at the umbilicus.

FIGURE 29-7
Stages in the descent of the testis into the scrotal sac.

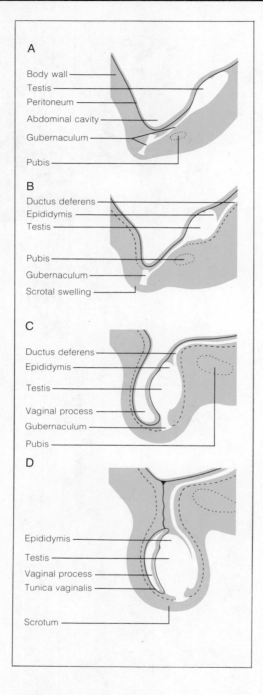

A

Body wall
Testis
Peritoneum
Abdominal cavity
Gubernaculum
Pubis

B

Ductus deferens
Epididymis
Testis

Pubis
Gubernaculum
Scrotal swelling

C

Ductus deferens
Epididymis
Testis
Vaginal process
Gubernaculum
Pubis

D

Epididymis
Testis
Vaginal process
Tunica vaginalis
Scrotum

The location of the testes outside the abdominal cavity seems to have some significance, since as long as a testis remains in the abdominal cavity and fails to descend, it does not produce viable sperm. The sterility associated with the condition, known as *cryptorchidism,* may be due to the fact that the higher temperature in the abdominal cavity does not permit maturation of sperm. There are smooth muscle fibers in the subcutaneous tissue of the scrotum that serve to regulate testicular temperature by contracting and bringing the testes closer to the body when the environmental temperature is low and by relaxing at higher temperatures.

Transport and Delivery of Sperm

Epididymis   When the sperm have reached a certain stage of maturation, they detach from their supporting cells and move along the seminiferous tubules to the tiny **efferent ductules** that lead to the **epididymis** for storage. The epididymis consists of a mass of narrow, coiled tubules sitting like a cap atop the testis. Its epithelium has long microvilli resembling cilia and contains secretory cells that provide a fluid medium for storage of the nonmotile sperm. The microvilli and smooth muscle in the walls of the epididymal tubules is important in moving the sperm from the epididymis. While in the epididymis, however, sperm become motile and are capable of propelling themselves. The epididymis undoubtedly performs other important functions, but they are still not understood.

Ductus Deferens   The tubules of the epididymis converge to form the **ductus (vas) deferens,** which carries sperm up through the inguinal canal into the abdominal cavity, and around behind the urinary bladder (see Figure 29-8). The walls of the ductus deferens are lined with ciliated epithelium and contain smooth muscle which is capable of peristaltic contractions; both aid the movement of the sperm.

Glands and Related Ducts   The ductus deferens remains retroperitoneal as it circles around to the posterior inferior surface of the urinary bladder. There it enlarges to form the *ampulla* just before it receives the duct of the **seminal vesicle,** an irregularly shaped sac with a greatly folded mucous membrane lining. The seminal vesicle produces a viscous substance that constitutes the largest share of the semen and has a high content of fructose.

The junction of the ductus deferens and the duct of the seminal vesicle forms a short **ejaculatory duct** which penetrates the **prostate gland** to enter the **urethra** as it too passes through the prostate gland. The openings of the ejaculatory ducts are near those of the prostate gland. The prostatic secretion contains a number of enzymes plus high concentrations of citric acid and zinc. The prostate may hypertrophy in later life, and obstruct the urethra interfering with urination. The tiny **bulbourethral glands,** found below the prostate, deposit their mucous secretions into the urethra as it enters the penis.

Other than fructose, which provides a nutrient source for the sperm, the specific roles of most of the constituents of semen are not known. It has been suggested, however, that the buffers present help protect the sperm against the acidity of the female reproductive tract and that some of the enzymes help maintain the proper consistency of the seminal fluid.

Urethra   The male urethra may be divided into three parts as shown in Figure 29-9. The first is the **prostatic urethra,** that portion passing from the bladder and through the prostate gland. The second portion is the very short **membranous urethra,** which passes through the membranous diaphragm (urogenital diaphragm) that lies between the rami of the pubic bones (across the pubic angle). The third and longest portion is the **cavernous urethra,** which is surrounded by spongy or cavernous tissue as it passes along the length of the penis.

Penis   The **penis** is the copulatory organ. It terminates distally in the enlarged **glans,** and its skin is only loosely attached to the underlying connective tissue. The flap of skin over the glans, the **prepuce** or **foreskin,** is often removed surgically shortly after birth (*circumcision*).

Besides the urethra, the penis contains three cavernous bodies, the paired **corpora cavernosum,** and the single slender midline **corpus spongiosum** (which contains the urethra).

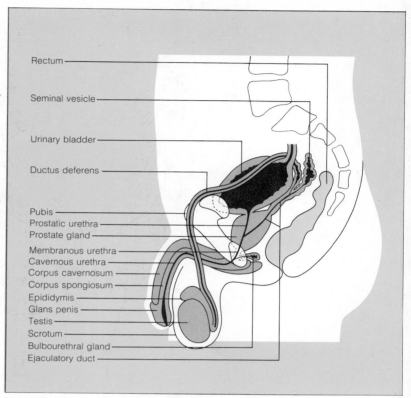

Rectum

Seminal vesicle

Urinary bladder

Ductus deferens

Pubis
Prostatic urethra
Prostate gland
Membranous urethra
Cavernous urethra
Corpus cavernosum
Corpus spongiosum
Epididymis
Glans penis
Testis
Scrotum
Bulbourethral gland
Ejaculatory duct

FIGURE 29-8   Midsagittal section through the male pelvis.

The cavernous bodies extend well back into the perineal region, where they are attached to the urogenital diaphragm. They are vascular sinusoid structures surrounded by a thick, inelastic connective tissue sheath. Upon proper stimulation they become greatly engorged with blood, causing the penis to enlarge and grow rigid. This is the phenomenon of **erection.**

Erection   Erection is brought about by a spinal reflex initiated primarily by touch receptors in the external genitalia, especially those in the glans penis, which is richly supplied with tactile receptors. Stimuli arising from other parts of the body, especially the brain, also impinge upon motoneurons of the erection reflex; psychic or emotional factors are, in fact, highly important in initiating the reflexes involved in the male sexual act.

The immediate cause of erection is the abrupt vasodilatation of the penile blood vessels, brought about by stimulation of sacral parasympathetic fibers

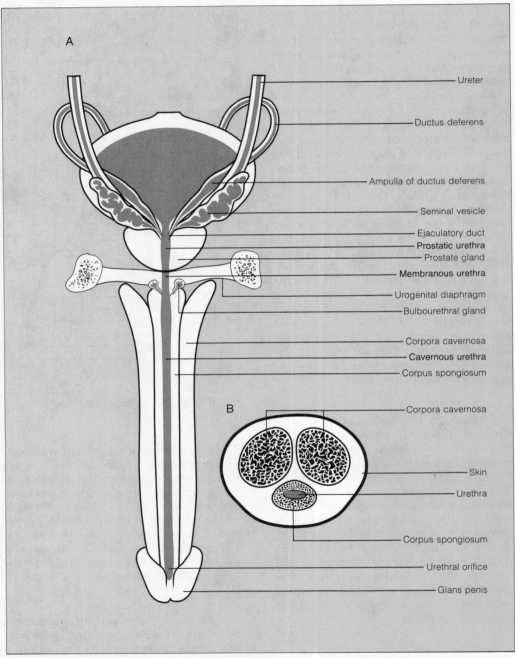

A

Ureter

Ductus deferens

Ampulla of ductus deferens

Seminal vesicle

Ejaculatory duct

Prostatic urethra

Prostate gland

Membranous urethra

Urogenital diaphragm

Bulbourethral gland

Corpora cavernosa

Cavernous urethra

Corpus spongiosum

B

Corpora cavernosa

Skin

Urethra

Corpus spongiosum

Urethral orifice

Glans penis

FIGURE 29-9   A. Posterior view of the bladder and related structures, with the urethra and related structures shown in longitudinal section. B. Cross section of the penis.

rather than by inhibition of sympathetic fibers. As the vessels dilate, local resistance falls and blood rapidly flows into them, filling the vascular spaces of the cavernous bodies. The penis becomes somewhat enlarged as it becomes engorged with blood, but since the connective tissue surrounding it does not stretch, the organ cannot balloon out to great size. The rising pressure within it tends to occlude the thin-walled veins, and the venous outflow is reduced. The penis then becomes quite firm and rigid, which facilitates the deposition of sperm in the female reproductive tract. Erection lasts until the stimulation ceases, whereupon the tone of the arteries and arterioles returns, the inflow of blood decreases and the cavernous bodies are partially drained.

Ejaculation Sperm remain nonmotile until the time of ejaculation, when the action of the cilia and smooth muscle of the epididymis and ductus deferens starts them on their way toward the urethra. The same stimuli that provoke erection are responsible for this activity and for eliciting the secretions of the epididymal and ductal secretory cells, the seminal vesicles and prostate gland, and the bulbourethral glands. Contraction of smooth muscle cells in the prostate and seminal vesicle and in the capsules surrounding them empties their secretions into the ducts. These secretions together constitute the semen, which provides the sperm with fructose for energy and gives them an alkaline environment of the proper consistency, both of which support survival in the female tract.

When the smooth muscle of the internal genital organs contracts, small amounts of sperm-containing semen enter the urethra. The skeletal muscles near the base of the penis then contract too, resulting in the ejaculation of fluid and sperm from the urethra. As the culmination of the sexual act, this ejaculation is accompanied by sensations of extreme pleasure, due at least in part to stimulation of receptors in the glans by the walls of the vagina. This emotional peak is the sexual orgasm.

The material expelled in a normal ejaculation consists of about 3 ml of the viscous milky semen, containing about 200–300 million sperm. Only a single spermatozoon is needed to fertilize an ovum, but fertilization is much less likely to occur when the sperm count is very low. When it falls toward 20 million sperm per milliliter, fertilization becomes increasingly rare. The need for large numbers of sperm is rather puzzling, but it may be explainable in terms of the great disparity of size between sperm and ovum and the basic improbability that a single sperm will even find an ovum. In addition, the female reproductive tract is a hostile environment for sperm. The vagina is relatively acidic, and the mucus in the cervix of the uterus is an additional obstacle. Furthermore, the ovum is surrounded by several layers of cells. Many sperm, with the enzymes they carry in their heads, are probably required to break a path to the ovum itself, even though only one sperm actually penetrates (see Chapter 30).

The number of sperm in the ejaculate is important, but it is not as critical as it was once thought to be, nor is it the only factor in male fertility. The motility of the sperm and their morphology are also important. Normally a certain percent of the sperm do not demonstrate "forward progression" and a certain percent of them exhibit abnormal structure. As the fractions of these nonfunctional sperm increase in the ejaculate, the fertility decreases. Abnormally formed sperm usually show poor motility, and both are associated with low sperm counts. Motility, however, seems to be the critical factor, since if the quality of the sperm is high, it can compensate to a considerable degree for other deficiencies.

ENDOCRINE FUNCTION OF THE TESTES

Besides sperm, the testes also produce androgens, the male sex hormones. A number of substances, including some of the adrenal cortical steroids, have similar actions, but the most important of the androgens is **testosterone** produced by the interstitial cells of the testes.

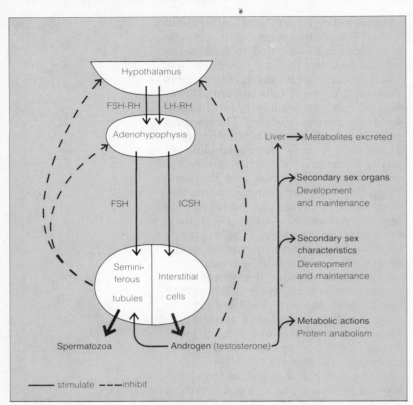

FIGURE 29–10   Hormonal relationships of the male reproductive system.

Control of Testicular Function

Testicular function is under the control of the pituitary gonadotropins. Follicle-stimulating hormone (FSH) in the male is identical with that in the female and serves a comparable function, namely stimulation of the production of gametes (Figure 29–10). Luteinizing hormone (LH) is known as interstitial cell-stimulating hormone (ICSH) in the male because it stimulates the interstitial cells of the testes to secrete androgen (testosterone). Testosterone inhibits the release of LH (ICSH) by a negative feedback effect exerted on the hypothalamus (on the LH-releasing hormone). Testosterone also reduces FSH secretion, but the complete mechanism of regulation of FSH in the male is not clear—unidentified substances from the seminiferous tubules appear to be involved in some way.

Although the interstitial cells of the fetal testes produce testosterone (due to stimulation by maternal gonadotropins) and provoke differentiation of the external genitalia, these cells disappear soon after birth, and testosterone production is virtually nonexistent until the pituitary gonadotropins are secreted at the time of puberty. During the prepubertal period, the gonads are perfectly capable of functioning, but they fail to do so because there is no stimulus. How gonadotropin secretion is turned on at the time of puberty is not understood, but it involves a neural mechanism and the hypothalamic releasing hormones. One suggestion is that until puberty something inhibits the hypothalamus. Another is that before puberty the brain is extremely sensitive to certain inhibitory steroids, and as it becomes less sensitive the hypothalamus is no longer blocked.

Actions of Testicular Hormones

Testosterone and other androgens exert their effects upon all sexual structures—the primary and secondary organs and the secondary sex characteristics. In general, they stimulate the growth of their target organs and these organs are dependent upon the testosterone. Since growth involves the synthesis of protein, testosterone and other androgens are recognized as promoting protein anabolism and reducing protein catabolism.

Testosterone increases the size of the testes and seminiferous tubules and stimulates spermatogenesis. It is apparently able to maintain a certain degree of spermatogenesis in a hypophysectomized animal, but its role in the normal individual is not entirely clear. (In the intact animal, testosterone inhibits secretion of pituitary gonadotropins and may thus actually reduce spermatogenesis.) The action of testosterone upon the seminiferous tubules, however, is rather unusual, since hormones usually do not act upon the gland that produces them. The effect is probably a local action, due to diffusion of testosterone into the tubules.

Androgens are necessary for the development and maintenance of the accessory sex organs, including their enlargement and secretory functions, and for the increase in size of the tubular structures and external genitalia. They are responsible for the growth and functional development of these structures at puberty and for their maintenance in the adult. Without androgens, these organs atrophy and cease to function properly.

The secondary sex characteristics are also totally dependent upon androgens for their development and maintenance. At puberty, androgens stimulate the larynx to enlarge, the vocal cords to become heavier, and the voice to become deeper. There is a general increase in body hair, but there is also a tendency to baldness in men with a genetic predisposition to it. The general male body configuration appears as the shoulders become broader, while the hips stay narrow, and as muscles grow heavier. The sebaceous glands of the skin are stimulated to secrete their oily secretion (which may result in acne).

The metabolic effects of androgens are mainly upon skin, bone, and muscle, and some of these effects have been mentioned above. Because androgens favor protein anabolism (and a positive nitrogen balance), they contribute to the growth spurt of males at puberty. They also, however, help terminate the growth spurt, since they bring about a closure of the epiphyses in the long bones. In addition, they are associated with certain mental and behavioral patterns, such as greater aggressiveness, increased interest in the opposite sex, and greater sex drive (libido). Behavioral and attitudinal changes are difficult to evaluate in humans, but androgens have marked effects of this nature in experimental animals.

Abnormal Testicular Function

Hypogonadism may result from deficiencies in the hypothalamus, the pituitary, or the testes. The first two would result in failure of spermatogenesis, causing sterility, and in reduced secretion of testosterone. Failure in the testes themselves could involve either the seminiferous tubules or the interstitial cells. The inhibition of spermatogenesis by increased temperature when the testes remain in the abdominal cavity has already been mentioned.

The actual effects produced by any deficiency depend largely upon when the lack develops. If an inadequate gonadotropin production or poor response by the testes develops before puberty, or if the testes are removed (castration), the changes associated with puberty cannot occur. The seminiferous tubules do not begin to form spermatozoa, the external genitalia remain juvenile, and the internal organs do not mature and become capable of performing their roles. The secondary sex characteristics also fail to appear. The individual may be taller than average because of delayed closure of the epiphyses. The shoulders are rather narrow, the muscle mass is small, and fat is deposited around the hips, giving the body a typically female contour. The voice remains high-pitched, and the growth of facial and body hair is scant or absent.

If hypogonadism develops after puberty, it does not affect the development or maturation of the reproductive organs and structures, but it does affect those organs and characteristics that depend upon androgens for maintenance. The muscle mass decreases and fat distribution changes, but the pitch of the voice stays low. The regression of androgen-dependent structures is not abrupt following removal of the testes. Castrated humans often retain for long periods of time their libido and the ability to copulate.

Hypergonadism results from excessive androgen production, usually due to a tumor of the hypothalamus, pituitary, testes, or adrenal cortex. The symptoms might be described as excessive masculinity. Unless it is very marked, however, or unless there are other symptoms, such a condition is usually not presented for treatment.

CHAPTER 29 SUMMARY

The elements of the reproductive system fit into three categories: (1) *Primary sex organs* or *gonads*, the *testis* in the male and the *ovary* in the female. They produce the *gametes* (*spermatozoa* and *ova*) and secrete the sex hormones. (2) The *secondary sex organs*, which store, transport, and care for the gametes, and (3) the *secondary sex characteristics.*

Origin of Gametes and the Individual

Typical cells have 46 chromosomes, including two sex chromosomes. In cells of females the sex chromosomes are XX and in males they are XY. Gametes have 23 chromosomes including one sex chromosome. Ova have an X and sperm have either an X or a Y chromosome. The reduction in the number of chromosomes comes about by *meiosis. Spermatogenesis* begins when a spermatogonium forms a primary spermatocyte which, by meiotic cell division, produces four spermatids, half with X and half with Y chromosomes. *Oogenesis* follows a similar course, except that a primary oocyte leads to only one *ootid (ovum).* Conception occurs when a sperm penetrates an ovum forming a cell with 46 chromosomes. If both sex chromosomes are X the new individual is female; if one is Y, it is male.

During the first few weeks of development the reproductive structures are the same in both sexes, but if the individual is to be a male, the gonads differentiate into testes and begin to secrete testosterone and a regression factor. They cause development of male sex organs and regression of structures that would become female organs. In the absence of these factors, female organs develop.

The Male Reproductive Organs

The testes are suspended outside the body cavity in the *scrotum.* They consist of *seminiferous tubules* which produce sperm, and *interstitial cells* which produce testosterone. The testes develop in the abdominal cavity and descend into the scrotum carrying vessels and nerves and the ductus deferens with it. These structures, with their fascial sheaths, make up the *spermatic cord.* A testis that remains in the abdominal cavity will not produce sperm;

the temperature in the abdominal cavity is apparently too high.

After puberty, spermatogenesis is a continuous process. The spermatids become associated with supporting cells in the seminiferous tubules where they develop into spermatozoa, then move to the epididymis for storage and further maturation.

The *epididymis* is continuous with the *ductus deferens,* which carries the sperm back into the abdominal cavity. Ducts of the *seminal vesicles* open into the ductus deferens forming the *ejaculatory ducts* which penetrate the *prostate gland* and open into the urethra. The urethra is divided into *prostatic, membranous,* and *cavernous* portions. The *penis* consists largely of the *corpus spongiosum,* which contains the urethra, and the paired *corpora cavernosa.*

Erection is a reflex phenomenon that results from engorgement of the cavernous tissue with blood. This makes the penis relatively firm and rigid and facilitates its entry into the female reproductive tract. Erection is brought about by tactile stimulation of the tip of the penis and psychic stimuli, which lead to dilation of the vessels of the cavernous tissue. With continued stimulation peristaltic contraction of the smooth muscle of the epididymis, ductus deferens, and other ducts moves the sperm toward the urethral orifice. Secretions, particularly those of the seminal vesicle and prostate gland, are released. These secretions make up the *semen,* which provides a fluid medium for the sperm. It contains fructose for energy and suitable buffers and other substances to aid their survival in the female tract. The muscle contractions lead to ejaculation, in which a few ml of semen containing 200–300 million sperm are expelled from the urethra.

Endocrine Function of the Testes

The testes are controlled by pituitary gonadotropins. *Follicle stimulating hormone (FSH)* stimulates spermatogenesis, and *interstitial cell stimulating hormone (ICSH or LH)* stimulates secretion of testosterone by the interstitial cells. Gonadotropin secretion is caused by hypothalamic releasing hormones which are responsible for both initiating their secretion at puberty and for maintaining it.

Testosterone stimulates spermatogenesis and, as the male sex hormone, it is necessary for the growth and development of secondary sex organs and the secondary sex characteristics at puberty, and for their maintenance throughout life. Without testosterone, spermatogenesis will not begin, the genital organs remain juvenile, and secondary sex characteristics do not develop. Absence of testosterone in adults results in failure of spermatogenesis, inadequate maintenance of secondary sex organs, and regression of some of the sex characteristics.

STUDY QUESTIONS

1   How does meiosis differ from mitosis?

2   How does the development of sperm (spermatogenesis) differ from the development of ova (oogenesis)?

3   At what stage of development is the sex of an individual determined? What determines whether that individual will be male or female? whether that individual will develop male or female sex organs?

4   What is the source of semen? What are some of its functions?

5   What processes are involved in erection, and what mechanisms make it possible for erection to occur?

6   What is ejaculation, and what brings it about?

7   What are the sources of androgens, and what controls their secretion?

8   What initiates the changes that occur at puberty?

9   Compare the effects of castration performed before puberty with those of castration performed after puberty.

10   Can the testicular functions of spermatogenesis and hormone secretion be separated? (Can one be blocked without interfering with the other?) Explain.

30

The Female Reproductive Organs

Endocrine Function of the Ovary

The Menstrual Cycle and Its Control

Pregnancy

Mammary Glands and Lactation

Control of Fertility

The Female Reproductive System

As in the male, the structures and activities of the female reproductive system are geared to the care and nourishment of ova, the female germ cells, which are produced before birth and await their turn to develop and mature. The female reproductive system has an additional role, however, for every time an ovum matures the reproductive organs must be prepared for the possibility that the ovum might be fertilized. If it is, it will need a suitable environment in which to develop, and a connection with the maternal circulation will have to be established quickly.

The discussion of the female reproductive system therefore includes not only the organs of the system and the periodic development of a mature ovum, but also the preparations that are made repeatedly to receive a fertilized ovum, and the changes that occur when a fertilized ovum becomes implanted in the female reproductive tract.

As in the male, these processes are hormone controlled. Hormones of the anterior pituitary control the development of ova and secretion of the female hormones. The female sex hormones control the preparations by the other organs for growth and development of a fertilized ovum (fetus).

The primary female sex organs, the ovaries, produce the female gametes, or ova, and the female sex hormones, estrogen and progesterone. These hormones are responsible for the female secondary sex characteristics and the development and function of the accessory sex organs—the uterine tubes, the uterus, and the vagina.

The organs of the female reproductive system are very different from those of the male, but they perform comparable functions and there are numerous similarities between the two systems. One of the major differences is that where the male produces gametes and hormones relatively continuously, the female has a cyclic pattern. Ova mature one at a time, and the hormones—pituitary gonadotropins as well as those produced by the ovary—are secreted in a cyclic manner. The result is a complex of rhythmic interdependent changes affecting the female's primary and secondary sex organs and her secondary sex characteristics throughout her reproductive life. This pattern of changes constitutes what is known as the **menstrual cycle** in the human female and other primates.

THE FEMALE REPRODUCTIVE ORGANS

When an embryo begins to show sexual differentiation in its sixth or seventh week of life, the cortex of the indifferent gonad develops into the ovary and the Müllerian ducts develop into the uterine tubes, uterus, and vagina. In the human female the organs of reproduction are located in the pelvic portion of the abdominopelvic cavity, since the ovary does not migrate like the testis. It remains in the pelvic cavity near, but not connected to, the uterine tubes that transport the ova to the uterus. Like the organs of the urinary and male reproductive systems, the female reproductive organs are retroperitoneal. However, in the female, the urinary and reproductive organs are entirely separate. There are separate openings to the outside for the urinary tract (the urethra) and the reproductive system (the vagina).

The **uterus** or **womb** is an unpaired muscular organ located between the rectum and the urinary bladder (Figure 30–1). Extending laterally from either side of the uterus is a slender duct known as the **uterine tube,** or **fallopian tube** or **oviduct,** whose lateral ends open into the peritoneal cavity near the ovary. Inferior to the uterus is the **vagina,** or birth canal, which leads to the external opening, the **vaginal orifice.**

These organs are supported by ligamentous folds of peritoneum (Figures 30–1, 30–2, 30–4). From the floor of the pelvic cavity the peritoneum passes over the top of the urinary bladder, up over the front of the uterus, down behind it, and then over the rectum and up the posterior body wall. The peritoneum thus forms a fold that encloses the uterus and uterine tubes and extends to the body wall on either side. This fold is known as the **broad ligament of the uterus.** The ovaries, on either side of the uterus and under the uterine tubes, are suspended from the posterior surface of the broad ligament by a small fold of peritoneum, the **mesovarium.** The ovaries themselves are not actually covered by peritoneum, but rather by a layer of epithelium that is continuous with the epithelium on the peritoneal surface. The lateral portions of the broad ligament attach to the body wall in such a way as to support and suspend the ovaries and the ends of the uterine tubes. They form the **suspensory ligaments** through which nerves and blood vessels reach the ovaries and uterine tubes. A thickening of the broad ligament extending from the uterus to the ovary is known as the **ovarian ligament.**

The **round ligament** of the uterus is an embryonic remnant rather than a supporting peritoneal fold. It lies between the layers of the broad ligament and extends from the uterus near where the uterine tubes attach, and then passes laterally and anteriorly to enter the inguinal canal; it eventually terminates in the labium majus. The round ligament is comparable to the gubernaculum in the male.

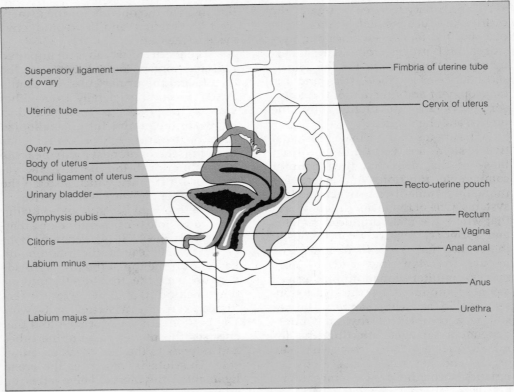

Suspensory ligament
of ovary

Uterine tube

Ovary
Body of uterus
Round ligament of uterus
Urinary bladder

Symphysis pubis

Clitoris

Labium minus

Labium majus

Fimbria of uterine tube

Cervix of uterus

Recto-uterine pouch

Rectum
Vagina
Anal canal

Anus

Urethra

FIGURE 30–1   Midsagittal section through the female pelvis.

The Ovary

The ovary, roughly the size and shape of an almond, is somewhat smaller than the testis. Its surface is covered with a layer of low cuboidal epithelium. Underlying the epithelium is the connective tissue **stroma** of the ovary, containing blood vessels and numerous follicles in all stages of development. Each follicle consists of a rather large cell, an oocyte, surrounded by one or more layers of relatively undifferentiated cells known as **granulosa** or **follicle cells.** The majority of follicles have but a single layer of flattened granulosa cells around the oocyte, and they are known as **primordial follicles.**

Several million oocytes are present in the two fetal ovaries, but a kind of degeneration known as **atresia** is the fate of most of them. At birth there are about two million oocytes. No new germ cells are formed after birth and, because of continuing atresia, only about 400,000 remain at the time of puberty. All the follicles remain as pri-

mordial follicles, immature and undeveloped, until puberty when an increase in the secretion of pituitary gonadotropins stimulates the growth of some of them. Thereafter a few follicles begin to develop each month (each menstrual cycle), but only one normally completes the process and releases its ovum. Thus only about 400 of the 400,000 or so primordial follicles ever reach maturity in the 30–40 reproductive years of the human female. Although we know that most follicles regress and degenerate, we do not know the mechanism by which just one follicle is selected to be the one to reach maturity in each cycle. A microscopic section of an adult ovary shows follicles in many stages of development and atresia.

Development of a Follicle The primordial follicle is but the first of several stages in development of the follicle. As shown in Figure 30–3, the process begins with growth and proliferation of the granulosa cells, and growth of the oocyte as well. Soon a distinct noncellular membrane, the **zona pellucida,** comes to separate the ovum from the granulosa cells. As the follicle continues to grow, scattered spaces between granulosa cells appear, enlarge, and merge to form the **antrum,** a cavity that is filled with fluid produced by the granulosa cells. The ovum, surrounded by several layers of cells (the *corona radiata*), remains attached to one side of the follicle.

While the follicle is growing, nearby cells of the stroma become oriented around it as a loose, poorly organized capsule known as the theca. Its inner portion, the **theca interna,** is quite vascular and soon takes on a secretory function, while the outer **theca externa** is more fibrous. A thin basement membrane separates the theca interna from the granulosa cells. As the follicle enlarges it works its way to the surface of the ovary where it forms a prominent translucent bump. It is now a **mature (graafian) follicle,** and it is about a centimeter in diameter.

Finally the mature follicle begins to accumulate fluid and enlarge rather rapidly, but there is no increase in its interior pressure. The follicle wall becomes thin and bulges even more at one point, which eventually breaks, releasing the follicular fluid and the ovum with its corona radiata out into the abdominal cavity. The rupture of the follicle and release of the ovum constitute **ovulation,** marking the end of the **follicular phase** of the ovarian cycle. It has been suggested that an enzyme may be released that weakens the connective tissue of the follicle wall, but the actual means by which ovulation comes about is not understood. Normally, a single follicle ovulates, but occasionally two follicles may rupture, and sometimes no follicle ruptures (an *anovulatory* cycle).

So far we have referred to the female germ cell as an oocyte rather than an ovum. The formation of a mature ovum (oogenesis), like spermatogenesis, involves several stages, including meiosis, but it is not a continuous process as is the formation of spermatozoa. The germ cell in an ovarian follicle throughout its development is a prima-

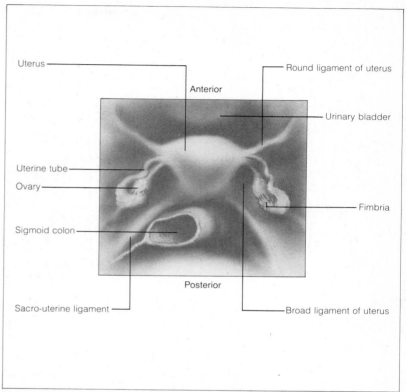

FIGURE 30–2   Organs of the female pelvis, viewed from above.

ry oocyte. The first meiotic division, which gives rise to a secondary oocyte (and the first polar body), is completed just before ovulation. The second meiotic division, which produces a mature ovum (and the second polar body), begins immediately, but it is arrested in metaphase and is completed only when (and if) fertilization has taken place. Strictly speaking, very few female germ cells ever become mature ova. The student should recognize, however, that in practice the female germ cell is commonly called an ovum at any stage.

The Corpus Luteum After the ovum is released, the remaining portion of the follicle undergoes marked changes. Its walls collapse, thicken, and become infolded to form the **corpus luteum.** The granulosa cells hypertrophy to fill the antrum, and they acquire granules of lipid and yellow pigment (which provide the yellowish color for which the structure is named). These

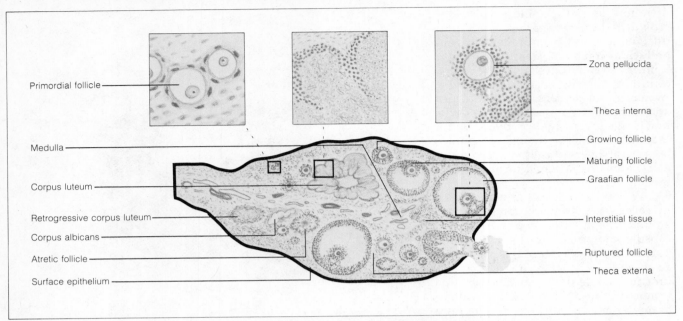

Primordial follicle

Medulla

Corpus luteum

Retrogressive corpus luteum

Corpus albicans

Atretic follicle

Surface epithelium

Zona pellucida

Theca interna

Growing follicle

Maturing follicle

Graafian follicle

Interstitial tissue

Ruptured follicle

Theca externa

FIGURE 30–3   Diagram of a cross section of the ovary, to show the development of follicles.

cells are now known as **granulosa lutein** cells. Blood vessels from the theca interna invade the expanding cell mass and the thecal cells (now **theca lutein** cells) form clusters around the periphery. The corpus luteum attains considerable size during the week following ovulation and marks the **luteal phase** of the ovarian cycle. During this time both the granulosa lutein cells and the theca lutein cells have important endocrine functions. After a week or ten days the corpus luteum begins to degenerate and is eventually replaced by connective tissue. All that remains several weeks later is a small scar, the *corpus albicans*. If the ovum is fertilized, however, the life of the corpus luteum is prolonged. It then reaches a still greater diameter of 2–3 cm, and it remains large and functional for several months before it slowly degenerates.

Internal Genital Organs

Uterine Tubes   The uterine or fallopian tube transports the newly released ovum to the uterus and provides the environment that is essential for the viability of the ovum before and after fertilization. Each tube is quite narrow, with numerous longitudinal folds that

almost obliterate the lumen. Laterally, it widens a bit and curves so as to open immediately over the ovary. The wall contains smooth muscle which is capable of peristalsis, and many cells of the epithelium are ciliated, especially near the open end. The opening is characterized by a fringed (or *fimbriated*) margin, whose long finger-like processes, or **fimbriae,** are virtually draped over the surface of the ovary. Both muscular and ciliary activity of the tube and fimbriae are increased around the time of ovulation. Some granulosa cells (the corona radiata) adhere to the ovum when it is released, so that the cell mass containing the ovum tends to stick to any part of the fimbria with which it comes in contact. The movements of the cilia are also believed to generate currents toward the entrance to the tubes, so that when an ovum is released into the peritoneal cavity, it is almost invariably swept into the uterine tubes. Peristaltic contractions of the smooth muscle in the tubal walls and the current created by the cilia move the ovum toward the uterus.

The ovum usually remains in the uterine tubes for three or four days. It rapidly moves about halfway along the tube then comes to rest, and this is where fertilization usually occurs. If the

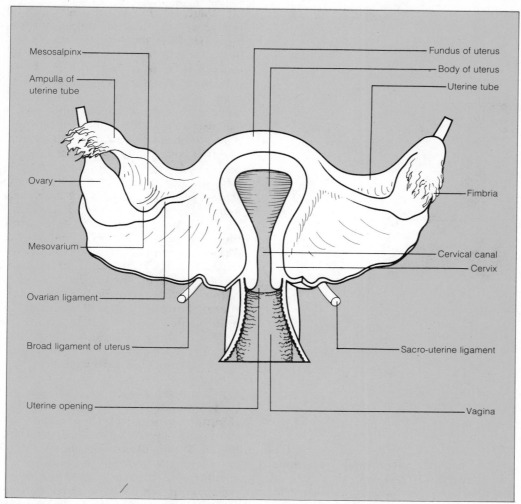

Mesosalpinx

Ampulla of
uterine tube

Ovary

Mesovarium

Ovarian ligament

Broad ligament of uterus

Uterine opening

Fundus of uterus

Body of uterus

Uterine tube

Fimbria

Cervical canal

Cervix

Sacro-uterine ligament

Vagina

FIGURE 30-4   Posterior view of the female reproductive organs.

ovum is fertilized, it then continues toward the uterus where it becomes implanted for further development. If the ovum is not fertilized, it soon begins to degenerate. It may continue on to the uterus, perhaps to be absorbed or phagocytosed along the way.

Uterus The uterus is the somewhat pear-shaped structure to which the uterine tubes are attached (Figure 30–4). It is commonly divided into two portions, the larger **body** superiorly, and the narrower, more cylindrical **cervix** inferiorly. That portion of the body above the entry of the uterine tubes is known as the **fundus.** The cavity of the uterus is relatively small due to the thickness of the muscular walls. In fron-

tal section, it is triangular in shape, with the angles marked by the openings of the uterus—the uterine tubes laterally and the vagina inferiorly. The inferior portion of the cavity is quite narrow, as it extends into the cervix and terminates at the tiny opening into the vagina.

Normally the uterus is tipped slightly forward, so that there is a small angle between the cervix and the vagina. However, since the pelvic organs are all soft structures held in place by rather loose ligaments, their positions are not fixed (the position of the uterus, for example, varies with the fullness of the urinary bladder).

The wall of the uterus consists of three layers: the outer serosa (peritoneum) covering most of the organ; a

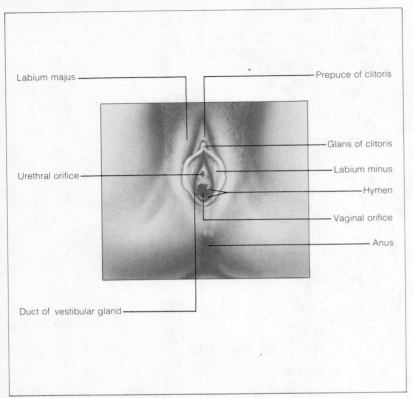

Labium majus

Prepuce of clitoris

Glans of clitoris

Urethral orifice

Labium minus

Hymen

Vaginal orifice

Anus

Duct of vestibular gland

FIGURE 30-5   Inferior view of the external genitalia (vulva) and perineum.

thick muscular layer; and the inner mucosal lining. The serosal covering of the uterus is the broad ligament, the fold of peritoneum that passes over the anterior and posterior surfaces of both the uterus and the uterine tubes. Adventitia is the outer layer of some parts. The muscular layer is known as the **myometrium.** It consists of bundles of smooth muscle fibers passing in several directions. It is the thickest layer and is capable of independent rhythmic contractions. The innermost layer, the **endometrium,** consists of epithelium, lamina propria with numerous blood vessels, and it undergoes marked proliferation and development during the menstrual cycle. It is the endometrium in which the fertilized ovum becomes embedded and which supplies all the needs of the developing embryo. Also, it constitutes the most prominent aspect of the menstrual flow, since its epithelium, blood vessels, and connective tissue are periodically sloughed off if a fertilized ovum does not implant.

Vagina The vagina is a canal whose walls are much thinner and less muscular than those of the uterus. Its lumen has a larger diameter, but its mucosa is more folded and the whole tube is collapsed, so that it may actually seem narrower. The smooth muscle layer is capable of constriction, but it can also be greatly stretched. The vaginal epithelium is stratified squamous, like that covering the external surfaces of the body, but it too undergoes periodic proliferation as part of the menstrual cycle. The changes are much less marked than those of the endometrium, however, and there is no bleeding from the vaginal epithelium. The vagina serves as the birth canal and is the female organ of copulation, since it receives the sperm. The vagina provides a rather hostile environment for the sperm, however, since it is normally quite acidic.

External Genital Organs

The female external genitalia include structures associated with the external openings of the vagina and urethra. Collectively these organs are known as the **vulva** or *pudendum.*

Over the symphysis pubis is a cushion of adipose tissue (the *mons pubis*) from which two fat-containing folds of skin extend posteriorly. They are the **labia majora,** which are comparable to the scrotum in the male; the round ligaments of the uterus terminate in them. They enclose the structures in the region of the urethral and vaginal openings. The **labia minora** are two smaller folds of skin lying between the labia majora, and at their anterior junction is the small rounded **clitoris** (see Figure 30-5). It is comparable to the male penis and, like the penis, it contains erectile tissue and is extremely sensitive to tactile stimulation.

The space enclosed by the labia minora is known as the **vestibule.** The urethra opens into it just posterior to the clitoris, and the vaginal opening is posterior to that. The vaginal orifice is partially covered by a fold of highly vascular mucous membrane, the **hymen.** It may completely close the vaginal opening, or it may be totally absent, and it may persist after sexual intercourse. Its presence or absence, therefore, cannot prove or disprove virginity, as was once believed. There are glands in the vestibule *(vestibular glands)* that produce a mucous secretion that facilitates the entry and movement of the penis during sexual intercourse.

On either side of the vagina, deep to the labia, is a mass of erectile tissue, the bulb of the vestibule, which corresponds to the erectile tissue at the base of the penis (the bulb of the penis). It is attached to the urogenital diaphragm.

The **perineum** is the region of the pelvic outlet. It consists of the soft tissues between the symphysis pubis and the coccyx in the sagittal plane and between the inferior rami of the pubic bones and the ischia laterally. Because the perineal structures, particularly those between the vaginal orifice and the anus, can be torn during childbirth, an incision (or *episiotomy*) is sometimes made to permit the passage of the infant. An incision, after all, is easier to repair than a tear.

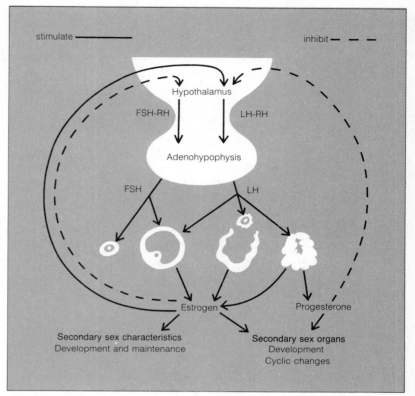

FIGURE 30–6   Hormonal relationships of the female reproductive system.

ENDOCRINE FUNCTION OF THE OVARY

The hormones produced by the ovary are estrogen, progesterone, relaxin, and even a little androgen, the first two being the most important. Before ovulation, the developing follicle secretes estrogen; afterward, the corpus luteum secretes both estrogen and progesterone.

Control of Ovarian Secretion

The ovary is controlled by the pituitary gonadotropins, follicle-stimulating hormone (FSH) and luteinizing hormone (LH), whose actions are well described by their names. They not only stimulate the development of the follicle and the formation of the corpus luteum, but they also provoke secretion of the ovarian hormones. The secretion of gonadotropins is stimulated by hypothalamic releasing hormones (FSH-RH and LH-RH) and is inhibited by the

ovarian hormones through the familiar negative feedback mechanism (see Figure 30–6). The result of this hormonal interplay is a sequence of events involving not only the development and maturation of ova and the secretion of hormones, but repeated intermittent stimulation of all reproductive structures.

Estrogen

Estrogen, as the counterpart of the androgens, has actions in the female comparable to those of androgens in the male. It is responsible for the stimulation, development, and function of the accessory sex organs (the uterus, uterine tubes, and vagina) and for the development and maintenance of the secondary sex characteristics. There is also some evidence that estrogen may affect follicle growth.

Naturally occurring estrogens are steroids characterized by 18 carbon

FIGURE 30-7
Structural formulas of
some natural and
synthetic ovarian
hormones.

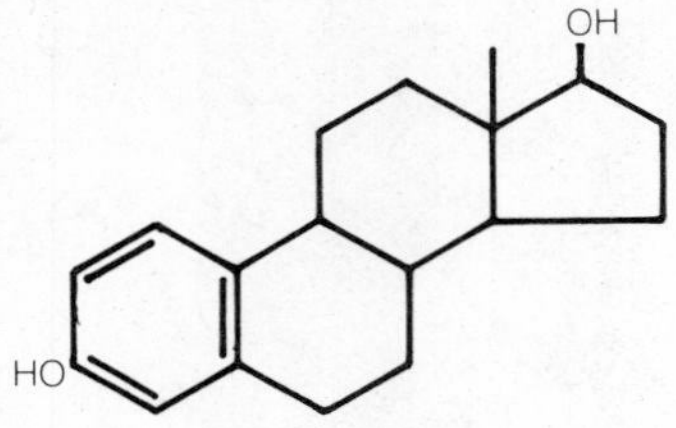

Estradiol (the major naturally occurring estrogen)

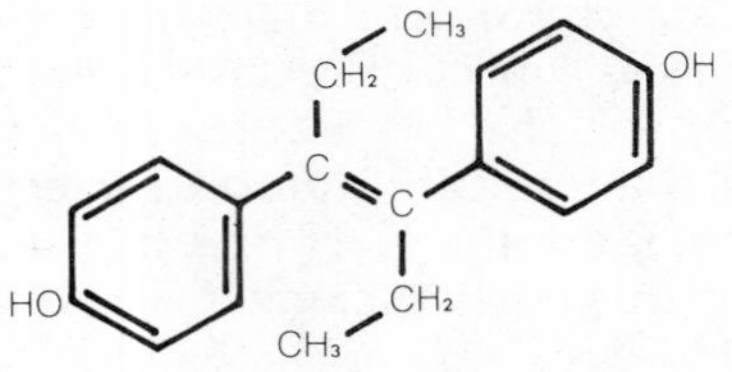

Diethylstilbestrol (a synthetic estrogen)

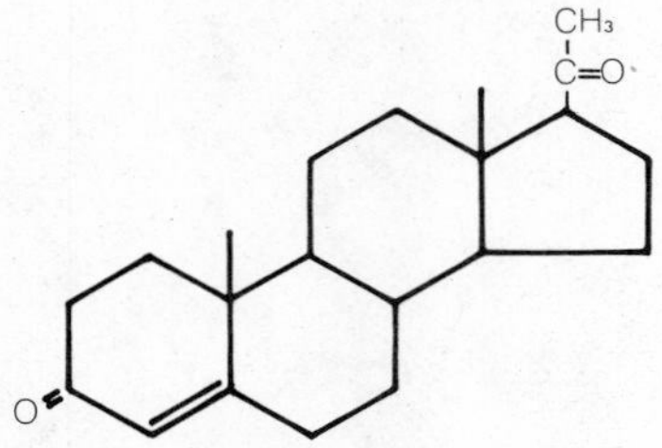

Progesterone

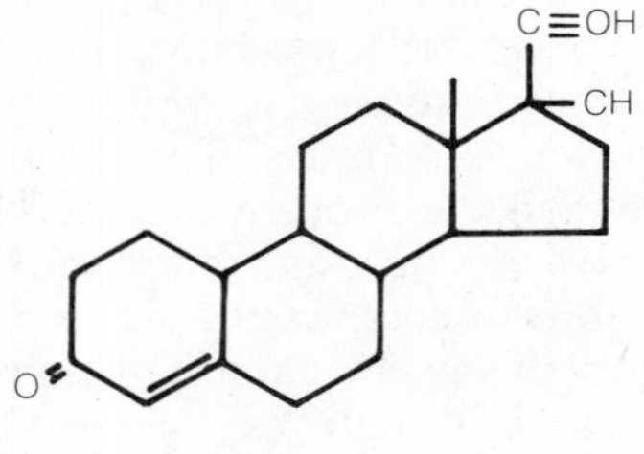

Norethindrone (a synthetic progesterone
and oral contraceptive)

atoms and three double bonds in the first ring (see Figure 30-7). Three compounds, *estradiol*, *estrone*, and *estriol*, found in the female bloodstream have been shown to have estrogenic activity, but estradiol is the major estrogen secreted by the human ovary. A number of synthetic estrogens have been made available, but they are not steroids. (The synthetic estrogens are effective orally, but naturally occurring estrogens are not.)

The secretion of estrogen varies throughout life and throughout the menstrual cycle. Very little is produced before puberty, but at that time the ovaries are stimulated to put out increased amounts of estrogen. The hormone then causes development of the reproductive organs and the secondary sex characteristics. Thereafter, it is produced by both the follicle (cells of theca interna) and the corpus luteum (theca lutein cells). Its concentration varies with the phase of the menstrual cycle. During pregnancy, estrogen is also produced by the placenta. After the reproductive years, that is, after *menopause,* estrogen secretion declines markedly.

The effects of estrogen are brought about primarily by an increase in cell proliferation, particularly of the epithelium of the sex organs. These effects are most conspicuous on the uterus, where it causes proliferation of the endometrial epithelium, glands, and blood vessels. Its action is not limited to the endometrium, however, since estrogen also causes increased contractility of the myometrium and of the smooth muscle of the uterine tubes. In addition, it causes a greater uptake of water and electrolytes by all the uterine cells, which increases the weight of that organ.

The vaginal epithelium is also very sensitive to estrogen, and its cells respond with characteristic changes. This response is so marked that microscopic examination of cells obtained from the vaginal epithelium (a vaginal smear) provides a practical method of identifying the stages of the menstrual cycle and has been used as a bioassay for estrogen.

Estrogen causes the enlargement of the breasts and the development of the duct system of these glands at puberty. Similarly, it dictates the broader hips and narrower shoulders of women and causes their tendency to deposit fat around the hips, abdomen, and buttocks.

Progesterone

Progesterone is secreted by the granulosa lutein cells of the corpus luteum. Since the corpus luteum does not form until after ovulation, there is very little circulating progesterone present before that time. There is evidence, however, that the follicle cells (which become granulosa lutein cells) do secrete some progesterone into the follicular fluids, and progesterone has also been found in the circulation a few hours before ovulation. Progesterone is a 21-carbon steroid which resembles the adrenal cortical steroids more closely than estrogen (see Figure 13–7). In fact, the formation of progesterone is an intermediate step in the synthesis of adrenal steroids.

The administration of large doses of progesterone has much less striking effects than similar doses of estrogen. But if progesterone is given to an animal previously treated with estrogen, the effects of the progesterone are more marked. This suggests that the estrogens produced by the pre-ovulatory follicle may prime the tissues for progesterone. For this reason, progesterone can be considered to exert a "finishing" action, as it continues what estrogen starts. Where estrogen causes growth and proliferation of new tissue, progesterone tends to cause differentiation and development of that tissue.

Under the influence of progesterone, the estrogen-primed endometrium tends to become secretory; its simple tubular glands and arteries become progressively more twisted and coiled, and glycogen accumulates in the epithelial cells of the glands. Progesterone also causes secretory changes in the epithelium of the vagina and uterine tubes, and it stimulates the development of ducts and alveoli in the mammary glands. It reduces the contractility of the myometrium (antagonistic to the action of estrogen), and it is believed to be responsible for the slight increase in body temperature that occurs after ovulation and lasts until menstruation.

Estrogen begins the preparation of the uterus to receive a fertilized ovum, while progesterone completes that preparation and creates a suitable site for its implantation and development. Together these two hormones play a vital role in maintaining a favorable environment throughout pregnancy.

THE MENSTRUAL CYCLE AND ITS CONTROL

It should now be readily apparent that the reproductive cycle involves the close coordination of several organs and a number of hormones. In primates, including humans, this cycle is known as the menstrual cycle. It has two notable events: *ovulation* midway in the cycle, and *menstruation* at the end of it. By convention, however, the menstrual cycle is considered to begin with the first day of uterine bleeding because that is the most easily identified event of the cycle. The "normal" human menstrual cycle is said to be 28 days, but that is merely a convenient average, since its length is highly variable and notoriously unpredictable.

Perhaps the best way to get a picture of the interrelationships between the various hormones and organs involved might be to follow through the sequence of events in the various organs during a "typical" menstrual cycle, as illustrated diagrammatically in Figure 30–8. The most dramatic changes occur in the ovary and the uterus. In the ovary each cycle brings the development and release of an ovum, followed by the formation of a corpus luteum. In the uterus there is the build-up of the endometrium and, if the ovum is not fertilized, the breakdown and loss of that build-up.

We know that events in the ovary (follicle development and hormone secretion) are controlled by the pituitary gonadotropins, which are controlled by hypothalamic releasing hormones and negative feedback of the ovarian hormones. We also know that events in other organs such as the endometrium are controlled by the ovarian hormones. By presenting these and other events of the menstrual cycle in a schematic diagram (Figure 30–8), showing the time courses and relative hormone concentrations, it may seem that the cycle is a

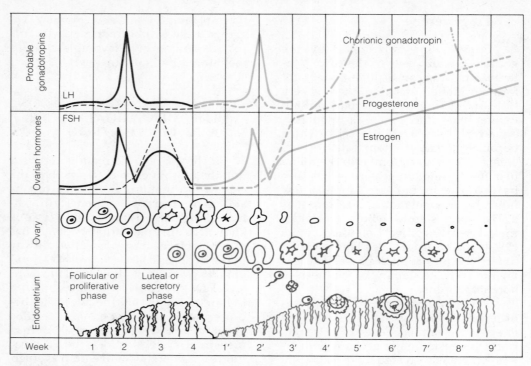

FIGURE 30–8 Cyclic changes in gonadotropins, ovarian hormones, the ovary, and endometrium during a menstrual cycle and early stages of pregnancy.

tidy self-contained package whose workings are well-defined and understood. Such a diagram, however, is no more than a description of what happens, and when; it is not a map of cause and effect. From it we can say that event A is followed immediately by event B, but much as we might wish, many times we cannot say with assurance that the occurrence of event A *causes* event B. Both logic and circumstantial evidence may support a causal relationship, but so often the hard evidence is either lacking, inconclusive, or conflicting. Reproductive functions show great species variations, and one must be extremely cautious in applying to humans what is known for other species.

If one assumes a cycle of 28 days beginning with menstruation, the blood loss usually lasts for the first four or five days. At the end of this time the endometrium is reduced to its minimal state. Much of it has been lost, leaving a relatively thin covering whose exposed surface may show protruding stubs of torn blood vessels and fragments of glands.

At the beginning of the cycle, follicle-stimulating hormone (FSH) from the anterior pituitary is the dominant hormone and it stimulates initial growth of several follicles. A small amount of luteinizing hormone (LH) is also present, since the output of gonadotropin is normally a mixture of the two. The LH is necessary, in conjunction with FSH, to bring about the secretion by the growing follicle.

Toward the latter part of the first week one follicle begins to grow rapidly, and the others regress and become atretic. The "chosen" follicle produces significant amounts of estrogen, and the endometrium shows signs of estrogen stimulation. The remaining endometrial epithelium quickly begins to proliferate and soon establishes a complete epithelial cover over the endometrial surface. Glands develop as straight tubes extending down from the surface. The torn blood vessels sprout from deeper layers and invade the growing endometrium. As the estrogen content of the blood slowly rises, the production of FSH declines and the proliferation continues. The endometrium thickens until it is several millimeters thick, highly vascular, and with numerous glands. This is the **proliferative phase** of the uterine cycle, and it coincides with the follicular phase of the ovarian cycle.

Near the end of the second week of the menstrual cycle, there is a sharp but brief rise in the estrogen concentration, which causes a great surge in gonadotropin output. It is not understood why estrogen increases gonadotropin secretion by positive feedback at this time, while it reduces such secretion by negative feedback at other times. Many mechanisms have been postulated, such as changes in sensitivity, or involvement of two different centers in the hypothalamus. Whatever the mechanism, there is marked increase in both FSH and LH. The increase in LH is much greater, and it is LH that is responsible for rupture of the ovarian follicle and release of the ovum (ovulation) on about the fourteenth day of a 28 day cycle. It also has been suggested that LH causes the release of some progesterone by granulosa cells of the follicle just before ovulation, and that it causes the production of an enzyme that weakens the wall of the follicle just prior to ovulation. It is not clear why there is a peak in FSH secretion along with the LH. In any event, LH is now the dominant gonadotropin, and it causes conversion of the ruptured follicle into a corpus luteum that soon begins to secrete estrogen and progesterone in significant amounts.

With the added influence of progesterone, the ovarian hormones cause conspicuous differentiation and development of the newly formed endometrial tissue. The endometrial glands continue to grow but they now become convoluted and begin to secrete a scant amount of a thick fluid that is rich in glycogen. The arteries that grow into the endometrium lengthen and become tightly coiled, and are now called spiral arteries. The entire endometrium becomes edematous, and a week or so after ovulation it has attained its full development; it is then ready to receive a fertilized ovum. Other structures, including the accessory organs, and sex

characteristics have also reached the peak of their cyclic stimulation. This period after ovulation marks the **secretory phase** of the uterine cycle, and it coincides with the luteal phase of the ovarian cycle.

Perhaps the concentrations of the ovarian hormones in the blood is great enough to inhibit the release of gonadotropins needed to maintain the corpus luteum, or perhaps the corpus luteum is simply programmed to function for only a certain period of time. Whatever the reason, in the fourth week the corpus luteum begins to regress and ceases to produce the estrogen and progesterone needed to maintain the endometrial development. Soon there are marked changes in the endometrial blood vessels. Here and there a few superficial coiled arteries constrict and maintain a spasm for prolonged intervals, restricting the blood flow to those parts of the endometrium they serve and producing a local *ischemia* (lack of blood). When the vessels finally relax, some of them rupture, blood oozes from them, and some of the ischemic tissue begins to deteriorate and break loose. As more vessels rupture, more blood is lost, and more endometrial tissue breaks away. The blood that drains from the endometrium and uterus into the vagina does not clot readily. This is the **menstrual flow,** and it usually amounts to about 25–60 ml of blood, containing the disintegrated endometrial tissue and the secretions of the uterine glands. At the end of the four or five days of breakdown, the endometrium is almost completely denuded. All that remains is the deepest (basal) layer, with a few stubs of blood vessels protruding from the surface and the deep end of some of the glands. This is the source of the epithelium that will soon cover the whole surface once more.

The decrease of estrogen and progesterone that precipitates menstruation also means the end to their inhibition of the pituitary and hypothalamus. The secretion of gonadotropins, especially of FSH, rises during menstrual flow, and another batch of follicles is stimulated to begin to grow. Thus the cycle continues.

Although the length of the menstrual cycle is notoriously variable, most of the variation is in the first half, before ovulation. A two-week interval between ovulation and menstruation is much more dependable.

Puberty and the Onset of Menstruation

Until the time of puberty, the ovary is quiescent and produces no mature ova. It does produce a very small amount of estrogen, but not enough to stimulate any of the estrogen-dependent structures. The uterus, vagina, and other accessory sex organs remain undeveloped and nonfunctional, and there is no appearance of the secondary sex characteristics. The ovary is quite capable of responding to gonadotropin, so its inactivity at this time must be due to a lack of stimulation. It has been suggested that the prepubertal hypothalamus is very sensitive to estrogen and that the small amount present is sufficient to inhibit the production of gonadotropin releasing hormones. Presumably puberty is marked by a changing sensitivity to estrogen of the hypothalamus and more estrogen becomes necessary to block the release of releasing hormones, and hence of gonadotropins. When the prepubertal estrogen levels can no longer prevent it, releasing hormones are produced which elicit the gonadotropins, and the ovary produces ova and hormones. To date, there is no suitable explanation for either the prepubertal inhibition of the hypothalamus or its release from inhibition at puberty, but it is likely that these changes are of neural (brain) origin.

The first menstrual period (menarche) usually occurs between the ages of twelve and fourteen, but the maturation process begins a year or two earlier, as indicated by a gradually increasing secretion of estrogen. It may be several years, however, before regular cycles are established, as early cycles may vary greatly in length and amount of hormone secreted, and some cycles may be anovulatory. Once established, however, menstrual cycles continue for 30–40 years, after which ovarian function declines. During this time, how-

ever, the menstrual cycle is influenced and altered by many factors (including pregnancy) which affect the release of the gonadotropins, usually by acting through the nervous system and the hypothalamic releasing hormones. This neural tie accounts for the effects of emotional and physical stresses upon the menstrual cycle. It is not uncommon, for example, for a woman who goes away to college to have no menstrual periods until she goes home for vacation.

Menopause

After several decades the human ovaries become less responsive to the gonadotropins, and the secretion of ovarian hormones declines. Menstrual cycles become irregular and eventually cease altogether, usually around the age of fifty. Although estrogen secretion declines, gonadotropin secretion, lacking the negative feedback provided by estrogen, is at a much higher level than before. There is some atrophy of estrogen-dependent structures (such as the mammary glands), but there is not necessarily a cessation of sexual interest. The "hot flashes" caused by dilation of cutaneous blood vessels are apparently related to hormone changes and they can be prevented by estrogen treatment. There may be serious emotional problems at this time, but they probably are not directly due to the estrogen lack.

PREGNANCY

Fertilization and Implantation

The ovum remains fertile for about 12 hours after ovulation. Sperm live for up to about 48 hours in the female reproductive tract. Thus in order for fertilization to occur, sperm must be placed in the female tract between 48 hours before and 12 hours after ovulation. Upon introduction into the vagina, sperm rapidly move upward toward the uterine tubes. The means by which this is accomplished is not well understood, but the motility of the sperm and the contractions of the smooth muscle in the vagina and uterus play important roles. The composition and consistency of the mucus in the cervix have a critical effect upon the passage of sperm through the cervix into the uterus. It is less viscous, more watery, and more alkaline around the time of ovulation. These and other properties of the cervical mucus probably also affect the survival of the sperm and their acquisition of the ability to fertilize an ovum (capacitation).

The ovum, which is released into the abdominal cavity at ovulation, does not progress so rapidly, however, and it actually pauses for several days in the uterine tube. Fertilization therefore normally occurs in the uterine tube, although very rarely it may take place in the abdominal cavity.

Upon reaching the ovum a few sperm push through the cells of the corona radiata, but only one enters the ovum. Enzymes released from the acrosome in the head of the sperm make this possible. The disruption that occurs when the first spermatozoon penetrates the zona pellucida initiates rapid changes in the zona that prevent additional spermatozoa from crossing it. The head of the successful spermatozoon enters the ovum (its tail is lost), and its nucleus fuses with that of the ovum to form a fertilized ovum with a complete set of 46 chromosomes, including 2 sex chromosomes. Mitotic cell divisions begin immediately, and the newly fertilized ovum resumes its journey toward the uterus (Figure 30–9).

About a week after ovulation (third week of the menstrual cycle), the fertilized ovum is ready to implant in the endometrium. By this time it has already undergone several cell divisions and consists of a fluid-filled ball of cells known as a **blastocyst.** Along one side of the blastocyst is a thickening, the **inner cell mass,** from which the embryo will develop. The outer layer of cells develops into a specialized structure the **trophoblast.** Upon contact with the endometrium, the cells of the trophoblast release enzymes that digest the endometrial tissue and cause a breakdown at that site. Thus the blastocyst erodes the surface and burrows into

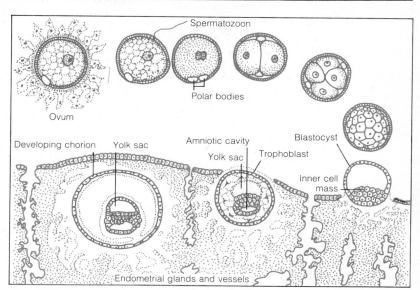

FIGURE 30–9   Fertilization and implantation of an ovum.

the endometrium, which responds by growing around it. In only a few days it is securely implanted.

Development of the Placenta

The further development of the embryo depends critically upon establishing an immediate path for obtaining nutrients from the endometrial tissues. Fortunately, the contact between trophoblast and endometrium develops rapidly into a complex but effective connection. The endometrium near the site of implantation becomes quite thick and highly vascular; many large blood sinuses form where the endometrium has eroded. The trophoblast soon develops a thin cellular membrane over its entire surface, which has projections that extend into the blood sinuses. This membrane is the **chorion,** and the projections are the **chorionic villi.** Before long the chorionic connections become concentrated on the side of the trophoblast nearest to the endometrium, and the mass of tissue formed by the interdigitation of chorionic villi and endometrium becomes the **placenta.** The communication between embryo and the mother is, therefore, partly of embryonic and partly of maternal origin.

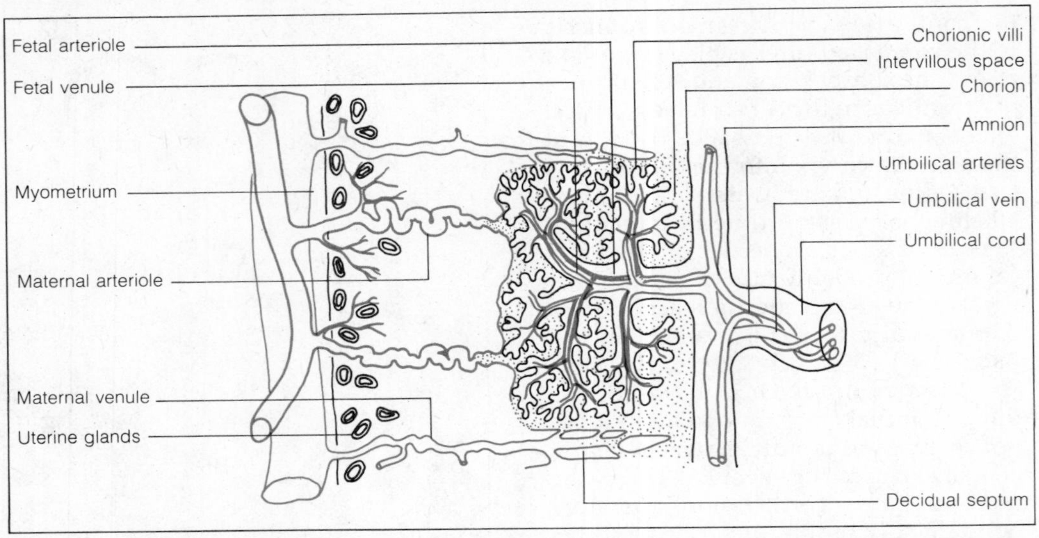

FIGURE 30-10   Vascular organization of the placenta, showing the relationship between fetal and maternal blood vessels.

As development continues, blood vessels extend from the growing embryo into the chorionic villi (Figure 30-10). These vessels form the terminal branches of the umbilical arteries and vein, the major vascular connections between the placenta and the embryo— or *fetus*, as the embryo is called after the third month. Blood from the fetus flows into the chorionic villi and exchanges gases, nutrients, and wastes with the maternal blood across the chorionic membrane. The maternal and infant blood do not mix, however; all exchanges must take place across this membrane, subject to the limitations of membrane transport and diffusion.

Maintenance of the Endometrium

Implantation occurs in the second week after ovulation, the fourth week of the menstrual cycle, which is when menstruation is about to occur. It is clear that survival and further development of the embryo depends upon preventing menstruation and the loss of the endometrium. Further follicle development and ovulations should also be postponed; in fact, the entire menstrual cycle needs to be interrupted for the duration of the pregnancy.

The normal cycle is interrupted by the same mechanism that sustains it:

hormones. Where the nonpregnant endometrium is maintained as long as estrogen and progesterone are secreted, so is the endometrium in pregnancy. In pregnancy the corpus luteum does not degenerate as usual; it remains and enlarges and continues to produce estrogen and progesterone for several weeks at a rate considerably in excess of that seen in a nonpregnant cycle. Its importance is shown by the fact that if the corpus luteum is removed (by ovariectomy) during the first two months, the pregnancy is terminated. The rising concentrations of estrogen and progesterone, of course, sustain the endometrium and cause further growth and development. The accompanying inhibition of FSH release (by negative feedback) prevents further follicle development as well.

The major factor in maintaining a functional corpus luteum seems to be the cells of the chorionic villi (derived from the trophoblast), which begin to secrete a hormone within a few days after implantation (and thus before menstruation would begin). It is known as **human chorionic gonadotropin (HCG),** and it can be detected in the maternal blood and urine soon after implantation (see Figure 30-8). Its concentration rises dramatically, to a very high level by the end of the second month, and then it declines sharply but does not disappear completely. The action of HCG is similar to that of LH, so it maintains the corpus luteum and drives it to produce great amounts of estrogen and progesterone during the first two or three months of pregnancy.

The corpus luteum is essential only in the first months of pregnancy, for a later ovariectomy does not terminate the pregnancy. The need for estrogen and progesterone to maintain the endometrium continues, however, and the role of the corpus luteum as a hormone source is taken over by the placenta itself (the chorion), which develops the ability to produce estrogen and progesterone in sufficient amounts. The concentrations of these hormones continue to rise throughout pregnancy, even though the corpus luteum becomes nearly nonfunctional during the latter part of pregnancy (see Figure 30–11).

The presence of HCG is the basis of tests to determine whether or not a woman is pregnant. Within a few weeks of conception (shortly after the first missed menstrual period), the secretion of this hormone is great enough that detectable amounts are excreted in the urine. The excreted HCG retains its gonadotropic properties, and this was used in the classical pregnancy tests. They depend upon the response of an immature animal (mouse, rat, rabbit, frog, etc.) to the injection of an extract of urine. After a suitable interval the test animal is examined for signs of the LH-like effects of HCG. A positive response would be ovulation and/or the presence of corpora lutea, or, in male frogs, the release of sperm. If the woman was not pregnant, her urine would contain no HCG and the test animal would show no response. Today quicker tests are used which are based on certain antigenic properties of HCG, in which its presence in very small amounts is detected by *immunoassay* rather than by its biological activity. These tests do not take as long to perform and do not involve sacrifice of an animal.

Gestation

Shortly after development of the chorion, there appears another important fetal membrane, the **amnion,** which develops from the inner cell

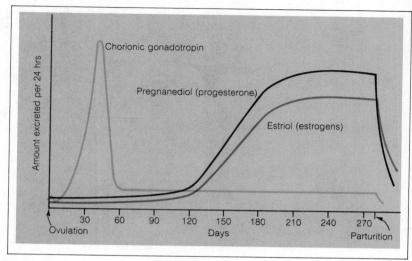

FIGURE 30–11   Urinary excretion of estrogen, progesterone, and chorionic gonadotropin during pregnancy.

mass of the blastocyst. It grows and spreads until it lines the chorion and surrounds the developing embryo, forming the **amniotic cavity.** This cavity is filled with the **amniotic fluid,** in which the embryo is literally suspended. The fluid serves as a protective covering and cushion for the embryo, much as does the cerebrospinal fluid for the brain and spinal cord.

The period of **gestation** (pregnancy) is considered to be 40 weeks (280 days), give or take a week or two, from the beginning of the last menstrual period. During this time the fetus increases its size many times and develops all the organ systems needed to sustain it in the outside world (see Figure 30–12). To provide for such growth the mother must increase her food intake and the rate of her metabolic processes. The increase in basal metabolic rate and changes in metabolism of the various foodstuffs are brought about by endocrine changes, including increased production of thyroxin, cortisol, and a placental hormone with actions similar to growth hormone. To accommodate the growth of the fetus, the uterus must increase greatly in size (and the other abdominal organs must adjust accordingly; this sometimes causes complications). There is an increase in uterine blood flow, both to the growing fetus and to the uterine muscle, and this increase requires an increase of about 30

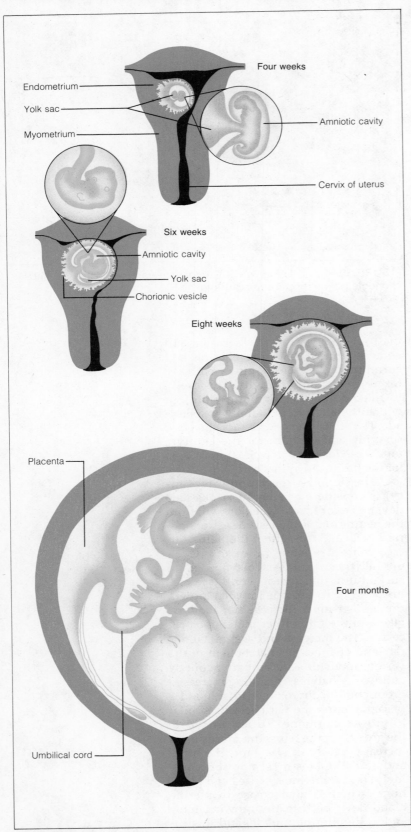

Endometrium

Yolk sac

Myometrium

Four weeks

Amniotic cavity

Cervix of uterus

Six weeks

Amniotic cavity

Yolk sac

Chorionic vesicle

Eight weeks

Placenta

Four months

Umbilical cord

FIGURE 30–12   Development of the embryo.

percent in both cardiac output and blood volume. The increase in blood and extracellular fluid volume is probably associated with a rise in aldosterone production.

Parturition

As pregnancy nears its end, the uterine musculature becomes increasingly sensitive to almost any stimulus, and the occasional spontaneous contractions become stronger and more frequent. These contractions, which bring about birth or **parturition,** constitute *labor.* The specific initiator of labor is not known, but a number of factors have been suggested. It is difficult, however, to distinguish between those factors that initiate labor and those that ensure completion once it has begun. Estrogen is said to increase myometrial contractility, and progesterone to decrease it, and both are present in increasing concentrations throughout pregnancy. Until shortly before the onset of labor, however, progesterone is believed to dominate and block the stimulating effect of estrogen. As labor approaches, there is a decline in both hormones, but indications are that progesterone declines first and more. The response to estrogen is then no longer blocked and the highly excitable myometrium responds with greater contractions. Contractility of the myometrium, like that of smooth muscle elsewhere, is increased by stretch, but this factor is probably less important in initiating labor than in continuing it.

Regardless of the triggering mechanism, however, at some particular time peristaltic contractions of the uterus begin to sweep toward the vagina with increasing frequency and intensity. The gradual buildup of contractions over a period of several hours is an important preparation for birth, since the repeated contractions cause the infant's head to act as a wedge and dilate the narrow cervix and its tiny opening. If the fetus is not positioned with its head down, parturition is more difficult (as in a *breech delivery*).

Once labor is under way, it is also aided by a neuroendocrine reflex. Impulses arising from mechanical stimulation of the cervix by the head of the

Focus

Amniocentesis

Congenital disorders are diseases or defects that a person is born with. They may be the product of injury in the process of birth, of imperfect development of a body part, or of a genetic problem such as a new mutation, a chance combination of genes donated by the parents, or damage to one or more chromosomes. There is very little that can be done to prevent these genetic problems, which include sickle cell anemia, Huntington's chorea, Down's syndrome, and many other debilitating or fatal diseases. However, they can often be detected early in the fetus' existence by a process known as amniocentesis.

The key to amniocentesis is the fact that the fetus sheds cells from its skin and respiratory tract into the amniotic fluid that surrounds it within the uterus. These cells can be collected and examined by a physician. He or she inserts a hypodermic needle through the mother's abdomen and into the uterus and amniotic sac, being careful not to touch the fetus. The doctor then withdraws a sample of amniotic fluid, isolates the fetal cells from it, and lets them grow for a day or two in a cell culture system. The cells can then be examined in two ways. They can be looked at with a microscope; in nondividing cells, the presence of a Barr body indicates a female fetus; in dividing cells, such chromosomal abnormalities as extra and missing chromosomes and translocations, deletions, and inversions of chromosomal segments can be detected. In addition, the composition of the cells can be analyzed for the presence or absence of biochemical compounds that indicate the existence of metabolic (and hence genetic) defects.

Usually performed early in the second trimester of pregnancy, amniocentesis can detect fetal problems early. This is, in fact, the usual reason for amniocentesis—especially with parents who have already had one or more defective children. Amniocentesis has been shown to be generally safe for fetus and mother, although there is some indication that it causes a slight increase in spontaneous abortions, or miscarriages. There is always a possibility of other complications. It is quite accurate, leading to mistaken diagnoses in only 0.4 percent of 3000 cases in one study.

However, amniocentesis sometimes makes it possible to cure certain problem fetuses. In one case, a woman had already borne one child whose inherited disease meant a buildup of acid in the body, developmental retardation, failure to thrive, and death. Amniocentesis showed that her current fetus suffered from the same problem. Fortunately, the disease (methylmalonic acidemia) was known to result from a defect in vitamin B_{12} synthesis. Her doctors therefore gave the mother large doses of B_{12} in the hope that it would reach the fetus through the placenta. It did, and at birth the baby was normal. Without amniocentesis, the outcome could not have been so happy.

fetus cause the release of **oxytocin** from the posterior pituitary gland. By the end of pregnancy, a very small amount of this hormone is sufficient to make the myometrium contract.

Relaxin is the name applied to a substance probably produced by the corpus luteum during pregnancy. Its effects include a "relaxation" or softening of structures such as the symphysis pubis and the pelvic ligaments. It also presumably softens the endometrium of the uterus and cervix and enhances the response of the uterine muscle to estrogen and progesterone mixtures. All these actions would be very helpful in parturition, but they are, unfortunately, more pronounced and better documented in animals than in the human female. Relaxin has been found in the blood of pregnant women, but its exact role in humans is not known.

The contractions of the myometrium raise the intrauterine pressure, and in time the fetal membranes burst, releasing the amniotic fluid which escapes through the vagina. As the infant's head moves into the cervix, it further dilates the passageway and triggers the release of oxytocin. The uterine contractions become stronger and more frequent, and the infant is forced through the birth canal into the outside world. The umbilical cord (umbilical arteries and vein) is still attached.

Within a few minutes after delivery, the uterus resumes rhythmic contractions and expels the placenta as the **afterbirth.** This involves tearing the highly developed vascular connections of the placenta away from the uterus. Excessive hemorrhage is normally prevented by strong sustained contractions of the myometrium and by powerful constriction of the uterine blood vessels. The uterine contractions also help reduce the size of the greatly distended uterus.

MAMMARY GLANDS AND LACTATION

The mammary glands (Figure 30–13) do not begin to develop until they are exposed to periodic stimulation by estrogen and progesterone at

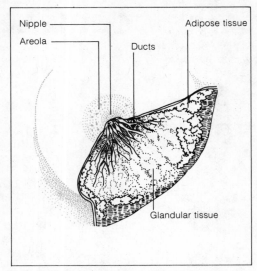

FIGURE 30–13   Structure of the mammary gland.

puberty. Estrogen causes growth and branching of the previously rudimentary duct system and the deposition of fat, which is responsible for the increased size of the breasts. Progesterone causes the formation of lobules and some small alveoli. Growth hormone and prolactin, however, are also needed for normal development of the mammary glands.

The adult nonpregnant mammary gland is exposed to cyclic fluctuations of estrogen and progesterone, but the changes produced in the gland are small compared to those that occur during pregnancy. Because the levels of estrogen and progesterone are then considerably elevated, the entire gland enlarges, the duct system proliferates further, and the secretory apparatus develops enormously. Alveoli, whose walls consist of a single layer of secretory cells, form at the ends of the ducts and enlarge to a functional status. In spite of the great stimulation of the glands during this time, however, they do not normally produce significant amounts of milk until one to three days after parturition. The reason for this is not completely clear, but it is likely that high levels of progesterone cause the release of a prolactin inhibitory hormone (PIH) that blocks the release of prolactin. After delivery and the removal of the placenta, the progesterone level falls and prolactin secretion increases dramatically. There are undoubtedly other factors, but the increase in prolactin is particularly important in the initiation of *lactation*.

After the milk is secreted into the alveoli, it is forced into the ducts by specialized cells surrounding the alveoli. These **myoepithelial cells** have the ability to contract and literally squeeze the newly secreted milk from the alveoli into the ducts, from which it can be removed by sucking. Tactile stimulation of the nipple by the sucking infant excites receptors there and sends impulses to the hypothalamus, resulting in the release of oxytocin from the posterior pituitary. When oxytocin reaches the mammary glands, it causes contraction of the myoepithelial cells and the ejection of milk into the ducts. This is sometimes called the milk "let-down." Since the reflex requires a little time, the sucking infant gets no milk until oxytocin reaches the myoepithelial cells, usually within a minute.

Sucking also stimulates the further release of prolactin, which ensures the continued secretion of milk. In some societies the children are nursed for several years, and lactation continues throughout this time. Continued prolactin secretion associated with nursing an infant tends to inhibit ovulation by preventing the release of FSH and LH through a similar neuroendocrine reflex. It reduces the likelihood of an early second pregnancy, since about half of all nursing mothers do not begin to ovulate while they are still nursing.

CONTROL OF FERTILITY

In recent years the problem of an increasing population in a world of unchanging dimensions and diminishing resources has received much attention. The world population now doubles every fifty years or less, in part due to our own history. High birth rates were once desirable because infant mortality was high and life expectancy was low, only about 40 years. But now, with low infant mortality and a longer life expectancy, and with relatively unchanged birth rates, the problem of overpopulation is much more pressing. It is, in fact, already with us, and the control of human fertility is of great importance.

Early methods of preventing pregnancy consisted of avoiding coitus from shortly before ovulation until after ovulation. This was not, however, a very reliable method, since there was, and is, no good way to determine the time of ovulation. It presumably occurs midway in the menstrual cycle and is accompanied by a slight rise in body temperature due to progesterone, but the rise does not occur until *after* ovulation. The rhythm method thus depends upon predicting according to the average of past cycles, but many women are inconsistent in cycle length and ovulation time. Other and still current methods of birth control depend upon preventing sperm from reaching the ovum, but they have disadvantages of inconvenience, and they are not totally effective either.

Fertility in women could theoretically be controlled by interfering at any of several sites: the release of gonadotropins, the ovary, the uterine tubes, the endometrium, or the cervical secretions. Experimental efforts have been directed at all these possibilities.

The development of oral contraceptives ("the pill") was hailed as a major breakthrough in preventing unwanted pregnancies. Usually containing an estrogen-progesterone combination, these contraceptives are taken daily for about the first 20 days of the cycle. They raise the estrogen and progesterone levels enough to inhibit the release of the pituitary gonadotropins. Without the preovulatory surge of LH there is no ovulation, and without ovulation there is no ovum available to be fertilized. When treatment is stopped, the estrogen and progesterone levels fall and menstruation occurs due to withdrawal of the hormones needed to maintain the endometrium. This is an effective method (if the pills are taken regularly and on schedule), but a small percentage of women suffer serious vascular side effects.

Another reasonably effective method is the intrauterine device (IUD). It is a small plastic ring, coil, or other form that is placed in the uterus. It apparently causes a response by the endometrium that interferes with implantation of a blastocyst or hastens its passage.

The role of the uterus as a possible site of action has been investigated from several different approaches. One of these involves altering the uterine environment to reduce the survival of sperm and/or their progress toward an ovum. The properties of the cervical mucus provide a promising target for such investigations. Another approach involves rendering the uterine environment unsuitable for implantation of a fertilized ovum. Efforts have been made to develop a pill that would do this. One group of substances that has received considerable attention in this regard is the *prostaglandins*. They are found in many tissues, are extremely potent, and have a bewildering variety of effects, one of which is to increase contractility of the uterine muscle. On this basis they have been considered as a potential "morning after" pill, that is, one which prevents implantation. So far, these hopes have not been realized.

Attempts have also been made to produce an effective male pill, but this, too, has not yet met with notable success. In the meantime, **vasectomy** is one method of limiting male fertility. A procedure in which the ductus deferens is bilaterally sectioned and tied off, vasectomy blocks, usually irreversibly, the passage of sperm from the testes to the urethra. After a time there is some, but apparently not total, degeneration of the seminiferous tubules. If the blood supply has not been disrupted, the interstitial cells continue to produce testosterone. The vasectomized male is sterile and is unable to father children, but he is able to carry on sexual activities and does not show the signs of androgen lack that are characteristic of hypogonadal males.

Because of the increasing world population, most of the attention has gone toward limiting fertility. However, the matter of *infertility* is a serious problem for a sizable fraction of the population, as about 10–15 percent of all marriages are infertile. Infertility in the male, as indicated in Chapter 29, is most often related to the quality of the sperm and may be caused by any of a number of conditions or diseases. It can usually be alleviated if the cause can be corrected. Infertility in the female is more common, and may be due to failure to ovulate or to conditions that interfere with ovum transport. The former is more frequent and may result either from an ovarian failure or from secretion of too much or too little gonadotropin. Infertility due to a lack of gonadotropin can be treated with HCG, which acts like LH in stimulating ovulation and the formation of a corpus luteum. Sometimes the treatment is too successful, however, and causes several follicles to ovulate, resulting in multiple births.

CHAPTER 30 SUMMARY

The ovary is the female gonad, and the major secondary sex organs are the uterus, uterine tubes (oviducts, fallopian tubes), and the vagina. The reproductive organs are covered by a fold of peritoneum, the *broad ligament*, which has folds such as the mesovarium and suspensory ligament that support the ovary and ends of the uterine tubes.

The *ovary* contains numerous *primordial follicles*, each consisting of an oocyte surrounded by a few granulosa cells. Beginning at puberty one ovum matures each month (each menstrual cycle) for the next 30–40 years. The "chosen" oocyte develops a zona pellucida and the granulosa cells begin to proliferate. Cells around the follicle become organized into the vascular *theca interna* and the outer *theca externa*. An antrum containing follicular fluid appears in the follicle. It becomes quite large and soon the *mature* or *graafian follicle* is ready for *ovulation*. The follicle ruptures and the egg is released into the abdominal cavity. The remaining follicle develops into a *corpus luteum* which functions for 7–10 days and then degenerates.

Each *uterine tube* ends near an ovary, and the finger-like processes of their fringed openings sweep over the ovary. After its release the egg is normally carried into a uterine tube, which is where fertilization usually occurs.

The *uterus* is divided into the *fundus, body,* and *cervix*. Its wall is a thick layer of muscle, the *myometrium*, lined with *endometrium* and covered by the broad ligament. Its lumen is very small.

The thin-walled *vagina* is the birth canal. It encloses the cervix at one end and opens to the outside at the other. It lies between the rectum and the urethra.

The external genital organs include the *labia majora*, which enclose the *labia minora*, and they enclose a small space, the *vestibule*, into which the urethra and vagina open. The *clitoris* is located at the anterior junction of the two labia minora. It contains cavernous tissue and is capable of erection.

Endocrine Function of the Ovary

Hormone secretion by the ovary is controlled by the pituitary gonadotropins, and they are controlled by hypothalamic releasing hormones. *Estrogen* is secreted by theca interna cells in the developing follicle and by theca lutein cells in the corpus luteum. Estrogen causes growth, development, and maintenance of the secondary sex organs and secondary sex characteristics. Its action is mainly one of proliferation, which is particularly marked in the endometrium.

Progesterone, produced by granulosa lutein cells of the corpus luteum, causes differentiation and is most effective on tissues that have been previously acted upon by estrogen. In the endometrium, progesterone causes development and secretion by its glands and development of its blood vessels.

Both estrogen and progesterone have similar effects on other organs of the reproductive tract.

Menstrual Cycle and its Control

The menstrual cycle is the sequence of events by which the female reproductive system prepares to receive a fertilized ovum approximately every 28 days. The cycle is described as beginning with menstrual flow, with ovulation occurring about day 14. At the end of 4–5 days of menstruation the endometrium is very thin. FSH from the pituitary stimulates follicle growth and LH causes estrogen secretion. The first two weeks of the cycle are known as the *proliferative phase* in view of events in the endometrium, or the *follicular phase* in view of events in the ovary.

Late in the second week, the concentration of estrogen rises and is probably the cause of an outpouring of LH. It is the LH surge that triggers ovulation, formation of a corpus luteum, and its secretion of both estrogen and progesterone. The ovarian hormones cause further development of the endometrium and secretion by its glands. The post-ovulatory portion of the cycle is known as the *secretory* or *luteal phase*.

After 7–10 days the corpus luteum begins to deteriorate, its secretion of ovarian hormones declines, and soon is not enough to maintain the endometrium, which begins to disintegrate, and menstruation occurs. Meanwhile, the anterior pituitary has begun to secrete gonadotropins and another cycle begins.

Pregnancy

An ovum remains fertile for about 12 hours after ovulation, and sperm remain fertile about 48 hours in the female reproductive tract. Sperm move toward the uterine tubes, aided by muscular contractions of the vagina and uterus. Upon reaching an ovum, the sperm release enzymes which help penetrate the surrounding granulosa cells and the zona pellucida. The first spermatozoan to cross the zona pellucida causes changes that prevent additional sperm from entering. Mitotic cell divisions begin almost immediately after fertilization.

The fertilized ovum resumes its journey toward the uterus, and a week or so after ovulation it is ready to implant. By this time it is a *blastocyst*, consisting of an outer layer, the *trophoblast*, and an inner cell mass from which the embryo develops. Enzymes from the trophoblast erode the endometrial surface, the blastocyst burrows in, and the endometrium grows over it.

The endometrium becomes quite thick and vascular and develops blood sinuses near the site of implantation. The trophoblast forms a thin membrane, the *chorion*, with processes, the *chorionic villi*, that protrude into the endometrial sinuses. This connection and communication between the embryo and endometrium develops into the *placenta*, which is the site of all exchange between the developing embryo (and later, the growing fetus) and the maternal blood.

It is necessary to maintain the endometrium and prevent menstruation throughout pregnancy, and this requires estrogen and progesterone. The chorionic villi secrete a hormone, *chorionic gonadotropin (HCG)*, that stimulates the corpus luteum to remain and continue to secrete hormones. HCG appears very early in pregnancy and rises to a high concentration in a few weeks and then declines. Its presence is the basis of pregnancy tests. The corpus luteum grows and secretes estrogen and progesterone during the first three months; if it is removed, the pregnancy will terminate. After that time the chorion itself secretes estrogen and progesterone, and the corpus luteum is no longer necessary.

Early in development another membrane, the *amnion*, is formed. It encloses the embryo and is filled with *amniotic fluid*.

Toward the end of pregnancy the myometrium becomes increasingly sensitive to many chemicals, including estrogen. Rhythmic contractions begin to occur, which gradually become stronger and more frequent. The head of the fetus presses against the cervix and stimulates receptors whose afferents lead to the hypothalamus and cause release of *oxytocin*, a powerful stimulant to the myometrium. Repeated contractions gradually dilate the cervical opening. Eventually the amnion breaks, releasing the amniotic fluid, and a short time later the infant is forced through the birth canal. Subsequent contractions dislodge and expel the placenta (afterbirth), followed by strong myometrial contraction which helps prevent excessive loss of blood.

Mammary Glands and Lactation

The mammary gland, which begins to develop ducts and secretory alveoli at puberty, undergoes further growth during pregnancy. After parturition, prolactin is released, causing secretion of milk into the alveoli. The suckling infant stimulates receptors in the nipple resulting in the reflex release of oxytocin. The hormone causes contraction of myoepithelial cells which surround the alveoli, forcing milk from the alveoli and from the gland.

STUDY QUESTIONS

1 Describe the relationship between the peritoneum and the organs of the female reproductive system.

2 What is involved in the development of a mature ovum? What controls it?

3 In the first few days after ovulation, what happens to the egg? What happens to the ruptured follicle?

4 What is believed to be the immediate cause of ovulation? The beginning of a menstrual cycle? Formation of a corpus luteum? Secretion of estrogen? Secretion of progesterone?

5 What is menstruation? How does the endometrium prepare for it? What happens to the endometrium in menstruation, and what causes these events?

6 When and where is an ovum usually fertilized?

7 What is involved in implantation? How soon does it usually occur?

8 What is the source of human chorionic gonadotropin? What function does it serve? Why is it not important in later pregnancy?

9 What is the nature of the "connection" between the embryo and the endometrium?

10 Once labor is initiated, what are some of the events that occur? What causes and/or controls them?

11 What is the mechanism for causing release of oxytocin?

Further Readings

Most of the information contained in this book is based on observations and investigations reported in the scientific literature. Unfortunately, it was usually necessary to simply state the pertinent facts and describe the mechanisms in skeleton form, omitting background material and ramifications. Much of the original literature is quite detailed or technical, requiring a more extensive background in chemistry, mathematics, or biology than students in an introductory course are likely to possess.

There are, however, numerous sources of information on broad or specific topics in which you can pursue your interests with the background you have. Some of these are listed below. Included are a number of textbooks which also have reading lists. Some of the suggestions are articles on a specific topic, usually written by a person doing research in that area. Many of these readings should be quite comprehensible to you after you have studied the material in that particular chapter. Some are a little more detailed or difficult, but with a little effort on your part, you should be able to understand most of what is presented. These texts and papers are identified by an asterisk (*) in the following lists.

Certain textbooks cover broad areas of anatomy and physiology and are useful for supplementing discussions in all chapters. They are listed at the beginning rather than by chapters. The readings listed for each chapter are concerned primarily with material covered in that chapter, but of course, they often cut across topics discussed in other chapters.

Gross Anatomy

J. E. Anderson, *Grant's Atlas of Anatomy*, 7th ed., Williams and Wilkins, Baltimore, 1978.

* J. V. Basmajian, *Grant's Method of Anatomy*, 9th ed., Williams and Wilkins, Baltimore, 1975.

J. E. Crouch, *Functional Human Anatomy*, 3rd ed., Lea and Febiger, Philadelphia, 1978.

* C. M. Goss, *Gray's Anatomy of the Human Body*, 29th ed., Lea and Febiger, Philadelphia, 1973.

* R. T. Woodburne, *Essentials of Human Anatomy*, 6th ed., Oxford University Press, New York, 1978.

Microscopic and Submicroscopic Anatomy

* W. Bloom and D. W. Fawcett, *A Textbook of Histology*, 10th ed., W. B. Saunders, Philadelphia, 1975.

M. S. H. DiFiore, *An Atlas of Histology*, 4th ed., Lea and Febiger, Philadelphia, 1974.

* A. W. Ham, *Histology*, 7th ed., H. P. Lippincott, Philadelphia, 1974.

* L. C. Junqueiro, J. Carneiro, and A. N. Contopoulos, *Basic Histology*, 2nd ed., Lange Medical Publications, Los Altos, California, 1977.

R. G. Kessel and R. H. Kardon, *Tissues and Organs: A Text-Atlas of Scanning Electron Microscopy*, W. H. Freeman and Co., San Francisco, 1979.

K. A. Porter and M. A. Bonneville, *An Introduction to the Fine Structure of Cells and Tissues*, 4th ed., Lea and Febiger, Philadelphia, 1973.

W. F. Windle, *Textbook of Histology*, 5th ed., McGraw-Hill, New York, 1976.

Physiology

* J. R. Brobeck, *Best and Taylor's Physiological Basis of Medical Practice*, 10th ed., Williams and Wilkins, Baltimore, 1979.

* W. F. Ganong, *Review of Medical Physiology*, 9th ed., Lange Medical Publications, Los Altos, California, 1979.

* A. C. Guyton, *Textbook of Medical Physiology*, 5th ed., W. B. Saunders, Philadelphia, 1976.

* V. Mountcastle, *Medical Physiology*, 13th ed., C. V. Mosby, St. Louis, 1974.

A. J. Vander, J. H. Sherman, and D. S. Luciano, *Human Physiology—The Mechanisms of Body Function*, 2nd ed., McGraw-Hill, New York, 1975.

Chapter 1

* E. F. Adolph, "Early concepts of physiological regulations," *Physiological Reviews*, 41:737, (1961).

W. B. Cannon, *The Wisdom of the Body*, Norton, New York, 1939.

R. A. Capaldi, "A dynamic model of cell membranes," *Scientific American*, March, 1974.

J. F. Fulton, *Selected Readings in the History of Physiology*, 2nd ed., Charles C Thomas, Springfield, Illinois, 1966.

H. F. Ladish and J. E. Rothman, "The assembly of cell membranes," *Scientific American*, January, 1979.

The Living Cell, Readings from *Scientific American*, W. H. Freeman and Co., San Francisco, 1965.

A. G. Loewy and P. Siekevitz, *Cell Structure and Function*, 2nd ed., Holt, Rinehart and Winston, New York, 1970.

C. Singer, *A Short History of Anatomy and Physiology from the Greeks to Harvey*, 2nd ed., Dover, New York, 1957.

* S. J. Singer and G. L. Nicolson, "The fluid mosaic model of the structure of cell membranes," *Science*, 175:720, (1972).

Chapter 2

D. B. Brown, "Isolations of genes," *Scientific American,* August 1973.

A. Claude, "The coming of age of the cell," *Science,* 189: 433 (1975).

F. H. C. Crick, "The genetic code," *Scientific American,* April, 1966.

E. Frieden, "The chemical elements of life," *Scientific American,* July, 1972.

E. Frieden, "The enzyme-cell complex," *Scientific American,* August, 1969.

* H. A. Harper, V. W. Rodwell, and P. A. Mayes, *Review of Physiological Chemistry,* 17th ed., Lange Medical Publications, Los Altos, California, 1979.

* A. L. Lehninger, *Biochemistry,* 2nd ed., Worth, New York, 1975.

A. G. Loewy and P. Siekevitz, *Cell Structure and Function,* 2nd ed., Holt, Rinehart and Winston, New York, 1970.

D. Mazia, "The cell cycle," *Scientific American,* January, 1974.

O. L. Miller, Jr., "The visualization of genes in action," *Scientific American,* March, 1973.

E. Racker, "The membrane of mitochondria," *Scientific American,* February, 1968.

* E. Racker, "The inner mitochondrial membrane: Basic and applied concepts," *Hospital Practice,* February, 1974.

A. Rich and S. H. Kim, "The three-dimensional structure of transfer RNA," *Scientific American,* January, 1978.

Chapter 3

W. Montagna, "The skin," *Scientific American,* February, 1965.

R. Ross, "Wound Healing," *Scientific American,* June 1969.

R. Ross and P. Bornstein, "Elastic fibers in the body," *Scientific American,* June, 1971.

B. Satir, "The final steps in secretion," *Scientific American,* October, 1975.

Chapter 4

W. Platzer, *Locomotor System* Vol 1 in W. Kahle, H. Leonhardt and W. Platzer, *Color Atlas and Textbook of Human Anatomy,* Year Book Medical Publishers, Chicago, 1978.

* P. J. Rasche and R. K. Burke, *Kinesiology and Applied Anatomy: The Science of Human Movement,* 6th ed., Lea and Febiger, Philadelphia, 1978.

* C. Rosse and D. K. Clawson, *The Musculoskeletal System in Health and Disease,* Harper and Row, Philadelphia, 1980.

Chapter 5

J. Napier, "The antiquity of human walking," *Scientific American,* April, 1967.

W. Platzer, *Locomotor System,* Vol 1 in W. Kahle, H. Leonhardt, and W. Platzer, *Color Atlas and Textbook of Human Anatomy,* Year Book Medical Publishers, Chicago, 1978.

* P. J. Rasche and R. K. Burke, *Kinesiology and Applied Anatomy: The Science of Human Movement,* 6th ed., Lea and Febiger, Philadelphia, 1978.

* C. Rosse and D. K. Clawson, *The Musculoskeletal System in Health and Disease,* Harper and Row, Philadelphia, 1980.

D. A. Sonstegard, L. S. Matthews, and H. Kaufer "The surgical replacement of the human knee joint," *Scientific American,* January 1978.

Chapter 6

* J. V. Basmajian, *Muscles Alive,* 4th ed., Williams and Wilkins, Baltimore, 1978.

C. D. Clemente, *Anatomy: A Regional Atlas of the Human Body,* Lea and Febiger, Philadelphia, 1975.

* P. J. Rasche and R. K. Burke, *Kinesiology and Applied Anatomy: The Science of Human Movement,* 6th ed., Lea and Febiger, Philadelphia, 1978.

* C. Rosse and D. K. Clawson, *The Musculoskeletal System in Health and Disease,* Harper and Row, Philadelphia, 1980.

W. Platzer, *Locomotor System,* Vol 1 in W. Kahle, H. Leonhardt, and W. Platzer, *Color Atlas and Textbook of Human Anatomy,* Year Book Medical Publishers, Chicago, 1978.

Chapter 7

C. Cohen, "The protein switch in muscle," *Scientific American,* November 1975.

G. Hoyle, "How is muscle turned on and off?" *Scientific American,* April, 1970.

H. E. Huxley, "The mechanism of muscular contraction," *Scientific American,* December, 1965.

H. E. Huxley, "The mechanism of muscular contraction," *Science,* 164:1356, (1969).

R. B. Layzer and L. P. Rowland, "Cramps," *New England Journal of Medicine,* 285:31, 1971.

R. Margaria, "The sources of muscular energy," *Scientific American,* March, 1974.

P. A. Merton, "How we control contraction of our muscles," *Scientific American,* May, 1972.

J. M. Murray and A. Weber, "The cooperative action of muscle proteins," *Scientific American,* February, 1974.

K. R. Porter and C. Franzini-Armstrong, "The sarcoplasmic reticulum," *Scientific American,* 1965.

Note: The entire September 1979 issue of *Scientific American* is devoted to the brain. Many, but not all, of the articles in this issue are cited in the following chapters.

Chapter 8

J. Axelrod, "Neurotransmitters," *Scientific American,* June, 1974.

* M. L. Barr, *The Human Nervous System, an Anatomical Viewpoint,* 2nd ed., Harper and Row, New York, 1974.

* J. G. Chusid, *Correlative Neuroanatomy*, 17th ed., Lange Medical Publications, Los Altos, California, 1979.

J. C. Eccles, "The Synapse," *Scientific American*, January, 1965.

* C. Eyzaguirre and S. J. Fidone, *Physiology of the Nervous System*, 2nd ed., Year Book Medical Publishers, Chicago, 1975.

E. Gardner, *Fundamentals of Neurology: A Psychophysiological Approach*, 6th ed., W. B. Saunders, Philadelphia, 1975.

* A. C. Guyton, *Structure and Function of the Nervous System*, 2nd ed., W. B. Saunders, Philadelphia, 1976.

L. L. Iverson, The chemistry of the brain," *Scientific American*, September, 1979.

* B. Katz, *Nerve, Muscle, and Synapse*, McGraw-Hill, New York, 1966.

R. D. Keynes, "Ion channels in the nerve cell membrane," *Scientific American*, March 1979.

H. A. Lester, "The response to acetylcholine," *Scientific American*, February, 1977.

F. H. Netter, *The CIBA Collection of Medical Illustrations*, Vol 1, *The Nervous System*, CIBA Pharmaceutical Products, Inc., Summit, N. J., 1953.

* H. D. Patton, J. W. Sundsten, W. E. Crill, and P. D. Swanson, *Introduction to Basic Neurology*, W. B. Saunders, Philadelphia, 1976.

V. J. Wilson, "Inhibition in the central nervous system," *Scientific American*, May, 1966.

Chapter 9

G. Austin, *The Spinal Cord: Basic Aspects and Surgical Considerations*, 2nd ed., Charles C Thomas, Springfield, Illinois, 1972.

C. Eyzaguirre and S. J. Fidone, *Physiology of the Nervous System*, 2nd ed., Year Book Medical Publishers, Chicago, 1975.

E. Gardner, *Fundamentals of Neurology: A Psychophysiological Approach*, 6th ed., W. B. Saunders, Philadelphia, 1975.

A. C. Guyton, *Structure and Function of the Nervous System*, 2nd ed., W. B. Saunders, Philadelphia, 1976.

W. Kahle, *Nervous System and Sensory Organs*, Vol 3 in W. Kahle, H. Leonhardt and W. Platzer, *Color Atlas and Textbook of Human Anatomy*, Year Book Medical Publishers, Chicago, 1978.

E. R. Kandel, "Small systems of neurons," *Scientific American*, September, 1979.

F. H. Netter, *The CIBA Collection of Medical Illustrations*, Vol 1, *The Nervous System*, CIBA Pharmaceutical Products, Inc., Summit, N. J., 1953.

* H. D. Patton, J. W. Sundsten, W. E. Crill, and P. D. Swanson, *Introduction to Basic Neurology*, W. B. Saunders, Philadelphia, 1976.

G. M. Sheperd, "Microcircuits in the nervous system," *Scientific American*, February, 1978.

Chapter 10

* M. L. Barr, *The Human Nervous System, an Anatomical Viewpoint*, 2nd ed., Harper and Row, New York, 1974.

* M. B. Carpenter, *Core Text in Neuroanatomy*, 2nd ed., Williams and Wilkins, Baltimore, 1978.

* J. G. Chusid, *Correlative Neuroanatomy*, 17th ed., Lange Medical Publications, Los Altos, California, 1979.

W. M. Cowen, "The development of the brain," *Scientific American*, September, 1979.

W. Kahle, *Nervous System and Sensory Organs*, Vol 3 in W. Kahle, H. Leonhardt and W. Platzer, *Color Atlas and Textbook of Human Anatomy*, Year Book Medical Publishers, Chicago, 1978.

A. R. Luria, "The functional organization of the brain," *Scientific American*, March, 1970.

W. J. H. Nauta and M. Feirtag, "The organization of the brain," *Scientific American*, September, 1979.

F. H. Netter, *The CIBA Collection of Medical Illustrations*, Vol 1, *The Nervous System*, CIBA Pharmaceutical Products, Inc., Summit, N. J., 1953.

Chapter 11

* M. L. Barr, *The Human Nervous System, an Anatomical Viewpoint*, 2nd ed., Harper and Row, New York, 1974.

K. L. Casey, "Pain: A current view of neural mechanisms," *American Scientist*, 61:194, (1973).

* C. Eyzaguirre and S. J. Fidone, *Physiology of the Nervous System*, 2nd ed., Year Book Medical Publishers, Chicago, 1975.

E. Gardner, *Fundamentals of Neurology, a Psychophysiological Approach*, 6th ed., W. B. Saunders, Philadelphia, 1975.

E. H. Gombrich, "The visual image," *Scientific American*, September, 1972.

B. Gordon, "The superior colliculus of the brain," *Scientific American*, December 1972.

* A. C. Guyton, *Structure and Function of the Nervous System*, 2nd ed., W. B. Saunders, Philadelphia, 1976.

D. H. Hubel and T. N. Weisel, "Brain mechanisms of vision," *Scientific American*, September, 1979.

W. Kahle, *Nervous System and Sensory Organs*, Vol 3 in W. Kahle, H. Leonhardt, and W. Platzer, *Color Atlas and Textbook of Human Anatomy*, Year Book Medical Publishers, Chicago, 1978.

W. J. H. Nauta and M. Feirtag, "The organization of the brain," *Scientific American*, September, 1979.

* H. D. Patton, J. W. Sundsten, W. E. Crill, and P. D. Swanson, *Introduction to Basic Neurology*, W. B. Saunders, Philadelphia, 1976.

J. D. Pettigrew, "The neurophysiology of binocular vision," *Scientific American*, August 1972.

J. Ross, "The resources of binocular vision," *Scientific American*, March, 1976.

W. A. H. Rushton, "Visual pigments and color blindness," *Scientific American*, March, 1975.

G. von Békésy, "The ear," *Scientific American*, August, 1957.

F. Warshofsky and S. S. Stevens, *Sound and Hearing*, Life Science Library, Time-Life, New York, 1969.

F. S. Werblin, "The control of sensitivity of the retina," *Scientific American*, January, 1973.

R. W. Young, "Visual cells," *Scientific American*, October 1970.

Chapter 12

* M. L. Barr, *The Human Nervous System, an Anatomical Viewpoint*, 2nd ed., Harper and Row, New York, 1974.

L. V. DiCara, "Learning in the autonomic nervous system," *Scientific American*, January 1970.

E. V. Evarts, "Brain mechanisms of movement," *Scientific American*, September, 1979.

* C. Eyzaguirre and S. J. Fidone, *Physiology of the Nervous System*, 2nd ed., Year Book Medical Publishers, Chicago, 1975.

E. Gardner, *Fundamentals of Neurology: A Psychophysiological Approach*, 6th ed., W. B. Saunders, Philadelphia, 1975.

M. S. Gazzaniga, "The split-brain in man," *Scientific American*, August, 1967.

N. Geschwind, "Specializations of the human brain," *Scientific American*, September, 1979.

N. Geschwind, "Language and the brain," *Scientific American*, April, 1972.

* A. C. Guyton, *Structure and Function of the Nervous System*, 2nd ed., W. B. Saunders, Philadelphia, 1976.

L. L. Iverson, "The chemistry of the brain," *Scientific American*, September, 1979.

W. Kahle, *Nervous System and Sensory Organs*, Vol 3 in W. Kahle, H. Leonhardt and W. Platzer, *Color Atlas and Textbook of Human Anatomy*, Year Book Medical Publishers, Chicago, 1978.

R. R. Llinás, "The cortex of the cerebellum," *Scientific American*, January, 1975.

A. R. Luria, "Functional organization of the brain," *Scientific American*, March 1970.

P. A. Merton, "How we control the contraction of our muscles," *Scientific American*, May 1972.

N. E. Miller, B. R. Dworkin, "Effects of learning on visceral functions—Biofeedback," *New England Journal of Medicine*, 296:1274, (1977).

W. J. H. Nauta and M. Feirtag, "The organization of the brain," *Scientific American*, September, 1979.

* H. D. Patton, J. W. Sundsten, W. E. Crill and P. D. Swanson, *Introduction to Basic Neurology*, W. B. Saunders, Philadelphia, 1976.

K. Pearson, "The control of walking," *Scientific American*, December, 1976.

* J. Pick, *The Autonomic Nervous System*, Lippincott, Philadelphia, 1970.

K. H. Pribram, "The neurophysiology of remembering," *Scientific American*, January, 1969.

A. Routtenberg, "The reward system of the brain," *Scientific American*, November, 1978.

S. H. Snyder, "Opiate receptors and internal organs," *Scientific American*, January, 1977.

* S. H. Snyder, "Opiate receptors in the brain," *New England Journal of Medicine*, 296:266, (1977).

Chapter 13

R. L. Byyny, "Withdrawal from glucocorticoid therapy," *New England Journal of Medicine*, 295:30, (1976).

R. B. Gillie, "Endemic goiter," *Scientific American*, June, 1971.

* R. S. Goldsmith, "Hyperparathyroidism," *New England Journal of Medicine*, 281:367, (1969).

R. Guillemin and R. Burgus, "The hormones of the hypothalamus," *Scientific American*, November, 1972.

* R. J. Lefkowitz, "Isolated hormone receptors," *New England Journal of Medicine*, 288:1061, (1973).

S. Levine, "Stress and behavior," *Scientific American*, January, 1971.

J. P. Martin, "Neural regulation of growth hormone secretion," *New England Journal of Medicine*, 288:1384, (1973).

B. S. McEwen, "Interactions between hormones and nerve tissue," *Scientific American*, December, 1976.

J. A. Nathanson and P. Greengard "'Second messengers' in the brain," *Scientific American*, August, 1977.

F. H. Netter, *The CIBA Collection of Medical Illustrations*, Vol 4, *The Endocrine System and Selected Metabolic Diseases*, CIBA Pharmaceutical Products, Inc., Summit, N. J., 1953.

B. W. O'Malley and W. T. Schrader, "The receptors of steroid hormones," *Scientific American*, February, 1976.

I. Pastan, "Cyclic AMP," *Scientific American*, August 1972.

J. E. Pike, "Prostaglandins," *Scientific American*, November, 1971.

H. Rasmussen and M. M. Pechet, "Calcitonin," *Scientific American*, October, 1970.

* E. W. Sutherland, "Studies on the mechanism of hormone action," *Science*, 177:401, (1972).

* J. Tepperman, *Metabolic and Endocrine Physiology*, 3rd ed., Year Book Medical Publishers, Chicago, 1973.

C. D. Turner and J. T. Bagnara, *General Endocrinology*, 5th ed., W. B. Saunders, Philadelphia, 1976.

Chapter 14

* M. A. Beaven, "Histamine," *New England Journal of Medicine*, 294:30, and 294:320, (1976).

* A. C. Burton, *Physiology and Biophysics of the Circulation*, 2nd ed., Year Book Medical Publishers, Chicago, 1972.

D. J. Capra and A. B. Edmundson, "Antibody combining site," *Scientific American*, January 1977.

A. Cerami and C. M. Peterson, "Cyanate and sickle cell disease," *Scientific American*, April, 1975.

M. D. Cooper and A. R. Lawton, III, "The development of the immune system," *Scientific American*, November, 1974.

* H. W. Davenport, *The ABC of Acid-Base Chemistry*, 6th ed., University of Chicago Press, Chicago, 1974.

* E. W. Davie and O. D. Ratnoff, "Waterfall sequence for intrinsic blood clotting," *Science*, 145:1310, (1964).

* D. Deykin, "Emerging concepts of platelet function," *New England Journal of Medicine*, 290:144 (1974).

G. M. Edelman, "The structure and function of antibodies," *Scientific American*, August, 1970.

* V. J. Freda, W. Pollack, J. G. Gorman, "Rh disease: How near the end?" *Hosp. Pract.*, 13:61, June 1978.

N. K. Jerne, "The immune system," *Scientific American*, July, 1973.

R. A. Lerner and F. J. Dixon, "The human lymphocytes as an experimental animal," *Scientific American*, June, 1973.

M. M. Mayer, "The complement system," *Scientific American*, November, 1973.

Chapter 15

E. F. Adolph, "The heart's pacemaker," *Scientific American*, March, 1967.

* R. M. Berne and M. V. Levy, *Cardiovascular Physiology*, 3rd ed., C. V. Mosby, St. Louis, 1977.

* A. C. Burton, *Physiology and Biophysics of the Circulation*, 2nd ed., Year Book Medical Publishers, Chicago, 1972.

* B. Folkow and E. Neil, *Circulation*, Oxford University Press, New York, 1971.

* K. M. Kent and T. Cooper, "The denervated heart: A model for studying the autonomic control of the heart," *New England Journal of Medicine*, 291:1017, (1974).

* B. Lown and R. L. Verrier, "Neural activity and ventricular fibrillation," *New England Journal of Medicine*, 294:1165,(1976).

F. H. Netter, *The CIBA Collection of Medical Illustrations*, Vol. 5, *The Heart*, CIBA Pharmaceutical Products, Inc., Summit, N. J., 1969.

* T. C. Ruch and H. D. Patton, *Physiology and Biophysics*, Vol II *Circulation, Respiration and Fluid Balance*, 20th ed. W. B. Saunders, Philadelphia, 1974.

* R. F. Rushmer, *Structure and Function of the Cardiovascular System*, 2nd ed., W. B. Saunders, Philadelphia, 1976.

A. M. Scher, "The electrocardiogram," *Scientific American*, November, 1957.

C. J. Wiggers, "The heart," *Scientific American*, May 1957.

Chapter 16

C. D. Clemente, *Anatomy: A Regional Atlas of the Human Body*, Lea and Febiger, Philadelphia, 1975.

H. Leonhardt, *Internal Organs*, Vol 2 in W. Kahle, H. Leonhardt and W. Platzer, *Color Atlas and Textbook of Human Anatomy*, Year Book Medical Publishers, Chicago, 1978.

H. S. Mayerson, "The lymphatic system," *Scientific American*, June 1963.

J. E. Wood, "The venous system," *Scientific American*, January, 1968.

Chapter 17

* R. M. Berne and M. V. Levy, *Cardiovascular Physiology*, 3rd ed., C. V. Mosby, St. Louis, 1977.

* E. Braunwald, "Regulation of the circulation," *New England Journal of Medicine*, 290:1124 and 290:1420, (1974).

* A. C. Burton, *Physiology and Biophysics of the Circulation*, 2nd ed., Year Book Medical Publishers, Chicago, 1972.

* B. Folkow and E. Neil, *Circulation*, Oxford University Press, New York, 1971.

* T. C. Ruch and H. D. Patton, *Physiology and Biophysics*, Vol II, *Circulation, Respiration, and Fluid Balance*, 20th ed. W. B. Saunders, Philadelphia, 1974.

* R. F. Rushmer, *Structure and Function of the Cardiovascular System*, 2nd ed., W. B. Saunders, Philadelphia, 1976.

* L. Talbot and S. A. Berger, "Fluid-mechanical aspects of the human circulation," *American Scientist*, 62:671, (1974).

* S. F. Vatner and E. Braunwald, "Cardiovascular control mechanisms in the conscious state," *New England Journal of Medicine*, 293:970, (1976).

J. V. Warren, "The physiology of the giraffe," *Scientific American*, November, 1974.

B. W. Zweifach, "The microcirculation of the blood," *Scientific American*, January, 1959.

Chapter 18

E. P. Benditt, "The origin of atherosclerosis," *Scientific American*, February, 1977.

* R. M. Berne and M. V. Levy, *Cardiovascular Physiology*, 3rd ed., C. V. Mosby, St. Louis, 1977.

* A. C. Burton, *Physiology and Biophysics of the Circulation*, 2nd ed., Year Book Medical Publishers, Chicago, 1972.

C. B. Chapman and J. H. Mitchell, "The physiology of exercise," *Scientific American*, May, 1965.

* B. Folkow and E. Neil, *Circulation*, Oxford University Press, New York, 1971.

N. A. Lassen, D. H. Ingvar, and E. Skinhoj, "Brain function and blood flow," *Scientific American*, October, 1978.

* G. S. Moss and J. D. Saletta, "Traumatic shock in man," *New England Journal of Medicine*, 290:724, (1974).

N. Pace, "Weightlessness: A matter of gravity," *New England Journal of Medicine*, 297:32, (1977).

* T. C. Ruch and H. D. Patton, *Physiology and Biophysics*, Vol II, *Circulation, Respiration and Fluid Balance*, 20th ed., W. B. Saunders, Philadelphia, 1974.

* R. F. Rushmer, *Structure and Function of the Cardiovascular System*, 2nd ed., W. B. Saunders, Philadelphia, 1976.

* R. F. Rushmer, *Cardiovascular Dynamics*, 4th ed., W. B. Saunders, Philadelphia, 1976.

Chapter 19

M. E. Avery, N. Wang, and H. W. Taeusch, Jr., "The lung of the newborn infant," *Scientific American*, April, 1973.

C. D. Clemente, *Anatomy: A Regional Atlas of the Human Body*, Lea and Febiger, Philadelphia, 1975.

J. H. Comroe, Jr., "The lung," *Scientific American*, February, 1966.

* J. H. Comroe, Jr., *Physiology of Respiration*, 2nd ed., Year Book Medical Publishers, Chicago, 1974.

H. Leonhardt, *Internal Organs*, Vol 2 in W. Kahle, H. Leonhardt, and W. Platzer, *Color Atlas and Textbook of Human Anatomy*, Year Book Medical Publishers, Chicago, 1978.

* N. B. Slonim and L. H. Hamilton, *Respiratory Physiology*, 3rd ed., C. V. Mosby, St. Louis, 1976.

J. Sundberg, "The acoustics of the singing voice," *Scientific American*, March 1977.

Chapter 20

* P. T. Baker, "Human adaptation to high altitude," *Science*, 163:1149, (1969).

* J. H. Comroe, Jr., *Physiology of Respiration*, 2nd ed., Year Book Medical Publishers, Chicago, 1974.

* H. W. Davenport, *The ABC of Acid – Base Chemistry*, 6th ed., University of Chicago Press, 1974.

C. A. Finch and C. Lenfant, "Oxygen transport in man," *New England Journal of Medicine*, 286:407, (1972).

R. J. Hock, "The physiology of high altitude," *Scientific American*, February, 1970.

* C. Lenfant and K. Sullivan, "Adaptation to high altitude," *New England Journal of Medicine*, 284:1298, (1971).

M. F. Perutz, "Hemoglobin structure and respiratory transport," *Scientific American*, December, 1978.

* N. B. Slonim and L. H. Hamilton, *Respiratory Physiology*, 3rd ed., C. V. Mosby, St. Louis, 1976.

* J. B. West, *Respiratory Physiology – The Essentials*, 2nd ed., Williams and Wilkins, Baltimore, 1976.

Chapter 21

A. J. Berger, R. A. Mitchell and J. W. Severinghaus, "Regulation of the respiration," *New England Journal of Medicine*, 297:92, (1977).

* J. H. Comroe, Jr., *Physiology of Respiration*, 2nd ed., Year Book Medical Publishers, Chicago, 1974.

* R. H. Kellogg, "Oxygen and carbon dioxide in the regulation of respiration," *Federation Proceedings*, 36: 1658, (1977).

* T. C. Ruch and H. D. Patton, *Physiology and Biophysics*, Vol II, *Circulation, Respiration and Fluid Balance*, 20th ed., W. B. Saunders, Philadelphia, 1974.

* N. B. Slonim and L. H. Hamilton, *Respiratory Physiology*, 3rd ed., C. V. Mosby, St. Louis, 1976.

* J. B. West, *Respiratory Physiology – The Essentials*, 2nd ed., Williams and Wilkins, Baltimore, 1976.

P. M. Winter and E. Lowenstein, "Acute respiratory failure," *Scientific American*, November, 1969.

Chapter 22

H. Leonhardt, *Internal Organs*, Vol 2 in W. Kahle, H. Leonhardt and W. Platzer, *Color Atlas and Textbook of Human Anatomy*, Year Book Medical Publishers, Chicago, 1978.

F. H. Netter *The CIBA Collection of Medical Illustrations*, Vol 3, Part I *Upper Digestive System*, 1959 Part II *Lower Digestive System*, 1962 Part III *Liver, Biliary Tract and Pancreas*, 1959. CIBA Pharmaceutical Products, Inc., Summit, N. J.

Chapter 23

* F. P. Brooks, *Control of Gastrointestinal Function*, Macmillan, New York, 1970.

H. W. Davenport, *Digest of Digestion*, 2nd ed., Year Book Medical Publishers, Chicago, 1978.

* H. W. Davenport, *Physiology of the Digestive Tract*, 4th ed., Year Book Medical Publishers, Chicago, 1977.

H. W. Davenport, "Why the stomach does not digest itself," *Scientific American*, January, 1972.

* L. R. Johanson, *Gastrointestinal Physiology*, C. V. Mosby, St. Louis, 1977.

N. Kretchmer, "Lactose and lactase," *Scientific American*, October, 1972.

A. I. Mendeloff, "Dietary fiber and human health," *New England Journal of Medicine*, 297:811, (1977).

H. Neurath, "Protein digesting enzymes," *Scientific American*, December, 1974.

S. Phillips, "Fluid and electrolyte fluxes in the gut," *Hospital Practice*, March, 1973.

* L. H. Stahlgren, "The dumping syndrome: A study of its hemodynamics," *Hospital Practice*, December, 1970.

J. H. Walsh and M. I. Grossman, "Gastrin," *New England Journal of Medicine*, 292:1324, (1975).

R. C. N. Williamson, "Intestinal adaptation," *New England Journal of Medicine*, 298:1393, (1978).

Chapter 24

* S. S. Fajans and J. C. Floyd, "Fasting hypoglycemia in adults," *New England Journal of Medicine*, 294: 766,(1976).

P. Felig and J. Wahrer, "Fuel homeostasis in exercise," *New England Journal of Medicine*, 293:1078, (1975).

* H. A. Harper, V. W. Rodwell, and P. A. Mayes, *Review of Physiological Chemistry*, 17th ed., Lange Medical Publications, Los Altos, California, 1979.

P. C. Hinkle and R. E. McCart, "How cells make ATP," *Scientific American*, February, 1974.

G. O. Kermode, "Food additives," *Scientific American*, March, 1972.

*A. L. Lehninger, *Biochemistry*, 2nd ed., Worth, New York, 1975.

A. Kappas, and A. P. Alvares, "How the liver metabolizes foreign substances," *Scientific American*, June 1975.

C. S. Leiber, "The metabolism of alcohol," *Scientific American*, March 1976.

* D. Levine and D. Haft, "Carbohydrate homeostasis," *New England Journal of Medicine*, 283:175, (1970).

R. Margaria, "The sources of muscular energy," *Scientific American*, March, 1972.

* J. Tepperman, *Metabolic and Endocrine Physiology*, 3rd ed., Year Book Medical Publishers, Chicago, 1973.

Chapter 25

J. D. Fernstrom and R. J. Wurtman, Nutrition and the brain," *Scientific American*, February, 1974.

* R. J. Havel, "Caloric homeostasis and disorders of fuel transport," *New England Journal of Medicine*, 287: 1186, (1972).

* L. Landsberg and J. B. Young, "Fasting, feeding, and regulation of the sympathetic nervous system," *New England Journal of Medicine*, 298:1295, (1978).
J. Mayer, *Overweight, Causes, Cost, and Control*, Prentice-Hall, Englewood Cliffs, N. J., 1968.
* J. Mayer and D. W. Thomas, "Regulation of food intake and obesity," *Science*, 156:327, (1967).
* T. C. Ruch and H. D. Patton, *Physiology and Biophysics*, Vol III, *Digestion, Metabolism, Endocrine Function and Reproduction*, 20th ed., W. B. Saunders, Philadelphia, 1973.
N. S. Scrimshaw, and V. R. Young, "The requirements of human nutrition," *Scientific American*, September, 1976.
E. M. Stricker, "Hyperphagia," *New England Journal of Medicine*, 298:1010, (1978).
* J. Tepperman, *Metabolic and Endocrine Physiology*, 3rd ed., Year Book Medical Publishers, Chicago, 1973.
V. R. Young and N. S. Scrimshaw, "The physiology of starvation," *Scientific American*, October, 1971.

Chapter 26

E. Atkins and P. Bodel, "Fever," *New England Journal of Medicine*, 286:27, (1972).
M. A. Baker, "A brain cooling system in mammals," *Scientific American*, May 1979.
T. H. Benzinger, "The human thermostat," *Scientific American*, January 1961.
H. C. Heller, L. Crawshaw, and H. T. Hammel, "The thermostat of vertebrates," *Scientific American*, August 1978.
L. Irving, "Adaptation to cold," *Scientific American*, January 1966.
* T. C. Ruch and H. D. Patton, *Physiology and Biophysics*, Vol III, *Digestion, Metabolism, Endocrine Function and Reproduction*, 20th ed., W. B. Saunders, Philadelphia, 1973.
* K. Schmidt-Nielsen, *Animal Physiology*, 2nd ed., Cambridge University Press, New York, 1978.

Chapter 27

A. C. Barger, "The renal circulation," *New England Journal of Medicine*, 284:482, (1971).
* B. M. Brenner and R. Beeuwkes, "The renal circulation," *Hospital Practice*, 13:35, (1978).
* B. M. Brenner and H. D. Humes, "Mechanics of glomerular ultrafiltration," *New England Journal of Medicine*, 297:148, (1977).
* J. H. Hamburger, G. Richet, and J. P. Grundfeld, *Structure and Function of the Kidney*, W. B. Saunders, Philadelphia, 1971.
J. P. Merrill, "The artificial kidney," *Scientific American*, July 1961.
F. H. Netter, *The CIBA Collection of Medical Illustrations*, Vol 6, *Kidneys, Ureters, and Urinary Bladder*, CIBA Pharmaceutical Products, Inc., Summit, N. J., 1973.
* R. F. Pitts, *Physiology of the Kidney and Body Fluids*, 3rd ed., Year Book Medical Publishers, Chicago, 1974.
H. W. Smith, *From Fish to Philosopher*, Anchor Books, Doubleday, Garden City, New York, 1961.

H. W. Smith, "The kidney," *Scientific American*, August 1962.
A. J. Vander, *Renal Physiology*, McGraw-Hill, New York, 1975.

Chapter 28

P. J. Cannon, "The kidney in heart failure," *New England Journal of Medicine*, 296:32, (1977).
* H. W. Davenport, *The ABC of Acid–Base Chemistry*, 6th ed., University of Chicago Press, Chicago, 1974.
P. V. Feig, and D. K. McCurdy, "The hypertonic state," *New England Journal of Medicine*, 297:1444, (1977).
* H. S. Frazier, "Renal regulation of sodium balance," *New England Journal of Medicine*, 279:865, (1968).
F. J. Gennari, and J. P. Kassirer, "Osmotic diuresis," *New England Journal of Medicine*, 291:714, (1974).
* W. S. Peart, "Renin-angiotensin system," *New England Journal of Medicine*, 292:302, (1975).
* R. F. Pitts, *Physiology of the Kidney and Body Fluids*, 3rd ed., Year Book Medical Publishers, Chicago, 1974.
* R. F. Pitts, "The role of ammonia production in regulation of acid–base balance," *New England Journal of Medicine*, 284:32, (1971).
* J. Tepperman, *Metabolic and Endocrine Physiology*, 3rd ed., Year Book Medical Publishers, Chicago, 1973.
A. J. Vander, *Renal Physiology*, McGraw-Hill, New York, 1975.

Chapter 29

R. D. Allen, "The moment of fertilization," *Scientific American*, July, 1959.
* L. Chan and B. W. O'Malley, "Recent studies on the mechanisms of action of the sex steroid hormones," *New England Journal of Medicine*, 294:1322, (1976).
* J. M. Davidson, "Hormones and sexual behavior in the male," *Hospital Practice*, September, 1975.
T. Friedman, "Prenatal diagnosis of genetic disease," *Scientific American*, November, 1971.
F. H. Netter, *The CIBA Collection of Medical Illustrations*, Vol 2, *Reproductive System*, CIBA Pharmaceutical Products, Inc., Summit, N. J., 1954.
* E. W. Page, C. A. Villee, and D. B. Villee, *Human Reproduction: The Core Content of Obstetrics, Gynecology, and Perinatal Medicine*, 2nd ed., W. B. Saunders, Philadelphia, 1976.
* T. C. Ruch and H. D. Patton, *Physiology and Biophysics*, Vol III, *Digestion, Metabolism, Endocrine Function and Reproduction*, 20th ed., W. B. Saunders, Philadelphia, 1973.
S. J. Segal, "The physiology of human reproduction," *Scientific American*, March, 1974.
* J. Tepperman, *Metabolic and Endocrine Physiology*, 3rd ed., Year Book Medical Publishers, Chicago, 1973.
C. D. Turner and J. T. Bagnara, *General Endocrinology*, 5th ed., W. B. Saunders, Philadelphia, 1976.

Chapter 30

A. E. Beer and R. E. Billingham, "The embryo as a transplant," *Scientific American*, April, 1974.

D. Eppel, "The program of fertilization," *Scientific American*, November, 1977.

A. G. Franz, "Prolactin," *New England Journal of Medicine*, 298:201, (1978).

C. Grobstein, "External human fertilization," *Scientific American*, November, 1977.

P. H. Klopfer, "Mother love, What turns it on?" *American Scientist*, 59:404, (1971).

* S. M. McCann, "Luteinizing-hormone-releasing hormone," *New England Journal of Medicine*, 296:797, (1977).

* R. P. Michael, "Hormones and sexual behavior in the female," *Hospital Practice*, December, 1975.

F. H. Netter, *The CIBA Collection of Medical Illustrations*, Vol 2, *Reproductive System*, CIBA Pharmaceutical Products, Inc., Summit, N. J., 1954.

* J. Tepperman, *Metabolic and Endocrine Physiology*, 3rd ed., Year Book Medical Publishers, Chicago, 1973.

C. D. Turner and J. T. Bagnara, *General Endocrinology*, 5th ed., W. B. Saunders, Philadelphia, 1976.

* S. Y. C. Yen, "Neuroendocrine regulation of the menstrual cycle," *Hospital Practice*, March, 1979.

Some Commonly Used Metric Units and Conversions

Prefixes are used with the basic units of the metric system, as well as other basic units, including those for time (seconds), electricity (volts), and amounts or concentrations (mols, osmols, equivalents).

Prefix	Symbol	Magnitude
tera-	T	10^{12}
giga-	G	10^{9}
mega-	M	10^{6}
kilo-	k	10^{3}
deci-	d	10^{-1}
centi-	c	10^{-2}
milli-	m	10^{-3}
micro-	μ	10^{-6}
nano-	n	10^{-9}
pico-	p	10^{-12}

Temperature

Degrees Celsius (Centigrade), °C	Degrees Fahrenheit, °F
−20	−4
−10	14
0	32
10	50
20	68
30	86
37	98.6
40	104
50	122
60	140
70	158
80	176
90	194
100	212

Length

1 kilometer (km) = 1000 meters (m)
1 m = 100 centimeters (cm)
    = 1000 millimeters (mm)
1 mm = 1000 micrometers (μm)
1 μm = 1000 nanometers (nm)

1 inch = 2.54 cm
       = 25.4 mm
1 meter = 39.37 inches
   1 km = 0.62 (5/8) miles

Weight

1 kilogram (kg) = 1000 grams (g)
1 g = 1000 milligrams (mg)
1 mg = 1000 micrograms (μg)
1 μg = 1000 nanograms (ng)

1 pound (lb) = 453.6 g
1 ounce (oz) = 28.35 g
1 kg = 35.27 oz.
    2.2+ lb.

Volume

1 liter (1) = 1000 milliliters (ml)
        = 1000 cubic centimeters (cm)*
    1 ml = 1000 microliters (μl)
    1 μl = 1000 nanometers (nm)
    *(1 ml = 1.000027 cc)

    1 liter = 1.05 quarts
1 quart = 0.946 liters
1 gallon = 3.785 liters

Glossary

Acquisition of the vocabulary necessary for communication in the medical sciences can be a formidable task. It can be greatly simplified, however, by learning the meanings of some of the prefixes, suffixes, and stems from which these words are formed. Following are lists of a few of the word-parts which are commonly used in the basic sciences as well as clinically. Mastery of these terms should quickly provide a sizable working vocabulary.

Prefixes

a-, ab-	away from, away	hemi-	half	noci-	injurious
a-, an-	without, lack, not	hetero-	varied, unlike, different	oligo-	scant, sparse
ad-	to, toward	histo-	tissue	para-	beside, near
ante-	before	homeo-	similar, alike	peri-	around
anti-	against	hyper-	above, beyond, excess	phago-	eat
auto-	self	hypo-	below, under, deficient	poly-	many
bi-	two	in-	into or not	post-	behind, after
bio-	life (biology)	infra-	below, under	pre-	before, in front of
brady-	slow	inter-	between or among	pro-	before, giving rise to
chromo-	color	intra-	within or inside	proprio-	one's own
circum-	around	ipse-	same	retro-	backward, behind
contra-	opposed, against	iso-	equal, like	sclero-	hard
de-	remove, decrease	juxta-	next to	semi-	half
di-	two	leuko-	white	sub-	below
dia-	through	macro-	large	super-⎫	
dys-	bad, difficult	mal-	bad	supra-⎭	above, excess
ecto-	on, without, on outside	mega-⎫	large, great	syn-	with, binding together
endo-	in, within	megalo-⎭		tachy-	swift
epi-	upon, above	meso-	middle	trans-	across
erythro-	red	mono-	one, single	ultra-	beyond, excessive
eu-	good, well (euphoria)	morpho-	shape, form	uni-	one
ex-	out, away from	necro-	dead		

Suffixes

-algia	pain	-lysis	dissolving, destruction, separation	-rrhea	flow
-ase	enzyme			-scope	vision; instrument for visual examination
-cele	cyst	-meter	measure		
-cle	small	-oid	like, similar to	-soma, -some	body
-ectomy	cut out, removal	-ole	small	-stomy	make an opening
-emia	blood	-oma	tumor, swelling		
-gen	producing	-opia	vision	-tome, -tomy	cutting
-gnosis	knowledge	-osis	a condition, a process	-trope	to turn
-graph	writing	-plegia,⎫	stroke	-trophe	nourishment
-itis	inflammation	-plexy ⎭		-ule, -ulus	small
-ject	throw	-pnea	breathing	-uria	urine
		-rrhage	to burst forth, excessive discharge		

Combining Forms

adeno-	gland	cyto-	cell	nephros-	kidney
amyl-	starch	derm-	skin	neuro-	nerve
angio-	vessel	entero-	intestine (small)	ophthalmo-	eye
ano-	anus	erg-	work	osteo-	bone
arthro-	joint	gastro-	stomach	oto-	ear
bili-	bile	glosso-	tongue	ovi, ovo-	egg
brach-	arm	gluc-⎤		patho-	disease
broncho-	bronchus, trachea	glyc-⎦ glucose, sugar		phlebo-	vein
calor-	caloric, heat	hem-	blood	pneumo-	air, lungs
cardio-	heart	hepat-	liver	psycho-	mind
cephalo-	head	hydro-	water	pulmo-	lung
cerebro-	brain	hystero-	uterus	renal-	kidney
chole-	bile	ileo-	ileum	sarco-	flesh
chondro-	cartilage	ilio-	ilium	thermo-	heat
cor-	heart	myel-	marrow	thrombo-	clot
corpus-	body	myo-	muscle	ur-	urine
costo-	rib			vas-, vaso-	vessel

Definitions of Terms

A band   Anisotropic band of striated muscle cell.

abdomen   That part of the body between the diaphragm and pelvis.

abduction   Movement away from the midline of the body or part.

ablation   Removal of a part of the body.

absorption   Taking up of substances by cells or through mucous membrane, skin, or blood vessels.

accommodation   Adjusting the focus of the eye for objects at various distances.

acetabulum   Cup-shaped depression in the pelvic bone into which the head of the femur fits.

acetoacetic acid   One of the ketone bodies formed during the oxidation of fatty acids; found in the blood and urine of diabetics and during starvation.

acetone   A volatile, fragrant, colorless liquid. One of the ketones formed during oxidation of fatty acids (*see* acetoacetic acid).

acetonuria   Presence of acetone in the urine.

acetylcholine, ACh   Substance released by nerve endings of cholinergic fibers; transmits nerve impulses and then is rapidly destroyed.

acetylcholinesterase, AChE   The enzyme that inactivates acetylcholine.

acetyl coenzyme A, acetyl Co-A   Product of condensation of acetic acid and coenzyme A involved in the transfer of two-carbon fragments, notably in their entrance into the Krebs cycle.

ACh   Acetylcholine.

AChE   Acetylcholinesterase.

Achilles tendon   Strong tendon of the large muscles of the calf (gastrocnemius and soleus); inserts into the bone of the heel (calcaneus).

acid   A substance that dissociates (breaks up) in aqueous solution to yield hydrogen ions.

acidosis   A condition in which there is a relative decrease in alkali (particularly bicarbonate) in body fluids in proportion to acids.

acinus   (*pl.* acini) Cells arranged as a small grapelike sac, forming the secretory portion of certain glands.

acromegaly   A disease characterized by enlargement of the hands, feet, head, and face due to excessive secretion of growth hormone by the anterior pituitary gland.

acromion   Outer end of the spine of the scapula; forms the tip of the shoulder.

acrosome   A dense, caplike cover on the anterior portion of heads of sperm; derived from Golgi apparatus; its enzymes aid in penetration of ovum at fertilization.

actin   One of the contractile proteins in muscle; found in thin filaments of the sarcomere of striated muscle.

action potential   The change in electrical potential across the membrane of a nerve or muscle cell when it becomes active.

active state   The tension produced internally by the action of the muscle proteins; immediately follows excitation of the muscle cell and the release of calcium.

active transport   Carrier-mediated transfer of a substance across a membrane requiring the expenditure of energy; usually against a gradient.

ACTH   Adrenocorticotropic hormone.

actomyosin   A protein complex formed by the combination of actin and myosin.

adaptation   (1) The process by which an organism becomes fitted to its environment; evolution. (2) Decreased excitability of receptors to repeated stimuli of constant intensity.

Addison's disease   Chronic condition caused by insufficient secretion of hormones of the adrenal cortex.

adduction   Movement toward the midline of the body or part.

adenine   One of the nitrogen bases found in RNA and DNA, and in nucleotides such as ATP, NAD, and others.

adenohypophysis   Anterior lobe of the pituitary gland.

adenoids   Enlargement of the pharyngeal tonsils in the posterior wall of the nasopharynx.

adenosine diphosphate, ADP   The molecule formed when one phosphate is split from ATP with the release of energy—a reversible reaction.

adenosine triphosphate, ATP   An energy carrier molecule, consisting of the base *adenine*, the pentose *ribose*, and three *phosphate groups*; formed by adding one phosphate plus energy to ADP.

adenyl cyclase   An enzyme, found in cell membranes, that catalyzes the formation of cAMP from ATP.

ADH   Antidiuretic hormone.

adipose   Fatty; relating to fat.

ADP Adenosine diphosphate.

adrenal gland An endocrine organ located upon each kidney. Each consists of 2 glands of different origin and function: the outer *adrenal cortex* and the inner *adrenal medulla.*

adrenalectomy Removal of the adrenal glands.

adrenaline Epinephrine.

adrenergic fiber An axon whose terminal releases epinephrine or norepinephrine.

adrenocorticotropic hormone, ACTH A hormone released by the anterior lobe of the pituitary gland; stimulates growth of the adrenal cortex and secretion of the corticosteroid hormones.

adventitia The outermost covering of an organ; derived chiefly from connective tissue. The outer layer of the wall of a blood vessel.

aerobic Requiring, or occurring in the presence of, oxygen.

afferent arteriole The arteriole which brings blood to the glomerulus of the kidney.

afferent neuron A sensory nerve cell; conducts impulses toward the CNS.

afterbirth The placenta and membranes that are expelled after delivery of the fetus.

agglutination Clumping or sticking together, as of red blood cells or bacteria.

agglutinin An antibody that causes clumping or agglutination.

agglutinogen A substance that acts as an antigen causing the production of a specific agglutinin.

agonist A prime mover. A muscle whose action is opposed by an antagonist muscle.

agranulocyte A white blood cell lacking granules in the cytoplasm.

albino One who lacks pigments in skin, hair, or eyes.

albumin A protein that is widely distributed throughout the tissues of the body. The major protein of blood plasma.

aldosterone A hormone of the adrenal cortex; increases sodium retention; most potent mineralcorticoid.

alkaline Relating to or having the properties of a base; strongly basic.

alkalosis A condition in which there is a relative increase of alkali (particularly bicarbonate) in body fluids; pH usually rises.

allergy Excessive sensitivity to a substance that is harmless to most people in the same concentration.

alpha adrenergic receptors Adrenergic receptors that can be activated or blocked by specific drugs; their activation results in increased peripheral vascular resistance, dilation of pupils, and contraction of piloerectors.

alpha cells The pancreatic cells which secrete glucagon.

alveolus (*pl.* alveoli) (1) A terminal air sac in the lung. (2) Terminal secretory portion of many glands. (3) Tooth socket in the jaw bone.

amacrine cells Cells in retina of eye; have short branching dendrites that make lateral connections between ganglion cell dendrites.

amine A compound derived from ammonia (NH_3) by replacing hydrogen with other radicals. Epinephrine and norepinephrine are amines.

amino acid An organic acid with an NH_2 (amino) group.

amnion Innermost of the membranes that enclose the embryo. It encloses the amniotic cavity.

amniotic cavity Cavity enclosed by amnion; filled with the amniotic fluid in which the embryo is immersed.

ampulla A saclike dilation at the end of a tube or duct.

amygdaloid nucleus An almond-shaped nucleus in anterior portion of temporal lobe, anterior to inferior horn of lateral ventricle; part of limbic system.

amylase An enzyme that splits starches and related polysaccharides.

anabolism The processes by which living cells assimilate nutrients and convert simple substances to more complex substances; synthesis. Requires energy.

anaerobic Living or occurring in the absence of oxygen.

anaphase The stage of cell division in which the chromosomes move apart, to opposite poles of the cell.

anaphylactic Relating to an excessive sensitivity to foreign protein or other substance due to previous exposure.

anastomosis A natural connection between blood vessels or other tubular structures; the passage of nerve fibers from one nerve to another.

anatomy The science of the structure of organisms.

androgen A hormone, or other agent, with male sex hormone activity.

anemia A deficiency in the amount of hemoglobin in the blood.

anesthesia Total or partial loss of sensation due to injury, disease, or administration of a drug.

aneurysm A blood-filled sac or dilation in the wall of an artery.

angina A severe constrictive pain; especially *angina pectoris*, a suffocating chest pain, associated with heart disease.

angiotensin A vasoconstrictor substance in the blood produced by the action of renin on the plasma globulin angiotensinogen.

anion An ion that carries a negative charge. It is attracted to the positively charged anode.

anisotropic Referring to a property of myosin filaments in muscle; birefringent or doubly refractile.

annulus fibrosus The fibrous ring that forms the circumference of intervertebral discs.

ANS Autonomic nervous system.

antagonist Acting in opposition; especially a muscle acting in opposition to a prime mover or agonist.

anterior Front or ventral.

antibody A specific substance produced in the body in response to an antigen.

anticodon The three-base sequence on tRNA that is complementary to, and base-pairs with, a codon on mRNA.

antidiuretic hormone, ADH A hormone released by the posterior lobe of the pituitary gland; promotes conservation of water by increasing water reabsorption in the kidney.

antigen A foreign substance stimulates the immune response by the formation of antibodies and/or cell-mediated immunity.

antrum An open space; a cavity or sinus; the pyloric end of the stomach.

anus The inferior opening of the digestive tract.

aorta The major systemic artery; arises from the left ventricle.

aortic sinus (1) Region of pressoreceptors in the arch of the aorta. (2) Dilated pocket behind the cusps of the aortic semilunar valves.

apnea Temporary cessation of breathing.

apneusis A type of respiration characterized by prolonged inspiration and brief expiration.

aponeurosis A fibrous sheetlike tendon for attachment of muscles.

appendage A subordinate or subsidiary part attached to the main structure; the limbs.

appendix A small wormlike appendage attached to the cecum of the large intestine.

appetite A desire, or craving, to satisfy a particular need; specifically, a desire to take food.

aqueous humor The watery fluid in the anterior and posterior chambers of the eye.

arachnoid Resembling a cobweb. Specifically, the delicate weblike middle layer of the meninges.

ARAS Ascending reticular activating system.

arcuate Bent or curved in the form of a bow; arched or bowed.

areolar Characterized by having small openings or spaces. Loosely organized, as areolar connective tissue.

arrector pili Smooth muscle in skin attached to a hair follicle; a piloerector.

arrhythmia Irregularity; loss of rhythm.

arteriole Smallest artery with a muscular wall. It is continuous with capillary network.

arteriosclerosis Hardening of the arteries.

artery A vessel that carries blood away from the heart.

arthritis An inflammation of the joints.

arthrology The study of the joints.

articulation Joining or connecting together; specifically, the joints between bones or cartilages of the skeleton.

arytenoid cartilages A pair of small cartilages in the larynx.

ascites Accumulation of a serous fluid in the peritoneal cavity.

ascorbic acid Vitamin C. A water soluble vitamin; a strong reducing agent used as an antioxidant in foodstuffs.

asphyxiation Loss of consciousness from interruption of breathing, as in suffocation or drowning.

asthma A disease characterized by difficulty of breathing, due to spasmodic bronchial constriction.

astigmatism A visual defect due to unequal curvatures of the refractile surfaces of the eye. Light rays from a point do not come to focus at a point on the retina.

astrocyte A star-shaped glial cell.

atelectasis Failure of expansion of alveoli of lungs; pulmonary collapse.

atherosclerosis Arteriosclerosis due to lipid deposits in lining of arteries.

atlas First cervical vertebra. Articulates with the occipital bone of the skull.

atom The smallest particle of an element that still retains the properties of that element; composed of protons, neutrons, and electrons.

ATP Adenosine triphosphate.

atresia Closure of a passage or channel; degeneration of ovarian follicle.

atrioventricular, AV Pertaining to both the atrium and ventricle of the heart.

atrioventricular bundle Bundle of specialized fibers in the heart that conducts impulses from the atrioventricular node to the ventricular muscle.

atrioventricular (AV) node Small mass of specialized cardiac muscle tissue in right atrium near entry of coronary sinus; the atrioventricular bundle arises from it.

atrium A chamber or opening; particularly the upper (receiving) chamber on each side of the heart.

atrophy A decrease in size, or wasting away, of a tissue, organ, or part.

atropine A poisonous substance derived from the deadly nightshade plant. It blocks transmission at parasympathetic postganglionic nerve endings.

auditory Pertaining to the sense of hearing.

auditory tube Eustachian tube; a tube connecting the middle ear with the nasopharynx.

auricle An earlike appendage; formerly used synonymously with atrium in heart, and now used to describe only a portion of each atrium.

auscultation Act of listening to sounds within the body, as in blood vessels and in the chest and abdomen.

automaticity Ability of a structure to initiate its own activity.

autonomic Autonomous, independent, self-directed.

autonomic nervous system The involuntary portion of the nervous system which innervates smooth and cardiac muscle and glands.

autoregulation The ability of a part or area to regulate itself independently; particularly applied to blood flow to an organ.

axial skeleton The bones of the head and trunk, as distinguished from those of the appendages.

axilla The region of the armpit.

axis (1) The second cervical vertebra. (2) That imaginary straight line passing through a movable joint about which the movable bone revolves.

axon The nerve cell process which carries impulses away from the nerve cell body; sometimes used more generally as any long nerve cell process.

axon hillock Raised area of a neuron cell body from which the axon arises; it lacks Nissl substance.

axoplasm Cytoplasm in a nerve cell process.

azygos An unpaired anatomical structure; specifically, the veins draining the chest wall.

B lymphocyte Lymphocyte associated with humoral immunities; gives rise to plasma cells that produce and release antibodies.

bacteria A large group of typically one-celled microorganisms widely distributed in air, water, soil, and the bodies of living plants and animals.

Barr body Sex chromatin; a small compact mass of chromatin near the nuclear membrane of cells of females, representing an X chromosome.

basal ganglia Several distinct masses of gray matter (nuclei) within the cerebral hemispheres: includes the corpus striatum and other nuclei. They are part of the telencephalon.

basal metabolism Minimal metabolism required to maintain life in an awake individual.

base A substance capable of reacting with acid.

basement lamina Layer of extracellular material under the deep surface of epithelium.

basic electrical rhythm, BER Train of spontaneous depolarization of pacemaker cells in longitudinal muscle of stomach and small intestine; is conducted through longitudinal muscle; determines frequency of contraction.

basophil A white blood cell characterized by a rather pale nucleus and large cytoplasmic granules that stain readily with basophilic dye.

beta adrenergic receptors Adrenergic receptors that can be activated or blocked by specific drugs; their activation results in increased heart rate and force of contraction, bronchial dilation, glycogenolysis.

beta cells The pancreatic cells which secrete insulin.

β-hydroxybutyric acid One of the ketone bodies formed during the oxidation of fatty acids and found in the blood and urine of diabetics and during starvation.

beta oxidation Alternate oxidation. In the oxidation of fatty acid, the reaction occurs between the 2nd and 3rd carbons, splitting off two-carbon groups.

biaxial joint A joint with two axes; movement can occur in two planes.

biceps Having two heads, especially applied to certain muscles.

bicuspid valve Having two processes or cusps; the left atrioventricular (mitral) valve of the heart.

bifid Split or separated into two parts.

bifurcation Forming two branches, a forking.

bilateral Relating to, or having, two sides.

bile A fluid secreted by the liver, stored in the gall bladder, and released into the duodenum.

bilirubin   A red pigment found in bile.

binocular vision   Vision with two eyes, in which the image formed reflects the fact that the two eyes view the object from slightly different angles, creating a three-dimensional appearance.

bioassay   Determination of the concentration or potency of a substance by comparing its effect upon a tissue, animal, or microorganism with that of a standard preparation.

bipennate   An arrangement of muscle fibers in which fibers are attached to both sides of the tendon, as in a feather.

bipolar cells   Neurons having two processes.

blastocyst   Stage in early development of an embryo, consisting of a hollow ball of cells with an inner cell mass.

blind spot   That part of the retina from which the optic nerve leaves. It contains no receptors and so is insensitive to light.

blood   The fluid tissue of the body contained within the cardiovascular system.

blood-brain barrier   Indication of the existence of a modification of the membranes that separate blood from the cells in the brain; it blocks passage of certain substances that pass into tissues elsewhere.

BMR   Basal metabolic rate.

body   The principal mass of an organ or structure; corpus; soma.

bolus   A rounded mass, especially a soft mass of chewed food ready for swallowing.

bomb calorimeter   A special chamber designed to measure heat liberated upon combustion of the material inside it—specifically the heat (energy) content of foods.

bone   The hard connective tissue forming the skeleton. Its fibers are collagen and its ground substance contains salts, mainly of calcium and phosphate.

Bowman's capsule   Glomerular capsule; enlarged end of a nephron.

brachial   Pertaining to the arm.

brachial plexus   Nerve plexus formed of anterior primary divisions of fifth to eighth cervical and first thoracic nerves; provides innervation to upper extremity.

bradykinin   A potent vasodilator; a polypeptide formed in blood by action of a proteolytic enzyme.

brain   The mass of nervous tissue within the cranium; it consists of the brainstem, cerebral hemispheres, and cerebellum.

brainstem   The brain except for the cerebral hemispheres and cerebellum. That is, it consists of the medulla oblongata, pons, and midbrain.

broad ligament of uterus   The fold of peritoneum which covers the anterior and posterior surfaces of the uterus, enclosing the uterine tubes and blood vessels, and from whose posterior surface the ovary is suspended.

bronchiole   Small terminal branch of respiratory airway.

bronchitis   Inflammation of the mucous membrane lining the bronchial tubes.

bronchus   (*pl.* bronchi) The large branch of the trachea leading to the lungs. There are a right and a left bronchus, each of which divides several times.

buffer   A compound in solution that is able to combine with an acid or with a base to form a weaker acid or base. It is used to stabilize pH values.

bulbourethral glands   Small glands located on either side of the male urethra; they open into the cavernous urethra.

bundle of His   The atrioventricular bundle.

bursa   (*pl.* bursae) A fluid filled sac or cavity lined with synovial membrane. Found around joints or sites of friction or pressure.

bursa of Fabricius   Site in poultry where lymphocytes acquire the special properties to become B lymphocytes.

calcification   The process by which insoluble calcium salts are deposited in the matrix of bone, or cartilage; also in other tissues.

calcitonin   A hormone produced by the thyroid gland. It lowers blood calcium level by inhibiting the removal of calcium from bone.

callus   (1) The hard bonelike material that develops around the fragments of a broken bone. (2) A thickened area of skin, due to friction or pressure.

calorie   A unit of heat energy. The amount of heat required to raise 1 g water 1°C. 1000 calories = 1 kilocalorie.

calorimetry   The measurement of the amount of heat given off.

calyx   (*pl.* calyces) Cuplike extension of the renal pelvis, surrounding the tip of one or more papillae.

cAMP   Cyclic adenosine monophosphate.

canal   A duct or channel; a tubular structure.

canaliculus   (*pl.* canaliculi) A small canal.

cancellous bone   Spongy bone, with a latticework of interconnected spicules of bone; forms the inner, more spongy portion of bone.

capacitance vessels   Venules and veins that are capable of containing a large volume of blood.

capacitation   The process by which sperm acquire the ability to fertilize ova.

capillary   The smallest of blood vessels. They connect arterioles and venules. The site of exchange between blood and interstitial fluid.

capsule   A fibrous or membranous sac that encloses an organ or part.

carbaminohemoglobin   The compound formed in red blood cells by the combination of carbon dioxide and hemoglobin.

carbohydrates   Organic compounds containing carbon, hydrogen, and oxygen (the latter in the proportion of 2:1); includes starches, sugars, cellulose, etc.

carbon monoxide   A colorless, odorless, poisonous gas formed by the union of one atom of carbon and one atom of oxygen; CO.

carbonic anhydrase   An enzyme that catalyzes the reaction between carbon dioxide and water to carbonic acid, and also the reverse reaction.

carboxyl   The chemical group that is characteristic of organic

$$acids; - COOH (C - OH).$$

with structure $-COOH$ having C double-bonded to O.

cardiac   Pertaining to the heart.

cardiac centers   Areas in the brain for the control of heart rate.

cardiac muscle   Special type of muscle found in the heart.

cardiac output   The amount of blood pumped by the heart each minute.

carotene   A yellow pigment that can be converted to vitamin A (a provitamin).

carotid bodies   Location of chemoreceptors located at the bifurcation of the carotid arteries.

carotid sheath   Connective tissue sheath enclosing the carotid artery, jugular vein, and vagus nerve in the neck.

carotid sinus   Location of pressoreceptors in the bifurcation of the carotid arteries.

carpals   Pertaining to the wrist and eight bones of the wrist.

carrier   A chemical substance that can accept an atom, radical, or subatomic group from one compound and pass it on to another.

cartilage   A type of connective tissue in which cells are surrounded by a firm, resilient nonvascular matrix containing numerous fibers, mostly collagenous.

cartilage bone   Bone developed from and within cartilage, by cartilage replacement.

castration   Removal of the testes or ovaries.

catabolism   Processes by which complex compounds are broken down by living cells to simpler compounds; often accompanied by the release of energy.

catalyst   A substance that accelerates a chemical reaction, but is, itself, not consumed or permanently changed in the reaction. Enzymes are catalysts.

cataract   Opacity of the lens of the eye.

catecholamine   One of a group of chemical compounds that includes epinephrine, norepinephrine, and dopa (dihydroxyphenylalanine).

catheter   A narrow hollow tube designed to be passed into certain organs of the body, such as the urinary bladder or heart.

cation   An ion carrying a positive charge. It is attracted to the negatively charged cathode.

cauda equina   The bundle of spinal nerve roots extending below the end of the spinal cord; horse's tail.

caudal   Toward the tail end of an animal; inferior, in humans.

caudate nucleus   One of the basal ganglia in the brain. Characterized by a long taillike process.

cavernous body   Structure in the penis and clitoris that contains blood spaces or sinuses; erectile tissue.

CCK   Cholecystokinin-pancreozymin; CCK-PZ.

cecum   A blind end pouch; specifically, the sac at the beginning of the large intestine, into which the ileum empties.

cell   The structural and functional unit of living organisms; a mass of protoplasm enclosed in a membrane and usually containing a nucleus.

cell body   Soma; the mass of cytoplasm containing the nucleus of a nerve cell. Processes arise from the cell body.

cementum   Covering of the root of a tooth that is connected to the bone of the alveolar process.

center   (1) A group of nerve cell bodies in the central nervous system with a restricted, special function. It is the site of synaptic connections. (2) A region in which a particular process begins, as in ossification.

central inhibition   The reduction of the number of outgoing impulses from a reflex center.

central nervous system, CNS   The brain and spinal cord.

centriole   Two tiny structures near the nucleus of a cell. They move apart during mitosis, forming the spindle.

centromere   The constricted part of a chromosome to which spindle fibers attach during mitosis.

cephalic   Pertaining to the head; head end. It is superior in humans.

cerebellum   Part of the hindbrain, consisting of two hemispheres and a small central portion.

cerebral aqueduct   A narrow canal through the midbrain, connecting the third and fourth ventricles.

cerebrospinal fluid, CSF   The fluid produced in the ventricles of the brain. It fills the ventricles and surrounds the central nervous system.

cerebrum   Generally includes the structures of the telencephalon, such as the cerebral hemispheres, the basal ganglia, and rhinencephalon.

cervix   A neck; necklike constriction. The narrow inferior end of the uterus, leading into the vagina.

charge   A term referring to an excess or deficiency of electrons, as with ions.

chemoreceptors   Receptors that are stimulated by chemical changes, such as taste buds.

chemosensitive area, CSA   Area on the ventral surface of the medulla of the brain that is sensitive to changes in the composition of the cerebrospinal fluid, and influences respiration.

chemotaxis   Property by which chemical stimuli attract or repel organisms or protoplasm; especially migration of certain white blood cells.

chiasm   A crossing or intersection, as of the optic nerves (at the *optic chiasm*).

cholecystokinin-pancreozymin, CCK   A hormone, formerly believed to be two hormones, produced by the intestinal mucosa that causes contraction of the gall bladder (cholecystokinin) and secretion of an enzyme-rich pancreatic juice (pancreozymin).

cholesterol   The most abundant steroid in animal tissues, especially in bile.

cholinergic   A term applied to any nerve fiber whose terminals release acetylcholine upon stimulation.

cholinesterase, AChE   A widely distributed enzyme that inactivates acetylcholine.

chondroblast   A cell of growing cartilage; a cartilage-forming cell.

chondrocyte   A cartilage cell.

chordae tendineae   Tendinous strands extending from the margins of the cusps of the atrioventricular valves to papillary muscles in the ventricles of the heart.

chorion   Outermost of the fetal membranes.

chorionic gonadotropin   A hormone produced by trophoblastic cells of the placenta. It stimulates the corpus luteum to produce estrogen and progesterone during the first months of pregnancy.

chorionic villi   Processes that develop from the external part of the chorion and become embedded in the endometrium. They become vascularized and are the means of carrying out metabolic exchanges between the fetus and the mother.

choroid   A skinlike coat. The middle layer of the eyeball.

choroid plexus   Vascular projections into the ventricles of the brain which produce the cerebrospinal fluid.

chromatin   Deeply staining material in the nucleus of a cell which emerges as chromosomes during mitosis.

chromatolysis   The changes that occur in a nerve cell body as a result of damage to its peripheral process. Includes disintegration of the highly staining Nissl bodies.

chromosome   The gene-containing material in the nucleus of a cell. The chromosomes appear as recognizable units during mitosis.

chylomicron   Microscopic droplet of fat, especially in lymph vessels after a meal high in fat.

chyme   Semifluid mass of partly digested food leaving the stomach.

chymotrypsin   A protein-splitting enzyme secreted by the pancreas, as the inactive chymotrypsinogen.

cilia   Tiny hairlike processes on the free edge of certain epithelial cells. They often tend to "beat" in a coordinated wavelike fashion.

ciliary body   The thickened anterior edge of the choroid layer of the eye; contains the ciliary muscles which participate in accommodation.

circle of Willis   A series of arteries on the ventral surface of the brain, that forms a ring around the pituitary stalk. It provides a connection between the two major arteries to the brain.

circumcision   Removal of the foreskin of the penis.

circumduction   Movement of a part, such as an eye or limb, in a circular direction.

cisterna   A cavity or enclosed space serving as a reservoir; a cistern. Specifically, the cisterna chyli, the dilated sac at the lower end of the thoracic duct; receives lymph from the inferior half of the body.

citric acid cycle   Krebs cycle; the common pathway for oxidation of carbohydrate, fat, and some amino acids. A sequence of reactions in which two-carbon fragments (acetate) are metabolized to carbon dioxide with the release of energy.

clitoris The small erectile organ in the female, homologous to the penis in the male.

clone A colony or group of cells or organisms that have arisen from a single cell or individual by asexual reproduction.

CNS The central nervous system.

Cochlea The portion of the inner ear associated with the reception of sound waves. It lies in a cavity in the temporal bone and is shaped like a snail shell.

codon Sequence of three bases on a strand of DNA that specifies a particular amino acid.

coenzyme A nonprotein substance that is necessary for the action of an enzyme.

coenzyme A, CoA The coenzyme involved in transferring two-carbon units into the Krebs cycle.

collagen The protein found in the most widely distributed connective tissue fibers (collagenous fibers). It is strong and inelastic and yields gelatin upon boiling.

collateral Accessory to the main thing, as a collateral ganglion; or a side branch, as collateral blood vessels or nerves.

colloid (1) A dispersion of solid particles in a medium. The particles are larger than crystalloidal particles and do not pass through natural membranes. (2) The material in follicles of the thyroid gland.

colloid osmotic pressure, COP The osmotic pressure developed by particles of colloidal substances in solution. Generally refers to that osmotic pressure of blood plasma and body fluids that is due to the presence of protein.

collum The neck; cervix.

colon The large intestine, from the cecum to the rectum.

color blindness Inability to distinguish one or more of the primary colors.

commissure A bundle of nerve fibers connecting the right and left sides of the brain or spinal cord.

communicate To connect.

communicating ramus A branch of a spinal nerve carrying nerve fibers to the sympathetic chain.

compact bone Dense bone, consisting largely of haversian systems and interstitial lamellae; forms the surface portion of bones.

complement A group of proteins that participate in certain antigen-antibody reactions; they destroy certain bacteria or other cells which have been sensitized by specific complement fixing antibodies.

compliance The ease with which a structure or substance can be deformed or distended; the volume change per unit pressure, especially in the lung, as a measure of its mechanical properties.

compound A distinct substance formed by the union of two or more elements reacting in definite proportions by weight.

compound action potential The action potential in a nerve trunk; the sum of action potentials in the individual axons.

concave Having a depressed or hollowed, curved surface.

conception Fertilization of an ovum; becoming pregnant.

concha (*pl.* conchae) A shell-shaped structure, such as the nasal conchae on the medial wall of the nasal cavity.

conduction (1) The transmission or propagation of certain forms of energy; particularly electrical energy, as nerve impulses. (2) The transfer of heat by direct contact.

condyle A large, rounded articular surface at the end of a bone.

cones Receptor cells in the retina of the eye for color perception.

congenital Existing at birth.

conjunctiva The mucous membrane covering the anterior surface of the eyeball and lining the eyelids.

connective tissue One of the primary tissues. Characterized by scattered cells (chiefly fibroblasts and macrophages) in large amounts of intercellular substance composed of fibers (collagenous, elastic, and reticular) and amorphous ground substance.

contraceptive Any agent for the prevention of conception.

contractility The ability or property of a substance to shorten or develop tension, especially as applied to muscle.

contraction A shortening or increase in tension; the normal response of muscle.

contracture A permanent shortening or maintained contraction of a muscle, with impaired or retarded relaxation.

contralateral On the opposite side.

convection Transfer of heat by movement of heated particles, as when warm air or fluid rises.

convergence The coming together of two or more objects or lines at a common point; several neurons synapsing on a single neuron; inward deviation of the eyes for near vision.

convex A curved surface that bulges outward.

COP Colloid osmotic pressure.

copulation Coitus; sexual intercourse.

coracoid Shaped like a crow's beak; applied particularly to a process on the scapula.

core temperature The temperature of the internal parts of the body; deep temperature.

corium Dermis.

cornea The transparent epithelium and connective tissue membrane that covers the anterior portion of the eye.

cornified A "horny" layer; converted to keratin.

corona radiata (1) Arrangement of cells around a mature ovum. (2) Fan-shaped appearance of nerve fibers radiating from the upper portion of the internal capsule in the brain.

coronal Pertaining to any structure that resembles a crown.

coronal plane A frontal plane, dividing an object into anterior and posterior parts.

coronary Encircling in the manner of a crown; applied particularly to blood vessels to the heart.

coronoid Shaped like a crow's beak; applied to processes of certain bones.

corpora cavernosa Columns of erectile tissue that form the major portion of the penis.

corpora quadrigemini Four small rounded masses on the dorsal surface of the midbrain (the superior and inferior colliculi), associated with visual and auditory functions.

corpus Body or mass; the main portion of an organ or other structure.

corpus albicans The white scar remaining after a corpus luteum has degenerated.

corpus luteum Yellow body; a temporary endocrine gland formed of a ruptured ovarian follicle.

corpus spongiosum Column of cavernous tissue in the penis, traversed by the urethra.

corpus striatum Certain basal ganglia (caudate, globus pallidus, and putamen) considered as a unit. Its striated appearance is due to connecting bundles of white matter penetrating the gray matter.

cortex The bark; the outer portion of an organ, such as of the kidney, adrenal gland, or brain, as distinguished from the inner or medullary portion.

corticoids Substances having actions like those of the steroids of the adrenal cortex.

corticosteroids The steroid hormones released by the adrenal cortex.

corticotropin Adrenocorticotropic hormone; ACTH; a hormone of the anterior pituitary which stimulates secretion by the adrenal cortex.

cortisol The major glucocorticoid produced by the adrenal cortex in humans.

costal Pertaining to a rib.

countercurrent mechanism A mechanism based upon the effects produced by two currents flowing in opposite directions near one another, as in the kidney.

cranial Pertaining to the head or skull.

cranial nerve One of the twelve pairs of nerves that arise from the brain.

craniosacral division That part of the autonomic nervous system that arises from the brain and the sacral part of the spinal cord; the parasympathetic division.

cranium Skull; the bones of the head, specifically those enclosing the brain, excluding the bones of the face.

creatine phosphate, CP An energy-rich compound found particularly in skeletal muscle; a source of energy for resynthesis of ATP from ADP.

creatinine A component of urine; formed in the catabolism of creatine.

crenation Shriveling or shrinking of a cell due to withdrawal of water, e.g., red blood cells in a hypertonic solution.

crest A ridge, especially a bony ridge.

cretinism A condition of stunted growth and development, both physical and mental, caused by severe hypofunction of the thyroid gland in infancy.

cribriform Sievelike; containing many perforations.

cricoid cartilage A ring-shaped cartilage of the larynx.

crista A ridge, crest, or line projecting from a surface.

crossed extensor reflex Extension of the contralateral limb during a flexion reflex.

crown The exposed portion of a tooth; the part above the gum.

cruciate Shaped like a cross; crossing or overlapping.

cryptorchidism Condition in which the testes have not descended into the scrotum.

crystalloid Very small diameter particles which, in solution, can pass through a natural membrane.

crystalloid osmotic pressure The osmotic pressure that is due to the presence of crystalloids in a solution.

CSA Chemosensitive area.

CSF Cerebrospinal fluid.

cupula A cup-shaped or domelike structure, as in the semicircular canals of the inner ear.

curare A drug derived from various plants used in poison arrow tips; causes paralysis by blocking transmission at skeletal neuromuscular junctions.

Cushing's disease A disease caused by excessive secretion of adrenocorticotropic hormone; marked by symptoms of hyperfunction of the adrenal cortex.

cusp (1) A leaflet of one of the heart's valves. (2) A point on the crown of a tooth.

cutaneous Pertaining to the skin.

cyanosis A bluish or purplish coloration of the skin and mucous membranes caused by deficient oxygenation of blood; specifically, by an increased content of reduced hemoglobin in the blood.

cyclic adenosine monophosphate, cAMP Adenosine 3′, 5′ cyclic monophosphate. An intracellular mediator of many hormone actions; "second messenger."

cystic duct The duct leading from the gall bladder to the common bile duct.

cytology The study of cells.

cytoplasm The substance (protoplasm) in a cell, exclusive of the nucleus; contains organelles and inclusions within a colloidal protoplasm.

cytosine One of the nitrogenous bases found in DNA and RNA.

dark adaptation The process by which the receptors in the eye adjust to low levels of illumination. It involves regeneration of rhodopsin and dilation of the pupil.

dead space The space in the respiratory "tree" (nasal passages, trachea, bronchi, and bronchioles) in which there is not effective exchange of gases with the pulmonary blood.

deamination A chemical reaction in which an amino group (NH_2) is removed from an amino acid.

decerebrate rigidity The increase in tone of antigravity muscles that develops after decerebration.

decerebration Removal of the brain above the corpora quadrigemini.

deciduous teeth Teeth that are not permanent; the "baby teeth."

decompression sickness The "bends"; a condition resulting from a too rapid decrease in barometric pressure, or decompression; caused by nitrogen forming bubbles in the tissues as it comes out of solution when the pressure falls rapidly.

decortication Removal of the outer layer or cortex; particularly of the cerebrum.

deep Below the surface, as opposed to superficial.

defecation The discharge of the contents of the rectum.

deglutition The act of swallowing.

dendrites Branching cytoplasmic processes of nerve cells that receive impulses and transmit them to the nerve cell body.

denervate To cut the nerve supply to a part.

denervation hypersensitivity The increased sensitivity of a denervated structure to the transmitter substances normally released by its nerve supply.

dens Tooth; a toothlike structure.

dental caries Tooth cavity; localized destruction of a tooth; penetration of enamel and invasion of dentin and, eventually, the tooth pulp.

denticulate ligament Ligament extending laterally along each side of the spinal cord as narrow toothlike attachments to the dura mater.

dentin The main part of a tooth; the bonelike substance beneath the enamel.

deoxycorticosterone A steroid compound with mineralocorticoid actions; is less potent than aldosterone, and less prominent in adrenal secretion.

deoxyhemoglobin Hemoglobin that has given up its oxygen; reduced hemoglobin.

deoxyribonucleic acid, DNA The nucleic acid constituent of chromosomes; carries the genetic information and is the chemical basis of heredity.

depolarization The destruction or neutralization of polarity. In excitable tissues, it is the reduction of the negative charge inside the cell or cell process; associated with increased excitability.

depression Movement of the shoulder girdle so as to lower the tip of the shoulder (the acromion process); opposite of elevation.

depressor Any agent that causes inhibition, reduction, or slowing; as, for example, a reduction of blood pressure or slowing of the heart.

dermatome The area of skin supplied by sensory fibers of a single dorsal root.

dermis The deep thick inner layer of skin; corium; under the epidermis.

desmosome Site of adhesion between two cells; consisting of plate of dense material on the two cells, separated by a thin layer of extracellular material.

Dextrin A polysaccharide formed as an intermediate in the breakdown of starch.

dextrose Glucose; a sugar.

diabetes insipidus A condition marked by increased output of very dilute urine, due to inadequate production of antidiuretic hormone.

diabetes mellitus A condition marked by increased output of urine with a high content of sugar, due to inadequate production of insulin; characterized by severe problems of metabolism, particularly of carbohydrate and fat.

diabetogenic Any agent that tends to cause symptoms of diabetes mellitus, such as elevated blood glucose level.

dialysis Separation of crystalloids from colloids in a solution by placing a semipermeable membrane between the solution and water. Crystalloids will diffuse and colloids will not.

diaphragm A thin partition or wall, such as (1) the partition of skeletal muscle and tendon that separates the thoracic and abdominal cavities; or (2) the thin disc for controlling the amount of light entering a microscope.

diaphysis (*pl.* diaphyses) The shaft of a long bone.

diastole Period between contractions in the heart; relaxation, dilation, and filling of the chambers.

diastolic pressure The lowest pressure that occurs in an artery during diastole; also in the ventricle.

diencephalon That part of the forebrain between the telencephalon and mesencephalon. It contains primarily the thalamus, hypothalamus, subthalamic structures, and neurohypophysis.

differentiation Specialization; development of different properties or characteristics, as in tissues or cells.

diffusion The random movement of particles in a gas or solution that leads to a uniform distribution throughout the available volume.

digestion The mechanical and chemical breakdown of the foodstuffs to compounds that are suitable for absorption by the cells of the body.

digit A finger or toe.

diopter The unit of refractile power of a lens.

dipeptide A compound formed by two amino acids, joined together by a peptide bond.

diplopia Double vision; a single object is perceived as two objects.

direct calorimetry Actual measurement of the heat produced, as contrasted with indirect methods in which heat production is calculated from the measurement of other variables.

disaccharide A sugar formed by the union of two monosaccharides, such as sucrose (table sugar).

dissociation Separation; a complex chemical compound becomes a simpler one, as by ionization.

distal Farthest from the center or median line, as opposed to proximal.

diuresis Increased rate of urine production.

diurnal Daily; recurring every day.

divergence Movement or spreading apart in different directions; especially the spreading of branches of a neuron to synapse with several other neurons.

DNA Deoxyribonucleic acid.

dopamine Immediate precursor of norepinephrine; also a transmitter at some brain synapses.

dorsal Pertaining to, or lying near, the back; in humans, posterior.

dorsiflexion Movement at the ankle in which the foot and toes are turned upward.

dorsum The back or posterior surface of a part; the back of the body; the back of the hand (as opposed to the palmar surface); the upper surface of the foot (as opposed to the plantar surface).

Down's syndrome A congenital defect associated with the presence of an extra chromosome 21; associated with mental retardation and characteristic physical features; mongolism.

duct A canal or tubular passage.

ductus arteriosus Fetal vessel connecting the left pulmonary artery with the arch of the aorta. It closes a few weeks after birth and becomes the ligamentum arteriosum.

ductus deferens The duct that transports sperm from the epididymis to the ejaculatory duct; transports sperm from the testis.

ductus venosus Fetal vessel, a continuation of the umbilical vein, which empties into the inferior vena cava. After birth it becomes obliterated and remains as the ligamentum venosum.

duodenum The first portion of the small intestine, leading from the stomach. It receives secretions from the pancreas and liver.

dura mater Tough outer layer of the meninges enclosing the brain and spinal cord.

dysmenorrhea Difficult and painful menstruation.

dyspnea Difficult or labored breathing; "shortness of breath."

ECF Extracellular fluid.

ECG Electrocardiogram.

ectopic Displaced; not in the normal place; such as a heartbeat not arising from the sinoatrial node, or a pregnancy occurring outside of the uterus.

edema Excessive accumulation of fluids in the tissues.

EDV End-diastolic volume.

EEG Electroencephalogram.

effector The responding tissue; skeletal, smooth, or cardiac muscle or glands.

efferent Conveying away from; used of a vessel, duct, or nerve, but especially of nerve fibers that carry impulses away from the central nervous system; the motoneuron.

efferent ductules Small ducts leading from the testis to epididymis.

ejaculation The act of throwing out or ejecting a fluid from a duct; specifically the emission of seminal fluid.

elastin The major connective tissue protein in elastic connective tissue.

electrocardiogram, ECG, EKG Graphic record of electrical changes associated with each beat of the heart.

electrode One of the two terminals of an electrical circuit.

electroencephalogram, EEG A record of the electrical activity of the brain, obtained by electrodes on the scalp.

electrolyte A substance which ionizes in solution and conducts an electric current.

electromyogram, EMG A graphic record of the electrical activity associated with contraction of muscle.

electron A negatively charged particle; a unit of negative electricity.

electron microscope A microscope in which beams of electrons are used instead of beams of visual light. Because of the much shorter wavelength of the electron beams, much higher magnifications are possible.

element A substance composed of atoms of one particular kind. An element cannot be decomposed into two or more other substances.

elevation A movement of the shoulder girdle in which the tip of the shoulder (acromion) is raised; opposite of depression.

embolism Obstruction of a blood vessel by a moving blood clot or other mass carried in the bloodstream; may also be a bubble of gas (air) in the vessel.

embryo An organism in the early stages of development. In humans, from conception approximately to the end of the second month, after which it is commonly designated a fetus.

emphysema A condition in which the air spaces of the lungs are enlarged, either by dilation or destruction of the walls between alveoli.

enamel The hard material covering the exposed part of a tooth.

end-diastolic volume, EDV   Volume of blood in the ventricle just before systole; an indication of the degree of filling.

end-feet   Presynaptic endings; small terminal enlargements of nerve cell processes that make synaptic contact with another neuron.

endocardium   The innermost layer of the heart, primarily endothelial.

endocrine   Pertaining to glands that secrete internally, into the bloodstream; the hormone secretion of such glands.

endocytosis   Movement of materials into cells, particularly of large molecules, as in phagocytosis and pinocytosis.

endolymph   Fluid contained within the membranous labyrinth of the inner ear.

endometrium   Mucous membrane lining the uterus, composed of epithelium and its underlying connective tissue.

endomysium   Thin connective tissue sheath surrounding each muscle fiber.

endoneurium   Thin connective tissue surrounding each nerve fiber; holds the nerve fibers of a small bundle together.

endoplasmic reticulum, ER   A network of fine tubules and vesicles within the cytoplasm of a cell; an organelle.

endorphin   A small peptide found in the brain capable of producing effects similar to opiates (such as morphine), particularly easing pain.

endosteum   The membrane lining the inner surface of bone in marrow cavities.

endothelium   Layer of flat epithelial cells lining the heart, blood vessels, and lymphatics.

end-plate   Slightly raised area marking the junction of a somatic motoneuron and a skeletal muscle fiber.

end-plate potential, EPP   Local, nonpropagated potential developed in the end-plate region of a skeletal muscle cell when an impulse arrives on its motoneuron.

end-systolic volume, ESV   Volume of blood in the ventricle at the end of contraction; an indication of the degree of emptying.

energy   The capacity to do work.

enteric   Pertaining to the intestine.

enterohepatic circulation   The circuit by which constituents of the bile, secreted by the liver and entering the duodenum, are absorbed from the intestine into the portal vein and returned to the liver.

enterokinase   An enzyme produced by the intestinal mucosa that activates trypsinogen.

enzyme   An organic catalyst (protein) produced by a living cell.

eosin   A commonly used histological stain; an acid dye.

eosinophil   A granular leukocyte whose granules stain readily with acid dyes.

ependyma   The cellular membrane lining the ventricles of the brain and central canal of the spinal cord. Its cells are neuroglial cells.

epicardium   The visceral (inner) layer of pericardium that immediately covers the heart.

epicondyle   A projection on or above a condyle of a bone.

epidermis   The outer epithelial portion of the skin.

epididymis   The tortuously coiled tube attached to the testis and leading to the ductus deferens. Sperm are stored here.

epiglottis   Leaf-shaped elastic cartilage at the base of the tongue and superior end of the larynx.

epilepsy   A chronic disorder marked by recurring attacks of brain dysfunction; often associated with some alteration of consciousness; may be very slight, but ranges up to generalized convulsions.

epimysium   The connective tissue sheath surrounding a skeletal muscle.

epinephrine   The chief secretion of the adrenal medulla in humans and a component of the transmitter substance of adrenergic fibers.

epineurium   The connective tissue surrounding a nerve trunk, binding together bundles of fibers.

epiphysis   (pl. epiphyses) The end of a long bone and the last part of the bone to ossify. In early life it is separated from the shaft by cartilage.

epiploic   Pertaining to the omentum.

epithelium   One of the primary tissues; characterized by closely packed cells; covers internal and external surfaces, and lines ducts and tubes, and includes glands.

eponychium   The fold of skin over the root of the nail; its free edge forms the cuticle.

EPP   End plate potential.

EPSP   Excitatory postsynaptic potential.

equilibrium   Balance. A state when opposing reactions or events are occurring at equal rates.

equivalent   The weight in grams of an element that combines with, or replaces, 1 g of hydrogen.

ER   Endoplasmic reticulum.

erection   The hard and rigid state, particularly of the penis, produced by engorgement with blood of the erectile tissue.

erythroblastosis fetalis   Hemolytic disease of the newborn, usually caused by antibodies developed by an Rh-negative mother in response to antigens in the blood of an Rh-positive fetus.

erythrocyte   Red blood cell; RBC.

erythropoiesis   Formation of red blood cells.

erythropoietin   A substance that enhances or increases erythropoiesis; produced when the kidney is hypoxic.

esophagus   That part of the digestive tract between the pharynx and the stomach.

essential amino acids   Amino acids that are needed for proper bodily function but cannot be synthesized in the body. They must therefore be provided in the diet.

essential fatty acids   Fatty acids that are necessary for proper bodily function but cannot be produced in the body. They must be included in the diet.

essential hypertension   High arterial pressure, of unknown cause.

estrogen   Any substance with female sex hormone activity. Normally produced chiefly by the ovary and placenta. Stimulates the secondary sex organs and secondary sex characteristics.

ESV   End-systolic volume.

eupnea   Normal quiet breathing.

eustachian tube   The auditory tube.

evagination   A protrusion of some part or organ; outpocketing.

evaporation   Change from liquid to vapor form.

eversion   Turning outward; a movement at the tarsal joints in which the sole of the foot is turned outward.

excitability   Irritability; ability to respond to a stimulus.

excitation   Stimulation; increasing the rapidity or intensity of a process, especially of excitable tissue.

excitatory postsynaptic potential, EPSP   Local partial depolarization of the postsynaptic membrane, caused by the action of an excitatory transmitter at a synapse; raises the excitability of the neuron.

excretion   The process by which metabolic wastes and undigested substances are eliminated; the product or substance that is removed.

exocrine   Pertaining to glands that secrete outwardly, through a duct or to a surface; pertaining to the secretion of such glands.

exocytosis   Movement of materials, particularly large molecules, out of cells, as in secretion.

exophthalmos   A condition marked by protruding eyeballs.

expiration   Exhalation; breathing out.

expiratory reserve   The air that can be exhaled after or beyond a normal quiet expiration.

extension   A movement that increases the angle between two bones; to straighten a limb; opposite of flexion.

extensor   A muscle which, upon contraction, straightens a limb or part.

external   Exterior; on the outside, as opposed to internal.

exteroceptor   A receptor that is stimulated by changes in the external environment, particularly receptors in the skin.

extracellular   Within an organism, but outside the cells.

extracellular fluid compartment, ECF   Body water that is outside of the cells; composed mostly of intercellular, or interstitial, fluid between the cells and the blood plasma.

extrafusal   Outside of the muscle spindle; applied to skeletal muscle fibers.

extrapyramidal tracts   Those nerve tracts descending from the brain that do not pass through the pyramids of the medulla of the brain.

extremity   A limb; arm (superior extremity) or leg (inferior extremity).

facet   A small smooth area on a bone for articulation.

facilitated diffusion   Diffusion across a membrane with the aid of a carrier molecule. Transfer is "down" the gradient and energy is not required to cause transfer.

facilitation   A decrease in a neuron's resting potential due to incoming impulses and excitatory postsynaptic potentials (EPSPs); increased excitability of a neuron.

FAD   Flavin adenine dinucleotide.

fallopian tube   Uterine tube; oviduct; tube by which ovum is transported to the uterus.

falx   A sickle-shaped structure; as in falx cerebri, which is the fold of dura mater in the longitudinal fissure between the cerebral hemispheres of the brain.

fascia   Sheet of fibrous connective tissue which envelops the body under the skin (superficial fascia) or encloses muscles and groups of muscles (deep fascia).

fascicle   A small bundle of fibers, usually of nerve or muscle fibers.

fat   A compound formed by the union of a molecule of glycerol and three molecules of fatty acid; a neutral fat; triglyceride.

fatty acid   An organic acid consisting of a long hydrocarbon chain with a carboxyl (COOH) group on one end.

feces   The material expelled from the digestive tract during defecation; consists of epithelium, secretions, bacteria, undigested residue of food, etc.

ferritin   Iron-protein complex formed by the union of iron and apoferritin in the intestinal mucosa.

fertilization   The process by which a spermatozoon penetrates and unites with an ovum.

fetus   The unborn young after it has taken form in the uterus. In humans this is from about the end of the eighth week until birth.

fever   An elevated body temperature.

fiber   Slender thread or strand. It may be extracellular, as connective tissue fibers; or cell processes, as nerve fibers; or whole cells, as muscle fibers.

fibril   A minute fiber, often, but not always, intracellular, as myofibrils of muscle.

fibrillation   Uncoordinated contraction of muscle fibers, but not simultaneous contraction of the whole muscle.

fibrin   Insoluble protein formed during the clotting of blood; forms the substance of the clot.

fibrinogen   Soluble protein in the blood plasma that is converted to fibrin when blood clots.

fibroblast   Connective tissue cell responsible for formation of connective tissue fibers and ground substance.

filament   A very fine threadlike structure; a small fibril.

filtration   Passage of fluid and solute through a membrane due to a mechanical (hydrostatic or filtration) pressure gradient.

filum terminale   Thin nonneural thread extending from the inferior end of the spinal cord to the coccyx.

fimbria   A fringelike structure, especially that surrounding the abdominal opening of the uterine tubes.

final common path   The motoneuron to a skeletal muscle cell; the only path by which nerve impulses can reach that cell.

fissure   A furrow, cleft, or slit.

fixators   Muscles that contribute to a movement by fixing or stabilizing the origin of the prime movers so that their action can be exerted at the insertion, e.g., scapular muscles are fixators for those muscles which originate on the scapula and move the arm.

flaccid   Flabby; without tone.

flagellum   A hairlike process of certain cells; its whiplike action provides for locomotion of the cell; characteristic of spermatozoa.

flavin adenine dinucleotide   A riboflavin-containing coenzyme; a hydrogen carrier in oxidation-reduction reactions.

flexion   A movement that decreases the angle between two bones; bending of a joint.

flexion reflex   A reflex response characterized by withdrawal of a limb, usually by flexion, in response to a cutaneous stimulation of the limb.

flexor   A muscle which, upon contracting, causes flexion or binding at a joint.

focal distance (length)   The distance from a lens at which parallel rays of light come to focus.

follicle   A small sac.

follicle cells   Cells that surround the ovum and line the wall of the ovarian follicle.

follicle-stimulating hormone, FSH   The hormone of the anterior pituitary gland that stimulates the development and maturation of an ovarian follicle; in the male it stimulates the seminiferous tubules to produce sperm.

fontanel   One of several membranous areas at the junctions of cranial bones of infants and fetuses.

foramen   A hole or opening in a bone or membranous structure.

foramen magnum   Large opening in base of occipital bone through which the spinal cord passes.

forebrain   Anterior portion of the brain, including the telencephalon and diencephalon.

fornix   A paired structure of white fibers located beneath the corpus callosum of the brain; extends from hippocampus in the temporal lobe to the mammillary bodies.

fossa   A depression, cavity, or hollow.

fovea   A cup-shaped pit or depression.

fovea centralis   The shallow pit in the retina containing only cones. It is the area of greatest visual acuity.

frontal plane   A vertical plane that divides a part into anterior and posterior portions; a coronal plane.

fructose   A sugar; a hexose monosaccharide, found in sucrose.

FSH   Follicle-stimulating hormone.

fulcrum   The fixed support or hinge about which a lever turns.

fundus   The base of a hollow organ; the part farthest from the outlet.

fusimotor fibers   Small nerve fibers providing the efferent innervation to the intrafusal muscle fibers (those in the muscle spindles); gamma efferents.

galactose   A sugar; a hexose monosaccharide found in lactose.

gall bladder   A small sac located under the liver. It stores bile.

gametes   Sexual or germ cells; ova and spermatozoa.

gamma efferents   Fusimotor fibers; the motoneurons that innervate intrafusal fibers.

ganglion   A collection of nerve cell bodies located outside the central nervous system (except for the basal ganglia of the brain, which are really nuclei).

ganglion cells   (1) Parasympathetic postganglionic cells; their cell bodies are found in the myenteric and submucous plexuses of the digestive tract. (2) Cells of the retina; their central processes form the optic nerve.

gap junction   Intercellular junction characterized by low electrical resistance, which permits electrical coupling of adjacent cells; found in smooth muscle of digestive tract and in cardiac muscle.

gastric   Pertaining to the stomach.

gastrin   A hormone produced by the gastric mucosa which stimulates the secretion of gastric juice, particularly of hydrochloric acid.

gastrointestinal, GI   Pertaining to the stomach and intestine, but often used loosely to denote the digestive tract or digestive system.

gene   Functional unit of heredity. The part of a chromosome that transmits a particular hereditary characteristic.

genetics   The branch of science that deals with heredity.

genitalia   The reproductive organs.

germinal center   Area in lymphoid tissue containing lymphocytes, some of which are undergoing mitosis. Usually stains more lightly than surrounding tissue.

gestation   Pregnancy.

gigantism   Abnormal increase in size; overgrowth. Often due to hypersecretion of growth hormone.

gingiva   The gums; dense fibrous connective tissue covered by mucous membrane; envelops bony socket and surrounds neck of each tooth.

gland   A secretory organ; may be a single cell or a complex organ.

glans   A conical, acorn-shaped structure; the small mass of erectile tissue capping the clitoris and at the end of the penis.

glaucoma   A condition characterized by increased intraocular pressure.

glenoid   Resembling a socket; applied to depressions in articular surfaces of bones of shoulder joint and jaw.

glia   Neuroglia.

globulin   A type of protein found in blood plasma; includes the proteins that are important in immunities and antigen-antibody reactions.

globus pallidus   Part of the corpus striatum, medial to putamen.

glomerulus   A small tuft of capillaries, especially the tuft of capillary loops of the renal corpuscle at the beginning of a nephron.

glossal   Pertaining to the tongue.

glottis   The slitlike opening between the true vocal cords or folds.

glucagon   A hormone produced by the alpha cells of the pancreatic islands. It acts to raise the blood glucose level.

glucocorticoids   Hormones of the adrenal cortex whose major action is on metabolic processes, particularly on carbohydrate and fat metabolism. They also have an anti-inflammatory action.

gluconeogenesis   Formation of glucose from noncarbohydrate sources, particularly from protein.

glucose   Dextrose; a sugar, a hexose monosaccharide, found in sucrose. The principal sugar in the blood.

glucosuria   The presence of glucose in the urine.

glutamine   Source of ammonia produced in the kidney; derived from glutamic acid, which is an amino acid.

gluteal   Relating to the buttocks.

glycerol   A three-carbon alcohol. An important constituent of fat, along with fatty acids.

glycogen   Animal starch; a polysaccharide formed of glucose molecules. The storage form of carbohydrate in animal cells.

glycogenesis   The formation of glycogen from glucose.

glycogenolysis   The breakdown of glycogen to glucose.

glycolysis   The breakdown of glucose to lactic acid.

goblet cell   Mucus-secreting cell of some epithelial surfaces. Accumulated secretory material in the cell gives it its gobletlike appearance.

goiter   Chronic enlargement of the thyroid gland; may represent either a hypo- or hyper-thyroid condition.

Golgi apparatus   A cluster of flattened vesicles in the cytoplasm associated with preparation of material to be secreted from the cell; an organelle.

gonad   The organ that produces the gamete; the testis in the male and the ovary in the female.

gonadotropin   One of several hormones that promote the growth and/or function of the gonad; produced by the anterior pituitary gland and the placenta.

graafian follicle   A mature ovarian follicle. It is large and fluid filled, with the oocyte attached to one side.

gradient   The rate of change, as the slope or incline; the steepness. Also applies to temperature, pressure, concentration, and other variables.

gram molecular weight   The relative weight of a molecule; equal to the sum of the atomic weights of all atoms in the molecule; expressed in grams. One mole.

granulocyte   Granular leukocyte; includes neutrophils, eosinophils, and basophils, all of which have prominent cytoplasmic granules.

granulosa cells   Small cells in the ovarian follicle; follicular cells.

gray matter   Areas of the central nervous system containing cell bodies of neurons; lacks myelin and hence has grayish appearance.

greater omentum   An apronlike fold of mesentery suspended from the greater curvature of the stomach and transverse colon.

ground substance   Amorphous intercellular substance; a protein-polysaccharide complex containing the interstitial fluid. It varies greatly in consistency, being solid and firm in bone and cartilage.

growth hormone, GH   Somatotropin; a hormone of the anterior pituitary that promotes growth of the organism; favors protein anabolism and fat mobilization.

guanine   One of the nitrogenous bases found in DNA and RNA.

gubernaculum   A fibrous cord connecting the testis and the inside of the scrotal sac.

gut   The intestine.

gyrus   A convolution on the surface of the cerebrum.

hair   A slender filamentous outgrowth of the epidermis of animals; or fine processes on certain epithelial cells (hair cells) of the inner ear.

hallux   The great toe; the first digit of the foot.

hamstrings   Tendons of the posterior thigh muscles (the biceps femoris, semimembranosus, and semitendinosus) which pass on either side of the popliteal fossa; also, applied to these muscles.

haustra   Sacculations in the colon.

haversian system   The organization of the substance in compact bone, consisting of a central (haversian) canal and concentric lamellae around it.

Hb   Hemoglobin (deoxyhemoglobin).

HbO₂   Oxyhemoglobin.

HCG   Human chorionic gonadotropin.

head   (1) The superior or anterior end of the animal body, containing the brain. (2) The proximal end of a bone. (3) The end of a muscle that is attached to the less movable part of the skeleton. (4) The upper, anterior, or larger extremity of any body or structure.

hematocrit   The percent volume of red blood cells in a volume of blood.

heme   The iron-containing pigmented component of hemoglobin. Essential for oxygen transport.

hemodynamics   The study of the dynamics of the blood circulation or blood flow.

hemoglobin, Hb   The oxygen-carrying protein in red blood cells; forms a loose reversible combination with oxygen, composed of an iron-containing pigment, heme, and a protein, globin.

hemolysis   The destruction of red blood cells with the release of their hemoglobin.

hemophilia   An inherited disorder of the blood-clotting mechanism, marked by delayed clotting and the tendency to frequent and prolonged hemorrhage.

hemopoiesis   Formation of blood cells and other formed elements.

hemorrhage   Bleeding; especially when it is profuse.

hemostasis   The arrest or stopping of bleeding.

heparin   An anticoagulant; found in several tissues, especially the lungs and liver.

hepatic   Pertaining to the liver.

hernia   Rupture; the protrusion of an organ or part of an organ through an abnormal opening, such as a loop of intestine through a part of the abdominal wall.

hexose   A monosaccharide containing six carbon atoms, such as glucose.

hilus   The part of an organ where vessels and nerves enter and leave. Often there is a notch or depression at this site.

hindbrain   That part of the brain consisting of the pons, cerebellum, and medulla oblongata.

hippocampus   Specialized cerebral cortex in temporal lobe, on the floor of the inferior horn of the lateral ventricle; part of limbic system.

hirsutism   The presence of excessive bodily and facial hair.

histamine   A substance found in normal tissues and released when they are injured. It constricts bronchioles and is a powerful vasodilator, thus lowering blood pressure and producing shocklike effects.

histology   The science that deals with the minute structure of the tissues; microscopic anatomy.

homeostasis   A state of equilibrium in the body of the chemical composition and other properties of the fluids and tissues; maintenance of a constancy of the internal environment.

homeotherm   An animal that maintains a relatively constant body temperature; a warm-blooded animal.

horizontal cells   Cells of the retina whose processes extend roughly parallel to the surface; make interconnections between rods and cones.

hormone   A chemical substance produced in one part of the body (usually an endocrine gland) and carried in the blood to another organ or part, where it exerts a specific effect.

horn   A curved or pointed structure resembling a horn; as the anterior and posterior horns of gray matter of the spinal cord, or horn of the lateral ventricles of the brain.

HSP   Hydrostatic pressure.

human chorionic gonadotropin, HCG   Hormone produced by the placental trophoblastic cells. It is important in the first months of pregnancy because it stimulates secretion of estrogen and progesterone by the corpus luteum.

humoral   Pertaining to fluids in the body such as blood and lymph; or to chemicals produced in the body that act locally, such as acetylcholine; or to the fluids that fill the eyeball—the aqueous and vitreous humors.

hunger   A desire or need for food; viewed as a physiological need for food, as opposed to a psychological need (appetite).

hyaline   Having a glassy, homogeneous, translucent appearance.

hyaline membrane disease   A disease of the newborn, especially premature infants, characterized by respiratory distress; associated with reduced amounts of lung surfactant.

hydrocephalus   Excessive accumulation of fluid in the ventricles of the brain causing a thinning of brain tissue.

hydrogen transport system   A series of enzymes which transfer hydrogen ions from one to the next in oxidation-reduction reactions. Energy is released in several of these reactions.

hydrolysis   The splitting of larger molecules to smaller ones with the addition of water.

hydrostatic pressure, HSP   The pressure exerted by fluids; a mechanical or filtration pressure.

hymen   Fold of mucous membrane that partially closes off the external opening of the vagina in a virgin.

hypercapnia   Excessive amount of carbon dioxide in the blood.

hyperextension   Extension of a limb or joint beyond the normal limit.

hyperglycemia   Excessive amount of glucose in the blood.

hyperopia (hypermetropia)   Farsightedness. Parallel light rays come to focus beyond the retina because the refractile system is too weak or the eyeball is flattened (relatively).

hyperphagia   Overeating; gluttony.

hyperplasia   An increase in the number of cells in a tissue or organ.

hyperpnea   Increased pulmonary ventilation; increased rate and depth of breathing.

hyperpolarization   An increase in the polarization of membranes of nerve or muscle cells. It reduces the excitability of the cell.

hypertension   High arterial blood pressure.

hyperthermia   Elevated body temperature; fever.

hypertonic   Refers to a solution having a greater osmotic pressure than another solution to which it is compared.

hypertrophy   Growth, enlargement of a part or organ; often refers to an increase in the size of the elements rather than an increase in their number (*see also* hyperplasia).

hyperventilation   Increase of pulmonary ventilation beyond that needed; may result in "blowing off" an excessive amount of carbon dioxide.

hypocapnia   Abnormally low carbon dioxide tension in the blood.

hypodermic   Subcutaneous; beneath the skin.

hypogastric   Relating to the region of the hypogastrium; the lower middle region of the abdomen.

hypoglycemia   Abnormally low concentration of glucose in the blood.

hyponychium   The layer of cornified epithelium beneath the free border of the nail.

hypophysectomy   Removal or destruction of the pituitary gland.

hypophysis   The pituitary gland.

hypotension   Low arterial blood pressure.

hypothalamus   Portion of the diencephalon just below the thalamus, on either side of the third ventricle; contains centers that exert control over body temperature, water and electrolyte balance, the pituitary gland, and other processes.

hypothenar   Pertaining to the fleshy mass on the medial side of the palm of the hand.

hypothermia   Reduced or subnormal body temperature.

hypotonic Refers to a solution having an osmotic pressure less than the solution to which it is compared.

hypoxia Oxygen deficiency in organs and tissues.

I band Isotropic band of striated muscle cell.

ICF Intracellular fluid.

ICSH Interstitial cell stimulating hormone.

ileum The lower part of the small intestine, between the jejunum and the cecum.

ilium One of the bones of the pelvis.

immunoassay Identification of a substance, such as a protein, through its ability to act as an antigen.

immunoglobulin, Ig Proteins with specific antibody activity; associated with humoral immunities.

implantation Transplantation of an organ or part to a new site in the body; the embedding of a fertilized ovum in the endometrium of the uterus.

impulse The "message" conducted along a nerve or muscle fiber; the action potential.

inclusion A transient substance in a cell—such as metabolic products (secretions), stored products (fat), or engulfed materials.

incompetence Inadequate or insufficient; particularly valves of the heart that do not close properly, permitting regurgitation of blood.

incus The middle of the three ossicles of the middle ear; the anvil.

indifferent gonad The embryonic gonad, before it has differentiated into an ovary or a testis.

indirect calorimetry Determination of heat production in an oxidation reaction by measuring oxygen consumption or carbon dioxide production and calculating the heat produced.

inferior Lower; below, in relation to another structure.

inflammatory reaction The local response to a local injury; includes dilatation of blood vessels with loss of fluid into the area, invasion by leukocytes.

infundibulum Funnel-shaped structure or passage; stalk of the pituitary; free end of the uterine tubes, etc.

ingestion The introduction of food or drink into the stomach.

inguinal Pertaining to the region of the groin.

inhibition The depression or arrest of a function; in a neuron, increasing the membrane potential, thereby reducing the number of impulses discharged.

inhibitory postsynaptic potential, IPSP Local hyperpolarization of the postsynaptic membrane caused by the action of an inhibitory transmitter at a synapse; lowers excitability of the neuron.

innervation Nerve supply to a part.

insertion The attachment of a muscle at the more movable end, usually more distal.

inspiration Inhalation; breathing in.

inspiratory reserve The air that can be inhaled after or beyond a normal quiet inspiration.

insufficiency Incomplete function; inadequacy; incompetency, as of a heart valve; hypofunction, as in adrenal insufficiency.

insulin A hormone produced by the beta cells of the pancreatic islands. It affects many processes concerned with carbohydrate and fat metabolism; lowers blood glucose level.

intercalated disc Short line extending across a fiber of cardiac muscle; indicates the boundary between cells.

intercellular Between the cells.

intercellular fluid The fluid between the cells, excluding blood plasma; the interstitial fluid.

intermedin Melanocyte-stimulating hormone, produced by the intermediate lobe of the pituitary gland; causes dispersion of the pigment granules of melanocytes in the skin of frogs and some fishes, thus darkening the skin.

internal Interior; on the inside, as opposed to external.

internal capsule The white matter in the brain between the thalamus and caudate nucleus medially, and the globus pallidus and putamen laterally; made up of numerous ascending and descending fiber tracts.

internal environment The immediate fluid environment of the cells of the body, from which they obtain needed substances; the interstitial fluid.

interneuron A neuron interposed between a sensory and a motor neuron.

interoceptor A sensory nerve ending or receptor located in the viscera.

interphase The stage between two cell divisions, during which the cell performs its normal functions.

interstices Small spaces, gaps, or holes in the substance of an organ or tissue.

interstitial Pertaining to interspaces in the tissues.

interstitial cells Androgen-secreting cells of the testes, located in spaces between the seminiferous tubules of the testes.

interstitial fluid The intercellular fluid.

intracellular Within the cell.

intracellular fluid That portion of the body fluid contained within the cells of the tissues.

intrafusal fiber The small striated muscle cells within a muscle spindle.

intrapleural Within the pleura; between the visceral and parietal layers of pleura.

intrapulmonary Within the lungs.

intrathoracic Within the chest cavity.

intrauterine device, IUD A plastic or metal coil or loop, etc., inserted into the uterus as a contraceptive agent.

intrinsic Belonging entirely to a part.

intrinsic factor A substance present in gastric mucosa and necessary for adequate absorption of vitamin B_{12}.

intrinsic muscles Muscles that lie totally within a part, such as the intrinsic muscles of the eye, as contrasted with extrinsic muscles which move the eye, or the intrinsic muscles of the hand or foot.

invagination Infolding; one part passing or pushing into or within another part.

inversion Turning inward; a movement at the tarsal joints in which the sole of the foot is turned inward.

ion An atom or group of atoms, with either a positive or a negative electrical charge due to loss or gain of an electron.

ionization The process of dissociation or separation of atoms or molecules into ions.

ipsilateral On the same side.

IPSP Inhibitory postsynaptic potential.

iris A thin muscular diaphragm in front of the lens of the eye. It controls the amount of light entering the eye. It is the pigmented portion.

irritability The ability to respond to a stimulus.

ischemia Temporary local lack of blood to an area due to an obstruction.

islands (islets) of langerhans Pancreatic islands; the endocrine portion of the pancreas.

isometric Of equal length; having the same length.

isometric contraction A muscle contraction in which the ends of the muscle are fixed and the muscle develops tension but does not shorten.

isotonic Of equal tension; having the same tension or pressure.

isotonic contraction A muscle contraction in which the muscle is allowed to shorten and lift a load and thus do work.

isotonic solution A solution with the same osmotic pressure as the solution to which it is compared.

isotope One of a group of atoms having the same number of protons but different numbers of neutrons in the nucleus. The isotopes of a given element have the same atomic numbers, but their atomic masses are different.

isotropic Having equal refractile power in all directions; as the I band of striated muscle.

isthmus A narrow neck or constriction connecting two parts of an organ or other structure.

IUD Intrauterine device.

jaundice A yellowish color of the skin and mucous membranes due to the presence of bile in the blood.

jejunum The portion of the small intestine between the duodenum and the ileum.

joint Articulation; the place of union of two or more bones; may or may not be movable.

juxtaglomerular apparatus The juxtaglomerular cells of the afferent arteriole plus the macula densa of the distal tubules in the kidney.

juxtaglomerular cells Modified cells in the wall of the afferent arteriole near its entry into the glomerulus of the nephron; associated with the release of renin.

kcal kilocalorie.

keratin A tough, insoluble, indigestible protein present in skin structures such as hair and nails.

keto acid An acid with a general formula $R\text{-}\overset{\displaystyle O}{\overset{\|}{C}}\text{-OH}$.

ketogenic Leading to the production of ketone bodies.

ketone Chemical compound containing a ketone group.

ketone bodies Ketones produced during fat metabolism, specifically acetone and acetoacetic and beta-hydroxybutyric acids.

ketonuria Elevated excretion of ketone bodies in the urine.

ketosis A condition characterized by increased production of ketone bodies.

kilocalorie A unit of heat; the amount of heat required to raise 1 kg water 1°C; 1000 calories.

kinase (1) An enzyme that catalyzes the transfer of phosphate groups to form triphosphates (ATP). (2) An enzyme that catalyzes the conversion of a proenzyme to an active enzyme (enterokinase).

kinesthesia The sense of perception of movement; the muscular sense.

kinetic energy The energy of motion.

kinins Certain peptides formed from globulins in the blood. They are potent vasodilators, but cause contraction of visceral smooth muscle, as well as other actions.

Krebs cycle The common pathway for oxidation of carbohydrates, fats, and some amino acids. A sequence of reactions in which two-carbon fragments (acetate) are metabolized to carbon dioxide and hydrogen atoms, which are eventually transferred to oxygen with the release of energy; known also as the citric acid cycle or the tricarboxylic acid cycle.

kyphosis Abnormal, concave forward curvature of the spine; hunchback.

labia Lips, or lip-shaped structures.

labia majora Large folds of skin that form outer lips on either side of the labia minora.

labia minora Smaller folds of mucous membrane, between the labia majora, which enclose the vestibule and the urethral and vaginal openings.

labor Delivery; childbirth; the process of expulsion of the fetus at the end of a normal pregnancy.

labyrinth A maze; an intricate system of interconnecting passages. Particularly those in the inner ear: *bony labyrinth*, the series of canals and cavities in the temporal bone that contain the cochlea, vestibule, and semicircular canals of the inner ear; *membranous labyrinth*, the system of membranous sacs within the bony labyrinth, the utricle, saccule, and cochlear and semicircular ducts.

lacrimal Pertaining to tears.

lactation The secretion of milk.

lacteal Tiny lymphatic vessels in the small intestine.

lactic acid An organic acid formed in glycolysis; formed in greatest amounts during muscle contraction.

lactogenic hormone Prolactin, a hormone of the anterior pituitary that stimulates secretion by the mammary glands.

lactose Milk sugar, a disaccharide.

lacuna (*pl.* lacunae) A small space, cavity, or depression. In cartilage and bone, lacunae are occupied by cells.

lambdoidal Resembling the Greek letter lambda (λ) in shape.

lamellae Thin sheets or layers, as in compact bone.

lamina A thin plate or flat layer.

lamina propria Connective tissue layer underlying epithelium in mucous membranes.

laryngopharynx That part of the pharynx that lies behind the larynx, just above the esophagus.

larynx The voicebox; lies between the pharynx and trachea.

latent period The period of time between the application of a stimulus and the beginning of the response.

lateral On the side; away from the midline.

lateral rotation Movement of a part about a longitudinal axis, such that the anterior surface moves laterally.

lateral sacs Vesicles of sarcoplasmic reticulum; adjacent to transverse tubules; release calcium ions upon stimulation.

lens Transparent convex structure in the eye, just behind the pupil. Its curvature can be altered to focus on near or far objects.

lesion A wound or injury.

lesser omentum A fold of peritoneum extending between the lesser curvature of the stomach and the inferior surface of the liver.

lesser peritoneal sac The omental bursa.

leukemia Pathological condition marked by a great and progressive increase in the number of white cells in the blood.

leukocyte A white blood cell; WBC.

leukocytosis Great increase in the number of leukocytes, as in an acute infection.

Leydig cells Interstitial secretory cells of the testes.

LH Luteinizing hormone.

ligament A band or sheet of fibrous connective tissue that binds two bones at a joint; also binds cartilage or other structures.

ligamentum arteriosum The remains of the ductus arteriosus of the fetus.

ligamentum nuchae Strong ligamentous band at the back of the neck.

ligamentum teres Round ligament: of the liver, the remains of the umbilical vein of the fetus; of the uterus, extends from either side of the uterus through the inguinal canal to end in the labium majus; of the femur, extends from the head of the femur to the acetabulum.

ligamentum venosum The remains of the ductus venosus of the fetus.

light reflex Reflex reduction in the size of the pupil when light is shined into the eye.

limbic system A portion of the brain concerned with emotions and autonomic activity. Generally medial structures surrounding the corpus callosum; the "old" brain; rhinencephalon.

liminal Threshold.

linea alba White line; the fibrous band running vertically along the midline of the anterior abdominal wall. Continuous with the fascia of the abdominal muscles.

lingual Pertaining to the tongue.

lipase A fat-splitting enzyme.

lipid Fat and fatlike substances.

lipogenesis Production of fat, either the normal deposition of fat, or the conversion of carbohydrate or protein to fat.

lipolysis The splitting or chemical breakdown of fat.

lobe A subdivision of an organ or part; bounded by fissures, septa, or other structural elements.

lobectomy Excision or removal of a lobe of an organ or gland.

lobotomy Incision into a lobe; cutting one or more of the nerve tracts in a lobe of the cerebrum, e.g., frontal lobotomy.

lobule A small lobe, or a subdivision of a lobe.

local response A localized response limited to a part or area; a nonpropagated change in membrane potential of excitable tissue.

lordosis Abnormal, concave backward curvature of the spine; hollow or swing back; swayback.

lumbosacral plexus Plexus formed by lumbar and sacral nerves; provides innervation chiefly to the lower extremity.

lumen The interior space of a tubular or hollow structure (such as a blood vessel).

luteinizing hormone, LH Hormone of the anterior pituitary that stimulates the final maturation of ovarian follicles, ovulation, and formation of corpora lutea. Stimulates secretion by interstitial cells of testes (as interstitial cell stimulating hormone).

luteotropic Having a stimulating effect on the development and function of the corpus luteum.

lymph Fluid in the lymphatic vessels; collected from the intercellular fluid in the tissues.

lymph node Small organ along the course of the lymphatic vessels; consists of lymphoid tissue and contains lymphocytes and macrophages which remove foreign substances from the lymph.

lymphocyte A granular leukocyte; important in immune responses.

lysosome Small membrane-bound organelle that contains hydrolytic enzymes (digestive enzymes).

macromolecule A large molecule, notably a protein, nucleic acid, or polysaccharide.

macrophage A large cell with phagocytic properties. Widely distributed, particularly in connective tissue.

macula densa Area of modified cells of distal convoluted tubule of kidney, where it comes into contact with the afferent arteriole; part of the juxtaglomerular apparatus.

malleolus A small hammer; a rounded prominence, such as those on either side of the ankle joint.

malleus A hammer; one of the small bones of the middle ear.

maltose Malt sugar; a disaccharide formed from two molecules of glucose.

mammal Any member of a vertebrate species that suckles its young.

mammary glands Milk-producing glands on the ventral surface of female mammals.

mammillary body One of a pair of small rounded nuclei bulging from the inferior surface of the hypothalamus. Receives fibers from the fornix, and sends fibers to thalamus.

manubrium Upper portion of the sternum. Sometimes it remains as a separate bone; a handle.

marrow A soft gelatinous fatty substance found in bones. Red marrow is found in adult cancellous bone and generates red blood cells. Yellow marrow is found in the medullary cavity of adult long bones and does not.

mast cells Connective tissue cells whose granules are believed to contain histamine and heparin.

mastication Chewing.

mastoid Pertaining to the mastoid process of the temporal bone; behind the ear.

matrix (1) The intercellular substance of a tissue, such as of bone or cartilage; (2) basic material from which a structure, such as a tooth or nail, develops.

mean arterial pressure The average pressure in an artery during the course of a cardiac cycle. The mean between systolic and diastolic pressure (considering the relative time at each pressure).

meatus A passage or canal.

medial (mesial) Toward the middle; near to the midline.

medial rotation Movement of a part about a longitudinal axis, such that the anterior surface moves medially.

median Central; in the middle.

mediastinum The portion of the thoracic cavity between the lungs; contains the heart and blood vessels, trachea and airways, esophagus, and nerves, forming a septum between the lungs.

medulla Central or inner portion of an organ or part; core; marrow.

meiosis Type of cell division that occurs in gametes, in which new cells are produced with the number of chromosomes reduced by half.

melanin A dark brown to black pigment normally found to varying degrees in skin, hair, and some other sites.

melanocyte Melanin-containing cell.

melanotropin Melanocyte-stimulating hormone; intermedin; MSH. Hormone of the intermediate lobe of the pituitary gland; causes dispersion of granules of melanocytes, thus darkening the skin of certain species.

membrane (1) A thin sheet of pliable tissue which covers or encloses a part, lines a cavity, provides a partition or connection; (2) the surface layer of a cell (the plasma membrane), or of intracellular organelles.

membrane potential The potential difference between the inside and outside of a resting cell, especially a nerve or muscle cell.

menarche The time of the first menstrual flow; establishment of menstrual function.

meninges The membranes that enclose the brain and spinal cord.

menopause Termination of menstrual function; cessation of menstrual cycles.

menstruation Periodic (monthly) discharge of fluid and blood from the uterus.

mental Pertaining to the chin.

mesencephalon The midbrain, including the cerebral peduncles, corpora quadrigemini, and the cerebral aqueduct.

mesenchymal cells Primitive cells with great potential to develop into one of several more specialized cells (depending upon their location).

mesenchyme A primitive tissue, containing mesenchymal cells in a jellylike matrix.

mesentery   A double layer of peritoneum, attached to the posterior wall of the abdominal cavity, which encloses in its folds such viscera as the stomach, intestine, and spleen.

mesocolon   The fold of peritoneum attaching the colon to the posterior abdominal wall.

mesonephros   One of the embryonic excretory organs. It regresses, but its duct system remains in the human male to become the epididymis and ductus deferens.

mesovarium   A short fold of peritoneum that connects the ovary with the posterior layer of the broad ligament of the uterus.

messenger RNA, mRNA   The RNA sequence that carries the coded "message" from DNA in the nucleus to the cytoplasmic sites of protein synthesis.

metabolism   The sum of all chemical processes, constructive and destructive, in living cells involving energy exchange.

metabolite   Any product of metabolism, especially of catabolism; foodstuffs, intermediates, or waste products.

metacarpals   Pertaining to the five bones of the hand, between the carpal bones and the digits.

metaphase   The stage of mitosis in which the chromosomes become aligned along the equatorial plate of the cell.

metarteriole   A small vessel between an arteriole and a true capillary; contains scattered groups of smooth muscle fibers in its wall.

metatarsals   Pertaining to the five bones of the foot, between the tarsal bones and the digits.

metencephalon   The anterior part of the hindbrain; includes the pons and cerebellum.

microcephaly   Condition marked by an abnormally small head or skull.

microcirculation   Circulation in the smallest of blood vessels, e.g., the arterioles, capillaries, and venules.

microelectrode   A very fine electrode, usually a fine wire or glass capillary tube, that can be inserted into a single cell without excessive damage in order to study its electrical activity.

microglia   Small phagocytic cells found in gray and white matter of the central nervous system. Often considered with neuroglia, although they do not arise from neural tissue as neuroglia do.

microtubules   Cylindrical cytoplasmic organelles found in many cells; become more prominent during mitosis when they form the mitotic spindle.

microvilli   Very tiny protoplasmic projections on the free surface of epithelial cells which greatly increase the surface. With the light microscope they appear as striate or brush borders.

micturition   Urination.

midbrain   Mesencephalon; the cerebral peduncles, corpora quadrigemini, and cerebral aqueduct.

mineralocorticoids   Hormones of the adrenal cortex whose major action is on sodium and potassium metabolism.

mitochondria   Organelles of the cytoplasm that contain the enzymes for fatty acid oxidation, the Krebs cycle, electron transport, and oxidative phosphorylation; thus they are the principal sites of energy release in the cells.

mitosis   Usual process of cell division, in which the two daughter cells have exactly the same chromosomes as the original cell.

mitral valve   Valve between the left atrium and left ventricle of the heart; the bicuspid valve.

mixture   With regard to chemistry, a combination of two or more ingredients without a chemical reaction, that is, without a permanent loss or gain of electrons. The substances retain their original properties (*see also* compound).

modality   (1) One of the main types of sensation, e.g., touch, sight, taste, etc. (2) Any mode used in physical therapy such as heat or electrical methods.

molar solution   A solution in which 1 liter of solution contains an amount of solute equal to its molecular weight in grams.

mole   Molecular weight of a compound in grams; gram molecular weight.

molecule   The smallest possible quantity of a substance that retains the chemical properties of that substance and is composed of one or more atoms.

monocyte   A large agranular leukocyte with phagocytic properties.

monosaccharide   A simple sugar, including hexoses and pentoses; it cannot be further broken down by hydrolysis.

monosynaptic   Pertaining to a nerve pathway with only one synapse.

mons pubis   A pad of fatty tissue forming a prominence in front of the symphysis pubis in the female.

motility   The power of spontaneous movement.

motoneuron   A nerve cell with motor function; the nerve supplying an effector; motor neuron.

motor   Concerned with motion, especially activity of effectors, such as motor cortex (of brain), motor end plate, motoneuron, motor unit, etc.

motor unit   A motoneuron together with the muscle cells it innervates.

mRNA   Messenger RNA.

mucin   A glycoprotein found in many mucous secretions, such as those of goblet cells and salivary glands; it is also present in ground substance.

mucosa   Mucous membrane lining cavities and tubular structures; consists of epithelium, lamina propria, and (in digestive system) muscularis mucosa.

mucous membrane   The membrane lining the cavities and tubes that communicate with the surface of the body and producing a mucous secretion.

mucus   The secretion of mucous glands.

multiaxial joint   A joint capable of movement about more than two axes; or of movement in more than two planes, such as the joints at the hip and shoulder.

multipennate   Muscle fiber arrangement in which the fibers converge to many tendons.

murmur   Sound heard from within the body, particularly from the heart valves, lungs, or blood vessels.

muscle spindle   Muscle proprioceptor; a spindle shaped structure in skeletal muscle. Consists of several small muscle fibers (intrafusal fibers) enclosed in a capsule. There are sensory endings on these fibers, and efferent fibers to them.

muscularis   The muscular coat in the wall of a hollow organ or tubular structure.

muscularis mucosa   Thin layer of smooth muscle of the mucosa of most of the digestive tract.

myelencephalon   The inferior portion of the hindbrain, chiefly the medulla oblongata.

myelin   Lipid material with protein arranged in layers around many axons. It gives white matter its color.

myelinated   Having a myelin sheath.

myenteric   Relating to the muscular layer of the intestine.

myocardium   The heart muscle.

myoepithelial cells   Cells of epithelial origin which are also contractile. Found around the secretory alveoli of mammary glands; also in sweat, lacrimal, and salivary glands.

myofibrils   Fine longitudinal fibrils or strands in muscle cells. They are contractile.

myofilaments   Extremely fine strands found in myofibrils of striated muscles; contain actin or myosin.

myometrium   Muscular wall of the uterus.

myoneural junction   Neuromuscular junction; the site at which branches of a motoneuron terminate at a muscle cell.

myopia Nearsightedness. Parallel light rays come to focus in front of the retina because the refractile system is too strong, or the eyeball is elongated.

myosin A protein found in the thick filaments of a myofibril. In combination with actin, it forms the contractile unit of muscle.

myotatic reflex A stretch reflex.

myxedema Hypothyroid condition in the adult. A disease characterized by accumulation of a thick fluid in the subcutaneous tissues, as well as reduced metabolic functions.

NAD Nicotinamide adenine dinucleotide.

narcosis A state of stupor, insensibility, or unconsciousness produced by the action of some narcotic drug.

nares Nostrils; openings of the nasal cavity, especially the paired anterior openings (external nares), but also the posterior openings (internal nares) into the nasopharynx.

nasal cavity Cavity of the nose, above the hard palate; divided medially by a septum with three conchae on each side; lined with mucous membrane.

nasolacrimal duct Passage leading from the lacrimal sac to the nasal cavity. It drains tears from the eye.

nasopharynx The upper portion of the pharynx; behind the nasal cavity.

near point The nearest point at which an object can be seen distinctly, using maximum accommodation.

neck Any constricted portion, bearing some resemblance to the neck of an animal; often below the head, as in a bone or an organ.

negative feedback The mechanism in which the output of a system (such as nerve impulses or a secretion) inhibits or depresses the activity of the source, thus reducing the output.

nephron The functional unit of the kidney, consisting of the renal corpuscle and tubule.

neural Relating to any part of the nervous system.

neurilemma The sheath of Schwann.

neuroendocrine reflex A reflex in which the afferent limb is a sensory neuron and the efferent limb is a hormone.

neuroglia Nonnervous cellular elements of nervous tissue; believed to perform important metabolic functions.

neurohypophysis The posterior lobe of the pituitary gland.

neuromuscular junction Myoneural junction; the site where a branch of a motoneuron terminates at a muscle cell.

neuron A nerve cell.

neurosecretion A chemical substance released by a nerve cell; includes transmitters such as acetylcholine, secretions of the neurohypophysis, and releasing hormones of the hypothalamus.

neutral fat Triglyceride; glycerol combined with three molecules of fatty acid.

neutron Electrically neutral particle found in the nucleus of an atom.

neutrophil A granular leukocyte whose numerous small granules stain readily with neutral dyes; most abundant leukocyte; has phagocytic properties.

niacin Nicotinic acid; a water-soluble vitamin; one of the B vitamins.

nicotinamide adenine dinucleotide, NAD An important hydrogen carrier for oxidation-reduction reactions of cellular metabolism.

Nissl bodies, Nissl substance or granules. Granules formed by clumps of rough endoplasmic reticulum in nerve cell bodies and dendrites; contain RNA.

nitrogen balance The difference between the total nitrogen ingested and the total nitrogen excreted by an organism. Should be approximately zero in the adult. It is positive during growth and negative during starvation and certain illnesses.

nodes of Ranvier Interruptions in the myelin sheath of a nerve fiber, occurring at fairly regular intervals.

nodule A small node, or closely packed mass of cells, distinct from surrounding cells.

nonelectrolyte A substance whose molecules do not dissociate into ions when in solution and hence do not conduct an electric current.

nonrapid eye movement sleep, NREM Typical normal sleep, with slower, larger waves on the electroencephalogram; slow wave sleep; characterized by neither rapid eye movements nor the small, rapid EEG waves of paradoxical or REM sleep.

noradrenaline Norepinephrine.

norepinephrine The chief transmitter released by adrenergic nerve endings; also present in the secretion of the adrenal medulla.

normal solution A solution containing one equivalent weight (the weight in grams reacting with 1 g hydrogen) per liter of solution.

normoblast An immature red blood cell that still contains a nucleus.

notch An indentation at the edge of a structure.

NREM sleep Nonrapid eye movement sleep.

nuchal Relating to the back or nape of the neck.

nuclease An enzyme that catalyzes the hydrolysis of nucleic acids to nucleotides.

nucleic acid A substance of high molecular weight found in the chromosomes, nucleoli, mitochondria, and cytoplasm of all cells. DNA and RNA are nucleic acids.

nucleolus Small mass in nucleus of a cell; contains RNA.

nucleoplasm The protoplasm within the nucleus of a cell.

nucleotidase An enzyme that catalyzes the hydrolysis of nucleotides into phosphoric acid and nucleotides.

nucleotide A component of DNA and RNA; composed of a sugar, a nitrogenous base, and a phosphate group.

nucleus (1) The central portion of an atom, containing protons and neutrons. (2) The rounded mass of protoplasm within the cytoplasm of a cell, containing the chromosomes. (3) A collection of nerve cell bodies in the central nervous system; gray matter.

nucleus pulposus The cushion of gelatinous material in the center of an intervertebral disc.

nystagmus Rapid involuntary oscillations of the eyeball; may be horizontal, vertical, or rotary.

obesity An excessive accumulation of fat.

obturator A structure that closes or occludes a bodily opening.

occipital Pertaining to the region of the back of the head.

occlusion (1) The act of closing, or the state of being closed, as of an artery. (2) A form of indirect inhibition in the nervous system, as a result of convergence.

ohm A unit of electrical resistance.

olfactory Relating to the sense of smell.

oligodendroglia A neuroglial cell characterized by having few processes. The processes form sheaths around nerve fibers in the central nervous system.

omental bursa The lesser peritoneal sac, generally behind the stomach and liver; opens into the greater sac (peritoneal cavity) at the epiploic foramen.

oncotic pressure Osmotic pressure exerted by colloids in solution; colloid osmotic pressure.

oocyte An immature ovum.

oogenesis The process of formation and development of an ovum.

optic   Pertaining to the eye, vision, or the properties of light and its refraction.

optic chiasm   Structure on the ventral surface of the brain, anterior to the pituitary gland, where some of the fibers of the optic nerve cross. Those fibers from the nasal half of each retina cross to form the optic tract.

optic disc   The area on the retina where the fibers of the optic nerve leave the eye; the blind spot.

orbit   Eye socket; the bony cavity occupied by the eyeball and its muscles, nerves, etc.

organ   Any part or structure in the body carrying out a specific function or functions, such as the heart, liver, kidney, etc.

organ of Corti   Spiral organ; the sense organ for hearing.

organ system   A group of organs having a common function, such as respiration, digestion, etc.

organelle   An intracellular organ; a specialized living part of a cell that carries out a specific individual function. Examples are mitochondria, lysosomes, etc.

orgasm   The peak or culmination of the sexual act.

orifice   An aperture or opening.

origin   (1) The less movable attachment of a muscle, usually the proximal attachment. (2) The point from which a nerve arises.

oropharynx   The central part of the pharynx, that portion posterior to the oral cavity.

oscilloscope   An electronic instrument for amplifying and displaying on a fluorescent screen electrical fluctuations (oscillations), such as action potentials.

osmol   The molecular weight of a solute, in grams, divided by the number of ions or particles into which it dissociates in solution.

osmolarity   An expression of osmotic concentration of a solution, taking into account the number of osmotically active particles in solution.

osmoreceptors   Receptors in the central nervous system (hypothalamus) that respond to changes in the osmotic pressure of the blood.

osmosis   The movement of solvent across a membrane due to a concentration gradient caused by the presence of a nondiffusible solute. Solvent moves toward the nondiffusible solute.

osmotic pressure   The pressure generated by osmosis.

ossicles   Small bones; specifically the tiny bones of the middle ear.

ossification   The formation of bone: endochondral ossification is the formation of bone within cartilage, as in long bones; intramembranous ossification is the formation of bone within a connective tissue layer, as the bones of the skull.

osteoblast   A bone-forming cell; forms bone matrix.

osteoclast   A large, multinucleated cell that dissolves or erodes bone tissue.

osteocyte   A bone cell. It occupies a lacuna and has processes that extend through canaliculi in the bone matrix.

osteoid   Bone tissue before it has calcified.

osteon   Haversian system; a central canal and concentric bony lamellae found in compact bone.

otoliths   Crystalline particles of calcium carbonate in the gelatinous material over hair cells of the utricle and saccule of the inner ear; statoconia.

oval window   An oval opening in the bony wall between the middle and inner ear; occupied by the foot of the stapes.

ovary   The female primary reproductive organ; produces ova and hormones.

oviduct   Uterine tube; fallopian tube; the tube by which the ovum is transported to the uterus.

ovulation   The rupture of an ovarian follicle with release of an ovum.

ovum   (*pl.* ova) A mature unfertilized female sex cell; female gamete; egg.

oxidation   A reaction involving the combination with oxygen, or the loss of hydrogen, or one or more electrons, from an atom.

oxidative phosphorylation   The process of forming energy-rich bonds (i.e., producing ATP) from the energy released in oxidation reactions.

oxygen debt   The quantity of oxygen required by contracting muscles in excess of that actually supplied to them during activity. The extra oxygen consumed after activity is used to oxidize the lactic acid produced.

oxyhemoglobin   Hemoglobin in combination with oxygen.

oxytocin   A hormone produced by the neurohypophysis which increases contractions in the pregnant uterus and causes milk release during lactation.

pacemaker   A cell or a center that discharges rhythmically and spontaneously and thus "drives" other cells; particularly the cells of the sinoatrial node, which initiate the heart beat; also an electronic device that substitutes for the normal cardiac pacemaker.

pacinian corpuscle   A small oval receptor composed of concentric layers of connective tissue, sensitive to deformation caused by pressure; found in subcutaneous tissue and other areas.

palate   The roof of the mouth.

palmar   Pertaining to the palm of the hand.

palpation   Touching; feeling or perceiving by the sense of touch.

pancreas   A glandular organ in the abdominal cavity. It produces both exocrine and endocrine secretions.

pancreatic islands (islets)   The endocrine portion of the pancreas; consists of clumps of cells that produce insulin and glucagon.

panniculus adiposus   Superficial fascia that contains adipose tissue.

pantothenic acid   One of the B vitamins. It is an essential growth factor, and is part of coenzyme A.

papilla   Any small nipplelike elevation.

papillary muscle   Portions of cardiac muscle that protrude from the inner surface of the ventricles. The chordae tendineae attach to them.

paradoxical sleep   Sleep with EEG patterns resembling those of wakefulness and an alert brain; rapid eye movement (REM) sleep.

parathormone   Hormone of the parathyroid gland; participates in calcium and phosphorus metabolism; raises blood calcium level.

parathyroid glands   Several tiny endocrine glands embedded in the posterior surface of the thyroid gland.

parasympathetic   Pertaining to that part of the autonomic nervous system that arises from the brain and the sacral portion of the central nervous system; craniosacral division.

parasympathomimetic   Any drug or chemical whose action resembles that of the parasympathetic nervous system.

paraventricular nucleus   Hypothalamic nucleus. Gives rise to nerve fibers that end in the neurophyophysis; associated with release of oxytocin.

parietal   Pertaining to the wall of any cavity.

parotid gland   Largest of the salivary glands; located below and in front of the ear.

partial pressure   The pressure or tension exerted by one gas in a mixture of gases, such as the partial pressure of oxygen (P_{O_2}) in the air.

parturition   Childbirth.

pathology   The science that deals with all aspects of disease, its nature, cause, and development, as well as structural and functional changes produced.

pectoral girdle The incomplete ring that provides attachment and support for the upper extremity; consists of the sternum, clavicles, and scapulae.

peduncle A foot, stalk, or stem. Especially a large mass of nerve fibers connecting certain parts of the brain with other parts of the nervous system, such as cerebral or cerebellar peduncles.

pelvic girdle The bony ring formed by the innominate bone (os coxae) and the sacrum, to which the lower extremity is attached.

pelvis (1) The pelvic girdle. (2) The cavity enclosed by the pelvic girdle; the lower portion of the abdominopelvic cavity. (3) Any basinlike or cup-shaped cavity, such as the renal pelvis.

penis The male copulatory organ.

pepsin A protein-splitting enzyme, secreted by the stomach in an inactive form as pepsinogen.

peptidase An enzyme capable of splitting a peptide.

peptide bond The common form of linkage between amino acids. The acid (carboxyl) group of one amino acid reacts with the amino (NH_2) group of the next acid.

perfusion The artificial passage of fluid through the blood vessels.

pericardium The fibrous membrane that forms a closed sac around the heart.

perichondrium The connective tissue membrane around cartilage.

perilymph The clear fluid within the osseous labyrinth and around the membranous labyrinth.

perimysium The fibrous sheath enveloping bundles of skeletal muscle fibers.

perineum The region of the pelvic outlet, between the anus and scrotum or urethral opening.

perineurium The connective tissue sheath around a bundle of nerve fibers in a peripheral nerve.

periosteum The fibrous connective tissue membrane covering a bone entirely, except for its articular surfaces.

peripheral Related to the periphery; outer part; away from the center.

peripheral inhibition Inhibition that occurs at the neuroeffector junction.

peripheral nervous system That part of the nervous system outside of the brain and spinal cord; consists of nerves and ganglia (including those of the autonomic nervous system).

peripheral resistance Resistance to the flow of blood in the systemic circuit; particularly that caused by the resistance vessels.

peristalsis Wavelike rings of contraction that proceed along tubular structures, such as the intestine, by which the contents are moved along the tube.

peritoneum The serous membrane that lines the abdominopelvic cavity and encloses the pelvic viscera.

pernicious anemia A type of anemia related to a lack of vitamin B_{12}, often because of faulty absorption; characterized by decreased number of red blood cells and low hemoglobin.

peroneal Relating to the fibula or lateral side of the leg.

petrous Of stony hardness.

pH The negative logarithm of the hydrogen ion concentration. An indication of the degree of acidity or alkalinity of a solution. On a scale of 0–14, 7 is neutrality, below 7 is progressively more acid, and above is progressively more alkaline.

phagocytosis Process by which certain scavenger cells, called phagocytes, engulf and digest foreign particles in their environment.

phalanx (*pl.* phalanges) The bone (or bones) of the digits (fingers and toes).

pharmacology The science that deals with drugs, their sources, appearance, chemistry, actions, and uses.

pharynx The throat. That part of the digestive system between the oral cavity and esophagus.

phospholipid A lipid containing phosphoric acid; found in plant and animal tissues, including myelin.

phosphorylation The addition of phosphate to an organic molecule, such as glucose upon its entry to a cell, or the formation of ATP from ADP.

photoperiod Varying length (diurnal or seasonal) of exposure of a living organism to light; causes physiological and behavioral changes.

photosynthesis The process by which green plants, using chlorophyll and the energy from the sun, are able to produce carbohydrate from carbon dioxide and water.

phrenic Relating to the diaphragm.

phylogenetic Relating to the evolutionary development of a species.

physiology The science that deals with the normal processes that occur in living things.

pia mater The delicate fibrous membrane that closely covers the brain and spinal cord; the innermost layer of the meninges.

piloerection Erection of the hairs of the skin by action of smooth muscles.

pineal gland A small glandlike structure on the dorsal surface of the brain, near the corpora quadrigemini.

pinna The outer portion of the ear; the external ear, exclusive of the external acoustic meatus.

pinocytosis The process by which cells engulf droplets of fluid from their environment, forming fluid-filled vesicles in the cytoplasm.

pituitary gland The hypophysis. Composed of an anterior portion, the adenohypophysis; a posterior portion, the neurohypophysis; and an intermediate lobe.

placenta The organ of metabolic interchange between the fetus and mother. It is partly of embryonic origin and partly of maternal origin.

plantar Pertaining to the sole of the foot.

plantar flexion Movement at the ankle in which the foot and toes are moved downward, toward the sole of the foot.

plasma The fluid portion of circulating (unclotted) blood.

plasma cell Cell derived from B lymphocytes; active in antibody production.

plasma membrane The cell membrane; the lipid and protein boundary that encloses the cell substance.

platelets Thrombocytes. Small fragments of a large bone marrow cell; found in circulating blood and involved in blood clotting.

pleura The serous membrane that lines the thoracic cavity and covers the lungs.

pleural space The intrapleural space or cavity; the potential space between the visceral and parietal layers of the pleura.

plexus A network of interlacing nerves or blood vessels.

plicae circulares Numerous transverse folds in mucous membrane of the small intestine.

pneumonia Inflammation of the lungs.

pneumothorax The presence of air or other gas in the pleural space; in the thoracic cavity but outside of the lungs.

podocyte Epithelial cells of the inner layer of the glomerular capsule; characterized by footlike processes that wrap around the glomerular capillaries.

poikilothermic Having a variable body temperature, such as cold-blooded animals.

polar body The minute cell, formed in the meiotic division of an oocyte, that gets virtually none of the cytoplasm. The cytoplasm goes to the other daughter cell, the one destined to become a mature ovum.

polarization The development of differences in electrical potential between two points in living tissue; specifically the unequal distribution of positive and negative charges between the inside and outside of a cell.

poliomyelitis Inflammation of cells of anterior horn of spinal cord (motoneurons). Causes flaccid paralysis followed by atrophy of affected muscles.

pollex The thumb or first finger.

polycythemia An abnormal increase in the number of red blood cells in the blood.

polydipsia Frequent drinking due to extreme thirst.

polymerization Process by which a high molecular weight compound is produced by successive additions or condensations of a simpler compound.

polymorphonuclear leukocytes Leukocytes whose nuclei are of varied shapes. The granular leukocytes, especially neutrophils.

polypeptide Small protein or protein fragment; chain composed of a number of amino acids.

polysaccharide Carbohydrate composed of long chains of simple sugars; includes starch and cellulose.

polyuria Excretion of an excessive volume of urine.

pons That part of the brain, or brainstem, just above the medulla. It is an important connection (bridge) to the cerebellum.

popliteal Pertaining to the posterior surface of the knee.

porta hepatis The fissure on the under surface of the liver, lodging the vessels, nerves, ducts, and lymphatics of the liver.

portal circulation Vessels that carry blood from one set of capillaries to another. Specifically, the blood from the intestinal capillaries to the liver; also the circulation from the hypothalamus to the anterior pituitary.

postabsorptive After absorption is completed. Nutrients in the digestive tract have been absorbed, and energy needs are being provided from the body's stores; a fasting state.

posterior Behind, or toward the rear; in humans, dorsal.

postganglionic neuron The second neuron in an autonomic efferent pathway. Its cell body is outside the central nervous system, and it terminates at an effector.

postsynaptic membrane Portion of the membrane of cell body and dendrites at the site of a synapse; sensitive to (contains receptors for) neurotransmitter substance.

potential The difference in electrical charge or the electrical gradient; especially that across the membrane of certain cells (nerve and muscle cells).

potential energy The energy of position or existence which is not being exerted at that time.

precursor Anything that precedes another, or from which another is derived; especially an inactive substance that is converted to an active form.

preganglionic neuron The first neuron in an autonomic efferent pathway. Its cell body is in the central nervous system, and it terminates on a postganglionic neuron.

pregnancy The state of a female after conception until the birth of the child; gestation.

prepuce Foreskin; the free fold of skin that covers, more or less completely, the glans penis.

presbyopia Loss of the eyes' ability to accommodate; the loss of elasticity of the lens of the eye that causes the near point to recede.

pressoreceptor Baroreceptor; a sense organ that responds to changes in pressure.

pressor response A response characterized by a rise in pressure, especially of blood pressure.

presynaptic membrane Portion of the cell membrane of an axon at the site of a synapse; site of release of neurotransmitter.

primary bronchus The main right or left bronchus; arises directly from the trachea.

primary follicle Early stage of development of an ovarian follicle, in which the oocyte is surrounded by one or more layers of follicular cells.

primate A member of the highest order of mammals, which includes humans, apes, and monkeys.

prime mover A muscle directly responsible for a particular movement.

process (1) A projection or outgrowth, as of a bone or certain cells. (2) A method or mode of action.

progesterone A steroid hormone produced by the corpus luteum and placenta. Responsible for preparing the reproductive organs for pregnancy.

projection fibers Nerve fibers connecting the cerebral cortex with other centers in the brain or spinal cord.

prolactin Hormone of the anterior pituitary that stimulates the secretion of milk by the mammary gland.

prolapse To fall or sink down; said of an organ or part, especially of the uterus.

pronation Rotation of the forearm so that the palm of the hand faces backward (in the anatomical position) or downward; or, in reference to the whole body, the face and abdomen downward.

prophase The first stage of mitosis during which the chromosomes appear within the nucleus.

proprioceptors Receptors in muscles, tendons, or joints that are concerned with giving information about movement and position of the body parts.

prostaglandins A class of widely distributed physiologically active substances derived from certain essential fatty acids: They are probably produced and act locally; their varied actions include vasodilatation and stimulation of intestinal and uterine smooth muscle, and many others.

prostate A gland that surrounds the male urethra just below the urinary bladder; secretes a milky fluid as a contribution to semen.

protein Large nitrogen-containing molecules, consisting of long chains of amino acids.

proteinuria Excretion, in the urine, of protein.

proteolytic Pertaining to the ability to decompose or break down protein.

prothrombin A protein in the plasma that is converted to thrombin to cause clotting of the blood.

proton A unit of positive charge found in the nucleus of an atom.

protoplasm Living matter; the substance of which animal and vegetable tissues are formed.

protraction The movement of the shoulder girdle in which the tip of the shoulder (acromion) moves forward.

proximal Nearest the body or point of attachment.

pterygoid Wing-shaped; particularly the structures associated with the winglike parts of the sphenoid bone.

ptyalin Salivary amylase.

puberty The sequence of events by which a child is transformed into a young adult and in which the reproductive organs become functional.

pudendum The external genital organs, especially of the female; the vulva.

pulmonary Relating to the lungs or pulmonary vessels.

pulmonary ventilation The amount of air delivered to the lung each minute.

pulp cavity The central cavity of a tooth, containing vessels and nerve fibers.

pulse The rhythmic dilation of an artery caused by the increased volume of blood forced into it with each heart beat; a wave of distension which travels over the artery and can be felt.

pulse pressure   The variation in pressure in an artery during the cardiac cycle; the difference between systolic and diastolic pressure.

pupil   The circular opening in the center of the iris through which light rays enter the eye.

Purkinje fibers   Terminal ramifications, in the ventricles, of the specialized conducting system of the heart.

putamen   Part of the basal ganglia; lateral portion of corpus striatum, lateral to globus pallidus.

pylorus   The junction between the stomach and the duodenum; the sphincter at this junction.

pyramid   Any of several anatomical structures with a more or less pyramidal shape, including the medullary pyramids on the ventral surface of the brain, and the pyramids in the renal medulla.

pyramidal tracts   Descending nerve fiber tracts that pass through the pyramids of the medulla; the corticospinal tracts.

pyridoxine   Vitamin B_6; associated with utilization of unsaturated fatty acids.

pyrogen   Any agent that causes a rise in body temperature.

pyruvic acid   An important intermediate compound in the metabolism of carbohydrates.

ramus   A branch, as of an artery or nerve, or a part of an irregularly shaped bone.

raphé   A ridge or seamlike structure at the junction of two similar adjacent structures.

rapid-eye movement sleep, REM   Paradoxical sleep. Intervals during sleep in which brain waves resemble those of wakefulness. Characterized by symmetrical jerky movements of the eyes.

RAS   Reticular activating system.

RBC   Red blood cells.

receptor   (1) A sense organ; the peripheral endings of sensory nerves. They are especially sensitive to a particular type of stimulus. (2) A specific configuration of a cell membrane with which a specific substance (neural transmitter, hormone, or drug) may combine in a way that leads to altered cell function. Receptors for steroid hormones are inside cells. Many substances, including transmitters, are believed to act by combining with a receptor site.

reciprocal innervation   The neural connections whereby contraction of a muscle is accompanied by relaxation of its antagonist. Similar relations are exhibited by many reflex centers in the central nervous system.

rectum   The terminal portion of the digestive tract, from the sigmoid colon to the anal canal.

rectus   Straight; associated especially with muscles that follow a straight course.

red nucleus   A mass of gray matter with a reddish color in the upper part of the midbrain. Gives rise to the rubrospinal tract.

red pulp   That part of the splenic tissue consisting of blood sinuses and intervening tissue.

reduced hemoglobin   Deoxyhemoglobin.

reduction   A chemical reaction in which an ion or compound gains one or more electrons. It is always accompanied by oxidation of the compound that gives up the electrons.

referred pain   Pain perceived as originating at a site remote from its actual origin.

reflex   An involuntary movement or response of a part in response to a stimulus applied to a receptor.

refraction   The deflection or bending of light rays when they pass from a medium of one density to a medium of another density, as from air to glass and glass to air in passing through a lens.

refractory period   Period immediately following a stimulus, during which a cell does not respond to a stimulus.

relaxin   A hormone produced by the corpus luteum during pregnancy. Its role in humans is unclear, but in such animals as guinea pig and mouse, it softens structures of the symphysis pubis and pelvic ligaments.

releasing hormones   Substances of hypothalamic origin that are capable of increasing the rate of secretion of individual hormones of the anterior pituitary.

REM sleep   Rapid eye-movement sleep.

renal   Pertaining to the kidney.

renin   A substance secreted by the kidney which causes a rise in arterial blood pressure by activating angiotensin.

rennin   An enzyme produced by the gastric mucosa which curdles milk through its action on milk protein.

residual volume   That volume of air remaining in the lungs after a maximal expiration.

resistance vessels   Mainly muscular arterioles and small arteries that regulate regional blood flow to capillary beds, and are responsible for most of the peripheral resistance.

resorption   The taking up again of that which has been excreted or put out, such as the removal (dissolving) of the intercellular material of bone, or the removal of a blood clot.

respiration   (1) The exchange of gases (oxygen and carbon dioxide) between an organism and its environment. (2) Breathing; the movements by which air in the lungs is exchanged.

response   The change in an organism produced by a stimulus.

resting potential   Membrane potential; the electrical potential difference between the inside and outside of a cell.

reticular   Pertaining to a reticulum; a network, especially a fine network of cells and fine (reticular) connective tissue fibers.

reticular activating system   A system of fibers and synapses in the reticular formation of the brainstem which, when activated, results in a generalized alerting or arousal of the electrical activity of the cerebral cortex. Its destruction results in a permanent coma.

reticular formation   Area of the brainstem composed of diffuse neurons and synapses. Has numerous connections to other areas and ascending and descending pathways. Contains many important centers.

reticulocyte   An immature red blood cell which contains a network formation that can be shown by use of certain stains.

reticuloendothelial system   The system of macrophages—the phagocytic cells found in the lining of sinuses and the reticular tissue of many organs and tissues.

retina   The innermost layer of the eye; the neural layer, containing the receptors for light.

retinene   Retinal; visual yellow; split off when rhodopsin (visual purple) of the rods of the eye is bleached by light.

retraction   The movement of the shoulder girdle in which the tip of the shoulder (acromion) is moved backward.

retrograde   Moving backward; retracing original course; reversing normal order and development; degenerating.

retroperitoneal   Behind the peritoneum.

Rh factor   Antigen associated with red blood cells of most individuals. Those lacking this factor (Rh negative) produce antibodies when exposed to it.

rhinencephalon   The olfactory (sense of smell) brain. Phylogenetically old portion of the brain.

rhodopsin   Visual purple; a protein found in rods of the retina; it is bleached by the action of light, and is restored in the absence of light; consists of retinene and a protein, scotopsin.

riboflavin   Vitamin B_2; a coenzyme for certain enzymes important in metabolic reactions.

ribosomes   Particles consisting largely of ribonucleic acid (RNA) found in cytoplasm, often in association with endoplasmic reticulum.

ribonucleic acid, RNA   Nucleic acid found in nuclei and cytoplasm of cells; plays an important role in protein synthesis.

rickets   A calcium deficiency disease in infants and young children; characterized by softening of bones and skeletal deformities.

RNA   Ribonucleic acid.

rod   One of the light receptors in the retina. Sensitive to low light intensity and provides for black and white vision.

root   The base, foundation, or beginning of a part; the bundles of nerve fibers emerging from the spinal cord to form spinal nerves; the part of a tooth embedded in the bony alveolus.

rotation   Movement of a bone or part about its own (longitudinal) axis, or an axis of another bone or part.

round window   A round opening in the bony wall between the middle and inner ear. In life it is covered by a membrane.

rugae   Folds or creases; especially those of the lining of the stomach.

saccule   (1) A small sac. (2) The smaller of two chambers of the vestibular portion of the membranous labyrinth in the inner ear.

sagittal   In an antero-posterior direction; a plane that divides a given part into right and left parts.

salivary glands   The three pairs of saliva-producing exocrine glands located near the mouth.

saltatory conduction   Conduction in nerve fibers in which the impulse "jumps" from one node of Ranvier to the next.

SA node   Sinoatrial node.

saphenous   Relating to, or associated with, saphenous veins in the leg; denotes various structures in the leg.

sarcolemma   The plasma membrane of a muscle fiber.

sarcomere   A segment or unit of a myofibril of a striated muscle cell. Specifically, the region between two adjacent Z lines.

sarcoplasm   The fluid nonfibrillar portion of the cytoplasm of a muscle fiber.

sarcoplasmic reticulum   The network of vesicles and tubules which surrounds each myofibril in striated muscle. It is involved in transmission of excitation from the cell membrane to the contractile protein in the myofibril.

satiety center   A center in the hypothalamus involved in regulation of food intake. Stimulation causes a decrease in food intake.

saturated fatty acid   A fatty acid whose carbon chain has no double bonds.

scala tympani   The lower portion of the divided canal of the cochlea. The round window is at its lower end.

scala vestibuli   The upper portion of the divided canal of the cochlea. The oval window with the stapes is at its lower end.

Schwann cell   Neurilemma cell; found around the nerve cell processes of peripheral nerve fibers. Its membrane forms the myelin sheath.

sciatic   Relating to the hip or ischium, or structures in the vicinity.

sclera   The tough fibrous outermost layer of the eyeball; the white of the eye.

scoliosis   Lateral curvature of the spine.

scotopsin   Protein component of the pigment in rods of the retina (rhodopsin).

scrotum   The sac or pouch that contains the testes.

sebaceous glands   Glands in the skin that secrete an oily substance (sebum); usually open into a hair follicle.

sebum   The secretion of sebaceous glands.

secondary sex characteristics   Characteristics peculiar to male or female which develop at puberty; such as distribution of body hair, voice changes, etc.

secondary sex organs   Accessory sex organs; these organs and structures that provide for protection, transport, storage, or nutrition of ova or sperm while in the body.

secretin   A hormone produced by the intestinal mucosa when there is acid in the duodenum. It stimulates pancreatic secretion and bile secretion.

secretion   The process by which a substance elaborated (manufactured) by a cell is transferred outside of the cell; also the substance that is secreted.

section   (1) The act of cutting. (2) A cut or division. (3) A cut surface, as a cross section. (4) A thin slice of material for examination under the microscope.

segmentation   The division or splitting into segments or parts; an alternate contraction and relaxation of adjacent segments of the small intestine; a rhythmic, nonpropulsive mixing movement.

sella turcica   A saddlelike structure on the supper surface of the sphenoid bone; contains the pituitary gland.

semen   The secretion of the male reproductive organs (testes, seminal vesicles, prostate, and bulbourethral glands); contains sperm and seminal fluid.

semicircular canals   Three canals in the temporal bone that lie approximately at right angles to one another. They contain receptors for equilibrium, specifically for rotation.

semilunar   Half-moon shaped; crescentic.

seminal vesicles   Small, convoluted, glandular structures that are outgrowths of the ductus deferens. Their secretion contributes to the semen.

seminiferous tubules   Coiled tubules that make up the bulk of the testes. Site of production of sperm.

sensory   Pertaining to sensation; as a sensory neuron, which conveys impulses from a receptor to the central nervous system.

septum   (*pl.* septa) A partition or wall separating two cavities or masses of soft tissue.

septum pellucidum   Partition between anterior portion of lateral ventricles (between fornix and corpus callosum).

serosa   Serous membrane; the surface portion of the lining of body cavities and outermost layer of structures that line body cavities; consists of epithelium and connective tissue.

serous   Relating to or resembling serum (the watery portion of animal fluid).

serous membrane   Surface layer of lining of body cavities.

serrated   Having a notched or saw-toothed edge.

Sertoli cell   Supporting cell in seminiferous tubules of testes.

serum   The liquid part of animal fluid after coagulation; particularly the fluid portion of clotted blood.

sesamoid bone   A bone formed in a tendon where it passes over a joint.

sex chromatin   A small compact mass of chromatin near the nuclear membrane of cells of females; represents an X chromosome; Barr body.

shaft   A long rodlike structure; the part of a long bone between the epiphyses; the diaphysis.

sheath of Schwann   Thin membrane around processes of peripheral nerve cells; formed of Schwann cells; neurilemma.

sickle cell anemia   A type of anemia characterized by the presence of sickle-shaped red blood cells and excessive destruction of red blood cells. A hereditary condition caused by an abnormal hemoglobin molecule.

sigmoid   Shaped like the letter S.

simple goiter   An enlarged thyroid gland; commonly due to iodine lack.

sinoatrial node, SA node   A collection of specialized cells in the right atrium which initiate the heartbeat; the pacemaker.

sinus A cavity or recess, as in a bone; a channel for the passage of blood or lymph, as venous sinuses in the cranium.

sinusoid A small blood-filled space, as in the liver.

slow-wave sleep Normal sleep, characterized by relatively slow and larger brain waves (synchronized); NREM sleep.

sodium—potassium pump An active transport system, located in the cell membrane, which maintains the resting membrane potential by pumping sodium ions out of and potassium ions into the cell.

solute The dissolved substance in a solution.

solvent A substance, usually liquid, capable of dissolving something. Water is the solvent in body fluids.

soma The cell body of a neuron; the body, exclusive of head and limbs; the trunk.

somatic Pertaining to the body; especially to the nonvisceral parts of the body, as somatic vs. autonomic.

somatostatin Growth hormone releasing hormone (GH-RH); also found in the stomach, small intestine, and pancreatic islets.

somatotropin, STH Growth hormone, secreted by the anterior pituitary; promotes growth of the organism and enhances protein anabolism and fat mobilization.

somesthetic Relating to bodily sensations; awareness.

spasticity A state of increased muscular tone, with exaggerated stretch reflexes.

spatial summation An increase in excitability of a neuron due to an increase in the number of active excitatory presynaptic terminals.

specific dynamic action, SDA The increase in metabolic rate following ingestion of food. It is related to the metabolism of the foods, rather than their digestion and absorption.

specific gravity Density; the weight of any body, compared with that of a standard of the same volume. Usually compared to distilled water, which has a specific gravity of 1.000.

spermatic cord The structure formed by the ductus deferens, testicular arteries, veins, nerves, lymphatics, and fuscia, extending from the scrotum to the inguinal ligament.

spermatogenesis The formation and development of spermatozoa.

spermatozoa The male gametes; sperm.

sphincter Muscle fibers arranged around a tubular structure, such as the opening to an organ of visceral cavity, which, by contracting, can close that opening.

sphygmomanometer An instrument for measuring blood pressure.

spicule A small needle-shaped body.

spinal ganglion Site of cell bodies of sensory neurons of spinal nerves; located on posterior roots; posterior root ganglion.

spinal nerve Nerves that emerge from the spinal cord. There are 31 pairs.

spinal shock Period after injury to spinal cord marked by loss of spinal reflexes in muscles innervated by nerves arising from the cord below the cut. Lasts for a few weeks in humans.

spine (1) A short, sharp process. (2) Spinal column.

spiral organ The auditory sense organ; contains the receptor cells that are ultimately stimulated by sound waves; organ of Corti.

spirometer An instrument for measuring the volumes of air breathed.

splanchnic nerve An autonomic nerve carrying sympathetic preganglionic fibers to collateral ganglia, and which supplies abdominal and pelvic viscera.

spongy bone Cancellous bone.

squamous Scalelike; a thin plate.

stapes The smallest of the ossicles of the middle ear; the stirrup; its footplate fits against the oval window.

starch A polysaccharide, formed by the union of many monosaccharides.

statoconia Otoliths.

stem cell Precursor of all blood cells; a primitive cell of bone marrow.

stenosis A narrowing of any canal, but especially of one of the cardiac valves.

stereognosis The appreciation of the form of an object by means of cutaneous and muscle senses.

steroid One of a large family of chemical substances, including the adrenal cortical hormones, sex hormones, vitamin D, cholesterol, and some components of bile.

sterol Steroids containing —OH (alcohol) groups.

stimulus Any change in the environment that in some way modifies the activity of protoplasm.

STPD Standard temperature and pressure, dry ($0°C$, 760 mm Hg, no water vapor).

strabismus A condition in which the optical axes of the two eyes are not parallel; the two eyes do not "look at" the same object.

stratum A layer, especially of differentiated tissues.

stretch reflex Contraction of a muscle in response to stretch of the tendon of that muscle, which stimulates its proprioceptors; myotatic reflex; tendon reflex.

striated Striped.

Stroke volume The volume per stroke; in the heart, the amount of blood pumped by a ventricle each beat.

stroma Framework, especially of connective tissue in an organ, gland, or other tissue.

styloid A slender bony process, resembling a stylus.

subarachnoid space The space between the arachnoid and pia mater layers of the meninges; contains cerebrospinal fluid.

subcutaneous Beneath the skin; hypodermic.

subliminal Subthreshold; below the limit of conscious perception.

sublingual gland One of the salivary glands; located in the floor of the mouth.

submandibular gland One of the salivary glands; located near the angle of the mandible.

submucosa A layer of connective tissue beneath a mucous membrane.

substantia nigra A mass of gray matter containing pigmented cells; functionally related to the basal ganglia.

substrate The substance acted upon or changed by an enzyme.

subthreshold Said of a stimulus that is insufficient to cause a sensation or a response of an effector.

sucrose A disaccharide; table sugar; composed of one molecule of fructose and one of glucose.

sugar A saccharide; one of many small carbohydrates with a general formula of $(CH_2O)_n$; name usually ends in -ose. Important energy sources in the animal body.

sulcus A furrow or groove, especially on the surface of the brain.

summation An accumulation or adding of effects; the accumulation of the local effects of subliminal stimuli at a synapse, leading to threshold and discharge of that neuron; the increased height of contraction of a muscle when a second response is elicited before the first completed summation of twitches.

superficial On or near the surface.

superior Above; higher in relation to another structure; in humans, cephalic.

supination Rotation of the forearm so that the palm of the hand faces forward (in the anatomical position) or upward; or, in reference to the whole body, the face and abdomen upward; opposite of pronation.

supporting cell Cell in seminiferous tubules of the testes with which spermatids associate as they mature to spermatozoa.

supramaximal stimulus   A stimulus whose intensity is greater than that required to produce a maximal response.

supraoptic nucleus   Hypothalamic nucleus located above the optic chiasm. Origin of fibers to neurohypophysis; associated with release of antidiuretic hormone.

suprarenal gland   The adrenal gland.

surae   The calf of the leg.

surfactant   A surface-active substance that reduces surface tension; used to describe such agents in the alveoli of the lung.

suture   An immovable fibrous joint, such as between the bones of the skull.

sympathetic   Pertaining to that part of the autonomic nervous system that arises from the thoracic and lumbar portions of the spinal cord; the thoraco-lumbar division.

sympatho-adrenal   Pertaining to the close relationship between the actions of the sympathetic nervous system and the adrenal medulla.

sympathomimetic   Any drug or chemical whose actions resemble those of the sympathetic nervous system.

symphysis   (*pl.* symphyses) A slightly movable cartilaginous joint, with a fibrocartilage pad or disc between the two bones, e.g., the symphysis pubis and intervertebral discs.

synapse   The junction between two neurons, where an impulse is transmitted from one neuron to another, and where excitation and inhibition occur.

synaptic cleft   The space between the pre- and postsynaptic membranes in a synapse.

synchondrosis   (*pl.* synchondroses) A joint in which two bones are joined by cartilage, as between the ribs and sternum.

syncytium   A group of cells whose separate cytoplasms have fused; the cells are not separated by cell membranes.

syndesmosis   (*pl.* syndesmoses) A joint in which two bones are held together by fibrous connective tissue; a ligamentous junction.

syndrome   The signs and symptoms that appear together in a disease or condition.

synergist   A muscle that assists another (prime mover) in its action by fixing one of its attachments, or by stabilizing the intermediate joint when the prime mover crosses more than one joint.

synovial joint   A joint characterized by a fibrous capsule enclosing a fluid-filled cavity around the joint. Generally freely movable.

systemic   Generally relating to the whole organism, as distinguished from any of its individual parts.

systole   The rhythmic contraction of the heart, especially the ventricles.

systolic pressure   The highest pressure that occurs in an artery during contraction of the ventricles.

T_3, T_4   Tri-iodothyronine and tetra-iodothyronine (thyroxin); thyroid hormones.

T lymphocytes   Lymphocytes associated with cell-mediated immunities. Acquire their properties in the thymus gland.

tactile   Pertaining to touch.

taenia coli   The three longitudinal bands of muscle in the colon.

target organ   The tissue or organ upon which a hormone exerts its action.

tarsals   Relating to the region of the instep and the seven bones of the foot between the leg and the metatarsals.

taste buds   Sense organs for taste located in papillae of the mucous membrane of the tongue.

tectorial membrane   A gelatinous membrane that overlies the hair cells of the spiral organ in the inner ear.

telencephalon   The most anterior portion of the brain, including the cerebral hemispheres, lateral ventricles, and part of the third ventricle.

telereceptor   A receptor that responds to stimuli from a distance, such as the retina.

telophase   The final stage of mitosis, in which the cytoplasm divides and the nuclei re-form.

temporal summation   An increase in response of a nerve due to an increased frequency of excitation.

tendon   A band or cord of fibrous connective tissue that connects a muscle to a bone.

tendon organ   Golgi tendon organ; tension receptors located near the junction of muscle and tendon. Stimulated by both active and passive stretch of the muscle.

tendon sheath   A double-layered synovial sheath surrounding certain tendons.

tentorium cerebelli   A fold of dura mater roofing over the cerebellum, separating it from the cerebral hemispheres.

terminal ganglion   Location of the cell bodies of parasympathetic postganglionic neurons; may be either tiny ganglia or scattered cells in or close to the wall of the organ innervated.

testis   (*pl.* testes) The primary sex organ of the male; produces sperm and secretes androgens; located in the scrotal sac.

testosterone   The chief hormone secreted by the testes; an androgen, and a steroid.

tetanus   (1) A smooth, sustained contraction of muscle, due to rapidly repeated stimulation. (2) A disease (lockjaw) marked by painful tonic contraction, produced by a toxin.

tetany   A disorder characterized by involuntary contraction of skeletal muscles; an increased irritability of motor and sensory neurons.

thalamus   Mass of gray matter in the diencephalon; serves as a sensory relay station and has important integrative and nonspecific effects.

theca   A case or sheath, such as around a tendon; also, the theca interna and theca externa, which are the vascular and fibrous layers around a developing ovarian follicle.

thenar   The fleshy mass on the lateral side of the palm.

thermoreceptor   A receptor that is sensitive to heat.

thiamin   Vitamin B_1, an important coenzyme in metabolic functions.

thoracic duct   The largest lymph vessel in the body. Extends from the cisterna chyli in the lumbar region of the abdominal cavity to the junction of the left subclavian and jugular veins; drains lymph from the lower half and the left side of the body.

thoracolumbar division   That part of the autonomic nervous system that arises from the thoracic and lumbar portions of the spinal cord; the sympathetic division.

thorax   The chest; the upper part of the trunk.

threshold   (1) The least stimulus that will produce a sensation, or a response of an effector. (2) The concentration of a substance in the plasma above which that substance appears in the urine.

thrombin   An enzyme that is present in the blood during clotting; formed from prothrombin and converts fibrinogen to fibrin.

thrombocyte   Blood platelet.

thromboplastin   A factor whose release initiates blood clotting via the extrinsic pathway.

thrombus   A blood clot in a blood vessel and adhering to its wall.

thymine   One of the nitrogenous bases commonly found in DNA.

thymus   A lymphoid organ in the thoracic cavity above the heart. It is where lymphocytes become T lymphocytes by acquiring the ability to carry out cell-mediated immunity. It regresses after puberty.

thyrocalcitonin   Calcitonin.

thyroid cartilage   The largest single cartilage in the larynx; contains vocal cords.

thyroid gland   An endocrine gland in the neck; secretes thyroxin and calcitonin.

thyrotropin   Thyroid-stimulating hormone; TSH; a hormone of the anterior pituitary gland that stimulates secretion by the thyroid gland.

thyroxin T_4; tetra-iodothyronine; the major hormone secreted by the thyroid gland; increases oxygen consumption by cells.

tidal volume   The volume of air that is inspired or expired in normal quiet breathing.

tight junction   Intercellular junction in which adjacent cell membranes appear to fuse, leaving no intercellular space between them.

tissue   A collection of similar cells and the intercellular substances surrounding them. The four basic tissues are: epithelium, connective tissue (including blood, bone, and cartilage), muscle tissue, and nerve tissue.

tissue fluid   The interstitial fluid; fluid that is outside of cells and blood vessels.

tone, tonus   (1) Normal state of tension of a muscle or healthy state or organs. (2) In skeletal muscles, the resistance to stretch, and is dependent upon an intact innervation.

tonic   Pertaining to a state of continuous activity, especially of muscles, but also of neurons and other cells, and reflexes and other responses.

tonofibrils   A system of tiny fibrils found in the cytoplasm of epithelial cells; often prominent near desmosome.

tonsil   A collection of lymphoid tissue, specifically several aggregates in the pharynx (palatine, pharyngeal, and lingual tonsils).

total lung capacity   The volume of air in the lungs after a maximal inspiration; includes vital capacity and residual volume.

toxin   A poisonous substance derived from animal or plant sources.

trabecula   (*pl.* trabeculae) Supporting bundle of fibers crossing the substance of a structure; septum; small piece or spicule of spongy bone.

trachea   Airway leading from the larynx to the bronchi; has cartilage rings to prevent collapse; windpipe.

tract   A path or track. (1) A collection of nerve fibers of the same origin, termination, and function. (2) A system of organs arranged in series that contribute to a common function (e.g., digestive tract).

transamination   The process by which an amino group is transferred from an amino acid to a keto acid.

transcription   In protein synthesis, the process of transferring genetic code information from DNA to mRNA.

transection   Cutting across; a cross section.

transfer RNA, tRNA   The short chain RNA molecules of several varieties, each of which combines with a specific amino acid and brings it to the site of protein synthesis.

translation   In protein synthesis, the process by which genetic code information contained in mRNA is used to bring about a specific sequence of amino acids.

transmitter   A substance that aids in transmission of impulses from a nerve cell to another nerve cell or a muscle; a neurotransmitter; produced and released by the nerve cell.

transmural pressure   The pressure gradient across the wall of an organ; the across-the-wall pressure difference.

transverse   Crosswise; across the longitudinal axis of the body or part. A transverse plane is a horizontal plane.

transverse tubules   Tubular invaginations of the sarcolemma that pass around the myofibrils, between the lateral sacs of the sarcoplasmic reticulum; carry the impulse into the muscle cell.

trauma   A wound or injury.

tricarboxylic acid cycle   The Krebs cycle; citric acid cycle.

tricuspid valve   Having three cusps or processes; the right atrioventricular valve of the heart.

triglyceride   Neutral fat; glycerol combined with three molecules of fatty acid.

trigone   A triangle; specifically, the small triangular area in the urinary bladder between the openings of the ureters and the urethra.

tRNA   Transfer RNA

trochanter   A large rounded bony process, specifically on the femur.

trochlea   A pulleylike structure.

trophoblast   Outer embryonic layer of the blastocyst; erodes the uterine mucosa to establish connections with the maternal tissues; develops the chorionic villi.

tropocollagen   The fundamental unit of collagen; consists of three polypeptide chains wound together.

tropomyosin   A strand of protein associated with the actin molecules in muscle. One of the "relaxing" proteins; helps regulate contraction.

troponin   A globular protein associated with the actin molecules in muscle. One of the "relaxing" proteins; helps regulate contraction.

trypsin   A protein-splitting enzyme; secreted by the pancreas as inactive trypsinogen.

tubercle   Usually a small rounded enlargement.

tuberculosis   A disease caused by a particular type of bacteria that may affect almost any tissue or organ, but most commonly the lungs.

tuberosity   Usually a large rounded enlargement.

tubular maximum, T_m   The maximal rate of transfer (reabsorption or secretion) of a substance by the renal tubules.

tunica adventitia   The outer coat of a blood vessel or other tubular structure, mostly connective tissue.

tunica intima   The innermost coat of a blood vessel, consisting of endothelium and subendothelial connective tissue.

tunica media   The middle coat in the wall of a blood vessel, consisting chiefly of muscle in some vessels, and connective tissue (collagenous or elastic) in others.

tunica vaginalis   A connective tissue sheath, derived from the peritoneum, that partially surrounds most of the testis and epididymis.

turbinate   Concha; one of the bony processes on the lateral wall of the nasal cavity.

twitch   The response of a muscle to a single stimulus; a single contraction.

tympanic membrane   Eardrum; lies between the external and middle ear.

umbilical cord   A cord containing the umbilical arteries and vein, connecting the fetus with the placenta.

umbilicus   The navel; the site where the umbilical cord left the fetus.

undifferentiated cell   A primitive cell that has not taken on any specialized structural or functional characteristics.

uniaxial joint   A joint capable of movement about a single axis, or in a single plane, such as the elbow joint.

unilateral   Occurring on only one side.

unipolar cell   A nerve cell that has only a single process, such as a sensory neuron, whose single process divides into a central and a peripheral branch.

unit membrane   The typical membrane of a cell and many organelles, as seen under the electron microscope.

unmyelinated   Having no myelin sheath.

unsaturated fatty acid A fatty acid whose carbon chain contains one or more double or triple bonds.

uracil One of the nitrogenous bases found in RNA.

urea The chief nitrogenous waste product excreted in the urine; formed in the deamination of amino acids.

uremia A toxic condition caused by an excess of urea and other nitrogenous wastes in blood.

ureter Duct transporting urine from the kidney to the urinary bladder.

urethra Duct transporting urine from the urinary bladder to the exterior. In the male it also transports seminal fluid and sperm.

uric acid A substance excreted in the urine; is formed from metabolism of certain nitrogenous bases (purine bases).

urinary bladder Readily distensible sac in which urine is stored.

urine The fluid containing dissolved solutes that is excreted by the kidneys; is stored in the bladder and excreted through the urethra.

urogenital diaphragm A muscular and fibrous layer between the pubic rami and below the symphysis pubis. In the male the urethra passes through it.

uterine tubes Fallopian tubes; oviducts; ducts by which the ovum is transported to the uterus.

uterus The womb; the hollow muscular organ in which the fertilized ovum implants and develops.

utricle The larger of two chambers of the vestibule of the membranous labyrinth in the inner ear; contains receptors for stimulation by static position and by linear acceleration.

uvula A grape-shaped structure; specifically, the soft rounded projection of the soft palate.

vagina Any sheathlike structure, such as around some tendons; the canal leading from the uterus to the vestibule.

vagotomy Cutting the vagus nerves.

vas deferens The ductus deferens; leads from the epididymis to the ejaculatory duct.

vasa recta Straight vessels; said of capillaries in the renal pyramids.

vasa vasorum "Vessels of the vessels"; small arteries in the outer layers of the walls of large arteries and veins.

vascular Related to, or containing vessels, especially blood vessels.

vasectomy Removal of a small portion of the ductus (vas) deferens.

vasoconstriction Narrowing of the blood vessels.

vasodilatation Dilation (relaxation) of the blood vessels.

vasopressin Antidiuretic hormone; ADH. In large doses the antidiuretic hormone causes constriction of blood vessels.

vein A blood vessel that carries blood toward the heart.

vena comitantes Deep veins that follow the same course as their corresponding arteries.

ventilation The cyclic process of moving air; specifically, the exchange of air in the lungs with that of the atmosphere.

ventral Pertaining to the belly or abdomen; in humans, anterior.

ventricle A small cavity; especially those in the brain which contain cerebrospinal fluid, and those in the heart where they contain blood to be pumped from the heart; also in the larynx.

ventricular folds False vocal cords; a pair of folds in the larynx, just above the true vocal cords.

venule A small vein; lies between capillaries and veins.

vermis Wormlike; the midline portion of the cerebellum.

vertigo Dizziness; a sensation of whirling, either of oneself or of the environment.

vesicles Small membrane-bound sacs in cell cytoplasm, such as synaptic vesicles or pinocytotic vesicles; a sac or distensible organ, such as the urinary bladder or gall bladder.

vestibule A small cavity or space at the entrance to a canal, such as that in the inner ear, and at the vaginal opening.

vestigeal Related to a rudimentary structure; often the remnant of an embryonic structure.

villi (*sing.* villus) Tiny projections into the lumen of the small intestine.

viscera (*sing.* viscus) Internal organs.

visual acuity Acuteness of vision; the power of the visual apparatus to distinguish visual detail, such as recognizing a letter of the alphabet.

visual purple Rhodopsin.

visual yellow Retinene.

vital capacity The greatest amount of air that can be exhaled after a maximal inspiration.

vital center Essential to life; usually includes cardiovascular and respiratory centers in the medulla of the brainstem.

vitamin An organic substance that is necessary for normal metabolism but not produced in the body; thus it must be obtained in food to prevent a deficiency disease.

vitreous humor (body) The colorless, transparent gel filling the cavity of the eye behind the lens.

vocal folds True vocal cords; folds of mucous membrane in the larynx. Sound is produced when they vibrate.

volts Unit of electromotive force; the electrical "driving force."

volume receptor Low-pressure receptors, such as in great veins, pulmonary vessels, and right and left atria which monitor fullness of the vascular system.

vulva The female external genitalia.

white matter The portion of the brain and spinal cord composed of nerve fibers.

white pulp That part of splenic tissue consisting of nodules and aggregations of lymphocytes.

withdrawal reflex A flexion reflex.

womb Uterus.

Z line Narrow cross-striation of striated muscle; delineates the sarcomere.

zona fasciculata The middle layer of the adrenal cortex; cells arranged in cords.

zona glomerulosa Outer layer of adrenal cortex; cells arranged in clumps.

zona pellucida A homogeneous translucent layer of material surrounding the oocyte in an ovarian follicle.

zona reticularis The innermost layer of the adrenal cortex, adjacent to the adrenal medulla.

zygote Single cell formed by the union of sperm and ovum; a fertilized ovum.

Index

Page numbers in **boldface** type refer to pages on which the entry is illustrated.

A band of striated muscle, 193, **194, 195, 196**
Abdomen,
  arteries, 439–42
  muscles, 163–64, 165, **165, 166**
  veins, 447
Abdominal,
  aorta, descending, 439, **441, 442**
  cavity, 75, 77, **77, 550,** 559–60, **561**
Abducens nerve, **264,** 273, 274
Abduction, 139, **141**
Absorption,
  electrolytes, 585
  food, 582–84
  water, 584–85
Accommodation, of lens, 300–301
Acetabulum, 121, **125,** 126, **127**
Acetylcholine, 233
  action on pacemaker cells, 420
  pepsinogen production, 575
  release by cholinergic fibers, 343
Achilles, tendon of, **155,** 183, **186, 187**
Acid-base balance, 404–6, 673–75
Acidity, 24–25
Acidosis, 525, 535, 612
Acids, 23
Acoustic,
  meatus,
    external, 100, **101, 102,** 105, **107, 309,** 310
    internal, **103, 104,** 106, **107**
  nerve. *See* Nerve, statoacoustic
Acromegaly, 364
Acromion process of scapula, 116, **120**
ACTH,
  action, 360, 380
  effect on glucocorticoids, 370–71
  role in stress response, 371
  secretion, 371
Actin, 193, **194**
Action potential, 51–52, **52,** 196, **201,** 229–32

Active transport processes, 33, 36–38
Acupuncture, 293
Adam's apple, 498
Adaptation, 7
  of receptors, 239–40, **241**
  visual, light and dark, 308
Addison's disease, 371, 375
Adduction, 139, **141**
Adductor muscles, 181, 184, **185**
Adenine, 43, **43**
Adenohypophysis, 267, 358–60, **358,** 363–65
Adenoids, 452, 498
Adenosine,
  diphosphate, 41, **42**
  triphosphate, 40–41, **41,** 206
    storage of energy, 590–93
Adipose,
  cell, 63
  tissue, 64, **65**
ADP. *See* Adenosine diphosphate
Adrenal,
  cortex, **365,** 367, 369–71
    layers, **365,** 366
  glands, 365–71, **365,** 650, **651**
  medulla, 341, **365,** 366–67
    epinephrine secretion, 341, 366–67
Adrenergic fibers, release of transmitter substance, 343, **343**
Adventitia, 542, **544**
Afferent neurons, 251, **251**
Afterbirth, 714
Agglutination, 403, **405**
Agglutinins, 403
Agglutinogens, 403
Agonist muscle, 152
Agranulocytes, 392–93
Air,
  breathing mechanism, 505–10
  composition, 516
  movements, nonrespiratory, 511–12
  tidal, 509, **510**
  volume in lungs, 509–10
Albumin, 398
Aldosterone, 369, 380, 613, 668

  reabsorptive action, 369, 668–69
  secretion, 369
Alkalosis, 525, 535
Allergy, 396
All-or-none law, 229, 415
Alveolar,
  ducts, 502, **502**
  membrane, 502, **503**
  processes, 545
  ventilation, 517, **517,** 531
Alveoli, **501,** 502, **502,** 503
Alveolus, of bone, 91
Amines, 353
Amino acids, 27
  essential, 600, 602
  from protein breakdown, 600
  oxidation of, 602
Aminopeptidase, 566, 580
Ammonia,
  formed by deamination, 602
  tubular secretion, 675, **675**
Amniocentesis, 713
Amnion, 711
Amnionic,
  cavity, **709,** 711, **712**
  fluid, 711
Amoeba, 7–8, **7**
Ampulla, of semicircular canal, **309,** 315, **315**
Amylase,
  pancreatic, 566, 576
  salivary, 566, 570
Anabolism, 40
  and catabolism, 590
  protein, 363
    stimulation by growth hormone, 363
Anal,
  canal, 553
  sphincters, 553–54, 581
Anaphase, 47, **47**
Anastomosis, between coronary arteries, 414
Androgens, 371, 607, 691–93
Anemia,
  hemoglobin deficiency, 389
  hemolytic, 390
  pernicious, 390
  sickle cell, 390
Aneurysm, 457

Angiotensin, role in hypertension, 490, 675–76
Anions, 22, **22**
Annulus fibrosus, 136, **136**
Antagonist muscle, 152
Anterior,
  pituitary,
    gland, 358–61, **358, 359**
    hormones and action, 363–65
  plane of body, 75, **76**
Antibodies,
  blood groups, 403–4
  immune response, 394–96
Antidiuretic hormone,
  action, 360, 364, 380
  role in reabsorption of water, 669–70
Antigens,
  blood groups, 403–4
  immune response, 394–96
Antrum of ovum, 699
Aorta,
  abdominal, 439, **441, 442**
  arch, 437, **437, 438, 441**
  ascending, 437, **441**
  branches, 437
  thoracic, 439, **441**
Aortic,
  arch, 437, **437, 438, 441**
  body, 437, 466, **466**
  and carotid chemoreceptors, 466
  sinus, 426, 437, 466, **466**
  valves, 414
Apnea, 529
Apneustic center, 529, **529, 530, 533**
Aponeurosis, 150
  palmar, **175,** 179
  plantar, 185, **191**
Appendicular skeleton, 96, 99
Appendix, **543,** 553, **554**
Appetite, regulatory centers in hypothalamus, 624
Aqueduct, cerebral, 266, 276, **277, 278**
Aqueous humor, 298
Arachnoid, 277, **278, 279**
  membrane, 245, **245**
  villi, 279, **279**
Arbor vitae of cerebellum, 266

Arch(es),
  aortic, 437, **437, 438, 441**
  foot, 129, 131, **131**
  glossopalatine, **545,** 547
  metatarsal, 131, **131**
  neural, 110, **113, 115, 116**
  palmar, **438, 439, 440**
  pharyngopalatine, **545,** 547, **549**
  plantar, 443
  vertebral, 110, **113, 115, 116**
  zygomatic, 99, 100, **102, 105**
Arcuate line, of ilium, 121, 124, **125, 127**
Areolar tissue, 64, **65**
Arterial blood pressure. *See* Blood pressure, arterial
Arterioles, 411, **434,** 435, 436
  kidney, 653, **654, 655**
Arteriosclerosis, 460, 490
Artery(ies), **434,** 435, 436-43
  of abdomen, 439-42
  anterior cerebral, **280, 281,** 282, 437
  anterior intercostal, 438
  anterior tibial, **438,** 443, **444**
  aorta, 414, 437, **437, 439, 441**
  arcuate, **438,** 443, 653, **655**
  axillary, 438, **438, 440**
  basilar, 280, 281-82, **281,** 438
  brachial, 438, **438, 440**
  brachiocephalic trunk, 437, **438, 441**
  of brain, **280, 281,** 282
  bronchial, **438,** 439
  carotid, common, 437, **438, 441**
    external, 437, **438, 439,** 557
    internal, **280, 281,** 282, 437, **438, 439**
  celiac, **438,** 439, 440, **442**
  cerebral, 437
  communicating, **280, 281,** 282
  coronary, **413,** 414, **414,** 483-84, **483**
  costocervical trunk, 438
  deep femoral, **438,** 443, **444**
  deep palmar arch, **438,** 439, **440**
  digital, **438, 439, 440,** 443
  dorsalis pedis, 443, **444**
  elastic, 435, 436
  esophageal, 439
  facial, 437, **439**
  femoral, **438,** 443, **444**
  gastric, 439, **441**
  gastroepiploics, 439, 440, **442**
  gluteal, 442
  of head and neck, 437, **439**
  hepatic, 439, 440, **441**
  hypogastric. *See* Artery, iliac, internal
  iliac, common, **438,** 439, **442, 543, 550, 553, 554**

  external, **438,** 441, 443, **443**
  internal, **438, 441, 442, 443,** 558
  inferior alveolar, 438
  inferior mesenteric, **438,** 439, 442, **442,** 557
  inferior pudendal, 443
  interlobar, 653
  interlobular, 653, **655**
  internal thoracic, 438, 439, **439**
  lingual, 437, **439**
  lumbar, 439, **441**
  maxillary, 437, **439**
  middle cerebral, **280,** 282, 437
  middle sacral, **438,** 439, **441**
  muscular, 435, 436
  obturator, 442, **443**
  occipital, 437, **439**
  ovarian, 439, 441, **441**
  of pelvic region, 442-43
  peroneal, **438,** 443, **444**
  plantar arch, 443
  popliteal, **438,** 443, **444**
  posterior auricular, 437, **439**
  posterior cerebral, **280, 281,** 282, 438
  posterior intercostal, 439, **441**
  posterior tibial, **438,** 443, **444**
  pulmonary, 436, **437**
  radial, 438-39, **438, 440**
  rectal, 442
  renal, **438,** 439, 441, **441** 653
  splenic, **438,** 439
  subclavian, 437, 438, **438, 440, 441**
  superficial palmar arch, **438, 439, 440**
  superficial temporal, 437-38, **439,** 557
  superior alveolar, 438
  superior mesenteric, **438,** 439, 440, **441, 442,** 557
  superior thyroid, 437, **439**
  systemic, 437-43, **438**
  testicular, 439, 441-42, **441**
  of thorax, 439
  thyrocervical trunk, 438, **439**
  ulnar, 438-39, **438, 440**
  umbilical, 478, **479,** 480
  vertebral, **280,** 281-82, **281,** 438, **439, 440, 441**
Arthritis, 144
Articular,
  capsule, 136
  cartilage, 136
  cavity, 136
  processes of vertebrae, 112, **115, 116, 117**
  surface, 89
Articulations. *See* Joints
Artificial,
  kidney, 659
  knee joint, 145

Ascending,
  aorta, 437, **441**
  colon, **543, 550,** 553, **554**
Ascorbic acid, 617
Asphyxia, 536
Association areas of cerebral cortex, 268, **270,** 290, **290**
Asthma, 536
Astigmatism, 302, **303**
Astrocytes, 226, **227**
Atelectasis, 537
Atherosclerosis, 399, 460, 490
Atlas, 112, **113**
Atom, 18
Atomic,
  number, 18
  structure, 18-20, **19**
  weight, 19
ATP. *See* Adenosine triphosphate
Atresia, 698
Atrioventricular,
  bundle, 416, **416**
  groove, 414
  node, 415-16, **416**
  valve, 412, **414**
Atrium of heart, 410, **410, 411,** 412, **413, 415**
Auditory
  area of cerebral cortex, 268
  ossicles, 110, **309,** 310
  pathway, 313-14, **313**
  tube, 309, **310**
Auscultation, 423
Autonomic,
  nervous system, 220, 335-46
    divisions, 337-39
    functions, 339-46
    structure, 336-39
  reflex, 336
    arc, 336
    centers, 336
Autoregulation, of kidney, 657
Axial skeleton, 96
Axis, **113,** 114
Axon, 221, **222**
Axoplasm, 224

Back muscles, 161, 162, **163, 169, 171**
Ball-and-socket joint, 140
Bands, A, H, and I, of striated muscle, 193, **194, 195, 196**
Barometric pressure, effect on respiration, 535-36
Barr body, 684
Basal,
  ganglion, 263, 268, 270, **270,** 330-32
  metabolic rate, 627, 629-31 631, **631**
  size, age, and sex factors, 630-31

Base (chem.), 23
Basic electrical rhythm (BER), 567
Basilar,
  artery, **280, 281,** 282
  membrane, **311,** 312
Basophils, **391,** 392
Bends, decompression sickness, 536
Beriberi, 616
Bernard, Claude, 9-10
Bicarbonate,
  buffer system, 405-6, 524
  CO_2 transport, 522-24
Bicipital groove, 118-19
Bicuspid,
  tooth, 545
  valve, 414, **414**
Bile,
  canaliculi, 556-57
  digestive function, 577-78
  duct, 550, **555, 556,** 557
  formation process, **578**
  secretion, 577-78, 582
Bilirubin, 389, 577
Binocular vision, 303
Bioassay, 354
Birth, skull at, 92, 104-5, **106**
  control, 715-16
Bladder, urinary, 651, **651, 652,** 662-63
  and male reproductive organs, 688-89, **689**
Blastocyst, 709, **709**
Blind spot, 298, **299**
Blood, 385-406
  cells, 386, 387-98, **391**
    red. *See* Erythrocyte
    white. *See* Leukocyte
  clotting, 400-402, **400, 401,** 616
    mechanism, 400-401
      failure of, 402
    pharmaceutical agents that retard, 402
  composition, 386
  fetal, 478, **479,** 480
  flow, 455-60, **455**
    resistances, 456-58, **456, 457**
    velocity, 458-59, **458**
    volume per minute, 455
  glucose level, 357, **357**
  groups, 403-4, **405**
  oxygen, 519-20
  pH, 404-5
    body fluids, 404-5
    control,
      buffer system, 405-6, 473-75
  plasma, 398-99
    composition, 398
    electrolyte concentration, 399
    proteins, 398-99

platelets, **391**, 392, 398
pressure,
    arterial, 420–22, 460–68
        effect of changes on
            cerebral blood flow,
            484, **484**
        gravity and, 485–86, **485**,
            486
        maintenance, 461–64, **463**
        measurement, 460–61,
            461
    diastolic, 422, 461, **461, 462**
    gradient, 456, **456**
        and peripheral resistance,
            457
    influence of kidney, 675–76
    pulmonary, 480–82
    systolic, 422, 461, **461, 462**
reservoir, function of spleen,
    451
serum, 400, **400**
storage, function of spleen,
    451
sugar, 399
temperature, hypothalamic
    centers, 642
transfusion, 403–4
transport of gases, 517–20,
    522–24
types, 403–4
universal donor and recipient,
    403
velocity, 458–59, **458**
venous, 388, 434
vessels, 433–50
    disorders, 460
    layers, 434, 435
    structure, 434–36, **434**
    velocity, 457
    volume, 386
        determinant of arterial
            blood pressure, 462,
            463, 464
Blue babies, 481
B lymphocyte, 394–95, 450, 451
Body(ies),
    anterior view, 75, **76**
    Barr, 684
    cavities, 75, 77, **77**
    ciliary, 297, **297, 298**
    fluids, 666–67
    ketone, 599
    lateral view, 75
    Nissl, 221
    pineal, 267, **267, 379**
    planes, 75, **76**
    temperature, 635–44, **637, 638,**
        641
    vertebrae, 110, **113, 115, 116**
Bolus, 570–71, **571**
Bomb calorimeter, 622, **622**
Bone(s), 87–133
    arm, 95, 118–20, **122, 123**
    atlas, 112, **113**

axis, **113,** 114
calcaneus, 128, **130**
canaliculi, **90,** 91
cancellous, 89
carpal, 120, **124**
cells, 66
classification, 88–91
clavicle, 95, 118, **119**
coccyx, **111,** 114, **117**
collar. *See* clavicle
compact, 89, **90**
conchae, 99, **100, 102, 104**
cranial, 99–106
definition, 88
diaphysis, 89, **90**
ear, 110, **309,** 310
elbow, 138, 139
epiphysis, 89, **90**
ethmoid, 99, **103, 104,** 106
face, **100, 101,** 106–8
femur, 89, **90,** 124, 126, **128**
fibula, 126, **129**
finger, 120, **124**
flat, 89, **89**
foot, 126, 128, 129, **130**
formation, 92–96, **93**
fracture, 95–96, **96**
frontal, 99, **100, 101, 104,** 106,
    106
function, 88
growth, 92–95
hand, 120, **124**
heelbone, 128, **130**
humerus, 118–19, **119, 122**
hyoid, 110
ilium, 121, 124, **125, 127**
incus, 110, **309,** 310
irregular, 89, **89**
ischium, 124, **125, 127**
jaw, 108, **109**
kneecap, 126
lacrimal, 100, **101,** 108
lacunae, 65, **90,** 91
lamellae, **90,** 91
long, 88, **89**
malleus, 110, **309,** 310
mandible, 99, **100, 101,** 108,
    109
marrow, 66, 89, **90**
maxillary, **100, 101, 104,** 106–8
medullary cavity, 89
metacarpal, 120, **124**
metatarsal, 128, **130**
nasal, 99, **100, 101, 104,** 108
number, 96, 99
occipital, **101, 102, 104,** 106
orbits, 99, **100**
os coxae, 121, **125**
palatine, **102,** 108
parietal, **100, 101, 103, 104,**
    105, 106
patella, 126
pelvic, 121, 124, **125, 127**
phalanges, 120, **124,** 129, **130**

processes, 91
pubis, 124, **125, 127**
radius, 119–20, **123**
ribs, 114, **118**
sacrum, 95, **111, 114, 117**
scapula, 95, 116, **120, 121**
sesamoid, 89
short, 88–89, **89**
sinusus, **103, 104, 105,** 106,
    107
skull, 99–110
sphenoid, **100, 101, 102, 104,**
    106, **108**
spongy, 89, **90**
stapes, 110, **309,** 310
sternum, 95, 114–15, **118**
structure, 88–89, **90,** 91
talus, 126, 128, **130**
tarsal, 128, **130**
temporal, **100, 101, 103, 104,**
    105, **107**
of thorax, 114, **118**
tibia, 126, **129**
tissue, 66
toe, 129, **130**
ulna, 119, **123**
vertebrae, 95, 110–15
vomer, 100, **102, 104,** 108
wrist, 120, **124**
zygomatic, 99, **100, 101,** 107
Bony labyrinth, **309,** 310
Bowman's capsule, 652
Brachial plexus, 247–48, **249**
Brain, 261–85
    death, 295
    development, 262–63, **262**
    hemispheres, 262, 263, 267–71
    lobes, **264, 265,** 268
    meninges, 244–45, **245,**
        276–77, **278**
    metabolism, 348–49
    motor area, 268, 325–26, **327**
    sensory area, 287–90, **290**
    structure, 263–71
    ventricles, **262,** 263, 276, **277,**
        278
Brainstem, 263, 265–66, **267,** 334
Breast. *See* Mammary glands
Broad ligaments of uterus, 697,
    699, 701
Bronchi, 301–2, **301, 302**
Bonchioles, 501–2, **501, 502**
Bronchitis, 536
Buffer(s),
    action, 405–6
    mechanism for control of pH
        of body fluid, 405–6,
        473–75
    pairs in fluid, 405
    role in pH control, 524–25
Bulbourethral glands, 689, **689**
Bursa(ae), 137, **138**
    of Fabricius, 394

Calcaneus, 128, **130**
Calcification, 93
Calcitonin, 95
Calcium,
    level maintenance by
        parathormone, 376–77,
        377
    metabolism of, 612–13
Callus, 96, **96**
Calorie, 622
    content in foods, 622–23
Calorimetry, direct and indirect,
    631, **631, 632**
Calyces, renal, 652, **653**
Canal(s),
    anal, 553
    bone, 91
    carotid, 101, **102**
    central, of spinal cord, **262,**
        263
    ear, **309,** 310
    haversian, **90,** 91
    inguinal, 688, **688**
    semicircular, **309,** 310, 315–17,
        315, 316
Canaliculi,
    of bone, **90,** 91
    lacrimal, 295, **295**
Cancellous bone, 89
Cancer, 473
Canine tooth, 545, **546**
Cannon, Walter B., 10
Capillary(ies), 411, **434,** 435, 436
    and alveolar membranes,
        diffusion of gases, 517–18
    equilibrium, 470–71
    exchange, 468–70, **470**
    function, 468–70
    glomerular, 653–54, **654, 655**
    walls, structure, 468
Capsule,
    Bowman's, 652
    glomerular, 652, **654**
    internal, of cerebrum, 270, **270**
Carbaminohemoglobin, 522, **523**
Carbohydrate(s), 25
    absorption, 583
    metabolism, 591–97
Carbon dioxide,
    exchange, 520, 522–24, **523**
    and O_2, diffusion of, 516–18
    partial pressure in air and
        body, 518
    respiratory stimulant, 531
    tension, 522–23
    transport, 520, 522–24, **523**
Carbon monoxide poisoning,
    534–35
Carbonic acid, buffering action,
    405–6, 524
Carbonic anhydrase, 522
Carboxypeptidase, 566, 577

Cardiac,
  catherization, 421
  center of medulla, 426–27, **426**
  cycle, 420–23, **424**
  decompensation, 490
  muscle, 79, 209, 212–13, **212**
  output, 423–24, 426–31, **424,** 455, 456, 464
Cardioaccelerator center (CA), 426, **426**
Cardioinhibitor center (CI), 426, **426**
Cardiopulmonary resuscitation (CPR), 425
Cardiovascular,
  system, 79–80, 409–31
  shock, 488–89
Caries, dental, 545
Carotid,
  and aortic chemoreceptors, **530,** 532, **533**
  arteries, 280–82, 437–41
  body, 437, 466
  canal, 101, **102**
  sheath, 445
  sinus, 426, **426,** 437, **439**
    pressoreceptor mechanism, 466, **466**
Carpal bones, 120, **124**
Cartilage, 65–66
  articular, 136
  arytenoid, 499, **499, 500**
  corniculate, 499, **499, 500**
  costal, 114
  cricoid, 498, **499**
  cuneiform, 499, **500**
  elastic, 66
  fibrous, 66
  hyaline, 66
  laryngeal, 499
  semilunar, 143, **146**
  thyroid, 498, **499**
  types, 66
Catabolism and anabolism, 40, 590
Cataract, 297
Catecholamines, 366
Catheterization, cardiac, 421
Cations, 22, **22**
Cauda equina, 244
Caudal, 75, **76**
Caudate,
  lobe of liver, 555, **556**
  nucleus of cerebrum, 268, **269,** 270, **270**
Cavernous,
  sinus of cranial cavity, 282, **284**
  urethra, 689, **690**
Cavity(ies),
  abdominal, 75, 77, **77, 550,** 559–60, **561**
  articular, 136

body, 75, 77, **77**
cranial, 75, **77**
dental, 545
dorsal, 75, **77**
joint, 136, 137
medullary, of bone, 89
nasal, 99, **100, 105,** 496–98, **497, 498**
oral, **543,** 544–47, **545, 549**
pelvic, 75, 77, **77**
pericardial, 412
peritoneal, 560–61
synovial, 137, **138**
thoracic, 75, 77, **77,** 503–4, **507**
ventral, 75, 77, **77**
vertebral, 75, **77**
Cecum, **550,** 553, **554**
Cell(s),
  blood,
    red. *See* Erythrocyte
    white. *See* Leukocyte
  bone, 66
  chief, of stomach, 549, **552**
  cytoplasm, 11, 12
  definition, 6–7
  division, 46–47, **47**
  growth, 46
  hair, **311,** 312, **314, 315**
  Kupffer, 555–56
  membrane, 11–12
  muscle, 193, 210
  nerve, 221–22, **222**
  neuroglia, 226–27, **227**
  neuron, 221–22, **222**
  nucleus, 11, **11, 13**
  pacemaker, 576
  parietal, of stomach, 549, **552**
  permeability, 11
  pigments, 15
  Schwann, **222,** 224–26, **225**
  stem, 388
  structure and function, 10–16
  theory, 6–7
Cementum, 545, **547**
Center, reflex, 251
Central canal of spinal cord, 263, **278**
Central nervous system, 220, **220,** 244, **263,** 287
Centriole, **13,** 15, 46–47
Centrosome, 15
Cephalic phase of gastric juice secretion, 575
Cerebellar,
  cortex, 266
  peduncles, **264,** 266, **267**
Cerebellum, 262, **262,** 263, **264,** 266, 329–30
  function, 266
  structure, 266
Cerebral,
  aqueduct, 266, 276, **277, 278**
  arteries, 437

circulation, 484, **484,** 485
cortex, 263, 267–68, **269,** 287, 335
hemispheres, 262, **262,** 263, 267–71
nucleus(i), 263
peduncles, 266, **267, 269**
Cerebrospinal fluid, 245, 277, **278,** 279, **279**
Cerebrum, **262,** 263, **265,** 267–71
  basal ganglia, 268, 270, **271**
  convolutions of, 268, **271**
  lobes, **265,** 268, **270**
Cervical,
  enlargement, 244, **244**
  plexus, 247, **249**
  vertebrae, 112, **113,** 114
Chemical,
  compound, 18
  mixture, 18
  reactions,
    citric acid cycle, 41–42, 591, 593–94, **594**
    glycogenesis, 593
    thermoregulation, 638
    transmitters, 343
Chemoreceptors, **465,** 466
  aortic, **530,** 532, **533**
  carotid, **530,** 532, **533**
Chemosensitive areas, 531–32
Chemotaxis, 393
Chest, 503
  cavity, 503–504, **505, 507**
Chewing, 570
Chiasm, optic, **264, 265,** 267, **268,** 305, **306**
Chief cell, of stomach, 549, **552**
Childbirth. *See* Labor and Parturition
Chlorine, metabolism, 613
Choking, Heimlich maneuver, 513
Cholecystokinin, 381, **478,** 568, 573, 577, 578
Cholesterol, 27, 399, 579
Cholinergic fibers, release of acetylcholine, 343, **343**
Chondroblast, 65, 92
Chondrocyte, 65, 92
Chordae tendineae, 414, **414**
Chorion, 709, **709, 710**
Chorionic villi, 478, **479,** 709, **710**
Choroid,
  coat of eye, 296–97, **297**
  plexus, **278,** 279
Chromatin, 11, 15
Chromatolysis, 225, **226**
Chromosomes, 15, 46–47
  sex determination, 682–83
Chylomicrons, 583, **584,** 597
Chyme, 550, 573
Chymotrypsin, 566, 577

Cilia, 58
Ciliary,
  body, 297, **297, 298**
  muscle, 297, **297, 298**
Circle of Willis, **280,** 282, 438
Circulation, 409–31, **410, 482**
  cerebrospinal fluid, 279
  cerebral, 484
  coronary, 483–84, **483**
  enterohepatic, of bile salts, 578, **578**
  fetal, 478, **479,** 480
  lymphatic, **450,** 472, 474
  portal, 447, **447**
  pulmonary, 410, **410,** 436, **437,** 480–82
  systemic, 410, 436–37
  venous, 471–72
Circulatory system, 10, 409–31
Circumduction, 140
Circumvallate papillae of tongue, 318, **318,** 547, **547**
Cisterna chyli, **449,** 450
Citric acid cycle, 41–42, 591, 593–94, **594**
Clavicle, 95, 118, **119**
Cleft palate, 544
Clitoris, **698, 701,** 702, **702**
Clot, blood, 400–401
  retraction, 400, **400**
Clotting, blood. *See* Blood, clotting
Coagulation, blood. *See* Blood, clotting
Coats of eyeball, 296–97, **297**
Cobalamin, 617
Cobalt, 615
Coccygeal vertebrae, 114
Coccyx, **111,** 114, **117**
Cochlea of ear, **309,** 310, **311**
Cochlear,
  duct, **311,** 312, **312**
  nerve, 273, **309, 311,** 312
Coenzyme A, 591
Collagen, 62
Collagenous fibers, 62, 63
Collateral ganglia, 337, 339
Collecting tubule, renal, 653, **655**
Colloid,
  osmotic pressure, 469
    effective, 469–71, **469**
  thyroid gland, 372
Colon, **543, 550,** 553, **554**
  digestive process, 580–81
  secretions, 580–81
Color, vision and blindness, 308
Column(s),
  spinal cord, 246
  vertebral, 110, **111**
Columnal epithelium, 60–61, **60**
Common,
  bile duct, 550, **555,** 557
  carotid artery, 437, **438, 441**

iliac,
    artery, **438, 439, 442**
    vein, **441, 445,** 447
    path, final, 253, **253**
Communicating,
    arteries, **280, 281,** 282
    rami, 247, 337
Compact bone, 89, **90**
Compliance, of lungs, 507–509
Compound, 18
Compound action potential,
    232, **233**
Concave lens, 299, **300**
Concentration, 22
Conchae, nasal, 99, **100, 102,
    104,** 106, 108, 496, **497, 498**
Conduction,
    heat regulation, 639
    nerve impulse, 232
    saltatory, 232
    system, of heart, 415–17, **416**
Condyles, 91
    of femur, 126, **128**
    of humerus, 119
    occipital, 100, **102,** 106
    of tibia, 126, **129**
Condyloid,
    joint, 139
    process of mandible, 108, **109**
Cones of retina, 297, 306–308,
    307
Conjunctiva, 295, **298**
Connective tissue, 62–66, **65**
    adipose, 64, **65**
    cells, 63
    classification, 63–64
    dense, 64, **65**
    elastic, 64–65, **65**
    fibers, 62–63
    loose, 64, **65**
    reticular, 64, **65**
Consciousness, 346
Constipation, 581–82
Contraception, 715–16
    intrauterine device, 715–16
    pill, 715
Contraction,
    atrial, 418
    muscle, 192–215
        isometric, 204
        isotonic, 204
        tetanic, 202–203, **203**
        twitch, 200
    uterine, 712, 714
Contracture, muscle, 203, **204**
Convection, heat regulation, 639
Convergence, 236, **237**
Convex lens, 299, **300**
Convoluted tubule, 653, **655**
Copper, 614
Coracoid process of scapula,
    116, **120, 121**
Cord,
    spermatic, **686,** 687

spinal, 244–49, 262–63, **262,**
    334
    of brain, **262, 263, 264**
    transection, 334
    umbilical, 478, **479**
Core temperature, 636–37, **637**
Cornea of eye, 296, **297, 298**
Coronal,
    plane, 75
    suture, **100, 101,** 103, **104**
Corona radiata, 699
Coronary,
    arteries, **413,** 414, **414,** 483–84,
        483
    circulation, 483–84, **483**
    occlusion, 483–84
    sinus, 412, **413,** 415
    veins, 415
Coronoid process,
    of mandible, **101,** 108, **109**
    of ulna, 119, **123**
Corpora quadrigemini, **265,** 266
Corpus,
    albicans, 700, **700**
    callosum, **265,** 268, **269, 270**
    cavernosum of penis, 689,
        689, 690
    luteum, 699–700, **700,** 710–11
    spongiosum of penis, 689,
        689, 690
    striatum, 270
Corpuscle, renal, 652, **654**
Corresponding points of vision,
    304, **305**
Cortex,
    adrenal, **365,** 367, 369–71
        layers, **365,** 366
    cerebellar, 266
    cerebral, 263, 267–68, **269,**
        287, 335
    kidney, 651–52, **653**
Corticospinal tracts, 246, **246,**
    326, **326**
Corticosterone, **368,** 370
Cortisol, **368,** 370, 607, 608
Cortisone, 370
Costal,
    breathing, 506
    cartilage, 114
    pits, 114
Coughing, 511
Countercurrent mechanism, 670,
    671
Covalent bonds, 20
Cranial, 75, **76**
    bones, 99–106
    cavity, 75, **77**
    fossa, 101
    nerves, 220, **220,** 263, **264,**
        271–76
Craniosacral division of
    autonomic nervous system,
    337
Cranium, 99–110
Creatine phosphate, 206

Crenation, 387
Cretinism, 374
Cribriform plate, **103, 104,** 106
Cricoid cartilage, 498, **499**
Cristae of mitochondria, 13, **13**
Crown of tooth, 545, **547**
Cryptorchidism, 688
Cuboidal epithelium, 60, **60**
Cupula, 315, **315**
Curare, 234
Curves,
    of stomach, 549, **552**
    of vertebral column, 110, **111**
Cushing's disease, 371
Cyanosis, 534
Cycle(s),
    citric acid, 41–42, 591,
        593–94, **594**
    endometrial, 706–707, **706**
    menstrual, 697, 705–708, **706**
Cyclic AMP, 355–56
Cystic duct, **555, 556,** 557
Cytoplasm, 7, 11, 12–13
    organelles, 13–15
Cytosine, 43, **43**

Daylight vision, 308
Dead space, 510, **510**
Deamination, 602
Debt, oxygen, 207–208, 595
Decerebrate rigidity, 334
Deciduous teeth, 545
Defecation, 581
Deglutition, 571–72, **571**
Deltoid tuberosity, 119, **122**
Dendrite, 221, **222**
Denervation hypersensitivity,
    334
Dens, **113,** 114
Denticulate ligaments, 245, **245**
Dentin, of tooth, 66, 545, **547**
Deoxycorticosterone, 369
Deoxyhemoglobin, 387, 524
Deoxyribonuclease, 466, 577
Deoxyribonucleic acid, 28,
    43–44, **43**
Depolarization, 229
Depression of joints, 141
Depth perception, 304
Dermatome, 247, **248**
Dermis, 69–70
Desmosomes, 59, **59**
Descending,
    aorta, 439, **441**
    colon, **543, 550,** 553
Desoxycorticosterone—*see*
Diabetes,
    insipidus, 364, 669
    mellitis, 378, 605, 611–12
Diabetic coma, 612
Diaphragm, 161, 164, **181**
    role during inspiration, 506
    urogenital, 651
Diaphragmatic breathing,
    605–607

Crenation, 387
Diaphysis, 89, **90**
Diarrhea, 581
Diastole, 421, **421**
Diastolic blood pressure, 422,
    461, **461, 462**
Dicumarol, anticoagulant, 402
Diencephalon, **262,** 263, 266–67,
    268
Diethylstilbestrol, **704**
Differential white count, 398
Diffusion, 33–34
    capillary exchange, 469
    facilitated, 37, **37**
    gases, 31–33, **32,** 516–18, **519**
Digestion, 516–18, **519,** 564–85
Digestive,
    enzymes, 565, 566, 573, 574,
        576–77
    juices, 573–78, 580
    organs, 544–57
    system, 10, 80, 541–61
    tract, 542, **543**
Dipeptidase, 566, 580
Diplopia, 304
Disaccharides, 25, 566, 580
Discs,
    intercalated, 212
    intervertebral, 110, 136
Dislocation, 143
Distal, 75
    convoluted tubule, 653, **655**
Diuresis, 669, **669**
Diuretics, 671
Divergence, 237, **237**
Division, cell, 46–47, **47**
DNA, 28, 43–44, **43**
Dopamine, 366
Dorsal, 75, **76**
    cavity, 75, **77**
Dorsiflexion, 141
Down's syndrome, 684–85
Duct(s),
    alveolar, 502, **502**
    bile, 550, **555, 556,** 557
    cochlear, **311,** 312, **312**
    cyctic, **555, 556,** 557
    ejaculatory, 689, **689**
    hepatic, **555, 556,** 557
    lymphatic, **449,** 450
    nasolacrimal, 99, 108, 295, 496
    pancreatic, 551, 554
    thoracic, **449,** 450
Ductus,
    arteriosus, 478, **479**
    deferens, **686,** 688, **688, 689**
    venosus, 478, **479**
Duodenum, **543,** 550–51, **550,**
    552
    digestive process, 573
Dura mater, 245, **245,** 277, **278**
Dwarf, pituitary, 363
Dyspnea, 536

Ear, 309–13
  ossicles, 110, **309**, 310
  structure, 309–10, **309, 311,** 312
Eardrum, 310
Edema, 464
  pulmonary, 437
Effectors, 219, 251
Efferent neurons, 251, **251**
Ejaculation of sperm, 691
Ejaculatory duct, 689, **689**
Elastic,
  fibers, 62, 63
  cartilage, 66
  membrane, external and internal, 434–35, **434**
Elastin, 62
Elbow, 138, 139
Electroanesthesia, 235
Electrocardiogram (ECG), 418–19, **419,** 420
Electroencephalogram (EEG), 294–95, **294**
Electrolytes, 22
  absorption, 585
  blood plasma, 399
  constituent of urine, 661
  and fluid balance, 667–69, 672–73
  reabsorption and secretion, 654–58, 660
Electron, 18
Electroneutrality, 35
Element, 18
Elevation of joints, 141
Embolus, 402
Embryo, 709, 710, **712**
Emotions, 347–48
  role of hypothalamus and limbic system, 347
Emphysema, 536–37
Enamel of tooth, 66, 545, **547**
End-diastolic volume, **424,** 427–28, 429
End-feet, 223, 236
Endocardium, 412
Endochondral ossification, 92–94, **93**
Endocrine,
  glands, 61, 352–82
  system, 61, 352–82
  and nervous system, hypothalamus link between, 362
Endocytosis, 38
Endolymph, 310
Endometrium, 702
  cyclical changes in, 706–707, **706**
  implantation of fertilized ovum, 709–10, **709, 712**
  layers composing, 702
Endomysium, 149, **149**
Endoneurium, 221, **221**

Endoplasmic reticulum, **11, 13,** 14
Endothelium, 434, **434,** 436
End plate potential, 234
End-systolic volume, **424,** 428–29
Energy, 39–42
  balance, 621–22
  caloric values of foods, 622–23
  cost of activity, 627–29
  expenditure, 626–33
  heat as by-product, 621
  heat, measurement, 621
  for muscle contraction, 206–207, **207**
  photosynthesis, 622
  release, 594
  sources, 622–24, 626
  for work, 621
Enterogastric reflex, 573
Enterohepatic circulation, 578, **578**
Enterokinase, 566, 576, 580
Enzyme(s), 27
  action, 29–31, **30**
  digestive, 565, 566, 573, 574, 576–77
  fat metabolism, 597–98
Eosinophils, 391–92, **391**
Ependymal cells, 226–27, **227**
Epicardium, 412
Epicondyles, 91, 119, **122**
Epidermis, 67–69
Epididymis, **686,** 688, **688, 689**
Epiglottis, **497,** 498, **499, 500**
Epimysium, 149, **149**
Epinephrine,
  action, 341, 343, 606–607
  blood glucose level increase, 608
  cardiac accelerator, 429
  effect on muscle cells, 467
  secretion, 366–67
Epineurium, 220, **221**
Epiphyseal centers of ossification, 94
Epiphysis, 89, **90, 93**
Epiploic,
  appendages of colon, 553, **554**
  foramen, 561, **561**
Episiotomy, 703
Epithelium, 58–62
  columnal, 60–61, **60**
  cuboidal, 60, **60**
  glandular, 61–62, **61**
  of mucosa, 542
  simple, 60–61, **60**
  squamous, 60–61, **60**
  stratified, **60,** 61
  transitional, **60,** 61
EPP, 234
EPSP, 237

Equilibrium, 314–17
  molecular, 32
  potential, 49, **49**
Erection, of penis, 689, 691
Erythroblastosis fetalis, 404
Erythrocyte, 387–90, **391,** 392
  count, 387
  formation, 388
  life-span, 388
  removal by spleen, 451
Erythropoiesis, 388
Erythropoietin, 389
Esophagus, 543, 548, **549**
Estradiol, 704, **704**
Estrogen, 381, 703–704
  functions, 703
  secretion,
    cyclical changes, 703, 706–708, **706**
    stimulation,
      follicle-stimulating hormone, 706–707
      luteinizing hormone, 706–707
Eustachian tube, 310
Evaporation, and heat loss, 639–40
Eversion of foot, 141, **142**
Excitatory postsynaptic potential, 237
Excretion and excretory organs, 649–63
Excretory system, 10, 80, 649–63
Exercise,
  application of Starling's law of heart, 429–30
  cardiovascular adjustments, 487–88
  oxygen debt, 207–208, 595
  pulmonary ventilation, 533–34
Exocrine glands, 61, 353
Exocytosis, 38
Exophthalmic goiter, 374
Expiration, 507, **508,** 510
Expiratory reserve volume, 509, **510**
Extension, 139, **140, 141**
Extensors, 151, **151**
External, 75
  acoustic meatus, 100, **101, 102,** 105, **107, 309,** 310
  ear, **309,** 310
  iliac,
    arteries, **438,** 441, 443, **443**
    veins, **445,** 447
Exteroceptors, 223
  cutaneous receptors, 223
  telereceptors, 223
Extracellular fluid, 9, 666–76
  pH regulation, 673–75
  role of antidiuretic hormone, 360, 364, 380, 669–70
Extrapyramidal tracts, 327, **328**
Extremity(ies), 77, **78**

Extrinsic innervation,
  of gut, 559, **559**
  of heart, 420
Eye(s), 295–308
  accessory structures, 295–99
  anatomy, 295–99
  disorders, 301–302
  extrinsic muscles, 295–96, **296**
  focusing mechanism, 300–301
  movements during sleep, 347
Eyeball, 296–98
  coats, 296–97, **297**
  convergence of, 301
  receptors, 297–98

Facial,
  arteries, 437, **439**
  expression muscles, 156, **157**
  nerve, **264,** 273, 274, 337
Facilitated diffusion, 37, **37**
Facilitation, 238, **238**
Facilitory area, 329, **329, 330**
Fainting, 485
Falciform ligament of liver, 550, 555, **555, 556**
Fallopian tube, 697, **698, 701**
Falx cerebri, 277, **279, 284**
Farsightedness, 301
Fascia lata, 179, **182**
Fat(s), 26
  cell, 63
  metabolism, 597–99
    hormonal control, 363
  oxidation of, 598–99
  storage, 597
  synthesis, 599
Fatty acid, 26
  oxidation, 598–99
  synthesis, 599
Fauces, isthmus of, 547
Feedback, control of hormone secretion, 357–58, **357,** 703
Female,
  genitals,
    external, 702–3, **702**
    internal, **698,** 700–2
  pelvis, 124, **127**
  reproductive organs, 696–702, **698, 699**
  sexual cycles, 697, 705–8, **706**
Femoral,
  artery, **438, 443, 444**
  vein, 448, **448**
Femur, 89, **90,** 124, 126, **128**
  origin of muscles, 179–80
Ferritin, 585, 614
Fertility,
  control of, 715–16
  male, 691, 716
Fertilization of ovum, 684, 708–9, **709**
Fetal circulation, 478, **479,** 480
Fetus, 710
Fever, 644

Fiber(s),
  adrenergic, acetylcholine
    release, 343, **343**
  cholinergic, acetylcholine
    release, 343, **343**
  collagenous, 62
  elastic, 62, 63
  fusimotor, 324, **324**
  intrafusal, 323–24, **323, 324**
  muscle, 150, **150**
  nerve, 220–27, **221, 225, 226**
  reticular, 62
  tracts, 246, **246,** 268
Fibrillation, atrial and
    ventricular, 417
Fibrils, 15
Fibrin, 400
Fibrinogen, 399
Fibroblasts, 63
Fibrocartilage, 66
Fibrous,
  cartilage, 66
  joints, 135, **135**
Fibula, 126, **129**
  articulation of, 135
  insertion of muscles, 184
Filiform papillae of tongue, **318,**
    547, **547**
Filtration, 35, **35**
  glomerular, 656–58, **656, 657**
  pressure, effective, 470, **470**
Filum terminale, 244, **244**
Fimbriae of uterine tubes, **698,**
    699, 700, **701**
Final common path, 253, **253**
Fissure(s), 91
  of brain, 268
    lateral, 268, **270, 271**
    longitudinal, 268, **270, 271**
    orbital, 99, **100, 102, 103**
  of spinal cord, 244, **245**
Fixator, 153
Flaccid paralysis, 233
Flat,
  bones, 89, **89**
  feet, 131
Flavin adenine dinucleotide
    (FAD), 591, 594
Flexion, 139, **140, 141, 151**
Flexors, 151, **151**
Floating ribs, 114
Fluid(s),
  body, 666–67
  cerebrospinal, 245, 277, **278,**
    279, **279**
  and electrolyte balance,
    667–69, 672–73
  extracellular, 9, 666–76
  interstitial, 9, 667
  intracellular, 9, 667
  pericardial, 412
  pleural, 503
  seminal. *See* Semen
  synovial, 136
  water reabsorption, 669

Fluorine, 614
Focal length and point, 299, **300**
Follicle(s),
  graafian, 699, **700**
  hair, 70, **72**
  primordial, 698–99, **700**
  thyroid gland, 372
Follicle-stimulating hormone,
    360, 380, 692, 703, 706–7
Follicular phase, 699, **706**
Fontanels, 92, 104–5, **106**
Food(s),
  absorption of, 582–84
  energy-supplying, 622–23
  intake, 624, 626
  metabolism, function of liver,
    629
  specific dynamic action, 629
Foot,
  arches, 129, 131, **131**
  bones, 126, 128, 129, **130**
  muscles, 182–85, **186, 187,**
    188, 189, 190, 191
Foramen, 91
  epiploic, 561, **561**
  interventricular, 276, **277, 278**
  intervertebral, 110
  jugular, 101, **102, 103, 104**
  magnum, 100, **102, 103, 104,**
    106
  mandibular, 108, **109**
  mental, **100,** 110
  obturator, **125,** 126, **127**
  optic, 99
  ovale, 478, **479**
  transverse, 112, **113**
  vertebral, 112, **113, 115, 116**
Forebrain, 262–63, **262**
Fornix, 270, **271**
Fovea centralis, **297,** 298
Fracture, 95–96, **96**
Frictional drag, 457, **457**
Frontal,
  bone, 99, **100, 101, 104,** 106,
    106
  plane, 75, **76**
Fulcrum, 151–52, **152**
Fundus,
  stomach, 549, **552**
  uterus, 701, **701**
Fungiform papillae of tongue,
    318, **318,** 547, **547**

Gall bladder, **543, 550, 555,** 557,
    578
Gallstones, 578, 579
Gametes, 682
Gamma,
  efferents, 324, **325**
  globulin, 395
  loop, 324, **325**
Ganglion,
  basal, 263, 268, 270, **270,**
    330–32

collateral, 337, 339
  semilunar, 272, **273**
  spinal, **245,** 246, **247**
  spiral, of cochlea, **311,** 312
  sympathetic, 337, **338**
  terminal, 337
Gas(es),
  blood transport, 517–20,
    522–24
  diffusion, 31–33, **32,** 516–18,
    519
  exchange, 515–25
    in lungs, 510
  laws, 516
  properties, 516
Gastric,
  glands, 549
  inhibitory peptide, 568
  juice, 549, 573–75
    composition, 573–74
    role of HCL in, 574
    secretion, 573–75, **573,** 582
  motility, 572–73, **573**
  phase of juice secretion, 575
  pits, 549, **552**
Gastrin, 381, 568
Gastrointestinal tract, 542, **543**
Generator potential, 239, **240**
Genes, 15, 43
Geniculate body, 305, **306**
Genitalia,
  female,
    external, 702–3, **702**
    internal, **698,** 700–2
  male, 685–89, **689**
Germinal center of lymph
    nodes, 449–50, **450**
Gestation, 711–12
Gigantism, 363
Gingiva of tooth, 545, **547**
Girdle,
  pectoral, 115–16, **119**
  pelvic, 115, 120–21, **125**
Gland(s),
  adrenal, 365–71, **365,** 650, **651**
  bulbourethral, 689, **689**
  digestive, 575–77
  endocrine, 61, 352–82
  exocrine, 61, 353
  lacrimal, 99, 295, **295**
  mammary, 714–15, **714**
  parathyroid, 376–78, 380
  parotid, 547, **548**
  pineal, 267, **267,** 379
  pituitary, 266–67, **268,**
    358–61, **358, 359,** 363–65
  prostate, 689, **689**
  salivary, 547, **548,** 570–71
  sublingual, 547, **548**
  submandibular, 547, **548**
  tear, 295
  thyroid, 371–74, **372**
  vestibular, **702,** 703
Glandular epithelium, 61–62, **61**
Glans penis, 689, **689, 690**

Glaucoma, 298
Glenohumeral joint, **137,** 140,
    141
Glenoid fossa, 116, **121**
Glial cell. *See* Neuroglia
Gliding joint, 141
Globin, 388
Globulins, 399
Globus pallidus of cerebrum,
    268
Glomerular filtration, 656–58,
    656, 657
Glomerulus, 652–54
  Bowman's capsule, 653
  role in urine formation,
    654–55
Glossopalatine arch, **545,** 547
Glossopharyngeal nerve, **264,**
    273, 275, 337
Glottis, 499, **500**
Glucagon, 378–79, 381, 605, 608
Glucocorticoid(s), 367, 369–71,
    380, 607
  blood glucose level increase,
    357
  inflammatory reaction, 370
  secretion, effect of ACTH, 370
Gluconeogenesis, 597
Glucose,
  absorption, 660
  blood,
    level in, 596
    regulation by insulin, 378
  brain metabolism, 348–49
  metabolism, 591–97
  phosphorylation, 592
  reabsorption, 660
Glucostatic theory, 626
Glucosuria, 596, 662
Gluteal tuberosity, 126, **128**
Gluteus muscles, 179, 180, **182,**
    183
Glycerol, 26
Glycogen, 26, 593
Glycogenesis, 593
Glycogenolysis, 593
Glycolipids, 27
Glycolysis, 41, **592,** 593
  anaerobic, 595
Gnotobiosis, 397
Goiter, toxic and exophthalmic,
    374
Golgi apparatus, **11, 13,** 14
Gonad(s), 381
  female, 682, 684, **685**
  indifferent, 684, **684**
  male, 682, 684, **685**
Gonadotropin, chorionic,
    710–11, **711**
Gout, 144
Graafian follicle, 699, **700**
Gradients, concentration, 32
Granulocytes, 391, **391**
Gravity, effect on cardiovascular
    system, 484–86, **485, 486**

Gray matter,
  of brain, 263
  of spinal cord, 245, **245**
Groove,
  bicipital, of humerus, 118–19
  intertubercular, of humerus,
    119, **122**
Ground substance, 62
Growth, 7
  cause of positive nitrogen
    balance, 602
  effect of estrogens, 703–4
  hormone, 95, 606
    abnormal secretion, 363–64
    action, 363, 381
    blood glucose elevation, 608
  puberty,
    female, 704, 708
    male, 692–93
Guanine, 43, **43**
Gubernaculum of testes, 687,
  688
Gyri of cerebral cortex, 268, **270**

Hair, 70–71, **72**
  cells, **311**, 312, **314, 315**
Hard palate, 100, 106, 544, **545**
Harelip, 545
Haustra of colon, **550,** 553, **554**
Haversian,
  canal, **90,** 91
  system, **90,** 91, 94–95
H band and zone of striated
    muscle, 193, **194, 195, 196**
Head,
  femur, 89, **89,** 124, **128**
  humerus, 118, **122**
  and neck,
    arteries, 437, **439**
    muscles, 156–60, **160**
    veins, 445, **446**
  radius, 119, **123**
Hearing,
  defects, 314
  mechanism, 308–14
Heart, 409–31
  abnormalities, 416–17
  atria, 410, **410, 411,** 412, **413,
    415**
  beat, premature, 417, **418**
  block, 417, **418, 419**
  blood supply, 410, **410**
  chambers, 412
  conduction system, 415–17,
    416
  murmurs, 423
  muscle, 209, 212–13, **212**
  nerve supply, 420
  pacemaker mechanism,
    416–17, **417, 418**
  pericardial covering, 412
  rate, 416, 423
    control of, 424, 426–27
    and stroke volume, 423

sounds during cycle, 422–23
Starling's law, 429–30, **430**
valves, 412, 414, **414**
walls, layers, 412, **413**
Heat,
  energy, measurement, 621
  of muscle contraction, 208
  production and loss, 638–41,
    640
Heimlich anti-choking
    maneuver, 513
Hematocrit, 386, **387**
Heme, 388, **388**
Hemiplegia, 255
Hemispheres of cerebrum, 262,
  262, 263, 267–71
Hemodynamics, 455–60
Hemoglobin, 387, **388**
  as buffer, 406
  catabolism of, 388, **389**
  fetal, 521
  hypoxia, 389
  levels, normal, 387
  and oxygen, 387, 534
    transport, 519
  saturation, 519–20
Hemolysis, 387, 662
Hemolytic anemia, 390
Hemophilia, 402
Hemorrhagic shock, 489
Hemostasis, 400
Heparin, 402
Hepatic,
  artery, 555, **556**
  duct, 555, **556,** 557
  vein, **441, 445,** 447, **447,** 557
Hering-Breuer reflex, 530
Hernia, 687
Hexoses, 25
Hiccup, 511
Hilus,
  kidney, 651
  lung, 502
Hindbrain, 262–63, **262**
Hinge joint, 139
Hip joint, 143, **144**
Hippocampal gyrus, 270, **271**
Histamine, 393
Homeostasis, 8–10
  arterial blood pressure
      maintenance, 461–62
  blood,
    pH, 673
    sugar, 399
  of body fluids, 673–75
  cardiovascular, 484–85
  excretory mechanisms, 650
Homeotherm, 636
Horizontal plane, 75
Hormone(s), 352–82
  adrenal,
    cortex, 367, 369–71, 380, 607
    medulla, 366–67, 380
  anterior pituitary, 360,
    363–64, 380

antidiuretic, 360, 364, 380
  role in reabsorption of
      water, 669–70
control,
  fat metabolism, 363
  pancreatic secretion, 577,
    577
  protein metabolism, 363
digestive action, 568–69
follicle-stimulating, 360, 380,
  703, 706–7
growth, 360, 363–64
  abnormal secretion, 363–64
interstitial cell-stimulating,
  360, 380, 692
luteinizing, 360, 380, 692, 703,
  706–7
mechanisms of action, 354–57
melanocyte-stimulating, 360,
  365
ovarian, 381, 703–8
  cyclical changes in, 705–8
  feedback mechanism, 703
posterior pituitary, 360, 364
  380
prolactin, 360
releasing, 359, 360–61
secretion, control of, 357–58
thyroid, 358, 372–74, 607
thyroid-stimulating, 358,
  372–74
Horns of spinal cord, 245–46,
  245
Human chorionic gonadotropin,
  381, 710–11, **711**
Humerus, 118–19, **119, 122**
  muscles, 166–70, **177**
Humor, aqueous and vitreous,
  298
Humoral immunity, 394, 395
Hunchback, 110
Hunger, regulatory centers in
    hypothalamus, 624
Hyaline,
  cartilage, 66
  membrane disease, 509
Hydrocephalus, 279
Hydrochloric acid,
  in buffer system, 405–6
  gastric juice, 573, 574, **574**
Hydrogen,
  ions, tubular secretion, 673–75,
    673, 674
  transport system, 594, **595**
Hydrostatic pressure, 469–70,
  470
Hymen, 702-3, **702**
Hyoid bone, 110
  muscle attachment, **159,** 160
Hypercalcemia, 377
Hypercapnia, 535
Hyperextension, 139
Hyperglycemic effect, 606
Hypergonadism, 693
Hyperopia, 301, **302**

Hyperoxia, 534
Hyperpnea, 533–34
Hypersensitivity, 396
Hypertension, 460, 489–90
  causes, 490
  renal, 490
Hypertonic solution, 36
Hyperventilation and acidosis,
  525
Hypocapnia, 535
Hypoglossal nerve, **264, 267,**
  275, 276
Hypoglycemia, 605
Hypogonadism, 693
Hypophyseal portal system, 359,
  359
Hypophysectomy, 363
Hypophysis. See Pituitary gland
Hypothalamus, 263, 266, **268**
  body temperature regulation,
    641–42, **641,** 644, **644**
  control of pituitary secretion,
    359–61
  derivation, 263, 266
  effect on emotions, 347–48
  feeding center, 624, **624,** 626,
    626
  function, 266
  visceral control, 344, 346
Hypothenar eminence, 178
Hypothermia, 644
Hypotonic solution, 36
Hypoxia, 389, 534
  myocardial, 483

I band of striated muscle, 193,
  194, 195, 196
Ileocecal valve, 551
Ileum, **550,** 551
Iliac,
  arteries, **438,** 439, **441, 442,**
    443, **443**
  crest, 121, **125**
  fossa, 124, **125**
  spine, 121, **125, 127**
  veins, **441, 445,** 447
Iliotibial tract, 179, **182, 183, 185**
Ilium, 121, 124, **125, 127**
Image formation, 299, **300**
Immunity,
  active, 395
  humoral, 395
  passive, 395
Immunoglobulin, 395
Implantation, of ovum, 709, **709**
Impulse,
  conduction through heart,
    416–17
  nerve, 227–32
    conduction, 232
    facilitory, 238
    inhibitory, 238
Inclusions of cell, 13, 15
Incus of ear, 110, **309,** 310

Indirect calorimetry, 631
Inferior, region of body, 75, **76**
Infertility,
    female, 716
    male, 691, 716
Inflammation, decrease induced
    by glucocorticoids, 370
Inflammatory response, 393
Inflation, reflex, 531
Infundibulum, of pituitary, 267,
    358
Inguinal,
    canal, 688, **688**
    ligament, 164, **165**
Inhibitory,
    area, 329, **329, 330**
    postsynaptic potential, 238
Insertion of muscle, 150
Inspiration, 505–6, **508, 510**
Inspiratory reserve volume, 509,
    510
Insula of cerebrum, 268, **271**
Insulin,
    action, 378, 381, 603–5, 608
    deficiency in diabetes
        mellitus, 378, 605
    hypoglycemic effect, 378
    secretion, 357, **357**
    shock, 612
Intelligence, 348
Intercalated discs, 212, **212**
Intercostal muscles, 163, **163,** 164
Intermediate lobe, of pituitary,
    358, 359, 360, 365
Internal, 75
    acoustic meatus, **103, 104,** 106,
        107
    capsule of cerebrum, 270, **270**
    ear, **309,** 310
Interneurons, 222, **222**
Interoceptors, 223–24
Interosseous membrane, 120
Interphase, 46, **47**
Interstitial,
    cells, 686, **687**
    fluid, 667, 670, 672–73
Interstitial-cell stimulating
    hormone, 360, 380, 692
Intertubercular groove of
    humerus, 118–19, **122**
Interventricular,
    foramen, 276, **277, 278**
    septum, 412, **414**
Intervertebral,
    discs, 110, 136, **136**
    foramina, 110
Intestinal,
    mucosa, 542, **544, 552, 553**
    phase of gastric juice
        secretion, 575
    secretion, 580, 582
Intestine,
    large, **543, 550, 553, 554**
        modifications of mucosa
            and muscle, 553

small, **543,** 550–52, **550, 553**
    digestive process, 575–80
    divisions, 550–51
    functions, 550
Intracellular fluid, 9, 667
Intrafusal fibers, **223,** 224
Intramembranous ossification,
    92
Intrauterine device (IUD),
    715–16
Intrinsic,
    factor, 390
    innervation of gut, 567, 559,
        559
Inversion of foot, 141, **142**
Involuntary muscle, 209
Iodine,
    radioactive, 372–73
    in thyroid hormones, 372, 614
Ionic bond, 21
Ionization, 20–21, **21**
Ions, 21, 399
IPSP, 238
Iris, 297, **297, 298**
Iron,
    deficiency, 614
    lung, 511
    metabolism, 614
Irradiation, 256
Irregular bones, 89, **89**
Irritability, 7
Ischial,
    rami, 124, **125, 127**
    spine, 124, **125,** 126, **127**
    tuberosity, 124, **125**
Ischium, 124, **125, 127**
Islets of Langerhans, 378, 554
Isometric contraction, 204
Isotonic,
    contraction, 204
    solution, 36
Isotopes, 19
Isthmus of the fauces, 547

Jejunum, **550,** 551
Joint(s), 79, 134–47
    ankle, 141, **142**
    artificial, 145
    ball and socket, 140
    biaxial, 138, 139–40, **140**
    capsule, 136
    cartilaginous, 135–36
    cavity, 136, 137
    condyloid, 139
    dislocation, 143
    disorders, 143–44
    elbow, 138, 139
    fibrous, 135, **135**
    of foot, 141–42
    glenohumeral, **137,** 140, 141,
        142
    gliding, 141
    hinge, 139

hip, 143, **144**
    knee, 143, **146**
    movements, 138–41
    multiaxial, 138, 140, **141**
    pivot, 139
    radioulnar, 139
    rib, **112,** 114
    sacroiliac, 124, 142
    saddle, 140
    shoulder, **137**
    suture, 135, **135**
    symphyses, 135–36, **136**
    synchondrosis, 135
    syndesmosis, 135, **135**
    synovial, 136–37, **137**
    temporomandibular, 99, 142,
        143
    tibiofibular, **135**
    uniaxial, 138, 139, **140**
    vertebral, 136, **136**
Jugular,
    foramen, 101, **102, 103, 104**
    notch, 115, **118**
    vein, **441,** 445, **445, 446, 447**
Juice(s),
    digestive, 573–78, 580
    gastric, 549, 573–75
        composition, 573–75
        role of HCL in, 574
        secretion, 573–75, **573,** 583
    intestinal, 580, 582
    pancreatic, 551, **554**
        functions, 576
        secretion, 576–77, 582
Junctions,
    heuroeffector, 222
    neuromuscular, 222, 233
Juxtaglomerular complex, 653,
    655

Keratin, 68, 71
Ketone bodies, 599
Kidney, 80
    artificial, 659
    autoregulation mechanism,
        657
    function, 650
    influence on blood pressure,
        675–76
    internal structure, 650–51, **653**
    role of angiotensin, 675–76
Kilocalorie, 622
Kinetic energy, 39
Kinin, 467–68
Knee,
    jerk, 257
    joint, 143, 145, **146**
Kneecap, 126
Krebs cycle, 41–42, 591, 592–93,
    593
Kupffer cells, of liver, 555–56
Kyphosis, 110

Labia,
    majora, **698, 702, 702**
    minora, **698, 701,** 702, **702**
Labor, 712–13
    function of uterus, 714
    oxytocin role, 364, 714
Labyrinth, bony and
    membranous, **309,** 310
Lacrimal,
    bones, **100, 101,** 108
    canaliculi, 295, **295**
    gland, 99, 295, **295**
Lactation, 714–15
Lacteal, 474, 552, **553**
Lactic acid, 206–7
Lacunae of bone, 65, **90,** 91
Lambdoidal suture, **101, 102,**
    103, **104**
Lamellae of bone, **90,** 91
Lamina,
    basement, 58
    propria, 542, **544**
Laminar flow, 457, **457**
Langerhans, islets of, 378, 554
Large intestine, **543, 550,** 553,
    554
    modifications of mucosa and
        muscle, 553
Laryngeal,
    aperture, 499
    cartilages, 498–99, **499**
    prominence, 498
Laryngopharynx, **497,** 498, 543,
    548, **549**
Larynx, 498–500, **499**
Latent period, 200, **200**
Lateral, 75
Law(s),
    of conservation of energy, 621
    of gases, 516
    of mass action, 28–29
Leukemia, 398
Learning, 348
Lens, of retina, 297, **297, 298,**
    300
Lesions of brain, 334–35
Lesser peritoneal sac, 560–61,
    561
Leukocyte(s), 390–96, 398
    classification, 391–93
    count, 390
    phagocytic, 393
Leukocytosis, 396
Leukopenia, 396
Lever, 151–52, **152**
Ligament(s),
    cruciate, 143, **146**
    denticulate, 245, **245**
    inguinal, 164, **165**
    ovarian, 697, **701**
    palmar, **178,** 179
    round, of uterus, 697, **698, 701**
    suspensory,
        of eyeball, 297, **297, 298**
        of ovaries, 697, **698**

Ligaments (cont.)
  transverse, 185, **186, 187, 189**
  uterine, 697, **698, 701**
Ligamentum,
  arteriosum, 480, **482**
  nuchae, 160
  teres of liver, 480, **482, 550,**
    555, **555, 556**
  venosum, 480, **482**
Light,
  adaptation, 308
  reflex, 305
Limbic system, 270, **272,** 344,
  347
Liminal stimulus and response,
  200
Linea,
  alba, 164, **165, 166**
  aspera, 126, **128**
Line, Z, of striated muscle, 193,
  194, 195, 196
Lingual tonsils, 547
Lipase, pancreatic, 566, 576
Lipids, 26
  absorption, 583–84
  metabolism of, 597–98
Lipogenesis, 597
Lipolysis, 597
Liver, **543,** 550, 555–57, **555,**
  556, 557
  carbohydrate metabolism,
    592-93
  cell function, 555-56
  deamination process, 602
  digestive process, 577–78
  gluconeogenesis, 597
  glycogenesis, 597
  glycogenolysis, 597
Lobes,
  of brain, **264, 265,** 268
  liver, **550,** 555, **555, 556**
  lung, 501–2, **501**
Lobules,
  liver, 555, **557**
  splenic, 450
Loewi, Otto, 342
Long bones, 88, **89**
Longitudinal,
  arch of foot, 129, **131**
  fissure of brain, 268, **270, 271**
Loop of Henle, 653, **655**
Loose connective tissue, 64, **65**
Lordosis, 110
Lung(s),
  air volume in, 509–10, **510**
  breathing mechanism, 505–9
  collapse, 511
  heat loss mechanism, 641
  structure, 501–3, **501, 502**
  substances excreted, 650
Lumbar,
  enlargement, 244, **244**
  vertebrae, **111,** 114, **116**
Lumbosacral plexus, 248, **250**
Luteal phase of ovarian cycle,
  700, **706,** 707

Luteinizing hormone, 360, 380,
  692, 703, 706–7
Lymph,
  flow, 472, 474
  formation, 448
  nodes, 449–50, **450,** 474
  nodules, 552, **553**
  vessels, 448, **450,** 472, 474
Lymphatic,
  circulation, 472, 474
  drainage, 450
  duct, **449,** 450
  system, 448–52, **449**
    cancer of, 473
Lymphocytes, **391,** 392–93, 474
  T and B cells, 394–95, 450,
    451
Lymphoid,
  organs, 392, 450–52, **451**
  tissue, 66
Lysosomes, **11, 13,** 14–15

Macrophages, 63
Magnesium, metabolism of,
  613–14
Male,
  fertility, 691, 716
  reproductive organs, 681–82,
    686, 689
  urethra, 689, **689, 690**
Malleolus,
  fibular, 126, **128**
  tibial, 126, **128**
Malleus of ear, 110, **309,** 310
Mammary glands, 714–15, **714**
  development, 714
  milk ejection, oxytocin
    function, 715
  role of estrogen and
    progesterone, 714
Mammillary bodies, 270
Mandible, 99, **100, 101,** 108, **109**
Mandibular,
  canal, 108
  foramen, 108, **109**
  fossa, 105, **107**
  nerve, 272–73, **273**
  ramus, 100, 108, **109**
Manganese, 615
Manubrium, 114, **118**
Marrow,
  bone, 66
    red, 89
    yellow, 89, **90**
  cavity of long bone, 89, **90**
Mast cell, 63
Mastication, 570
  muscles, 157, 158, **158**
Mastoid process, 100, **100, 101,**
  102, 104, 105, **107**
Matter,
  gray, 245, **245,** 263
  white, 245, **245, 279**
Maxillary nerve, 272–73, **273**
Maximal stimulus, 201, **201**

Meatus,
  acoustic,
    external, 100, **101, 102,** 105,
      107
    internal, **103, 104,** 106, **107**
  nasal, 496, **497, 498**
Medial, 75
Mediastinum, 411, 503, 504, **505,**
  506
Medulla,
  adrenal, 341, 651–52, **653**
    epinephrine secretion, 341,
      606–7
  oblongata, **264,** 265, **265, 267**
Medullary,
  cavity of bone, 89
  vasomotor centers, 465
Meiosis, 683, **683, 684,** 699
Melanin, 69, 71
Melanocyte, 69, 71
Melanocyte-stimulating,
  hormone, 360, 365
Membrane(s),
  alveolar and capillary,
    diffusion of gases,
    517–18
  arachnoid, 245, **245,** 277
  basilar, **311,** 312
  cell, 11–12
  elastic, of blood vessels,
    434–35, **434**
  interosseous, 120
  nuclear, **13**
  plasma, 11, **12, 13**
  potential, 49–51, **49**
  selectively permeable, 34
  synovial, 136
  tectorial, **311,** 312
  tympanic, **309,** 310
  vestibular, **311,** 312
Membranous,
  labyrinth, **309,** 310
  urethra, 689, **689, 690**
Memory, 348
Menarche, 708
Meninges of brain, 244–45, **245,**
  276–77, **278**
Menopause, 704, 708
Menstrual cycle, 697, 705–8, **706**
Menstruation, 705–7
Mental foramen, **100,** 110
Mesencephalon, **262,** 266
Mesenchyme, 62
Mesentery, **550,** 551, 553, 560,
  561
Mesial, 75
Messenger RNA, 44, **44**
Metacarpal bones, 120, **124**
Metatarsal,
  arches, 131, **131**
  bones, 128, **130**
Metencephalon, 263
Microglia, 227, **227**
Microscope, electron, 10–11
Microtubules, 15
Microvilli, 58

Micturition, 662–63, **663**
Midbrain, 262–63, **262,** 266
Middle ear, 310
Midsagittal plane, 75, **76**
Mineralocorticoids, 367, 369
  action, 380
Minerals, metabolism of,
  612–15
Mitochondria, **11,** 13–14, **13**
  action of thyroid hormone,
    373
Mitosis, 46–47, **47**
Mitral valve, 414
Mixture, chemical, 18
Molar tooth, 545, **546**
Molarity, 22–23
Mole (chem), 22
Molecular,
  structure, 20, **20**
  weight, 23
Molecule, 18
Mongolism, 684–85
Motoneurons, 222, **222,** 250–54,
  256–58
  alpha, 323, **324**
Motor,
  area of brain, 268, 325–26,
    327
  end plate, 222, 233, **234**
  pathways, 325–27, **326**
  tracts, 246, 325–27, **326**
  unit, 204
Mouth, 544, **545**
  digestive process, 570–72, **571**
Mucosa, intestinal, 542, **544, 552,**
  553
Müllerian ducts, 684
Muscle(s),
  abdominal, 163–64, **165, 165,**
    166
  adductor group, 181, 184, **185**
  agonist, 152
  antagonist, 152
  antigravity, 153
  of arm, 166–77
  attachments, 149–50
  of back, 161, 162, **163,** 169, **171**
  biceps brachii, 169–72, **173,**
    174, 175
  biceps femoris, 181, **182, 183,**
    184
  brachialis, 169, **172, 173, 174,**
    175
  brachioradialis, 171, 172, **173**
    174, 175, 176
  buccinator, 156, **157**
  cardiac, 209, 212–13, **212,** 415
  cell, 193, 210
  ciliary, 297, **297, 298**
  coccygeus, 164
  contraction, 192–215
    uterine, 712, 714
  coracobrachialis, 170, **173**
  cramps, 205
  deltoid, 168, **169,** 170, **171,**
    173, 174

depressor anguli oris, 156, **157**
depressor labii inferioris, 156, **157**
diaphragm, 161, 164, **181**
digastric, 159, **160**
epicranius frontalis, 156, **157**
   occipitalis, 156
erector spinae, 161, 162
extensor, 151, **151**
extensor carpi radialis, 172, **175, 176,** 177, **177**
   ulnaris, 172, **176,** 177, **177**
extensor digitorum communis, 172, **176**
   longus, 188, 190
extensor hallucis longus, 188, 190
extensor pollicis longus, 172, **176**
external oblique of abdomen, 163, 165, **165**
extrinsic, of eye, 156–57, 295–96, **296**
of facial expression, 156, **157**
fascia, 179, 180, **182**
fibers, 150, **150**
flexor, 151, **151**
flexor carpi radialis, 172, **175,** 177, **177**
   ulnaris, 172, **175,** 177, **177**
flexor digitorum longus, **186, 187,** 188, **190**
   profundus, 171, 172, **175, 178**
   superficialis, 171, 172, **175, 178**
flexor hallucis longus, **186, 187,** 188, **189**
flexor pollicis longus, 172, **175, 178**
of foot, 182–83, 185, **186, 187,** 188, **189, 190, 191**
gastrocnemius, 183, **186, 187,** 188, **189**
gemellus, 180, **182, 183**
geniohyoid, 159
gluteus maximus, 179, 180, **182, 183**
   medius, 179, 180, **182, 183**
   minimus, 179, 180, **183**
gracilis, **182, 183,** 184, **185, 186**
gross structure, 193
hallucis, 186, 188, **189, 191**
hamstrings, 181, 184
of hand, 177–79, **177, 178**
of head, 156–60
heart. *See* Cardiac.
of hip, 179–81, **182, 183**
iliacus, 179, 180, **181**
iliocostalis, 161, 162, **163**
iliopsoas, 179, 180, **181**
infrahyoid, 157, 159, **160**
infraspinatus, 168, 170, **174**
insertion, 150

intercostals, 163, **163,** 164
internal oblique of abdomen, 163, 165, **165**
interossei, of foot, **191**
   of hand, 178, **178**
intertransversarii, 162
involuntary, 209
of larynx, 157–58, **161**
latent period, 200, **200**
latissimus dorsi, 167, 168, **169,** 170, **171**
of leg, 181–90
levator ani, 164, **167**
   labii superioris, 156, **157**
   scapulae, 168, **171, 173**
longissimus, 161, 162, **163**
longus capitis and colli, 159
lumbricales, of foot, **191**
   of hand, 178, **178**
masseter, 157, **157,** 158
of mastication, 157, 158, **158**
multifidus, 161, 162
mylohyoid, 159, **160**
names, 153–54
of neck, 158–60, **160**
oblique, of eye, 163, 165, **165, 166,** 296, **296**
obturator externus, 180, **183**
   internus, 180, **182**
omohyoid, 159, **160**
orbicularis oculi, 156, **157**
   oris, 156, **157**
origin, 150
palmaris longus, 172, **175**
papillary, 414, **414**
pectineus, 184, **185**
pectoralis major, 167–68, 170, **173, 174**
   minor, **165,** 166, 168, **173**
of pelvic floor, 164, **167**
of perineum, 164, **167**
peroneus brevis, 183, **188, 189,** 190
   longus, 183, 188, **189,** 190
   tertius, 183, 188, **189,** 190
of pharynx, 157–58, **161**
piriformis, 180, **182, 183**
platysma, 156, **157**
pollicis, 177, **178**
pronator quadratus, 172, **175**
   teres, 172, **175**
psoas major, 160, 180, **181**
pterygoids, 157, 158, **158**
pump, 486
quadratus femoris, 180, **182, 183**
   lumborum, 165, **181**
quadriceps femoris, 182, **185**
rectus abdominus, 163, 165, **165, 166**
rectus, of eye, 296, **296**
rectus femoris, 182, 184, **185**
rhomboids, 166, 168, **171, 173**
risorius, 156, **157**

rotatores, 161, 162
sartorius, 181, **182, 183,** 184, **185, 186**
scalene, 159, 160, **160**
semimembranosus, 181, **182, 183, 184**
semispinalis, 161, 162, **163**
semitendinosus, 181, **182, 183,** 184
serratus anterior, 166, 168, **171**
   posterior, 164
skeletal, 79, 193–208, **193,** 323–35
smooth, 79, 209–12, **210**
soleus, 183, **186, 187, 188, 189**
spinalis, 161, 162, **163**
spindles, **223,** 224, 324, **325**
splenius capitis, 162, **163**
   cervicis, 162, **163**
sternocleidomastoid, 158, 159, **160,** 165
sternohyoid, 159, **160**
sternothyroid, 159, **160**
striated, 79, 193
stylohyoid, 159, **160**
suboccipital, 159
subscapularis, 168, 170, **173**
supinator, 172, **175, 176**
suprahyoid, 157, 159, **160**
supraspinatus, 168, 170, **173, 174**
temporalis, 157, 158, **158**
tensor fascia lata, 179, 180, **185**
teres major, 168, 170, **173, 174**
   minor, 168, 170, **173, 174**
of thigh, 179–81, **182, 183, 185**
of thorax, 161, 163, 164
thyrohyoid, 159, **160**
tibialis anterior, 185, 188, **189, 190**
   posterior, **187,** 188
of tongue, 157–58, **161**
transversus abdominus, 163, 165, **166**
trapezius, 158, 159, **160, 165,** 166, 168
triceps brachii, 169, 170, 172, **174, 176**
   surae, 183
twitch, 200, 202–3, **202, 203**
vastus intermedius, 182, 184, **185**
   lateralis, 182, **182, 183,** 184
   medialis, 182, 184, **185**
of vertebral column, 160–62, **163,** 168
visceral, 209
voluntary, 193
zygomaticus major, 156, **157**
   minor, 156, **157**
Muscular,
   system, 79, 148–91
   work, 627, 629
Muscularis, 542, **544**

mucosa, 542, **552, 553, 544**
Myelencephalon, 263
Myelin, 222, 225, **225**
Myocardium, 412, **414, 415**
Myofibrils, 15, 193, **194, 196**
Myofilaments, 15, 193, **194**
Myometrium, 702, **712**
   contractions during labor, 712, 714
Myopia, 301–2, **302**
Myosin, 193, **194, 195**
Myxedema, 374

NAD, 591, 594, **595**
Nails, 70, **71**
Nasal,
   bones, 99, **100, 101, 104,** 108
   cavity, 99, **100, 105,** 317, **317,** 496–98, **497, 498**
   conchae, 99, **100, 102, 104,** 106, 108, 496, **497, 498**
   septum, 99, **100, 104**
Nasolacrimal duct, 99, 108, 295
Nasopharynx, **497, 498, 543, 547, 548, 549**
Near point of vision, 301, **301**
Nearsightedness, 301–302
Neck,
   anatomical, of humerus, 118, **122**
   femur, 126, **128**
   and head,
      arteries, 437, **439**
      muscles, 156–60, **160**
      veins, 445, **446**
   radius, **123**
   of tooth, 545, **547**
Negative feedback mechanism, of hormones, 357, **357**
Nephron,
   blood supply, 653–54, **655**
   role in urine formation, 654–55
   structure, 652–53, **655**
Nerve(s),
   abducens, **264,** 273, 274
   acoustic, *See* statoacoustic.
   axillary, 248, **249**
   cranial, 220, **220,** 263, **264,** 271–76
   cochlear, 273, 275, **309, 311,** 312
   common peroneal, 249, **250**
   endings, 222–24
   facial, **264,** 273, 274, 337
   femoral, 248, **250**
   fiber, 220–27, **221, 225, 226**
   glossopharyngeal, **264,** 273, 275, 337
   hypoglossal, **264,** 275, 276
   impulse, 227–32
   mandibular, 272–73, **273,** 274
   maxillary, 272–73, **273,** 274

Nerves (cont.)
median, 248, **249**
musculocutaneous, 248, **249**
oculomotor, **264,** 272, 274, 337
olfactory, **264,** 271, 274
ophthalmic, 272–73, **273,** 274
optic, **264,** 272, 274, 305, **306**
pelvic, 337
peripheral, 220, **220,** 247–49
phrenic, 247, **249**
plexus, 247, **249, 250**
radial, 248, **249**
saphenous, 249, **250**
sciatic, 249, **250**
spinal, 220, **220, 245,** 246–47, 247
    accessory, **264, 267,** 275, 276, **280**
splanchnic, 339, 558
statoacoustic, **264,** 273, 275
tibial, 249, **250**
trigeminal, **264,** 272–73, **273,** 274
trochlear, **264,** 272, 274
ulnar, 248, **249**
vagus, **264,** 273, 275, **276,** 337
Nervous system, 80–81
autonomic, 81, 220, 335–46
    divisions, 337–39
    functions, 339–46
    structure, 336–39
cells, 221–22, **222,** 226–27, **227**
central, 220, **220,** 244, **263,** 287
and endocrine system, hypothalamus link between, 361–62
parasympathetic, 337, **338,** 341–42
peripheral, 220, **220,** 247–49
somatic, 220
sympathetic, 337, 339, 341
Neural,
arch, 110, **113, 115, 116**
tube, 262
Neurilemma, 225
Neuroeffector junction, 222
Neuroendocrine reflex, 362
Neuroglia, 226–27, **227**
Neurohypophysis, 266–67, 358–60, **358,** 364–80
Neuromuscular,
junction, 222, 233
transmission, 233–34, **234,** 236
Neuron(s), 221–22, **222.**
afferent, 251, **251, 252**
axon, 221, **222**
classification, 222
dendrite, 221, **222**
efferent, 251, **251**
interneuron, 222, **222**

motoneuron, 222, **222**
preganglionic and postganglionic, 336, **336**
sensory, 222, **222**
structure, 220–21
types, 222, **222**
Neutron, 18
Neutrophils, 391, **391, 392**
Niacin,
deficiency, 616
role in cellular respiration, 616
Nicotinamide adenine dinucleotide (NAD), 591, 594
Night blindness, 308
Nipples, oxytocin secretion, 364, 715
Nissl bodies, 221
Nitrogen,
balance, negative and positive, 602
in blood plasma, 399
narcosis, 536
partial pressure, 516
Node(s),
lymph, 449–50, **450,** 474
heart, 415–16, **416**
of Ranvier, **222,** 225, **225**
Nodules, lymph, 552, **553**
Nonelectrolytes, 22
Non-rapid eye movement (NREM), 347
Norepinephrine,
secretion, 366–67, 380
transmitter substance, 343, 467
Normality, 23–24
Normoblast, 388
Notch,
radial, 120, **123**
sciatic, 121, **125**
semilunar, 119, **123**
Nucleic acids, 28
Nucleolus, 11, 15–16
Nucleotidase, 566, 580
Nucleotide, 28
Nucleus,
of atom, 18
caudate, 268, **269,** 270, **270**
cell, 11, **11, 13,** 15–16
cerebral, 263
neuron, 221
pulposus, 136, **136**
Nystagmus, 316

Oblique muscles of eye, 163, 165, **165, 166,** 296, **296**
Obturator foramen, 121, **125, 127**
Occipital,
bone, **101, 102, 104,** 106
condyles, 100, **102,** 106
lobe of brain, **265,** 268, **270**

Oculomotor nerve, **264,** 272, 274, 337
Olecranon,
fossa, 119, **122**
process, 119, **123**
Olfaction, 317
Olfactory,
bulb, **264,** 270, **317**
nerve, **264,** 271, 274, **317**
tract, **264,** 270, **317**
Oligodendroglia, 226, **227**
Omentum, greater and lesser, **550, 560, 561**
Oncotic pressure, 469, 470
Oocytes, 683, 698
Oogenesis, 683
Ophthalmic nerve, 272–73, **273**
Optic,
chiasm, **264, 265, 267, 268,** 305, **306**
foramina, 99
nerve, **264,** 272, 274, 305, **306, 307**
tract, 305, **306**
Orbital fissures, 99, **100, 102, 103**
Orbits, 99
Organ(s),
of Corti, **311,** 312
digestive system, 544–57
nervous, 244–49
reproductive,
    female, 696–702, **698, 699**
    male, 681–82, **686, 689**
respiratory system, 496–504
retroperitoneal, 560
spiral, **311,** 312
Organelles, 13–14, **13**
Origin of muscle, 150
Oropharynx, **497,** 498, **543, 545,** 548, **549**
Os coxae, 121, **125**
Oscilloscope, 228, **228**
Osmoreceptors, 669
Osmosis, 34–35, **34**
Osmotic pressure, 34–36, 469–71, **469**
Ossicles of ear, 110, **309,** 310
Ossification, 92–96
centers, 92, **93**
endochondral, 92–94, **93**
intramembranous, 92
Osteoblasts, 91, 92
Osteoclasts, 94
Osteocytes, 66, 91
Osteoid, 92
Osteology, 88
Otoliths, 314
Oval window of ear, **309,** 310
Ovarian,
artery, 439, 441, **441**
hormones, 381, 703–705
    cyclical changes in, 705–708
    feedback mechanism, 703

ligament, 697, **701**
vein, **441,** 447
Ovary(ies), 682, 698–701, **698, 699, 700, 701**
Oviduct, 697, **698, 699, 700**
Ovulation, 699, 705
Ovum, 682
cyclical changes, 705, **706**
development, 698–99, **700**
fertilization, 708–709, **709**
Oxidation, 29
energy release, 41–42
Oxygen,
blood, 519–20
caloric value, 631–32
and carbon dioxide, transport, 520, 522–24, **523**
combination with hemoglobin, 519
consumption, metabolic rate determination, 631–32, **632**
debt, 207–208, 595
diffusion, 516–18
and hemoglobin saturation, 519–20
partial pressure in air and body, 518
saturation of blood, 519–20
tension, 517–20
Oxyhemoglobin, 387, 519, 524
Oxytocin, 360, 364, 380, 714, 715
action, 360, 364, 380
milk ejection role, 364, 715
release during labor, 364, 714

Pacemaker, 416, **416, 418**
cells, 576
Palate,
cleft, 544
hard, 100, 106, 544, **545**
soft, 544, **545**
Palatine,
bones, **102,** 108
process of maxilla, **102, 104,** 106
tonsils, 452, 545, **547**
Palmar,
aponeurosis, **175,** 179
arches, deep and superficial, **438, 439, 440**
ligament, **178,** 179
Pancreas, 378–79, **550,** 554–55
Pancreatic,
amylase, 566, 576
duct, 551, 554
islands, 378, 554
juice, 551, **554**
    function, 576
    secretion, 576–77, 582
lipase, 566, 576
Pancreozymin, 381, 568
Panniculus adiposus, 164

Pantothenic acid, 616
Papilla(ae),
    duodenal, 551
    renal, 652, **653**
    tongue, 318, **318,** 547, **547**
Paralysis, flaccid, 333
Paranasal sinuses, 103, **103, 104,**
    105, 496, **497**
Paraplegia, 255
Parasympathetic division of
    autonomic nervous system,
    337, **338,** 341–42
Parathormone, action, 95,
    376–77, **377,** 380
Parietal,
    bones, **100, 101, 103, 104,** 105,
        106
    cells, of stomach, 549, **552**
    peritoneum, 559–60
    pleura, 503, **505**
Parkinson's disease, 331
Parotid gland, 547, **548**
Partial pressure, 516
Parturition, 712, 714
    function of uterus, 714
    oxytocin role, 714
Passive transport processes,
    33–36
Patella, 126
Patellar tendon, 182, **189, 190**
Pathways,
    auditory, 313–14, **313**
    motor, 325–27, **326**
        extrapyramidal, 327, **328**
        pyramidal, 326, **326**
    sympathetic, 337, 339, **339**
Pectoral girdle, 115–16, **119**
Peduncles,
    cerebellar, **264,** 266, **267**
    cerebral, 263, **265,** 266, **267,**
        269
Pellagra, 616
Pelvic,
    bones, 121, 124, **125, 127**
    cavity, 75, 77, **77**
    floor, 164, **167**
    girdle, 115, 120, 121, **125**
    nerve, 337
Pelvis,
    female, 124, 126, **127**
    greater and lesser, 124, 126
    male, 124, 126, **127**
    renal, 651, **653**
    true and false, 124, 126
Penis, 689, **689, 690**
Pepsin, 566, 573
Pepsinogen, 573
Peptide(s), 353
    gastric inhibitory, 568
Pericardial cavity, 412
Pericardium, 412
Perichondrium, 65–66
Perilymph, 310
Perimysium, 149, **149**

Perineurium, 221, **221**
Periosteal,
    bone collar, 93, **93**
    bud, 94
Periosteum, 66, 89, **90,** 93
Peripheral,
    inhibition, 343
    nervous system, 220, **220,**
        247–49
    receptors, **530,** 532, **533**
    resistance, 457–58
        determinant of arterial
            blood pressure, **463,**
            464–65
        neural control, 465–67, **465**
    run-off blood flow, 459
Perirenal fat, 651
Peristalsis, 569, **569,** 576
Peritoneal,
    cavity, 560
    sac, 560
Peritoneum, 67, 542, **544,**
    559–61, 561
    parietal, 559–60
    visceral, 559–60
Permanant teeth, 545, **546**
Permeability of cell, 11
Pernicious anemia, 390
Perpendicular plate, **104,** 106
Perspiration, 640–41
Peyer's patches, 552
pH, 24–25
    blood, 404–405
    control, action of buffers,
        405–406
    regulation in extracellular
        fluid, 673–75
    urine, 674–75
Phagocytic leukocytes, 393
Phagocytosis, 38, **38**
Phalanges, 120, **124,** 129, **130**
Pharyngeal tonsils, 452
Pharyngopalatine arch, **545,** 547,
    549
Pharynx, **497,** 498, **543,** 547, **549**
Phenylketonuria(PKU), 43, 662
Phosphate,
    blood level maintenance by
        parathormone, 376–77
    buffer pair, 405–406
Phospholipids, 26–27
Phosphorus, metabolism of, 612
Phosphorylation, 41, 592
    oxidative, 594
Photoreceptors, 297–98
Photosynthesis, 622
Physical thermoregulation, 638
Physical transport processes,
    33–36
Physiological transport
    processes, 33, 36–38
Pia mater, 245, **245,** 276–77, **278,**
    279
Pigmentation of skin, function

    of melanocyte-stimulating
        hormone, 360, 365
Pineal gland, 267, **267,** 379
Pinna of external ear, 309, **309**
Pinocytosis, 38, **38**
Pituitary,
    dwarf, 363
    gland, 266–67, **268,** 358–61,
        358, 359
        anterior lobe, 358–61, **358,**
            359, 380
            hormones and actions,
                363–65, 380, 606
        intermediate lobe, **358,** 359,
            365
        location, 267, 358
        posterior lobe, 358, **358**
            hormones and action,
                364–65
    portal system, 359, **359**
    stalk, 267, 358, **358**
Pivot joint, 139
Placenta, 478, **479,** 480, 709–10,
    710
Planes of body, 75, **76**
Plantar,
    aponeurosis, 185, **191**
    arch, 443
    flexion, 141
Plasma, 66
    blood. *See* Blood plasma
    membrane, 11, **12, 13**
    proteins, 398–99
        as buffers, 406
        capillary exchange, 469–71
Plate, perpendicular, **104,** 106
Platelet(s), **391,** 392, 398
    role in blood clotting, 400
Pleura, 67, 503
    parietal, 503, **505**
    visceral, 503, **505**
Pleural space, 503, **507**
Plexus, 247
    brachial, 247–48, **249**
    cardiac, 420
    cervical, 247, **249**
    choroid, **278,** 279
    lumbrosacral, 248, **250**
Pneumonia, 536
Pneumotaxic center, 529, **529,**
    530, 533
Pneumothorax, 511
Podocytes, 652
Poikilotherm, 636
Polycythemia, 390
Polypeptide, 353
Polysaccharides, 26
Pons, **264,** 265–66, **265, 267, 268,**
    269
Popliteal,
    artery, **438,** 443, **444**
    fossa, 181, **182, 183**
    surface, 126, **128**
    vein, 448, **448**

Portal,
    circulation, 447, **447**
    system, hypophyseal, 359, **359**
    vein, 447, **447, 556,** 558
        route, 555, 556
Posterior, 75, **76**
    median sulcus, 244, **245**
    pituitary gland, 358, **358**
    white columns, 246, 287, **289**
Postganglionic,
    fibers of autonomic nervous
        system, 337, **338,** 339
    neurons, 336, **336**
Post-junctional membrane, 233,
    234
Posture, 332–33
Potassium,
    reabsorption, 667–69
    metabolism, 613
Potential,
    action, 51–52, **52,** 196, **201,**
        229–32
    compound action, 232, **233**
    difference, 48
    electrical, 48
    end plate, 234
    energy, 39
    equilibrium, 49, **49**
    excitatory postsynaptic, 237
    generator, 239, **240**
    inhibitory postsynaptic, 238
    membrane, 49–51, 416, **417**
    resting, 49–51, **50, 51,** 196
Precapillary sphincter, 435, 468,
    468
Preganglionic,
    fibers of autonomic nervous
        system, 337, **338,** 339
    neurons, 336, **336**
Pregnancy, 708–12, 714
    influence on basal metabolic
        rate, 631
    Rh-negative mother with Rh-
        positive fetus, 403–404
Premature beat, 417, **418**
Prepuce, male, 689, **689**
Presbyopia, 301
Pressoreceptor, **465,** 466
Pressure(s),
    atmospheric, 516
    blood. *See* Blood pressure
    filtration, effective, 470, **470**
    glomerular filtration, 656–58,
        656, 657
    hydrostatic, 469–70, **470**
    intraocular, 298
    in lungs during inspiration,
        505
    osmotic, 469, 470
    partial, 516
    pulse, 461
    tension, 516
    transmural, 428
    tympanic membrane, **309,** 310

Presynaptic terminals, 223, 236, **236**
Progesterone, 381, **704, 705**
  action with estrogen, 705
  secretion,
    cyclical changes, **705,** 706
    stimulation by luteinizing hormone, 707
Projection tracts, 268, **269**
Prolactin, 360
Proliferative phase of menstrual cycle, **706,** 707
Pronation, 139, **143**
Prophase, 46–47
Proprioceptors, 224
Prostaglandins, 356–57, 716
Prostate gland, 689, **689**
Prostatic urethra, 689, **689**
Protein, 27
  absorption, 584
  anabolism, 363
    stimulation by growth hormone, 363
  capillary exchange, 469
  catabolism, 370
  kinase, 355
  metabolism, 600, 602
  plasma, 398–99
    buffer pairs in, 406
  specific dynamic action, 629
  synthesis, 42–45, **44,** 600, 602
    promotion by insulin, 378
Proteinuria, 662
Prothrombin, 401, 616
Proton, 18
Protoplasm, 7
Protraction, of joints, 141
Proximal, 75
  convoluted tubule, renal, 653, **655**
Pterygoid processes of sphenoid bone, 100, **104,** 106, **108**
Ptyalin, 570
Puberty,
  female, 704, 708
  male, 692–93
Pubic,
  arch, 126
  rami, 124, **125, 127**
  tubercle, 124, **125, 127**
Pubis, 124, **125, 127**
Pulmonary,
  arteries, 414, 436, **437, 438**
  circulation, 410, **410,** 436, **437,** 480–82
  space, 503, **507**
  trunk, 436, **437**
  valves, 414
  veins, 436, **437, 445**
  ventilation, 510, 531–34
  vessels, 480–82
Pulp cavity of tooth, 545, **547**
Pulse,
  pressure, 461

waves, 459–60
Pupil of eye, 297, **297**
  constriction mechanism, 301
Putamen of cerebrum, 268, **269,** 270
Pyloric,
  antrum, 549
  canal, 549
  sphincter, 549, **552**
Pyramidal tracts, 326, **326**
Pyramids,
  of medulla oblongata, **264,** 265, **269**
  renal, 652, **653**
Pyridoxine, 616
Pyruvic acid, 42, **42,** 206, 593–96

Quadrate lobe of liver, 555, **556**

Radial,
  artery, 438–39, **438, 440**
  notch, 119, **123**
  tuberosity, 119, **123**
Radiation and heat loss, 639
Radioimmunoassay, 354
Radio-ulnar joint, 139, **143**
Radius, 119–20, **123**
Ramus(i),
  communicating, 227, 247
  ischial, 124, **125, 127**
  pubic, **124, 125,** 127
  spinal nerves, 247, **247**
Ranvier, nodes of, 222, **222, 225**
Rapid eye movement (REM), 347
Reabsorption, tubular, 655, 658, 660, **661**
Receptive relaxation, 572
Receptor(s), 81, 219, 223–24, **223,** 251, **251**
  cutaneous, 223, 333, **465,** 466–67
  hair cells, **311,** 312, **314, 315**
  for light, 297–98, 306–307
  peripheral, **530,** 532, **533**
  properties of, 239–40
  skin, 67, 74
  of smell, 317
  muscle spindles, **223,** 224
  taste, 319
  tendon organs, 324, **325**
  vestibular, 333
Reciprocal innervation, 256, **256**
Rectum, **543, 551,** 553–54, **554,** 581
Rectus,
  abdominus, 163, 165, **165, 166**
  femoris, 182, 184, **185**
  muscles of eye, 296, **296**

Red,
  blood cell. *See* Erythrocyte
  bone marrow, 89
  pulp of spleen, 450–51, **451**
Reduction, 29
Reflex(es), 81
  accommodation, 300–301
  action, 253–54, 256
  arc, 251, **251**
  contralateral, 256
  crossed extensor, **257,** 258
  enterogastric, 573
  extensor, 257
  flexor, **257,** 258
  grasp, 258
  Hering-Breuer, 530
  inflation, 530, **533**
  intersegmental, 258
  ipsilateral, 256
  knee jerk, 257
  light, 305
  multineuronal, 258
  multisynaptic, 258
  myotatic, 257
  postural, 258
  proprioceptive, 257
  pupil, 301
  righting, 258, 333
  spinal, 250–58
  stretch, 257, **257**
  tendon, 257
  visceral, 258
Refraction, 299, **299**
Refractory period, **200,** 203, 230
Regeneration of nerves, 225–26
Relaxation period, 200, **200**
Relaxin, 381, 714
Renal,
  arteries, **438,** 439, 441, **441,** 653
  calyces, 652, **653**
  capsule, 652, **654**
  circulation, 654
  corpuscle, 652, **654**
  hypertension, 490
  papillae, 652, **653**
  pelvis, 651, **653**
  pyramids, 652, **653**
  sinus, 651, **653**
  tubule, 653, **655**
    role in urine tonicity, 670
  veins, **441,** 447, 654
Renin, 381, 675–76
  angiotensin system, 675–76, **676**
  role in hypertension, 490
Rennin, 574
Replication, DNA, 46
Repolarization, 230
Reproduction, 5
  of cells, 46–47
Reproductive,
  cells, 682

organs,
  female, 696–702, **698, 699**
  male, 681–89, **686, 689**
  system, 80, 681–716
Residual air volume, 509, **510**
Resistance,
  blood flow, 456–58, **456, 457**
  peripheral, 457–58
  variable, **456,** 457
Respiration, 504–10
  artificial, 511
  controlling mechanism, 527–34
  effect of barometric pressure, 535–36
  Hering-Breuer reflex, 530
  mechanisms, 504–10
  muscles, 161, 163, 164
  types, 506
Respiratory,
  centers, medullary, 528–30, **529, 530, 533**
  cycle, 528–31
  diseases, 536–37
  distress syndrome, 508–509
  minute volume, 510
  problems, 534–537
  system, 10, 80, 504–10
Response, muscle to stimulus, 198–204
Resting potential, 49–51, **50, 51,** 196, 229, **229**
Resuscitation, mouth-to-mouth, 511
Reticular,
  activating system, 292, **292,** 294
  fibers, 62
  formation, 265, 290–91, **291,** 328–29, **330**
  tissue, 64, **65**
Reticuloendothelial system, 388
Reticulolytes, 388
Reticulum,
  endoplasmic, **13,** 14
  sarcoplasmic, 193, **196**
Retina, 297, **297, 298,** 306–8, **307**
Retinal grain, 302
Retinene, 308
Retraction, of joints, 141
Rh factor, 403–4
Rhinencephalon, 263, 270–71, **272**
Rhodopsin, 308
Ribonuclease, 566, 577
Rib(s), 114
  articulation, **112,** 114
  cage, 114, **118**
  facet, 114, **115**
  floating, 114
  origin of muscles, 164
Riboflavin, 616

Ribonucleic acid, 14, 28, 43–45
  messenger, 44, **44**
  ribosomal, 44
  transfer, **44**, 45
Ribosomes, **13**, 14
Rickets, 377, 615
Righting reflexes, 258, 333
RNA, 14, 28, 43–45
  messenger, 44, **44, 45**
  ribosomal, 44
  transfer, **44**, 45, **45**
Rods of retina, 297, 306–8, **307**
Root(s),
  canal, 545, **547**
  lung, 502
  spinal cord, **245**, 246, **247**
  tooth, 545, **547**
Rotation, 140, **141**
Round, ligament, 697, **698, 701**
  window of ear, **309**, 310
Rubrospinal tract, 327
Rugae, of stomach, 549, **552**

Saccule, **309**, 314
Sacral vertebrae, **111**, 114, **117**
Sacroiliac joint, 124, 142
Sacrum, 95, **111**, 114, **117**
Saddle joint, 140
Sagittal,
  plane, 75, **76**
  sinus, 282, **284**
  suture, 103
Saliva, 570, 571
Salivary,
  amylase, 570–71
  glands, 547, **548**, 570–71
  juice, 582
  secretion, 547, **548**
Saltatory conduction, 232
Sarcolemma, 193, **193, 196**
Sarcomere, 193
Sarcoplasm, 193
Sarcoplasmic reticulum, 193, **196**
Satiety center, 624, **626**
Scala,
  tympani, **311**, 312
  vestibuli, **311**, 312
Scapula, 95, 116, **120, 121**
  insertion of muscles, 164, 166, 168
  origin of muscles, 164, 166, 168
  spine, 116, **120**
Schwann,
  cell, 222, 224–26, **225**
  sheath, 225
Sciatic,
  nerve, **245**, 249, **250**
  notch, **125**, 126, **127**
Sclera of eyeball, 296, **297**
Scoliosis, 110

Scrotum, 685–86, **686, 687, 688, 689**
Scurvy, 617
Sebaceous gland, 71–72, **72**
Sebum, 71
Secretin, 381, 568
Secretion,
  bile, 577–78, 582
  digestive glands, 573–78, 580
  gastric juice, 573–75, **573**, 582
  granule, 14, 15
  hormone control, 357–58
  intestinal juice, 580, 582
  saliva, 570, 571
  tubular, 655, 660, **661**
Segmentation, 569, **569**, 576
Sella turcica, 101, **103, 104**
Semen, 687
Semicircular canals, 309, 310, 315–17, **315, 316**
Semilunar,
  cartilages, 143, **146**
  ganglion, 272, **273**
  notch of ulna, 119, **123**
  valves, 414, **414**
Seminal,
  duct. *See* Vas deferens
  vesicles, 688, **689**
Seminiferous tubules, 686, **687**
Sense organs, 286–319
Sensory,
  areas of cerebral cortex, 268
  fibers of cranial nerves, 271
  neurons, 222, **222**
  pathways, 287–90
  tracts, 287–88
Septum,
  interventricular, 412, **414**
  nasal, 99, **100, 104**
Serosa, 67, 542, **544**
Sertoli cells, 686
Serum, 400, **400**
Sesamoid bone, 89
Sex,
  basal metabolic rate
    determinant, 630, **631**
  cells, meiosis of, 683, **683**, 699
  characteristics, 682, 693
  chromatin, 684
  chromosomes, 682–83
  determination, 684–85
  differentiation, 684, **685**, 697
  organs,
    female, **685**, 697–703, **698, 699**
    male, **685**, 685–91, **689**
Sheath,
  myelin, **222**, 225, **225**
  Schwann, 225
  tendon, 137, 138
Shock,
  cardiovascular, 488–89
  hemorrhagic, 489
  spinal, 253, 334

Short bones, 88–89, **89**
Shoulder,
  blades. *See* Scapula
  girdle. *See* Pectoral
  muscles, 168, **169**
Sickle cell anemia, 390
Sigmoid colon, **543, 550**, 553, **554**
Sinoatrial node of heart, 415, **416**
Sinus(es), 102
  aortic, 426, 437, 466, **466**
  carotid, 426, **426**, 437, **439**
  cavernous, 282, **284**
  ethmoid, 106
  frontal, **104**, 106
  maxillary, 107
  paranasal, 103, **103, 104**, 105, 496, **497**
  renal, 651, **653**
  sagittal, 282, **284**
  sphenoid, **104**, 106
  straight, 282, **284**
  transverse, 282, **284**
  venous, 282, **284**
Sinusoids, 445
Skeletal muscle, 193–208, **193**
Skeleton, 86, **98**
  appendicular, 96, 115–31
  axial, 96, 99–115
Skin, 66–70
  blood vessels, 74
  derivatives, 70–72
  heat loss mechanism, 640–41, **640**
  receptors, 67, 74
Skull, 99–110
  anterior aspect, 99, **100**
  bones, 99–110
  infant, 92, 104–5, **106**
  inferior aspect, 100–1, **102**
  interior base aspect, 101–2, **103**
  lateral aspect, 99–100, **101**
  sagittal section, 102–3
Sleep, 346–47
Small intestine, **543**, 550–52, **550, 553**
  digestive process, 575–80
  divisions, 550–51
  functions, 550
Smell, sense of, 317
Smooth muscle, 79, 209–12, **210**
Sneezing, 511
Sodium,
  bicarbonate buffer system, 405–6, 524
  hydroxide, in buffer system, 406
  metabolism, 613
  reabsorption, 369, 380, 660, 667–68
Soft, palate, 544, **545**
  spots, 92, 104–5, **106**

Solute, 21
Solution, 21, 36
Solvent, 21
Somatic nervous system, 220
Somatosensory area of cerebral cortex, 268, 290, **290**
Somesthetic cortex, 290
Sound, frequency and intensity, 309
Spatial summation, 237
Specific dynamic action, 629
Sperm, 682, **683**, 685–87, **687**
  count, 691
  ejaculation, 691
Spermatic cord, **686**, 687
Spermatid, 683, 686, **687**
Spermatocyte, 683, 686, **687**
Spermatogonium, 683, 686, **687**
Sphincter(s),
  anal, 553–54
  duodenal, 557
  precapillary, 435
  pyloric, 549, **552**
Sphygmomanometer, 460–61, **461**
Spinal,
  cord, 244–49, 262–63, **262, 334**
    transection of, 334
  ganglion, **245**, 246, **247**
  meninges, 244–45, **245**, 276–77, **278**
  nerves, 220, **220, 245**, 246–47, **247**
    accessory, **264, 267**, 275, 276, **280**
  reflexes, 250–58
  shock, 253, 334
  tracts, 246, **246**
Spine(s),
  iliac, 121, **125, 127**
  ischial, 124, **125**, 126, **127**
  scapula, 116, **120**
Spinocerebellar tracts, 246, 288, **289**
Spinothalamic tracts, 246, **246**, 287–88, **288**
Spinous process of vertebra, 112, **113, 115, 116**
Spiral,
  ganglion, **311**, 312
  organ, **311**, 312
Spirometer, 632, **632**
Splanchnic nerves, 339, 558
Spleen, 450–51, **451**
Sprain, 143–44
Squamous,
  epithelium, 60–61, **60**
  portion of temporal bone, 105, **107**
  suture, **101**, 103, **104**
Stalk, pituitary, 267, 358
Stapes, of ear, 110, **309**, 310

Starches, 26
Starling's,
  curve, 429–30, **430**
  law of the heart, 429
Starvation, body's response to,
  610
Statoacoustic nerve, **264,** 273,
  275
Statoconia, 314, **314**
Stem cell, 388
Stenosis, 423
Stereognosis, 291
Steroids, 354, 356, 367, 368–71
Sternum, 95, 114–15, **119**
Sterols, 27
Stimulus, 198–99
  liminal, 200
  maximal, 201, **201**
  nerve impulse conduction,
    227–32
  skeletal muscle contraction,
    203–4
  subliminal, 200, 201–2, **201**
  subthreshold, 200, **201**
  supramaximal, 201, **201**
  threshold, 200, **201**
Strabismus, 304
Straight sinus, 282, **284**
Stratified epithelium, **60,** 61
Stratum,
  corneum, 68–69, **68**
  germinativum, 68, **68**
  granulosum, 68, **68**
  lucidum, 68, **68**
  spinosum, 68, **68**
Stress response, induced by
  glucocorticoids, 371
Stretch reflex, 257, **257**
Striated muscle, 79, 193
Stroke volume of heart, 423, **424,**
  427–31, **430**
Stroma of ovary, 698
Styloid process,
  of radius, 119, **123**
  of temporal bone, 100, **101,**
    102, 104, 105, **107**
  of ulna, 119, **123**
Subarachnoid space, 245, **245,**
  277, 278
Subliminal stimulus, 200,
  201–2, **201**
Sublingual gland, 547, **548**
Submandibular gland, 547, **548**
Submucosa, 542, **544, 552, 553**
Substrate, of enzyme, 29–30, **30**
Subthreshold stimulus, 200, **201**
Sulcus(i),
  of cerebral cortex, 268, **270**
  posterior median, 244, **245**
Sulfur, metabolism, 614
Summation, 202
  spatial, 237
  of stimuli, 202, **202,** 203
  temporal, 237

Sunburn, 73
Superficial veins, 443
Superior plane of body, 75, **76**
Supination, 139, **143**
Supporting cells, 686, **687**
Supramaximal stimulus, 201,
  201
Suprarenal glands. *See* Adrenal
  glands
Surgical neck of humerus, 118,
  122
Suspensory ligaments,
  of eye, 297, **297, 298**
  of ovaries, 697, **698**
Suture(s),
  coronal, 100, **101,** 103, **104**
  lambdoidal, **101, 102,** 103, **104**
  sagittal, 103
  squamosal, **101,** 103, **104**
Swallowing, 571–72, **571**
Sweat gland, **68,** 72
Sweating, 640–41
Sympathectomy, 341
Sympathetic division of
    autonomic nervous system,
    337, 339, 341
  ganglia, 337, **338**
  pathways, **338,** 339, **339**
Sympathoadrenal system, 341
Symphyses, 135–36, **136**
  pubis, 124, **125,** 126, **127**
Synapses, 223, 236–39
Synaptic,
  excitation, 237–38
  knobs, 236, **236**
  transmission, 236–37
Synchondroses, 135
Synergist, 153
Synovial,
  cavity, 137, **138**
  fluid, 136
  folds, 136–37
  joints, 136–37, **137**
  membrane, 136
Systemic,
  arteries, 437–43, **438**
  circulation, 410, 436–37
  veins, 443–48
Systole, 421, **421**
Systolic blood pressure, 422,
  461, **461, 462**

Taenia coli of colon, 550, 553,
  554
Talus, 126, 128, **130**
Target cell, 355
Tarsal bones, 126, 128–29, **130**
Taste, 318–19
  buds, 318, **318**
  classification, 318
Tectorial membrane, **311,** 312
Teeth, 545, **546, 547**

Telencephalon, 262, **263**
Telophase, 47
Temperature,
  blood, hypothalamic centers,
    642
  body, 635–44, **637, 638, 641**
    energy expenditure, and
      heat loss, 629
    regulation and
      hypothalamus, 641–42,
      641, 644
  core, 636, **637**
  sensory pathway, 287–88
  skin, heat loss mechanism,
    640–41, **640**
Temporal,
  bone, **100, 101, 103, 104,** 105,
    107
  lobe of brain, **264, 265,** 268
  summation, 237
Temporary teeth, 545
Temporomandibular joint, 142,
  143
Tendon(s),
  Achilles, 183, **186, 187**
  organs, 324, **325**
  reflex, 257
  sheath, 137, **138**
Tension pressure, 516
Tentorium cerebelli, 277, 284
Terminal ganglia, 337
Testis(es), 682, 685–86, **686**
  descent, 687–88, **688**
  development, 684, **685**
  endocrine function, 691–93
  undescended, 688
Testosterone, 381, 684
Tetanus, muscle, 202–3, **203**
Thalamus, **265,** 266, 290
Theca interna and externa, 699,
  700
Thenar eminence, 178
Thermoregulation, chemical and
    physical, 638
Thiamin,
  deficiency, 616
  metabolic role, 616
Thoracic,
  aorta, 439, **441**
  cavity, 75, 77, **77,** 503–4, **507**
  duct, **449,** 450
  veins, 446–47
  vertebrae, **112,** 114
Thoracolumbar division of
    autonomic nervous system,
    337, **338**
Thorax, 503, **506**
  bones, 114, **118**
  cavity, 503–4, **507**
  muscles, 161, 164, **164**
Threshold,
  contraction, 200, ∠01
  stimulus, 200, **201**
  of urine, 660

Thrombin, 401
Thromboplastin, 401
Thrombus, 402
Thymine, 43, **43**
Thymus, 394, 451–52
Thyrocalcitonin, 377
Thyroglobulin, 372, 373
Thyroid,
  cartilages, 498, **499**
  gland, 371–74, **372**
  hormone, 358, 372–74
    action of mitochondria, 373
    hyposecretion and
      hypersecretion, 374
    secretion, stimulation by
      thyrotropin, 373
    synthesis, 372–73, **373, 374**
Thyroid-stimulating hormone,
  358, 360
Thyrotropin, 373
Thyrotropin-releasing hormone,
  373
Thyroxin, 372–74, 380
  blood glucose level increase,
    357–58, **357**
  heat production increase, 638
Tibia, 126, **129**
  articulation of, **135**
  insertion of muscles, 184
  origin of muscles, 184
Tibial tuberosity, 126, **129**
Tibiofibular joint, **135**
Tidal air, 509, **510**
Tissue(s), 58–66
  adipose, 64, **65**
  areolar, 64, **65**
  bone, 88
  cartilage, 65–66, **66**
  classification, 58
  connective, 62–66, **65**
  deep fascia, 64
  epithelial, 58–62, **60**
  hemopoietic, 88
  loose connective, 64, **65**
  lymph, 66
  reticular, 64, **65**
T lymphocyte, 394–95, 450, 451
Tongue, 318, **318,** 545–47, **547,**
  549
  muscles, 157–58, **161**
  papillae, 318, **318,** 547, **547**
Tonsils,
  lingual, 452, 547
  palatine, 452, **545,** 547, **549**
  pharyngeal, 452
Tonus, 256–57
Trabeculae, **90,** 91
Trachea, **497,** 500–1, **501, 502**
Tract(s),
  association, 268, **270**
  cerebellar, 287–88, **289**
  cerebral, 268, **270**
  corticospinal, 326, **326**
  digestive, 542, **543**

extrapyramidal, 327, **328**
fiber, 246, **246**
gastrointestinal, 542, **543**
optic, 305, **306**
projection, 268, **269**
pyramidal, 226, **226**
respiratory, upper and lower, 496–503, **501**
reticulospinal, 327
rubrospinal, 327
spinocerebellar, **246**, 288, **289**
spinothalamic, 246, **246**, 287–88, **288**
vestibulospinal, 327
Transamination, 600
Transfer RNA, **44, 45, 45**
Transfusion, blood, 403–4
Transitional epithelium, **60**, 61
Transmitter substance, 234, 236–39, 360
Transmural pressure, 428
Transport, 33–38
active, 33, 36–38
passive, 33–36
Transverse,
arch of foot, 131, **131**
colon, **543, 550**, 553, **554**
foramen, 112, **113**
ligament, 185, **186, 187, 189**
mesocolon, 560, **561**
plane of body, 75, **76**
processes of vertebrae, 112, **113, 115, 116**
sinus, 282, **284**
tubules, 195, **196**
Tricarboxylic acid cycle, 41–42
Tricuspid valve, 414, **414**
Trigeminal nerve, **264**, 272–73, **273**, 274
Triglycerides, 26
Trigone, 651, **652**
Tri-iodothyronine, 372, 380
Trochanter, 91
of femur, 126, **128**
Trochlea of eye, 296
Trochlear nerve, **264**, 272, 274
Trophoblast, 709, **709**
Tropocollagen, 63
Tropomyosin, **194**, 195
Troponin, **194**, 195
Trunk of body, 75
Trypsin, 566, 576
Trypsinogen, 566, 576
T-system of striated muscle cells, 195, **196**
Tube(s),
auditory, **309**, 310
fallopian, 697, **698, 701**
uterine, 697, 700–1, **698, 699, 701**
Tubercle, 91
of humerus, 118, **122**
Tuberculosis, 536

Tuberosity, 91
deltoid, 119, **122**
gluteal, 126, **128**
ischial, 124, **125**
radial, 119, **123**
tibial, 126, **129**
Tubular,
maximum, 660
reabsorption, 655, 658, 660, **661**
Tubule(s),
convoluted, 653, **655**
renal, 653, **655**
seminiferous, 686, **687**
T-system, 195, **196**
Tunica,
adventitia, 434, **434**, 435, 436
intima, 434, **434**, 435, 436
media, 434, **434**, 435, 436
vaginalis, 687, **688**
Twitch, 200
Tympanic membrane, **309**, 310
Tyrosine, 372, **373**

Ulna, 119, **123**
articulation, 139
insertion of muscles, 172, **177**
origin of muscles, 172, **177**
Umbilical,
arteries, 478, **479**, 480
cord, 478, **479**
ligaments, 480, **482**
veins, 478, **479**, 480
Umbilicus, 478
Universal donor and recipient, 403
Urea, 602, 658, 661
Ureters, 651, **651**, 652
Urethra, 651, **652, 653**
female, 698
male, 689, **689, 690**
Urinary,
bladder, 651, **651, 652**, 662–63
system, 649–63
Urination, 662–63
Urine,
composition, 660–61
abnormal constituents, 661–62
formation, 654–58, 660
pH, 674–75
tonicity, 670
volume, 661
Urogenital diaphragm, 651
Uterine,
contractions, 712, 714
tube, 697, 700–1, **698, 699, 701**
Uterus, 697, **698, 699**, 701–2
Utricle, **309**, 314, **314**
Uvula, **497**, 498, 544, **545**

Vagal mechanism, 530–31
Vagina, 697, **698, 701**, 702
Vagus nerve, **264**, 273, 275, **276**, 337
Valence, 21
Valve(s),
aortic, 414
atrioventricular, 412, **414**
cuspid, 414, **414**
heart, 412, 414, **414**, 421, 423
iliocecal, 551
mitral, 414
pulmonary, 414
semilunar, 414, **414**
venous, 435
Valvular defects, 423
Vas deferens, **686**, 688, **688, 689**
Vasa,
rectae, 653–54, **655**
vasorum, 434
Vasectomy, 716
Vasoconstriction, 411, 435
Vasoconstrictor center, 465
Vasodilatation, 411, 435
Vasodilator center, 465
Vasomotor, pressoreceptor mechanism, 465
Vasopressin, 364
Vein(s),
of abdomen, 447
axillary, 445, **445, 446**
azygos, **441**, 446, **447**
basilic, 445–46, **445, 446**
brachial, **445**, 446, **446**
brachiocephalic, **441**, 445, **445**
cephalic, 445, **445, 446**
coronary, 415
deep, 443
dorsal venous arch, 445, 448, **448**
femoral, 448, **448**
gastric, 447, **447**
gastroepiploic, 447, **447**
of head and neck, 445, **446**
hemiazygos, **441**, 446, **447**
hepatic, **441, 445**, 447, **447**
iliac, common, **441, 445**, 447
external, **445**, 447
internal, **445**, 447
inferior mesenteric, 447, **447**
inferior vena cava, 412, **413**
intercostal, **441**, 446–47, **447**
jugular, external, **445**, 447
internal, **445**, 447
lumbar, **441**, 447
median cubital, **445**, 446, **446**
ovarian, **441**, 447
of pelvic region, 447
popliteal, 448, **448**
portal, 447, **447, 556**, 558
pulmonary, 436, **437, 445**
renal, **441**, 447
saphenous, 448, **448**

splenic, 447, **447**
subclavian, 445, **445, 446, 447**
superficial, 443
superior mesenteric, 447, **447**
superior vena cava, 412, **413**, **441**, 445, **445, 446, 447**
systemic, 443–48
testicular, **441**, 447
of thoracic cavity, 446–47
umbilical, 478, **479**
of upper extremity, 445–46
Vena cava,
inferior, **441, 447, 447**, 412, **413**
superior, 412, **413, 441**, 445, **445, 446, 447**
Venae comitantes, 443
Venous,
blood, 388, 434
circulation, 471–72
return, 428, 472
sinus, 282, **284**, 445
Ventilation, 504–12
alveolar, 510, 531
pulmonary, 531–34
Ventral, 75, **76**
cavity, 75, 77, **77**
Ventricles,
of brain, **262**, 263, 276, **277**, **278**
of heart, 410, **410**, 411, 412, **412, 413**
role in cardiac output, 423–24, 426–31
Ventricular folds, **497**, 499, **500**
Venule, 435, 436
Vermis of cerebellum, 266
Vertebra(ae), 95, 110–15
cervical, 112, **113**, 114
coccygeal, **111**, 114
lumbar, **111**, 114, **116**
muscle origins, 160–62
sacral, **111**, 114, **117**
thoracic, **112**, 114
Vertebral,
arch, 110, **112, 113, 115, 116**
arteries, **280**, 281–82, **281**
cavity, 75, **77**
column, 110, **111**
foramen, 112, **113, 115, 116**
joints, 136, **136**
Vertigo, 316
Vesicles, seminal, 688, **689**
Vessel(s),
blood. *See* Blood vessels
lymphatic, 448–50
afferent, 450, **450**
efferent, 450, **450**
pulmonary, 480–82
Vestibular,
apparatus, 314, 333
glands, **702**, 703
membrane, **311**, 312
nerve, 273, 275
receptors, 333

Vestibule,
    of inner ear, **309,** 310
    of mouth, 544
    of vagina, 702
Vestibulospinal tract, 327
Villi, intestinal, 552, **553**
Visceral
    muscle, 209
    peritoneum, 559–60
    pleura, 503, **505, 561**
    reflexes, 258
Viscosity, blood, 457
    determinant of arterial blood
        pressure, **463,** 464
Vision, 295–308
    binocular, 303
    color, 308
    daylight, 308
    night, 308
Vital capacity of lung, 509
Vitamin(s), 615–17
    A, and vision, 308, 615
        and epithelium, 615
    absorption of, 585
    B_{12}, 390
        role in red blood cell
            formation, 390, 617
    C, deficiency, 617
    D, deficiency, 615
        roles, 377–78, 615
    E, possible roles, 615–16
    fat-soluble, 615–16
    K, role in blood clotting, 402,
        616
    niacin, 616
    pantothenic acid, 616–17
    pyridoxine (B_6), 616
    riboflavin (B_2), 616
    thiamin (B_1), 616
    water-soluble, 616–17
Vitreous humor, 298
Vocal cords, 499, **500**
Voltage, 48, 228
Voluntary,
    movement, 332
    muscle, 193
Vomer, 100, **102, 104,** 108
Vomiting, 582
Vulva, 702, **702**

Walking, 131, 328
Water, 21–22
    absorption of, 584–85
    evaporation and heat loss,
        639–40
    reabsorption, 669
        role of antidiuretic
            hormone, 669–70
Wave, pulse, 459–60
White,
    blood cell. *See* Leukocyte
    matter, 245, **245**
    pulp of spleen, 451, **451**
Willis, circle of, **280,** 282
Windpipe. *See* Trachea
Wisdom tooth, 545
Womb, 697
Wrist bones, 120, **124**

X and Y chromosomes, 682–83,
    683
Xiphoid process, 114–15, **118**

Yawning, 511
Yellow bone marrow, 89

Zinc, 615
Z line of striated muscle, 193,
    194, 195, 196
Zona,
    fasciculata, **365,** 366
    glomerulosa, **365,** 366
    pellucida, 699, **700**
    reticularis, **365,** 366
Zone, H, of striated muscle, 193,
    194, 195, 196
Zygomatic,
    arch, 99, 100, **102, 105**
    bone, 99, **100, 101,** 107
    process, **100,** 105, **107**